Elementary Algeb

MW01702287

Fourth Edition

Alan S. Tussy | R. David Gustafson

CENGAGE
Learning™

Australia • Brazil • Japan • Korea • Mexico • Singapore • Spain • United Kingdom • United States

CENGAGE
Learning™

Elementary Algebra: Fourth Edition

Alan S. Tussy | R. David Gustafson

Executive Editors:
Maureen Staudt
Michael Stranz

Senior Project Development Manager:
Linda DeStefano

Marketing Specialist:
Sara Mercurio
Lindsay Shapiro

Senior Production / Manufacturing Manager:
Donna M. Brown

PreMedia Supervisor:
Joel Brennecke

Rights & Permissions Specialist:
Kalina Hintz
Todd Osborne

Cover Image:
Getty Images*

* Unless otherwise noted, all cover images used by Custom Solutions, a part of Cengage Learning, have been supplied courtesy of Getty Images with the exception of the Earthview cover image, which has been supplied by the National Aeronautics and Space Administration (NASA).

ISBN-13: 978-1-4240-6843-2

ISBN-10: 1-4240-6843-6

Cengage Learning
5191 Natorp Boulevard
Mason, Ohio 45040
USA

Cengage Learning is a leading provider of customized learning solutions with office locations around the globe, including Singapore, the United Kingdom, Australia, Mexico, Brazil, and Japan. Locate your local office at: **international.cengage.com/region**

Cengage Learning products are represented in Canada by Nelson Education, Ltd.

For your lifelong learning solutions, visit **www.cengage.com/custom**

Visit our corporate website at **www.cengage.com**

Printed in the United States of America

Brief Contents

PREFACE

Elementary Algebra, Fourth Edition, is more than a simple upgrade of the third edition. Substantial changes have been made to the example structure, the Study Sets, and the pedagogy. Throughout the process, the objective has been to ease teaching challenges and meet students' educational needs.

Algebra, for many of today's developmental math students, is like a foreign language. They have difficulty translating the words, their meanings, and how they apply to problem solving. With these needs in mind (and as educational research suggests), the fundamental goal is to have students read, write, think, and speak using the *language of algebra*. Instructional approaches that include vocabulary, practice, and well-defined pedagogy, along with an emphasis on reasoning, modeling, communication, and technology skills have been blended to address this need.

The most common student question as they watch their instructors solve problems and as they read the textbook is . . . *Why?* The new fourth edition addresses this question in a unique way. Experience teaches us that it's not enough to know *how* a problem is solved. Students gain a deeper understanding of algebraic concepts if they know *why* a particular approach is taken. This instructional truth was the motivation for adding a **Strategy** and **Why** explanation to the solution of each worked example. The fourth edition now provides, on a consistent basis, a concise answer to that all-important question: *Why?*

This is just one of several changes in this revision, and we trust that all of them will make the course a better experience for both instructor and student.

NEW TO THIS EDITION

- New Example Structure
- New Chapter Opening Applications
- New *Study Skills Workshops*
- New Chapter Objectives
- New *Guided Practice* and *Try It Yourself* sections in the *Study Sets*
- New End-of-Chapter Organization

Chapter Openers Answering The Question: When Will I Use This?

Have you heard this question before? Instructors are asked this question time and again by students. In response, we have written chapter openers called *From Campus to Careers*. This feature highlights vocations that require various algebraic skills. Designed to inspire career exploration, each includes job outlook, educational requirements, and annual earnings information. Careers presented in the openers are tied to an exercise found later in the *Study Sets*.

Systems of Linear Equations and Inequalities

© Peter Steiner/Alamy

***from* Campus to Careers**
Portrait Photographer

Portrait photographers take pictures of individuals or groups of people and often work in their own studios. Some specialize in weddings, religious ceremonies, or

JOB TITLE: Portrait Photographer
EDUCATION: A well-rounded education including art and business cours... is preferred.

3

Examples That Tell Students Not Just How, But WHY

Why? That question is often asked by students as they watch their instructor solve problems in class and as they are working on problems at home. It's not enough to know how a problem is solved. Students gain a deeper understanding of the algebraic concepts if they know why a particular approach was taken. This instructional truth was the motivation for adding a *Strategy* and *Why* explanation to each worked example.

EXAMPLE 4 Solve the system: $\begin{cases} 4a + 7b = -8 \\ 5a + 6b = 1 \end{cases}$.

Strategy We will use the elimination method to solve this system.

Why Since none of the variables has coefficient 1 or −1, it would be difficult to solve this system using substitution.

Solution

Step 1: Both equations are written in standard $Ax + By = C$ form.

Step 2: In this example, we must write *both* equations in equivalent forms to obtain like terms that are opposites. To eliminate a, we can multiply the first equation by 5 to create the term $20a$, and we can multiply the second equation by -4 to create the term $-20a$.

$$\begin{cases} 4a + 7b = -8 \\ 5a + 6b = 1 \end{cases} \begin{matrix} \xrightarrow{\text{Multiply by 5}} \\ \xrightarrow[\text{Multiply by } -4]{} \end{matrix} \begin{matrix} 5(4a + 7b) = 5(-8) \\ -4(5a + 6b) = -4(1) \end{matrix} \begin{matrix} \xrightarrow{\text{Simplify}} \\ \xrightarrow[\text{Simplify}]{} \end{matrix} \begin{cases} 20a + 35b = -40 \\ -20a - 24b = -4 \end{cases}$$

Step 3: When we add the resulting equations, a is eliminated.

$$\begin{array}{r} 20a + 35b = -40 \\ -20a - 24b = -4 \\ \hline 11b = -44 \end{array} \quad \text{In the left column: } 20a + (-20a) = 0.$$

Step 4: Solve the resulting equation for b.

$$\begin{aligned} 11b &= -44 \\ b &= -4 \quad \text{Divide both sides by 11. This is the } b\text{-value of the solution.} \end{aligned}$$

Step 5: To find a, we can substitute -4 for b in any equation that contains both variables. It appears the computations will be simplest if we use $5a + 6b = 1$.

$$\begin{aligned} 5a + 6b &= 1 && \text{This is the second equation of the original system.} \\ 5a + 6(-4) &= 1 && \text{Substitute } -4 \text{ for } b. \\ 5a - 24 &= 1 && \text{Multiply.} \\ 5a &= 25 && \text{Add 24 to both sides.} \\ a &= 5 && \text{Divide both sides by 5. This is the } a\text{-value of the solution.} \end{aligned}$$

Step 6: Written in (a, b) form, the solution is $(5, -4)$. Check it in the original equations.

Self Check 4 Solve the system: $\begin{cases} 5a + 3b = -7 \\ 3a + 4b = 9 \end{cases}$.

Now Try **Problem 45**

Examples That Offer Immediate Feedback

Each example includes a *Self Check*. These can be completed by students on their own or as classroom lecture examples, which is how Alan Tussy uses them. Alan asks selected students to read aloud the *Self Check* problems as he writes what the student says on the board. The other students, with their books open to that page, can quickly copy the *Self Check* problem to their notes. This speeds up the note-taking process and encourages student participation in his lectures. It also teaches students how to read mathematical symbols. Each *Self Check* answer is printed adjacent to the corresponding problem in the Annotated Instructor's Edition for easy reference. *Self Check* solutions can be found at the end of each section in the student edition before the *Study Sets* begin.

Examples That Ask Students To Try

Each example ends with a *Now Try* problem. These are the final step in the learning process. Each one is linked to similar problems found within the *Guided Practice* section of the *Study Sets*.

Emphasis on Study Skills

Each chapter begins with a *Study Skills Workshop*. Instead of simple suggestions printed in the margins, each workshop contains a *Now Try This* section offering students actionable skills, assignments, and projects that will impact their study habits throughout the course.

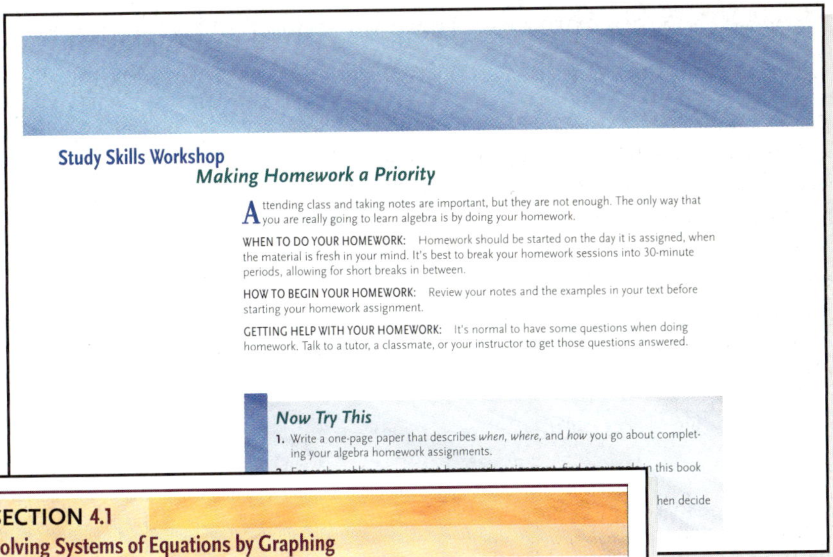

Study Skills Workshop

Making Homework a Priority

Attending class and taking notes are important, but they are not enough. The only way that you are really going to learn algebra is by doing your homework.

WHEN TO DO YOUR HOMEWORK: Homework should be started on the day it is assigned, when the material is fresh in your mind. It's best to break your homework sessions into 30-minute periods, allowing for short breaks in between.

HOW TO BEGIN YOUR HOMEWORK: Review your notes and the examples in your text before starting your homework assignment.

GETTING HELP WITH YOUR HOMEWORK: It's normal to have some questions when doing homework. Talk to a tutor, a classmate, or your instructor to get those questions answered.

Now Try This

1. Write a one-page paper that describes *when, where,* and *how* you go about completing your algebra homework assignments.

Useful Objectives Help Keep Students Focused

Objectives are now numbered at the start of each section to focus students' attention on the skills that they will learn as they work through the section. When each objective is introduced, the number and heading will appear again to remind them of the objective at hand.

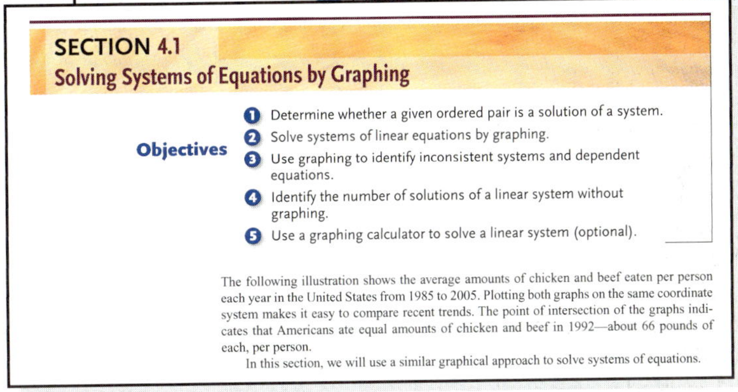

SECTION 4.1

Solving Systems of Equations by Graphing

Objectives

1. Determine whether a given ordered pair is a solution of a system.
2. Solve systems of linear equations by graphing.
3. Use graphing to identify inconsistent systems and dependent equations.
4. Identify the number of solutions of a linear system without graphing.
5. Use a graphing calculator to solve a linear system (optional).

The following illustration shows the average amounts of chicken and beef eaten per person each year in the United States from 1985 to 2005. Plotting both graphs on the same coordinate system makes it easy to compare recent trends. The point of intersection of the graphs indicates that Americans ate equal amounts of chicken and beef in 1992—about 66 pounds of each, per person.

In this section, we will use a similar graphical approach to solve systems of equations.

Heavily Revised Study Sets

The *Study Sets* have been thoroughly revised to ensure every concept is covered even if the instructor traditionally assigns every other problem. Particular attention was paid to developing a gradual level of progression.

Guided Practice

All of the problems in the *Guided Practice* portion of the *Study Sets* are linked to an associated worked example from that section. This feature will promote student success by referring them to the proper example(s) if they encounter difficulties solving homework problems.

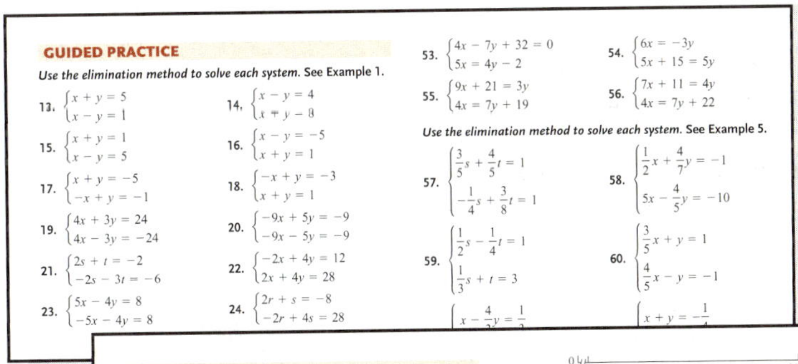

GUIDED PRACTICE

Use the elimination method to solve each system. See Example 1.

13. $\begin{cases} x + y = 5 \\ x - y = 1 \end{cases}$
14. $\begin{cases} x - y = 4 \\ x + y = 8 \end{cases}$
15. $\begin{cases} x + y = 1 \\ x - y = 5 \end{cases}$
16. $\begin{cases} x - y = -5 \\ x + y = 1 \end{cases}$
17. $\begin{cases} x + y = -5 \\ -x + y = -1 \end{cases}$
18. $\begin{cases} -x + y = -3 \\ x + y = 1 \end{cases}$
19. $\begin{cases} 4x + 3y = 24 \\ 4x - 3y = -24 \end{cases}$
20. $\begin{cases} -9x + 5y = -9 \\ -9x - 5y = -9 \end{cases}$
21. $\begin{cases} 2s + t = -2 \\ -2s - 3t = -6 \end{cases}$
22. $\begin{cases} -2x + 4y = 12 \\ 2x + 4y = 28 \end{cases}$
23. $\begin{cases} 5x - 4y = 8 \\ -5x - 4y = 8 \end{cases}$
24. $\begin{cases} 2r + s = -8 \\ -2r + 4s = 28 \end{cases}$

53. $\begin{cases} 4x - 7y + 32 = 0 \\ 5x = 4y - 2 \end{cases}$
54. $\begin{cases} 6x = -3y \\ 5x + 15 = 5y \end{cases}$
55. $\begin{cases} 9x + 21 = 3y \\ 4x = 7y + 19 \end{cases}$
56. $\begin{cases} 7x + 11 = 4y \\ 4x = 7y + 22 \end{cases}$

Use the elimination method to solve each system. See Example 5.

57. $\begin{cases} \frac{3}{5}s + \frac{4}{5}t = 1 \\ -\frac{1}{4}s + \frac{3}{8}t = 1 \end{cases}$
58. $\begin{cases} \frac{1}{2}x + \frac{4}{7}y = -1 \\ 5x - \frac{4}{5}y = -10 \end{cases}$
59. $\begin{cases} \frac{1}{2}s - \frac{1}{4}t = 1 \\ \frac{1}{3}s + t = 3 \end{cases}$
60. $\begin{cases} \frac{3}{5}x + y = 1 \\ \frac{4}{5}x - y = -1 \end{cases}$

Try It Yourself

To promote problem recognition, some *Study Sets* now include a collection of *Try It Yourself* problems that do not have the example linking. The problem types are thoroughly mixed and are not linked, giving students an opportunity to practice decision making and strategy selection as they would when taking a test or quiz.

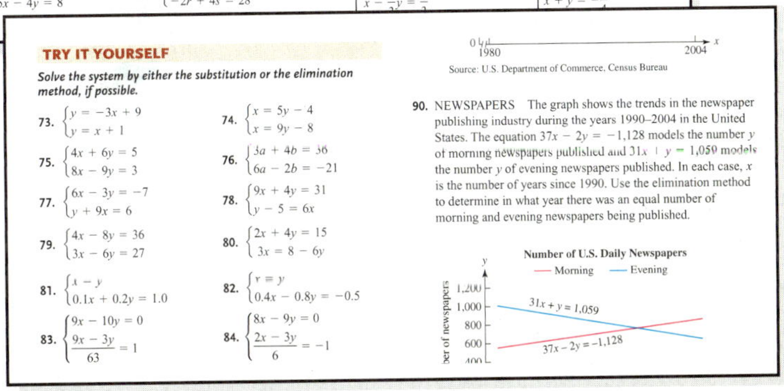

TRY IT YOURSELF

Solve the system by either the substitution or the elimination method, if possible.

73. $\begin{cases} y = -3x + 9 \\ y = x + 1 \end{cases}$
74. $\begin{cases} x = 5y - 4 \\ x = 9y - 8 \end{cases}$
75. $\begin{cases} 4x + 6y = 5 \\ 8x - 9y = 3 \end{cases}$
76. $\begin{cases} 3a + 4b = 36 \\ 6a - 2b = -21 \end{cases}$
77. $\begin{cases} 6x - 3y = -7 \\ y + 9x = 6 \end{cases}$
78. $\begin{cases} 9x + 4y = 31 \\ y - 5 = 6x \end{cases}$
79. $\begin{cases} 4x - 8y = 36 \\ 3x - 6y = 27 \end{cases}$
80. $\begin{cases} 2x + 4y = 15 \\ 3x = 8 - 6y \end{cases}$
81. $\begin{cases} x = y \\ 0.1x + 0.2y = 1.0 \end{cases}$
82. $\begin{cases} x = y \\ 0.4x - 0.8y = -0.5 \end{cases}$
83. $\begin{cases} 9x - 10y = 0 \\ \frac{9x - 3y}{63} = 1 \end{cases}$
84. $\begin{cases} 8x - 9y = 0 \\ \frac{2x - 3y}{6} = -1 \end{cases}$

Source: U.S. Department of Commerce, Census Bureau

90. NEWSPAPERS The graph shows the trends in the newspaper publishing industry during the years 1990–2004 in the United States. The equation $37x - 2y = -1{,}128$ models the number y of morning newspapers published and $31x + y = 1{,}059$ models the number y of evening newspapers published. In each case, x is the number of years since 1990. Use the elimination method to determine in what year there was an equal number of morning and evening newspapers being published.

Comprehensive End-of-Chapter Summary with Integrated Chapter Review

The end-of-chapter material has been redesigned to function as a complete study guide for students. New Chapter Summaries that include definitions, concepts, and examples, by section, have been written. Review problems for each section have been placed after each section summary.

CHAPTER 4
Summary & Review

SECTION 4.1 Solving Systems of Equations by Graphing

DEFINITIONS AND CONCEPTS

When two equations are considered at the same time, we say that they form a **system of equations.**

A **solution of a system** of equations in two variables is an ordered pair that satisfies both equations of the system.

EXAMPLES

Is (4, 3) a solution of the system $\begin{cases} x + y = 7 \\ x - y = 5 \end{cases}$?

To answer this question, we substitute 4 for x and 3 for y in each equation.

$$x + y = 7 \qquad\qquad x - y = 5$$
$$4 + 3 \stackrel{?}{=} 7 \qquad\qquad 4 - 3 \stackrel{?}{=} 5$$
$$7 = 7 \quad \text{True} \qquad\qquad 1 = 5 \quad \text{False}$$

Although (4, 3) satisfies the first equation, it does not satisfy the second. Because it does not satisfy both equations, it is not a solution of the system.

To **solve a system graphically:**

1. Graph each equation on the s system.
2. Determine the coordinates of section of the graphs. That or solution.
3. Check the solution in each eq nal system.

SECTION 4.3 Solving Systems of Equations by Elimination (Addition)

DEFINITIONS AND CONCEPTS

To solve a system of equations in x and y using **elimination (addition):**

1. Write each equation in the standard $Ax + By = C$ form.
2. Multiply one (or both) equations by nonzero quantities to make the coefficients of x (or y) opposites.
3. Add the equations to eliminate the terms involving x (or y).
4. Solve the equation obtained in step 3.
5. Find the value of the other variable by substituting the value of the variable found in step 4 into any equation containing both variables.
6. Check the solution in the equations of the original system.

With the elimination method, the basic objective is to obtain two equations whose sum will be one equation in one variable.

EXAMPLES

Use elimination to solve $\begin{cases} 2x - 3y = 4 \\ 3x + y = -5 \end{cases}$.

Step 1: Both equations are written in $Ax + By = C$ form.

Step 2: Multiply the second equation by 3 so that the coefficients of y are opposites.

Step 3:

$$2x - 3y = 4$$
$$\underline{9x + 3y = -15}$$
$$11x \quad = -11 \quad \text{Add the like terms, column by column.}$$

Step 4: Solve for x.

$$11x = -11$$
$$x = -1 \quad \text{Divide both sides by 11.}$$

Step 5: Find y.

$$3x + y = -5 \quad \text{This is the second equation.}$$
$$3(-1) + y = -5 \quad \text{Substitute } -1 \text{ for x.}$$
$$y = -2$$

TRUSTED FEATURES

- **The Study Sets** found in each section offer a multifaceted approach to practicing and reinforcing the concepts taught in each section. They are designed for students to methodically build their knowledge of the section concepts, from basic recall to increasingly complex problem solving, through reading, writing, and thinking mathematically.

  Vocabulary—Each Study Set begins with the important Vocabulary discussed in that section. The fill-in-the-blank vocabulary problems emphasize the main concepts taught in the chapter and provide the foundation for learning and communicating the language of algebra.

  Concepts—In Concepts, students are asked about the specific subskills and procedures necessary to successfully complete the practice problems that follow.

  Notation—In Notation, the students review the new symbols introduced in a section. Often, they are asked to fill in steps of a sample solution. This helps to strengthen their ability to read and write mathematics and prepares them for the practice problems by modeling solution formats.

  Guided Practice—The problems in Guided Practice are linked to an associated worked example from that section. This feature will promote student success by referring them to the proper examples if they encounter difficulties solving homework problems.

  Try It Yourself—To promote problem recognition, the Try It Yourself problems are thoroughly mixed and are not linked, giving students an opportunity to practice decision-making and strategy selection as they would when taking a test or quiz.

  Applications—The Applications provide students the opportunity to apply their newly acquired algebraic skills to relevant and interesting real-life situations.

  Writing—The Writing problems help students build mathematical communication skills.

  Review—The Review problems consist of randomly selected problems from previous chapters. These problems are designed to keep students' successfully mastered skills fresh and at the forefront of their minds before moving on to the next section.

  Challenge Problems—The Challenge Problems provide students with an opportunity to stretch themselves and develop their skills beyond the basics. Instructors often find these to be useful as extra-credit problems.

- **Detailed Author Notes** guide students along in a step-by-step process continue to be found in the solutions to every example.
- **The Language of Algebra** boxes draw connections between mathematical terms and everyday references to reinforce the language of algebra thread that runs throughout the text.
- **The Notation, Success Tips, Caution,** and **Calculators** boxes offer helpful tips to reinforce correct mathematical notation, improve students' problem-solving abilities, warn students of potential pitfalls and increase clarity, and offer tips on using scientific calculators.
- **Strategic use of color** has been implemented within the new design to help the visual learner.
- **Chapter Tests** are available at the end of every chapter as preparation for the class exam.

- **The Cumulative Review** following the end-of-chapter material keeps students' skills sharpened before moving on to the next chapter. Each problem is now linked to the associated section from which the problem came for ease of reference. The final Cumulative Review, found at the end of the last chapter, is often used by instructors as a Final Exam Review.

CHANGES TO THE TABLE OF CONTENTS

Based on feedback from colleagues and users of the third edition, the following changes have been made to the table of contents in an effort to further streamline the text and make it even easier to use.

- Chapter 2 topics have been reorganized and the section *Simplifying Algebraic Expressions Using Properties of Real Numbers* has been moved from Chapter 2 to Section 1.9.

  2.1 *Solving Equations Using Properties of Equality*
  2.2 *More about Solving Equations*
  2.3 *Applications of Percent* (Commission and discount problems were added)
  2.4 *Formulas*
  2.5 *Problem Solving* (Consecutive integer, commission, and set-up fee/cost per item problems were added)
  2.6 *More about Problem Solving*
  2.7 *Solving Inequalities*

- Parallel and perpendicular lines are now introduced in Section 3.4 *Slope and Rate of Change.*
- The section *An Introduction to Functions* has been moved from Chapter 9 to Section 3.8. This topic is a natural fit after studying linear equations in two variables.
- Former Chapter 7 *Systems of Linear Equations and Inequalities* has been moved up to Chapter 4, promoting the logical progression from the linear equations and inequalities in two variables of Chapter 3 to systems of equations in two variables in Chapter 4. The built-in flexibility of the content allows instructors who prefer to cover this toward the end of the course to continue to do so.
- The section *Variation* has been moved from Chapter 9 to Chapter 7. The new order of topics for Chapter 9 is:

  9.1 *Solving Quadratic Equations: The Square Root Property*
  9.2 *Solving Quadratic Equations: Completing the Square*
  9.3 *Solving Quadratic Equations: The Quadratic Formula*
  9.4 *Complex Numbers*
  9.5 *Graphing Quadratic Equations*

GENERAL REVISIONS AND OVERALL DESIGN

- We have edited the prose so that it is even more clear and concise.
- Strategic use of color has been implemented within the new design to help the visual learner.
- Added color in the solutions highlight strategic steps and improve readability.
- We have updated all data and graphs and have added scaling to all axes in all graphs.

- We have added more real-world applications and deleted some of the more "contrived" problems.
- We have included more problem-specific photographs.

INSTRUCTOR RESOURCES

Print Ancillaries

INSTRUCTOR'S RESOURCE BINDER (0-495-55472-3)
Maria H. Andersen, *Muskegon Community College*

NEW! Offered exclusively with Tussy/Gustafson. Each section of the main text is discussed in uniquely designed Teaching Guides containing instruction tips, examples, activities, worksheets, overheads, assessments, and solutions to all worksheets and activities.

COMPLETE SOLUTIONS MANUAL (0-495-38965-X)
Alexander H. Lee, *Hinds Community College*

The Complete Solutions Manual provides worked-out solutions to all of the problems in the text.

TEST BANK (0-495-38966-8)
Carol M. Walker & David J. Walker, *Hinds Community College*

Drawing from hundreds of text-specific questions, an instructor can easily create tests that target specific course objectives. The Test Bank includes multiple tests per chapter, as well as final exams. The tests are made up of a combination of multiple-choice, free-response, true/false, and fill-in-the-blank questions.

ANNOTATED INSTRUCTOR'S EDITION (0-495-38962-5)

The Instructor's Edition provides the complete student text with answers next to each respective exercise.

Electronic Ancillaries

ENHANCED WebAssign ENHANCED WEBASSIGN (0-495-38972-2)

Instant feedback and ease of use are just two reasons why WebAssign is the most widely used homework system in higher education. WebAssign's homework delivery system allows you to assign, collect, grade, and record homework assignments via the web. And now, this proven system has been enhanced to include links to textbook sections, video examples, and problem-specific tutorials. Enhanced WebAssign is more than a homework system—it is a complete learning system for math students.

ThomsonNOW™ THOMSONNOW™ (0-495-39456-4)

ThomsonNOW™ is an online teaching and learning resource that gives you more control in less time and delivers the results you want—NOW.

POWERLECTURE™: A 1-STOP MICROSOFT® POWERPOINT® TOOL (0-495-55649-1)

NEW! The ultimate multimedia manager for your course needs. The PowerLecture CD-ROM includes the Complete Solutions Manual, ExamView®, JoinIn™, and custom PowerPoint® lecture slides authored by Richard D. Townsend, North Carolina Central University.

TEXT SPECIFIC DVDs (0-495-38967-6)

These text specific DVDs provide additional guidance and support to students when they are preparing for an upcoming quiz or exam.

STUDENT RESOURCES

Print Ancillaries

STUDENT WORKBOOK (0-495-55468-5)
Maria H. Andersen, *Muskegon Community College*

NEW! Get a head start. The Student Workbook contains all of the Assessments, Activities, and Worksheets from the Instructor's Resource Binder for classroom discussions, in-class activities, and group work.

STUDENT SOLUTIONS MANUAL (0-495-38964-1)
Alexander H. Lee, *Hinds Community College*

The Student Solutions Manual provides worked-out solutions to the odd-numbered problems in the text.

Electronic Ancillaries

ENHANCED WebAssign **ENHANCED WEBASSIGN (0-495-38972-2)**

Get instant feedback on your homework assignments with Enhanced WebAssign (assigned by your instructor). This online homework system is easy to use and includes helpful links to textbook sections, video examples, and problem-specific tutorials.

INSTANT ACCESS CODE, THOMSONNOW™ (0-495-39458-0)

Instant Access gives students without a new copy of Tussy/Gustafson's *Elementary Algebra, Fourth Edition*, one access code to all available technology associated with this textbook. ThomsonNOW, a powerful and fully integrated teaching and learning system, provides instructors and students with unsurpassed control, variety, and all-in-one utility. ThomsonNOW ties together the fundamental learning activities: diagnostics, tutorials, homework, personalized study, quizzing, and testing. Personalized Study is a learning companion that helps students gauge their unique study needs and makes the most of their study time by building focused personalized learning plans that reinforce key concepts. Pre-Tests give students an initial assessment of their knowledge. Personalized study plans, based on the students' answers to the Pre-Test questions, outline key elements for review. Post-Tests assess student mastery of core chapter concepts. Results can even be e-mailed to the instructor!

PRINTED ACCESS CARD, THOMSONNOW™ (0-495-39457-2)

This printed access card provides entrance to all the content that accompanies Tussy/Gustafson's *Elementary Algebra, Fourth Edition*, within ThomsonNOW.

WEBSITE *www.thomsonedu.com/math/tussy*
Visit us on the web for access to a wealth of free learning resources, including tutorials, final exams, chapter outlines, chapter reviews, web links, videos, flashcards, and more!

ACKNOWLEDGMENTS

We want to express our gratitude to Steve Odrich, Maria H. Andersen, Diane Koenig, Alexander Lee, Ed Kavanaugh, Karl Hunsicker, Cathy Gong, Dave Ryba, Terry Damron, Marion Hammond, Lin Humphrey, Doug Keebaugh, Robin Carter, Tanja Rinkel, Bob Billups, Jeff Cleveland, Jo Morrison, Sheila White, Jim McClain, Paul Swatzel, and the Citrus College Library staff (including Barbara Rugeley) for their help with this project. Your encouragement, suggestions, and insight have been invaluable to us.

We would also like to express our thanks to the Brooks/Cole editorial, marketing, production and design staff for helping us craft this new edition: Charlie Van Wagner, Danielle Derbenti, Greta Kleinert, Laura Localio, Lynh Pham, Cassandra Cummings, Donna Kelley, Sam Subity, Cheryll Linthicum, Vernon Boes, and Graphic World.

Additionally, we would like to say that authoring a textbook is a tremendous undertaking. A revision of this scale would not have been possible without the thoughtful feedback and support from the following colleagues listed below. Their contributions to this edition have shaped this revision in countless ways.

Alan S. Tussy
R. David Gustafson

Advisory Board

Kim Caldwell, Volunteer State Community College
Peter Embalabala, Lincoln Land Community College
John Garlow, Tarrant Community College–Southeast Campus
Becki Huffman, Tyler Junior College
Mary Legner, Riverside Community College
Ann Loving, J. Sargeant Reynolds Community College
Trudy Meyer, El Camino College
Carol Ann Poore, Hinds Community College
Jill Rafael, Sierra College
Pamelyn Reed, Cy-Fair College
Patty Sheeran, McHenry Community College
Valerie Wright, Central Piedmont Community College
Loris Zucca, Kingwood College

Reviewers

Maria Andersen, Muskegon Community College
Scott Barnett, Henry Ford Community College
David Behrman, Somerset Community College
Jeanne Bowman, University of Cincinnati
Carol Cheshire, Macon State College
Suzanne Doviak, Old Dominion University
Peter Embalabala, Lincoln Land Community College
Joan Evans, Texas Southern University
Rita Fielder, University of Central Arkansas
Anissa Florence, Jefferson Community and Technical College
Pat Foard, South Plains College
Tom Fox, Cleveland State Community College
Heng Fu, Thomas Nelson Community College
Kim Gregor, Delaware Technical Community College–Wilmington
Haile Kebede Haile, Minneapolis Community and Technical College
Jennifer Hastings, Northeast Mississippi Community College
Kristy Hill, Hinds Community College
Laura Hoye, Trident Technical College
Becki Huffman, Tyler Junior College
Angela Jahns, North Idaho College
Cynthia Johnson, Heartland Community College

Ann Loving, J. Sargeant Reynolds Community College
Lynette King, Gadsden State Community College
Mike Kirby, Tidewater Community College
Mary Legner, Riverside Community College
Wayne (Paul) Lee, Saint Philip's College
Yixia Lu, South Suburban College
Keith Luoma, Augusta State University
Susan Meshulam, Indiana University/Purdue University Indianapolis
Trudy Meyer, El Camino College
Molly Misko, Gadsden State Community College
Elsie Newman, Owens Community College
Charlotte Newsom, Tidewater Community College
Randy Nichols, Delta College
Stephen Nicoloff, Paradise Valley Community College
Charles Odion, Houston Community College
Jason Pallett, Longview Community College
Mary Beth Pattengale, Sierra College
Naeemah Payne, Los Angeles Community College
Carol Ann Poore, Hinds Community College
Jill Rafael, Sierra College
Pamela Reed, North Harris Montgomery Community College
Nancy Ressler, Oakton Community College
Emma Sargent, Tennessee State University
Ned Schillow, Lehigh Carbon Community College
Debra Shafer, University of North Carolina
Hazel Shedd, Hinds Community College
Donald Solomon, University of Wisconsin
John Squires, Cleveland State Community College
Robin Steinberg, Pima Community College
Eden Thompson, Utah Valley State College
Carol Walker, Hinds Community College
Diane Williams, Northern Kentucky University
Loris Zucca, Kingwood College

Class Testers

Candace Blazek, Anoka Ramsey Community College
Jennifer Bluth, Anoka Ramsey Community College
Vicki Gearhart, San Antonio College
Megan Goodwin, Anoka Ramsey Community College
Haile Haile, Minneapolis Community and Technical College
Vera Hu-Hyneman, SUNY–Suffolk Community College
Marlene Kutesky, Virginia Commonwealth University
Richard Leedy, Polk Community College
Wendiann Sethi, Seton Hall University
Eleanor Storey, Frontrange Community College
Cindy Thore, Central Piedmont Community College
Gowribalan "Ana" Vamadeva, University of Cincinnati
Cynthia Wallin, Central Virginia Community College
John Ward, Jefferson Community and Technical College

Focus Groups

Khadija Ahmed, Monroe Community College
Maria Andersen, Muskegon Community College
Chad Bemis, Riverside Community College
A. Elena Bogardus, Camden Community College
Carilynn Bouie, Cuyahoga Community College
Kim Brown, Tarrant Community College
Carole Carney, Brookdale Community College
Joe Castillo, Broward Community College
John Close, Salt Lake Community College
Chris Copple, Northwest State Community College
Mary Deas, Johnson County Community College
Maggie Flint, Northeast State

Douglas Furman, SUNY Ulster Community College
Abel Gage, Skagit Valley College
Amy Hoherz, Johnson County Community College
Pete Johnson, Eastern Connecticut State University
Ed Kavanaugh, Schoolcraft College
Leonid Khazanov, Borough of Manhattan Community College
MC Kim, Suffolk County Community College
Fred Lang, Art Institute of Washington
Hoat Le, San Diego Community College
Richard Leedy, Polk Community College
Daniel Lopez, Brookdale Community College
Ann Loving, J. Sargeant Reynolds Community College
Charles Odion, Houston Community College
Maggie Pasqua Viz, Brookdale Community College
Fred Peskoff, Borough of Manhattan Community College
Sheila Pisa, Riverside Community College–Moreno Valley
Jill Rafael, Sierra College
Christa Solheid, Santa Ana College
Jim Spencer, Santa Rosa Junior College
Teresa Sutcliffe, Los Angeles Valley College
Rose Toering, Kilian Community College
Judith Wood, Central Florida Community College
Mary Young, Brookdale Community College

Workshops

Andrea Adlman, Ventura College
Rodney Alford, Calhoun Community College
Maria Andersen, Muskegon Community College
Hamid Attarzadeh, Jefferson Community and Technical College
Victoria Baker, University of Houston–Downtown
Betty Barks, Lansing Community College
Susan Beane, University of Houston–Downtown
Barbara Blass, Oakland Community College
Charles A. Bower, St. Philip's College
Tony Craig, Paradise Valley Community College
Patrick Cross, University of Oklahoma
Archie Earl, Norfolk State University
Melody Eldred, State University of New York at Cobleskill
Joan Evans, Texas Southern University
Mike Everett, Santa Ana College
Betsy Farber, Bucks County Community College
Nancy Forrest, Grand Rapids Community College
Radu Georgescu, Prince George's Community College
Rebecca Giles, Jefferson State Community College
Thomas Grogan, Cincinnati State
Paula Jean Haigis, Calhoun Community College
Haile Haile, Minneapolis Community and Technical College
Kelli Jade Hammer, Broward Community College
Julia Hassett, Oakton Community College
Alan Hayashi, Oxnard College
Joel Helms, University of Cincinnati
Jim Hodge, Mountain State University
Jeffrey Hughes, Hinds Community College
Leslie Johnson, John C. Calhoun State Community College
Cassandra Johnson, Robeson Community College
Ed Kavanaugh, Schoolcraft College
Alex Kolesnik, Ventura College
Marlene Kustesky, Virginia Commonwealth University
Lider-Manuel Lamar, Seminole Community College
Roger Larson, Anoka Ramsey Community College
Alexander Lee, Hinds Community College, Rankin Campus
Richard Leedy, Polk Community College
Marcus McGuff, Austin Community College
Owen Mertens, Missouri State University

James Metz, Kapi'olani Community College
Pam Miller, Phoenix College
Tania Munding, Ohlone College
Charlie Naffziger, Central Oregon Community College
Oscar Neal, Grand Rapids Community College
Doug Nelson, Central Oregon Community College
Katrina Nichols, Delta College
Megan Nielsen, St. Cloud State University
Nancy Ressler, Oakton Community College
Elaine Richards, Eastern Michigan University
Harriette Roadman, New River Community College
Lilia Ruvalcaba, Oxnard College
Wendiann Sethi, Seton Hall University
Karen Smith, Nicholls State University
Donald Solomon, University of Wisconsin–Milwaukee
Frankie Solomon, University of Houston–Downtown
Michael Stack, South Suburban College
Kristen Starkey, Rose State College
Kristin Stoley, Blinn College
Eleanor Storey, Front Range Community College–Westminster Campus
Fariheh Towfiq, Palomar College
Gowribalan Vamadeva, University of Cincinnati
Beverly Vredevelt, Spokane Falls Community College
Andreana Walker, Calhoun Community College
Cynthia Wallin, Central Virginia Community College
John Ward, Kentucky Community and Technical College–Jefferson Community College
Richard Watkins, Tidewater Comunity College
Antoinette Willis, St. Philip's College
Nazar Wright, Guilford Technical Community College
Shishen Xie, University of Houston–Downtown
Catalina Yang, Oxnard College
Heidi Young, Bryant and Stratton College
Ghidei Zedingle, Normandale Community College

APPLICATIONS INDEX

Examples that are applications are shown with boldface page numbers.
Exercises that are applications are shown with lightface page numbers.

CHAPTER 1

An Introduction to Algebra

© AP/Wide World Photo

from Campus to Careers

Lead Transportation Security Officer

Since 9/11, Homeland Security is one of the fastest-growing career choices in the United States. A lead transportation security officer works in an airport where he or she searches passengers, screens baggage, reviews tickets, and determines staffing requirements. The job description calls for the ability to perform arithmetic computations correctly and solve practical problems by choosing from a variety of mathematical techniques such as formulas and percentages.

In **Problem 93** of **Study Set 1.5,** we will make some arithmetic computations using data from two of the busiest airports in the United States, Orlando International and New York La Guardia.

JOB TITLE:
Lead Transportation Security Officer
EDUCATION:
High school diploma or GED, some college helpful
JOB OUTLOOK:
Excellent in many locations
ANNUAL EARNINGS:
$31,100 to $46,700
FOR MORE INFORMATION:
www.tsa.gov

Study Skills Workshop

Committing to the Course

Starting a new course is exciting, but it might also make you a bit nervous. In order to be successful in your algebra class, you need a plan.

MAKE TIME FOR THE COURSE: As a general guideline, 2 hours of independent study time is recommended for every hour in the classroom.

KNOW WHAT IS EXPECTED: Read your instructor's syllabus thoroughly. It lists class policies about attendance, homework, tests, calculators, grading, and so on.

BUILD A SUPPORT SYSTEM: Know where to go for help. Take advantage of your instructor's office hours, your school's tutorial services, the resources that accompany this textbook, and the assistance that you can get from classmates.

Now Try This

Each of the forms referred to below can be found online at: http://www.thomsonedu.com/math/tussy.

1. To help organize your schedule, fill out the *Weekly Planner Form.*
2. Review the class policies by completing the *Course Information Sheet.*
3. Use the *Support System Worksheet* to build your course support system.

SECTION 1.1
Introducing the Language of Algebra

Objectives

1. Read tables and graphs.
2. Use the basic vocabulary and notation of algebra.
3. Identify expressions and equations.
4. Use equations to construct tables of data.

Algebra is the result of contributions from many cultures over thousands of years. The word *algebra* comes from the title of the book *Ihm Al-jabr wa'l muqābalah,* written by an Arabian mathematician around A.D. 800. Using the vocabulary and notation of algebra, we can mathematically **model** many situations in the real world. In this section, we begin to explore the language of algebra by introducing some of its basic components.

1 Read Tables and Graphs.

In algebra, we use tables to show relationships between quantities. For example, the following table lists the number of bicycle tires a production planner must order when a given number of bicycles is to be manufactured. For a production run of, say, 300 bikes, we locate 300 in the left column and then scan across the table to see that the company must order 600 tires.

© Andersen Ross/Getty Images

Bicycles to be manufactured	Tires to order
100	200
200	400
300	600
400	800

The Language of Algebra
Horizontal is a form of the word *horizon*. Think of the sun setting over the *horizon*. *Vertical* means in an upright position. Pro basketball player LeBron James' *vertical* leap measures more than 40 inches.

The information in the table can also be presented in a **bar graph,** as shown below. The **horizontal axis,** labeled "Number of bicycles to be manufactured," is scaled in units of 100 bicycles. The **vertical axis,** labeled "Number of tires to order," is scaled in units of 100 tires. The height of a bar indicates the number of tires to order. For example, if 200 bikes are to be manufactured, we see that the bar extends to 400, meaning 400 tires are needed.

Another way to present this information is with a **line graph.** Instead of using a bar to represent the number of tires to order, we use a dot drawn at the correct height. After drawing the data points for 100, 200, 300, and 400 bicycles, we connect them with line segments to create the following graph, on the right.

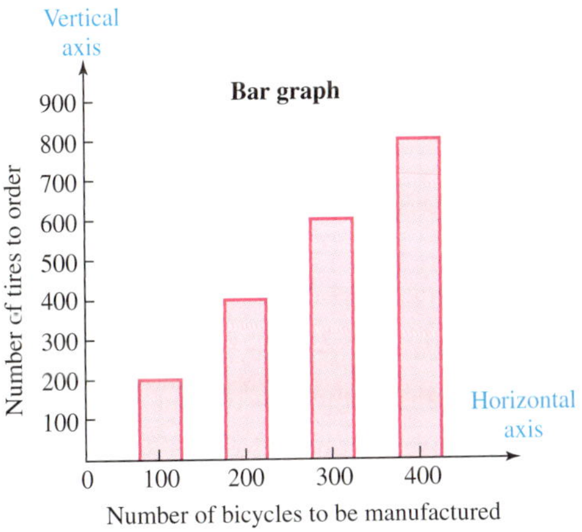

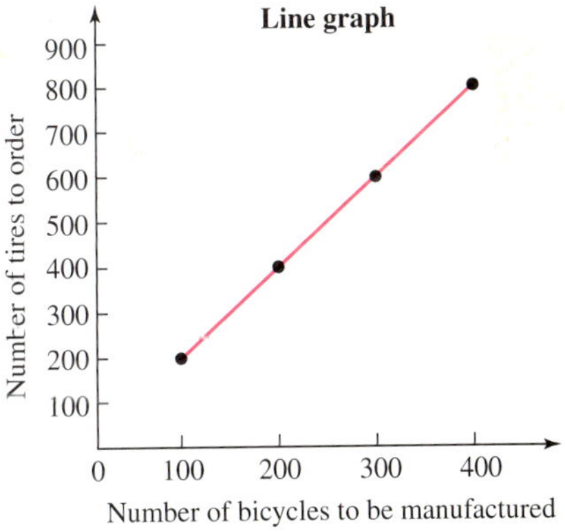

EXAMPLE 1 Use the line graph to find the number of tires needed if 250 bicycles are to be manufactured.

Strategy Since we know the number of bicycles to be manufactured, we will begin on the horizontal axis of the graph and scan up and over to read the answer on the vertical axis.

Why We scan up and over because the number of tires is given by the scale on the vertical axis.

Solution We locate 250 between 200 and 300 on the horizontal axis and draw a dashed line upward to intersect the graph. From the point of intersection, we draw a dashed horizontal line to the left that intersects the vertical axis at 500. This means that 500 tires should be ordered if 250 bicycles are to be manufactured.

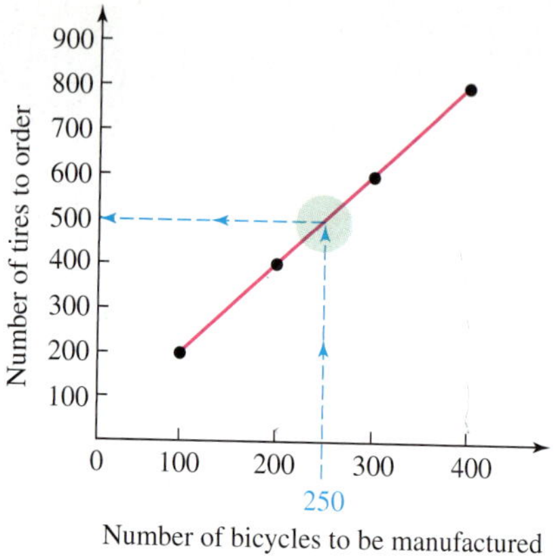

Success Tip
Answers to the Self Check problems are given at the end of each section, before each Study Set.

Self Check 1 Use the graph to find the number of tires needed if 350 bicycles are to be manufactured.

Now Try **Problem 31**

2 Use the Basic Vocabulary and Notation of Algebra.

From the table and graphs, we see that there is a relationship between the number of tires to order and the number of bicycles to be manufactured. Using words, we can express this relationship as a **verbal model:**

"The number of tires to order is two times the number of bicycles to be manufactured."

Since the word **product** indicates the result of a multiplication, we can write:

"The number of tires to order is the *product* of 2 and the number of bicycles to be manufactured."

To indicate other arithmetic operations, we will use the following words.

- A **sum** is the result of an addition: the sum of 5 and 6 is 11.
- A **difference** is the result of a subtraction: the difference of 3 and 2 is 1.
- A **quotient** is the result of a division: the quotient of 6 and 3 is 2.

Many symbols used in arithmetic are also used in algebra. For example, a $+$ symbol is used to indicate addition, a $-$ symbol is used to indicate subtraction, and an $=$ symbol means *is equal to.*

Since the letter x is often used in algebra and could be confused with the multiplication symbol $\times$, we usually write multiplication using a **raised dot** or **parentheses.**

Symbols for Multiplication

$\times$	Times symbol	$6 \times 4 = 24$
$\cdot$	Raised dot	$6 \cdot 4 = 24$
( )	Parentheses	$(6)4 = 24$ or $6(4) = 24$ or $(6)(4) = 24$

In algebra, the symbol most often used to indicate division is the *fraction bar.*

Symbols for Division

$\div$	Division symbol	$24 \div 4 = 6$
$\overline{)\ }$	Long division	$4\overline{)24}$ with quotient 6
—	Fraction bar	$\frac{24}{4} = 6$

EXAMPLE 2 Write each statement in words, using one of the words *sum, product, difference,* or *quotient:* **a.** $\frac{22}{11} = 2$ **b.** $22 + 11 = 33$

Strategy We will examine each statement to determine whether addition, subtraction, multiplication, or division is being performed.

Why The word that we should use (*sum, product, difference,* or *quotient*) depends on the arithmetic operation that we have to describe.

Solution

a. Since the fraction bar indicates division, we have: The quotient of 22 and 11 equals 2.

b. The symbol indicates addition: The sum of 22 and 11 equals 33.

Self Check 2 Write the following statement in words: $22 - 10 = 12$

Now Try **Problems 33 and 35**

3 Identify Expressions and Equations.

Another way to describe the tires–to–bicycles relationship uses *variables.* **Variables** are letters (or symbols) that stand for numbers. If we let the letter b represent the number of bicycles to be manufactured, then the number of tires to order is two times b, written $2b$. In the notation, the number 2 is an example of a **constant** because it does not change value.

When multiplying a variable by a number, or a variable by another variable, we can omit the symbol for multiplication. For example,

$$2b \text{ means } 2 \cdot b \qquad xy \text{ means } x \cdot y \qquad 8abc \text{ means } 8 \cdot a \cdot b \cdot c$$

We call $2b$, xy, and $8abc$ *algebraic expressions.*

The Language of Algebra Since the number of bicycles to be manufactured can *vary,* or change, it is represented using a *variable.*

Algebraic Expressions

Variables and/or numbers can be combined with the operations of addition, subtraction, multiplication, and division to create **algebraic expressions.**

Here are some other examples of algebraic expressions.

$4a + 7$ — This expression is a combination of the numbers 4 and 7, the variable a, and the operations of multiplication and addition.

$\frac{10 - y}{3}$ — This expression is a combination of the numbers 10 and 3, the variable y, and the operations of subtraction and division.

$15mn(2m)$ — This expression is a combination of the numbers 15 and 2, the variables m and n, and the operation of multiplication.

The Language of Algebra We often refer to *algebraic expressions* as simply *expressions.*

The Language of Algebra
The equal symbol = can be represented by verbs such as:

is *are* *gives* *yields*

The symbol ≠ is read as *"is not equal to."*

In the bicycle manufacturing example, if we let the letter t stand for the number of tires to order, we can translate the **verbal model** to mathematical symbols.

The number of tires to order	is	two	times	the number of bicycles to be manufactured.
t	$=$	2	$\cdot$	b

The statement $t = 2 \cdot b$, or more simply, $t = 2b$, is called an *equation.* An **equation** is a mathematical sentence that contains an = symbol. The = symbol indicates that the expressions on either side of it have the same value. Other examples of equations are

$3 + 5 = 8 \qquad x + 5 = 20 \qquad 17 - 2r = 14 + 3r \qquad p = 100 - d$

EXAMPLE 3 Translate the verbal model into an equation.

The number of decades	is	the number of years	divided by	10.

Strategy We will represent the unknown quantities using variables and we will use symbols to represent the words *is* and *divided by.*

Why To translate a verbal (word) model into an equation means to write it using mathematical symbols.

Solution We can represent the two unknown quantities using variables: Let d = the number of decades and y = the number of years. Then we have:

The number of decades	is	the number of years	divided by	10.
d	$=$	y	$\div$	10

If we write the division using a fraction bar, then the verbal model translates to the equation $d = \frac{y}{10}$.

Self Check 3 Translate into an equation: The number of unsold tickets is the difference of 500 and the number of tickets that have been purchased.

Now Try **Problems 41 and 45**

In the bicycle manufacturing example, using the equation $t = 2b$ to describe the relationship has one major advantage over the other methods. It can be used to determine the number of tires needed for a production run of any size.

EXAMPLE 4 Use the equation $t = 2b$ to find the number of tires needed for a production run of 178 bicycles.

Strategy In $t = 2b$, we will replace b with 178. Then we will multiply 178 by 2 to obtain the value of t.

Why The equation $t = 2b$ indicates that the number of tires is found by multiplying the number of bicycles by 2.

Solution

$t = 2b$ — This is the describing equation.

$t = 2(178)$ — Replace b, which stands for the number of bicycles, with 178. Use parentheses to show the multiplication. We could also write $2 \cdot 178$.

$t = 356$ — Multiply.

If 178 bicycles are manufactured, 356 tires will be needed.

Self Check 4 Use the equation $t = 2b$ to find the number of tires needed if 604 bicycles are to be manufactured.

Now Try **Problem 53**

4 Use Equations to Construct Tables of Data.

Equations such as $t = 2b$, which express a relationship between two or more variables, are called **formulas.** Some applications require the repeated use of a formula.

EXAMPLE 5 Find the number of tires to order for production runs of 233 and 852 bicycles. Present the results in a table.

Strategy We will use the equation $t = 2b$ twice.

Why There are two different-sized production runs: one of 233 bikes and another of 852 bikes.

Solution

Step 1: We construct a two-column table. Since b represents the number of bicycles to be manufactured, we use it as the heading of the first column. Since t represents the number of tires needed, we use it as the heading of the second column. Then we enter the size of each production run in the first column, as shown.

The Language of Algebra
To *substitute* means to put or use in place of another, as with a *substitute* teacher. Here, we *substitute* 233 and 852 for b.

Bicycles to be manufactured b	Tires to order t
233	466
852	1,704

Step 2: We **substitute** 233 and 852 for b in $t = 2b$ and find each corresponding value of t. The results are entered in the second column.

$$t = 2b \qquad t = 2b$$
$$t = 2(233) \qquad t = 2(852)$$
$$t = 466 \qquad t = 1{,}704$$

Self Check 5 Find the number of tires needed for production runs of 87 and 487 bicycles. Present the results in a table.

Now Try **Problem 55**

Success Tip
Answers to the odd-numbered problems in each Study Set can be found at the back of the book in Appendix 3, beginning on page A-7.

ANSWERS TO SELF CHECKS **1.** 700 **2.** The difference of 22 and 10 equals 12. **3.** $u = 500 - p$ **4.** 1,208 **5.**

b	t
87	174
487	974

STUDY SET 1.1

VOCABULARY

Fill in the blanks.

1. A _____ is the result of an addition. A __________ is the result of a subtraction. A _________ is the result of a multiplication. A _________ is the result of a division.

2. __________ are letters (or symbols) that stand for numbers.

3. A number, such as 8, is called a _________ because it does not change.

4. Variables and numbers can be combined with the operations of addition, subtraction, multiplication, and division to create algebraic ___________.

5. An _________ is a mathematical sentence that contains an = symbol.

6. An equation such as $t = 2b$, which expresses a relationship between two or more variables, is called a ________.

7. The __________ axis of a graph extends left and right and the vertical axis extends up and down.

8. The word ________ comes from the title of a book written by an Arabian mathematician around A.D. 800.

CONCEPTS

Classify each item as an algebraic expression or an equation.

9. a. $m + 18 = 23$ **b.** $m + 18$

10. a. $30x$ **b.** $30x = 600$

11. a. $\frac{c-7}{5}$ **b.** $\frac{c-7}{5} = 7c$

12. a. $r = \frac{2}{3}$ **b.** $\frac{2}{3}r$

13. What arithmetic operations does the expression $\frac{12 + 9t}{25}$ contain? What variable does it contain?

14. What arithmetic operations does the equation $4y - 14 = 5(6)$ contain? What variable does it contain?

15. Construct a line graph using the data in the following table.

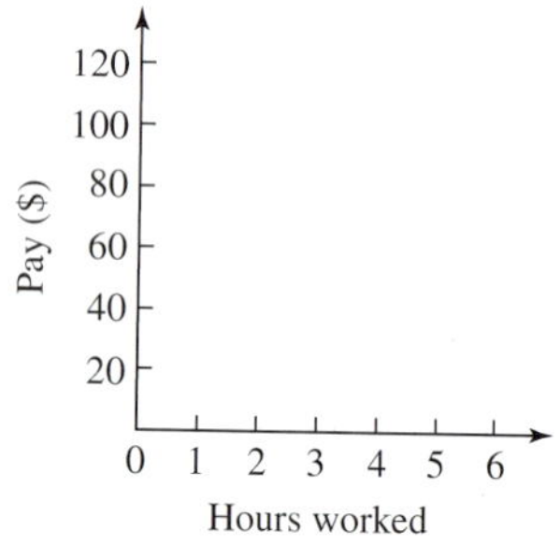

Hours worked	Pay (dollars)
1	20
2	40
3	60
4	80
5	100

16. Use the data in the graph to complete the table.

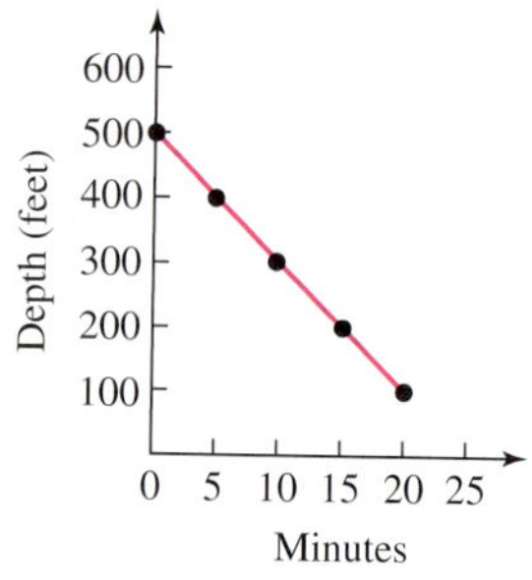

Minutes	Depth (feet)

NOTATION

Fill in the blanks.

17. The symbol ≠ means ___ ____ ______ ___.

18. The symbols ( ) are called ___________.

19. Write the multiplication 5×6 using a raised dot and then using parentheses.

20. Give four verbs that can be represented by an equal symbol =.

Write each expression without using a multiplication symbol or parentheses.

21. $4 \cdot x$ **22.** $P \cdot r \cdot t$

23. $2(w)$ **24.** $(x)(y)$

Write each division using a fraction bar.

25. $32 \div x$ **26.** $30\overline{)90}$

27. $5\overline{)55}$ **28.** $h \div 15$

GUIDED PRACTICE

***Use the line graph in Example 1 to find the number of tires needed to build the following number of bicycles.* See Example 1.**

29. 150

30. 400

31. Explain what the dashed lines help us find in the graph.

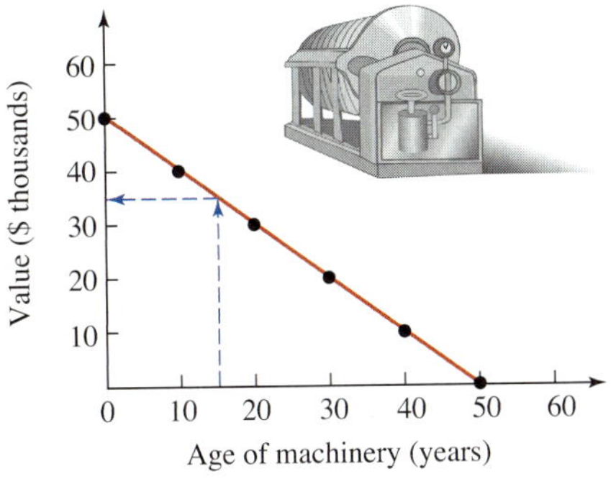

32. Use the line graph to find the income received from 30, 50, and 70 customers.

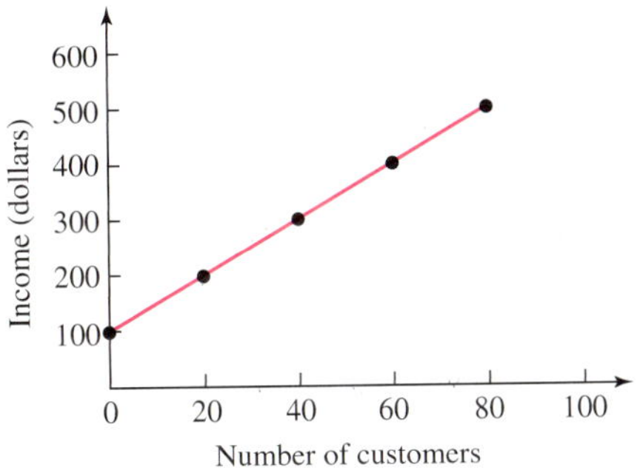

***Express each statement using one of the words* sum, product, difference, *or* quotient. See Example 2.**

33. $8(2) = 16$

34. $45 \cdot 12 = 540$

35. $11 - 9 = 2$

36. $65 + 89 = 154$

37. $x + 2 = 10$

38. $16 - t = 4$

39. $\frac{66}{11} = 6$

40. $12 \div 3 = 4$

***Translate each verbal model into an equation. (Answers may vary, depending on the variables chosen.)* See Example 3.**

41.

The sale price	is	$100	minus	the discount.

42.

The cost of dining out	equals	the cost of the meal	plus	$7 for parking.

43.

7	times	the age of a dog in years	gives	the dog's equivalent human age.

44.

The number of centuries	is	the number of years	divided by	100.

45. The amount of sand that should be used is the product of 3 and the amount of cement used.

46. The number of waiters needed is the quotient of the number of customers and 10.

47. The weight of the truck is the sum of the weight of the engine and 1,200.

48. The number of classes still open is the difference of 150 and the number of classes that are closed.

49. The profit is the difference of the revenue and 600.

50. The distance is the product of the rate and 3.

51. The quotient of the number of laps run and 4 gives the number of miles run.

52. The sum of the tax and 35 gives the total cost.

***Use the formula to complete each table.* See Examples 4 and 5.**

53. $d = 360 + L$

Lunch time (minutes) L	School day (minutes) d
30	
40	
45	

54. $b = 1{,}024k$

Kilobytes k	Bytes b
1	
5	
10	

55. $t = 1{,}500 - d$

Deductions d	Take-home pay t
200	
300	
400	

56. $w = \frac{s}{12}$

Inches of snow s	Inches of water w
12	
24	
72	

Use the data in the table to complete the formula.

57. $d = \frac{e}{\square}$

Eggs e	Dozens d
24	2
36	3
48	4

58. $p = \square c$

Canoes c	Paddles p
6	12
7	14
8	16

59. $I = \square c$

Couples c	Individuals I
20	40
100	200
200	400

60. $t = \frac{p}{\square}$

Players p	Teams t
5	1
10	2
15	3

APPLICATIONS

61. TRAFFIC SAFETY As the railroad crossing guard drops, the measure of angle 1 (written ∠1) increases while the measure of ∠2 decreases. At any instant the *sum* of the measures of the two angles is 90°. Complete the table. Then use the data to construct a line graph. Scale each axis in units of 15°.

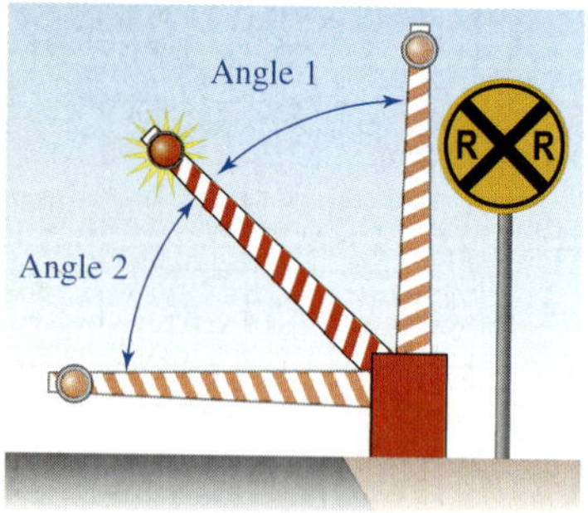

Angle 1 (degrees)	Angle 2 (degrees)
0	
15	
30	
45	
60	
75	
90	

62. U.S. CRIME STATISTICS Property crimes include burglary, theft, and motor vehicle theft. Graph the following property crime rate data using a bar graph. Scale the vertical axis in units of 50.

Year	Crimes per 1,000 households
1992	325
1994	310
1996	266
1998	217

Year	Crimes per 1,000 households
2000	178
2002	159
2004	161

Source: Bureau of Justice Statistics.

WRITING

63. Many students misuse the word *equation* when discussing mathematics. What is an equation? Give an example.

64. Explain the difference between an algebraic expression and an equation. Give an example of each.

65. In this section, four methods for describing numerical relationships were discussed: tables, words, graphs, and equations. Which method do you think is the most useful? Explain why.

66. In your own words, define *horizontal* and *vertical.*

CHALLENGE PROBLEMS

67. Complete the table and the formula.

$t = \square$

s	t
10	
18	19
	34
47	48

68. Suppose $h = 4n$ and $n = 2g$. Complete the following formula: $h = \square g$.

SECTION 1.2
Fractions

Objectives

1. Factor and prime factor natural numbers.
2. Recognize special fraction forms.
3. Multiply and divide fractions.
4. Build equivalent fractions.
5. Simplify fractions.
6. Add and subtract fractions.
7. Simplify answers.
8. Compute with mixed numbers.

In arithmetic, we add, subtract, multiply, and divide **natural numbers:** 1, 2, 3, 4, 5, and so on. Assuming that you have mastered those skills, we will now review the arithmetic of fractions.

1 Factor and Prime Factor Natural Numbers.

To compute with fractions, we need to know how to *factor* natural numbers. To **factor** a number means to express it as a product of two or more numbers. For example, some ways to factor 8 are

$$1 \cdot 8, \qquad 4 \cdot 2, \qquad \text{and} \qquad 2 \cdot 2 \cdot 2$$

The Language of Algebra
When we say "factor 8," we are using the word *factor* as a verb. When we say "2 is a *factor* of 8," we are using the word *factor* as a noun.

The numbers 1, 2, 4, and 8 that were used to write the products are called *factors* of 8. In general, a **factor** is a number being multiplied.

Sometimes a number has only two factors, itself and 1. We call such numbers *prime numbers.*

Prime Numbers and Composite Numbers

A **prime number** is a natural number greater than 1 that has only itself and 1 as factors. The first ten prime numbers are 2, 3, 5, 7, 11, 13, 17, 19, 23, and 29.

A **composite number** is a natural number, greater than 1, that is not prime. The first ten composite numbers are 4, 6, 8, 9, 10, 12, 14, 15, 16, and 18.

Every composite number can be factored into the product of two or more prime numbers. This product of these prime numbers is called its **prime factorization.**

EXAMPLE 1 Find the prime factorization of 210.

Strategy We will use a series of steps to express 210 as a product of only prime numbers.

Why To *prime factor* a number means to write it as a product of prime numbers.

Solution First, write 210 as the product of two natural numbers other than 1.

The Language of Algebra
Prime factors are often written in *ascending* order. To *ascend* means to move upward.

$210 = 10 \cdot 21$ The resulting prime factorization will be the same no matter which two factors of 210 you begin with.

Neither 10 nor 21 are prime numbers, so we factor each of them.

$$210 = 2 \cdot 5 \cdot 3 \cdot 7$$ Factor 10 as $2 \cdot 5$ and factor 21 as $3 \cdot 7$.

Writing the factors in ascending order, the **prime-factored form** of 210 is $2 \cdot 3 \cdot 5 \cdot 7$. Two other methods for prime factoring 210 are shown below.

> ### Success Tip
> A whole number is divisible by
> - 2 if it ends in 0, 2, 4, 6, or 8
> - 3 if the sum of the digits is divisible by 3
> - 5 if it ends in 0 or 5
> - 10 if it ends in 0

Work downward. Factor each number as a product of two numbers (other than 1 and itself) until all factors are prime. Circle prime numbers as they appear at the end of a branch. ↓

Factor tree

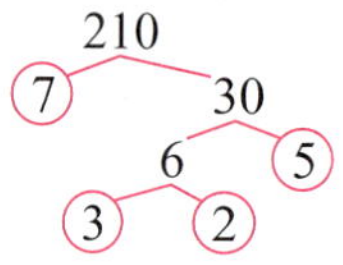

Division ladder

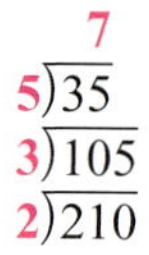

↑ Work upward. Perform repeated division until the final quotient is a prime number. It is helpful to start with the smallest prime, 2, as a trial divisor. Then, in order, try larger primes as divisors: 3, 5, 7, 11, and so on.

Either way, the factorization is $2 \cdot 3 \cdot 5 \cdot 7$. To check it, multiply the prime factors. The product should be 210.

Self Check 1 Find the prime factorization of 189.

Now Try **Problem 15**

2 Recognize Special Fraction Forms.

In a fraction, the number above the **fraction bar** is called the **numerator,** and the number below is called the **denominator.**

> ### The Language of Algebra
> The word *fraction* comes from the Latin word *fractio* meaning "breaking in pieces."

Fraction bar ⟶ $\frac{5}{6}$ ⟵ numerator / ⟵ denominator

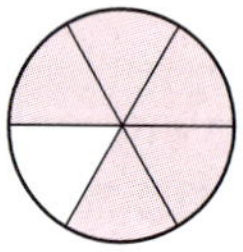

Fractions can describe the number of equal parts of a whole. For example, consider the circle with 5 of 6 equal parts colored red. We say that $\frac{5}{6}$ (five-sixths) of the circle is shaded.

Fractions are also used to indicate division. For example, $\frac{8}{2}$ indicates that the numerator, 8, is to be divided by the denominator, 2:

$$\frac{8}{2} = 8 \div 2 = 4$$ We know that $\frac{8}{2} = 4$ because of its related multiplication statement: $2 \cdot 4 = 8$.

If the numerator and denominator of a fraction are the same nonzero number, the fraction indicates division of a number by itself, and the result is 1. Each of the following fractions is, therefore, a **form of 1.**

$$1 = \frac{1}{1} = \frac{2}{2} = \frac{3}{3} = \frac{4}{4} = \frac{5}{5} = \frac{6}{6} = \frac{7}{7} = \frac{8}{8} = \frac{9}{9} = \ldots$$

If a denominator is 1, the fraction indicates division by 1, and the result is simply the numerator. For example, $\frac{5}{1} = 5$ and $\frac{24}{1} = 24$.

Special Fraction Forms

For any nonzero number a,

$$\frac{a}{a} = 1 \quad \text{and} \quad \frac{a}{1} = a$$

3 Multiply and Divide Fractions.

The rule for multiplying fractions can be expressed in words and in symbols as follows.

Multiplying Fractions

To multiply two fractions, multiply the numerators and multiply the denominators.

For any two fractions $\frac{a}{b}$ and $\frac{c}{d}$,

$$\frac{a}{b} \cdot \frac{c}{d} = \frac{a \cdot c}{b \cdot d}$$

EXAMPLE 2 Multiply: $\frac{7}{8} \cdot \frac{3}{5}$

Strategy To find the product, we will multiply the numerators, 7 and 3, and multiply the denominators, 8 and 5.

Why This is the rule for multiplying two fractions.

Solution

$$\frac{7}{8} \cdot \frac{3}{5} = \frac{7 \cdot 3}{8 \cdot 5}$$ Multiply the numerators. Multiply the denominators.

$$= \frac{21}{40}$$

Self Check 2 Multiply: $\frac{5}{9} \cdot \frac{2}{3}$

Now Try **Problem 27**

One number is called the **reciprocal** of another if their product is 1. To find the reciprocal of a fraction, we invert its numerator and denominator.

$\frac{3}{4}$ is the reciprocal of $\frac{4}{3}$, because $\frac{3}{4} \cdot \frac{4}{3} = \frac{12}{12} = 1$.

$\frac{1}{10}$ is the reciprocal of 10, because $\frac{1}{10} \cdot 10 = \frac{10}{10} = 1$.

We use reciprocals to divide fractions.

Success Tip

Every number, except 0, has a reciprocal. Zero has no reciprocal, because the product of 0 and a number cannot be 1.

Dividing Fractions

To divide two fractions, multiply the first fraction by the reciprocal of the second.

For any two fractions $\frac{a}{b}$ and $\frac{c}{d}$, where $c \neq 0$,

$$\frac{a}{b} \div \frac{c}{d} = \frac{a}{b} \cdot \frac{d}{c}$$

EXAMPLE 3 Divide: $\frac{1}{3} \div \frac{4}{5}$

Strategy We will multiply the first fraction, $\frac{1}{3}$, by the reciprocal of the second fraction, $\frac{4}{5}$.

Why This is the rule for dividing two fractions.

Solution

$$\frac{1}{3} \div \frac{4}{5} = \frac{1}{3} \cdot \frac{5}{4}$$ Multiply $\frac{1}{3}$ by the reciprocal of $\frac{4}{5}$. The reciprocal of $\frac{4}{5}$ is $\frac{5}{4}$.

$$= \frac{1 \cdot 5}{3 \cdot 4}$$ Multiply the numerators. Multiply the denominators.

$$= \frac{5}{12}$$

Self Check 3 Divide: $\frac{6}{25} \div \frac{1}{2}$

Now Try **Problem 31**

4 Build Equivalent Fractions.

The two rectangles on the right are the same size. The first rectangle is divided into 10 equal parts. Since 6 of those parts are red, $\frac{6}{10}$ of the figure is shaded.

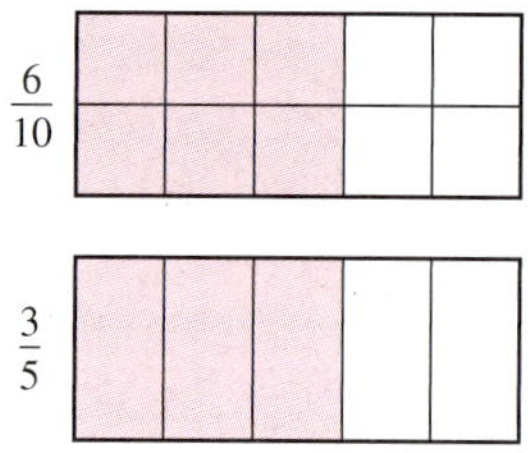

The second rectangle is divided into 5 equal parts. Since 3 of those parts are red, $\frac{3}{5}$ of the figure is shaded. We can conclude that $\frac{6}{10} = \frac{3}{5}$ because $\frac{6}{10}$ and $\frac{3}{5}$ represent the same shaded part of the rectangle. We say that $\frac{6}{10}$ and $\frac{3}{5}$ are *equivalent fractions.*

Equivalent Fractions

Two fractions are **equivalent** if they represent the same number.

Writing a fraction as an equivalent fraction with a larger denominator is called **building** the fraction. To build a fraction, we multiply it by a form of 1. Since any number multiplied by 1 remains the same (identical), 1 is called the **multiplicative identity element.**

Multiplication Property of 1

The product of 1 and any number is that number.

For any number a,

$$1 \cdot a = a \quad \text{and} \quad a \cdot 1 = a$$

EXAMPLE 4 Write $\frac{3}{5}$ as an equivalent fraction with a denominator of 35.

Strategy We will compare the given denominator to the required denominator and ask, "By what must we multiply 5 to get 35?"

Success Tip
Multiplying $\frac{3}{5}$ by $\frac{7}{7}$ changes its appearance, but does not change its value, because we are multiplying it by a form of 1.

Why The answer to that question helps us determine the form of 1 to be used to build an equivalent fraction.

Solution We need to multiply the denominator of $\frac{3}{5}$ by 7 to obtain a denominator of 35. It follows that $\frac{7}{7}$ should be the form of 1 that is used to build $\frac{3}{5}$. Multiplying $\frac{3}{5}$ by $\frac{7}{7}$ changes its appearance but does not change its value, because we are multiplying it by 1.

$$\frac{3}{5} = \frac{3}{5} \cdot \frac{7}{7} \quad \frac{7}{7} = 1$$

$$= \frac{3 \cdot 7}{5 \cdot 7} \quad \text{Multiply the numerators. Multiply the denominators.}$$

$$= \frac{21}{35}$$

Self Check 4 Write $\frac{5}{8}$ as an equivalent fraction with a denominator of 24.

Now Try **Problem 35**

Building Fractions

To build a fraction, multiply it by 1 in the form of $\frac{c}{c}$, where c is any nonzero number.

5 Simplify Fractions.

Every fraction can be written in infinitely many equivalent forms. For example, some equivalent forms of $\frac{10}{15}$ are:

$$\frac{2}{3} = \frac{4}{6} = \frac{6}{9} = \frac{8}{12} = \frac{10}{15} = \frac{12}{18} = \frac{14}{21} = \frac{16}{24} = \frac{18}{27} = \frac{20}{30} = \ldots$$

The Language of Algebra
The word *infinitely* is a form of the word *infinite,* which means endless.

Of all of the equivalent forms in which we can write a fraction, we often need to determine the one that is in *simplest form.*

Simplest Form of a Fraction

A fraction is in **simplest form,** or **lowest terms,** when the numerator and denominator have no common factors other than 1.

To **simplify a fraction,** we write it in simplest form by removing a factor equal to 1. For example, to simplify $\frac{10}{15}$, we note that the greatest factor common to the numerator and denominator is 5 and proceed as follows:

$$\frac{10}{15} = \frac{2 \cdot 5}{3 \cdot 5} \quad \text{Factor 10 and 15.}$$

$$= \frac{2}{3} \cdot \frac{5}{5} \quad \text{Use the rule for multiplying fractions in reverse: write } \frac{2 \cdot 5}{3 \cdot 5} \text{ as the product of two fractions, } \frac{2}{3} \text{ and } \frac{5}{5}.$$

$$= \frac{2}{3} \cdot 1 \quad \text{A nonzero number divided by itself is equal to 1: } \frac{5}{5} = 1.$$

$$= \frac{2}{3} \quad \text{Use the multiplication property of 1: any number multiplied by 1 remains the same.}$$

To simplify $\frac{10}{15}$, we removed a factor equal to 1 in the form of $\frac{5}{5}$. The result, $\frac{2}{3}$, is equivalent to $\frac{10}{15}$.

We can easily identify the greatest common factor of the numerator and the denominator of a fraction if we write them in prime-factored form.

EXAMPLE 5 Simplify each fraction, if possible: **a.** $\frac{63}{42}$ **b.** $\frac{33}{40}$

Strategy We will begin by prime factoring the numerator and denominator of the fraction. Then, to simplify it, we will remove a factor equal to 1.

Why We need to make sure that the numerator and denominator have no common factors other than 1. If that is the case, then the fraction is in *simplest form.*

Solution

a. After prime factoring 63 and 42, we see that the greatest common factor of the numerator and the denominator is $3 \cdot 7 = 21$.

$$\frac{63}{42} = \frac{3 \cdot 3 \cdot 7}{2 \cdot 3 \cdot 7}$$ Write 63 and 42 in prime-factored form.

$$= \frac{3}{2} \cdot \frac{3 \cdot 7}{3 \cdot 7}$$ Write $\frac{3 \cdot 3 \cdot 7}{2 \cdot 3 \cdot 7}$ as the product of two fractions, $\frac{3}{2}$ and $\frac{3 \cdot 7}{3 \cdot 7}$.

$$= \frac{3}{2} \cdot 1$$ A nonzero number divided by itself is equal to 1: $\frac{3 \cdot 7}{3 \cdot 7} = 1$.

$$= \frac{3}{2}$$ Any number multiplied by 1 remains the same.

b. Prime factor 33 and 40.

$$\frac{33}{40} = \frac{3 \cdot 11}{2 \cdot 2 \cdot 2 \cdot 5}$$

Since the numerator and the denominator have no common factors other than 1, the fraction $\frac{33}{40}$ is in simplest form (lowest terms).

The Language of Algebra What do Calvin Klein, Queen Latifah, and Tom Hanks have in common? They all attended a community college. The word *common* means shared by two or more. In this section, we will work with *common* factors and *common* denominators.

Self Check 5 Simplify each fraction, if possible:
a. $\frac{24}{56}$ **b.** $\frac{16}{125}$

Now Try **Problem 45**

To streamline the simplifying process, we can replace pairs of factors common to the numerator and denominator with the equivalent fraction $\frac{1}{1}$.

EXAMPLE 6 Simplify: $\frac{90}{105}$

Strategy We will begin by prime factoring the numerator, 90, and denominator, 105. Then we will look for any factors common to the numerator and denominator and remove them.

Why When the numerator and/or denominator of a fraction are large numbers, such as 90 and 105, writing their prime factorizations is helpful in identifying any common factors.

Solution

$$\frac{90}{105} = \frac{2 \cdot 3 \cdot 3 \cdot 5}{3 \cdot 5 \cdot 7}$$ Write 90 and 105 in prime-factored form.

$$= \frac{2 \cdot \overset{1}{\cancel{3}} \cdot 3 \cdot \overset{1}{\cancel{5}}}{\underset{1}{\cancel{3}} \cdot \underset{1}{\cancel{5}} \cdot 7}$$ Slashes and 1's are used to show that $\frac{3}{3}$ and $\frac{5}{5}$ are replaced by the equivalent fraction $\frac{1}{1}$. A factor equal to 1 in the form of $\frac{3 \cdot 5}{3 \cdot 5} = \frac{15}{15}$ was removed.

$$= \frac{6}{7}$$ Multiply the remaining factors in the numerator: $2 \cdot 1 \cdot 3 \cdot 1 = 6$. Multiply the remaining factors in the denominator: $1 \cdot 1 \cdot 7 = 7$.

Self Check 6 Simplify: $\frac{126}{70}$

Now Try Problem 53

We can use the following steps to simplify a fraction.

Simplifying Fractions

1. Factor (or prime factor) the numerator and denominator to determine their common factors.
2. Remove factors equal to 1 by replacing each pair of factors common to the numerator and denominator with the equivalent fraction $\frac{1}{1}$.
3. Multiply the remaining factors in the numerator and in the denominator.

The procedure for simplifying fractions is based on the following property.

The Fundamental Property of Fractions

If $\frac{a}{b}$ is a fraction and c is a nonzero real number,

$$\frac{ac}{bc} = \frac{a}{b}$$

Caution When all common factors of the numerator and/or the denominator of a fraction are removed, forgetting to write 1's above the slashes can lead to a common mistake.

Correct

$$\frac{15}{45} = \frac{\overset{1}{\cancel{3}} \cdot \overset{1}{\cancel{5}}}{\underset{1}{\cancel{3}} \cdot 3 \cdot \underset{1}{\cancel{5}}} = \frac{1}{3}$$

Incorrect

$$\frac{15}{45} = \frac{\cancel{3} \cdot \cancel{5}}{\cancel{3} \cdot 3 \cdot \cancel{5}} = \frac{0}{3} = 0$$

6 Add and Subtract Fractions.

In algebra as in everyday life, we can only add or subtract objects that are similar. For example, we can add dollars to dollars, but we cannot add dollars to oranges. This concept is important when adding fractions.

Consider the problem $\frac{2}{5} + \frac{1}{5}$. When we write it in words, it is apparent we are adding similar objects.

two-**fifths** + one-**fifth**

└ Similar objects ┘

Because the denominators of $\frac{2}{5}$ and $\frac{1}{5}$ are the same, we say that they have a **common denominator.**

Adding and Subtracting Fractions that Have the Same Denominator

To add (or subtract) fractions that have the same denominator, add (or subtract) their numerators and write the sum (or difference) over the common denominator.

For any fractions $\frac{a}{d}$ and $\frac{b}{d}$,

$$\frac{a}{d} + \frac{b}{d} = \frac{a+b}{d} \quad \text{and} \quad \frac{a}{d} - \frac{b}{d} = \frac{a-b}{d}$$

Caution

We **do not** add fractions by adding the numerators and adding the denominators!

$$\frac{2}{5} + \frac{1}{5} = \frac{2+1}{5+5} = \frac{3}{10} \quad \text{(crossed out)}$$

The same caution applies when subtracting fractions.

For example,

$$\frac{2}{5} + \frac{1}{5} = \frac{2+1}{5} = \frac{3}{5} \quad \text{and} \quad \frac{18}{23} - \frac{9}{23} = \frac{18-9}{23} = \frac{9}{23}$$

Caution Be careful when adding (or subtracting) numerators and writing the result over the common denominator. Only *factors* common to the numerator and the denominator of a fraction can be removed. For example, it is incorrect to remove the 5's in $\frac{5+8}{5}$ because 5 is not used as a factor in the expression 5 + 8. This error leads to an incorrect answer of 9.

Correct	***Incorrect***
$\frac{5+8}{5} = \frac{13}{5}$	$\frac{5+8}{5} = \frac{\overset{1}{\cancel{5}}+8}{\underset{1}{\cancel{5}}} = \frac{9}{1} = 9$ (crossed out)

Success Tip

To determine the LCD of two fractions, list the multiples of one of the denominators. The first number in the list that is exactly divisible by the other denominator is their LCD. For $\frac{2}{5}$ and $\frac{1}{3}$, the multiples of the first denominator, 5, are

5, 10, (15), 20, 25, . . .

Since 15 is the first number in the list that is exactly divisible by the second denominator, 3, the LCD is 15.

Now we consider the problem $\frac{2}{5} + \frac{1}{3}$. Since the denominators are not the same, we cannot add these fractions in their present form.

two-**fifths** + one-**third**

└ Not similar objects ┘

To add (or subtract) fractions with different denominators, we express them as equivalent fractions that have a common denominator. The smallest common denominator, called the **least** or **lowest common denominator,** is usually the easiest common denominator to use.

Least Common Denominator (LCD)

The **least** or **lowest common denominator (LCD)** for a set of fractions is the smallest number each denominator will divide exactly (divide with no remainder).

The denominators of $\frac{2}{5}$ and $\frac{1}{3}$ are 5 and 3. The numbers 5 and 3 divide many numbers exactly (30, 45, and 60, to name a few), but the smallest number that they divide exactly is 15. Thus, 15 is the LCD for $\frac{2}{5}$ and $\frac{1}{3}$.

To find $\frac{2}{5} + \frac{1}{3}$, we find equivalent fractions that have denominators of 15 and we use the rule for adding fractions.

$$\frac{2}{5} + \frac{1}{3} = \frac{2}{5} \cdot \frac{3}{3} + \frac{1}{3} \cdot \frac{5}{5}$$ Multiply $\frac{2}{5}$ by 1 in the form of $\frac{3}{3}$. Multiply $\frac{1}{3}$ by 1 in the form of $\frac{5}{5}$.

$$= \frac{6}{15} + \frac{5}{15}$$ Multiply the numerators and multiply the denominators. Note that the denominators are now the same.

$$= \frac{6 + 5}{15}$$ Add the numerators. Write the sum over the common denominator.

$$= \frac{11}{15}$$

When adding (or subtracting) fractions with unlike denominators, the least common denominator is not always obvious. Prime factorization is helpful in determining the LCD.

Finding the LCD Using Prime Factorization

1. Prime factor each denominator.
2. The LCD is a product of prime factors, where each factor is used the greatest number of times it appears in any one factorization found in step 1.

EXAMPLE 7 Subtract: $\frac{3}{10} - \frac{5}{28}$

Strategy We will begin by expressing each fraction as an equivalent fraction that has the LCD for its denominator. Then we will use the rule for subtracting fractions with *like* denominators.

Why To add or subtract fractions, the fractions must have like denominators.

Solution To find the LCD, we find the prime factorization of both denominators and use each prime factor the *greatest* number of times it appears in any one factorization:

$$\left.\begin{aligned} 10 &= 2 \cdot 5 \\ 28 &= 2 \cdot 2 \cdot 7 \end{aligned}\right\} \text{LCD} = 2 \cdot 2 \cdot 5 \cdot 7 = 140$$

2 appears twice in the factorization of 28.
5 appears once in the factorization of 10.
7 appears once in the factorization of 28.

Since 140 is the smallest number that 10 and 28 divide exactly, we write $\frac{3}{10}$ and $\frac{5}{28}$ as fractions with the LCD 140.

$$\frac{3}{10} - \frac{5}{28} = \frac{3}{10} \cdot \frac{14}{14} - \frac{5}{28} \cdot \frac{5}{5}$$ We must multiply 10 by 14 to obtain 140. We must multiply 28 by 5 to obtain 140.

$$= \frac{42}{140} - \frac{25}{140}$$ Multiply the numerators and multiply the denominators. Note that the denominators are now the same.

$$= \frac{42 - 25}{140}$$ Subtract the numerators. Write the difference over the common denominator.

$$= \frac{17}{140}$$

Self Check 7 Subtract: $\frac{11}{48} - \frac{7}{40}$

Now Try **Problem 65**

We can use the following steps to add or subtract fractions with different denominators.

Adding and Subtracting Fractions that Have Different Denominators

1. Find the LCD.
2. Rewrite each fraction as an equivalent fraction with the LCD as the denominator. To do so, build each fraction using a form of 1 that involves any factors needed to obtain the LCD.
3. Add or subtract the numerators and write the sum or difference over the LCD.
4. Simplify the result, if possible.

7 Simplify Answers.

When adding, subtracting, multiplying, or dividing fractions, remember to express the answer in simplest form.

EXAMPLE 8 Perform the operations and simplify:

a. $45\left(\frac{4}{9}\right)$ **b.** $\frac{5}{12} + \frac{3}{2} - \frac{1}{4}$

Strategy We will perform the indicated operations and then make sure that the answer is in simplest form (lowest terms).

Why Fractional answers should always be given in simplest form.

Solution

a. $45\left(\frac{4}{9}\right) = \frac{45}{1}\left(\frac{4}{9}\right)$ Write 45 as a fraction: $45 = \frac{45}{1}$.

$= \frac{45 \cdot 4}{1 \cdot 9}$ Multiply the numerators. Multiply the denominators.

$= \frac{5 \cdot \overset{1}{\cancel{9}} \cdot 4}{1 \cdot \underset{1}{\cancel{9}}}$ To simplify the result, factor 45 as $5 \cdot 9$. Then remove the common factor 9 of the numerator and denominator.

$= 20$ Multiply the remaining factors in the numerator. Multiply the remaining factors in the denominator. $\frac{20}{1} = 20$.

b. Since the smallest number that 12, 2, and 4 divide exactly is 12, the LCD is 12.

$\frac{5}{12} + \frac{3}{2} - \frac{1}{4} = \frac{5}{12} + \frac{3}{2} \cdot \frac{6}{6} - \frac{1}{4} \cdot \frac{3}{3}$ $\frac{5}{12}$ already has a denominator of 12. Build $\frac{3}{2}$ and $\frac{1}{4}$ so that their denominators are 12.

$= \frac{5}{12} + \frac{18}{12} - \frac{3}{12}$ Multiply the numerators and multiply the denominators. The denominators are now the same.

$= \frac{20}{12}$ Add the numerators, 5 and 18, to get 23. From that sum, subtract 3. Write that result, 20, over the common denominator.

$= \frac{\overset{1}{\cancel{4}} \cdot 5}{3 \cdot \underset{1}{\cancel{4}}}$ To simplify $\frac{20}{12}$, factor 20 and 12, using their greatest common factor, 4. Then remove $\frac{4}{4} = 1$.

$= \frac{5}{3}$

Caution
Remember that an LCD is **not needed** when multiplying or dividing fractions.

Self Check 8 Perform the operations and simplify: **a.** $24\left(\frac{7}{6}\right)$
b. $\frac{1}{15} + \frac{31}{30} - \frac{3}{10}$

Now Try **Problems 67 and 71**

8 Compute with Mixed Numbers.

A **mixed number** represents the sum of a whole number and a fraction. For example, $5\frac{3}{4}$ means $5 + \frac{3}{4}$.

EXAMPLE 9 Divide: $5\frac{3}{4} \div 2$

Strategy We begin by writing the mixed number $5\frac{3}{4}$ and the whole number 2 as fractions. Then we use the rule for dividing two fractions.

Why To multiply (or divide) with mixed numbers, we first write them as fractions, and then multiply (or divide) as usual.

Solution

$$5\frac{3}{4} \div 2 = \frac{23}{4} \div \frac{2}{1}$$

Write $5\frac{3}{4}$ as an improper fraction by multiplying its whole-number part by the denominator: $5 \cdot 4 = 20$. Then add the numerator to that product: $3 + 20 = 23$. Finally, write the result, 23, over the denominator 4. Write 2 as a fraction: $2 = \frac{2}{1}$.

$$= \frac{23}{4} \cdot \frac{1}{2}$$

Multiply by the reciprocal of $\frac{2}{1}$, which is $\frac{1}{2}$.

$$= \frac{23}{8}$$

Multiply the numerators.
Multiply the denominators.

$$= 2\frac{7}{8}$$

Write $\frac{23}{8}$ as a mixed number by dividing the numerator, 23, by the denominator, 8. The quotient, 2, is the whole-number part; the remainder, 7, over the divisor, 8, is the fractional part.

The Language of Algebra

Fractions such as $\frac{23}{4}$, with a numerator greater than or equal to the denominator, are called **improper fractions.** In algebra, such fractions are often preferable to their equivalent mixed number form.

Self Check 9 Multiply: $1\frac{1}{8} \cdot 9$

Now Try **Problem 77**

EXAMPLE 10 ***Freeway Signs.*** How far apart are the Downtown San Diego and Sea World Drive exits?

Strategy We can find the distance between exits by finding the difference in the mileages on the freeway sign: $6\frac{1}{2} - 1\frac{3}{4}$.

Why The word *difference* indicates subtraction.

Success Tip

This problem could also be solved by writing the mixed numbers $6\frac{1}{2}$ and $1\frac{3}{4}$, as improper fractions and subtracting them.

Solution

$$\begin{array}{rcrcrcr} 6\frac{1}{2} & = & 6\frac{2}{4} & = & 5\frac{2}{4}+\frac{4}{4} & = & 5\frac{6}{4} \\ -1\frac{3}{4} & = & -1\frac{3}{4} & = & -1\frac{3}{4} & = & -1\frac{3}{4} \\ \hline & & & & & & 4\frac{3}{4} \end{array}$$

Using vertical form, express $\frac{1}{2}$ as an equivalent fraction with denominator 4. Then, borrow 1 in the form of $\frac{4}{4}$ from 6 to subtract the fractional parts of the mixed numbers.

The Downtown San Diego and Sea World Drive exits are $4\frac{3}{4}$ miles apart.

Now Try Problem 107

ANSWERS TO SELF CHECKS **1.** $189 = 3 \cdot 3 \cdot 3 \cdot 7$ **2.** $\frac{10}{27}$ **3.** $\frac{12}{25}$ **4.** $\frac{15}{24}$ **5. a.** $\frac{3}{7}$ **b.** In simplest form **6.** $\frac{9}{5}$ **7.** $\frac{13}{240}$ **8. a.** 28 **b.** $\frac{4}{5}$ **9.** $\frac{81}{8} = 10\frac{1}{8}$

STUDY SET 1.2

VOCABULARY

Fill in the blanks.

1. A factor is a number being ________.

2. Numbers that have only 1 and themselves as factors, such as 23, 37, and 41, are called ______ numbers.

3. When we write 60 as $2 \cdot 2 \cdot 3 \cdot 5$, we say that we have written 60 in ________ form.

4. The ________ of the fraction $\frac{3}{4}$ is 3, and the ________ is 4.

5. Two fractions that represent the same number, such as $\frac{1}{2}$ and $\frac{2}{4}$, are called ________ fractions.

6. $\frac{2}{3}$ is the ________ of $\frac{3}{2}$, because their product is 1.

7. The ________ common denominator for a set of fractions is the smallest number each denominator will divide exactly.

8. The ______ number $7\frac{1}{3}$ represents the sum of a whole number and a fraction: $7 + \frac{1}{3}$.

CONCEPTS

Complete each fact about fractions.

9. a. $\frac{a}{a} =$ ____ **b.** $\frac{a}{1} =$ ____

c. $\frac{a}{b} \cdot \frac{c}{d} =$ ____ **d.** $\frac{a}{b} \div \frac{c}{d} =$ ____

e. $\frac{a}{d} + \frac{b}{d} =$ ____ **f.** $\frac{a}{d} - \frac{b}{d} =$ ____

10. What two equivalent fractions are shown?

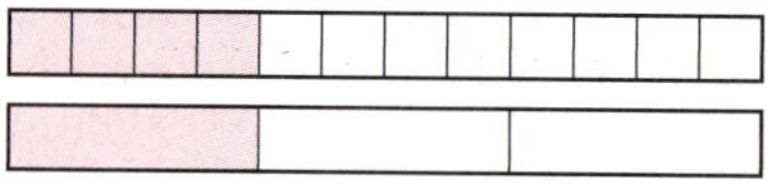

11. Complete each statement.

a. To simplify a fraction, we remove factors equal to ____ in the form of $\frac{2}{2}$, $\frac{3}{3}$, or $\frac{4}{4}$, and so on.

b. To build a fraction, we multiply it by ____ in the form of $\frac{2}{2}$, $\frac{3}{3}$, or $\frac{4}{4}$, and so on.

12. What is the LCD for fractions having denominators of 24 and 36?

NOTATION

Fill in the blanks.

13. a. Multiply $\frac{5}{6}$ by a form of 1 to build an equivalent fraction with denominator 30.

$$\frac{5}{6} \cdot \square = \square$$

b. Remove common factors to simplify $\frac{12}{42}$.

$$\frac{12}{42} = \frac{2 \cdot \square \cdot 3}{2 \cdot 3 \cdot \square} = \square$$

14. a. Write $2\frac{15}{16}$ as an improper fraction.

b. Write $\frac{49}{12}$ as a mixed number.

GUIDED PRACTICE

Find the prime factorization of each number. **See Example 1.**

15. 75
16. 20
17. 28
18. 54
19. 81
20. 125
21. 117
22. 147
23. 220
24. 270
25. 1,254
26. 1,144

Perform each operation. **See Examples 2 and 3.**

27. $\frac{5}{6} \cdot \frac{1}{8}$
28. $\frac{2}{3} \cdot \frac{1}{5}$
29. $\frac{7}{11} \cdot \frac{3}{5}$
30. $\frac{13}{9} \cdot \frac{2}{3}$
31. $\frac{3}{4} \div \frac{2}{5}$
32. $\frac{7}{8} \div \frac{6}{13}$
33. $\frac{6}{5} \div \frac{5}{7}$
34. $\frac{4}{3} \div \frac{3}{2}$

Build each fraction or whole number to an equivalent fraction with the indicated denominator. **See Example 4.**

35. $\frac{1}{3}$, denominator 9
36. $\frac{3}{8}$, denominator 24
37. $\frac{4}{9}$, denominator 54
38. $\frac{9}{16}$, denominator 64
39. 7, denominator 5
40. 12, denominator 3
41. 5, denominator 7
42. 6, denominator 8

Simplify each fraction, if possible. **See Examples 5 and 6.**

43. $\frac{6}{18}$
44. $\frac{6}{9}$
45. $\frac{24}{28}$
46. $\frac{35}{14}$
47. $\frac{15}{40}$
48. $\frac{22}{77}$
49. $\frac{33}{56}$
50. $\frac{26}{21}$
51. $\frac{26}{39}$
52. $\frac{72}{64}$
53. $\frac{36}{225}$
54. $\frac{175}{490}$

Perform the operations and, if possible, simplify. **See Objective 6 and Example 7.**

55. $\frac{3}{5} + \frac{3}{5}$
56. $\frac{4}{9} - \frac{1}{9}$
57. $\frac{6}{7} - \frac{2}{7}$
58. $\frac{5}{13} + \frac{6}{13}$
59. $\frac{1}{6} + \frac{1}{24}$
60. $\frac{17}{25} - \frac{2}{5}$
61. $\frac{7}{10} - \frac{1}{14}$
62. $\frac{9}{8} - \frac{5}{6}$
63. $\frac{2}{15} + \frac{7}{9}$
64. $\frac{7}{25} + \frac{3}{10}$
65. $\frac{21}{56} - \frac{9}{40}$
66. $\frac{13}{24} - \frac{3}{40}$

Perform the operations and, if possible, simplify. **See Example 8.**

67. $16\left(\frac{3}{2}\right)$
68. $30\left(\frac{5}{6}\right)$
69. $18\left(\frac{2}{9}\right)$
70. $14\left(\frac{3}{7}\right)$
71. $\frac{2}{3} - \frac{1}{6} + \frac{5}{18}$
72. $\frac{3}{5} - \frac{7}{10} + \frac{7}{20}$
73. $\frac{5}{12} + \frac{1}{3} - \frac{2}{5}$
74. $\frac{7}{15} + \frac{1}{5} - \frac{4}{9}$

Perform the operations and, if possible, simplify. **See Examples 9 and 10.**

75. $4\frac{2}{3} \cdot 7$
76. $7 \cdot 1\frac{3}{28}$
77. $8 \div 3\frac{1}{5}$
78. $15 \div 3\frac{1}{3}$
79. $8\frac{2}{9} - 7\frac{2}{3}$
80. $3\frac{4}{5} - 3\frac{1}{10}$
81. $3\frac{3}{16} + 2\frac{5}{24}$
82. $15\frac{5}{6} + 11\frac{5}{8}$

TRY IT YOURSELF

Perform the operations and, if possible, simplify.

83. $\frac{3}{5} + \frac{2}{3}$
84. $\frac{4}{3} + \frac{7}{2}$
85. $21\left(\frac{10}{3}\right)$
86. $28\left(\frac{4}{7}\right)$
87. $6 \cdot 2\frac{7}{24}$
88. $3\frac{1}{2} \cdot \frac{1}{5}$
89. $\frac{2}{3} - \frac{1}{4} + \frac{1}{12}$
90. $\frac{3}{7} - \frac{2}{5} + \frac{2}{35}$
91. $\frac{21}{35} \div \frac{3}{14}$
92. $\frac{23}{25} \div \frac{46}{5}$
93. $\frac{4}{3}\left(\frac{6}{5}\right)$
94. $\frac{21}{8}\left(\frac{2}{15}\right)$
95. $\frac{4}{63} + \frac{1}{45}$
96. $\frac{5}{18} + \frac{1}{99}$
97. $3 - \frac{3}{4}$
98. $4 - \frac{7}{3}$
99. $\frac{1}{2} \cdot \frac{3}{5}$
100. $\frac{3}{4} \cdot \frac{5}{7}$
101. $3\frac{1}{3} \div 1\frac{5}{6}$
102. $2\frac{1}{2} \div 1\frac{5}{8}$
103. $\frac{11}{21} - \frac{8}{21}$
104. $\frac{19}{35} - \frac{12}{35}$

APPLICATIONS

105. FORESTRY A ranger cut down a pine tree and measured the widths of the outer two growth rings.

a. What was the growth over this 2-year period?

b. What is the difference in the widths of the rings?

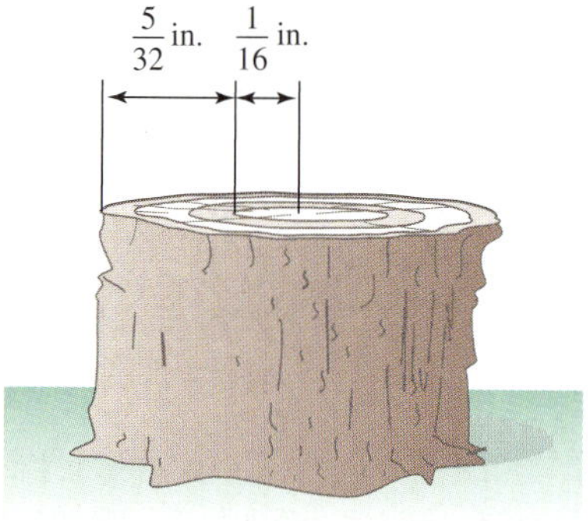

106. HARDWARE To secure the bracket to the stock, a bolt and a nut are used. How long should the threaded part of the bolt be?

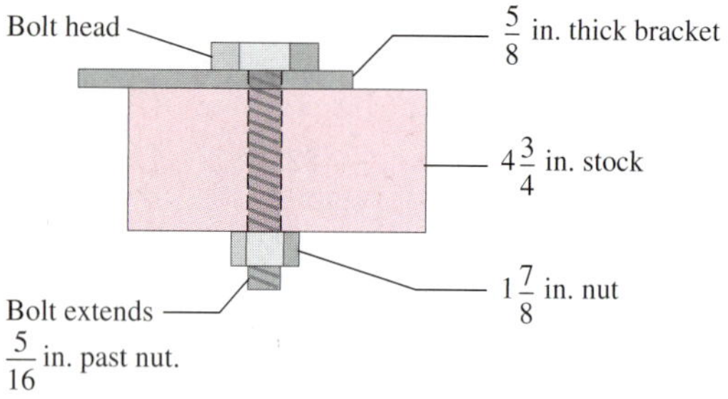

107. COOKING How much butter is left in a $10\frac{1}{2}$-pound tub of butter if $4\frac{3}{4}$ pounds are used to make a wedding cake?

108. CALORIES A company advertises that its mints contain only 3 calories a piece. What is the calorie intake if you eat an entire package of 20 mints?

109. FRAMES How many inches of molding are needed to make the square picture frame?

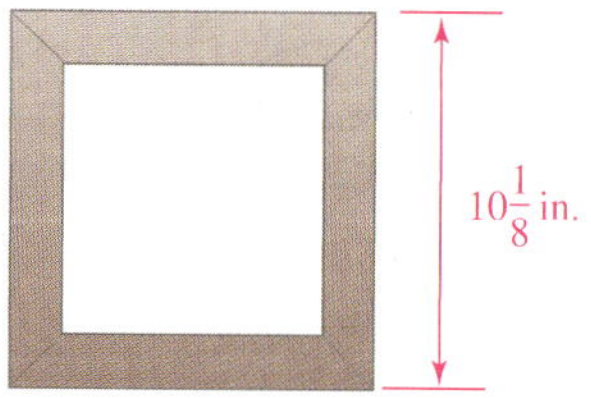

110. DECORATING The materials used to make a pillow are shown in the next column. Examine the inventory list to decide how many pillows can be manufactured in one production run with the materials in stock.

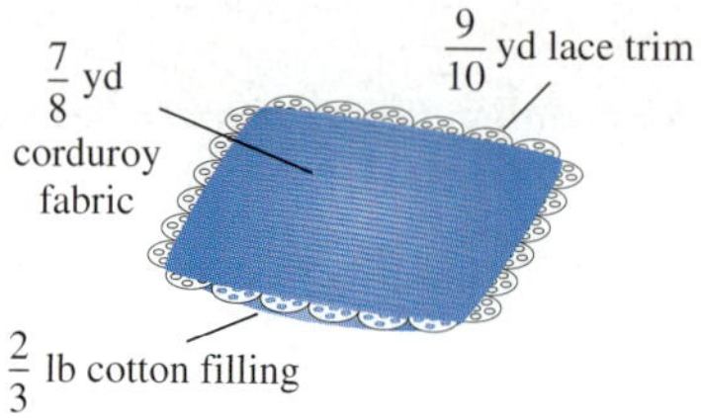

Factory Inventory List

Materials	*Amount in stock*
Lace trim	135 yd
Corduroy fabric	154 yd
Cotton filling	98 lb

WRITING

111. Explain how to add two fractions having unlike denominators.

112. To multiply two fractions, must they have like denominators? Explain.

113. What are equivalent fractions?

114. Explain the error in the following addition.

$$\frac{4}{3} + \frac{3}{2} = \frac{4+3}{3+2} = \frac{7}{5}$$

REVIEW

Use the formula to complete each table.

115. $T = 15g$

Number of gears g	Number of teeth T
10	
12	

116. $p = r - 200$

Revenue r	Profit p
1,000	
5,000	

CHALLENGE PROBLEMS

117. Which is larger: $\frac{11}{12}$ or $\frac{8}{9}$?

118. If the circle represents a whole, find the missing value.

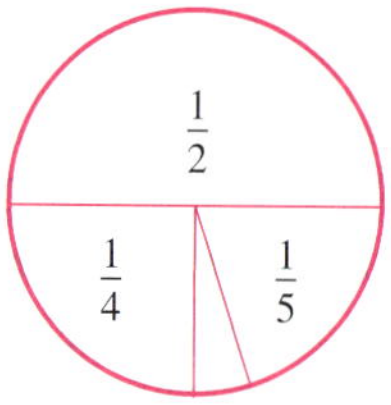

SECTION 1.3
The Real Numbers

Objectives

1. Define the set of integers.
2. Define the set of rational numbers.
3. Define the set of irrational numbers.
4. Classify real numbers.
5. Graph sets of numbers on the number line.
6. Find the absolute value of a real number.

A **set** is a collection of objects, such as a set of golf clubs or a set of dishes. In this section, we will define some important sets of numbers that are used in algebra.

1 Define the Set of Integers.

Natural numbers are the numbers that we use for counting. To write this set, we list its **members** (or **elements**) within **braces** { }.

Natural Numbers

The set of **natural numbers** is $\{1, 2, 3, 4, 5, \ldots\}$. Read as "the set containing one, two, three, four, five, and so on."

The natural numbers, together with 0, form the set of **whole numbers.**

Whole Numbers

The set of **whole numbers** is $\{0, 1, 2, 3, 4, 5, \ldots\}$.

Notation

The symbol . . . used in the previous definitions is called an **ellipsis** and it indicates that the established pattern continues forever.

Whole numbers are not adequate for describing many real-life situations. For example, if you write a check for more than what's in your account, the account balance will be less than zero.

We can use the **number line** below to visualize numbers less than zero. A number line is straight and has uniform markings. The arrowheads indicate that it extends forever in both directions. For each natural number on the number line, there is a corresponding number, called its *opposite,* to the left of 0. In the diagram, we see that 3 and -3 (negative three) are opposites, as are -5 (negative five) and 5. Note that 0 is its own opposite.

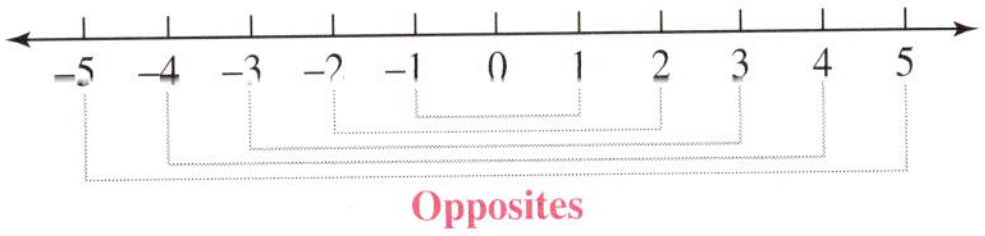

Opposites

Two numbers that are the same distance from 0 on the number line, but on opposite sides of it, are called **opposites.**

The whole numbers, together with their opposites, form the set of **integers.**

Integers The set of **integers** is $\{\ldots, -4, -3, -2, -1, 0, 1, 2, 3, 4, \ldots\}$.

The Language of Algebra
The *positive integers* are:
$1, 2, 3, 4, 5, \ldots$
The *negative integers* are:
$-1, -2, -3, -4, -5, \ldots$

On the number line, numbers greater than 0 are to the right of 0. They are called **positive numbers.** Positive numbers can be written with or without a **positive sign** $+$. For example, $2 = +2$ (positive two). They are used to describe such quantities as an elevation above sea level ($+3000$ ft) or a stock market gain (25 points).

Numbers less than 0 are to the left of 0 on the number line. They are called **negative numbers.** Negative numbers are always written with a **negative sign** $-$. They are used to describe such quantities as an overdrawn checking account ($-\$75$) or a below-zero temperature ($-12°$).

Positive and negative numbers are called **signed numbers.**

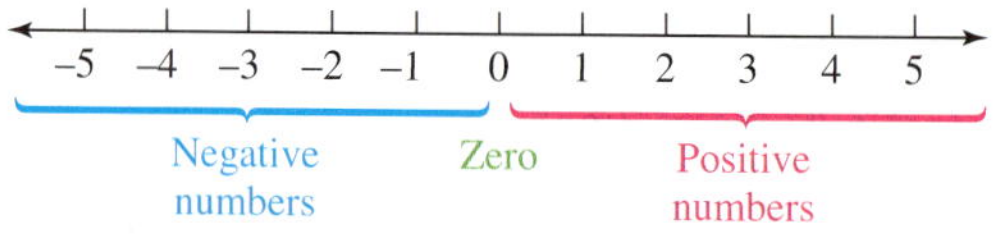

2 Define the Set of Rational Numbers.

We use fractions to describe many situations in daily life. For example, a morning commute might take $\frac{1}{4}$ hour or a recipe might call for $\frac{2}{3}$ cup of sugar. Fractions such as $\frac{1}{4}$ and $\frac{2}{3}$, that are quotients of two integers, are called *rational numbers.*

Rational Numbers A **rational number** is any number that can be expressed as a fraction with an integer numerator and a nonzero integer denominator.

The Language of Algebra
Rational numbers are so named because they can be expressed as the *ratio* (quotient) of two integers.

Some other examples of rational numbers are

$$\frac{3}{8}, \quad \frac{41}{100}, \quad \frac{25}{25}, \quad \text{and} \quad \frac{19}{12}$$

To show that negative fractions are rational numbers, we use the following fact.

Negative Fractions For any numbers a and b where b is not 0,

$$-\frac{a}{b} = \frac{-a}{b} = \frac{a}{-b}$$

To illustrate this rule, we consider $-\frac{11}{16}$. It is a rational number because it can be written as $\frac{-11}{16}$ or as $\frac{11}{-16}$.

Positive and negative mixed numbers are also rational numbers because they can be expressed as fractions. For example,

$$7\frac{5}{8} = \frac{61}{8} \quad \text{and} \quad -6\frac{1}{2} = -\frac{13}{2} = \frac{-13}{2}$$

Any natural number, whole number, or integer can be expressed as a fraction with a denominator of 1. For example, $5 = \frac{5}{1}$, $0 = \frac{0}{1}$, and $-3 = \frac{-3}{1}$. Therefore, every natural number, whole number, and integer is also a rational number.

The Language of Algebra
To *terminate* means to bring to an end. In the movie *The Terminator,* actor Arnold Schwarzenegger plays a heartless machine sent to Earth to bring an end to his enemies.

Many numerical quantities are written in decimal notation. For instance, a candy bar might cost \$0.89, a dragster might travel at 203.156 mph, or a business loss might be −\$4.7 million. These decimals are called **terminating decimals** because their representations terminate (stop). As shown below, terminating decimals can be expressed as fractions. Therefore, terminating decimals are rational numbers.

$$0.89 = \frac{89}{100} \qquad 203.156 = 203\frac{156}{1000} = \frac{203{,}156}{1{,}000} \qquad -4.7 = -4\frac{7}{10} = \frac{-47}{10}$$

Decimals such as 0.3333 . . . and 2.8167167167 . . . , which have a digit (or block of digits) that repeats, are called **repeating decimals.** Since any repeating decimal can be expressed as a fraction, repeating decimals are rational numbers.

The set of rational numbers cannot be listed as we listed other sets in this section. Instead, we use **set-builder** notation.

Rational Numbers

The set of rational numbers is

$$\left\{ \frac{a}{b} \,\middle|\, a \text{ and } b \text{ are integers, with } b \neq 0. \right\}$$

Read as "the set of all numbers of the form $\frac{a}{b}$, such that a and b are integers, with $b \neq 0$."

The Language of Algebra
A fraction and its *decimal equivalent* are different notations that represent the same value.

To find the *decimal equivalent* for a fraction, we divide its numerator by its denominator. For example, to write $\frac{1}{4}$ and $\frac{5}{22}$ as decimals, we proceed as follows:

$$\begin{array}{r} 0.25 \\ 4\overline{)1.00} \\ \underline{8} \\ 20 \\ \underline{20} \\ 0 \end{array}$$

Write a decimal point and additional zeros to the right of 1.

The remainder is 0.

$$\begin{array}{r} 0.22727\ldots \\ 22\overline{)5.00000} \\ \underline{4\,4} \\ 60 \\ \underline{44} \\ 160 \\ \underline{154} \\ 60 \\ \underline{44} \\ 160 \end{array}$$

Write a decimal point and additional zeros to the right of 5.

60 and 160 continually appear as remainders. Therefore, 2 and 7 will continually appear in the quotient.

The Language of Algebra
Since every natural number belongs to the set of whole numbers, we say the set of natural numbers is a *subset* of the set of whole numbers. Similarly, the set of whole numbers is a subset of the set of integers, and the set of integers is a subset of the set of rational numbers.

The decimal equivalent of $\frac{1}{4}$ is 0.25 and the decimal equivalent of $\frac{5}{22}$ is 0.2272727 . . . . We can use an **overbar** to write repeating decimals in more compact form: $0.2272727\ldots = 0.2\overline{27}$. Here are more fractions and their decimal equivalents.

Terminating decimals	***Repeating decimals***
$\frac{1}{2} = 0.5$	$\frac{1}{6} = 0.166666\ldots$ or $0.1\overline{6}$
$\frac{5}{8} = 0.625$	$\frac{1}{3} = 0.333333\ldots$ or $0.\overline{3}$
$\frac{3}{4} = 0.75$	$\frac{5}{11} = 0.454545\ldots$ or $0.\overline{45}$

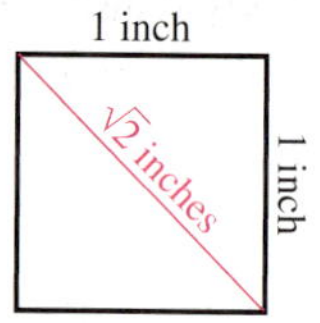

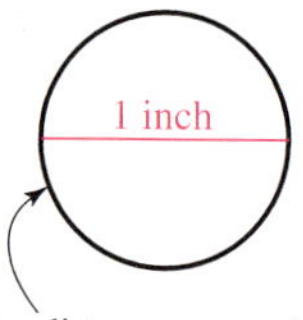

The distance around the circle is π inches.

3 Define the Set of Irrational Numbers.

Not all numbers are rational numbers. One example is the square root of 2, written $\sqrt{2}$. It is the number that, when multiplied by itself, gives 2: $\sqrt{2} \cdot \sqrt{2} = 2$. It can be shown that $\sqrt{2}$ *cannot* be written as a fraction with an integer numerator and an integer denominator. Therefore, it is not rational; it is an irrational number. It is interesting to note that a square with sides of length 1 inch has a diagonal that is $\sqrt{2}$ inches long.

The number represented by the Greek letter π (pi) is another example of an irrational number. A circle, with a 1-inch diameter, has a circumference of π inches.

Expressed in decimal form,

$$\sqrt{2} = 1.414213562\ldots \qquad \text{and} \qquad \pi = 3.141592654\ldots$$

These decimals neither terminate nor repeat.

Irrational Numbers

An **irrational number** is a nonterminating, nonrepeating decimal. An irrational number cannot be expressed as a fraction with an integer numerator and an integer denominator.

The Language of Algebra

Since π is irrational, its decimal representation has an infinite number of *decimal places.* In 2002, a University of Tokyo mathematician used a super computer to calculate π to over one trillion *decimal places.*

Other examples of irrational numbers are:

$$\sqrt{3} = 1.732050808\ldots \qquad -\sqrt{5} = -2.236067977\ldots$$

$$-\pi = -3.141592654\ldots \qquad 3\pi = 9.424777961\ldots$$

3π means $3 \cdot \pi$.

We can use a calculator to approximate the decimal value of an irrational number. To approximate $\sqrt{2}$ using a scientific calculator, we use the square root key $\sqrt{\ }$. To approximate π, we use the *pi* key π.

$$\sqrt{2} \approx 1.414213562 \qquad \text{and} \qquad \pi \approx 3.141592654$$

Rounded to the nearest thousandth, $\sqrt{2} \approx 1.414$ and $\pi \approx 3.142$.

4 Classify Real Numbers.

The set of **real numbers** is formed by combining the set of rational numbers and the set of irrational numbers. Every real number has a decimal representation. If it is rational, its corresponding decimal terminates or repeats. If it is irrational, its decimal representation is nonterminating and nonrepeating.

The Real Numbers

A **real number** is any number that is a rational number or an irrational number.

The following diagram shows how various sets of numbers are related. Note that a number can belong to more than one set. For example, -6 is an integer, a rational number, and a real number.

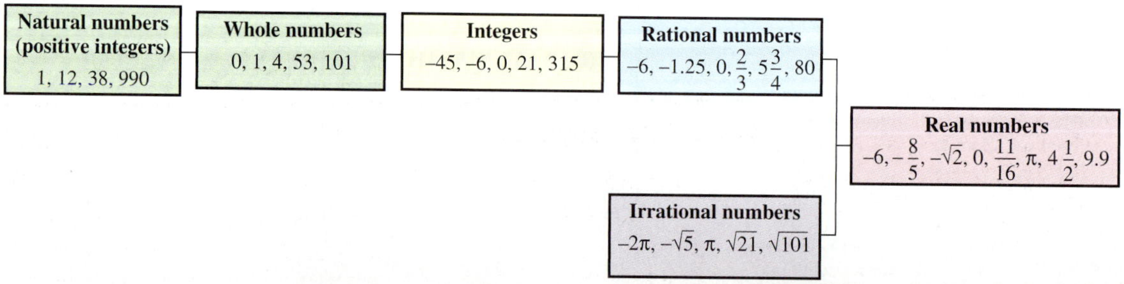

EXAMPLE 1 Which numbers in the following set are natural numbers, whole numbers, integers, rational numbers, irrational numbers, real numbers? $\left\{-3.4, \frac{2}{5}, 0, -6, 1\frac{3}{4}, \pi, 16\right\}$

Strategy We begin by scanning the given set, looking for any natural numbers. Then we scan it five more times, looking for whole numbers, for integers, for rational numbers, for irrational numbers, and finally, for real numbers.

Why We need to scan the given set of numbers six times, because numbers in that set can belong to more than one classification.

Solution

Natural numbers: 16 — 16 is a member of $\{1, 2, 3, 4, 5, \ldots\}$.

Whole numbers: 0, 16 — 0 and 16 are members of $\{0, 1, 2, 3, 4, 5, \ldots\}$.

Integers: 0, -6, 16 — 0, -6, and 16 are members of $\{\ldots, -3, -2, -1, 0, 1, 2, 3, \ldots\}$.

Rational numbers:
$-3.4, \frac{2}{5}, 0, -6, 1\frac{3}{4}, 16$ — A rational number can be expressed as a fraction of two integers: $-3.4 = \frac{-34}{10}$, $0 = \frac{0}{1}$, $-6 = \frac{-6}{1}$, $1\frac{3}{4} = \frac{7}{4}$, and $16 = \frac{16}{1}$.

Irrational numbers: π — $\pi = 3.1415\ldots$ is a nonterminating, nonrepeating decimal.

Real numbers:
$-3.4, \frac{2}{5}, 0, -6, 1\frac{3}{4}, \pi, 16$ — Every natural number, whole number, integer, rational number, and irrational number is a real number.

Self Check 1 Use the instructions for Example 1 with: $\left\{0.1, \sqrt{2}, -\frac{2}{7}, 45, -2, \frac{13}{4}, -6\frac{7}{8}\right\}$

Now Try **Problem 27**

5 Graph Sets of Numbers on the Number Line.

Every real number corresponds to a point on the number line, and every point on the number line corresponds to exactly one real number. As we move right on the number line, the values of the numbers increase. As we move left, the values decrease. On the number line below, we see that 5 is greater than -3, because 5 lies to the right of -3. Similarly, -3 is less than 5, because it lies to the left of 5.

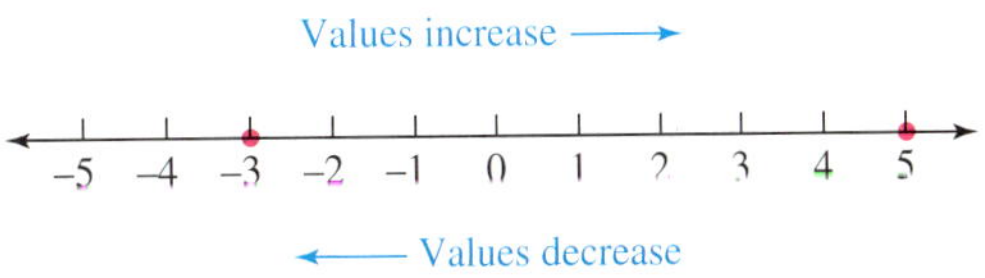

The Language of Algebra
The prefix *in* means *not*. For example:

inaccurate ↔ not accurate
inexpensive ↔ not expensive
inequality ↔ not equal

The **inequality symbol** $>$ means "is greater than." It is used to show that one number is greater than another. The inequality symbol $<$ means "is less than." It is used to show that one number is less than another. For example,

$5 > -3$ — Read as "5 is greater than -3."

$-3 < 5$ — Read as "-3 is less than 5."

To distinguish between these inequality symbols, remember that each one points to the smaller of the two numbers involved.

$5 > -3$ $\qquad\qquad$ $-3 < 5$

Points to the smaller number.

EXAMPLE 2 Use one of the symbols $>$ or $<$ to make each statement true:

a. $-4 \square 4$ **b.** $-2 \square -3$ **c.** $4.47 \square 12.5$ **d.** $\frac{3}{4} \square \frac{5}{8}$

Strategy To pick the correct inequality symbol to place between a given pair of numbers, we need to determine the position of each number on a number line.

Why For any two numbers on a number line, the number to the *left* is the smaller number and the number to the *right* is the larger number.

Solution

a. Since -4 is to the left of 4 on the number line, we have $-4 < 4$.

b. Since -2 is to the right of -3 on the number line, we have $-2 > -3$.

c. Since 4.47 is to the left of 12.5 on the number line, we have $4.47 < 12.5$.

d. To compare fractions, express them in terms of the same denominator, preferably the LCD. If we write $\frac{3}{4}$ as an equivalent fraction with denominator 8, we see that $\frac{3}{4} = \frac{3 \cdot 2}{4 \cdot 2} = \frac{6}{8}$. Therefore, $\frac{3}{4} > \frac{5}{8}$.

To compare the fractions, we could also convert each to its decimal equivalent. Since $\frac{3}{4} = 0.75$ and $\frac{5}{8} = 0.625$, we know that $\frac{3}{4} > \frac{5}{8}$.

Self Check 2 Use one of the symbols $<$ or $>$ to make each statement true:

a. $1 \_ -1$ **b.** $-5 \_ -4$ **c.** $6.7 \_ 4.999$ **d.** $\frac{3}{5} \_ \frac{2}{3}$

Now Try **Problems 37 and 44**

The Language of Algebra
To state that a number x is positive, we can write $x > 0$. To state that a number x is negative, we can write $x < 0$.

To **graph a number** means to mark its position on the number line.

EXAMPLE 3 Graph each number in the set: $\left\{-2.43, \sqrt{2}, 1, -0.\overline{3}, 2\frac{5}{6}, -\frac{3}{2}\right\}$

Strategy We locate the position of each number on the number line, draw a bold dot, and label it.

Why To *graph a number* means to make a drawing that represents the number.

Solution

- To locate -2.43, we round it to the nearest tenth: $-2.43 \approx -2.4$.
- To locate $\sqrt{2}$, we use a calculator: $\sqrt{2} \approx 1.4$.
- To locate $-0.\overline{3}$, we recall that $0.\overline{3} = 0.333\ldots = \frac{1}{3}$. Therefore, $-0.\overline{3} = -\frac{1}{3}$.
- In mixed-number form, $-\frac{3}{2} = -1\frac{1}{2}$. This is midway between -1 and -2.

Success Tip
It is often helpful to approximate the value of a number or to write the number in an equivalent form to determine its location on a number line.

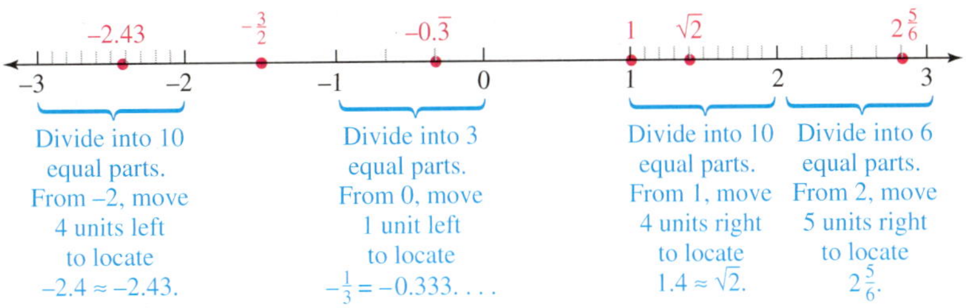

Self Check 3 Graph each number in the set: $\left\{1.7,\ \pi,\ -1\frac{3}{4},\ 0.\overline{6},\ \frac{5}{2},\ -3\right\}$

Now Try **Problem 57**

6 Find the Absolute Value of a Real Number.

A number line can be used to measure the distance from one number to another. For example, in the following figure we see that the distance from 0 to −4 is 4 units and the distance from 0 to 3 is 3 units.

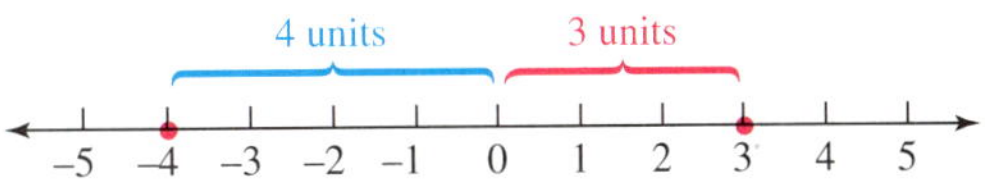

To express the distance that a number is from 0 on a number line, we can use absolute values.

Absolute Value The **absolute value** of a number is its distance from 0 on the number line.

Success Tip
Since absolute value expresses distance, the absolute value of a number is always positive or zero, but never negative.

To indicate the absolute value of a number, we write the number between two vertical bars. From the figure above, we see that $|-4| = 4$. This is read as "the absolute value of negative 4 is 4" and it tells us that the distance from 0 to −4 is 4 units. It also follows from the figure that $|3| = 3$.

EXAMPLE 4 Find each absolute value: **a.** $|18|$ **b.** $\left|-\frac{7}{8}\right|$ **c.** $|98.6|$ **d.** $|0|$

Strategy We need to determine the distance that the number within the vertical absolute value bars is from 0.

Why The absolute value of a number is the distance between 0 and the number on a number line.

Solution

a. Since 18 is a distance of 18 from 0 on the number line, $|18| = 18$.

b. Since $-\frac{7}{8}$ is a distance of $\frac{7}{8}$ from 0 on the number line, $\left|-\frac{7}{8}\right| = \frac{7}{8}$.

c. Since 98.6 is a distance of 98.6 from 0 on the number line, $|98.6| = 98.6$.

d. Since 0 is a distance of 0 from 0 on the number line, $|0| = 0$.

Self Check 4 Find each absolute value:

a. $|100|$ **b.** $|-4.7|$ **c.** $|\sqrt{2}|$

Now Try **Problems 61 and 67**

ANSWERS TO SELF CHECKS **1.** Natural numbers: 45; whole numbers: 45; integers: 45, -2; rational numbers: 0.1, $-\frac{2}{7}$, 45, -2, $\frac{13}{4}$, $-6\frac{7}{8}$; irrational numbers: $\sqrt{2}$; real numbers: all **2. a.** > **b.** < **c.** > **d.** < **3.**

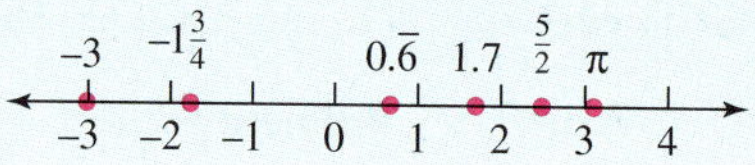

4. a. 100 **b.** 4.7 **c.** $\sqrt{2}$

STUDY SET 1.3

VOCABULARY

Fill in the blanks.

1. The set of _______ numbers is {0, 1, 2, 3, 4, 5, . . .}.
2. The set of _______ numbers is {1, 2, 3, 4, 5, . . .}.
3. The set of _______ is {. . . , -2, -1, 0, 1, 2, . . .}.
4. Positive and negative numbers are called _______ numbers.
5. 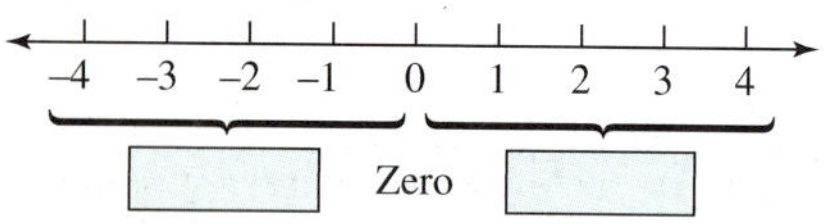

6. The symbols < and > are _________ symbols.
7. A _______ number is any number that can be expressed as a fraction with an integer numerator and a nonzero integer denominator.
8. 0.25 is called a _________ decimal and 0.333 . . . is called a _________ decimal.
9. An _________ number cannot be expressed as a quotient of two integers.
10. An irrational number is a nonterminating, nonrepeating _______.
11. Every point on the number line corresponds to exactly one _____ number.
12. The _________ ______ of a number is the distance on the number line between the number and 0.

CONCEPTS

13. Represent each situation using a signed number.
    a. A loss of \$15 million
    b. A building foundation $\frac{5}{16}$ inch above grade
14. Show that each of the following numbers is a rational number by expressing it as a fraction with an integer numerator and a nonzero integer denominator: 6, -9, $-\frac{7}{8}$, $3\frac{1}{2}$, -0.3, 2.83.
15. Give the opposite of each number.
    a. 20 **b.** $-\frac{2}{3}$
16. What two numbers are a distance of 8 away from 5 on the number line?
17. What two numbers are a distance of 5 away from -9 on the number line?
18. Refer to the graph below. Use an inequality symbol, < or >, to make each statement true.
    a. a ☐ b **b.** b ☐ a
    c. b ☐ 0 and a ☐ 0 **d.** $|a|$ ☐ $|b|$

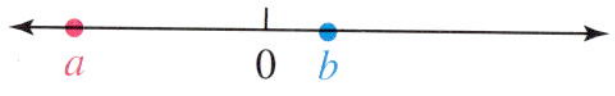

NOTATION

Fill in the blanks.

19. $\sqrt{2}$ is read "the _______ _____ of 2."
20. $|-15|$ is read "the _________ _______ of -15."
21. The symbol $\approx$ means ___ ___________ ______ ___.
22. The symbols { } are called _______.
23. The symbol π is a letter from the _______ alphabet.
24. To find the decimal equivalent for the fraction $\frac{2}{3}$ we divide: ☐$\overline{)☐}$
25. Fill in the blanks: $-\frac{4}{5} = \frac{☐}{5} = \frac{4}{☐}$

26. Write each repeating decimal using an overbar.
  a. 0.666 . . . b. 0.2444 . . .
  c. 0.717171 . . . d. 0.456456456 . . .

GUIDED PRACTICE

Place check marks in the table to show the set or sets to which each number belongs. For example, the check shows that $\sqrt{2}$ is irrational. **See Example 1.**

27.

	5	0	-3	$\frac{7}{8}$	0.17	$-9\frac{1}{4}$	$\sqrt{2}$	π
Real								
Irrational							✓	
Rational								
Integer								
Whole								
Natural								

28. Which numbers in the following set are natural numbers, whole numbers, integers, rational numbers, irrational numbers, real numbers? $\left\{67, \frac{4}{13}, -5.9, 11\frac{2}{3}, \sqrt{2}, 0, -3, \pi\right\}$

Determine whether each statement is true or false. **See Example 1.**

29. Every whole number is an integer.
30. Every integer is a natural number.
31. Every integer is a whole number.
32. Irrational numbers are nonterminating, nonrepeating decimals.
33. Irrational numbers are real numbers.
34. Every whole number is a rational number.
35. Every rational number can be written as a fraction of two integers.
36. Every rational number is a whole number.

Use one of the symbols < or > to make each statement true. **See Example 2.**

37. $0 \quad -4$ 38. $27 \quad 115$
39. $5 \quad 4$ 40. $0 \quad 32$
41. $917 \quad 971$ 42. $898 \quad 889$
43. $-2 \quad -3$ 44. $-5 \quad -4$
45. $-\frac{5}{8} \quad -\frac{3}{8}$ 46. $-19\frac{2}{3} \quad -19\frac{1}{3}$
47. $\frac{2}{3} \quad \frac{3}{5}$ 48. $-2.27 \quad -5.25$

Write each fraction as a decimal. If the result is a repeating decimal, use an overbar. **See Objective 2.**

49. $\frac{5}{8}$ 50. $\frac{3}{32}$
51. $\frac{1}{30}$ 52. $\frac{7}{9}$
53. $\frac{1}{60}$ 54. $\frac{5}{11}$
55. $\frac{21}{50}$ 56. $\frac{2}{125}$

Graph each set of numbers on a number line. **See Example 3.**

57. $\left\{-\pi, 4.25, -1\frac{1}{2}, -0.333\ldots, \sqrt{2}, -\frac{35}{8}, 3\right\}$
58. $\left\{-2\frac{1}{8}, \pi, 2.75, -\sqrt{2}, \frac{17}{4}, 0.666\ldots, -3\right\}$
59. The integers between -5 and 2
60. The whole numbers less than 4

Find each absolute value. **See Example 4.**

61. $|83|$ 62. $|29|$
63. $\left|\frac{4}{3}\right|$ 64. $\left|\frac{9}{16}\right|$
65. $|-11|$ 66. $|-14|$
67. $|-6.1|$ 68. $|-25.3|$

Insert one of the symbols >, <, or = in the blank. **See Examples 2 and 4.**

69. $|3.4| \quad -3$ 70. $0.08 \quad 0.079$
71. $|-1.1| \quad 1.2$ 72. $-5.5 \quad -5\frac{1}{2}$
73. $\left|-\frac{15}{2}\right| \quad 7.5$ 74. $\sqrt{2} \quad \pi$
75. $\frac{99}{100} \quad 0.99$ 76. $|2| \quad |-2|$
77. $0.3 \quad 0.333\ldots$ 78. $\left|-2\frac{2}{3}\right| \quad \frac{7}{3}$
79. $1 \quad \left|-\frac{15}{16}\right|$ 80. $-0.666\ldots \quad -0.6$

APPLICATIONS

81. DRAFTING Which dimensions of the aluminum bracket shown below are natural numbers, whole numbers, integers, rational numbers, irrational numbers, and real numbers?

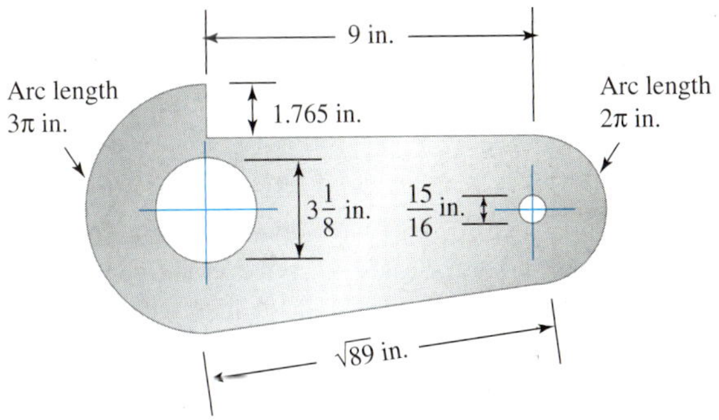

82. HISTORY Refer to the time line shown on the next page.
  a. What basic unit was used to scale the time line?
  b. What symbolism is used to represent zero?
  c. Which numbers could be thought of as positive and which as negative?
  d. Express the dates for the Maya civilization using positive and negative numbers.

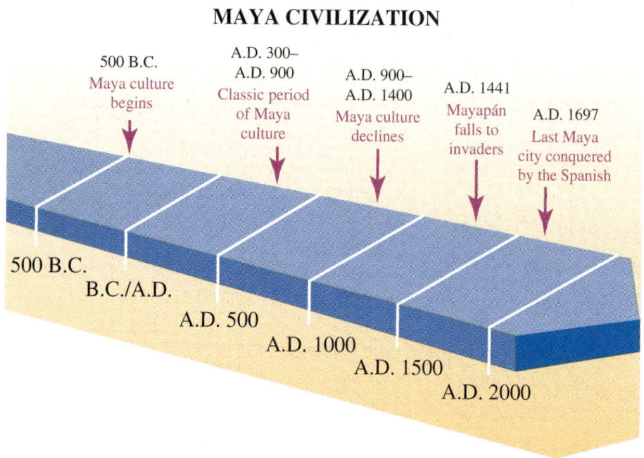

Based on data from *People in Time and Place, Western Hemisphere* (Silver Burdett & Ginn, 1991), p. 129.

83. ARCHERY Which arrow landed farther from the target? How does the concept of absolute value apply here?

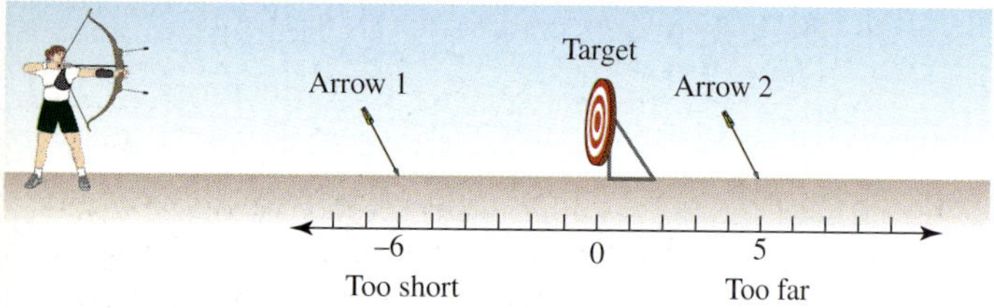

84. DRAFTING On an architect's scale, the edge marked 16 divides each inch into 16 equal parts. Find the decimal form for each fractional part of one inch that is highlighted on the scale.

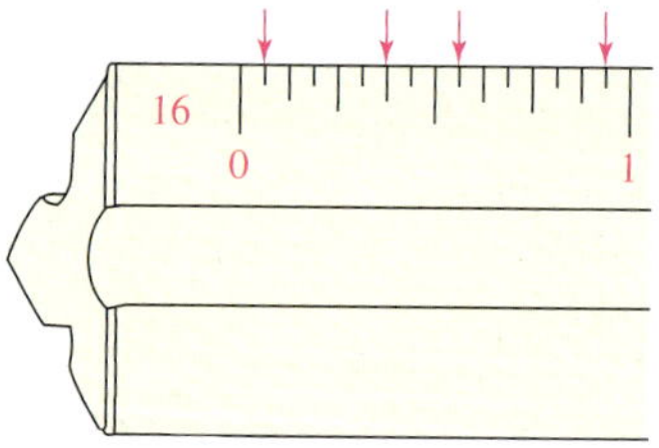

85. TRADE Each year from 1990 through 2005, the United States imported more goods and services from Japan than it exported to Japan. This caused trade deficits, which are represented by negative numbers on the following graph.

a. In which year was the deficit the worst? Express that deficit using a signed number.

b. In which year was the deficit the smallest? Express that deficit using a signed number.

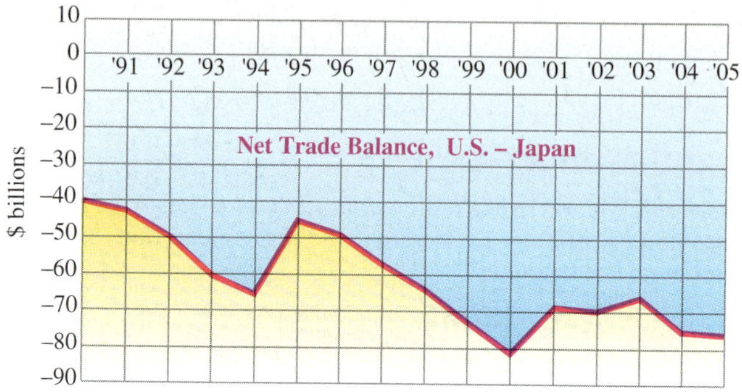

Source: U.S. Bureau of the Census

86. U.S. BUDGET A budget *deficit* is a negative number that indicates the government spent more money than it took in that year. A budget *surplus* is a positive number that indicates the government took in more money than it spent that year. Refer to the graph.

a. In which year was the federal budget deficit the worst? Express that deficit using a signed number.

b. In which year was the federal budget surplus the greatest? Estimate that surplus.

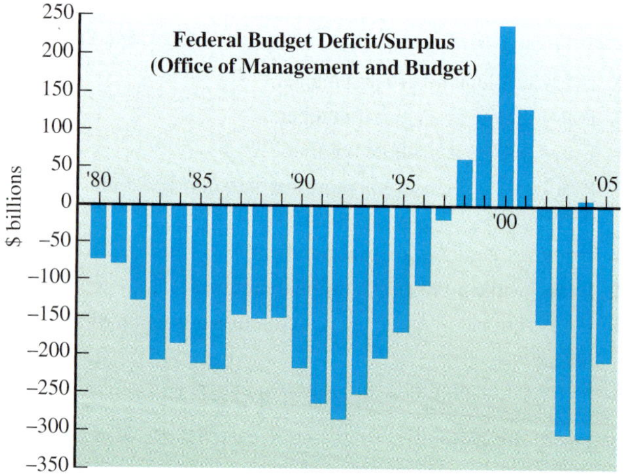

Source: U.S. Bureau of the Census

WRITING

87. Explain the difference between a rational and an irrational number.

88. Can two different numbers have the same absolute value? Explain.

89. Explain how to find the decimal equivalent of a fraction.

90. What is a real number?

91. *Pi Day* (or *Pi Approximation Day*) is an unofficial holiday held to celebrate π. Why do you think Pi Day is observed each year on March 14?

92. Explain why $0.1\overline{333}$ is not the simplest way to represent $0.1333\ldots$.

REVIEW

93. Simplify: $\frac{24}{54}$

94. Find: $\frac{3}{4}\left(\frac{8}{5}\right)$

95. Find: $5\frac{2}{3} \div 2\frac{5}{9}$

96. Find: $\frac{3}{10} + \frac{2}{15}$

CHALLENGE PROBLEMS

97. What is the set of nonnegative integers?

98. Is 0.10100100010000 . . . a repeating decimal? Explain.

Find a rational number between each pair of numbers.

99. $\frac{1}{8}$ and $\frac{1}{9}$

100. $1.7\overline{1}$ and $1.7\overline{2}$

SECTION 1.4
Adding Real Numbers; Properties of Addition

Objectives

1. Add two numbers that have the same sign.
2. Add two numbers that have different signs.
3. Use properties of addition.
4. Identify opposites (additive inverses).

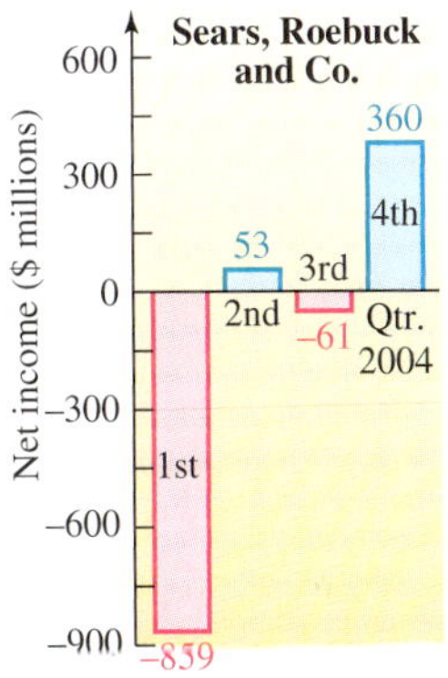

Source: 2004 Sears Annual Report

In the graph on the left, signed numbers are used to show the financial performance of Sears, Roebuck and Company for the year 2004. Positive numbers indicate *profits* and negative numbers indicate *losses*. To find Sears' net income (in millions of dollars), we need to calculate the following sum:

$$\text{Net income} = -859 + 53 + (-61) + 360$$

In this section, we discuss how to perform this addition and others involving signed numbers.

1 Add Two Numbers That Have the Same Sign.

A number line can be used to explain the addition of signed numbers. For example, to compute $5 + 2$, we begin at 0 and draw an arrow five units long that points right. It represents 5. From the tip of that arrow, we draw a second arrow two units long that points right. It represents 2. Since we end up at 7, it follows that $5 + 2 = 7$.

The Language of Algebra

The names of the parts of an addition fact are:

Addend *Addend* *Sum*

$5 + 2 = 7$

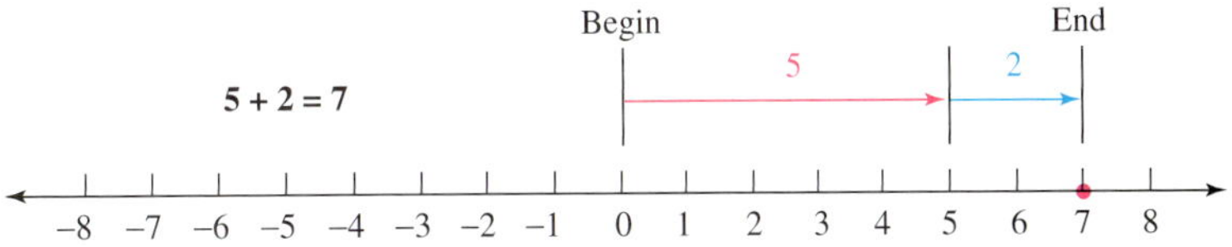

To compute $-5 + (-2)$, we begin at 0 and draw an arrow five units long that points left. It represents -5. From the tip of that arrow, we draw a second arrow two units long that points left. It represents -2. Since we end up at -7, it follows that $-5 + (-2) = -7$.

Notation

To avoid confusion, we write negative numbers within parentheses to separate the negative sign − from the addition symbol +.

$-5 + (-2)$

To check this result, think of the problem in terms of money. If you lost \$5 (-5) and then lost another \$2 (-2), you would have lost a total of \$7 (-7).

When we use a number line to add numbers with the same sign, the arrows point in the same direction and they build upon each other. Furthermore, the answer has the same sign as the numbers that we added. These observations suggest the following rules.

Adding Two Numbers That Have the Same (Like) Signs

1. To add two positive numbers, add them as usual. The final answer is positive.
2. To add two negative numbers, add their absolute values and make the final answer negative.

EXAMPLE 1 Add: **a.** $-20 + (-15)$ **b.** $-7.89 + (-0.6)$ **c.** $-\frac{1}{3} + \left(-\frac{1}{2}\right)$

Strategy We will use the rule for adding two numbers that have the same sign.

Why In each case, we are asked to add two negative numbers.

Solution

a. $-20 + (-15) = -35$ Add their absolute values, 20 and 15, to get 35. Then make the final answer negative.

b. Add their absolute values, 7.89 and 0.6.

$$\begin{array}{r} 7.89 \\ +0.6 \\ \hline 8.49 \end{array}$$ Remember to align the decimal points when adding decimals.

Then make the final answer negative: $-7.89 + (-0.6) = -8.49$.

c. Add their absolute values, $\frac{1}{3}$ and $\frac{1}{2}$.

$$\frac{1}{3} + \frac{1}{2} = \frac{2}{6} + \frac{3}{6}$$ The LCD is 6. Build each fraction: $\frac{1}{3} \cdot \frac{2}{2} = \frac{2}{6}$ and $\frac{1}{2} \cdot \frac{3}{3} = \frac{3}{6}$.

$$= \frac{5}{6}$$ Add the numerators and write the sum over the LCD.

Then make the final answer negative: $-\frac{1}{3} + \left(-\frac{1}{2}\right) = -\frac{5}{6}$.

The Language of Algebra
Two negative numbers, as well as two positive numbers, are said to have *like* signs.

Success Tip
The sum of two positive numbers is *always* positive. The sum of two negative numbers is *always* negative.

Self Check 1 Add: **a.** $-51 + (-9)$ **b.** $-12.3 + (-0.88)$ **c.** $-\frac{1}{4} + \left(-\frac{2}{3}\right)$

Now Try **Problems 15, 21, and 25**

2 Add Two Numbers That Have Different Signs.

To compute $5 + (-2)$, we begin at 0 and draw an arrow five units long that points right. From the tip of that arrow, we draw a second arrow two units long that points left. Since we end up at 3, it follows that $5 + (-2) = 3$. In terms of money, if you won \$5 and then lost \$2, you would have \$3 left.

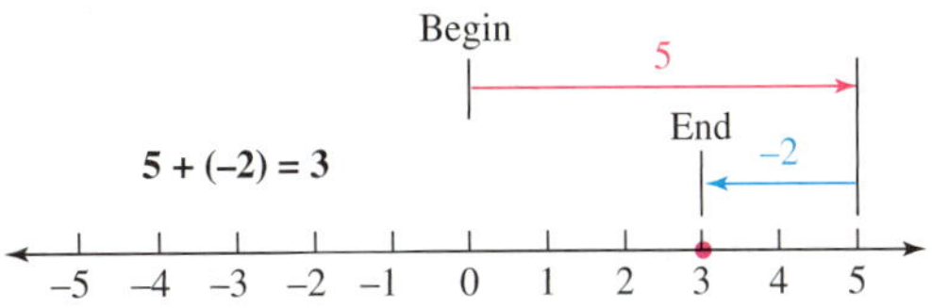

The Language of Algebra
A positive number and a negative number are said to have *unlike* signs.

To compute $-5 + 2$, we begin at 0 and draw an arrow five units long that points left. From the tip of that arrow, we draw a second arrow two units long that points right. Since we end up at -3, it follows that $-5 + 2 = -3$. In terms of money, if you lost \$5 and then won \$2, you have lost \$3.

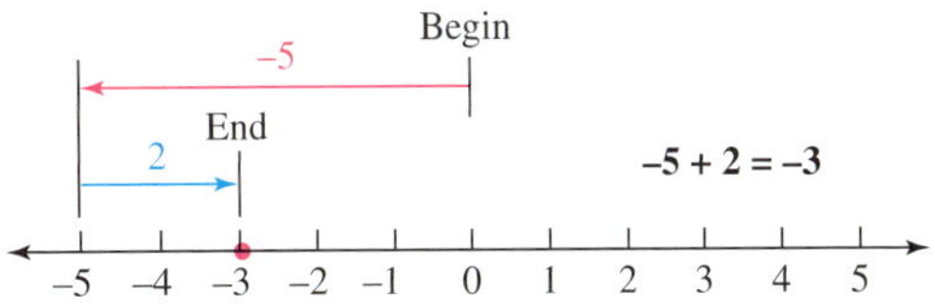

When we use a number line to add numbers with different signs, the arrows point in opposite directions and the longer arrow determines the sign of the answer. If the longer arrow represents a positive number, the sum is positive. If it represents a negative number, the sum is negative. These observations suggest the following rules.

Adding Two Numbers That Have Different (Unlike) Signs

To add a positive number and a negative number, subtract the smaller absolute value from the larger.

1. If the positive number has the larger absolute value, the final answer is positive.
2. If the negative number has the larger absolute value, make the final answer negative.

EXAMPLE 2 Add: **a.** $-20 + 32$ **b.** $5.7 + (-7.4)$ **c.** $-\frac{19}{25} + \frac{2}{5}$

Strategy We will use the rule for adding two numbers that have different (unlike) signs.

Why In each case, we are asked to add a positive number and a negative number.

Solution

a. $-20 + 32 = 12$ Subtract the smaller absolute value from the larger: $32 - 20 = 12$. The positive number, 32, has the larger absolute value, so the final answer is positive.

b. Subtract the smaller absolute value, 5.7, from the larger, 7.4.

$$\begin{array}{r} 7.4 \\ -5.7 \\ \hline 1.7 \end{array}$$

Remember to align the decimal points when subtracting decimals.

Since the negative decimal, -7.4, has the larger absolute value, make the final answer negative: $5.7 + (-7.4) = -1.7$.

c. Since $\frac{2}{5} = \frac{10}{25}$, the fraction $-\frac{19}{25}$ has the larger absolute value. We subtract the smaller absolute value from the larger:

$$\frac{19}{25} - \frac{2}{5} = \frac{19}{25} - \frac{10}{25}$$ Replace $\frac{2}{5}$ with the equivalent fraction $\frac{10}{25}$.

$$= \frac{9}{25}$$ Subtract the numerators and write the difference over the LCD.

Success Tip
The sum of two numbers with different signs may be positive or negative. The sign of the sum is the sign of the number with the greater absolute value.

Calculators
Entering negative numbers
We don't do anything special to enter positive numbers on a calculator. To enter a negative number, say -7.4, some calculators require the $-$ sign to be entered before entering 7.4 while others require the $-$ sign to be entered after entering 7.4. Consult your owner's manual to determine the proper keystrokes.

Since the negative fraction $-\frac{19}{25}$ has the larger absolute value, make the final answer negative: $-\frac{19}{25} + \frac{10}{25} = -\frac{9}{25}$.

Self Check 2 Add: **a.** $63 + (-87)$ **b.** $-6.27 + 8$ **c.** $-\frac{1}{10} + \frac{1}{2}$

Now Try **Problems 29, 33, and 35**

EXAMPLE 3 ***Accounting.*** Find the net earnings of Sears, Roebuck and Company for the year 2004 using the data in the graph on page 35.

Strategy To find the net income, we will add the quarterly profits and losses (in millions of dollars), performing the additions as they occur from left to right.

Why The phrase *net income* means that we should combine (add) the quarterly profits and losses to determine whether there was an overall profit or loss that year.

Solution

$$
\begin{aligned}
-859 + 53 + (-61) + 360 &= -806 + (-61) + 360 && \text{Add: } -859 + 53 = -806. \\
&= -867 + 360 && \text{Add: } -806 + (-61) = -867. \\
&= -507
\end{aligned}
$$

In 2004, Sears' net income was −$507 million.

The Language of Algebra *Net* refers to what remains after all the deductions (losses) have been accounted for. *Net income* is a term used in business that often is referred to as the *bottom line.* Net income indicates what a company has earned (or lost) in a given period of time (usually one year).

Self Check 3 Add: $650 + (-13) + 87 + (-155)$

Now Try **Problem 43**

3 Use Properties of Addition.

The addition of two numbers can be done in any order and the result is the same. For example, $8 + (-1) = 7$ and $-1 + 8 = 7$. This example illustrates that addition is **commutative.**

The Commutative Property of Addition

Changing the order when adding does not affect the answer.

For any real numbers a and b,

$$a + b = b + a$$

The Language of Algebra *Commutative* is a form of the word *commute,* meaning to go back and forth. *Commuter* trains take people to and from work.

In the following example, we add $-3 + 7 + 5$ in two ways. We will use grouping symbols ( ), called **parentheses,** to show this. Standard practice requires that the operation within the parentheses be performed first.

Method 1: Group −3 and 7

$$
\begin{aligned}
(-3 + 7) + 5 &= 4 + 5 \\
&= 9
\end{aligned}
$$

Method 2: Group 7 and 5

$$
\begin{aligned}
-3 + (7 + 5) &= -3 + 12 \\
&= 9
\end{aligned}
$$

It doesn't matter how we group the numbers in this addition; the result is 9. This example illustrates that addition is **associative.**

The Associative Property of Addition

Changing the grouping when adding does not affect the answer.
For any real numbers a, b, and c,

$$(a + b) + c = a + (b + c)$$

Sometimes, an application of the associative property can simplify a computation.

The Language of Algebra
Associative is a form of the word *associate,* meaning to join a group. The NBA (National Basketball Association) is a group of professional basketball teams.

EXAMPLE 4 Find the sum: $98 + (2 + 17)$

Strategy We will use the associative property to group 2 with 98. Then, we evaluate the expression by following the rules for the order of operations.

Why It is helpful to regroup because 98 and 2 are a pair of numbers that are easily added.

Solution

$98 + (2 + 17) = (98 + 2) + 17$ Use the associative property of addition to regroup.

$= 100 + 17$ Do the addition within the parentheses first.

$= 117$

Self Check 4 Find the sum: $(39 + 25) + 75$

Now Try Problem 53

EXAMPLE 5 ***Game Shows.*** A contestant on *Jeopardy!* correctly answered the first question to win \$100, missed the second to lose \$200, correctly answered the third to win \$300, and missed the fourth to lose \$400. What is her score after answering four questions?

Strategy We can represent money won by a positive number and money lost by a negative number. Her score is the sum of 100, −200, 300, and −400. Instead of doing the additions from left to right, we will use another approach. Applying the commutative and associative properties, we will add the positives, add the negatives, and then add those results.

Why It is easier to add numbers that have the same sign than numbers that have different signs. This method minimizes the possibility of an error, because we only have to add numbers that have different signs once.

Solution

$100 + (-200) + 300 + (-400)$

$= (100 + 300) + [(-200) + (-400)]$ Reorder the numbers. Group the positives together. Group the negatives together using brackets [ ].

$= 400 + (-600)$ Add the positives. Add the negatives.

$= -200$ Add the results.

After four questions, her score was −\$200, which represents a loss of \$200.

Self Check 5 Add: $-6 + 1 + (-4) + (-5) + 9$

Now Try Problem 49

The Language of Algebra
Identity is a form of the word *identical*, meaning the same. You have probably seen *identical* twins.

Whenever we add 0 to a number, the result is the number. Therefore, $8 + 0 = 8$, $2.3 + 0 = 2.3$, and $0 + (-16) = -16$. These examples illustrate the **addition property of 0.** Since any number added to 0 remains the same, 0 is called the **identity element** for addition.

Addition Property of 0 (Identity Property of Addition)

When 0 is added to any real number, the result is the same real number.

For any real number a,

$$a + 0 = a \quad \text{and} \quad 0 + a = a$$

4 Identify Opposites (Additive Inverses).

The Language of Algebra
Don't confuse the words *opposite* and *reciprocal*. The opposite of 4 is -4. The reciprocal of 4 is $\frac{1}{4}$.

Recall that two numbers that are the same distance from 0 on a number line, but on opposite sides of it, are called **opposites.** To develop a property for adding opposites, we will find $-4 + 4$ using a number line. We begin at 0 and draw an arrow four units long that points left, to represent -4. From the tip of that arrow, we draw a second arrow, four units long that points right, to represent 4. We end up at 0; therefore, $-4 + 4 = 0$.

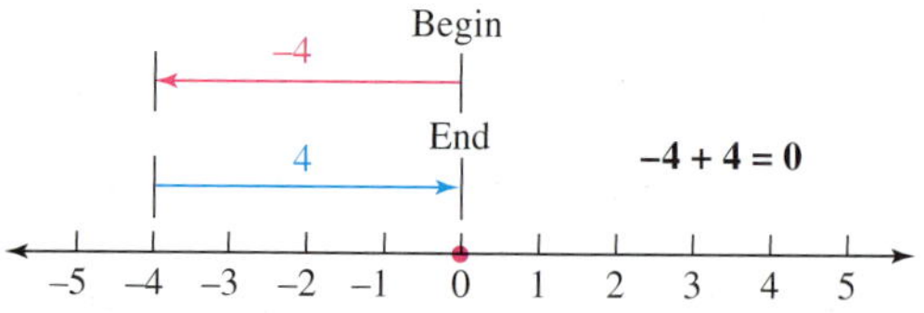

This example illustrates that when we add opposites, the result is 0. Therefore, $1.6 + (-1.6) = 0$ and $-\frac{3}{4} + \frac{3}{4} = 0$. Also, whenever the sum of two numbers is 0, those numbers are opposites. For these reasons, opposites are also called **additive inverses.**

Addition Property of Opposites (Inverse Property of Addition)

The sum of a number and its opposite (additive inverse) is 0.

For any real number a and its opposite or additive inverse $-a$,

$$a + (-a) = 0$$ Read $-a$ as "the opposite of a."

EXAMPLE 6 Add: $12 + (-5) + 6 + 5 + (-12)$

Strategy Instead of working from left to right, we will use the commutative and associative properties of addition to add pairs of opposites.

Why Since the sum of a number and its opposite is 0, it is helpful to identify such pairs in an addition.

Solution

opposites (12 and −12)

$$12 + (-5) + 6 + 5 + (-12) = 0 + 0 + 6$$
$$= 6$$

opposites (−5 and 5)

Self Check 6 Add: $8 + (-1) + 6 + 5 + (-8) + 1$

Now Try **Problem 61**

ANSWERS TO SELF CHECKS **1. a.** -60 **b.** -13.18 **c.** $-\frac{11}{12}$ **2. a.** -24 **b.** 1.73 **c.** $\frac{2}{5}$ **3.** 569 **4.** 139 **5.** -5 **6.** 11

STUDY SET 1.4

VOCABULARY

Fill in the blanks.

1. In the addition statement $-2 + 5 = 3$, the result, 3, is called the _____.
2. Two numbers that are the same distance from 0 on a number line, but on opposite sides of it, are called _________ or additive ________.
3. The ___________ property of addition states that changing the order when adding does not affect the answer. The ___________ property of addition states that changing the grouping when adding does not affect the answer.
4. Since any number added to 0 remains the same (is identical), the number 0 is called the ________ element for addition.

CONCEPTS

5. For each pair of numbers, which one has the larger absolute value?
   a. 6 or 5 **b.** 8.9 or -9.2
6. Determine whether each statement is true or false.
   a. The sum of a number and its opposite is always 0.
   b. The sum of two negative numbers is always negative.
   c. The sum of two numbers with different signs is always negative.
7. For each addition, just determine the sign of the answer.
   a. $39 + (-64)$ **b.** $-189 + 198$
8. Complete each property of addition. Then give its name.
   a. $a + (-a) =$
   b. $a + 0 =$
   c. $a + b = b +$
   d. $(a + b) + c = a +$
9. Use the commutative property of addition to complete each statement.
   a. $-5 + 1 =$ ________
   b. $15 + (-80.5) =$ ________
   c. $-20 + (4 + 20) = -20 + ($ $)$
   d. $(2.1 + 3) + 6 = ($ $) + 6$
10. Use the associative property of addition to complete each statement.
    a. $(-6 + 2) + 8 =$ ________
    b. $-7 + (7 + 3) =$ ________
11. What properties were used in Step 1 and Step 2 of the solution?
    $$(99 + 4) + 1 = (4 + 99) + 1 \quad \text{Step 1}$$
    $$= 4 + (99 + 1) \quad \text{Step 2}$$
    $$= 4 + 100$$
    $$= 104$$
12. Consider: $-3 + 6 + (-9) + 8 + (-4)$
    a. Add all the positives in the expression.
    b. Add all of the negatives.
    c. Add the results from parts **a** and **b.**

NOTATION

13. a. Express the commutative property of addition using the variables x and y.

b. Express the associative property of addition using the variables x, y, and z.

14. Fill in the blank: We read $-a$ as "the ________ of a."

GUIDED PRACTICE

***Add.* See Example 1.**

15. $-8 + (-1)$

16. $-3 + (-2)$

17. $-5 + (-12)$

18. $-4 + (-14)$

19. $-29 + (-45)$

20. $-23 + (-31)$

21. $-4.2 + (-6.1)$

22. $-5.1 + (-5.1)$

23. $-\frac{3}{4} + \left(-\frac{2}{3}\right)$

24. $-\frac{1}{5} + \left(-\frac{3}{4}\right)$

25. $-\frac{1}{4} + \left(-\frac{1}{10}\right)$

26. $-\frac{3}{8} + \left(-\frac{1}{3}\right)$

***Add.* See Example 2.**

27. $-7 + 4$

28. $-9 + 7$

29. $50 + (-11)$

30. $27 + (-30)$

31. $15.84 + (-15.84)$

32. $9.19 + (-9.19)$

33. $-6.25 + 8.5$

34. $21.37 + (-12.1)$

35. $-\frac{7}{15} + \frac{3}{15}$

36. $-\frac{8}{11} + \frac{3}{11}$

37. $\frac{1}{2} + \left(-\frac{1}{8}\right)$

38. $\frac{5}{6} + \left(-\frac{1}{4}\right)$

***Add.* See Examples 3 and 5.**

39. $8 + (-5) + 13$

40. $17 + (-12) + (-23)$

41. $21 + (-27) + (-9)$

42. $-32 + 12 + 17$

43. $-27 + (-3) + (-13) + 22$

44. $53 + (-27) + (-32) + (-7)$

45. $-20 + (-16) + 10$

46. $-13 + (-16) + 4$

47. $19.35 + (-20.21) + 1.53$

48. $33.12 + (-35.7) + 2.98$

49. $-60 + 70 + (-10) + (-10) + 205$

50. $-100 + 200 + (-300) + (-100) + 200$

***Apply the associative property of addition to find the sum.* See Example 4.**

51. $-99 + (99 + 215)$

52. $67 + (-67 + 127)$

53. $(-112 + 56) + (-56)$

54. $(-67 + 5) + (-5)$

55. $\frac{1}{8} + \left(\frac{7}{8} + \frac{2}{3}\right)$

56. $\left(\frac{1}{2} + \frac{9}{16}\right) + \frac{7}{16}$

57. $(12.4 + 1.9) + 1.1$

58. $87.6 + (2.4 + 1.7)$

***Add.* See Example 6.**

59. $-1 + 9 + 1$

60. $5 + 8 + (-5)$

61. $-7 + 5 + (-10) + 7$

62. $-3 + 6 + (-9) + (-6)$

63. $-8 + 11 + (-11) + 8 + 1$

64. $2 + 15 + (-15) + 8 + (-2)$

65. $-2.1 + 6.5 + (-8.2) + 2.1$

66. $0.9 + 0.5 + (-0.2) + (-0.9)$

TRY IT YOURSELF

Add.

67. $-9 + 81 + (-2)$

68. $11 + (-21) + (-13)$

69. $0 + (-6.6)$

70. $0 + (-2.14)$

71. $-\frac{9}{16} + \frac{7}{16}$

72. $-\frac{3}{4} + \frac{1}{4}$

73. $-6 + (-8)$

74. $-4 + (-3)$

75. $-167 + 167$

76. $-25 + 25$

77. $19.2 + (-41.3)$

78. $57.93 + (-93.27)$

79. $2{,}345 + (-178)$

80. $-4{,}061 + 5{,}000$

81. $3 + (-6) + (-3) + 74$

82. $4 + (-3) + (-4) + 5$

83. $-\frac{1}{4} + \left(-\frac{2}{7}\right)$

84. $-\frac{3}{32} + \left(-\frac{1}{2}\right)$

85. $-0.2 + (-0.3) + (-0.4)$

86. $-0.9 + (-1.9) + (-2.9)$

APPLICATIONS

87. MILITARY SCIENCE During a battle, an army retreated 1,500 meters, regrouped, and advanced 2,400 meters. The next day, it advanced another 1,250 meters. Find the army's net gain.

88. HEALTH Find the point total for the six risk factors (in blue) on the medical questionnaire. Then use the table to determine the patient's risk of contracting heart disease in the next 10 years.

Age	Points	Total Cholesterol	Points
34	−1	150 reading	−3
HDL Cholesterol	**Points**	**Blood Pressure**	**Points**
62 reading	−2	124/100	3
Diabetic	**Points**	**Smoker**	**Points**
Yes	2	Yes	2

10-Year Heart Disease Risk

Total Points	Risk	Total Points	Risk
−2 or less	1%	5	4%
−1 to 1	2%	6	6%
2 to 3	3%	7	6%
4	4%	8	7%

89. GOLF The leaderboard below shows the top finishers from the 1997 Masters Golf Tournament. Scores for each round are compared to *par,* the standard number of strokes necessary to complete the course. A score of -2, for example, indicates that the golfer used two strokes less than par to complete the course. A score of 5 indicates five strokes more than par. Determine the tournament total for each golfer.

Leaderboard

	Round 1	2	3	4	Total
Tiger Woods	–2	–6	–7	–3	
Tom Kite	+5	–3	–6	–2	
Tommy Tolles	0	0	0	–5	
Tom Watson	+3	–4	–3	0	

90. SUBMARINES A submarine was cruising at a depth of 1,250 feet. The captain gave the order to climb 550 feet. Compared to sea level, find the new depth of the sub.

91. CREDIT CARDS Refer to the monthly statement. What is the new balance?

Previous Balance	New Purchases, Fees, Advances & Debits	Payments & Credits	New Balance
3,660.66	1,408.78	3,826.58	

Billing Date	Date Payment Due	Credit Line
04/21/08	05/16/08	9,100

92. POLITICS The following proposal to limit campaign contributions was on the ballot in a state election, and it passed. What will be the net fiscal impact on the state government?

Proposition 212 Campaign Spending Limits

YES ☐
NO ☐

Limits contributions to $200 in state campaigns. Financial impact: Cost of $4.5 million for enforcement. Increases state revenue by $6.7 million by eliminating tax deductions for lobbying.

93. MOVIE LOSSES According to the Numbers Box Office Data website, the movie *Stealth,* released in 2005 by Sony Pictures, cost about $176,350,000 to produce, promote, and distribute. It reportedly earned back just $76,700,000 worldwide. Express the dollar loss suffered by Sony as a signed number.

94. STOCKS The last entry on the line for June 12 indicates that one share of Walt Disney Co. stock lost $0.81 in value that day. How much did the value of a share of Disney stock rise or fall over the 5-day period?

June 12	43.88	23.38	Disney	.21	0.5	87	–43	40.75	–.81
June 13	43.88	23.38	Disney	.21	0.5	86	–15	40.19	–.56
June 14	43.88	23.38	Disney	.21	0.5	87	–50	41.00	+.81
June 15	43.88	23.38	Disney	.21	0.5	89	–28	41.81	+.81
June 16	43.88	23.38	Disney				–15	41.19	–.63

Based on data from the *Los Angeles Times*

95. CHEMISTRY An atom is composed of protons (with a charge of $+1$), neutrons (with no charge), and electrons (with a charge of -1). Two simple models of atoms are shown. What is the overall charge of each atom?

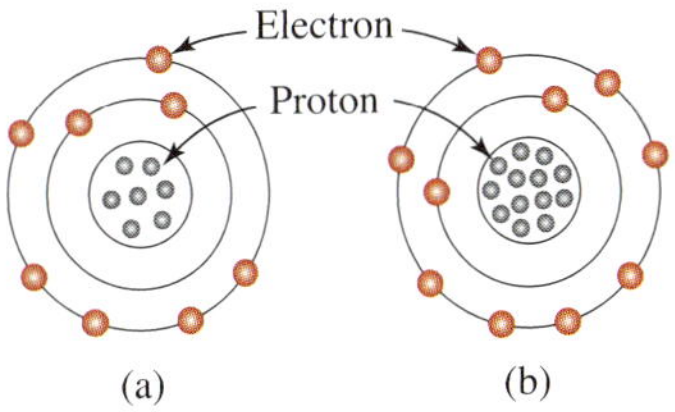

96. PHYSICS In the illustration, arrows show the two forces acting on a lamp hanging from a ceiling. What is the sum of the forces?

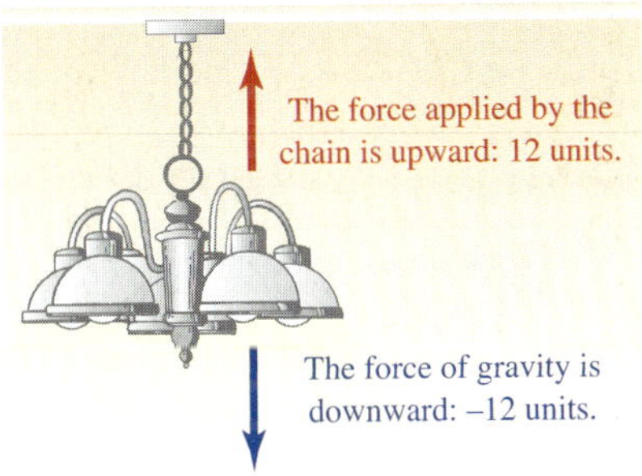

97. THE BIG EASY The city of New Orleans lies, on average, 6 feet below sea level. What is the elevation of the top of an 85-foot tall building in New Orleans?

98. ELECTRONICS A closed circuit contains two batteries and three resistors. The sum of the voltages in the loop must be 0. Is it?

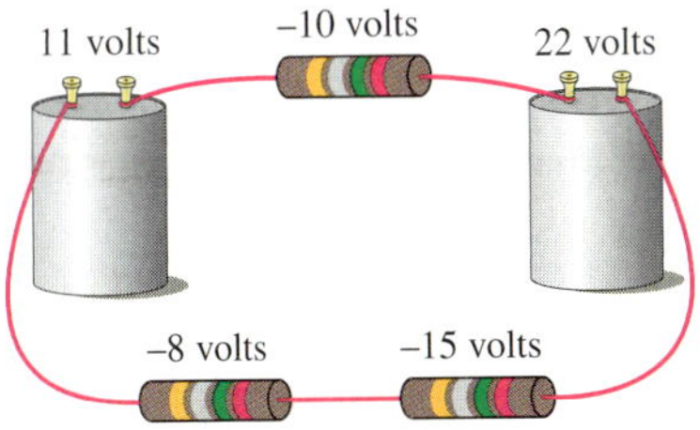

99. ACCOUNTING The 2004 quarterly profits and losses of Greyhound Bus Lines are shown in the table. Losses are denoted using parentheses. Calculate the company's total net income for 2004.

Quarter	Net income ($ million)
1st	(21.1)
2nd	(4.3)
3rd	2.6
4th	(0.4)

Source: www.greyhound.com

100. POLITICS Six months before an election, the incumbent trailed the challenger by 18 points. To overtake her opponent, the incumbent decided to use a four-part strategy. Each part of the plan is shown below, with the expected point gain. With these gains, will the incumbent overtake the challenger on election day?

- TV ads $+10$
- Voter mailing $+3$
- Union endorsement $+2$
- Telephone calls $+1$

WRITING

101. Explain why the sum of two positive numbers is always positive and the sum of two negative numbers is always negative.

102. Explain why the sum of a negative number and a positive number is sometimes positive, sometimes negative, and sometimes zero.

REVIEW

103. True or false: Every real number can be expressed as a decimal.

104. Multiply: $\frac{1}{3} \cdot \frac{1}{3}$

105. What two numbers are a distance of 6 away from -3 on the number line?

106. Graph: $\left\{-2.5,\ \sqrt{2},\ \frac{11}{3},\ -0.333\ldots,\ 0.75\right\}$

CHALLENGE PROBLEMS

107. A set is said to be *closed under addition* if the sum of any two of its members is also a member of the set. Is the set $\{-1, 0, 1\}$ a closed set under addition? Explain.

108. Think of two numbers. First, add the absolute value of the two numbers, and write your answer. Second, add the two numbers, take the absolute value of that sum, and write that answer. Do the two answers agree? Can you find two numbers that produce different answers? When do you get answers that agree, and when don't you?

SECTION 1.5 Subtracting Real Numbers

Objectives

1. Use the definition of subtraction.
2. Solve application problems using subtraction.

In this section, we discuss a rule to use when subtracting signed numbers.

1 Use the Definition of Subtraction.

A minus symbol $-$ is used to indicate subtraction. However, this symbol is also used in two other ways, depending on where it appears in an expression.

$5 - 18$ This is read as "five minus eighteen."

-5 This is usually read as "negative five." It could also be read as "the additive inverse of five" or "the opposite of five."

$-(-5)$ This is usually read as "the opposite of negative five." It could also be read as "the additive inverse of negative five."

In $-(-5)$, parentheses are used to write the opposite of a negative number. When such expressions are encountered in computations, we simplify them by finding the opposite of the number within the parentheses.

$-(-5) = 5$ Read as "the opposite of negative five is five."

This observation illustrates the following rule.

Opposite of an Opposite

The opposite of the opposite of a number is that number.

For any real number a,

$$-(-a) = a$$ Read as "the opposite of the opposite of a is a."

EXAMPLE 1

Simplify each expression: **a.** $-(-45)$ **b.** $-(-h)$ **c.** $-|-10|$

Strategy To simplify each expression, we will use the concept of opposite.

Why In each case, the outermost $-$ symbol is read as "the opposite."

Solution

a. The number within the parentheses is -45. Its opposite is 45. Therefore, $-(-45) = 45$.

b. The opposite of the opposite of h is h. Therefore, $-(-h) = h$.

c. The notation $-|-10|$ means "the opposite of the absolute value of negative ten." Since $|-10| = 10$, we have:

$$-|\mathbf{-10}| = \mathbf{-10}$$ The absolute value bars is do not affect the $-$ symbol outside them. Therefore, the result is negative.

Self Check 1

Simplify each expression: **a.** $-(-1)$ **b.** $-(-y)$ **c.** $-|-500|$

Now Try **Problems 13, 15, and 17**

To develop a rule for subtraction, we consider the following illustration. It represents the subtraction $5 \quad 2 = 3$.

The Language of Algebra

The names of the parts of a subtraction fact are:

Minuend — 5; *Subtrahend* — 2; *Difference* — 3

$$5 - 2 = 3$$

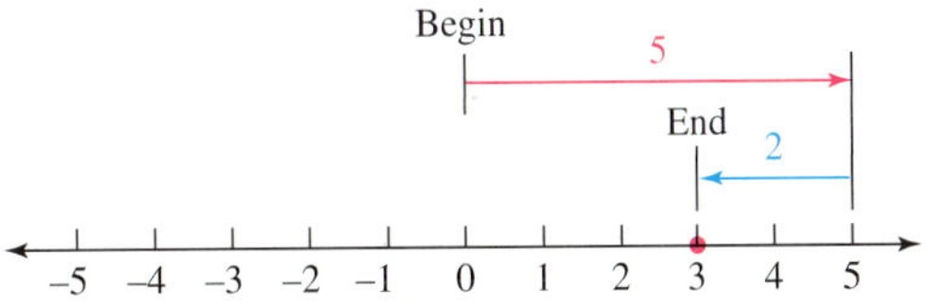

The illustration above also represents the addition $5 + (-2) = 3$. We see that

Subtracting 2 from 5 is the same as adding the opposite of 2 to 5.

$$5 - 2 = 3 \qquad\qquad 5 + (-2) = 3$$

The results are the same.

This observation suggests the following definition.

Subtraction of Real Numbers

To subtract two real numbers, add the first number to the opposite (additive inverse) of the number to be subtracted.

For any real numbers a and b,

$$a - b = a + (-b)$$

EXAMPLE 2 Subtract and check the result.

a. $-13 - 8$ **b.** $-7 - (-45)$ **c.** $\frac{1}{4} - \left(-\frac{1}{8}\right)$

Strategy To find each difference, we will apply the rule for subtraction: Add the first number to the opposite of the number to be subtracted.

Why It is easy to make an error when subtracting signed numbers. We will probably be more accurate if we write each subtraction as addition of the opposite.

Solution

a. We read $-13 - 8$ as "negative thirteen *minus* eight." Subtracting 8 is the same as adding -8.

Change the subtraction to addition.

$$-13 - 8 = -13 + (-8) = -21$$

Change the number being subtracted to its opposite.

To check, we add the *difference,* -21, and the *subtrahend,* 8, to obtain the *minuend,* -13.

Check: $-21 + 8 = -13$

b. We read $-7 - (-45)$ as "negative seven *minus* negative forty-five." Subtracting -45 is the same as adding 45.

Add

$$-7 - (-45) = -7 + 45 = 38$$

the opposite

Check: $38 + (-45) = -7$

c. $\frac{1}{4} - \left(-\frac{1}{8}\right) = \frac{2}{8} - \left(-\frac{1}{8}\right)$ Express $\frac{1}{4}$ in terms of the LCD 8: $\frac{1}{4} \cdot \frac{2}{2} = \frac{2}{8}$.

$= \frac{2}{8} + \frac{1}{8}$ To subtract, add the opposite.

$= \frac{3}{8}$

Check: $\frac{3}{8} + \left(-\frac{1}{8}\right) = \frac{2}{8} = \frac{1}{4}$

Self Check 2 Subtract: **a.** $-32 - 25$ **b.** $17 - (-12)$ **c.** $-\frac{1}{3} - \left(-\frac{3}{4}\right)$

Now Try **Problems 25, 37, and 53**

The Language of Algebra
When we change a number to its opposite, we say we have *changed* (or *reversed*) its sign.

The Language of Algebra
The rule for subtracting real numbers is often summarized as: *Subtracting a number is the same as adding its opposite.*

Calculators
The subtraction key
When using a calculator to subtract signed numbers, be careful to distinguish between the *subtraction* key $-$ and the keys that are used to enter negative values: $+/-$ on a scientific calculator and $(-)$ on a graphing calculator.

EXAMPLE 3 **a.** Subtract 0.5 from 4.6 **b.** Subtract 4.6 from 0.5

Strategy We will translate each phrase to mathematical symbols and then perform the subtraction. We must be careful when translating the instruction to subtract one number *from* another number.

Why The order of the numbers in each word phrase must be reversed when we translate it to mathematical symbols.

Solution

a. The number to be subtracted is 0.5.

Subtract 0.5 from 4.6

$4.6 - 0.5 = 4.1$ To translate, reverse the order in which 0.5 and 4.6 appear in the sentence.

b. The number to be subtracted is 4.6.

Subtract 4.6 from 0.5

$0.5 - 4.6 = 0.5 + (-4.6)$ To translate, reverse the order in which 4.6 and 0.5 appear in the sentence. Add the opposite of 4.6.

$= -4.1$

Caution Notice from parts **a** and **b** that $4.6 - 0.5 \neq 0.5 - 4.6$. This result illustrates an important fact: subtraction is *not* commutative. When subtracting two numbers, it is important that we write them in the correct order, because, in general, $a - b \neq b - a$.

Self Check 3 **a.** Subtract 2.2 from 4.9 **b.** Subtract 4.9 from 2.2

Now Try Problem 57

EXAMPLE 4 Perform the operations: $-9 - 15 + 20 - (-6)$

Strategy This expression contains addition and subtraction. We will write each subtraction as addition of the opposite and then evaluate the expression.

Why It is easy to make an error when subtracting signed numbers. We will probably be more accurate if we write each subtraction as addition of the opposite.

Solution

$$-9 - 15 + 20 - (-6) = -9 + (-15) + 20 + 6$$

$$= -24 + 26$$ Add the negatives. Add the positives. Add the results.

$$= 2$$

Self Check 4 Perform the operations: $-40 - (-10) + 7 - (-15)$

Now Try Problem 63

2 Solve Application Problems Using Subtraction.

Subtraction finds the *difference* between two numbers. When we find the difference between the maximum value and the minimum value of a collection of measurements, we are finding the **range** of the values.

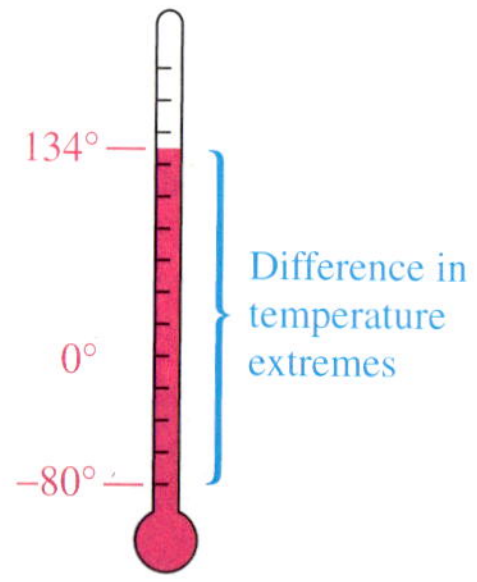

EXAMPLE 5 ***U.S. Temperatures.*** The record high temperature in the United States was 134°F in Death Valley, California, on July 10, 1913. The record low was −80°F at Prospect Creek, Alaska, on January 23, 1971. Find the temperature range for these extremes.

Strategy We will subtract the lowest temperature from the highest temperature.

Why The *range* of a collection of data indicates the spread of the data. It is the difference between the largest and smallest values.

Solution

$$134 - (-80) = 134 + 80 \quad \text{134° is the higher temperature and −80° is the lower.}$$
$$= 214$$

The temperature range for these extremes is 214°F.

Now Try Problem 89

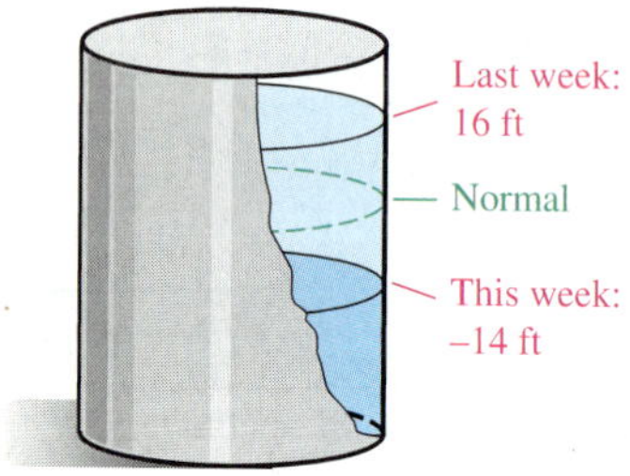

EXAMPLE 6 ***Water Levels.*** In one week, the water level in a storage tank went from 16 feet above normal to 14 feet below normal. Find the change in the water level.

Strategy We can represent a water level above normal using a positive number and a water level below normal using a negative number. To find the change in the water level, we will subtract.

Why In general, *to find the change in a quantity, we subtract the earlier value from the later value.*

Solution

$$-14 - 16 = -14 + (-16) \quad \text{The earlier water level, 16, is subtracted from the later water level, −14.}$$
$$= -30$$

The negative result indicates that the water level fell 30 feet that week.

Now Try Problem 91

Caution

When applying the subtraction rule, *do not* change the first number:

$$-14 - 16 = -14 + (-16)$$

ANSWERS TO SELF CHECKS **1. a.** 1 **b.** y **c.** −500 **2. a.** −57 **b.** 29 **c.** $\frac{5}{12}$ **3. a.** 2.7 **b.** −2.7 **4.** −8

STUDY SET 1.5

VOCABULARY

Fill in the blanks.

1. ________ finds the difference between two numbers.

2. In the subtraction $-2 - 5 = -7$, the result of -7 is called the ________.

3. The difference between the maximum and the minimum value of a collection of measurements is called the ______ of the values.

4. To find the _______ in a quantity, subtract the earlier value from the later value.

CONCEPTS

5. Find the opposite, or additive inverse, of each number.

a. 12 **b.** $-\frac{1}{5}$

c. 2.71 **d.** 0

6. Complete each statement.

a. $a - b = a +$

To subtract two numbers, add the first number to the ________ of the number to be subtracted.

b. $-(-a) =$

The opposite of the opposite of a number is that ________.

7. In each case, determine what number is being subtracted.

a. $5 - 8$ **b.** $-5 - (-8)$

8. Apply the rule for subtraction and fill in the blanks.

$$1 - (-9) = 1 \quad \square \quad = \quad \square$$

9. Use addition to check this subtraction: $15 - (-8) = 7$. Is the result correct?

10. Write each subtraction in the following expression as addition of the opposite.

$$-10 - 8 + (-23) + 5 - (-34)$$

NOTATION

11. Write each phrase using symbols. Then find its value.

a. One minus negative seven

b. The opposite of negative two

c. The opposite of the absolute value of negative three

d. Subtract 6 from 2

12. Write each expression in words.

a. $-(-m)$

b. $-2 - (-3)$

c. $x - (-y)$

GUIDED PRACTICE

***Simply each expression.* See Example 1.**

13. $-(-55)$
14. $-(-27.2)$
15. $-(-x)$
16. $-(-t)$
17. $-|-25|$
18. $-|-100|$
19. $-\left|-\frac{3}{16}\right|$
20. $-\left|-\frac{4}{3}\right|$

***Subtract.* See Example 2.**

21. $4 - 7$
22. $1 - 6$
23. $2 - 15$
24. $3 - 14$
25. $-6 - 4$
26. $-3 - 4$
27. $8 - (-3)$
28. $17 - (-21)$
29. $0 - 6$
30. $0 - 9$
31. $0 - (-1)$
32. $0 - (-8)$
33. $-1 - (-3)$
34. $-1 - (-7)$
35. $20 - (-20)$
36. $30 - (-30)$
37. $-2 - (-7)$
38. $-9 - (-1)$
39. $-14 - 55$
40. $-13 - 47$
41. $-44 - 44$
42. $-33 - 33$
43. $0 - (-12)$
44. $0 - 12$
45. $-0.9 - 0.2$
46. $-0.3 - 0.2$
47. $6.3 - 9.8$
48. $2.1 - 9.4$
49. $-1.5 - 0.81$
50. $-1.57 - (-0.8)$
51. $-\frac{1}{8} - \frac{3}{8}$
52. $-\frac{3}{4} - \frac{1}{4}$
53. $-\frac{9}{16} - \left(-\frac{1}{4}\right)$
54. $-\frac{1}{2} - \left(-\frac{1}{4}\right)$
55. $\frac{1}{3} - \frac{3}{4}$
56. $\frac{1}{6} - \frac{5}{8}$

***Perform the indicated operation.* See Example 3.**

57. Subtract -5 from 17.

58. Subtract 45 from -50.

59. Subtract 12 from -13.

60. Subtract -11 from -20.

***Perform the operations.* See Example 4.**

61. $8 - 9 - 10$
62. $1 - 2 - 3$
63. $-25 - (-50) - 75$
64. $-33 - (-22) - 44$
65. $-6 + 8 - (-1) - 10$
66. $-4 + 5 - (-3) - 13$
67. $61 - (-62) + (-64) - 60$
68. $93 - (-92) + (-94) - 95$

TRY IT YOURSELF

Perform the operations.

69. $244 - (-12)$

70. $354 - (-29)$

71. $-20 - (-30) - 50 + 40$

72. $-24 - (-28) - 48 - 44$

73. $-1.2 - 0.9$

74. $-2.52 - 1.72$

75. $\frac{1}{8} - \left(-\frac{5}{7}\right)$

76. $\frac{5}{8} - \left(-\frac{2}{9}\right)$

77. $-62 - 71 - (-37) + 99$

78. $-17 - 32 - (-85) - 51$

79. Subtract 47.5 from 0.

80. Subtract 30.3 from 0.

81. Subtract -137 from 12.

82. Subtract 512 from -47.

83. $-1{,}903 - (-1{,}732)$

84. $-300 - (-11)$

85. $2.83 - (-1.8)$

86. $4.75 - (-1.9)$

87. $-\frac{5}{6} - \frac{3}{4}$

88. $-\frac{3}{7} - \frac{2}{5}$

APPLICATIONS

89. THE EMPIRE STATE New York state's record high temperature of 108°F was set in 1926, and the record low of −52°F was set in 1979. What is the range of these temperature extremes?

90. EYESIGHT Nearsightedness, the condition where near objects are clear and far objects are blurry, is measured using negative numbers. Farsightedness, the condition where far objects are clear and near objects are blurry, is measured using positive numbers. Find the range in the measurements shown.

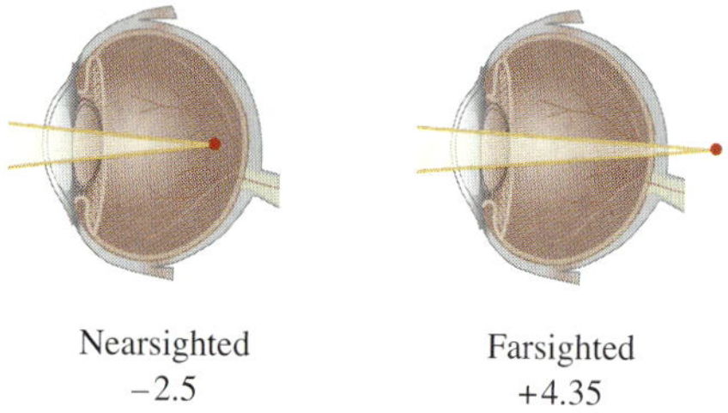

91. LAW ENFORCEMENT A burglar scored −18 on a lie detector test, a score that indicates deception. However, on a second test, he scored +3, a score that is inconclusive. Find the change in the scores.

92. RACING To improve handling, drivers often adjust the angle of the wheels of their car. When the wheel leans out, the degree measure is considered positive. When the wheel leans in, the degree measure is considered negative. Find the change in the position of the wheel shown in the next column.

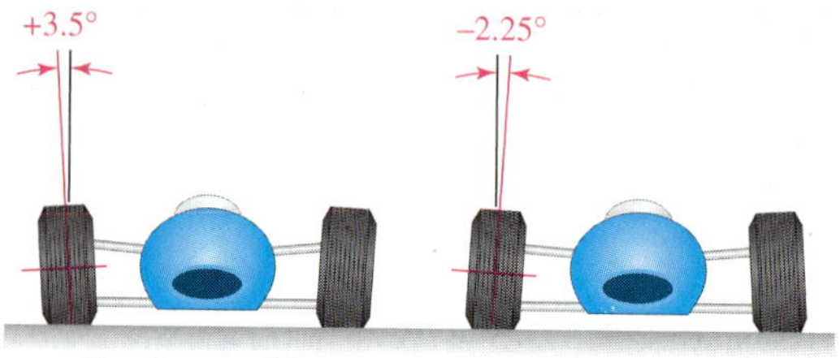

93. ***from Campus to Careers***

Lead Transportation Security Officer

Determine the change in the number of passengers using each airport in 2006 compared with 2005.

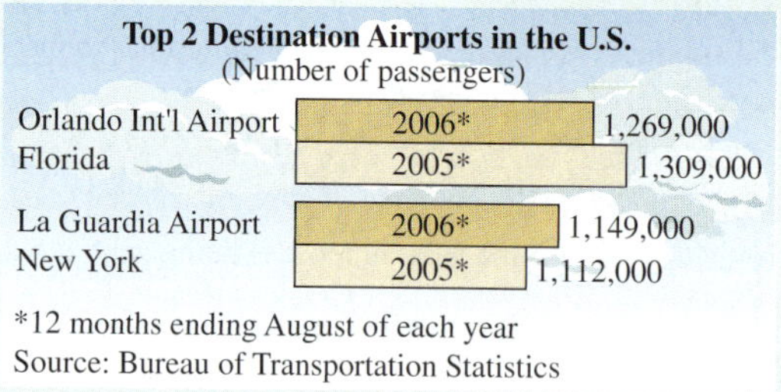

94. U.S. JOBS The table lists the three occupations that are predicted to have the largest job declines from 2004–2014. Complete the column labeled "Change."

Occupation	Number of jobs 2004	2014	Change
Farmers/ranchers	1,065,000	910,000	
Stock clerks	1,566,000	1,451,000	
Sewing machine operators	256,000	163,000	

Source: Bureau of Labor Statistics

95. GEOGRAPHY The elevation of Death Valley, California, is 282 feet below sea level. The elevation of the Dead Sea in Israel is 1,312 feet below sea level. Find the difference in their elevations.

96. CARD GAMES Gonzalo won the second round of a card game and earned 50 points. Matt and Hydecki had to deduct the value of each of the cards left in their hands from their score on the first round. Use this information to update the score sheet on the next page. (Face cards are counted as 10 points, aces as 1 point, and all others have the value of the number printed on the card.)

Running point total	Round 1	Round 2
Matt	+50	
Gonzalo	−15	
Hydecki	−2	

97. FOREIGN POLICY In 2004, Congress forgave \$4.1 billion of Iraqi debt owed to the United States. Before that, Iraq's total debt was estimated to be \$120.2 billion.

a. Which expression below can be used to find Iraq's total debt after getting debt relief from the United States?

i. $120.2 + 4.1$ **ii.** $120.2 - (-4.1)$

iii. $-120.2 - (-4.1)$ **iv.** $-120.2 - 4.1$

b. Find Iraq's total debt after getting the debt relief.

98. HISTORY Plato, a famous Greek philosopher, died in 347 B.C. at the age of 81. When was he born?

99. NASCAR Complete the table below to determine how many points the third, fourth, and fifth place finishers were behind the leader.

2006 Final Driver Standings			
Rank	Driver	Points	Points behind leader
1	Jimmie Johnson	6,475	. . .
2	Matt Kenseth	6,419	−56
3	Denny Hamlin	6,407	
4	Kevin Harvick	6,397	
5	Dale Earnhardt, Jr	6,328	

100. GAUGES With the engine off, the ammeter on a car reads 0. If the headlights, which draw a current of 7 amps, and the radio, which draws a current of 6 amps, are both turned on, what will be the new reading?

WRITING

101. Explain what it means when we say that subtraction is *not commutative.*

102. Why is addition of signed numbers taught before subtraction of signed numbers?

103. Explain why we know that the answer to $4 - 10$ is negative without having to do any computation.

104. Is the following statement true or false? Explain.

Having a debt of \$100 forgiven is equivalent to gaining \$100.

REVIEW

105. Find the prime factorization of 30.

106. Write the set of integers.

107. True or false: $-4 > -5$?

108. Use the associative property of addition to simplify the calculation: $-18 + (18 + 89)$.

CHALLENGE PROBLEMS

109. Suppose x is positive and y is negative. Determine whether each statement is true or false.

a. $x - y > 0$ **b.** $y - x < 0$

c. $-x < 0$ **d.** $-y < 0$

110. Find:

$$1 - 2 + 3 - 4 + 5 - 6 + . . . + 99 - 100$$

SECTION 1.6
Multiplying and Dividing Real Numbers; Multiplication and Division Properties

Objectives

1. Multiply signed numbers.
2. Use properties of multiplication.
3. Divide signed numbers.
4. Use properties of division.

In this section, we will develop rules for multiplying and dividing positive and negative numbers.

1 Multiply Signed Numbers.

Multiplication represents repeated addition. For example, 4(3) is equal to the sum of four 3's.

$$4(3) = 3 + 3 + 3 + 3$$
$$= 12$$

The Language of Algebra
The names of the parts of a multiplication fact are:

Factor, Factor, Product
$4(3) = 12$

This example illustrates that *the product of two positive numbers is positive.*

To develop a rule for multiplying a positive number and a negative number, we will find $4(-3)$, which is equal to the sum of four -3's.

$$4(-3) = -3 + (-3) + (-3) + (-3)$$
$$= -12$$

We see that the result is negative. As a check, think in terms of money. If you lose \$3 four times, you have lost a total of \$12, which is written $-\$12$. This example illustrates that *the product of a positive number and a negative number is negative.*

Multiplying Two Numbers That Have Different (Unlike) Signs

To multiply a positive number and a negative number, multiply their absolute values. Then make the final answer negative.

EXAMPLE 1 Multiply: **a.** $8(-12)$ **b.** $-151 \cdot 5$ **c.** $(-0.6)(1.2)$ **d.** $\frac{3}{4}\left(-\frac{4}{15}\right)$

Success Tip
The product of two numbers with unlike signs is *always* negative.

Strategy We will use the rule for multiplying two numbers that have different signs.

Why In each case, we are asked to multiply a positive number and a negative number.

Solution

a. $8(-12) = -96$ Multiply the absolute values, 8 and 12, to get 96. Since the signs are unlike, make the final answer negative.

b. $-151 \cdot 5 = -755$ Multiply the absolute values, 151 and 5, to get 755. Since the signs are unlike, make the final answer negative.

c. To find the product of these two decimals with unlike signs, first multiply their absolute values, 0.6 and 1.2.

$$\begin{array}{r} 1.2 \\ \times 0.6 \\ \hline 0.72 \end{array}$$

Place the decimal point in the result so that the answer has the same number of decimal places as the sum of the number of decimal places in the factors.

Then make the final answer negative: $(-0.6)(1.2) = -0.72$.

d. $\frac{3}{4}\left(-\frac{4}{15}\right) = -\frac{\overset{1}{\cancel{3}} \cdot \overset{1}{\cancel{4}}}{\underset{1}{\cancel{4}} \cdot \underset{1}{\cancel{3}} \cdot 5}$ Multiply the absolute values $\frac{3}{4}$ and $\frac{4}{15}$. Since the signs are unlike, make the final answer negative.

$= -\frac{1}{5}$ To simplify the fraction, factor 15 as $3 \cdot 5$. Remove the common factors 3 and 4 in the numerator and denominator.

Self Check 1 Multiply: **a.** $20(-3)$ **b.** $-3 \cdot 5$ **c.** $4.3(-2.6)$ **d.** $-\frac{5}{8} \cdot \frac{16}{25}$

Now Try **Problems 19, 25, and 27**

To develop a rule for multiplying two negative numbers, consider the following list, where we multiply -4 by factors that decrease by 1. We know how to find the first four products. Graphing those results on a number line is helpful in determining the last three products.

Factor decreases by 1 each time → Look for a pattern here

$$-4(3) = -12$$
$$-4(2) = -8$$
$$-4(1) = -4$$
$$-4(0) = 0$$
$$-4(-1) = ?$$
$$-4(-2) = ?$$
$$-4(-3) = ?$$

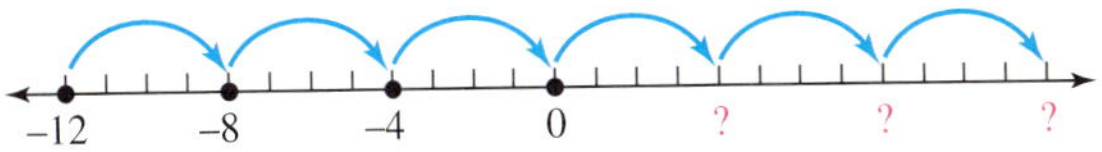

A graph of the products

From the pattern, we see that the product increases by 4 each time. Thus,

$$-4(-1) = 4, \quad -4(-2) = 8, \quad \text{and} \quad -4(-3) = 12$$

These results illustrate that *the product of two negative numbers is positive.* As a check, think of losing four debts of \$3. This is equivalent to gaining \$12. Therefore, $-4(-\$3) = \$12$.

Since the product of two positive numbers is positive, and the product of two negative numbers is also positive, we can summarize the multiplication rule as follows.

Multiplying Two Numbers That Have the Same (Like) Signs

To multiply two real numbers that have the same sign, multiply their absolute values. The final answer is positive.

EXAMPLE 2

Multiply: **a.** $-5(-6)$ **b.** $\left(-\frac{1}{2}\right)\left(-\frac{5}{8}\right)$

Strategy We will use the rule for multiplying two numbers that have the same sign.

Why In each case, we are asked to multiply two negative numbers.

Solution

a. $-5(-6) = 30$ Multiply the absolute values, 5 and 6, to get 30. Since both factors are negative, the final answer is positive.

b. $\left(-\frac{1}{2}\right)\left(-\frac{5}{8}\right) = \frac{5}{16}$ Multiply the absolute values, $\frac{1}{2}$ and $\frac{5}{8}$, to get $\frac{5}{16}$. Since the two factors have the same sign, the final answer is positive.

Success Tip
The product of two numbers with like signs is *always* positive.

Self Check 2 Multiply: **a.** $-15(-8)$ **b.** $-\frac{1}{4}\left(-\frac{1}{3}\right)$

Now Try **Problems 31 and 39**

2 Use Properties of Multiplication.

The multiplication of two numbers can be done in any order; the result is the same. For example, $-9(4) = -36$ and $4(-9) = -36$. This illustrates that multiplication is **commutative.**

The Commutative Property of Multiplication

Changing the order when multiplying does not affect the answer.

For any real numbers a and b,

$$ab = ba$$

In the following example, we multiply $-3 \cdot 7 \cdot 5$ in two ways. Recall that the operation within the parentheses should be performed first.

Method 1: Group −3 and 7	***Method 2: Group 7 and 5***
$(-3 \cdot 7)5 = (-21)5$	$-3(7 \cdot 5) = -3(35)$
$= -105$	$= -105$

It doesn't matter how we group the numbers in this multiplication; the result is -105. This example illustrates that multiplication is **associative.**

The Associative Property of Multiplication

Changing the grouping when multiplying does not affect the answer.

For any real numbers a, b, and c,

$$(ab)c = a(bc)$$

EXAMPLE 3 Multiply: **a.** $-5(-37)(-2)$ **b.** $-4(-3)(-2)(-1)$

Strategy First, we will use the commutative and associative properties of multiplication to reorder and regroup the factors. Then we will perform the multiplications.

Why Applying of one or both of these properties before multiplying can simplify the computations and lessen the chance of a sign error.

Solution Using the commutative and associative properties of multiplication, we can reorder and regroup the factors to simplify computations.

a. Since it is easy to multiply by 10, we will find $-5(-2)$ first.

$$-5(-37)(-2) = -5(-2)(-37)$$ Use the commutative property of multiplication.

$$= 10(-37)$$

$$= -370$$

b. $-4(-3)(-2)(-1) = 12(2)$ Multiply the first two factors and multiply the last two factors.

$$= 24$$

Self Check 3 Multiply: **a.** $-25(-3)(-4)$ **b.** $-1(-2)(-3)(-3)$

Now Try **Problems 43 and 47**

In Example 3a, we multiplied three negative numbers. In Example 3b, we multiplied four negative numbers. The results illustrate the following fact.

Multiplying Negative Numbers

The product of an even number of negative numbers is positive. The product of an odd number of negative numbers is negative.

Recall that the product of 0 and any whole number is 0. The same is true for any real number. Therefore, $-6 \cdot 0 = 0$, $\frac{7}{16} \cdot 0 = 0$, and $0(4.51) = 0$.

Multiplication Property of 0

The product of 0 and any real number is 0.
For any real number a,

$$0 \cdot a = 0 \quad \text{and} \quad a \cdot 0 = 0$$

Whenever we multiply a number by 1, the number remains the same. Therefore, $1 \cdot 6 = 6$, $4.57 \cdot 1 = 4.57$, and $1(-9) = -9$. Since any number multiplied by 1 remains the same (is identical), the number 1 is called the **identity element** for multiplication.

Multiplication Property of 1 (Identity Property of Multiplication)

The product of 1 and any number is that number.
For any real number a,

$$1 \cdot a = a \quad \text{and} \quad a \cdot 1 = a$$

Two numbers whose product is 1 are **reciprocals** or **multiplicative inverses** of each other. For example, 8 is the multiplicative inverse of $\frac{1}{8}$, and $\frac{1}{8}$ is the multiplicative inverse of 8, because $8 \cdot \frac{1}{8} = 1$. Likewise, $-\frac{3}{4}$ and $-\frac{4}{3}$ are multiplicative inverses because $-\frac{3}{4}\left(-\frac{4}{3}\right) = 1$. All real numbers, except 0, have a multiplicative inverse.

Multiplicative Inverses (Inverse Property of Multiplication)

The product of any number and its multiplicative inverse (reciprocal) is 1.
For any nonzero real number a,

$$a\left(\frac{1}{a}\right) = 1$$

EXAMPLE 4 Find the reciprocal of each number: **a.** $\frac{2}{3}$ **b.** $-\frac{2}{3}$ **c.** -11

Strategy To find the reciprocal of a fraction, we invert the numerator and the denominator.

Why We want the product of the given number and its reciprocal to be 1.

Solution

a. The reciprocal of $\frac{2}{3}$ is $\frac{3}{2}$ because $\frac{2}{3}\left(\frac{3}{2}\right) = 1$. *To find the reciprocal of a fraction, invert the numerator and denominator.*

b. The reciprocal of $-\frac{2}{3}$ is $-\frac{3}{2}$ because $-\frac{2}{3}\left(-\frac{3}{2}\right) - 1$.

c. The reciprocal of -11 is $-\frac{1}{11}$ because $-11\left(-\frac{1}{11}\right) = 1$. *Think of -11 as $\frac{-11}{1}$ to find its reciprocal.*

Caution
Do not change the sign of a number when finding its reciprocal.

Self Check 4 Find the reciprocal of each number: **a.** $-\frac{15}{16}$ **b.** $\frac{15}{16}$ **c.** -27

Now Try **Problems 53 and 55**

3 Divide Signed Numbers.

Every division fact can be written as an equivalent multiplication fact.

Division

For any real numbers a, b, and c, where $b \neq 0$,

$$\frac{a}{b} = c \quad \text{provided that} \quad c \cdot b = a$$

The Language of Algebra
The names of the parts of a division fact are:

Dividend → 15, *Divisor* → 5, *Quotient* → 3: $\frac{15}{5} = 3$

We can use this relationship between multiplication and division to develop rules for dividing signed numbers. For example,

$$\frac{15}{5} = 3 \qquad \text{because} \qquad 3(5) = 15$$

From this example, we see that *the quotient of two positive numbers is positive.*

To determine the quotient of two negative numbers, we consider $\frac{-15}{-5}$.

$$\frac{-15}{-5} = 3 \qquad \text{because} \qquad 3(-5) = -15$$

From this example, we see that the *quotient of two negative numbers is positive.*

To determine the quotient of a positive number and a negative number, we consider $\frac{15}{-5}$.

$$\frac{15}{-5} = -3 \qquad \text{because} \qquad -3(-5) = 15$$

From this example, we see that *the quotient of a positive number and a negative number is negative.*

To determine the quotient of a negative number and a positive number, we consider $\frac{-15}{5}$.

$$\frac{-15}{5} = -3 \qquad \text{because} \qquad -3(5) = -15$$

From this example, we see that *the quotient of a negative number and a positive number is negative.*

We summarize the rules from the previous examples and note that they are similar to the rules for multiplication.

Dividing Two Real Numbers

To divide two real numbers, divide their absolute values.

1. The quotient of two numbers that have the same *(like)* signs is positive.
2. The quotient of two numbers that have different *(unlike)* signs is negative.

EXAMPLE 5 Divide and check the result: **a.** $\frac{-81}{-9}$ **b.** $\frac{45}{-9}$

c. $-2.87 \div 0.7$ **d.** $-\frac{5}{16} \div \left(-\frac{1}{2}\right)$

Strategy We will use the rules for dividing signed numbers. In each case, we need to ask, "Is it a quotient of two numbers with the same sign or different signs?"

Why The signs of the numbers that we are dividing determine the sign of the result.

Solution

a. $\frac{-81}{-9} = 9$ Divide the absolute values, 81 by 9, to get 9. Since the signs are like, the final answer is positive.

Multiply to check the result: $9(-9) = -81$.

b. $\frac{45}{-9} = -5$ Divide the absolute values, 45 by 9, to get 5. Since the signs are unlike, make the final answer negative.

Check: $-5(-9) = 45$

c. $-2.87 \div 0.7 = -4.1$ Since the signs are unlike, make the final answer negative.

Check: $-4.1(0.7) = -2.87$

d. $-\frac{5}{16} \div \left(-\frac{1}{2}\right) = -\frac{5}{16}\left(-\frac{2}{1}\right)$ Multiply the first fraction by the reciprocal of the second fraction. The reciprocal of $-\frac{1}{2}$ is $-\frac{2}{1}$.

$= \frac{5 \cdot 2}{16 \cdot 1}$ Multiply the absolute values $\frac{5}{16}$ and $\frac{2}{1}$. Since the signs are like, the final answer is positive.

$= \frac{5 \cdot \overset{1}{\cancel{2}}}{\underset{1}{\cancel{2}} \cdot 8 \cdot 1}$ To simplify the fraction, factor 16 as $2 \cdot 8$. Then remove the common factor 2.

$= \frac{5}{8}$

Check: $\frac{5}{8}\left(-\frac{1}{2}\right) = -\frac{5}{16}$

Success Tip

To perform this decimal division, move each decimal point one place to the right.

$0.7\overline{)2.87}$

Self Check 5 Find each quotient: **a.** $\frac{-28}{-4}$ **b.** $\frac{75}{-25}$

c. $0.32 \div (-1.6)$ **d.** $\frac{3}{4} \div \left(-\frac{5}{8}\right)$

Now Try Problems 57, 61, 69, and 77

EXAMPLE 6 ***Depreciation.*** Over an 8-year period, the value of a \$150,000 house fell at a uniform rate to \$110,000. Find the amount of depreciation per year.

Strategy The phrase *uniform rate* means that the value of the house fell the same amount each year, for 8 straight years. We can determine the amount it depreciated in one year (per year) by dividing the total change in value of the house by 8.

Why The process of separating a quantity into equal parts (in this case, the change in the value of the house) indicates division.

Solution First, we find the change in the value of the house.

$110{,}000 - 150{,}000 = -40{,}000$ Subtract the previous value from the current value.

The result represents a drop in value of \$40,000. Since the depreciation occurred over 8 years, we divide −40,000 by 8.

$\frac{-40{,}000}{8} = -5{,}000$ Divide the absolute values, 40,000 by 8, to get 5,000, and make the quotient negative.

The house depreciated \$5,000 per year.

Now Try **Problem 105**

The Language of Algebra *Depreciation* is a form of the word *depreciate*, meaning to lose value. You've probably heard that the minute you drive a new car off the lot, it has *depreciated*.

4 Use Properties of Division.

Whenever we divide a number by 1, the quotient is that number. Therefore, $\frac{12}{1} = 12$, $\frac{-80}{1} = -80$, and $7.75 \div 1 = 7.75$. Furthermore, whenever we divide a nonzero number by itself, the quotient is 1. Therefore, $\frac{35}{35} = 1$, $\frac{-4}{-4} = 1$, and $0.9 \div 0.9 = 1$. These observations suggest the following properties of division.

Division Properties

Any number divided by 1 is the number itself. Any number (except 0) divided by itself is 1. For any real number a,

$$\frac{a}{1} = a \quad \text{and} \quad \frac{a}{a} = 1 \quad (\text{where } a \neq 0).$$

Caution Division is *not commutative*. For example, $\frac{6}{3} \neq \frac{3}{6}$ and $\frac{-12}{4} \neq \frac{4}{-12}$. In general, $\frac{a}{b} \neq \frac{b}{a}$.

The Language of Algebra When we say a division by 0, such as $\frac{2}{0}$, is *undefined*, we mean that $\frac{2}{0}$ does not represent a real number.

We will now consider division that involves zero. First, we examine division of zero. Let's look at two examples. We know that

$$\frac{0}{2} = 0 \quad \text{because} \quad 0 \cdot 2 = 0 \quad \text{and} \quad \frac{0}{-5} = 0 \quad \text{because} \quad 0(-5) = 0$$

These examples illustrate that *0 divided by a nonzero number is 0.*

The Language of Algebra Division of 0 by 0, written $\frac{0}{0}$, is called *indeterminate*. This form is studied in advanced mathematics classes.

To examine division by zero, let's look at $\frac{2}{0}$ and its related multiplication statement.

$$\frac{2}{0} = ? \quad \text{because} \quad ? \cdot 0 = 2$$

There is no number that can make $0 \cdot ? = 2$ true because any number multiplied by 0 is equal to 0, not 2. Therefore, $\frac{2}{0}$ does not have an answer. We say that such a division is **undefined.**

These results suggest the following division facts.

Division Involving 0

For any nonzero real number a,

$$\frac{0}{a} = 0 \quad \text{and} \quad \frac{a}{0} \text{ is undefined.}$$

EXAMPLE 7 Find each quotient, if possible: **a.** $\frac{0}{8}$ **b.** $\frac{-24}{0}$

Strategy In each case, we need to determine if we have division *of* 0 or division *by* 0.

Why *Division of 0* by a nonzero number is defined, and the result is 0. However, *division by 0* is undefined; there is no result.

Solution

a. $\frac{0}{8} = 0$ because $0 \cdot 8 = 0$. This is division of 0 by 8.

b. $\frac{-24}{0}$ is undefined. This is division of -24 by 0.

Self Check 7 Find each quotient, if possible: **a.** $\frac{4}{0}$ **b.** $\frac{0}{17}$

Now Try Problems 73 and 75

ANSWERS TO SELF CHECKS **1. a.** -60 **b.** -15 **c.** -11.18 **d.** $-\frac{2}{5}$ **2. a.** 120 **b.** $\frac{1}{12}$ **3. a.** -300 **b.** 18 **4. a.** $-\frac{16}{15}$ **b.** $\frac{16}{15}$ **c.** $-\frac{1}{27}$ **5. a.** 7 **b.** -3 **c.** -0.2 **d.** $-\frac{6}{5}$ **7. a.** Undefined **b.** 0

STUDY SET 1.6

VOCABULARY

Fill in the blanks.

1. The answer to a multiplication problem is called a ________. The answer to a division problem is called a ________.
2. The ____________ property of multiplication states that changing the order when multiplying does not affect the answer.
3. The ____________ property of multiplication states that changing the grouping when multiplying does not affect the answer.
4. Division of a nonzero number by 0 is __________.

CONCEPTS

Fill in the blanks.

5. **a.** The product or quotient of two numbers with like signs is ________.
   b. The product or quotient of two numbers with unlike signs is ________.
6. **a.** The product of an even number of negative numbers is ________.
   b. The product of an odd number of negative numbers is ________.
7. **a.** $\frac{-9}{3} = -3$ because $\quad \cdot \quad =$
   b. $\frac{0}{8} = 0$ because $\quad \cdot \quad =$
8. Complete each property of multiplication.
   a. $a \cdot b = b \cdot$
   b. $(ab)c =$
   c. $0 \cdot a =$
   d. $1 \cdot a =$
   e. $a\left(\frac{1}{a}\right) =$
9. Complete each property of division.
   a. $\frac{a}{1} =$
   b. $\frac{a}{a} =$
   c. $\frac{0}{a} =$
   d. $\frac{a}{0} -$
10. Which property justifies each statement?
    a. $-5(2 \cdot 17) = (-5 \cdot 2)17$
    b. $-5\left(-\frac{1}{5}\right) = 1$
    c. $-5 \cdot 2 = 2(-5)$
    d. $-5(1) = -5$
    e. $-5 \cdot 0 = 0$

11. Complete each statement using the given property.

a. Commutative property of multiplication

$5 \cdot 8 =$ ____

b. Associative property of multiplication

$-2(6 \cdot 9) =$ ____

c. Inverse property of multiplication

$5(\ \ ) = 1$

d. Multiplication property of 1

____ $(-20) = -20$

12. Complete the table.

Number	Opposite (additive inverse)	Reciprocal (multiplicative inverse)
2		
$-\frac{4}{5}$		
1.75		

Let POS stand for a positive number and NEG stand for a negative number. Determine the sign of each result, if possible.

13. a. POS · NEG b. POS + NEG

c. POS − NEG d. $\frac{\text{POS}}{\text{NEG}}$

14. a. NEG · NEG b. NEG + NEG

c. NEG − NEG d. $\frac{\text{NEG}}{\text{NEG}}$

NOTATION

Write each sentence using symbols.

15. The product of negative four and negative five is twenty.

16. The quotient of sixteen and negative eight is negative two.

GUIDED PRACTICE

***Multiply.* See Example 1.**

17. $4(-1)$
18. $6(-1)$
19. $-2 \cdot 8$
20. $-3 \cdot 4$
21. $12(-5)$
22. $(-9)(11)$
23. $3(-22)$
24. $-8 \cdot 9$
25. $1.2(-0.4)$
26. $(-3.6)(0.9)$
27. $\frac{1}{3}\left(-\frac{3}{4}\right)$
28. $\left(-\frac{3}{4}\right)\left(\frac{4}{5}\right)$

***Multiply.* See Example 2.**

29. $(-1)(-7)$
30. $(-2)(-5)$
31. $(-6)(-9)$
32. $(-8)(-7)$
33. $-3(-3)$
34. $-1(-1)$
35. $63(-7)$
36. $43(-6)$
37. $-0.6(-4)$
38. $-0.7(-8)$
39. $\left(-\frac{7}{8}\right)\left(-\frac{2}{21}\right)$
40. $\left(-\frac{5}{6}\right)\left(-\frac{2}{15}\right)$

***Multiply.* See Example 3.**

41. $-3(-4)(0)$
42. $15(0)(-22)$
43. $3.3(-4)(-5)$
44. $(-2.2)(-4)(-5)$
45. $-2(-3)(-4)(-5)(-6)$
46. $-9(-7)(-5)(-3)(-1)$
47. $(-41)(3)(-7)(-1)$
48. $56(-3)(-4)(-1)$
49. $(-6)(-6)(-6)$
50. $(-5)(-5)(-5)$
51. $(-2)(-2)(-2)(-2)$
52. $(-3)(-3)(-3)(-3)$

***Find the reciprocal of each number. Then find the product of the given number and its reciprocal.* See Example 4.**

53. $\frac{7}{9}$
54. $-\frac{8}{9}$
55. -13
56. $\frac{1}{8}$

***Divide.* See Example 5.**

57. $-30 \div (-3)$
58. $-12 \div (-2)$
59. $-6 \div (-2)$
60. $-36 \div (-9)$
61. $\frac{24}{-6}$
62. $\frac{-78}{6}$
63. $\frac{85}{-5}$
64. $\frac{-84}{7}$
65. $\frac{17}{-17}$
66. $\frac{-24}{24}$
67. $\frac{-110}{-110}$
68. $\frac{-200}{-200}$
69. $\frac{-10.8}{1.2}$
70. $\frac{-13.5}{-1.5}$
71. $\frac{0.5}{-100}$
72. $\frac{-1.7}{10}$
73. $\frac{0}{150}$
74. $\frac{0}{-12}$
75. $\frac{-17}{0}$
76. $\frac{225}{0}$
77. $-\frac{1}{3} \div \frac{4}{5}$
78. $-\frac{2}{3} \div \frac{7}{8}$
79. $-\frac{9}{16} \div \left(-\frac{3}{20}\right)$
80. $-\frac{4}{5} \div \left(-\frac{8}{25}\right)$

TRY IT YOURSELF

Perform the operations.

81. $\frac{-23.5}{5}$

82. $\frac{-337.8}{6}$

83. $-5.2 \cdot 100$

84. $-1.17 \cdot 1{,}000$

85. $\frac{1}{2}\left(-\frac{1}{3}\right)\left(-\frac{1}{4}\right)$

86. $\frac{1}{3}\left(-\frac{1}{5}\right)\left(-\frac{1}{7}\right)$

87. $\frac{550}{-50}$

88. $\frac{440}{-20}$

89. $-3\frac{3}{8} \div \left(-2\frac{1}{4}\right)$

90. $-3\frac{4}{15} \div \left(-2\frac{1}{10}\right)$

91. $7.2(-2.1)(-2)$

92. $4.6(-5.4)(-2)$

93. $\frac{1}{2}\left(-\frac{3}{4}\right)$

94. $\frac{1}{3}\left(-\frac{5}{16}\right)$

95. $-\frac{16}{25} \div \frac{64}{15}$

96. $-\frac{15}{16} \div \frac{25}{8}$

97. $\frac{-24.24}{-0.8}$

98. $\frac{-55.02}{-0.7}$

99. $-1\frac{1}{4}\left(-\frac{3}{4}\right)$

100. $-1\frac{1}{8}\left(-\frac{3}{8}\right)$

Use the associative property of multiplication to find each product.

101. $-\frac{1}{2}(2 \cdot 67)$

102. $\left(-\frac{5}{16} \cdot \frac{1}{7}\right)7$

103. $-0.2(-10 \cdot 3)$

104. $-1.5(-100 \cdot 4)$

APPLICATIONS

105. REAL ESTATE Over a 5-year period, the value of a \$200,000 lot fell at a uniform rate to \$160,000. Find the amount of depreciation per year.

106. TOURISM The ocean liner Queen Mary cost \$22,500,000 to build in 1936. The ship was purchased by the city of Long Beach, California, in 1967 for \$3,450,000. It now serves as a convention center. What signed number indicates the annual average depreciation of the ship over the 31-year period from 1936 to 1967? Round to the nearest dollar.

107. FLUID FLOW In a lab, the temperature of a fluid was decreased 6° per hour for 12 hours. What signed number indicates the drop in temperature?

108. STRESS ON THE JOB A health care provider for a company estimates that 75 hours per week are lost by employees suffering from stress-related illness. In one year, how many hours are lost? Use a signed number to answer.

109. WEIGHT LOSS As a result of a diet, Tom has been steadily losing $4\frac{1}{2}$ pounds per month.

a. Which expression below can be used to determine how much heavier Tom was 8 months ago?

i. $-4\frac{1}{2} \cdot 8$ **ii.** $-4\frac{1}{2}(-8)$

iii. $4\frac{1}{2}(-8)$ **iv.** $-4\frac{1}{2} - 8$

b. How much heavier was Tom 8 months ago?

110. PLANETS The temperature on Pluto gets as low as −386°F. This is twice as low as the lowest temperature reached on Jupiter. What is the lowest temperature on Jupiter?

111. CAR RADIATORS The instructions on a container of antifreeze state, "A 50/50 mixture of antifreeze and water protects against freeze-ups down to −34°F, while a 60/40 mix protects against freeze-ups down to one and one-half times that temperature." To what temperature does the 60/40 mixture protect?

112. ACCOUNTING For 2004, the net income for Martha Stewart Living Omnimedia, Inc., was about −\$60,000,000. The company's losses for 2005 were even worse, by a factor of about 1.25. What signed number indicates the company's net income that year?

113. AIRLINES In the 2005 income statement for Delta Air Lines, numbers within parentheses represent a loss. Complete the statement given these facts. The second and fourth quarter *losses* were approximately the same and totaled \$2,200 million. The third quarter loss was about $\frac{1}{3}$ of the first quarter loss.

DELTA INCOME STATEMENT				2005
All amounts in millions of dollars	1st Qtr (1,200)	2nd Qtr (?)	3rd Qtr (?)	4th Qtr (?)

Source: Yahoo! Finance

114. COMPUTERS The formula = A1*B1*C1 in cell D1 of the spreadsheet instructs the computer to multiply the values in cells A1, B1, and C1 and to print the result *in place of the formula* in cell D1. (The symbol * represents multiplication.) What value will be printed in the cell D1? What values will be printed in cells D2 and D3?

Microsoft Excel - Book 1

File Edit View Insert Format Tools Data Window

	A	B	C	D
1	4	−5	−17	= A1*B1*C1
2	22	−30	14	= A2*B2*C2
3	−60	−20	−34	= A3*B3*C3
4				
5				

115. PHYSICS An oscilloscope displays electrical signals as wavy lines on a screen. See the next page. By switching the magnification dial to ×2, for example, the height of the "peak" and the depth of the "valley" of a graph will be doubled. Use signed numbers to indicate the height and depth of the display for each setting of the dial.

a. normal **b.** ×0.5

c. ×1.5 **d.** ×2

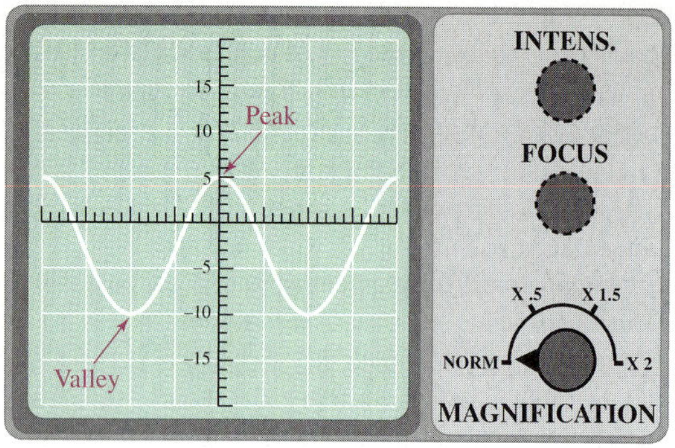

116. LIGHT Water acts as a selective filter of light. In the illustration, we see that red light waves penetrate water only to a depth of about 5 meters. How many times deeper does

a. yellow light penetrate than red light?

b. green light penetrate than orange light?

c. blue light penetrate than yellow light?

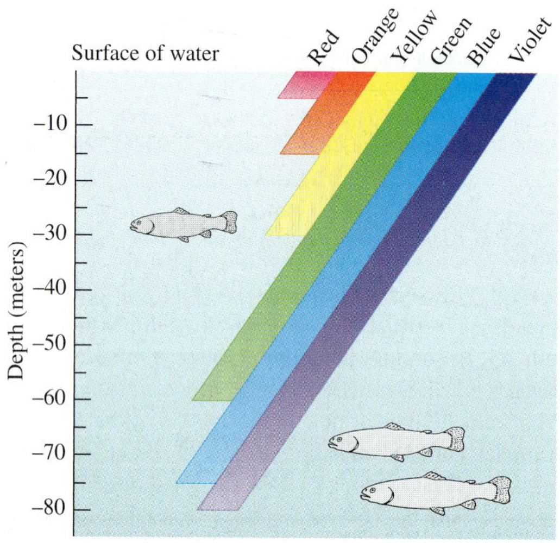

WRITING

117. Explain why $\frac{16}{0}$ is undefined.

118. The commutative property states that changing the order when multiplying does not change the answer. Are the following activities commutative? Explain.

a. Washing a load of clothes; drying a load of clothes

b. Putting on your left sock; putting on your right sock

119. What is wrong with the following statement?

A negative and a positive is a negative.

120. If we multiply two different numbers and the answer is 0, what must be true about one of the numbers? Explain your answer.

REVIEW

121. Add: $-3 + (-4) + (-5) + 4 + 3$

122. Write $-3 - (-5)$ as addition of the opposite.

123. Find $\frac{1}{2} + \frac{1}{4} + \frac{1}{3}$. Answer in decimal form.

124. Which integers have an absolute value equal to 45?

CHALLENGE PROBLEMS

125. If the product of five numbers is negative, how many of them could be negative? Explain.

126. Suppose a is a positive number and b is a negative number. Determine whether the given expression is positive or negative.

a. $-a(-b)$

b. $\frac{-a}{b}$

c. $\frac{-a}{a}$

d. $\frac{1}{b}$

SECTION 1.7
Exponents and Order of Operations

Objectives

1. Evaluate exponential expressions.
2. Use the order of operations rules.
3. Evaluate expressions containing grouping symbols.
4. Find the mean (average).

In algebra, we often have to find the value of expressions that involve more than one operation. In this section, we introduce an order-of-operations rule to follow in such cases. But first, we discuss a way to write repeated multiplication using *exponents*.

1 Evaluate Exponential Expressions.

In the expression $3 \cdot 3 \cdot 3 \cdot 3 \cdot 3$, the number 3 repeats as a factor five times. We can use **exponential notation** to write this product in a more compact form.

Exponent and Base

An **exponent** is used to indicate repeated multiplication. It is how many times the **base** is used as a factor.

The Language of Algebra

5^2 represents the area of a square with sides 5 units long. 4^3 represents the volume of a cube with sides 4 units long.

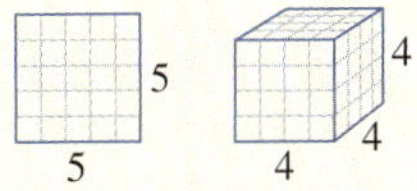

$$\underbrace{3 \cdot 3 \cdot 3 \cdot 3 \cdot 3}_{\text{Five repeated factors of 3.}} = 3^5$$

The exponent is 5.

The base is 3.

In the **exponential expression** 3^5, 3 is the base, and 5 is the exponent. The expression is called a power of 3. Some other examples of exponential expressions are:

5^2 Read as "5 to the second power" or "5 squared."

4^3 Read as "4 to the third power" or "4 cubed."

$(-2)^5$ Read as "−2 to the fifth power."

EXAMPLE 1 Write each product using exponents: **a.** $7 \cdot 7 \cdot 7$ **b.** $(-5)(-5)(-5)(-5)(-5)$ **c.** $8 \cdot 8 \cdot 15 \cdot 15 \cdot 15 \cdot 15$ **d.** $a \cdot a \cdot a \cdot a \cdot a \cdot a$ **e.** $4 \cdot \pi \cdot r \cdot r$

Strategy We need to determine the number of repeated factors in the expression.

Why An exponent can be used to represent repeated multiplication.

Solution

a. The factor 7 is repeated 3 times. We can represent this repeated multiplication with an exponential expression having a base of 7 and an exponent of 3: $7 \cdot 7 \cdot 7 = 7^3$.

b. The factor −5 is repeated five times: $(-5)(-5)(-5)(-5)(-5) = (-5)^5$.

c. $8 \cdot 8 \cdot 15 \cdot 15 \cdot 15 \cdot 15 = 8^2 \cdot 15^4$

d. $a \cdot a \cdot a \cdot a \cdot a \cdot a = a^6$

e. $4 \cdot \pi \cdot r \cdot r = 4\pi r^2$

Self Check 1 Write each product using exponents:

a. $(12)(12)(12)(12)(12)(12)$ **b.** $2 \cdot 9 \cdot 9 \cdot 9$

c. $(-30)(-30)(-30)$ **d.** $y \cdot y \cdot y \cdot y$

e. $12 \cdot b \cdot b \cdot b \cdot c$

Now Try **Problems 15 and 21**

To **evaluate** (find the value of) an exponential expression, we write the base as a factor the number of times indicated by the exponent. Then we multiply the factors, working left to right.

EXAMPLE 2 Evaluate each expression: **a.** 5^3 **b.** 10^1 **c.** $\left(-\frac{2}{3}\right)^3$ **d.** $(0.6)^2$ **e.** $(-3)^4$ **f.** $(-3)^5$

Strategy We will rewrite each exponential expression as a product of repeated factors, and then perform the multiplication. This requires that we identify the base and the exponent.

Why The exponent tells the number of times the base is to be written as a factor.

Solution

a. $5^3 = 5 \cdot 5 \cdot 5$ Write the base, 5, as a factor 3 times.

$= 125$ Multiply, working left to right. We say 125 is the *cube of 5.*

b. $10^1 = 10$ The base is 10. Since the exponent is 1, we write the base once.

c. $\left(-\frac{2}{3}\right)^3 = \left(-\frac{2}{3}\right)\left(-\frac{2}{3}\right)\left(-\frac{2}{3}\right)$ Since $-\frac{2}{3}$ is the base and 3 is the exponent, we write $-\frac{2}{3}$ as a factor three times.

$= \frac{4}{9}\left(-\frac{2}{3}\right)$ Work from left to right. $\left(-\frac{2}{3}\right)\left(-\frac{2}{3}\right) = \frac{4}{9}$

$= -\frac{8}{27}$

d. $(0.6)^2 = (0.6)(0.6)$ Since 0.6 is the base and 2 is the exponent, we write 0.6 as a factor two times.

$= 0.36$ We say 0.36 is the *square of 0.6.*

e. $(-3)^4 = (-3)(-3)(-3)(-3)$ Write the base, -3, as a factor 4 times.

$= 9(-3)(-3)$ Work from left to right.

$= -27(-3)$

$= 81$

f. $(-3)^5 = (-3)(-3)(-3)(-3)(-3)$ Write the base, -3, as a factor 5 times.

$= 9(-3)(-3)(-3)$ Work from left to right.

$= -27(-3)(-3)$

$= 81(-3)$

$= -243$

Self Check 2 Evaluate: **a.** 2^5 **b.** 9^1 **c.** $\left(-\frac{3}{4}\right)^3$ **d.** $(-0.3)^2$ **e.** $(-6)^2$ **f.** $(-5)^3$

Now Try **Problems 23, 29, and 33**

Caution

Don't make the common mistake of multiplying the base and the exponent:

$$5^3 \neq 5 \cdot 3$$

Calculators

Finding a power

The squaring key x^2 can be used to find the square of a number. To raise a number to a power, we use the y^x key on a scientific calculator and the ^ key on a graphing calculator.

The Language of Algebra

A number or a variable has an *understood* exponent of 1. For example,

$$8 = 8^1 \quad \text{and} \quad x = x^1$$

In Example 2e, we raised -3 to an even power; the result was positive. In part f, we raised -3 to an odd power; the result was negative. These results illustrate the following rule.

Even and Odd Powers of a Negative Number

When a negative number is raised to an even power, the result is positive.

When a negative number is raised to an odd power, the result is negative.

The Language of Algebra
Read $(-4)^2$ as "negative four squared" and -4^2 as "the opposite of the square of four."

Although the expressions $(-4)^2$ and -4^2 look alike, they are not. When we find the value of each expression, it becomes clear that they are not equivalent.

$$(-4)^2 = (-4)(-4) = 16$$ The base is −4, the exponent is 2.

$$-4^2 = -(4 \cdot 4) = -16$$ The base is 4, the exponent is 2.

Different results

EXAMPLE 3 Evaluate: -2^4

Strategy We will rewrite the expression as a product of repeated factors and then perform the multiplication. We must be careful when identifying the base. It is 2, not -2.

Why Since there are no parentheses around -2, the base is 2.

Solution

$-2^4 = -(2 \cdot 2 \cdot 2 \cdot 2)$ Read as "the opposite of the fourth power of two."

$= -16$ Do the multiplication within the parentheses to get 16. Then write the opposite of that result.

Self Check 3 Evaluate: -5^4

Now Try **Problem 35**

2 Use the Order of Operations Rules.

Suppose you have been asked to contact a friend if you see a Rolex watch for sale when you are traveling in Europe. While in Switzerland, you find the watch and send the text message shown on the left. The next day, you get the response shown on the right.

Something is wrong. The first part of the response (No price too high!) says to buy the watch at any price. The second part (No! Price too high.) says not to buy it, because it's too expensive. The placement of the exclamation point makes us read the two parts of the response differently, resulting in different meanings. When reading a mathematical statement, the same kind of confusion is possible. For example, consider the expression

$$2 + 3 \cdot 6$$

We can evaluate this expression in two ways. We can add first, and then multiply. Or we can multiply first, and then add. However, the results are different.

$2 + 3 \cdot 6 = 5 \cdot 6$	Add 2 and 3 first.	$2 + 3 \cdot 6 = 2 + 18$	Multiply 3 and 6 first.
$= 30$	Multiply 5 and 6.	$= 20$	Add 2 and 18.

Different answers

If we don't establish a uniform order of operations, the expression has two different values. To avoid this possibility, we will always use the following set of priority rules.

Order of Operations

1. Perform all calculations within parentheses and other grouping symbols following the order listed in Steps 2–4 below, working from the innermost pair of grouping symbols to the outermost pair.
2. Evaluate all exponential expressions.
3. Perform all multiplications and divisions as they occur from left to right.
4. Perform all additions and subtractions as they occur from left to right.

When grouping symbols have been removed, repeat Steps 2–4 to complete the calculation.

If a fraction is present, evaluate the expression above and the expression below the bar separately. Then simplify the fraction, if possible.

It isn't necessary to apply all of these steps in every problem. For example, the expression $2 + 3 \cdot 6$ does not contain any parentheses, and there are no exponential expressions. So we look for multiplications and divisions to perform and proceed as follows:

$2 + 3 \cdot 6 = 2 + 18$	Do the multiplication first.
$= 20$	Do the addition.

EXAMPLE 4 Evaluate: **a.** $3 \cdot 2^3 - 4$ **b.** $-30 - 4 \cdot 5 + 9$ **c.** $24 \div 6 \cdot 2$ **d.** $160 - 4 + 6(-2)(-3)$

Strategy We will scan the expression to determine what operations need to be performed. Then we will perform those operations, one-at-a-time, following the order of operations rules.

Why If we don't follow the correct order of operations, the expression can have more than one value.

The Language of Algebra
Sometimes, for problems like these, the instruction *simplify* is used instead of *evaluate*.

Solution

a. Three operations need to be performed to evaluate this expression: multiplication, raising to a power, and subtraction. By the order of operations rules, we evaluate 2^3 first.

$3 \cdot 2^3 - 4 = 3 \cdot 8 - 4$	Evaluate the exponential expression: $2^3 = 8$.
$= 24 - 4$	Do the multiplication: $3 \cdot 8 = 24$.
$= 20$	Do the subtraction.

b. This expression involves subtraction, multiplication, and addition. The order of operations rules tell us to multiply first.

$$-30 - 4 \cdot 5 + 9 = -30 - 20 + 9$$ Do the multiplication: $4 \cdot 5 = 20$.

$$= -50 + 9$$ Working from left to right, do the subtraction: $-30 - 20 = -30 + (-20) = -50$.

$$= -41$$ Do the addition.

Caution
A common mistake is to forget to work from left to right and incorrectly perform the multiplication before the division. This produces the wrong answer, 2.

~~$24 \div 6 \cdot 2 = 24 \div 12 = 2$~~

c. Since there are no calculations within parentheses nor are there exponents, we perform the multiplications and divisions as they occur from left to right. The division occurs before the multiplication, so it must be performed first.

$$24 \div 6 \cdot 2 = 4 \cdot 2$$ Working left to right, do the division: $24 \div 6 = 4$.

$$= 8$$ Do the multiplication.

d. Although this expression contains parentheses, there are no operations to perform within them. Since there are no exponents, we will perform the multiplications as they occur from left to right.

$$160 - 4 + 6(-2)(-3) = 160 - 4 + (-12)(-3)$$ Do the multiplication, working left to right: $6(-2) = -12$.

$$= 160 - 4 + 36$$ Complete the multiplication: $(-12)(-3) = 36$.

$$= 156 + 36$$ Working left to right, the subtraction occurs before the addition. The subtraction must be performed first: $160 - 4 = 156$.

$$= 192$$

Self Check 4 Evaluate: **a.** $2 \cdot 3^2 + 17$ **b.** $-40 - 9 \cdot 4 + 10$ **c.** $18 \div 2 \cdot 3$ **d.** $240 - 8 + 3(-2)(-4)$

Now Try Problems 39, 45, 49, and 51

3 Evaluate Expressions Containing Grouping Symbols.

Grouping symbols serve as mathematical punctuation marks. They help determine the order in which an expression is to be evaluated. Examples of grouping symbols are parentheses ( ), brackets [ ], braces { }, absolute value symbols | |, and the fraction bar —.

EXAMPLE 5 Evaluate each expression: **a.** $(6 - 3)^2$ **b.** $5^3 + 2(-8 - 3 \cdot 2)$

Strategy We will perform the operation(s) within the parentheses first. When there is more than one operation to perform within the parentheses, we follow the order of operations rules.

Why This is the first step of the order of operations rule.

Solution

a. $(6 - 3)^2 = 3^2$ Do the subtraction within the parentheses: $6 - 3 = 3$.

$= 9$ Evaluate the exponential expression.

Notation
Multiplication is indicated when a number is next to a parenthesis or bracket.

$5^3 + 2(-8 - 3 \cdot 2)$

b. We begin by performing the operations within the parentheses in the proper order: multiplication first, and then subtraction.

$$\begin{aligned} 5^3 + 2(-8 - 3 \cdot 2) &= 5^3 + 2(-8 - 6) && \text{Do the multiplication: } 3 \cdot 2 = 6. \\ &= 5^3 + 2(-14) && \text{Do the subtraction: } -8 - 6 = -14. \\ &= 125 + 2(-14) && \text{Evaluate } 5^3. \\ &= 125 + (-28) && \text{Do the multiplication: } 2(-14) = -28. \\ &= 97 && \text{Do the addition.} \end{aligned}$$

Self Check 5 Evaluate: **a.** $(12 - 6)^3$ **b.** $1^3 + 6(-6 - 3 \cdot 0)$

Now Try **Problem 73**

Expressions can contain two or more pairs of grouping symbols. To evaluate the following expression, we begin within the innermost pair of grouping symbols, the parentheses. Then we work within the outermost pair, the brackets.

Innermost pair

$$-4[2 + 3(4 - 8^2)] - 2$$

Outermost pair

EXAMPLE 6 Evaluate: $-4[2 + 3(4 - 8^2)] - 2$

Strategy We will work within the parentheses first and then within the brackets. At each stage, we follow the order of operations rules.

Why By the order of operations, we must work from the *innermost* pair of grouping symbols to the *outermost.*

The Language of Algebra
When one pair of grouping symbols is inside another pair, we say that those grouping symbols are *nested,* or *embedded.*

Solution

$$\begin{aligned} &-4[2 + 3(4 - 8^2)] - 2 \\ &= -4[2 + 3(4 - 64)] - 2 && \text{Evaluate the exponential expression within the parentheses: } 8^2 = 64. \\ &= -4[2 + 3(-60)] - 2 && \text{Do the subtraction within the parentheses: } 4 - 64 = 4 + (-64) = -60. \\ &= -4[2 + (-180)] - 2 && \text{Do the multiplication within the brackets: } 3(-60) = -180. \\ &= -4[-178] - 2 && \text{Do the addition within the brackets: } 2 + (-180) = -178. \\ &= 712 - 2 && \text{Do the multiplication: } -4[-178] = 712. \\ &= 710 && \text{Do the subtraction.} \end{aligned}$$

Self Check 6 Evaluate: $-5[4 + 2(5^2 - 15)] - 10$

Now Try **Problem 67**

EXAMPLE 7 Evaluate: $\frac{-3(3+2)+5}{17-3(-4)}$

Strategy We will evaluate the expression above and the expression below the fraction bar separately. Then we will simplify the fraction, if possible.

Why Fraction bars are grouping symbols. They group the numerator and denominator. The expression could be written $[-3(3+2)+5] \div [17-3(-4)]$.

Solution

$$\frac{-3(3+2)+5}{17-3(-4)} = \frac{-3(5)+5}{17-(-12)}$$

In the numerator, do the addition within the parentheses. In the denominator, do the multiplication.

$$= \frac{-15+5}{17+12}$$

In the numerator, do the multiplication. In the denominator, write the subtraction as the addition of the opposite of −12, which is 12.

$$= \frac{-10}{29}$$

Do the additions.

$$= -\frac{10}{29}$$

Write the − sign in front of the fraction: $\frac{-10}{29} = -\frac{10}{29}$. The fraction does not simplify.

Self Check 7 Evaluate: $\frac{-4(-2+8)+6}{8-5(-2)}$

Now Try Problem 81

Calculators

Order of operations

Calculators have the order of operations built in. A left parenthesis key (and a right parenthesis key) should be used when grouping symbols, including a fraction bar, are needed.

EXAMPLE 8 Evaluate: $10|9-15|-2^5$

Strategy The absolute value bars are grouping symbols. We will perform the calculation within them first.

Why By the order of operations, we must perform all calculations within parentheses and other grouping symbols (such as absolute value bars) first.

Solution

$10|9-15|-2^5 = 10|-6|-2^5$ Subtract: $9 - 15 = 9 + (-15) = -6$.

$= 10(6) - 2^5$ Find the absolute value: $|-6| = 6$.

$= 10(6) - 32$ Evaluate the exponential expression: $2^5 = 32$.

$= 60 - 32$ Do the multiplication: $10(6) = 60$.

$= 28$ Do the subtraction.

Self Check 8 Evaluate: $10^3 + 3|24-25|$

Now Try Problem 85

Notation

Multiplication is indicated when a number is next to an absolute value symbol.

↓

$10|9-15|-2^5$

4 Find the Mean (Average).

The **arithmetic mean** (or **average**) of a set of numbers is a value around which the values of the numbers are grouped.

Finding an Arithmetic Mean

To find the **mean** of a set of values, divide the sum of the values by the number of values.

EXAMPLE 9 ***Hotel Reservations.*** In an effort to improve customer service, a hotel electronically recorded the number of times the reservation desk telephone rang before it was answered by a receptionist. The results of the week-long survey are shown in the table. Find the average number of times the phone rang before a receptionist answered.

Number of rings	Number of calls
1	11
2	46
3	45
4	28
5	20

Strategy First, we will determine the total number of times the reservation desk telephone rang during the week. Then we will divide that result by the total number of calls received.

Why To find the *average* value of a set of values, we divide the sum of the values by the number of values.

Solution To find the total number of rings, we multiply each *number of rings* (1, 2, 3, 4, and 5 rings) by the respective number of occurrences and add those subtotals.

$$\text{Total number of rings} = 11(1) + 46(2) + 45(3) + 28(4) + 20(5)$$

The total number of calls received was $11 + 46 + 45 + 28 + 20$. To find the average, we divide the total number of rings by the total number of calls.

$$\text{Average} = \frac{11(1) + 46(2) + 45(3) + 28(4) + 20(5)}{11 + 46 + 45 + 28 + 20}$$

$$= \frac{11 + 92 + 135 + 112 + 100}{150}$$ In the numerator, do the multiplications. In the denominator, do the additions.

$$= \frac{450}{150}$$ Do the addition.

$$= 3$$ Simplify the fraction.

The average number of times the phone rang before it was answered was 3.

Now Try **Problem 115**

ANSWERS TO SELF CHECKS **1. a.** 12^6 **b.** $2 \cdot 9^3$ **c.** $(-30)^3$ **d.** y^4 **e.** $12b^3c$ **2. a.** 32 **b.** 9 **c.** $-\frac{27}{64}$ **d.** 0.09 **e.** 36 **f.** -125 **3.** -625 **4. a.** 35 **b.** -66 **c.** 27 **d.** 256 **5. a.** 216 **b.** -35 **6.** -130 **7.** -1 **8.** 1,003

STUDY SET 1.7

VOCABULARY

Fill in the blanks.

1. In the exponential expression 7^5, 7 is the ______, and 5 is the __________. 7^5 is the fifth _______ of seven.
2. 10^2 can be read as ten _________, and 10^3 can be read as ten _______.
3. An _________ is used to represent repeated multiplication.
4. To _________ the expression $2(-1 + 4^2)$ means to find its value.
5. The rules for the ______ of operations guarantee that an evaluation of a numerical expression will result in a single answer.
6. To find the arithmetic ______ or average of a set of values, divide the sum of the values by the number of values.

CONCEPTS

7. To evaluate each expression, what operation should be performed first?
   a. $24 - 4 + 2$
   b. $32 \div 8 \cdot 4$
   c. $8 - (3 + 5)^2$
   d. $65 \cdot 3^3$
8. To evaluate $\frac{36 - 4(7)}{2(10 - 8)}$, what operation should be performed first in the numerator? In the denominator?

NOTATION

9. a. Give the name of each grouping symbol: ( ), [ ], { }, | |, and —.
   b. In the expression $-8 + 2[15 - (-6 + 1)]$, which grouping symbols are innermost, and which are outermost?
10. What operation is indicated?

$$2 + 9\underset{\uparrow}{}|5 - (2 + 4)|$$

11. a. In the expression $(-5)^2$, what is the base?
    b. In the expression -5^2, what is the base?
12. Write each expression using symbols. Then evaluate it.
    a. Negative two squared
    b. The opposite of the square of two

Complete the evaluation of each expression.

13. $-19 - 2[(1 + 2)^2 \cdot 3] = -19 - 2[\quad^2 \cdot 3]$
    $= -19 - 2[\quad \cdot 3]$
    $= -19 - 2[\quad]$
    $= -19 - \quad$
    $= \quad$

14. $\dfrac{46 - 2^3}{-3(5) - 4} = \dfrac{46 - \quad}{\quad}$
    $= \dfrac{\quad}{\quad}$
    $= -2$

GUIDED PRACTICE

Write each product using exponents. **See Example 1.**

15. $8 \cdot 8 \cdot 8$
16. $(-4)(-4)(-4)(-4)$
17. $7 \cdot 7 \cdot 7 \cdot 12 \cdot 12$
18. $5 \cdot 5 \cdot 5 \cdot 5 \cdot 5 \cdot 5 \cdot 7 \cdot 7 \cdot 7$
19. $x \cdot x \cdot x$
20. $b \cdot b \cdot b \cdot b$
21. $r \cdot r \cdot r \cdot r \cdot s \cdot s$
22. $m \cdot m \cdot m \cdot n \cdot n \cdot n \cdot n$

Evaluate each expression. **See Example 2.**

23. 7^2
24. 9^2
25. 6^3
26. 6^4
27. $(-5)^4$
28. $(-5)^3$
29. $(-0.1)^2$
30. $(-0.8)^2$
31. $\left(-\frac{1}{4}\right)^3$
32. $\left(-\frac{1}{3}\right)^4$
33. $\left(\frac{2}{3}\right)^3$
34. $\left(\frac{3}{4}\right)^3$

Evaluate each expression. **See Example 3.**

35. $(-6)^2$ and -6^2
36. $(-4)^2$ and -4^2
37. $(-8)^2$ and -8^2
38. $(-9)^2$ and -9^2

Evaluate each expression. **See Example 4.**

39. $3 - 5 \cdot 4$
40. $-4 \cdot 6 + 5$
41. $32 - 16 \div 4 + 2$
42. $60 - 20 \div 10 + 5$
43. $9 \cdot 5 - 6 \div 3$
44. $8 \cdot 5 - 4 \div 2$
45. $12 \div 3 \cdot 2$
46. $18 \div 6 \cdot 3$
47. $-22 - 15 + 3$
48. $-33 - 8 + 10$
49. $2(9) - 2(5)(10)$
50. $-6(7) - 3(-4)(-2)$
51. $2 \cdot 5^2 + 4 \cdot 3^2$
52. $5 \cdot 3^3 - 4 \cdot 2^3$
53. $-2(-1)^2 + 3(-1) - 3$
54. $-4(-3)^2 + 3(-3) - 1$

Evaluate each expression. **See Examples 5 and 6.**

55. $-4(6 + 5)$
56. $-3(5 - 4)$
57. $(9 - 3)(9 - 9)^2$
58. $-(-8 - 6)(6 - 6)^2$

59. $(-1 - 3^2 \cdot 4)2^2$

60. $-1(28 - 5^2 \cdot 2)3^2$

61. $1 + 5(10 + 2 \cdot 5) - 1$

62. $14 + 3(7 - 5 \cdot 3)$

63. $-(2 \cdot 3 - 2^2)^5$

64. $-(3 \cdot 5 - 2 \cdot 6)^4$

***Evaluate each expression.* See Example 6.**

65. $(-1)^9[-7^2 - (-2)^2]$

66. $[-9^2 - (-8)^2](-1)^{10}$

67. $64 - 6[15 + 2(-3 + 8)]$

68. $4 - 2[26 + 2(5 - 3)]$

69. $-2[2 + 4^2(8 - 9)]^2$

70. $-3[5 + 3^2(4 - 5)]^2$

71. $3 + 2[-1 - (4 - 5)]$

72. $4 + 2[-7 - (3 - 9)]$

73. $8 - 3[5^2 - (7 - 3)^2]$

74. $3 - [3^3 + (3 - 1)^3]$

***Evaluate each expression.* See Example 7.**

75. $\dfrac{-2 - 5}{-7 + (-7)}$

76. $\dfrac{-3 - (-1)}{-2 + (-2)}$

77. $\dfrac{4 \cdot 2^4 - 60 + (-4)}{5^4 - (-4)(-5)}$

78. $\dfrac{(6 - 5)^8 - 1}{(-9)(-3) - 4}$

79. $\dfrac{2(-4 - 2 \cdot 2)}{3(-3)(-2)}$

80. $\dfrac{3(-3^2 + 2 \cdot 2^2)}{(5 - 8)(7 - 9)}$

81. $\dfrac{72 - (2 - 2 \cdot 4)}{10^2 - (9 \cdot 10 + 2^2)}$

82. $\dfrac{13^2 - 5^2}{-3(5 - 3^2)}$

***Evaluate each expression.* See Example 8.**

83. $-2|4 - 8|$

84. $-5|1 - 8|$

85. $-|7 - 2^3(4 - 7)|$

86. $-|9 - 5(1 - 2^3)|$

87. $\dfrac{(3 + 5)^2 + |-2|}{-2(5 - 8)}$

88. $\dfrac{|-25| - 8(-5)}{2^4 - 29}$

89. $\dfrac{|6 - 4| + 2|-4|}{26 - 2^4}$

90. $\dfrac{4|9 - 7| + |-7|}{3^2 - 2^2}$

TRY IT YOURSELF

Evaluate each expression.

91. $[6(5) - 5(5)]^3(-4)$

92. $5 - 2 \cdot 3^4 - (-6 + 5)^3$

93. $8 - 6[(130 - 4^3) - 2]$

94. $91 - 5[(150 - 3^3) - 1]$

95. $-2\left(\dfrac{15}{-5}\right) - \dfrac{6}{2} + 9$

96. $-6\left(\dfrac{25}{-5}\right) - \dfrac{36}{9} + 1$

97. $-5(-2)^3 - |-2 + 1|$

98. $-6(-3)^3 - |-6 + 5|$

99. $\dfrac{18 - [2 + (1 - 6)]}{16 - (-4)^2}$

100. $\dfrac{6 - [6(-1) - 88]}{4 - 2^2}$

101. $-|-5 \cdot 2^4| - 30$

102. $2 + |-3 \cdot 2^2 \cdot 2^2 \cdot 1^2|$

103. $(-3)^3\left(\dfrac{-4}{2}\right)(-1)$

104. $(-2)^3\left(\dfrac{-6}{2}\right)(-1)$

105. $\dfrac{1}{2}\left(\dfrac{1}{8}\right) + \left(-\dfrac{1}{4}\right)^2$

106. $-\dfrac{1}{9}\left(\dfrac{1}{4}\right) + \left(-\dfrac{1}{6}\right)^2$

107. $\dfrac{-5^2 \cdot 10 + 10 \cdot 2^4}{-5 - 3 - 1}$

108. $\dfrac{(-6^2 - 2^4 \cdot 2) + 5}{-4 - 3}$

109. $-\left(\dfrac{40 - 1^3 - 2^4}{3(2 + 5) + 2}\right)$

110. $-\left(\dfrac{8^2 - 10}{2(3)(4) - 5(3)}\right)$

APPLICATIONS

111. LIGHT As light energy passes through the first unit of area, 1 yard away from the bulb, it spreads out. How much area does that light energy cover 2 yards, 3 yards, and 4 yards from the bulb? Express each answer using exponents.

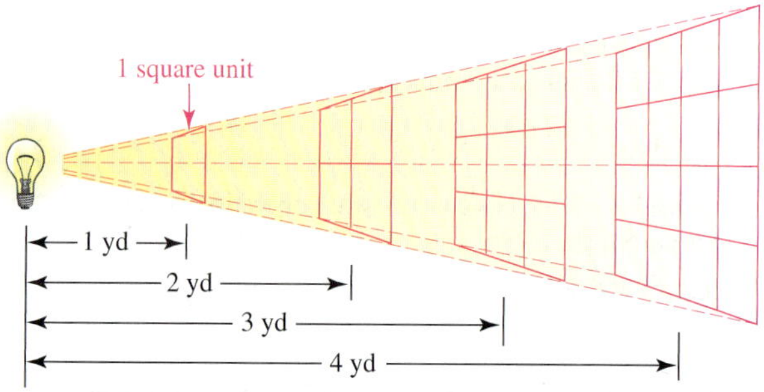

112. CHAIN LETTERS A woman sent two friends a letter with the following request: "Please send a copy of this letter to two of your friends." Assume that all those receiving letters responded and that everyone in the chain received just one letter. Complete the table and then determine how many letters will be circulated in the 10th level?

Level	Number of letters circulated
1st	$2 = 2^1$
2nd	$= 2$
3rd	$= 2$
4th	$= 2$

113. HURRICANES The table lists the number of major hurricanes to strike the mainland United States by decade. Find the average number per decade.

Decade	Number	Decade	Number
1901–1910	4	1951–1960	8
1911–1920	7	1961–1970	6
1921–1930	5	1971–1980	4
1931–1940	8	1981–1990	5
1941–1950	10	1991–2000	5

Source: National Hurricane Center

114. ENERGY USAGE Find the average number of therms of natural gas used per month.

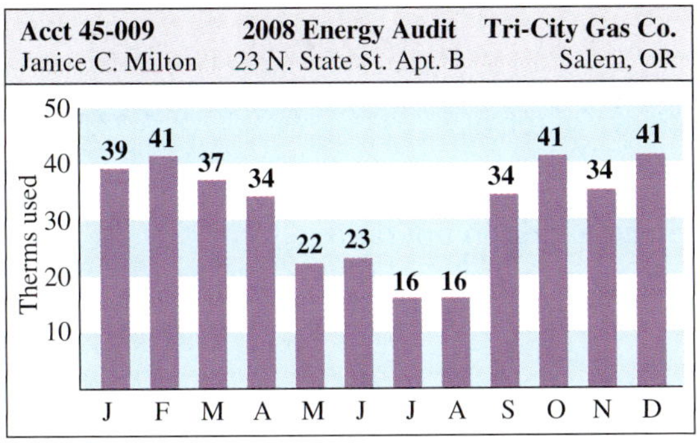

115. CASH AWARDS A contest is to be part of a promotional kickoff for a new children's cereal. The prizes to be awarded are shown.

a. How much money will be awarded in the promotion?

b. What is the average cash prize?

Coloring Contest

Grand prize: Disney World vacation plus $2,500
Four 1st place prizes of $500
Thirty-five 2nd place prizes of $150
Eighty-five 3rd place prizes of $25

116. SURVEYS Some students were asked to rate their college cafeteria food on a scale from 1 to 5. The responses are shown on the tally sheet. Find the average rating.

<table>
<tr><th>Poor</th><th></th><th>Fair</th><th></th><th>Excellent</th></tr>
<tr><td>1</td><td>2</td><td>3</td><td>4</td><td>5</td></tr>
<tr><td></td><td>|||</td><td>|||</td><td>𝍸</td><td>𝍸 ||||</td></tr>
</table>

117. WRAPPING GIFTS How much ribbon is needed to wrap the package if 15 inches of ribbon are needed to make the bow?

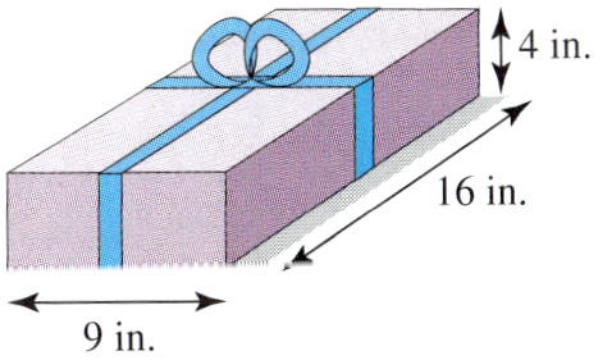

118. SCRABBLE Write an expression to determine the number of points received for playing the word QUARTZY and then evaluate it. (The number on each tile gives the point value of the letter.)

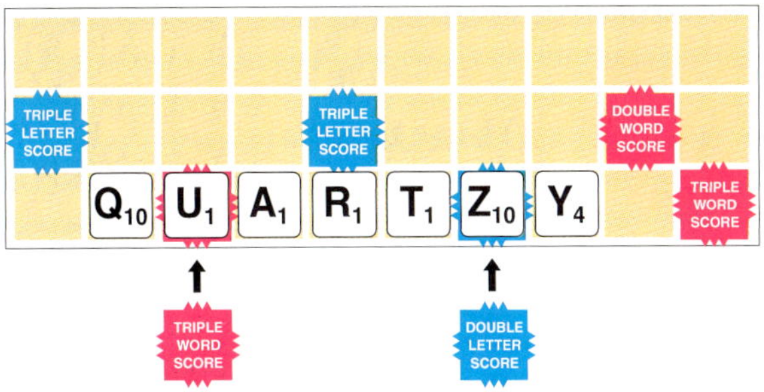

WRITING

119. Explain the difference between 2^3 and 3^2.

120. Why are the order of operations rules necessary?

121. Explain the error. What is the correct answer?

$$\begin{aligned} 40 \div 4 \cdot 2 &= 40 \div 8 \\ &= 5 \end{aligned}$$

122. Explain the error. What is the correct answer?

$$\begin{aligned} 5 + 3(2 - 6) &= 5 + 3(-4) \\ &= 8(-4) \\ &= -32 \end{aligned}$$

REVIEW

123. What numbers are a distance of 6 away from -11 on a number line?

124. Fill in the blank with $>$ or $<$: $0.3 \;\_\_\; \frac{1}{3}$

CHALLENGE PROBLEMS

125. Using each of the numbers 2, 3, and 4 only once, what is the greatest value that the following expression can have?

$$(\square^{\square})^{\square}$$

126. Insert a pair of parentheses into $4 \cdot 3^2 - 4 \cdot 2$ so that it has a value of 40.

Translate the set of instructions to an expression and then evaluate it.

127. Subtract the sum of -9 and 8 from the product of the cube of -3 and the opposite of 4.

128. Increase the square the reciprocal of -2 by the difference of -0.25 and -1.

SECTION 1.8
Algebraic Expressions

Objectives

1. Identify terms and coefficients of terms.
2. Write word phrases as algebraic expressions.
3. Analyze problems to determine hidden operations.
4. Evaluate algebraic expressions.

Since problems in algebra are often presented in words, the ability to interpret what you read is important. In this section, we will introduce several strategies that will help you translate words into algebraic expressions.

1 Identify Terms and Coefficients of Terms.

Recall that variables and/or numbers can be combined with the operations of arithmetic to create **algebraic expressions.** Addition symbols separate expressions into parts called *terms.* For example, the expression $x + 8$ has two terms.

$$x \quad + \quad 8$$

First term Second term

Since subtraction can be written as addition of the opposite, the expression $a^2 - 3a - 9$ has three terms.

$$a^2 - 3a - 9 = a^2 + (-3a) + (-9)$$

First term Second term Third term

In general, a **term** is a product or quotient of numbers and/or variables. A single number or variable is also a term. Examples of terms are:

$$4, \quad y, \quad 6r, \quad -w^3, \quad 3.7x^5, \quad \frac{3}{n}, \quad -15ab^2$$

Notation

By the commutative property of multiplication, $r6 = 6r$ and $-15b^2a = -15ab^2$. However, we usually write the numerical factor first and the variable factors in alphabetical order.

The numerical factor of a term is called the **coefficient** of the term. For instance, the term $6r$ has a coefficient of 6 because $6r = 6 \cdot r$. The coefficient of $-15ab^2$ is -15 because $-15ab^2 = -15 \cdot ab^2$. More examples are shown below.

A term such as 4, that consists of a single number, is called a **constant term.**

The Language of Algebra

Terms such as x and y have *implied* coefficients of 1. *Implied* means suggested without being precisely expressed.

Term	Coefficient	
$8y^2$	8	
$-0.9pq$	-0.9	
$\frac{3}{4}b$	$\frac{3}{4}$	This term could be written $\frac{3b}{4}$.
$-\frac{x}{6}$	$-\frac{1}{6}$	Because $-\frac{x}{6} = -\frac{1x}{6} = -\frac{1}{6} \cdot x$
x	1	Because $x = 1x$
$-t$	-1	Because $-t = -1t$
27	27	

EXAMPLE 1 Identify the coefficient of each term in the expression: $7x^2 - x + 6$

Strategy We will begin by writing the subtraction as addition of the opposite. Then we will determine the numerical factor of each term.

Why Addition symbols separate expressions into terms.

Solution If we write $7x^2 - x + 6$ as $7x^2 + (-x) + 6$, we see that it has three terms: $7x^2$, $-x$, and 6. The numerical factor of each term is its coefficient.

The coefficient of $7x^2$ is **7** because $7x^2$ means $7 \cdot x^2$.

The coefficient of $-x$ is **−1** because $-x$ means $-1 \cdot x$.

The coefficient of the constant 6 is 6.

Self Check 1 Identify the coefficient of each term in the expression: $p^3 - 12p^2 + 3p - 4$

Now Try **Problem 19**

It is important to be able to distinguish between the *terms* of an expression and the *factors* of a term.

EXAMPLE 2 Is m used as a *factor* or a *term* in each expression?

a. $m + 6$ **b.** $8m$

Strategy We will begin by determining whether m is involved in an addition or a multiplication.

Why Addition symbols separate expressions into *terms*. A *factor* is a number being multiplied.

Solution

a. Since m is added to 6, m is a term of $m + 6$.

b. Since m is multiplied by 8, m is a factor of $8m$.

Self Check 2 Is b used as a *factor* or a *term* in each expression?

a. $-27b$ **b.** $5a + b$

Now Try **Problems 21 and 23**

2 Write Word Phrases as Algebraic Expressions.

The tables on the next page show how key phrases can be translated into algebraic expressions.

Caution
Be careful when translating subtraction. Order is important. For example, when a translation involves the phrase *less than*, note how the terms are reversed.

18 less than w

$w - 18$

Addition	
the sum of a and 8	$a + 8$
4 plus c	$4 + c$
16 added to m	$m + 16$
4 more than t	$t + 4$
20 greater than F	$F + 20$
T increased by r	$T + r$
exceeds y by 35	$y + 35$

Subtraction	
the difference of 23 and P	$23 - P$
550 minus h	$550 - h$
18 less than w	$w - 18$
7 decreased by j	$7 - j$
M reduced by x	$M - x$
12 subtracted from L	$L - 12$
5 less f	$5 - f$

Caution
Be careful when translating division. As with subtraction, order is important. For example, s divided by d is *not* written $\frac{d}{s}$.

Multiplication	
the product of 4 and x	$4x$
20 times B	$20B$
twice r	$2r$
double the amount a	$2a$
triple the profit P	$3P$
three-fourths of m	$\frac{3}{4}m$

Division	
the quotient of R and 19	$\frac{R}{19}$
s divided by d	$\frac{s}{d}$
the ratio of c to d	$\frac{c}{d}$
k split into 4 equal parts	$\frac{k}{4}$

EXAMPLE 3 Write each phrase as an algebraic expression:
a. one-half of the profit P **b.** 5 less than the capacity c
c. the product of the weight w and 2,000, increased by 300

Strategy We will begin by identifying any key phrases.

Why Key phrases can be translated to mathematical symbols.

Solution

a. Key phrase: *One-half of* **Translation:** multiplication by $\frac{1}{2}$

The algebraic expression is: $\frac{1}{2}P$.

b. Key phrase: *less than* **Translation:** subtraction

Sometimes thinking in terms of specific numbers makes translating easier. Suppose the capacity was 100. Then 5 *less than* 100 would be $100 - 5$. If the capacity is c, then we need to make it 5 less. The algebraic expression is: $c - 5$.

c. Key phrase: *product of* **Translation:** multiplication

Key phrase: *increased by* **Translation:** addition

In the given wording, the comma after 2,000 means w is first multiplied by 2,000; then 300 is added to that product. The algebraic expression is: $2{,}000w + 300$.

Caution
$5 < c$ is the translation of the statement 5 *is less than the capacity c* not 5 *less than the capacity c*.

Self Check 3 Write each phrase as an algebraic expression:
a. 80 less than the total t **b.** $\frac{2}{3}$ of the time T
c. the difference of twice a and 15, squared

Now Try **Problems 25, 31, and 35**

To solve application problems, we often let a variable stand for an unknown quantity.

EXAMPLE 4 ***Swimming.*** A pool is to be sectioned into 8 equally wide swimming lanes. Write an algebraic expression that represents the width of each lane.

Strategy We will begin by letting $x =$ the width of the swimming pool in feet. Then we will identify any key phrases.

Why The width of the pool is unknown.

Solution The key phrase, *sectioned into 8 equally wide lanes,* indicates division.

Therefore, the width of each lane is $\frac{x}{8}$ feet.

Self Check 4 It takes Val m minutes to get to work by bus. If she drives her car, her travel time exceeds this by 15 minutes. How long does it take her to get to work by car?

Now Try **Problem 61**

EXAMPLE 5 ***Painting.*** A 10-inch-long paintbrush has two parts: a handle and bristles. Choose a variable to represent the length of one of the parts. Then write an expression to represent the length of the other part.

Strategy There are two approaches. We can let $h =$ the length of the handle or we can let $b =$ the length of the bristles.

Why Both the length of the handle and the length of the bristles are unknown.

Solution Refer to the drawing on the top. If we let $h =$ the length of the handle (in inches), then the length of the bristles is $10 - h$.

Now refer to the drawing on the bottom. If we let $b =$ the length of the bristles (in inches), then the length of the handle is $10 - b$.

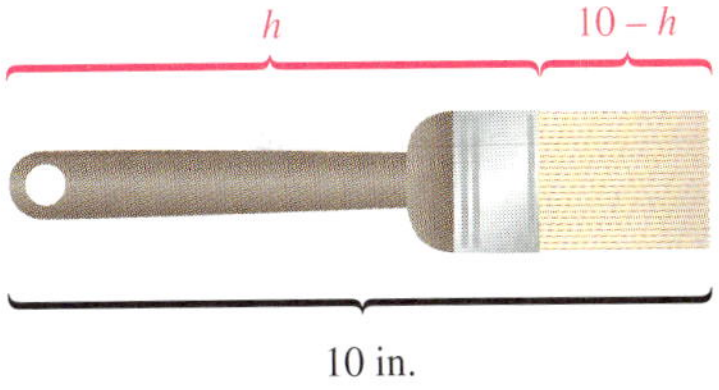

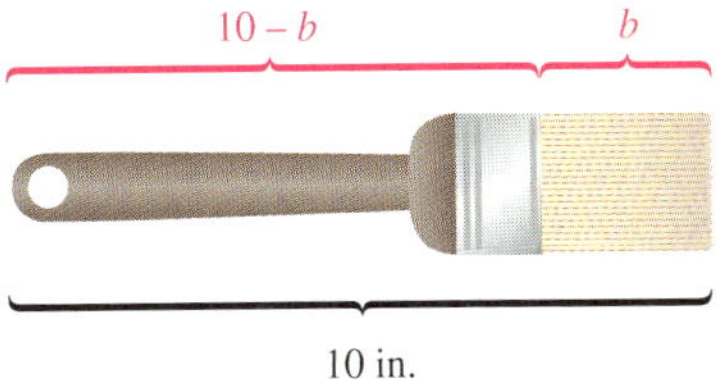

Self Check 5 Part of a \$900 donation to a college went to the scholarship fund, the rest to the building fund. Choose a variable to represent the amount donated to one of the funds. Then write an expression that represents the amount donated to the other fund.

Now Try **Problem 13**

EXAMPLE 6 ***Enrollments.*** Second semester enrollment in a nursing program was 32 more than twice that of the first semester. Let x represent the enrollment for one of the semesters. Write an expression that represents the enrollment for the other semester.

Strategy We will begin by letting $x =$ the enrollment for the first semester.

Why Because the second-semester enrollment is related to the first-semester enrollment.

Solution

Key phrase: *more than* **Translation:** addition

Key phrase: *twice that* **Translation:** multiplication by 2

The second semester enrollment was $2x + 32$.

Self Check 6 In an election, the incumbent received 55 fewer votes than three times the challenger's votes. Let x represent the number of votes received by one candidate. Write an expression that represents the number of votes received by the other.

Now Try **Problem 105**

3 Analyze Problems to Determine Hidden Operations.

Many applied problems require insight and analysis to determine which mathematical operations to use.

EXAMPLE 7 ***Vacations.*** Disneyland, in California, was in operation 16 years before the opening of Disney World in Florida. Euro Disney, in France, was constructed 21 years after Disney World. Write algebraic expressions to represent the ages (in years) of each Disney attraction.

Attraction	Age
Disneyland	$x + 16$
Disney World	x
Euro Disney	$x - 21$

Strategy We will begin by letting $x =$ the age of Disney World.

Why The ages of Disneyland and Euro Disney are both related to the age of Disney World.

Solution In carefully reading the problem, we see that Disneyland was built 16 years before Disney World. That makes its age 16 years more than that of Disney World. The key phrase *more than* indicates addition.

$x + 16 =$ the age of Disneyland

Euro Disney was built 21 years *after* Disney World. That makes its age 21 years less than that of Disney World. The key phrase *less than* indicates subtraction.

$x - 21 =$ the age of Euro Disney

Now Try **Problem 107**

Number of years	Number of months
1	12
2	24
3	36
x	$12x$

We multiply the number of years by 12 to find the number of months.

EXAMPLE 8 How many months are in x years?

Strategy There are no key phrases so we must carefully analyze the problem. We will begin by considering some specific cases.

Why It's often easier to work with specifics first to get a better understanding of the relationship between the two quantities. Then we can generalize using a variable.

Solution Let's calculate the number of months in 1 year, 2 years, and 3 years. When we write the results in a table, a pattern is apparent.

The number of months in x years is $12 \cdot x$ or $12x$.

Self Check 8 How many days is h hours?

Now Try **Problems 7 and 67**

In some problems, we must distinguish between *the number of* and *the value of* the unknown quantity. For example, to find the value of 3 quarters, we multiply the number of quarters by the value (in cents) of one quarter. Therefore, the value of 3 quarters is $3 \cdot 25$ cents $= 75$ cents.

The same distinction must be made if the number is unknown. For example, the value of n nickels is not n cents. The value of n nickels is $n \cdot 5$ cents $= 5n$ cents. For problems of this type, we will use the relationship

Number · value = total value

EXAMPLE 9 Find the total value of: **a.** five dimes **b.** q quarters **c.** $x + 1$ half-dollars

Strategy To find the total value (in cents) of each collection of coins, we multiply the number of coins by the value (in cents) of one coin, as shown in the table.

Why Number · value = total value

Solution

Type of coin	Number	Value	Total value	
Dime	5	10	50	Multiply: $5 \cdot 10 = 50$.
Quarter	q	25	$25q$	Multiply: $q \cdot 25$ can be written $25q$.
Half-dollar	$x + 1$	50	$50(x + 1)$	Multiply: $(x + 1) \cdot 50$ can be written $50(x + 1)$.

Self Check 9 Find the value of: **a.** six \$50 savings bonds **b.** t \$100 savings bonds **c.** $x - 4$ \$1,000 savings bonds

Now Try **Problems 14 and 69**

4 Evaluate Algebraic Expressions.

To evaluate an algebraic expression, we substitute given numbers for each variable and perform the necessary calculations in the proper order.

EXAMPLE 10 Evaluate each expression for $x = 3$ and $y = -4$: **a.** $y^3 + y^2$ **b.** $-y - x$ **c.** $|5xy - 7|$ **d.** $\frac{y - 0}{x - (-1)}$

Strategy We will replace each x and y in the expression with the given value of the variable, and evaluate the expression using the order of operation rules.

Why To *evaluate an expression* means to find its numerical value, once we know the value of its variable(s).

Solution

a. $y^3 + y^2 = (-4)^3 + (-4)^2$ Substitute −4 for each y. We must write −4 within parentheses so that it is the base of each exponential expression.

$= -64 + 16$ Evaluate each exponential expression.

$= -48$

Caution
When replacing a variable with its numerical value, we must often write the replacement number within parentheses to convey the proper meaning.

b. $-y - x = -(-4) - 3$ Substitute −4 for y and 3 for x. Don't forget to write the − sign in front of (−4).

$= 4 - 3$ Simplify: $-(-4) = 4$.

$= 1$

c. $|5xy - 7| = |5(3)(-4) - 7|$ Substitute 3 for x and −4 for y.

$= |-60 - 7|$ Do the multiplication: $5(3)(-4) = -60$.

$= |-67|$ Do the subtraction: $-60 - 7 = -60 + (-7) = -67$.

$= 67$ Find the absolute value of −67.

d. $\frac{y - 0}{x - (-1)} = \frac{-4 - 0}{3 - (-1)}$ Substitute 3 for x and −4 for y.

$= \frac{-4}{4}$ In the denominator, do the subtraction: $3 - (-1) = 3 + 1 = 4$.

$= -1$ Simplify the fraction.

Self Check 10 Evaluate each expression for $a = -2$ and $b = 5$:
a. $|a^3 + b^2|$ **b.** $-a + 2ab$ **c.** $\frac{a + 2}{b - 3}$

Now Try **Problems 79 and 91**

EXAMPLE 11 ***Rocketry.*** If a toy rocket is shot into the air with an initial velocity of 80 feet per second, its height (in feet) after t seconds in flight is given by $-16t^2 + 80t$. How many seconds after the launch will it hit the ground?

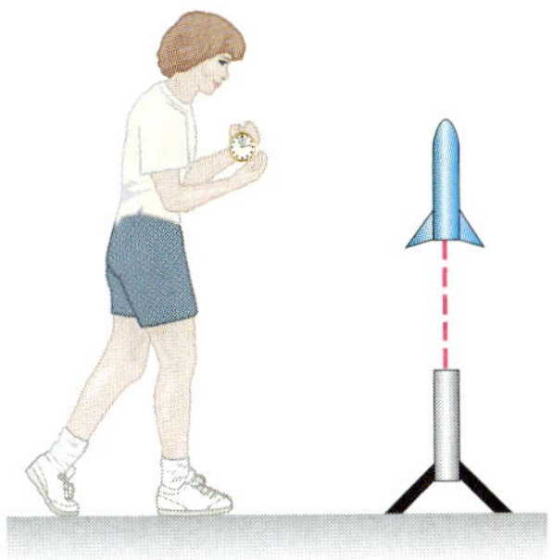

Strategy We can substitute positive values for t, the time in flight, until we find the one that gives a height of 0.

Why When the height of the rocket is 0, it is on the ground.

Solution We begin by finding the height after the rocket has been in flight for 1 second ($t = 1$).

$$\begin{aligned} -16t^2 + 80t &= -16(1)^2 + 80(1) \quad \text{Substitute 1 for } t. \\ &= 64 \end{aligned}$$

As we evaluate $-16t^2 + 80t$ for several more values of t, we record each result in a table. The columns of the table can also be headed with the terms **input** and **output.** The values of t are the inputs into the expression $-16t^2 + 80t$, and the resulting values are the outputs.

t	$-16t^2 + 80t$
1	64
2	96
3	96
4	64
5	0

Evaluate for $t = 2$: $-16t^2 + 80t = -16(2)^2 + 80(2) = 96$

Evaluate for $t = 3$: $-16t^2 + 80t = -16(3)^2 + 80(3) = 96$

Evaluate for $t = 4$: $-16t^2 + 80t = -16(4)^2 + 80(4) = 64$

Evaluate for $t = 5$: $-16t^2 + 80t = -16(5)^2 + 80(5) = 0$

Input	Output
1	64
2	96
3	96
4	64
5	0

The height of the rocket is 0 when $t = 5$. The rocket will hit the ground 5 seconds after being launched.

Self Check 11 In Example 11, suppose the height of the rocket is given by $-16t^2 + 112t$. What will be the height of the rocket 6 seconds after launch?

Now Try **Problem 97**

ANSWERS TO SELF CHECKS **1.** 1, −12, 3, −4 **2. a.** Factor **b.** Term **3. a.** $t - 80$ **b.** $\frac{2}{3}T$ **c.** $(2a - 15)^2$ **4.** $(m + 15)$ minutes **5.** s = amount donated to scholarship fund in dollars; $900 - s$ = amount donated to building fund **6.** x = number of votes received by the challenger; $3x - 55$ = number of votes received by the incumbent **8.** $\frac{h}{24}$ **9. a.** \$300 **b.** \$100t **c.** \$1,000$(x - 4)$ **10. a.** 17 **b.** −18 **c.** 0 **11.** 96 ft

STUDY SET 1.8

VOCABULARY

Fill in the blanks.

1. Variables and/or numbers can be combined with the operations of arithmetic to create algebraic ___________.
2. A ______ is a product or quotient of numbers and/or variables. Examples are: $8x$, $\frac{t}{2}$, and $-cd^3$.
3. Addition symbols separate algebraic expressions into parts called _______.
4. A term, such as 27, that consists of a single number is called a _________ term.
5. The ___________ of the term $10x$ is 10.
6. To _________ $4x - 3$ for $x = 5$, we substitute 5 for x and perform the necessary calculations.

CONCEPTS

7. Complete the table below on the left to determine the number of days in w weeks.
8. Complete the table below on the right to determine the number of minutes in s seconds.

Number of weeks	Number of days
1	
2	
3	
w	

Number of seconds	Number of minutes
60	
120	
180	
s	

9. The knife shown below is 12 inches long. Write an expression that represents the length of the blade.

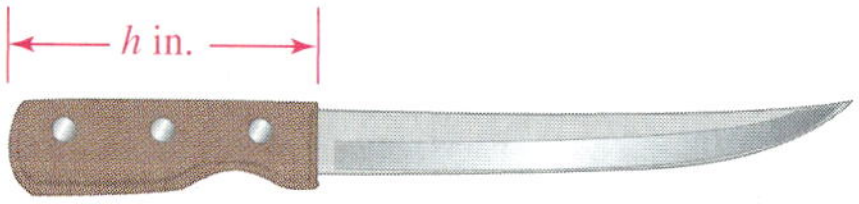

10. A student inherited \$5,000 and deposits x dollars in American Savings. Write an expression that represents the amount of money left to deposit in a City Mutual account.

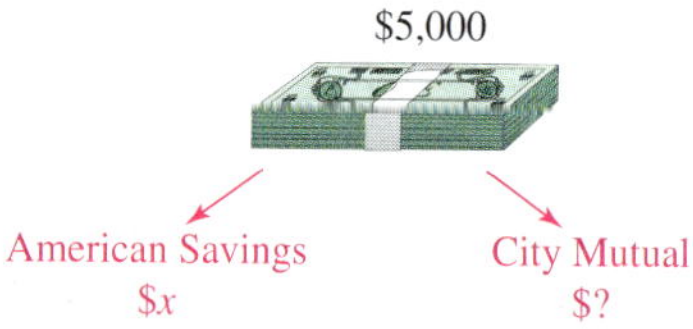

11. Solution 1 is poured into solution 2. Write an expression that represents the number of ounces in the mixture.

12. Peanuts were mixed with p pounds of cashews to make 100 pounds of a mixture. Write an expression that represents the number of pounds of peanuts that were used.

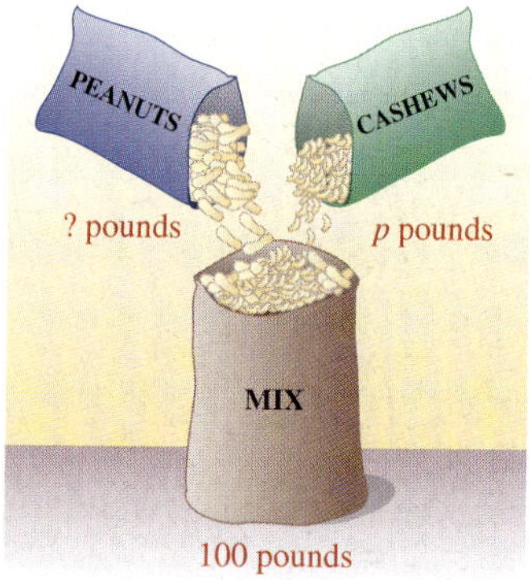

13. **a.** Let b = the length of the beam shown below (in feet). Write an expression that represents the length of the pipe.
    b. Let p = the length of the pipe (in feet). Write an expression that represents the length of the beam.

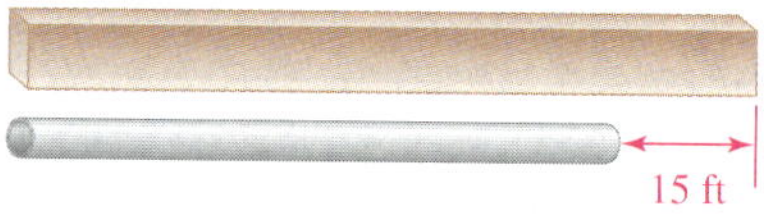

14. Complete the table. Give each value in cents.

Coin	Number	· Value	= Total value
Nickel	6		
Dime	d		
Half-dollar	$x + 5$		

NOTATION

Complete each solution. Evaluate each expression for $a = 5$, $x = -2$, and $y = 4$.

15. $9a - a^2 = 9(\quad) - (5)^2$
$= 9(5) - \quad$
$= \quad - 25$
$= 20$

16. $-x + 6y = -(\quad) + 6(\quad)$
$= \quad + 24$
$= 26$

17. Write each term in standard form.
a. $y8$ **b.** $d2c$
c. What property of multiplication did you use?

18. Fill in the blanks.
a. $\frac{w}{2} = \quad w$ **b.** $\frac{2}{3}m = \frac{\quad}{3}$

GUIDED PRACTICE

19. Consider the expression $3x^3 + 11x^2 - x + 9$. See Example 1.
a. How many terms does the expression have?
b. What is the coefficient of each term?

20. Complete the following table.

Term	$6m$	$-75t$	w	$\frac{1}{2}bh$	$\frac{x}{5}$	t
Coefficient						

***Determine whether the variable c is used as a factor or as a term.* See Example 2.**

21. $c + 32$ **22.** $-24c + 6$
23. $5c$ **24.** $a + b + c$

***Translate each phrase to an algebraic expression. If no variable is given, use x as the variable.* See Example 3.**

25. The sum of the length l and 15
26. The difference of a number and 10
27. The product of a number and 50
28. Three-fourths of the population p
29. The ratio of the amount won w and lost l
30. The tax t added to c
31. P increased by two-thirds of p
32. 21 less than the total height h
33. The square of k minus 2,005
34. s subtracted from S
35. 1 less than twice the attendance a
36. J reduced by 500
37. 1,000 split n equal ways
38. Exceeds the cost c by 25,000
39. 90 more than twice the current price p
40. 64 divided by the cube of y
41. 3 times the total of 35, h, and 300
42. Decrease x by -17
43. 680 fewer than the entire population p
44. Triple the number of expected participants
45. The product of d and 4, decreased by 15
46. The quotient of y and 6, cubed
47. Twice the sum of 200 and t
48. The square of the quantity 14 less than x
49. The absolute value of the difference of a and 2
50. The absolute value of a, decreased by 2
51. One-tenth of the distance d
52. Double the difference of x and 18

***Translate each algebraic expression into an English phrase. (Answers may vary.)* See Example 3.**

53. $\frac{3}{4}r$
54. $\frac{2}{3}d$
55. $t - 50$
56. $c + 19$
57. xyz
58. $10ab$
59. $2m + 5$
60. $2s - 8$

***Answer with an algebraic expression.* See Example 4.**

61. A model's skirt is x inches long. The designer then lets the hem down 2 inches. What is the length of the altered skirt?
62. A soft drink manufacturer produced c cans of cola during the morning shift. Write an expression for how many six-packs of cola can be assembled from the morning shift's production.
63. The tag on a new pair of 36-inch-long jeans warns that after washing, they will shrink x inches in length. What is the length of the jeans after they are washed?
64. A caravan of b cars, each carrying 5 people, traveled to the state capital for a political rally. How many people were in the caravan?

***Answer with an algebraic expression.* See Example 8.**

65. How many minutes are there in h hours?
66. How many feet are in y yards?
67. How many feet are in i inches?
68. How many centuries in y years?

***Answer with an algebraic expression.* See Example 9.**

69. A sales clerk earns \$$x$ an hour; how much does he earn in an 8-hour day?

70. A cashier earns \$$d$ an hour; how much does she earn in a 40-hour week?

71. If a car rental agency charges 49¢ a mile, what is the rental fee if a car is driven x miles?

72. If one egg is worth e cents, find the value (in cents) of one dozen eggs.

73. A ticket to a concert costs \$$t$. What would a pair of concert tickets cost?

74. If one apple is worth a cents, find the value (in cents) of 20 apples.

75. Tickets to a circus cost \$25 each. What will tickets cost for a family of x people if they also pay for two of their neighbors?

76. A certain type of office desk that used to sell for \$$x$ is now on sale for \$50 off. What will a company pay if it purchases 80 of the desks?

Evaluate each expression, for x = 3, y = −2, and z = −4. **See Example 10.**

77. $-y$

78. $-z$

79. $-z + 3x$

80. $-y - 5x$

81. $3y^2 - 6y - 4$

82. $-z^2 - z - 12$

83. $(3 + x)y$

84. $(4 + z)y$

85. $(x + y)^2 - |z + y|$

86. $[(z - 1)(z + 1)]^2$

87. $-\dfrac{2x + y^3}{y + 2z}$

88. $-\dfrac{2z^2 - x}{2x - y^2}$

Evaluate each expression. **See Example 10.**

89. $b^2 - 4ac$ for $a = -1$, $b = 5$, and $c = -2$

90. $(x - a)^2 + (y - b)^2$ for $x = -2$, $y = 1$, $a = 5$, and $b = -3$

91. $a^2 + 2ab + b^2$ for $a = -5$ and $b = -1$

92. $\dfrac{a - x}{y - b}$ for $x = -2$, $y = 1$, $a = 5$, and $b = 2$

93. $\dfrac{n}{2}[2a + (n - 1)d]$ for $n = 10$, $a = -4.2$, and $d = 6.6$

94. $\dfrac{a(1 - r^n)}{1 - r}$ for $a = -5$, $r = 2$, and $n = 3$

95. $(27c^2 - 4d^2)^3$ for $c = \frac{1}{3}$ and $d = \frac{1}{2}$

96. $\dfrac{-b^2 + 16a^2 + 1}{2}$ for $a = \frac{1}{4}$ and $b = -10$

Complete each table. **See Example 11.**

97.

x	$x^3 - 1$
0	
−1	
−3	

98.

g	$g^2 - 7g + 1$
0	
7	
−10	

99.

s	$\frac{5s + 36}{s}$
1	
6	
−12	

100.

a	$2{,}500a + a^3$
2	
4	
−5	

101.

Input x	Output $2x - \frac{x}{2}$
100	
−300	

102.

Input x	Output $\frac{x}{3} + \frac{x}{4}$
12	
−36	

103.

x	$(x + 1)(x + 5)$
−1	
−5	
−6	

104.

x	$\frac{1}{x + 8}$
−7	
−9	
−8	

APPLICATIONS

105. VEHICLE WEIGHTS A Hummer H2 weighs 340 pounds less than twice a Honda Element.

a. Let x represent the weight of one of the vehicles. Write an expression for the weight of the other vehicle.

b. If the weight of the Element is 3,370 pounds, what is the weight of the Hummer?

106. SOD FARMS The expression $20{,}000 - 3s$ gives the number of square feet of sod that are left in a field after s strips have been removed. Suppose a city orders 7,000 strips of sod. Evaluate the expression and explain the result.

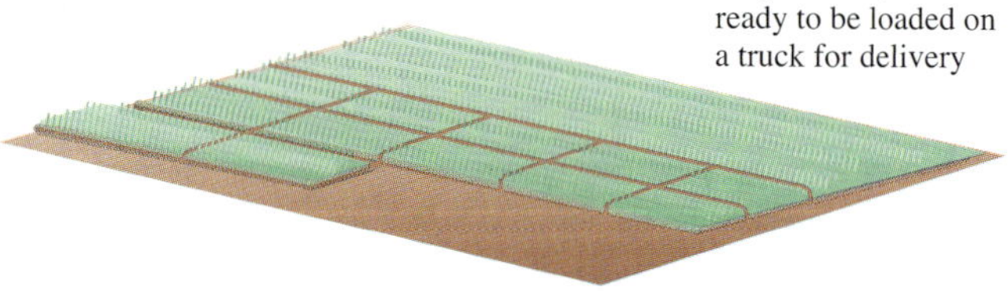

107. COMPUTER COMPANIES IBM was founded 80 years before Apple Computer. Dell Computer Corporation was founded 9 years after Apple.

a. Let x represent the age (in years) of one of the companies. Write expressions to represent the ages (in years) of the other two companies.

b. On April 1, 2008, Apple Computer Company was 32 years old. How old were the other two computer companies then?

108. THRILL RIDES The distance in feet that an object will fall in t seconds is given by the expression $16t^2$. Find the distance that riders on "Drop Zone" will fall during the times listed in the table.

© Joel Rogers, www.coastergallery.com

Time (seconds)	Distance (feet)
1	
2	
3	
4	

WRITING

109. What is an algebraic expression? Give some examples.

110. Explain why 2 *less than* x does not translate to $2 < x$.

111. In this section, we substituted a number for a variable. List some other uses of the word *substitute* that you encounter in everyday life.

112. Explain why d dimes are not worth d¢.

REVIEW

113. Find the LCD for $\frac{5}{12}$ and $\frac{1}{15}$.

114. Simplify: $\frac{3 \cdot 3 \cdot 5}{3 \cdot 5 \cdot 5 \cdot 11}$

115. Evaluate: $\left(\frac{2}{3}\right)^3$

116. Find the result when $\frac{7}{8}$ is multiplied by its reciprocal.

CHALLENGE PROBLEMS

117. Evaluate: $(8 - 1)(8 - 2)(8 - 3) \ldots (8 - 49)(8 - 50)$

118. Translate to an expression: The sum of a number decreased by six, and seven more than the quotient of triple the number and five.

SECTION 1.9
Simplifying Algebraic Expressions Using Properties of Real Numbers

Objectives

1. Simplify products.
2. Use the distributive property.
3. Identify like terms.
4. Combine like terms.

In algebra, we frequently replace one algebraic expression with another that is equivalent and simpler in form. That process, called *simplifying an algebraic expression,* often involves the use of one or more properties of real numbers.

1 Simplify Products.

The commutative and associative properties of multiplication can be used to simplify certain products. For example, let's simplify $8(4x)$.

$8(4x) = 8 \cdot (4 \cdot x)$ Rewrite $4x$ as $4 \cdot x$.

$= (8 \cdot 4) \cdot x$ Use the associative property of multiplication to group 4 with 8.

$= 32x$ Do the multiplication within the parentheses.

We have found that $8(4x) = 32x$. We say that $8(4x)$ and $32x$ are **equivalent expressions** because for each value of x, they represent the same number.

Success Tip

By the commutative property of multiplication, we can *change* the order of factors. By the associative property of multiplication, we can change the *grouping* of factors.

If $x = 10$

$8(4x) = 8[4(10)]$ $\quad$ $32x = 32(10)$
$= 8(40)$ $\quad$ $= 320$
$= 320$

If $x = -3$

$8(4x) = 8[4(-3)]$ $\quad$ $32x = 32(-3)$
$= 8(-12)$ $\quad$ $= -96$
$= -96$

EXAMPLE 1 Simplify: **a.** $-9(3b)$ **b.** $15a(6)$ **c.** $3(7p)(-5)$ **d.** $\frac{8}{3} \cdot \frac{3}{8}r$ **e.** $35\left(\frac{4}{5}x\right)$

Strategy We will use the commutative and associative properties of multiplication to reorder and regroup the factors in each expression.

Why We want to group all of the numerical factors of an expression together so that we can find their product.

Solution

a. $-9(3b) = (-9 \cdot 3)b$ Use the associative property of multiplication to regroup the factors.
$= -27b$ Do the multiplication within the parentheses.

b. $15a(6) = 15(6)a$ Use the commutative property of multiplication to reorder the factors.
$= 90a$ Do the multiplication, working from left to right: $15(6) = 90$.

c. $3(7p)(-5) = [3(7)(-5)]p$ Use the commutative and associative properties of multiplication to reorder and regroup the factors.
$= -105p$ Do the multiplication within the brackets.

d. $\frac{8}{3} \cdot \frac{3}{8}r = \left(\frac{8}{3} \cdot \frac{3}{8}\right)r$ Use the associative property of multiplication to regroup the factors.
$= 1r$ Multiply within the parentheses. The product of a number and its reciprocal is 1.
$= r$ The coefficient 1 need not be written.

e. $35\left(\frac{4}{5}x\right) = \left(35 \cdot \frac{4}{5}\right)x$ Use the associative property of multiplication to regroup the factors.
$= \left(\frac{\overset{1}{\cancel{5}} \cdot 7 \cdot 4}{\underset{1}{\cancel{5}}}\right)x$ Factor 35 as $5 \cdot 7$ and then remove the common factor 5.
$= 28x$

Self Check 1 Multiply: **a.** $9 \cdot 6s$ **b.** $-4(6u)(-2)$ **c.** $\frac{2}{3} \cdot \frac{3}{2}m$ **d.** $36\left(\frac{2}{9}y\right)$

Now Try **Problems 15, 25, 29, and 31**

2 Use the Distributive Property.

Another property that is often used to simplify algebraic expressions is the **distributive property.** To introduce it, we will evaluate $4(5 + 3)$ in two ways.

The Language of Algebra
To *distribute* means to give from one to several. You have probably *distributed* candy to children coming to your door on Halloween.

Method 1
Use the order of operations:

$$4(5 + 3) = 4(8)$$
$$= 32$$

Method 2
Distribute the multiplication:

$$4(5 + 3) = 4(5) + 4(3)$$
$$= 20 + 12$$
$$= 32$$

Each method gives a result of 32. This observation suggests the following property.

The Distributive Property

For any real numbers a, b, and c,

$$a(b + c) = ab + ac$$

The Language of Algebra
Formally, it is called the *distributive property of multiplication over addition.* When we use it to write a product, such as $5(x + 2)$, as a sum, $5x + 10$, we say that we have *removed* or *cleared* the parentheses.

To illustrate one use of the distributive property, let's consider the expression $5(x + 3)$. Since we are not given the value of x, we cannot add x and 3 within the parentheses. However, we can distribute the multiplication by the factor of 5 that is outside the parentheses to x and to 3 and add those products.

$5(x + 3) = 5(x) + 5(3)$ Distribute the multiplication by 5.
$= 5x + 15$ Do the multiplications.

EXAMPLE 2 Multiply: **a.** $8(m + 9)$ **b.** $-12(4t + 1)$ **c.** $6\left(\frac{x}{3} + \frac{9}{2}\right)$

Strategy In each case, we will distribute the multiplication by the factor *outside* the parentheses over each term *within* the parentheses.

Why In each case, we cannot simplify the expression within the parentheses. To multiply, we must use the distributive property.

Solution

The Language of Algebra
We read $8(m + 9)$ as "eight times the *quantity* of m plus nine." The word *quantity* alerts us to the grouping symbols in the expression.

a. $8(m + 9) = 8 \cdot m + 8 \cdot 9$ Distribute the multiplication by 8.
$= 8m + 72$ Do the multiplications.

b. $-12(4t + 1) = -12(4t) + (-12)(1)$ Distribute the multiplication by −12.
$= -48t + (-12)$ Do the multiplications.
$= -48t - 12$ Write the result in simpler form. Recall that adding −12 is the same as subtracting 12.

c. $6\left(\frac{x}{3} + \frac{9}{2}\right) = 6 \cdot \frac{x}{3} + 6 \cdot \frac{9}{2}$ Distribute the multiplication by 6.

$= \frac{2 \cdot \overset{1}{\cancel{3}} \cdot x}{\underset{1}{\cancel{3}}} + \frac{\overset{1}{\cancel{2}} \cdot 3 \cdot 9}{\underset{1}{\cancel{2}}}$ Factor 6 as $2 \cdot 3$ and then remove the common factors 3 and 2.

$= 2x + 27$

Self Check 2 Multiply: **a.** $7(m + 2)$ **b.** $-80(8x + 3)$ **c.** $24\left(\frac{y}{6} + \frac{3}{8}\right)$

Now Try **Problems 35, 37, and 39**

Since subtraction is the same as adding the opposite, the distributive property also holds for subtraction.

$$a(b - c) = ab - ac$$

EXAMPLE 3

Multiply: **a.** $3(3b - 4)$ **b.** $-6(-3y - 8)$ **c.** $-1(t - 9)$

Strategy In each case, we will distribute the multiplication by the factor *outside* the parentheses over each term *within* the parentheses.

Why In each case, we cannot simplify the expression within the parentheses. To multiply, we must use the distributive property.

Solution

a.

$3(3b - 4) = 3(3b) - 3(4)$	Distribute the multiplication by 3.
$= 9b - 12$	Do the multiplications.

Caution
A common mistake is to forget to distribute the multiplication over each of the terms within the parentheses.

~~$3(3b - 4) = 9b - 4$~~

b.

$-6(-3y - 8) = -6(-3y) - (-6)(8)$	Distribute the multiplication by -6.
$= 18y - (-48)$	Do the multiplications.
$= 18y + 48$	Write the result in simpler form. Add the opposite of -48.

Another approach is to write the subtraction within the parentheses as addition of the opposite. Then we distribute the multiplication by -6 over the addition.

$-6(-3y - 8) = -6[-3y + (-8)]$	Add the opposite of 8.
$= -6(-3y) + (-6)(-8)$	Distribute the multiplication by -6.
$= 18y + 48$	Do the multiplications.

Success Tip
Notice that distributing the multiplication by -1 *changes the sign* of each term within the parentheses.

c.

$-1(t - 9) = -1(t) - (-1)(9)$	Distribute the multiplication by -1.
$= -t - (-9)$	Do the multiplications.
$= -t + 9$	Write the result in simpler form. Add the opposite of -9.

Self Check 3 Multiply: **a.** $5(2x - 1)$ **b.** $-9(-y - 4)$ **c.** $-1(c - 22)$

Now Try **Problems 43, 47, and 49**

Caution The distributive property does not apply to every expression that contains parentheses—only those where multiplication is distributed over addition (or subtraction). For example, to simplify $6(5x)$, we do not use the distributive property.

Correct	Incorrect
$6(5x) = (6 \cdot 5)x = 30x$	$6(5x) = 30 \cdot 6x = 180x$

The distributive property can be extended to several other useful forms. Since multiplication is commutative, we have:

$$(b + c)a = ba + ca \qquad (b - c)a = ba - ca$$

For situations in which there are more than two terms within parentheses, we have:

$$a(b + c + d) = ab + ac + ad \qquad a(b - c - d) = ab - ac - ad$$

EXAMPLE 4 Multiply: **a.** $(6x + 4)\frac{1}{2}$ **b.** $2(a - 3b)8$ **c.** $-0.3(3a - 4b + 7)$

Strategy We will multiply each term within the parentheses by the factor (or factors) outside the parentheses.

Why In each case, we cannot simplify the expression within the parentheses. To multiply, we must use the distributive property.

Solution

a. $(6x + 4)\frac{1}{2} = (6x)\frac{1}{2} + (4)\frac{1}{2}$ Distribute the multiplication by $\frac{1}{2}$.

$= 3x + 2$ Do the multiplications.

b. $2(a - 3b)8 = 2 \cdot 8(a - 3b)$

$= 16(a - 3b)$ Multiply 2 and 8 to get 16.

$= 16a - 48b$ Distribute the multiplication by 16.

c. $-0.3(3a - 4b + 7) = -0.3(3a) - (-0.3)(4b) + (-0.3)(7)$

$= -0.9a + 1.2b - 2.1$ Do each multiplication.

Self Check 4 Multiply: **a.** $(-6x - 24)\frac{1}{3}$

b. $6(c - 2d)9$

c. $-0.7(2r + 5s - 8)$

Now Try **Problems 53, 55, and 57**

We can use the distributive property to find the opposite of a sum. For example, to find $-(x + 10)$, we interpret the $-$ symbol as a factor of -1, and proceed as follows:

$-(x + 10) = -1(x + 10)$ Replace the $-$ symbol with -1.

$= -1(x) + (-1)(10)$ Distribute the multiplication by -1.

$= -x - 10$

In general, we have the following property of real numbers.

The Opposite of a Sum

The opposite of a sum is the sum of the opposites.

For any real numbers a and b,

$$-(a + b) = -a + (-b).$$

EXAMPLE 5 Simplify: $-(-9s - 3)$

Strategy We will multiply each term within the parentheses by -1.

Why The $-$ outside the parentheses represents a factor of -1 that is to be distributed.

Solution

$-(-9s - 3) = -1(-9s - 3)$ Replace the $-$ symbol in front of the parentheses with -1.

$= -1(-9s) - (-1)(3)$ Distribute the multiplication by -1.

$= 9s + 3$

Self Check 5 Simplify: $-(-5x + 18)$

Now Try Problem 59

3 Identify Like Terms.

Before we can discuss methods for simplifying algebraic expressions involving addition and subtraction, we need to introduce some new vocabulary.

Like Terms

Like terms are terms containing exactly the same variables raised to exactly the same powers. Any constant terms in an expression are considered to be like terms. Terms that are not like terms are called **unlike terms.**

Success Tip
When looking for like terms, don't look at the coefficients of the terms. Consider only the variable factors of each term. If two terms are like terms, only their coefficients may differ.

Here are several examples.

Like terms	*Unlike terms*	
$4x$ and $7x$	$4x$ and $7y$	The variables are not the same.
$-10p^2$ and $25p^2$	$-10p$ and $25p^2$	Same variable, but different powers.
$\frac{1}{3}c^3d$ and c^3d	$\frac{1}{3}c^3d$ and c^3	The variables are not the same.

EXAMPLE 6 List the like terms in each expression: **a.** $7r + 5 + 3r$ **b.** $6x^4 - 6x^2 - 6x$ **c.** $-17m^3 + 3 - 2 + m^3$

Strategy First, we will identify the terms of the expression. Then we will look for terms that contain the same variables raised to exactly the same powers.

Why If two terms contain the same variables raised to the same powers, they are like terms.

Solution

a. $7r + 5 + 3r$ contains the like terms $7r$ and $3r$.

b. Since the exponents on x are different, $6x^4 - 6x^2 - 6x$ contains no like terms.

c. $-17m^3 + 3 - 2 + m^3$ contains two pairs of like terms: $-17m^3$ and m^3 are like terms, and the constant terms, 3 and -2, are like terms.

Self Check 6 List the like terms: **a.** $2x - 2y + 7y$ **b.** $5p^2 - 12 + 17p^2 + 2$

Now Try Problem 63

4 Combine Like Terms.

To add or subtract objects, they must be similar. For example, fractions that are to be added must have a common denominator. When adding decimals, we align columns to be sure to add tenths to tenths, hundredths to hundredths, and so on. The same is true when working with terms of an algebraic expression. They can be added or subtracted only if they are like terms.

This expression can be simplified because it contains like terms.	This expression cannot be simplified because its terms are not like terms.
$3x + 4x$	$3x + 4y$

Recall that the distributive property can be written in the following forms:

$$(b + c)a = ba + ca \qquad (b - c)a = ba - ca$$

We can use these forms of the distributive property in reverse to simplify a sum or difference of like terms. For example, we can simplify $3x + 4x$ as follows:

$$\begin{aligned} 3x + 4x &= (3 + 4)x \quad \text{Use } ba + ca = (b + c)a. \\ &= 7x \end{aligned}$$

The Language of Algebra Simplifying a sum or difference of like terms is called *combining like terms.*

We can simplify $15m^2 - 9m^2$ in a similar way:

$$\begin{aligned} 15m^2 - 9m^2 &= (15 - 9)m^2 \quad \text{Use } ba - ca = (b - c)a. \\ &= 6m^2 \end{aligned}$$

In each case, we say that we *combined like terms.* These examples suggest the following general rule.

Combining Like Terms Like terms can be combined by adding or subtracting the coefficients of the terms and keeping the same variables with the same exponents.

EXAMPLE 7 Simplify by combining like terms, if possible: **a.** $2x + 9x$ **b.** $-8p + (-2p) + 4p$ **c.** $0.5s^3 - 0.3s^3$ **d.** $4w + 6$ **e.** $\frac{4}{9}b + \frac{7}{9}b$

Strategy We will use the distributive property in reverse to add (or subtract) the coefficients of the like terms. We will keep the same variables raised to the same powers.

Why To *combine like terms* means to add or subtract the like terms in an expression.

Solution

Success Tip
Just as 2 apples plus 9 apples is 11 apples, $2x + 9x = 11x$.

a. Since $2x$ and $9x$ are like terms with the common variable x, we can combine them.

$2x + 9x = 11x$ Think: $(2 + 9)x = 11x$.

b. $-8p + (-2p) + 4p = -6p$ Think: $[-8 + (-2) + 4]p = -6p$.

c. $0.5s^3 - 0.3s^3 = 0.2s^3$ Think: $(0.5 - 0.3)s^3 = 0.2s^3$.

d. Since $4w$ and 6 are not like terms, they cannot be combined. $4w + 6$ doesn't simplify.

e. $\frac{4}{9}b + \frac{7}{9}b = \frac{11}{9}b$ Think: $\left(\frac{4}{9} + \frac{7}{9}\right)b = \frac{11}{9}b$.

Self Check 7 Simplify, if possible: **a.** $3x + 5x$ **b.** $-6y + (-6y) + 9y$ **c.** $4.4s^4 - 3.9s^4$ **d.** $4a - 2$ **e.** $\frac{10}{7}c - \frac{4}{7}c$

Now Try Problems 67, 71, 79, and 83

EXAMPLE 8 Simplify by combining like terms: **a.** $16t - 15t$ **b.** $16t - t$ **c.** $15t - 16t$ **d.** $16t + t$

Strategy As we combine like terms, we must be careful when working with the terms such as t and $-t$.

Why Coefficients of 1 and -1 are usually not written.

Solution

a. $16t - 15t = t$ Think: $(16 - 15)t = 1t = t$.

b. $16t - t = 15t$ Think: $16t - 1t = (16 - 1)t = 15t$.

c. $15t - 16t = -t$ Think: $(15 - 16)t = -1t = -t$.

d. $16t + t = 17t$ Think: $16t + 1t = (16 + 1)t = 17t$.

Self Check 8 Simplify: **a.** $9h - h$ **b.** $9h + h$ **c.** $9h - 8h$ **d.** $8h - 9h$

Now Try Problems 73 and 77

EXAMPLE 9 Simplify: $6a^2 + 54a - 4a - 36$

Strategy First, we will identify any like terms in the expression. Then we will use the distributive property in reverse to combine them.

Why To *simplify* an expression we use properties of real numbers to write an equivalent expression in simpler form.

Solution We can combine the like terms that involve the variable a.

$$6a^2 + 54a - 4a - 36 = 6a^2 + 50a - 36 \quad \text{Think: } (54 - 4)a = 50a.$$

Self Check 9 Simplify: $7y^2 + 21y - 2y - 6$

Now Try **Problem 93**

EXAMPLE 10 Simplify: $4(x + 5) - 5 - (2x - 4)$

Strategy First, we will remove the parentheses. Then we will identify any like terms and combine them.

Why To *simplify* an expression we use properties of real numbers, such as the distributive property, to write an equivalent expression in simpler form.

Solution

> ***Success Tip***
> Here, the distributive property is used both *forward* (to remove parentheses) and in *reverse* (to combine like terms).

$$4(x + 5) - 5 - (2x - 4) = 4(x + 5) - 5 - 1(2x - 4)$$
Replace the − symbol in front of $(2x - 4)$ with −1.

$$= 4x + 20 - 5 - 2x + 4$$
Distribute the multiplication by 4 and −1.

$$= 2x + 19$$
Think: $(4 - 2)x = 2x$.
Think: $(20 - 5 + 4) = 19$.

Self Check 10 Simplify: $6(3y - 1) + 2 - (-3y + 4)$

Now Try **Problem 99**

ANSWERS TO SELF CHECKS **1. a.** $54s$ **b.** $48u$ **c.** m **d.** $8y$ **2. a.** $7m + 14$ **b.** $-640x - 240$ **c.** $4y + 9$ **3. a.** $10x - 5$ **b.** $9y + 36$ **c.** $-c + 22$ **4. a.** $-2x - 8$ **b.** $54c - 108d$ **c.** $-1.4r - 3.5s + 5.6$ **5.** $5x - 18$ **6. a.** $-2y$ and $7y$ **b.** $5p^2$ and $17p^2$; -12 and 2 **7. a.** $8x$ **b.** $-3y$ **c.** $0.5s^4$ **d.** Does not simplify **e.** $\frac{6}{7}c$ **8. a.** $8h$ **b.** $10h$ **c.** h **d.** $-h$ **9.** $7y^2 + 19y - 6$ **10.** $21y - 8$

STUDY SET 1.9

VOCABULARY

Fill in the blanks.

1. To ________ the expression $5(6x)$ means to write it in simpler form: $5(6x) = 30x$.
2. $5(6x)$ and $30x$ are ________ expressions because for each value of x, they represent the same number.
3. To perform the multiplication $2(x + 8)$, we use the ________ property.
4. We call $-(c + 9)$ the ________ of a sum.

5. Terms such as $7x^2$ and $5x^2$, which have the same variables raised to exactly the same power, are called _____ terms.

6. When we write $9x + x$ as $10x$, we say we have _________ like terms.

CONCEPTS

7. a. Fill in the blanks to simplify the expression.

$$4(9t) = (\ \square \cdot \square\ )t = \square\, t$$

b. What property did you use in part a?

8. a. Fill in the blanks to simplify the expression.

$$-6y \cdot 2 = \square \cdot \square \cdot y = \square\, y$$

b. What property did you use in part a?

9. Fill in the blanks.

a. $2(x + 4) = 2x \ \square\ 8$ b. $2(x - 4) = 2x \ \square\ 8$

c. $-2(x + 4) = -2x \ \square\ 8$ d. $-2(-x - 4) = 2x \ \square\ 8$

10. Fill in the blanks to combine like terms.

a. $4m + 6m = (\ \square\ )m = \square\, m$

b. $30n^2 - 50n^2 = (\ \square\ )n^2 = \square\, n^2$

c. $12 + 32d + 15 = 32d + \square$

d. Like terms can be combined by adding or subtracting the __________ of the terms and keeping the same __________ with the same exponents.

11. Simplify each expression, if possible.

a. $5(2x)$ b. $5 + 2x$

c. $6(-7x)$ d. $6 - 7x$

e. $2(3x)(3)$ f. $2 + 3x + 3$

12. Fill in the blanks: Distributing multiplication by -1 changes the _____ of each term within the parentheses.

$$-(x + 10) = \square(x + 10) = -x \ \square\ 10$$

NOTATION

13. Translate to symbols.

a. Six times the quantity of h minus four.

b. The opposite of the sum of z and sixteen.

14. Write an equivalent expression for the given expression using fewer symbols.

a. $1x$ b. $-1d$ c. $0m$

d. $5x - (-1)$ e. $16t + (-6)$

GUIDED PRACTICE

Simplify each expression. **See Example 1.**

15. $3 \cdot 4t$
16. $9 \cdot 3s$
17. $9(7m)$
18. $12n(8)$
19. $5(-7q)$
20. $-7(5t)$
21. $5t \cdot 60$
22. $70a \cdot 10$
23. $(-5.6x)(-2)$
24. $(-4.4x)(-3)$
25. $5(4c)(3)$
26. $9(2h)(2)$
27. $-4(-6)(-4m)$
28. $-5(-9)(-4n)$
29. $\frac{5}{3} \cdot \frac{3}{5}g$
30. $\frac{9}{7} \cdot \frac{7}{9}k$
31. $12\left(\frac{5}{12}x\right)$
32. $15\left(\frac{4}{15}w\right)$
33. $8\left(\frac{3}{4}y\right)$
34. $27\left(\frac{2}{3}x\right)$

Multiply. **See Examples 2–5.**

35. $5(x + 3)$
36. $4(x + 2)$
37. $-3(4x + 9)$
38. $-5(8x + 9)$
39. $45\left(\frac{x}{5} + \frac{2}{9}\right)$
40. $35\left(\frac{y}{5} + \frac{8}{7}\right)$
41. $0.4(x - 4)$
42. $2.2(2q - 1)$
43. $6(6c - 7)$
44. $9(9d - 3)$
45. $-6(13c - 3)$
46. $-2(10s - 11)$
47. $-15(-2t - 6)$
48. $-20(-4z - 5)$
49. $-1(-4a + 1)$
50. $-1(-2x + 3)$
51. $(3t + 2)8$
52. $(2q + 1)9$
53. $(3w - 6)\frac{2}{3}$
54. $(2y - 8)\frac{1}{2}$
55. $4(7y + 4)2$
56. $8(2a - 3)4$
57. $25(2a - 3b + 1)$
58. $5(9s - 12t - 3)$
59. $-(x - 7)$
60. $-(y + 1)$
61. $-(-5.6y + 7)$
62. $-(-4.8a - 3)$

List the like terms in each expression, if any. **See Example 6.**

63. $3x + 2 - 2x$
64. $3y + 4 - 11y + 6$
65. $-12m^4 - 3m^3 + 2m^2 - m^3$
66. $6x^3 + 3x^2 + 6x$

Simplify by combining like terms. **See Examples 7 and 8.**

67. $3x + 7x$
68. $12y - 15y$
69. $-4x + 4x$
70. $-16y + 16y$
71. $-7b^2 + 27b^2$
72. $-2c^3 + 12c^3$
73. $13r - 12r$
74. $25s + s$
75. $36y + y - 9y$
76. $32a - a + 5a$
77. $43s^3 - 44s^3$
78. $8j^3 - 9j^3$
79. $-9.8c + 6.2c$
80. $-5.7m + 4.3m$
81. $-0.2r - (-0.6r)$
82. $-1.1m - (-2.4m)$
83. $\frac{3}{5}t + \frac{1}{5}t$
84. $\frac{3}{16}x - \frac{5}{16}x$
85. $-\frac{7}{16}x - \frac{3}{16}x$
86. $-\frac{5}{18}x - \frac{7}{18}x$

***Simplify by combining like terms.* See Example 9.**

87. $15y - 10 - y - 20y$

88. $9z - 7 - z - 19z$

89. $3x + 4 - 5x + 1$

90. $4b + 9 - 9b + 9$

91. $9m^2 - 6m + 12m - 4$

92. $6a^2 + 18a - 9a + 5$

93. $4x^2 + 5x - 8x + 9$

94. $10y^2 - 8y + y - 7$

***Simplify.* See Example 10.**

95. $2z + 5(z - 3)$

96. $12(m + 11) - 11$

97. $2(s^2 - 7) - (s^2 - 2)$

98. $4(d^2 - 3) - (d^2 - 1)$

99. $-9(3r - 9) - 7(2r - 7)$

100. $-6(3t - 6) - 3(11t - 3)$

101. $36\left(\frac{2}{9}x - \frac{3}{4}\right) + 36\left(\frac{1}{2}\right)$

102. $40\left(\frac{3}{8}y - \frac{1}{4}\right) + 40\left(\frac{4}{5}\right)$

TRY IT YOURSELF

Simplify each expression.

103. $6 - 4(-3c - 7)$

104. $10 - 5(-5g - 1)$

105. $-4r - 7r + 2r - r$

106. $-v - 3v + 6v + 2v$

107. $24\left(-\frac{5}{6}r\right)$

108. $\frac{3}{4} \cdot \frac{1}{2}g$

109. $a + a + a$

110. $t - t - t - t$

111. $60\left(\frac{3}{20}r - \frac{4}{15}\right)$

112. $72\left(\frac{7}{8}f - \frac{8}{9}\right)$

113. $5(-1.2x)$

114. $5(-6.4c)$

115. $-(c + 7) + 2(c - 3)$

116. $-(z + 2) + 5(3 - z)$

117. $a^3 + 2a^2 + 4a - 2a^2 - 4a - 8$

118. $c^3 - 3c^2 + 9c + 3c^2 - 9c + 27$

APPLICATIONS

In Exercises 119–122, recall that the perimeter of a figure is equal to the sum of the lengths of its sides.

119. THE RED CROSS In 1891, Clara Barton founded the Red Cross. Its symbol is a white flag bearing a red cross. If each side of the cross has length x, write an expression that represents the perimeter of the cross.

120. BILLIARDS Billiard tables vary in size, but all tables are twice as long as they are wide.

a. If the billiard table is x feet wide, write an expression that represents its length.

b. Write an expression that represents the perimeter of the table.

121. PING-PONG Write an expression that represents the perimeter of the Ping-Pong table.

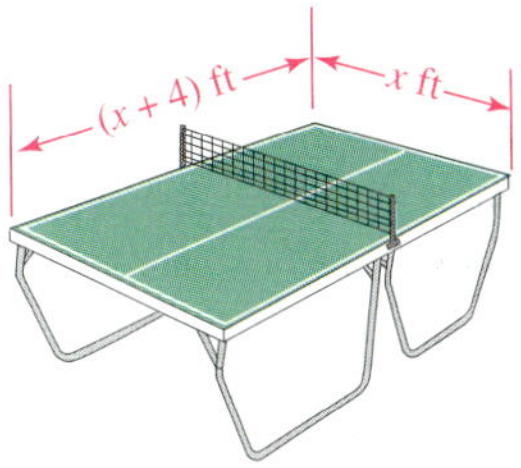

122. SEWING Write an expression that represents the length of the yellow trim needed to outline a pennant with the given side lengths.

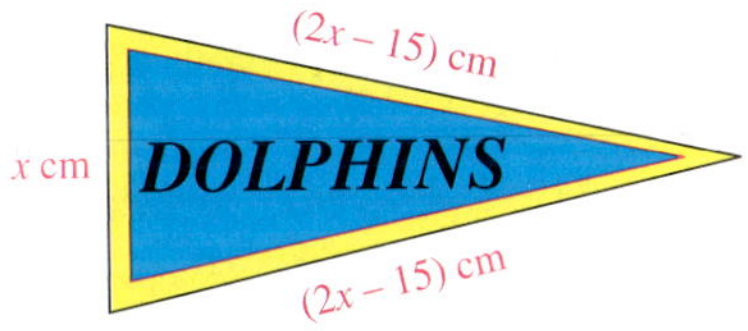

WRITING

123. Explain why the distributive property applies to $2(3 + x)$ but not to $2(3x)$.

124. Tell how to combine like terms.

REVIEW

Evaluate each expression for $x = -3$, $y = -5$, and $z = 0$.

125. $\dfrac{x - y^2}{2y - 1 + x}$

126. $\dfrac{2y + 1}{x} - x$

CHALLENGE PROBLEMS

127. Fill in the blanks: $\_\_\_(\_\_\_ - \_\_\_) = -75x + 40$

Simplify.

128. $-2[x + 4(2x + 1)] - 5[x + 2(3x + 4)]$

CHAPTER 1 Summary & Review

SECTION 1.1 Introducing the Language of Algebra

DEFINITIONS AND CONCEPTS	EXAMPLES
Tables, bar graphs, and **line graphs** are used to describe numerical relationships.	See page 3 for examples of tables and graphs.
A **sum** is the result of an addition. A **difference** is the result of a subtraction. A **product** is the result of a multiplication. A **quotient** is the result of a division.	$3 + 15 = 18$ (sum) $16 - 1 = 15$ (difference) $7 \cdot 8 = 56$ (product) $\frac{63}{9} = 7$ (quotient)
A **variable** is a letter (or symbol) that stands for a number.	Variables: x, a, and y
Algebraic expressions contain variables and numbers combined with the operations of addition, subtraction, multiplication, and division.	Expressions: $5y + 7$, $\frac{12 - x}{5}$, and $8a(b-3)$
An **equation** is a statement that two expressions are equal.	Equations: $3x-4 = 12$ and $\frac{t}{9} = 12$
Equations that express a relationship between two or more variables are called **formulas.**	$A = lw$ (The formula for the area of a rectangle)

REVIEW EXERCISES

The line graph shows the number of cars in a parking structure from 6 P.M. to 12 midnight on a Saturday.

1. What units are used to scale the horizontal and vertical axes?
2. How many cars were in the parking structure at 11 P.M.?
3. At what time did the parking structure have 500 cars in it?

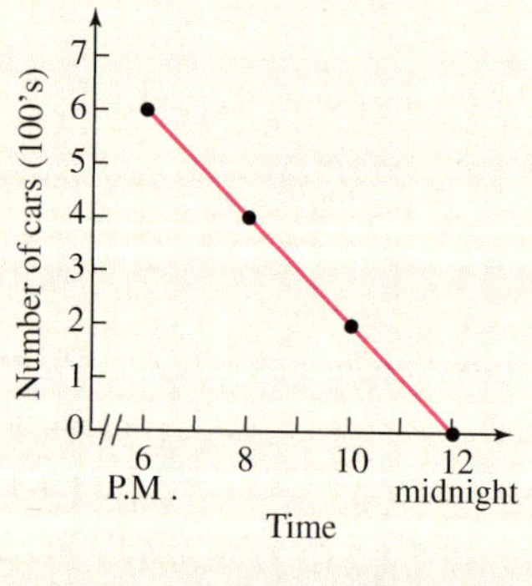

4. When was the structure empty of cars?

Express each statement in words, using one of the words sum, difference, product, or quotient.

5. $15-3 = 12$
6. $15 + 3 = 18$
7. $15 \div 3 = 5$
8. $15 \cdot 3 = 45$
9. **a.** Write the multiplication 4×9 with a raised dot and then with parentheses.
   b. Write the division $9 \div 3$ using a fraction bar.
10. Write each multiplication without a multiplication symbol.
    a. $8 \cdot b$ **b.** $P \cdot r \cdot t$
11. Classify each item as either an expression or an equation.
    a. $5 = 2x + 3$ **b.** $2x + 3$
12. Use the formula $n = b + 5$ to complete the table.

Brackets (b)	Nails (n)
5	
10	
20	

SECTION 1.2 Fractions

DEFINITIONS AND CONCEPTS	EXAMPLES
A **factor** is a number being multiplied.	$8 \cdot 9 = 72$ (8: Factor; 9: Factor)
A **prime number** is a natural number that is greater than 1 that has only itself and 1 as factors. A **composite number** is a natural number, greater than 1, that is not prime.	Primes: $\{2, 3, 5, 7, 11, 13, 17, 19, 23, \ldots\}$ Composites: $\{4, 6, 8, 9, 10, 12, 14, 15, \ldots\}$
Any composite numbers can be factored into the product of two or more prime factors.	Find the prime factorization of 98. $98 = 2 \cdot 49 = 2 \cdot 7 \cdot 7$
In a fraction, the number above the **fraction bar** is the **numerator** and the number below the faction bar is called the **denominator.**	$\frac{11}{15}$ (11 ← Numerator; 15 ← Denominator)
Two fractions are **equivalent** if they represent the same number.	Equivalent fractions: $\frac{1}{2} = \frac{2}{4} = \frac{3}{6} = \frac{4}{8} = \ldots$
To **multiply two fractions,** multiply their numerators and multiply their denominators.	Multiply: $\frac{5}{8} \cdot \frac{3}{4} = \frac{15}{32}$
One number is the **reciprocal** of another if their product is 1.	The reciprocal of $\frac{4}{5}$ is $\frac{5}{4}$ because $\frac{4}{5} \cdot \frac{5}{4} = 1$.
To **divide two fractions,** multiply the first fraction by the reciprocal of the second fraction.	Divide: $\frac{4}{7} \div \frac{5}{8} = \frac{4}{7} \cdot \frac{8}{5} = \frac{32}{35}$
Multiplication property of 1: the product of 1 and any number is that number.	$1 \cdot 5 = 5$ and $\frac{7}{8} \cdot 1 = \frac{7}{8}$
To **build a fraction,** multiply it by a form of 1 such as $\frac{2}{2}, \frac{3}{3}, \frac{4}{4}, \ldots$.	Write $\frac{3}{4}$ as an equivalent fraction with a denominator of 20. $\frac{3}{4} = \frac{3}{4} \cdot \frac{5}{5} = \frac{15}{20}$
To **simplify a fraction,** remove pairs of factors common to the numerator and the denominator. A fraction is in **simplest form,** or **lowest terms,** when the numerator and denominator have no common factors other than 1.	Simplify: $\frac{12}{18} = \frac{2 \cdot \overset{1}{\cancel{6}}}{3 \cdot \underset{1}{\cancel{6}}} = \frac{2}{3}$ $\frac{6}{6} = 1$
To find the **LCD** of two fractions, prime factor each denominator and find the product of the prime factors, using each factor the greatest number of times it appears in any one factorization.	Find the LCD of $\frac{5}{12}$ and $\frac{7}{8}$. $\left.\begin{array}{l} 12 = 2 \cdot 2 \cdot 3 \\ 8 = 2 \cdot 2 \cdot 2 \end{array}\right\}$ LCD $= 2 \cdot 2 \cdot 2 \cdot 3 = 24$
To **add (or subtract) fractions that have the same denominator,** add (or subtract) the numerators and keep the common denominator. Simplify, if possible. To **add (or subtract) fractions that have different denominators,** rewrite each fraction as an equivalent fraction with the LCD as the denominator. Then add (or subtract) as usual. Simplify, if possible.	Add: $\frac{5}{12} + \frac{7}{8} = \frac{5}{12} \cdot \frac{2}{2} + \frac{7}{8} \cdot \frac{3}{3}$ The LCD is 24. Build each fraction. $= \frac{10}{24} + \frac{21}{24}$ The denominators are now the same. $= \frac{31}{24}$ This result does not simplify.

SECTION 1.2 Fractions–*continued*

DEFINITIONS AND CONCEPTS	EXAMPLES
A **mixed number** represents the sum of a whole number and a fraction. In some computations, it is necessary to write mixed numbers as fractions.	Mixed numbers: $6\frac{1}{3} = 6 + \frac{1}{3}$ and $1\frac{3}{4} = \frac{7}{4}$

REVIEW EXERCISES

13. a. Write 24 as the product of two factors.

b. Write 24 as the product of three factors.

c. List the factors of 24.

14. What do we call fractions, such as $\frac{1}{8}$ and $\frac{2}{16}$, that represent the same number?

Give the prime factorization of each number, if possible.

15. 54 **16.** 147

17. 385 **18.** 41

Simplify each fraction.

19. $\frac{20}{35}$ **20.** $\frac{24}{18}$

Build each number to an equivalent fraction with the indicated denominator.

21. $\frac{5}{8}$, denominator 64 **22.** 12, denominator 3

What is the LCD for fractions having the following denominators?

23. 10 and 18 **24.** 21 and 70

Perform each operation and simplify, if possible.

25. $\frac{1}{8} \cdot \frac{7}{8}$ **26.** $\frac{16}{35} \cdot \frac{25}{48}$

27. $\frac{1}{3} \div \frac{15}{16}$ **28.** $16\frac{1}{4} \div 5$

29. $\frac{17}{25} - \frac{7}{25}$ **30.** $\frac{8}{11} - \frac{1}{2}$

31. $\frac{17}{24} + \frac{11}{40}$ **32.** $4\frac{1}{9} - 3\frac{5}{6}$

33. THE INTERNET A popular website averaged $1\frac{3}{4}$ million hits per day during a 30-day period. How many hits did it receive during that time?

34. MACHINE SHOPS How much must be milled off the $\frac{17}{24}$-inch-thick steel rod so that the collar will slip over it?

SECTION 1.3 The Real Numbers

DEFINITIONS AND CONCEPTS	EXAMPLES
To write a set, we list its **elements** within **braces** { }.	In the English alphabet, the set of vowels is {a, e, i, o, u}.
The **natural numbers** are the numbers we count with.	Natural numbers: {1, 2, 3, 4, 5, 6, . . .}
The **whole numbers** are the natural numbers together with 0.	Whole numbers: {0, 1, 2, 3, 4, 5, 6, . . .}
Two numbers are called **opposites** if they are the same distance from 0 on the number line but are on opposite sides of it.	Opposites: 3 and −3
The **integers** include the whole numbers and their opposites.	Integers: {. . . , −3, −2, −1, 0, 1, 2, 3, . . .}
The **rational numbers** are numbers that can be expressed as fractions with an integer numerator and a nonzero integer denominator.	Rational numbers: $-6, -3.1, -\frac{1}{2}, 0, \frac{11}{12}, 9\frac{4}{5},$ and 87
Terminating and **repeating decimals** can be expressed as fractions and are, therefore, rational numbers.	Rational numbers: $-0.25 = -\frac{1}{4}$ and $0.\overline{6} = \frac{2}{3}$

SECTION 1.3 The Real Numbers–*continued*

DEFINITIONS AND CONCEPTS	EXAMPLES
An **irrational number** is a nonterminating, nonrepeating decimal. An irrational number cannot be expressed as a fraction with an integer numerator and a nonzero integer denominator. A **real number** is any number that is either a rational or an irrational number. Every real number corresponds to a point on the **number line,** and every point on the number line corresponds to exactly one real number.	Irrational numbers: $\sqrt{5}$, π, and $-\sqrt{7}$ Graph the set $\left\{-2, -0.75, 1\frac{3}{4}, \pi\right\}$ on a number line. 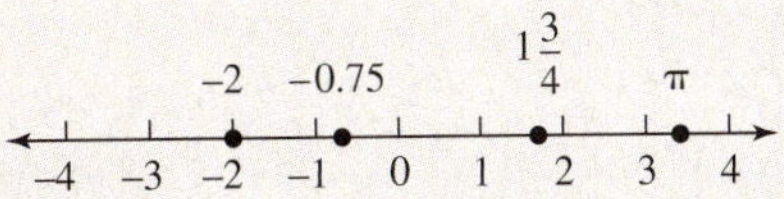
Inequality symbols: $>$ is greater than $<$ is less than	$25 > 15$ and $-2 > -7$ $3.3 < 9.7$ and $-10 < -9$
The **absolute value** of a number is the distance on the number line between the number and 0.	$\|5\| = 5$, $\|-7\| = 7$, and $-\left\|-\frac{5}{9}\right\| = -\frac{5}{9}$

REVIEW EXERCISES

35. **a.** Which number is a whole number but not a natural number?

b. Write the set of integers.

36. Represent 206 feet below sea level with a signed number.

37. Use one of the symbols $>$ or $<$ to make each statement true.

a. $0 \quad 5$ **b.** $-12 \quad -13$

38. Show that each of the following numbers is a rational number by expressing it as a ratio (quotient) of two integers.

a. 0.7 **b.** $4\frac{2}{3}$

Write each fraction as a decimal. Use an overbar if the result is a repeating decimal.

39. $\frac{1}{250}$ **40.** $\frac{17}{22}$

41. Graph each number on a number line:
$\{\pi, 0.33\overline{3}\ldots, 3.75, \sqrt{2}, -\frac{17}{4}, \frac{7}{8}, -2\}$

42. Determine which numbers in the given set are natural numbers, whole numbers, integers, rational numbers, irrational numbers, and real numbers. $\{-\frac{4}{5}, 99.99, 0, \sqrt{2}, -12, 4\frac{1}{2}, 0.66\overline{6}, \ldots, 8\}$

Determine whether each statement is true or false.

43. All integers are whole numbers.

44. π is a rational number.

45. The set of real numbers corresponds to all points on the number line.

46. A real number is either rational or irrational.

Insert one of the symbols >, <, or = in the blank to make each statement true.

47. $|-6| \quad |5|$ **48.** $-9 \quad |-10|$

SECTION 1.4 Adding Real Numbers; Properties of Addition

DEFINITIONS AND CONCEPTS	EXAMPLES
To **add two real numbers with like signs:**	
1. To add two positive numbers, add them as usual. The final answer is positive.	Add: $3 + 5 = 8$
2. To add two negative numbers, add their absolute values and make the final answer negative.	Add: $-5 + (-11) = -16$
To **add two real numbers with unlike signs:**	
1. Subtract their absolute values (the smaller from the larger).	Add: $-8 + 6 = -2$
2. To that result, attach the sign of the number with the larger absolute value.	Add: $12 + (-5) = 7$

SECTION 1.4 Adding Real Numbers; Properties of Addition—*continued*

DEFINITIONS AND CONCEPTS	EXAMPLES
Properties of Addition	
Commutative property: $a + b = b + a$ *Changing the order when adding does not affect the answer.*	$5 + (-9) = -9 + 5$ Reorder.
Associative property: $(a + b) + c = a + (b + c)$ *Changing the grouping when adding does not affect the answer.*	$(3 + 7) + 5 = 3 + (7 + 5)$ Regroup.
Addition property of 0: $a + 0 = a$ and $0 + a = a$	$-6 + 0 = -6$ 0 is the additive identity element.
Addition property of opposites: $a + (-a) = 0$ and $(-a) + a = 0$	$11 + (-11) = 0$ 11 and -11 are additive inverses.

REVIEW EXERCISES

Add.

49. $-45 + (-37)$

50. $25 + (-13)$

51. $0 + (-7)$

52. $-7 + 7$

53. $12 + (-8) + (-15)$

54. $-9.9 + (-2.4)$

55. $\frac{5}{16} + \left(-\frac{1}{2}\right)$

56. $35 + (-13) + (-17) + 6$

57. Determine what property of addition is shown.

a. $-2 + 5 = 5 + (-2)$

b. $(-2 + 5) + 1 = -2 + (5 + 1)$

c. $80 + (-80) = 0$

d. $-5.75 + 0 = -5.75$

58. TEMPERATURES Determine Washington State's record high temperature if it is 166° greater than the state's record low temperature of -48°F.

SECTION 1.5 Subtracting Real Numbers

DEFINITIONS AND CONCEPTS	EXAMPLES
The opposite of the opposite of a number is that number. For any real number a, $-(-a) = a$.	$-(-13) = 13$
To **subtract two real numbers,** add the first to the opposite (additive inverse) of the number to be subtracted. For any real numbers a and b, $a - b = a + (-b)$	Subtract: $4-7 = 4 + (-7) = -3$ $6 - (-8) = 6 + 8 = 14$ $-1 - (-2) = -1 + 2 = 1$
To **check** a subtraction, the difference plus the subtrahend should equal the minuend.	To check $-6 - 2 = -8$, verify that $-8 + 2 = -6$.

REVIEW EXERCISES

Write the expression in simpler form.

59. a. The opposite of 10

b. The additive inverse of -3

60. a. $-\left(-\frac{9}{16}\right)$ **b.** $-|-4|$

Perform the operations.

61. $45-64$

62. Subtract $\frac{1}{3}$ from $-\frac{3}{5}$

63. $-7 - (-12)$

64. $3.6 - (-2.1)$

65. $0-10$

66. $-33 + 7-5 - (-2)$

67. GEOGRAPHY The tallest peak on Earth is Mount Everest, at 29,028 feet, and the greatest ocean depth is the Mariana Trench, at $-36,205$ feet. Find the difference in these elevations. Check thc result.

68. HISTORY Plato, a famous Greek philosopher, died in 347 B.C. (-347) at the age of 81. When was he born? Check the result.

SECTION 1.6 Multiplying and Dividing Real Numbers; Multiplication and Division Properties

DEFINITIONS AND CONCEPTS	EXAMPLES
To **multiply two real numbers,** multiply their absolute values.	
1. If the numbers have **like signs,** the final answer is positive.	Multiply: $-5(-7) = 35$ and $14(3) = 42$
2. If the numbers have **unlike signs,** the final answer is negative.	Multiply: $6(-6) = -36$ and $-11(5) = -55$
Properties of multiplication	
Commutative property: $ab = ba$ *Changing the order when multiplying does not affect the answer.*	$-8(12) = 12(-8)$ Reorder.
Associative property: $(ab)c = a(bc)$ *Changing the grouping when multiplying does not affect the answer.*	$(-4 \cdot 9) \cdot 7 = -4(9 \cdot 7)$ Regroup.
Multiplication property of 0: $0 \cdot a = 0$ and $a \cdot 0 = 0$	$0 \cdot (-7) = 0$
Multiplication property of 1: $1 \cdot a = a$ and $a \cdot 1 = a$	$1 \cdot 32 = 32$ 1 is the multiplicative identity.
Multiplicative inverse property: $a\left(\frac{1}{a}\right) = 1$ and $\frac{1}{a}(a) = 1$	$4\left(\frac{1}{4}\right) = 1$ 4 and $\frac{1}{4}$ are multiplicative inverses.
To **divide two real numbers,** divide their absolute values.	
1. If the numbers have **like signs,** the final answer is positive.	Divide: $\frac{16}{8} = 2$ and $\frac{-25}{-5} = 5$
2. If the numbers have **unlike signs,** the final answer is negative.	Divide: $\frac{-36}{9} = -4$ and $\frac{56}{-7} = -8$
For any real number, $\frac{a}{1} = a$ and $\frac{a}{a} = 1$, where $a \neq 0$.	$\frac{25}{1} = 25$ and $\frac{-32}{-32} = 1$
Division of zero by a nonzero number is 0. **Division by zero** is undefined.	$\frac{0}{17} = 0$ but $\frac{17}{0}$ is undefined.
To **check** the division $\frac{a}{b} = c$, verify that $c \cdot b = a$.	To check $\frac{6}{-2} = -3$, verify that $-3(-2) = 6$.

REVIEW EXERCISES

Multiply.

69. $-8 \cdot 7$

70. $-9\left(-\frac{1}{9}\right)$

71. $2(-3)(-2)$

72. $(-4)(-1)(-3)$

73. $-1.2(-5.3)$

74. $0.002(-1{,}000)$

75. $-\frac{2}{3}\left(\frac{1}{5}\right)$

76. $-6(-3)(0)(-1)$

77. ELECTRONICS The picture on the screen can be magnified by switching a setting on the monitor. What would be the new high and low if every value changed by a factor of 1.5?

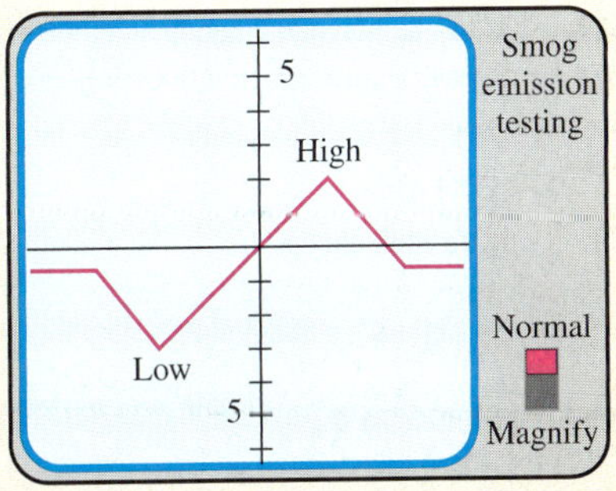

78. Determine what property of multiplication is shown.

a. $(2 \cdot 3)5 = 2(3 \cdot 5)$

b. $(-5)(-6) = (-6)(-5)$

c. $-6 \cdot 1 = -6$

d. $\frac{1}{2}(2) = 1$

Perform each division, if possible.

79. $\frac{44}{-44}$

80. $\frac{-272}{16}$

81. $\frac{-81}{-27}$

82. $-\frac{3}{5} \div \frac{1}{2}$

83. $\frac{-60}{0}$

84. $\frac{-4.5}{1}$

85. Fill in the blanks: $\frac{0}{18} = 0$ because ___ · ___ = ___.

86. GEMSTONES A 3-carat yellow sapphire stone valued at $3,000 five years ago is now worth $1,200. What signed number indicates the average annual depreciation of the sapphire?

SECTION 1.7 Exponents and Order of Operations

DEFINITIONS AND CONCEPTS	EXAMPLES
An **exponent** represents repeated multiplication.	$8^5 = 8 \cdot 8 \cdot 8 \cdot 8 \cdot 8$ *The exponent of 5 indicates that 8 is to be used as a factor 5 times.*
In a^n, a is the **base** and n is the **exponent.**	In 7^4, the base is 7 and 4 is the exponent.
Order of Operations 1. Perform all calculations within grouping symbols, working from the innermost to the outermost in the following order. 2. Evaluate all exponential expressions 3. Perform all multiplications and divisions as they occur from left to right. 4. Perform all additions and subtractions as they occur from left to right. In fractions, evaluate the numerator and denominator separately. Then simplify the fraction.	Evaluate: $\frac{3(6-4^3)-2^4+4}{8 \div 4 \cdot 3} = \frac{3(6-64)-2^4+4}{2 \cdot 3}$ *Evaluate: $4^3 = 64$.* $= \frac{3(-58)-2^4+4}{6}$ *Subtract within the parentheses.* $= \frac{3(-58)-16+4}{6}$ *Evaluate: $2^4 = 16$.* $= \frac{-174-16+4}{6}$ *Multiply.* $= \frac{-190+4}{6}$ *Subtract.* $= \frac{-186}{6}$ *Add.* $= -31$ *Divide.*
$\text{Mean} = \frac{\text{sum of values}}{\text{number of values}}$	Find the mean of the test scores of 74, 83, 79, 91, and 73. $\text{Mean} = \frac{74+83+79+91+73}{5} = 80$

REVIEW EXERCISES

87. Write each expression using exponents.

a. $8 \cdot 8 \cdot 8 \cdot 8 \cdot 8$

b. $9 \cdot \pi \cdot r \cdot r$

88. Evaluate each expression.

a. 9^2 **b.** $\left(-\frac{2}{3}\right)^3$

c. 2^5 **d.** 50^1

Evaluate each expression.

89. $2 + 5 \cdot 3$

90. $-24 \div 2 \cdot 3$

91. $-(16 - 3)^2$

92. $43 + 2(-6 - 2 \cdot 2)$

93. $10 - 5[-3 - 2(5 - 7^2)] - 5$

94. $\frac{-4(4+2)-4}{2|-18-4(5)|}$

95. $(-3)^3\left(\frac{-8}{2}\right) + 5$

96. $\frac{2^4 - (4-6)(3-6)}{12 + 4[(-1)^8 - 2^2]}$

97. Write each expression in symbols and then evaluate it.

a. Negative nine squared

b. The opposite of the square of nine

98. WALK-A-THONS Use the data in the table to find the average (mean) donation to a charity walk-a-thon.

Donation	\$5	\$10	\$20	\$50	\$100
Number received	20	65	25	5	10

SECTION 1.8 Algebraic Expressions

DEFINITIONS AND CONCEPTS	EXAMPLES
Addition symbols separate algebraic expressions into **terms.** In a term, the numerical factor is called the **coefficient.**	Since $a^2 + 3a - 5$ can be written as $a^2 + 3a + (-5)$, it has three terms. The coefficient of a^2 is 1, the coefficient of $3a$ is 3, and the coefficient of -5 is -5.
Key phrases can be translated to algebraic expressions.	*5 more than x* can be expressed as $x + 5$. *25 less than twice y* can be expressed as $2y - 25$. One-half of c can be expressed as $\frac{1}{2}c$.
Number · value = total value	The total value (in cents) of n nickels is $n \cdot 5 = 5n$ cents.
To **evaluate algebraic expressions,** we substitute the values of its variables and use the rules for the order of operations rule.	Evaluate $\frac{x^2 - y^2}{x + y}$ for $x = 2$ and $y = -3$. $\frac{x^2 - y^2}{x + y} = \frac{2^2 - (-3)^2}{2 + (-3)}$ Substitute 2 for x and −3 for y. $= \frac{4 - 9}{-1}$ $= \frac{-5}{-1}$ $= 5$

REVIEW EXERCISES

99. How many terms does each expression have?

a. $3x^2 + 2x - 5$ **b.** $-12xyz$

100. Identify the coefficient of each term of the given expression.

a. $16x^2 - 5x + 25$ **b.** $\frac{x}{2} + y$

Write each phrase as an algebraic expression.

101. 25 more than the height h

102. 15 less than triple the cutoff score s

103. 6 less than one-half of the time

104. The absolute value of the difference of 2 and the square of a

105. HARDWARE Let n represent the length of the nail. Write an algebraic expression that represents the length of the bolt (in inches).

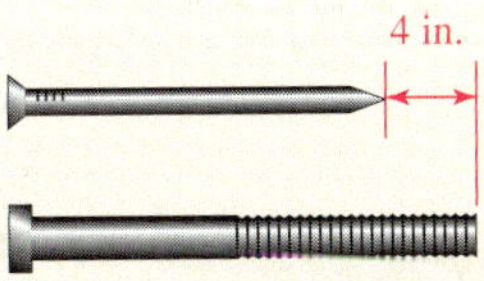

106. HARDWARE Let b represent the length of the bolt. Write an algebraic expression that represents the length of the nail (in inches).

107. How many years are in d decades?

108. Five years after a house was constructed, a patio was added. How old, in years, is the patio if the house is x years old?

109. Complete the table below. The units are cents.

Coin	Number	Value	Total value
Nickel	6	5	
Dime	d	10	

110. Complete the table below.

x	$20x - x^3$
0	
1	
−4	

Evaluate each algebraic expression for the given values of the variables.

111. $b^2 - 4ac$ for $b = -10$, $a = 3$, and $c = 5$

112. $\frac{x + y}{-x - z}$ for $x = 19$, and $y = 17$, and $z = -18$

SECTION 1.9 Simplifying Algebraic Expressions Using Properties of Real Numbers

DEFINITIONS AND CONCEPTS	EXAMPLES
We often use the *commutative property of multiplication* to reorder factors and the *associative property of multiplication* to regroup factors when **simplifying expressions.**	Simplify: $-5(3y) = (-5 \cdot 3)y = -15y$ $-45b\left(\frac{5}{9}\right) = -45\left(\frac{5}{9}b\right) = \left(-45 \cdot \frac{5}{9}\right)b = -25b$
The **distributive property** can be used to remove parentheses: $a(b + c) = ab + ac$ $a(b - c) = ab - ac$ $a(b + c + d) = ab + ac + ad$	Multiply: $7(x + 3) = 7 \cdot x + 7 \cdot 3 = 7x + 21$ $-0.2(4m - 5n - 7) = -0.2(4m) - (-0.2)(5n) - (-0.2)(7)$ $= -0.8m + n + 1.4$
Like terms are terms with exactly the same variables raised to exactly the same powers.	$3x$ and $-5x$ are like terms. $-4t^3$ and $3t^2$ are unlike terms because the variable t has different exponents. $0.5xyz$ and $3.7xy$ are unlike terms because they have different variables.
Simplifying the sum or difference of like terms is called **combining like terms.** Like terms can be combined by adding or subtracting the coefficients of the terms and keeping the same variables with the same exponents.	Simplify: $4a + 2a = 6a$ Think: $(4 + 2)a = 6a$. $5p^2 + p - p^2 - 9p = 4p^2 - 8p$ Think: $(5 - 1)p^2 = 4p^2$ and $(1 - 9)p = -8p$. $2(k - 1) - 3(k + 2) = 2k - 2 - 3k - 6 = -k - 8$

REVIEW EXERCISES

Simplify each expression.

113. $-4(7w)$

114. $3(-2x)(-4)$

115. $0.4(5.2f)$

116. $\frac{7}{2} \cdot \frac{2}{7}r$

Use the distribution property to remove parentheses.

117. $5(x + 3)$

118. $-(2x + 3 - y)$

119. $\frac{3}{4}(4c - 8)$

120. $-2(-3c - 7)(2.1)$

Simplify each expression by combining like terms.

121. $8p + 5p - 4p$

122. $-5m + 2 - 2m - 2$

123. $n + n + n + n$

124. $5(p - 2) - 2(3p + 4)$

125. $55.7k^2 - 55.6k^2$

126. $8a^3 + 4a^3 + 2a - 4a^3 - 2a - 1$

127. $\frac{3}{5}w - \left(-\frac{2}{5}w\right)$

128. $36\left(\frac{1}{9}h - \frac{3}{4}\right) + 36\left(\frac{1}{3}\right)$

129. GEOMETRY Write an algebraic expression in simplified form that represents the perimeter of the triangle.

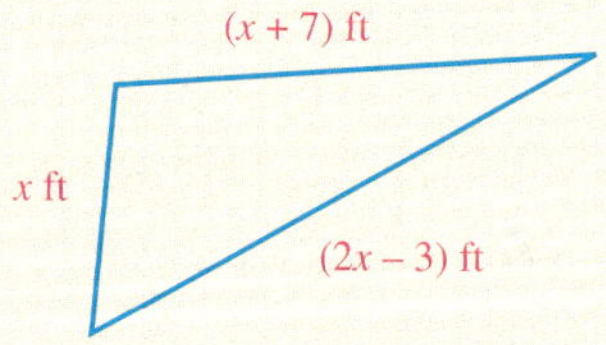

130. Write an equivalent expression for the given expression using fewer symbols.

a. $1x$

b. $-1x$

c. $4x - (-1)$

d. $4x + (-1)$

CHAPTER 1 Test

1. Fill in the blanks.

a. Two fractions, such as $\frac{1}{2}$ and $\frac{5}{10}$, that represent the same number are called ________ fractions.

b. The result of a multiplication is called a ________.

c. $\frac{8}{7}$ is the ________ of $\frac{7}{8}$ because $\frac{8}{7} \cdot \frac{7}{8} = 1$.

d. $9x^2$ and $7x^2$ are ____ ______ because they have the same variable raised to exactly the same power.

e. For any nonzero real number a, $\frac{a}{0}$ is ________.

2. SECURITY GUARDS The graph shows the cost to hire a security guard.

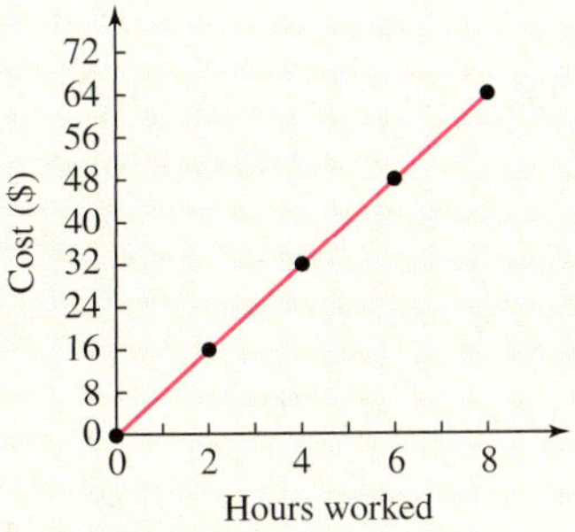

a. What will it cost to hire a security guard for 3 hours?

b. If a school was billed \$40 for hiring a security guard for a dance, for how long did the guard work?

3. Use the formula $f = \frac{a}{5}$ to complete the table.

Square miles (a)	Fire stations (f)
15	
100	
350	

4. Give the prime factorization of 180.

5. Simplify: $\frac{42}{105}$

6. Divide: $\frac{15}{16} \div \frac{5}{8}$

7. Add: $\frac{7}{10} + \frac{1}{14}$

8. Subtract: $8\frac{2}{5} - 1\frac{2}{3}$

9. SHOPPING Find the cost of the fruit on the scale.

10. Write $\frac{5}{6}$ as a decimal.

11. Graph each member of the set on a number line.

$$\left\{-1\frac{1}{4}, \sqrt{2}, -3.75, \frac{7}{2}, 0.5, -3\right\}$$

12. Determine whether each statement is true or false.

a. Every integer is a rational number.

b. Every rational number is an integer.

c. π is an irrational number.

d. 0 is a whole number.

13. Describe the set of real numbers.

14. Insert the proper symbol, $>$ or $<$, in the blank.

a. $-2 \quad \square \quad -3$

b. $-|-9| \quad \square \quad 8$

c. $|-4| \quad \square \quad -(-5)$

d. $\left|-\frac{7}{8}\right| \quad \square \quad 0.5$

15. TELEVISION During "sweeps week," networks try to gain viewers by showing flashy programs. Use the data to determine the average daily gain (or loss) of ratings points by a network for the 7-day "sweeps period."

Day	M	T	W	Thr	F	Sa	Su
Point loss/gain	0.6	−0.3	1.7	1.5	−0.2	1.1	−0.2

Perform the operations.

16. $(-6) + 8 + (-4)$

17. $-\frac{1}{2} + \frac{7}{8}$

18. **a.** $-10 - (-4)$

b. Show a check of the result.

19. **a.** $\frac{-126}{-9}$

b. Show a check of the result.

20. $(-2)(-3)(-5)$

21. $-6.1(0.4)$

22. $\frac{0}{-3}$

23. $\left(-\frac{3}{5}\right)^3$

24. $3 + (-3)$

25. $0 - 3$

26. $-30 + 50 - 10 - (-40)$

27. ASTRONOMY *Magnitude* is a term used in astronomy to describe the brightness of planets and stars. Negative magnitudes are associated with brighter objects. By how many magnitudes do a full moon and the sun differ?

Object	Magnitude
Sun	−26.5
Full moon	−12.5

28. What property of real numbers is illustrated?

a. $(-12 + 97) + 3 = -12 + (97 + 3)$

b. $2(x + 7) = 2x + 14$

c. $-2(m)5 = -2(5)m$

d. $\frac{1}{8}(8) = 1$

e. $0 + 15 = 15$

29. Write each product using exponents:

a. $9(9)(9)(9)(9)$ **b.** $3 \cdot x \cdot x \cdot z \cdot z \cdot z$

Evaluate each expression.

30. $8 + 2 \cdot 3^4$ **31.** $\dfrac{3(40 - 2^3)}{-2(6 - 4)^2}$

32. -10^2 **33.** $9 - 3[45 - 5^2(1^5 - 4)]$

34. Evaluate $3(x - y) - 5(x + y)$ for $x = 2$ and $y = -5$.

35. Complete the table.

x	$2x - \frac{30}{x}$
5	
10	
−30	

36. Translate to an algebraic expression: seven less than twice the width w.

37. a. MUSIC A band recorded x songs for a CD. However, two of the songs were not included in the album because of poor sound quality. Write an expression that represents the number of songs on the CD.

b. MONEY Find the value of q quarters in cents.

38. How many terms are in the expression $4x^2 + 5x - 7$? What is the coefficient of the second term?

Simply each expression.

39. $5(-4x)$ **40.** $-8(-7t)(4)$

41. $\frac{4}{5}(15a + 5) - 16a$ **42.** $-1.1d^3 - 3.8d^3 - d^3$

43. $9x^2 + 2(7x - 3) - 9(x^2 - 1)$

44. Write an expression that represents the perimeter of the rectangle.

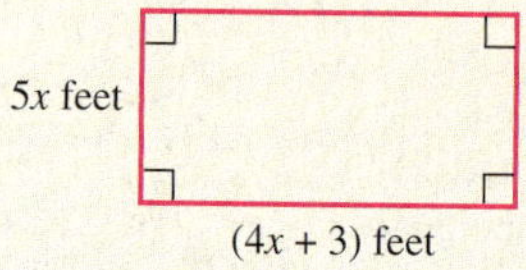

GROUP PROJECT

WRITING FRACTIONS AS DECIMALS

Overview: This is a good activity to try at the beginning of the course. You can become acquainted with other students in your class while you review the process for finding decimal equivalents of fractions.

Instructions: Form groups of 6 students. Select one person from your group to record the group's responses on the questionnaire. Express the results in fraction form and in decimal form.

What fraction (decimal) of the students in your group . . .	Fraction	Decimal
have the letter *a* in their first names?		
have a birthday in January or February?		
work full-time or part-time?		
have ever been on television?		
live more than 10 miles from the campus?		
say that summer is their favorite season of the year?		

CHAPTER 2

Equations, Inequalities, and Problem Solving

from Campus to Careers

Automotive Service Technician

Anyone whose car has ever broken down appreciates the talents of automotive service technicians. To work on today's high-tech cars and trucks, a person needs strong diagnostic and problem-solving skills. Courses in automotive repair, electronics, physics, chemistry, English, computers, and mathematics provide a good educational background for a career as a service technician.

Service technicians must be knowledgeable about the repair and maintenance of automobiles and the fuels that power them. In **Problem 75** of **Study Set 2.4,** you will see how the octane ratings of three familiar grades of gasoline, unleaded, unleaded plus, and premium, are calculated using a formula.

JOB TITLE:
Automotive Service Technician
EDUCATION:
Strongly recommended formal training at a vocational school or community college.
JOB OUTLOOK:
Demand for technicians will grow as the number of vehicles in operation increases.
ANNUAL EARNINGS:
\$37,000 to \$47,000
FOR MORE INFORMATION:
www.bls.gov/oco/home.htm

Study Skills Workshop

Preparing to Learn

Many students feel that there are two types of people—those who are good at math and those who are not—and that this cannot be changed. This isn't true! Here are some suggestions that can increase your chances for success in algebra.

DISCOVER YOUR LEARNING STYLE: Are you a visual, verbal, or audio learner? Knowing this will help you determine how best to study.

GET THE MOST OUT OF THE TEXTBOOK: This book and the software that comes with it contain many student support features. Are you taking advantage of them?

TAKE GOOD NOTES: Are your class notes complete so that they are helpful when doing your homework and studying for tests?

Now Try This

1. To determine what type of learner you are, take the *Learning Style Survey* found online at http://www.metamath.com/multiple/multiple_choice_questions.html. Then, write a one-page paper explaining what you learned from the survey results and how you will use the information to help you succeed in the class.
2. To learn more about the student support features of this book, take the *Textbook Tour* found online at http://www.thomsonedu.com/math/tussy.
3. Rewrite a set of your class notes to make them more readable and to clarify the concepts and examples covered. If they are not already, write them in outline form. Fill in any information you didn't have time to copy down in class and complete any phrases or sentence fragments.

SECTION 2.1
Solving Equations Using Properties of Equality

Objectives

1. Determine whether a number is a solution.
2. Use the addition property of equality.
3. Use the subtraction property of equality.
4. Use the multiplication property of equality.
5. Use the division property of equality.

In this section, we introduce four fundamental properties of equality that are used to solve equations.

1 Determine Whether a Number is a Solution.

An **equation** is a statement indicating that two expressions are equal. An example is $x + 5 = 15$. The equal symbol $=$ separates the equation into two parts: The expression $x + 5$

The Language of Algebra
It is important to know the difference between an equation and an expression. An equation contains an = symbol and an expression does not.

is the **left side** and 15 is the **right side.** The letter x is the **variable** (or the **unknown**). The sides of an equation can be reversed, so we can write $x + 5 = 15$ or $15 = x + 5$

- An equation can be true: $6 + 3 = 9$
- An equation can be false: $2 + 4 = 7$
- An equation can be neither true nor false. For example, $x + 5 = 15$ is neither true nor false because we don't know what number x represents.

An equation that contains a variable is made true or false by substituting a number for the variable. If we substitute 10 for x in $x + 5 = 15$, the resulting equation is true: $10 + 5 = 15$. If we substitute 1 for x, the resulting equation is false: $1 + 5 = 15$. A number that makes an equation true when substituted for the variable is called a **solution** and it is said to **satisfy** the equation. Therefore, 10 is a solution of $x + 5 = 15$, and 1 is not. The **solution set** of an equation is the set of all numbers that make the equation true.

EXAMPLE 1 Is 9 a solution of $3y - 1 = 2y + 7$?

Strategy We will substitute 9 for each y in the equation and evaluate the expression on the left side and the expression on the right side separately.

Why If a true statement results, 9 is a solution of the equation. If we obtain a false statement, 9 is not a solution.

Solution

Evaluate the expression on the left side.

$$3y - 1 = 2y + 7$$
$$3(9) - 1 \stackrel{?}{=} 2(9) + 7$$
$$27 - 1 \stackrel{?}{=} 18 + 7$$
$$26 = 25$$

Evaluate the expression on the right side.

The Language of Algebra
Read $\stackrel{?}{=}$ as "is possibly equal to."

Since $26 = 25$ is false, 9 is not a solution of $3y - 1 = 2y + 7$.

Self Check 1 Is 25 a solution of $10 - x = 35 - 2x$?

Now Try **Problem 19**

2 Use the Addition Property of Equality.

To **solve an equation** means to find all values of the variable that make the equation true. We can develop an understanding of how to solve equations by referring to the scales shown on the right.

The first scale represents the equation $x - 2 = 3$. The scale is in balance because the weights on the left side and right side are equal. To find x, we must add 2 to the left side. To keep the scale in balance, we must also add 2 to the right side. After doing this, we see that x grams is balanced by 5 grams. Therefore, x must be 5. We say that we have solved the equation $x - 2 = 3$ and that the solution is 5.

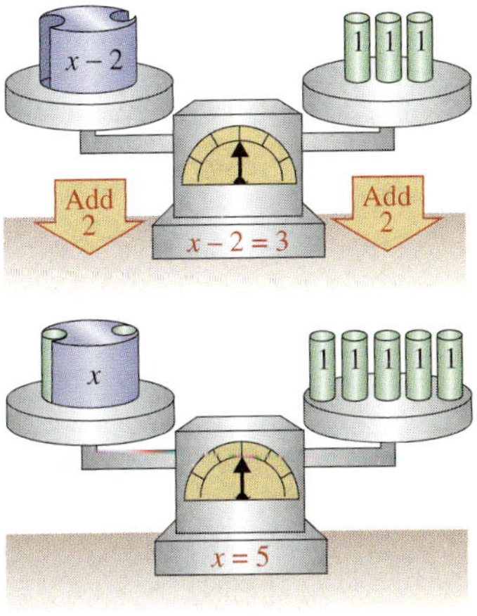

In this example, we solved $x - 2 = 3$ by transforming it to a simpler *equivalent equation*, $x = 5$.

Equivalent Equations

Equations with the same solutions are called **equivalent equations.**

The procedure that we used suggests the following property of equality.

Addition Property of Equality

Adding the same number to both sides of an equation does not change its solution.

For any real numbers a, b, and c,

if $a = b$, then $a + c = b + c$

When we use this property, the resulting equation is *equivalent to the original one.* We will now show how it is used to solve $x - 2 = 3$ algebraically.

The Language of Algebra
We solve equations by writing a series of steps that result in an equivalent equation of the form

$x = $ *a number*

or

a number $= x$

We say the variable is *isolated* on one side of the equation. *Isolated* means alone or by itself.

EXAMPLE 2

Solve: $x - 2 = 3$

Strategy We will use a property of equality to isolate the variable on one side of the equation.

Why To solve the original equation, we want to find a simpler equivalent equation of the form **$x =$ a number**, whose solution is obvious.

Solution We will use the addition property of equality to isolate x on the left side of the equation. We can undo the subtraction of 2 by adding 2 to both sides.

$x - 2 = 3$ This is the equation to solve.

$x - 2 + 2 = 3 + 2$ Add 2 to both sides.

$x + 0 = 5$ The sum of a number and its opposite is zero: $-2 + 2 = 0$.

$x = 5$ When 0 is added to a number, the result is the same number.

Since 5 is obviously the solution of the equivalent equation $x = 5$, the solution of the original equation, $x - 2 = 3$, is also 5. To check this result, we substitute 5 for x in the original equation and simplify.

$x - 2 = 3$

$5 - 2 \stackrel{?}{=} 3$ Substitute 5 for x.

$3 = 3$ True

Since the statement is true, 5 is the solution. A more formal way to present this result is to write the solution within braces as a solution set: $\{5\}$.

Self Check 2 Solve: $n - 16 = 33$

Now Try Problem 37

EXAMPLE 3 Solve: **a.** $-19 = y - 7$ **b.** $-27 + y = -3$

Strategy We will use a property of equality to isolate the variable on one side of the equation.

Why To solve the original equation, we want to find a simpler equivalent equation of the form **y = a number** or **a number = y**, whose solution is obvious.

Solution

a. To isolate y on the right side, we use the addition property of equality. We can undo the subtraction of 7 by adding 7 to both sides.

$$-19 = y - 7 \quad \text{This is the equation to solve.}$$
$$-19 + 7 = y - 7 + 7 \quad \text{Add 7 to both sides.}$$
$$-12 = y \quad \text{The sum of a number and its opposite is zero: } -7 + 7 = 0.$$

Check:
$$-19 = y - 7 \quad \text{This is the original equation.}$$
$$-19 \stackrel{?}{=} -12 - 7 \quad \text{Substitute } -12 \text{ for } y.$$
$$-19 = -19 \quad \text{True}$$

Since the statement is true, the solution is -12. The solution set is $\{-12\}$.

b. To isolate y, we use the addition property of equality. We can eliminate -27 on the left side by adding its opposite (additive inverse) to both sides.

$$-27 + y = -3 \quad \text{The equation to solve.}$$
$$-27 + y + 27 = -3 + 27 \quad \text{Add 27 to both sides.}$$
$$y = 24 \quad \text{The sum of a number and its opposite is zero: } -27 + 27 = 0.$$

Check:
$$-27 + y = -3 \quad \text{This is the original equation.}$$
$$-27 + 24 \stackrel{?}{=} -3 \quad \text{Substitute 24 for } y.$$
$$-3 = -3 \quad \text{True}$$

The solution is 24. The solution set is $\{24\}$.

Notation

We may solve an equation so that the variable is isolated on either side of the equation. Note that $-12 = y$ is equivalent to $y = -12$.

Caution

After checking a result, be careful when stating your conclusion. Here, it would be incorrect to say:

The solution is -3.

The number we were checking was 24, not -3.

Self Check 3 Solve: **a.** $-5 = b - 38$ **b.** $-20 + n = 29$

Now Try **Problems 39 and 43**

3 Use the Subtraction Property of Equality.

Since any subtraction can be written as an addition by adding the opposite of the number to be subtracted, the following property is an extension of the addition property of equality.

Subtraction Property of Equality

Subtracting the same number from both sides of an equation does not change its solution. For any real numbers a, b, and c,

$$\text{if } a = b, \text{ then } \quad a - c = b - c$$

When we use this property, the resulting equation is equivalent to the original one.

EXAMPLE 4 Solve: **a.** $x + \frac{1}{8} = \frac{7}{4}$ **b.** $54.9 + x = 45.2$

Strategy We will use a property of equality to isolate the variable on one side of the equation.

Why To solve the original equation, we want to find a simpler equivalent equation of the form **x = a number**, whose solution is obvious.

Solution

a. To isolate x, we use the subtraction property of equality. We can undo the addition of $\frac{1}{8}$ by subtracting $\frac{1}{8}$ from both sides.

> ***The Language of Algebra***
> We could also isolate x by adding the additive inverse of $\frac{1}{8}$, which is $-\frac{1}{8}$, to both sides:
> $$x + \frac{1}{8} + \left(-\frac{1}{8}\right) = \frac{7}{4} + \left(-\frac{1}{8}\right)$$

$x + \frac{1}{8} = \frac{7}{4}$ This is the equation to solve.

$x + \frac{1}{8} - \frac{1}{8} = \frac{7}{4} - \frac{1}{8}$ Subtract $\frac{1}{8}$ from both sides.

$x = \frac{7}{4} - \frac{1}{8}$ On the left side, $\frac{1}{8} - \frac{1}{8} = 0$.

$x = \frac{7}{4} \cdot \frac{2}{2} - \frac{1}{8}$ Build $\frac{7}{4}$ so that it has a denominator of 8.

$x = \frac{14}{8} - \frac{1}{8}$ Multiply the numerators and multiply the denominators.

$x = \frac{13}{8}$ Subtract the numerators. Write the result over the common denominator 8.

Verify that $\frac{13}{8}$ is the solution by substituting it for x in the original equation and simplifying.

b. To isolate x, we use the subtraction property of equality. We can undo the addition of 54.9 by subtracting 54.9 from both sides.

$54.9 + x = 45.2$ This is the equation to solve.

$54.9 + x - 54.9 = 45.2 - 54.9$ Subtract 54.9 from both sides.

$x = -9.7$ On the left side, $54.9 - 54.9 = 0$.

Check:

$54.9 + x = 45.2$ This is the original equation.

$54.9 + (-9.7) \stackrel{?}{=} 45.2$ Substitute -9.7 for x.

$45.2 = 45.2$ True

The solution is -9.7. The solution set is $\{-9.7\}$.

Self Check 4 Solve: **a.** $x + \frac{4}{15} = \frac{11}{5}$ **b.** $0.7 + a = 0.2$

Now Try **Problems 49 and 51**

4 Use the Multiplication Property of Equality.

The first scale shown below represents the equation $\frac{x}{3} = 25$. The scale is in balance because the weights on the left side and right side are equal. To find x, we must triple (multiply by 3) the weight on the left side. To keep the scale in balance, we must also triple the weight on the right side. After doing this, we see that x is balanced by 75. Therefore, x must be 75.

The procedure that we used suggests the following property of equality.

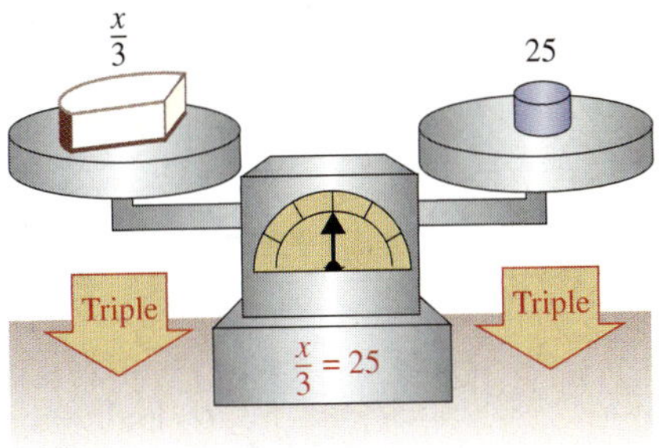

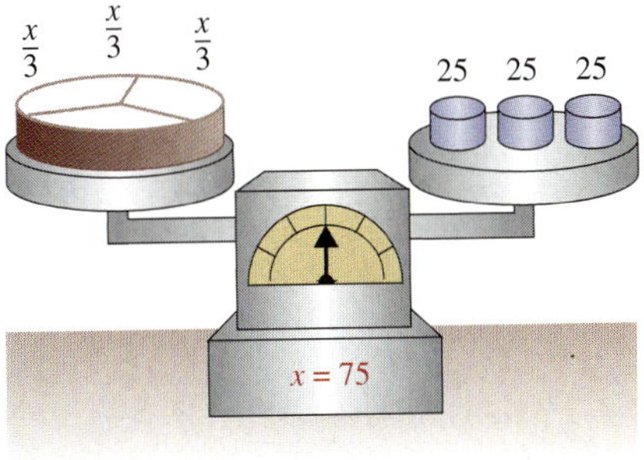

Multiplication Property of Equality

Multiplying both sides of an equation by the same nonzero number does not change its solution.

For any real numbers a, b, and c, where c is not 0,

$$\text{if } a = b, \text{ then } \quad ca = cb$$

When we use this property, the resulting equation is equivalent to the original one. We will now show how it is used to solve $\frac{x}{3} = 25$ algebraically.

EXAMPLE 5 Solve: $\frac{x}{3} = 25$

Strategy We will use a property of equality to isolate the variable on one side of the equation.

Why To solve the original equation, we want to find a simpler equivalent equation of the form **x = a number**, whose solution is obvious.

Solution To isolate x, we use the multiplication property of equality. We can undo the division by 3 by multiplying both sides by 3.

$\frac{x}{3} = 25$	This is the equation to solve.
$3 \cdot \frac{x}{3} = 3 \cdot 25$	Multiply both sides by 3.
$\frac{3x}{3} = 75$	Do the multiplications.
$1x = 75$	Simplify $\frac{3x}{3}$ by removing the common factor of 3 in the numerator and denominator: $\frac{3}{3} = 1$.
$x = 75$	The coefficient 1 need not be written since $1x = x$.

If we substitute 75 for x in $\frac{x}{3} = 25$, we obtain the true statement $25 = 25$. This verifies that 75 is the solution. The solution set is $\{75\}$.

Self Check 5 Solve: $\frac{b}{24} = 3$

Now Try **Problem 53**

Since the product of a number and its reciprocal (or multiplicative inverse) is 1, we can solve equations such as $\frac{2}{3}x = 6$, where the coefficient of the variable term is a fraction, as follows.

EXAMPLE 6

Solve: **a.** $\frac{2}{3}x = 6$ **b.** $-\frac{5}{4}x = 3$

Strategy We will use a property of equality to isolate the variable on one side of the equation.

Why To solve the original equation, we want to find a simpler equivalent equation of the form **x = a number**, whose solution is obvious.

Solution

a. Since the coefficient of x is $\frac{2}{3}$, we can isolate x by multiplying both sides of the equation by the reciprocal of $\frac{2}{3}$, which is $\frac{3}{2}$.

$$\frac{2}{3}x = 6$$ This is the equation to solve.

$$\frac{3}{2} \cdot \frac{2}{3}x = \frac{3}{2} \cdot 6$$ To undo the multiplication by $\frac{2}{3}$, multiply both sides by the reciprocal of $\frac{2}{3}$.

$$\left(\frac{3}{2} \cdot \frac{2}{3}\right)x = \frac{3}{2} \cdot 6$$ Use the associative property of multiplication to group $\frac{3}{2}$ and $\frac{2}{3}$.

$$1x = 9$$ On the left, $\frac{3}{2} \cdot \frac{2}{3} = 1$. On the right, $\frac{3}{2} \cdot 6 = \frac{18}{2} = 9$.

$$x = 9$$ The coefficient 1 need not be written since $1x = x$.

Check:

$$\frac{2}{3}x = 6$$ This is the original equation.

$$\frac{2}{3}(9) \stackrel{?}{=} 6$$ Substitute 9 for x in the original equation.

$$6 = 6$$ On the left side, $\frac{2}{3}(9) = \frac{18}{3} = 6$.

Since the statement is true, 9 is the solution. The solution set is $\{9\}$.

b. To isolate x, we multiply both sides by the reciprocal of $-\frac{5}{4}$, which is $-\frac{4}{5}$.

$$-\frac{5}{4}x = 3$$ This is the equation to solve.

$$-\frac{4}{5}\left(-\frac{5}{4}x\right) = -\frac{4}{5}(3)$$ To undo the multiplication by $-\frac{5}{4}$, multiply both sides by the reciprocal of $-\frac{5}{4}$.

Notation

Variable terms with fractional coefficients can be written in two ways. For example:

$\frac{2}{3}x = \frac{2x}{3}$ and $-\frac{5}{4}a = -\frac{5a}{4}$

$$1x = -\frac{12}{5}$$ On the left side, $-\frac{4}{5}\left(-\frac{5}{4}\right) = 1$.

$$x = -\frac{12}{5}$$ The coefficient 1 need not be written since $1x = x$.

The solution is $-\frac{12}{5}$. Verify that this is correct by checking.

Self Check 6 Solve: **a.** $\frac{7}{2}x = 21$ **b.** $-\frac{3}{8}b = 2$

Now Try **Problems 61 and 67**

5 Use the Division Property of Equality.

Since any division can be rewritten as a multiplication by multiplying by the reciprocal, the following property is a natural extension of the multiplication property.

Division Property of Equality

Dividing both sides of an equation by the same nonzero number does not change its solution. For any real numbers a, b, and c, where c is not 0,

$$\text{if } a = b, \text{ then } \frac{a}{c} = \frac{b}{c}$$

When we use this property, the resulting equation is equivalent to the original one.

EXAMPLE 7 Solve: **a.** $2t = 80$ **b.** $-6.02 = -8.6t$

Strategy We will use a property of equality to isolate the variable on one side of the equation.

Why To solve the original equation, we want to find a simpler equivalent equation of the form **t = a number** or **a number = t**, whose solution is obvious.

Solution

a. To isolate t on the left side, we use the division property of equality. We can undo the multiplication by 2 by dividing both sides of the equation by 2.

$$2t = 80$$ This is the equation to solve.

$$\frac{2t}{\mathbf{2}} = \frac{80}{\mathbf{2}}$$ Use the division property of equality: Divide both sides by 2.

$$1t = 40$$ Simplify $\frac{2t}{2}$ by removing the common factor of 2 in the numerator and denominator: $\frac{2}{2} = 1$.

$$t = 40$$ The product of 1 and any number is that number: $1t = t$.

If we substitute 40 for t in $2t = 80$, we obtain the true statement $80 = 80$. This verifies that 40 is the solution. The solution set is $\{40\}$.

The Language of Algebra
Since division by 2 is the same as multiplication by $\frac{1}{2}$, we can also solve $2t = 80$ using the multiplication property of equality. We could also isolate t by multiplying both sides by the *multiplicative inverse* of 2, which is $\frac{1}{2}$:

$$\frac{1}{2} \cdot 2t = \frac{1}{2} \cdot 80$$

Success Tip

It is usually easier to multiply on each side if the coefficient of the variable term is a *fraction*, and divide on each side if the coefficient is an *integer* or *decimal*.

b. To isolate t on the right side, we use the division property of equality. We can undo the multiplication by -8.6 by dividing both sides by -8.6.

$-6.02 = -8.6t$ This is the equation to solve.

$\frac{-6.02}{-8.6} = \frac{-8.6t}{-8.6}$ Use the division property of equality: Divide both sides by -8.6.

$0.7 = t$ Do the division: $8.6\overline{)6.02}$. The quotient of two negative numbers is positive.

The solution is 0.7. Verify that this is correct by checking.

Self Check 7 Solve: **a.** $16x = 176$ **b.** $10.04 = -0.4r$

Now Try Problems 69 and 79

EXAMPLE 8 Solve: $-x = 3$

Strategy The variable x is not isolated, because there is a $-$ sign in front of it. Since the term $-x$ has an understood coefficient of -1, the equation can be written as $-1x = 3$. We need to select a property of equality and use it to isolate the variable on one side of the equation.

Why To find the solution of the original equation, we want to find a simpler equivalent equation of the form **x = a number**, whose solution is obvious.

Solution To isolate x, we can either multiply or divide both sides by -1.

Multiply both sides by -1:

$-x = 3$ The equation to solve

$-1x = 3$ Write: $-x = -1x$

$(-1)(-1x) = (-1)3$

$1x = -3$

$x = -3$ $1x = x$

Divide both sides by -1:

$-x = 3$ The equation to solve

$-1x = 3$ Write: $-x = -1x$

$\frac{-1x}{-1} = \frac{3}{-1}$

$1x = -3$ On the left side, $\frac{-1}{-1} = 1$.

$x = -3$ $1x = x$

Check:

$-x = 3$ This is the original equation.

$-(-3) \stackrel{?}{=} 3$ Substitute -3 for x.

$3 = 3$ On the left side, the opposite of -3 is 3.

Since the statement is true, -3 is the solution. The solution set is $\{-3\}$.

Self Check 8 Solve: $-h = -12$

Now Try Problem 81

ANSWERS TO SELF CHECKS **1.** Yes **2.** 49 **3. a.** 33 **b.** 49 **4. a.** $\frac{29}{15}$ **b.** -0.5 **5.** 72 **6. a.** 6 **b.** $-\frac{16}{3}$ **7. a.** 11 **b.** -25.1 **8.** 12

STUDY SET 2.1

VOCABULARY

Fill in the blanks.

1. An ________, such as $x + 1 = 7$, is a statement indicating that two expressions are equal.
2. Any number that makes an equation true when substituted for the variable is said to ________ the equation. Such numbers are called ________.
3. To ________ an equation means to find all values of the variable that make the equation true.
4. To solve an equation, we ________ the variable on one side of the equal symbol.
5. Equations with the same solutions are called ________ equations.
6. To ________ the solution of an equation, we substitute the value for the variable in the original equation and determine whether the result is a true statement.

CONCEPTS

7. Given $x + 6 = 12$,
   a. What is the left side of the equation?
   b. Is this equation true or false?
   c. Is 5 the solution?
   d. Does 6 satisfy the equation?
8. For each equation, determine what operation is performed on the variable. Then explain how to undo that operation to isolate the variable.
   a. $x - 8 = 24$
   b. $x + 8 = 24$
   c. $\frac{x}{8} = 24$
   d. $8x = 24$
9. Complete the following properties of equality.
   a. If $a = b$, then

   $$a + c = b + \square \quad \text{and} \quad a - c = b - \square$$

   b. If $a = b$, then $ca = \square\, b$ and $\frac{a}{c} = \frac{b}{\square}$ $(c \neq 0)$
10. a. To solve $\frac{h}{10} = 20$, do we multiply both sides of the equation by 10 or 20?
    b. To solve $4k = 16$, do we subtract 4 from both sides of the equation or divide both sides by 4?
11. Simplify each expression.
    a. $x + 7 - 7$
    b. $y - 2 + 2$
    c. $\frac{5t}{5}$
    d. $6 \cdot \frac{h}{6}$
12. a. To solve $-\frac{4}{5}x = 8$, we can multiply both sides by the reciprocal of $-\frac{4}{5}$. What is the reciprocal of $-\frac{4}{5}$?
    b. What is $-\frac{5}{4}\left(-\frac{4}{5}\right)$?

NOTATION

Complete each solution to solve the equation.

13.
$$x - 5 = 45$$
$$x - 5 + \square = 45 + \square$$
$$x = \square$$

Check:
$$x - 5 = 45$$
$$\square - 5 \stackrel{?}{=} 45$$
$$\square = 45 \quad \text{True}$$

$\square$ is the solution.

14.
$$8x = 40$$
$$\frac{8x}{\square} = \frac{40}{\square}$$
$$x = \square$$

Check:
$$8x = 40$$
$$8(\square) \stackrel{?}{=} 40$$
$$\square = 40 \quad \text{True}$$

$\square$ is the solution.

15. a. What does the symbol $\stackrel{?}{=}$ mean?
    b. If you solve an equation and obtain $50 = x$, can you write $x = 50$?
16. Fill in the blank: $-x = \square\, x$

GUIDED PRACTICE

***Check to determine whether the given number is a solution of the equation.* See Example 1.**

17. $6,\ x + 12 = 28$
18. $110,\ x - 50 = 60$
19. $-8,\ 2b + 3 = -15$
20. $-2,\ 5t - 4 = -16$
21. $5,\ 0.5x = 2.9$
22. $3.5,\ 1.2 + x = 4.7$
23. $6,\ 33 - \frac{x}{2} = 30$
24. $-8,\ \frac{x}{4} + 98 = 100$
25. $-2,\ |c - 8| = 10$
26. $-45,\ |30 - r| = 15$
27. $12,\ 3x - 2 = 4x - 5$
28. $5,\ 5y + 8 = 3y - 2$
29. $-3,\ x^2 - x - 6 = 0$
30. $-2,\ y^2 + 5y - 3 = 0$
31. $1,\ \frac{2}{a + 1} + 5 = \frac{12}{a + 1}$
32. $4,\ \frac{2t}{t - 2} - \frac{4}{t - 2} = 1$
33. $\frac{3}{4},\ x - \frac{1}{8} = \frac{5}{8}$
34. $\frac{7}{3},\ -4 = a + \frac{5}{3}$
35. $-3,\ (x - 4)(x + 3) = 0$
36. $5,\ (2x + 1)(x - 5) = 0$

Use a property of equality to solve each equation. Then check the result. See Examples 2–4.

37. $a - 5 = 66$

38. $x - 34 = 19$

39. $9 = p - 9$

40. $3 = j - 88$

41. $x - 1.6 = -2.5$

42. $y - 1.2 = -1.3$

43. $-3 + a = 0$

44. $-1 + m = 0$

45. $d - \frac{1}{9} = \frac{7}{9}$

46. $\frac{7}{15} = b - \frac{1}{15}$

47. $x + 7 = 10$

48. $y + 15 = 24$

49. $s + \frac{1}{5} = \frac{4}{25}$

50. $\frac{1}{6} = h + \frac{4}{3}$

51. $3.5 + f = 1.2$

52. $9.4 + h = 8.1$

Use a property of equality to solve each equation. Then check the result. See Examples 5–8.

53. $\frac{x}{15} = 3$

54. $\frac{y}{7} = 12$

55. $0 = \frac{v}{11}$

56. $\frac{d}{49} = 0$

57. $\frac{d}{-7} = -3$

58. $\frac{c}{-2} = -11$

59. $\frac{y}{0.6} = -4.4$

60. $\frac{y}{0.8} = -2.9$

61. $\frac{4}{5}t = 16$

62. $\frac{11}{15}y = 22$

63. $\frac{2}{3}c = 10$

64. $\frac{9}{7}d = 81$

65. $-\frac{7}{2}r = 21$

66. $-\frac{4}{5}s = 36$

67. $-\frac{5}{4}h = -5$

68. $-\frac{3}{8}t = -3$

69. $4x = 16$

70. $5y = 45$

71. $63 = 9c$

72. $40 = 5t$

73. $23b = 23$

74. $16 = 16h$

75. $-8h = 48$

76. $-9a = 72$

77. $-100 = -5g$

78. $-80 = -5w$

79. $-3.4y = -1.7$

80. $-2.1x = -1.26$

81. $-x = 18$

82. $-y = 50$

83. $-n = \frac{4}{21}$

84. $-w = \frac{11}{16}$

TRY IT YOURSELF

Solve each equation. Then check the result.

85. $8.9 = -4.1 + t$

86. $7.7 = -3.2 + s$

87. $-2.5 = -m$

88. $1.8 = -b$

89. $-\frac{9}{8}x = 3$

90. $-\frac{14}{3}c = 7$

91. $\frac{3}{4} = d + \frac{1}{10}$

92. $\frac{5}{9} = r + \frac{1}{6}$

93. $-15x = -60$

94. $-14x = -84$

95. $-10 = n - 5$

96. $-8 = t - 2$

97. $\frac{h}{-40} = 5$

98. $\frac{x}{-7} = 12$

99. $a - 93 = 2$

100. $18 = x - 3$

APPLICATIONS

101. SYNTHESIZERS To find the unknown angle measure, which is represented by x, solve the equation $x + 115 = 180$.

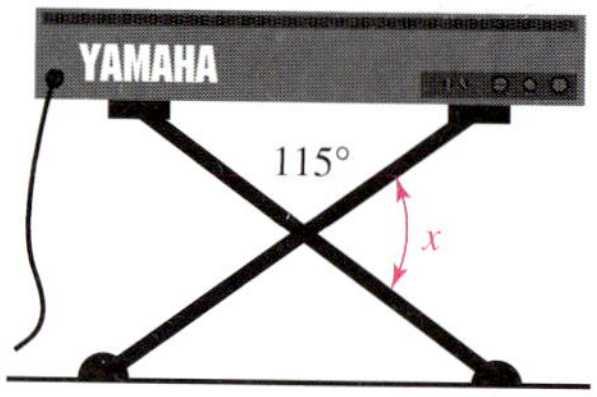

102. STOP SIGNS To find the measure of one angle of the stop sign, which is represented by x, solve the equation $8x = 1{,}080$.

103. SHARING THE WINNING TICKET When a 2006 Florida Lotto Jackpot was won by a group of 16 nurses employed at a Southwest Florida Medical Center, each received \$375,000. To find the amount of the jackpot, which is represented by x, solve the equation $\frac{x}{16} = 375{,}000$.

104. TENNIS Billie Jean King won 40 Grand Slam tennis titles in her career. This is 14 less than the all-time leader, Martina Navratilova. To find the number of titles won by Navratilova, which is represented by x, solve the equation $40 = x - 14$.

WRITING

105. What does it mean to solve an equation?

106. When solving an equation, we *isolate* the variable on one side of the equation. Write a sentence in which the word *isolate* is used in a different context.

107. Explain the error in the following work.

Solve: $x + 2 = 40$

$x + 2 - 2 = 40$

$x = 40$

108. After solving an equation, how do we check the result?

REVIEW

109. Evaluate $-9 - 3x$ for $x = -3$.

110. Evaluate: $-5^2 + (-5)^2$

111. Translate to symbols: Subtract x from 45

112. Evaluate: $\frac{2^3 + 3(5 - 3)}{15 - 4 \cdot 2}$

CHALLENGE PROBLEMS

113. If $a + 80 = 50$, what is $a - 80$?

114. Find two solutions of $|x + 1| = 100$.

SECTION 2.2
More about Solving Equations

Objectives

1. Use more than one property of equality to solve equations.
2. Simplify expressions to solve equations.
3. Clear equations of fractions and decimals.
4. Identify identities and contradictions.

We have solved simple equations by using properties of equality. We will now expand our equation-solving skills by considering more complicated equations. We want to develop a general strategy that can be used to solve any kind of *linear equation in one variable.*

Linear Equation in One Variable

A **linear equation in one variable** can be written in the form

$$ax + b = c$$

where a, b and c are real numbers and $a \neq 0$.

1 Use More Than One Property of Equality to Solve Equations.

Sometimes we must use several properties of equality to solve an equation. For example, on the left side of $2x + 6 = 10$, the variable x is multiplied by 2, and then 6 is added to that product. To isolate x, we use the order of operations rules in reverse. First, we undo the addition of 6, and then we undo the multiplication by 2.

$2x + 6 = 10$	This is the equation to solve.
$2x + 6 - 6 = 10 - 6$	To undo the addition of 6, subtract 6 from both sides.
$2x = 4$	Do the subtractions.
$\frac{2x}{2} = \frac{4}{2}$	To undo the multiplication by 2, divide both sides by 2.
$x = 2$	Do the divisions.

The solution is 2.

EXAMPLE 1 Solve: $-12x + 5 = 17$

Strategy First we will use a property of equality to isolate the *variable term* on one side of the equation. Then we will use a second property of equality to isolate the *variable* itself.

Why To solve the original equation, we want to find a simpler equivalent equation of the form **x = a number**, whose solution is obvious.

Solution On the left side of the equation, x is multiplied by -12, and then 5 is added to that product. To isolate x, we undo the operations in the opposite order.

- To isolate the variable term, $-12x$, we subtract 5 from both sides to undo the addition of 5.
- To isolate the variable, x, we divide both sides by -12 to undo the multiplication by -12.

The Language of Algebra
We subtract 5 from both sides to isolate the *variable term*, $-12x$. Then we divide both sides by -12 to isolate the *variable*, x.

$-12x + 5 = 17$	This is the equation to solve.
$-12x + 5 - 5 = 17 - 5$	Use the subtraction property of equality: Subtract 5 from both sides to isolate the variable term $-12x$.
$-12x = 12$	Do the subtractions: $5 - 5 = 0$ and $17 - 5 = 12$.
$\dfrac{-12x}{-12} = \dfrac{12}{-12}$	Use the division property of equality: Divide both sides by -12 to isolate x.
$x = -1$	Do the divisions.

Caution
When checking solutions, always use the original equation.

Check:

$-12x + 5 = 17$	This is the original equation.
$-12(-1) + 5 \stackrel{?}{=} 17$	Substitute -1 for x.
$12 + 5 \stackrel{?}{=} 17$	Do the multiplication on the left side.
$17 = 17$	True

The solution is -1. The solution set is $\{-1\}$.

Self Check 1 Solve: $8x - 13 = 43$

Now Try **Problem 15**

EXAMPLE 2 Solve: $\frac{5}{8}m - 2 = -12$

Strategy We will use properties of equality to isolate the variable on one side of the equation.

Why To solve the original equation, we want to find a simpler equivalent equation of the form **m = a number**, whose solution is obvious.

Solution We note that the coefficient of m is $\frac{5}{8}$ and proceed as follows.

- To isolate the variable term $\frac{5}{8}m$, we add 2 to both sides to undo the subtraction of 2.
- To isolate the variable, m, we multiply both sides by $\frac{8}{5}$ to undo the multiplication by $\frac{5}{8}$.

$\dfrac{5}{8}m - 2 = -12$	This is the equation to solve.
$\dfrac{5}{8}m - 2 + 2 = -12 + 2$	Use the addition property of equality: Add 2 to both sides to isolate the variable term $\frac{5}{8}m$.
$\dfrac{5}{8}m = -10$	Do the additions: $-2 + 2 = 0$ and $-12 + 2 = -10$.
$\dfrac{8}{5}\left(\dfrac{5}{8}m\right) = \dfrac{8}{5}(-10)$	Use the multiplication property of equality: Multiply both sides by $\frac{8}{5}$ (which is the reciprocal of $\frac{5}{8}$) to isolate m.
$m = -16$	On the left side: $\frac{8}{5}\left(\frac{5}{8}\right) = 1$ and $1m = m$. On the right side: $\frac{8}{5}(-10) = -\frac{8 \cdot 2 \cdot \overset{1}{\cancel{5}}}{\underset{1}{\cancel{5}}} = -16$.

The solution is -16. Verify this by substituting -16 into the original equation. The solution set is $\{-16\}$.

Self Check 2 Solve: $\frac{7}{12}a - 6 = -27$

Now Try Problem 21

EXAMPLE 3 Solve: $-0.2 = -0.8 - y$

Strategy First, we will use a property of equality to isolate the variable term on one side of the equation. Then we will use a second property of equality to isolate the variable itself.

Why To solve the original equation, we want to find a simpler equivalent equation of the form **a number** $= y$, whose solution is obvious.

Solution To isolate the variable term $-y$ on the right side, we eliminate -0.8 by adding 0.8 to both sides.

$$-0.2 = -0.8 - y \quad \text{This is the equation to solve.}$$

$$-0.2 + 0.8 = -0.8 - y + 0.8 \quad \text{Add 0.8 to both sides to isolate } -y.$$

$$0.6 = -y \quad \text{Do the additions.}$$

Since the term $-y$ has an understood coefficient of -1, the equation can be written as $0.6 = -1y$. To isolate y, we can either multiply both sides or divide both sides by -1. If we choose to divide both sides by -1, we proceed as follows.

$$0.6 = -1y$$

$$\frac{0.6}{-1} = \frac{-1y}{-1} \quad \text{To undo the multiplication by } -1\text{, divide both sides by } -1.$$

$$-0.6 = y$$

The solution is -0.6. Verify this by substituting -0.6 into the original equation.

Self Check 3 Solve: $-6.6 - m = -2.7$

Now Try Problem 35

2 Simplify Expressions to Solve Equations.

When solving equations, we should simplify the expressions that make up the left and right sides before applying any properties of equality. Often, that involves removing parentheses and/or combining like terms.

EXAMPLE 4 Solve: **a.** $3(k + 1) - 5k = 0$ **b.** $8a - 2(a - 7) = 68$

Strategy We will use the distributive property along with the process of combining like terms to simplify the left side of each equation.

Why It's best to simplify each side of an equation before using a property of equality.

Solution

a.

$3(k + 1) - 5k = 0$	This is the equation to solve.
$3k + 3 - 5k = 0$	Distribute the multiplication by 3.
$-2k + 3 = 0$	Combine like terms: $3k - 5k = -2k$.
$-2k + 3 - 3 = 0 - 3$	To undo the addition of 3, subtract 3 from both sides. This isolates the variable term $-2k$.
$-2k = -3$	Do the subtractions: $3 - 3 = 0$ and $0 - 3 = -3$.
$\frac{-2k}{-2} = \frac{-3}{-2}$	To undo the multiplication by -2, divide both sides by -2. This isolates the variable k.
$k = \frac{3}{2}$	Simplify: $\frac{-3}{-2} = \frac{3}{2}$.

> **Caution**
>
> To check a result, we evaluate each side of the equation following the order of operations rules. On the left side, perform the addition within parentheses first. *Don't distribute the multiplication by 3.*
>
> $3\left(\frac{3}{2} + 1\right)$
>
> ↑ Add first

Check:

$3(k + 1) - 5k = 0$	This is the original equation.
$3\left(\frac{3}{2} + 1\right) - 5\left(\frac{3}{2}\right) \stackrel{?}{=} 0$	Substitute $\frac{3}{2}$ for k.
$3\left(\frac{5}{2}\right) - 5\left(\frac{3}{2}\right) \stackrel{?}{=} 0$	Do the addition within the parentheses. Think of 1 as $\frac{2}{2}$ and then add: $\frac{3}{2} + \frac{2}{2} = \frac{5}{2}$.
$\frac{15}{2} - \frac{15}{2} \stackrel{?}{=} 0$	Do the multiplications.
$0 = 0$	True

The solution is $\frac{3}{2}$ and the solution set is $\left\{\frac{3}{2}\right\}$.

b.

$8a - 2(a - 7) = 68$	This is the equation to solve.
$8a - 2a + 14 = 68$	Distribute the multiplication by -2.
$6a + 14 = 68$	Combine like terms: $8a - 2a = 6a$.
$6a + 14 - 14 = 68 - 14$	To undo the addition of 14, subtract 14 from both sides. This isolates the variable term $6a$.
$6a = 54$	Do the subtractions.
$\frac{6a}{6} = \frac{54}{6}$	To undo the multiplication by 6, divide both sides by 6. This isolates the variable a.
$a = 9$	Do the divisions.

Self Check 4 Solve: **a.** $4(a + 2) - a = 11$ **b.** $9x - 5(x - 9) = 1$

Now Try Problems 45 and 47

When solving an equation, if variables appear on both sides, we can use the addition (or subtraction) property of equality to get all variable terms on one side and all constant terms on the other.

EXAMPLE 5 Solve: $3x - 15 = 4x + 36$

Strategy There are variable terms ($3x$ and $4x$) on both sides of the equation. We will eliminate $3x$ from the left side of the equation by subtracting $3x$ from both sides.

Why To solve for x, all the terms containing x must be on the same side of the equation.

Solution

$3x - 15 = 4x + 36$	This is the equation to solve.
$3x - 15 - 3x = 4x + 36 - 3x$	Subtract $3x$ from both sides to isolate the variable term on the right side.
$-15 = x + 36$	Combine like terms: $3x - 3x = 0$ and $4x - 3x = x$.
$-15 - 36 = x + 36 - 36$	To undo the addition of 36, subtract 36 from both sides.
$-51 = x$	Do the subtractions.

Check:

$3x - 15 = 4x + 36$	The original equation.
$3(-51) - 15 \stackrel{?}{=} 4(-51) + 36$	Substitute -51 for x.
$-153 - 15 \stackrel{?}{=} -204 + 36$	Do the multiplications.
$-168 = -168$	True

The solution is -51 and the solution set is $\{-51\}$.

> ### *Success Tip*
> We could have eliminated $4x$ from the right side by subtracting $4x$ from both sides:
>
> $$3x - 15 - 4x = 4x + 36 - 4x$$
> $$-x - 15 = 36$$
>
> However, it is usually easier to isolate the variable term on the side that will result in a *positive* coefficient.

Self Check 5 Solve: $30 + 6n = 4n - 2$

Now Try **Problem 57**

3 Clear Equations of Fractions and Decimals.

Equations are usually easier to solve if they don't involve fractions. We can use the multiplication property of equality to clear an equation of fractions by multiplying both sides of the equation by the least common denominator.

EXAMPLE 6 Solve: $\frac{1}{6}x + \frac{5}{2} = \frac{1}{3}$

Strategy To clear the equations of fractions, we will multiply both sides by their LCD.

Why It's easier to solve an equation that involves only integers.

Solution

$\frac{1}{6}x + \frac{5}{2} = \frac{1}{3}$	This is the equation to solve.
$6\left(\frac{1}{6}x + \frac{5}{2}\right) = 6\left(\frac{1}{3}\right)$	Multiply both sides by the LCD of $\frac{1}{6}$, $\frac{5}{2}$, and $\frac{1}{3}$, which is 6. Don't forget the parentheses.
$6\left(\frac{1}{6}x\right) + 6\left(\frac{5}{2}\right) = 6\left(\frac{1}{3}\right)$	On the left side, distribute the multiplication by 6.
$x + 15 = 2$	Do each multiplication: $6\left(\frac{1}{6}\right) = 1$, $6\left(\frac{5}{2}\right) = \frac{30}{2} = 15$, and $6\left(\frac{1}{3}\right) = \frac{6}{3} = 2$.
$x + 15 - 15 = 2 - 15$	To undo the addition of 15, subtract 15 from both sides.
$x = -13$	

Check the solution by substituting -13 for x in $\frac{1}{6}x + \frac{5}{2} = \frac{1}{3}$.

> ### *Caution*
> Before multiplying both sides of an equation by the LCD, enclose the left and right sides with parentheses or brackets.
>
> $$\left(\frac{1}{6}x + \frac{5}{2}\right) = \left(\frac{1}{3}\right)$$

Self Check 6 Solve: $\frac{1}{4}x + \frac{1}{2} = -\frac{1}{8}$

Now Try Problem 63

If an equation contains decimals, it is often convenient to multiply both sides by a power of 10 to change the decimals in the equation to integers.

EXAMPLE 7 Solve: $0.04(12) + 0.01x = 0.02(12 + x)$

Strategy To clear the equations of decimals, we will multiply both sides by a carefully chosen power of 10.

Why It's easier to solve an equation that involves only integers.

Solution The equation contains the decimals 0.04, 0.01, and 0.02. Since the greatest number of decimal places in any one of these numbers is two, we multiply both sides of the equation by 10^2 or 100. This changes 0.04 to 4, and 0.01 to 1, and 0.02 to 2.

$$0.04(12) + 0.01x = 0.02(12 + x)$$

$$100[0.04(12) + 0.01x] = 100[0.02(12 + x)]$$ Multiply both sides by 100. Don't forget the brackets.

$$100 \cdot 0.04(12) + 100 \cdot 0.01x = 100 \cdot 0.02(12 + x)$$ Distribute the multiplication by 100.

$$4(12) + 1x = 2(12 + x)$$ Multiply each decimal by 100 by moving its decimal point 2 places to the right.

$$48 + x = 24 + 2x$$ Distribute the multiplication by 2.

$$48 + x - 24 - x = 24 + 2x - 24 - x$$ Subtract 24 and x from both sides.

$$24 = x$$ Simplify each side.

$$x = 24$$

The solution is 24. Check by substituting 24 for x in the original equation.

Success Tip
Recall that multiplying a decimal by 10 moves the decimal point 1 place to the right, multiplying it by 100 moves it 2 places to the right, and so on.

Success Tip
When we write the decimals in the equation as fractions, it becomes more apparent why it is helpful to multiply both sides by the LCD, 100.

$$\frac{4}{100}(12) + \frac{1}{100}x = \frac{2}{100}(12 + x)$$

Self Check 7 Solve: $0.08x + 0.07(15{,}000 - x) = 1{,}110$

Now Try Problem 71

The previous examples suggest the following strategy for solving equations. It is important to note that not every step is needed to solve every equation.

Strategy for Solving Linear Equations in One Variable

1. **Clear the equation of fractions or decimals:** Multiply both sides by the LCD to clear fractions or multiply both sides by a power of 10 to clear decimals.
2. **Simplify each side of the equation:** Use the distributive property to remove parentheses, and then combine like terms on each side.
3. **Isolate the variable term on one side:** Add (or subtract) to get the variable term on one side of the equation and a number on the other using the addition (or subtraction) property of equality.
4. **Isolate the variable:** Multiply (or divide) to isolate the variable using the multiplication (or division) property of equality.
5. **Check the result:** Substitute the possible solution for the variable in the *original* equation to see if a true statement results.

EXAMPLE 8 Solve: $\frac{7m+5}{5} = -4m + 1$

Strategy We will follow the steps of the equation solving strategy to solve the equation.

Why This is the most efficient way to solve a linear equation in one variable.

Solution

$$\frac{7m+5}{5} = -4m + 1$$ This is the equation to solve.

Step 1 $$5\left(\frac{7m+5}{5}\right) = 5(-4m+1)$$ Clear the equation of the fraction by multiplying both sides by 5.

Step 2 $$7m + 5 = -20m + 5$$ On the left side, remove the common factor 5 in the numerator and denominator. On the right side, distribute the multiplication by 5.

Step 3 $$7m + 5 + 20m = -20m + 5 + 20m$$ To eliminate the term $-20m$ on the right side, add $20m$ to both sides.

$$27m + 5 = 5$$ Combine like terms: $7m + 20m = 27m$ and $-20m + 20m = 0$.

$$27m + 5 - 5 = 5 - 5$$ To isolate the term $27m$, undo the addition of 5 by subtracting 5 from both sides.

$$27m = 0$$ Do the subtractions.

Step 4 $$\frac{27m}{27} = \frac{0}{27}$$ To isolate m, undo the multiplication by 27 by dividing both sides by 27.

$$m = 0$$ 0 divided by any nonzero number is 0.

Step 5 Substitute 0 for m in $\frac{7m+5}{5} = -4m + 1$ to check that the solution is 0.

Success Tip

We remove the common factor 5 in this way:

$$\frac{\overset{1}{\cancel{5}}}{1}\left(\frac{7m+5}{\underset{1}{\cancel{5}}}\right)$$

Caution

Remember that when you multiply one side of an equation by a nonzero number, you must multiply the other side of the equation by the same number.

Self Check 8 Solve: $6c + 2 = \frac{18 - c}{9}$

Now Try Problem 79

4 Identify Identities and Contradictions.

Each of the equations in Examples 1 through 8 had exactly one solution. However, some equations have no solutions while others have infinitely many solutions.

An equation that is true for all values of its variable is called an **identity.** One example is

$x + x = 2x$ If we substitute −10 for x, we get the true statement −20 = −20. If we substitute 7 for x, we get 14 = 14, and so on.

Since we can replace x with any number and the equation will be true, all real numbers are solutions of $x + x = 2x$. This equation has infinitely many solutions. Its solution set is written as {all real numbers}.

An equation that is not true for any values of its variable is called a **contradiction.** One example is

$x = x + 1$ No number is 1 greater than itself.

Since $x = x + 1$ has no solutions, its solution set is the **empty set,** or **null set,** and is written as $\varnothing$.

EXAMPLE 9 Solve: $3(x + 8) + 5x = 2(12 + 4x)$

Strategy We will follow the steps of the equation solving strategy to solve the equation.

Why This is the most efficient way to solve a linear equation in one variable.

Solution

$3(x + 8) + 5x = 2(12 + 4x)$ This is the equation to solve.

$3x + 24 + 5x = 24 + 8x$ Distribute the multiplication by 3 and by 2.

$8x + 24 = 24 + 8x$ Combine like terms: 3x + 5x = 8x. Note that the sides of the equation are identical.

$8x - 8x + 24 = 24 + 8x - 8x$ To eliminate the term 8x on the right side, subtract 8x from both sides.

$24 = 24$ Combine like terms on both sides: 8x − 8x = 0.

In this case, the terms involving x drop out and the result is true. This means that any number substituted for x in the original equation will give a true statement. Therefore, *all real numbers* are solutions and this equation is an identity.

Success Tip
Note that at the step $8x + 24 = 24 + 8x$ we know that the equation is an identity.

Self Check 9 Solve: $3(x + 5) - 4(x + 4) = -x - 1$

Now Try Problem 87

EXAMPLE 10 Solve: $3(d + 7) - d = 2(d + 10)$

Strategy We will follow the steps of the equation solving strategy to solve the equation.

Why This is the most efficient way to solve a linear equation in one variable.

Solution

$3(d + 7) - d = 2(d + 10)$ This is the equation to solve.

$3d + 21 - d = 2d + 20$ Distribute the multiplication by 3 and by 2.

$2d + 21 = 2d + 20$ Combine like terms: 3d − d = 2d.

$2d + 21 - 2d = 2d + 20 - 2d$ To eliminate the term 2d on the right side, subtract 2d from both sides.

$21 = 20$ Combine like terms on both sides: 2d − 2d = 0.

The Language of Algebra
Contradiction is a form of the word *contradict,* meaning conflicting ideas. During a trial, evidence might be introduced that *contradicts* the testimony of a witness.

In this case, the terms involving d drop out and the result is false. This means that any number that is substituted for d in the original equation will give a false statement. Since this equation has *no solution,* it is a contradiction.

Self Check 10 Solve: $-4(c - 3) + 2c = 2(10 - c)$

Now Try **Problem 89**

ANSWERS TO SELF CHECKS **1.** 7 **2.** −36 **3.** −3.9 **4. a.** 1 **b.** −11 **5.** −16 **6.** $-\frac{5}{2}$ **7.** 6,000 **8.** 0 **9.** All real numbers; the equation is an identity **10.** No solution; the equation is a contradiction

STUDY SET 2.2

VOCABULARY

Fill in the blanks.

1. $3x + 8 = 10$ is an example of a linear ________ in one variable.

2. To solve $\frac{s}{3} + \frac{1}{4} = -\frac{1}{2}$, we can ______ the equation of the fractions by multiplying both sides by 12.

3. An equation that is true for all values of its variable is called an ________.

4. An equation that is not true for any values of its variable is called a ____________.

CONCEPTS

Fill in the blanks.

5. To solve $3x - 5 = 1$, we first undo the __________ of 5 by adding 5 to both sides. Then we undo the ______________ by 3 by dividing both sides by 3.

6. To solve $\frac{x}{2} + 3 = 5$, we can undo the ________ of 3 by subtracting 3 from both sides. Then we can undo the ________ by 2 by multiplying both sides by 2.

7. a. Combine like terms on the left side of $6x - 8 - 8x = -24$.

b. Distribute and then combine like terms on the right side of $20 = 4(3x - 4) - 9x$.

8. Is −2 a solution of the equation?

a. $6x + 5 = 7$ **b.** $8(x + 3) = 8$

9. Multiply.

a. $20\left(\frac{3}{5}x\right)$ **b.** $100 \cdot 0.02x$

10. By what must you multiply both sides of $\frac{2}{3} - \frac{1}{2}b = -\frac{4}{3}$ to clear it of fractions?

11. By what must you multiply both sides of $0.7x + 0.3(x - 1) = 0.5x$ to clear it of decimals?

12. a. Simplify: $3x + 5 - x$

b. Solve: $3x + 5 = 9$

c. Evaluate $3x + 5 - x$ for $x = 9$

d. Check: Is −1 a solution of $3x + 5 - x = 9$?

NOTATION

Complete the solution.

13. Solve:

$$2x - 7 = 21$$
$$2x - 7 \qquad = 21$$
$$2x = 28$$
$$\frac{2x}{\quad} = \frac{28}{\quad}$$
$$x = 14$$

Check:

$$2x - 7 = 21$$
$$2(\quad) - 7 \quad 21$$
$$- 7 \stackrel{?}{=} 21$$
$$= 21$$

is the solution.

14. A student multiplied both sides of $\frac{3}{4}t + \frac{5}{8} = \frac{1}{2}t$ by 8 to clear it of fractions, as shown below. Explain his error in showing this step.

$$8 \cdot \frac{3}{4}t + \frac{5}{8} = 8 \cdot \frac{1}{2}t$$

GUIDED PRACTICE

Solve each equation and check the result. **See Examples 1 and 2.**

15. $2x + 5 = 17$
16. $3x - 5 = 13$
17. $5q - 2 = 23$
18. $4p + 3 = 43$
19. $-33 = 5t + 2$
20. $-55 = 3w + 5$
21. $\frac{5}{6}k - 5 = 10$
22. $\frac{2}{5}c - 12 = 2$
23. $-\frac{7}{16}h + 28 = 21$
24. $-\frac{5}{8}h + 25 = 15$
25. $\frac{t}{3} + 2 = 6$
26. $\frac{x}{5} - 5 = -12$
27. $-3p + 7 = -3$
28. $-2r + 8 = -1$
29. $-5 - 2d = 0$
30. $-8 - 3c = 0$
31. $2(-3) + 4y = 14$
32. $4(-1) + 3y = 8$
33. $0.7 - 4y = 1.7$
34. $0.3 - 2x = -0.9$

Solve each equation and check the result. **See Example 3.**

35. $1.2 - x = -1.7$
36. $0.6 = 4.1 - x$
37. $-6 - y = -2$
38. $-1 - h = -9$

Solve each equation and check the result. **See Example 4.**

39. $3(2y - 2) - y = 5$
40. $2(-3a + 2) + a = 2$
41. $4(5b) + 2(6b - 1) = -34$
42. $9(x + 11) + 5(13 - x) = 0$
43. $-(4 - m) = -10$
44. $-(6 - t) = -12$
45. $10.08 = 4(0.5x + 2.5)$
46. $-3.28 = 8(1.5y - 0.5)$
47. $6a - 3(3a - 4) = 30$
48. $16y - 8(3y - 2) = -24$
49. $-(19 - 3s) - (8s + 1) = 35$
50. $2(3x) - 5(3x + 1) = 58$

Solve each equation and check the result. **See Example 5.**

51. $5x = 4x + 7$
52. $3x = 2x + 2$
53. $8y + 44 = 4y$
54. $9y + 36 = 6y$
55. $60r - 50 = 15r - 5$
56. $100f - 75 = 50f + 75$
57. $8y - 2 = 4y + 16$
58. $7 + 3w = 4 + 9w$
59. $2 - 3(x - 5) = 4(x - 1)$
60. $2 - (4x + 7) = 3 + 2(x + 2)$
61. $3(A + 2) = 2(A - 7)$
62. $9(T - 1) = 6(T + 2) - T$

Solve each equation and check the result. **See Example 6.**

63. $\frac{1}{8}y - \frac{1}{2} = \frac{1}{4}$
64. $\frac{1}{15}x - \frac{4}{5} = \frac{2}{3}$
65. $\frac{1}{3} = \frac{5}{6}x + \frac{2}{9}$
66. $\frac{2}{3} = -\frac{2}{3}x + \frac{3}{4}$
67. $\frac{1}{6}y + \frac{1}{4}y = -1$
68. $\frac{1}{3}x + \frac{1}{4}x = -2$
69. $\frac{2}{3}y + 2 = \frac{1}{5} + y$
70. $\frac{2}{5}x + 1 = \frac{1}{3} + x$

Solve each equation and check the result. **See Example 7.**

71. $0.06(s + 9) - 1.24 = -0.08s$
72. $0.08(x + 50) - 0.16x = 0.04(50)$
73. $0.09(t + 50) + 0.15t = 52.5$
74. $0.08(x - 100) = 44.5 - 0.07x$
75. $0.06(a + 200) + 0.1a = 172$
76. $0.03x + 0.05(6{,}000 - x) = 280$
77. $0.4b - 0.1(b - 100) = 70$
78. $0.105x + 0.06(20{,}000 - x) = 1{,}740$

Solve each equation and check the result. **See Example 8.**

79. $\frac{10 - 5s}{3} = s$
80. $\frac{40 - 8s}{5} = -2s$
81. $\frac{7t - 9}{16} = t$
82. $\frac{11r + 68}{3} = -3$
83. $\frac{5(1 - x)}{6} = -x + 1$
84. $\frac{3(14 - u)}{8} = -3u + 6$
85. $\frac{3(d - 8)}{4} = \frac{2(d + 1)}{3}$
86. $\frac{3(c - 2)}{2} = \frac{2(2c + 3)}{5}$

Solve each equation, if possible. **See Examples 9–10.**

87. $8x + 3(2 - x) = 5x + 6$
88. $5(x + 2) = 5x - 2$
89. $-3(s + 2) = -2(s + 4) - s$
90. $21(b - 1) + 3 = 3(7b - 6)$
91. $2(3z + 4) = 2(3z - 2) + 13$
92. $x + 7 = \frac{2x + 6}{2} + 4$
93. $4(y - 3) - y = 3(y - 4)$
94. $5(x + 3) - 3x = 2(x + 8)$

TRY IT YOURSELF

Solve each equation, if possible. Check the result.

95. $3x - 8 - 4x - 7x = -2 - 8$
96. $-6t - 7t - 5t - 1 = 12 - 3$
97. $0.05a + 0.01(90) = 0.02(a + 90)$
98. $0.03x + 0.05(2{,}000 - x) = 99.5$
99. $\frac{3(b + 2)}{2} = \frac{4b - 10}{4}$
100. $\frac{2(5a - 7)}{4} = \frac{9(a - 1)}{3}$
101. $4(a - 3) = -2(a - 6) + 6a$
102. $9(t + 2) = -6(t - 3) + 15t$
103. $10 - 2y = 8$
104. $7 - 7x = -21$
105. $2n - \frac{3}{4}n = \frac{1}{2}n + \frac{13}{3}$
106. $\frac{5}{6}n + 3n = -\frac{1}{3}n - \frac{11}{9}$
107. $-\frac{2}{3}z + 4 = 8$
108. $-\frac{7}{5}x + 9 = -5$
109. $-2(9 - 3s) - (5s + 2) = -25$
110. $4(x - 5) - 3(12 - x) = 7$

WRITING

111. To solve $3x - 4 = 5x + 1$, one student began by subtracting $3x$ from both sides. Another student solved the same equation by first subtracting $5x$ from both sides. Will the students get the same solution? Explain why or why not.

112. What does it mean to clear an equation such as $\frac{1}{4} + \frac{1}{2}x = \frac{3}{8}$ of the fractions?

113. Explain the error in the following solution.

Solve: $2x + 4 = 30$.

$$\frac{2x}{2} + 4 = \frac{30}{2}$$
$$x + 4 = 15$$
$$x + 4 - 4 = 15 - 4$$
$$x = 11$$

114. Write an equation that is an identity. Explain why every real number is a solution.

REVIEW

Name the property that is used.

115. $x \cdot 9 = 9x$

116. $4 \cdot \frac{1}{4} = 1$

117. $(x + 1) + 2 = x + (1 + 2)$

118. $2(30y) = (2 \cdot 30)y$

CHALLENGE PROBLEMS

119. In this section, we discussed equations that have no solution, one solution, and an infinite number of solutions. Do you think an equation could have exactly two solutions? If so, give an example.

120. The equation $4x - 3y = 5$ contains two different variables. Solve the equation by determining a value of x and a value for y that make the equation true.

SECTION 2.3
Applications of Percent

Objectives

1. Change percents to decimals and decimals to percents.
2. Solve percent problems by direct translation.
3. Solve applied percent problems.
4. Find percent of increase and decrease.
5. Solve discount and commission problems.

In this section, we will use translation skills from Chapter 1 and equation-solving skills from Chapter 2 to solve problems involving percents.

1 Change Percents to Decimals and Decimals to Percents.

The word **percent** means parts per one hundred. We can think of the percent symbol % as representing a denominator of 100. Thus, $93\% = \frac{93}{100}$. Since the fraction $\frac{93}{100}$ is equal to the decimal 0.93, it is also true that $93\% = 0.93$.

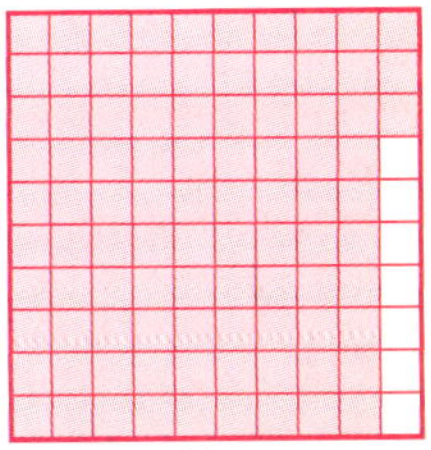

93% or $\frac{93}{100}$ or 0.93 of the figure is shaded.

When solving percent problems, we must often convert percents to decimals and decimals to percents. To change a percent to a decimal, we *divide by 100 by moving the decimal point 2 places to the left and dropping the % symbol.* For example,

$$31\% = 31.0\% = 0.31$$

To change a decimal to a percent, we *multiply the decimal by 100 by moving the decimal point 2 places to the right, and inserting a % symbol.* For example,

$$0.678 = 67.8\%$$

2 Solve Percent Problems by Direct Translation.

There are three basic types of percent problems. Examples of these are:

Type 1 What number is 8% of 215?

Type 2 102 is 21.3% of what number?

Type 3 31 is what percent of 500?

Every percent problem has three parts: the *amount,* the *percent,* and the *base.* For example, in the question *What number is 8% of 215?,* the words "what number" represent the **amount,** 8% represents the **percent,** and 215 represents the **base.** In these problems, the word "is" means "is equal to," and the word "of" means "multiplication."

What number	is	8%	of	215?
↓	↓	↓	↓	↓
Amount	=	**Percent**	·	**base**

EXAMPLE 1

What number is 8% of 215?

Strategy We will translate the words of this problem into an equation and then solve the equation.

Why The variable in the translation equation represents the unknown number that we are asked to find.

> ***The Language of Algebra***
> Translate the word
> - *is* to an equal symbol =
> - *of* to multiplication
> - *what* to a variable

Solution In this problem, the phrase "what number" represents the amount, 8% is the percent, and 215 is the base.

What number	is	8%	of	215?	
↓	↓	↓	↓	↓	
x	=	0.08	·	215	Change the percent to a decimal: 8% = 0.08.
x	=	17.2			Do the multiplication.

Thus, 8% of 215 is 17.2.

To check, we note that 17.2 out of 215 is $\frac{17.2}{215} = 0.08 = 8\%$.

Self Check 1 What number is 5.6% of 40?

Now Try **Problem 13**

We will illustrate the other two types of percent problems with application problems.

3 Solve Applied Percent Problems.

One method for solving applied percent problems is to use the given facts to write a **percent sentence** of the form

______ is ______ % of ______ ?

We enter the appropriate numbers in two of the blanks and the words "what" or "what number" in the remaining blank. As before, we translate the words into an equation and solve it.

EXAMPLE 2 ***Aging Populations.*** By the year 2075, the U.S. Bureau of the Census predicts that about 102 million residents will be age 65 or older. The **circle graph** (or **pie chart**) indicates that age group will make up 21.3% of the population. If this prediction is correct, find the population of the United States in 2075. (Round to the nearest million.)

Projection of the 2075 U.S. Population by Age

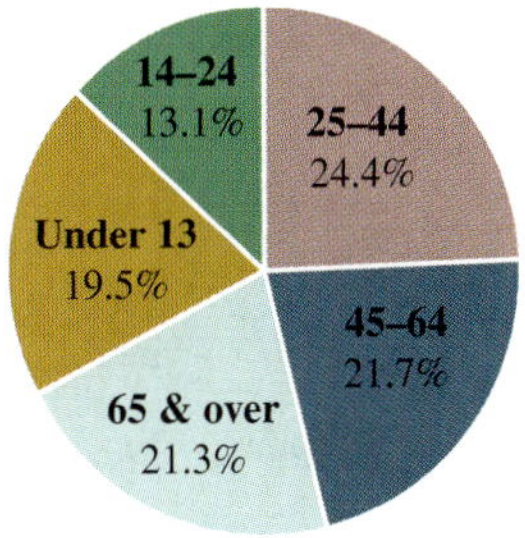

Source: U.S. Bureau of the Census (2000).

Strategy To find the predicted U.S. population in 2075, we will translate the words of the problem into an equation and then solve the equation.

Why The variable in the translation equation represents the unknown population in 2075 that we are asked to find.

Solution In this problem, 102 is the amount, 21.3% is the percent, and the words "what number" represent the base. The units are in millions.

> **Caution**
> Don't forget to change percents to decimal form (or fraction form) before performing any calculations.

102	is	21.3%	of	what number?
↓	↓	↓	↓	↓
102	=	0.213	·	x

$$\frac{102}{\mathbf{0.213}} = \frac{0.213x}{\mathbf{0.213}}$$ To undo the multiplication by 0.213, divide both sides by 0.213.

$$478.9 \approx x$$ Do the divisions.

$$479 \approx x$$ Round 478.9 to the nearest million.

The U.S. population is predicted to be about 479 million in the year 2075. We can check using estimation: 102 million out of a population of 479 million is approximately $\frac{100 \text{ million}}{500 \text{ million}}$, or $\frac{1}{5}$, which is 20%. Since this is close to 21.3%, the answer 479 seems reasonable.

Self Check 2 By the year 2100, it is predicted that 131 million, or 23%, of the U.S. residents will be age 65 or older. If the prediction is correct, find the population in 2100.

Now Try **Problem 17**

We pay many types of taxes in our daily lives, such as sales tax, gasoline tax, income tax, and Social Security tax. **Tax rates** are usually expressed as percents.

EXAMPLE 3 ***Taxes.*** A maid makes $500 a week. One of the deductions from her weekly paycheck is a Social Security tax of $31. Find her Social Security tax rate.

Strategy To find the tax rate, we will translate the words of the problem into an equation and then solve the equation.

Why The variable in the translation equation represents the unknown tax rate that we are asked to find.

Solution

31	is	what percent	of	500?
↓	↓	↓	↓	↓
31	=	x	·	500

31 is the amount, x is the percent, and 500 is the base.

$$\frac{31}{\mathbf{500}} = \frac{500x}{\mathbf{500}}$$ To undo the multiplication by 500, divide both sides by 500.

$$0.062 = x$$ Do the divisions.

$$6.2\% = x$$ Change the decimal 0.062 to a percent.

The Social Security tax rate is 6.2%

We can use estimation to check: \$31 out of \$500 is about $\frac{30}{500}$ or $\frac{6}{100}$, which is 6%. Since this is close to 6.2%, the answer seems reasonable.

Success Tip

One way to check an answer is using *estimation.* If your estimate and your answer are close, you can be reasonably sure that your answer is correct.

Self Check 3 The maid mentioned in Example 3 also has \$7.25 of Medicare tax deducted from her weekly paycheck. Find her Medicare tax rate.

Now Try **Problem 21**

4 Find Percent of Increase and Decrease.

Percents are often used to describe how a quantity has changed. For example, a health care provider might increase the cost of medical insurance by 3%, or a police department might decrease the number of officers assigned to street patrols by 10%. To describe such changes, we use **percent of increase** or **percent of decrease.**

EXAMPLE 4 ***Identity Theft.*** The Federal Trade Commission receives complaints involving the theft of someone's identity information, such as a credit card, Social Security number, or cell phone account. Refer to the data in the table. What was the percent of increase in the number of complaints from 2001 to 2005? (Round to the nearest percent.)

IDENTITY THEFT
Data Clearinghouse

Year	2001	2005
Number of Complaints	**86,000**	**256,000**

Strategy First, we will subtract to find the *amount of increase* in the number of complaints. Then we will translate the words of the problem into an equation and solve it.

Why A percent of increase problem involves finding the *percent of change,* and the change in a quantity is found using subtraction.

Solution To find the *amount of increase,* we subtract the earlier value, 86,000, from the later value, 256,000.

$$256{,}000 - 86{,}000 = 170{,}000$$

We know that an increase of 170,000 is some unknown percent of the number of complaints in 2001, which was 86,000.

Caution
The percent of increase (or decrease) is a percent of the *original* number, that is, the number before the change occurred.

170,000	is	what percent	of	86,000?	
↓	↓	↓	↓	↓	
170,000	=	x	·	86,000	170,000 is the amount, x is the percent, and 86,000 is the base.

$$\frac{170{,}000}{86{,}000} = \frac{86{,}000x}{86{,}000}$$ To undo the multiplication by 86,000, divide both sides by 86,000.

$$1.977 \approx x$$ Do the divisions.

$$197.7\% \approx x$$ Change 1.977 to a percent.

Rounding 197.7% to the nearest percent, we find that the number of identity theft complaints increased by about 198% from 2001 to 2005.

A 200% increase would be double the number of complaints: 2(86,000) = 172,000. It seems reasonable that 170,000 complaints is a 198% increase.

Self Check 4 In 2004, there were 247,000 complaints of identity theft. Find the percent increase from 2004 to 2005. (Round to the nearest percent.)

Now Try **Problem 43**

5 Solve Discount and Commission Problems.

When the price of an item is reduced, we call the amount of the reduction a **discount.** If a discount is expressed as a percent, it is called the **rate of discount.**

EXAMPLE 5 ***Health Club Discounts.*** A 30% discount on a 1-year membership for a fitness center amounted to a $90 savings. Find the cost of a 1-year membership before the discount.

Strategy We will translate the words of the problem into an equation and then solve the equation.

Why The variable in the translation equation represents the unknown cost of a 1-year membership before the discount that we are asked to find.

Solution We are told that $90 is 30% of some unknown membership cost.

90	is	30%	of	what number?	
↓	↓	↓	↓	↓	
90	=	0.30	·	x	90 is the amount, 30% is the percent, and x is the base.

$$\frac{90}{0.30} = \frac{0.30x}{0.30}$$ To undo the multiplication by 0.30, divide both sides by 0.30.

$$300 = x$$ Do the divisions.

A one-year membership cost $300 before the discount.

Self Check 5 A shopper saved \$6 on a pen that was discounted 5%. Find the original cost.

Now Try Problem 51

Instead of working for a salary or at an hourly rate, many salespeople are paid on **commission.** An employee who is paid a commission is paid a percent of the goods or services that he or she sells. We call that percent the **rate of commission.**

EXAMPLE 6 ***Commissions.*** A real estate agent earned \$14,025 for selling a house. If she received a $5\frac{1}{2}\%$ commission, what was the selling price?

Strategy We will translate the words of the problem into an equation and then solve the equation.

Why The variable in the translation equation represents the unknown selling price of the house that we are asked to find.

Solution We are told that \$14,025 is $5\frac{1}{2}\%$ of some unknown selling price of a house.

\$14,025	is	5.5%	of	what number?	Write $5\frac{1}{2}\%$ as 5.5%.
↓	↓	↓	↓	↓	
\$14,025	=	0.055	·	x	14,025 is the amount, 5.5% is the percent, and x is the base.

$$\frac{14{,}025}{0.055} = \frac{0.055x}{0.055}$$ To undo the multiplication by 0.055, divide both sides by 0.055.

$$255{,}000 = x$$ Do the divisions.

The selling price of the house was \$255,000.

Self Check 6 A jewelry store clerk receives a 4.5% commission on all sales. What was the price of a gold necklace sold by the clerk if his commission was \$15.75?

Now Try Problem 53

ANSWERS TO SELF CHECKS **1.** 2.24 **2.** 570 million **3.** 1.45% **4.** 4% **5.** \$120 **6.** \$350

STUDY SET 2.3

VOCABULARY

Fill in the blanks.

1. ________ means parts per one hundred.

2. In the statement "10 is 50% of 20," 10 is the ________, 50% is the percent, and 20 is the ______.

3. In percent questions, the word *of* means ____________, and ___ means equals.

4. An employee who is paid a ___________ is paid a percent of the goods or services that he or she sells.

CONCEPTS

5. Represent the amount of the figure that is shaded using a fraction, a decimal, and a percent.

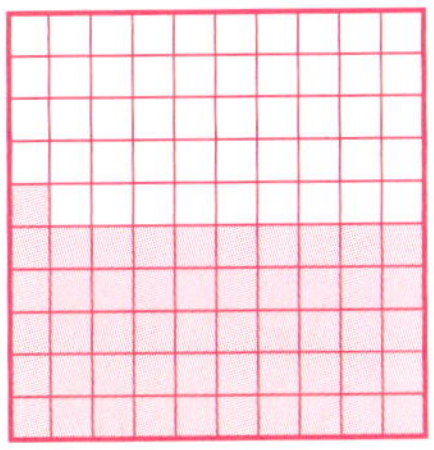

6. Fill in the blank: To solve a percent problem, we translate the words of the problem into an ________ and solve it.

7.

High School Sports Programs Girl's Water Polo—Number of Participants	
2001	**2005**
14,792	17,241

Source: National Federation of State High School Associations

a. Find the *amount* of increase in participation.

b. Fill in blanks to find the percent of increase in participation:
______ is _____% of ______?

8. Fill in the blanks using the words *percent, amount,* and *base.*

= ·

9. Translate each sentence into an equation. **Do not solve.**

a. 12 is 40% of what number?

b. 99 is what percent of 200?

c. What is 66% of 3?

10. Use estimation to determine if each statement is reasonable.

a. 18 is 48% of 93.

b. 47 is 6% of 206.

NOTATION

11. Change each percent to a decimal.

a. 35% **b.** 8.5%

c. 150% **d.** $2\frac{3}{4}\%$

12. Change each decimal to a percent.

a. 0.9 **b.** 9

c. 0.999

GUIDED PRACTICE

See Examples 1–3.

13. What number is 48% of 650?

14. What number is 60% of 200?

15. 78 is what percent of 300?

16. 143 is what percent of 325?

17. 75 is 25% of what number?

18. 78 is 6% of what number?

19. What number is 92.4% of 50?

20. What number is 2.8% of 220?

21. 0.42 is what percent of 16.8?

22. 199.92 is what percent of 2,352?

23. 128.1 is 8.75% of what number?

24. 1.12 is 140% of what number?

APPLICATIONS

25. ANTISEPTICS Use the facts on the label to determine the amount of pure hydrogen peroxide in the bottle.

26. DINING OUT Refer to the sales receipt. Compute the 15% tip (*rounded up* to the nearest dollar). Then find the total cost of the meal.

27. U.S. FEDERAL BUDGET The circle graph shows how the government spent \$2,500 billion in 2005. How much was spent on

a. Social Security/Medicare?

b. Defense/Veterans?

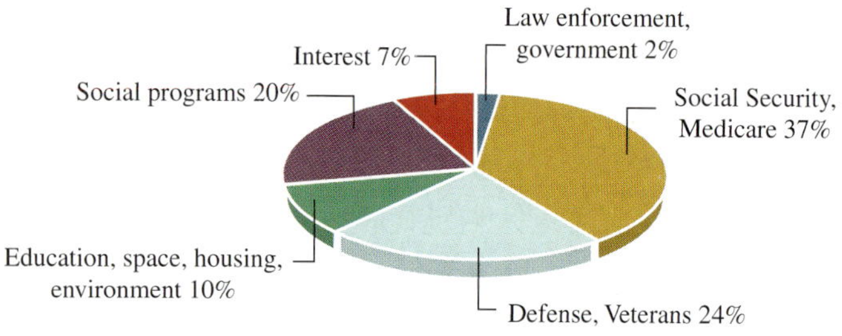

Based on 2006 Federal Income Tax Form 1040

28. TAX TABLES Use the table to compute the amount of federal income tax to be paid on an income of \$39,909.

If your income is over—	*But not over—*	Your income tax is—	*of the amount over—*
\$0	\$7,300	**......... 10%**	**\$0**
7,300	29,700	**\$730.00 + 15%**	**7,300**
29,700	71,950	**4,090.00 + 25%**	**29,700**

29. PAYPAL Many e-commerce businesses use PayPal to perform payment processing for them. For certain transactions, merchants are charged a fee of 2.9% of the selling price of the item plus \$0.30. What would PayPal charge an online art store to collect payment on a painting selling for \$350?

30. eBAY When a student sold an Xbox on eBay for \$153, she was charged a two-part final value fee: 5.25% of the first \$25 of the selling price plus 3.25% of the remainder of the selling price over \$25. Find the fee to sell the Xbox on eBay.

31. PRICE GUARANTEES Home Club offers a "10% Plus" guarantee: If the customer finds the same item selling for less somewhere else, he or she receives the difference in price plus 10% of the difference. A woman bought miniblinds at the Home Club for \$120 but later saw the same blinds on sale for \$98 at another store. How much can she expect to be reimbursed?

32. ROOM TAXES A guest at the San Antonio Hilton Airport Hotel paid \$180 for a room plus a 9% city room tax, a $1\frac{3}{4}\%$ county room tax, and a 6% state room tax. Find the total amount of tax that the guest paid on the room.

33. COMPUTER MEMORY The *My Computer* screen on a student's computer is shown in the next column. What percent of the memory on the hard drive Local Disk (C:) of his computer is used? What percent is free? (GB stands for gigabytes.)

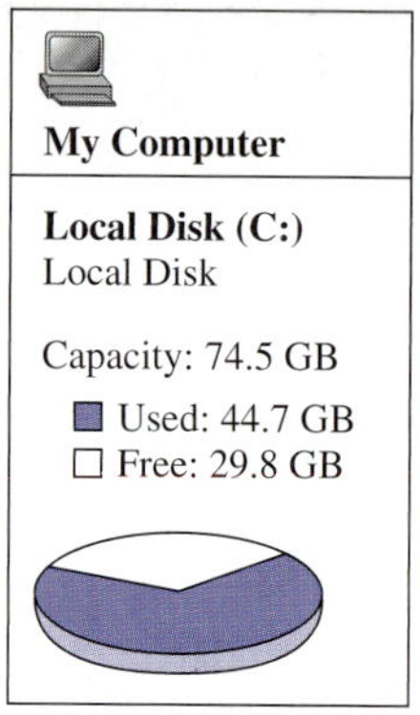

34. GENEALOGY Through an extensive computer search, a genealogist determined that worldwide, 180 out of every 10 million people had his last name. What percent is this?

35. DENTISTRY Refer to the dental record. What percent of the patient's teeth have fillings? Round to the nearest percent.

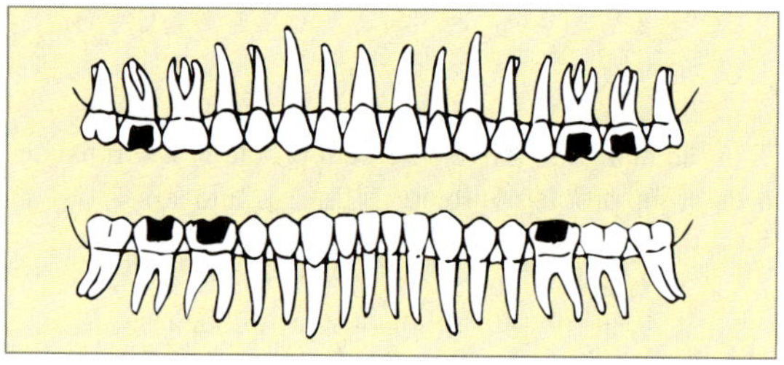

36. TEST SCORES The score 175/200 was written by an algebra instructor at the top of a student's test paper. Write the test score as a percent.

37. DMV WRITTEN TEST To obtain a learner's permit to drive in Nevada, a score of 80% (or better) on a 50-question multiple-choice test is required. If a teenager answered 33 questions correctly, did he pass the test?

38. iPODS The settings menu screen of an Apple iPod is shown. What percent of the memory capacity is still available? Round to the nearest percent. (GB stands for gigabytes.)

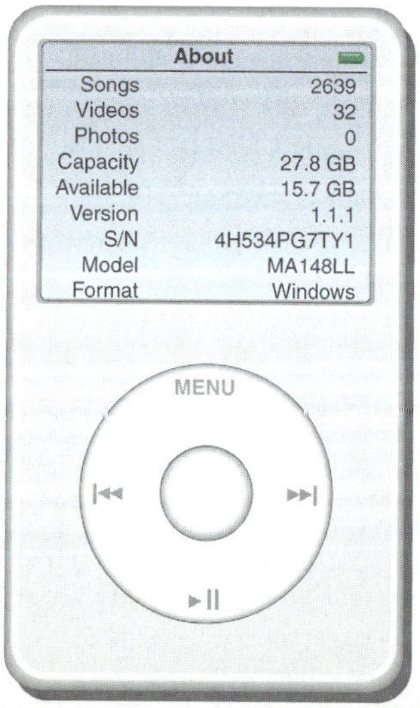

39. CHILD CARE After the first day of registration, 84 children had been enrolled in a day care center. That represented 70% of the available slots. Find the maximum number of children the center could enroll.

40. RACING PROGRAMS One month before a stock car race, the sale of ads for the official race program was slow. Only 12 pages, or just 30% of the available pages, had been sold. Find the total number of pages devoted to advertising in the program.

41. NUTRITION The Nutrition Facts label from a can of clam chowder is shown.

a. Find the number of grams of saturated fat in one serving. What percent of a person's recommended daily intake is this?

b. Determine the recommended number of grams of saturated fat that a person should consume daily.

Nutrition Facts	
Serving Size 1 cup (240mL) Servings Per Container about 2	
Amount per serving	
Calories 240	Calories from Fat 140
	% Daily Value*
Total Fat 15 g	**23%**
Saturated Fat 5 g	**25%**
Cholesterol 10 mg	**3%**
Sodium 980 mg	**41%**
Total Carbohydrate 21 g	**7%**
Dietary Fiber 2 g	**8%**
Sugars 1 g	
Protein 7 g	

42. COMMERCIALS Jared Fogle credits his tremendous weight loss to exercise and a diet of low-fat Subway sandwiches. His current weight (about 187 pounds) is 44% of his maximum weight (reached in March of 1998). What did he weigh then?

43. EXPORTS According to the graph, between what two years was there the greatest percent decrease in U.S. exports to Mexico? Find the percent of decrease.

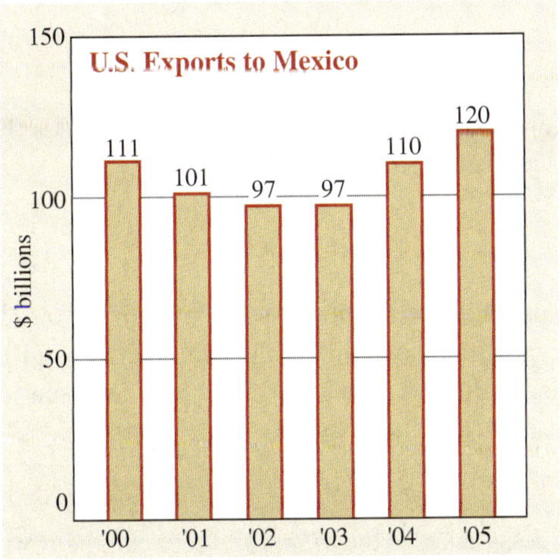

Based on data from www.census.gov/foreign-trade

44. AUCTIONS A pearl necklace of former First Lady Jacqueline Kennedy Onassis, originally valued at $700, was sold at auction in 1996 for $211,500. Find the percent of increase in the value of the necklace. (Round to the nearest percent.)

45. INSURANCE COSTS A college student's good grades earned her a student discount on her car insurance premium. Find the percent of decrease to the nearest percent if her annual premium was lowered from $1,050 to $925.

46. U.S. LIFE EXPECTANCY Use the following life expectancy data for 1900 and 2004 to find the percent of increase for males and for females. Round to the nearest percent.

Years of life expected at birth		
	Male	Female
1900	46.3	48.3
2004	75.2	80.4

Source: National Vital Statistics Reports

47. TALK RADIO Refer to the table and find the percent increase in the number of news/talk radio stations from 2004 to 2005. Round to the nearest percent.

Number of U.S. news/talk radio stations	
2004: 1,282	**2005:** 1,324

Source: *The World Almanac, 2006*

48. FOOD LABELS To be labeled "Reduced Fat," foods must contain at least 25% less fat per serving than the regular product. One serving of the original Jif peanut butter has 16 grams of fat per serving. The new Jif Reduced Fat product contains 12 grams of fat per serving. Does it meet the labeling requirement?

49. TV SHOPPING Jan bought a toy from the QVC home shopping network that was discounted 20%. If she saved $15, what was the original price of the toy?

50. DISCOUNTS A 12% discount on a watch saved a shopper $48. Find the price of the watch before the discount.

51. SALES The price of a certain model patio set was reduced 35% because it was being discontinued. A shopper purchased two of them and saved a total of $210. Find the price of a patio set before the discount.

52. TV SALES The price of a plasma screen television was reduced $800 because it was used as a floor model. If this was a 40% savings, find the original price of the TV.

53. REAL ESTATE The $3\frac{1}{2}$% commission paid to a real estate agent on the sale of a condominium earned her $3,325. Find the selling price of the condo.

54. CONSIGNMENT An art gallery agreed to sell an artist's sculpture for a commission of 45%. What must be the selling price of the sculpture if the gallery would like to make $13,500?

55. STOCKBROKERS A stockbroker charges a 2.5% commission to sell shares of a stock for a client. Find the value of stock sold by a broker if the commission was $640.

56. AGENTS An agent made one million dollars by charging a 12.5% commission to negotiate a long-term contract for a professional athlete. Find the amount of the contract.

WRITING

57. Explain the error:

What is 5% of 8?

$x = 5 \cdot 8$

$x = 40$

40 is 5% of 8.

58. Write a real-life situation that could be described by "9 is what percent of 20?"

59. Explain why 150% of a number is more than the number.

60. Why is the problem "What is 9% of 100?" easy to solve?

REVIEW

61. Divide: $-\frac{16}{25} \div \left(-\frac{4}{15}\right)$

62. What two numbers are a distance of 8 away from 4 on the number line?

63. Is -34 a solution of $x + 15 = -49$?

64. Evaluate: $2 + 3[24 - 2(2 - 5)]$

CHALLENGE PROBLEMS

65. SOAPS A soap advertises itself as $99\frac{44}{100}\%$ pure. First, determine what percent of the soap is impurities. Then express your answer as a decimal.

66. Express $\frac{1}{20}$ of 1% as a percent using decimal notation.

SECTION 2.4 Formulas

Objectives

1. Use formulas from business.
2. Use formulas from science.
3. Use formulas from geometry.
4. Solve for a specified variable.

A **formula** is an equation that states a relationship between two or more variables. Formulas are used in fields such as business, science, and geometry.

1 Use Formulas from Business.

A formula for retail price: To make a profit, a merchant must sell an item for more than he or she paid for it. The price at which the merchant sells the product, called the **retail price,** is the *sum* of what the item cost the merchant plus the **markup.** Using r to represent the retail price, c the cost, and m the markup, we can write this formula as

$$r = c + m \quad \text{Retail price = cost + markup}$$

A formula for profit: The **profit** a business makes is the *difference* between the **revenue** (the money it takes in) and the cost. Using p to represent the profit, r the revenue, and c the cost, we can write this formula as

$$p = r - c \quad \text{Profit = revenue − cost}$$

If we are given the values of all but one of the variables in a formula, we can use our equation solving skills to find the value of the remaining variable.

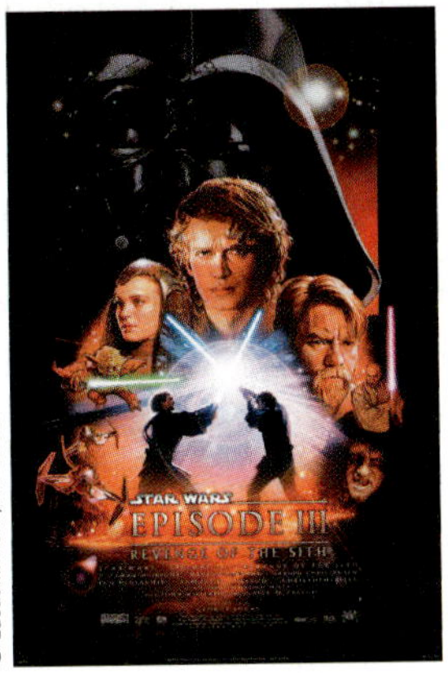

EXAMPLE 1 ***Films.*** Estimates are that 20th Century Fox made a \$309 million profit on the movie *Star Wars: Revenge of the Sith.* If the studio received \$424 million in worldwide box office revenue, find the cost to make and distribute the film. (Source: www.the-numbers.com, August 2006)

Strategy To find the cost to make and distribute the film, we will substitute the given values in the formula $p = r - c$ and solve for c.

Why The variable c represents the unknown cost.

Solution The movie made \$309 million (the profit p) and the studio took in \$424 million (the revenue r). To find the cost c, we proceed as follows.

$$p = r - c \quad \text{This is the formula for profit.}$$

$$309 = 424 - c \quad \text{Substitute 309 for } p \text{ and 424 for } r.$$

$$309 - 424 = 424 - c - 424 \quad \text{To eliminate 424 on the right side, subtract 424 from both sides.}$$

$$-115 = -c \quad \text{Do the subtractions.}$$

$$\frac{-115}{-1} = \frac{-c}{-1} \quad \text{To solve for } c\text{, divide (or multiply) both sides by } -1.$$

$$115 = c \quad \text{The units are millions of dollars.}$$

It cost \$115 million to make and distribute the film.

Self Check 1 A PTA spaghetti dinner made a profit of \$275.50. If the cost to host the dinner was \$1,235, how much revenue did it generate?

Now Try **Problem 11**

The Language of Algebra
The word *annual* means occurring once a year. An *annual* interest rate is the interest rate paid per year.

A formula for simple interest: When money is borrowed, the lender expects to be paid back the amount of the loan plus an additional charge for the use of the money, called **interest.** When money is deposited in a bank, the depositor is paid for the use of the money. The money the deposit earns is also called interest.

Interest is computed in two ways: either as **simple interest** or as **compound interest.** Simple interest is the *product* of the principal (the amount of money that is invested, deposited, or borrowed), the annual interest rate, and the length of time in years. Using I to represent the simple interest, P the principal, r the annual interest rate, and t the time in years, we can write this formula as

$$I = Prt \quad \text{Interest} = \text{principal} \cdot \text{rate} \cdot \text{time}$$

EXAMPLE 2 ***Retirement Income.*** One year after investing \$15,000, a retired couple received a check for \$1,125 in interest. Find the interest rate their money earned that year.

Strategy To find the interest rate, we will substitute the given values in the formula $I = Prt$ and solve for r.

Why The variable r represents the unknown interest rate.

Caution
When using the formula $I = Prt$, always write the interest rate r (which is given as a percent) as a decimal (or fraction) before performing any calculations.

Solution The couple invested \$15,000 (the principal P) for 1 year (the time t) and made \$1,125 (the interest I). To find the annual interest rate r, we proceed as follows.

$I = Prt$ — This is the formula for simple interest.

$1{,}125 = 15{,}000r(1)$ — Substitute 1,125 for I, 15,000 for P, and 1 for t.

$1{,}125 = 15{,}000r$ — Simplify the right side.

$\frac{1{,}125}{15{,}000} = \frac{15{,}000r}{15{,}000}$ — To solve for r, undo the multiplication by 15,000 by dividing both sides by 15,000.

$0.075 = r$ — Do the divisions.

$7.5\% = r$ — To write 0.075 as a percent, multiply 0.075 by 100 by moving the decimal point two places to the right and inserting a % symbol.

The couple received an annual rate of 7.5% that year on their investment. We can display the facts of the problem in a table.

	P	$\cdot\ r$	$\cdot\ t =$	I
Investment	15,000	0.075	1	1,125

Self Check 2 A father loaned his daughter \$12,200 at a 2% annual simple interest rate for a down payment on a house. If the interest on the loan amounted to \$610, for how long was the loan?

Now Try **Problem 15**

2 Use Formulas from Science.

A formula for distance traveled: If we know the average rate (of speed) at which we will be traveling and the time we will be traveling at that rate, we can find the distance traveled. Using d to represent the distance, r the average rate, and t the time, we can write this formula as

$d = rt$ — Distance = rate · time

EXAMPLE 3 ***Whales.*** As they migrate from the Bering Sea to Baja California, gray whales swim for about 20 hours each day, covering a distance of approximately 70 miles. Estimate their average swimming rate in miles per hour (mph).

Strategy To find the swimming rate, we will substitute the given values in the formula $d = rt$ and solve for r.

Why The variable r represents the unknown average swimming rate.

Solution The whales swam 70 miles (the distance d) in 20 hours (the time t). To find their average swimming rate r, we proceed as follows.

Caution
When using the formula $d = rt$, make sure the units are consistent. For example, if the rate is given in miles per hour, the time must be expressed in hours.

$d = rt$ — This is the formula for distance traveled.

$70 = r(20)$ — Substitute 70 for d and 20 for t.

$\frac{70}{20} = \frac{20r}{20}$ — To solve for r, undo the multiplication by 20 by dividing both sides by 20.

$3.5 = r$ — Do the divisions.

The whales' average swimming rate is 3.5 mph. The facts of the problem can be shown in a table.

	r	$\cdot\ t$	$= d$
Gray whale	3.5	20	70

Self Check 3 An elevator travels at an average rate of 288 feet per minute. How long will it take the elevator to climb 30 stories, a distance of 360 feet?

Now Try **Problem 19**

A formula for converting temperatures: In the American system, temperature is measured on the Fahrenheit scale. The Celsius scale is used to measure temperature in the metric system. The formula that relates a Fahrenheit temperature F to a Celsius temperature C is:

$$C = \frac{5}{9}(F - 32)$$

EXAMPLE 4

Convert the temperature shown on the City Savings sign to degrees Fahrenheit.

Strategy To find the temperature in degrees Fahrenheit, we will substitute the given Celsius temperature in the formula $C = \frac{5}{9}(F - 32)$ and solve for F.

Why The variable F represents the unknown temperature in degrees Fahrenheit.

Solution The temperature in degrees Celsius is 30°. To find the temperature in degrees Fahrenheit F, we proceed as follows.

$$C = \frac{5}{9}(F - 32)$$ This is the formula for temperature conversion.

$$30 = \frac{5}{9}(F - 32)$$ Substitute 30 for C, the Celsius temperature.

$$\frac{9}{5} \cdot 30 = \frac{9}{5} \cdot \frac{5}{9}(F - 32)$$ To undo the multiplication by $\frac{5}{9}$, multiply both sides by the reciprocal of $\frac{5}{9}$.

$$54 = F - 32$$ Do the multiplications.

$$54 + 32 = F - 32 + 32$$ To isolate F, undo the subtraction of 32 by adding 32 to both sides.

$$86 = F$$

30°C is equivalent to 86°F.

Self Check 4 Change −175°C, the temperature on Saturn, to degrees Fahrenheit.

Now Try **Problem 25**

The Language of Algebra
In 1724, Daniel Gabriel *Fahrenheit*, a German scientist, introduced the temperature scale that bears his name. The Celsius scale was invented in 1742 by Swedish astronomer Anders *Celsius*.

3 Use Formulas from Geometry.

The Language of Algebra When you hear the word *perimeter*, think of the distance around the "rim" of a flat figure.

To find the **perimeter** of a plane (two-dimensional, flat) geometric figure, such as a rectangle or triangle, we find the distance around the figure by computing the sum of the lengths of its sides. Perimeter is measured in American units, such as inches, feet, yards, and in metric units such as millimeters, meters, and kilometers.

EXAMPLE 5 ***Flags.*** The largest flag ever flown was an American flag that had a perimeter of 1,520 feet and a length of 505 feet. It was hoisted on cables across Hoover Dam to celebrate the 1996 Olympic Torch Relay. Find the width of the flag.

Strategy To find the width of the flag, we will substitute the given values in the formula $P = 2l + 2w$ and solve for w.

Why The variable w represents the unknown width of the flag.

Solution The perimeter P of the rectangular-shaped flag is 1,520 ft and the length l is 505 ft. To find the width w, we proceed as follows.

Perimeter formulas

$P = 2l + 2w$ (rectangle)
$P = 4s$ (square)
$P = a + b + c$ (triangle)

$P = 2l + 2w$	This is the formula for the perimeter of a rectangle.
$1{,}520 = 2(505) + 2w$	Substitute 1,520 for P and 505 for l.
$1{,}520 = 1{,}010 + 2w$	Do the multiplication.
$510 = 2w$	To undo the addition of 1,010, subtract 1,010 from both sides.
$255 = w$	To isolate w, undo the multiplication by 2 by dividing both sides by 2.

The width of the flag is 255 feet. If its length is 505 feet and its width is 255 feet, its perimeter is $2(505) + 2(255) = 1{,}010 + 510 = 1{,}520$ feet, as given.

Self Check 5 The largest flag that consistently flies is the flag of Brazil in Brasilia, the country's capital. It has perimeter 1,116 feet and length 328 feet. Find its width.

Now Try **Problem 27**

Area formulas

$A = lw$ (rectangle)
$A = s^2$ (square)
$A = \frac{1}{2}bh$ (triangle)
$A = \frac{1}{2}h(B + b)$ (trapezoid)

The **area** of a plane (two-dimensional, flat) geometric figure is the amount of surface that it encloses. Area is measured in square units, such as square inches, square feet, square yards, and square meters (written as in.^{2}, ft^2, yd^2, and m^2, respectively).

EXAMPLE 6 **a.** What is the circumference of a circle with diameter 14 feet? Round to the nearest tenth of a foot. **b.** What is the area of the circle? Round to the nearest tenth of a square foot.

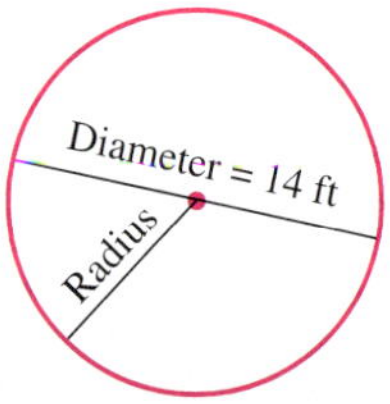

Strategy To find the circumference and area of the circle, we will substitute the proper values into the formulas $C = \pi D$ and $A = \pi r^2$ and find C and A.

Why The variable C represents the unknown circumference of the circle and A represents the unknown area.

Solution

a. Recall that the circumference of a circle is the distance around it. To find the circumference C of a circle with diameter D equal to 14 ft, we proceed as follows.

$C = \pi D$	This is the formula for the circumference of a circle. πD means $\pi \cdot D$.
$C = \pi(14)$	Substitute 14 for D, the diameter of the circle.
$= 14\pi$	The exact circumference of the circle is 14π.
≈ 43.98229715	To use a scientific calculator to approximate the circumference, enter π × 14 =. If you do not have a calculator, use 3.14 as an approximation of π. (Answers may vary slightly depending on which approximation of π is used.)

The circumference is exactly 14π ft. Rounded to the nearest tenth, this is 44.0 ft.

b. The radius r of the circle is one-half the diameter, or 7 feet. To find the area A of the circle, we proceed as follows.

$A = \pi r^2$	This is the formula for the area of a circle. πr^2 means $\pi \cdot r^2$.
$A = \pi(7)^2$	Substitute 7 for r, the radius of the circle.
$= 49\pi$	Evaluate the exponential expression: $7^2 = 49$. The exact area is 49π ft^2.
≈ 153.93804	To use a calculator to approximate the area, enter 49 × π =.

The area is exactly 49π ft^2. To the nearest tenth, the area is 153.9 ft^2.

Circle formulas

$D = 2r$ (diameter)

$r = \frac{1}{2}D$ (radius)

$C = 2\pi r = \pi D$ (circumference)

$A = \pi r^2$ (area)

Notation

When an approximation of π is used in a calculation, it produces an approximate answer. Remember to use an *is approximately equal to* symbol $\approx$ in your solution to show that.

Self Check 6 Find the circumference of a circle with radius 10 inches. Round to the nearest hundredth of an inch.

Now Try **Problem 28**

The **volume** of a three-dimensional geometric solid is the amount of space it encloses. Volume is measured in cubic units, such as cubic inches, cubic feet, and cubic meters (written as in.^{3}, ft^3, and m^3, respectively).

EXAMPLE 7

Find the volume of the cylinder. Round to the nearest tenth of a cubic centimeter.

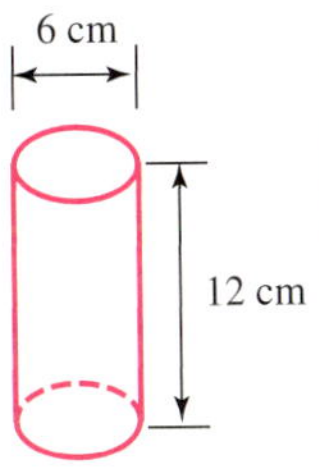

Strategy To find the volume of the cylinder, we will substitute the proper values into the formula $V = \pi r^2 h$ and find V.

Why The variable V represents the unknown volume.

Solution Since the radius of a circle is one-half its diameter, the radius r of the circular base of the cylinder is $\frac{1}{2}(6 \text{ cm}) = 3$ cm. The height h of the cylinder is 12 cm. To find volume V of the cylinder, we proceed as follows.

Volume formulas

$V = lwh$ (rectangular solid)

$V = s^3$ (cube)

$V = \frac{4}{3}\pi r^3$ (sphere)

$V = \pi r^2 h$ (cylinder)

$V = \frac{1}{3}\pi r^2 h$ (cone)

$V = \pi r^2 h$	This is the formula for the volume of a cylinder. $\pi r^2 h$ means $\pi \cdot r^2 \cdot h$.
$V = \pi(3)^2(12)$	Substitute 3 for r and 12 for h.
$= \pi(9)(12)$	Evaluate the exponential expression.
$= 108\pi$	Multiply. The exact volume is 108π cm^3.
≈ 339.2920066	Use a calculator to approximate the volume.

To the nearest tenth, the volume is 339.3 cubic centimeters. This can be written as 339.3 cm^3.

Self Check 7 Find the volume of a cone whose base has radius 12 meters and whose height is 9 meters. Round to the nearest tenth of a cubic meter. Use the formula $V = \frac{1}{3}\pi r^2 h$.

Now Try Problem 29

4 Solve for a Specified Variable.

The Language of Algebra The word *specified* is a form of the word *specify*, which means to select something for a purpose. Here, we select a variable for the purpose of solving for it.

Suppose a shopper wishes to calculate the markup m on several items, knowing their retail price r and their cost c to the merchant. It would take a lot of time to substitute values for r and c into the formula for retail price $r = c + m$ and then repeatedly solve for m. A better way is to solve the formula for m first, substitute values for r and c, and then compute m directly.

To **solve a formula for a specified variable** means to isolate that variable on one side of the equation, with all other variables and constants on the opposite side.

EXAMPLE 8 Solve the formula for retail price, $r = c + m$ for m.

Strategy To solve for m, we will focus on it as if it is the only variable in the equation. We will use a strategy similar to that used to solve linear equations in one variable to isolate m on one side. (See page 125 if you need to review the strategy.)

Why We can solve the formula as if it were an equation in one variable because all the other variables are treated as if they were numbers (constants).

Solution

The Language of Algebra We say that the formula is *solved for m* because m is alone on one side of the equation and the other side does not contain m.

$r = c + m$	To solve for m, we will isolate m on this side of the equation.
$r - c = c + m - c$	To isolate m, undo the addition of c by subtracting c from both sides.
$r - c = m$	Simplify the right side: $c - c = 0$.
$m = r - c$	Reverse the sides of the equation so that m is on the left.

Self Check 8 Solve the formula for profit, $p = r - c$, for r.

Now Try Problem 31

EXAMPLE 9

Solve $A = \frac{1}{2}bh$ for b.

Strategy To solve for b, we will treat b as the only variable in the equation and use properties of equality to isolate it on one side. We will treat the other variables as if they were numbers (constants).

Why To solve for a specified variable means to isolate it on one side of the equation.

Solution We use the same steps to solve an equation for a specified variable that we use to solve equations with only one variable.

Success Tip

To solve for b, think of it as the only variable in the equation. Treat A and h as if they were numbers.

$$A = \frac{1}{2}bh$$ To solve for b, we will isolate b on this side of the equation.

$$2 \cdot A = 2 \cdot \frac{1}{2}bh$$ To clear the equation of the fraction, multiply both sides by 2.

$$2A = bh$$ Simplify.

$$\frac{2A}{h} = \frac{bh}{h}$$ To isolate b, undo the multiplication by h by dividing both sides by h.

$$\frac{2A}{h} = b$$ On the right side, remove the common factor of h: $\frac{b\cancel{h}^{1}}{\cancel{h}_{1}} = b$.

$$b = \frac{2A}{h}$$ Reverse the sides of the equation so that b is on the left.

Self Check 9 Solve $A = \frac{1}{2}r^2a$ for a.

Now Try Problem 37

EXAMPLE 10

Solve $P = 2l + 2w$ for l.

Strategy To solve for l, we will treat l as the only variable in the equation and use properties of equality to isolate it on one side. We will treat the other variables as if they were numbers (constants).

Why To solve for a specified variable means to isolate it on one side of the equation.

Solution

$$P = 2l + 2w$$ To solve for l, we will isolate l on this side of the equation.

$$P - 2w = 2l + 2w - 2w$$ To undo the addition of $2w$, subtract $2w$ from both sides.

$$P - 2w = 2l$$ Combine like terms: $2w - 2w = 0$.

$$\frac{P - 2w}{2} = \frac{2l}{2}$$ To isolate l, undo the multiplication by 2 by dividing both sides by 2.

$$\frac{P - 2w}{2} = l$$ Simplify the right side.

We can write the result as $l = \frac{P - 2w}{2}$.

Caution

Do not try to simplify the result this way:

$$l = \frac{P - \cancel{2}^{1}w}{\cancel{2}_{1}}$$ (crossed out)

This step is incorrect because 2 is not a factor of the entire numerator.

Self Check 10 Solve $P = 2l + 2w$ for w.

Now Try Problem 45

EXAMPLE 11 In Chapter 3, we will work with equations that involve the variables x and y, such as $3x + 2y = 4$. Solve this equation for y.

Strategy To solve for y, we will treat y as the only variable in the equation and use properties of equality to isolate it on one side.

Why To solve for a specified variable means to isolate it on one side of the equation.

Solution

To solve for y, we will isolate y on this side of the equation.

$3x + 2y = 4$

$3x + 2y - 3x = 4 - 3x$ To eliminate 3x on the left side, subtract 3x from both sides.

$2y = 4 - 3x$ Combine like terms: $3x - 3x = 0$.

$\frac{2y}{2} = \frac{4 - 3x}{2}$ To isolate y, undo the multiplication by 2 by dividing both sides by 2.

$y = \frac{4}{2} - \frac{3x}{2}$ Write $\frac{4-3x}{2}$ as the difference of two fractions with like denominators, $\frac{4}{2}$ and $\frac{3x}{2}$.

$y = 2 - \frac{3}{2}x$ Simplify: $\frac{4}{2} = 2$. Write $\frac{3x}{2}$ as $\frac{3}{2}x$.

$y = -\frac{3}{2}x + 2$ On the right side, write the x term first.

Success Tip
When solving for a specified variable, there is often more than one way to express the result. Keep this in mind when you are comparing your answers with those in the back of the text.

Self Check 11 Solve $x + 3y = 12$ for y.

Now Try Problem 47

EXAMPLE 12 Solve $V = \pi r^2 h$ for r^2.

Strategy To solve for r^2, we will treat it as the only variable expression in the equation and isolate it on one side.

Why To solve for a specified variable means to isolate it on one side of the equation.

Solution

To solve for r^2, we will isolate r^2 on this side of the equation.

$V = \pi r^2 h$

$\frac{V}{\pi h} = \frac{\pi r^2 h}{\pi h}$ $\pi r^2 h$ means $\pi \cdot r^2 \cdot h$. To isolate r^2, undo the multiplication by π and h on the right side by dividing both sides by πh.

$\frac{V}{\pi h} = r^2$ On the right side, remove the common factors of π and h: $\frac{\overset{1}{\cancel{\pi}} r^2 \overset{1}{\cancel{h}}}{\underset{1}{\cancel{\pi}}\underset{1}{\cancel{h}}} = r^2$.

$r^2 = \frac{V}{\pi h}$ Reverse the sides of the equation so that r^2 is on the left.

Self Check 12 Solve $V = lwh$ for w.

Now Try Problem 55

ANSWERS TO SELF CHECKS 1. \$1,510.50 2. 2.5 yr 3. 1.25 min 4. −283°F 5. 230 ft 6. 62.83 in. 7. 1,357.2 m^3 8. $r = p + c$ 9. $a = \frac{2A}{r^2}$ 10. $w = \frac{P - 2l}{2}$ 11. $y = 4 - \frac{1}{3}x$ or $y = -\frac{1}{3}x + 4$ 12. $w = \frac{V}{lh}$

STUDY SET 2.4

VOCABULARY

Fill in the blanks.

1. A ________ is an equation that is used to state a known relationship between two or more variables.
2. The distance around a plane geometric figure is called its ________, and the amount of surface that it encloses is called its ______.
3. The ________ of a three-dimensional geometric solid is the amount of space it encloses.
4. The formula $a = P - b - c$ is ______ for a because a is isolated on one side of the equation and the other side does not contain a.

CONCEPTS

5. Use variables to write the formula relating:
   a. Time, distance, rate
   b. Markup, retail price, cost
   c. Costs, revenue, profit
   d. Interest rate, time, interest, principal
6. Complete the table.

	Principal ·	rate ·	time =	interest
Account 1	\$2,500	5%	2 yr	
Account 2	\$15,000	4.8%	1 yr	

7. Complete the table to find how far light and sound travel in 60 seconds. (*Hint:* mi/sec means miles per second.)

	Rate ·	time =	distance
Light	186,282 mi/sec	60 sec	
Sound	1,088 ft/sec	60 sec	

8. Determine which concept (perimeter, area, or volume) should be used to find each of the following. Then determine which unit of measurement, ft, ft^2, or ft^3, would be appropriate.
   a. The amount of storage in a freezer
   b. The amount of ground covered by a sleeping bag lying on the floor
   c. The distance around a dance floor

NOTATION

Complete the solution.

9. Solve $Ax + By = C$ for y.

$$Ax + By = C$$
$$Ax + By - \square = C - \square$$
$$By = C - Ax$$
$$\frac{By}{\square} = \frac{C - Ax}{\square}$$
$$y = \frac{C - Ax}{\square}$$

10. a. Approximate 98π to the nearest hundredth.
    b. In the formula $V = \pi r^2 h$, what does r represent? What does h represent?
    c. What does 45°C mean?
    d. What does 15°F mean?

GUIDED PRACTICE

***Use a formula to solve each problem.* See Example 1.**

11. HOLLYWOOD As of 2006, the movie *Titanic* had brought in \$1,835 million worldwide and made a gross profit of \$1,595 million. What did it cost to make the movie?

12. VALENTINE'S DAY Find the markup on a dozen roses if a florist buys them wholesale for \$12.95 and sells them for \$47.50.

13. SERVICE CLUBS After expenses of \$55.15 were paid, a Rotary Club donated \$875.85 in proceeds from a pancake breakfast to a local health clinic. How much did the pancake breakfast gross?

14. NEW CARS The factory invoice for a minivan shows that the dealer paid \$16,264.55 for the vehicle. If the sticker price of the van is \$18,202, how much over factory invoice is the sticker price?

See Example 2.

15. ENTREPRENEURS To start a mobile dog-grooming service, a woman borrowed \$2,500. If the loan was for 2 years and the amount of interest was \$175, what simple interest rate was she charged?

16. SAVINGS A man deposited \$5,000 in a credit union paying 6% simple interest. How long will the money have to be left on deposit to earn \$6,000 in interest?

17. LOANS A student borrowed some money from his father at 2% simple interest to buy a car. If he paid his father \$360 in interest after 3 years, how much did he borrow?

18. BANKING Three years after opening an account that paid simple interest of 6.45% annually, a depositor withdrew the \$3,483 in interest earned. How much money was left in the account?

See Example 3.

19. SWIMMING In 1930, a man swam down the Mississippi River from Minneapolis to New Orleans, a total of 1,826 miles. He was in the water for 742 hours. To the nearest tenth, what was his average swimming rate?

20. PARADES Rose Parade floats travel down the 5.5-mile-long parade route at a rate of 2.5 mph. How long will it take a float to complete the route if there are no delays?

21. HOT-AIR BALLOONS If a hot-air balloon travels at an average of 37 mph, how long will it take to fly 166.5 miles?

22. AIR TRAVEL An airplane flew from Chicago to San Francisco in 3.75 hours. If the cities are 1,950 miles apart, what was the average speed of the plane?

See Example 4.

23. FRYING FOODS One of the most popular cookbooks in U.S. history, *The Joy of Cooking,* recommends frying foods at 365°F for best results. Convert this to degrees Celsius.

24. FREEZING POINTS Saltwater has a much lower freezing point than freshwater does. For saltwater that is as saturated as much it can possibly get (23.3% salt by weight), the freezing point is −5.8°F. Convert this to degrees Celsius.

25. BIOLOGY Cryobiologists freeze living matter to preserve it for future use. They can work with temperatures as low as −270°C. Change this to degrees Fahrenheit.

26. METALLURGY Change 2,212°C, the temperature at which silver boils, to degrees Fahrenheit. Round to the nearest degree.

See Examples 5–7. If you do not have a calculator, use 3.14 as an approximation of π. Answers may vary slightly depending on which approximation of π is used.

27. ENERGY SAVINGS One hundred inches of foam weather stripping tape was placed around the perimeter of a rectangular-shaped window. If the length of the window is 30 inches, what is its width?

28. RUGS Find the amount of floor area covered by a circular throw rug that has a radius of 15 inches. Round to the nearest square inch.

29. STRAWS Find the volume of a 150 millimeter-long drinking straw that has an inside diameter of 4 millimeters. Round to the nearest cubic millimeter.

30. RUBBER BANDS The world's largest rubber band ball is $5\frac{1}{2}$ ft tall and was made in 2006 by Steve Milton of Eugene, Oregon. Find the volume of the ball. Round to the nearest cubic foot. (*Hint:* The formula for the volume of a sphere is $V = \frac{4}{3}\pi r^3$.)

***Solve each formula for the specified variable (or expression).* See Examples 8–12.**

31. $r = c + m$ for c

32. $p = r - c$ for c

33. $P = a + b + c$ for b

34. $a + b + c = 180$ for a

35. $E = IR$ for R

36. $d = rt$ for t

37. $V = lwh$ for l

38. $I = Prt$ for r

39. $C = 2\pi r$ for r

40. $V = \pi r^2 h$ for h

41. $V = \frac{1}{3}Bh$ for h

42. $C = \frac{1}{7}Rt$ for R

43. $w = \frac{s}{f}$ for f

44. $P = \frac{ab}{c}$ for c

45. $T = 2r + 2t$ for r

46. $y = mx + b$ for x

47. $Ax + By = C$ for x

48. $A = P + Prt$ for t

49. $K = \frac{1}{2}mv^2$ for m

50. $V = \frac{1}{3}\pi r^2 h$ for h

51. $A = \frac{a + b + c}{3}$ for c

52. $x = \frac{a + b}{2}$ for b

53. $2E = \frac{T - t}{9}$ for t

54. $D = \frac{C - s}{n}$ for s

55. $s = 4\pi r^2$ for r^2

56. $E = mc^2$ for c^2

57. $Kg = \dfrac{wv^2}{2}$ for v^2

58. $c^2 = a^2 + b^2$ for a^2

59. $V = \dfrac{4}{3}\pi r^3$ for r^3

60. $A = \dfrac{\pi r^2 S}{360}$ for r^2

61. $\dfrac{M}{2} - 9.9 = 2.1B$ for M

62. $\dfrac{G}{0.5} + 16r = -8t$ for G

63. $S = 2\pi rh + 2\pi r^2$ for h

64. $c = bn + 16t^2$ for t^2

65. $3x + y = 9$ for y

66. $-5x + y = 4$ for y

67. $-x + 3y = 9$ for y

68. $5y - x = 25$ for y

69. $4y + 16 = -3x$ for y

70. $6y + 12 = -5x$ for y

71. $A = \dfrac{1}{2}h(b + d)$ for b

72. $C = \dfrac{1}{4}s(t - d)$ for t

73. $\dfrac{7}{8}c + w = 9$ for c

74. $\dfrac{3}{4}m - t = 5b$ for m

APPLICATIONS

75. ***from Campus to Careers***
Automotive Service Technician

If your automobile engine is making a knocking sound, a service technician will probably tell you that the octane rating of the gasoline that you are using is too low. Octane rating numbers are printed on the yellow decals on gas pumps. The formula used to calculate them is

© Jeremy Hardie/Getty Images

$$\text{Pump octane number} = \frac{(R + M)}{2}$$

where R is the *research octane number*, which is determined with a test engine running at a low speed and M is the *motor octane number*, which is determined with a test engine running at a higher speed. Calculate the octane rating for the following three grades of gasoline.

Gasoline grade	R	M	Octane rating
Unleaded	92	82	
Unleaded plus	95	83	
Premium	97	85	

76. PROPERTIES OF WATER The boiling point and the freezing point of water are to be given in both degrees Celsius and degrees Fahrenheit on the thermometer. Find the missing degree measures.

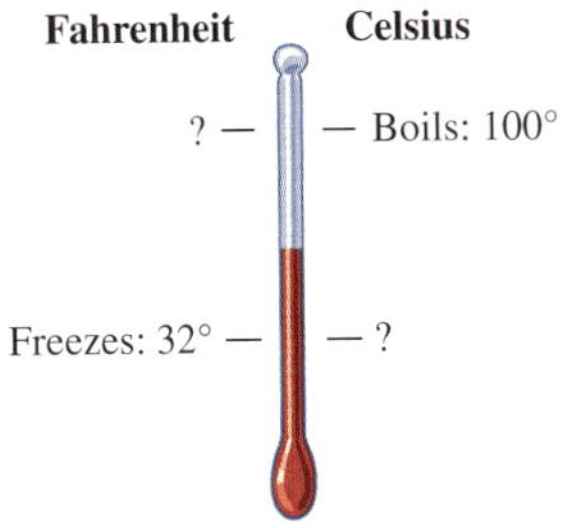

77. AVON PRODUCTS Complete the financial statement.

Income statement (dollar amounts in millions)	Quarter ending Sep 04	Quarter ending Sep 05
Revenue	1,806.2	1,886.0
Cost of goods sold	1,543.4	1,638.9
Operating profit		

Source: Avon Products, Inc.

78. CREDIT CARDS The finance charge that a student pays on his credit card is 19.8% APR (annual percentage rate). Determine the finance charges (interest) the student would have to pay if the account's average balance for the year was $2,500.

79. CAMPERS The perimeter of the window of the camper shell is 140 in. Find the length of one of the shorter sides of the window.

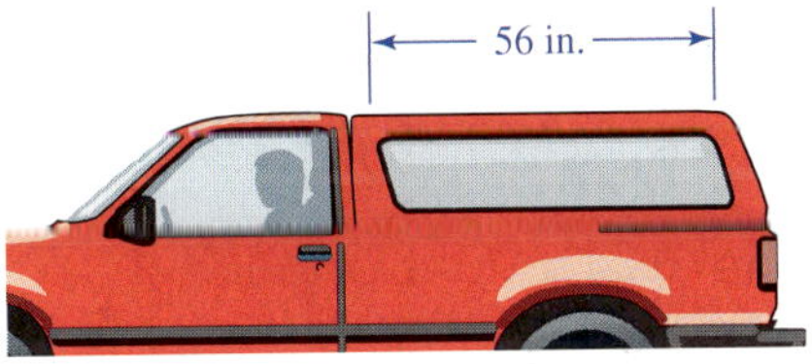

80. FLAGS The flag of Eritrea, a country in east Africa, is shown. The perimeter of the flag is 160 inches.

a. What is the width of the flag?

b. What is the area of the red triangular region of the flag?

81. KITES 650 in.^2 of nylon cloth were used to make the kite shown. If its height is 26 inches, what is the wingspan?

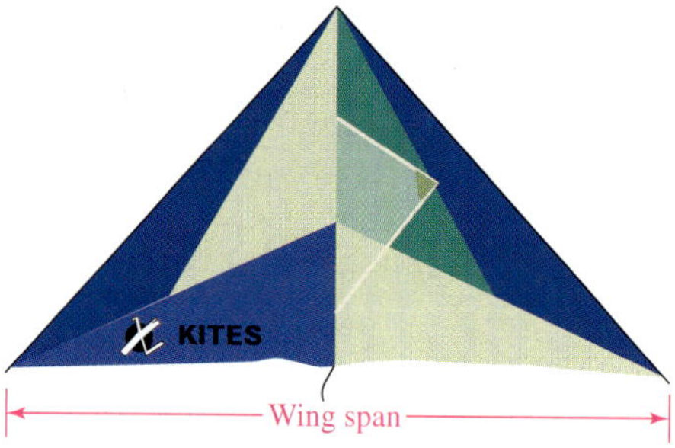

82. MEMORIALS The Vietnam Veterans Memorial is a black granite wall recognizing the more than 58,000 Americans who lost their lives or remain missing. Find the total area of the two triangular-shaped surfaces on which the names are inscribed.

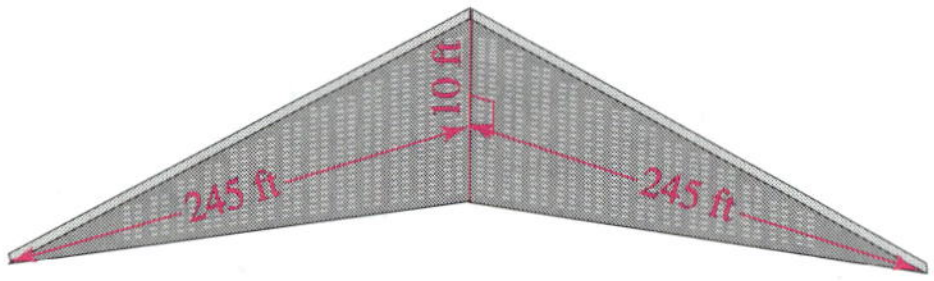

83. WHEELCHAIRS Find the diameter of the rear wheel and the radius of the front wheel.

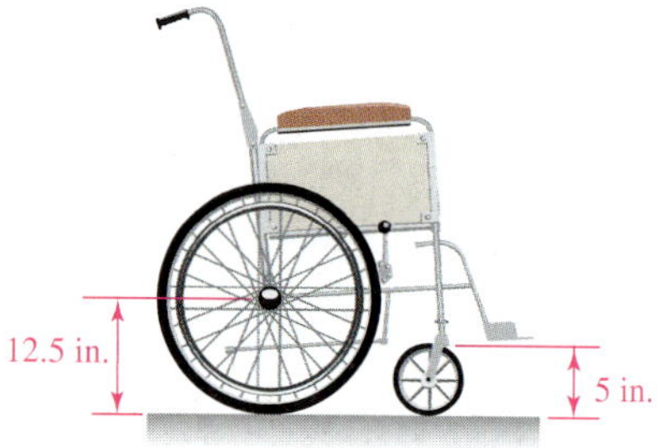

84. ARCHERY The diameter of a standard archery target used in the Olympics is 48.8 inches. Find the area of the target. Round to the nearest square inch.

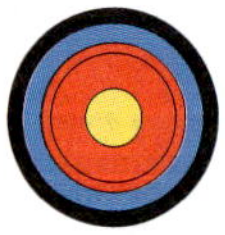

85. BULLS-EYE See Exercise 84. The diameter of the center yellow ring of a standard archery target is 4.8 inches. What is the area of the bulls-eye? Round to the nearest tenth of a square inch.

86. GEOGRAPHY The circumference of the Earth is about 25,000 miles. Find its diameter to the nearest mile.

87. HORSES A horse trots in a circle around its trainer at the end of a 28-foot-long rope. Find the area of the circle that is swept out. Round to the nearest square foot.

88. YO-YOS How far does a yo-yo travel during one revolution of the "around the world" trick if the length of the string is 21 inches?

89. WORLD HISTORY The Inca Empire (1438–1533) was centered in what is now called Peru. A special feature of Inca architecture was the trapezoid-shaped windows and doorways. A standard Inca window was 70 cm (centimeters) high, 50 cm at the base and 40 cm at the top. Find the area of a window opening. [*Hint:* The formula for the area of a trapezoid is $A = \frac{1}{2}(\text{height})(\text{upperbase} + \text{lowerbase})$].

© DIOMEDIA/Alamy

90. HAMSTER HABITATS Find the amount of space in the tube.

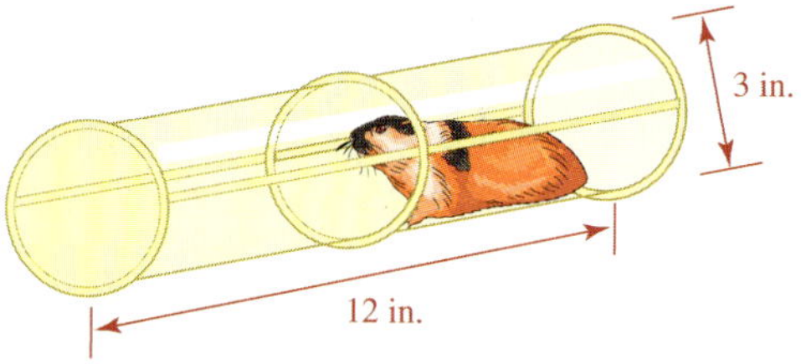

91. TIRES The road surface footprint of a sport truck tire is approximately rectangular. If the area of the footprint is 45 in.^2, about how wide is the tire?

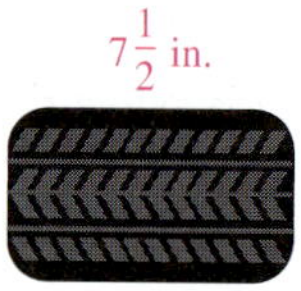

92. SOFTBALL The strike zone in fast-pitch softball is between the batter's armpit and the top of her knees, as shown. If the area of the strike zone for this batter is 442 in.^2, what is the width of home plate?

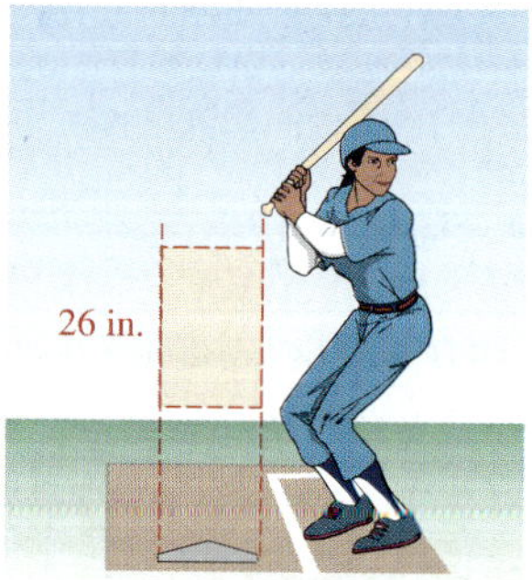

93. FIREWOOD The cord of wood shown occupies a volume of 128 ft^3. How long is the stack?

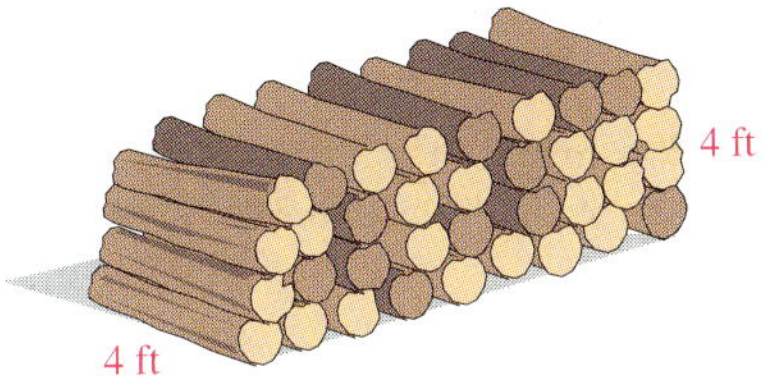

94. TEEPEES The teepees constructed by the Blackfoot Indians were cone-shaped tents about 10 feet high and about 15 feet across at the ground. Estimate the volume of a teepee with these dimensions, to the nearest cubic foot.

95. IGLOOS During long journeys, some Canadian Eskimos built winter houses of snow blocks stacked in the dome shape shown. Estimate the volume of an igloo having an interior height of 5.5 feet to the nearest cubic foot.

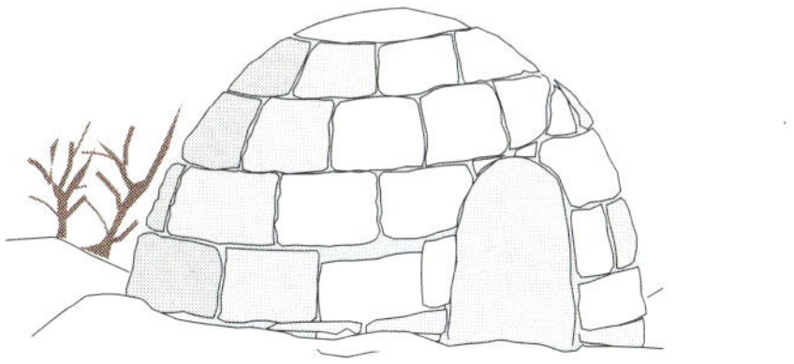

96. PYRAMIDS The Great Pyramid at Giza in northern Egypt is one of the most famous works of architecture in the world. Find its volume to the nearest cubic foot.

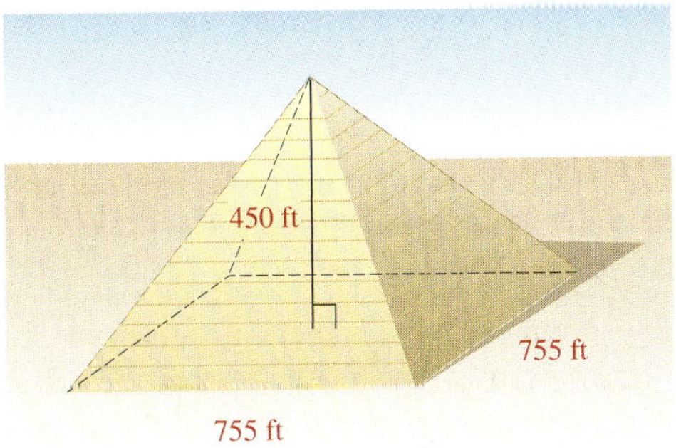

97. COOKING If the fish shown in the illustration in the next column is 18 inches long, what is the area of the grill? Round to the nearest square inch.

98. SKATEBOARDING A half-pipe ramp is in the shape of a semicircle with a radius of 8 feet. To the nearest tenth of a foot, what is the length of the arc that the rider travels on the ramp?

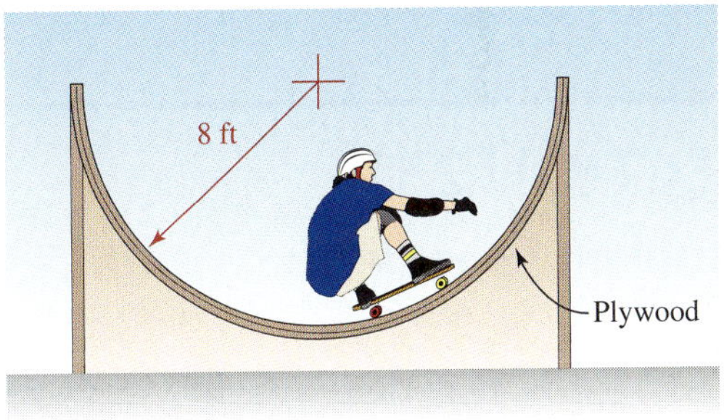

99. PULLEYS The approximate length L of a belt joining two pulleys of radii r and R feet with centers D feet apart is given by the formula $L = 2D + 3.25(r + R)$. Solve the formula for D.

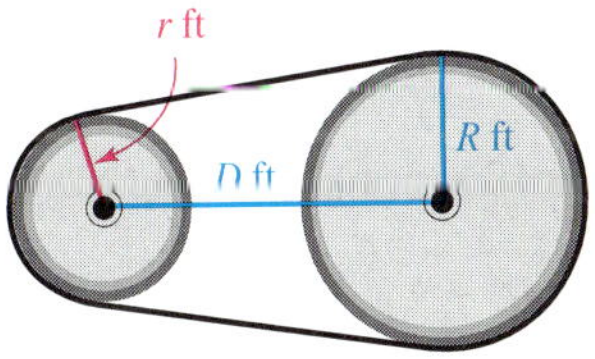

100. THERMODYNAMICS The Gibbs free-energy function is given by $G = U - TS + pV$. Solve this formula for the pressure p.

WRITING

101. After solving $A = B + C + D$ for B, a student compared her answer with that at the back of the textbook. Could this problem have two different-looking answers? Explain why or why not.

Student's answer: $B = A - C - D$

Book's answer: $B = A - D - C$

102. A student solved $x + 5c = 3c + a$ for c. His answer was $c = \frac{3c + a - x}{5}$ for c. Explain why the equation is not solved for c.

103. Explain the difference between what perimeter measures and what area measures.

104. Explain the error made below.

$$y = \frac{3x + \overset{1}{\cancel{2}}}{\underset{1}{\cancel{2}}}$$

REVIEW

105. Find 82% of 168.

106. 29.05 is what percent of 415?

107. What percent of 200 is 30?

108. A woman bought a coat for \$98.95 and some gloves for \$7.95. If the sales tax was 6%, how much did the purchase cost her?

CHALLENGE PROBLEMS

109. In mathematics, letters from the Greek alphabet are often used as variables. Solve the following equation for α (read as "alpha"), the first letter of the Greek alphabet.

$$-7(\alpha - \beta) - (4\alpha - \theta) = \frac{\alpha}{2}$$

110. When a car of mass collides with a wall, the energy of the collision is given by the formula $E = \frac{1}{2}mv^2$. Compare the energy of two collisions: a car striking a wall at 30 mph, and at 60 mph. Then complete this sentence: Although the speed is only twice as fast, the energy is ___ times greater.

SECTION 2.5
Problem Solving

Objectives

1. Apply the steps of a problem-solving strategy.
2. Solve consecutive integer problems.
3. Solve geometry problems.

In this section, you will see that algebra is a powerful tool that can be used to solve a wide variety of real-world problems.

1 Apply the Steps of a Problem-Solving Strategy.

To become a good problem solver, you need a plan to follow, such as the following five-step strategy.

Strategy for Problem Solving

1. **Analyze the problem** by reading it carefully to understand the given facts. What information is given? What are you asked to find? What vocabulary is given? Often, a diagram or table will help you visualize the facts of the problem.
2. **Form an equation** by picking a variable to represent the numerical value to be found. Then express all other unknown quantities as expressions involving that variable. Key words or phrases can be helpful. Finally, translate the words of the problem into an equation.
3. **Solve the equation.**
4. **State the conclusion.**
5. **Check the result** using the original wording of the problem, not the equation that was formed in step 2.

EXAMPLE 1 ***California Coastline.*** The first part of California's magnificent 17-Mile Drive begins at the Pacific Grove entrance and continues to Seal Rock. It is 1 mile longer than the second part of the drive, which extends from Seal Rock to the Lone Cypress. The final part of the tour winds through the Monterey Peninsula, eventually returning to the entrance. This part of the drive is 1 mile longer than four times the length of the second part. How long is each part of 17-Mile Drive?

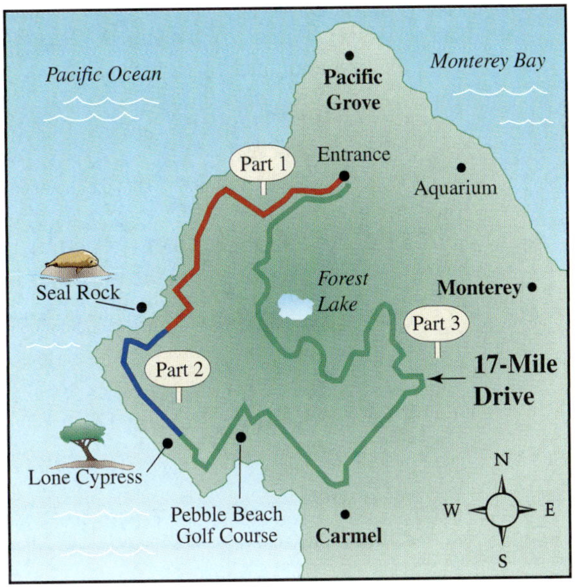

Analyze the Problem The drive is composed of three parts. We need to find the length of each part. We can straighten out the winding 17-Mile Drive and model it with a line segment.

Form an Equation Since the lengths of the first part and of the third part of the drive are related to the length of the second part, we will let x represent the length of that part. We then express the other lengths in terms of x. Let

x = the length of the second part of the drive
$x + 1$ = the length of the first part of the drive
$4x + 1$ = the length of the third part of the drive

Caution

For this problem, one common mistake is to let

x = the length of ~~each~~ part of the drive

The three parts of the drive have different lengths; x cannot represent three different distances.

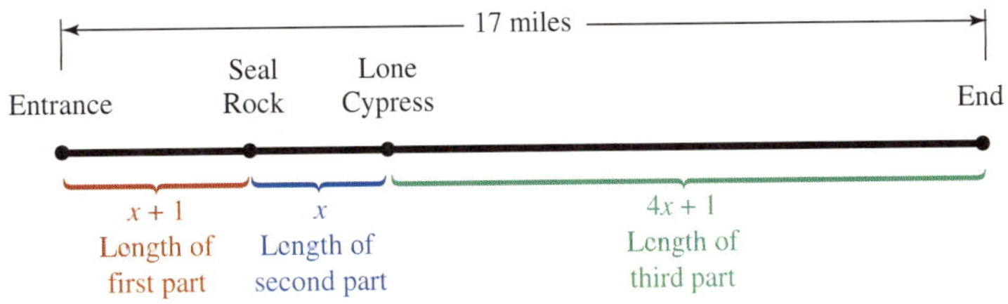

The sum of the lengths of the three parts must be 17 miles.

The length of part 1	plus	the length of part 2	plus	the length of part 3	equals	the total length.
$x + 1$	$+$	x	$+$	$4x + 1$	$=$	17

Solve the Equation

$$x + 1 + x + 4x + 1 = 17$$

$$6x + 2 = 17 \quad \text{Combine like terms: } x + x + 4x = 6x \text{ and } 1 + 1 = 2.$$

$$6x = 15 \quad \text{To undo the addition of 2, subtract 2 from both sides.}$$

$$\frac{6x}{6} = \frac{15}{6} \quad \text{To isolate } x\text{, undo the multiplication by 6 by dividing both sides by 6.}$$

$$x = 2.5 \quad \text{Do the divisions.}$$

Recall that x represents the length of the second part of the drive. To find the lengths of the first and third parts, we evaluate $x + 1$ and $4x + 1$ for $x = 2.5$.

Caution
Check the result using the original wording of the problem, not by substituting it into the equation. Why? The equation may have been solved correctly, but the danger is that you may have formed it incorrectly.

First part of drive

$$x + 1 = 2.5 + 1$$
$$= 3.5$$

Third part of drive

$$4x + 1 = 4(2.5) + 1 \quad \text{Substitute 2.5 for } x.$$
$$= 11$$

State the Conclusion The first part of the drive is 3.5 miles long, the second part is 2.5 miles long, and the third part is 11 miles long.

Check the Result Since 3.5 mi + 2.5 mi + 11 mi = 17 mi, the answers check.

Now Try Problem 13

EXAMPLE 2

Computer Logos. A trucking company had their logo embroidered on the front of baseball caps. They were charged \$8.90 per hat plus a one time set up fee of \$25. If the project cost \$559, how many hats were embroidered?

Analyze the Problem

- It cost \$8.90 to have a logo embroidered on a hat.
- The set up charge was \$25.
- The project cost \$559.
- We are to find the number of hats that were embroidered.

Success Tip
The *Form an Equation* step is often the hardest. To help, write a **verbal model** of the situation (shown here in blue) and then translate it into an equation.

Form an Equation Let x = the number of hats that were embroidered. If x hats are embroidered, at a cost of \$8.90 per hat, the cost to embroider all of the hats is $x \cdot \$8.90$ or $\$8.90x$. Now we translate the words of the problem into an equation.

The cost to embroider one hat	times	the number of hats	plus	the set up charge	equals	the total cost.
8.90	$\cdot$	x	+	25	=	559

Solve the Equation

$$8.90x + 25 = 559$$

$$8.90x = 534 \quad \text{To undo the addition of 25, subtract 25 from both sides.}$$

$$\frac{8.90x}{8.90} = \frac{534}{8.90} \quad \text{To isolate } x\text{, undo the multiplication by 8.90 by dividing both sides by 8.90.}$$

$$x = 60 \quad \text{Do the divisions.}$$

State the Conclusion The company had 60 hats embroidered.

Check the Result The cost to embroider 60 hats is 60($8.90) = $534. When the $25 set up charge is added, we get $534 + $25 = $559. The answer checks.

Now Try Problem 21

EXAMPLE 3 ***Auctions.*** A classic car owner is going to sell his 1960 Chevy Impala at an auction. He wants to make $46,000 after paying an 8% commission to the auctioneer. For what selling price (called the "hammer price") will the car owner make this amount of money?

Analyze the Problem When the commission is subtracted from the selling price of the car, the owner wants to have $46,000 left.

The Language of Algebra
Phrases such as *should be* or *will be* translate to an equal symbol =.

Form an Equation Let x = the selling price of the car. The amount of the commission is 8% of x, or $0.08x$. Now we translate the words of the problem to an equation.

The selling price of the car	minus	the auctioneer's commission	should be	$46,000.
x	$-$	$0.08x$	$=$	46,000

Solve the Equation

$$x - 0.08x = 46{,}000$$

$$0.92x = 46{,}000 \quad \text{Combine like terms: } 1.00x - 0.08x = 0.92x.$$

$$\frac{0.92x}{0.92} = \frac{46{,}000}{0.92} \quad \text{To isolate } x\text{, undo the multiplication by 0.92 by dividing both sides by 0.92.}$$

$$x = 50{,}000 \quad \text{Do the divisions.}$$

State the Conclusion The owner will make $46,000 if the car sells for $50,000.

Check the Result An 8% commission on $50,000 is 0.08($50,000) = $4,000. The owner will keep $50,000 − $4,000 = $46,000. The answer checks.

Now Try Problem 27

2 Solve Consecutive Integer Problems.

Integers that follow one another, such as 15 and 16, are called **consecutive integers.** They are 1 unit apart. **Consecutive even integers** are even integers that differ by 2 units, such as 12 and 14. Similarly, **consecutive odd integers** differ by 2 units, such as 9 and 11. When solving consecutive integer problems, if we let x = the first integer, then

- two consecutive integers are x and $x + 1$
- two consecutive even integers are x and $x + 2$
- two consecutive odd integers are x and $x + 2$

EXAMPLE 4 ***U.S. History.*** The year George Washington was chosen president and the year the Bill of Rights went into effect are consecutive odd integers whose sum is 3,580. Find the years.

Analyze the Problem We need to find two consecutive odd integers whose sum is 3,580. From history, we know that Washington was elected president first and the Bill of Rights went into effect later.

The Language of Algebra
Consecutive means following one after the other in order. Elton John holds the record for the most *consecutive* years with a song on the Top 50 music chart: 31 years (1970 to 2000).

Form an Equation Let x = the first odd integer (the date when Washington was chosen president). The next odd integer is 2 *greater than* x, therefore $x + 2$ = the next larger odd integer (the date when the Bill of Rights went into effect).

The first odd integer	plus	the second odd integer	is	3,580.
x	$+$	$x + 2$	$=$	3,580

Solve the Equation

$$x + x + 2 = 3{,}580$$

$$2x + 2 = 3{,}580 \quad \text{Combine like terms: } x + x = 2x.$$

$$2x = 3{,}578 \quad \text{To undo the addition of 2, subtract 2 from both sides.}$$

$$x = 1{,}789 \quad \text{To isolate } x\text{, undo the multiplication by 2 by dividing both sides by 2.}$$

State the Conclusion George Washington was chosen president in the year 1789. The Bill of Rights went into effect in $1789 + 2 = 1791$.

Check the Result 1789 and 1791 are consecutive odd integers whose sum is $1789 + 1791 = 3{,}580$. The answers check.

Now Try **Problem 33**

3 Solve Geometry Problems.

EXAMPLE 5 ***Crime Scenes.*** Police used 400 feet of yellow tape to fence off a rectangular-shaped lot for an investigation. Fifty less feet of tape was used for each width as for each length. Find the dimensions of the lot.

The Language of Algebra
Dimensions are measurements of length and width. We might speak of the *dimensions* of a dance floor or a TV screen.

Success Tip
When solving geometry problems, a sketch is often helpful.

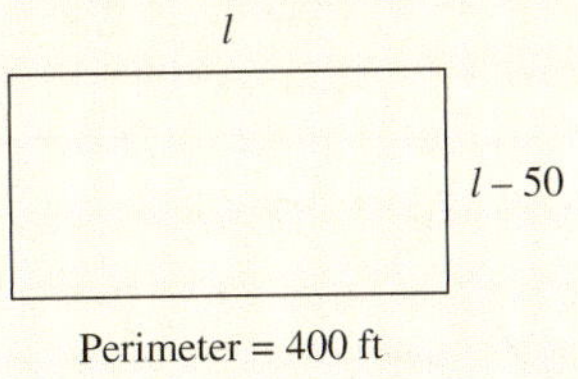

Analyze the Problem Since the yellow tape surrounded the lot, the concept of perimeter applies. Recall that the formula for the perimeter of a rectangle is $P = 2l + 2w$. We also know that the width of the lot is 50 feet less than the length.

Form an Equation Since the width of the lot is given in terms of the length, we let l = the length of the lot. Then $l - 50$ = the width. Using the perimeter formula, we have:

2	times	the length	plus	2	times	the width	is	the perimeter.
2	$\cdot$	l	$+$	2	$\cdot$	$(l - 50)$	$=$	400

Solve the Equation

$2l + 2(l - 50) = 400$ Write the parentheses so that the entire expression $l - 50$ is multiplied by 2.

$2l + 2l - 100 = 400$ Distribute the multiplication by 2.

$4l - 100 = 400$ Combine like terms: $2l + 2l = 4l$.

$4l = 500$ To undo the subtraction of 100, add 100 to both sides.

$l = 125$ To isolate l, undo the multiplication by 4 by dividing both sides by 4.

State the Conclusion The length of the lot is 125 feet and width is $125 - 50 = 75$ feet.

Check the Result The width (75 feet) is 50 less than the length (125 feet). The perimeter of the lot is $2(125) + 2(75) = 250 + 150 = 400$ feet. The answers check.

Now Try Problem 39

EXAMPLE 6

Isosceles Triangles. If the vertex angle of an isosceles triangle is 56°, find the measure of each base angle.

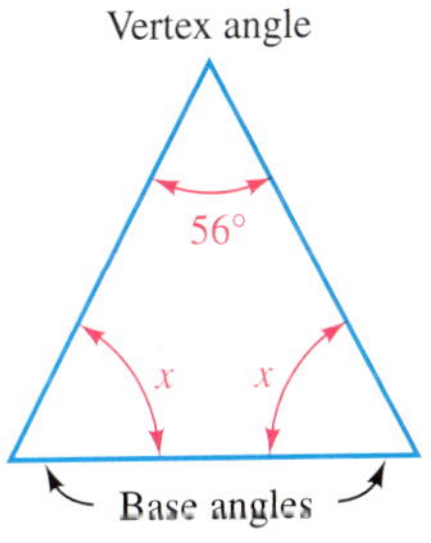

Analyze the Problem An **isosceles triangle** has two sides of equal length, which meet to form the **vertex angle.** In this case, the measurement of the vertex angle is 56°. We can sketch the triangle as shown. The **base angles** opposite the equal sides are also equal. We need to find their measure.

Form an Equation If we let x = the measure of one base angle, the measure of the other base angle is also x. Since the sum of the angles of any triangle is 180°, the sum of the base angles and the vertex angle is 180°. We can use this fact to form the equation.

One base angle	plus	the other base angle	plus	the vertex angle	is	180°.
x	$+$	x	$+$	56	$=$	180

Solve the Equation

$x + x + 56 = 180$

$2x + 56 = 180$ Combine like terms: $x + x = 2x$.

$2x = 124$ To undo the addition of 56, subtract 56 from both sides.

$x = 62$ To isolate x, undo the multiplication by 2 by dividing both sides by 2.

State the Conclusion The measure of each base angle is 62°.

Check the Result Since $62° + 62° + 56° = 180°$, the answer checks.

Now Try Problem 43

STUDY SET 2.5

VOCABULARY

Fill in the blanks.

1. Integers that follow one another, such as 7 and 8, are called ________ integers.
2. An ________ triangle is a triangle with two sides of the same length.
3. The equal sides of an isosceles triangle meet to form the ______ angle. The angles opposite the equal sides are called _____ angles, and they have equal measures.
4. When asked to find the dimensions of a rectangle, we are to find its ______ and ______.

CONCEPTS

5. A 17-foot pipe is cut into three sections. The longest section is three times as long as the shortest, and the middle-sized section is 2 feet longer than the shortest. Complete the diagram.

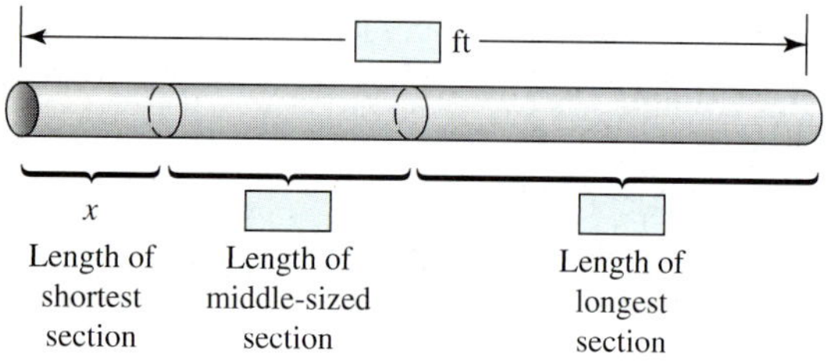

6. It costs \$28 per hour to rent a trailer. Write an expression that represents the cost to rent the trailer for x hours.
7. A realtor is paid a 3% commission on the sale of a house. Write an expression that represents the amount of the commission if a house sells for \$$x$.
8. The perimeter of the rectangle is 15 feet. Fill in the blanks: $2(\quad) + 2x = \quad$

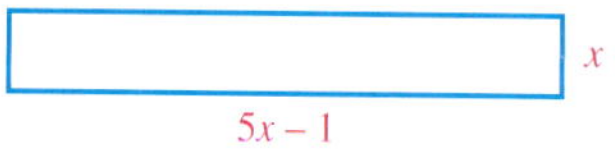

9. What is the sum of the measures of the angles of any triangle?
10. What is x?

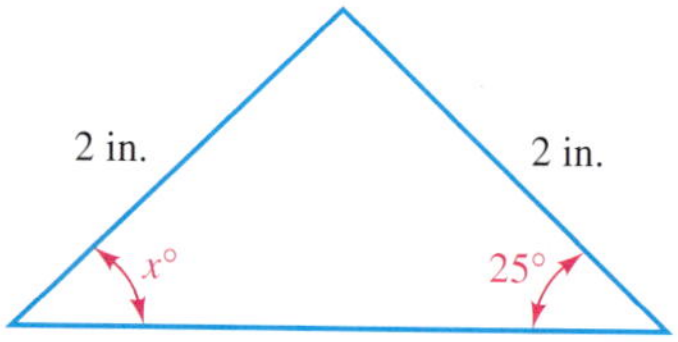

NOTATION

11. a. If x represents an integer, write an expression for the next largest integer.
    b. If x represents an odd integer, write an expression for the next largest odd integer.
12. What does 45° mean?

GUIDED PRACTICE

See Example 1.

13. A 12-foot board has been cut into two sections, one twice as long as the other. How long is each section?

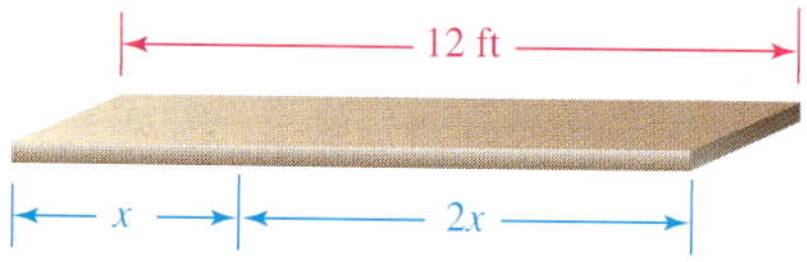

14. The robotic arm will extend a total distance of 18 feet. Find the length of each section.

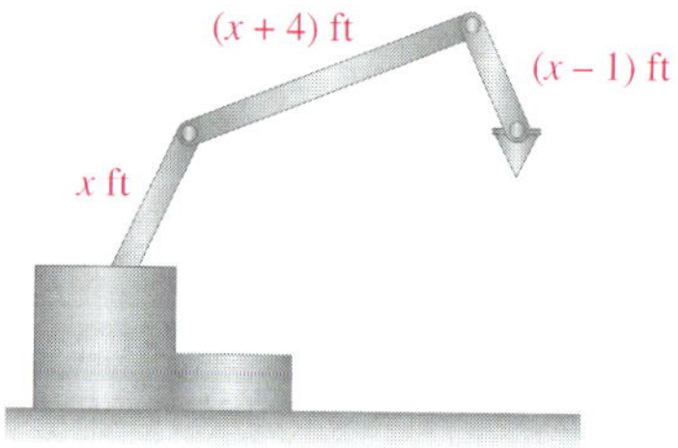

APPLICATIONS

15. NATIONAL PARKS The Natchez Trace Parkway is a historical 444-mile route from Natchez, Mississippi, to Nashville, Tennessee. A couple drove the Trace in four days. Each day they drove 6 miles more than the previous day. How many miles did they drive each day?

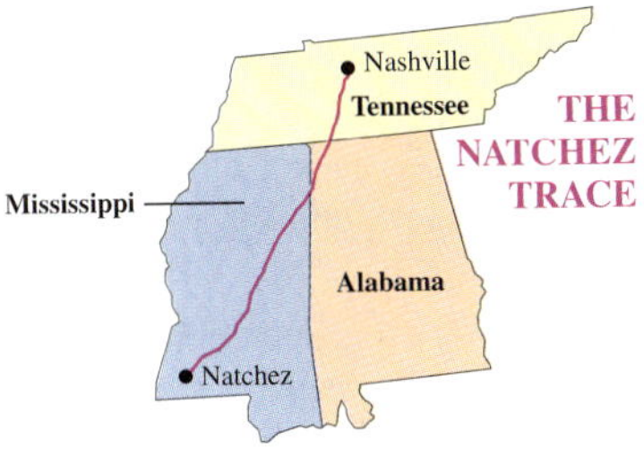

16. TOURING A rock group plans to travel for a total of 38 weeks, making three concert stops. They will be in Japan for 4 more weeks than they will be in Australia. Their stay in Sweden will be 2 weeks shorter than that in Australia. How many weeks will they be in each country?

17. SOLAR HEATING One solar panel is 3.4 feet wider than the other. Find the width of each panel.

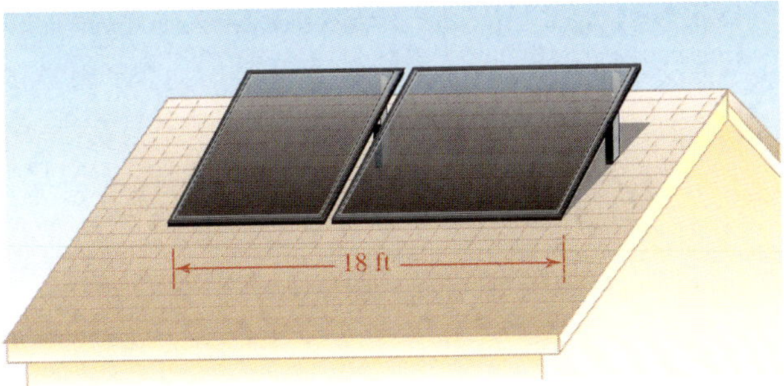

18. ACCOUNTING Determine the 2005 income of Abercrombie & Fitch Company for each quarter from the data in the graph.

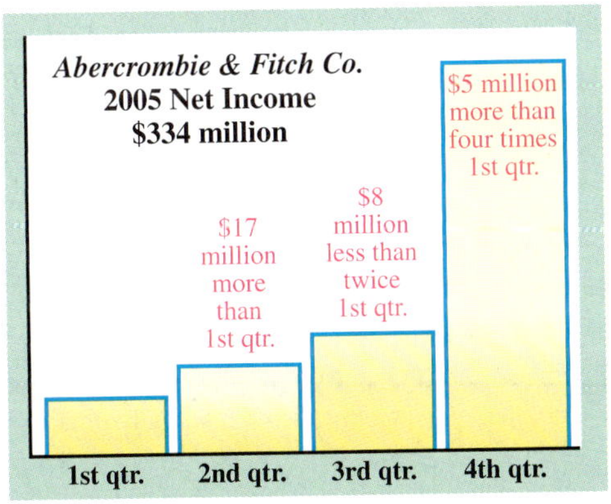

19. COUNTING CALORIES A slice of pie with a scoop of ice cream has 850 calories. The calories in the pie alone are 100 more than twice the calories in the ice cream alone. How many calories are in each food?

20. WASTE DISPOSAL Two tanks hold a total of 45 gallons of a toxic solvent. One tank holds 6 gallons more than twice the amount in the other. How many gallons does each tank hold?

21. CONCERTS The fee to rent a concert hall is $2,250 plus $150 per hour to pay for the support staff. For how many hours can an orchestra rent the hall and stay within a budget of $3,300?

22. TRUCK MECHANICS An engine repair cost a truck owner $1,185 in parts and labor. If the parts were $690 and the mechanic charged $45 per hour, how many hours did the repair take?

23. FIELD TRIPS It costs a school $65 a day plus $0.25 per mile to rent a 15-passenger van. If the van is rented for two days, how many miles can be driven on a $275 budget?

24. DECORATIONS A party supply store charges a set-up fee of $80 plus 35¢ per balloon to make a balloon arch. A business has $150 to spend on decorations for their grand opening. How many balloons can they have in the arch? (*Hint:* 35¢ = $0.35.)

25. TUTORING High school students enrolling in a private tutoring program must first take a placement test (cost $25) before receiving tutoring (cost $18.75 per hour). If a family has set aside $400 to get their child extra help, how many hours of tutoring can they afford?

26. DATA CONVERSION The *Books2Bytes* service converts old print books to Microsoft Word electronic files for $20 per book plus $2.25 per page. If it cost $1,201.25 to convert a novel, how many pages did the novel have?

27. CATTLE AUCTIONS A cattle rancher is going to sell one of his prize bulls at an auction and would like to make $45,500 after paying a 9% commission to the auctioneer. For what selling price will the rancher make this amount of money?

28. LISTING PRICE At what price should a home be listed if the owner wants to make $567,000 on its sale after paying a 5.5% real estate commission?

29. SAVINGS ACCOUNTS The balance in a savings account grew by 5% in one year, to $5,512.50. What was the balance at the beginning of the year?

30. AUTO INSURANCE Between the years 2000 and 2006, the average cost for auto insurance nationwide grew 27%, to $867. What was the average cost in 2000? Round to the nearest dollar.

Consecutive integer problems

31. SOCCER Ronaldo of Brazil and Gerd Mueller of Germany rank 1 and 2, respectively, with the most goals scored in World Cup play. The number of goals Ronaldo and Mueller have scored are consecutive integers that total 29. Find the number of goals scored by each man.

32. DICTIONARIES The definitions of the words *job* and *join* are on back-to-back pages in a dictionary. If the sum of those page numbers is 1,411, on what page can the definition of *job* be found?

33. TV HISTORY *Friends* and *Leave It to Beaver* are two of the most popular television shows of all time. The number of episodes of each show are consecutive even integers whose sum is 470. If there are more episodes of *Friends,* how many episodes of each were there?

34. VACATIONS The table shows the average number of vacation days an employed adult receives for selected countries. Complete the table. (The numbers of days are listed in descending order.)

Average Number of Vacation Days per Year	
Country	Days
Italy	42
France	
Germany	
U.S.	13

Consecutive odd integers whose sum is 72.

Source: The World Almanac, 2006.

35. CELEBRITY BIRTHDAYS Elvis Presley, George Foreman, and Kirstie Alley have birthdays (in that order) on consecutive even-numbered days in January. The sum of the calendar dates of their birthdays is 30. Find each birthday.

36. LOCKS The three numbers of the combination for a lock are consecutive integers, and their sum is 81. Find the combination.

Geometry problems

37. TENNIS The perimeter of a regulation singles tennis court is 210 feet and the length is 3 feet less than three times the width. What are the dimensions of the court?

38. SWIMMING POOLS The seawater Orthlieb Pool in Casablanca, Morocco, is the largest swimming pool in the world. With a perimeter of 1,110 meters, this rectangular-shaped pool is 30 meters longer than 6 times its width. Find its dimensions.

39. ART The *Mona Lisa* was completed by Leonardo da Vinci in 1506. The length of the picture is 11.75 inches shorter than twice the width. If the perimeter of the picture is 102.5 inches, find its dimensions.

© Réunion des Musées Nationaux/Art Resource, NY

40. NEW YORK CITY Central Park, which lies in the middle of Manhattan, is rectangular-shaped and has a 6-mile perimeter. The length is 5 times the width. What are the dimensions of the park?

41. ENGINEERING A truss is in the form of an isosceles triangle. Each of the two equal sides is 4 feet shorter than the third side. If the perimeter is 25 feet, find the lengths of the sides.

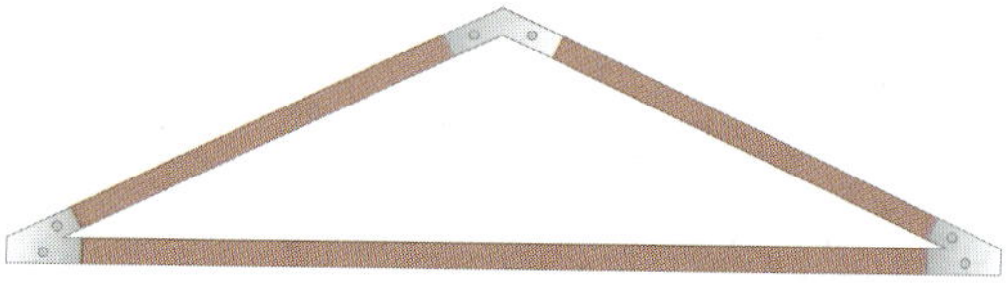

42. FIRST AID A sling is in the shape of an isosceles triangle with a perimeter of 144 inches. The longest side of the sling is 18 inches longer than either of the other two sides. Find the lengths of each side.

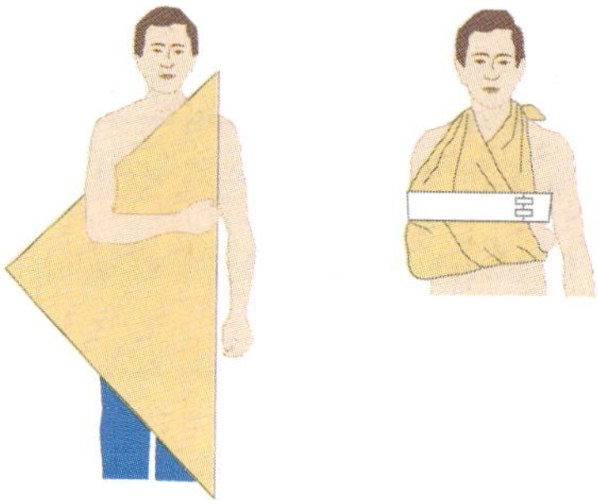

43. TV TOWERS The two guy wires supporting a tower form an isosceles triangle with the ground. Each of the base angles of the triangle is 4 times the third angle (the vertex angle). Find the measure of the vertex angle.

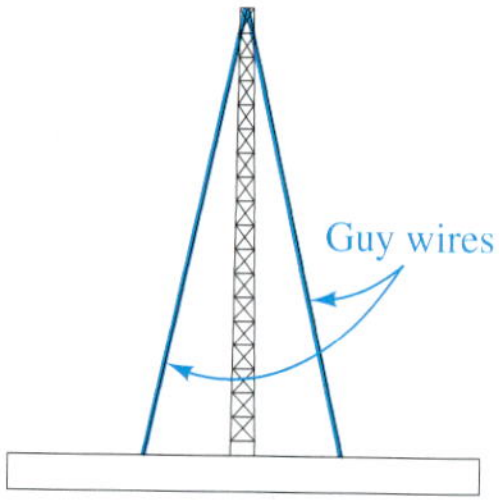

44. CLOTHESLINES A pair of damp jeans are hung in the middle of a clothesline to dry. (See the next page.) Find $x°$, the angle that the clothesline makes with the horizontal.

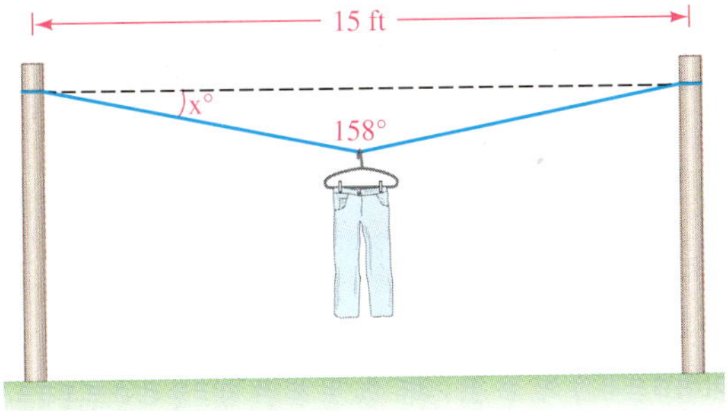

45. MOUNTAIN BICYCLES For the bicycle frame shown, the angle that the horizontal crossbar makes with the seat support is 15° less than twice the angle at the steering column. The angle at the pedal gear is 25° more than the angle at the steering column. Find these three angle measures.

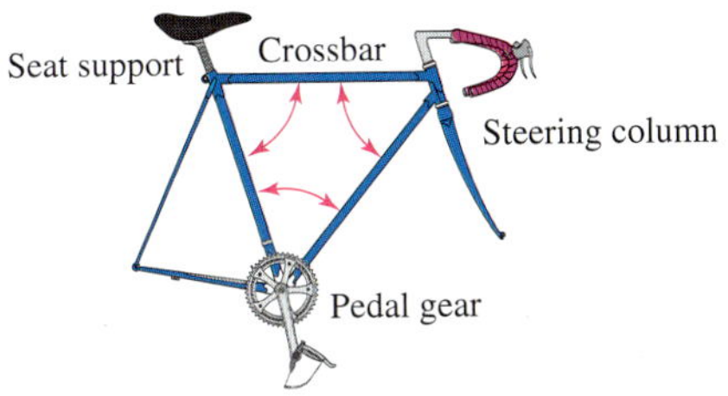

46. TRIANGLES The measure of $\angle 1$ (read as angle 1) of a triangle is one-half that of $\angle 2$. The measure of $\angle 3$ is equal to the sum of the measures of $\angle 1$ and $\angle 2$. Find each angle measure.

47. COMPLEMENTARY ANGLES Two angles are called ***complementary angles*** when the sum of their measures is 90°. Find the measures of the complementary angles shown in the illustration.

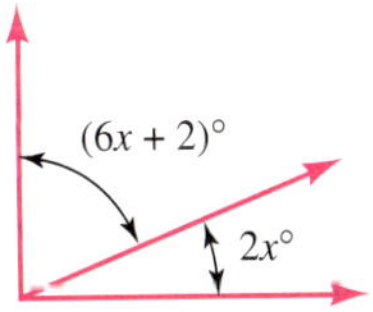

48. SUPPLEMENTARY ANGLES Two angles are called ***supplementary angles*** when the sum of their measures is 180°. Find the measures of the supplementary angles shown in the illustration.

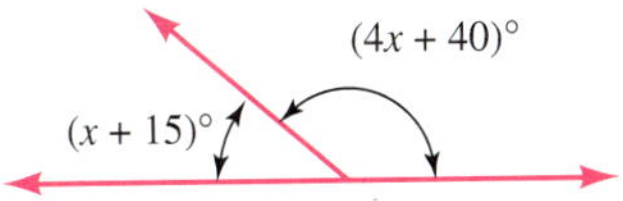

WRITING

49. Create a geometry problem that could be answered by solving the equation $2w + 2(w + 5) = 26$.

50. What information do you need to know to answer the following question?

A business rented a copy machine for \$85 per month plus 4¢ for every copy made. How many copies can be made each month?

51. Make a list of words and phrases that translate to an equal symbol =.

52. Define the word *strategy.*

REVIEW

Solve.

53. $\frac{5}{8}x = -15$

54. $\frac{12x + 24}{13} = 36$

55. $\frac{3}{4}y = \frac{2}{5}y - \frac{3}{2}y - 2$

56. $6 + 4(1 - x) = 3(x + 1)$

57. $4.2(y - 4) - 0.6y = -13.2$

58. $16 - 8(b + 4) = 24b + 64$

CHALLENGE PROBLEMS

59. What concept discussed in this section is illustrated by the following day and time?

Two minutes and three seconds past 1 A.M. on the 5th day of April, 2006

60. MANUFACTURING A company has two machines that make widgets. The production costs are listed below.

Machine 1: Setup cost \$400 and \$1.70 per widget

Machine 2: Setup cost \$500 and \$1.20 per widget

Find the number of widgets for which the cost to manufacture them on either machine is the same.

SECTION 2.6
More about Problem Solving

Objectives

1. Solve investment problems.
2. Solve uniform motion problems.
3. Solve liquid mixture problems.
4. Solve dry mixture problems.
5. Solve number-value problems.

In this section, we will solve problems that involve money, motion, and mixtures. Tables are a helpful way to organize the information given in these problems.

1 Solve Investment Problems.

To find the amount of *simple interest* I an investment earns, we use the formula $I = Prt$, where P is the principal (the amount invested), r is the annual interest rate, and t is the time in years.

EXAMPLE 1 ***Paying Tuition.*** A college student invested the \$12,000 inheritance he received and decided to use the annual interest earned to pay his tuition cost of \$945. The highest rate offered by a bank at that time was 6% annual simple interest. At this rate, he could not earn the needed \$945, so he invested some of the money in a riskier, but more profitable, investment offering a 9% return. How much did he invest at each rate?

Analyze the Problem We know that \$12,000 was invested for 1 year at two rates: 6% and 9%. We are asked to find the amount invested at each rate so that the total return would be \$945.

Form an Equation Let $x =$ the amount invested at 6%. Then $12{,}000 - x =$ the amount invested at 9%. To organize the facts of the problem, we enter the principal, rate, time, and interest earned in a table.

Step 1: List each investment in a row of the table.

Bank				
Riskier Investment				

Step 2: Label the columns using $I = Prt$ reversed and also write Total:.

	P	$\cdot\ r$	$\cdot\ t =$	I
Bank				
Riskier Investment				
				Total:

Step 3: Enter the rates, times, and total interest.

	P	$\cdot\ r$	$\cdot\ t =$	I
Bank		**0.06**	**1**	
Riskier Investment		**0.09**	**1**	
				Total: **945**

Step 4: Enter each unknown principal.

	P	$\cdot\ r$	$\cdot\ t =$	I
Bank	x	0.06	1	
Riskier Investment	**$12{,}000 - x$**	0.09	1	
				Total: 945

Step 5: In the last column, multiply P, r, and t to obtain expressions for the interest earned.

	P	$\cdot\ r$	$\cdot\ t =$	I
Bank	x	0.06	1	$\mathbf{0.06x}$
Riskier Investment	$12{,}000 - x$	0.09	1	$\mathbf{0.09(12{,}000 - x)}$
				Total: 945

← This is $x \cdot 0.06 \cdot 1$.

← This is $(12{,}000 - x) \cdot 0.09 \cdot 1$.

Use the information in this column to form an equation.

The interest earned at 6%	plus	the interest earned at 9%	equals	the total interest.
$0.06x$	$+$	$[0.09(12{,}000 - x)]$	$=$	945

Solve the Equation

$$0.06x + 0.09(12{,}000 - x) = 945$$

$$100[0.06x + 0.09(12{,}000 - x)] = 100(945)$$ Multiply both sides by 100 to clear the equation of decimals.

$$100(0.06x) + 100(0.09)(12{,}000 - x) = 100(945)$$ Distribute the multiplication by 100.

$$6x + 9(12{,}000 - x) = 94{,}500$$ Do the multiplications by 100.

$$6x + 108{,}000 - 9x = 94{,}500$$ Use the distributive property.

$$-3x + 108{,}000 = 94{,}500$$ Combine like terms.

$$-3x = -13{,}500$$ Subtract 108,000 from both sides.

$$x = 4{,}500$$ To isolate x, divide both sides by -3.

Success Tip

We can *clear an equation of decimals* by multiplying both sides by a power of 10. Here, we multiply 0.06 and 0.09 by 100 to move each decimal point two places to the right:

$100(0.06) = 6 \qquad 100(0.09) = 9$

Caution

On the left side of the equation, do not incorrectly distribute the multiplication by 100 over addition **and** multiplication.

$100[0.06x + 0.09(12{,}000 - x)]$

State the Conclusion The student invested \$4,500 at 6% and \$12,000 − \$4,500 = \$7,500 at 9%.

Check the Result The first investment earned 0.06(\$4,500), or \$270. The second earned 0.09(\$7,500), or \$675. Since the total return was \$270 + \$675 = \$945, the answers check.

Now Try Problem 17

2 Solve Uniform Motion Problems.

If we know the rate r at which we will be traveling and the time t we will be traveling at that rate, we can find the distance d traveled by using the formula $d = rt$.

EXAMPLE 2 ***Rescues at Sea.*** A cargo ship, heading into port, radios the Coast Guard that it is experiencing engine trouble and that its speed has dropped to 3 knots (this is 3 sea miles per hour). Immediately, a Coast Guard cutter leaves port and speeds at a rate of 25 knots directly toward the disabled ship, which is 56 sea miles away. How long will it take the Coast Guard to reach the ship? (Sea miles are also called nautical miles.)

Success Tip
A sketch is helpful when solving uniform motion problems.

Analyze the Problem We know the *rate* of each ship (25 knots and 3 knots), and we know that they must close a *distance* of 56 sea miles between them. We don't know the *time* it will take to do this.

Form an Equation Let $t =$ the time it takes the Coast Guard to reach the cargo ship. During the rescue, the ships don't travel at the same rate, but they do travel for the same amount of time. Therefore, t also represents the travel time for the cargo ship.

We enter the rates, the variable t for each time, and the total distance traveled by the ships (56 sea miles) in the table. To fill in the last column, we use the formula $r \cdot t = d$ twice to find an expression for each distance traveled: $25 \cdot t = 25t$ and $3 \cdot t = 3t$.

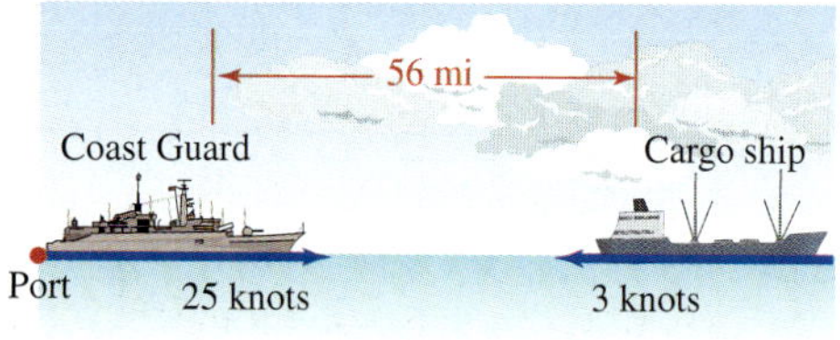

	r	t	d
Coast Guard cutter	25	t	$25t$
Cargo ship	3	t	$3t$
			Total: 56

Multiply $r \cdot t$ to obtain an expression for each distance traveled.

Use the information in this column to form an equation.

The distance the cutter travels	plus	the distance the ship travels	equals	the original distance between the ships.
$25t$	$+$	$3t$	$=$	56

Solve the Equation

$$25t + 3t = 56$$

$$28t = 56 \quad \text{Combine like terms: } 25t + 3t = 28t.$$

$$t = \frac{56}{28} \quad \text{To isolate } t \text{, divide both sides by 28.}$$

$$t = 2 \quad \text{Do the division.}$$

Caution
A common error is to enter 56 miles as the distance traveled for each ship. However, neither ship traveled 56 miles. Together, they travel 56 miles.

	r	t	d
Cutter			~~56~~
Cargo			~~56~~

State the Conclusion The ships will meet in 2 hours.

Check the Result In 2 hours, the Coast Guard cutter travels $25 \cdot 2 = 50$ sea miles, and the cargo ship travels $3 \cdot 2 = 6$ sea miles. Together, they travel $50 + 6 = 56$ sea miles. Since this is the original distance between the ships, the answer checks.

Now Try Problem 27

EXAMPLE 3 ***Concert Tours.*** While on tour, a country music star travels by bus. Her musical equipment is carried in a truck. How long will it take her bus, traveling 60 mph, to overtake the truck, traveling at 45 mph, if the truck had a $1\frac{1}{2}$-hour head start to her next concert location?

Analyze the Problem We know the rate of each vehicle (60 mph and 45 mph) and that the truck began the trip $1\frac{1}{2}$ or 1.5 hours earlier than the bus. We need to determine how long it will take the bus to catch up to the truck.

Form an Equation Let $t =$ the time it takes the bus to overtake the truck. With a 1.5-hour head start, the truck is on the road longer than the bus. Therefore, $t + 1.5 =$ the truck's travel time.

We enter each rate and time in the table, and use the formula $r \cdot t = d$ twice to fill in the distance column.

	r ·	t =	d
Bus	60	t	$60t$
Truck	45	$t + 1.5$	$45(t + 1.5)$

Multiply $r \cdot t$ to obtain an expression for each distance traveled.

Enter this information first.

Use the information in this column to form an equation.

When the bus overtakes the truck, they will have traveled the same distance.

The distance traveled by the bus	is the same as	the distance traveled by the truck.
$60t$	=	$45(t + 1.5)$

Solve the Equation

$$60t = 45(t + 1.5)$$

$$60t = 45t + 67.5 \quad \text{Distribute the multiplication by 45: } 45(1.5) = 67.5.$$

$$15t = 67.5 \quad \text{Subtract } 45t \text{ from both sides: } 60t - 45t = 15t.$$

$$t = 4.5 \quad \text{To isolate } t\text{, divide both sides by 15: } \frac{67.5}{15} = 4.5.$$

> ***Success Tip***
> We used 1.5 hrs for the head start because it is easier to solve $60t = 4.5(t + 1.5)$ than $60t = 45\left(t + 1\frac{1}{2}\right)$.

State the Conclusion The bus will overtake the truck in 4.5 or $4\frac{1}{2}$ hours.

Check the Result In 4.5 hours, the bus travels $60(4.5) = 270$ miles. The truck travels for $1.5 + 4.5 = 6$ hours at 45 mph, which is $45(6) = 270$ miles. Since the distance traveled are the same, the answer checks.

Now Try **Problem 31**

3 Solve Liquid Mixture Problems.

We now discuss how to solve mixture problems. In the first type, a liquid mixture of a desired strength is made from two solutions with different concentrations.

EXAMPLE 4 ***Mixing Solutions.*** A chemistry experiment calls for a 30% sulfuric acid solution. If the lab supply room has only 50% and 20% sulfuric acid solutions, how much of each should be mixed to obtain 12 liters of a 30% acid solution?

Analyze the Problem The 50% solution is too strong and the 20% solution is too weak. We must find how much of each should be combined to obtain 12 liters of a 30% solution.

Success Tip
The strength *(concentration)* of a mixture is always between the strengths of the two solutions used to make it.

Form an Equation If x = the number of liters of the 50% solution used in the mixture, the remaining $(12 - x)$ liters must be the 20% solution. The amount of pure sulfuric acid in each solution is given by

Amount of solution · strength of the solution = amount of pure sulfuric acid

A table and sketch are helpful in organizing the facts of the problem.

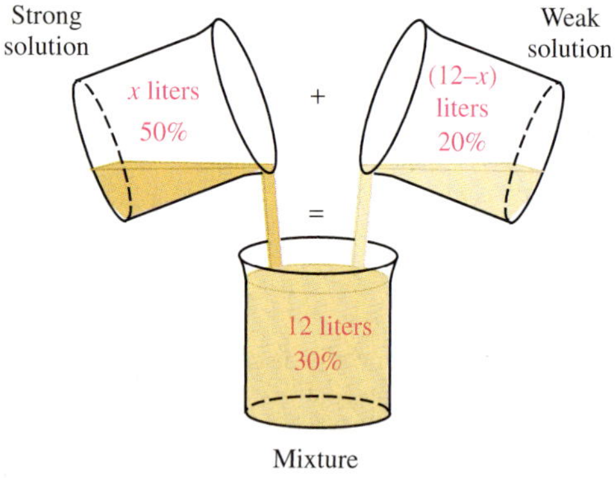

	Amount ·	Strength =	Amount of pure sulfuric acid
Strong	x	0.50	$0.50x$
Weak	$12 - x$	0.20	$0.20(12 - x)$
Mixture	12	0.30	$12(0.30)$

Enter this information first. (Amount and Strength columns)

Multiply amount · strength three times to fill in this column.

Use the information in this column to form an equation.

The sulfuric acid in the 50% solution	plus	the sulfuric acid in the 20% solution	equals	the sulfuric acid in the mixture.
$0.50x$	$+$	$0.20(12 - x)$	$=$	$12(0.30)$

Solve the Equation

$$0.50x + 0.20(12 - x) = 12(0.30)$$

$$0.5x + 2.4 - 0.2x = 3.6$$ Distribute the multiplication by 0.20.

$$0.3x + 2.4 = 3.6$$ Combine like terms: $0.5x - 0.2x = 0.3x$.

$$0.3x = 1.2$$ Subtract 2.4 from both sides.

$$x = 4$$ To isolate x, undo the multiplication by 0.3 by dividing both sides by 0.3: $\frac{1.2}{0.3} = 4$.

Success Tip
We could begin by multiplying both sides of the equation by 10 to clear it of the decimals.

State the Conclusion 4 liters of 50% solution and $12 - 4 = 8$ liters of 20% solution should be used.

Check the Result The amount of acid in 4 liters of the 50% solution is $0.50(4) = 2.0$ liters and the amount of acid in 8 liters of the 20% solution is $0.20(8) = 1.6$ liters. Thus, the amount of acid in these two solutions is $2.0 + 1.6 = 3.6$ liters. The amount of acid in 12 liters of the 30% mixture is also $0.30(12) = 3.6$ liters. Since the amounts of acid are equal, the answers check.

Now Try **Problem 39**

4 Solve Dry Mixture Problems.

In another type of mixture problem, a dry mixture of a specified value is created from two differently priced ingredients.

EXAMPLE 5 ***Snack Foods.*** Because cashews priced at \$9 per pound were not selling, a produce clerk decided to combine them with less expensive peanuts and sell the mixture for \$7 per pound. How many pounds of peanuts, selling at \$6 per pound, should be mixed with 50 pounds of cashews to obtain such a mixture?

Analyze the Problem We need to determine how many pounds of peanuts to mix with 50 pounds of cashews to obtain a mixture worth \$7 per pound.

Form an Equation Let $x =$ the number of pounds of peanuts to use in the mixture. Since 50 pounds of cashews will be combined with the peanuts, the mixture will weigh $50 + x$ pounds. The value of the mixture and of each of its ingredients is given by

Amount · the price = the total value

We can organize the facts of the problem in a table.

	Amount	· Price	= Total value
Peanuts	x	6	$6x$
Cashews	50	9	450
Mixture	$50 + x$	7	$7(50 + x)$

Enter this information first. (Amount and Price columns)

Multiply amount · price three times to fill in this column.

Use the information in this column to form an equation.

The value of the peanuts	plus	the value of the cashews	equals	the value of the mixture.
$6x$	$+$	450	$=$	$7(50 + x)$

Solve the Equation

$6x + 450 = 7(50 + x)$

$6x + 450 = 350 + 7x$ Distribute the multiplication by 7.

$450 = 350 + x$ To eliminate the term $6x$ on the left side, subtract $6x$ from both sides: $7x - 6x = x$.

$100 = x$ To isolate x, subtract 350 from both sides.

State the Conclusion 100 pounds of peanuts should be used in the mixture.

Check the Result The value of 100 pounds of peanuts, at \$6 per pound, is $100(6) = \$600$ and the value of 50 pounds of cashews, at \$9 per pound, $50(9) = \$450$. Thus, the value of these two amounts is \$1,050. Since the value of 150 pounds of the mixture, at \$7 per pound, is also $150(7) = \$1,050$, the answer checks.

Now Try **Problem 45**

5 Solve Number–Value Problems.

When problems deal with collections of different items having different values, we must distinguish between the *number of* and the *value of* the items. For these problems, we will use the fact that

Number · value = total value

EXAMPLE 6 ***Dining Area Improvements.*** A restaurant owner needs to purchase some tables, chairs, and dinner plates for the dining area of her establishment. She plans to buy four chairs and four plates for each new table. She also plans to buy 20 additional plates in case of breakage. If a table costs \$100, a chair \$50, and a plate \$5, how many of each can she buy if she takes out a loan for \$6,500 to pay for the new items?

Analyze the Problem We know the *value* of each item: Tables cost \$100, chairs cost \$50, and plates cost \$5 each. We need to find the *number* of tables, chairs, and plates she can purchase for \$6,500.

Form an Equation The number of chairs and plates she needs depends on the number of tables she buys. So we let t = the number of tables to be purchased. Since every table requires four chairs and four plates, she needs to order $4t$ chairs. Because 20 additional plates are needed, she should order $(4t + 20)$ plates. We can organize the facts of the problem in a table.

	Number	· Value	= Total value
Tables	t	100	$100t$
Chairs	$4t$	50	$50(4t)$
Plates	$4t + 20$	5	$5(4t + 20)$
			Total: 6,500

Multiply number · value three times to fill in this column.

Enter this information first.

Use the information in this column to form an equation.

The value of the tables	plus	the value of the chairs	plus	the value of the plates	equals	the value of the purchase.
$100t$	$+$	$50(4t)$	$+$	$5(4t + 20)$	$=$	6,500

Solve the Equation

$$100t + 50(4t) + 5(4t + 20) = 6{,}500$$

$$100t + 200t + 20t + 100 = 6{,}500 \quad \text{Do the multiplications and distribute.}$$

$$320t + 100 = 6{,}500 \quad \text{Combine like terms: } 100t + 200t + 20t = 320t.$$

$$320t = 6{,}400 \quad \text{Subtract 100 from both sides.}$$

$$t = 20 \quad \text{To isolate } t\text{, divide both sides by 320.}$$

To find the number of chairs and plates to buy, we evaluate $4t$ and $4t + 20$ for $t = 20$.

Chairs: $4t = 4(20) = 80$ ***Plates:*** $4t + 20 = 4(20) + 20 = 100$ Substitute 20 for t.

State the Conclusion The owner needs to buy 20 tables, 80 chairs, and 100 plates.

Check the Result The total value of 20 tables is 20(\$100) = \$2,000, the total value of 80 chairs is 80(\$50) = \$4,000, and the total value of 100 plates is 100(\$5) = \$500. Because the total purchase is \$2,000 + \$4,000 + \$500 = \$6,500, the answers check.

Now Try **Problem 53**

STUDY SET 2.6

VOCABULARY

Fill in the blanks.

1. Problems that involve depositing money are called __________ problems, and problems that involve moving vehicles are called uniform ________ problems.
2. Problems that involve combining ingredients are called ________ problems, and problems that involve collections of different items having different values are called __________ problems.

CONCEPTS

3. Complete the *principal column* given that part of $30,000 is invested in stocks and the rest in art.

	P ·	r ·	t =	I
Stocks	x			
Art	?			

4. A man made two investments that earned a combined annual interest of $280. Complete the table and then form an equation for this investment problem.

	P	· r	· t =	I
Bank	x	0.04	1	
Stocks	$6{,}000 - x$	0.06	1	
				Total:

5. Complete the *rate column* given that the east-bound plane flew 150 mph slower than the west-bound plane.

	r ·	t =	d
West	r		
East	?		

6. a. Complete the *time column* given that a runner wants to overtake a walker and the walker had a $\frac{1}{2}$-hour head start.

	r ·	t =	d
Runner		t	
Walker		?	

b. Complete the *time column* given that part of a 6-hour drive was in fog and the other part was in clear conditions

	r ·	t =	d
Foggy		t	
Clear		?	

7. A husband and wife drive in opposite directions to work. Their drives last the same amount of time and their workplaces are 80 miles apart. Complete the table and then form an equation for this distance problem.

	r ·	t =	d
Husband	35	t	
Wife	45		
			Total:

8. a. How many gallons of acetic acid are there in barrel 2?

b. Suppose the contents of the two barrels are poured into an empty third barrel. How many gallons of liquid will the third barrel contain?

c. Estimate the strength of the solution in the third barrel: 15%, 35%, or 60% acid?

Barrel 1
x gallons of a 20% acetic acid solution

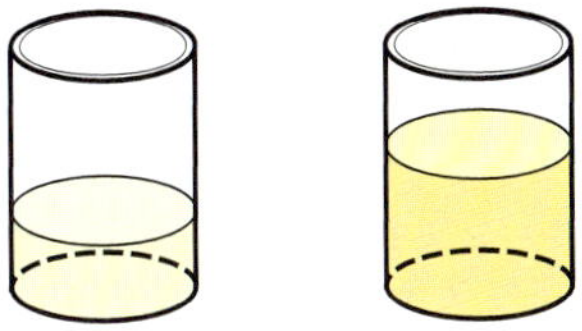

Barrel 2
42 gallons of a 40% acetic acid solution

9. a. Two antifreeze solutions are combined to form a mixture. Complete the table and then form an equation for this mixture problem.

	Amount ·	Strength =	Pure antifreeze
Strong	6	0.50	
Weak	x	0.25	
Mixture		0.30	

b. Two oil-and-vinegar salad dressings are combined to make a new mixture. Complete the table and then form an equation for this mixture problem.

	Amount ·	Strength =	Pure vinegar
Strong	x	0.06	
Weak		0.03	
Mixture	10	0.05	

10. The value of all the nylon brushes that a paint store carries is $670. Complete the table and then form an equation for this number-value problem.

	Number ·	Value =	Total value
1-inch	$2x$	4	
2-inch	x	5	
3-inch	$x + 10$	7	
			Total:

NOTATION

11. Write 6% and 15.2% in decimal form.

12. By what power of 10 should each decimal be multiplied to make it a whole number?

a. 0.08 **b.** 0.162

GUIDED PRACTICE

***Solve each equation.* See Example 1.**

13. $0.18x + 0.45(12 - x) = 0.36(12)$

14. $0.12x + 0.20(4 - x) = 0.6$

15. $0.08x + 0.07(15{,}000 - x) = 1{,}110$

16. $0.108x + 0.07(16{,}000 - x) = 1{,}500$

APPLICATIONS

Investment problems

17. CORPORATE INVESTMENTS The financial board of a corporation invested $25,000 overseas, part at 4% and part at 7% annual interest. Find the amount invested at each rate if the first-year combined income from the two investments was $1,300.

18. LOANS A credit union loaned out $50,000, part at an annual rate of 5% and the rest at an annual rate of 8%. They collected combined simple interest of $3,400 from the loans that year. How much was loaned out at each rate?

19. OLD COINS A salesperson used her $3,500 year-end bonus to purchase some old coins, with hopes of earning 15% annual interest on the gold coins and 12% annual interest on the silver coins. If she saw return on her investment of $480 the first year, how much did she invest in each type of coin?

20. HIGH-RISK COMPANIES An investment club used funds totaling $200,000 to invest in a bio-tech company and in an ethanol plant, with hopes of earning 11% and 14% annual interest, respectively. Their hunch paid off. The club made a total of $24,250 interest the first year. How much was invested at each rate?

21. RETIREMENT A professor wants to supplement her pension with investment interest. If she invests $28,000 at 6% interest, how much would she have to invest at 7% to achieve a goal of $3,500 per year in supplemental income?

22. EXTRA INCOME An investor wants to receive $1,000 annually from two investments. He has put $4,500 in a money market account paying 4% annual interest. How much should he invest in a stock fund that pays 10% annual interest to achieve his goal?

23. 1099 FORMS The form shows the interest income Terrell Washington earned in 2008 from two savings accounts. He deposited a total of $15,000 at the first of that year, and made no further deposits or withdrawals. How much money did he deposit in account 822 and in account 721?

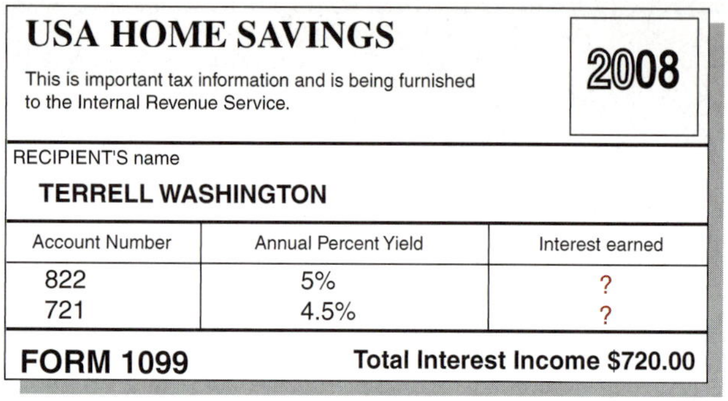
USA HOME SAVINGS

This is important tax information and is being furnished to the Internal Revenue Service.

2008

RECIPIENT'S name

TERRELL WASHINGTON

Account Number	Annual Percent Yield	Interest earned
822	5%	?
721	4.5%	?

FORM 1099 Total Interest Income $720.00

24. INVESTMENT PLANS A financial planner recommends a plan for a client who has $65,000 to invest. (See the chart.) At the end of the presentation, the client asks, "How much will be invested at each rate?" Answer this question using the given information.

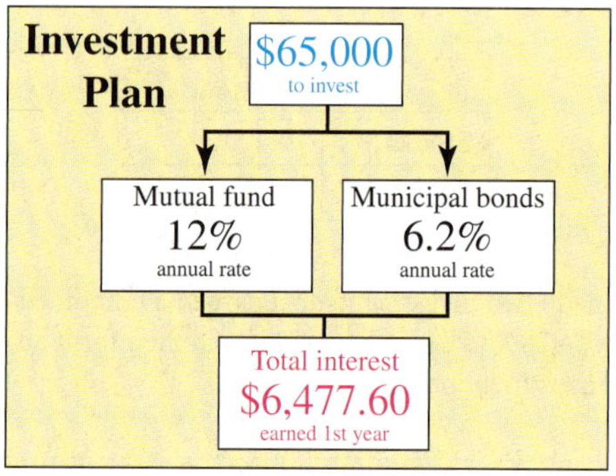

25. INVESTMENTS Equal amounts are invested in each of three accounts paying 7%, 8%, and 10.5% annually. If one year's combined interest income is $1,249.50, how much is invested in each account?

26. PERSONAL LOANS Maggy lent her brother some money at 2% annual interest. She lent her sister twice as much money at half of the interest rate. In one year, Maggy collected combined interest of $200 from her brother and sister. How much did she lend each of them?

Uniform motion problems

27. TORNADOES During a storm, two teams of scientists leave a university at the same time in vans to search for tornadoes. The first team travels east at 20 mph and the second travels west at 25 mph. If their radios have a range of up to 90 miles, how long will it be before they lose radio contact?

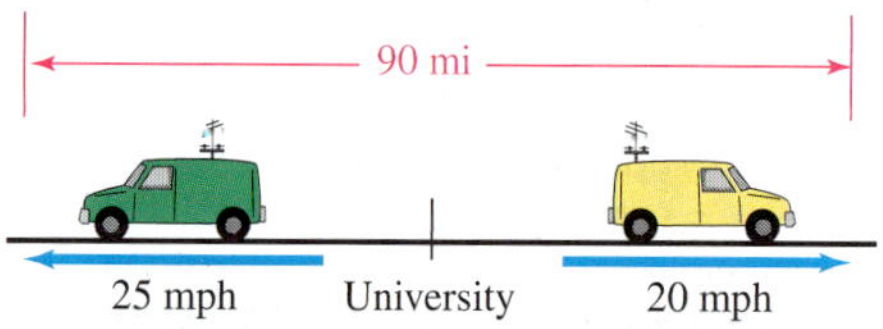

28. UNMANNED AIRCRAFT Two remotely controlled unmanned aircraft are launched in opposite directions. One flies east at 78 mph and the other west at 82 mph. How long will it take the aircraft to fly a combined distance of 560 miles?

29. HELLO/GOODBYE A husband and wife work different shifts at the same plant. When the husband leaves from work to make the 20-mile trip home, the wife leaves their home and drives to work. They travel on the same road. The husband's driving rate is 45 mph and the wife's is 35 mph. How long into their drives can they wave at each when passing on the road?

30. AIR TRAFFIC CONTROL An airliner leaves Berlin, Germany, headed for Montreal, Canada, flying at an average speed of 450 mph. At the same time, an airliner leaves Montreal headed for Berlin, averaging 500 mph. If the airports are 3,800 miles apart, when will the air traffic controllers have to make the pilots aware that the planes are passing each other?

31. CYCLING A cyclist leaves his training base for a morning workout, riding at the rate of 18 mph. One and one-half hours later, his support staff leaves the base in a car going 45 mph in the same direction. How long will it take the support staff to catch up with the cyclist?

32. PARENTING How long will it take a mother, running at 4 feet per second, to catch up with her toddler, running down the sidewalk at 2 feet per second, if the child had a 5-second head start?

33. ROAD TRIPS A car averaged 40 mph for part of a trip and 50 mph for the remainder. If the 5-hour trip covered 210 miles, for how long did the car average 40 mph?

34. CROSS-TRAINING An athlete runs up a set of stadium stairs at a rate of 2 stairs per second, immediately turns around, and then descends the same stairs at a rate of 3 stairs per second. If the workout takes 90 seconds, how long does it take him to run up the stairs?

35. WINTER DRIVING A trucker drove for 4 hours before he encountered icy road conditions. He reduced his speed by 20 mph and continued driving for 3 more hours. Find his average speed during the first part of the trip if the entire trip was 325 miles.

36. SPEED OF TRAINS Two trains are 330 miles apart, and their speeds differ by 20 mph. Find the speed of each train if they are traveling toward each other and will meet in 3 hours.

Liquid mixture problems

37. SALT SOLUTIONS How many gallons of a 3% salt solution must be mixed with 50 gallons of a 7% solution to obtain a 5% solution?

38. PHOTOGRAPHY A photographer wishes to mix 2 liters of a 5% acetic acid solution with a 10% solution to get a 7% solution. How many liters of 10% solution must be added?

39. MAKING CHEESE To make low-fat cottage cheese, milk containing 4% butterfat is mixed with milk containing 1% butterfat to obtain 15 gallons of a mixture containing 2% butterfat. How many gallons of each milk must be used?

40. ANTIFREEZE How many quarts of a 10% antifreeze solution must be mixed with 16 quarts of a 40% antifreeze solution to make a 30% solution?

41. PRINTING A printer has ink that is 8% cobalt blue color and ink that is 22% cobalt blue color. How many ounces of each ink are needed to make 1 gallon (64 ounces) of ink that is 15% cobalt blue color?

42. FLOOD DAMAGE One website recommends a 6% chlorine bleach-water solution to remove mildew. A chemical lab has 3% and 15% chlorine bleach-water solutions in stock. How many gallons of each should be mixed to obtain 100 gallons of the mildew spray?

43. INTERIOR DECORATING The colors on the paint chip card below are created by adding different amounts of orange tint to a white latex base. How many gallons of Desert Sunrise should be mixed with 1 gallon of Bright Pumpkin to obtain Cool Cantaloupe?

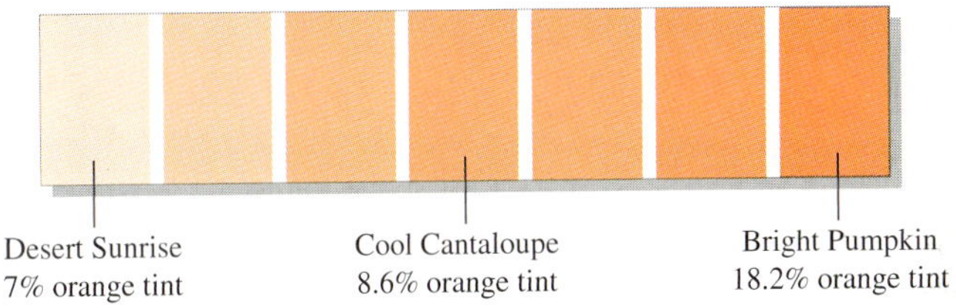

44. ANTISEPTICS A nurse wants to add water to 30 ounces of a 10% solution of benzalkonium chloride to dilute it to an 8% solution. How much water must she add? (*Hint:* Water is 0% benzalkonium chloride.)

Dry mixture problems

45. LAWN SEED A store sells bluegrass seed for $6 per pound and ryegrass seed for $3 per pound. How much ryegrass must be mixed with 100 pounds of bluegrass to obtain a blend that will sell for $5 per pound?

46. COFFEE BLENDS A store sells regular coffee for $8 a pound and gourmet coffee for $14 a pound. To get rid of 40 pounds of the gourmet coffee, a shopkeeper makes a blend to put on sale for $10 a pound. How many pounds of regular coffee should he use?

47. RAISINS How many scoops of natural seedless raisins costing $3.45 per scoop must be mixed with 20 scoops of golden seedless raisins costing $2.55 per scoop to obtain a mixture costing $3 per scoop?

48. FERTILIZER Fertilizer with weed control costing $38 per 50-pound bag is to be mixed with a less expensive fertilizer costing $6 per 50-pound bag to make 16 bags of fertilizer that can be sold for $28 per bag. How many bags of cheaper fertilizer should be used?

49. PACKAGED SALAD How many 10-ounce bags of Romaine lettuce must be mixed with fifty 10-ounce bags of Iceberg lettuce to obtain a blend that sells for \$2.50 per ten-ounce bag?

50. MIXING CANDY Lemon drops worth \$3.80 per pound are to be mixed with jelly beans that cost \$2.40 per pound to make 100 pounds of a mixture worth \$2.96 per pound. How many pounds of each candy should be used?

51. BRONZE A pound of tin is worth \$1 more than a pound of copper. Four pounds of tin are mixed with 6 pounds of copper to make bronze that sells for \$3.65 per pound. How much is a pound of tin worth?

52. SNACK FOODS A bag of peanuts is worth \$.30 less than a bag of cashews. Equal amounts of peanuts and cashews are used to make 40 bags of a mixture that sells for \$1.05 per bag. How much is a bag of cashews worth?

Number-value problems

53. RENTALS The owners of an apartment building rent equal numbers of 1-, 2-, and 3-bedroom units. The monthly rent for a 1-bedroom is \$550, a 2-bedroom is \$700, and a 3-bedroom is \$900. If the total monthly income is \$36,550, how many of each type of unit are there?

54. WAREHOUSING A store warehouses 40 more portables than big-screen TV sets, and 15 more consoles than big-screen sets. The monthly storage cost for a portable is \$1.50, a console is \$4.00, and a big-screen is \$7.50. If storage for all the televisions costs \$276 per month, how many big-screen sets are in stock?

55. SOFTWARE Three software applications are priced as shown. Spreadsheet and database programs sold in equal numbers, but 15 more word processing applications were sold than the other two combined. If the three applications generated sales of \$72,000, how many spreadsheets were sold?

Software	Price
Spreadsheet	**\$150**
Database	**\$195**
Word processing	**\$210**

56. INVENTORIES With summer approaching, the number of air conditioners sold is expected to be double that of stoves and refrigerators combined. Stoves sell for \$350, refrigerators for \$450, and air conditioners for \$500, and sales of \$56,000 are expected. If stoves and refrigerators sell in equal numbers, how many of each appliance should be stocked?

57. PIGGY BANKS When a child emptied his coin bank, he had a collection of pennies, nickels, and dimes. There were 20 more pennies than dimes and the number of nickels was triple the number of dimes. If the coins had a value of \$5.40, how many of each type coin were in the bank?

58. WISHING WELLS A scuba diver, hired by an amusement park, collected \$121 in nickels, dimes, and quarters at the bottom of a wishing well. There were 500 nickels, and 90 more quarters than dimes. How many quarters and dimes were thrown into the wishing well?

59. BASKETBALL Epiphanny Prince, of New York, scores 113 points in a high school game on February 1, 2006, breaking a national prep record that was held by Cheryl Miller. Prince made 46 more 2-point baskets than 3-point baskets, and only 1 free throw. How many 2-point and 3-point baskets did she make?

60. MUSEUM TOURS The admission prices for the Coca-Cola Museum in Atlanta are shown. A family purchased 3 more children's tickets than adult tickets, and 1 less senior ticket than adult tickets. The total cost of the tickets was \$73. How many of each type of ticket did they purchase?

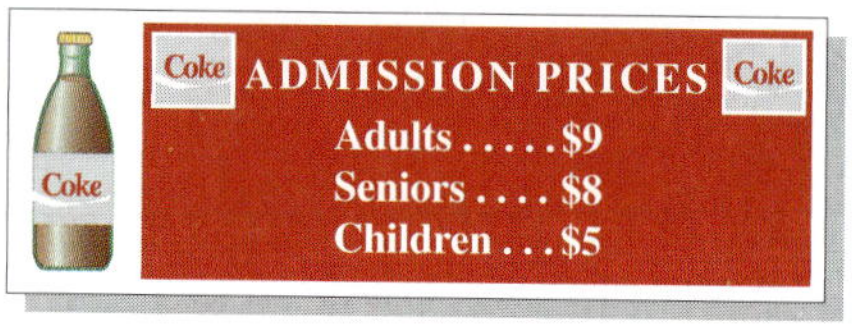

WRITING

61. Create a mixture problem of your own, and solve it.

62. Is it possible to mix a 10% sugar solution with a 20% sugar solution to get a 30% sugar solution? Explain.

REVIEW

Multiply.

63. $-25(2x - 5)$

64. $-12(3a + 4b - 32)$

65. $-(3x - 3)$

66. $\frac{1}{2}(4b - 8)$

67. $(4y - 4)4$

68. $3(5t + 1)2$

CHALLENGE PROBLEMS

69. EVAPORATION How much water must be boiled away to increase the concentration of 300 milliliters of a 2% salt solution to a 3% salt solution?

70. TESTING A teacher awarded 4 points for each correct answer and deducted 2 points for each incorrect answer when grading a 50-question true-false test. A student scored 56 points on the test and did not leave any questions unanswered. How many questions did the student answer correctly?

71. FINANCIAL PLANNING A plumber has a choice of two investment plans:

- An insured fund that pays 11% interest
- A risky investment that pays a 13% return

If the same amount invested at the higher rate would generate an extra $150 per year, how much does the plumber have to invest?

72. INVESTMENTS The amount of annual interest earned by $8,000 invested at a certain rate is $200 less than $12,000 would earn at a rate 1% lower. At what rate is the $8,000 invested?

SECTION 2.7 Solving Inequalities

Objectives

1. Determine whether a number is a solution of an inequality.
2. Graph solution sets and use interval notation.
3. Solve linear inequalities.
4. Solve compound inequalities.
5. Solve inequality applications.

In our daily lives, we often speak of one value being *greater than* or *less than* another. For example, a sick child might have a temperature *greater than* 98.6°F or a granola bar might contain *less than* 2 grams of fat.

In mathematics, we use *inequalities* to show that one expression is greater than or is less than another expression.

1 Determine Whether a Number is a Solution of an Inequality.

An **inequality** is a statement that contains one or more of the following symbols.

Inequality Symbols

$<$ is less than	$>$ is greater than	$\neq$ is not equal to
$\leq$ is less than or equal to	$\geq$ is greater than or equal to	

An inequality can be true, false, or neither true nor false. For example,

- $9 \geq 9$ is true because $9 = 9$.
- $37 < 24$ is false.
- $x + 1 > 5$ is neither true nor false because we don't know what number x represents.

The Language of Algebra

Because $<$ requires one number to be strictly less than another number and $>$ requires one number to be strictly greater than another number, $<$ and $>$ are called *strict inequalities*.

An inequality that contains a variable can be made true or false depending on the number that is substituted for the variable. If we substitute 10 for x in $x + 1 > 5$, the resulting inequality is true: $10 + 1 > 5$. If we substitute 1 for x, the resulting inequality is false: $1 + 1 > 5$. A number that makes an inequality true is called a **solution** of the inequality, and we say that the number *satisfies* the inequality. Thus, 10 is a solution of $x + 1 > 5$ and 1 is not.

In this section, we will find the solutions of *linear inequalities in one variable.*

Linear Inequality in One Variable

A **linear inequality in one variable** can be written in one of the following forms where a, b, and c are real numbers and $a \neq 0$.

$$ax + b > c \qquad ax + b \geq c \qquad ax + b < c \qquad ax + b \leq c$$

EXAMPLE 1 Is 9 a solution of $2x + 4 \leq 21$?

Strategy We will substitute 9 for x and evaluate the expression on the left side.

Why If a true statement results, 9 is a solution of the inequality. If we obtain a false statement, 9 is not a solution.

Solution

$$2x + 4 \leq 21$$

$$2(9) + 4 \overset{?}{\leq} 21 \quad \text{Substitute 9 for x. Read } \overset{?}{\leq} \text{ as "is possibly less than or equal to."}$$

$$18 + 4 \overset{?}{\leq} 21$$

$$22 \leq 21 \quad \text{This inequality is false.}$$

The statement $22 \leq 21$ is false because neither $22 < 21$ nor $22 = 21$ is true. Therefore, 9 is not a solution.

Self Check 1 Is 2 a solution of $3x - 1 \geq 0$?

Now Try Problem 13

2 Graph Solution Sets and Use Interval Notation.

The **solution set** of an inequality is the set of all numbers that make the inequality true. Some solution sets are easy to determine. For example, if we replace the variable in $x > -3$ with a number greater than -3, the resulting inequality will be true. Because there are infinitely many real numbers greater than -3, it follows that $x > -3$ has infinitely many solutions. Since there are too many solutions to list, we use **set-builder notation** to describe the solutions set.

$$\{x \mid x > -3\} \quad \text{Read as "the set of all x such that x is greater than −3."}$$

We can illustrate the solution set by **graphing the inequality** on a number line. To graph $x > -3$, a **parenthesis** or **open circle** is drawn on the endpoint -3 to indicate that -3 is not part of the graph. Then we shade all of the points on the number line to the right of -3. The right arrowhead is also shaded to show that the solutions continue forever to the right.

Notation
The parenthesis (opens in the direction of the shading and indicates that an endpoint is not included in the shaded interval.

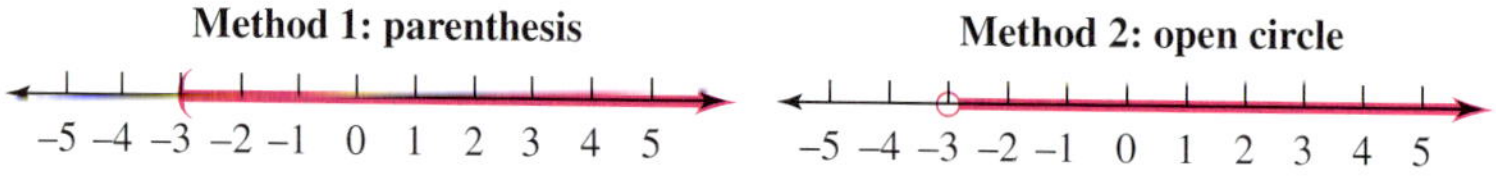

All real numbers greater than –3

The graph of $x > -3$ is an example of an **interval** on the number line. We can write intervals in a compact form called **interval notation.**

> **Notation**
> The *infinity* symbol ∞ does not represent a number. It indicates that an interval extends to the right without end.

The interval notation that represents the graph of $x > -3$ is $(-3, \infty)$. As on the number line, a left parenthesis is written next to -3 to indicate that -3 is not included in the interval. The **positive infinity symbol** ∞ that follows indicates that the interval continues without end to the right. With this notation, *a parenthesis is always used next to an infinity symbol.*

The illustration below shows the relationship between the symbols used to graph an interval and the corresponding interval notation. If we begin at -3 and move to the right, the shaded arrowhead on the graph indicates that the interval approaches positive infinity ∞.

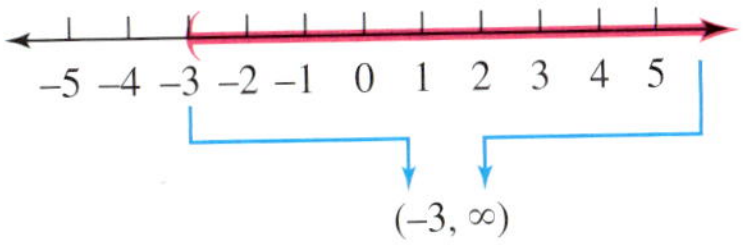

We now have three ways to describe the solution set of an inequality.

Set-builder notation	***Number line graph***	***Interval notation***
$\{x \mid x > -3\}$	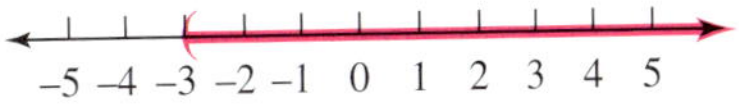	$(-3, \infty)$

EXAMPLE 2

Graph: $x \leq 2$

Strategy We need to determine which real numbers, when substituted for x, would make $x \leq 2$ a true statement.

Why To graph $x \leq 2$ means to draw a "picture" of all of the values of x that make the inequality true.

Solution If we replace x with a number less than or equal to 2, the resulting inequality will be true. To graph the solution set, a **bracket** or a **closed circle** is drawn at the endpoint 2 to indicate that 2 is part of the graph. Then we shade all of the points on the number line to the left of 2 and the left arrowhead.

> **Notation**
> The *bracket*] opens in the direction of the shading and indicates that an endpoint is included in the shaded interval.

Method 1: bracket

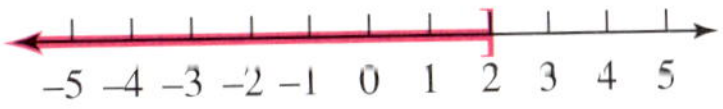

Method 2: closed circle

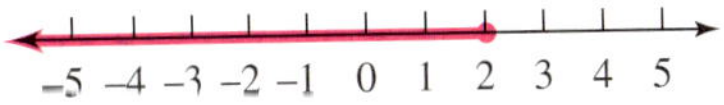

All real numbers less than or equal to 2

The interval is written as $(-\infty, 2]$. The right bracket indicates that 2 is included in the interval. The **negative infinity symbol** $-\infty$ shows that the interval continues forever to the left. The illustration below shows the relationship between the symbols used to graph the interval and the corresponding interval notation.

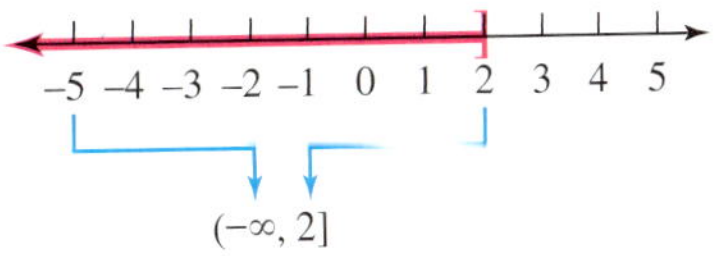

Self Check 2 Graph: $x \geq 0$

Now Try **Problem 17**

3 Solve Linear Inequalities.

To **solve an inequality** means to find all values of the variable that make the inequality true. As with equations, there are properties that we can use to solve inequalities.

Addition and Subtraction Properties of Inequality

Adding the same number to, or subtracting the same number from, both sides of an inequality does not change its solutions.

For any real numbers a, b, and c,

If $a < b$, then $a + c < b + c$.

If $a < b$, then $a - c < b - c$.

Similar statements can be made for the symbols $\leq$, $>$, and $\geq$.

After applying one of these properties, the resulting inequality is equivalent to the original one. **Equivalent inequalities** have the same solution set.

Like equations, inequalities are solved by isolating the variable on one side.

EXAMPLE 3

Solve $x + 3 > 2$. Write the solution set in interval notation and graph it.

Strategy We will use a property of inequality to isolate the variable on one side.

Why To solve the original inequality, we want to find a simpler equivalent inequality of the form **$x >$ a number** or **$x <$ a number**, whose solution is obvious.

Solution We will use the subtraction property of inequality to isolate x on the left side of the inequality. We can undo the addition of 3 by subtracting 3 from both sides.

$x + 3 > 2$ This is the inequality to solve.

$x + 3 - 3 > 2 - 3$ Subtract 3 from both sides.

$x > -1$

All real numbers greater than -1 are solutions of $x + 3 > 2$. The solution set can be written in set-builder notation as $\{x \mid x > -1\}$ and in interval notation as $(-1, \infty)$. The graph of the solution set is shown below.

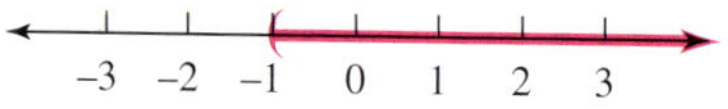

Since there are infinitely many solutions, we cannot check all of them.

As an informal check, we can pick some numbers in the graph, say 0 and 30, substitute each number for x in the original inequality, and see whether true statements result.

Check:

$x + 3 > 2$

$0 + 3 \overset{?}{>} 2$ Substitute 0 for x.

$3 > 2$ True.

$x + 3 > 2$

$30 + 3 \overset{?}{>} 2$ Substitute 30 for x.

$33 > 2$ True.

The solution set appears to be correct.

Success Tip

We solve linear inequalities by writing a series of steps that result in an equivalent inequality of the form

$x >$ *a number*

or

$x <$ *a number*

Similar statements apply to linear inequalities containing $\leq$ and $\geq$.

Notation

Since we use parentheses and brackets in interval notation, we will use them to graph inequalities. Note that parentheses, not brackets, are written next to ∞ and $-\infty$ because there is no endpoint.

$(-3, \infty)$ $(-\infty, 2]$

Self Check 3 Solve $x - 3 < -2$. Write the solution set in interval notation and graph it.

Now Try Problem 25

As with equations, there are properties for multiplying and dividing both sides of an inequality by the same number. To develop what is called *the multiplication property of inequality,* we consider the true statement $2 < 5$. If both sides are multiplied by a positive number, such as 3, another true inequality results.

$$2 < 5 \quad \text{This inequality is true.}$$
$$3 \cdot 2 < 3 \cdot 5 \quad \text{Multiply both sides by 3.}$$
$$6 < 15 \quad \text{This inequality is true.}$$

However, if we multiply both sides of $2 < 5$ by a negative number, such as -3, the direction of the inequality symbol is reversed to produce another true inequality.

$$2 < 5 \quad \text{This inequality is true.}$$
$$-3 \cdot 2 > -3 \cdot 5 \quad \text{Multiply both sides by } -3 \text{ and reverse the direction of the inequality.}$$
$$-6 > -15 \quad \text{This inequality is true.}$$

The inequality $-6 > -15$ is true because -6 is to the right of -15 on the number line.

Dividing both sides of an inequality by the same negative number also requires that the direction of the inequality symbol be reversed.

$$-4 < 6 \quad \text{This inequality is true.}$$
$$\frac{-4}{-2} > \frac{6}{-2} \quad \text{Divide both sides by } -2 \text{ and change } < \text{ to } >.$$
$$2 > -3 \quad \text{This inequality is true.}$$

These examples illustrate the multiplication and division properties of inequality.

Caution

If the inequality symbol is not reversed when both sides of a true inequality are multiplied by a negative number, the result is a false inequality. For example,

$$3 < 6 \quad \text{True}$$
$$-3 \cdot 3 < -3 \cdot 6$$
$$-9 < -18 \quad \text{False}$$

The same is true if we divide both sides of a true inequality by a negative number. Divide both sides of $3 < 6$ by -3 to see for yourself.

Multiplication and Division Properties of Inequality

Multiplying or dividing both sides of an inequality by the same positive number does not change its solutions.

For any real numbers a, b, and c, where c is positive,

$$\text{If } a < b, \text{ then } ac < bc. \qquad \text{If } a < b, \text{ then } \frac{a}{c} < \frac{b}{c}.$$

If we multiply or divide both sides of an inequality by a negative number, the direction of the inequality symbol must be reversed for the inequalities to have the same solutions.

For any real numbers a, b, and c, where c is negative,

$$\text{If } a < b, \text{ then } ac > bc. \qquad \text{If } a < b, \text{ then } \frac{a}{c} > \frac{b}{c}.$$

Similar statements can be made for the symbols $\leq$, $>$, and $\geq$.

EXAMPLE 4 Solve each inequality. Write the solution set in interval notation and graph it. **a.** $-\frac{3}{2}t \geq -12$ **b.** $-5t < 55$

Strategy We will use a property of inequality to isolate the variable on one side.

Why To solve the original inequality, we want to find a simpler equivalent inequality, whose solution is obvious.

Solution

a. To undo the multiplication by $-\frac{3}{2}$, we multiply both sides by the reciprocal, which is $-\frac{2}{3}$.

$$-\frac{3}{2}t \geq -12 \quad \text{This is the inequality to solve.}$$

$$-\frac{2}{3}\left(-\frac{3}{2}t\right) \leq -\frac{2}{3}(-12) \quad \text{Multiply both sides by } -\tfrac{2}{3}\text{. Since we are multiplying both sides by a negative number, reverse the direction of the inequality.}$$

$$t \leq 8 \quad \text{Do the multiplications.}$$

The solution set is $(-\infty, 8]$ and it is graphed as shown.

b. To undo the multiplication by -5, we divide both sides by -5.

$$-5t < 55 \quad \text{This is the inequality to solve.}$$

$$\frac{-5t}{-5} > \frac{55}{-5} \quad \text{To isolate } t\text{, undo the multiplication by } -5 \text{ by dividing both sides by } -5\text{. Since we are dividing both sides by a negative number, reverse the direction of the inequality.}$$

$$t > -11$$

The solution set is $(-11, \infty)$ and it is graphed as shown.

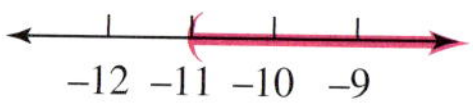

Self Check 4 Solve each inequality. Write the solution set in interval notation and graph it. **a.** $-\frac{h}{20} \leq 10$ **b.** $-12a > -144$

Now Try Problems 31 and 33

EXAMPLE 5

Solve $-5 > 3x + 7$. Write the solution set in interval notation and graph it.

Strategy First we will use a property of inequality to isolate the *variable term* on one side. Then we will use a second property of inequality to isolate the *variable* itself.

Why To solve the original inequality, we want to find a simpler equivalent inequality of the form **$x >$ a number** or **$x <$ a number**, whose solution is obvious.

Solution

$$-5 > 3x + 7 \quad \text{This is the inequality to solve.}$$

$$-5 - 7 > 3x + 7 - 7 \quad \text{To isolate the variable term, } 3x\text{, undo the addition of 7 by subtracting 7 from both sides.}$$

$$-12 > 3x \quad \text{Do the subtractions.}$$

$$\frac{-12}{3} > \frac{3x}{3} \quad \text{To isolate } x\text{, undo the multiplication by 3 by dividing both sides by 3.}$$

$$-4 > x \quad \text{Do the divisions.}$$

Caution

Don't be confused by the negative number on the left side. We didn't reverse the > symbol because we divided both sides by *positive* 3.

$$\frac{-12}{3} > \frac{3x}{3}$$

To determine the solution set, it is useful to rewrite the inequality $-4 > x$ in an equivalent form with the variable on the left side. If -4 is greater than x, it follows that x must be less than -4.

$$x < -4$$

The solution set is $(-\infty, -4)$ whose graph is shown below.

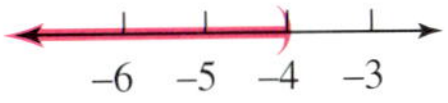

Self Check 5 Solve $-13 < 2r - 7$. Write the solution set in interval notation and graph it.

Now Try **Problem 39**

EXAMPLE 6 Solve $5.1 - 3k < 19.5$. Write the solution set in interval notation and graph it.

Strategy We will use properties of inequality to isolate the variable on one side.

Why To solve the original inequality, we want to find a simpler equivalent inequality of the form **$k >$ a number** or **$k <$ a number**, whose solution is obvious.

Solution

$5.1 - 3k < 19.5$ — This is the inequality to solve.

$5.1 - 3k - 5.1 < 19.5 - 5.1$ — To isolate $-3k$ on the left side, subtract 5.1 from both sides.

$-3k < 14.4$ — Do the subtractions.

$\frac{-3k}{-3} > \frac{14.4}{-3}$ — To isolate k, undo the multiplication by -3 by dividing both sides by -3 and reverse the direction of the $<$ symbol.

$k > 4.8$ — Do the divisions.

The solution set is $(-4.8, \infty)$, whose graph is shown below.

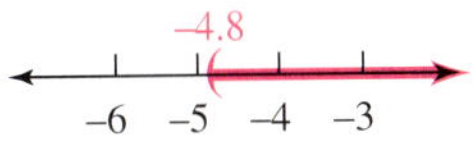

Self Check 6 Solve $-9n + 1.8 > -17.1$. Write the solution set in interval notation and graph it.

Now Try **Problem 47**

The equation solving strategy on page 125 can be applied to inequalities. However, when solving inequalities, we must remember to *change the direction of the inequality symbol when multiplying or dividing both sides by a negative number.*

EXAMPLE 7 Solve $8(y + 1) \geq 2(y - 4) + y$. Write the solution set in interval notation and graph it.

Strategy We will follow the steps of the equation solving strategy (adapted to inequalities) to solve the inequality.

Why This is the most efficient way to solve a linear inequality in one variable.

Solution

$8(y + 1) \geq 2(y - 4) + y$ — This is the inequality to solve.

$8y + 8 \geq 2y - 8 + y$ — Distribute the multiplication by 8 and by 2.

$8y + 8 \geq 3y - 8$ — Combine like terms: $2y + y = 3y$.

$8y + 8 - 3y \geq 3y - 8 - 3y$ — To eliminate 3y from the right side, subtract 3y from both sides.

$5y + 8 \geq -8$ — Combine like terms on both sides.

$5y + 8 - 8 \geq -8 - 8$ — To isolate 5y, undo the addition of 8 by subtracting 8 from both sides.

$5y \geq -16$ — Do the subtractions.

$\frac{5y}{5} \geq \frac{-16}{5}$ — To isolate y, undo the multiplication by 5 by dividing both sides by 5. Do not reverse the direction of the $\geq$ symbol.

$y \geq -\frac{16}{5}$

The solution set is $\left[-\frac{16}{5}, \infty\right)$. To graph it, we note that $-\frac{16}{5} = -3\frac{1}{5}$.

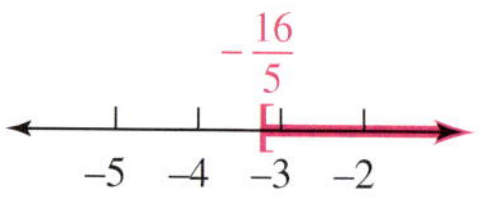

Success Tip

As an informal check, substitute a number on the graph that is shaded, such as 0, into $8(y + 1) \geq 2(y - 4) + y$. A true statement should result. Then substitute a number on the graph that is not shaded, such as -4, into the inequality. A false statement should result.

Self Check 7 Solve $5(b - 2) \geq -(b - 3) + 2b$. Write the solution set in interval notation and graph it.

Now Try **Problem 53**

4 Solve Compound Inequalities.

The Language of Algebra

The word *compound* means made up of two or more parts. For example, a *compound* inequality has three parts. Other examples are: a *compound* sentence, a *compound* fracture, and a chemical *compound*.

Two inequalities can be combined into a **compound inequality** to show that an expression lies between two fixed values. For example, $-2 < x < 3$ is a combination of

$$-2 < x \quad \text{and} \quad x < 3$$

It indicates that x is greater than -2 and that x is also less than 3. The solution set of $-2 < x < 3$ consists of all numbers that lie between -2 and 3, and we write it as $(-2, 3)$. The graph of the compound inequality is shown below.

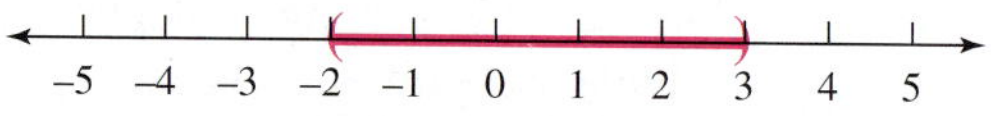

EXAMPLE 8 Graph: $-4 \le x < 0$

Strategy We need to determine which real numbers, when substituted for x, would make $-4 \le x < 0$ a true statement.

Why To graph $-4 \le x < 0$ means to draw a "picture" of all of the values of x that make the compound inequality true.

Solution If we replace the variable in $-4 \le x < 0$ with a number between -4 and 0, including -4, the resulting compound inequality will be true. Therefore, the solution set is the interval $[-4, 0)$. To graph the interval, we draw a bracket at -4, a parenthesis at 0, and shade in between.

Notation

Note that the two inequality symbols in $-4 \le x < 0$ point in the same direction and point to the smaller number.

To check, we pick a number in the graph, such as -2, and see whether it satisfies the inequality. Since $-4 \le -2 < 0$ is true, the answer appears to be correct.

Self Check 8 Graph $-2 \le x < 1$ and write the solution set in interval notation.

Now Try **Problem 61**

To solve compound inequalities, we isolate the variable in the middle part of the inequality. To do this, we apply the properties of inequality to all *three* parts of the inequality.

EXAMPLE 9 Solve $-4 < 2(x - 1) \le 4$. Write the solution set in interval notation and graph it.

Strategy We will use properties of inequality to isolate the variable by itself as the middle part of the inequality.

Why To solve the original inequality, we want to find a simpler equivalent inequality of the form **a number $<$ x $\le$ a number**, whose solution is obvious.

Solution

$-4 < 2(x - 1) \le 4$	This is the compound inequality to solve.
$-4 < 2x - 2 \le 4$	Distribute the multiplication by 2.
$-4 + 2 < 2x - 2 + 2 \le 4 + 2$	To isolate $2x$, undo the subtraction of 2 by adding 2 to all three parts.
$-2 < 2x \le 6$	Do the additions.
$\frac{-2}{2} < \frac{2x}{2} \le \frac{6}{2}$	To isolate x, we undo the multiplication by 2 by dividing all three parts by 2.
$-1 < x \le 3$	Do the divisions.

Success Tip

Think of interval notation as a way to tell someone how to draw the graph, from left to right, giving them only a "start" and a "stop" instruction.

The solution set is $(-1, 3]$ and its graph is shown.

Self Check 9 Solve $-6 \le 3(t + 2) \le 6$. Write the solution set in interval notation and graph it.

Now Try **Problem 69**

5 Solve Inequality Applications.

When solving problems, phrases such as "not more than," or "should exceed" suggest that the problem involves an inequality rather than an equation.

The Language of Algebra
Some phrases that suggest an inequality are:

surpass: $>$	*at least:* $\ge$
not exceed: $\le$	*at most:* $\le$

between: $\square < \square < \square$

EXAMPLE 10 ***Grades.*** A student has scores of 72%, 74%, and 78% on three exams. What percent score does he need on the last exam to earn a grade of no less than B (80%)?

Analyze the Problem We know three scores. We are to find what the student must score on the last exam to earn a grade of B or higher.

Form an Inequality We can let x = the score on the fourth (and last) exam. To find the average grade, we add the four scores and divide by 4. To earn a grade of *no less than* B, the student's average must be *greater than or equal to* 80%.

The average of the four grades	must be no less than	80.
$\dfrac{72 + 74 + 78 + x}{4}$	$\ge$	80

Solve the Inequality

$$\frac{224 + x}{4} \ge 80 \quad \text{Combine like terms in the numerator: } 72 + 74 + 78 = 224.$$

$$4\left(\frac{224 + x}{4}\right) \ge 4(80) \quad \text{To clear the inequality of the fraction, multiply both sides by 4.}$$

$$224 + x \ge 320 \quad \text{Simplify each side.}$$

$$x \ge 96 \quad \text{To isolate } x\text{, undo the addition of 224 by subtracting 224 from both sides.}$$

State the Conclusion To earn a B, the student must score 96% or better on the last exam. Assuming the student cannot score higher than 100% on the exam, the solution set is written as [96, 100]. The graph is shown below.

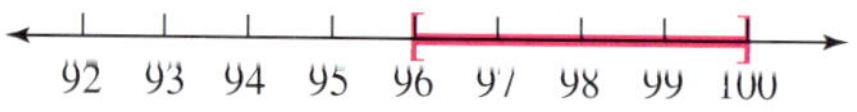

Check the Result Pick some numbers in the interval, and verify that the average of the four scores will be 80% or greater.

ANSWERS TO SELF CHECKS

1. Yes **2.** [0, ∞)
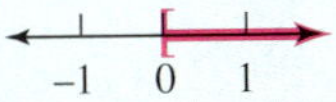

3. (−∞, 1)
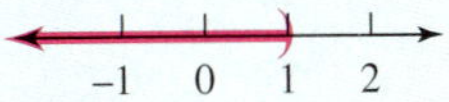

4. a. [−200, ∞)
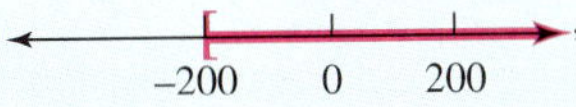

b. (−∞, 12)
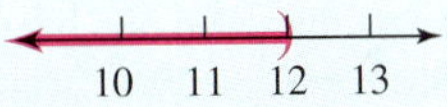

5. (−3, ∞)
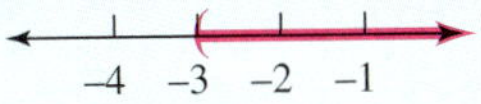

6. (−∞, 2.1)
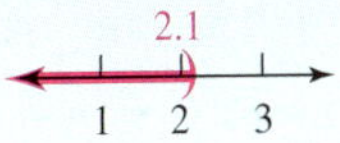

7. $\left[\frac{13}{4}, \infty\right)$
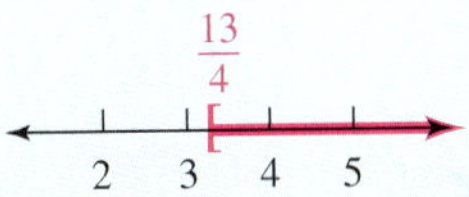

8. [−2, 1)
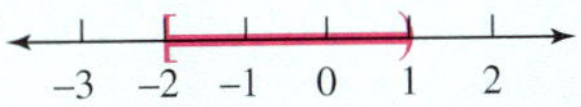

9. [−4, 0]
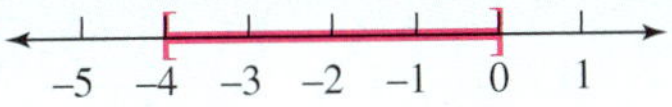

STUDY SET 2.7

VOCABULARY

Fill in the blanks.

1. An __________ is a statement that contains one of the symbols: $>$, $\geq$, $<$, or $\leq$.

2. To ______ an inequality means to find all the values of the variable that make the inequality true.

3. The solution set of $x > 2$ can be expressed in ________ notation as $(2, \infty)$.

4. The inequality $-4 < x \leq 10$ is an example of a __________ inequality.

CONCEPTS

Fill in the blanks.

5. a. Adding the ______ number to both sides of an inequality does not change the solutions.

b. Multiplying or dividing both sides of an inequality by the same ________ number does not change the solutions.

c. If we multiply or divide both sides of an inequality by a ________ number, the direction of the inequality symbol must be reversed for the inequalities to have the same solutions.

6. To solve $-4 \leq 2x + 1 < 3$, properties of inequality are applied to all ______ parts of the inequality.

7. Rewrite the inequality $32 < x$ in an equivalent form with the variable on the left side.

8. The solution set of an inequality is graphed below. Which of the four numbers, 3, −3, 2, and 4.5, when substituted for the variable in that inequality, would make it true?

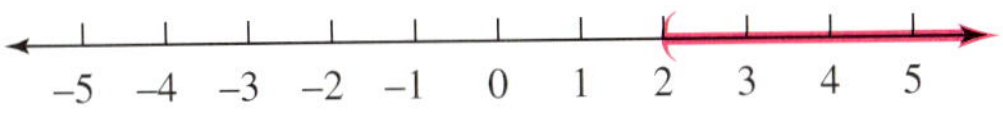

NOTATION

9. Write each symbol.

a. is less than or equal to **b.** infinity

c. bracket **d.** is greater than

10. Consider the graph of the interval [4, 8).

a. Is the endpoint 4 included or not included in the graph?

b. Is the endpoint 8 included or not included in the graph?

Complete the solution to solve each inequality.

11.
$$4x - 5 \geq 7$$
$$4x - 5 + \quad \geq 7 + $$
$$4x \geq $$
$$\frac{4x}{\quad} \geq \frac{12}{\quad}$$
$$x \geq 3 \qquad \text{Solution set: } [\quad, \infty)$$

12.
$$-6x > 12$$
$$\frac{-6x}{\quad} \quad \frac{12}{-6}$$
$$x < \qquad \text{Solution set: } (\quad, -2)$$

GUIDED PRACTICE

See Example 1.

13. Determine whether each number is a solution of $3x - 2 > 5$.

a. 5 **b.** −4

14. Determine whether each number is a solution of $3x + 7 < 4x - 2$.

a. 12 **b.** 9

15. Determine whether each number is a solution of $-5(x - 1) \geq 2x + 12$.

a. 1 **b.** -1

16. Determine whether each number is a solution of $\frac{4}{5}a \geq -2$.

a. $-\dfrac{5}{4}$ **b.** -15

Graph each inequality and describe the graph using interval notation. See Example 2.

17. $x < 5$ **18.** $x \geq -2$

19. $-3 < x \leq 1$ **20.** $-4 \leq x \leq 2$

Write the inequality that is represented by each graph. Then describe the graph using interval notation.

21. 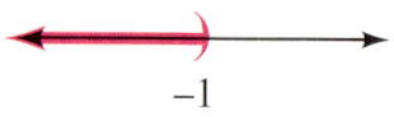

22. 

23. 

24. 

Solve each inequality. Write the solution set in interval notation and graph it. See Examples 3–4.

25. $x + 2 > 5$ **26.** $x + 5 \geq 2$

27. $g - 30 \geq -20$ **28.** $h - 18 \leq -3$

29. $8h < 48$ **30.** $2t > 22$

31. $-\dfrac{3}{16}x \geq -9$ **32.** $-\dfrac{7}{8}x \leq 21$

33. $-3y \leq -6$ **34.** $-6y \geq -6$

35. $\dfrac{2}{3}x \geq 2$ **36.** $\dfrac{3}{4}x < 3$

Solve each inequality. Write the solution set in interval notation and graph it. See Examples 5–6.

37. $9x + 1 > 64$ **38.** $4x + 8 < 32$

39. $0.5 \geq 2x - 0.3$ **40.** $0.8 > 7x - 0.04$

41. $\dfrac{x}{8} - (-9) \geq 11$ **42.** $\dfrac{x}{6} - (-12) > 14$

43. $\dfrac{m}{-42} - 1 > -1$ **44.** $\dfrac{a}{-25} + 3 < 3$

45. $-x - 3 \leq 7$ **46.** $-x - 9 > 3$

47. $-3x - 7 > -1$ **48.** $-5x + 7 \leq 12$

Solve each inequality. Write the solution set in interval notation and graph it. See Example 7.

49. $9a + 4 > 5a - 16$ **50.** $8t + 1 < 4t - 19$

51. $0.4x \leq 0.1x + 0.45$ **52.** $0.9s \leq 0.3s + 0.54$

53. $8(5 - x) \leq 10(8 - x)$ **54.** $17(3 - x) \geq 3 - 13x$

55. $8x + 4 > -(3x - 4)$ **56.** $7x + 6 \geq -(x - 6)$

57. $\dfrac{1}{2} + \dfrac{n}{5} > \dfrac{3}{4}$ **58.** $\dfrac{1}{3} + \dfrac{c}{5} > -\dfrac{3}{2}$

59. $\dfrac{6x + 1}{4} \leq x + 1$ **60.** $\dfrac{3x - 10}{5} \leq x + 4$

Solve each compound inequality. Write the solution set in interval notation and graph it. See Examples 8–9.

61. $2 < x - 5 < 5$ **62.** $-8 < t - 8 < 8$

63. $0 \leq x + 10 \leq 10$ **64.** $-9 \leq x + 8 < 1$

65. $-3 \leq \dfrac{c}{2} \leq 5$ **66.** $-12 < \dfrac{b}{3} < 0$

67. $3 \leq 2x - 1 < 5$ **68.** $4 < 3x - 5 \leq 7$

69. $-9 < 6x + 9 \leq 45$ **70.** $-30 \leq 10d + 20 < 90$

71. $6 < -2(x - 1) < 12$ **72.** $4 \leq -4(x - 2) < 20$

TRY IT YOURSELF

Solve each inequality or compound inequality. Write the solution set in interval notation and graph it.

73. $6 - x \leq 3(x - 1)$ **74.** $3(3 - x) \geq 6 + x$

75. $\dfrac{y}{4} + 1 \leq -9$ **76.** $\dfrac{r}{8} - 7 \geq -8$

77. $0 < 5(x + 2) \leq 15$ **78.** $-18 \leq 9(x - 5) < 27$

79. $-1 \leq -\dfrac{1}{2}n$ **80.** $-3 \geq -\dfrac{1}{3}t$

81. $-m - 12 > 15$ **82.** $-t + 5 < 10$

83. $-\frac{2}{3} \geq \frac{2y}{3} - \frac{3}{4}$

84. $-\frac{2}{9} \geq \frac{5x}{6} - \frac{1}{3}$

85. $9x + 13 \geq 2x + 6x$

86. $7x - 16 < 2x + 4x$

87. $7 < \frac{5}{3}a + (-3)$

88. $5 < \frac{7}{2}a + (-9)$

89. $-8 \leq \frac{y}{8} - 4 \leq 2$

90. $6 < \frac{m}{16} + 7 < 8$

91. $-2(2x - 3) > 17$

92. $-3(x + 0.2) < 0.3$

93. $\frac{5}{3}(x + 1) \geq -x + \frac{2}{3}$

94. $\frac{5}{2}(7x - 15) \geq \frac{11}{2}x - \frac{3}{2}$

95. $2x + 9 \leq x + 8$

96. $3x + 7 \leq 4x - 2$

97. $-7x + 1 < -5$

98. $-3x - 10 \geq -5$

APPLICATIONS

99. GRADES A student has test scores of 68%, 75%, and 79% in a government class. What must she score on the last exam to earn a B (80% or better) in the course?

100. OCCUPATIONAL TESTING An employment agency requires applicants average at least 70% on a battery of four job skills tests. If an applicant scored 70%, 74%, and 84% on the first three exams, what must he score on the fourth test to maintain a 70% or better average?

101. GAS MILEAGE A car manufacturer produces three models in equal quantities. One model has an economy rating of 17 miles per gallon, and the second model is rated for 19 mpg. If government regulations require the manufacturer to have a fleet average that exceeds 21 mpg, what economy rating is required for the third model?

102. SERVICE CHARGES When the average daily balance of a customer's checking account falls below \$500 in any week, the bank assesses a \$5 service charge. The table shows the daily balances of one customer. What must Friday's balance be to avoid the service charge?

Day	Balance
Monday	\$540.00
Tuesday	\$435.50
Wednesday	\$345.30
Thursday	\$310.00

103. GEOMETRY The perimeter of an equilateral triangle is at most 57 feet. What could the length of a side be? (*Hint:* All three sides of an equilateral triangle are equal.)

104. GEOMETRY The perimeter of a square is no less than 68 centimeters. How long can a side be?

105. COUNTER SPACE A rectangular counter is being built for the customer service department of a store. Designers have determined that the outside perimeter of the counter (shown in red) needs to exceed 30 feet. Determine the acceptable values for x.

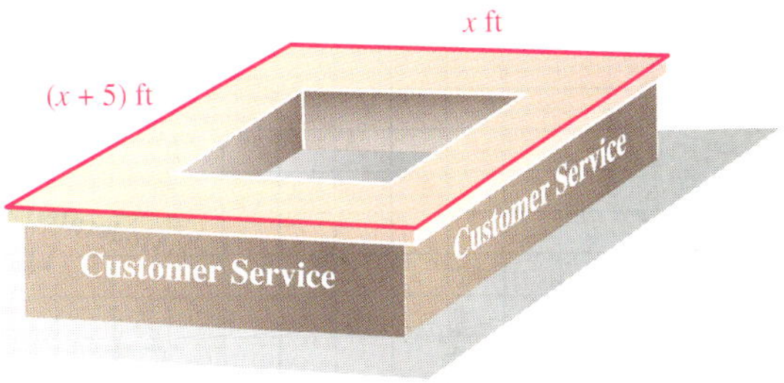

106. NUMBER PUZZLES What numbers satisfy the condition: Four more than three times the number is at most 10?

107. GRADUATIONS It costs a student \$18 to rent a cap and gown and 80 cents for each graduation announcement that she orders. If she doesn't want her spending on these graduation costs to exceed \$50, how many announcements can she order?

108. TELEPHONES A cellular telephone company has currently enrolled 36,000 customers in a new calling plan. If an average of 1,200 people are signing up for the plan each day, in how many days will the company surpass their goal of having 150,000 customers enrolled?

109. WINDOWS An architect needs to design a triangular-shaped bathroom window that has an area no greater than 100 in.^{2}. If the base of the window must be 16 inches long, what window heights will meet this condition?

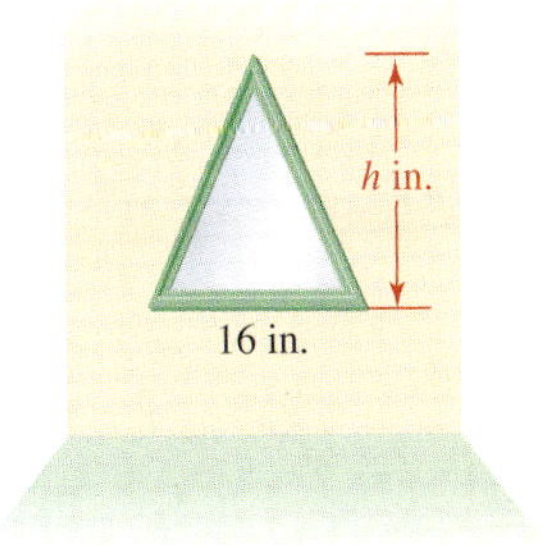

110. ROOM TEMPERATURES To hold the temperature of a room between 19° and 22° Celsius, what Fahrenheit temperatures must be maintained? *Hint:* Use the formula $C = \frac{5}{9}(F - 32)$.

SECTION 2.2 More about Solving Equations

DEFINITIONS AND CONCEPTS	EXAMPLES
A five-step **strategy for solving linear equations:** 1. *Clear* the equation of fractions or decimals. 2. *Simplify* each side. Use the distributive property and combine like terms when necessary. 3. *Isolate the variable term.* Use the addition and subtraction properties of equality. 4. *Isolate the variable.* Use the multiplication and division properties of equality. 5. *Check* the result in the original equation.	Solve: $2(y + 2) + 4y = 11 - y$ $2y + 4 + 4y = 11 - y$ Distribute the multiplication by 2. $6y + 4 = 11 - y$ Combine like terms: $2y + 4y = 6y$. $6y + 4 + y = 11 - y + y$ To eliminate $-y$ on the right, add y to both sides. $7y + 4 = 11$ Combine like terms. $7y + 4 - 4 = 11 - 4$ To isolate the variable term $7y$, subtract 4 from both sides. $7y = 7$ Simplify each side of the equation. $\frac{7y}{7} = \frac{7}{7}$ To isolate y, divide both sides by 7. $y = 1$
To clear an equation of fractions, multiply both sides of an equation by the LCD.	To solve $\frac{1}{2} + \frac{x}{3} = \frac{3}{4}$, first multiply both sides by 12: $12\left(\frac{1}{2} + \frac{x}{3}\right) = 12\left(\frac{3}{4}\right)$
To clear an equation of decimals, multiply both sides by a power of 10 to change the decimals in the equation to integers.	To solve $0.5(x - 4) = 0.1x + 0.2$, first multiply both sides by 10: $10[0.5(x - 4)] = 10(0.1x + 0.2)$
An equation that is true for all values of its variable is called an **identity.** An equation that is not true for any value of its variable is called a **contradiction.**	When we solve $x + 5 + x = 2x + 5$, the variables drop out and we obtain a true statement $5 = 5$. All real numbers are solutions. When we solve $y + 2 = y$, the variables drop out and we obtain a false statement $2 = 0$. The equation has no solutions.

REVIEW EXERCISES

Solve each equation. Check the result.

19. $5x + 4 = 14$

20. $98.6 - t = 129.2$

21. $\frac{n}{5} + (-2) = 4$

22. $\frac{b - 5}{4} = -6$

23. $5(2x - 4) - 5x = 0$

24. $-2(x - 5) = 5(-3x + 4) + 3$

25. $\frac{3}{4} = \frac{1}{2} + \frac{d}{5}$

26. $\frac{5(7 - x)}{4} = 2x - 3$

27. $\frac{3(2 - c)}{2} = \frac{-2(2c + 3)}{5}$

28. $\frac{b}{3} + \frac{11}{9} + 3b = -\frac{5}{6}b$

29. $0.15(x + 2) + 0.3 = 0.35x - 0.4$

30. $0.5 - 0.02(y - 2) = 0.16 + 0.36y$

31. $3(a + 8) = 6(a + 4) - 3a$

32. $2(y + 10) + y = 3(y + 8)$

83. $-\frac{2}{3} \geq \frac{2y}{3} - \frac{3}{4}$

84. $-\frac{2}{9} \geq \frac{5x}{6} - \frac{1}{3}$

85. $9x + 13 \geq 2x + 6x$

86. $7x - 16 < 2x + 4x$

87. $7 < \frac{5}{3}a + (-3)$

88. $5 < \frac{7}{2}a + (-9)$

89. $-8 \leq \frac{y}{8} - 4 \leq 2$

90. $6 < \frac{m}{16} + 7 < 8$

91. $-2(2x - 3) > 17$

92. $-3(x + 0.2) < 0.3$

93. $\frac{5}{3}(x + 1) \geq -x + \frac{2}{3}$

94. $\frac{5}{2}(7x - 15) \geq \frac{11}{2}x - \frac{3}{2}$

95. $2x + 9 \leq x + 8$

96. $3x + 7 \leq 4x - 2$

97. $-7x + 1 < -5$

98. $-3x - 10 \geq -5$

APPLICATIONS

99. GRADES A student has test scores of 68%, 75%, and 79% in a government class. What must she score on the last exam to earn a B (80% or better) in the course?

100. OCCUPATIONAL TESTING An employment agency requires applicants average at least 70% on a battery of four job skills tests. If an applicant scored 70%, 74%, and 84% on the first three exams, what must he score on the fourth test to maintain a 70% or better average?

101. GAS MILEAGE A car manufacturer produces three models in equal quantities. One model has an economy rating of 17 miles per gallon, and the second model is rated for 19 mpg. If government regulations require the manufacturer to have a fleet average that exceeds 21 mpg, what economy rating is required for the third model?

102. SERVICE CHARGES When the average daily balance of a customer's checking account falls below \$500 in any week, the bank assesses a \$5 service charge. The table shows the daily balances of one customer. What must Friday's balance be to avoid the service charge?

Day	Balance
Monday	\$540.00
Tuesday	\$435.50
Wednesday	\$345.30
Thursday	\$310.00

103. GEOMETRY The perimeter of an equilateral triangle is at most 57 feet. What could the length of a side be? (*Hint:* All three sides of an equilateral triangle are equal.)

104. GEOMETRY The perimeter of a square is no less than 68 centimeters. How long can a side be?

105. COUNTER SPACE A rectangular counter is being built for the customer service department of a store. Designers have determined that the outside perimeter of the counter (shown in red) needs to exceed 30 feet. Determine the acceptable values for x.

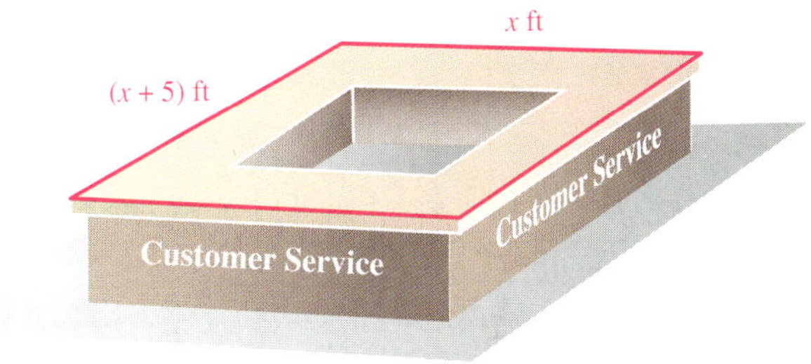

106. NUMBER PUZZLES What numbers satisfy the condition: Four more than three times the number is at most 10?

107. GRADUATIONS It costs a student \$18 to rent a cap and gown and 80 cents for each graduation announcement that she orders. If she doesn't want her spending on these graduation costs to exceed \$50, how many announcements can she order?

108. TELEPHONES A cellular telephone company has currently enrolled 36,000 customers in a new calling plan. If an average of 1,200 people are signing up for the plan each day, in how many days will the company surpass their goal of having 150,000 customers enrolled?

109. WINDOWS An architect needs to design a triangular-shaped bathroom window that has an area no greater than 100 in.^{2}. If the base of the window must be 16 inches long, what window heights will meet this condition?

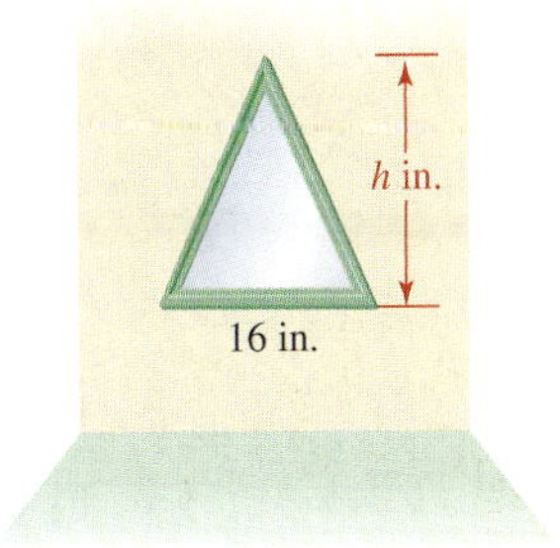

110. ROOM TEMPERATURES To hold the temperature of a room between 19° and 22° Celsius, what Fahrenheit temperatures must be maintained? *Hint:* Use the formula $C = \frac{5}{9}(F - 32)$.

111. INFANTS The graph is used to classify the weight of a baby boy from birth to 1 year. Estimate the weight range for boys in the following classifications, using a compound inequality:

a. 10 months old, "heavy"

b. 5 months old, "light"

c. 8 months old, "average"

d. 3 months old, "moderately light"

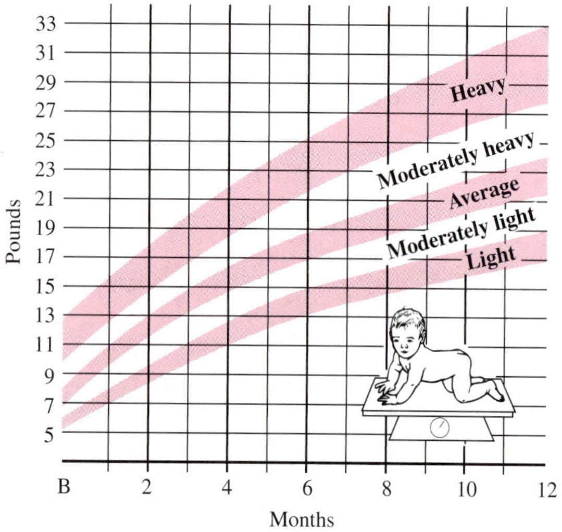

Based on data from *Better Homes and Gardens Baby Book* (Meredith Corp., 1969).

112. NUMBER PUZZLES What *whole* numbers satisfy the condition: Twice the number decreased by 1 is between 50 and 60?

WRITING

113. Explain why multiplying both sides of an inequality by a negative number reverses the direction of the inequality.

114. Explain the use of parentheses and brackets for graphing intervals.

REVIEW

Complete each table.

115.

x	$x^2 - 3$
−2	
0	
3	

116.

x	$\frac{x}{3} + 2$
−6	
0	
12	

CHALLENGE PROBLEMS

117. Solve the inequality. Write the solution set in interval notation and graph it.

$$3 - x < 5 < 7 - x$$

118. Use a guess-and-check approach to solve $\frac{1}{x} > 1$. Write the solution set in interval notation and graph it.

CHAPTER 2
Summary & Review

SECTION 2.1 Solving Equations Using Properties of Equality

DEFINITIONS AND CONCEPTS	EXAMPLES
An **equation** is a statement indicating that two expressions are equal. The equal symbol = separates an equation into two parts: the left side and the right side.	$2x + 4 = 10$ $\quad -5(a + 4) = -11a$ $\quad \frac{3}{2}t + 6 = t - \frac{1}{3}$
A number that makes an equation a true statement when substituted for the variable is called a **solution** of the equation.	Determine whether 2 is a solution of $x + 4 = 3x$. ***Check:*** $x + 4 = 3x$ $2 + 4 \stackrel{?}{=} 3(2)$ Substitute 2 for each x. $6 = 6$ True Since the resulting statement is true, 2 is a solution.
Equivalent equations have the same solutions.	$x - 2 = 6$ and $x = 8$ are equivalent equations because they have the same solution, 8.
To **solve an equation** isolate the variable on one side of the equation by undoing the operations performed on it using properties of equality. **Addition (Subtraction) property of equality:** If the same number is added to (or subtracted from) both sides of an equation, the result is an equivalent equation.	Solve: $x - 5 = 7$ $x - 5 + 5 = 7 + 5$ $x = 12$ Solve: $c + 9 = 16$ $c + 9 - 9 = 16 - 9$ $c = 7$
Multiplication (Division) property of equality: If both sides of an equation are multiplied (or divided) by the same nonzero number, the result is an equivalent equation.	Solve: $\frac{1}{3}m = 2$ $3\left(\frac{1}{3}m\right) = 3(2)$ $m = 6$ Solve: $10y = 50$ $\frac{10y}{10} = \frac{50}{10}$ $y = 5$

REVIEW EXERCISES

Determine whether the given number is a solution of the equation.

1. 84, $x - 34 = 50$
2. 3, $5y + 2 = 12$
3. -30, $\frac{x}{5} = 6$
4. 2, $a^2 - a - 1 = 0$
5. -3, $5b - 2 = 3b - 8$
6. 1, $\frac{2}{y + 1} = \frac{12}{y + 1} - 5$

Fill in the blanks.

7. An ________ is a statement indicating that two expressions are equal.
8. To solve $x - 8 = 10$ means to find all the values of the variable that make the equation a _____ statement.

Solve each equation and check the result.

9. $x - 9 = 12$
10. $-y = -32$
11. $a + 3.7 = -16.9$
12. $100 = -7 + r$
13. $120 = 5c$
14. $t - \frac{1}{2} = \frac{3}{2}$
15. $\frac{4}{3}t = -12$
16. $3 = \frac{q}{-2.6}$
17. $6b = 0$
18. $\frac{15}{16}s = -3$

SECTION 2.2 More about Solving Equations

DEFINITIONS AND CONCEPTS	EXAMPLES
A five-step **strategy for solving linear equations:** 1. *Clear* the equation of fractions or decimals. 2. *Simplify* each side. Use the distributive property and combine like terms when necessary. 3. *Isolate the variable term.* Use the addition and subtraction properties of equality. 4. *Isolate the variable.* Use the multiplication and division properties of equality. 5. *Check* the result in the original equation.	Solve: $2(y+2)+4y=11-y$ $2y+4+4y=11-y$ Distribute the multiplication by 2. $6y+4=11-y$ Combine like terms: $2y+4y=6y$. $6y+4+y=11-y+y$ To eliminate $-y$ on the right, add y to both sides. $7y+4=11$ Combine like terms. $7y+4-4=11-4$ To isolate the variable term $7y$, subtract 4 from both sides. $7y=7$ Simplify each side of the equation. $\frac{7y}{7}=\frac{7}{7}$ To isolate y, divide both sides by 7. $y=1$
To clear an equation of fractions, multiply both sides of an equation by the LCD.	To solve $\frac{1}{2}+\frac{x}{3}=\frac{3}{4}$, first multiply both sides by 12: $12\left(\frac{1}{2}+\frac{x}{3}\right)=12\left(\frac{3}{4}\right)$
To clear an equation of decimals, multiply both sides by a power of 10 to change the decimals in the equation to integers.	To solve $0.5(x-4)=0.1x+0.2$, first multiply both sides by 10: $10[0.5(x-4)]=10(0.1x+0.2)$
An equation that is true for all values of its variable is called an **identity.**	When we solve $x+5+x=2x+5$, the variables drop out and we obtain a true statement $5=5$. All real numbers are solutions.
An equation that is not true for any value of its variable is called a **contradiction.**	When we solve $y+2=y$, the variables drop out and we obtain a false statement $2=0$. The equation has no solutions.

REVIEW EXERCISES

Solve each equation. Check the result.

19. $5x+4=14$

20. $98.6-t=129.2$

21. $\frac{n}{5}+(-2)=4$

22. $\frac{b-5}{4}=-6$

23. $5(2x-4)-5x=0$

24. $-2(x-5)=5(-3x+4)+3$

25. $\frac{3}{4}=\frac{1}{2}+\frac{d}{5}$

26. $\frac{5(7-x)}{4}=2x-3$

27. $\frac{3(2-c)}{2}=\frac{-2(2c+3)}{5}$

28. $\frac{b}{3}+\frac{11}{9}+3b=-\frac{5}{6}b$

29. $0.15(x+2)+0.3=0.35x-0.4$

30. $0.5-0.02(y-2)=0.16+0.36y$

31. $3(a+8)=6(a+4)-3a$

32. $2(y+10)+y=3(y+8)$

SECTION 2.3 Applications of Percent

DEFINITIONS AND CONCEPTS	EXAMPLES
To solve **percent problems,** use the facts of the problem to write a sentence of the form: ☐ is ☐ % of ☐ ? Translate the sentence to mathematical symbols: *is* translates to an = symbol and *of* means multiply. Then solve the equation.	648 is 30% of what number? $648 = 30\% \cdot x$ Translate. $648 = 0.30x$ Change 30% to a decimal: 30% = 0.30. $\frac{648}{0.30} = x$ To isolate x, divide both sides by 0.30. $2{,}160 = x$ Do the division. Thus, 648 is 30% of 2,160.
To find the **percent of increase** or **the percent of decrease,** find what percent the increase or decrease is of the original amount.	SALE PRICES To find the percent of decrease when ground beef prices are reduced from \$4.89 to \$4.59 per pound, we first find the amount of decrease: $4.89 - 4.59 = 0.30$. Then we determine what percent 0.30 is of 4.89 (the original price). 0.30 is what% of 4.89? $0.30 = x \cdot 4.89$ Translate. $0.30 = 4.89x$ $\frac{0.30}{4.89} = x$ To isolate x, divide both sides by 4.89. $0.061349693 \approx x$ Do the division. $006.1349693\% \approx x$ Write the decimal as a percent. To the nearest tenth of a percent, the percent of decrease is 6.1%.

REVIEW EXERCISES

33. Fill in the blanks.

a. ________ means parts per one hundred.

b. When the price of an item is reduced, we call the amount of the reduction a ________.

c. An employee who is paid a ________ is paid a percent of the goods or services that he or she sells.

34. 4.81 is 2.5% of what number?

35. What number is 15% of 950?

36. What percent of 410 is 49.2?

37. U.S. ONLINE DATA The circle graph to the right shows Internet usage in the United States by the approximately 288.5 million people, ages 3 and over, in 2007. Determine the number of broadband users and the number of dial-up users. Round to the nearest tenth of one million.

38. COST OF LIVING A retired trucker receives a monthly Social Security check of \$764. If she is to receive a 3.5% cost-of-living increase soon, how much larger will her check be?

39. FAMILY BUDGETS It is recommended that a family pay no more than 30% of its monthly income (after taxes) on housing. If a family has an after-tax income of \$1,890 per month and pays \$625 in housing costs each month, are they within the recommended range?

U.S. Internet Usage,* 2007

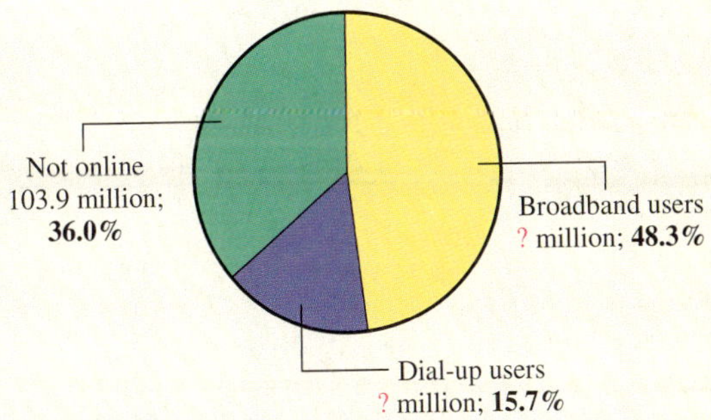

Source: *Advertising Age 2007 Fact Pack*

*An Internet user is defined as someone who uses the Internet at least once per month

40. DISCOUNTS A shopper saved \$148.50 on a food processor that was discounted 33%. What did it originally cost?

41. TUPPERWARE The hostess of a Tupperware party is paid a 25% commission on her in-home party's sales. What would the hostess earn if sales totaled \$600?

42. COLLECTIBLES A collector of football trading cards paid \$6 for a 1984 Dan Marino rookie card several years ago. If the card is now worth \$100, what is the percent of increase in the card's value? (Round to the nearest percent.)

SECTION 2.4 Formulas

DEFINITIONS AND CONCEPTS	EXAMPLES
A **formula** is an equation that states a relationship between two or more variables.	Retail price: $r = c + m$ Profit: $p = r - c$ Simple Interest: $I = Prt$ Distance: $d = rt$ Temperature: $C = \frac{5}{9}(F - 32)$
The **perimeter** of a plane geometric figure is the distance around it. The **area** of a plane geometric figure is the amount of surface that it encloses. The **volume** of a three-dimensional geometric solid is the amount of space it encloses.	Rectangle: $P = 2l + 2w$ Circle: $C = \pi D = 2\pi r$ $A = lw$ $A = \pi r^2$ Rectangular solid: $V = lwh$ Cylinder: $V = \pi r^2 h$ *See inside the back cover of the text for more geometric formulas.
If we are given the values of all but one of the variables in a formula, we can use our equation solving skills to find the value of the remaining variable.	BEDDING The area of a standard queen-size bed sheet is 9,180 in.^{2}. If the width is 102 inches, what is the length? $A = lw$ This is the formula for the area of a rectangle. $9{,}180 = 102w$ Substitute 9,180 for the area A and 102 for the width w. $\frac{9{,}180}{102} = w$ To isolate w, divide both sides by 102. $90 = w$ Do the division. The length of a standard queen-size bed sheet is 90 inches.
To solve a formula for a specific variable means to isolate that variable on one side of the equation, with all other variables and constants on the opposite side. Treat the specified variable as if it is the only variable in the equation. Treat the other variables as if they were numbers (constants).	Solve the formula for the volume of a cone for h. $V = \frac{1}{3}\pi r^2 h$ This is the formula for the volume of a cone. $3(V) = 3\left(\frac{1}{3}\pi r^2 h\right)$ To clear the equation of the fraction, multiply both sides by 3. $3V = \pi r^2 h$ Simplify. $\frac{3V}{\pi r^2} = \frac{\pi r^2 h}{\pi r^2}$ To isolate h, divide both sides by πr^2. $\frac{3V}{\pi r^2} = h$ or $h = \frac{3V}{\pi r^2}$

REVIEW EXERCISES

43. SHOPPING Find the markup on a CD player whose wholesale cost is $219 and whose retail price is $395.

44. RESTAURANTS One month, a restaurant had sales of $13,500 and made a profit of $1,700. Find the expenses for the month.

45. SNAILS A typical garden snail travels at an average rate of 2.5 feet per minute. How long would it take a snail to cross a 20-foot long flower bed?

46. CERTIFICATES OF DEPOSIT A $26,000 investment in a CD earned $1,170 in interest the first year. What was the annual interest rate?

47. JEWELRY Gold melts at about 1,065°C. Change this to degrees Fahrenheit.

48. CAMPING

a. Find the perimeter of the air mattress.

b. Find the amount of sleeping area on the top surface of the air mattress.

c. Find the approximate volume of the air mattress if it is 3 inches thick.

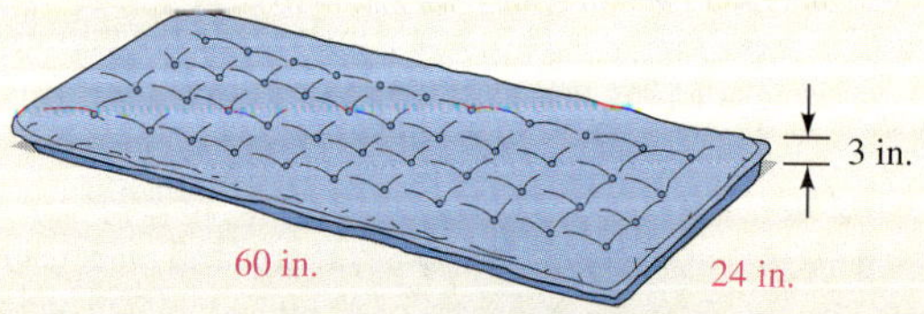

49. Find the area of a triangle with a base 17 meters long and a height of 9 meters.

50. Find the area of a trapezoid with bases 11 inches and 13 inches long and a height of 12 inches.

51. a. Find the circumference of a circle with a radius of 8 centimeters. Round to the nearest hundredth of one centimeter.

b. Find the area of the circle. Round to the nearest square centimeter.

52. Find the volume of a 12-foot cylinder whose circular base has a radius of 0.5 feet. Give the result to the nearest tenth.

53. Find the volume of a pyramid that has a square base, measuring 6 feet on a side, and a height of 10 feet.

54. HALLOWEEN After being cleaned out, a spherical-shaped pumpkin has an inside diameter of 9 inches. To the nearest hundredth, what is its volume?

Solve each formula for the specified variable.

55. $A = 2\pi rh$ for h

56. $A - BC = \frac{G - K}{3}$ for G

57. $C = \frac{1}{4}s(t - d)$ for t

58. $4y - 3x = 16$ for y

SECTION 2.5 Problem Solving

DEFINITIONS AND CONCEPTS	EXAMPLES
To solve application problems, use the five-step problem-solving strategy. 1. Analyze the problem. 2. Form an equation. 3. Solve the equation. 4. State the conclusion. 5. Check the result.	INCOME TAXES After taxes, an author kept \$85,340 of her total annual earnings. If her earnings were taxed at a 15% rate, how much did she earn that year?

Analyze the Problem The author earned some unknown amount of money. On that amount, she paid 15% in taxes. The difference between her total earnings and the taxes paid was \$85,340.

Form an Equation If we let x = the author's total earnings, the amount of taxes that she paid was 15% of x or $0.15x$. We can use the words of the problem to form an equation.

Her total earnings	minus	the taxes that she paid	equals	the money that she kept.
x	$-$	$0.15x$	$=$	85,340

Solve the Equation

$$x - 0.15x = 85{,}340$$

$$0.85x = 85{,}340 \quad \text{Combine like terms.}$$

$$x = 100{,}400 \quad \text{To isolate } x \text{, divide both sides by 0.85.}$$

State the Conclusion The author earned \$100,400 that year.

Check the Result The taxes were 15% of \$100,400 or \$15,060. If we subtract the taxes from her total earnings, we get \$100,400 − \$15,060 = \$85,340. The answer checks.

REVIEW EXERCISES

59. SOUND SYSTEMS A 45-foot-long speaker wire is to be cut into three pieces. One piece is to be 15 feet long. Of the remaining pieces, one must be 2 feet less than 3 times the length of the other. Find the length of the shorter piece.

60. SIGNING PETITIONS A professional signature collector is paid \$50 a day plus \$2.25 for each verified signature he gets from a registered voter. How many signatures are needed to earn \$500 a day?

61. LOTTERY WINNINGS After taxes, a lottery winner was left with a lump sum of \$1,800,000. If 28% of the original prize was withheld to pay federal income taxes, what was the original cash prize?

62. NASCAR The car numbers of drivers Bobby Labonte and Kyle Petty are consecutive odd integers whose sum is 88. If Labonte's number is the smaller, find the numbers of each car.

63. ART HISTORY *American Gothic* was painted in 1930 by Grant Wood. The length of the rectangular painting is 5 inches more than the width. Find the dimensions of the painting if it has a perimeter of $109\frac{1}{2}$ inches.

64. GEOMETRY Find the missing angle measures of the triangle.

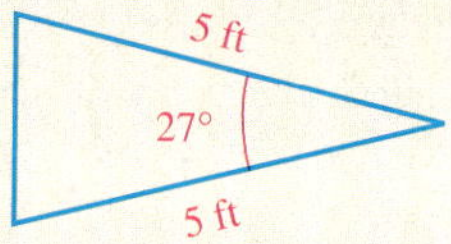

SECTION 2.6 More on Problem Solving

DEFINITIONS AND CONCEPTS

To solve application problems, use the five-step problem solving strategy.

1. Analyze the problem.
2. Form an equation.
3. Solve the equation.
4. State the conclusion.
5. Check the result.

Tables are a helpful way to organize the facts of a problem.

EXAMPLES

TRUCKING Two trucks leave from the same place at the same time traveling in opposite directions. One travels at a rate of 60 mph and the other at 50 mph. How long will it take them to be 165 miles apart?

Analyze the Problem We know that one truck travels at 60 mph and the other at 50 mph. Together, the trucks will travel a distance of 165 miles.

Form an Equation We enter each rate in the table under the heading r. Since the trucks travel for the same length of time, say t hours, we enter t for each truck under the heading t. Since $d = r \cdot t$, the first truck will travel $60t$ miles and the second will travel $50t$ miles. We enter the distances traveled under the heading d in the table.

	r	$\cdot\ t$	$= d$
Truck 1	60	t	$60t$
Truck 2	50	t	$50t$
			Total: 165

Use the information in this column to form an equation.

The distance the first truck travels	plus	the distance the second truck travels	is	165 miles.
$60t$	$+$	$50t$	$=$	165

Solve the Equation $60t + 50t = 165$

$110t = 165$ Combine like terms.

$\frac{110t}{110} = \frac{165}{110}$ To isolate t, divide both sides by 110.

$t = 1.5$

State the Conclusion The trucks will be 165 miles apart in 1.5 hours.

Check the Result If the first truck travels 60 mph for 1.5 hours, it will go $60(1.5) = 90$ miles. If the second truck travels 50 mph for 1.5 hours, it will go $50(1.5) = 75$ miles. Since 90 miles + 75 miles = 165 miles, the result checks.

REVIEW EXERCISES

65. INVESTMENT INCOME A woman has \$27,000. Part is invested for 1 year in a certificate of deposit paying 7% interest, and the remaining amount in a cash management fund paying 9%. After 1 year, the total interest on the two investments is \$2,110. How much is invested at each rate?

66. WALKING AND BICYCLING A bicycle path is 5 miles long. A man walks from one end at the rate of 3 mph. At the same time, a friend bicycles from the other end, traveling at 12 mph. In how many minutes will they meet?

67. AIRPLANES How long will it take a jet plane, flying at 450 mph, to overtake a propeller plane, flying at 180 mph, if the propeller plane had a $2\frac{1}{2}$-hour head start?

68. AUTOGRAPHS Kesha collected the autographs of 8 more television celebrities than she has of movie stars. Each TV celebrity autograph is worth \$75 and each movie star autograph is worth \$250. If her collection is valued at \$1,900, how many of each type of autograph does she have?

69. MIXTURES A store manager mixes candy worth 90¢ per pound with gumdrops worth \$1.50 per pound to make 20 pounds of a mixture worth \$1.20 per pound. How many pounds of each kind of candy does he use?

70. MILK Cream is about 22% butterfat and low-fat milk is about 2% butterfat. How many gallons of cream must be mixed with 18 gallons of low-fat milk to make whole milk that contains 4% butterfat?

SECTION 2.7 Solving Inequalities

DEFINITIONS AND CONCEPTS	EXAMPLES
An **inequality** is a mathematical statement that contains an $>$, $<$, $\geq$, or $\leq$ symbol.	$3x < 8$ $\quad \frac{1}{2}y - 4 \geq 12$ $\quad 2z + 4 \leq z - 5$
A **solution of an inequality** is any number that makes the inequality true.	Determine whether 3 is a solution of $2x - 7 < 5$. ***Check:*** $2x - 7 < 5$ $2(3) - 7 \overset{?}{<} 5$ Substitute 3 for x. $-1 < 5$ True Since the resulting statement is true, 3 is a solution.
We **solve inequalities** as we solve equations. However, if we multiply or divide both sides by a negative number, we must reverse the inequality symbol.	Solve: $-3(z - 1) \geq -6$ $-3z + 3 \geq -6$ Distribute the multiplication by -3. $-3z \geq -9$ To isolate the variable term $-3z$, subtract 3 from both sides. $\frac{-3z}{-3} \leq \frac{-9}{-3}$ To isolate z, divide both sides by -3. Reverse the inequality symbol. $z \leq 3$ Do the divisions.
Interval notation can be used to describe the solution set of an inequality. A **parenthesis** indicates that a number is not in the solution set of an inequality. A **bracket** indicates that a number is included in the solution set.	In interval notation, the solution set is $(-\infty, 3]$, whose graph is shown. 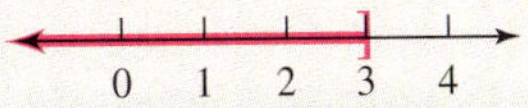

REVIEW EXERCISES

Solve each inequality. Write the solution set in interval notation and graph it.

71. $3x + 2 < 5$

72. $-\frac{3}{4}x \geq -9$

73. $\frac{3}{4} < \frac{d}{5} + \frac{1}{2}$

74. $5(3 - x) \leq 3(x - 3)$

75. $\frac{t}{-5} - (-1.8) \geq -6.2$

76. $63 < 7a$

77. $8 < x + 2 < 13$

78. $0 \leq 3 - 2x < 10$

79. SPORTS EQUIPMENT The acceptable weight w of Ping-Pong balls used in competition can range from 2.40 to 2.53 grams. Express this range using a compound inequality.

80. SIGNS A large office complex has a strict policy about signs. Any sign to be posted in the building must be rectangular in shape, its width must be 18 inches, and its perimeter is not to exceed 132 inches. What possible sign lengths meet these specifications?

CHAPTER 2 Test

1. Fill in the blanks.

a. To ______ an equation means to find all of the values of the variable that make the equation true.

b. ________ means parts per one hundred.

c. The distance around a circle is called its ____________.

d. An _________ is a statement that contains one of the symbols $>$, $\geq$, $<$, or $\leq$.

e. The ____________ property of ________ says that multiplying both sides of an equation by the same nonzero number does not change its solution.

2. Is 3 a solution of $5y + 2 = 12$?

Solve each equation.

3. $3h + 2 = 8$

4. $\frac{4}{5}t = -4$

5. $-22 = -x$

6. $\frac{11b - 11}{5} = 3b - 2$

7. $0.8(x - 1{,}000) + 1.3 = 2.9 + 0.2x$

8. $2(y - 7) - 3y = -(y - 3) - 17$

9. $\frac{m}{2} - \frac{1}{3} = \frac{1}{4} + \frac{m}{6}$

10. $9 - 5(2x + 10) = -1$

11. $24t = -6(8 - 4t)$

12. $6a + (-7) = 3a - 7 + 2a$

13. What is 15.2% of 80?

14. DOWN PAYMENTS To buy a house, a woman was required to make a down payment of $11,400. What did the house sell for if this was 15% of the purchase price?

15. BODY TEMPERATURES Suppose a person's body temperature rises from 98.6°F to a dangerous 105°F. What is the percent increase? Round to the nearest percent.

16. COMMISSIONS An appliance store salesperson receives a commission of 5% of the price of every item that she sells. What will she make if she sells a $599.99 refrigerator?

17. GRAND OPENINGS On its first night of business, a pizza parlor brought in $445. The owner estimated his profits that night to be $150. What were the costs?

18. Find the Celsius temperature reading if the Fahrenheit reading is 14°.

19. PETS The spherical fishbowl is three-quarters full of water. To the nearest cubic inch, find the volume of water in the bowl. (*Hint:* The volume of a sphere is given by $V = \frac{4}{3}\pi r^3$.)

20. Solve $A = P + Prt$ for r.

21. IRONS Estimate the area of the soleplate of the iron.

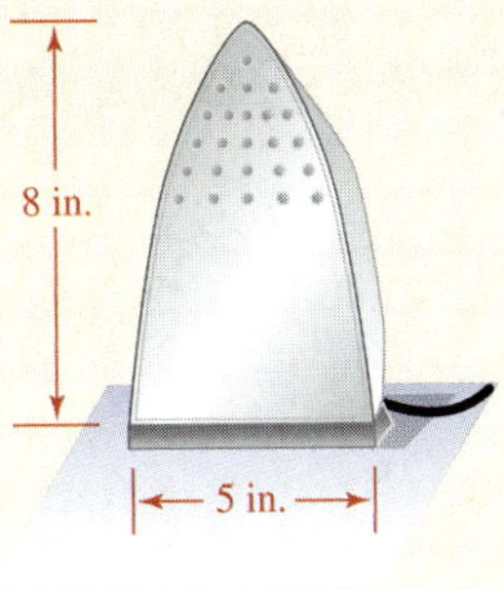

22. TELEVISION In a typical 30-minute block of time on TV, the number of programming minutes are 2 less than three times the number of minutes of commercials. How many minutes of programming and commercials are there?

23. HOME SALES A condominium owner cleared \$114,600 on the sale of his condo, after paying a 4.5% real estate commission. What was the selling price?

24. COLORADO The state of Colorado is approximately rectangular-shaped with perimeter 1,320 miles. Find the length (east to west) and width (north to south), if the length is 100 miles longer than the width.

25. TEA How many pounds of green tea, worth \$40 a pound, should be mixed with herbal tea, worth \$50 a pound, to produce 20 pounds of a blend worth \$42 a pound?

26. READING A bookmark is inserted between two page numbers whose sum is 825. What are the page numbers?

27. TRAVEL TIMES A car leaves Rockford, Illinois, at the rate of 65 mph, bound for Madison, Wisconsin. At the same time, a truck leaves Madison at the rate of 55 mph, bound for Rockford. If the cities are 72 miles apart, how long will it take for the car and the truck to meet?

28. PICKLES To make pickles, fresh cucumbers are soaked in a salt water solution called *brine.* How many liters of a 2% brine solution must be added to 30 liters of a 10% brine solution to dilute it to an 8% solution?

29. GEOMETRY If the vertex angle of an isosceles triangle is 44°, find the measure of each base angle.

30. INVESTMENTS Part of \$13,750 is invested at 9% annual interest, and the rest is invested at 8%. After one year, the accounts paid \$1,185 in interest. How much was invested at the lower rate?

Solve each inequality. Write the solution set in interval notation and graph it.

31. $-8x - 20 \le 4$

32. $-8.1 > \frac{t}{2} + (-11.3)$

33. $-12 \le 2(x + 1) < 10$

34. AWARDS A city honors its citizen of the year with a framed certificate. An artist charges \$15 for the frame and 75 cents per word for writing out the proclamation. If a city regulation does not allow gifts in excess of \$150, what is the maximum number of words that can be written on the certificate?

GROUP PROJECT

TRANSLATING KEY WORDS AND PHRASES

Overview: Students often say that the most challenging step of the five-step problem-solving strategy is forming an equation. This activity is designed to make that step easier by improving your translating skills.

Instructions: Form groups of 3 or 4 students. Select one person from your group to record the group's responses. Determine whether addition, subtraction, multiplication, or division is suggested by each of the following words or phrases. Then use the word or phrase in a sentence to illustrate its meaning.

deflate	recede	partition	evaporate	amplify
bisect	augment	hike	erode	boost
annexed	diminish	plummet	upsurge	wane
quadruple	corrode	taper off	trisect	broaden

COMPUTER SPREADSHEETS

Overview: In this activity, you will get some experience working with a spreadsheet.

Instructions: Form groups of 3 or 4 students. Examine the following spreadsheet, which consists of cells named by column and row. For example, 7 is entered in cell B3. In any cell you may enter data or a formula. For each formula in cells D1–D4 and E1–E4, the computer performs a calculation using values entered in other cells and prints the result in place of the formula. Find the value that will be printed in each formula cell. The symbol * means multiply, / means divide, and ^ means raise to a power.

	A	B	C	D	E
1	−8	20	−6	= 2*B1 − 3*C1 + 4	= B1 − 3*A1^2
2	39	2	−1	= A2/(B2 − C2)	= B3*B2*C2*2
3	50	7	3	= A3/5 + C3^3	= 65 − 2*(B3 − 5)^5
4	6.8	−2.8	−0.5	= 100*A4 + B4*C4	= A4/10 + A3/2*5

CHAPTER 3

Graphing Linear Equations and Inequalities in Two Variables; Functions

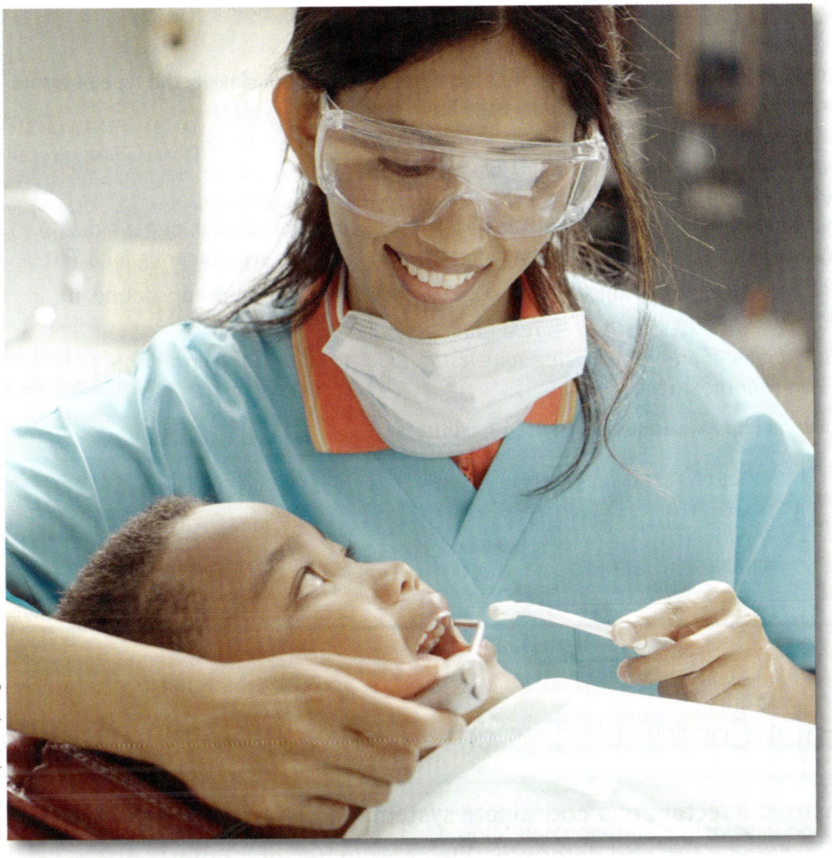

from Campus to Careers
Dental Assistant

A dental assistant is a valuable member of the dental health team who prepares patients for treatment, takes x-rays, sterilizes instruments, and keeps records. Part of the training of a dental assistant includes learning about a *coordinate system* that is used to identify the location of teeth in the mouth. This coordinate system is much like one used in algebra to graph points, lines, and curves. In **Problem 33** of **Study Set 3.1** you will see how the coordinate system used by dentists divides the mouth into four parts called *quadrants*.

JOB TITLE:
Dental Assistant

EDUCATION:
An associate's degree from an accredited program is required.

JOB OUTLOOK:
Excellent—One of the fastest growing occupations in the health care industry.

ANNUAL EARNINGS:
Median salary $31,739

FOR MORE INFORMATION:
www.bls.gov/oco/ocos163.htm

In the figure, red arrows are used to show how to plot the point $(-4, 3)$. We start at the origin, move 4 units to the *left* along the x-axis, then move *up* 3 units and draw a dot. This locates the point $(-4, 3)$.

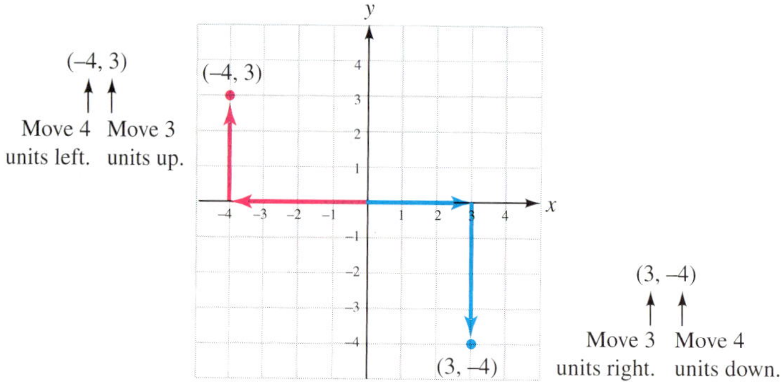

The Language of Algebra
Note that the points $(3, -4)$ and $(-4, 3)$ have different locations. Since the order of the coordinates of a point is important, we call them **ordered pairs.**

EXAMPLE 1

Plot each point. Then state the quadrant in which it lies or the axis on which it lies. **a.** $(4, 4)$ **b.** $\left(-1, -\frac{7}{2}\right)$ **c.** $(0, 2.5)$ **d.** $(-3, 0)$ **e.** $(0, 0)$

Strategy After identifying the x- and y-coordinates of the ordered pair, we will move the corresponding number of units left, right, up, or down to locate the point.

Why The coordinates of a point determine its location on the coordinate plane.

Solution

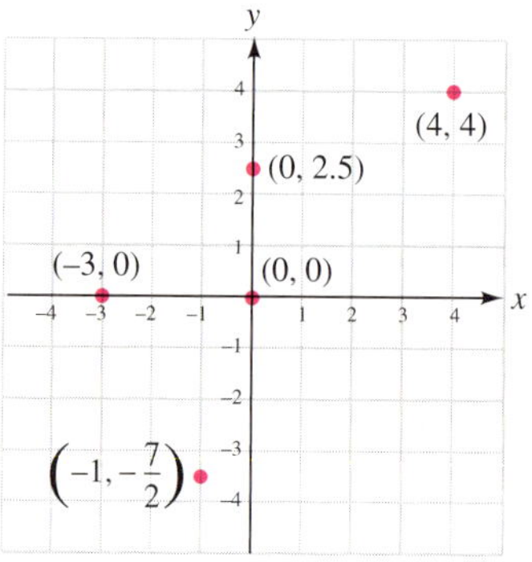

a. Since the x-coordinate, 4, is positive, we start at the origin and move 4 units to the *right* along the x-axis. Since the y-coordinate, 4, is positive, we then move *up* 4 units and draw a dot. This locates the point $(4, 4)$. The point lies in quadrant I.

b. To plot $\left(-1, -\frac{7}{2}\right)$, we begin at the origin and move 1 unit to the *left,* because the x-coordinate is -1. Then, since the y-coordinate is negative, we move $\frac{7}{2}$ units, or $3\frac{1}{2}$ units, *down.* The point lies in quadrant III.

Success Tip
Points with an x-coordinate that is 0 lie on the y-axis. Points with a y-coordinate that is 0 lie on the x-axis. Points that lie on an axis are not considered to be in any quadrant.

c. To plot $(0, 2.5)$, we begin at the origin and do not move right or left, because the x-coordinate is 0. Since the y-coordinate is positive, we move 2.5 units *up.* The point lies on the y-axis.

d. To plot $(-3, 0)$, we begin at the origin and move 3 units to the *left,* because the x-coordinate is -3. Since the y-coordinate is 0, we do not move up or down. The point lies on the x-axis.

e. To plot $(0, 0)$, we begin at the origin, and we remain there because both coordinates are 0. The point with coordinates $(0, 0)$ is the origin.

Self Check 1 Plot the points $(2, -2)$, $(-4, 0)$, $\left(1.5, \frac{5}{2}\right)$, and $(0, 5)$.

Now Try **Problem 17**

EXAMPLE 2 Find the coordinates of points A, B, C, D, E, and F plotted in figure (a) below.

Notation
Points are often labeled with capital letters. For example, the notation $A(2, 3)$ indicates that point A has coordinates $(2, 3)$.

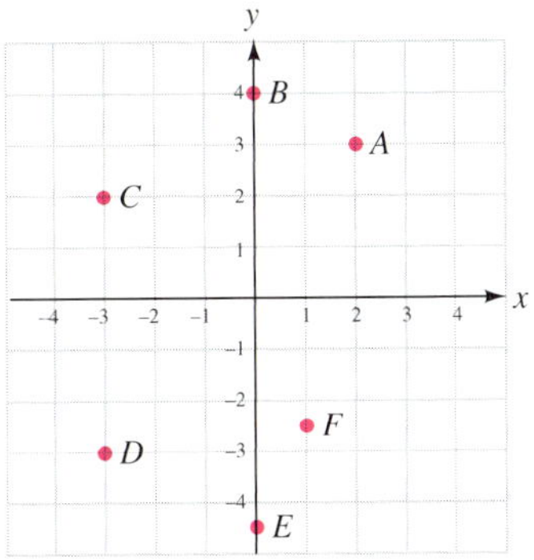

(a)

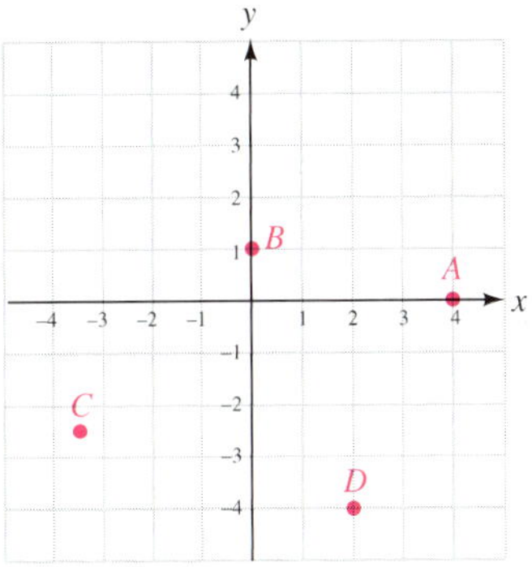

(b)

Strategy We will start at the origin and count to the left or right on the x-axis, and then up or down to reach each point.

Why The movement left or right gives the x-coordinate of the ordered pair and the movement up or down gives the y-coordinate.

Solution To locate point A, we start at the origin, move 2 units to the right on the x-axis, and then 3 units up. Its coordinates are $(2, 3)$. The coordinates of the other points are found in the same manner.

$B(0, 4)$ $C(-3, 2)$ $D(-3, -3)$ $E(0, -4.5)$ $F(1, -2.5)$

Self Check 2 Find the coordinates of each point in Figure (b) above.

Now Try **Problem 20**

3 Graph Paired Data.

Every day, we deal with quantities that are related:

- The time it takes to cook a roast depends on the weight of the roast.
- The money we earn depends on the number of hours we work.
- The sales tax that we pay depends on the price of the item purchased.

We can use graphs to visualize such relationships. For example, suppose a tub is filling with water, as shown on the next page. Obviously, the amount of water in the tub depends on how long the water has been running. To graph this relationship, we can use the measurements that were taken as the tub began to fill.

The Language of Algebra
Data that can be represented as an ordered pair is called **paired data**.

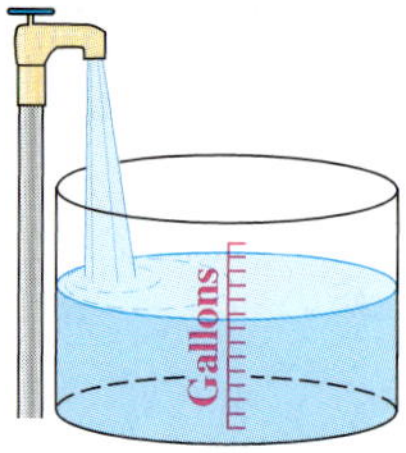

Time (mins)	Water in tub (gal)	
0	0	→ (0, 0)
1	8	→ (1, 8)
3	24	→ (3, 24)
4	32	→ (4, 32)

↑ x-coordinate ↑ y-coordinate

The data in the table can be expressed as ordered pairs (x, y).

The data in each row of the table can be written as an ordered pair and plotted on a rectangular coordinate system. Since the first coordinate of each ordered pair is a time, we label the x-axis *Time (min)*. The second coordinate is an amount of water, so we label the y-axis *Amount of water (gal)*. The y-axis is scaled in larger units (multiples of 4 gallons) because the size of the data ranges from 0 to 32 gallons.

After plotting the ordered pairs, we use a straightedge to draw a line through the points. As expected, the completed graph shows that the amount of water in the tub increases steadily as the water is allowed to run.

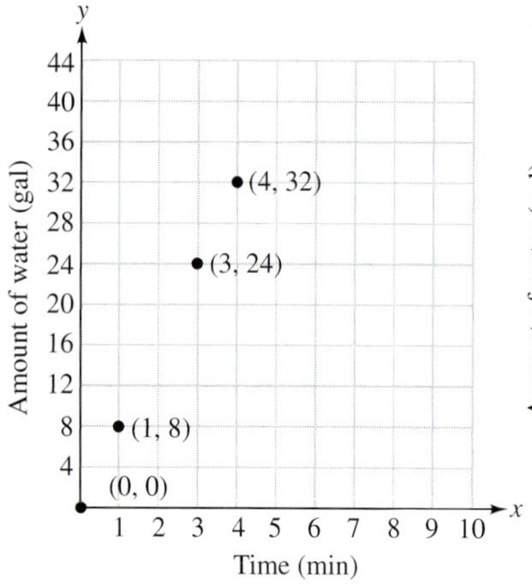

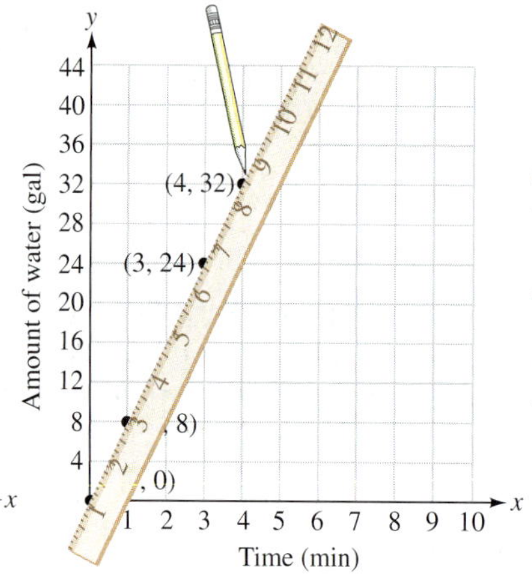

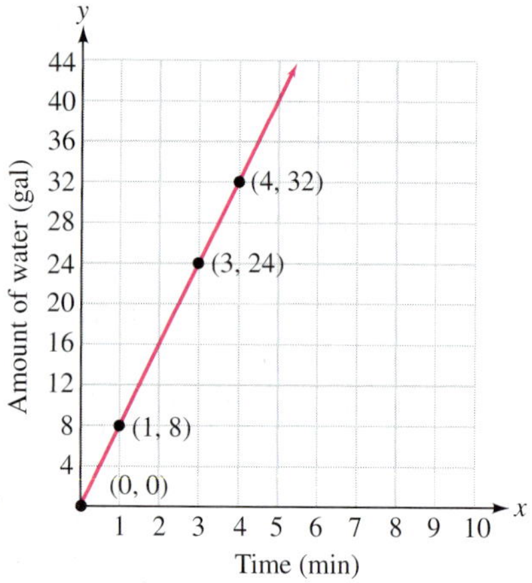

We can use the graph to determine the amount of water in the tub at various times. For example, the green dashed line on the graph shows that in 2 minutes, the tub will contain 16 gallons of water. This process, called **interpolation,** uses known information to predict values that are not known but are *within* the range of the data. The blue dashed line on the graph shows that in 5 minutes, the tub will contain 40 gallons of water. This process, called **extrapolation,** uses known information to predict values that are not known and are *outside* the range of the data.

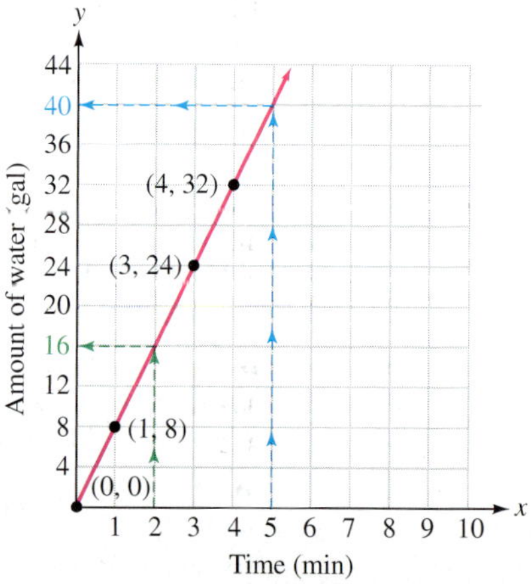

4 Read Line Graphs.

Since graphs are a popular way to present information, the ability to read and interpret them is very important.

The Language of Algebra
A rectangular coordinate system is a *grid*—a network of uniformly spaced perpendicular lines. At times, some large U.S. cities have such horrible traffic congestion that vehicles can barely move, if at all. The condition is called *gridlock*.

EXAMPLE 3 ***TV Shows.*** The following graph shows the number of people in an audience before, during, and after the taping of a television show. Use the graph to answer the following questions.

a. How many people were in the audience when the taping began?

b. At what times were there exactly 100 people in the audience?

c. How long did it take the audience to leave after the taping ended?

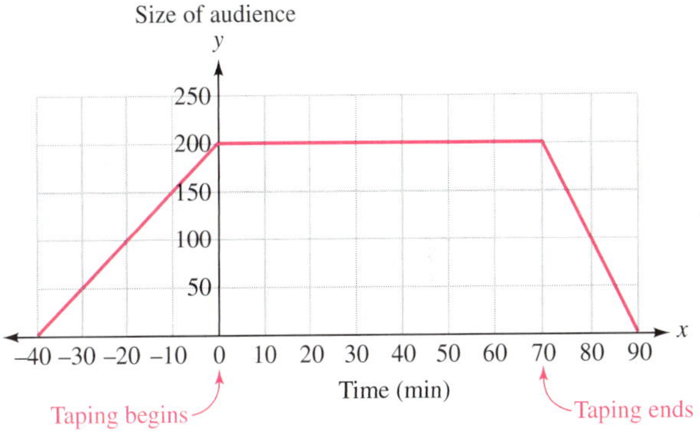

Strategy We will use an ordered pair of the form *(time, size of audience)* to describe each situation mentioned in parts (a), (b), and (c).

Why The coordinates of specific points on the graph can be used to answer each of these questions.

Solution

a. The time when the taping began is represented by 0 on the x-axis. The point on the graph directly above 0 is (0, 200). The y-coordinate indicates that 200 people were in the audience when the taping began.

b. We can draw a horizontal line passing through 100 on the y-axis. Since the line intersects the graph twice, at (−20, 100) and at (80, 100), there are two times when 100 people were in the audience. The x-coordinates of the points tell us those times: 20 minutes before the taping began, and 80 minutes after.

c. The x-coordinate of the point (70, 200) tells us when the audience began to leave. The x-coordinate of (90, 0) tells when the exiting was completed. Subtracting the x-coordinates, we see that it took $90 - 70 = 20$ minutes for the audience to leave.

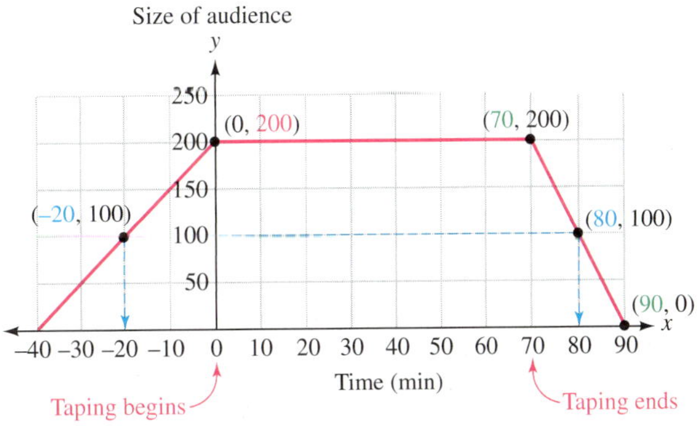

Self Check 3 Use the graph in Example 3 to answer the following questions.

a. At what times were there exactly 50 people in the audience?

b. How many people were in the audience when the taping took place?

c. When were the first audience members allowed into the taping session?

Now Try **Problems 21 and 23**

ANSWERS TO SELF CHECKS **1.**

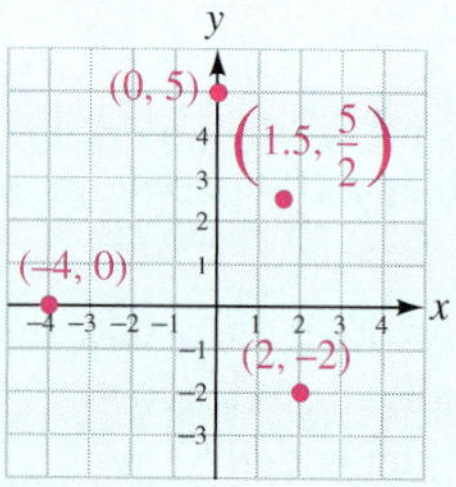

2. $A(4, 0)$; $B(0, 1)$; $C(-3.5, -2.5)$; $D(2, -4)$ **3. a.** 30 min before and 85 min after taping began **b.** 200 **c.** 40 min before taping began

STUDY SET 3.1

VOCABULARY

Fill in the blanks.

1. (7, 1) is called an ________ pair.

2. In the ordered pair (2, −5), the y-__________ is −5.

3. A rectangular coordinate system is formed by two perpendicular number lines called the x-______ and the y-______. The point where the axes cross is called the _______.

4. The x- and y-axes divide the coordinate plane into four regions called ___________.

5. The point with coordinates (4, 2) can be graphed on a ___________ coordinate system.

6. The process of locating the position of a point on a coordinate plane is called _______________ the point.

CONCEPTS

7. Fill in the blanks.

a. To plot (−5, 4), we start at the _______ and move 5 units to the _____ and then move 4 units _____.

b. To plot $\left(6, -\frac{3}{2}\right)$, we start at the _______ and move 6 units to the ______ and then move $\frac{3}{2}$ units _______.

8. In which quadrant is each point located?

a. (−2, 7) **b.** $\left(\frac{1}{2}, \frac{15}{16}\right)$

c. (−1, −2.75) **d.** (50, −16)

9. a. In which quadrants are the second coordinates of points positive?

b. In which quadrants are the first coordinates of points negative?

c. In which quadrant do points with a positive x-coordinate and a negative y-coordinate lie?

10. FARMING Write each row of data in the table as an ordered pair. Then plot the ordered pairs on the following graph and draw a straight line through the points. Use the graph to determine how many bushels will be produced if

a. 6 inches of rain fall. **b.** 10 inches of rain fall.

Rain (inches)	Bushels produced	
2	10	(,)
4	15	(,)
8	25	(,)

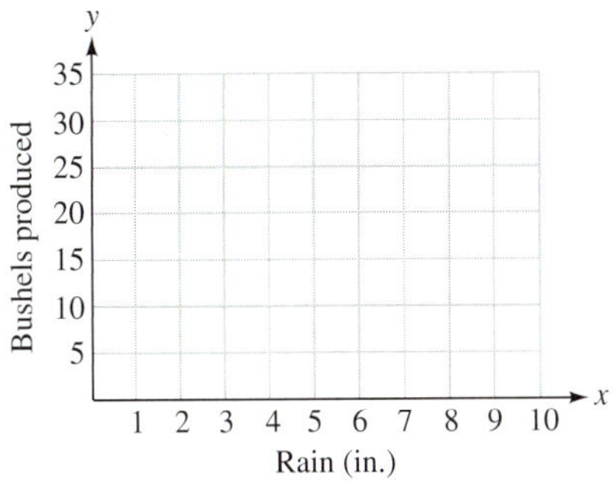

NOTATION

11. Explain the difference between (3, 5) and 3(5).

12. In the table, which column contains values associated with the vertical axis of a graph?

x	y
2	0
5	3
−1	−3

13. Do these ordered pairs name the same point?

$$\left(2.5, -\tfrac{7}{2}\right), \left(2\tfrac{1}{2}, -3.5\right), \left(2.5, -3\tfrac{1}{2}\right)$$

14. Do (3, 2) and (2, 3) represent the same point?

15. In the ordered pair (4, 5), is the number 4 associated with the horizontal or the vertical axis?

16. Fill in the blank: In the notation $P(4, 5)$, the capital letter P is used to name a ______.

GUIDED PRACTICE

See Examples 1 and 2.

17. Graph each point:
$(-3, 4), (4, 3.5), \left(-2, -\frac{5}{2}\right), (0, -4), \left(\frac{3}{2}, 0\right), (3, -4)$

18. Graph each point:
$(4, 4), (0.5, -3), (-4, -4), (0, -1), (0, 0), (0, 3), (-2, 0)$

19. Complete the coordinates for each point in Figure (a) below.

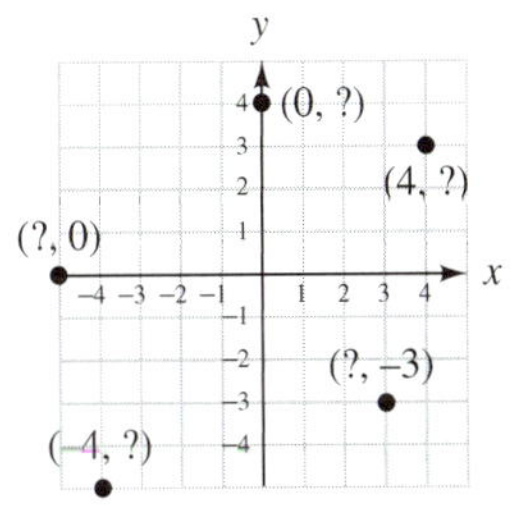

(a)

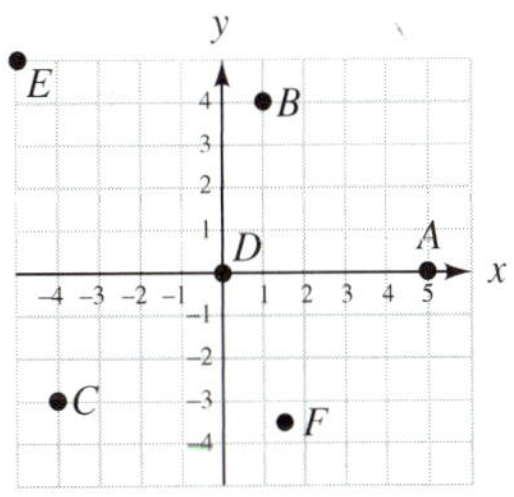

(b)

20. Find the coordinates of points A, B, C, D, E, and F in Figure (b) above.

***The following graph gives the heart rate of a woman before, during, and after an aerobic workout. Use it to answer Problems 21–24.* See Example 3.**

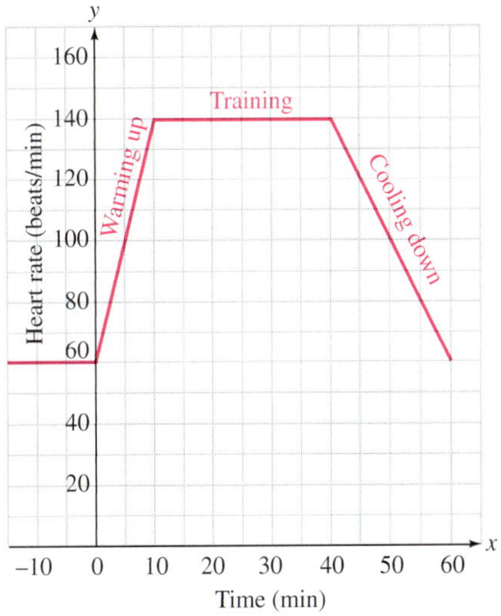

21. a. What was her heart rate before beginning the workout?

b. After beginning her workout, how long did it take the woman to reach her training-zone heart rate?

22. a. What was the woman's heart rate half an hour after beginning the workout?

b. For how long did the woman work out at her training zone?

23. a. At what time was her heart rate 100 beats per minute?

b. How long was her cool-down period?

24. a. What was the difference in the woman's heart rate before the workout and after the cool-down period?

b. What was her approximate heart rate 8 minutes after beginning?

***The following graph shows the depths of a submarine at certain times after it leaves port. Use the graph to answer Problems 25–28.* See Example 3.**

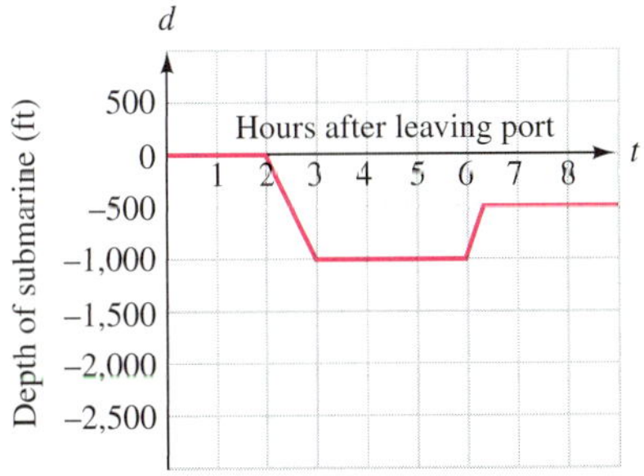

25. **a.** For how long does the sub travel at sea level?

b. What is the depth of the sub 5 hours after leaving port?

26. **a.** Once the sub begins to dive, how long does it take to reach −1,000 feet in depth?

b. For how long does the sub travel at a depth of 1,000 feet?

27. **a.** Explain what happens 6 hours after the sub leaves port.

b. What is the depth of the sub 8 hours after leaving port?

28. **a.** How long does it take the sub to first reach −500 feet in depth?

b. Approximate the time when the sub reaches −500 feet in depth for the second time.

APPLICATIONS

29. BRIDGE CONSTRUCTION Find the coordinates of each rivet, weld, and anchor.

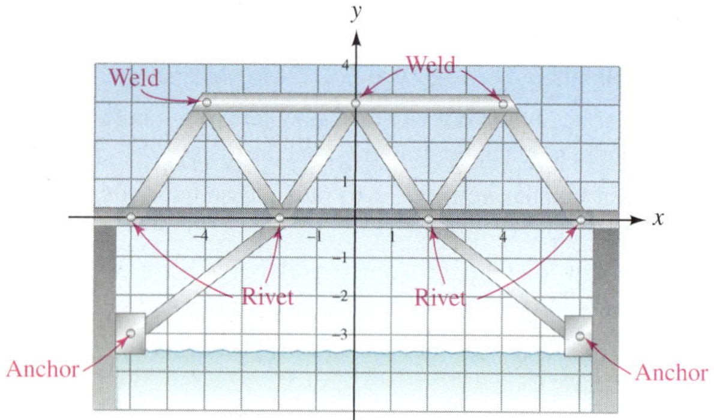

Scale: 1 unit = 8 ft

30. GOLF A golfer is videotaped and then has her swing displayed on a computer monitor so that it can be analyzed. Give the coordinates of the highlighted points.

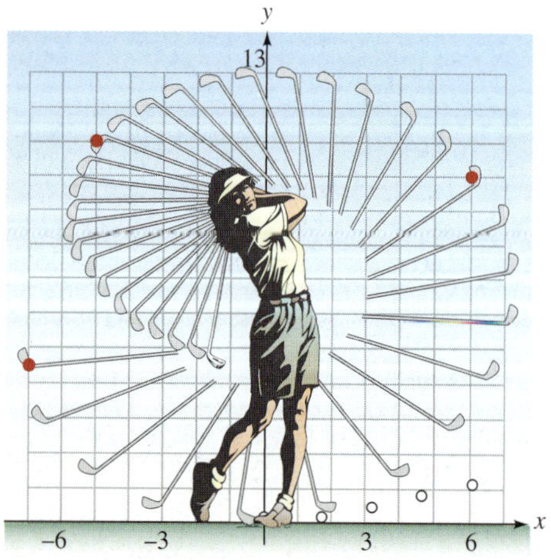

31. GAMES In the game Battleship, coordinates are used to locate ships. What are the coordinates of the ship shown? Express each answer in the form (letter, number).

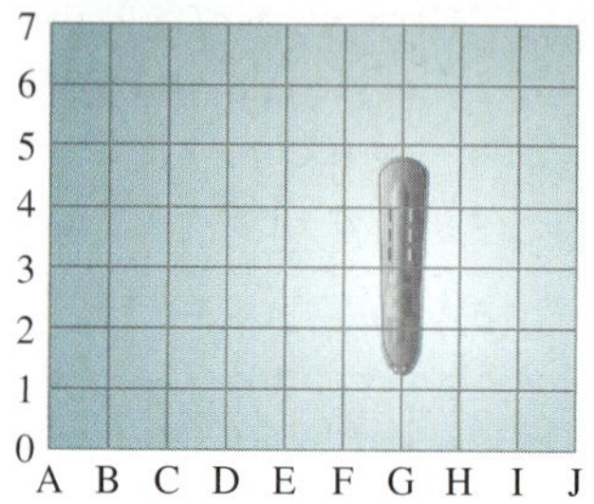

32. MAPS Use coordinates of the form (number, letter) to locate each of the following on the map: Rockford, Mount Carroll, Harvard, and the intersection of state Highway 251 and U.S. Highway 30.

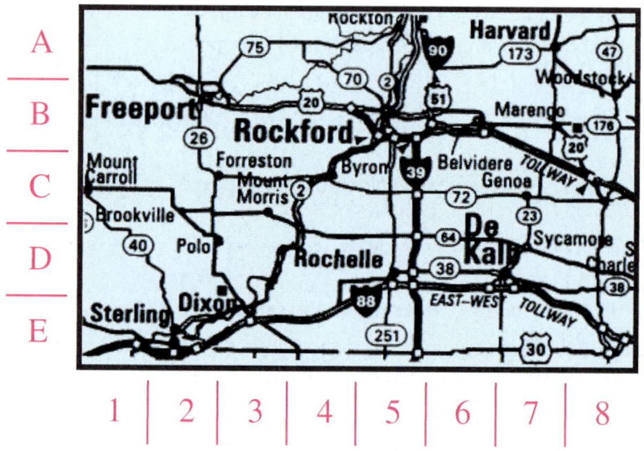

33. ***from Campus to Careers***

Dental Assistant

Dentists describe teeth as being located in one of four *quadrants* as shown below.

a. How many teeth are in the *upper left quadrant?*

b. Why would the upper left quadrant appear on the right in the illustration?

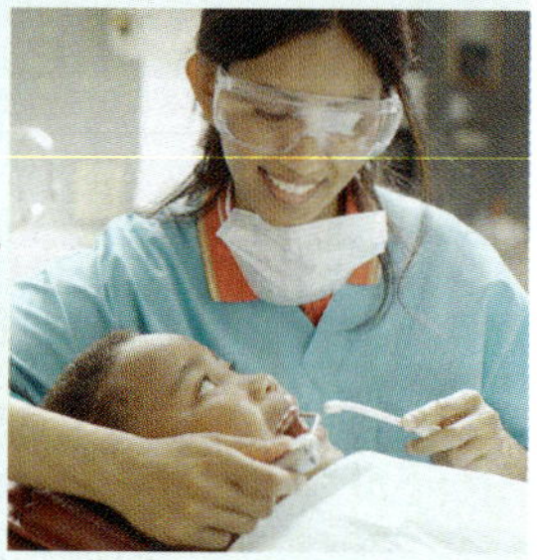

© Karin Dreyer/Getty Images

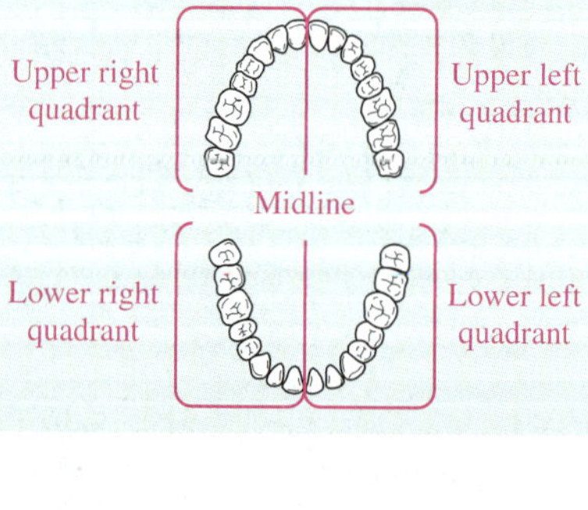

34. WATER PRESSURE The graphs show how the path of a stream of water changes when the hose is held at two different angles.

a. At which angle does the stream of water shoot up higher? How much higher?

b. At which angle does the stream of water shoot out farther? How much farther?

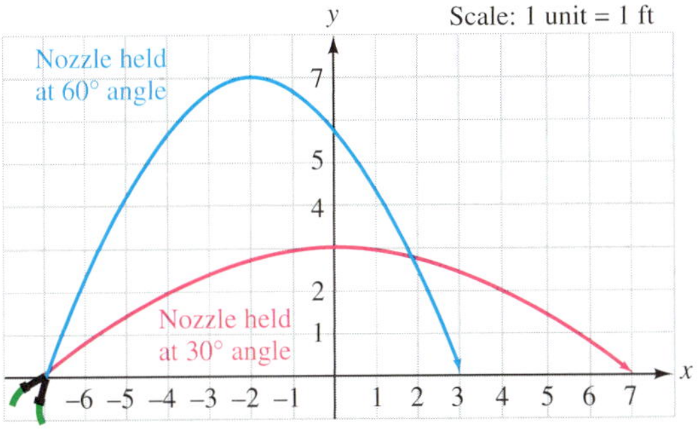

35. GEOMETRY Three vertices (corners) of a rectangle are (2, 1), (6, 1), and (6, 4). Find the coordinates of the fourth vertex. Then find the area of the rectangle.

36. GEOMETRY Three vertices (corners) of a right triangle are (−1, −7), (−5, −7), and (−5, −2). Find the area of the triangle.

37. TRUCKS The table below shows the number of miles that an 18-wheel truck can be driven on a given number of gallons of diesel fuel. Plot the data in the table as ordered pairs. Then draw a straight line through the points.

a. How far can the truck go on 4 gallons of fuel?

b. How many gallons of fuel are needed to travel a distance of 30 miles?

c. How far can the truck go on 7 gallons of fuel?

Fuel (gal)	Distance (mi)
2	10
3	15
5	25

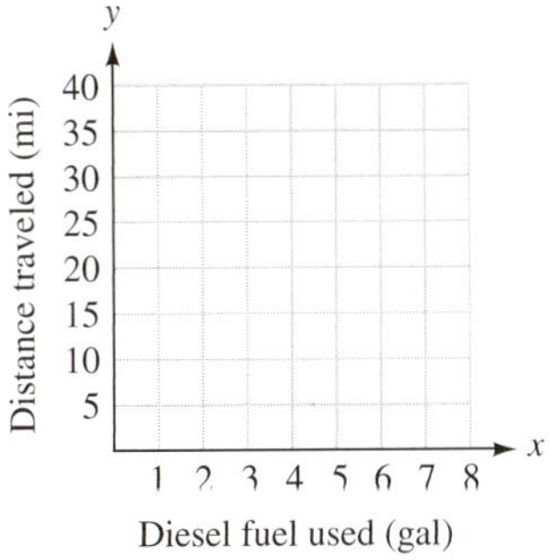

38. BOATING The table below shows the cost to rent a sailboat for a given number of hours. Plot the data in the table as ordered pairs. Then draw a straight line through the points.

a. What does it cost to rent the boat for 3 hours?

b. For how long can the boat be rented for $60?

c. What does it cost to rent the boat for 9 hours?

Rental time (hr)	Cost ($)
2	20
4	30
6	40

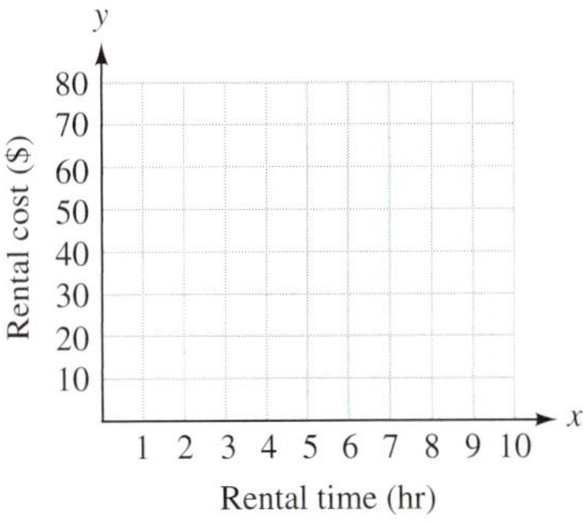

39. DEPRECIATION The table below shows the value (in thousands of dollars) of a color copier at various lengths of time after its purchase. Plot the data in the table as ordered pairs. Then draw a straight line passing through the points.

a. What does the point (3, 7) on the graph tell you?

b. Find the value of the copier when it is 7 years old.

c. After how many years will the copier be worth $2,500?

Age (yr)	Value ($1,000)
3	7
4	5.5
5	4

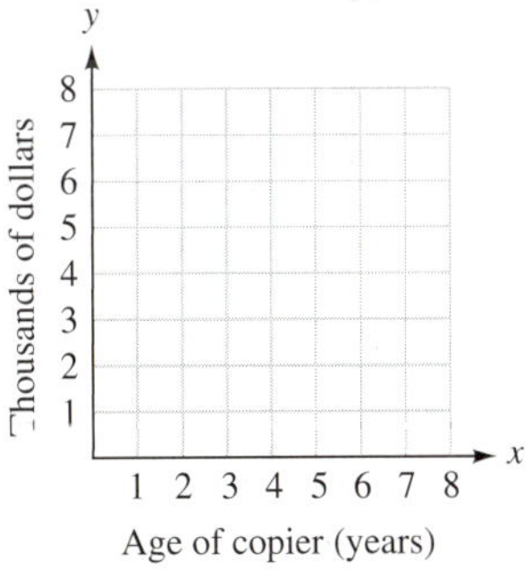

40. SWIMMING The table below shows the number of people at a public swimming pool at various times during the day. Plot the data in the table as ordered pairs. Then draw a straight line passing through the points.

a. How many people will be at the pool at 6 P.M.?

b. At what time will there be 250 people at the pool?

c. At what time will the number of people at the pool be half of what it was at noon?

Time	Number of people
0	350
3	200
5	100

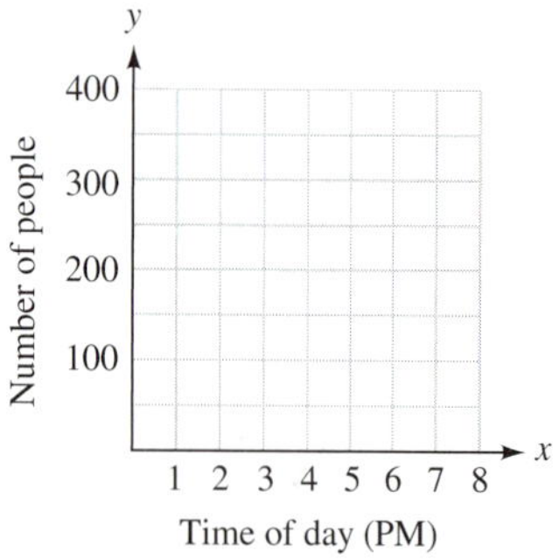

44. Explain this diagram.

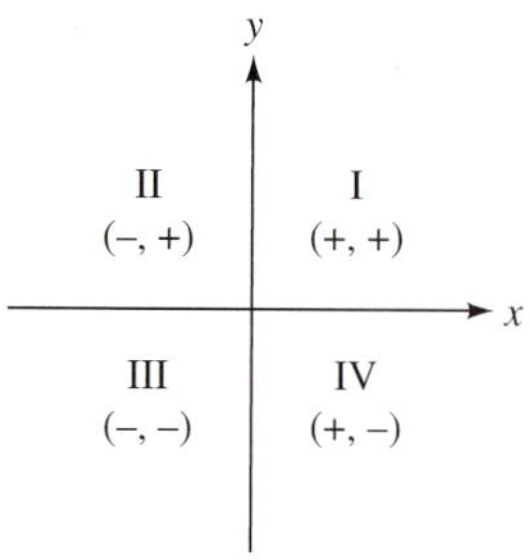

WRITING

41. Explain why the point $(-3, 3)$ is not the same as the point $(3, -3)$.

42. Explain how to plot the point $(-2, 5)$.

43. Explain why the coordinates of the origin are $(0, 0)$.

REVIEW

45. Solve $AC = \frac{2}{3}h - T$ for h.

46. Solve $5(x + 1) \le 2(x - 3)$. Write the solution set in interval notation and graph it.

47. Evaluate: $\dfrac{-4(4 + 2) - 2^3}{|-12 - 4(5)|}$

48. Simplify: $\dfrac{24}{54}$

CHALLENGE PROBLEMS

49. In what quadrant does a point lie if the *sum* of its coordinates is negative and the *product* of its coordinates is positive?

50. Draw line segment $\overline{AB}$ with endpoints $A(6, 5)$ and $B(-4, 5)$. Suppose that the x-coordinate of a point C is the average of the x-coordinates of points A and B, and the y-coordinate of point C is the average of the y-coordinates of points A and B. Find the coordinates of point C. Why is C called the midpoint of $\overline{AB}$?

SECTION 3.2
Graphing Linear Equations

Objectives

1. Determine whether an ordered pair is a solution of an equation.
2. Complete ordered-pair solutions of equations.
3. Construct a table of solutions.
4. Graph linear equations by plotting points.
5. Use graphs of linear equations to solve applied problems.

In this section, we will discuss equations that contain two variables. Such equations are often used to describe algebraic relationships between two quantities. To see a mathematical picture of these relationships, we will construct graphs of their equations.

1 Determine Whether an Ordered Pair Is a Solution of an Equation.

We have previously solved **equations in one variable.** For example, $x + 3 = 9$ is an equation in x. If we subtract 3 from both sides, we see that 6 is the solution. To verify this, we replace x with 6 and note that the result is a true statement: $9 = 9$.

In this chapter, we extend our equation-solving skills to find solutions of **equations in two variables.** To begin, let's consider $y = x - 1$, an equation in x and y.

A solution of $y = x - 1$ is a pair of values, one for x and one for y, that make the equation true. To illustrate, suppose x is 5 and y is 4. Then we have:

Notation
Equations in two variables often involve the variables x and y. However, other letters can be used. For example, $a - 3b = 5$ and $n = 4m + 6$ are equations in two variables.

$y = x - 1$ This is the given equation.

$4 \stackrel{?}{=} 5 - 1$ Substitute 5 for x and 4 for y.

$4 = 4$ True

Since the result is a true statement, $x = 5$ and $y = 4$ is a solution of $y = x - 1$. We write the solution as the ordered pair (5, 4), with the value of x listed first. We say that (5, 4) *satisfies* the equation.

In general, a **solution of an equation in two variables** is an ordered pair of numbers that makes the equation a true statement.

EXAMPLE 1 Is $(-1, -3)$ a solution of $y = x - 1$?

Strategy We will substitute -1 for x and -3 for y and see whether the resulting equation is true.

Why An ordered pair is a *solution* of $y = x - 1$ if replacing the variables with the values of the ordered pair results in a true statement.

Solution

$y = x - 1$ This is the given equation.

$-3 \stackrel{?}{=} -1 - 1$ Substitute -1 for x and -3 for y.

$-3 = -2$ False

Since $-3 = -2$ is false, $(-1, -3)$ is not a solution of $y = x - 1$.

Self Check 1 Is (9, 8) a solution of $y = x - 1$?

Now Try Problem 17

2 Complete Ordered-Pair Solutions of Equations.

If only one of the values of an ordered-pair solution is known, we can substitute it into the equation to determine the other value.

EXAMPLE 2 Complete the solution $(-5, \quad)$ of the equation $y = -2x + 3$.

Strategy We will substitute the known x-coordinate of the solution into the given equation.

Why We can use the resulting equation in one variable to find the unknown y-coordinate of the solution.

Solution In the ordered pair $(-5, \quad)$, the x-value is -5; the y-value is not known. To find y, we substitute -5 for x in the equation and evaluate the right side.

$y = -2x + 3$ This is the given equation.

$y = -2(-5) + 3$ Substitute -5 for x.

$y = 10 + 3$ Do the multiplication.

$y = 13$ This is the missing y-coordinate of the solution.

The completed ordered pair is $(-5, 13)$.

Self Check 2 Complete the solution $(-2, \_\_)$ of the equation $y = 4x - 2$.

Now Try Problems 29 and 31

Solutions of equations in two variables are often listed in a **table of solutions** (or **table of values**).

EXAMPLE 3 Complete the table of solutions for $3x + 2y = 5$.

x	y	(x, y)
7		(7,)
	4	(, 4)

Strategy In each case we will substitute the known coordinate of the solution into the given equation.

Why We can solve the resulting equation in one variable to find the unknown coordinate of the solution.

Solution In the first row, we are given an x-value of 7. To find the corresponding y-value, we substitute 7 for x and solve for y.

x	y	(x, y)
7	−8	(7, −8)

$3x + 2y = 5$ This is the given equation.

$3(7) + 2y = 5$ Substitute 7 for x.

$21 + 2y = 5$ Do the multiplication.

$2y = -16$ To isolate the variable term, 2y, subtract 21 from both sides.

$y = -8$ To isolate y, divide both sides by 2. This is the missing y-coordinate of the solution.

A solution of $3x + 2y = 5$ is $(7, -8)$. It is entered in the table on the left.

In the second row, we are given a y-value of 4. To find the corresponding x-value, we substitute 4 for y and solve for x.

x	y	(x, y)
7	−8	(7, −8)
−1	4	(−1, 4)

$3x + 2y = 5$ This is the given equation.

$3x + 2(4) = 5$ Substitute 4 for y.

$3x + 8 = 5$ Do the multiplication.

$3x = -3$ To isolate the variable term, 3x, subtract 8 from both sides.

$x = -1$ To isolate x, divide both sides by 3. This is the missing x-coordinate of the solution.

Another solution is $(-1, 4)$. It is entered in the table on the left.

Self Check 3 Complete the table of solutions for $3x + 2y = 5$.

x	y	(x, y)
__	-2	$(\_, -2)$
5	__	$(5, \_\_)$

Now Try **Problem 37**

3 Construct a Table of Solutions.

To find a solution of an equation in two variables, we can select a number, substitute it for one of the variables, and find the corresponding value of the other variable. For example, to find a solution of $y = x - 1$, we can select a value for x, say, -4, substitute -4 for x in the equation, and find y.

x	y	(x, y)
-4	-5	$(-4, -5)$

$y = x - 1$

$y = -4 - 1$ Substitute -4 for x.

$y = -5$

The ordered pair $(-4, -5)$ is a solution. We list it in the table on the left.

To find another solution of $y = x - 1$, we select another value for x, say, -2, and find the corresponding y-value.

x	y	(x, y)
-4	-5	$(-4, -5)$
-2	-3	$(-2, -3)$

$y = x - 1$

$y = -2 - 1$ Substitute -2 for x.

$y = -3$

A second solution is $(-2, -3)$, and we list it in the table of solutions.

If we let $x = 0$, we can find a third ordered pair that satisfies $y = x - 1$.

x	y	(x, y)
-4	-5	$(-4, -5)$
-2	-3	$(-2, -3)$
0	-1	$(0, -1)$

$y = x - 1$

$y = 0 - 1$ Substitute 0 for x.

$y = -1$

A third solution is $(0, -1)$, which we also add to the table of solutions.

We can find a fourth solution by letting $x = 2$, and a fifth solution by letting $x = 4$.

x	y	(x, y)
-4	-5	$(-4, -5)$
-2	-3	$(-2, -3)$
0	-1	$(0, -1)$
2	1	$(2, 1)$
4	3	$(4, 3)$

$y = x - 1$

$y = 2 - 1$ Substitute 2 for x.

$y = 1$

$y = x - 1$

$y = 4 - 1$ Substitute 4 for x.

$y = 3$

A fourth solution is $(2, 1)$ and a fifth solution is $(4, 3)$. We add them to the table.

Since we can choose any real number for x, and since any choice of x will give a corresponding value of y, it is apparent that the equation $y = x - 1$ has *infinitely many solutions*. We have found five of them: $(-4, -5)$, $(-2, -3)$, $(0, -1)$, $(2, 1)$, and $(4, 3)$.

4 Graph Linear Equations by Plotting Points.

It is impossible to list the infinitely many solutions of the equation $y = x - 1$. However, to show all of its solutions, we can draw a mathematical "picture" of them. We call this picture the *graph of the equation*.

> ***Notation***
> The graph only shows a part of the line. The arrowheads indicate that it extends indefinitely in both directions.

To graph $y = x - 1$, we plot the ordered pairs shown in the table on a rectangular coordinate system. Then we draw a straight line through the points, because the graph of any solution of $y = x - 1$ will lie on this line. Furthermore, every point on this line represents a solution. We call the line the **graph of the equation.** It represents all of the solutions of $y = x - 1$.

$y = x - 1$

x	y	(x, y)
-4	-5	$(-4, -5)$
-2	-3	$(-2, -3)$
0	-1	$(0, -1)$
2	1	$(2, 1)$
4	3	$(4, 3)$

↑ Select x ↑ Find y ↑ Plot (x, y)

Construct a table of solutions.

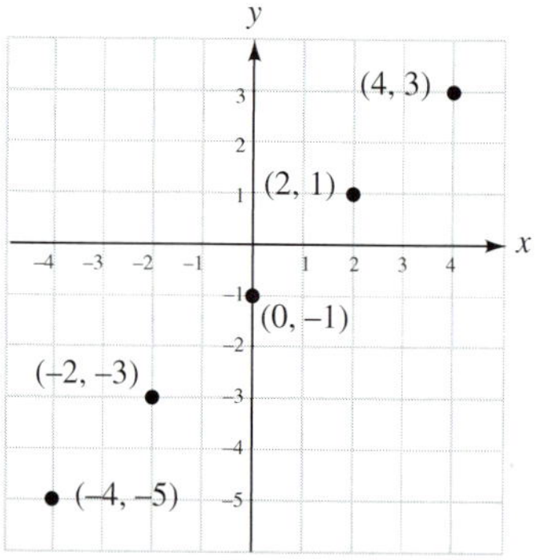

Plot the ordered pairs.

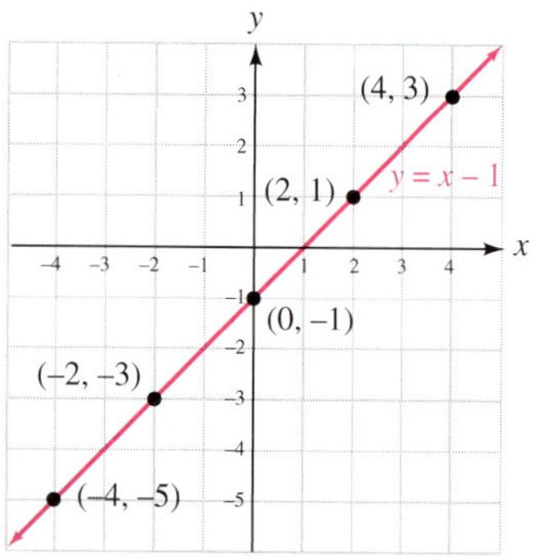

Draw a straight line through the points. This is the *graph of the equation.*

The equation $y = x - 1$ is said to be *linear* and its graph is a line. By definition, a linear equation in two variables is any equation that can be written in the following **standard** or **general form,** where the variable terms appears on one side of an equal symbol and a constant appears on the other.

Linear Equations

A **linear equation in two variables** is an equation that can be written in the form

$$Ax + By = C$$

where A, B, and C are real numbers and A and B are not both 0.

> ***Success Tip***
> The exponent on each variable of a linear equation is an understood 1. For example, $y = 2x + 4$ can be thought of as $y^1 = 2x^1 + 4$.

Some more examples of linear equations are

$y = 2x + 4$, $\quad 2x + 3y = 12$, $\quad$ and $\quad 3x = 5y$

Linear equations can be graphed in several ways. Generally, the form in which an equation is written determines the method that we use to graph it. To graph linear equations solved for y, such as $y = 2x + 4$, we can use the following method.

Graphing Linear Equations Solved for y by Plotting Points

1. Find three ordered pairs that are solutions of the equation by selecting three values for x and calculating the corresponding values of y.
2. Plot the solutions on a rectangular coordinate system.
3. Draw a straight line passing through the points. If the points do not lie on a line, check your computations.

EXAMPLE 4

Graph: $y = 2x + 4$

Strategy We will find three solutions of the equation, plot them on a rectangular coordinate system, and then draw a straight line passing through the points.

Why To *graph* a linear equation in two variables means to make a drawing that represents all of its solutions.

Solution To find three solutions of this linear equation, we select three values for x that will make the computations easy. Then we find each corresponding value of y.

Success Tip
When selecting x-values for a table of solutions, a rule of thumb is to choose a negative number, a positive number, and 0. When $x = 0$, the computations to find y are usually quite simple.

If $x = -2$	***If*** $x = 0$	***If*** $x = 2$
$y = 2x + 4$	$y = 2x + 4$	$y = 2x + 4$
$y = 2(-2) + 4$	$y = 2(0) + 4$	$y = 2(2) + 4$
$y = -4 + 4$	$y = 0 + 4$	$y = 4 + 4$
$y = 0$	$y = 4$	$y = 8$
$(-2, 0)$ is a solution.	$(0, 4)$ is a solution.	$(2, 8)$ is a solution.

We enter the results in a table of solutions and plot the points. Then we draw a straight line through the points and label it $y = 2x + 4$.

Success Tip
Since two points determine a line, only two points are needed to graph a linear equation. However, we should plot a third point as a check. If the three points do not lie on a straight line, then at least one of them is in error.

$y = 2x + 4$

x	y	(x, y)
-2	0	$(-2, 0)$
0	4	$(0, 4)$
2	8	$(2, 8)$

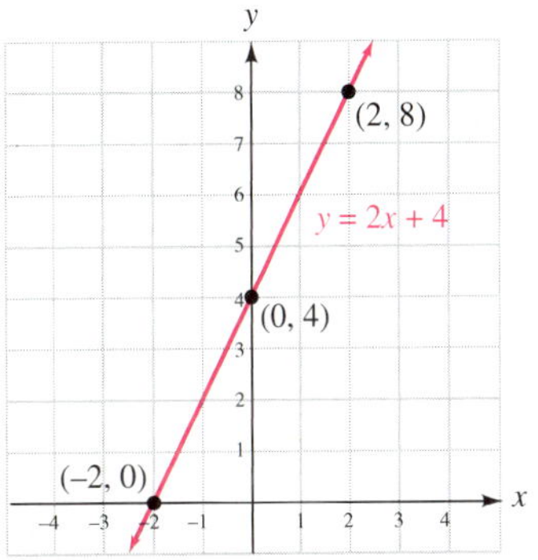

As a check, we can pick two points that the line appears to pass through, such as (1, 6) and (−1, 2). When we substitute their coordinates into the given equation, the two true statements that result indicate that (1, 6) and (−1, 2) are solutions and that the graph of the line is correctly drawn.

Check:

$y = 2x + 4$		$y = 2x + 4$	
$6 \stackrel{?}{=} 2(1) + 4$		$2 \stackrel{?}{=} 2(-1) + 4$	
$6 \stackrel{?}{=} 2 + 4$		$2 \stackrel{?}{=} -2 + 4$	
$6 = 6$	True	$2 = 2$	True

Self Check 4 Graph: $y = 2x - 2$

Now Try **Problem 41**

EXAMPLE 5 Graph: $y = -3x$

Strategy We will find three solutions of the equation, plot them on a rectangular coordinate system, and then draw a straight line passing through the points.

Why To *graph* a linear equation in two variables means to make a drawing that represents all of its solutions.

Solution To find three solutions, we begin by selecting three x-values: $-1, 0,$ and 1. Then we find the corresponding values of y. If $x = -1$, we have

$y = -3x$ This is the equation to graph.

$y = -3(-1)$ Substitute -1 for x.

$y = 3$

$(-1, 3)$ is a solution.

In a similar manner, we find the y-values for x-values of 0 and 1, and record the results in a table of solutions. After plotting the ordered pairs, we draw a straight line through the points and label it $y = -3x$.

$y = -3x$

x	y	(x, y)
-1	3	$(-1, 3)$
0	0	$(0, 0)$
1	-3	$(1, -3)$

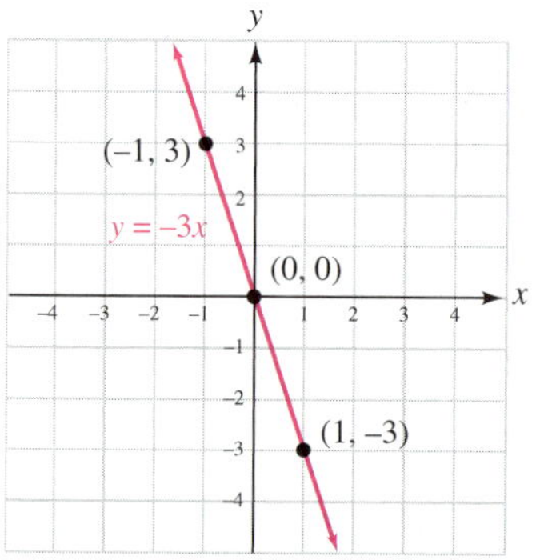

Self Check 5 Graph: $y = -4x$

Now Try **Problem 45**

To graph linear equations in x and y using the method discussed in this section, the variable y must be isolated on one side of the equation.

EXAMPLE 6

Graph $2x + 3y = -12$ by first solving for y.

Strategy We will use properties of equality to solve the given equation for y. Then we will use the point-plotting method of this section to graph the resulting equivalent equation.

Why The calculations to find several solutions of a linear equation in two variables are usually easier when the equation is solved for y.

Solution To solve for y, we proceed as follows.

$2x + 3y = -12$ This is the given equation.

$2x + 3y - 2x = -2x - 12$ To isolate the variable term $3y$ on the left side, subtract $2x$ from both sides. When solving for y, it is common practice to write the subtraction (or addition) of a variable term *before* the constant term.

$3y = -2x - 12$ On the left side, combine like terms, $2x - 2x = 0$.

Notation

Note that the division by 3 on the right side of the equation is done term-by-term instead of with a single fraction bar. It is common practice to write

$$\frac{-2x}{3} - \frac{12}{3} \quad \text{not} \quad \frac{-2x - 12}{3}$$

Success Tip

When we chose x-values that are multiples of the denominator 3, the corresponding y-values are integers, and not difficult to plot fractions.

$$\frac{3y}{3} = \frac{-2x}{3} - \frac{12}{3}$$ To isolate y, undo the multiplication by 3 by dividing both sides by 3.

$$y = -\frac{2}{3}x - 4$$ Write $\frac{-2x}{3}$ as $-\frac{2}{3}x$. Simplify: $\frac{12}{3} = 4$.

Since $y = -\frac{2}{3}x - 4$ is equivalent to $2x + 3y = -12$, we can use it to draw the graph of $2x + 3y = -12$.

To find solutions of $y = -\frac{2}{3}x - 4$, each value of x must be multiplied by $-\frac{2}{3}$. This computation is made easier if we select x-values that are *multiples of the denominator 3,* such as -3, 0, and 6. For example, if $x = -3$, we have

$$y = -\frac{2}{3}x - 4$$ This is the equation to graph.

$$y = -\frac{2}{3}(-3) - 4$$ Substitute -3 for x.

$$y = 2 - 4$$ Multiply: $-\frac{2}{3}(-3) = 2$. This step is simpler if we select x-values that are multiples of 3.

$$y = -2$$

Thus, $(-3, -2)$ is a solution.

Two more solutions, one for $x = 0$ and one for $x = 6$, can be found in a similar way, and entered in a table. We plot the ordered pairs, draw a straight line through the points, and label the line as $y = -\frac{2}{3}x - 4$ or as $2x + 3y = -12$.

$2x + 3y = -12$
or
$y = -\frac{2}{3}x - 4$

x	y	(x, y)
-3	-2	$(-3, -2)$
0	-4	$(0, -4)$
6	-8	$(6, -8)$

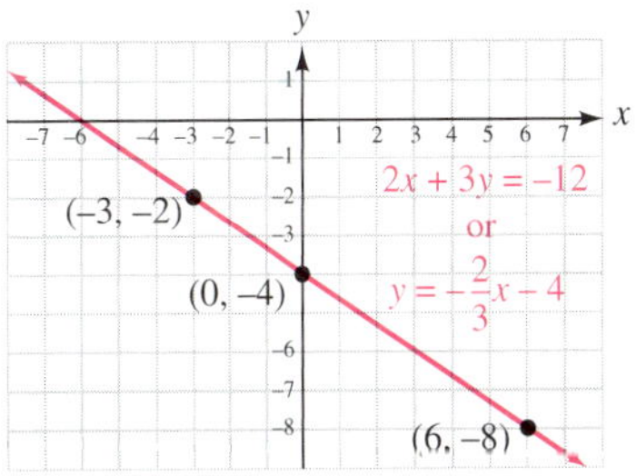

Self Check 6 Graph $5x - 2y = -2$ by first solving for y.

Now Try **Problem 67**

5 Use Graphs of Linear Equations to Solve Applied Problems.

When linear equations are used to model real-life situations, they are often written in variables other than x and y. In such cases, we must make the appropriate changes when labeling the table of solutions and the graph of the equation.

EXAMPLE 7 ***Cleaning Windows.*** The linear equation $A = -0.03n + 32$ estimates the amount A of glass cleaning solution (in ounces) that is left in the bottle after the sprayer trigger has been pulled a total of n times. Graph the equation and use the graph to estimate the amount of solution that is left after 500 sprays.

Strategy We will find three solutions of the equation, plot them on a rectangular coordinate system, and then draw a straight line passing through the points.

Why We can use the graph to estimate the amount of solution left after any number of sprays.

Solution Since A depends on n in the equation $A = -0.03n + 32$, solutions will have the form (n, A). To find three solutions, we begin by selecting three values of n. Because the number of trials cannot be negative, and the computations to find A involve decimal multiplication, we select 0, 100, and 1,000. For example, if $n = 100$, we have

$$A = -0.03n + 32 \quad \text{This is the equation to graph.}$$
$$A = -0.03(100) + 32$$
$$A = -3 + 32 \quad \text{Multiply: } -0.03(100) = -3.$$
$$A = 29$$

Thus, (100, 29) is a solution. It indicates that after 100 sprays, 29 ounces of cleaner will be left in the bottle.

In the same way, solutions are found for $n = 0$ and $n = 1{,}000$ and listed in the table. Then the ordered pairs are plotted and a straight line is drawn through the points.

To graphically estimate the amount of solution that is left after 500 sprays, we draw the dashed blue lines, as shown. Reading on the vertical A-axis, we see that after 500 sprays, about 17 ounces of glass cleaning solution would be left.

Success Tip

Since we selected large n-values such as 100 and 1,000, the horizontal n-axis was scaled in units of 100. Since the corresponding A-values range from 2 to 32, the vertical A-axis was scaled in units of 4.

$A = -0.03n + 32$

n	A	(n, A)
0	32	(0, 32)
100	29	(100, 29)
1,000	2	(1,000, 2)

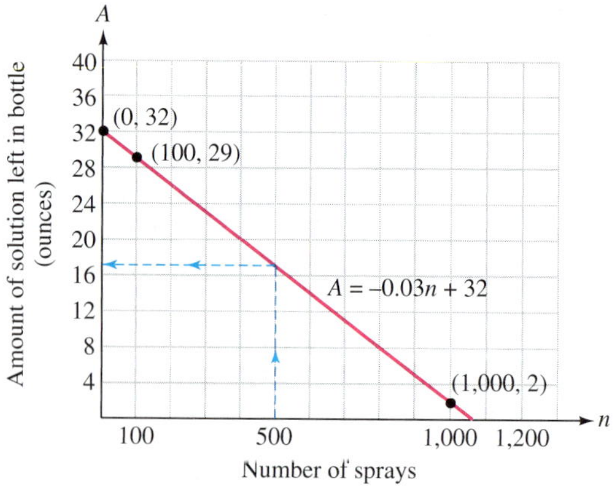

Now Try Problem 77

ANSWERS TO SELF CHECKS 1. Yes 2. $(-2, -10)$ 3.

x	y	(x, y)
3	-2	$(3, -2)$
5	-5	$(5, -5)$

4.

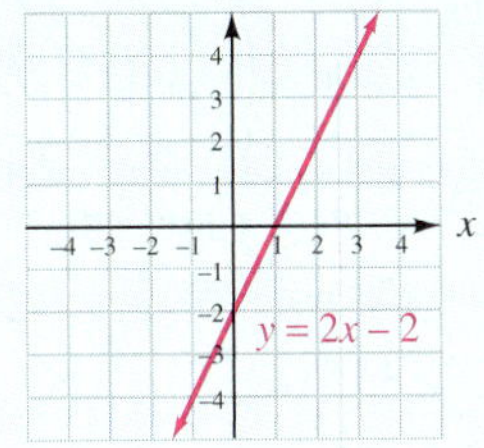

5.

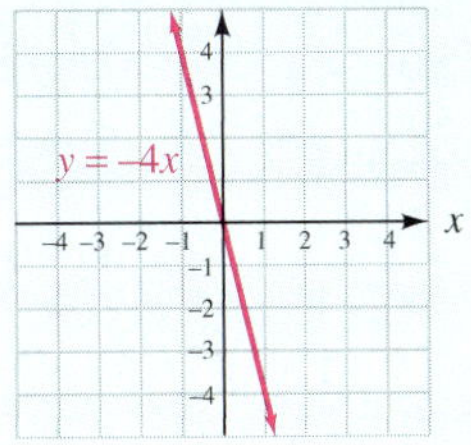

6.

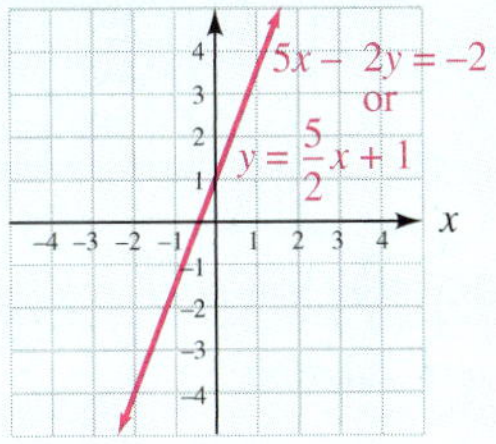

STUDY SET 3.2

VOCABULARY

Fill in the blanks.

1. $y = 9x + 5$ is an equation in _____ variables, x and y.
2. A ________ of an equation in two variables is an ordered pair of numbers that makes the equation a true statement.
3. Solutions of equations in two variables are often listed in a ______ of solutions.
4. The line that represents all of the solutions of a linear equation is called the _______ of the equation.
5. $y = 3x + 8$ is called a _______ equation because its graph is a line.
6. The ________________ form of a linear equation in two variables is $Ax + By = C$.

CONCEPTS

7. Consider: $y = -3x + 6$
   a. How many variables does the equation contain?
   b. Does $(4, -6)$ satisfy the equation?
   c. Is $(-2, 0)$ a solution?
   d. How many solutions does this equation have?
8. To graph a linear equation, three solutions were found, they were plotted (in black), and a straight line was drawn through them, as shown in the next column.
   a. Looking at the graph, complete the table of solutions.
   b. From the graph, determine three other solutions of the equation.

x	y	(x, y)
-4		(,)
-1		(,)
1		(,)

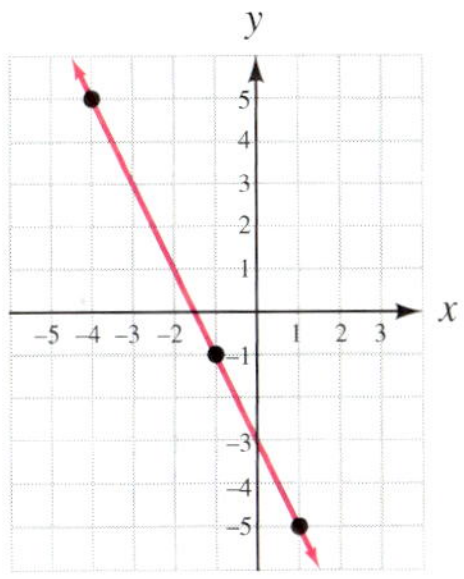

9. The graph of $y = -2x - 3$ is shown in Problem 8. Fill in the blanks: Every point on the graph represents an ordered-pair ________ of $y = -2x - 3$ and every ordered-pair solution is a ______ on the graph.
10. The graph of a linear equation is shown.
    a. If the coordinates of point M are substituted into the equation, will the result be true or false?
    b. If the coordinates of point N are substituted into the equation, will the result be true or false?

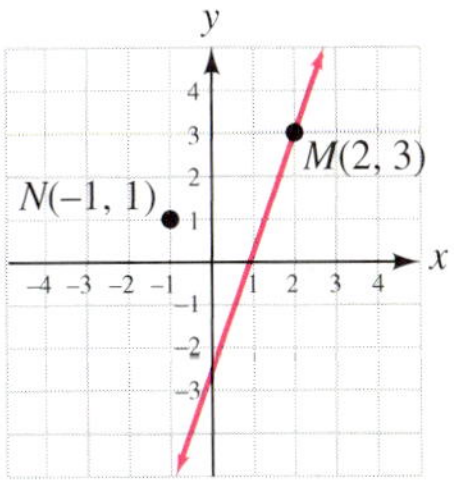

11. Suppose you are making a table of solutions for each given equation. What three x-values would you select to make the calculations for finding the corresponding y-values the easiest?
    a. $y = \frac{4}{5}x + 2$
    b. $y = 0.6x + 500$

12. A table of solutions for a linear equation is shown below. When constructing the graph of the equation, how would you scale the x-axis and the y-axis?

x	y	(x, y)
-20	600	$(-20, 600)$
5	100	$(5, 100)$
35	-500	$(35, -500)$

NOTATION

Complete each solution.

13. Verify that $(-2, 6)$ is a solution of $y = -x + 4$.

$$y = -x + 4$$
$$\quad \stackrel{?}{=} -(\quad) + 4$$
$$6 \stackrel{?}{=} \quad + 4$$
$$6 = \quad$$

14. Solve $5x + 3y = 15$ for y.

$$5x + 3y - 5x = \quad + 15$$
$$\quad = -5x + 15$$
$$\frac{3y}{\quad} = \frac{-5x}{\quad} + \frac{15}{\quad}$$
$$y = \quad x + \quad$$

15. **a.** Rewrite the linear equation $y = \frac{1}{2}x + 7$ showing the understood exponents on the variables.

$$y^{\quad} = \tfrac{1}{2}x^{\quad} + 7$$

b. Explain why $y = x^2 + 2$ and $y = x^3 - 4$ are not linear equations.

16. Complete the labeling of the table of solutions and graph of $c = -a + 4$.

		(,)
-1	5	$(-1, 5)$
0	4	$(0, 4)$
2	2	$(2, 2)$

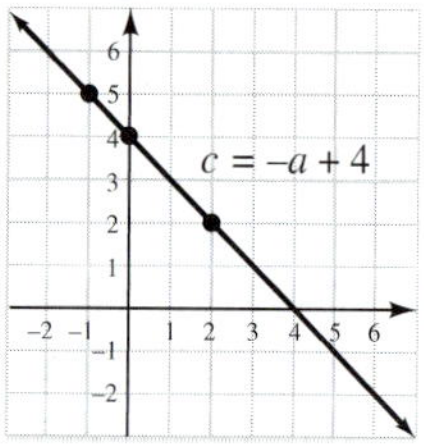

GUIDED PRACTICE

***Determine whether each equation has the given ordered pair as a solution.* See Example 1.**

17. $y = 5x - 4; (1, 1)$

18. $y = -2x + 3; (2, -1)$

19. $7x - 2y = 3; (2, 6)$

20. $10x - y = 10; (0, 0)$

21. $x + 12y = -12; (0, -1)$

22. $-2x + 3y = 0; (-3, -2)$

23. $3x - 6y = 12; (-3.6, -3.8)$

24. $8x + 4y = 10; (-0.5, 3.5)$

25. $y - 6x = 12; \left(\frac{5}{6}, 7\right)$

26. $y + 8x = 4; \left(\frac{3}{4}, 2\right)$

27. $y = -\frac{3}{4}x + 8; (-8, 12)$

28. $y = \frac{1}{6}x - 2; (-12, 4)$

***For each equation, complete the solution.* See Example 2.**

29. $y = -5x - 4; (-3, \quad)$

30. $y = 8x + 30; (-6, \quad)$

31. $4x - 5y = -4; (\quad, 4)$

32. $7x + y = -12; (\quad, 2)$

33. $y = \frac{x}{4} + 9; (16, \quad)$

34. $y = \frac{x}{6} - 8; (48, \quad)$

35. $7x = 4y; \left(\quad, -2\right)$

36. $11x = 16y; \left(\quad, -3\right)$

***Complete each table of solutions.* See Example 3.**

37. $y = 2x - 4$

x	y	(x, y)
8		
	8	

38. $y = 3x + 1$

x	y	(x, y)
-3		
	-2	

39. $3x - y = -2$

x	y	(x, y)
-5		
	-1	

40. $5x - 2y = -15$

x	y	(x, y)
5		
	0	

***Construct a table of solutions and then graph each equation.* See Examples 4 and 5.**

41. $y = 2x - 3$

42. $y = 3x + 1$

43. $y = 5x - 4$

44. $y = 6x - 3$

45. $y = x$

46. $y = 4x$

47. $y = -x - 1$

48. $y = -x + 2$

49. $y = -2x + 1$

50. $y = -3x + 2$

51. $y = \frac{x}{3}$

52. $y = -\frac{x}{3} - 1$

53. $y = -\frac{1}{2}x$

54. $y = \frac{3}{4}x$

55. $y = \frac{3}{8}x - 6$

56. $y = -\frac{3}{2}x + 2$

57. $y = \frac{2}{3}x - 2$

58. $y = \frac{5}{6}x - 5$

59. $y = 1.5x - 4$

60. $y = 0.5x + 3$

***Solve each equation for y and then graph it.* See Example 6.**

61. $3y = 12x + 15$

62. $5y = 20x - 30$

63. $-6y = 30x + 12$

64. $-3y = 9x - 15$

65. $8x + 4y = 16$

66. $14x + 7y = 28$

67. $5y - x = 20$

68. $4y - x = 8$

69. $7x - y = 1$

70. $2x - y = -3$

71. $7y = -2x$

72. $6y = -4x$

APPLICATIONS

73. BILLIARDS The path traveled by the black 8-ball is described by the equations $y = 2x - 4$ and $y = -2x + 12$. Construct a table of solutions for $y = 2x - 4$ using the x-values 1, 2, and 4. Do the same for $y = -2x + 12$, using the x-values 4, 6, and 8. Then graph the path of the 8-ball.

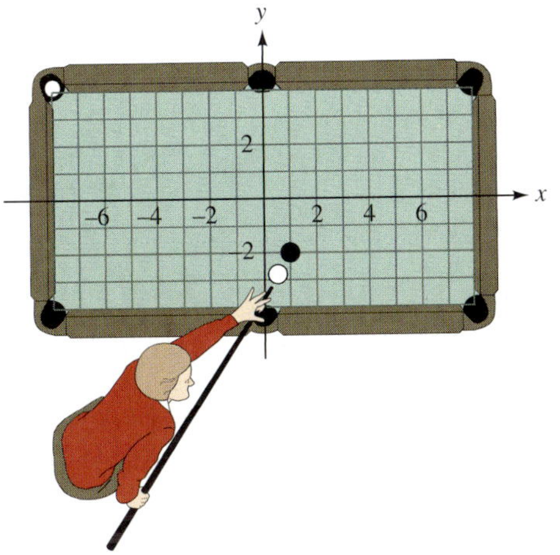

74. PING-PONG The path traveled by a Ping-Pong ball is described by the equations $y = \frac{1}{2}x + \frac{3}{2}$ and $y = -\frac{1}{2}x - \frac{3}{2}$. Construct a table of solutions for $y = \frac{1}{2}x + \frac{3}{2}$ using the x-values 7, 3, and −3. Do the same for $y = -\frac{1}{2}x - \frac{3}{2}$, using the x-values −3, −5, and −7. Then graph the path of the ball.

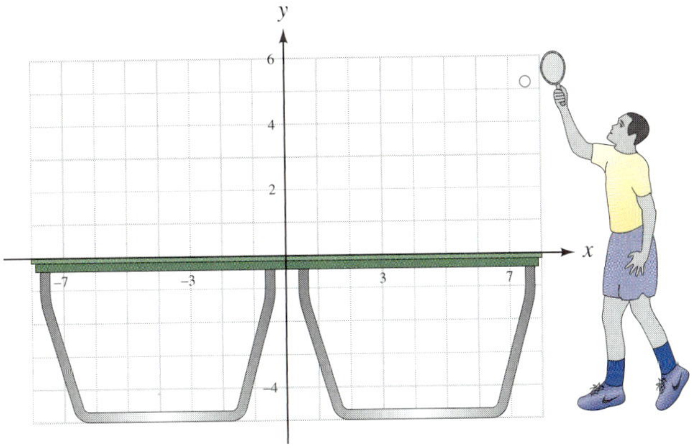

75. DEFROSTING POULTRY The number of hours h needed to defrost a turkey weighing p pounds in the refrigerator can be estimated by $h = 5p$. Graph the equation and use the graph to estimate the time needed to defrost a 25-pound turkey. (Source: helpwithcooking.com.)

76. OWNING A CAR In 2006, the average cost c (in dollars) to own and operate a car was estimated by $c = 0.52m$, where m represents the number of miles driven. Graph the equation and use the graph to estimate the cost of operating a car that is driven 25,000 miles. (Source: Automobile Association of America.)

77. HOUSEKEEPING The linear equation $A = -0.02n + 16$ estimates the amount A of furniture polish (in ounces) that is left in the bottle after the sprayer trigger has been pulled a total of n times. Graph the equation and use the graph to estimate the amount of polish that is left after 650 sprays.

78. SHARPENING PENCILS The linear equation $L = -0.04t + 8$ estimates the length L (in inches) of a pencil after it has been inserted into a sharpener and the handle turned a total of t times. Graph the equation and use the graph to estimate the length of the pencil after 75 turns of the handle.

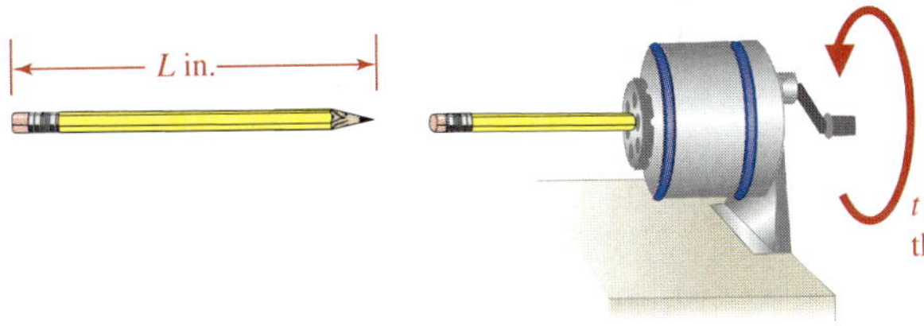

79. NFL TICKETS The average ticket price p to a National Football League game during the years 1990–2005 is approximated by $p = \frac{12}{5}t + 22$, where t is the number of years after 1990. Graph this equation and use the graph to predict the average ticket price in 2020. (Source: Team Marketing Report, NFL.)

80. U.S. AUTOMOBILE ACCIDENTS The number n of lives saved by seat belts during the years 1995–2004 is estimated by $n = 615t + 9{,}900$, where t is the number of years after 1995. Graph this equation and use the graph to predict the number of lives that will be saved by seat belts in 2015. (Source: Bureau of Transportation Statistics.)

81. RAFFLES A private school is going to sell raffle tickets as a fund raiser. Suppose the number n of raffle tickets that will be sold is predicted by the equation $n = -20p + 300$, where p is the price of a raffle ticket in dollars. Graph the equation and use the graph to predict the number of raffle tickets that will be sold at a price of \$6.

82. CATS The number n of cat owners (in millions) in the United States during the years 1995–2004 is estimated by $n = \frac{13}{20}t + 32$, where t is the number of years after 1995. Graph this equation and use the graph to predict the number of cat owners in the United States in 2015. (Source: Pet Food Institute.)

WRITING

83. When we say that $(-2, -6)$ is a solution of $y = x - 4$, what do we mean?

84. What is a table of solutions?

85. What does it mean when we say that a linear equation in two variables has infinitely many solutions?

86. A linear equation and a graph are two ways of describing a relationship between two quantities. Which do you think is more informative and why?

87. From geometry, we know that two points determine a line. Why is it a good practice when graphing linear equations to find and plot three solutions instead of just two?

88. A student found three solutions of a linear equation in two variables and plotted them as shown. What conclusion can be made about the location of the points?

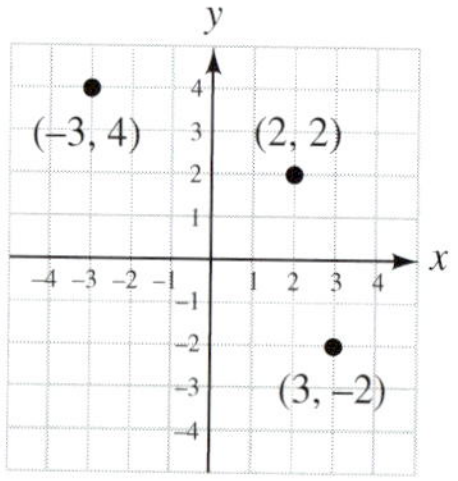

89. Two students were asked to graph $y = 3x - 1$. One made the table of solutions on the left. The other made the table on the right. The tables are completely different. Could they both be correct? Explain.

x	y	(x, y)
0	−1	(0, −1)
2	5	(2, 5)
3	8	(3, 8)

x	y	(x, y)
−2	−7	(−2, −7)
−1	−4	(−1, −4)
1	2	(1, 2)

90. Both graphs below are of the same linear equation $y = 10x$. Why do the graphs have a different appearance?

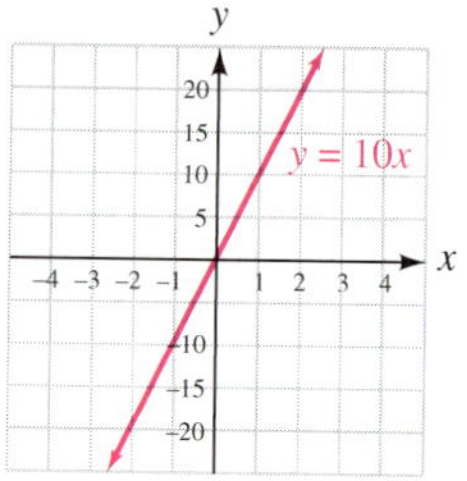

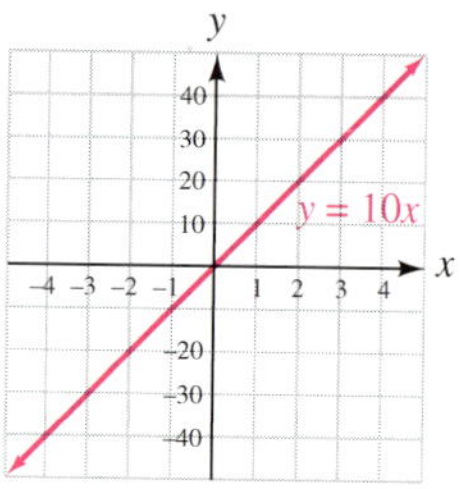

REVIEW

91. Simplify: $-(-5 - 4c)$

92. Write the set of integers.

93. Find the volume, to the nearest tenth of a cubic foot, of a sphere with radius 6 feet.

94. Solve: $-2(a + 3) = 3(a - 5)$

CHALLENGE PROBLEMS

Graph each of the following nonlinear equations in two variables by constructing a table of solutions consisting of seven ordered pairs. These equations are called nonlinear, because their graphs are not straight lines.

95. $y = x^2 + 1$

96. $y = x^3 - 2$

97. $y = |x| - 2$

98. $y = (x + 2)^2$

SECTION 3.3
Intercepts

Objectives

1. Identify intercepts of a graph.
2. Graph linear equations by finding intercepts.
3. Identify and graph horizontal and vertical lines.
4. Obtain information from intercepts.
5. (Optional) Use a calculator to graph linear equations.

In this section, we will graph linear equations by determining the points where their graphs intersect the x-axis and the y-axis. These points are called the *intercepts* of the graph.

1 Identify Intercepts of a Graph.

The graph of $y = 2x - 4$ is shown below. We see that the graph crosses the y-axis at the point $(0, -4)$; this point is called the ***y*-intercept** of the graph. The graph crosses the x-axis at the point $(2, 0)$; this point is called the ***x*-intercept** of the graph.

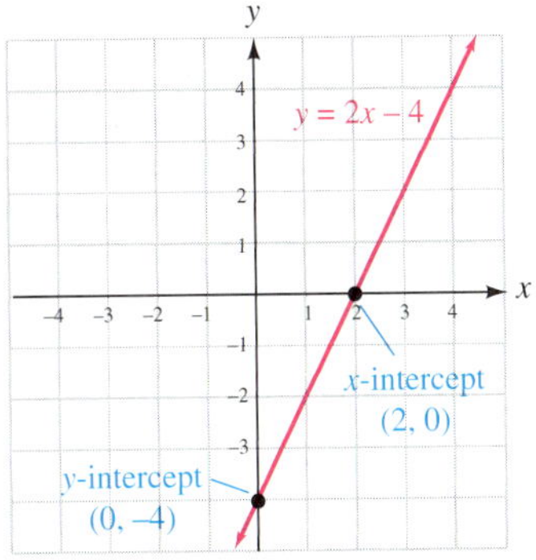

The Language of Algebra
The point where a line *intersects* the x- or y-axis is called an *intercept*.

EXAMPLE 1 For the graphs in figures (a) and (b), give the coordinates of the x- and y-intercepts.

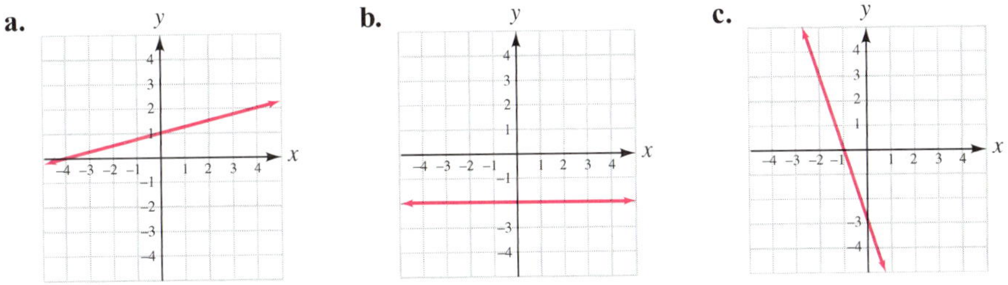

Strategy We will determine where each graph (shown in red) crosses the x-axis and the y-axis.

Why The point at which a graph crosses the x-axis is the x-intercept and the point at which a graph crosses the y-axis is the y-intercept.

Solution

a. In figure (a), the graph crosses the x-axis at $(-4, 0)$. This is the x-intercept. The graph crosses the y-axis at $(0, 1)$. This is the y-intercept.

b. In figure (b), the horizontal line does not cross the x-axis; there is no x-intercept. The graph crosses the y-axis at $(0, -2)$. This is the y-intercept.

Self Check 1 Give the coordinates of the x- and y-intercept of the graph in Figure (c).

Now Try **Problem 11**

From the previous examples, we see that a y-intercept has an x-coordinate of 0, and an x-intercept has a y-coordinate of 0. These observations suggest the following procedures for finding the intercepts of a graph from its equation.

Finding Intercepts

To find the y-intercept, substitute 0 for x in the given equation and solve for y.

To find the x-intercept, substitute 0 for y in the given equation and solve for x.

2 Graph Linear Equations by Finding Intercepts.

Plotting the x- and y-intercepts of a graph and drawing a line through them is called the **intercept method of graphing a line.** This method is useful when graphing linear equations written in the standard (general) form $Ax + By = C$.

EXAMPLE 2

Graph $x - 3y = 6$ by finding the y- and x-intercepts.

Strategy We will let $x = 0$ to find the y-intercept of the graph. We will then let $y = 0$ to find the x-intercept.

Why Since two points determine a line, the y-intercept and x-intercept are enough information to graph this linear equation.

Solution

***y*-intercept: $x = 0$**

$x - 3y = 6$

$0 - 3y = 6$ Substitute 0 for x.

$-3y = 6$

$y = -2$ To isolate y, divide both sides by −3.

The y-intercept is $(0, -2)$.

***x*-intercept: $y = 0$**

$x - 3y = 6$

$x - 3(0) = 6$ Substitute 0 for y.

$x - 0 = 6$

$x = 6$

The x-intercept is $(6, 0)$.

Since each intercept of the graph is a solution of the equation, we enter the intercepts in the table of solutions below.

As a check, we find one more point on the line. We select a convenient value for x, say, 3, and find the corresponding value of y.

Success Tip
The check point should lie on the same line as the x- and y-intercepts. If it does not, check your work to find the incorrect coordinate or coordinates.

$x - 3y = 6$

$3 - 3y = 6$ Substitute 3 for x.

$-3y = 3$ To isolate the variable term, −3y, subtract 3 from both sides.

$y = -1$ To isolate y, divide both sides by −3.

Therefore, $(3, -1)$ is a solution. It is also entered in the table.

We plot the intercepts and the check point, draw a straight line through them, and label the line as $x - 3y = 6$.

$x - 3y = 6$

x	y	(x, y)	
0	−2	(0, −2)	← y-intercept
6	0	(6, 0)	← x-intercept
3	−1	(3, −1)	← Check point

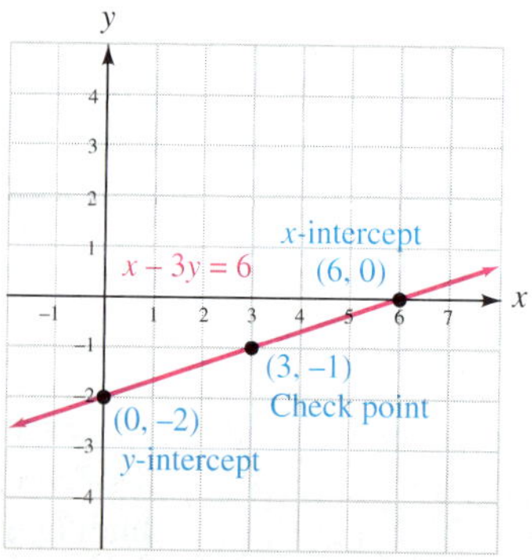

Self Check 2 Graph $x - 2y = 2$ by finding the intercepts.

Now Try **Problem 27**

The computations for finding intercepts can be simplified if we realize what occurs when we substitute 0 for y or 0 for x in an equation written in the form $Ax + By = C$.

EXAMPLE 3

Graph $40x + 3y = -120$ by finding the y- and x-intercepts.

Strategy We will let $x = 0$ to find the y-intercept of the graph. We will then let $y = 0$ to find the x-intercept.

Why Since two points determine a line, the y-intercept and x-intercept are enough information to graph this linear equation.

Solution When we substitute 0 for x, it follows that the term $40x$ will be equal to 0. Therefore, to find the y-intercept, we can cover the $40x$ and solve the remaining equation for y.

The Language of Algebra
This method to find the intercepts of the graph of a linear equation is commonly referred to as the *cover-over method*.

$$40x + 3y = -120 \quad \text{If } x = 0\text{, then } 40x = 40(0) = 0\text{. Cover the } 40x \text{ term.}$$
$$y = -40 \quad \text{To solve } 3y = -120\text{, divide both sides by 3.}$$

The y-intercept is $(0, -40)$.

When we substitute 0 for y, it follows that the term $3y$ will be equal to 0. Therefore, to find the x-intercept, we can cover the $3y$ and solve the remaining equation for x.

$$40x + 3y = -120 \quad \text{If } y = 0\text{, then } 3y = 3(0) = 0\text{. Cover the } 3y \text{ term.}$$
$$x = -3 \quad \text{To solve } 40x = -120\text{, divide both sides by 40.}$$

The x-intercept is $(-3, 0)$.

Caution
When using the cover-over method to find the y-intercept, be careful not to cover the sign in front of the y-term.

We can find a third solution by selecting a convenient value for x and finding the corresponding value for y. If we choose $x = -6$, we find that $y = 40$. The solution $(-6, 40)$ is entered in the table, and the equation is graphed as shown.

Success Tip
To fit y-values of 40 and -40 on the graph, the y-axis was scaled in units of 10.

$40x + 3y = -120$

x	y	(x, y)	
0	-40	$(0, -40)$	← y-intercept
-3	0	$(-3, 0)$	← x-intercept
-6	40	$(-6, 40)$	← Check point

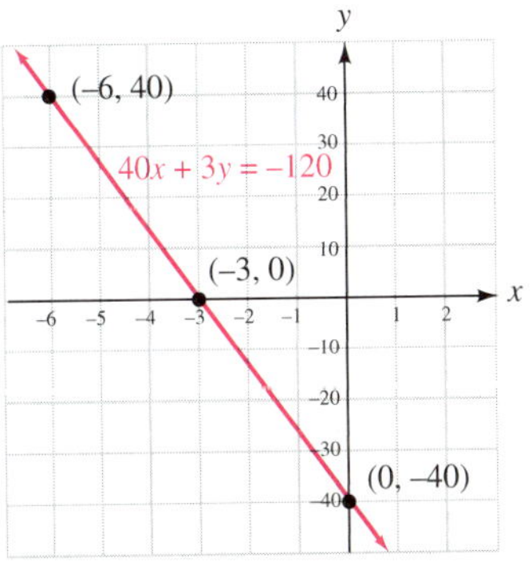

Self Check 3 Graph $32x + 5y = -160$ by finding the intercepts.

Now Try **Problem 35**

EXAMPLE 4

Graph $3x = -5y + 8$ by finding the intercepts.

Strategy We will let $x = 0$ to find the y-intercept of the graph. We will then let $y = 0$ to find the x-intercept.

Why Since two points determine a line, the y-intercept and x-intercept are enough information to graph this linear equation.

Solution We find the intercepts and select $x = 1$ to find a check point.

***y*-intercept: *x* = 0**

$$3x = -5y + 8$$
$$3(0) = -5y + 8$$
$$0 = -5y + 8$$
$$-8 = -5y$$
$$\frac{8}{5} = y$$
$$1\tfrac{3}{5} = y$$

The y-intercept is $\left(0, 1\tfrac{3}{5}\right)$.

***x*-intercept: *y* = 0**

$$3x = -5y + 8$$
$$3x = -5(0) + 8$$
$$3x = 8$$
$$x = \frac{8}{3}$$
$$x = 2\tfrac{2}{3}$$

The x-intercept is $\left(2\tfrac{2}{3}, 0\right)$.

***Check point:* *x* = 1**

$$3x = -5y + 8$$
$$3(1) = -5y + 8$$
$$3 = -5y + 8$$
$$-5 = -5y$$
$$1 = y$$

A check point is (1, 1).

The ordered pairs are plotted as shown, and a straight line is then drawn through them.

> ***Success Tip***
> When graphing, it is often helpful to write any coordinates that are improper fractions as mixed numbers. For example:
> $\left(\frac{8}{3}, 0\right) = \left(2\frac{2}{3}, 0\right)$

$3x = -5y + 8$

x	y	(x, y)	
0	$\frac{8}{5} = 1\frac{3}{5}$	$\left(0, 1\frac{3}{5}\right)$	← y-intercept
$\frac{8}{3} = 2\frac{2}{3}$	0	$\left(2\frac{2}{3}, 0\right)$	← x-intercept
1	1	(1, 1)	← Check point

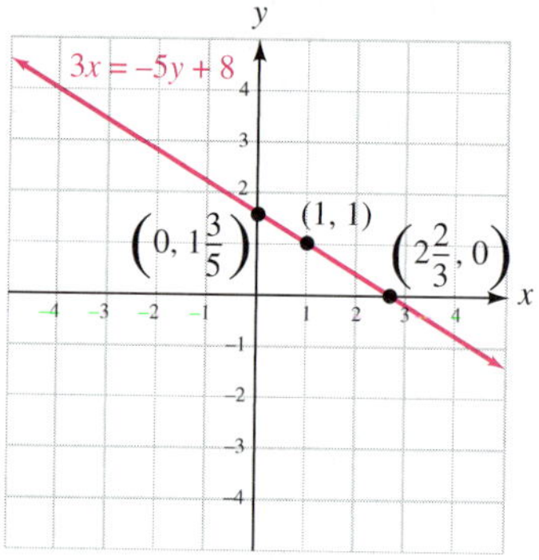

Self Check 4 Graph $8x = -4y + 15$ by finding the intercepts.

Now Try **Problem 39**

EXAMPLE 5

Graph $2x + 3y = 0$ by finding the intercepts.

Strategy We will let $x = 0$ to find the y-intercept of the graph. We will then let $y = 0$ to find the x-intercept.

Why Since two points determine a line, the y-intercept and x-intercept are enough information to graph this linear equation.

Solution When we find the y- and x-intercepts (shown on the next page), we see that they are both (0, 0). In this case, the line passes through the origin. Since we are using two points and a check point to graph lines, we need to find two more ordered-pair solutions.

If $x = 3$, we see that $(3, -2)$ is a solution. And if $x = -3$, we see that $(-3, 2)$ is also a solution. These two solutions and the origin are plotted and a straight line is drawn through them to give the graph of $2x + 3y = 0$.

***y*-intercept: $x = 0$**

$$2x + 3y = 0$$
$$2(0) + 3y = 0$$
$$3y = 0$$
$$y = 0$$

The y-intercept is $(0, 0)$.

***x*-intercept: $y = 0$**

$$2x + 3y = 0$$
$$2x + 3(0) = 0$$
$$2x = 0$$
$$x = 0$$

The x-intercept is $(0, 0)$.

Let $x = 3$

$$2x + 3y = 0$$
$$2(3) + 3y = 0$$
$$6 + 3y = 0$$
$$3y = -6$$
$$y = -2$$

$(3, -2)$ is a solution.

Let $x = -3$

$$2x + 3y = 0$$
$$2(-3) + 3y = 0$$
$$-6 + 3y = 0$$
$$3y = 6$$
$$y = 2$$

$(-3, 2)$ is a solution.

$2x + 3y = 0$

x	y	(x, y)	
0	0	$(0, 0)$	← The x-intercept and y-intercept.
3	-2	$(3, -2)$	← A solution.
-3	2	$(-3, 2)$	← This solution serves as a check point.

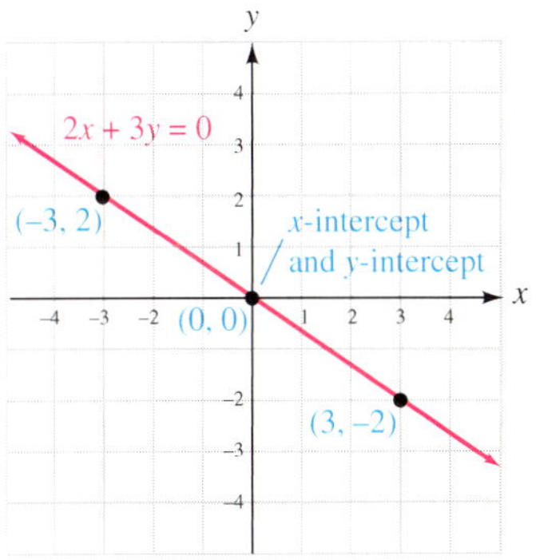

Self Check 5 Graph $5x - 2y = 0$ by finding the intercepts.

Now Try Problem 47

3 Identify and Graph Horizontal and Vertical Lines.

Equations such as $y = 4$ and $x = -3$ are linear equations, because they can be written in the general form $Ax + By = C$. For example, $y = 4$ is equivalent to $0x + 1y = 4$ and $x = -3$ is equivalent to $1x + 0y = -3$. We now discuss how to graph these types of linear equations.

EXAMPLE 6 Graph: $y = 4$

Strategy To find three ordered-pair solutions of this equation to plot, we will select three values for x and use 4 for y each time.

Why The given equation requires that $y = 4$.

Solution We can write the equation in general form as $0x + y = 4$. Since the coefficient of x is 0, the numbers chosen for x have no effect on y. The value of y is always 4. For example, if $x = 2$, we have

$0x + y = 4$ — This is the given equation, $y = 4$, written in standard (general) form.

$0(2) + y = 4$ — Substitute 2 for x.

$y = 4$ — Simplify the left side.

81. LANDSCAPING A developer is going to purchase x trees and y shrubs to landscape a new office complex. The trees cost \$50 each and the shrubs cost \$25 each. His budget is \$5,000. This situation is modeled by the equation $50x + 25y = 5{,}000$. Use the intercept method to graph it.

a. What information is given by the y-intercept?

b. What information is given by the x-intercept?

82. EGGS The number of eggs eaten by an average American in one year has remained almost constant since the year 2000. See the graph below. Draw a straight line that passes through, or near, the data points. What is the equation of the line?

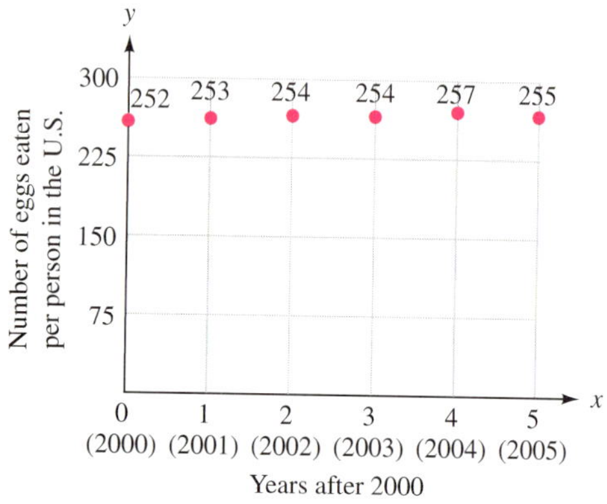

Source: United Egg

WRITING

83. To graph $3x + 2y = 12$, a student found the intercepts and a check point, and graphed them, as shown in figure (a). Instead of drawing a crooked line through the points, what should he have done?

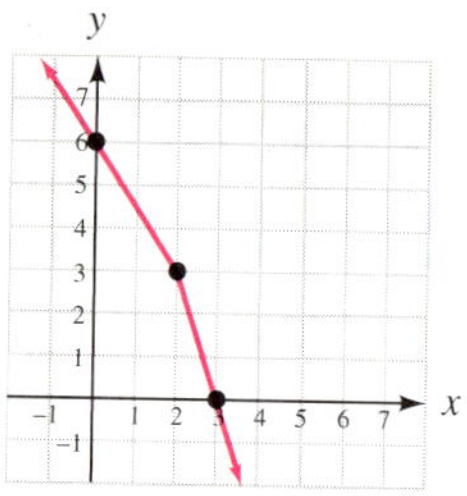

(a)

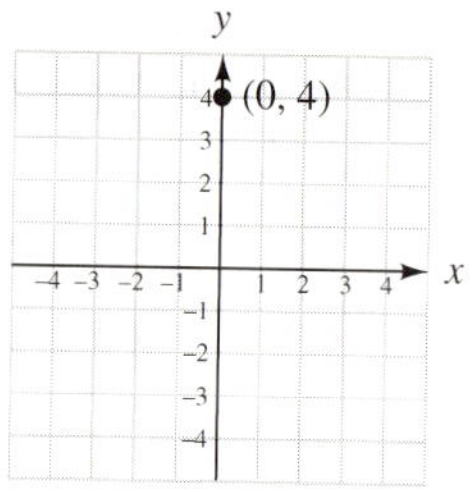

(b)

84. A student graphed the linear equation $y = 4$, as shown above in figure (b). Explain her error.

85. How do we find the intercepts of the graph of an equation without having to graph the equation?

86. In Section 3.2, we discussed a method to graph $y = 2x - 3$. In Section 3.3, we discussed a method to graph $2x + 3y = 6$. Briefly explain the steps involved in each method.

REVIEW

87. Simplify: $\frac{3 \cdot 5 \cdot 5}{3 \cdot 5 \cdot 5 \cdot 5}$

88. Simplify: $4\left(\frac{d}{2} - 3\right) - 5\left(\frac{2}{5}d - 1\right)$

89. Translate: Six less than twice x

90. Is -5 a solution of $2(3x + 10) = 5x + 6$?

CHALLENGE PROBLEMS

91. Where will the line $y = b$ intersect the line $x = a$?

92. Write an equation of the line that has an x-intercept of $(4, 0)$ and a y-intercept of $(0, 3)$.

93. What is the least number of intercepts a line can have? What is the greatest number a line can have?

94. On a rectangular coordinate system, draw a circle that has exactly two intercepts.

SECTION 3.4
Slope and Rate of Change

Objectives

1. Find the slope of a line from its graph.
2. Find the slope of a line given two points.
3. Find slopes of horizontal and vertical lines.
4. Solve applications of slope.
5. Calculate rates of change.
6. Determine whether lines are parallel or perpendicular using slope.

In this section, we introduce a means of measuring the steepness of a line. We call this measure the *slope of the line,* and it can be found in several ways.

1 Find the Slope of a Line from Its Graph.

The **slope of a line** is a ratio that compares the vertical change with the corresponding horizontal change as we move along the line from one point to another.

As an example, let's find the slope of the line graphed below. To begin, we select two points on the line, P and Q. One way to move from P to Q is to start at point P and count upward 5 grid squares. Then, moving to the right, we count 6 grid squares to reach point Q. The vertical change in this movement is called the **rise.** The horizontal change is called the **run.** Notice that a right triangle, called a **slope triangle,** is created by this process.

The slope of a line is defined to be *the ratio of the vertical change to the horizontal change.* So we have

$$\text{slope} = \frac{\text{vertical change}}{\text{horizontal change}} = \frac{\text{rise}}{\text{run}} = \frac{5}{6}$$

This ratio is a comparison of the rise and the run using a quotient.

The slope of the line is $\frac{5}{6}$. This indicates that there is a rise (vertical change) of 5 units for each run (horizontal change) of 6 units.

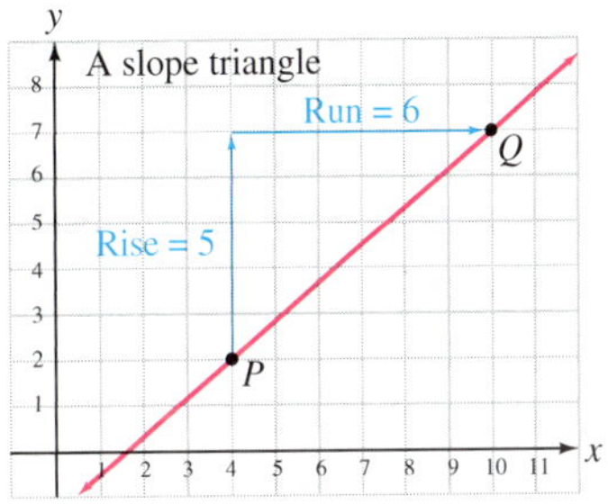

EXAMPLE 1 Find the slope of the line graphed in figure (a) below.

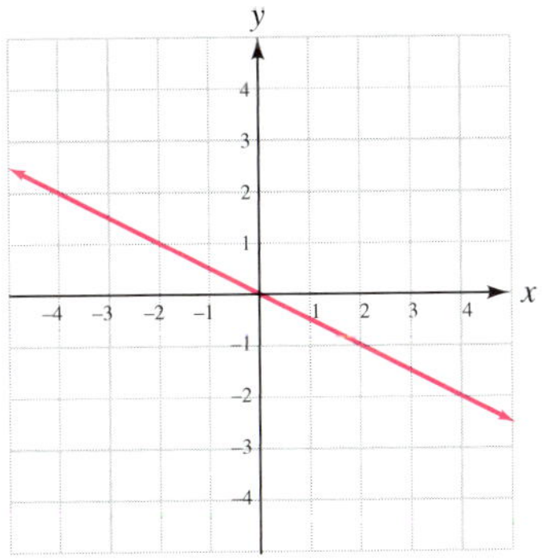

(a)

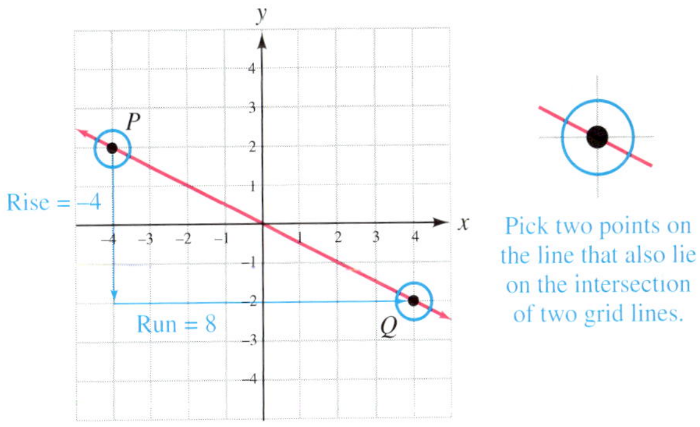

(b)

The Language of Algebra
A *ratio* is a comparison of two numbers using a quotient. Ratios are used in many settings. Mechanics speak of gear ratios. Colleges like to advertise their low student-to-teacher ratios.

Strategy We will pick two points on the line, construct a slope triangle, and find the rise and run. Then we will write the ratio of the rise to the run.

Why The slope of a line is the ratio of the rise to the run.

The Language of Algebra
Many historians believe that *m* was chosen to represent the slope of a line because it is the first letter of the French word *monter,* meaning to ascend or climb.

Solution We begin by choosing two points on the line, P and Q, as shown in figure (b), on the previous page. One way to move from P to Q is to start at point P and count *downward* 4 grid squares. Because this movement is downward, the rise is -4. Then, moving right, we count 8 grid squares to reach Q. This indicates that the run is 8.

To find the slope of the line, we write a ratio of the rise to the run in simplified form. Usually the letter m is used to denote slope, so we have

$$m = \frac{\text{rise}}{\text{run}} = \frac{-4}{8} = -\frac{1}{2}$$

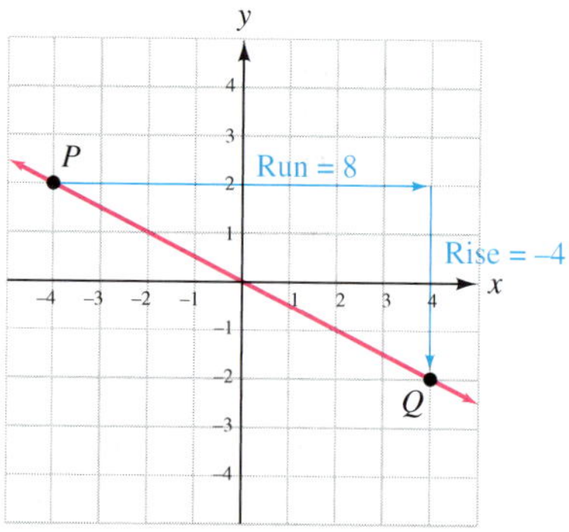

The slope of the line is $-\frac{1}{2}$.

Success Tip
When drawing a slope triangle, movement upward or to the right is positive. Movement downward or to the left is negative.

The movement from P to Q can be reversed. Starting at P, we can move to the right, a run of 8; and then downward, a rise of -4, to reach Q. With this approach, the slope triangle is above the line. When we form the ratio to find the slope, we get the same result as before:

$$m = \frac{\text{rise}}{\text{run}} = \frac{-4}{8} = -\frac{1}{2}$$

Self Check 1 Find the slope of the line shown above using two points different from those used in the solution of Example 1.

Now Try **Problem 21**

The identical answers from Example 1 and its Self Check illustrate an important fact: *For any line, the same value will be obtained no matter which two points on the line are used to find the slope.*

2 Find the Slope of a Line Given Two Points.

The Language of Algebra
The prefix *sub* means below or beneath, as in submarine or subway. In x_1, x_2, y_1, and y_2, the *subscripts* 1 and 2 are written lower than the variable. They are not exponents.

We can generalize the graphic method for finding slope to develop a slope formula. To begin, we select points P and Q on the line shown in the figure below. To distinguish between the coordinates of these points, we use **subscript notation.** Point P has coordinates (x_1, y_1), which are read as "x sub 1 and y sub 1." Point Q has coordinates (x_2, y_2), which are read as "x sub 2 and y sub 2."

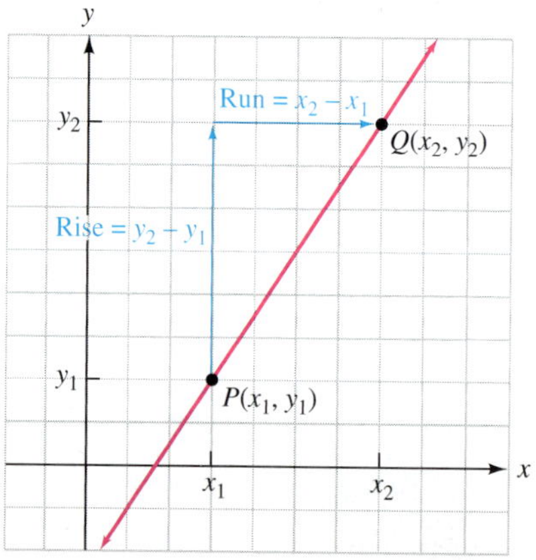

Success Tip
Recall that subtraction is used to measure change.

As we move from point P to point Q, the rise is the difference of the y-coordinates: $y_2 - y_1$. We call this difference the **change in y.** The run is the difference of the x-coordinates: $x_2 - x_1$. This difference is called the **change in x.** Since the slope is the ratio $\frac{\text{rise}}{\text{run}}$, we have the following formula for calculating slope.

Slope of a Line

The **slope** of a line passing through points (x_1, y_1) and (x_2, y_2) is

$$m = \frac{\text{vertical change}}{\text{horizontal change}} = \frac{\text{rise}}{\text{run}} = \frac{\text{change in } y}{\text{change in } x} = \frac{y_2 - y_1}{x_2 - x_1} \quad \text{if } x_2 \neq x_1$$

EXAMPLE 2

Find the slope of the line passing through (1, 2) and (3, 8).

Strategy We will use the slope formula to find the slope of the line.

Why We know the coordinates of two points on the line.

Solution When using the slope formula, it makes no difference which point you call (x_1, y_1) and which point you call (x_2, y_2). If we let (x_1, y_1) be (1, 2) and (x_2, y_2) be (3, 8), then

$$m = \frac{y_2 - y_1}{x_2 - x_1}$$ This is the slope formula.

$$m = \frac{8 - 2}{3 - 1}$$ Substitute 8 for y_2, 2 for y_1, 3 for x_2, and 1 for x_1.

$$m = \frac{6}{2}$$ Do the subtractions.

$$m = 3$$ Simplify. Think of this as a $\frac{3}{1}$ rise-to-run ratio.

Success Tip
The slope formula is a valuable tool because it allows us to calculate the slope of a line without having to view its graph.

The slope of the line is 3. The graph of the line, including the slope triangle, is shown here. Note that we obtain the same value for the slope if we let $(x_1, y_1) = (3, 8)$ and $(x_2, y_2) = (1, 2)$.

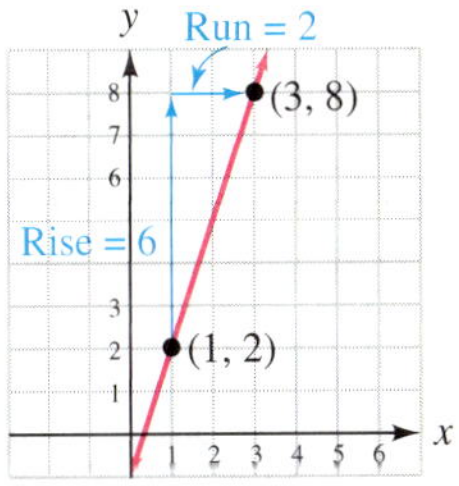

$$m = \frac{y_2 - y_1}{x_2 - x_1} = \frac{2 - 8}{1 - 3} = \frac{-6}{-2} = 3$$

Self Check 2 Find the slope of the line passing through (2, 1) and (4, 11).

Now Try Problem 33

Caution When using the slope formula, always subtract the y-coordinates and their corresponding x-coordinates in the same order. Otherwise, your answer will have the wrong sign.

$$m \neq \frac{y_2 - y_1}{x_1 - x_2} \quad \text{and} \quad m \neq \frac{y_1 - y_2}{x_2 - x_1}$$

EXAMPLE 3 Find the slope of the line that passes through $(-2, 4)$ and $(5, -6)$.

Strategy We will use the slope formula to find the slope of the line.

Why We know the coordinates of two points on the line.

Solution Since we know the coordinates of two points on the line, we can find its slope. If we let (x_1, y_1) be $(-2, 4)$ and (x_2, y_2) be $(5, -6)$, then

$$m = \frac{y_2 - y_1}{x_2 - x_1} \quad \text{This is the slope formula.}$$

$$m = \frac{-6 - 4}{5 - (-2)} \quad \text{Substitute } -6 \text{ for } y_2\text{, } 4 \text{ for } y_1\text{, } 5 \text{ for } x_2\text{, and } -2 \text{ for } x_1.$$

$$m = -\frac{10}{7} \quad \text{Do the subtractions. We can write the result as } \frac{-10}{7} \text{ or } -\frac{10}{7}.$$

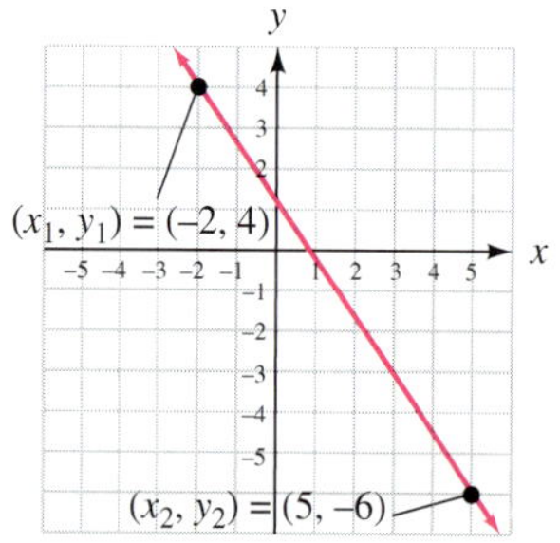

The slope of the line is $-\frac{10}{7}$.

If we graph the line by plotting the two points, we see that the line falls from left to right—a fact indicated by its negative slope.

Caution
Slopes are normally written as fractions, sometimes as decimals, but never as mixed numbers.
As with any fractional answer, always express slope in simplified form (lowest terms).

Self Check 3 Find the slope of the line that passes through $(-1, -2)$ and $(1, -7)$.

Now Try **Problem 39**

In Example 2, the slope of the line was positive. In Examples 1 and 3, the slopes of the lines were negative. In general, lines that rise from left to right have a positive slope. Lines that fall from left to right have a negative slope.

Success Tip
To classify the slope of a line as positive or negative, follow it from left to right, as you would read a sentence in a book.

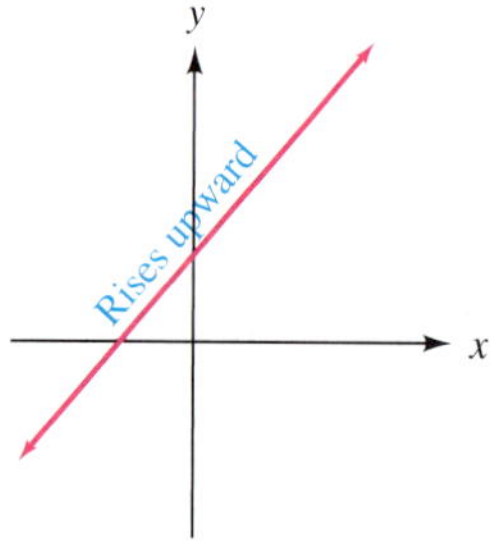

Positive slope

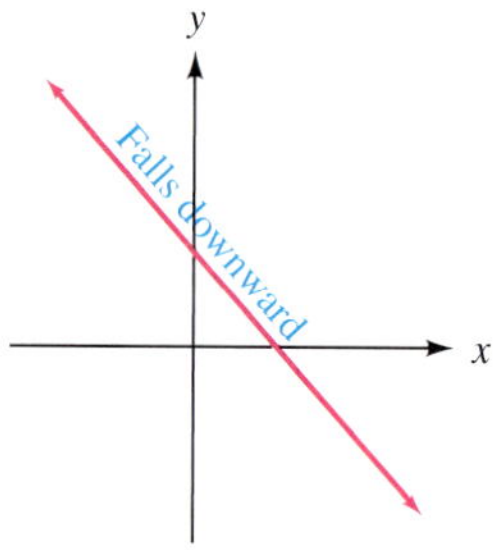

Negative slope

In the following illustration, we see a line with slope 3 is steeper than a line with slope of $\frac{5}{6}$, and a line with slope of $\frac{5}{6}$ is steeper than a line with slope of $\frac{1}{4}$. In general, *the larger the absolute value of the slope, the steeper the line.*

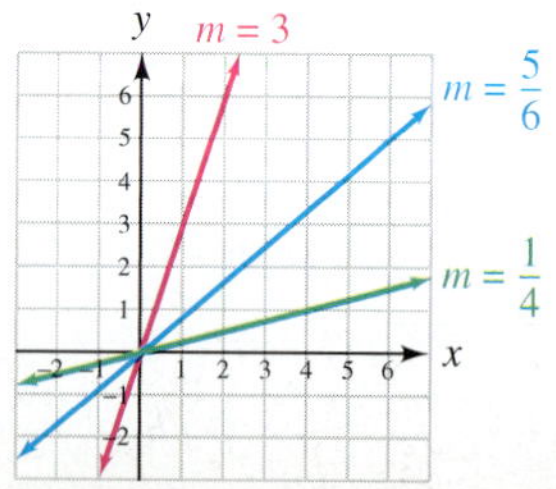

3 Find Slopes of Horizontal and Vertical Lines.

In the next two examples, we calculate the slope of a horizontal line and we show that a vertical line has no defined slope.

EXAMPLE 4 Find the slope of the line $y = 3$.

Strategy We will find the coordinates of two points on the line.

Why We can then use the slope formula to find the slope of the line.

Solution The graph of $y = 3$ is a horizontal line. To find its slope, we select two points on the line: $(-2, 3)$ and $(3, 3)$. If (x_1, y_1) is $(-2, 3)$ and (x_2, y_2) is $(3, 3)$, we have

$$m = \frac{y_2 - y_1}{x_2 - x_1}$$ This is the slope formula.

$$m = \frac{3 - 3}{3 - (-2)}$$ Substitute 3 for y_2, 3 for y_1, 3 for x_2, and -2 for x_1.

$$m = \frac{0}{5}$$ Simplify the numerator and the denominator.

$$m = 0$$

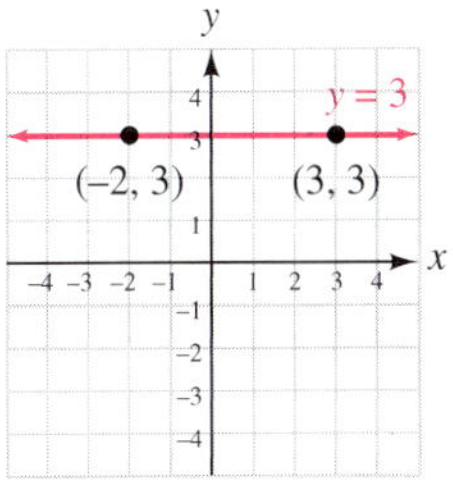

The rise = 0 for these two points.

The slope of the line $y = 3$ is 0.

Self Check 4 Find the slope of the line $y = 10$.

Now Try **Problem 61**

The y-coordinates of any two points on a horizontal line will be the same, and the x-coordinates will be different. Thus, the numerator of $\frac{y_2 - y_1}{x_2 - x_1}$ will always be zero, and the denominator will always be nonzero. Therefore, the slope of a horizontal line is 0.

EXAMPLE 5 If possible, find the slope of the line $x = -2$.

Strategy We will find the coordinate of two points on the line.

Why We can then use the slope formula to find the slope of the line, if it exits.

Solution The graph of $x = -2$ is a vertical line. To find its slope, we select two points on the line: $(-2, 3)$ and $(-2, -1)$. If (x_2, y_2) is $(-2, 3)$ and (x_1, y_1) is $(-2, -1)$, we have

$$m = \frac{y_2 - y_1}{x_2 - x_1}$$ This is the slope formula.

$$m = \frac{3 - (-1)}{-2 - (-2)}$$ Substitute 3 for y_2, -1 for y_1, -2 for x_2, and -2 for x_1.

$$m = \frac{4}{0}$$ Note that $x_1 = x_2$. Simplify the numerator and the denominator.

Since division by zero is undefined, $\frac{4}{0}$ has no meaning. The slope of the line $x = -2$ is undefined.

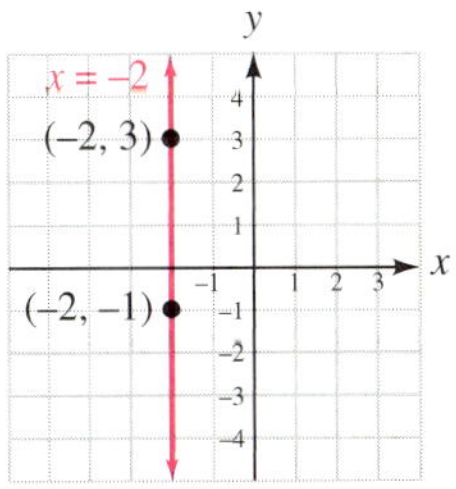

The run = 0 for these two points.

Notation
This example explains why the definition of slope includes the restriction that $x_1 \neq x_2$.

Self Check 5 If possible, find the slope of the line $x = 12$.

Now Try **Problem 67**

The y-coordinates of any two points on a vertical line will be different, and the x-coordinates will be the same. Thus, the numerator of $\frac{y_2 - y_1}{x_2 - x_1}$ will always be nonzero, and the denominator will always be 0. Therefore, the slope of a vertical line is undefined.

We now summarize the results from Examples 4 and 5.

Slopes of Horizontal and Vertical Lines

Horizontal lines (lines with equations of the form $y = b$) have slope 0.

Vertical lines (lines with equations of the form $x = a$) have undefined slope.

The Language of Algebra
Undefined and *0* do not mean the same thing. A horizontal line has a defined slope; it is 0. A vertical line does not have a defined slope; we say its slope is *undefined.*

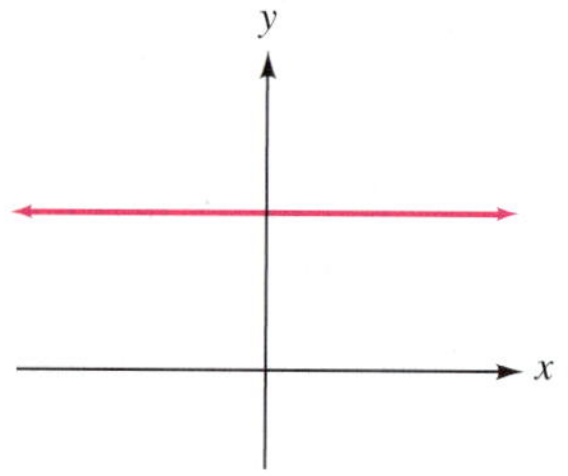

Horizontal line: 0 slope

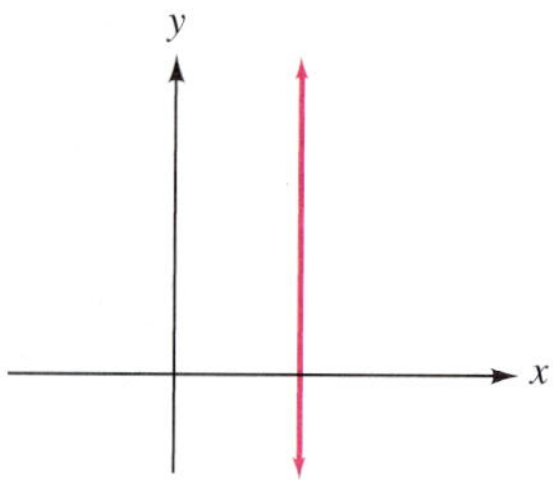

Vertical line: undefined slope

4 Solve Applications of Slope.

The concept of slope has many applications. For example, architects use slope when designing ramps and roofs. Truckers must be aware of the slope, or *grade,* of the roads they travel. Mountain bikers ride up rocky trails and snow skiers speed down steep slopes.

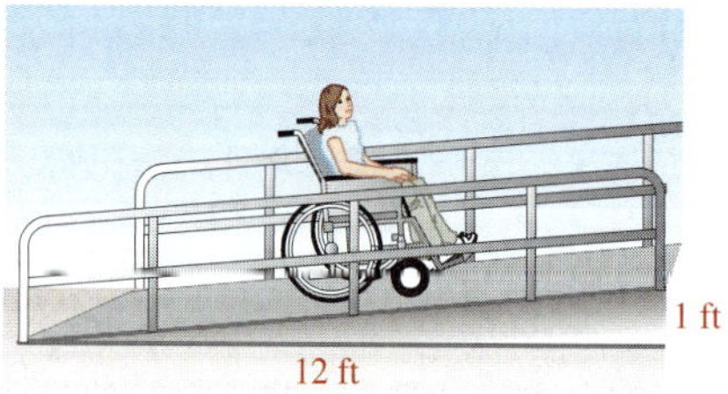

The Americans with Disabilities Act provides a guideline for the steepness of a ramp. The maximum slope for a wheelchair ramp is 1 foot of rise for every 12 feet of run: $m = \frac{1}{12}$.

The grade of an incline is its slope expressed as a percent: A 15% grade means a rise of 15 feet for every run of 100 feet: $m = \frac{15}{100}$, which simplifies to $\frac{3}{20}$.

EXAMPLE 6 ***Architecture.*** **Pitch** is the incline of a roof expressed as a ratio of the vertical rise to the horizontal run. Find the pitch of the roof shown in the illustration.

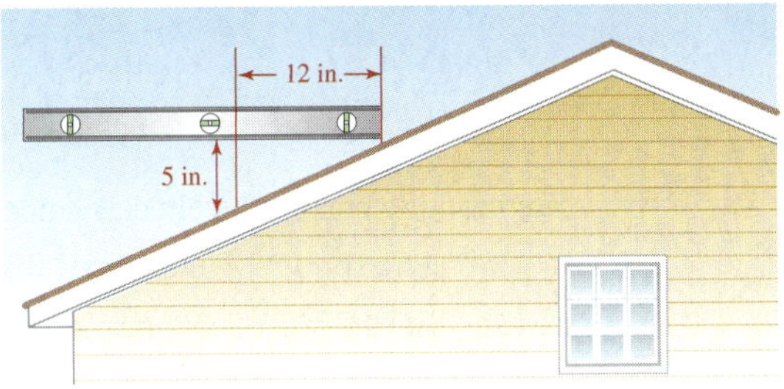

Strategy We will determine the rise and the run of the roof from the illustration. Then we will write the ratio of the rise to the run.

Why The pitch of a roof is its slope, and the slope of a line is the ratio of the rise to the run.

Solution In the illustration, a level is used to create a slope triangle. From the triangle, we see that the rise is 5 and the run is 12.

$$m = \frac{\text{rise}}{\text{run}} = \frac{5}{12}$$ The roof has a $\frac{5}{12}$ pitch.

Now Try **Problem 99**

5 Calculate Rates of Change.

We have seen that the slope of a line is a ratio of two numbers. If units are attached to a slope calculation, the result is called a **rate of change.** In general, a rate of change describes how much one quantity changes with respect to another. For example, we might speak of snow melting at the rate of 6 inches per day or a tourist exchanging money at the rate of 12 pesos per dollar.

EXAMPLE 7 ***Banking.*** A bank offers a business account with a fixed monthly fee, plus a service charge for each check written. The relationship between the monthly cost y and the number x of checks written is graphed below. At what rate does the monthly cost change?

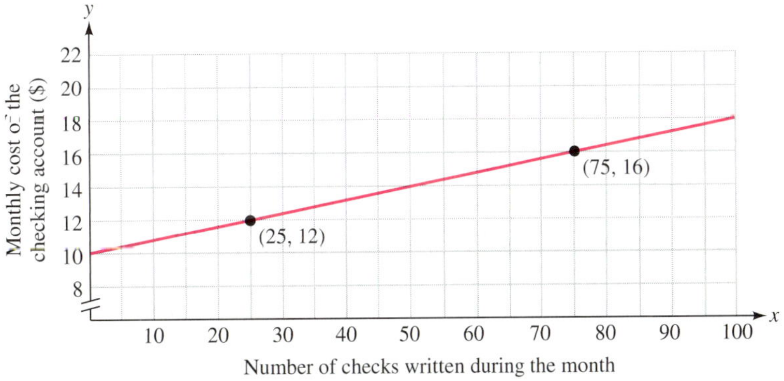

Notation

In the graph, the symbol ≑ indicates a break in the labeling of the vertical axis. The break enables us to omit a large portion of the grid that would not be used.

Strategy We will find the slope of the line and attach the proper units.

Why If units are attached to a slope calculation, the result is a rate of change.

Solution From the graph, we see that two points on the line are (25, 12) and (75, 16). If we let $(x_1, y_1) = (25, 12)$ and $(x_2, y_2) = (75, 16)$, we have

$$\text{Rate of change} = \frac{(y_2 - y_1)\text{ dollars}}{(x_2 - x_1)\text{ checks}} = \frac{(16 - 12)\text{ dollars}}{(75 - 25)\text{ checks}} = \frac{4\text{ dollars}}{50\text{ checks}} = \frac{2\text{ dollars}}{25\text{ checks}}$$

The Language of Algebra
The preposition *per* means for each, or for every. When we say the rate of change is 8¢ *per* check, we mean 8¢ for each check.

The monthly cost of the checking account increases \$2 for every 25 checks written.

We can express $\frac{2}{25}$ in decimal form by dividing the numerator by the denominator. Then we can write the rate of change in two other ways, using the word *per,* which indicates division.

Rate of change = \$0.08 per check or Rate of change = 8¢ per check

Now Try **Problem 103**

6 Determine Whether Lines Are Parallel or Perpendicular Using Slope.

Two lines that lie in the same plane but do not intersect are called **parallel lines.** Parallel lines have the same slope and different y-intercepts. For example, the lines graphed in figure (a) are parallel because they both have slope $-\frac{2}{3}$.

The Language of Algebra
The words *parallel* and *perpendicular* are used in many settings. For example, the gymnast on the *parallel* bars is in a position that is *perpendicular* to the floor.

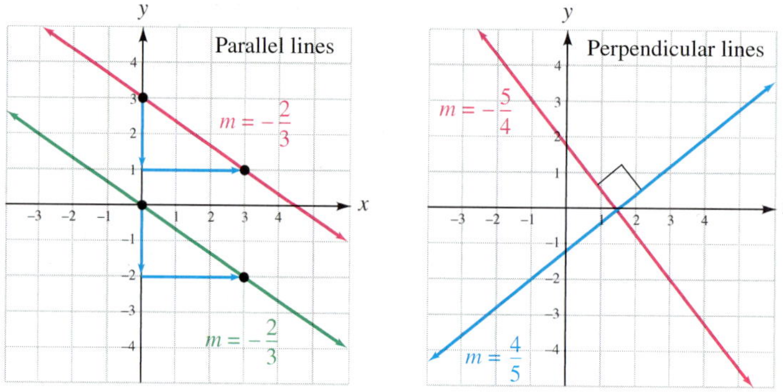

(a) (b)

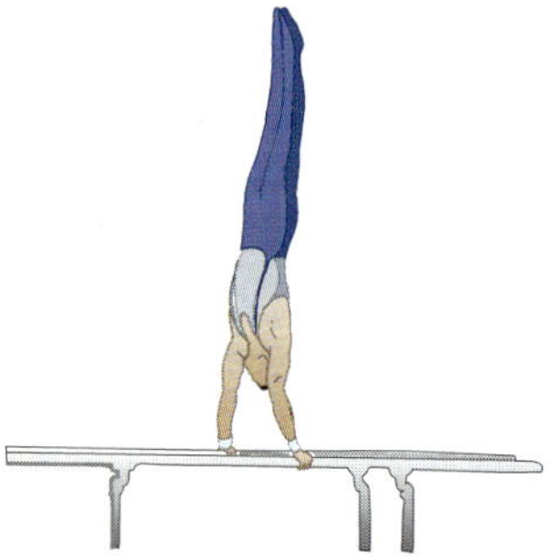

Lines that intersect to form four right angles (angles with measure 90°) are called **perpendicular lines.** If the product of the slopes of two lines is -1, the lines are perpendicular. This means that the slopes are **negative** (or **opposite**) **reciprocals.** In figure (b), we know that the lines with slopes $\frac{4}{5}$ and $-\frac{5}{4}$ are perpendicular because

$$\frac{4}{5}\left(-\frac{5}{4}\right) = -\frac{20}{20} = -1 \qquad \frac{4}{5}\text{ and }-\frac{5}{4}\text{ are negative reciprocals.}$$

Slopes of Parallel and Perpendicular Lines

1. Two lines with the same slope are parallel.
2. Two lines are perpendicular if the product of the slopes is -1; that is, if their slopes are negative reciprocals.
3. Any horizontal line and any vertical line are perpendicular.

EXAMPLE 8 Determine whether the line that passes through (7, −9) and (10, 2) and the line that passes through (0, 1) and (3, 12) are parallel, perpendicular, or neither.

Strategy We will use the slope formula to find the slope of each line.

Why If the slopes are equal, the lines are parallel. If the slopes are negative reciprocals, the lines are perpendicular. Otherwise, the lines are neither parallel nor perpendicular.

Solution To calculate the slope of each line, we use the slope formula.

***The line through* (7, −9) *and* (10, 2):**

$$m = \frac{y_2 - y_1}{x_2 - x_1} = \frac{2 - (-9)}{10 - 7} = \frac{11}{3}$$

***The line through* (0, 1) *and* (3, 12):**

$$m = \frac{y_2 - y_1}{x_2 - x_1} = \frac{12 - 1}{3 - 0} = \frac{11}{3}$$

Since the slopes are the same, the lines are parallel.

Self Check 8 Determine whether the line that passes through (2, 1) and (6, 8) and the line that passes through (−1, 0) and (4, 7) are parallel, perpendicular, or neither.

Now Try **Problems 73 and 75**

EXAMPLE 9 Find the slope of a line perpendicular to the line passing through (1, −4) and (8, 4).

Strategy We will use the slope formula to find the slope of the line passing through (1, −4) and (8, 4).

Why We can then form the negative reciprocal of the result to produce the slope of a line perpendicular to the given line.

Solution The slope of the line that passes through (1, −4) and (8, 4) is

$$m = \frac{y_2 - y_1}{x_2 - x_1} = \frac{4 - (-4)}{8 - 1} = \frac{8}{7}$$

The slope of a line perpendicular to the given line has slope that is the negative (or opposite) reciprocal of $\frac{8}{7}$, which is $-\frac{7}{8}$.

Self Check 9 Find the slope of a line perpendicular to the line passing through (−4, 1) and (9, 5).

Now Try **Problem 85**

ANSWERS TO SELF CHECKS **1.** $-\frac{1}{2}$ **2.** 5 **3.** $-\frac{5}{2}$ **4.** 0 **5.** Undefined slope **8.** Neither **9.** $-\frac{13}{4}$

STUDY SET 3.4

VOCABULARY

Fill in the blanks.

1. The ______ of a line is a measure of the line's steepness. It is the ______ of the vertical change to the horizontal change.

2. $m = \dfrac{\rule{2cm}{0.4pt}}{\text{horizontal change}} = \dfrac{\text{rise}}{\rule{1cm}{0.4pt}} = \dfrac{\text{change in } y}{\rule{2cm}{0.4pt}}$

3. The rate of ________ of a linear relationship can be found by finding the slope of the graph of the line and attaching the proper units.

4. ________ lines do not intersect. ____________ lines intersect to form four right angles.

CONCEPTS

5. Which line graphed has

a. a positive slope? **b.** a negative slope?

c. zero slope? **d.** undefined slope?

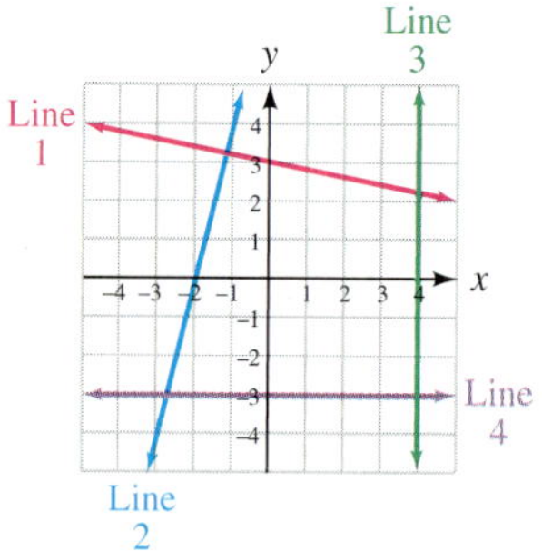

6. Consider each graph of a line and the slope triangle. What is the rise? What is the run? What is the slope of the line?

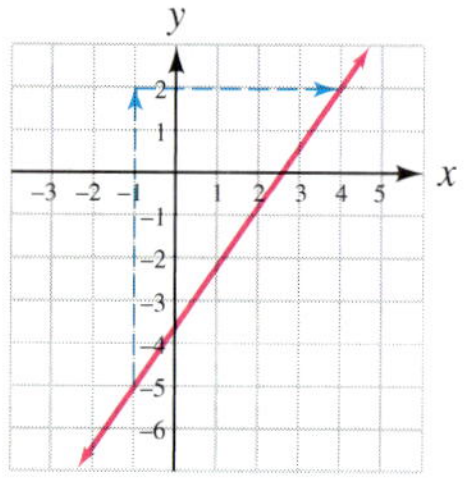
(a)

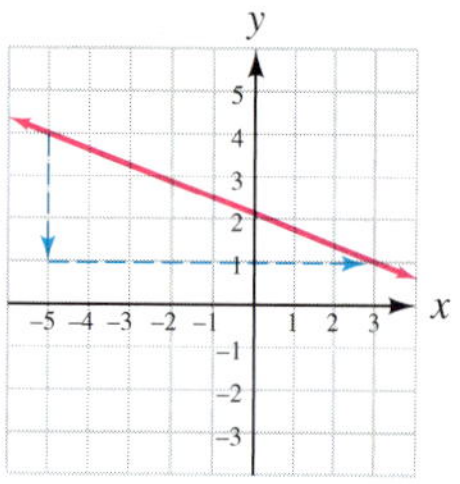
(b)

7. For each graph, determine which line has the greater slope.

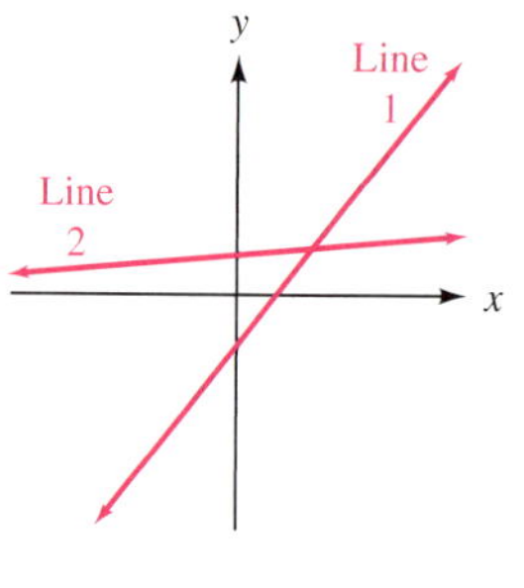

(a)

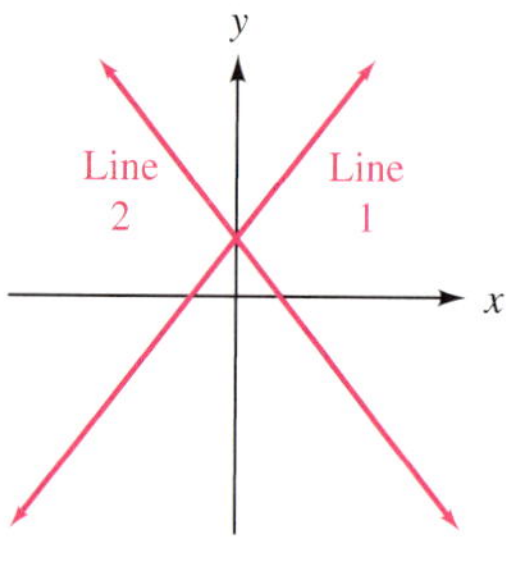

(b)

8. Which two labeled points should be used to find the slope of the line?

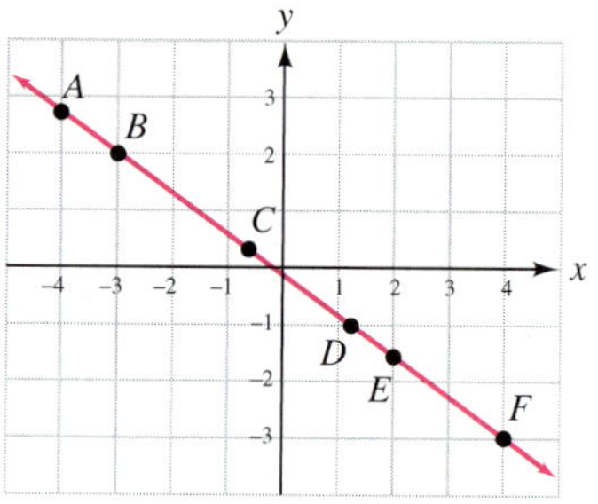

9. Fill in the blank: When calculating the slope of a line, the ______ value will be obtained no matter which two points on the line are used.

10. Evaluate each expression.

a. $\dfrac{10-4}{6-5}$ **b.** $\dfrac{-1-1}{-2-(-7)}$

11. Write each slope in a better way.

a. $m = \dfrac{0}{6}$ **b.** $m = \dfrac{8}{0}$

c. $m = \dfrac{3}{12}$ **d.** $m = \dfrac{-10}{-5}$

12. Fill in the blanks: __________ lines have a slope of 0. Vertical lines have __________ slope.

13. The *grade* of an incline is its slope expressed as a percent. Express the slope $\frac{2}{5}$ as a grade.

14. GROWTH RATES The graph on the next page shows how a child's height increased from ages 2 through 5. Fill in the correct units to find the rate of change in the child's height.

$$\text{Rate of change} = \frac{(40-31)\ \square}{(5-2)\ \square}$$

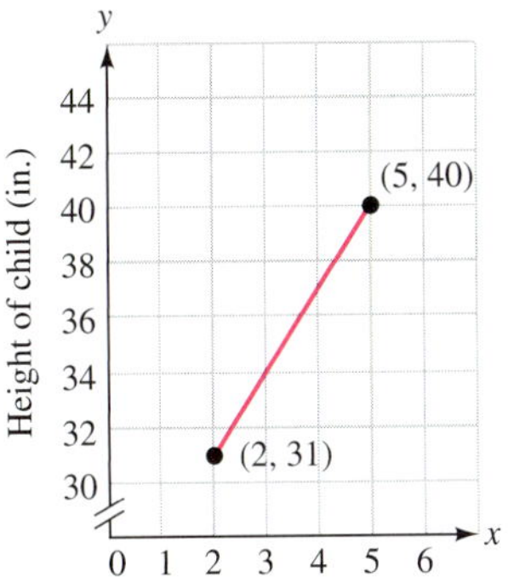

15. Find the negative reciprocal of each number.

a. 6 **b.** $-\frac{7}{8}$ **c.** -1

16. Fill in the blanks.

a. Two different lines with the same slope are ________.

b. If the slopes of two lines are negative reciprocals, the lines are ______________.

c. The product of the slopes of perpendicular lines is ____.

NOTATION

17. a. What is the formula used to find the slope of a line passing through (x_1, y_1) and (x_2, y_2)?

b. Fill in the blanks to state the slope formula in words: m equals y _____ two minus y _____ one ________________ x sub _____ minus x sub _____.

18. Explain the difference between y^2 and y_2.

19. Consider the points (7, 2) and (−4, 1). If we let $x_1 = 7$, then what is y_2?

20. The symbol ≑ is used when graphing to indicate a _______ in the labeling of an axis.

GUIDED PRACTICE

Find the slope of each line, if possible. **See Example 1.**

21.

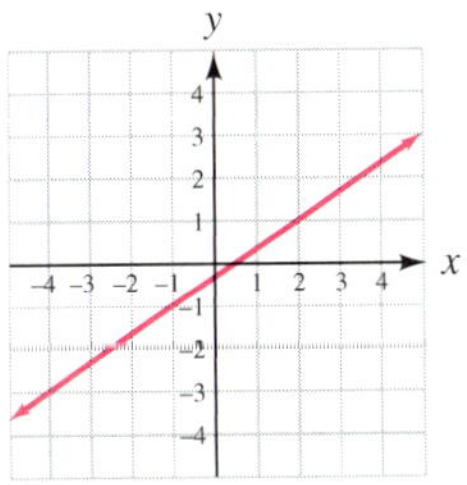

22.

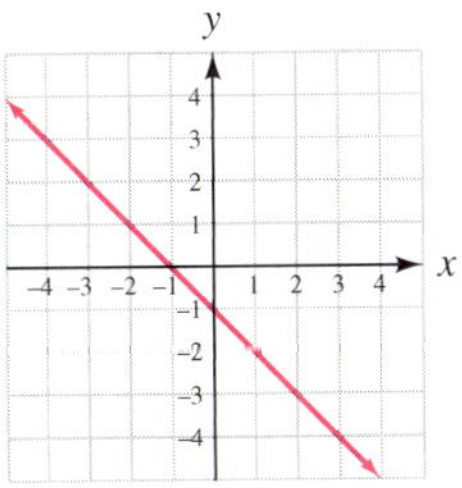

23.

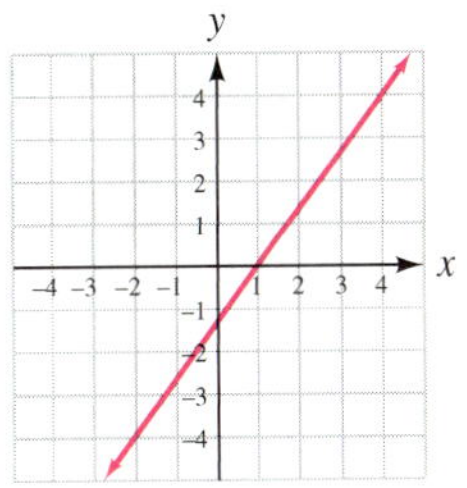

24.

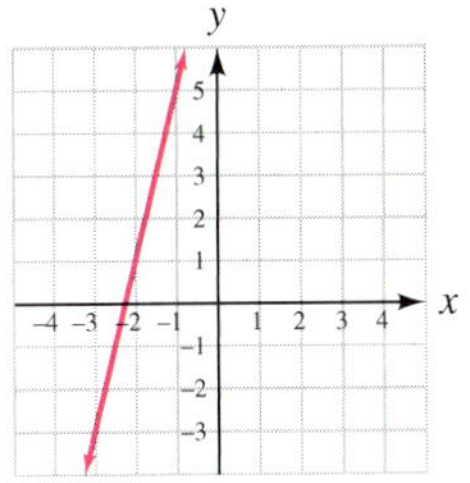

25.

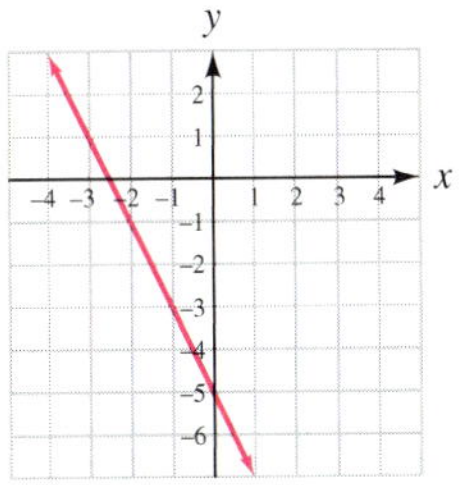

26.

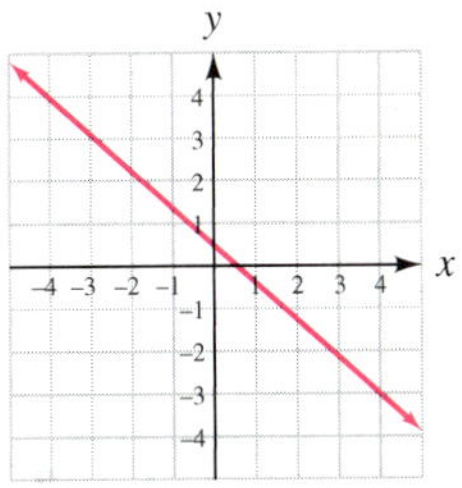

27.

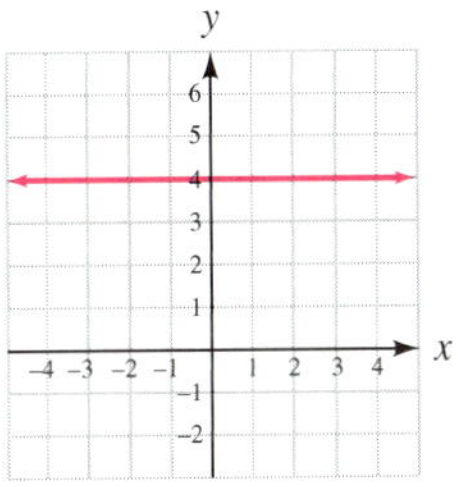

28.

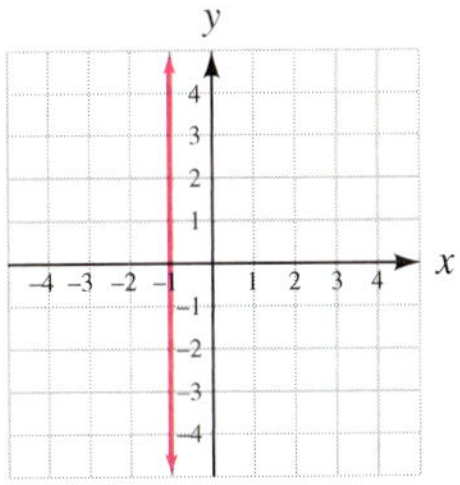

29.

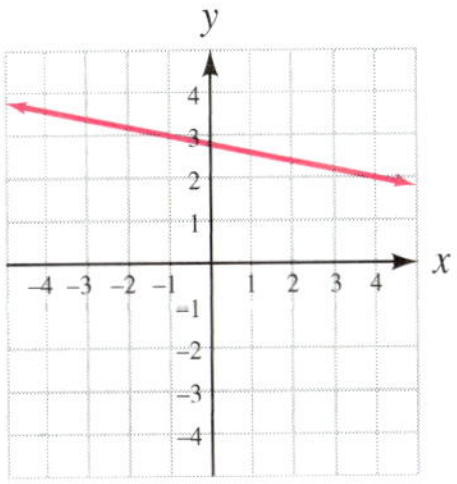

30.

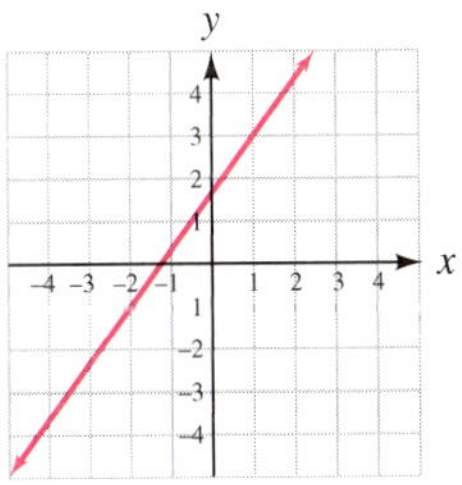

31.

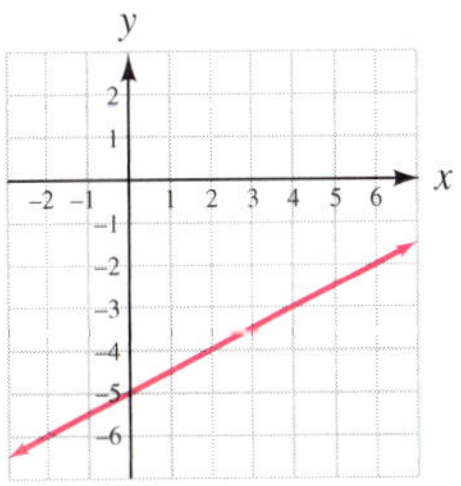

32.

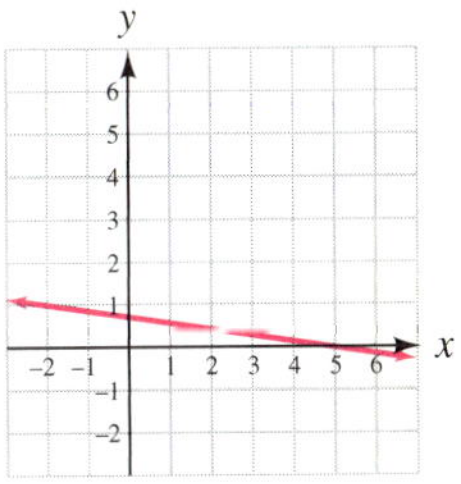

Find the slope of the line passing through the given points, when possible. **See Examples 2 and 3.**

33. (1, 3) and (2, 4)

34. (1, 3) and (2, 5)

35. (3, 4) and (2, 7)

36. (3, 6) and (5, 2)

37. (0, 0) and (4, 5)
38. (4, 3) and (7, 8)
39. $(-3, 5)$ and $(-5, 6)$
40. $(6, -2)$ and $(-3, 2)$
41. $(-2, -2)$ and $(-12, -8)$
42. $(-1, -2)$ and $(-10, -5)$
43. (5, 7) and $(-4, 7)$
44. $(-1, -12)$ and $(6, -12)$
45. $(8, -4)$ and $(8, -3)$
46. $(-2, 8)$ and $(-2, 15)$
47. $(-6, 0)$ and $(0, -4)$
48. $(0, -9)$ and $(-6, 0)$
49. $(-2.5, 1.75)$ and $(-0.5, -7.75)$
50. $(6.4, -7.2)$ and $(-8.8, 4.2)$
51. $(-2.2, 18.6)$ and $(-1.7, 18.6)$
52. $(4.6, 3.2)$ and $(4.6, -4.8)$
53. $\left(-\frac{4}{7}, -\frac{1}{5}\right)$ and $\left(\frac{3}{7}, \frac{6}{5}\right)$
54. $\left(-\frac{4}{9}, -\frac{1}{8}\right)$ and $\left(\frac{5}{9}, \frac{3}{8}\right)$
55. $\left(-\frac{3}{4}, \frac{2}{3}\right)$ and $\left(\frac{4}{3}, -\frac{1}{6}\right)$
56. $\left(\frac{1}{2}, \frac{3}{4}\right)$ and $\left(-\frac{11}{16}, -\frac{1}{2}\right)$

Determine the slope of the graph of the line that has the given table of solutions. **See Examples 2 and 3.**

57.

x	y	(x, y)
-3	-1	$(-3, -1)$
1	2	(1, 2)

58.

x	y	(x, y)
-3	6	$(-3, 6)$
0	2	(0, 2)

59.

x	y	(x, y)
-3	6	$(-3, 6)$
0	6	(0, 6)

60.

x	y	(x, y)
4	-5	$(4, -5)$
4	0	(4, 0)

Find the slope of each line, if possible. **See Examples 4 and 5.**

61. $y = -11$
62. $y = -2$
63. $y = 0$
64. $x = 0$
65. $x = 6$
66. $x = 6$
67. $x = -10$
68. $y = 8$
69. $y - 9 = 0$
70. $x + 14 = 0$
71. $3x = -12$
72. $2y + 2 = -6$

Determine whether the lines through each pair of points are parallel, perpendicular, or neither. **See Example 8.**

73. (5, 3) and (1, 4)
$(-3, -4)$ and $(1, -5)$
74. (2, 4) and $(-1, -1)$
(8, 0) and (11, 5)
75. $(-4, -2)$ and $(2, -3)$
(7, 1) and (8, 7)
76. $(-2, 4)$ and $(6, -7)$
$(-6, 4)$ and (5, 12)
77. (2, 2) and $(4, -3)$
$(-3, 4)$ and $(-1, 9)$
78. $(-1, -3)$ and (2, 4)
(5, 2) and $(8, -5)$
79. $(-1, 8)$ and $(-6, 8)$
(3, 3) and (3, 7)
80. (11, 0) and $(11, -5)$
(14, 6) and (25, 6)
81. (6, 4) and (2, 5)
$(-2, -3)$ and $(2, -4)$
82. $(-3, -1)$ and $(3, -2)$
(8, 2) and (9, 8)
83. (4, 2) and $(5, -3)$
$(-5, 3)$ and $(-2, 9)$
84. $(8, -3)$ and $(8, -8)$
(11, 3) and (22, 3)

Find the slope of a line perpendicular to the line passing through the given two points. **See Example 9.**

85. (0, 0) and $(5, -9)$
86. (0, 0) and (5, 12)
87. $(-1, 7)$ and (1, 10)
88. $(-7, 6)$ and (0, 4)
89. $\left(-2, \frac{1}{2}\right)$ and $\left(-1, \frac{3}{2}\right)$
90. $\left(\frac{1}{3}, -1\right)$ and $\left(\frac{4}{3}, -2\right)$
91. $(-1, 2)$ and $(-3, 6)$
92. $(5, -4)$ and $(-1, -7)$

APPLICATIONS

93. POOLS Find the slope of the bottom of the swimming pool as it drops off from the shallow end to the deep end.

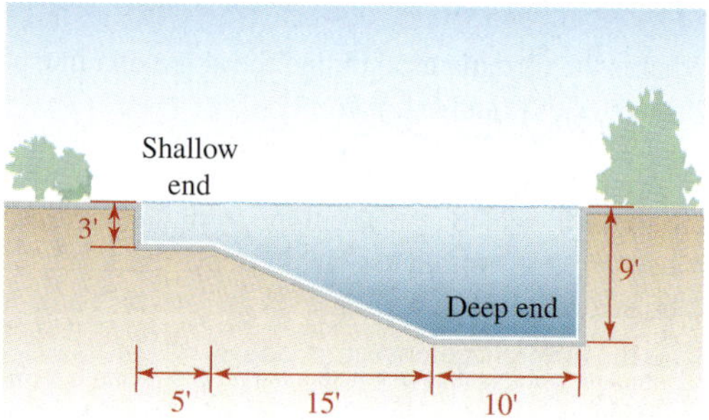

94. DRAINAGE Find the slope of the concrete patio slab using the 1-foot ruler, level, and 10-foot-long board shown in the illustration. (*Hint:* 10 feet = 120 in.)

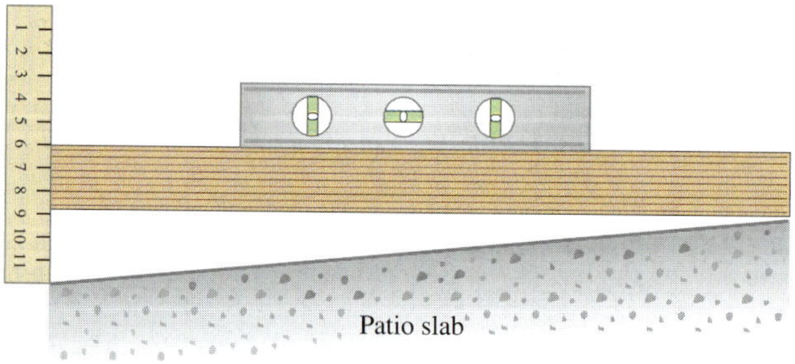

95. GRADE OF A ROAD Refer to the illustration on the next page. Find the slope of the decline and use that information to find the grade of the road.

96. STREETS One of the steepest streets in the United States is Eldred Street in Highland Park, California (near Los Angeles). It rises approximately 220 feet over a horizontal distance of 665 feet. What is the grade of the street?

Vince Compagnone/© LAT

97. TREADMILLS For each height setting listed in the table, find the resulting slope of the jogging surface of the treadmill. Then express each incline as a percent.

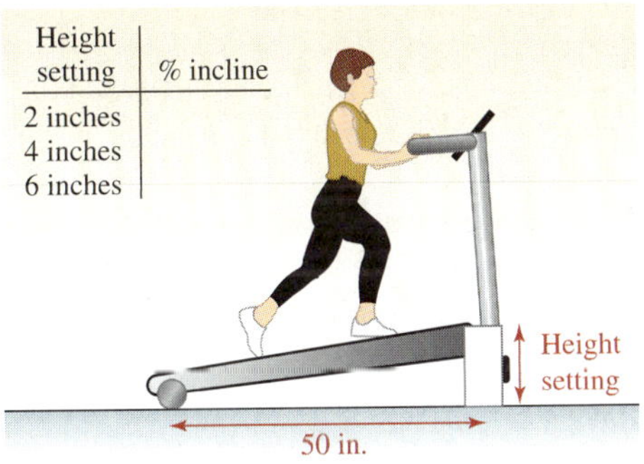

98. ARCHITECTURE Locate the coordinates of the peak of the roof if it is to have a pitch of $\frac{2}{5}$ and the roof line is to pass through the two given points in black.

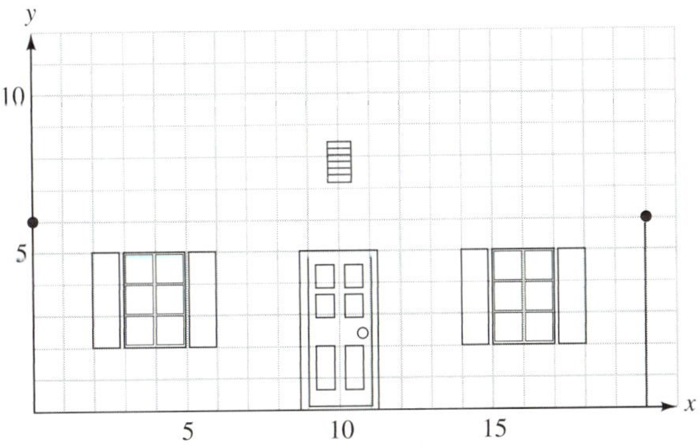

99. CARPENTRY Find the pitch of each roof.

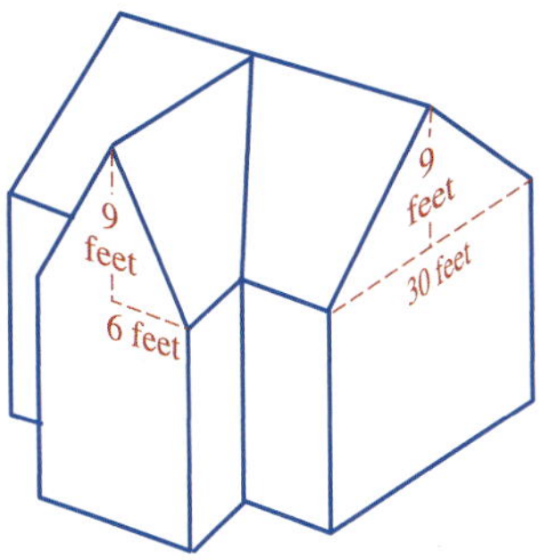

100. DOLL HOUSES Find x so that the pitch of the roof of the doll house is $\frac{4}{3}$.

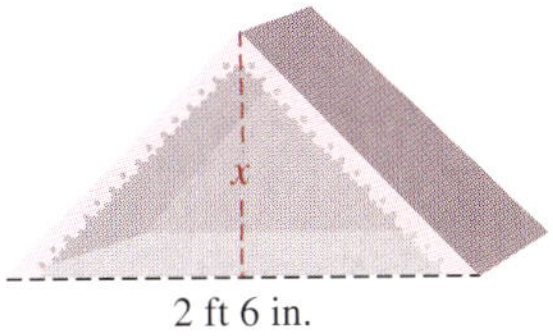

101. IRRIGATION The graph shows the number of gallons of water remaining in a reservoir as water is used from it to irrigate a field. Find the rate of change in the number of gallons of water in the reservoir.

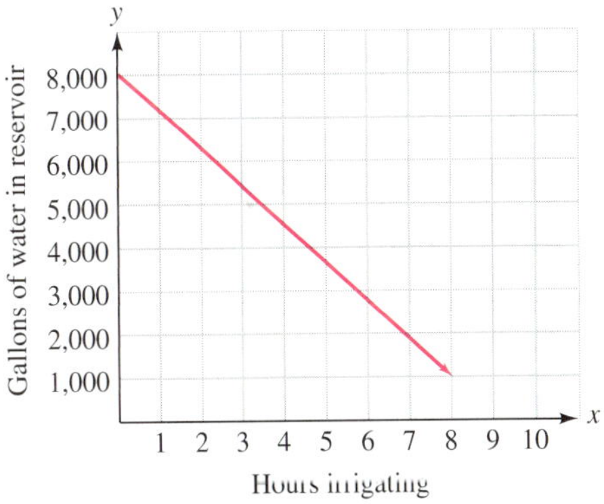

102. COMMERCIAL JETS Examine the graph and consider trips of more than 7,000 miles by a Boeing 777. Use a rate of change to estimate how the maximum payload decreases as the distance traveled increases.

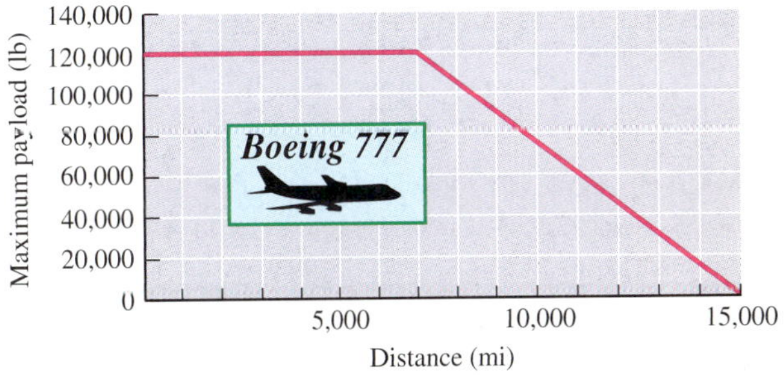

Based on data from Lawrence Livermore National Laboratory and *Los Angeles Times* (October 22, 1998).

103. MILK PRODUCTION The following graph approximates the amount of milk produced per cow in the United States for the years 1996-2005. Find the rate of change.

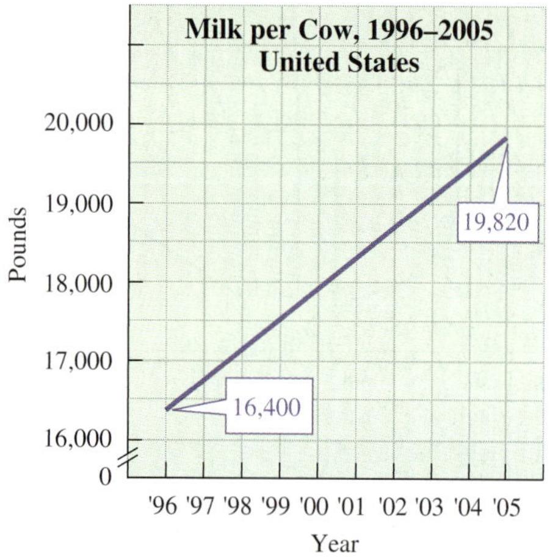

Source: United States Department of Agriculture

104. WAL-MART The graph below approximates the net sales of Wal-Mart for the years 1991–2006. Find the rate of change in sales for the years

a. 1991–1998

b. 1998–2006

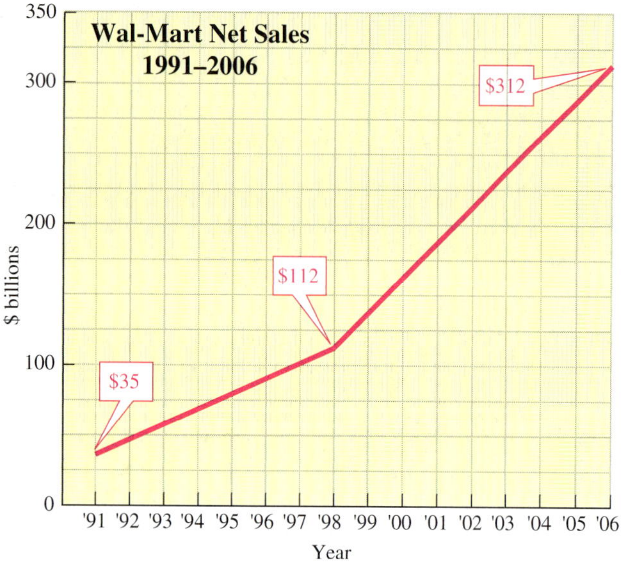

Based on data from the Wal-Mart 2006 Financial Summary

WRITING

105. Explain why the slope of a vertical line is undefined.

106. How do we distinguish between a line with positive slope and a line with negative slope?

107. Explain the error in the following solution: *Find the slope of the line that passes through* (6, 4) *and* (3, 1).

$$m = \frac{1-4}{6-3} = \frac{-3}{3} = -1$$

108. Explain the difference between a rate of change that is positive and one that is negative. Give an example of each.

REVIEW

109. HALLOWEEN CANDY A candy maker wants to make a 60-pound mixture of two candies to sell for \$2 per pound. If black licorice bits sell for \$1.90 per pound and orange gumdrops sell for \$2.20 per pound, how many pounds of each should be used?

110. MEDICATIONS A doctor prescribes an ointment that is 2% hydrocortisone. A pharmacist has 1% and 5% concentrations in stock. How many ounces of each should the pharmacist use to make a 1-ounce tube?

CHALLENGE PROBLEMS

111. Use the concept of slope to determine whether $A(-50, -10)$, $B(20, 0)$, and $C(34, 2)$ all lie on the same straight line.

112. A line having slope $\frac{2}{3}$ passes through the point $(10, -12)$. What is the y-coordinate of another point on the line whose x-coordinate is 16?

113. Subscripts are used in other disciplines besides mathematics. In what disciplines are the following symbols used?

a. H_2O and CO_2

b. C_7 and G_7

c. B_6 and B_{12}

114. Evaluate $2a_2^2 + 3a_3^3 + 4a_4^4$ for $a_2 = 2$, $a_3 = 3$, and $a_4 = 4$.

SECTION 3.5
Slope–Intercept Form

Objectives

1. Use slope–intercept form to identify the slope and y-intercept of a line.
2. Write a linear equation in slope–intercept form.
3. Write an equation of a line given its slope and y-intercept.
4. Use the slope and y-intercept to graph a linear equation.
5. Recognize parallel and perpendicular lines.
6. Use slope–intercept form to write an equation to model data.

Of all of the ways in which a linear equation can be written, one form, called *slope–intercept form,* is probably the most useful. When an equation is written in this form, two important features of its graph are evident.

1 Use Slope–Intercept Form to Identify the Slope and y-Intercept of a Line.

To explore the relationship between a linear equation and its graph, let's consider $y = 2x + 1$. We can graph this equation using the point-plotting method discussed in Section 3.2.

$y = 2x + 1$

x	y	(x, y)
-1	-1	$(-1, -1)$
0	1	$(0, 1)$
1	3	$(1, 3)$

To find the slope of the line, we pick two points on the line, $(-1, -1)$ and $(0, 1)$, and draw a slope triangle and count grid squares:

$$\text{Slope} = \frac{\text{rise}}{\text{run}} = \frac{2}{1} = 2$$

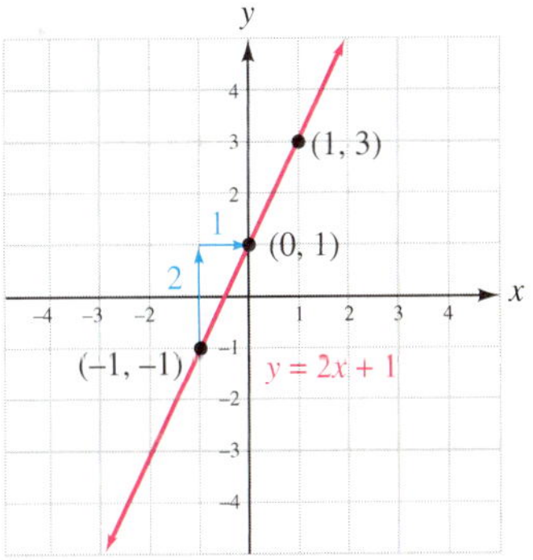

From the equation and the graph, we can make two observations:

- The graph crosses the y-axis at 1. This is the same as the constant term in $y = 2x + 1$.
- The slope of the line is 2. This is the same as the coefficient of x in $y = 2x + 1$.

This illustrates that the slope and y-intercept of the graph of $y = 2x + 1$ can be determined from the equation.

$$y = 2x + 1$$

The slope of the line is 2. (↑ under $2x$)

The y-intercept is $(0, 1)$. (↑ under 1)

These observations suggest the following form of an equation of a line.

Slope–Intercept Form of the Equation of a Line

If a linear equation is written in the form

$$y = mx + b$$

the graph of the equation is a line with slope m and y-intercept $(0, b)$.

When an equation of a line is written in slope–intercept form, the coefficient of the x-term is the line's slope and the constant term gives the y-coordinate of y-intercept.

$$y = mx + b$$

Slope (m) — y-intercept: $(0, b)$

Linear equation	*Equation written in slope–intercept form*	*Slope*	*y-intercept*
$y = 6x - 2$	$y = 6x + (-2)$	6	$(0, -2)$
$y = -\frac{5}{4}x$	$y = -\frac{5}{4}x + 0$	$-\frac{5}{4}$	$(0, 0)$
$y = \frac{x}{2} + 3$	$y = \frac{1}{2}x + 3$	$\frac{1}{2}$	$(0, 3)$
$y = -\frac{7}{8} - x$	$y = -x + \left(-\frac{7}{8}\right)$	-1	$\left(0, -\frac{7}{8}\right)$

Caution

For equations in $y = mx + b$ form, the slope of the line is the *coefficient* of x, not the x-term. For example, the graph of $y = 6x - 2$ has slope 6, *not* $6x$.

2 Write a Linear Equation in Slope–Intercept Form.

The equation of any nonvertical line can be written in slope–intercept form. To do so, we apply the properties of equality to solve the equation for y.

EXAMPLE 1

Find the slope and y-intercept of the line with the given equation.
a. $8x + y = 9$ **b.** $x + 4y = 16$ **c.** $-9x - 3y = 11$

Strategy We will write each equation in slope–intercept form, $y = mx + b$.

Why When the equations are written in slope–intercept form, the slope and y-intercept of their graphs become apparent.

Solution

a. The slope and y-intercept of the graph of $8x + y = 9$ are not obvious because the equation is not in slope–intercept form. To write it in $y = mx + b$ form, we isolate y on the left side.

$$8x + y = 9$$

$$8x + y - 8x = -8x + 9$$ To eliminate the term $8x$ on the left side, subtract $8x$ from both sides.

$$y = -8x + 9$$ On the left side, combine like terms: $8x - 8x = 0$.

$$y = -8x + 9$$

$m = -8$ — The slope is -8.
$b = 9$ — The y-intercept is $(0, 9)$.

Success Tip

Since we want the right side of the equation to have the form $mx + b$, we show the subtraction of $8x$ from that side as $-8x + 9$ rather than $9 - 8x$.

Success Tip

Since we want the right side of the equation to have the form $mx + b$, we show the division by 4 of that side as $\frac{-x}{4} + \frac{16}{4}$ rather than $\frac{-x + 16}{4}$.

b. To write the equation in slope–intercept form, we solve for y.

$$x + 4y = 16$$

$$x + 4y - x = -x + 16$$ To eliminate the x term on the left side, subtract x from both sides.

$$4y = -x + 16$$ Simplify the left side.

$$\frac{4y}{4} = \frac{-x + 16}{4}$$ To isolate y, undo the multiplication by 4 by dividing both sides by 4.

$$y = \frac{-x}{4} + \frac{16}{4}$$ On the right side, write $\frac{-x + 16}{4}$ as the sum of two fractions with like denominators.

$$y = -\frac{1}{4}x + 4$$ Write $\frac{-x}{4}$ as $-\frac{1}{4}x$. Simplify: $\frac{16}{4} = 4$.

Since $m = -\frac{1}{4}$ and $b = 4$, the slope is $-\frac{1}{4}$ and the y-intercept is (0, 4).

c. To write the equation in $y = mx + b$ form, we isolate y on the left side.

$$-9x - 3y = 11$$

$$-3y = 9x + 11$$ To eliminate the term $-9x$ on the left side, add $9x$ to both sides: $-9x + 9x = 0$.

$$\frac{-3y}{-3} = \frac{9x}{-3} + \frac{11}{-3}$$ To isolate y, undo the multiplication by -3 by dividing both sides by -3.

$$y = -3x - \frac{11}{3}$$ Simplify.

Since $m = -3$ and $b = -\frac{11}{3}$, the slope is -3 and the y-intercept is $\left(0, -\frac{11}{3}\right)$.

Self Check 1 Find the slope and y-intercept of the line with the given equation.
a. $9x + y = -4$ **b.** $x + 11y = -22$ **c.** $-10x - 2y = 7$

Now Try Problems 11, 35, and 45

3 Write an Equation of a Line Given Its Slope and y-Intercept.

If we are given the slope and y-intercept of a line, we can write an equation of the line by substituting for m and b in the slope–intercept form.

EXAMPLE 2 Write an equation of the line with slope -1 and y-intercept (0, 9).

Strategy We will use the slope–intercept form, $y = mx + b$, to write an equation of the line.

Why We know the slope of the line and its y-intercept.

Solution If the slope is -1 and the y-intercept is (0, 9), then $m = -1$ and $b = 9$.

$$y = mx + b$$ This is the slope–intercept form.

$$y = -1x + 9$$ Substitute -1 for m and 9 for b.

$$y = -x + 9$$ Simplify: $-1x = -x$.

The equation of the line with slope -1 and y-intercept (0, 9) is $y = -x + 9$.

Write an equation for each line shown. **See Example 3.**

59.

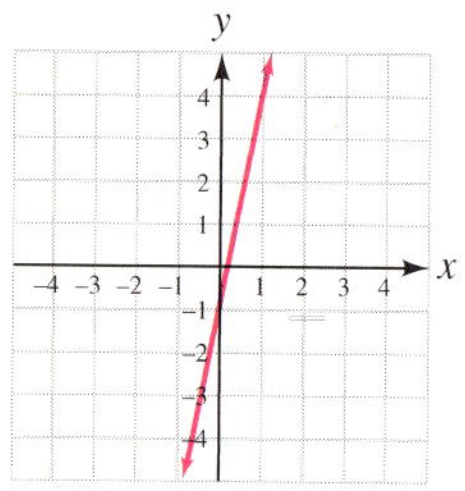

60.

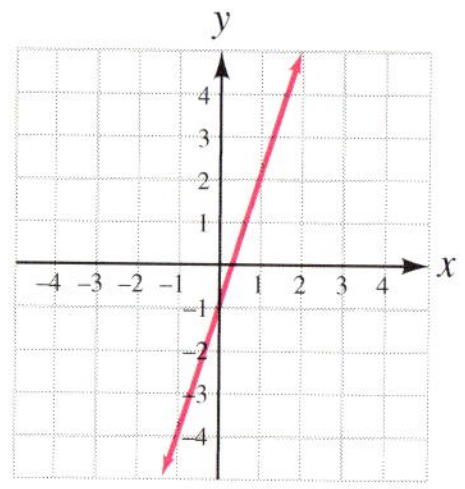

61.

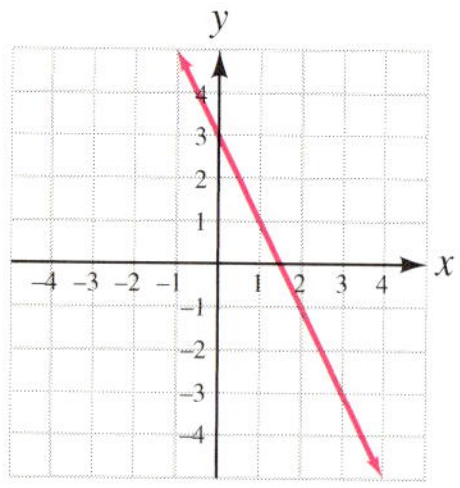

62.

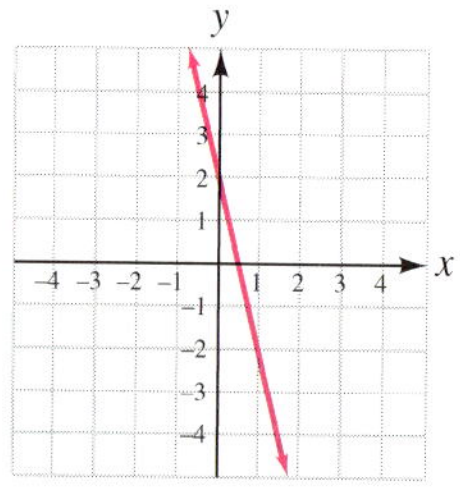

63.

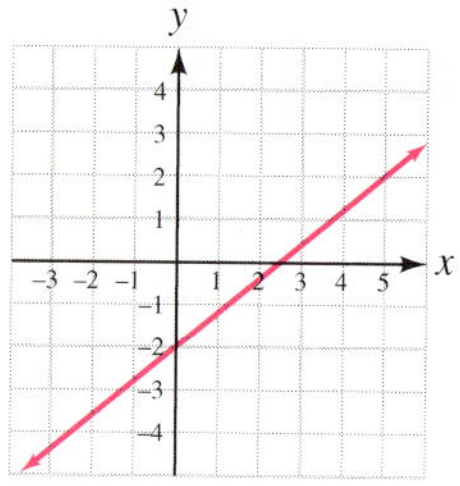

64.

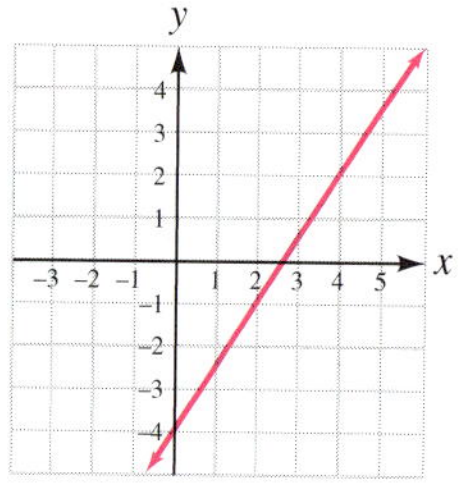

65.

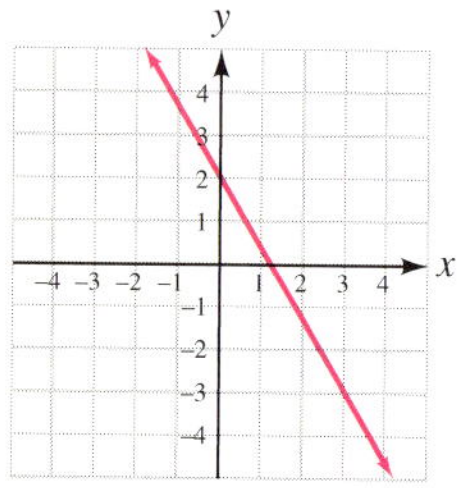

66.

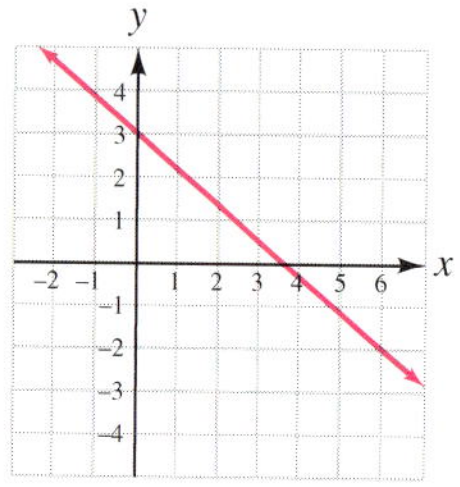

Find the slope and the y-intercept of the graph of each equation and graph it. **See Examples 4 and 5.**

67. $y = 3x + 3$

68. $y = -3x + 5$

69. $y = -\frac{1}{2}x + 2$

70. $y = \frac{x}{3}$

71. $y = -3x$

72. $y = -4x$

73. $4x + y = -4$

74. $2x + y = -6$

75. $3x + 4y = 16$

76. $2x + 3y = 9$

77. $10x - 5y = 5$

78. $4x - 2y = 6$

For each pair of equations, determine whether their graphs are parallel, perpendicular, or neither. **See Example 6.**

79. $y = 6x + 8$
$y = 6x$

80. $y = 3x - 15$
$y = -\frac{1}{3}x + 4$

81. $y = x$
$y = -x$

82. $y = \frac{1}{2}x - \frac{4}{5}$
$y = 0.5x + 3$

83. $y = -2x - 9$
$2x - y = 9$

84. $y = \frac{3}{4}x + 1$
$4x - 3y = 15$

85. $3x = 5y - 10$
$5x = 1 - 3y$

86. $-2y = 2 - x$
$2x - 3 = 4y$

87. $x - y = 12$
$-2x + 2y = -23$

88. $y = -3x + 1$
$3y = x - 5$

89. $x = 9$
$y = 8$

90. $-x + 4y = 10$
$2y + 16 = -8x$

APPLICATIONS

91. PRODUCTION COSTS A television production company charges a basic fee of \$5,000 and then \$2,000 an hour when filming a commercial.

a. Write a linear equation that describes the relationship between the total production costs c and the hours h of filming.

b. Use your answer to part a to find the production costs if a commercial required 8 hours of filming.

92. COLLEGE FEES Each semester, students enrolling at a community college must pay tuition costs of \$20 per unit as well as a \$40 student services fee.

a. Write a linear equation that gives the total fees t to be paid by a student enrolling at the college and taking x units.

b. Use your answer to part a to find the enrollment cost for a student taking 12 units.

93. CHEMISTRY A portion of a student's chemistry lab manual is shown below. Use the information to write a linear equation relating the temperature F (in degrees Fahrenheit) of the compound to the time t (in minutes) elapsed during the lab procedure.

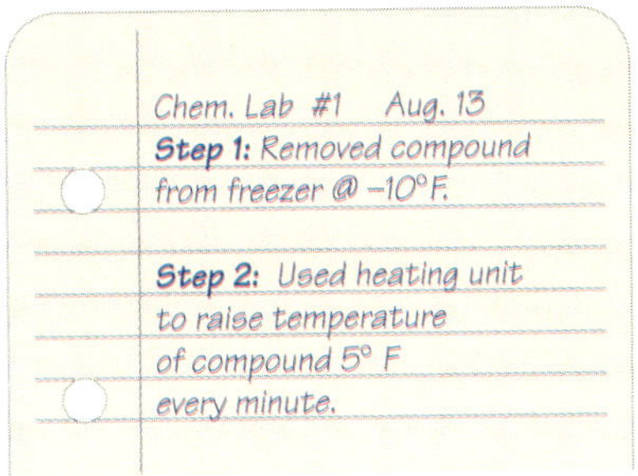

94. RENTALS Use the information in the newspaper advertisement to write a linear equation that gives the amount of income A (in dollars) the apartment owner will receive when the unit is rented for m months.

95. EMPLOYMENT SERVICE A policy statement of LIZCO, Inc., is shown below. Suppose a secretary had to pay an employment service $500 to get placed in a new job at LIZCO. Write a linear equation that tells the secretary the actual cost c of the employment service to her m months after being hired.

Policy no. 23452—A new hire will be reimbursed by LIZCO for any employment service fees paid by the employee at the rate of $20 per month.

96. VIDEOTAPES A VHS videocassette contains 800 feet of tape. In the long-play mode (LP), it plays 10 feet of tape every 3 minutes. Write a linear equation that relates the number of feet f of tape yet to be played and the number of minutes m the tape has been playing.

97. SEWING COSTS A tailor charges a basic fee of $20 plus $5 per letter to sew an athlete's name on the back of a jacket.

a. Write a linear equation that will find the cost c to have a name containing x letters sewn on the back of a jacket.

b. Graph the equation.

c. Suppose the tailor raises the basic fee to $30. On your graph from part b, draw the new graph showing the increased cost.

98. SALAD BARS For lunch, a delicatessen offers a "Salad and Soda" special where customers serve themselves at a well-stocked salad bar. The cost is $1.00 for the drink and 20¢ an ounce for the salad.

a. Write a linear equation that will find the cost c of a "Salad and Soda" lunch when a salad weighing x ounces is purchased.

b. Graph the equation.

c. How would the graph from part b change if the delicatessen began charging $2.00 for the drink?

d. How would the graph from part b change if the cost of the salad changed to 30¢ an ounce?

99. BASEBALL Use the following facts to write a linear equation in slope–intercept form that approximates the average price of a Major League Baseball ticket for the years 2000–2006.

- Let t represent the number of years since 2000 and c the average cost of a ticket in dollars.
- In 2000, the average ticket price was $16.63.
- From 2000 to 2006, the average ticket price increased 89¢ per year.

(Source: Team Marketing Report, MLB)

100. NAVIGATION The graph shows the recommended speed at which a ship should proceed into head waves of various heights.

a. What information does the y-intercept of the line give?

b. What is the rate of change in the recommended speed of the ship as the wave height increases?

c. Write the equation of the line.

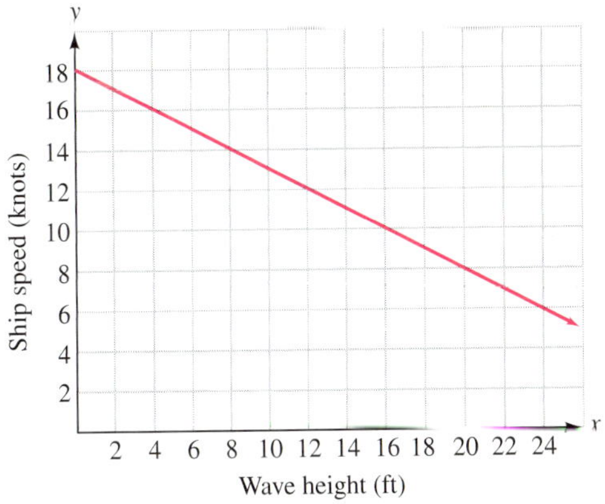

WRITING

101. Why is $y = mx + b$ called the slope–intercept form of the equation of a line?

102. On a quiz, a student was asked to find the slope of the graph of $y = 2x + 3$. She answered: $m = 2x$. Her instructor marked it wrong. Explain why the answer is incorrect.

REVIEW

103. CABLE TV A 186-foot television cable is to be cut into four pieces. Find the length of each piece if each successive piece is 3 feet longer than the previous one.

104. INVESTMENTS Joni received $25,000 as part of a settlement in a class action lawsuit. She invested some money at 10% and the rest at 9% simple interest rates. If her total annual income from these two investments was $2,430, how much did she invest at each rate?

CHALLENGE PROBLEMS

105. If the graph of $y = mx + b$ passes through quadrants I, II, and IV, what can be known about the constants m and b?

106. The equation $y = \frac{3}{4}x - \frac{5}{2}$ is in slope–intercept form. Write it in standard (general) form, $Ax + By = C$, where $A > 0$.

SECTION 3.6 Point–Slope Form

Objectives

1. Use point–slope form to write an equation of a line.
2. Write an equation of a line given two points on the line.
3. Write equations of horizontal and vertical lines.
4. Use a point and the slope to graph a line.
5. Write linear equations that model data.

If we know the slope of a line and its y-intercept, we can use the slope–intercept form to write the equation of the line. The question that now arises is, can *any* point on the line be used in combination with its slope to write its equation? In this section, we answer this question.

1 Use Point–Slope Form to Write an Equation of a Line.

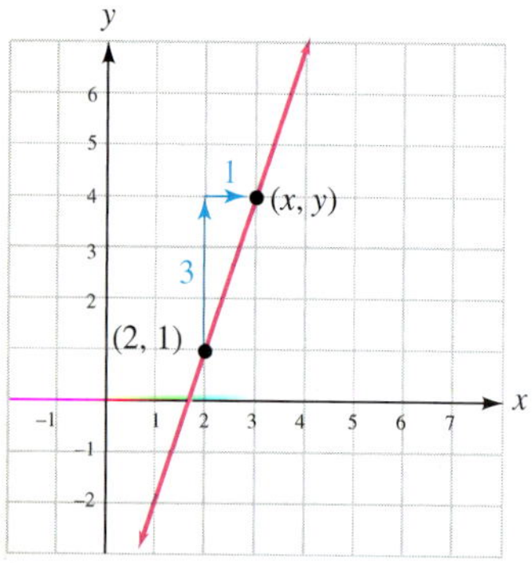

Refer to the line graphed on the left, with slope 3 and passing through the point (2,1). To develop a new form for the equation of a line, we will find the slope of this line in another way.

If we pick another point on the line with coordinates (x, y), we can find the slope of the line by substituting the coordinates of the points (x, y) and $(2, 1)$ into the slope formula.

$$\frac{y_2 - y_1}{x_2 - x_1} = m$$

$$\frac{y - 1}{x - 2} = m$$ Let (x_1, y_1) be $(2, 1)$ and (x_2, y_2) be (x, y). Substitute y for y_2, 1 for y_1, x for x_2, and 2 for x_1.

Since the slope of the line is 3, we can substitute 3 for m in the previous equation.

$$\frac{y - 1}{x - 2} = 3$$

We then multiply both sides by $x - 2$ to clear the equation of the fraction.

$$\frac{y-1}{x-2}(x-2) = 3(x-2)$$

$$y - 1 = 3(x - 2)$$ Simplify the left side. Remove the common factor $x - 2$ in the numerator and denominator: $\frac{y-1}{\cancel{x-2}_1} \cdot \frac{\cancel{x-2}^1}{1}$.

The resulting equation displays the slope of the line and the coordinates of one point on the line:

Slope of the line

$$y - 1 = 3(x - 2)$$

y-coordinate of the point — x-coordinate of the point

In general, suppose we know that the slope of a line is m and that the line passes through the point (x_1, y_1). Then if (x, y) is any other point on the line, we can use the definition of slope to write

$$\frac{y - y_1}{x - x_1} = m$$

If we multiply both sides by $x - x_1$ to clear the equation of the fraction, we have

$$y - y_1 = m(x - x_1)$$

This form of a linear equation is called **point–slope form.** It can be used to write the equation of a line when the slope and one point on the line are known.

Point–Slope Form of the Equation of a Line

If a line with slope m passes through the point (x_1, y_1), the equation of the line is

$$y - y_1 = m(x - x_1)$$

EXAMPLE 1 Find an equation of a line that has slope -8 and passes through $(-1, 5)$. Write the answer in slope–intercept form.

Strategy We will use the point–slope form, $y - y_1 = m(x - x_1)$, to write an equation of the line.

Why We are given the slope of the line and the coordinates of a point that it passes through.

Solution Because we are given the coordinates of a point on the line and the slope of the line, we begin by writing the equation of the line in the point–slope form. Since the slope is -8 and the given point is $(-1, 5)$, we have $m = -8$, $x_1 = -1$, and $y_1 = 5$.

$y - y_1 = m(x - x_1)$ This is the point–slope form.

$y - 5 = -8[x - (-1)]$ Substitute -8 for m, -1 for x_1, and 5 for y_1.

$y - 5 = -8(x + 1)$ Simplify within the brackets.

Notation
After writing an equation in point–slope form, it is common practice to solve it for y and write an equivalent equation in slope–intercept form.

To write this equation in slope–intercept form, we solve for y.

$$y - 5 = -8(x + 1)$$

$$y - 5 = -8x - 8 \quad \text{Distribute the multiplication by } -8.$$

$$y - 5 + 5 = -8x - 8 + 5 \quad \text{To isolate } y \text{, undo the subtraction of 5 by adding 5 to both sides.}$$

$$y = -8x - 3$$

In slope–intercept form, the equation is $y = -8x - 3$.

To verify this result, we note that $m = -8$. Therefore, the slope of the line is -8, as required. To see whether the line passes through $(-1, 5)$, we substitute -1 for x and 5 for y in the equation. If this point is on the line, a true statement should result.

$$y = -8x - 3$$

$$5 \stackrel{?}{=} -8(-1) - 3$$

$$5 \stackrel{?}{=} 8 - 3$$

$$5 = 5 \quad \text{True}$$

Self Check 1 Find an equation of the line that has slope -2 and passes through $(4, -3)$. Write the answer in slope–intercept form.

Now Try Problems 13 and 19

2 Write an Equation of a Line Given Two Points on the Line.

In the next example, we show that it is possible to write the equation of a line when we know the coordinates of two points on the line.

EXAMPLE 2 Find an equation of the line that passes through $(-2, 6)$ and $(4, 7)$. Write the equation in slope–intercept form.

Strategy We will use the point–slope form, $y - y_1 = m(x - x_1)$, to write an equation of the line.

Why We know the coordinates of a point that the line passes through and we can calculate the slope of the line using the slope formula.

Solution To find the slope of the line, we use the slope formula.

$$m = \frac{y_2 - y_1}{x_2 - x_1} = \frac{7 - 6}{4 - (-2)} = \frac{1}{6} \quad \text{Substitute 7 for } y_2\text{, 6 for } y_1\text{, 4 for } x_2\text{, and } -2 \text{ for } x_1.$$

Either point on the line can serve as (x_1, y_1). If we choose $(4, 7)$, we have

$$y - y_1 = m(x - x_1) \quad \text{This is the point–slope form.}$$

$$y - 7 = \frac{1}{6}(x - 4) \quad \text{Substitute } \tfrac{1}{6} \text{ for } m\text{, 7 for } y_1\text{, and 4 for } x_1.$$

Success Tip
In Example 2, either of the given points can be used as (x_1, y_1) when writing the point–slope equation. The results will be the same.

Looking ahead, we usually choose the point whose coordinates will make the computations the easiest.

To write this equation in slope–intercept form, we solve for y.

$$y - 7 = \frac{1}{6}x - \frac{2}{3}$$ Distribute the multiplication by $\frac{1}{6}$.

$$y - 7 + 7 = \frac{1}{6}x - \frac{2}{3} + 7$$ To isolate y, add 7 to both sides.

$$y = \frac{1}{6}x - \frac{4}{6} + \frac{42}{6}$$ Simplify the left side. Write 7 as $\frac{42}{6}$ to prepare to add the fractions.

$$y = \frac{1}{6}x + \frac{19}{3}$$ Simplify: $\frac{38}{6} = \frac{\overset{1}{\cancel{2}} \cdot 19}{\underset{1}{\cancel{2}} \cdot 3} = \frac{19}{3}$. This is slope–intercept form.

Success Tip
To check this result, verify that $(-2, 6)$ and $(4, 7)$ satisfy $y = \frac{1}{6}x + \frac{19}{3}$.

The equation of the line that passes through $(-2, 6)$ and $(4, 7)$ is $y = \frac{1}{6}x + \frac{19}{3}$.

Self Check 2 Find an equation of the line that passes through $(-5, 4)$ and $(8, -6)$. Write the equation in slope–intercept form.

Now Try Problem 29

3 Write Equations of Horizontal and Vertical Lines.

We have previously graphed horizontal and vertical lines. We will now discuss how to write their equations.

EXAMPLE 3 Write an equation of each line and graph it. **a.** A horizontal line passing through $(-2, -4)$ **b.** A vertical line passing through $(1, 3)$

Strategy We will use the appropriate form, either $y = b$ or $x = a$, to write an equation of each line.

Why These are the standard forms for the equations of a horizontal and a vertical line.

Solution

a. The equation of a horizontal line can be written in the form $y = b$. Since the y-coordinate of $(-2, -4)$ is -4, the equation of the line is $y = -4$. The graph is shown in the figure.

b. The equation of a vertical line can be written in the form $x = a$. Since the x-coordinate of $(1, 3)$ is 1, the equation of the line is $x = 1$. The graph is shown in the figure.

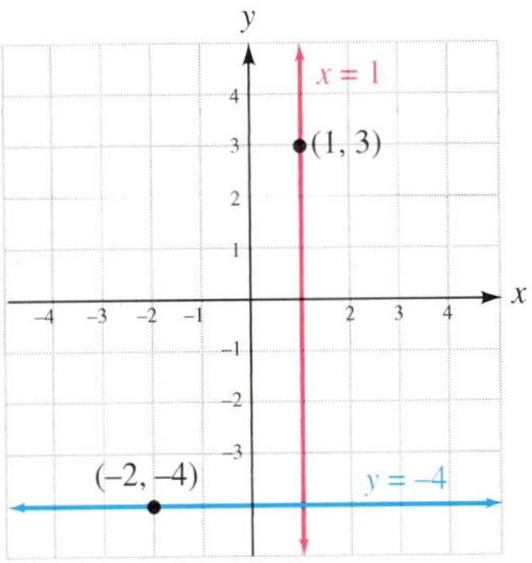

Self Check 3 Write an equation of each line. **a.** A horizontal line passing through $(3, 2)$ **b.** A vertical line passing through $(-1, -3)$

Now Try Problems 41 and 43

4 Use a Point and the Slope to Graph a Line.

If we know the coordinates of a point on a line, and if we know the slope of the line, we can use the slope to determine a second point on the line.

EXAMPLE 4 Graph the line with slope $\frac{2}{5}$ that passes through $(-1, -3)$.

Strategy First, we will plot the given point $(-1, -3)$. Then we will use the slope to find a second point that the line passes through.

Why Once we determine two points that the line passes through, we can draw the graph of the line.

Solution We begin by plotting the point $(-1, -3)$. From there, we move 2 units up and then 5 units to the right, since the slope is $\frac{2}{5}$. This puts us at a second point on the line, $(4, -1)$. We then draw a line through the two points.

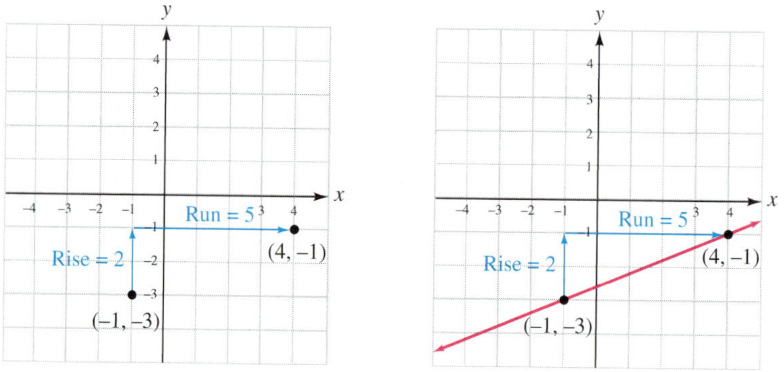

Self Check 4 Graph the line with slope -4 that passes through $(-4, 2)$.

Now Try **Problem 45**

5 Write Linear Equations That Model Data.

Many situations can be described by a linear equation. Quite often, these equations are written using variables other than x and y. In such cases, it is helpful to determine what an ordered-pair solution of the equation would look like.

EXAMPLE 5 ***Men's Shoe Sizes.*** The length (in inches) of a man's foot is not his shoe size. For example, the smallest adult men's shoe size is 5, and it fits a 9-inch-long foot. There is, however, a linear relationship between the two. It can be stated this way: Shoe size increases by 3 sizes for each 1-inch increase in foot length.

a. Write a linear equation that relates shoe size s to foot length L.

b. Shaquille O'Neal, a famous basketball player, has a foot that is about 14.6 inches long. Find his shoe size.

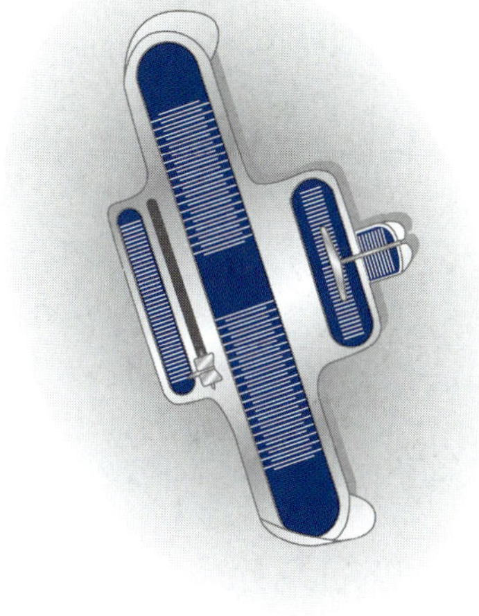

Strategy We will first find the slope of the of the line that describes the linear relationship between shoe size and the length of a foot. Then we will determine the coordinates of a point on that line.

Why Once we know the slope and the coordinates of one point on the line, we can use the point–slope form to write the equation of the line.

Solution

a. Since shoe size s depends on the length L of the foot, ordered pairs have the form (L, s). Because the relationship is linear, the graph of the desired equation is a line.

- The line's slope is the rate of change: $\frac{3 \text{ sizes}}{1 \text{ inch}}$. Therefore, $m = 3$.
- A 9-inch-long foot wears size 5, so the line passes through $(9, 5)$.

We substitute 3 for m and the coordinates of the point into the point–slope form and solve for s.

$$s - s_1 = m(L - L_1)$$ This is the point–slope form using the variables L and s.

$$s - 5 = 3(L - 9)$$ Substitute 3 for m, 9 for L_1, and 5 for s_1.

$$s - 5 = 3L - 27$$ Distribute the multiplication by 3.

$$s = 3L - 22$$ To isolate s, add 5 to both sides.

The equation relating men's shoe size and foot length is $s = 3L - 22$.

b. To find Shaquille's shoe size, we substitute 14.6 inches for L in the equation.

$$s = 3L - 22$$
$$s = 3(14.6) - 22$$
$$s = 43.8 - 22$$
$$s = 21.8$$

Since men's shoes only come in full- and half-sizes, we round 21.8 up to 22. Shaquille O'Neal wears size 22 shoes.

Now Try **Problem 73**

EXAMPLE 6 ***Studying Learning.*** In a series of trials, a rat was released in a maze to search for food. Researchers recorded the time that it took the rat to complete the maze on a **scatter diagram.** After the 40th trial, they drew a line through the data to obtain a model of the rat's performance. Write an equation of the line in slope–intercept form.

The Language of Algebra

The term *scatter diagram* is somewhat misleading. Often, the data points are not scattered loosely about. In this case, they fall, more or less, along an imaginary straight line, indicating a linear relationship.

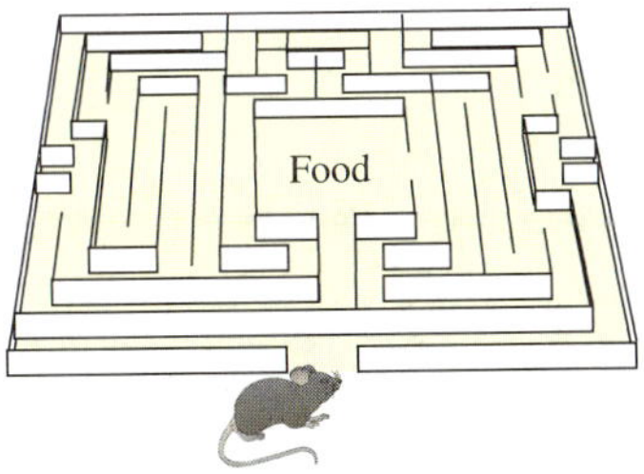

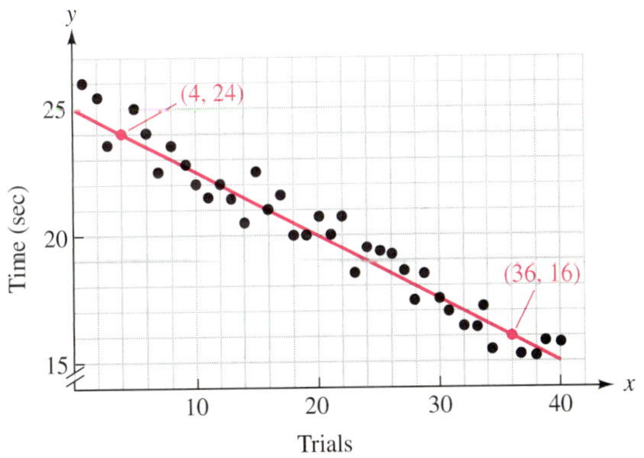

Strategy From the graph, we will determine the coordinates of two points on the line.

Why We can write an equation of a line when we know the coordinates of two points on the line. (See Example 2.)

Solution We begin by writing a point–slope equation. The line passes through several points; we will use (4, 24) and (36, 16) to find the slope.

$$m = \frac{y_2 - y_1}{x_2 - x_1} = \frac{16 - 24}{36 - 4} = \frac{-8}{32} = -\frac{1}{4}$$

Any point on the line can serve as (x_1, y_1). We will use (4, 24).

$y - y_1 = m(x - x_1)$ This is the point–slope form.

$y - 24 = -\frac{1}{4}(x - 4)$ Substitute $-\frac{1}{4}$ for m, 4 for x_1, and 24 for y_1.

To write this equation in slope–intercept form, solve for y.

$y - 24 = -\frac{1}{4}x + 1$ Distribute the multiplication by $-\frac{1}{4}$: $-\frac{1}{4}(-4) = 1$.

$y = -\frac{1}{4}x + 25$ To isolate y, add 24 to both sides.

A linear equation that models the rat's performance on the maze is $y = -\frac{1}{4}x + 25$, where x is the number of the trial and y is the time it took, in seconds.

Now Try **Problem 81**

ANSWERS TO SELF CHECKS

1. $y = -2x + 5$
2. $y = -\frac{10}{13}x + \frac{2}{13}$
3. a. $y = 2$ b. $x = -1$
4.

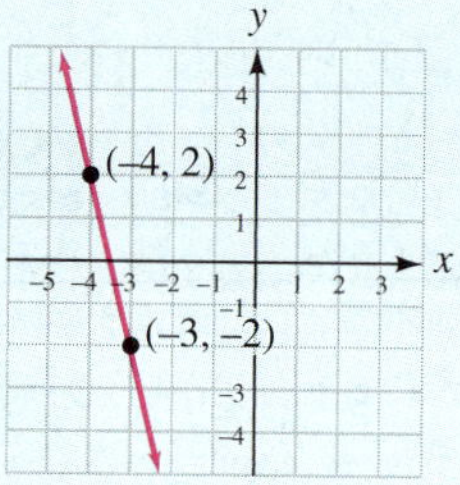

STUDY SET 3.6

VOCABULARY

Fill in the blanks.

1. $y - y_1 = m(x - x_1)$ is called the ____________ form of the equation of a line. In words, we read this as y minus y _____ one equals m ______ the quantity of x _______ x sub _____.

2. $y = mx + b$ is called the ______________ form of the equation of a line.

CONCEPTS

3. Determine in what form each equation is written.
   a. $y - 4 = 2(x - 5)$
   b. $y = 2x + 15$

4. What point does the graph of the equation pass through, and what is the line's slope?
   a. $y - 2 = 6(x - 7)$
   b. $y + 3 = -8(x + 1)$

5. Refer to the following graph of a line.
   a. What highlighted point does the line pass through?
   b. What is the slope of the line?
   c. Write an equation of the line in point–slope form.

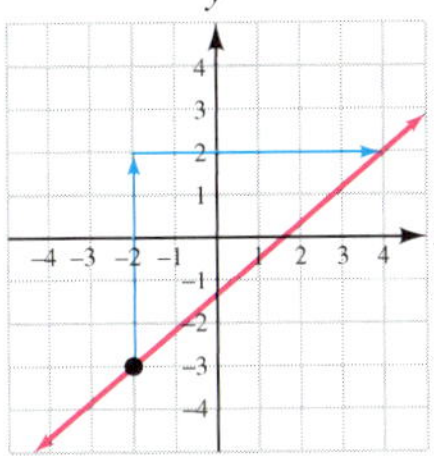

6. On a quiz, a student was asked to write the equation of a line with slope 4 that passes through $(-1, 3)$. Explain how the student can check her answer, $y = 4x + 7$.

7. Suppose you are asked to write an equation of the line in the scatter diagram below. What two points would you use to write the point–slope equation?

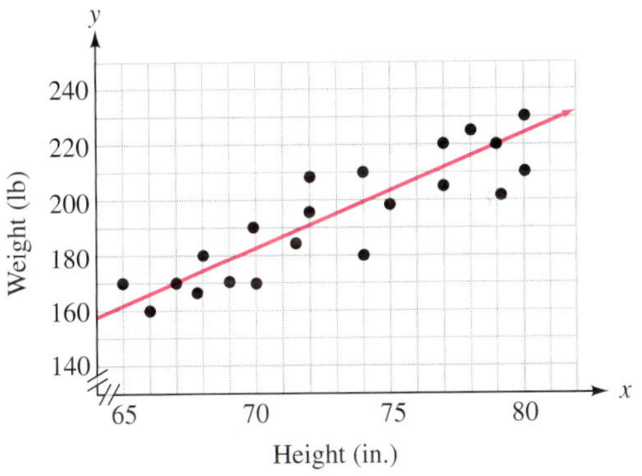

8. In each case, a linear relationship between two quantities is described. If the relationship were graphed, what would be the slope of the line?
   a. The sales of new cars increased by 15 every 2 months.
   b. There were 35 fewer robberies for each dozen police officers added to the force.
   c. One acre of forest is being destroyed every 30 seconds.

NOTATION

Complete the solution.

9. Write an equation of the line with slope -2 that passes through the point $(-1, 5)$. Write the answer in slope–intercept form.

$$y - y_1 = m(x - x_1)$$
$$y - \underline{\quad} = -2[x - (\underline{\quad})]$$
$$y - 5 = -2[x \underline{\quad} 1]$$
$$y - 5 = -2x - \underline{\quad}$$
$$y = -2x + \underline{\quad}$$

10. What is the point–slope form of the equation of a line?

11. Consider the steps below and then fill in the blanks:

$$y - 3 = 2(x + 1)$$
$$y - 3 = 2x + 2$$
$$y = 2x + 5$$

The original equation was in ____________ form. After solving for y, we obtain an equation in ____________ form.

12. Fill in the blanks: The equation of a horizontal line has the form $\underline{\quad} = b$ and the equation of a vertical line has the form $\underline{\quad} = a$.

GUIDED PRACTICE

Use the point–slope form to write an equation of the line with the given slope and point. Leave the equation in that form. See Example 1.

13. Slope 3, passes through $(2, 1)$
14. Slope 2, passes through $(4, 3)$
15. Slope $\frac{4}{5}$, passes through $(-5, -1)$
16. Slope $\frac{7}{8}$, passes through $(-2, -9)$

Use the point–slope form to write an equation of the line with the given slope and point. Then write the equation in slope–intercept form. See Example 1.

17. Slope 2, passes through $(3, 5)$
18. Slope 8, passes through $(2, 6)$
19. Slope -5, passes through $(-9, 8)$
20. Slope -4, passes through $(-2, 10)$
21. Slope -3, passes through the origin
22. Slope -1, passes through the origin
23. Slope $\frac{1}{5}$, passes through $(10, 1)$
24. Slope $\frac{1}{4}$, passes through $(8, 1)$
25. Slope $-\frac{4}{3}$,

x	y
6	-4

26. Slope $-\frac{3}{2}$,

x	y
-2	1

27. Slope $-\frac{11}{6}$, passes through $(2, -6)$
28. Slope $-\frac{5}{4}$, passes through $(2, 0)$

Find an equation of the line that passes through the two given points. Write the equation in slope–intercept form, if possible. See Example 2.

29. Passes through $(1, 7)$ and $(-2, 1)$
30. Passes through $(-2, 2)$ and $(2, -8)$

31.

x	y
−4	3
2	0

32.

x	y
−1	−4
1	−2

33. Passes through (5, 5) and (7, 5)

34. Passes through (−2, 1) and (−2, 15)

35. Passes through (5, 1) and (−5, 0)

36. Passes through (−3, 0) and (3, 1)

37. Passes through (−8, 2) and (−8, 17)

38. Passes through $\left(\frac{2}{3}, 2\right)$ and (0, 2)

39. Passes through $\left(\frac{2}{3}, \frac{1}{3}\right)$ and (0, 0)

40. Passes through $\left(\frac{1}{2}, \frac{3}{4}\right)$ and (0, 0)

***Write an equation of the line with the given characteristics.* See Example 3.**

41. Vertical, passes through (4, 5)

42. Vertical, passes through (−2, −5)

43. Horizontal, passes through (4, 5)

44. Horizontal, passes through (−2, −5)

***Graph the line that passes through the given point and has the given slope.* See Example 4.**

45. (1, −2), slope −1

46. (−4, 1), slope −3

47. (5, −3), $m = \frac{3}{4}$

48. (2, −4), $m = \frac{2}{3}$

49. (−2, −3), slope 2

50. (−3, −3), slope 4

51. (4, −3), slope $-\frac{7}{8}$

52. (4, 2), slope $-\frac{1}{5}$

TRY IT YOURSELF

Find an equation of the line with the following characteristics. Write the equation in slope–intercept form, if possible.

53. Passes through (5, 0) and (−11, −4)

54. Passes through (7, −3) and (−5, 1)

55. Horizontal, passes through (−8, 12)

56. Horizontal, passes through (9, −32)

57. Slope $-\frac{2}{3}$, passes through (3, 0)

58. Slope $-\frac{2}{5}$, passes through (15, 0)

59. Slope 8, passes through (2, 20)

60. Slope 6, passes through (1, −2)

61. Vertical, passes through (−3, 7)

62. Vertical, passes through (12, −23)

63. Slope 7 and y-intercept (0, 0)

64. Slope 3 and y-intercept (0, 4)

65. Passes through (−2, −1) and (−1, −5)

66. Passes through (−3, 6) and (−1, −4)

67. x-intercept (7, 0) and y-intercept (0, −2)

68. x-intercept (−3, 0) and y-intercept (0, 7)

69. Slope $\frac{1}{10}$, passes through the origin

70. Slope $\frac{9}{8}$, passes through the origin

71. Undefined slope, passes through $\left(-\frac{1}{8}, 12\right)$

72. Undefined slope, passes through $\left(\frac{2}{5}, -\frac{5}{6}\right)$

APPLICATIONS

73. ANATOMY There is a linear relationship between a woman's height and the length of her radius bone. It can be stated this way: Height increases by 3.9 inches for each 1-inch increase in the length of the radius. Suppose a 64-inch-tall woman has a 9-inch-long radius bone. Use this information to find a linear equation that relates height h to the length r of the radius. Write the equation in slope–intercept form.

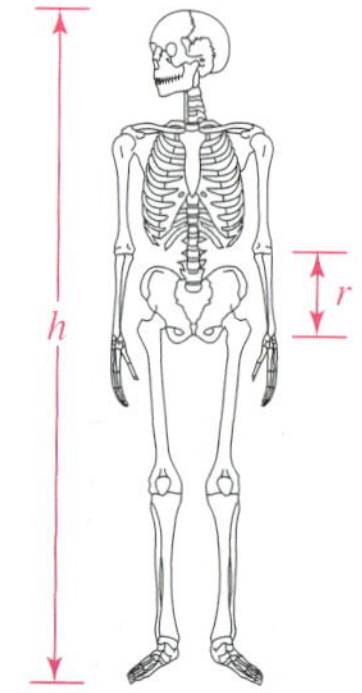

74. AUTOMATION An automated production line uses distilled water at a rate of 300 gallons every 2 hours to make shampoo. After the line had run for 7 hours, planners noted that 2,500 gallons of distilled water remained in the storage tank. Find a linear equation relating the time t in hours since the production line began and the number g of gallons of distilled water in the storage tank. Write the equation in slope–intercept form.

75. POLE VAULTING Find the equations of the lines that describe the positions of the pole for parts 1, 3, and 4 of the jump. Write the equations in slope–intercept form, if possible.

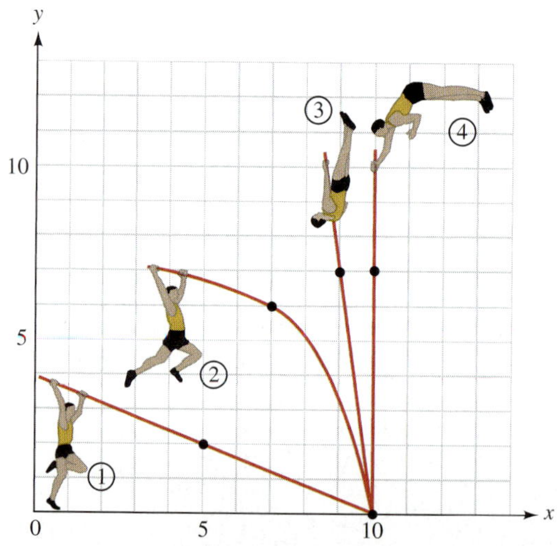

76. FREEWAY DESIGN The graph below shows the route of a proposed freeway.

a. Give the coordinates of the points where the proposed freeway will join Interstate 25 and Highway 40.

b. Write the equation of the line that describes the route of the proposed freeway. Give the answer in slope–intercept form.

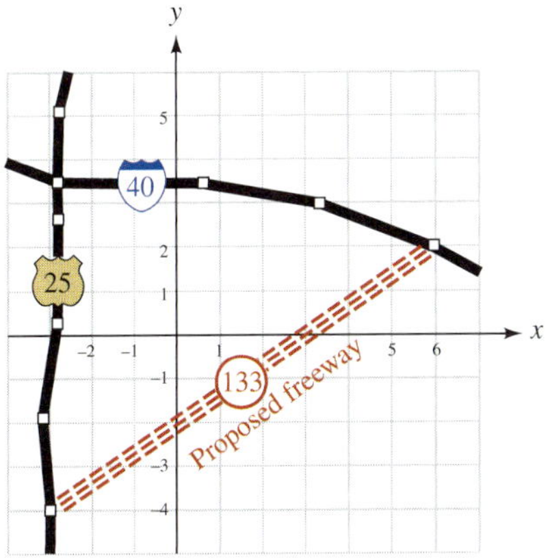

77. TOXIC CLEANUP Three months after cleanup began at a dump site, 800 cubic yards of toxic waste had yet to be removed. Two months later, that number had been lowered to 720 cubic yards.

a. Find an equation that describes the linear relationship between the length of time m (in months) the cleanup crew has been working and the number of cubic yards y of toxic waste remaining. Write the equation in slope–intercept form.

b. Use your answer to part (a) to predict the number of cubic yards of waste that will still be on the site one year after the cleanup project began.

78. DEPRECIATION To lower its corporate income tax, accountants of a company depreciated a word processing system over several years using a linear model, as shown in the worksheet.

a. Find a linear equation relating the years since the system was purchased, x, and its value, y, in dollars. Write the equation in slope–intercept form.

b. Find the purchase price of the system.

Tax Worksheet

Method of depreciation: *Linear*

Property	Years after purchase	Value
Word processing system	2	$60,000
"	4	$30,000

79. TRAMPOLINES There is a linear relationship between the length of the protective pad that wraps around a trampoline and the radius of the trampoline. Use the data in the table to find an equation that gives the length l of pad needed for any trampoline with radius r. Write the equation in slope–intercept form. Use units of feet for both l and r.

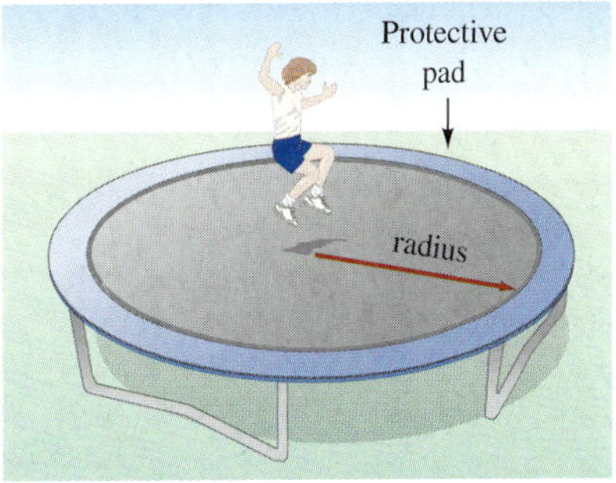

Radius	Pad length
3 ft	19 ft
7 ft	44 ft

80. CONVERTING TEMPERATURES The relationship between Fahrenheit temperature, F, and Celsius temperature, C, is linear.

a. Use the data in the illustration to write two ordered pairs of the form (C, F).

b. Use your answer to part (a) to find a linear equation relating the Fahrenheit and Celsius scales. Write the equation in slope–intercept form.

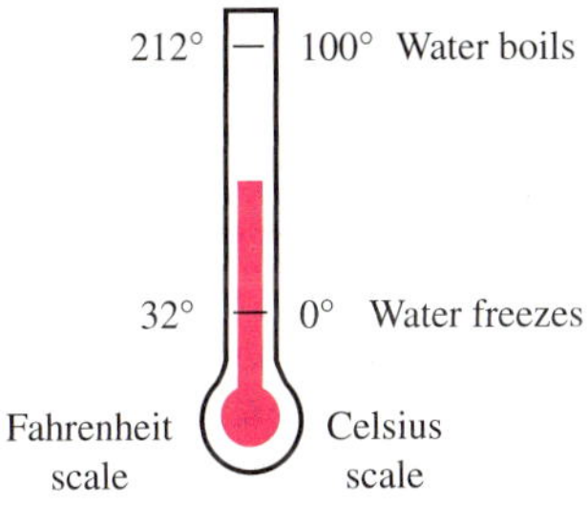

81. GOT MILK The scatter diagram shows the amount of milk that an average American drank in one year for the years 1980–2004. A straight line can be used to model the data.

a. Use two points on the line to find its equation. Write the equation in slope–intercept form.

b. Use your answer to part (a) to predict the amount of milk that an average American will drink in 2020.

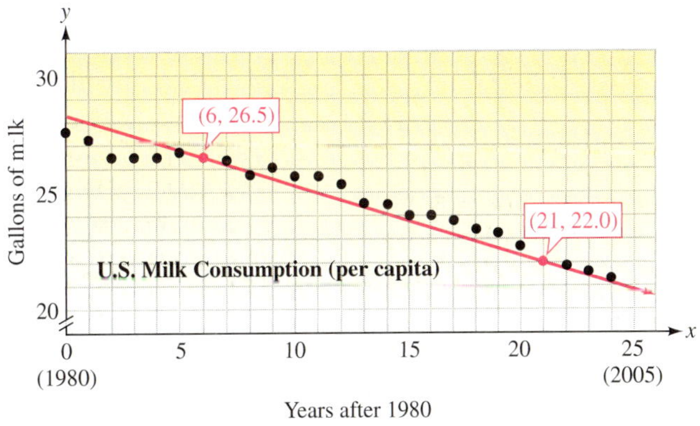

Source: United States Department of Agriculture

82. ENGINE OUTPUT The horsepower produced by an automobile engine was recorded for various engine speeds in the range of 2,400 to 4,800 revolutions per minute (rpm). The data were recorded on the following scatter diagram. Find an equation of the line that models the relationship between engine speed s and horsepower h. Write the equation in slope–intercept form.

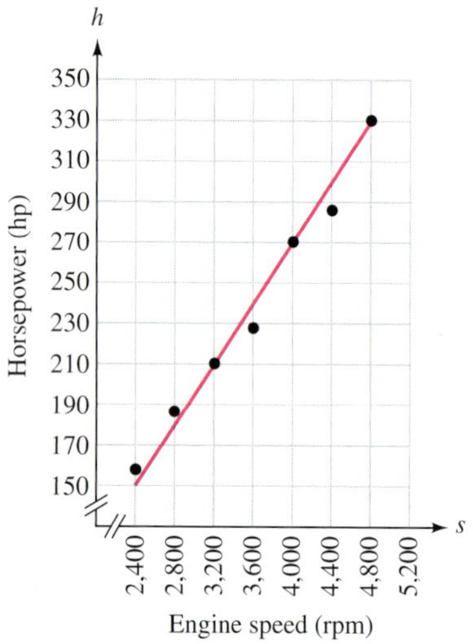

WRITING

83. Why is $y - y_1 = m(x - x_1)$ called the point–slope form of the equation of a line?

84. If we know two points that a line passes through, we can write its equation. Explain how this is done.

85. Explain the steps involved in writing $y - 6 = 4(x - 1)$ in slope–intercept form.

86. Think of the points on the graph of the horizontal line $y = 4$. What do the points have in common? How do they differ?

REVIEW

87. FRAMES The length of a rectangular picture is 5 inches greater than twice the width. If the perimeter is 112 inches, find the dimensions of the frame.

88. SPEED OF AN AIRPLANE Two planes are 6,000 miles apart, and their speeds differ by 200 mph. They travel toward each other and meet in 5 hours. Find the speed of the slower plane.

CHALLENGE PROBLEMS

89. Find an equation of the line that passes through (2, 5) and is parallel to the line $y = 4x - 7$. Write the equation in slope–intercept form.

90. Find an equation of the line that passes through (−6, 3) and is perpendicular to the line $y = -3x - 12$. Write the equation in slope–intercept form.

SECTION 3.7
Graphing Linear Inequalities

Objectives

1. Determine whether an ordered pair is a solution of an inequality.
2. Graph a linear inequality in two variables.
3. Graph inequalities with a boundary through the origin.
4. Solve applied problems involving linear inequalities in two variables.

Graph: $x + 6 < 8$

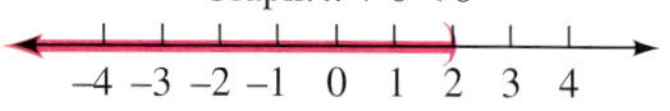

Graph: $5x + 3 \geq 4x$

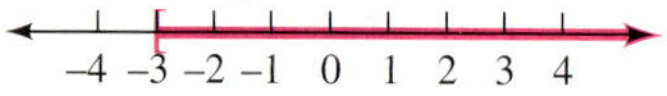

Recall that an **inequality** is a statement that contains one of the symbols $<$, $\leq$, $>$, or $\geq$. Inequalities in one variable, such as $x + 6 < 8$ and $5x + 3 \geq 4x$, were solved in Section 2.7. Because they have an infinite number of solutions, we represented their solution sets graphically, by shading intervals on a number line.

We now extend that concept to linear inequalities *in two variables,* as we introduce a procedure that is used to graph their solution sets.

1 Determine Whether an Ordered Pair Is a Solution of an Inequality.

If the = symbol in a linear equation in two variables is replaced with an inequality symbol, we have a **linear inequality in two variables.** Some examples are

$x - y \leq 5, \quad 4x + 3y < -6, \quad$ and $\quad y > 2x$

As with linear equations, a **solution of a linear inequality** in two variables is an ordered pair of numbers that makes the inequality true.

EXAMPLE 1 Determine whether each ordered pair is a solution of $x - y \leq 5$. Then graph each solution: **a.** (4, 2) **b.** (0, −6) **c.** (1, −4)

Strategy We will substitute each ordered pair of coordinates into the inequality.

Why If the resulting statement is true, the ordered pair is a solution.

Solution

a. For (4, 2):

$x - y \leq 5$ This is the given inequality.

$4 - 2 \overset{?}{\leq} 5$ Substitute 4 for x and 2 for y.

$2 \leq 5$ True

Because $2 \leq 5$ is true, (4, 2) is a solution of $x - y \leq 5$. We say that (4, 2) *satisfies* the inequality. This solution is graphed as shown, on the right.

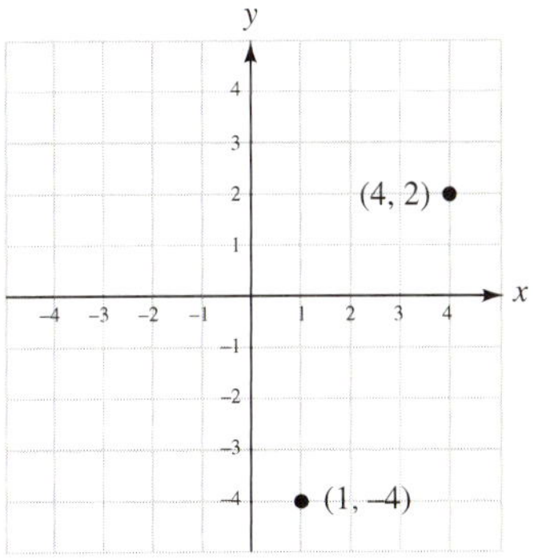

Two solutions of $x - y \leq 5$.

Notation

The symbol $\overset{?}{\leq}$ is read as "is possibly less than or equal to."

b. For (0, −6):

$x - y \leq 5$ This is the given inequality.

$0 - (-6) \overset{?}{\leq} 5$ Substitute 0 for x and −6 for y.

$6 \leq 5$ False

Because $6 \leq 5$ is false, (0, −6) is not a solution.

c. For (1, −4):

$x - y \leq 5$ This is the given inequality.

$1 - (-4) \overset{?}{\leq} 5$ Substitute 1 for x and −4 for y.

$5 \leq 5$ True

Because $5 \leq 5$ is true, (1, −4) is a solution, and we graph it as shown.

Self Check 1 Using the inequality in Example 1, determine whether each ordered pair is a solution: **a.** (8, 2) **b.** (4, −1) **c.** (−2, 4) **d.** (−3, −5)

Now Try **Problem 19**

In Example 1, we graphed two of the solutions of $x - y \leq 5$. Since there are infinitely more ordered pairs (x, y) that make the inequality true, it would not be reasonable to plot them. Fortunately, there is an easier way to show all of the solutions.

2 Graph a Linear Inequality in Two Variables.

The graph of a linear inequality is a picture that represents the set of all points whose coordinates satisfy the inequality. In general, such graphs are regions bounded by a line. We call those regions **half-planes,** and we use a two-step procedure to find them.

CONCEPTS

7. Determine whether $(-3, -5)$ is a solution of $5x - 3y \geq 0$.

8. Determine whether $(3, -1)$ is a solution of $x + 4y < -1$.

9. Fill in the blanks: A ________ line indicates that points on the boundary are not solutions and a ______ line indicates that points on the boundary are solutions.

10. The boundary for the graph of a linear inequality is shown. Why can't the origin be used as a test point to decide which side to shade?

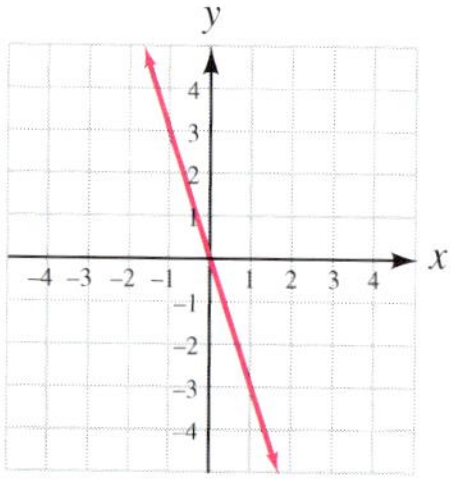

11. If a false statement results when the coordinates of a test point are substituted into a linear inequality, which half-plane should be shaded to represent the solution of the inequality?

12. A linear inequality has been graphed. Determine whether each point satisfies the inequality.

a. $(1, -3)$

b. $(-2, -1)$

c. $(2, 3)$

d. $(3, -4)$

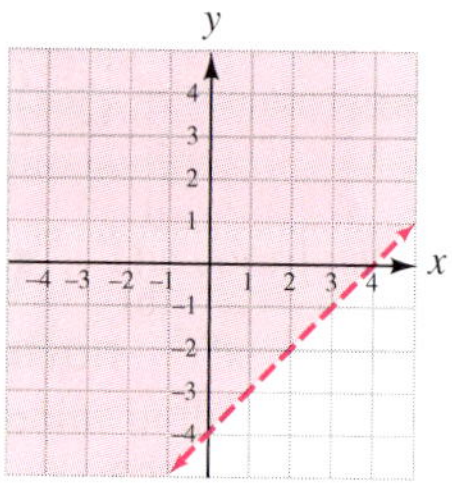

13. A linear inequality has been graphed. Determine whether each point satisfies the inequality.

a. $(2, 1)$

b. $(-2, -4)$

c. $(4, -2)$

d. $(-3, 4)$

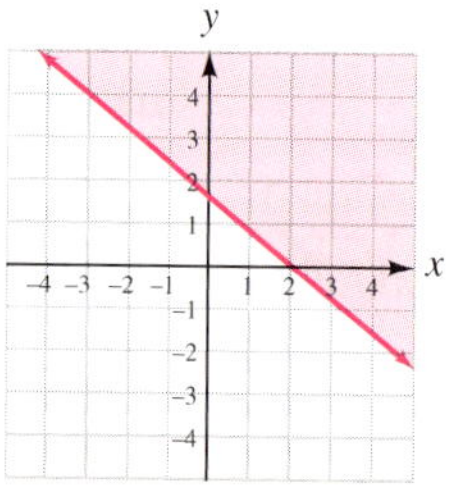

14. To graph linear inequalities, we must be able to graph boundary lines. Complete the table of solutions for each given boundary line.

a. $5x - 3y = 15$

x	y	(x, y)
0		
	0	
1		

b. $y = 3x - 2$

x	y	(x, y)
−1		
0		
2		

NOTATION

15. Write the meaning of each symbol in words.

a. $<$

b. $\geq$

c. $\leq$

d. $\overset{?}{>}$

16. **a.** When graphing linear inequalities, which inequality symbols are associated with a dashed boundary line?

b. When graphing linear inequalities, which inequality symbols are associated with a solid boundary line?

17. Fill in the blanks: The inequality $4x + 2y \leq 9$ means $4x + 2y$ ___ 9 or $4x + 2y$ ___ 9.

18. Fill in the blanks: The inequality $-x + 8y \geq 1$ means $-x + 8y$ ___ 1 or $-x + 8y$ ___ 1.

GUIDED PRACTICE

***Determine whether each ordered pair is a solution of the given inequality.* See Example 1.**

19. $2x + y > 6; (3, 2)$

20. $4x - 2y \geq -6; (-2, 1)$

21. $-5x - 8y < 8; (-8, 4)$

22. $x + 3y > 14; (-3, 8)$

23. $4x - y \leq 0; \left(\frac{1}{2}, 1\right)$

24. $9x - y \leq 2; \left(\frac{1}{3}, 1\right)$

25. $-5x + 2y > -4; (0.8, 0.6)$

26. $6x - 2y < -7; (-0.2, 1.5)$

***Complete the graph by shading the correct side of the boundary.* See Example 2.**

27. $x - y \geq -2$

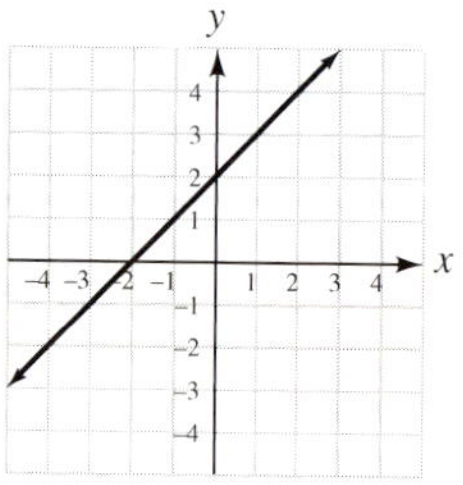

28. $x - y < 3$

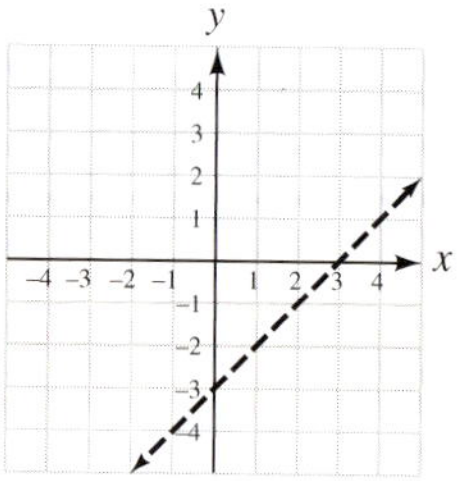

29. $y > 2x - 4$

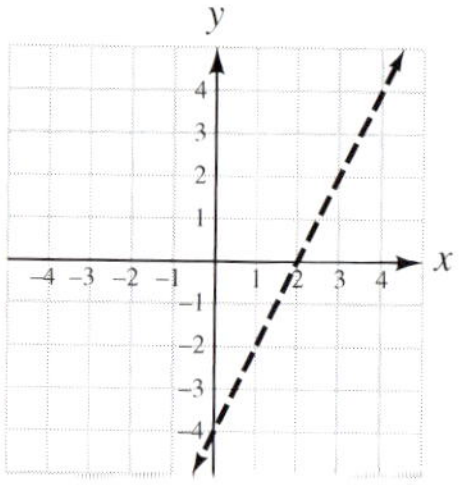

30. $y \leq -x + 1$

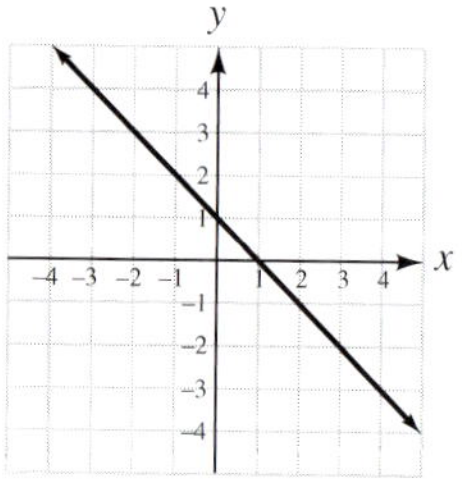

31. $x - 2y \geq 4$

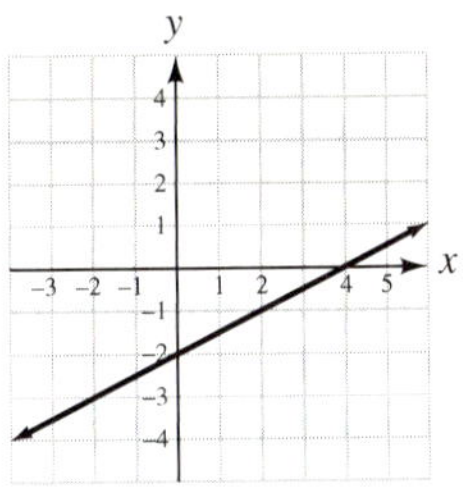

32. $3x + 2y > 12$

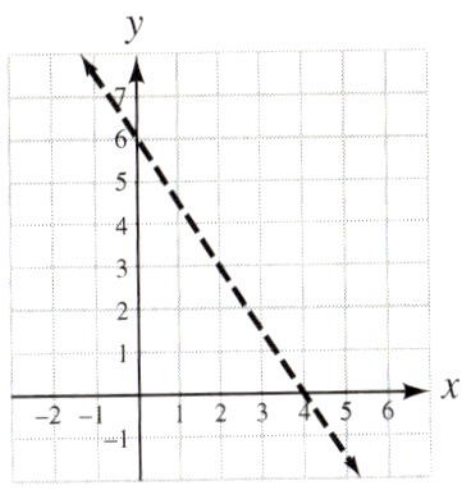

33. $y \leq 4x$

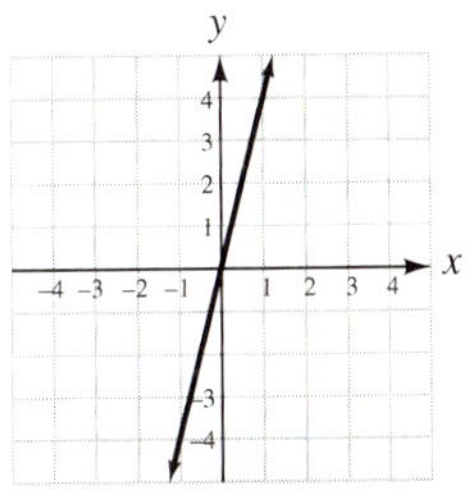

34. $y + 2x < 0$

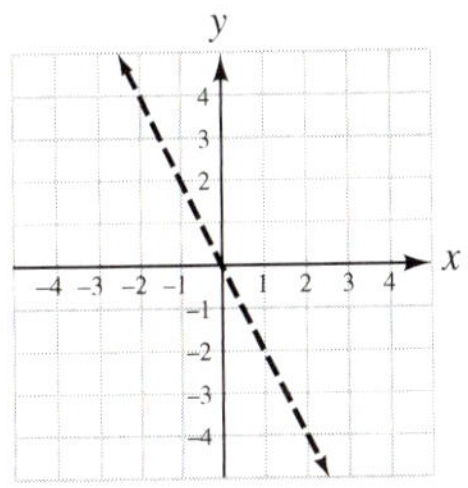

***Graph each inequality.* See Examples 2 and 3.**

35. $x + y \geq 3$

36. $x + y < 2$

37. $3x - 4y > 12$

38. $5x + 4y \geq 20$

39. $2x + 3y \leq -12$

40. $3x - 2y > 6$

41. $y < 2x - 1$

42. $y > x + 1$

43. $y < -3x + 2$

44. $y \geq -2x + 5$

45. $y \geq -\frac{3}{2}x + 1$

46. $y < \frac{x}{3} - 1$

47. $x - 2y \geq 4$

48. $4x + y \geq -4$

49. $2y - x < 8$

50. $y + 9x \geq 3$

51. $7x - 2y < 21$

52. $3x - 3y \geq -10$

53. $2x - 3y \geq 4$

54. $4x + 3y < 6$

***Graph each inequality.* See Example 4.**

55. $y \geq 2x$

56. $y < 3x$

57. $y < -\frac{x}{2}$

58. $y \geq x$

59. $y + x < 0$

60. $y - x < 0$

61. $5x + 3y < 0$

62. $2x + 5y > 0$

***Graph each inequality.* See Example 5.**

63. $x < 2$

64. $y > -3$

65. $y \leq 1$

66. $x \geq -4$

67. $y + 2.5 > 0$

68. $x - 1.5 \leq 0$

69. $x \leq 0$

70. $y < 0$

APPLICATIONS

71. DELIVERIES To decide the number x of pallets and the number y of barrels that a truck can hold, a driver refers to the graph below. Can a truck make a delivery of 4 pallets and 10 barrels in one trip?

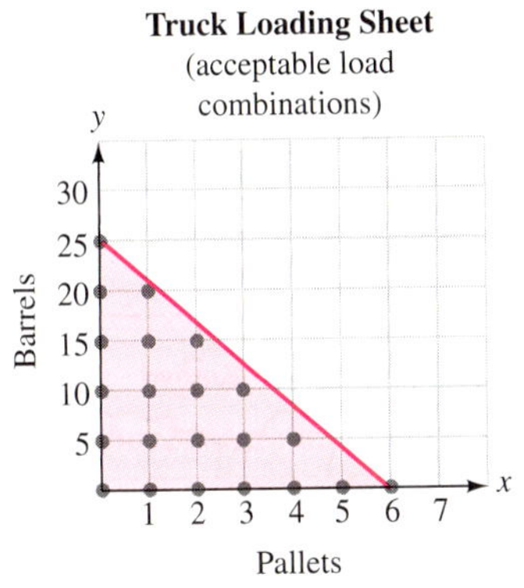

72. ZOOS To determine the allowable number of juvenile chimpanzees x and adult chimpanzees y that can live in an enclosure, a zookeeper refers to the graph. Can 6 juvenile and 4 adult chimps be kept in the enclosure?

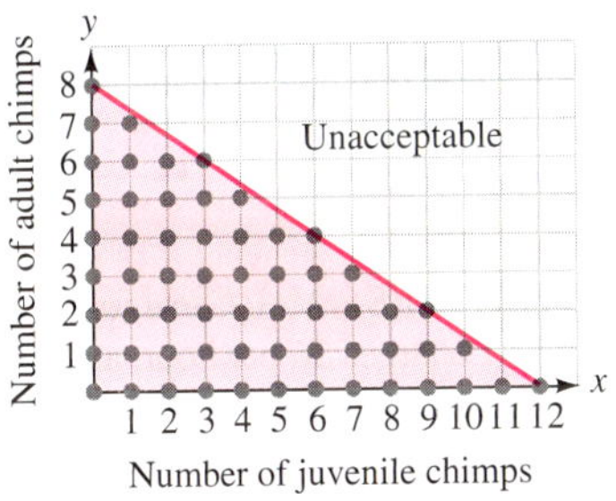

73. ROLLING DICE The points on the graph represent all of the possible outcomes when two fair dice are rolled a single time. For example, (5, 2), shown in red, represents a 5 on the first die and a 2 on the second. Which of the following sentences best describes the outcomes that lie in the shaded area?

(i) Their sum is at most 6.

(ii) Their sum exceeds 6.

(iii) Their sum does not exceed 6.

(iv) Their sum is at least 6.

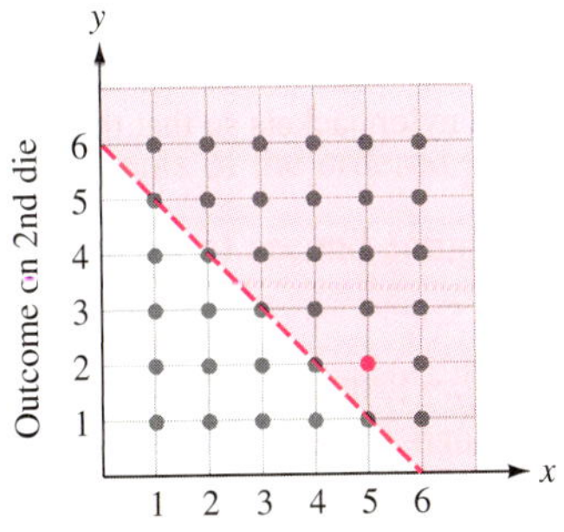

1st die

2nd die

REVIEW EXERCISES

Find the slope and the y-intercept of each line.

33. $y = \frac{3}{4}x - 2$

34. $y = -4x$

35. $y = \frac{x}{8} + 10$

36. $7x + 5y = -21$

37. Graph the line with slope 4 and y-intercept $(0, -1)$. Write an equation of the line.

38. Write an equation for the line shown here.

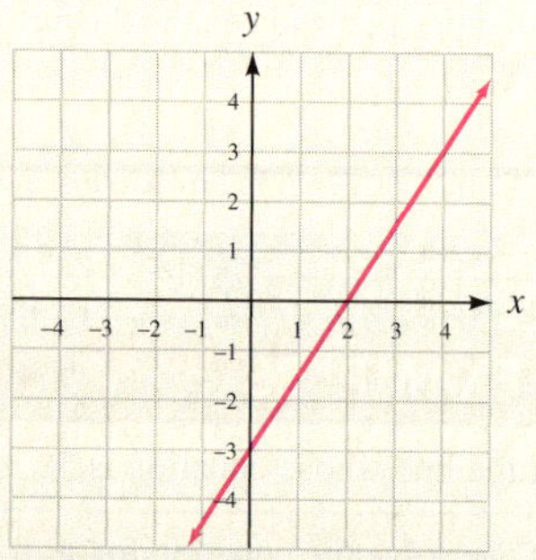

39. Find the slope and the y-intercept of the line whose equation is $9x - 3y = 15$. Then graph it.

40. COPIERS A business buys a used copy machine that has already produced 75,000 copies.

a. If the business plans to run 300 copies a week, write a linear equation that would find the number of copies c the machine has made in its lifetime after the business has used it for w weeks.

b. Use your result in part a to predict the total number of copies that will have been made on the machine 1 year, or 52 weeks, after being purchased by the business.

Without graphing, determine whether graphs of the given pairs of lines are parallel, perpendicular, or neither.

41. $y = -\frac{2}{3}x + 6$
$y = -\frac{2}{3}x - 6$

42. $x + 5y = -10$
$y - 5x = 0$

SECTION 3.6 Point–Slope Form

DEFINITIONS AND CONCEPTS

If a line with slope m passes through the point with coordinates (x_1, y_1), the equation of the line in **point–slope form** is

$$y - y_1 = m(x - x_1)$$

EXAMPLES

Find an equation of the line with slope -3 that passes through $(-2, 4)$. Write the equation in slope–intercept form.

We substitute the slope and the coordinates of the point into the point–slope form.

$y - y_1 = m(x - x_1)$ This is point–slope form.

$y - 4 = -3[x - (-2)]$ Substitute.

$y - 4 = -3(x + 2)$ Simplify within the brackets.

$y - 4 = -3x - 6$ Distribute.

$y = -3x - 2$ To isolate y, add 4 to both sides. This is slope–intercept form.

DEFINITIONS AND CONCEPTS

If we know **two points that a line passes through,** we can write its equation.

EXAMPLES

Find an equation of the line that passes through (2, 5) and (3, 7). Write the equation in slope–intercept form.

The slope of the line is

$$m = \frac{y_2 - y_1}{x_2 - x_1} = \frac{7 - 5}{3 - 2} = 2$$

Either point on the line can serve as (x_1, y_1). If we choose (2, 5), we have:

$y - y_1 = m(x - x_1)$ This is point–slope form.

$y - 5 = 2(x - 2)$ Substitute: $x_1 = 2$, $y_1 = 5$, and $m = 2$.

$y - 5 = 2x - 4$ Distribute.

$y = 2x + 1$ To isolate y, add 5 to both sides. This is the slope–intercept form.

REVIEW EXERCISES

Find an equation of the line with the given slope that passes through the given point. Write the equation in slope–intercept form and graph the equation.

43. $m = 3, (1, 5)$

44. $m = -\frac{1}{2}, (-4, -1)$

Find an equation of the line with the following characteristics. Write the equation in slope–intercept form.

45. passing through $(3, 7)$ and $(-6, 1)$

46. horizontal, passing through $(6, -8)$

47. CAR REGISTRATION When it was 2 years old, the annual registration fee for a Dodge Caravan was \$380. When it was 4 years old, the registration fee dropped to \$310. If the relationship is linear, write an equation that gives the registration fee f in dollars for the van when it is x years old.

48. U.S. WEDDINGS The scatter diagram shows the estimated average cost of a wedding for the years 2000–2006. A straight line can be used to model the data.

a. Use the two highlighted points in red to write the equation of the line. Write the answer in slope–intercept form.

b. Use your answer to part a to predict the average cost of a wedding in 2020.

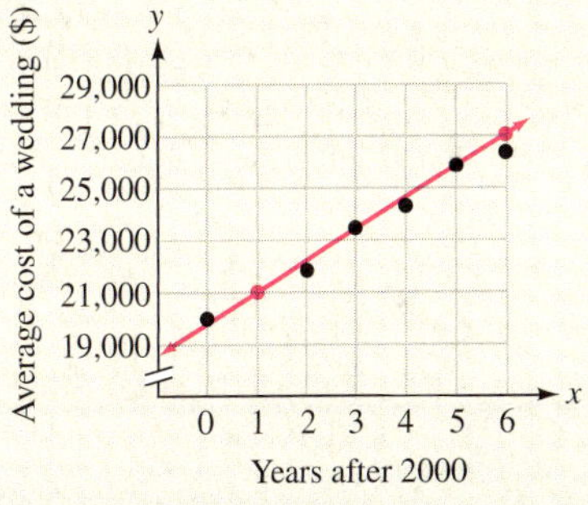

Source: U.S. Wedding Statistics and Marketing website

SECTION 3.7 Graphing Linear Inequalities

DEFINITIONS AND CONCEPTS	EXAMPLES
An ordered pair (x, y) is a **solution of an inequality** in x and y if a true statement results when the variables are replaced by the coordinates of the ordered pair.	Determine whether $(-2, 5)$ is a solution of $x + 3y > -6$. We substitute the coordinates into the equation and see if a true statement results. $x + 3y > -6$ $-2 + 3(5) \overset{?}{>} -6$ $13 > -6$ True Since the result is true, $(-2, 5)$ is a solution.
To graph a linear inequality: **1.** Replace the inequality symbol with an = symbol and graph the boundary line. Draw a solid line if the inequality contains $\le$ or $\ge$ and a dashed line if it contains $<$ or $>$. **2.** Pick a test point not on the boundary. Substitute its coordinates into the inequality. If the inequality is satisfied, shade the side that contains the test point. If the inequality is not satisfied, shade the other side.	Graph: $2x - y \le 4$ **1.** Graph the boundary line $2x - y = 4$ and draw it as a solid line because the inequality symbol is $\le$. **2.** Test the point (0, 0): $2x - y \le 4$ $2(0) - 0 \overset{?}{\le} 4$ $0 \le 4$ True Since the coordinates of the test point satisfy the inequality, we shade the side of the boundary line that contains (0, 0). 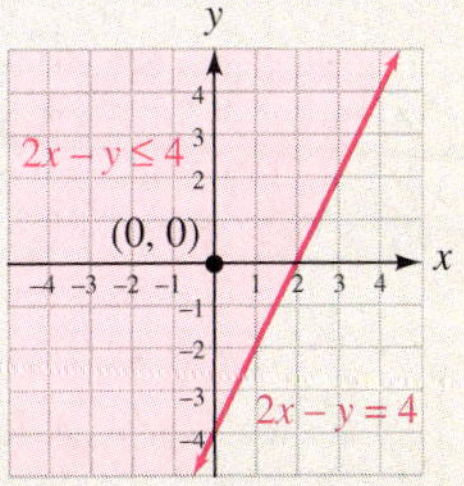

REVIEW EXERCISES

49. Determine whether each ordered pair is a solution of $2x - y \leq -4$.

a. $(0, 5)$ **b.** $(2, 8)$

c. $(-3, -2)$ **d.** $\left(\frac{1}{2}, -5\right)$

50. Fill in the blanks: $2x - 3y \geq 6$ means $2x - 3y$ ____ 6 or $2x - 3y$ ____ 6.

Graph each inequality.

51. $x - y < 5$ **52.** $2x - 3y \geq 6$

53. $y \leq -2x$ **54.** $y < -4$

55. The graph of a linear inequality is shown on the right. Would a true or a false statement result if the coordinates of

a. point A were substituted into the inequality?

b. point B were substituted into the inequality?

c. point C were substituted into the inequality?

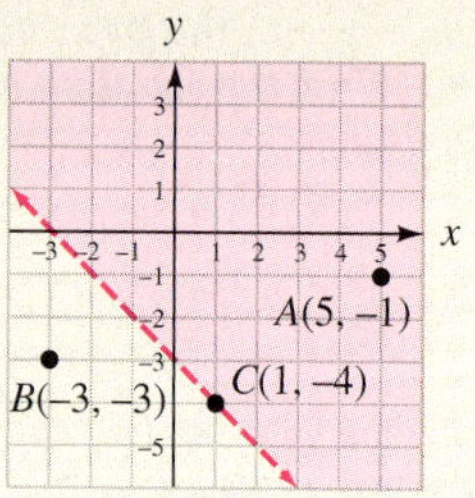

56. WORK SCHEDULES A student told her employer that during the school year, she would be available for up to 30 hours a week, working either 3- or 5-hour shifts. If x represents the number of 3-hour shifts she works and y represents the number of 5-hour shifts she works, the inequality $3x + 5y \leq 30$ shows the possible combinations of shifts she can work. Graph the inequality and find three possible combinations.

SECTION 3.8 An Introduction to Functions

DEFINITIONS AND CONCEPTS	EXAMPLES
A **relation** is a set of ordered pairs. The set of all **first components** is called the **domain** of the relation and the set of all **second components** is called the **range** of a relation.	The relation $\{(4, 7), (0, -3), (-3, 8), (1, 7)\}$ has the domain $\{-3, 0, 1, 4\}$ and the range is $\{-3, 7, 8\}$.

A **function** is a set of ordered pairs (a relation) in which to each first component there corresponds exactly one second component.

If to each value of x in the domain there is assigned exactly one value of y in the range, then ***y* is a function of *x*.**

x	y
1	2
5	3
6	8
5	9

y is not a function of x: $5 \rightarrow 3$ and $5 \rightarrow 9$

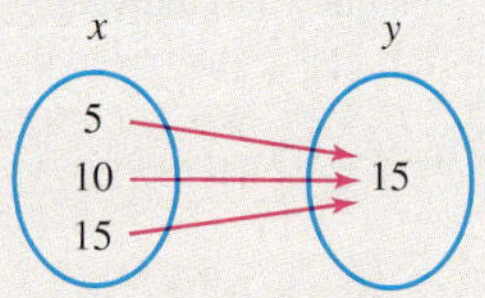

y is a function of x

A function can be defined by an equation. The notation $y = f(x)$ indicates that the variable y is a function of x. It is read as "f of x."

We can think of a function as a machine that takes some **input** x and turns it into some **output** $f(x)$, called a **function value.**

For the function $f(x) = 8x + 5$, $f(-2)$ is the value of $f(x)$ when $x = -2$.

$$f(x) = 8x + 5$$
$$f(-2) = 8(-2) + 5 \quad \text{Substitute } -2 \text{ for each } x.$$
$$= -16 + 5 \quad \text{Evaluate the right side.}$$
$$= -11$$

Thus, $f(-2) = -11$

The **vertical line test:** If a vertical line intersects a graph in more than one point, the graph is not the graph of a function.

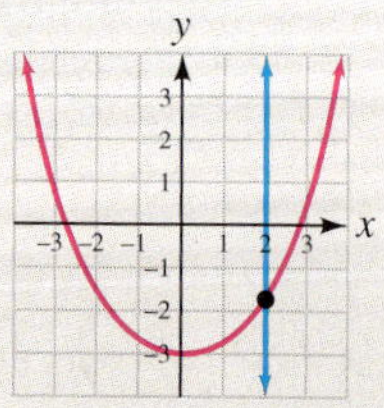

A function

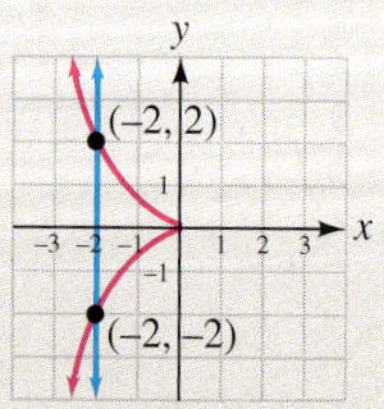

Not a function

SECTION 3.8 An Introduction to Functions–*continued*

DEFINITIONS AND CONCEPTS	EXAMPLES
The input-output pairs that a function generates can be written as ordered pairs and plotted on a rectangular coordinate system to give the **graph of a function.**	Graph the function: $f(x) = -\frac{2}{3}x + 3$ We make a table of solutions, plot the points, and draw the graph.

$f(x) = -\frac{2}{3}x + 3$

x	$f(x)$	
−3	5	→ (−3, 5)
0	3	→ (0, 3)
3	1	→ (3, 1)

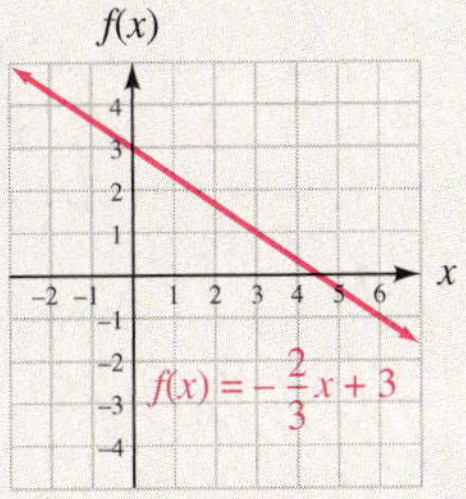

REVIEW EXERCISES

Find the domain and range of each relation.

57. $\{(7, -3), (-5, 9), (4, 4), (0, -11)\}$

58. $\{(2, -2), (15, -8), (-6, 9), (1, -8)\}$

Determine whether each arrow diagram or table defines y to be function of x. If a function is defined, give its domain and range. If it does not define a function, find ordered pairs that show a value of x that corresponds to more than one value of y.

59.

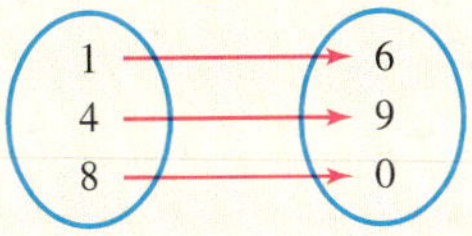

60.

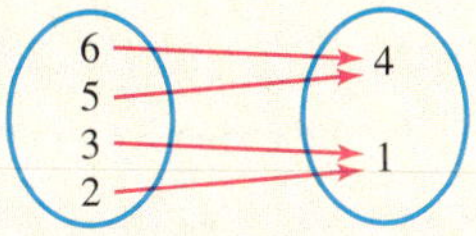

61.

x	y
9	81
7	49
5	25
3	9

62.

x	y
−1	2
0	3
−1	4
1	5

Fill in the blanks.

63. The set of all input values for a function is called the ________, and the set of all output values is called the ________.

64. Fill in the blank: Since $y =$ ____, the equations $y = 2x - 8$ and $f(x) = 2x - 8$ are equivalent.

For $f(x) = x^2 - 4x$, find each of the following function values.

65. $f(1)$

66. $f(0)$

67. $f(-3)$

68. $f\left(\frac{1}{2}\right)$

For $g(x) = 1 - 6x$, find each of the following function values.

69. $g(1)$

70. $g(-6)$

71. $g(0.5)$

72. $g\left(\frac{3}{2}\right)$

Determine whether each graph is the graph of a function. If it is not, find two ordered pairs that show a value of x that corresponds to more than one value of y.

73.

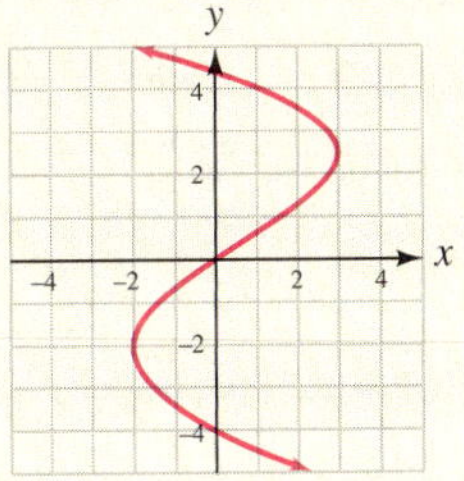

74.

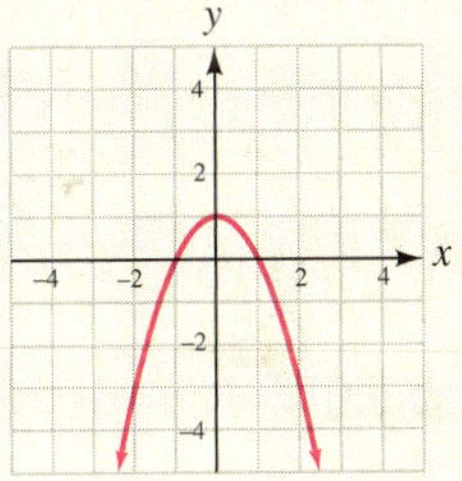

75. Complete the table of function values. Then graph the function.

$f(x) = 1 - |x|$

x	$f(x)$
0	
1	
2	
−1	
−2	
−3	

76. ALUMINUM CANS The function $V(r) = 15.7r^2$ estimates the volume in cubic inches of a can 5 inches tall with a radius of r inches. Find the volume of the can shown in the illustration. Round to the nearest tenth.

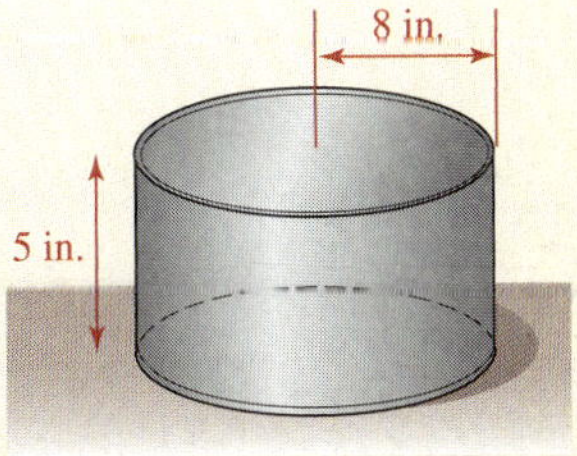

CHAPTER 3 Test

1. Fill in the blanks.
   a. A rectangular coordinate system is formed by two perpendicular number lines called the x-_____ and the y-_____.
   b. A ________ of an equation in two variables is an ordered pair of numbers that makes the equation a true statement.
   c. $3x + y = 10$ is called a ______ equation in two variables because its graph is a line.
   d. The ______ of a line is a measure of steepness.
   e. A ________ is a set of ordered pairs in which to each first component there corresponds exactly one second component.

The graph shows the number of dogs being boarded in a kennel over a 3-day holiday weekend.

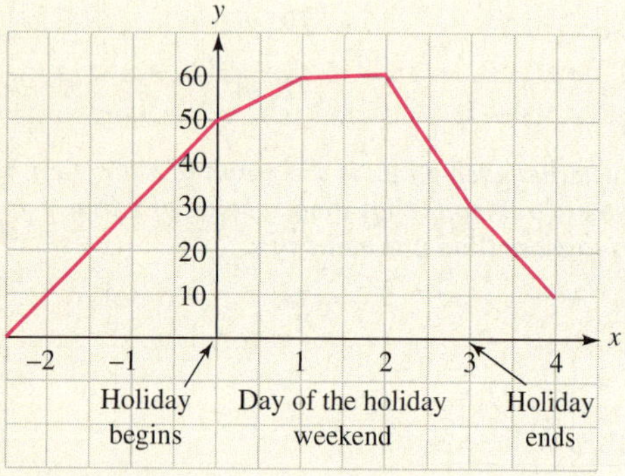

2. How many dogs were in the kennel 2 days before the holiday?

3. What is the maximum number of dogs that were boarded on the holiday weekend?

4. When were there 30 dogs in the kennel?

5. What information does the y-intercept of the graph give?

6. Plot each point on a rectangular coordinate system: $(1, 3)$, $(-2, 4)$, $(-3, -2)$, and $(3, -2)$.

7. Find the coordinates of each point shown in the graph.

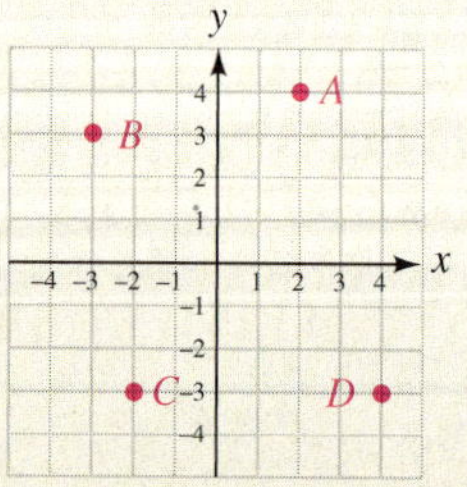

8. In which quadrant is each point located?
   a. $(-1, -5)$ b. $\left(6, -2\frac{3}{4}\right)$

9. Is $(-3, -4)$ a solution of $3x - 4y = 7$?

10. Complete the table of solutions for the linear equation.

$x + 4y = 6$

x	y	(x, y)
2		
	3	

11. The graph of a linear equation is shown.
    a. If the coordinates of point C are substituted into the equation, will the result be true or false?
    b. If the coordinates of point D are substituted into the equation, will the result be true or false?

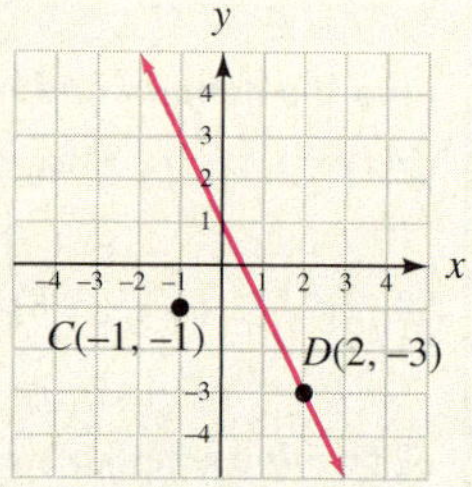

12. Graph: $y = \frac{x}{3}$

13. What are the x- and y-intercepts of the graph of $2x - 3y = 6$?

14. Graph: $8x + 4y = -24$

15. Find the slope of the line.

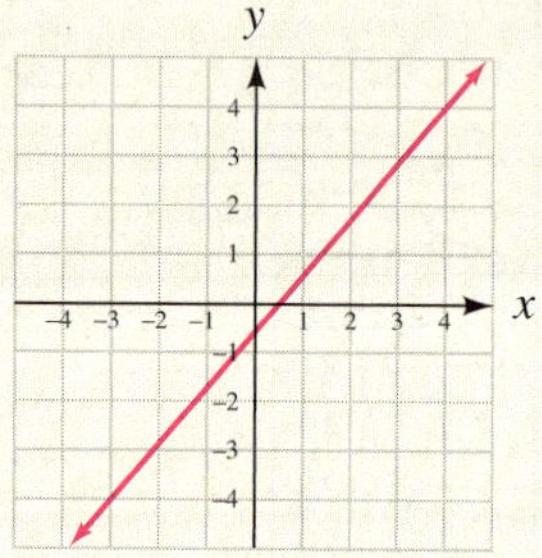

16. Find the slope of the line passing through $(-1, 3)$ and $(3, -1)$.

17. What is the slope of a horizontal line?

18. RAMPS Find the grade of a ramp that rises 2 feet over a horizontal distance of 20 feet.

19. One line passes through $(9, 2)$ and $(6, 4)$. Another line passes through $(0, 7)$ and $(2, 10)$. Without graphing, determine whether the lines are parallel, perpendicular, or neither.

20. When graphed, are the lines $y = 2x + 6$ and $2x - y = 0$ parallel, perpendicular, or neither?

In Problems 21 and 22, refer to the illustration that shows the elevation changes in a 26-mile marathon course.

21. Find the rate of change of the decline on which the woman is running.

22. Find the rate of change of the incline on which the man is running.

23. Graph: $x = -4$

24. Graph the line passing through $(-2, -4)$ having slope $\frac{2}{3}$.

25. Find the slope and the y-intercept of the graph of $x + 2y = 8$.

26. Find an equation of the line passing through $(-2, 5)$ with slope 7. Write the equation in slope–intercept form.

27. Find an equation for the line shown. Write the equation in slope–intercept form.

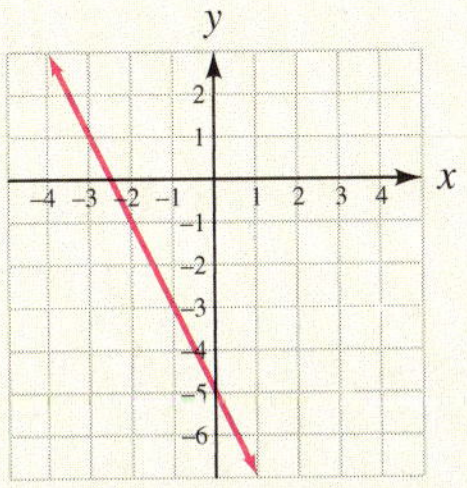

28. DEPRECIATION After it is purchased, a \$15,000 computer loses \$1,500 in resale value every year.

a. Write a linear equation that gives the resale value v of the computer x years after being purchased.

b. Use your answer to part (a) to predict the value of the computer 8 years after it is purchased.

29. Determine whether (6, 1) is a solution of $2x - 4y \geq 8$.

30. WATER HEATERS The scatter diagram shows how excessively high temperatures affect the life of a water heater. Write an equation of the line that models the data for water temperatures between 140° and 180°. Let T represent the temperature of the water in degrees Fahrenheit and y represent the expected life of the heater in years. Give the answer in slope–intercept form.

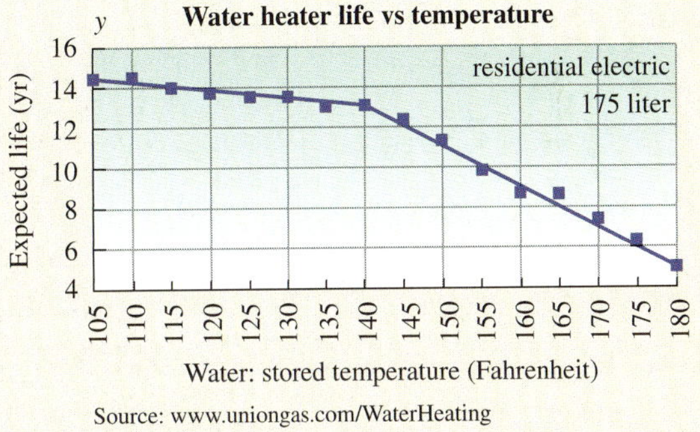

31. Graph the inequality: $x - y > -2$

32. Find the domain and range of the relation: $\{(5, 3), (1, 12), (-4, 3), (0, -8)\}$

Determine whether the table, arrow diagram, or graph define y to be function of x. If a function is defined, give its domain and range. If it does not define a function, find ordered pairs that show a value of x that corresponds to more than one value of y.

33.

x	y
1	4
2	3
3	2
4	1

34.

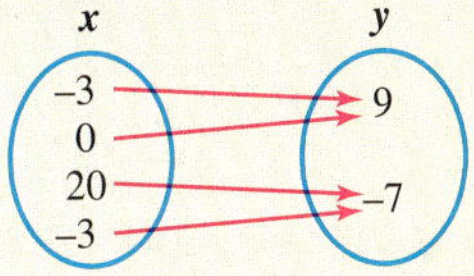

35.

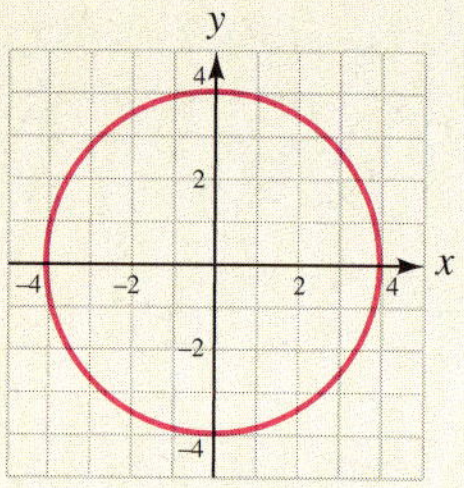

36.

x	y
5	12
10	12
15	12
20	12
25	12

37. If $f(x) = 2x - 7$, find: $f(-3)$

38. If $g(x) = 3.5x^3$ find: $g(6)$

39. TELEPHONE CALLS The function $C(n) = 0.30n + 15$ gives the cost C per month in dollars for making n phone calls. Find $C(45)$ and explain what it means.

40. Graph: $f(x) = |x| - 1$

GROUP PROJECT

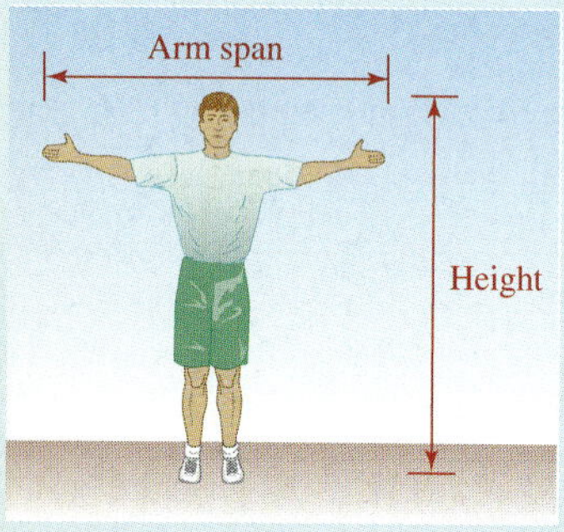

Overview: In this activity, you will explore the relationship between a person's height and arm span. Arm span is defined to be the distance between the tips of a person's fingers when their arms are held out to the side.

Instructions: Form groups of 5 or 6 students. Measure the height and arm span of each person in your group, and record the results in a table like the one shown below.

Name	Height (in.)	Arm span (in.)
1.		
2.		
3.		
4.		
5.		
6.		

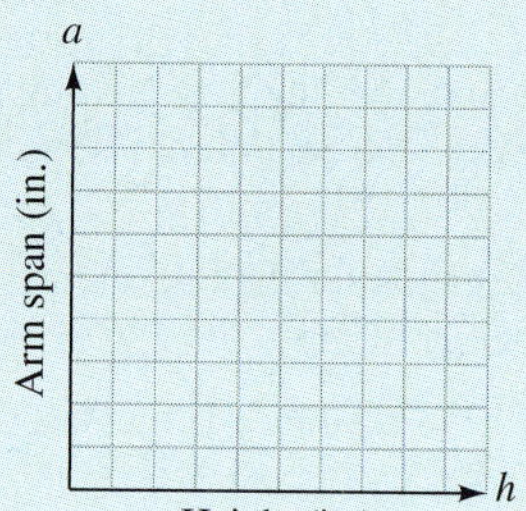

Plot the data in the table as ordered pairs of the form (height, arm span) on a graph like the one shown above. Then draw a straight-line model that best fits the data points.

Pick two convenient points on the line and find its slope. Use the point–slope form $a - a_1 = m(h - h_1)$ to find an equation of the line. Then, write the equation in slope–intercept form.

Ask a person from another group for his or her height measurement. Substitute that value into your linear model to predict that person's arm span. How close is your prediction to the person's actual arm span?

(From *Activities for Beginning and Intermediate Algebra* by Debbie Garrison, Judy Jones, and Jolene Rhodes)

CUMULATIVE REVIEW
Chapters 1–3

1. Find the prime factorization of 108. [Section 1.2]
2. Write $\frac{1}{250}$ as a decimal. [Section 1.3]
3. Determine whether each statement is true or false. [Section 1.3]
   a. Every whole number is an integer.
   b. Every integer is a real number.
   c. 0 is a whole number, an integer, and a rational number.

Perform the operations.

4. $-27 + 21 + (-9)$ [Section 1.4]
5. $-1.57 - (-0.8)$ [Section 1.5]
6. $-9(-7)(5)(-3)$ [Section 1.6]
7. $\frac{-180}{-6}$ [Section 1.6]
8. Evaluate: $\left|\frac{(6-5)^4 - (-21)}{-27 + 4^2}\right|$ [Section 1.7]
9. Evaluate $b^2 - 4ac$ for $a = 2$, $b = -8$, and $c = 4$. [Section 1.8]
10. Suppose x sheets from a 500-sheet ream of paper have been used. How many sheets are left? [Section 1.8]
11. How many terms does the algebraic expression $3x^2 - 2x + 1$ have? What is the coefficient of the second term? [Section 1.8]
12. Use the distributive property to remove parentheses. [Section 1.9]
   a. $2(x + 4)$
   b. $-2(x - 4)$

Simplify each expression. [Section 1.9]

13. $5a + 10 - a$
14. $-7(9t)$
15. $-2b^2 + 6b^2$
16. $5(-17)(0)(2)$
17. $(a + 2) - (a - 2)$
18. $-4(-5)(-8a)$
19. $-y - y - y$
20. $\frac{3}{2}(4x - 8) + x$

Solve each equation. [Sections 2.1 and 2.2]

21. $3x - 5 = 13$
22. $1.2 - x = -1.7$
23. $\frac{2x}{3} - 2 = 4$
24. $\frac{y - 2}{7} = -3$
25. $-3(2y - 2) - y = 5$
26. $9y - 3 = 6y$
27. $\frac{1}{3} + \frac{c}{5} = -\frac{3}{2}$
28. $5(x + 2) = 5x - 2$
29. $-x = 99$
30. $3c - 2 = \frac{11(c - 1)}{5}$

31. PENNIES A 2006 telephone survey of adults asked whether the penny should be discontinued from the national currency. The results are shown in the circle graph. If 869 people favored keeping the penny, how many took part in the survey? [Section 2.3]

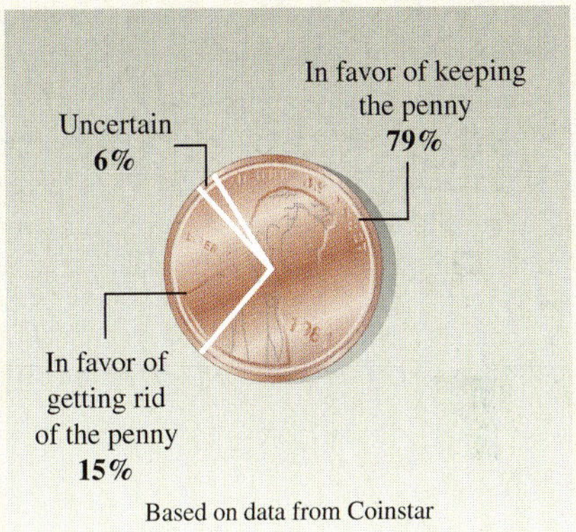

32. Solve for h: $S = 2\pi rh + 2\pi r^2$ [Section 2.4]
33. BAND AIDS Find the perimeter and the area of the gauze pad of the bandage. [Section 2.4]

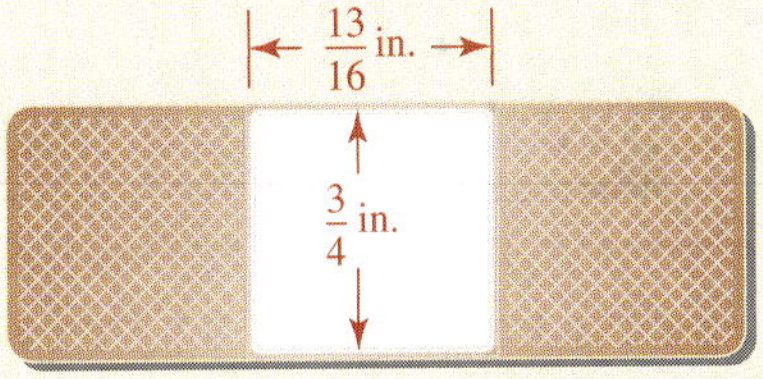

34. HIGH HEELS Find x. [Section 2.5]

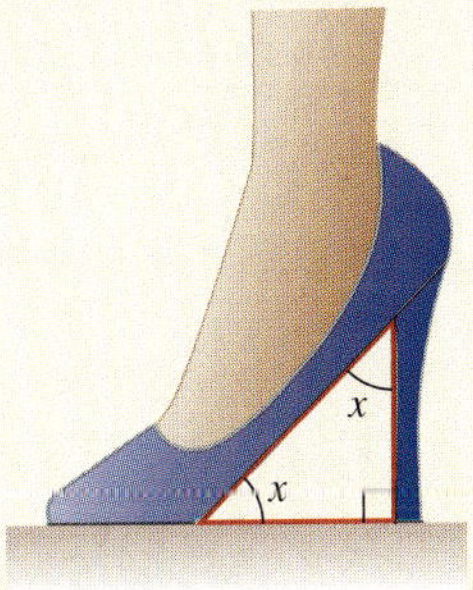

35. Complete the table. [Section 2.6]

	% acid	Liters	Amount of acid
50% solution	0.50	x	
25% solution	0.25	$13 - x$	
30% mixture	0.30	13	

36. ROAD TRIPS A bus, carrying the members of a marching band, and a truck, carrying their instruments, leave a high school at the same time. The bus travels at 60 mph and the truck at 50 mph. In how many hours will they be 75 miles apart? [Section 2.6]

37. MIXING CANDY Candy corn worth \$2.85 per pound is to be mixed with black gumdrops that cost \$1.80 per pound to make 200 pounds of a mixture worth \$2.22 per pound. How many pounds of each candy should be used? [Section 2.6]

Solve each inequality. Write the solution set in interval notation and graph it. [Section 2.7]

38. $-\frac{3}{16}x \geq -9$

39. $8x + 4 > 3x + 4$

40. In which quadrants are the second coordinates of ordered pairs positive? [Section 3.1]

41. Is $(-2, 4)$ a solution of $y = 2x - 8$? [Section 3.2]

Graph each equation.

42. $y = x$ [Section 3.2]

43. $4y + 2x = -8$ [Section 3.3]

44. What is the slope of the graph of the line $y = 5$? [Section 3.4]

45. What is the slope of the line passing through $(-2, 4)$ and $(5, -6)$? [Section 3.4]

46. ROOFING Find the pitch of the roof. [Section 3.4]

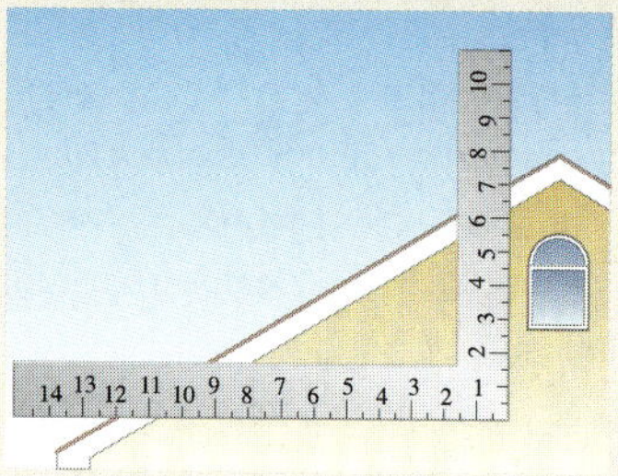

47. Find the slope and the y-intercept of the graph of the line described by $4x - 6y = -12$. [Section 3.5]

48. Write an equation of the line that has slope -2 and y-intercept $(0, 1)$. [Section 3.5]

49. Find an equation of the line that has slope $-\frac{7}{8}$ and passes through $(2, -9)$. Write the equation in point–slope form and in slope–intercept form. [Section 3.6]

50. Is $(-2, -4)$ a solution of $x + y \leq -6$? [Section 3.7]

51. Graph: $y \geq x + 1$ [Section 3.7]

52. Graph $x < 4$ on a rectangular coordinate system. [Section 3.7]

53. If $f(x) = x^4 + x$, find: $f(-3)$ [Section 3.8]

54. Is this the graph of a function? [Section 3.8]

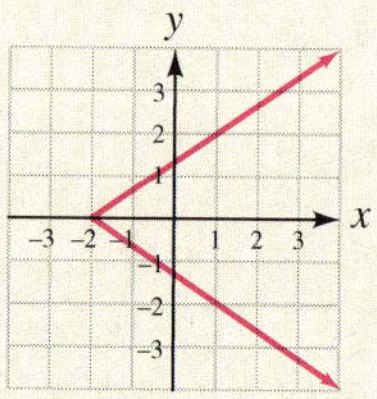

CHAPTER 4

Systems of Linear Equations and Inequalities

© Peter Steiner/Alamy

from Campus to Careers

Portrait Photographer

Portrait photographers take pictures of individuals or groups of people and often work in their own studios. Some specialize in weddings, religious ceremonies, or school photographs, and many work on location. Their job responsibilities require a variety of mathematical skills such as scheduling appointments, keeping financial records, pricing photographs, purchasing supplies, billing customers, and operating digital equipment.

Photographers often make packets of pictures available to their customers. In **Problem 29** of **Study Set 4.4,** we will find the costs of two sizes of photographs that are part of a wedding picture packet.

JOB TITLE:
Portrait Photographer
EDUCATION:
A well-rounded education including art and business courses is preferred.
JOB OUTLOOK:
Employment is expected to increase between 9% to 17% through the year 2014.
ANNUAL EARNINGS:
From $40,000, to an average of $50,000, up to $62,000 or more
FOR MORE INFORMATION:
www.bls.gov/oco/ocos264.htm

Study Skills Workshop

Making Homework a Priority

Attending class and taking notes are important, but they are not enough. The only way that you are really going to learn algebra is by doing your homework.

WHEN TO DO YOUR HOMEWORK: Homework should be started on the day it is assigned, when the material is fresh in your mind. It's best to break your homework sessions into 30-minute periods, allowing for short breaks in between.

HOW TO BEGIN YOUR HOMEWORK: Review your notes and the examples in your text before starting your homework assignment.

GETTING HELP WITH YOUR HOMEWORK: It's normal to have some questions when doing homework. Talk to a tutor, a classmate, or your instructor to get those questions answered.

Now Try This

1. Write a one-page paper that describes *when, where,* and *how* you go about completing your algebra homework assignments.
2. For each problem on your next homework assignment, find an example in this book that is similar. Write the example number next to the problem.
3. Make a list of questions that you have while doing your next assignment. Then decide whom you are going to ask to get those questions answered.

SECTION 4.1
Solving Systems of Equations by Graphing

Objectives

1. Determine whether a given ordered pair is a solution of a system.
2. Solve systems of linear equations by graphing.
3. Use graphing to identify inconsistent systems and dependent equations.
4. Identify the number of solutions of a linear system without graphing.
5. Use a graphing calculator to solve a linear system (optional).

The following illustration shows the average amounts of chicken and beef eaten per person each year in the United States from 1985 to 2005. Plotting both graphs on the same coordinate system makes it easy to compare recent trends. The point of intersection of the graphs indicates that Americans ate equal amounts of chicken and beef in 1992—about 66 pounds of each, per person.

In this section, we will use a similar graphical approach to solve systems of equations.

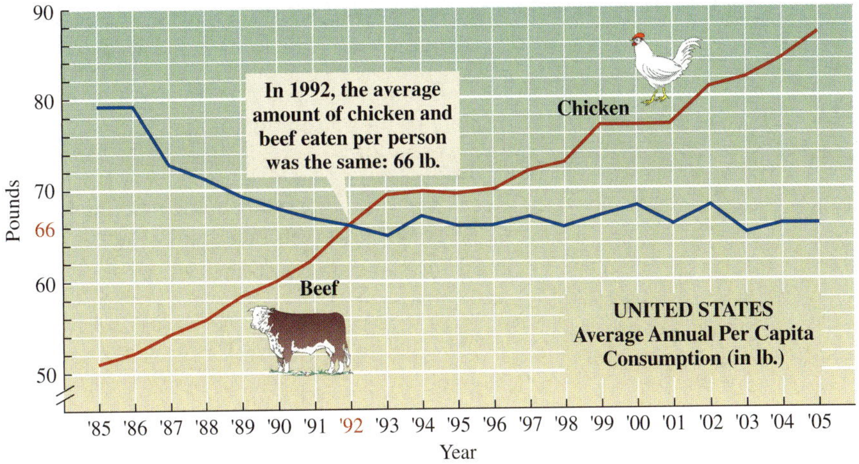

Source: U.S. Department of Agriculture

1 Determine Whether a Given Ordered Pair Is a Solution of a System.

We have previously discussed equations in two variables, such as $x + y = 3$. Because there are infinitely many pairs of numbers whose sum is 3, there are infinitely many pairs (x, y) that satisfy this equation. Some of these pairs are listed in Table (a).

Now consider the equation $x - y = 1$. Because there are infinitely many pairs of numbers whose difference is 1, there are infinitely many pairs (x, y) that satisfy $x - y = 1$. Some of these pairs are listed in the Table (b).

> ***The Language of Algebra***
> We say that (2, 1) *satisfies* $x + y = 3$, because the x-coordinate, 2, and the y-coordinate, 1, make the equation true when substituted for x and y: $2 + 1 = 3$. To *satisfy* means to make content, as in *satisfy* your thirst or a *satisfied* customer.

$x + y = 3$

x	y	(x, y)
0	3	(0, 3)
1	2	(1, 2)
2	1	(2, 1)
3	0	(3, 0)

(a)

$x - y = 1$

x	y	(x, y)
0	−1	(0, −1)
1	0	(1, 0)
2	1	(2, 1)
3	2	(3, 2)

(b)

From the two tables, we see that (2, 1) satisfies both equations.

When two equations with the same variables are considered simultaneously (at the same time), we say that they form a **system of equations.** Using a left brace { , we can write the equations from the previous example as a system:

$$\begin{cases} x + y = 3 \\ x - y = 1 \end{cases}$$

Read as "the system of equations $x + y = 3$ and $x - y = 1$."

Because the ordered pair (2, 1) satisfies both of these equations, it is called a **solution of the system.** In general, a system of linear equations can have exactly one solution, no solution, or infinitely many solutions.

EXAMPLE 1 Determine whether $(-2, 5)$ is a solution of each system of equations.

a. $\begin{cases} 3x + 2y = 4 \\ x - y = -7 \end{cases}$

b. $\begin{cases} 4y = 18 - x \\ y = 2x \end{cases}$

Strategy We will substitute the x- and y-coordinates of $(-2, 5)$ for the corresponding variables in both equations of the system.

Why If both equations are satisfied (made true) by the x- and y-coordinates, then the ordered pair is a solution of the system.

Solution

a. Recall that in an ordered pair, the first number is the x-coordinate and the second number is the y-coordinate. To determine whether $(-2, 5)$ is a solution, we substitute -2 for x and 5 for y in each equation.

The Language of Algebra A system of equations is two (or more) equations that we consider *simultaneously*—at the same time. Some professional sports teams *simulcast* their games. That is, the announcer's play-by-play description is broadcast on radio and television at the same time.

Check:

$3x + 2y = 4$ The first equation.
$3(-2) + 2(5) \stackrel{?}{=} 4$
$-6 + 10 \stackrel{?}{=} 4$
$4 = 4$ True

$x - y = -7$ The second equation.
$-2 - 5 \stackrel{?}{=} -7$
$-7 = -7$ True

Since $(-2, 5)$ satisfies both equations, it is a solution of the system.

b. Again, we substitute -2 for x and 5 for y in each equation.

Check:

$4y = 18 - x$ The first equation.
$4(5) \stackrel{?}{=} 18 - (-2)$
$20 \stackrel{?}{=} 18 + 2$
$20 = 20$ True

$y = 2x$ The second equation.
$5 \stackrel{?}{=} 2(-2)$
$5 = -4$ False

Although $(-2, 5)$ satisfies the first equation, it does not satisfy the second. Because it does not satisfy both equations, $(-2, 5)$ is not a solution of the system.

Self Check 1 Determine whether $(4, -1)$ is a solution of: $\begin{cases} x - 2y = 6 \\ y = 3x - 11 \end{cases}$

Now Try **Problem 15**

2 Solve Systems of Linear Equations by Graphing.

To **solve a system of equations** means to find all of the solutions of the system. One way to solve a system of linear equations is to graph the equations on the same set of axes.

EXAMPLE 2 Solve the system of equations by graphing: $\begin{cases} 2x + 3y = 2 \\ 3x - 2y = 16 \end{cases}$

Strategy We will graph both equations on the same coordinate system.

Why Recall that the graph of a linear equation is a "picture" of its solutions. If both equations are graphed on the same coordinate system, we can see whether they have any common solutions.

Solution The intercept-method is a convenient way to graph equations such as $2x + 3y = 2$ and $3x - 2y = 16$, because they are in standard $Ax + By = C$ form.

$2x + 3y = 2$

x	y	(x, y)
0	$\frac{2}{3}$	$\left(0, \frac{2}{3}\right)$
1	0	$(1, 0)$
-2	2	$(-2, 2)$

$3x - 2y = 16$

x	y	(x, y)
0	-8	$(0, -8)$
$\frac{16}{3}$	0	$\left(\frac{16}{3}, 0\right)$
2	-5	$(2, -5)$

To find the y-intercept, let x = 0 and solve for y.
To find the x-intercept, let y = 0 and solve for x.
As a check, pick another x-value, such as −2 or 2, and find y.

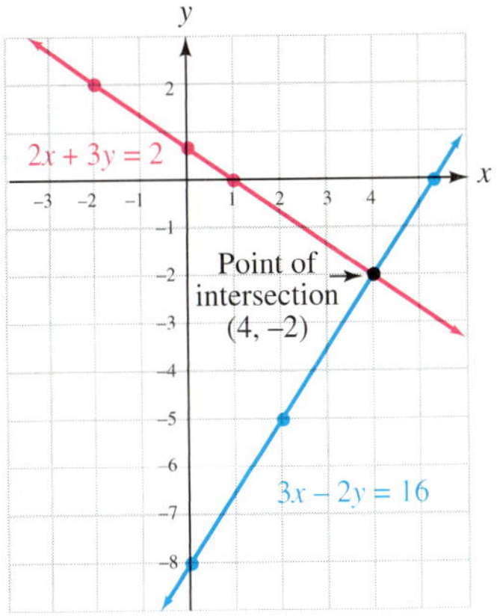

Success Tip
Accuracy is crucial when using the graphing method to solve a system. Here are some suggestions for improving your accuracy:
- Use graph paper.
- Use a sharp pencil.
- Use a ruler or straightedge.

The coordinates of each point on the line graphed in red satisfy $2x + 3y = 2$ and the coordinates of each point on the line graphed in blue satisfy $3x - 2y = 16$. Because the point of intersection is on both graphs, its coordinates satisfy both equations.

It appears that the graphs intersect at the point $(4, -2)$. To verify that it is the solution of the system, we substitute 4 for x and -2 for y in each equation.

Check:

$2x + 3y = 2$	The first equation.	$3x - 2y = 16$	The second equation.
$2(4) + 3(-2) \stackrel{?}{=} 2$		$3(4) - 2(-2) \stackrel{?}{=} 16$	
$8 + (-6) \stackrel{?}{=} 2$		$12 - (-4) \stackrel{?}{=} 16$	
$2 = 2$	True	$16 = 16$	True

Since $(4, -2)$ makes both equations true, it is the solution of the system. The solution set is written as $\{(4, -2)\}$.

Self Check 2 Solve the system of equations by graphing: $\begin{cases} 2x - y = -5 \\ x + y = -1 \end{cases}$

Now Try **Problem 25**

To solve a system of linear equations in two variables by graphing, follow these steps.

The Graphing Method

1. Carefully graph each equation on the same rectangular coordinate system.
2. If the lines intersect, determine the coordinates of the point of intersection of the graphs. That ordered pair is the solution of the system.
3. Check the proposed solution in each equation of the original system.

3 Use Graphing to Identify Inconsistent Systems and Dependent Equations.

A system of equations that has at least one solution, like that in Example 2, is called a **consistent system.** A system with no solution is called an **inconsistent system.**

EXAMPLE 3 Solve the system of equations by graphing: $\begin{cases} y = -2x - 6 \\ 4x + 2y = 8 \end{cases}$

Strategy We will graph both equations on the same coordinate system.

Why If both equations are graphed on the same coordinate system, we can see whether they have any common solutions.

Solution Since $y = -2x - 6$ is written in slope–intercept form, we can graph it by plotting the y-intercept $(0, -6)$ and then drawing a slope triangle whose rise is -2 and whose run is 1. We can graph $4x + 2y = 8$ using the intercept method.

$y = -2x - 6$

$m = -2$ $b = -6$

Slope: $\frac{-2}{1}$

y-intercept: $(0, -6)$

$4x + 2y = 8$

x	y	(x, y)
0	4	(0, 4)
2	0	(2, 0)
1	2	(1, 2)

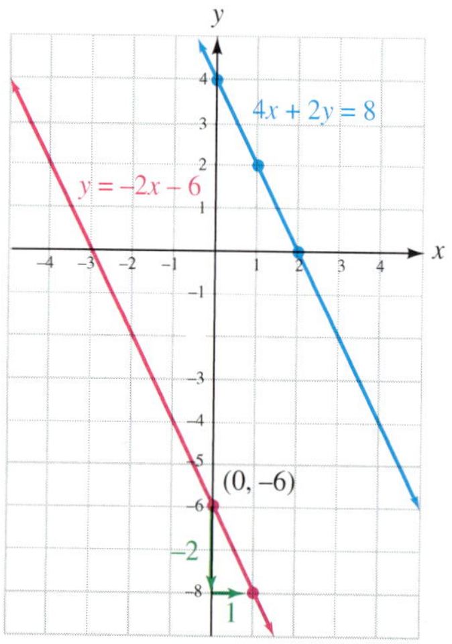

The lines in the graph appear to be parallel. We can verify this by writing the second equation in slope–intercept form and observing that the lines have the same slope, -2, and different y-intercepts, $(0, -6)$ and $(0, 4)$.

$y = -2x - 6$ $\qquad 4x + 2y = 8$

$2y = -4x + 8$ Subtract $4x$ from both sides.

$y = -2x + 4$ To isolate y, divide both sides by 2.

Different y-intercepts

Same slope

Caution
A common error is to graph the parallel lines, but forget to answer with the words *no solution*.

Because the lines are parallel, there is no point of intersection. Such a system has *no solution*. The solution set is the empty set, which is written $\varnothing$.

Self Check 3 Solve the system of equations by graphing: $\begin{cases} y = \frac{3}{2}x \\ 3x - 2y = 6 \end{cases}$

Now Try Problem 33

Some systems of equations have infinitely many solutions.

EXAMPLE 4 Solve the system of equations by graphing: $\begin{cases} y = 2x + 4 \\ 4x + 8 = 2y \end{cases}$

Strategy We will graph both equations on the same coordinate system.

Why If both equations are graphed on the same coordinate system, we will be able to see if they have any solutions in common.

Solution To graph $y = 2x + 4$, we use the slope and y-intercept, and to graph $4x + 8 = 2y$, we use the intercept method.

The Language of Algebra

The graphs of these lines *coincide.* That is, they occupy the same location. To illustrate this concept, think of a clock. At noon and at midnight, the hands of the clock *coincide.*

$y = 2x + 4$

$m = 2 \quad b = 4$

Slope: $\frac{2}{1}$

y-intercept: $(0, 4)$

$4x + 8 = 2y$

x	y	(x, y)
0	4	(0, 4)
−2	0	(−2, 0)
−3	−2	(−3, −2)

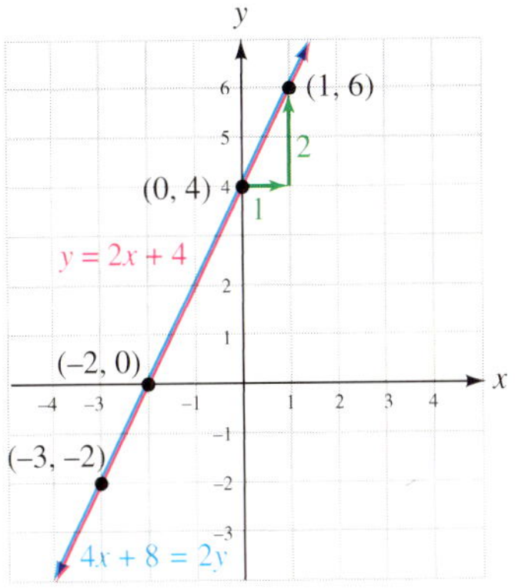

The graphs appear to be identical. We can verify this by writing the second equation in slope–intercept form and observing that it is the same as the first equation.

$y = 2x + 4$ The first equation.

$$4x + 8 = 2y \quad \text{The second equation.}$$
$$2y = 4x + 8$$
$$\frac{2y}{2} = \frac{4x}{2} + \frac{8}{2} \quad \text{Divide both sides by 2.}$$
$$y = 2x + 4$$

This confirms that $4x + 8 = 2y$ and $y = 2x + 4$ are different forms of the same equation. Thus, their graphs are identical.

Since the graphs are the same line, they have infinitely many points in common. The coordinates of each of those points satisfy both equations of the system. In cases like this, we say that there are *infinitely many solutions.*

From the graph, it appears that four of the infinitely many solutions are $(-3, -2)$, $(-2, 0)$, $(0, 4)$, and $(1, 6)$. Checks for two of these ordered pairs follow.

Caution

A common error is to graph the identical lines, but forget to answer with the words *infinitely many solutions.*

Check **(−3, −2):**

$$4x + 8 = 2y \qquad y = 2x + 4$$
$$4(-3) + 8 \stackrel{?}{=} 2(-2) \qquad -2 \stackrel{?}{=} 2(-3) + 4$$
$$-12 + 8 \stackrel{?}{=} -4 \qquad -2 \stackrel{?}{=} -6 + 4$$
$$-4 = -4 \qquad -2 = -2$$

Check **(0, 4):**

$$4x + 8 = 2y \qquad y = 2x + 4$$
$$4(0) + 8 \stackrel{?}{=} 2(4) \qquad 4 \stackrel{?}{=} 2(0) + 4$$
$$0 + 8 \stackrel{?}{=} 8 \qquad 4 \stackrel{?}{=} 0 + 4$$
$$8 = 8 \qquad 4 = 4$$

Self Check 4 Solve the system of equations by graphing: $\begin{cases} 6x - 4 = 2y \\ y = 3x - 2 \end{cases}$

Now Try **Problem 31**

In Examples 2 and 3, the graphs of the equations of the system were different lines. We call equations with different graphs **independent equations.** The equations in Example 4 have the same graph and are equivalent. Because they are different forms of the same equation, they are called **dependent equations.**

1 Solve Systems of Linear Equations by Substitution.

One algebraic method for solving a system of equations is the **substitution method.** It is introduced in the following example.

EXAMPLE 1

Solve the system: $\begin{cases} y = 3x - 2 \\ 2x + y = 8 \end{cases}$

Strategy Note that the first equation is solved for y. Because y and $3x - 2$ are equal (represent the same value), we will substitute $3x - 2$ for y in the second equation.

Why The objective is to obtain one equation containing only one unknown. When $3x - 2$ is substituted for y in the second equation, the result will be just that—an equation in one variable, x.

Solution Since the right side of $y = 3x - 2$ is used to make a substitution, $y = 3x - 2$ is called the **substitution equation.**

$$\begin{cases} y = 3x - 2 \\ 2x + y = 8 \end{cases}$$

To find the solution of the system, we proceed as follows:

$2x + y = 8$ — This is the second equation of the system.

$2x + 3x - 2 = 8$ — Substitute $3x - 2$ for y.

The resulting equation has only one variable and can be solved for x.

$2x + 3x - 2 = 8$

$5x - 2 = 8$ — Combine like terms: $2x + 3x = 5x$.

$5x = 10$ — To isolate the variable term, $5x$, add 2 to both sides.

$x = 2$ — Divide both sides by 5. This is the x-value of the solution.

We can find the y-value of the solution by substituting 2 for x in either equation of the original system. We will use the substitution equation because it is already solved for y.

$y = 3x - 2$ — This is the substitution equation.

$y = 3(2) - 2$ — Substitute 2 for x.

$y = 6 - 2$

$y = 4$ — This is the y-value of the solution. We would have obtained the same result if we had substituted 2 for x in $2x + y = 8$ and solved for y.

The ordered pair (2, 4) appears to be the solution of the system. To check, we substitute 2 for x and 4 for y in each equation.

Check:

$y = 3x - 2$	The first equation.	$2x + y = 8$	The second equation.
$4 \stackrel{?}{=} 3(2) - 2$		$2(2) + 4 \stackrel{?}{=} 8$	
$4 \stackrel{?}{=} 6 - 2$		$4 + 4 \stackrel{?}{=} 8$	
$4 = 4$	True	$8 = 8$	True

Since (2, 4) satisfies both equations, it is the solution. The solution set is written as $\{(2, 4)\}$. A graph of the equations of the system shows an intersection point of (2, 4). This illustrates an important fact: *The solution found using the substitution method will be the same as the solution found using the graphing method.*

Caution

When using the substitution method, a common error is to find the value of one of the variables, say x, and forget to find the value of the other. Remember that a solution of a linear system of two equations is an ordered pair (x, y).

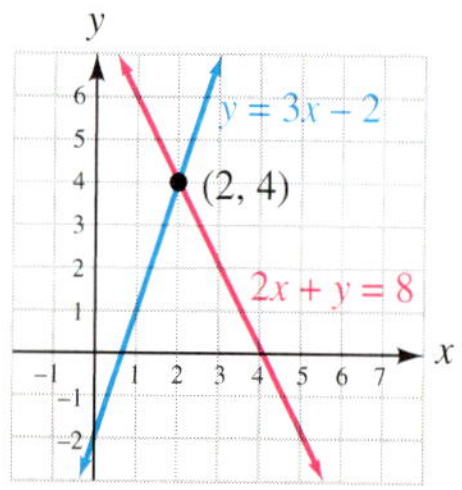

Self Check 1 Solve the system: $\begin{cases} x + 4y = 7 \\ x = 6y - 3 \end{cases}$

Now Try Problem 15

The substitution method works well for solving systems where one equation is solved, or can be easily solved, for one of the variables. To solve a system of equations in x and y by the substitution method, follow these steps.

The Substitution Method

1. Solve one of the equations for either x or y. If this is already done, go to step 2. (We call this equation the **substitution equation.**)
2. Substitute the expression for x or for y obtained in step 1 into the other equation and solve that equation.
3. Substitute the value of the variable found in step 2 into the substitution equation to find the value of the remaining variable.
4. Check the proposed solution in each equation of the original system. Write the solution as an ordered pair.

EXAMPLE 2

Solve the system: $\begin{cases} 4x + 27 = 7y \\ x = -5y \end{cases}$

Strategy We will use the substitution method to solve this system.

Why The substitution method works well when one of the equations of the system (in this case, $x = -5y$) is solved for a variable.

Solution

Step 1: Because x and $-5y$ represent the same value, we can substitute $-5y$ for x in the first equation.

$$\begin{cases} 4x + 27 = 7y \\ x = -5y \end{cases}$$ This is the substitution equation.

Step 2: When we substitute $-5y$ for x in the first equation, the resulting equation contains only one variable, and it can be solved for y.

$4x + 27 = 7y$ This is the first equation of the system.

$4(-5y) + 27 = 7y$ Substitute $-5y$ for x. Don't forget the parentheses.

$-20y + 27 = 7y$ Do the multiplication.

$27 = 27y$ To eliminate $-20y$ on the left side, add $20y$ to both sides.

$1 = y$ Divide both sides by 27. This is the y-value of the solution.

Step 3: To find x, substitute 1 for y in the equation $x = -5y$.

$x = -5y$ This is the substitution equation.

$x = -5(1)$ Substitute 1 for y.

$x = -5$ This is the x-value of the solution.

Step 4: The following check verifies that the solution is $(-5, 1)$.

Success Tip

The basic objective of this method is to use an appropriate substitution to obtain one equation in one variable.

Caution

We don't have to find the values of the variables in alphabetical order. Here, we found y first.

Check:

$4x + 27 = 7y$ The first equation.

$4(-5) + 27 \stackrel{?}{=} 7(1)$

$-20 + 27 \stackrel{?}{=} 7$

$7 = 7$ True

$x = -5y$ The second equation.

$-5 \stackrel{?}{=} -5(1)$

$-5 = -5$ True

Self Check 2 Solve the system: $\begin{cases} 3x + 40 = 8y \\ x = -4y \end{cases}$

Now Try **Problem 19**

2 Find a Substitution Equation.

Sometimes neither equation of a system is solved for a variable. In such cases, we can find a substitution equation by solving one of the equations for one of its variables.

EXAMPLE 3 Solve the system: $\begin{cases} 4x + y = 3 \\ 3x + 5y = 15 \end{cases}$

Strategy Since the system does not contain an equation solved for x or y, we must choose an equation and solve it for x or y. We will solve for y in the first equation, because y has a coefficient of 1. Then we will use the substitution method to solve the system.

Why Solving $4x + y = 3$ for x or solving $3x + 5y = 15$ for x or y would involve working with cumbersome fractions.

Solution

Step 1: To find a substitution equation, we proceed as follows:

$4x + y = 3$ This is the first equation of the system.

$y = 3 - 4x$ To isolate y, subtract 4x from both sides. This is the substitution equation.

Because y and $3 - 4x$ are equal, we can substitute $3 - 4x$ for y in the second equation of the system.

$\begin{cases} 4x + y = 3 & \rightarrow y = 3 - 4x \\ 3x + 5y = 15 \end{cases}$

Step 2: When we substitute for y in the second equation, the resulting equation contains only one variable and can be solved for x.

$3x + 5y = 15$ This is the second equation of the system.

$3x + 5(3 - 4x) = 15$ Substitute 3 − 4x for y. Don't forget the parentheses.

$3x + 15 - 20x = 15$ Distribute the multiplication by 5.

$15 - 17x = 15$ Combine like terms.

$-17x = 0$ To isolate the variable term, −17x, subtract 15 from both sides.

$x = 0$ Divide both sides by −17. This is the x-value of the solution.

Step 3: To find y, substitute 0 for x in the equation $y = 3 - 4x$.

$y = 3 - 4x$ This is the substitution equation.

$y = 3 - 4(0)$ Substitute 0 for x.

> ***Success Tip***
> To find a substitution equation, solve one of the equations of the system for one of its variables. If possible, solve for a variable whose coefficient is 1 or −1 to avoid working with fractions.

> ***Caution***
> Here, use parentheses when substituting $3 - 4x$ for y so that the multiplication by 5 is distributed over both terms of $3 - 4x$.
>
> $3x + 5(3 - 4x) = 15$

$y = 3 - 0$

$y = 3$ This is the y-value of the solution.

Step 4: The solution appears to be (0, 3). Check it in the original equations.

Check:

$4x + y = 3$	The first equation.	$3x + 5y = 15$	The second equation.
$4(0) + 3 \stackrel{?}{=} 3$		$3(0) + 5(3) \stackrel{?}{=} 15$	
$0 + 3 \stackrel{?}{=} 3$		$0 + 15 \stackrel{?}{=} 15$	
$3 = 3$	True	$15 = 15$	True

Self Check 3 Solve the system: $\begin{cases} 2x - 3y = 10 \\ 3x + y = 15 \end{cases}$

Now Try **Problem 31**

EXAMPLE 4

Solve the system: $\begin{cases} 3a - 3b = 5 \\ 3 - a = -2b \end{cases}$

Strategy Since the coefficient of a in the second equation is -1, we will solve that equation for a. Then we will use the substitution method to solve the system.

Why If we solve for the variable with a numerical coefficient of -1, we can avoid having to work with fractions.

Solution

Step 1: To find a substitution equation, we proceed as follows:

$3 - a = -2b$ This is the second equation of the system.

$-a = -2b - 3$ To isolate the variable term, $-a$, subtract 3 from both sides.

To obtain a on the left side, multiply both sides of the equation by -1.

$-1(-a) = -1(-2b - 3)$ Multiply both sides by -1. Don't forget the parentheses.

$a = 2b + 3$ Do the multiplications. This is the substitution equation.

Because a and $2b + 3$ represent the same value, we can substitute $2b + 3$ for a in the first equation.

$$\begin{cases} 3a - 3b = 5 \\ 3 - a = -2b \rightarrow a = 2b + 3 \end{cases}$$

Step 2: Substitute $2b + 3$ for a in the first equation and solve for b.

$3a - 3b = 5$ This is the first equation of the system.

$3(2b + 3) - 3b = 5$ Substitute $2b + 3$ for a. Don't forget the parentheses.

$6b + 9 - 3b = 5$ Distribute the multiplication by 3.

$3b + 9 = 5$ Combine like terms: $6b - 3b = 3b$.

$3b = -4$ To isolate the variable term, $3b$, subtract 9 from both sides.

$b = -\frac{4}{3}$ Divide both sides by 3. This is the b-value of the solution.

Step 3: To find a, substitute $-\frac{4}{3}$ for b in the equation $a = 2b + 3$.

Notation
Unless told otherwise, list the values of the variables of a solution in *alphabetical* order. For example, if the equations of a system involve the variables a and b, write the solution in the form (a, b).

$a = 2b + 3$ This is the substitution equation.

$a = 2\left(-\frac{4}{3}\right) + 3$ Substitute $-\frac{4}{3}$ for b.

$a = -\frac{8}{3} + \frac{9}{3}$ Do the multiplication and write 3 as a fraction with a denominator of 3.

$a = \frac{1}{3}$ Add. This is the a-value of the solution.

Step 4: The solution is $\left(\frac{1}{3}, -\frac{4}{3}\right)$. Check it in the original equations.

Self Check 4 Solve the system: $\begin{cases} 2s - t = 4 \\ 3s - 5t = 2 \end{cases}$

Now Try Problem 43

3 Solve Systems of Linear Equations that Contain Fractions.

It is usually helpful to clear any equations of fractions and combine any like terms before performing a substitution.

EXAMPLE 5 Solve the system: $\begin{cases} \frac{y}{4} = -\frac{x}{2} - \frac{3}{4} \\ 2x - y = -1 + y - x \end{cases}$

Strategy We will use properties of algebra to write each equation of the system in simpler form. Then we will use the substitution method to solve the resulting equivalent system.

Why The first equation will be easier to work with if we clear it of fractions. The second equation will be easier to work with if we eliminate the variable terms on the right side.

Solution We can clear the first equation of fractions by multiplying both sides by the LCD.

$$\frac{y}{4} = -\frac{x}{2} - \frac{3}{4}$$

$4\left(\frac{y}{4}\right) = 4\left(-\frac{x}{2} - \frac{3}{4}\right)$ Multiply both sides by the LCD, 4. Don't forget the parentheses.

$4\left(\frac{y}{4}\right) = 4\left(-\frac{x}{2}\right) - 4\left(\frac{3}{4}\right)$ Distribute the multiplication by 4.

(1) $y = -2x - 3$ Simplify. Call this equation 1.

We can write the second equation of the system in standard $Ax + By = C$ form by adding x and subtracting y from both sides.

$$2x - y = -1 + y - x$$

$$2x - y + x - y = -1 + y - x + x - y$$

(2) $3x - 2y = -1$ Combine like terms. Call this equation 2.

Notation
When a linear equation is written in standard $Ax + By = C$ form, the *variable terms* are on the left side of the equal symbol and the *constant* term is on the right side.

Step 1: Equations 1 and 2 form an equivalent system, which has the same solution as the original one. To find x, we proceed as follows:

$$\begin{cases} y = -2x - 3 & (1) \\ 3x - 2y = -1 & (2) \end{cases}$$ This is the substitution equation.

Step 2: To find x, substitute $-2x - 3$ for y in equation 2 and proceed as follows:

$$3x - 2y = -1$$

$$3x - 2(-2x - 3) = -1$$ Substitute $-2x - 3$ for y. Don't forget the parentheses.

$$3x + 4x + 6 = -1$$ Distribute the multiplication by -2.

$$7x + 6 = -1$$ Combine like terms: $3x + 4x = 7x$.

$$7x = -7$$ To isolate the variable term, $7x$, subtract 6 from both sides.

$$x = -1$$ Divide both sides by 7. This is the x-value of the solution.

Step 3: To find y, we substitute -1 for x in equation 1.

$$y = -2x - 3$$

$$y = -2(-1) - 3$$ Substitute -1 for x.

$$y = 2 - 3$$ Do the multiplication.

$$y = -1$$ Subtract. This is the y-value of the solution.

Step 4: The solution is $(-1, -1)$. Check it in the original system.

Caution

Always use the original equations when checking a solution. Do not use a substitution equation or an equivalent equation that you found algebraically. If an error was made, a proposed solution that would not satisfy the original system might appear to be correct.

Self Check 5 Solve the system: $\begin{cases} \dfrac{y}{6} = \dfrac{x}{3} + \dfrac{1}{2} \\ 2x - y = -3 + y - x \end{cases}$

Now Try Problem 45

4 Use Substitution to Identify Inconsistent Systems and Dependent Equations.

In the previous section, we solved inconsistent systems and systems of dependent equations graphically. We can also solve these systems using the substitution method.

EXAMPLE 6 Solve the system $\begin{cases} 4y - 12 = x \\ y = \dfrac{1}{4}x \end{cases}$ if possible.

Strategy We will use the substitution method to solve this system.

Why The substitution method works well when one of the equations of the system $\left(\text{in this case, } y = \frac{1}{4}x\right)$ is solved for a variable.

Solution To try to solve this system, substitute $\frac{1}{4}x$ for y in the first equation and solve for x.

$$4y - 12 = x$$

$$4\left(\frac{1}{4}x\right) - 12 = x$$ Substitute $\frac{1}{4}x$ for y.

$$x - 12 = x$$ Do the multiplication.

$$x - 12 - x = x - x$$ To eliminate x on the right side, subtract x from both sides.

$$-12 = 0$$ False

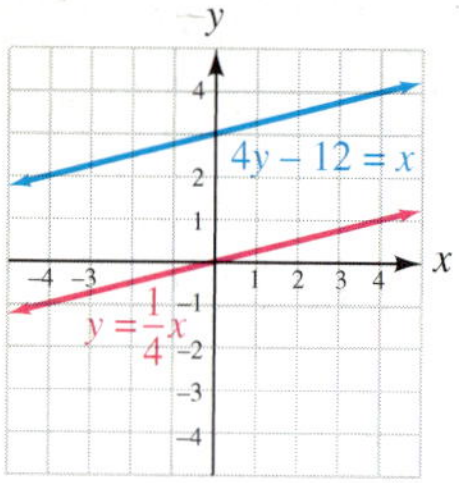

Here, the terms involving x drop out, and we get $-12 = 0$. This false statement indicates that the system has no solution and is inconsistent. The solution set is the empty set, $\varnothing$. The graphs of the equations of the system help to verify this; they are parallel lines.

Self Check 6 Solve the system $\begin{cases} x - 4 = y \\ -2y = 4 - 2x \end{cases}$ if possible.

Now Try **Problem 57**

EXAMPLE 7

Solve the system: $\begin{cases} x = -3y + 6 \\ 2x + 6y = 12 \end{cases}$

Strategy We will use the substitution method to solve this system.

Why The substitution method works well when one of the equations of the system (in this case, $x = -3y + 6$) is solved for a variable.

Solution To solve this system, substitute $-3y + 6$ for x in the second equation and solve for y.

$$2x + 6y = 12$$

$$2(-3y + 6) + 6y = 12 \quad \text{Substitute } -3y + 6 \text{ for } x. \text{ Don't forget the parentheses.}$$

$$-6y + 12 + 6y = 12 \quad \text{Distribute the multiplication by 2.}$$

$$12 = 12 \quad \text{True}$$

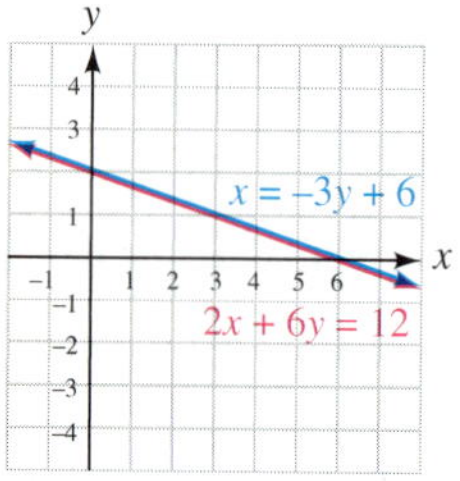

Here, the terms involving y drop out, and we get $12 = 12$. This true statement indicates that the two equations of the system are equivalent. Therefore, they are dependent equations and the system has infinitely many solutions. The graphs of the equations help to verify this; they are the same line.

Any ordered pair that satisfies one equation of this system also satisfies the other. To find several of the infinitely many solutions, we can substitute some values of x, say 0, 3, and 6, in either equation and solve for y. The results are: (0, 2), (3, 1), and (6, 0).

Self Check 7 Solve the system: $\begin{cases} y = 2 - x \\ 3x + 3y = 6 \end{cases}$

Now Try **Problem 59**

ANSWERS TO SELF CHECKS **1.** (3, 1) **2.** (−8, 2) **3.** (5, 0) **4.** $\left(\frac{18}{7}, \frac{8}{7}\right)$ **5.** (−3, −3) **6.** No solution **7.** Infinitely many solutions

STUDY SET 4.2

VOCABULARY

Fill in the blanks.

1. To solve the system $\begin{cases} x = y + 1 \\ 3x + 2y = 8 \end{cases}$ using the method discussed in this section, we begin by ________ $y + 1$ for x in the second equation.

2. We say that the equation $y = 2x + 4$ is solved for ___.

CONCEPTS

3. If the substitution method is used to solve $\begin{cases} 5x + y = 2 \\ y = -3x \end{cases}$, which equation should be used as the substitution equation?

4. Suppose the substitution method will be used to solve $\begin{cases} x - 2y = 2 \\ 2x + 3y = 11 \end{cases}$. Find a substitution equation by solving one of the equations for one of the variables.

5. Suppose $x - 4$ is substituted for y in the equation $x + 3y = 8$. Insert parentheses in $x + 3x - 4 = 8$ to show the substitution.

6. Fill in the blank. With the substitution method, the objective is to use an appropriate substitution to obtain one equation in _____ variable.

7. A student uses the substitution method to solve the system $\begin{cases} 4a + 5b = 2 \\ b = 3a - 11 \end{cases}$ and finds that $a = 3$. What is the easiest way for her to determine the value of b?

8. **a.** Clear $\frac{x}{5} + \frac{2y}{3} = 1$ of fractions.
b. Write $2x + y = x - 5y + 3$ in the form $Ax + By = C$.

9. Suppose $-2 = 1$ is obtained when a system is solved by the substitution method.
a. Does the system have a solution?
b. Which of the following is a possible graph of the system?

i.
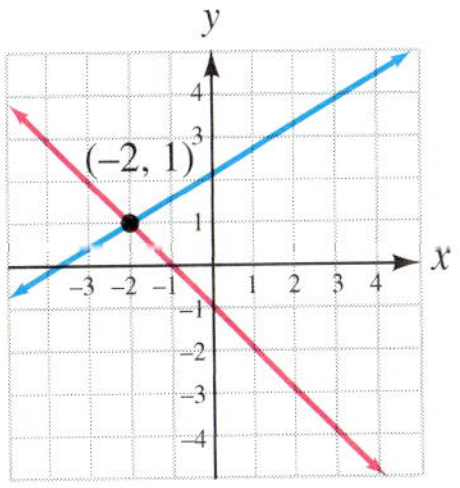

ii.
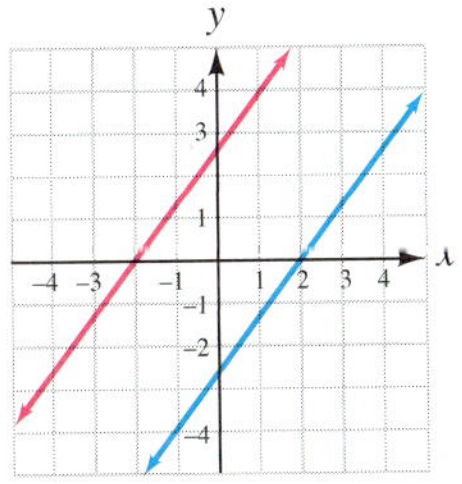

10. Suppose $2 = 2$ is obtained when a system is solved by the substitution method.
a. Does the system have a solution?
b. Which graph is a possible graph of the system?

i.
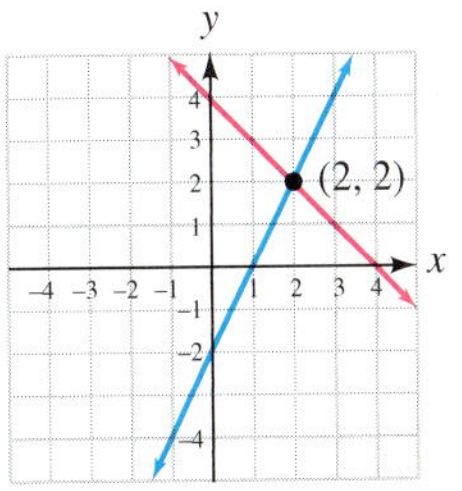

ii.
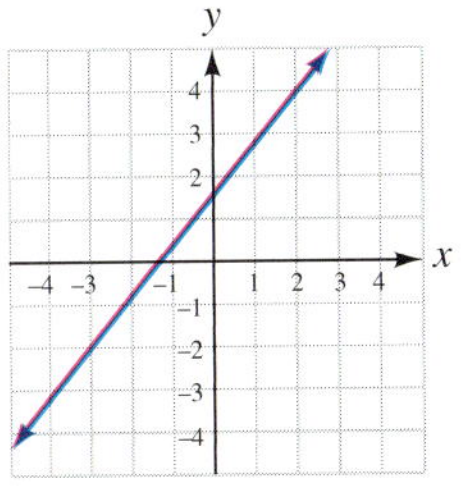

NOTATION

Complete the solution to solve the system.

11. Solve: $\begin{cases} y = 3x \\ x - y = 4 \end{cases}$

$$x - y = 4 \qquad \text{This is the second equation.}$$
$$x - (\quad) = 4$$
$$-2x = \quad$$
$$x = \quad$$
$$y = 3x \qquad \text{This is the first equation.}$$
$$y = 3(\quad)$$
$$y = \quad$$

The solution is $(\quad, \quad)$.

12. The system $\begin{cases} a = 3b + 2 \\ a + 3b = 8 \end{cases}$ was solved, and it was found that $b = 1$ and $a = 5$. Write the solution as an ordered pair.

GUIDED PRACTICE

Solve each system by substitution. **See Examples 1 and 2.**

13. $\begin{cases} y = 2x \\ x + y = 6 \end{cases}$

14. $\begin{cases} y = 3x \\ x + y = 4 \end{cases}$

15. $\begin{cases} y = 2x - 6 \\ 2x + y = 6 \end{cases}$

16. $\begin{cases} y = 2x - 9 \\ x + 3y = 8 \end{cases}$

17. $\begin{cases} 3x + y = -4 \\ x = y \end{cases}$

18. $\begin{cases} x + 2y = -6 \\ x = y \end{cases}$

19. $\begin{cases} x + 3y = -4 \\ x = -5y \end{cases}$

20. $\begin{cases} x + 5y = -3 \\ x = -4y \end{cases}$

21. $\begin{cases} 2x - y = -5 \\ x = -2y - 5 \end{cases}$

22. $\begin{cases} y = -2x \\ 3x + 2y = -1 \end{cases}$

23. $\begin{cases} b = \frac{2}{3}a \\ 8a - 3b = 3 \end{cases}$

24. $\begin{cases} a = \frac{2}{3}b \\ 9a + 4b = 5 \end{cases}$

25. $\begin{cases} 2x + 5y = -2 \\ y = -\frac{x}{2} \end{cases}$

26. $\begin{cases} y = -\frac{x}{2} \\ 2x - 3y = -7 \end{cases}$

27. $\begin{cases} x = \frac{1}{3}y - 1 \\ x = y + 5 \end{cases}$

28. $\begin{cases} x = \frac{1}{2}y + 2 \\ x = y - 6 \end{cases}$

Solve each system by substitution. **See Examples 3 and 4.**

29. $\begin{cases} r + 3s = 9 \\ 3r + 2s = 13 \end{cases}$

30. $\begin{cases} x - 2y = 2 \\ 2x + 3y = 11 \end{cases}$

31. $\begin{cases} 4x + y = -15 \\ 2x + 3y = 5 \end{cases}$

32. $\begin{cases} 4x + y = -5 \\ 2x - 3y = -13 \end{cases}$

33. $\begin{cases} 6x - 3y = 5 \\ x + 2y = 0 \end{cases}$

34. $\begin{cases} 5s + 10t = 3 \\ 2s + t = 0 \end{cases}$

35. $\begin{cases} 2x + 3 = -4y \\ x - 6 = -8y \end{cases}$

36. $\begin{cases} 5y + 2 = -4x \\ x + 2y = -2 \end{cases}$

37. $\begin{cases} 2a - 3b = -13 \\ -b = -2a - 7 \end{cases}$

38. $\begin{cases} a - 3b = -1 \\ -b = -2a - 2 \end{cases}$

39. $\begin{cases} 8x - 6y = 4 \\ 2x - y = -2 \end{cases}$

40. $\begin{cases} 5x + 4y = 0 \\ 2x - y = 0 \end{cases}$

41. $\begin{cases} 4x + 5y = 2 \\ 3x - y = 11 \end{cases}$

42. $\begin{cases} 5u + 3v = 5 \\ 4u - v = 4 \end{cases}$

43. $\begin{cases} 3x + 4y = -19 \\ 2y - x = 3 \end{cases}$

44. $\begin{cases} 5x - 2y = -7 \\ 5 - y = -3x \end{cases}$

Solve each system by substitution. **See Example 5.**

45. $\begin{cases} \frac{x}{2} + \frac{y}{2} = -1 \\ \frac{x}{3} - \frac{y}{2} = -4 \end{cases}$

46. $\begin{cases} \frac{2}{3}a + \frac{b}{5} = 1 \\ \frac{a}{3} - \frac{2}{3}b = \frac{13}{3} \end{cases}$

47. $\begin{cases} 5x = \frac{1}{2}y - 1 \\ \frac{1}{4}y = 10x - 1 \end{cases}$

48. $\begin{cases} \frac{x}{4} + y = \frac{1}{4} \\ \frac{y}{2} + \frac{11}{20} = \frac{x}{10} \end{cases}$

49. $\begin{cases} x - \frac{4}{5}y = 4 \\ \frac{y}{3} = \frac{x}{2} - \frac{5}{2} \end{cases}$

50. $\begin{cases} 3x - 2y = \frac{9}{2} \\ \frac{x}{2} - \frac{3}{4} = 2y \end{cases}$

51. $\begin{cases} y + x = 2x + 2 \\ 6x - 4y = 21 - y \end{cases}$

52. $\begin{cases} y - x = 3x \\ 2x + 2y = 14 - y \end{cases}$

53. $\begin{cases} 4x + 5y + 1 = -12 + 2x \\ x - 3y + 2 = -3 - x \end{cases}$

54. $\begin{cases} 6x + y = -8 + 3x - y \\ 3x - y = 2y + x - 1 \end{cases}$

55. $\begin{cases} 3(x - 1) + 3 = 8 + 2y \\ 2(x + 1) = 8 + y \end{cases}$

56. $\begin{cases} 4(x - 2) = 19 - 5y \\ 3(x - 2) - 2y = -y \end{cases}$

Solve each system by substitution, if possible. **See Examples 6 and 7.**

57. $\begin{cases} 2a + 4b = -24 \\ a = 20 - 2b \end{cases}$

58. $\begin{cases} 3a + 6b = -15 \\ a = -2b - 5 \end{cases}$

59. $\begin{cases} y - 3x = -5 \\ 21x = 7y + 35 \end{cases}$

60. $\begin{cases} 8y = 15 - 4x \\ x + 2y = 4 \end{cases}$

61. $\begin{cases} 6 - y = 4x \\ 2y = -8x - 20 \end{cases}$

62. $\begin{cases} 9x = 3y + 12 \\ 4 = 3x - y \end{cases}$

63. $\begin{cases} x = -3y + 6 \\ 2x + 4y = 6 + x + y \end{cases}$

64. $\begin{cases} 2x - y = x + y \\ -2x + 4y = 6 \end{cases}$

TRY IT YOURSELF

Solve each system by substitution, if possible. If a system has no solution or infinitely many solutions, so state.

65. $\begin{cases} -y = 11 - 3x \\ 2x + 5y = -4 \end{cases}$

66. $\begin{cases} -x = 10 - 3y \\ 2x + 8y = -6 \end{cases}$

67. $\begin{cases} \frac{x}{2} + \frac{y}{6} = \frac{2}{3} \\ \frac{x}{3} - \frac{y}{4} = \frac{1}{12} \end{cases}$

68. $\begin{cases} \frac{c}{2} + \frac{d}{14} = 1 \\ \frac{c}{5} - \frac{d}{2} = -\frac{33}{10} \end{cases}$

69. $\begin{cases} y - 4 = 2x \\ y = 2x + 2 \end{cases}$

70. $\begin{cases} x + 3y = 6 \\ x = -3y + 6 \end{cases}$

71. $\begin{cases} a + b = 1 \\ a - 2b = -1 \end{cases}$

72. $\begin{cases} 2b - a = -1 \\ 3a + 10b = -1 \end{cases}$

73. $\begin{cases} x = 7y - 10 \\ 2x - 14y + 20 = 0 \end{cases}$

74. $\begin{cases} y - 1 = 5x \\ 10x - 2y = 2 \end{cases}$

75. $\begin{cases} 4x + 1 = 2x + 5 + y \\ 2x + 2y = 5x + y + 6 \end{cases}$

76. $\begin{cases} 6x = 2(y + 20) + 5x \\ 5(x - 1) = 3y + 4(x + 10) \end{cases}$

77. $\begin{cases} 2a + 3b = 7 \\ 6a - b = 1 \end{cases}$

78. $\begin{cases} 3a + 5b = -6 \\ 5b - a = -3 \end{cases}$

79. $\begin{cases} 2x - 3y = -4 \\ x = -\frac{3}{2}y \end{cases}$

80. $\begin{cases} x = -\frac{3}{8}y \\ 8x - 3y = 4 \end{cases}$

APPLICATIONS

81. OFFROADING The *angle of approach* indicates how steep of an incline a vehicle can drive up without damaging the front bumper. The *angle of departure* indicates a vehicle's ability to exit an incline without damaging the rear bumper. The angle of approach a and the departure angle d for an H3 Hummer are described by the system $\begin{cases} a + d = 77 \\ a = d + 3 \end{cases}$. Use substitution to solve the system. (Each angle is measured in degrees.)

82. HIGH SCHOOL SPORTS The equations shown model the number of boys and girls taking part in high school soccer programs. In both models, x is the number of years after 2000, and y is the number of participants. If the trends continue, the graphs will intersect. Use the substitution method to predict the year when the number of boys and girls participating in high school soccer will be the same.

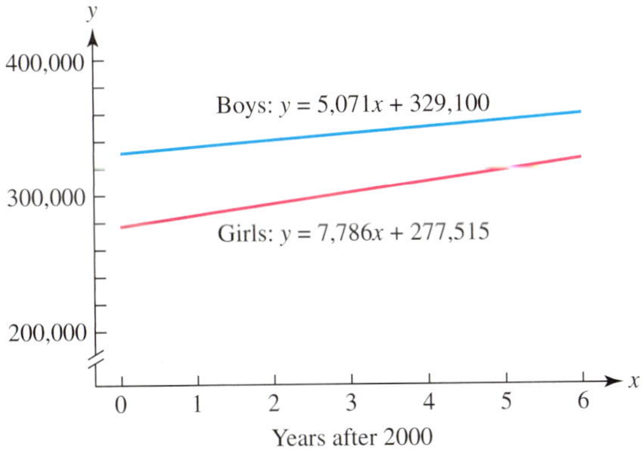

Source: National Federation of State High School Associations

WRITING

83. What concept does this diagram illustrate?

$$\begin{cases} 6x + 5y = 11 \\ y = 3x - 2 \end{cases}$$

84. When using the substitution method, how can you tell whether

a. a system of linear equations has no solution?

b. a system of linear equations has infinitely many solutions?

85. When solving a system, what advantages are there with the substitution method compared with the graphing method?

86. Consider the equation $5x + y = 12$. Explain why it is easier to solve for y than it is for x.

REVIEW

87. Find the prime factorization of 189.

88. Complete each statement. For any nonzero number a,

a. $\frac{a}{a} =$ **b.** $\frac{a}{1} =$

c. $\frac{0}{a} =$ **d.** $\frac{a}{0} =$

89. Simplify: $\frac{30}{36}$

90. Add: $\frac{5}{12} + \frac{1}{4}$

91. Multiply: $\frac{7}{8} \cdot \frac{3}{5}$

92. Divide: $\frac{1}{3} \div \frac{4}{5}$

CHALLENGE PROBLEMS

Use the substitution method to solve each system.

93. $\begin{cases} \frac{6x - 1}{3} - \frac{5}{3} = \frac{3y + 1}{2} \\ \frac{1 + 5y}{4} + \frac{x + 3}{4} = \frac{17}{2} \end{cases}$

94. $\begin{cases} 0.5x + 0.5y = 6 \\ 0.001x - 0.001y = -0.004 \end{cases}$

95. The system $\begin{cases} \frac{1}{2}x = y + 3 \\ x - 2y = 6 \end{cases}$ has infinitely many solutions. Find three of them.

96. Could the substitution method be used to solve the following system? Explain why or why not. If not, what method could be used?

$$\begin{cases} y = -2 \\ x = 5 \end{cases}$$

SECTION 4.3
Solving Systems of Equations by Elimination (Addition)

Objectives

1. Solve systems of linear equations by the elimination method.
2. Use multiplication to eliminate a variable.
3. Use the elimination method twice to solve a system.
4. Use elimination to identify inconsistent systems and dependent equations.
5. Determine the most efficient method to use to solve a linear system.

In the first step of the substitution method for solving a system of equations, we solve one of the equations for one of the variables. At times, this can be difficult, especially if none of the variables has a coefficient of 1 or -1. This is the case for the system

$$\begin{cases} 2x + 5y = 11 \\ 7x - 5y = 16 \end{cases}$$

Solving either equation for x or y involves working with cumbersome fractions. Fortunately, we can solve systems like this one using an easier algebraic method called the **elimination** or the **addition method.**

1 Solve Systems of Linear Equations by the Elimination Method.

The elimination method for solving a system is based on the **addition property of equality:** *When equal quantities are added to both sides of an equation, the results are equal.* In symbols, if $A = B$ and $C = D$, then adding the left sides and the right sides of these equations, we have $A + C = B + D$. This procedure is called *adding the equations.*

Add the terms on the left sides.

$$\begin{array}{r} A = B \\ C = D \\ \hline A + C = B + D \end{array}$$

Add the terms on the right sides.

EXAMPLE 1 Solve the system: $\begin{cases} 2x + 5y = 11 \\ 6x - 5y = 13 \end{cases}$

Strategy Since the coefficients of the y-terms are opposites, we will add the left sides and the right sides of the given equations.

Why When we add the equations in this way, the result will be an equation that contains only one variable, x.

Solution Since $6x - 5y$ and 13 are equal quantities, we can add $6x - 5y$ to the left side and 13 to the right side of the first equation, $2x + 5y = 11$.

The Language of Algebra
The *elimination* method, or *addition* method as it is also known, is so named because one of the variables is eliminated using addition.

$$\begin{aligned} 2x + 5y &= 11 \\ 6x - 5y &= 13 \\ \hline 8x \quad &= 24 \end{aligned}$$

To add the equations, add the like terms, column by column.

$11 + 13 = 24$

$5y + (-5y) = 0$

$2x + 6x = 8x$

Because the sum of the terms $5y$ and $-5y$ is 0, we say that the variable y has been eliminated. Since the resulting equation has only one variable, we can solve it for x.

$8x = 24$

$x = 3$ Divide both sides by 8. This is the x-value of the solution.

To find the y-value of the solution, substitute 3 for x in either equation of the original system.

$2x + 5y = 11$ This is the first equation of the system.

$2(3) + 5y = 11$ Substitute 3 for x.

$6 + 5y = 11$ Multiply.

$5y = 5$ Subtract 6 from both sides.

$y = 1$ Divide both sides by 5. This is the y-value of the solution.

Now we check the proposed solution (3, 1) in the equations of the original system.

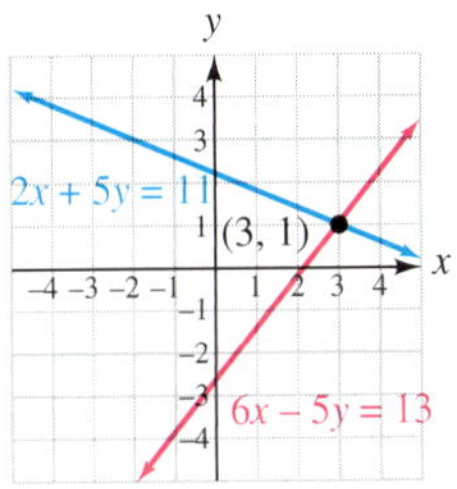

Check:

The first equation.		The second equation.	
$2x + 5y = 11$		$6x - 5y = 13$	
$2(3) + 5(1) \stackrel{?}{=} 11$		$6(3) - 5(1) \stackrel{?}{=} 13$	
$6 + 5 \stackrel{?}{=} 11$		$18 - 5 \stackrel{?}{=} 13$	
$11 = 11$	True	$13 = 13$	True

Since (3, 1) satisfies both equations, it is the solution. The graph on the left helps to verify this. The solution set is written $\{(3, 1)\}$.

Self Check 1 Solve the system: $\begin{cases} -4x + 3y = 4 \\ 4x + 5y = 28 \end{cases}$

Now Try **Problem 19**

To solve a system of equations in x and y by the elimination method, follow these steps.

The Elimination (Addition) Method

1. Write both equations of the system in standard $Ax + By = C$ form.
2. If necessary, multiply one or both of the equations by a nonzero number chosen to make the coefficients of x (or the coefficients of y) opposites.
3. Add the equations to eliminate the terms involving x (or y).
4. Solve the equation resulting from step 3.
5. Find the value of the remaining variable by substituting the solution found in step 4 into any equation containing both variables. Or, repeat steps 2–4 to eliminate the other variable.
6. Check the proposed solution in each equation of the original system. Write the solution as an ordered pair.

2 Use Multiplication to Eliminate a Variable.

In Example 1, the coefficients of the terms $5y$ in the first equation and $-5y$ in the second equation were opposites. When we added the equations, the variable y was eliminated. For many systems, however, we are not able to immediately eliminate a variable by adding. In such cases, we use the multiplication property of equality to create coefficients of x or y that are opposites.

EXAMPLE 2 Solve the system: $\begin{cases} 2x + 7y = -18 \\ 2x + 3y = -10 \end{cases}$

Strategy We will use the elimination method to solve this system.

Why Since none of the variables has a coefficient of 1 or -1, it would be difficult to solve this system using substitution.

Solution

Step 1: Both equations are in standard $Ax + By = C$ form. We see that neither the coefficients of x nor the coefficients of y are opposites. Adding these equations as written does not eliminate a variable.

Step 2: To eliminate x, we can multiply both sides of the second equation by -1. This creates the term $-2x$, whose coefficient is opposite that of the $2x$ term in the first equation.

$$\begin{cases} 2x + 7y = -18 \\ 2x + 3y = -10 \end{cases} \xrightarrow[\text{Multiply by } -1]{\text{Unchanged}} \begin{matrix} 2x + 7y = -18 \\ -1(2x + 3y) = -1(-10) \end{matrix} \xrightarrow[\text{Simplify}]{\text{Unchanged}} \begin{cases} 2x + 7y = -18 \\ -2x - 3y = 10 \end{cases}$$

Step 3: When the equations are added, x is eliminated.

$$\begin{array}{r} 2x + 7y = -18 \\ -2x - 3y = 10 \\ \hline 4y = -8 \end{array}$$

In the left column: $2x + (-2x) = 0$.

Step 4: Solve the resulting equation to find y.

$4y = -8$

$y = -2$ Divide both sides by 4. This is the y-value of the solution.

Step 5: To find x, we can substitute -2 for y in either of the equations of the original system, or in $-2x - 3y = 10$. It appears the computations will be simplest if we use $2x + 3y = -10$.

$2x + 3y = -10$ This is the second equation of the original system.

$2x + 3(-2) = -10$ Substitute -2 for y.

$2x - 6 = -10$ Multiply.

$2x = -4$ Add 6 to both sides.

$x = -2$ Divide both sides by 2. This is the x-value of the solution.

Step 6: The solution is $(-2, -2)$. Check this result in the original equations.

Success Tip

It doesn't matter which variable is eliminated first. We don't have to find the values of the variables in alphabetical order. Choose the one that is the easier to eliminate. The basic objective of this method is to obtain two equations whose sum will be *one* equation in *one* variable.

Self Check 2 Solve the system: $\begin{cases} x + 7y = -24 \\ 3x + 7y = -30 \end{cases}$

Now Try Problem 27

EXAMPLE 3 Solve the system: $\begin{cases} 7x + 2y - 14 = 0 \\ 9x = 4y - 28 \end{cases}$

Strategy We will use the elimination method to solve this system.

Why Since none of the variables has coefficient 1 or -1, it would be difficult to solve this system using substitution.

Solution

Step 1: To compare coefficients, write each equation in the standard $Ax + By = C$ form. Since each of the original equations will be written in an equivalent form, the resulting system will have the same solution as the original system.

$$\begin{cases} 7x + 2y = 14 & \text{Add 14 to both sides of } 7x + 2y - 14 = 0. \\ 9x - 4y = -28 & \text{Subtract 4y from both sides of } 9x = 4y - 28. \end{cases}$$

Step 2: Neither the coefficients of x nor the coefficients of y are opposites. To eliminate y, we can multiply both sides of the first equation by 2. This creates the term $4y$, whose coefficient is opposite that of the $-4y$ term in the second equation.

$$\begin{cases} 7x + 2y = 14 \\ 9x - 4y = -28 \end{cases} \xrightarrow[\text{Unchanged}]{\text{Multiply by 2}} \begin{matrix} 2(7x + 2y) = 2(14) \\ 9x - 4y = -28 \end{matrix} \xrightarrow[\text{Unchanged}]{\text{Simplify}} \begin{cases} 14x + 4y = 28 \\ 9x - 4y = -28 \end{cases}$$

Step 3: When the equations are added, y is eliminated.

$$\begin{aligned} 14x + 4y &= 28 \\ 9x - 4y &= -28 \\ \hline 23x \quad &= 0 \end{aligned} \qquad \text{In the middle column: } 4y + (-4y) = 0.$$

Step 4: Since the result of the addition is an equation in one variable, we can solve for x.

$$\begin{aligned} 23x &= 0 \\ x &= 0 && \text{Divide both sides by 23. This is the x-value of the solution.} \end{aligned}$$

Step 5: To find y, we can substitute 0 for x in any equation that contains both variables. It appears the computations will be simplest if we use $7x + 2y = 14$.

$$\begin{aligned} 7x + 2y &= 14 && \text{This is the first equation of the original system.} \\ 7(0) + 2y &= 14 && \text{Substitute 0 for x.} \\ 0 + 2y &= 14 && \text{Multiply.} \\ 2y &= 14 \\ y &= 7 && \text{Divide both sides by 2. This is the y-value of the solution.} \end{aligned}$$

Step 6: The solution is (0, 7). Check this result in the original equations.

Success Tip

We choose to eliminate y because the coefficient -4 is a *multiple* of the coefficient 2. The same cannot be said for 7 and 9, the coefficients of x.

Caution

When using the elimination method, don't forget to multiply both sides of an equation by the appropriate number.

Multiply both sides by 2

$2(7x + 2y) = 2(14)$

$$\begin{cases} 3(s + c) = 30 \\ 5(s - c) = 30 \end{cases} \xrightarrow{\text{Distribute}} \begin{cases} 3s + 3c = 30 \\ 5s - 5c = 30 \end{cases}$$

Solve the System To eliminate c, we proceed as follows.

$$\begin{aligned} 15s + 15c &= 150 && \text{Multiply both sides of } 3s + 3c = 30 \text{ by 5.} \\ 15s - 15c &= 90 && \text{Multiply both sides of } 5s - 5c = 30 \text{ by 3.} \\ \hline 30s \qquad &= 240 \\ s &= 8 && \text{Divide both sides by 30. This is the speed of the boat in still water.} \end{aligned}$$

To find c, it appears that the computations will be easiest if we use $3s + 3c = 30$.

$$\begin{aligned} 3s + 3c &= 30 \\ 3(8) + 3c &= 30 && \text{Substitute 8 for } s. \\ 24 + 3c &= 30 && \text{Multiply.} \\ 3c &= 6 && \text{Subtract 24 from both sides.} \\ c &= 2 && \text{Divide both sides by 3. This is the speed of the current.} \end{aligned}$$

State the Conclusion The speed of the boat in still water is 8 mph and the speed of the current is 2 mph.

Check the Results With a 2-mph current, the boat's downstream speed will be $8 + 2 = 10$ mph. In 3 hours, it will travel $10 \cdot 3 = 30$ miles. With a 2-mph current, the boat's upstream speed will be $8 - 2 = 6$ mph. In 5 hours, it will cover $6 \cdot 5 = 30$ miles. The results check.

Now Try **Problem 41**

EXAMPLE 7 ***Medical Technology.*** A laboratory technician has one batch of antiseptic that is 40% alcohol and a second batch that is 60% alcohol. She would like to make 8 fluid ounces of solution that is 55% alcohol. How many fluid ounces of each batch should she use?

Analyze the Problem Some 60% solution must be added to some 40% solution to make a 55% solution.

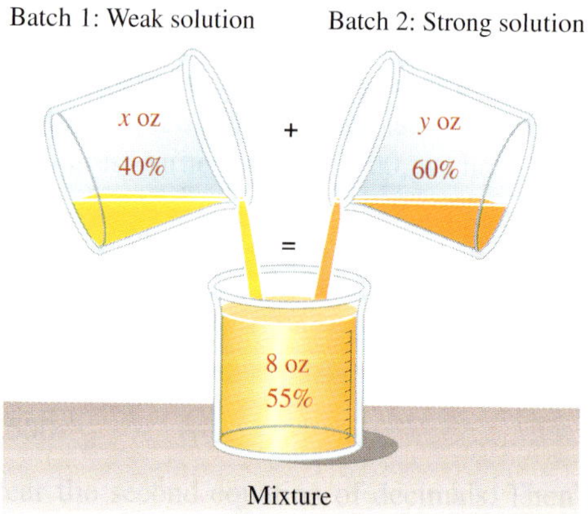

Form Two Equations Let $x =$ the number of ounces to be used from batch 1 and $y =$ the number of ounces to be used from batch 2. The amount of alcohol in each solution is given by

$$\text{Amount of solution} \cdot \text{strength of solution} = \text{amount of alcohol}$$

We can organize the facts of the problem in a table.

	Amount	· Strength	= Amount of alcohol
Batch 1 (too weak)	x	0.40	$0.40x$
Batch 2 (too strong)	y	0.60	$0.60y$
Mixture	8	0.55	0.55(8)

↑ One equation comes from information in this column. (Amount column)

↑ 40%, 60%, and 55% have been expressed as decimals. (Strength column)

↑ Another equation comes from information in this column. (Amount of alcohol column)

> **Success Tip**
> A *liter* is a basic unit of measurement of capacity in the metric system. A liter of liquid is slightly more than 1 quart.

The information in the table provides two equations.

$$\begin{cases} x + y = 8 \\ 0.40x + 0.60y = 0.55(8) \end{cases}$$

The number of ounces of batch 1 plus the number of ounces of batch 2 equals the total number of ounces in the mixture.

The amount of alcohol in batch 1 plus the amount of alcohol in batch 2 equals the amount of alcohol in the mixture.

Solve the System We can solve this system by elimination. To eliminate x, we proceed as follows.

$$-40x - 40y = -320$$ Multiply both sides of the first equation by -40.

$$40x + 60y = 440$$ Multiply both sides of the second equation by 100.

$$20y = 120$$

$$y = 6$$ Divide both sides by 20. This is the number of fluid ounces of batch 2 needed.

To find x, we substitute 6 for y in the first equation of the original system.

$$x + y = 8$$

$$x + 6 = 8$$ Substitute.

$$x = 2$$ Subtract 6 from both sides. This is the number of fluid ounces of batch 1 needed.

State the Conclusion The technician should use 2 fluid ounces of the 40% solution and 6 fluid ounces of the 60% solution.

Check the Results Note that 2 ounces + 6 ounces = 8 ounces, the required number. Also, the amount of alcohol in the two solutions is equal to the amount of alcohol in the mixture.

Alcohol in batch 1: $0.40x = 0.40(2) = 0.8$ ounces
Alcohol in batch 2: $0.60y = 0.60(6) = 3.6$ ounces
} Total: 4.4 ounces

Alcohol in the mixture: $0.55(8) = 4.4$ ounces

The results check.

Now Try Problem 45

EXAMPLE 8 ***Breakfast Cereal.*** One ounce of raisins (by weight) sells for 22¢ and one ounce of bran flakes (by weight) sells for 12¢. How many ounces of each should be used to create a 20-ounce box of raisin bran cereal that can be sold for 15¢ an ounce?

9. For each case below, write an algebraic expression that represents the speed of the canoe in miles per hour if its speed in still water is x mph.

Downstream

Upstream

10. Complete the table, which contains information about an airplane flying in windy conditions.

	Rate	· Time	= Distance
With wind	$x + y$	3	
Against wind	$x - y$	5	

11. a. If the contents of the two test tubes are poured into a third tube, how much solution will the third tube contain? (mL stands for milliliter. A milliliter is about 15 drops from an eyedropper.)

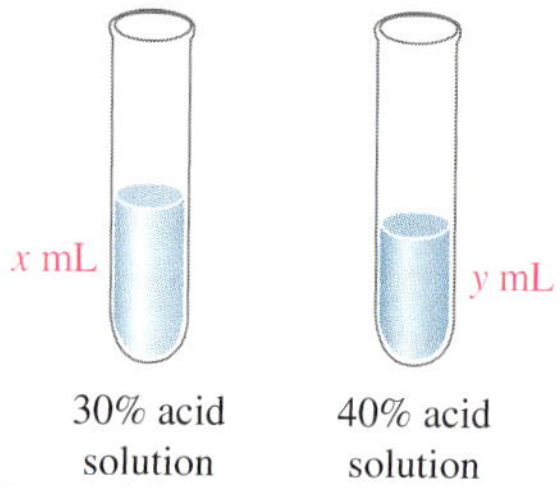

b. Which of the following strengths could the mixture possibly be: 27%, 33%, or 44% acid solution?

12. a. Complete the table, which contains information about mixing two salt solutions to get 12 gallons of a 3% salt solution.

	Amount	· Strength	= Amount of salt
Weak	x	0.01	
Strong	y	0.06	
Mix			

b. Use the information from the Amount column to write an equation.

c. Use the information from the Amount of salt column to write an equation.

GUIDED PRACTICE

See Example 2.

13. COMPLEMENTARY ANGLES Two angles are complementary. The measure of one angle is 10° more than three times the measure of the other. Find the measure of each angle.

14. SUPPLEMENTARY ANGLES Two angles are supplementary. The measure of one angle is 20° less than 19 times the measure of the other. Find the measure of each angle.

15. SUPPLEMENTARY ANGLES The difference of the measures of two supplementary angles is 80°. Find the measure of each angle.

16. COMPLEMENTARY ANGLES Two angles are complementary. The measure of one angle is 15° more than one-half of the measure of the other. Find the measure of each angle.

APPLICATIONS

Write a system of two equations in two variables to solve each problem.

17. TREE TRIMMING When fully extended, the arm on a tree service truck is 51 feet long. If the upper part of the arm is 7 feet shorter than the lower part, how long is each part of the arm?

18. ALASKA Most of the 1,422-mile-long Alaskan Highway is actually in Canada. Find the length of the highway that is in Alaska and the length of the highway that is in Canada if it is known that the difference in the lengths is 1,020 miles.

19. GOVERNMENT The salaries of the president and vice president of the United States total $608,100 a year. If the president makes $191,900 more than the vice president, find each of their salaries.

20. CAUSES OF DEATH According to the *National Vital Statistics Reports,* in 2004, the number of Americans who died from heart disease was about 6 times the number who died from accidents. If the total number of deaths from these two causes was approximately 763,000, how many Americans died from each cause in 2004?

21. MONUMENTS The Marine Corps War Memorial in Arlington, Virginia, portrays the raising of the U.S. flag on Iwo Jima during World War II. Find the measures of the two angles shown if the measure of $\angle 2$ is 15° less than twice the measure of $\angle 1$.

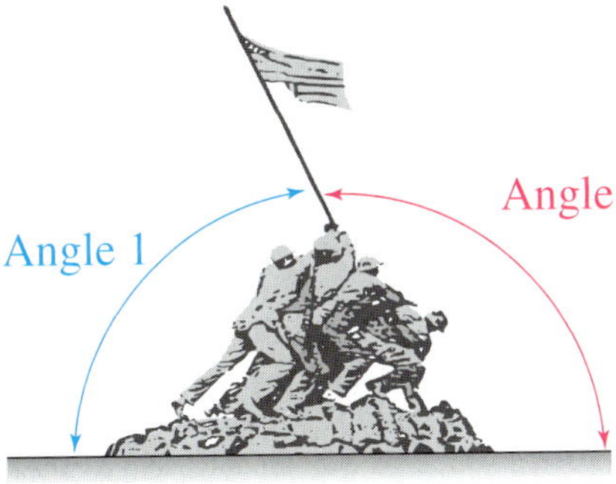

22. PHYSICAL THERAPY To rehabilitate her knee, an athlete does leg extensions. Her goal is to regain a full 90° range of motion in this exercise. Use the information in the illustration to determine her current range of motion in degrees and the number of degrees of improvement she still needs to make.

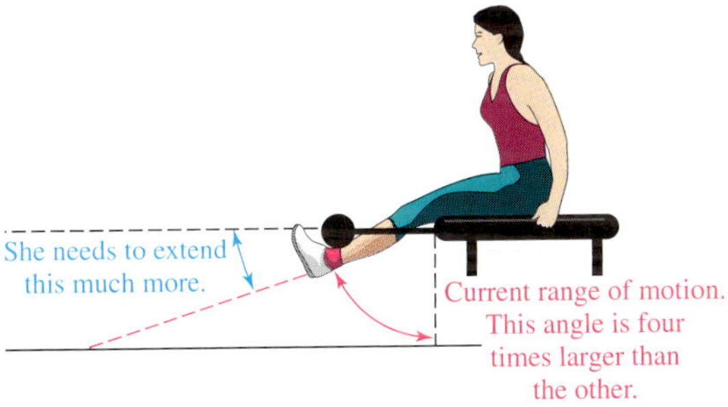

23. THEATER SCREENS At an IMAX theater, the giant rectangular movie screen has a width 26 feet less than its length. If its perimeter is 332 feet, find the length and the width of the screen.

24. ART In 1770, Thomas Gainsborough painted *The Blue Boy*. The sum of the length and width of the painting is 118 inches. The difference of the length and width is 22 inches. Find the length and width.

25. GEOMETRY A 50-meter path surrounds a rectangular garden. The width of the garden is two-thirds its length. Find the length and width.

26. BALLROOM DANCING A rectangular-shaped dance floor has a perimeter of 200 feet. If the floor were 20 feet wider, its width would equal its length. Find the length and width of the dance floor.

27. EMPTY CARTRIDGES A bank recycles its empty printer and copier cartridges. In January, the bank received $40 for recycling 5 printer and 2 copier cartridges. In February, the bank received $57 for recycling 6 printer and 3 copier cartridges. How much is the bank paid for an empty printer cartridge and for an empty copier cartridge?

28. THANKSGIVING DINNER There are a total of 510 calories in 6 ounces of turkey and one slice of pumpkin pie. There are a total of 580 calories in 4 ounces of turkey and two slices of pumpkin pie. How many calories are there in 1 ounce of turkey and in one slice of pumpkin pie?

29. ***from Campus to Careers***

Portrait Photographer

Suppose you are a wedding photographer and you sell:
Package 1: one 10×14 and ten 8×10 color photos for $239.50
Package 2: one 10×14 and five 8×10 color photos for $134.50
A newlywed couple buys Package 1 and decides that they want one more 10×14 and one more 8×10 photograph. At the same prices, what should you charge them for each additional photograph?

30. BUYING PAINTING SUPPLIES Two partial receipts for paint supplies are shown. (Assume no sales tax was charged.) Find the cost of one gallon of paint and the cost of one paint brush.

31. COLLECTING STAMPS Determine the price of an Elvis Presley stamp and a Statue of Liberty stamp given the following information.

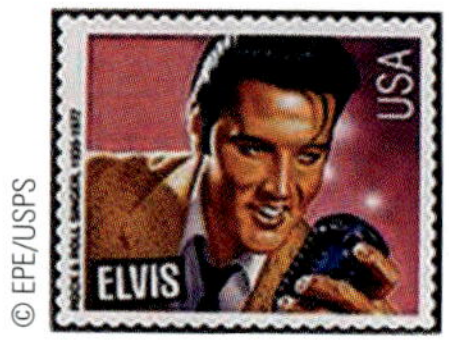

- One Elvis stamp and one Liberty stamp cost a total of 63¢.
- A sheet of 40 Elvis stamps and a sheet of 20 Liberty stamps cost a total of \$18.40. (*Hint:* 63¢ = \$0.63)

32. RECYCLING A boy scout troop earned \$24 by recycling a total of 330 beverage containers. The recycling rates are shown below. How many of the small capacity containers and how many of the large capacity containers did they recycle? (*Hint:* 5¢ = \$.05 and 10¢ = \$0.10)

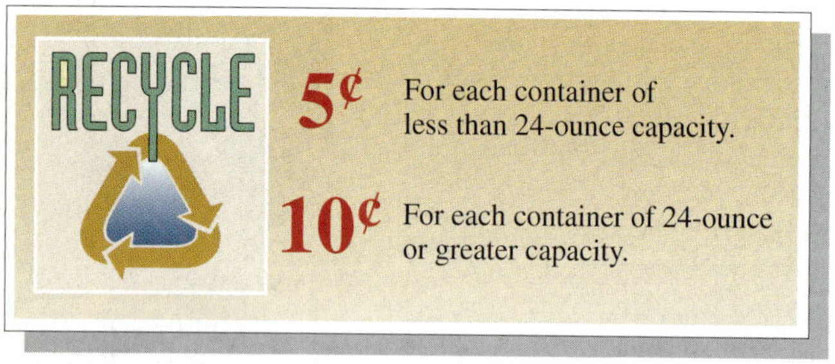

33. SELLING ICE CREAM At a store, ice cream cones cost \$1.80 and sundaes cost \$3.30. One day the receipts for a total of 148 cones and sundaes were \$360.90. How many cones were sold? How many sundaes?

34. BUYING TICKETS The ticket prices for a movie are shown below. Receipts for one showing were \$1,740 for an audience of 190 people. How many general admission tickets and how many senior citizen tickets were sold?

35. STUDENT LOANS A college used a \$5,000 gift from an alumnus to make two student loans. The first was at 5% annual interest to a nursing student. The second was at 7% to a business major. If the college collected \$310 in interest the first year, how much was loaned to each student?

36. FINANCIAL PLANNING In investing \$6,000 of a couple's money, a financial planner put some of it into a savings account paying 6% annual interest. The rest was invested in a riskier mini-mall development plan paying 12% annually. The combined interest earned for the first year was \$540. How much money was invested at each rate?

37. INVESTING A BONUS A businessman invested part of his \$40,000 end-of-the-year bonus in an international fund that paid an annual yield of 8%. The rest of the bonus was invested in an offshore bank that paid an annual yield of 9%. Find the amount of each investment if he made a total of \$3,415 in interest from them the first year.

38. PENSION FUNDS A state employees' pension fund invested a total of one million dollars in two accounts that earned 3.5% and 4.5% annual interest. At the end of the year, the total interest earned from the two investments was \$39,000. How much was invested at each rate?

39. LOSSES A CEO deposited part of \$22,000 in an account paying 4% interest annually. The rest of the money was invested in a biotech company that, after only one year, caused him to lose 3% of his initial investment in it. Find the amount of each investment if the net interest he earned the first year was only \$110.

40. LOTTERY WINNINGS After winning \$60,000 in the lottery, a retired teacher gave \$10,000 of it to her grandchildren. She invested part of the remainder in a growth fund that earned 4.4% annually and the rest in certificates of deposit paying a 5.8% annual percentage yield. The interest that she received on these two investments totaled \$2,732 at the end of the first year. Find the amount of each investment.

41. THE GULF STREAM The Gulf stream is a warm ocean current of the North Atlantic Ocean that flows northward, as shown below. Heading north with the Gulf Stream, a cruise ship traveled 300 miles in 10 hours. Against the current, it took 15 hours to make the return trip. Find the speed of the ship in still water and the speed of the current.

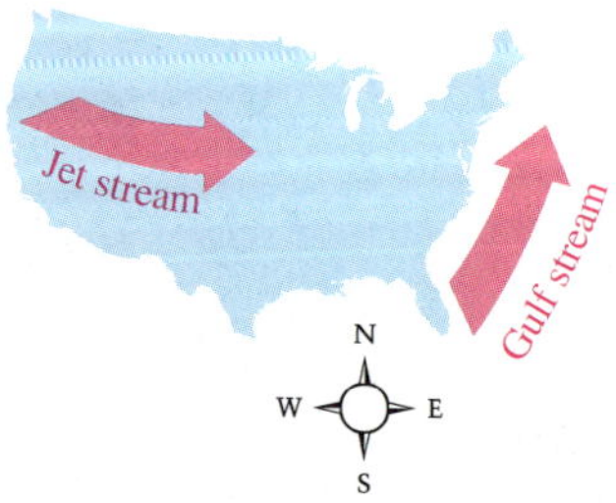

42. THE JET STREAM The jet stream is a strong wind current that flows across the United States, as shown on the previous page. Flying with the jet stream, a plane flew 3,000 miles in 5 hours. Against the same wind, the trip took 6 hours. Find the speed of the plane in still air and the speed of the wind current.

43. AVIATION An airplane can fly with the wind a distance of 800 miles in 4 hours. However, the return trip against the wind takes 5 hours. Find the speed of the plane in still air and the speed of the wind.

44. BOATING A boat can travel 24 miles downstream in 2 hours and can make the return trip in 3 hours. Find the speed of the boat in still water and the speed of the current.

45. MARINE BIOLOGY A marine biologist wants to set up an aquarium containing 3% salt water. He has two tanks on hand that contain 6% and 2% salt water. How much water from each tank must he use to fill a 32-gallon aquarium with a 3% saltwater mixture?

46. COMMEMORATIVE COINS A foundry has been commissioned to make souvenir coins. The coins are to be made from an alloy that is 40% silver. The foundry has on hand two alloys, one with 50% silver content and one with a 25% silver content. How many kilograms of each alloy should be used to make 20 kilograms of the 40% silver alloy?

47. CLEANING FLOORS A custodian is going to mix a 4% ammonia solution and a 12% ammonia solution to get 1 gallon (128 fluid ounces) of a 9% ammonia solution. How many fluid ounces of the 4% solution and the 12% solution should be used?

48. MOUTHWASH A pharmacist has a mouthwash solution that is 6% ethanol alcohol and another that is 18% ethanol alcohol. How many milliliters of each must be mixed to make 750 milliliters of a mouthwash that is 10% ethanol alcohol?

49. COFFEE SALES A coffee supply store waits until the orders for its special blend reach 100 pounds before making up a batch. Columbian coffee selling for $8.75 a pound is blended with Brazilian coffee selling for $3.75 a pound to make a product that sells for $6.35 a pound. How much of each type of coffee should be used to make the blend that will fill the orders?

50. MIXING NUTS A merchant wants to mix peanuts with cashews, as shown in the illustration, to get 48 pounds of mixed nuts that will be sold at $8 per pound. How many pounds of each should the merchant use?

51. GOURMET FOODS A New York delicatessen sells marinated mushrooms for $12 a pint and stuffed Kalamata olives for $9 a pint. How many pints of each should be used to get 20 pints of a mixture that will sell for $10 a pint?

52. HERBS Ginger root powder sells for $6.50 a pound and ginkgo leaf powder sells for $9.50 a pound. How many pounds of each should be used to make 15 pounds of a mixture that sells for $7 a pound?

WRITING

53. Explain why a table is helpful in solving uniform motion and mixture problems.

54. A man paid $89 for two shirts and four pairs of socks. If we let x = the cost of a shirt, in dollars, and y = the cost of a pair of socks, in dollars, an equation modeling the purchase is $2x + 4y = 89$. Explain why there is not enough information to determine the cost of a shirt or the cost of a pair of socks.

REVIEW

Graph each inequality. Then describe the graph using interval notation.

55. $x < 4$

56. $x \geq -3$

57. $-1 < x \leq 2$

58. $-2 \leq x \leq 0$

CHALLENGE PROBLEMS

59. On the last scale, how many nails will it take to balance 1 nut?

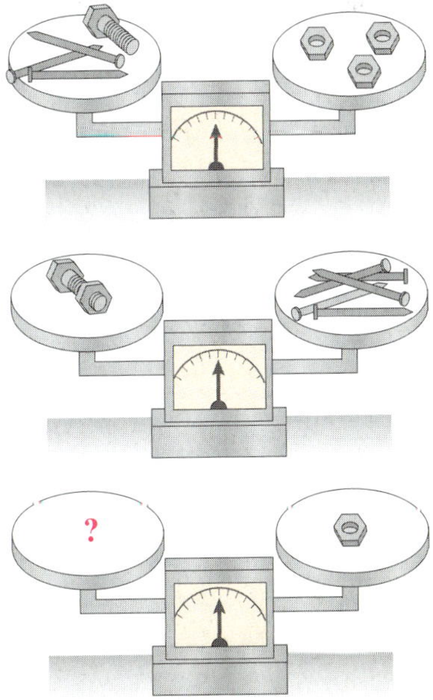

60. FARMING In a pen of goats and chickens, there are 40 heads and 130 feet. How many goats and chickens are in the pen?

SECTION 4.5
Solving Systems of Linear Inequalities

Objectives

1. Solve a system of linear inequalities by graphing.
2. Solve application problems involving systems of linear inequalities.

In Section 4.1, we solved systems of linear *equations* graphically by finding the point of intersection of two lines. Now we consider systems of linear *inequalities,* such as

$$\begin{cases} x + y \geq -1 \\ x - y \geq 1 \end{cases}$$

To solve systems of linear inequalities, we again find the points of intersection of graphs. In this case, however, we are not looking for an intersection of two lines, but an intersection of two regions.

1 Solve a System of Linear Inequalities by Graphing.

A solution of a **system of linear inequalities** is an ordered pair that satisfies each inequality. *To solve a system of linear inequalities* means to find all of its solutions. This can be done by graphing each inequality on the same set of axes and finding the points that are common to every graph in the system.

EXAMPLE 1 Graph the solutions of the system: $\begin{cases} x + y \geq -1 \\ x - y \geq 1 \end{cases}$

Strategy We will graph the solutions of $x + y \geq -1$ in one color and the solutions of $x - y \geq 1$ in another color on the same coordinate system.

Why We need to see where the graphs of the two inequalities intersect (overlap).

Solution To graph $x + y \geq -1$, we begin by graphing the boundary line $x + y = -1$. Since the inequality contains an $\geq$ symbol, the boundary is a solid line. Because the coordinates of the test point (0, 0) satisfy $x + y \geq -1$, we shade (in red) the side of the boundary that contains (0, 0). See part (a) of the figure on the next page.

Graph the boundary: The intercept method

$x + y = -1$

x	y	(x, y)
0	-1	$(0, -1)$
-1	0	$(-1, 0)$

***Shading: Check the test point* (0, 0)**

$x + y \geq -1$

$0 + 0 \overset{?}{\geq} -1$ Substitute.

$0 \geq -1$ True

(0, 0) is a solution of $x + y \geq -1$.

The Language of Algebra
To solve a system of linear inequalities, we *superimpose* the graphs of the inequalities. That is, we place one graph over the other. Most video camcorders can *superimpose* the date and time over the picture being recorded.

In part (b) of the figure, we superimpose the graph of $x - y \geq 1$ on the graph of $x + y \geq -1$ so that we can determine the points that the graphs have in common. To graph $x - y \geq 1$, we graph the boundary $x - y = 1$ as a solid line. Since the test point (0, 0) does not satisfy $x - y \geq 1$, we shade (in blue) the half-plane that does not contain (0, 0).

Graph the boundary: The intercept method

$x - y = 1$

x	y	(x, y)
0	-1	$(0, -1)$
1	0	$(1, 0)$

Shading: Check the test point (0, 0)

$x - y \geq 1$

$0 - 0 \overset{?}{\geq} 1$ Substitute.

$0 \geq 1$ False

$(0, 0)$ is not a solution of $x - y \geq 1$.

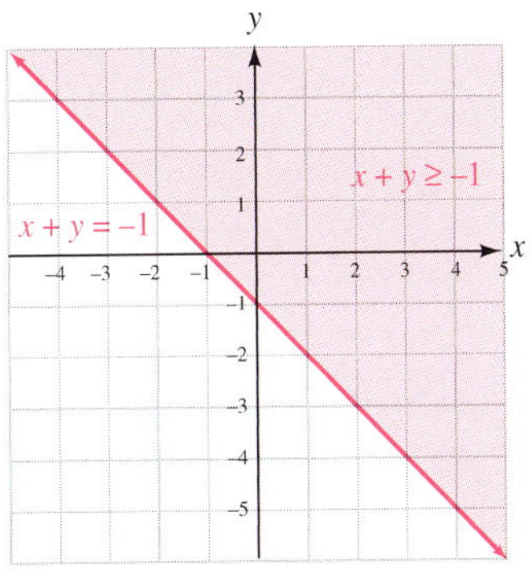

The graph of $x + y \geq -1$ is shaded in red.

(a)

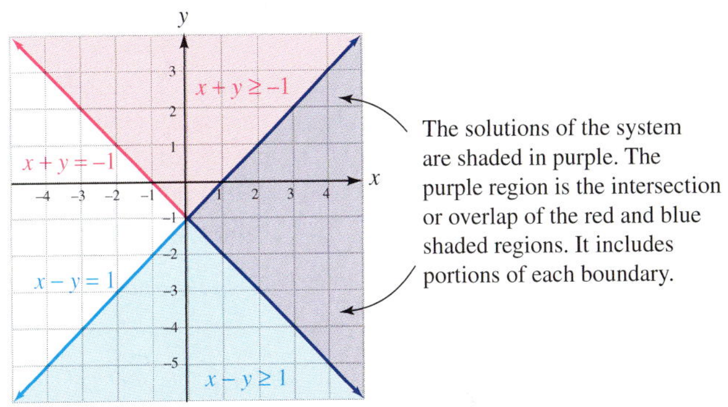

The graph of $x - y \geq 1$ is shaded in blue. It is drawn over the graph of $x + y \geq -1$.

(b)

> ***Success Tip***
> Colored pencils are often used to graph systems of inequalities. A standard pencil can also be used. Just draw different patterns of lines instead of shading.

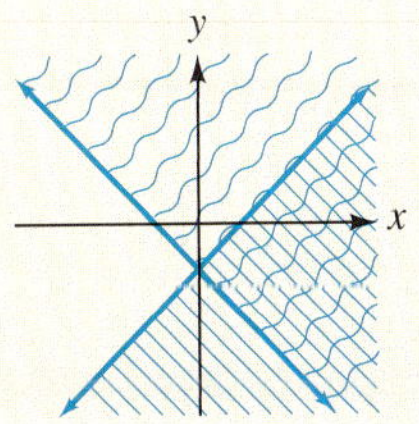

In part (b) of the figure, the area that is shaded twice represents the solutions of the given system. Any point in the doubly shaded region in purple (including the purple portions of each boundary) has coordinates that satisfy both inequalities.

Since there are infinitely many solutions, we cannot check each of them. However, as an informal check, we can select one ordered pair, say (4, 1), that lies in the doubly shaded region and show that its coordinates satisfy both inequalities of the system.

Check: $x + y \geq -1$ The first inequality. $x - y \geq 1$ The second inequality.

$4 + 1 \overset{?}{\geq} -1$ $4 - 1 \overset{?}{\geq} 1$

$5 \geq -1$ True $3 \geq 1$ True

The resulting true statements verify that (4, 1) is a solution of the system. If we pick a point that is not in the doubly shaded region, such as (1, 3), (−2, −2), or (0, −4), the coordinates of that point will fail to satisfy one or both of the inequalities.

Self Check 1 Graph the solutions of the system: $\begin{cases} x - y \leq 2 \\ x + y \geq -1 \end{cases}$

Now Try **Problem 15**

In general, to solve systems of linear inequalities, we will follow these steps.

Solving Systems of Linear Inequalities

1. Graph each inequality on the same rectangular coordinate system.
2. Use shading to highlight the intersection of the graphs (the region where the graphs overlap). The points in this region are the solutions of the system.
3. As an informal check, pick a point from the region where the graphs intersect and verify that its coordinates satisfy each inequality of the original system.

EXAMPLE 2 Graph the solutions of the system: $\begin{cases} y > 3x \\ 2x + y < 4 \end{cases}$

Strategy We will graph the solutions of $y > 3x$ in one color and the solutions of $2x + y < 4$ in another color on the same coordinate system to see where the graphs intersect.

Why The solution set of the system is the set of all points in the intersection of the two graphs.

Solution To graph $y > 3x$, we begin by graphing the boundary line $y = 3x$. Since the inequality contains an $>$ symbol, the boundary is a dashed line. Because the boundary passes through (0, 0), we use (2, 0) as the test point instead. Since (2, 0) does not satisfy $y > 3x$, we shade (in red) the half-plane that does not contain (2, 0). See part (a) of the following figure.

Graph the boundary: Slope and y-intercept

$$y = 3x + 0$$

$m = 3$ $b = 0$

Slope: $\frac{3}{1}$ y-intercept: (0, 0)

***Shading: Check the test point* (2, 0)**

$$y > 3x$$

$0 \overset{?}{>} 3(2)$ Substitute.

$0 > 6$ False

Since $0 > 6$ is false, (2, 0) is not a solution of $y > 3x$.

In part (b) of the figure, we superimpose the graph of $2x + y < 4$ on the graph of $y > 3x$ to determine the points that the graphs have in common. To graph $2x + y < 4$, we graph the boundary $2x + y = 4$ as a dashed line. Then we shade (in blue) the half-plane that contains (0, 0), because the coordinates of the test point satisfy $2x + y < 4$.

Success Tip

The ordered pairs that lie on a dashed boundary line are never part of the solution of a system of linear inequalities.

Graph the boundary: The intercept method

$$2x + y = 4$$

x	y	(x, y)
0	4	(0, 4)
2	0	(2, 0)

***Shading: Use the test point* (0, 0)**

$$2x + y < 4$$

$2(0) + 0 \overset{?}{<} 4$ Substitute.

$0 < 4$ True

Since $0 < 4$ is true, (0, 0) is a solution of $2x + y < 4$.

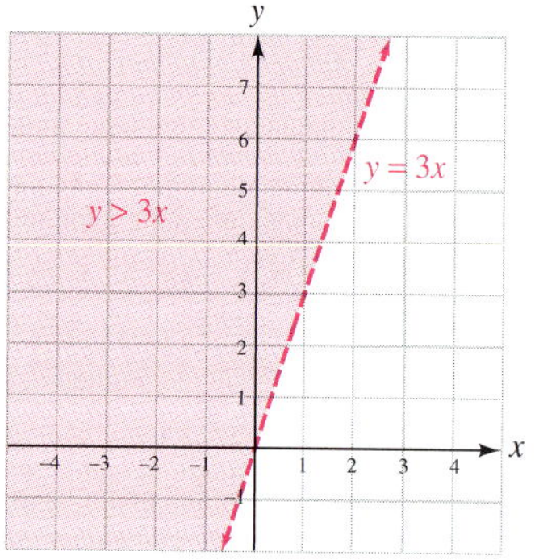

The graph of $y > 3x$ is shaded in red.

(a)

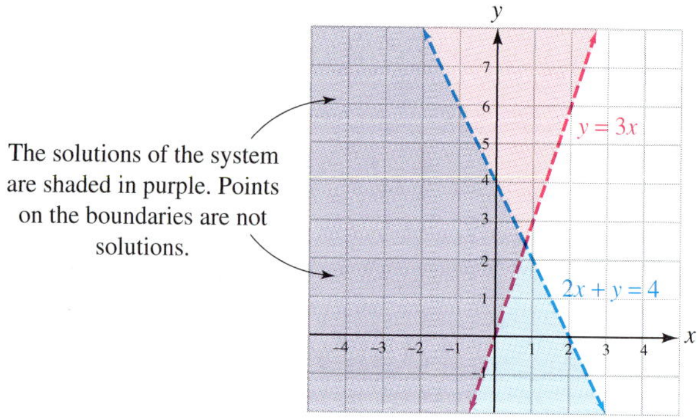

The graph of $2x + y < 4$ is shaded in blue. It is drawn over the graph of $y > 3x$.

(b)

In part (b) of the figure, the area that is shaded twice represents the solutions of the given system. Any point in the doubly shaded region in purple has coordinates that satisfy both inequalities. Pick a point in the region and show that this is true. Note that the region does not include either boundary; points on the boundaries are not solutions of the system.

Self Check 2 Graph the solutions of the system: $\begin{cases} x + 3y < 3 \\ y > \frac{1}{3}x \end{cases}$

Now Try **Problem 23**

EXAMPLE 3

Graph the solutions of the system: $\begin{cases} x \leq 2 \\ y > 3 \end{cases}$

Strategy We will graph the solutions of $x \leq 2$ in one color and the solutions of $y > 3$ in another color on the same coordinate system to see where the graphs of the two inequalities intersect.

Why The solution set of the system is the set of all points in the intersection of the two graphs.

Solution The boundary of the graph of $x \leq 2$ is the line $x = 2$. Since the inequality contains the symbol $\leq$, we draw the boundary as a solid line. The test point (0, 0) makes $x \leq 2$ true, so we shade the side of the boundary that contains (0, 0). See part (a) of the figure on the next page.

Graph the boundary: A table of solutions

$x = 2$

x	y	(x, y)
2	0	(2, 0)
2	2	(2, 2)
2	4	(2, 4)

***Shading: Check the test point* (0, 0)**

$x \leq 2$

$0 \leq 2$ True

Since $0 \leq 2$ is true, (0, 0) is a solution of $x \leq 2$.

In part (b) of the figure, the graph of $y > 3$ is superimposed over the graph of $x \leq 2$. The boundary of the graph of $y > 3$ is the line $y = 3$. Since the inequality contains the symbol $>$, we draw the boundary as a dashed line. The test point (0, 0) makes $y > 3$ false, so we shade the side of the boundary that does not contain (0, 0).

Graph the boundary: A table of solutions

$y = 3$

x	y	(x, y)
0	3	(0, 3)
1	3	(1, 3)
4	3	(4, 3)

***Shading: Check the test point* (0, 0)**

$y > 3$

$0 > 3$ False

Since $0 > 3$ is false, (0, 0) is not a solution of $y > 3$.

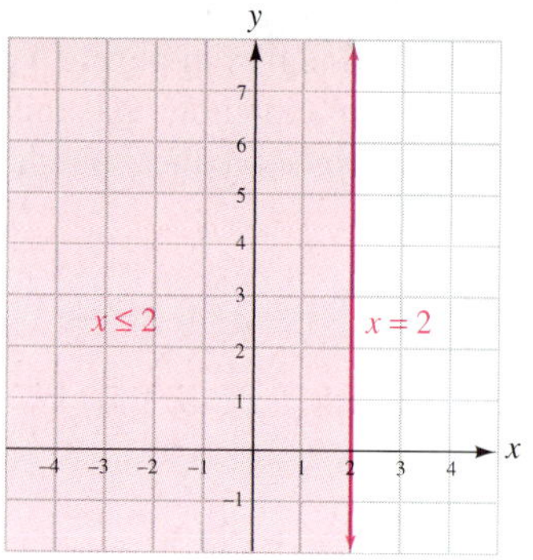

The graph of $x \leq 2$ is shaded in red.

(a)

The solutions of the system are shaded in purple. Points on the purple portion of $x = 2$ are solutions. Points on the dashed boundary line are not.

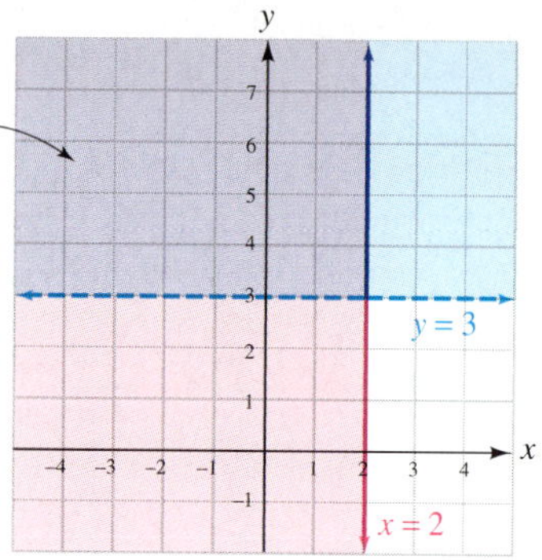

The graph of $y > 3$ is shaded in blue. It is drawn over the graph of $x \leq 2$.

(b)

The area that is shaded twice represents the solutions of the system of inequalities. Any point in the doubly shaded region in purple has coordinates that satisfy both inequalities, including the purple portion of the $x = 2$ boundary. Pick a point in the region and show that this is true.

Self Check 3 Graph the solutions of the system: $\begin{cases} y \leq 1 \\ x > 2 \end{cases}$

Now Try **Problem 25**

EXAMPLE 4

Graph the solutions of the system: $\begin{cases} x \geq 0 \\ y \geq 0 \\ x + 2y \leq 6 \end{cases}$

Strategy We will graph the solutions of $x \geq 0$, $y \geq 0$, and $x + 2y \leq 6$ on the same coordinate system to see where all three graphs intersect (overlap).

Why The solution set of the system is the set of all points in the intersection of the three graphs.

Solution This is a system of three linear inequalities. If shading is used to graph them on the same set of axes, it can become difficult to interpret the results. Instead, we can draw directional arrows attached to each boundary line in place of the shading.

- The graph of $x \geq 0$ has the boundary $x = 0$ and includes all points on the y-axis and to the right.
- The graph of $y \geq 0$ has the boundary $y = 0$ and includes all points on the x-axis and above.
- The graph of $x + 2y \leq 6$ has the boundary $x + 2y = 6$. Because the coordinates of the origin satisfy $x + 2y \leq 6$, the graph includes all points on and below the boundary.

The solutions of the system are the points that lie on triangle OPQ and the shaded triangular region that it encloses.

$x + 2y = 6$

x	y	(x, y)
0	3	(0, 3)
6	0	(6, 0)

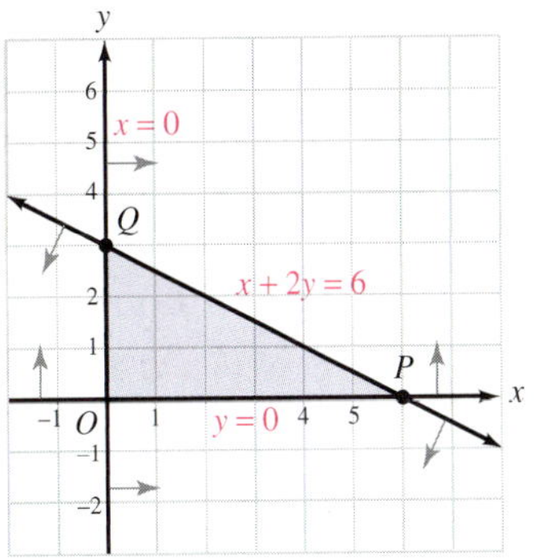

Self Check 4 Graph the solutions of the system: $\begin{cases} x \leq 1 \\ y \leq 2 \\ 2x - y \leq 4 \end{cases}$

Now Try **Problem 41**

2 Solve Application Problems Involving Systems of Linear Inequalities.

EXAMPLE 5 ***Landscaping.*** A homeowner budgets from \$300 to \$600 for trees and bushes to landscape his yard. After shopping around, he finds that good trees cost \$150 each and mature bushes cost \$75 each. What combinations of trees and bushes can he buy?

Analyze the Problem

- At least \$300 but not more than \$600 is to be spent for trees and bushes.
- Trees cost \$150 each and bushes cost \$75 each.
- What combination of trees and bushes can he buy?

Form Two Inequalities

Let x = the number of trees purchased and y = the number of bushes purchased. We then form the following system of inequalities:

The cost of a tree	times	the number of trees purchased	plus	the cost of a bush	times	the number of bushes purchased	should at least be	\$300.
\$150	$\cdot$	x	$+$	\$75	$\cdot$	y	$\geq$	\$300

The cost of a tree	times	the number of trees purchased	plus	the cost of a bush	times	the number of bushes purchased	should not be more than	\$600.
\$150	$\cdot$	x	$+$	\$75	$\cdot$	y	$\leq$	\$600

Solve the System

To solve the following system of linear inequalities

$$\begin{cases} 150x + 75y \geq 300 \\ 150x + 75y \leq 600 \end{cases}$$

we use the graphing methods discussed in this section. Neither a negative number of trees nor a negative number of bushes can be purchased, so we restrict the graph to Quadrant I.

The Language of Algebra
Phrases such as *must be at least* and *cannot go below* can be represented by the inequality symbol $\geq$. Phrases such as *is not more than* and *should not surpass* can be represented by the inequality symbol $\leq$.

State the Conclusion The coordinates of each point highlighted in the graph give a possible combination of the number of trees, x, and the number of bushes, y, that can be purchased. Written as ordered pairs, these possibilities are

(0, 4), (0, 5), (0, 6), (0, 7), (0, 8),
(1, 2), (1, 3), (1, 4), (1, 5), (1, 6),
(2, 0), (2, 1), (2, 2), (2, 3), (2, 4),
(3, 0), (3, 1), (3, 2), (4, 0)

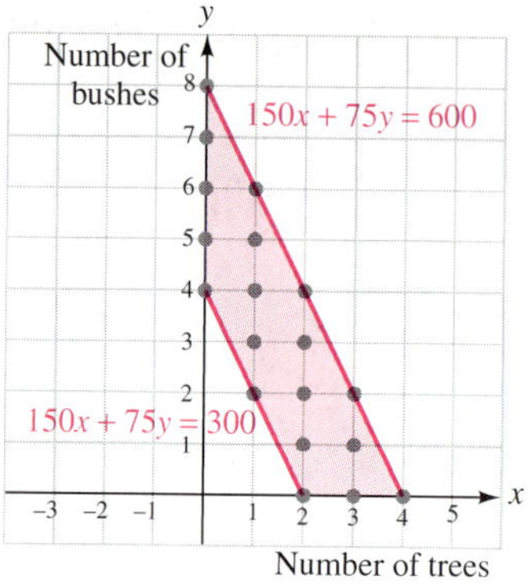

Check the Result Suppose the homeowner picks the combination of 3 trees and 2 bushes, as represented by (3, 2). Show that this point satisfies both inequalities of the system.

Now Try **Problem 47**

ANSWERS TO SELF CHECKS

1. 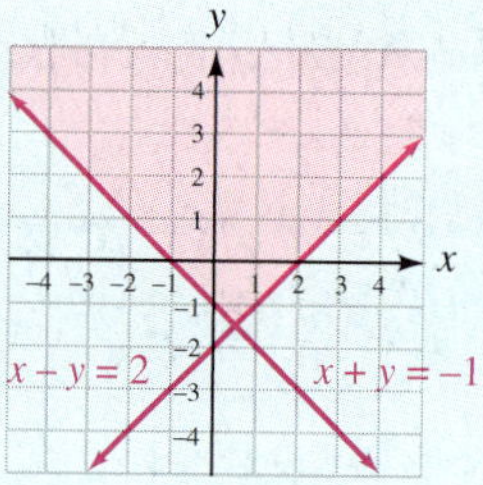

2. 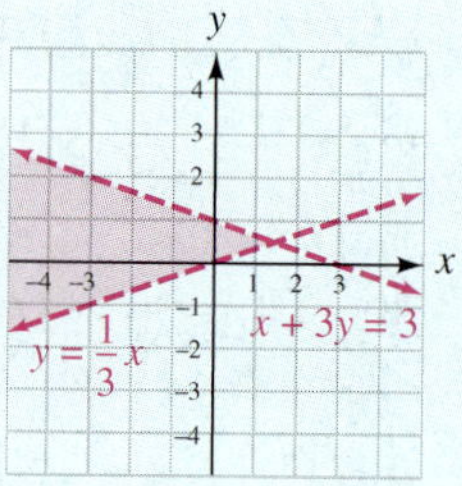

3. 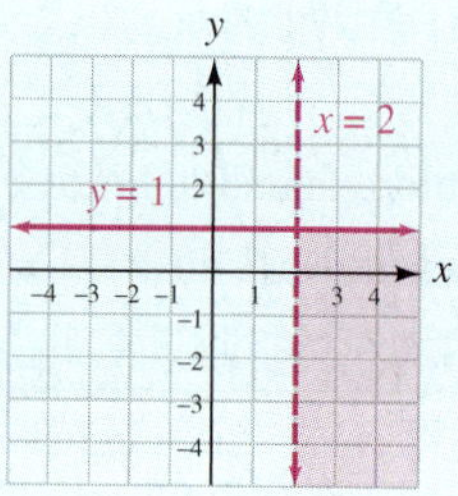

4. 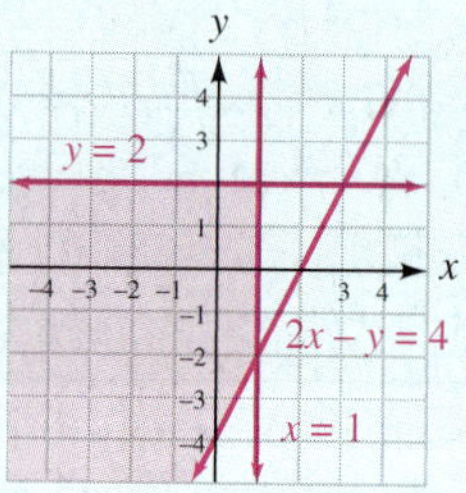

STUDY SET 4.5

VOCABULARY

Fill in the blanks.

1. $\begin{cases} x + y > 2 \\ x + y < 4 \end{cases}$ is a system of linear ___________.

2. To graph the linear inequality $x + y > 2$, first graph the __________ $x + y = 2$. Then pick the test ______ (0, 0) to determine which half-plane to shade.

3. To find the solutions of a system of two linear inequalities graphically, look for the ___________, or overlap, of the two shaded regions.

4. The phrase *should not surpass* can be represented by the inequality symbol ___ and the phrase *must be at least* can be represented by the inequality symbol ___.

CONCEPTS

5. **a.** What is the equation of the boundary line of the graph of $3x - y < 5$?

   b. Is the boundary a solid or dashed line?

6. a. What is the equation of the boundary line of the graph of $y \geq 4x$?
   b. Is the boundary a solid or dashed line?
   c. Why can't (0, 0) be used as a test point to determine what to shade?

7. Find the slope and the y-intercept of the line whose equation is $y = 4x - 3$.

8. Complete the table to find the x- and y-intercepts of the line whose equation is $3x - 2y = 6$.

x	y
0	
	0

9. The boundary of the graph of $2x + y > 4$ is shown.
   a. Does the point (0, 0) make the inequality true?
   b. Should the region above or below the boundary be shaded?

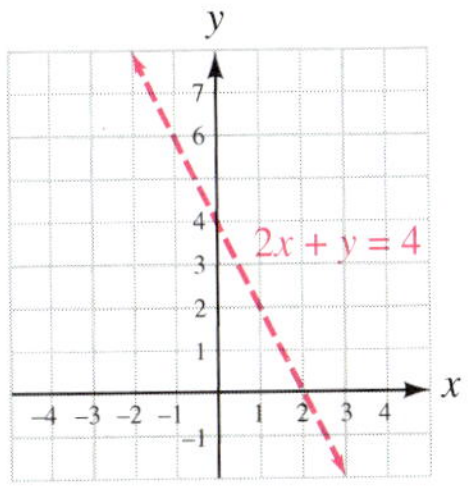

10. Linear inequality 1 is graphed in red and linear inequality 2 is graphed in blue. Determine whether a true or false statement results when
   a. The coordinates of point A are substituted into inequality 1
   b. The coordinates of point A are substituted into inequality 2
   c. The coordinates of point B are substituted into inequality 1
   d. The coordinates of point B are substituted into inequality 2
   e. The coordinates of point C are substituted into inequality 1
   f. The coordinates of point C are substituted into inequality 2

Inequality 1 solutions are in red

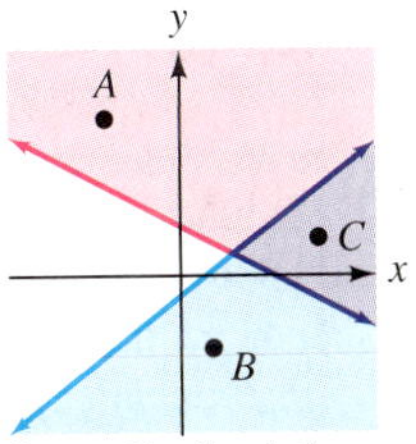

Inequality 2 solutions are in blue

11. The graph of a system of two linear inequalities is shown. Determine whether each point is a solution of the system.
   a. (4, −2)
   b. (1, 3)
   c. the origin

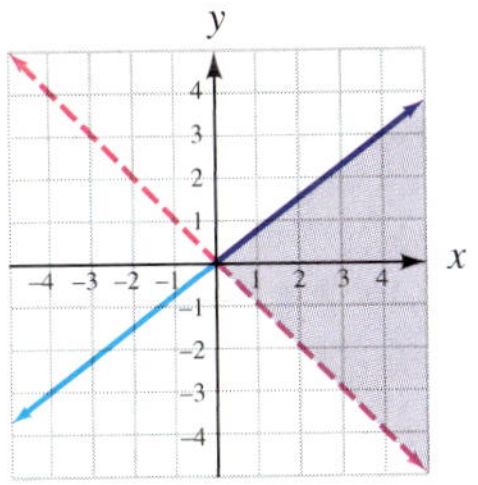

12. Use a check to determine whether each ordered pair is a solution of the system.

$$\begin{cases} x + 2y \geq -1 \\ x - y < 2 \end{cases}$$

a. (1, 4) b. (−2, 0)

13. Match each equation, inequality, or system with the graph of its solution.

a. $x + y = 2$ b. $x + y \geq 2$

c. $\begin{cases} x + y = 2 \\ x - y = 2 \end{cases}$ d. $\begin{cases} x + y \geq 2 \\ x - y \leq 2 \end{cases}$

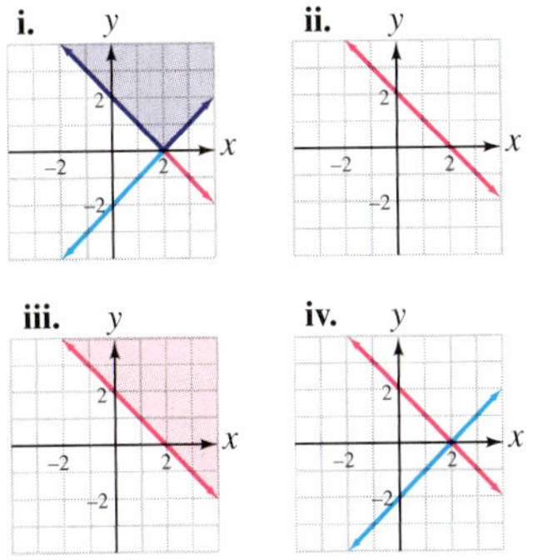

14. Match the system of inequalities with the correct graph.

a. $\begin{cases} x \geq 2 \\ y < 1 \end{cases}$ b. $\begin{cases} x > 2 \\ y \leq 1 \end{cases}$

c. $\begin{cases} x \geq 2 \\ y \geq 1 \end{cases}$ d. $\begin{cases} x > 2 \\ y > -1 \end{cases}$

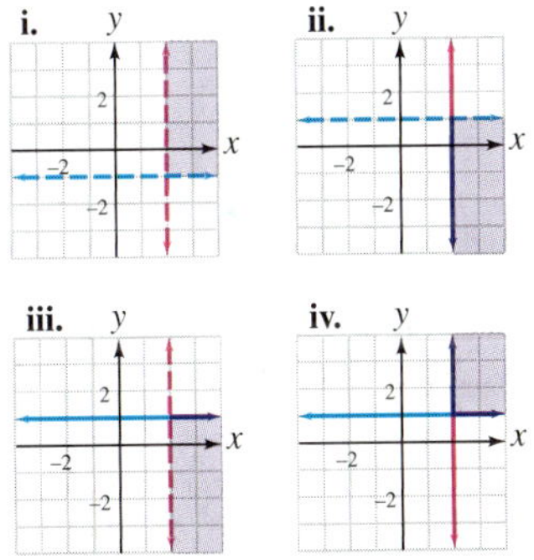

GUIDED PRACTICE

Graph the solutions of each system. **See Examples 1–3.**

15. $\begin{cases} x + 2y \leq 3 \\ 2x - y \geq 1 \end{cases}$

16. $\begin{cases} 2x + y \geq 3 \\ x - 2y \leq -1 \end{cases}$

17. $\begin{cases} x + y < -1 \\ x - y > -1 \end{cases}$

18. $\begin{cases} x + y > 2 \\ x - y < -2 \end{cases}$

19. $\begin{cases} 2x - 3y \leq 0 \\ y \geq x - 1 \end{cases}$

20. $\begin{cases} y > 2x - 4 \\ y \geq -x - 1 \end{cases}$

21. $\begin{cases} x + y < 2 \\ x + y \leq 1 \end{cases}$

22. $\begin{cases} y > -x + 2 \\ y < -x + 4 \end{cases}$

23. $\begin{cases} y > 2x \\ x + 2y < 6 \end{cases}$

24. $\begin{cases} y \leq 2x \\ x + y < 4 \end{cases}$

25. $\begin{cases} x \geq 2 \\ y \leq 3 \end{cases}$

26. $\begin{cases} x \geq -1 \\ y > -2 \end{cases}$

27. $\begin{cases} x > 0 \\ y > 0 \end{cases}$

28. $\begin{cases} x \leq 0 \\ y < 0 \end{cases}$

29. $\begin{cases} 3x + 4y \geq -7 \\ 2x - 3y \geq 1 \end{cases}$

30. $\begin{cases} 3x + y \leq 1 \\ 4x - y \geq -8 \end{cases}$

31. $\begin{cases} 2x + y < 7 \\ y > 2 - 2x \end{cases}$

32. $\begin{cases} 2x + y \geq 6 \\ y \leq 4x - 6 \end{cases}$

33. $\begin{cases} 2x - 4y > -6 \\ 3x + y \geq 5 \end{cases}$

34. $\begin{cases} 2x - 3y < 0 \\ 2x + 3y \geq 12 \end{cases}$

35. $\begin{cases} 3x - y + 4 \leq 0 \\ 3y > -2x - 10 \end{cases}$

36. $\begin{cases} 3x + 2y - 12 \geq 0 \\ x < -2 + y \end{cases}$

37. $\begin{cases} y \geq x \\ y \leq \frac{1}{3}x + 1 \end{cases}$

38. $\begin{cases} y > 3x \\ y \leq -x - 1 \end{cases}$

39. $\begin{cases} x + y > 0 \\ y - x < -2 \end{cases}$

40. $\begin{cases} y + 2x \leq 0 \\ y \leq \frac{1}{2}x + 2 \end{cases}$

***Graph the solutions of each system.* See Example 4.**

41. $\begin{cases} x \geq 0 \\ y \geq 0 \\ x + y \leq 3 \end{cases}$

42. $\begin{cases} x - y \leq 6 \\ x + 2y \leq 6 \\ x \geq 0 \end{cases}$

43. $\begin{cases} x - y < 4 \\ y \leq 0 \\ x \geq 0 \end{cases}$

44. $\begin{cases} 2x + y \leq 2 \\ y > x \\ x \geq 0 \end{cases}$

APPLICATIONS

45. BIRDS OF PREY Parts (a) and (b) of the illustration show the individual fields of vision for each eye of an owl. In part (c), shade the area where the fields of vision overlap—that is, the area that is seen by both eyes.

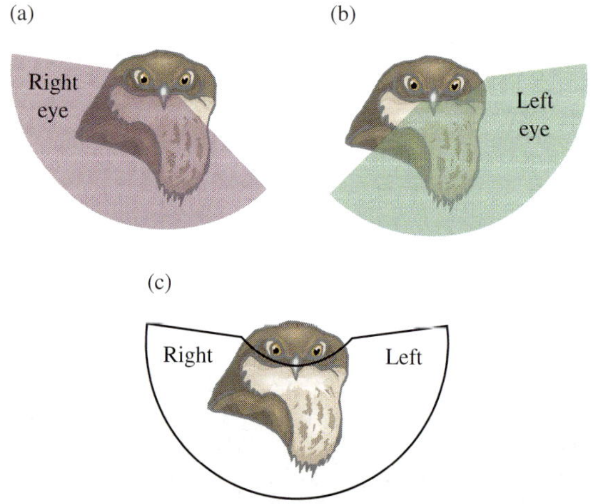

46. EARTH SCIENCE Shade the area of the earth's surface that is north of the Tropic of Capricorn and south of the Tropic of Cancer.

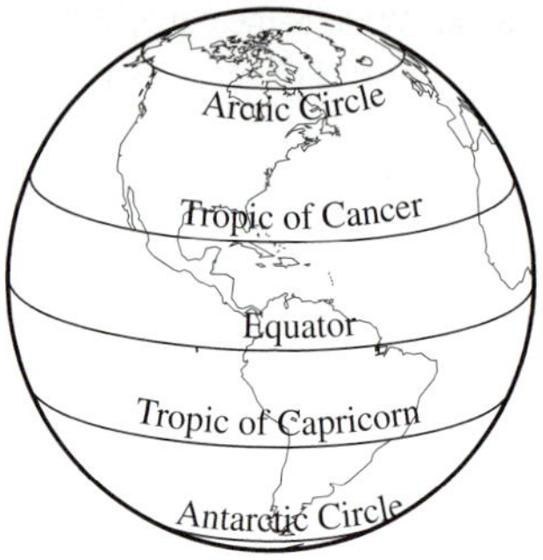

***In Problems 47–52, graph each system of inequalities and give two possible solutions.* See Example 5.**

47. BUYING COMPACT DISCS Melodic Music has compact discs on sale for either \$10 or \$15. If a customer wants to spend at least \$30 but no more than \$60 on CDs, graph a system of inequalities showing the possible combinations of \$10 CDs (x) and \$15 CDs (y) that the customer can buy.

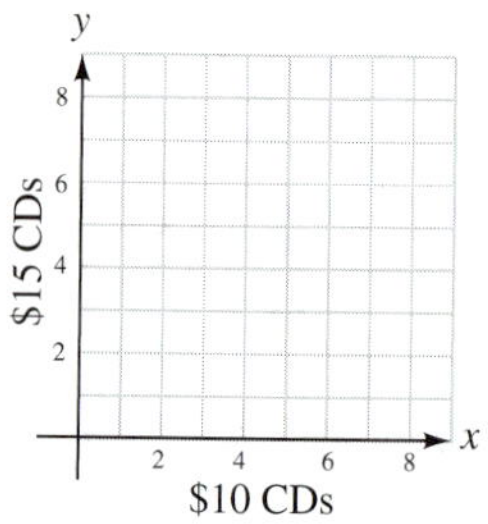

48. BUYING BOATS Boatworks wholesales aluminum boats for \$800 and fiberglass boats for \$600. Northland Marina wants to make a purchase totaling at least \$2,400 but no more than \$4,800. Graph a system of inequalities showing the possible combinations of aluminum boats (x) and fiberglass boats (y) that can be ordered.

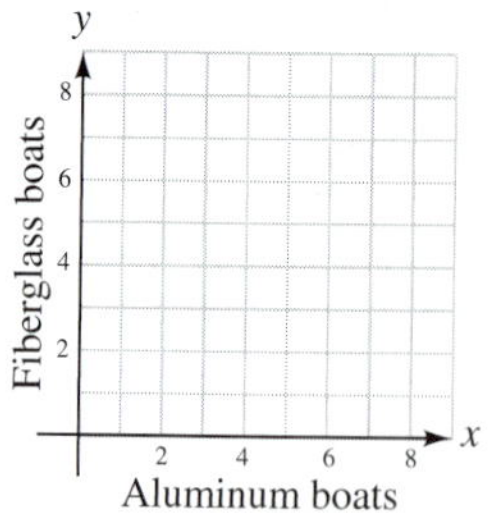

49. BUYING FURNITURE A distributor wholesales desk chairs for \$150 and side chairs for \$100. Best Furniture wants its order to total no more than \$900; Best also wants to order more side chairs than desk chairs. Graph a system of inequalities showing the possible combinations of desk chairs (x) and side chairs (y) that can be ordered.

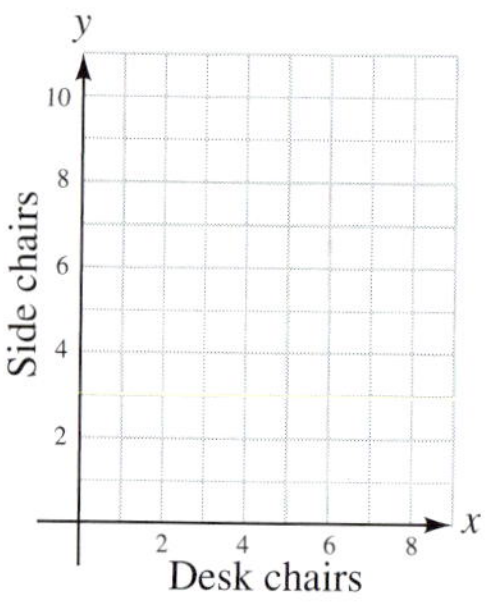

50. ORDERING FURNACE EQUIPMENT J. Bolden Heating Company wants to order no more than \$2,000 worth of electronic air cleaners and humidifiers from a wholesaler that charges \$500 for air cleaners and \$200 for humidifiers. If Bolden wants more humidifiers than air cleaners, graph a system of inequalities showing the possible combinations of air cleaners (x) and humidifiers (y) that can be ordered.

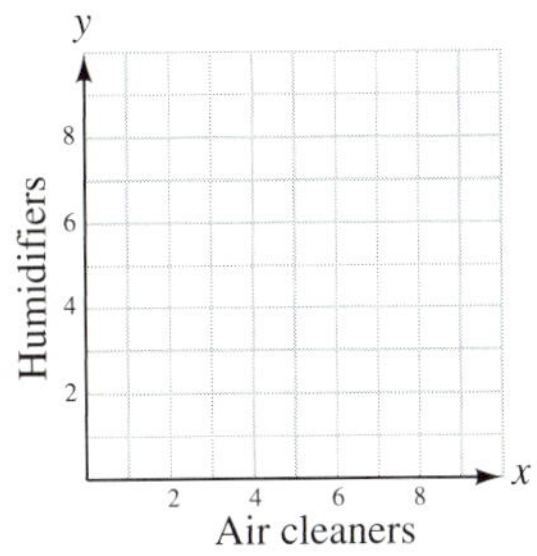

51. PESTICIDES To eradicate a fruit fly infestation, helicopters sprayed an area of a city that can be described by $y \geq -2x + 1$ (within the city limits). Two weeks later, more spraying was ordered over the area described by $y \geq \frac{1}{4}x - 4$ (within the city limits). Show the part of the city that was sprayed twice.

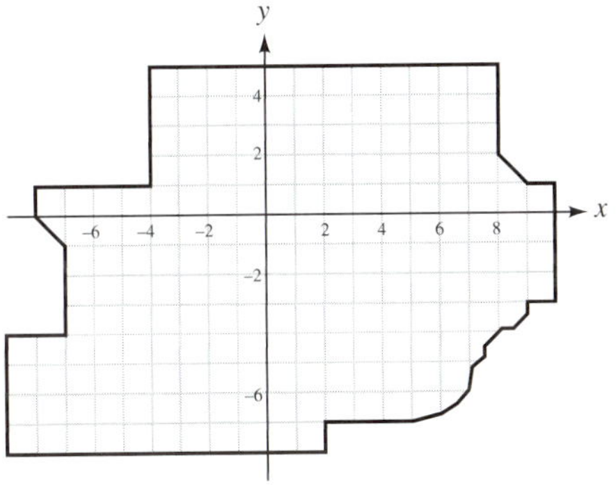

52. REDEVELOPMENT Refer to the following diagram. A government agency has declared an area of a city east of First Street, north of Second Avenue, south of Sixth Avenue, and west of Fifth Street as eligible for federal redevelopment funds. Describe this area of the city mathematically using a system of four inequalities, if the corner of Central Avenue and Main Street is considered the origin.

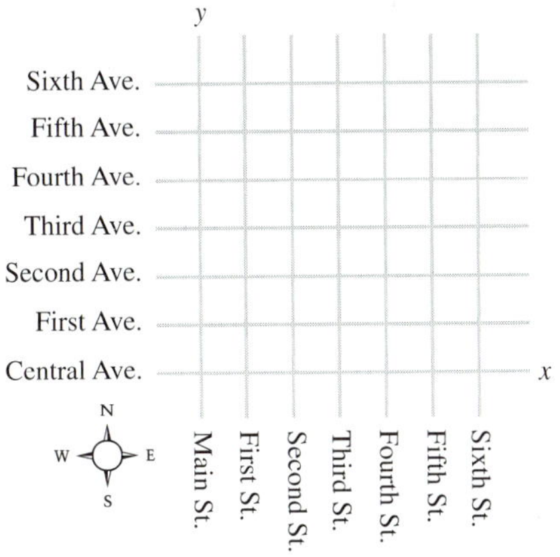

WRITING

53. Explain how to use graphing to solve a system of inequalities.

54. When a solution of a system of linear inequalities is graphed, what does the shading represent?

55. Describe how the graphs of the solutions of these systems are similar and how they differ.

$$\begin{cases} x + y = 4 \\ x - y = 4 \end{cases} \quad \text{and} \quad \begin{cases} x + y \geq 4 \\ x - y \geq 4 \end{cases}$$

56. Explain when a system of inequalities will have no solutions.

REVIEW

Simplify each expression.

57. $8\left(\frac{3}{4}t\right)$

58. $27\left(\frac{2}{3}m\right)$

59. $-\frac{2}{3}(3w - 6)$

60. $\frac{1}{2}(2y - 8)$

61. $-\frac{7}{16}x - \frac{3}{16}x$

62. $-\frac{5}{18}x - \frac{7}{18}x$

63. $60\left(\frac{3}{20}r - \frac{4}{15}\right)$

64. $72\left(\frac{7}{8}f - \frac{8}{9}\right)$

CHALLENGE PROBLEMS

Graph the solutions of each system.

65. $\begin{cases} \dfrac{x}{3} - \dfrac{y}{2} < -3 \\ \dfrac{x}{3} + \dfrac{y}{2} > -1 \end{cases}$

66. $\begin{cases} 3x + y < -2 \\ y > 3(1 - x) \end{cases}$

67. $\begin{cases} 2x + 3y \leq 6 \\ 3x + y \leq 1 \\ x \leq 0 \end{cases}$

68. $\begin{cases} x \geq 0 \\ y \geq 0 \\ 9x + 3y \leq 18 \\ 3x + 6y \leq 18 \end{cases}$

CHAPTER 4
Summary & Review

SECTION 4.1 Solving Systems of Equations by Graphing

DEFINITIONS AND CONCEPTS	EXAMPLES
When two equations are considered at the same time, we say that they form a **system of equations.** A **solution of a system** of equations in two variables is an ordered pair that satisfies both equations of the system.	Is (4, 3) a solution of the system $\begin{cases} x + y = 7 \\ x - y = 5 \end{cases}$? To answer this question, we substitute 4 for x and 3 for y in each equation. $x + y = 7$ $x - y = 5$ $4 + 3 \stackrel{?}{=} 7$ $4 - 3 \stackrel{?}{=} 5$ $7 = 7$ True $1 = 5$ False Although (4, 3) satisfies the first equation, it does not satisfy the second. Because it does not satisfy both equations, it is not a solution of the system.
To **solve a system graphically:** 1. Graph each equation on the same coordinate system. 2. Determine the coordinates of the point of intersection of the graphs. That ordered pair is the solution. 3. Check the solution in each equation of the original system.	Use graphing to solve the system: $\begin{cases} y = -2x + 3 \\ x - 2y = 4 \end{cases}$ **Step 1:** Graph each equation as shown below. $y = -2x + 3$: $m = \frac{-2}{1}$, $b = 3$ $x - 2y = 4$: x = 0, 4; y = −2, 0 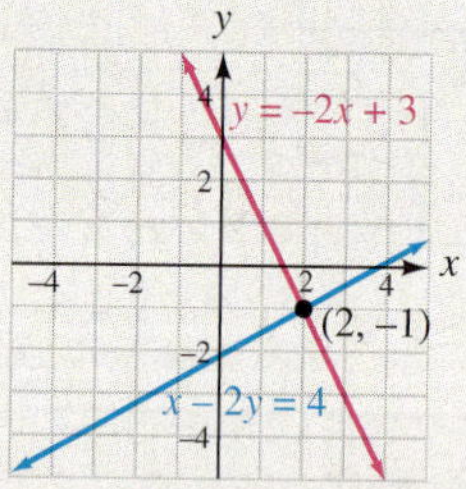 **Step 2:** It appears that the graphs intersect at the point (2, −1). To verify that this is the solution of the system, substitute 2 for x and −1 for y in each equation. **Step 3:** Check $x - 2y = 4$ $y = -2x + 3$ $2 - 2(-1) \stackrel{?}{=} 4$ $-1 \stackrel{?}{=} -2(2) + 3$ $4 = 4$ True $-1 = -1$ True Since (2, −1) makes both equations true, it is the solution of the system.

SECTION 4.1 Solving Systems of Equations by Graphing—*continued*

DEFINITIONS AND CONCEPTS

A system of equations that has at least one solution is called a **consistent system.** If the graphs of the equations of the system are parallel lines, the system has no solution and is called an **inconsistent system.**

Equations with different graphs are called **independent equations.** If the graphs of the equations in a system are the same line, the system has infinitely many solutions. The equations are called **dependent equations.**

We can determine the **number of solutions** that a system of two linear equations has by writing each equation in slope-intercept form, $y = mx + b$, and comparing the slopes and y-intercepts.

EXAMPLES

There are three possible outcomes when solving a system of two linear equations by graphing.

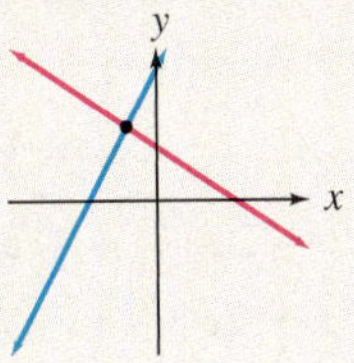

Consistent system
Independent equations
- Exactly one solution
- The lines have different slopes.

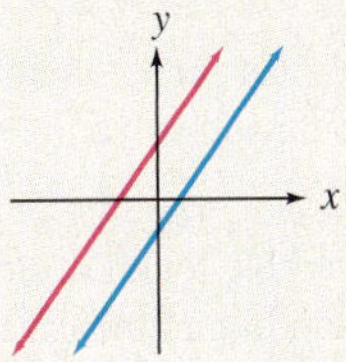

Inconsistent system
Independent equations
- No solution
- The lines have the same slope but different y-intercepts.

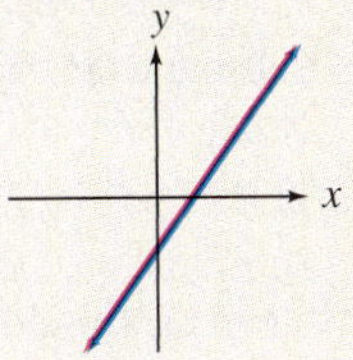

Consistent system
Dependent equations
- Infinitely many solutions
- The lines have the same slope and same y-intercept.

REVIEW EXERCISES

Determine whether the ordered pair is a solution of the system.

1. $(2, -3)$, $\begin{cases} 3x - 2y = 12 \\ 2x + 3y = -5 \end{cases}$

2. $\left(\frac{7}{2}, -\frac{2}{3}\right)$, $\begin{cases} 3y = 2x - 9 \\ 2x + 3y = 6 \end{cases}$

Use the graphing method to solve each system.

3. $\begin{cases} x + y = 7 \\ 2x - y = 5 \end{cases}$

4. $\begin{cases} 2x + y = 5 \\ y = -\frac{x}{3} \end{cases}$

5. $\begin{cases} 3x + 6y = 6 \\ x + 2y - 2 = 0 \end{cases}$

6. $\begin{cases} 6x + 3y = 12 \\ y = -2x + 2 \end{cases}$

7. Find the slope and the y-intercept of the graph of each line in the system $\begin{cases} y = -2x + 1 \\ 8x + 4y = 3 \end{cases}$. Then, use that information to determine the number of solutions of the system.

8. COLLEGE ENROLLMENT Estimate the point of intersection of the graphs. Explain its significance.

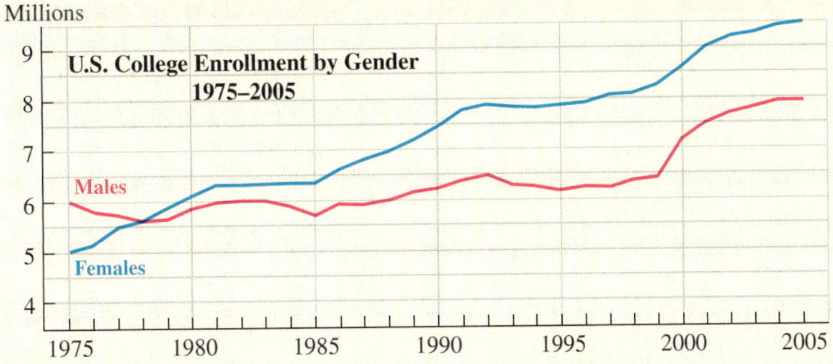

Source: *Digest of Education Statistics*

SECTION 4.2 Solving Systems of Equations by Substitution

DEFINITIONS AND CONCEPTS

To solve a system of equations in x and y by **substitution:**

1. Solve one of the equations for either x or y.
2. Substitute the expression for x (or for y) obtained in step 1 into the other equation and solve the equation.
3. Substitute the value of the variable found in step 2 into the substitution equation to find the value of the remaining variable.
4. Check the proposed solution in the equations of the original system.

With the substitution method, the objective is to use an appropriate substitution to obtain *one equation in one variable.*

If in step 2 the variable drops out and a false statement results, the system has **no solution.** If a true statement results, the system has **infinitely many solutions.**

EXAMPLES

Use substitution to solve the system: $\begin{cases} y = x + 7 \\ x + 2y = 5 \end{cases}$

Step 1: The first equation is already solved for y.

Step 2: Substitute $x + 7$ for y in the second equation.

$x + 2y = 5$

$x + 2(x + 7) = 5$ Substitute $x + 7$ for y.

$x + 2x + 14 = 5$ Distribute.

$3x + 14 = 5$ Combine like terms.

$3x = -9$ Subtract 14 from both sides.

$x = -3$ Divide both sides by 3.

Step 3: $y = x + 7$ This is the substitution equation.

$y = -3 + 7$ Substitute -3 for x.

$y = 4$ Simplify.

The following check verifies that the solution is $(-3, 4)$.

Step 4: Check

$y = x + 7$	$x + 2y = 5$
$4 \stackrel{?}{=} -3 + 7$	$-3 + 2(4) \stackrel{?}{=} 5$
$4 = 4$ True	$5 = 5$ True

REVIEW EXERCISES

Use the substitution method to solve each system.

9. $\begin{cases} y = 15 - 3x \\ 7y + 3x = 15 \end{cases}$

10. $\begin{cases} x = y \\ 5x - 4y = 3 \end{cases}$

11. $\begin{cases} 6x + 2y = 8 - y + x \\ 3x = 2 - y \end{cases}$

12. $\begin{cases} r = 3s + 7 \\ r = 2s + 5 \end{cases}$

13. $\begin{cases} 9x + 3y - 5 = 0 \\ 3x + y = \dfrac{5}{3} \end{cases}$

14. $\begin{cases} \dfrac{x}{2} + \dfrac{y}{2} = 11 \\ \dfrac{5x}{16} - \dfrac{3y}{16} = \dfrac{15}{8} \end{cases}$

15. When solving a system using the substitution method, suppose you obtain the result $8 = 9$.

   a. How many solutions does the system have?

   b. Describe the graph of the system.

   c. What term is used to describe the system?

16. Fill in the blank. With the substitution method, the objective is to use an appropriate substitution to obtain one equation in _____ variable.

SECTION 4.3 Solving Systems of Equations by Elimination (Addition)

DEFINITIONS AND CONCEPTS	EXAMPLES

DEFINITIONS AND CONCEPTS

To solve a system of equations in x and y using **elimination (addition):**

1. Write each equation in the standard $Ax + By = C$ form.
2. Multiply one (or both) equations by nonzero quantities to make the coefficients of x (or y) opposites.
3. Add the equations to eliminate the terms involving x (or y).
4. Solve the equation obtained in step 3.
5. Find the value of the other variable by substituting the value of the variable found in step 4 into any equation containing both variables.
6. Check the solution in the equations of the original system.

With the elimination method, the basic objective is to obtain two equations whose sum will be one equation in one variable.

If in step 3 both variables drop out and a false statement results, the system has **no solution.** If a true statement results, the system has **infinitely many solutions.**

EXAMPLES

Use elimination to solve: $\begin{cases} 2x - 3y = 4 \\ 3x + y = -5 \end{cases}$

Step 1: Both equations are written in $Ax + By = C$ form.

Step 2: Multiply the second equation by 3 so that the coefficients of y are opposites.

Step 3:

$$\begin{aligned} 2x - 3y &= 4 \\ 9x + 3y &= -15 \\ \hline 11x \quad &= -11 \end{aligned}$$

Add the like terms, column by column.

Step 4: Solve for x.

$$11x = -11$$
$$x = -1 \quad \text{Divide both sides by 11.}$$

Step 5: Find y.

$$3x + y = -5 \quad \text{This is the second equation.}$$
$$3(-1) + y = -5 \quad \text{Substitute } -1 \text{ for } x.$$
$$y = -2$$

The solution is $(-1, -2)$.

Step 6: Check

$$2x - 3y = 4 \qquad\qquad 3x + y = -5$$
$$2(-1) - 3(-2) \stackrel{?}{=} 4 \qquad\qquad 3(-1) + (-2) \stackrel{?}{=} -5$$
$$-2 + 6 \stackrel{?}{=} 4 \qquad\qquad -3 - 2 \stackrel{?}{=} -5$$
$$4 = 4 \quad \text{True} \qquad\qquad -5 = -5 \quad \text{True}$$

REVIEW EXERCISES

17. Write each equation of the system $\begin{cases} 4x + 2y - 7 = 0 \\ 3y = 5x + 6 \end{cases}$ in standard $Ax + By = C$ form.

18. Fill in the blank. With the elimination method, the basic objective is to obtain two equations whose sum will be one equation in ____ variable.

Solve each system using the elimination (addition) method.

19. $\begin{cases} 2x + y = 1 \\ 5x - y = 20 \end{cases}$

20. $\begin{cases} x + 8y = 7 \\ x - 4y = 1 \end{cases}$

21. $\begin{cases} 5a + b = 2 \\ 3a + 2b = 11 \end{cases}$

22. $\begin{cases} 11x + 3y = 27 \\ 8x + 4y = 36 \end{cases}$

23. $\begin{cases} 9x + 3y = 15 \\ 3x = 5 - y \end{cases}$

24. $\begin{cases} -\frac{a}{4} - \frac{b}{3} = \frac{1}{12} \\ \frac{a}{2} - \frac{5b}{4} = \frac{7}{4} \end{cases}$

25. $\begin{cases} 0.02x + 0.05y = 0 \\ 0.3x - 0.2y = -1.9 \end{cases}$

26. $\begin{cases} -\frac{1}{4}x = 1 - \frac{2}{3}y \\ 6x - 18y = 5 - 2y \end{cases}$

For each system, determine which method, substitution or elimination (addition), would be easier to use to solve the system and explain why.

27. $\begin{cases} 6x + 2y = 5 \\ 3x - 3y = -4 \end{cases}$

28. $\begin{cases} x = 5 - 7y \\ 3x - 3y = -4 \end{cases}$

SECTION 4.4 Problem Solving Using Systems of Equations

DEFINITIONS AND CONCEPTS

We can solve many types of problems using a system of two linear equations in two variables:

- Geometry problems
- Number-value problems
- Interest problems
- Uniform motion problems
- Liquid and dry mixture problems

To solve problems involving two unknown quantities:

1. **Analyze** the facts of the problem. Make a table or a diagram if it is helpful.
2. Pick different variables to represent the two unknown quantities. **Form two equations** involving those variables.
3. **Solve** the system of equations using graphing, substitution, or elimination.
4. **State** the conclusion.
5. **Check** the result.

Two angles are said to be **complementary** if the sum of their measures is 90°. Two angles are said to be **supplementary** if the sum of their measures is 180°.

EXAMPLES

The difference of the measures of two supplementary angles is 40°. Find the measure of each angle.

Analyze Since the angles are supplementary, the sum of their measures is 180°. If we subtract the smaller angle from the larger, the result is given to be 40°.

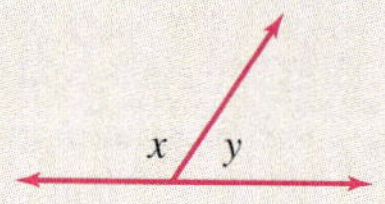

Form Let x = the measure (in degrees) of the larger angle and y = the measure (in degrees) of the smaller angle. Then we have the following two equations.

$$\begin{cases} x + y = 180 & \text{Their sum is } 180°. \\ x - y = 40 & \text{Their difference is } 40°. \end{cases}$$

Solve If we add the equations, we get

$$\begin{aligned} x + y &= 180 \\ x - y &= 40 \\ \hline 2x \quad &= 220 \\ x &= 110 \end{aligned}$$

Divide both sides by 2. This is the measure of the larger angle.

We can use the first equation of the system to find y.

$$\begin{aligned} x + y &= 180 \\ 110 + y &= 180 && \text{Substitute 110 for } x. \\ y &= 70 && \text{Subtract 110 from both sides. This is the measure of the smaller angle.} \end{aligned}$$

State The angles measure 110° and 70°.

Check Angles with measures of 110° and 70° are supplementary (their sum is 180°) and their difference is 40°. The results check.

REVIEW EXERCISES

Write a system of two equations in two variables to solve each problem.

29. ELEVATIONS The elevation of Las Vegas, Nevada, is 20 times greater than that of Baltimore, Maryland. The sum of their elevations is 2,100 feet. Find the elevation of each city.

30. PAINTING EQUIPMENT When fully extended, a ladder is 35 feet in length. If the extension is 7 feet shorter than the base, how long is each part of the ladder?

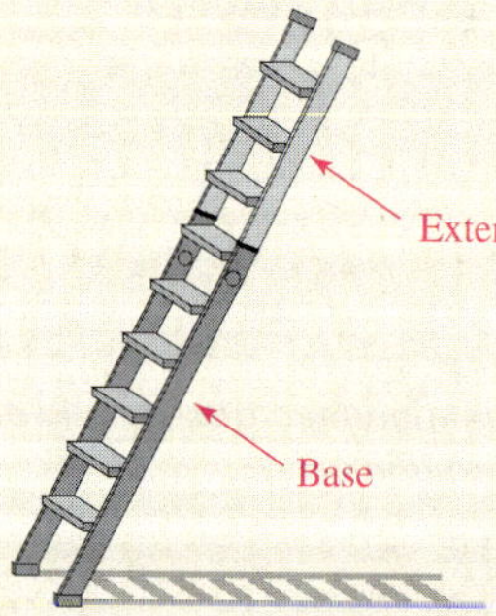

31. GEOMETRY Two angles are complementary. The measure of one is 15° more than twice the measure of the other. Find the measure of each angle.

32. CRASH INVESTIGATION In an effort to protect evidence, investigators used 420 yards of yellow "Police Line—Do Not Cross" tape to seal off a large rectangular-shaped area around an airplane crash site. How much area will the investigators have to search if the width of the rectangle is three-fourths of the length?

33. Complete each table.

a.

	Amount ·	Strength =	Amount of pesticide
Weak	x	0.02	
Strong	y	0.09	
Mixture	100	0.08	

b.

	Rate ·	Time =	Distance
With the wind	$s + w$	5	
Against the wind	$s - w$	7	

c.

	P ·	r ·	t =	I
Mack Financial	x	0.11	1	
Union Savings	y	0.06	1	

d.

	Amount	· Price	= Total value
Carmel corn	x	4	
Peanuts	y	8	
Mixture	10	5	

34. CANDY STORE A merchant wants to mix gummy worms worth \$6 per pound and gummy bears worth \$3 per pound to make 30 pounds of a mixture worth \$4.20 per pound. How many pounds of each type of candy should he use?

35. BOATING It takes a motorboat 4 hours to travel 56 miles down a river, and 3 hours longer to make the return trip. Find the speed of the current.

36. SHOPPING Packages containing two bottles of contact lens cleaner and three bottles of soaking solution cost \$63.40, and packages containing three bottles of cleaner and two bottles of soaking solution cost \$69.60. Find the cost of a bottle of cleaner and a bottle of soaking solution.

37. INVESTING Carlos invested part of \$3,000 in a 10% certificate account and the rest in a 6% passbook account. The total annual interest from both accounts is \$270. How much did he invest at 6%?

38. ANTIFREEZE How much of a 40% antifreeze solution must a mechanic mix with a 70% antifreeze solution if she needs 20 gallons of a 50% antifreeze solution?

SECTION 4.5 Solving Systems of Linear Inequalities

DEFINITIONS AND CONCEPTS

A solution of a **system of linear inequalities** is an ordered pair that satisfies each inequality.

To **solve a system of linear inequalities:**

1. Graph each inequality on the same coordinate system.
2. Use shading to highlight the intersection of the graphs. The points in this region are the solutions of the system.
3. As an informal check, pick a point from the region and verify that its coordinates satisfy each inequality of the original system.

EXAMPLES

Graph the solutions of the system: $\begin{cases} y \le x + 1 \\ y > -1 \end{cases}$

Step 1: Graph each inequality on the same coordinate system as shown.

Step 2: Use shading to highlight where the graphs intersect.

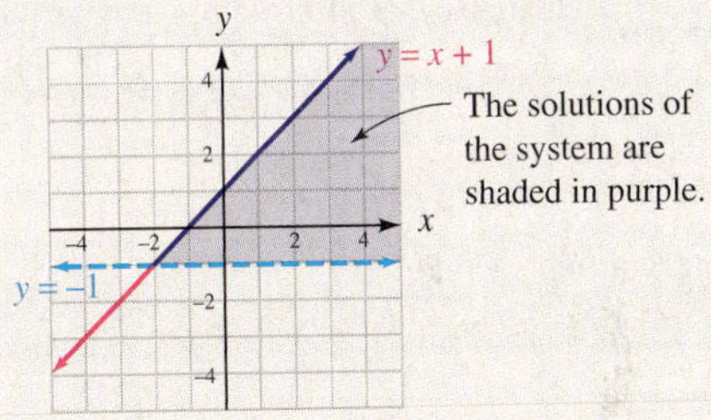

Step 3: Pick a point from the solution region such as (1, 0) and verify that it satisfies both inequalities.

REVIEW EXERCISES

Solve each system of inequalities.

39. $\begin{cases} 5x + 3y < 15 \\ 3x - y > 3 \end{cases}$

40. $\begin{cases} 3y \le x \\ y > 3x \end{cases}$

41. GIFT SHOPPING A grandmother wants to spend at least \$40 but no more than \$60 on school clothes for her grandson. If T-shirts sell for \$10 each and pants sell for \$20 each, write a system of inequalities that describes the possible numbers of T-shirts x and pairs of pants y that she can buy. Graph the system and give two possible solutions.

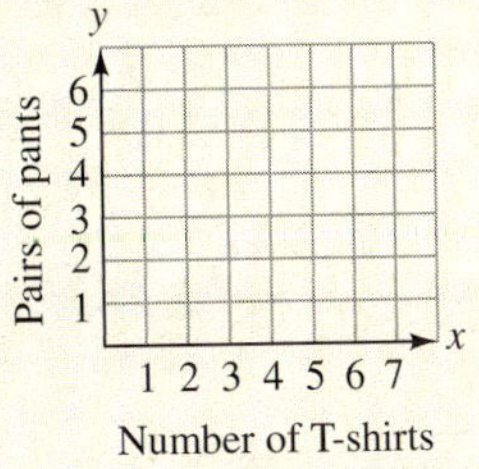

42. Use a check to determine whether each ordered pair is a solution of the system: $\begin{cases} x + 2y \le 3 \\ 2x - y > 1 \end{cases}$

a. (5, −4) **b.** (−1, −3)

CHAPTER 4 Test

Determine whether the ordered pair is a solution of the system.

1. $(5, 3)$, $\begin{cases} 3x + 2y = 21 \\ x + y = 8 \end{cases}$

2. $(-2, -1)$, $\begin{cases} 4x + y = -9 \\ 2x - 3y = -7 \end{cases}$

3. Fill in the blanks.

a. A ________ of a system of linear equations is an ordered pair that satisfies each equation.

b. A system of equations that has at least one solution is called a ________ system.

c. A system of equations that has no solution is called an ________ system.

d. Equations with different graphs are called ________ equations.

e. A system of ________ equations has an infinite number of solutions.

4. Find the slope and the y-intercept of the graph of each line in the system $\begin{cases} y = 4x - 10 \\ x - 2y = -16 \end{cases}$. Then, use that information to determine the *number of solutions* of the system. **Do not solve the system.**

Solve each system by graphing.

5. $\begin{cases} y = 2x - 1 \\ x - 2y = -4 \end{cases}$

6. $\begin{cases} x + y = 5 \\ y = -x \end{cases}$

The following graph shows two different ways in which a salesperson can be paid according to the number of items she sells each month.

7. What is the point of intersection of the graphs? Explain its significance.

8. Which plan is better for the salesperson if she feels that selling 30 items per month is almost impossible?

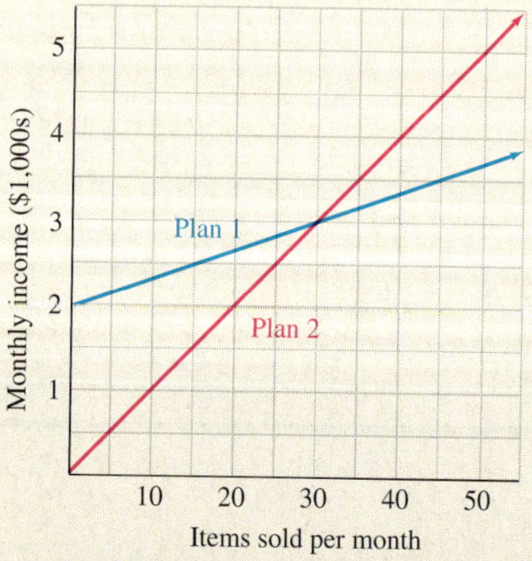

Solve each system by substitution.

9. $\begin{cases} y = x - 1 \\ 2x + y = -7 \end{cases}$

10. $\begin{cases} 3x + 6y = -15 \\ x + 2y = -5 \end{cases}$

Solve each system using elimination.

11. $\begin{cases} 3x - y = 2 \\ 2x + y = 8 \end{cases}$

12. $\begin{cases} 4x + 3y = -3 \\ -3x = -4y + 21 \end{cases}$

Solve each system using substitution or elimination.

13. $\begin{cases} 3x - 5y - 16 = 0 \\ \frac{x}{2} - \frac{5}{6}y = \frac{1}{3} \end{cases}$

14. $\begin{cases} 3a + 4b = -7 \\ 2b - a = -1 \end{cases}$

15. $\begin{cases} y = 3x - 1 \\ y = 2x + 4 \end{cases}$

16. $\begin{cases} 0.6c + 0.5c = 0 \\ 0.02c + 0.09d = 0 \end{cases}$

Write a system of two equations in two variables to solve each problem.

17. CHILD CARE On a mother's 22-mile commute to work, she drops her daughter off at a child care facility. The first part of the trip is 6 miles less than the second part. How long is each part of her morning commute?

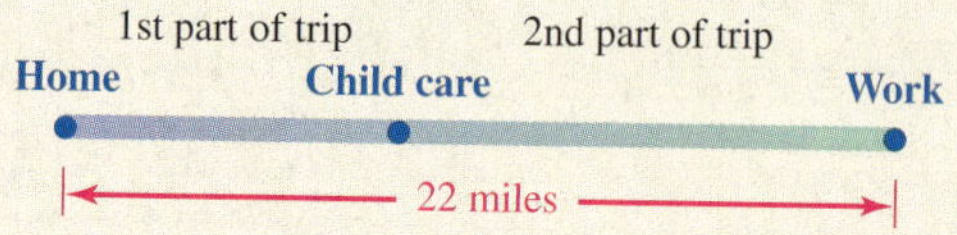

18. VACATIONING It cost a family of 7 a total of $187 for general admission tickets to the San Diego Zoo. How many adult tickets and how many child tickets were purchased?

19. FINANCIAL PLANNING A woman invested some money at 8% and some at 9%. The interest for 1 year on the combined investment of $10,000 was $840. How much was invested at each rate?

20. TAILWINDS/HEADWINDS Flying with a tailwind, a pilot flew an airplane 450 miles in 2.5 hours. Flying into a headwind, the return trip took 3 hours. Find the speed of the plane in calm air and the speed of the wind.

21. TETHER BALL The angles shown in the illustration are complementary. The measure of the larger angle is 10° more than three times the measure of the smaller angle. Find the measure of each angle.

22. ANTIFREEZE How many pints of a 5% antifreeze solution and how many pints of a 20% antifreeze solution must be mixed to obtain 12 pints of a 15% solution?

23. SUNSCREEN A sunscreen selling for \$1.50 per ounce is to be combined with another sunscreen selling for \$0.80 per ounce. How many ounces of each are needed to make 10 ounces of a sunscreen mix that sells for \$1.01 per ounce?

24. Solve the system by graphing.

$$\begin{cases} 2x + 3y \le 6 \\ x > 2 \end{cases}$$

25. CLOTHES SHOPPING This system of inequalities describes the number of \$20 shirts, x, and \$40 pairs of pants, y, a person can buy if he or she plans to spend not less than \$80 but not more than \$120. Graph the system. Then give three solutions.

$$\begin{cases} 20x + 40y \ge 80 \\ 20x + 40y \le 120 \end{cases}$$

26. Use a check to determine whether (3, 1) is a solution of the system: $\begin{cases} y \le 2x - 1 \\ x + 3y > 6 \end{cases}$

GROUP PROJECT

WRITING APPLICATION PROBLEMS

Overview: In Section 4.4, you solved application problems by translating the words of the problem into a system of two equations. In this activity, you will reverse these steps.

Instructions: Form groups of 2 or 3 students. For each type of application, write a problem that could be solved using the given equations. If you need help getting started, refer to the specific problem types in the text. When finished writing the five applications, pick one problem and solve it completely.

A rectangle problem:

$$\begin{cases} 2l + 2w = 320 \\ l = w + 40 \end{cases}$$

A number-value problem:

$$\begin{cases} 5x + 2y = 23 \\ 3x + 7y = 37 \end{cases}$$

An interest problem:

$$\begin{cases} x + y = 75{,}000 \\ 0.03x + 0.05y = 2{,}750 \end{cases}$$

A with–against the wind problem:

$$\begin{cases} 2(x + y) = 600 \\ 3(x - y) = 600 \end{cases}$$

A liquid mixture problem:

$$\begin{cases} x + y = 36 \\ 0.50x + 0.20y = 0.30(36) \end{cases}$$

CUMULATIVE REVIEW
Chapters 1–4

1. SPORTS CARS The graph shows the Porsche vehicle sales in the United States for the years 1986–2005.
   a. In what year were sales the lowest? [Section 1.1]
   b. In what year were sales the greatest? [Section 1.1]

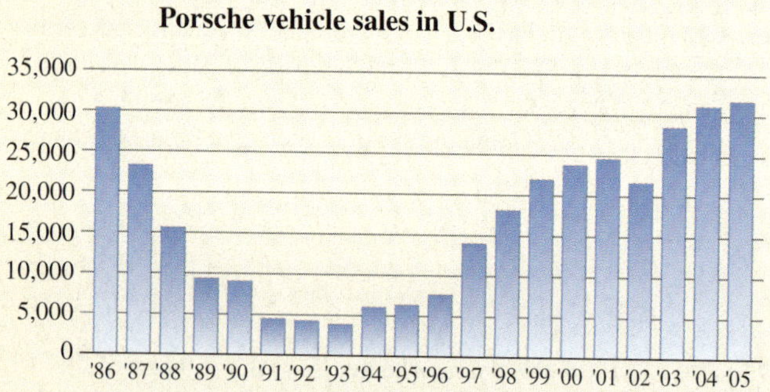

Source: Porsche Cars North America

2. Give the prime factorization of 100. [Section 1.2]
3. Divide: $\frac{3}{4} \div \frac{6}{5}$ [Section 1.2]
4. Subtract: $\frac{7}{10} - \frac{1}{14}$ [Section 1.2]
5. Is π a rational or irrational number? [Section 1.3]
6. Graph each member of the set on the number line. [Section 1.3]

$$\left\{-2\frac{1}{4}, \sqrt{2}, -1.75, \frac{7}{2}, 0.5\right\}$$

7. Write $\frac{2}{3}$ as a decimal. [Section 1.3]
8. What property of real numbers is illustrated? [Section 1.6]

$$3(2x) = (3 \cdot 2)x$$

Evaluate each expression.

9. $-3^2 + |4^2 - 5^2|$ [Section 1.7]
10. $(4 - 5)^{20}$ [Section 1.7]
11. $\frac{-3 - (-7)}{2^2 - 3}$ [Section 1.7]
12. $12 - 2[1 - (-8 + 2)]$ [Section 1.7]
13. RACING Suppose a driver has completed x laps of a 250-lap race. Write an expression for how many more laps he must make to finish the race. [Section 1.8]
14. What is the value of d dimes? [Section 1.8]

Simplify each expression.

15. $13r - 12r$ [Section 1.9]
16. $27\left(\frac{2}{3}x\right)$ [Section 1.9]
17. $4(d - 3) - (d - 1)$ [Section 1.9]
18. $(13c - 3)(-6)$ [Section 1.9]

Solve each equation. Check each result.

19. $3(x - 5) + 2 = 2x$ [Section 2.2]
20. $\frac{x - 5}{3} - 5 = 7$ [Section 2.2]
21. $\frac{2}{5}x + 1 = \frac{1}{3} + x$ [Section 2.2]
22. $-\frac{5}{8}h = 15$ [Section 2.2]
23. GYMNASTICS After the first day of registration, 119 children had been enrolled in a Gymboree class. That represented 85% of the available slots. Find the maximum number of children the center could enroll. [Section 2.3]
24. Solve $A = \frac{1}{2}h(b + B)$ for h. [Section 2.4]
25. MIXING CANDY The owner of a candy store wants to make a 30-pound mixture of two candies to sell for $4 per pound. If red licorice bits sell for $3.80 per pound and lemon gumdrops sell for $4.40 per pound, how many pounds of each should be used? [Section 2.6]
26. Solve $8(4 + x) > 10(6 + x)$. Write the solution set in interval notation and graph it. [Section 2.7]
27. In what quadrant does $(-3.5, 6)$ lie? [Section 3.1]
28. Is $(-2, 8)$ a solution of $y = -2x + 3$? [Section 3.2]

Graph each equation.

29. $x = 4$ [Section 3.3]
30. $4x - 3y = 12$ [Section 3.3]

Find the slope of the line with the given properties.

31. Passing through $(-2, 4)$ and $(6, 8)$ [Section 3.4]
32. A line that is horizontal [Section 3.4]
33. An equation of $2x - 3y = 12$ [Section 3.5]
34. Are the graphs of the lines parallel or perpendicular? [Section 3.5]

$$y = -\frac{3}{4}x + \frac{15}{4} \qquad 4x - 3y = 25$$

Find an equation of the line with the following properties. Write the equation in slope-intercept form.

35. Slope $= \frac{2}{3}$, y-intercept $= (0, 5)$ [Section 3.5]

36. Passing through $(-2, 4)$ and $(6, 10)$ [Section 3.6]

37. A horizontal line passing through $(2, 4)$ [Section 3.6]

38. Graph: $y < \frac{x}{3} - 1$ [Section 3.7]

39. If $f(x) = -2x^2 - 3x^3$, find $f(-1)$. [Section 3.8]

40. Is this the graph of a function? [Section 3.8]

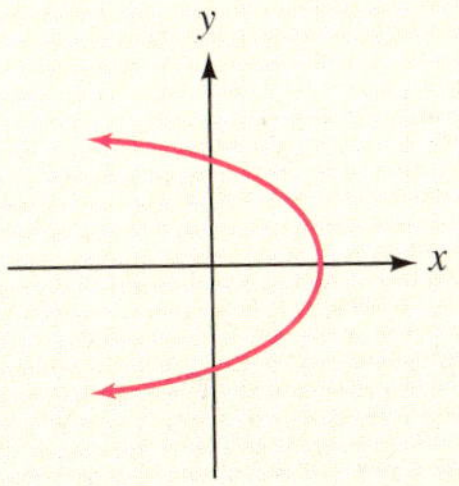

Solve each system by graphing.

41. $\begin{cases} x + 4y = -2 \\ y = -x - 5 \end{cases}$ [Section 4.1]

42. $\begin{cases} 2x - 3y < 0 \\ y > x - 1 \end{cases}$ [Section 4.5]

43. Solve $\begin{cases} x - 2y = 2 \\ 2x + 3y = 11 \end{cases}$ by substitution. [Section 4.2]

44. NUTRITION The table shows per serving nutritional information for egg noodles and rice pilaf. How many servings of each food should be eaten to consume exactly 22 grams of protein and 21 grams of fat? [Section 4.4]

	Protein (g)	Fat (g)
Egg noodles	5	3
Rice pilaf	4	5

CHAPTER 5

Exponents and Polynomials

from Campus to Careers

Sound Engineering Technician

Today's digitally recorded music is crystal clear thanks to the talents of sound engineering technicians. They operate console mixing boards and microphones to record the music, voices, and sound effects that are so much a part of our media-filled lives. The job requires strong mathematical skills and an aptitude for working with electronic equipment. Sound technicians constantly work with numbers as they read meters and graphs, adjust dials and switches, and keep written logs.

Many exciting advancements are taking place in the music industry. In **Problem 85** of **Study Set 5.4,** you will explore the rapid growth of iTunes, a computer program made by Apple Computer that plays, organizes, and enables us to buy music files and much more online.

JOB TITLE:
Sound Engineering Technician
EDUCATION:
Training from a technical school or community college is recommended.
JOB OUTLOOK:
Demand is expected to grow. Entry-level job prospects are very competitive.
ANNUAL EARNINGS:
Median annual salary $50,260
FOR MORE INFORMATION:
www.bls.gov/oco/ocos109.htm

Study Skills Workshop
Attendance

It's not uncommon for students' enthusiasm to lessen toward the middle of the term. Sometimes their effort and attendance begin to slip. Realize that missing even one class can have a great effect on your grade. Being tardy takes its toll as well. If you are just a few minutes late, or miss an entire class, you risk getting behind. So, keep the following tips in mind.

ARRIVE ON TIME, OR A LITTLE EARLY: When you arrive, get out your note-taking materials and homework. Identify any questions or comments that you plan to ask your instructor once the class starts.

IF YOU MUST MISS A CLASS: Get a set of notes, the homework assignments, and any handouts that the instructor may have provided for the day(s) that you missed.

STUDY THE MATERIAL YOU MISSED: Take advantage of the help that comes with this textbook, such as the CD and the videotapes. Watch the explanations of the material from the section(s) that you missed.

Now Try This

1. Plan ahead! List five possible situations that could cause you to be late to class or miss a class. (Some examples are parking/traffic delays, lack of a babysitter, oversleeping, or job responsibilities.) What can you do ahead of time so that these situations won't cause you to be tardy or absent?
2. Watch one section on the CD or the videotape series that accompanies this book. Take notes as you watch the explanations.

SECTION 5.1
Rules for Exponents

Objectives

1. Identify bases and exponents.
2. Multiply exponential expressions that have like bases.
3. Divide exponential expressions that have like bases.
4. Raise exponential expressions to a power.
5. Find powers of products and quotients.

In this section, we will use the definition of exponent to develop some rules for simplifying expressions that contain exponents.

1 Identify Bases and Exponents.

Recall that an **exponent** indicates repeated multiplication. It indicates how many times the **base** is used as a factor. For example, 3^5 represents the product of five 3's.

$$\underbrace{3^5}_{\text{Base}}{}^{\text{Exponent}} = \overbrace{3 \cdot 3 \cdot 3 \cdot 3 \cdot 3}^{\text{5 factors of 3}}$$

In general, we have the following definition.

Natural-Number Exponents

A natural-number exponent tells how many times its base is to be used as a factor. For any number x and any natural number n,

$$x^n = \overbrace{x \cdot x \cdot x \cdot \cdots \cdot x}^{n \text{ factors of } x}$$

Expressions of the form x^n are called **exponential expressions.** The base of an exponential expression can be a number, a variable, or a combination of numbers and variables. Some examples are:

$10^5 = 10 \cdot 10 \cdot 10 \cdot 10 \cdot 10$ The base is 10. The exponent is 5. Read as "10 to the fifth power."

$y^2 = y \cdot y$ The base is y. The exponent is 2. Read as "y squared."

$(-2s)^3 = (-2s)(-2s)(-2s)$ The base is $-2s$. The exponent is 3. Read as "negative $2s$ raised to the third power" or "negative $2s$ cubed."

$-8^4 = -(8 \cdot 8 \cdot 8 \cdot 8)$ Since the $-$ sign is not written within parentheses, the base is 8. The exponent is 4. Read as "the opposite (or the negative) of 8 to the fourth power."

When an exponent is 1, it is usually not written. For example, $4 = 4^1$ and $x = x^1$.

Notation

Bases that contain a $-$ sign *must* be written within parentheses.

$(-2s)^3$ ← Exponent; $-2s$: Base

EXAMPLE 1 Identify the base and the exponent in each expression:
a. 8^5 **b.** $7a^3$ **c.** $(7a)^3$

Strategy To identify the base and exponent, we will look for the form .

Why The exponent is the small raised number to the right of the base.

Solution

a. In 8^5, the base is 8 and the exponent is 5.

b. $7a^3$ means $7 \cdot a^3$. Thus, the base is a, not $7a$. The exponent is 3.

c. Because of the parentheses in $(7a)^3$, the base is $7a$ and the exponent is 3.

Self Check 1 Identify the base and the exponent:
a. $3y^4$ **b.** $(3y)^4$

Now Try **Problems 13 and 17**

EXAMPLE 2 Write each expression in an equivalent form using an exponent:

a. $\frac{p}{3} \cdot \frac{p}{3} \cdot \frac{p}{3} \cdot \frac{p}{3}$ **b.** $5 \cdot t \cdot t \cdot t$

Strategy We will look for repeated factors and count the number of times each appears.

Why We can use an exponent to represent repeated multiplication.

Solution

a. Since there are four repeated factors of $\frac{p}{3}$ in $\frac{p}{3} \cdot \frac{p}{3} \cdot \frac{p}{3} \cdot \frac{p}{3}$, the expression can be written $\left(\frac{p}{3}\right)^4$.

b. Since there are three repeated factors of t in $5 \cdot t \cdot t \cdot t$, the expression can be written $5t^3$.

Self Check 2 Write as an exponential expression:
$(x + y)(x + y)(x + y)(x + y)(x + y)$

Now Try **Problems 25 and 27**

2 Multiply Exponential Expressions That Have Like Bases.

To develop a rule for multiplying exponential expressions that have the same base, we consider the product $6^2 \cdot 6^3$. Since 6^2 means that 6 is to be used as a factor two times, and 6^3 means that 6 is to be used as a factor three times, we have

$$6^2 \cdot 6^3 = \underbrace{6 \cdot 6}_{\text{2 factors of 6}} \cdot \underbrace{6 \cdot 6 \cdot 6}_{\text{3 factors of 6}}$$

$$= \underbrace{6 \cdot 6 \cdot 6 \cdot 6 \cdot 6}_{\text{5 factors of 6}}$$

$$= 6^5$$

We can quickly find this result if we keep the common base of 6 and add the exponents on 6^2 and 6^3.

$$6^2 \cdot 6^3 = 6^{2+3} = 6^5$$

This example suggests the following rule for exponents.

Product Rule for Exponents

To multiply exponential expressions that have the same base, keep the common base and add the exponents.

For any number x and any natural numbers m and n,

$$x^m \cdot x^n = x^{m+n}$$ Read as "x to the mth power times x to the nth power equals x to the m plus nth power."

EXAMPLE 3 Simplify: **a.** $9^5(9^6)$ **b.** $x^3 \cdot x^4$ **c.** y^2y^4y **d.** $(x + 2)^8(x + 2)^7$ **e.** $(c^2d^3)(c^4d^5)$

Strategy In each case, we want to write an equivalent expression using one base and one exponent. We will use the product rule for exponents to do this.

Why The product rule for exponents is used to multiply exponential expressions that have the same base.

Caution
Don't make the mistake of multiplying the bases when using the product rule. Keep the *same* base.

$$9^5(9^6) \neq 81^{11}$$

Caution
Don't make the mistake of "distributing" an exponent over a sum (or difference). *There is no such rule.*

$$(x + 2)^{15} \neq x^{15} + 2^{15}$$

Solution

a. $9^5(9^6) = 9^{5+6} = 9^{11}$ Keep the common base, 9, and add the exponents. Since 9^{11} is a very large number, we will leave the answer in this form. We won't evaluate it.

b. $x^3 \cdot x^4 = x^{3+4} = x^7$ Keep the common base, x, and add the exponents.

c. $y^2y^4y = y^2y^4y^1$ Write y as y^1.
$= y^{2+4+1}$ Keep the common base, y, and add the exponents.
$= y^7$

d. $(x + 2)^8(x + 2)^7 = (x + 2)^{8+7}$ Keep the common base, x + 2, and add the exponents.
$= (x + 2)^{15}$

e. $(c^2d^3)(c^4d^5) = (c^2c^4)(d^3d^5)$ Use the commutative and associative properties of multiplication to group like bases together.
$= (c^{2+4})(d^{3+5})$ Keep the common base, c, and add the exponents. Keep the common base, d, and add the exponents.
$= c^6d^8$

Self Check 3 Simplify: **a.** $7^8(7^7)$ **b.** x^2x^3x **c.** $(y - 1)^5(y - 1)^5$ **d.** $(s^4t^3)(s^4t^4)$

Now Try **Problems 35 and 41**

Caution We cannot use the product rule to simplify expressions like $3^2 \cdot 2^3$, where the bases are not the same. However, we can simplify this expression by doing the arithmetic:

$$3^2 \cdot 2^3 = 9 \cdot 8 = 72$$

EXAMPLE 4 ***Geometry.*** Find an expression that represents the area of the rectangle.

x^3 feet

x^5 feet

Strategy We will multiply the length of the rectangle by its width.

Why The area of a rectangle is equal to the product of its length and width.

Solution

Area = **length** · **width** This is the formula for the area of a rectangle.
$= x^5 \cdot x^3$ Substitute x^5 for the length and x^3 for the width.
$= x^{5+3}$ Use the product rule: Keep the common base, x, and add the exponents.
$= x^8$

The area of the rectangle is x^8 square feet, which can be written as x^8 ft^2.

Now Try **Problem 45**

3 Divide Exponential Expressions That Have Like Bases.

To develop a rule for dividing exponential expressions that have the same base, we consider the quotient $\frac{4^5}{4^2}$, where the exponent in the numerator is greater than the exponent in the denominator. We can simplify this fraction by removing the common factors of 4 in the numerator and denominator:

$$\frac{4^5}{4^2} = \frac{4 \cdot 4 \cdot 4 \cdot 4 \cdot 4}{4 \cdot 4} = \frac{\overset{1}{\cancel{4}} \cdot \overset{1}{\cancel{4}} \cdot 4 \cdot 4 \cdot 4}{\underset{1}{\cancel{4}} \cdot \underset{1}{\cancel{4}}} = 4^3$$

We can quickly find this result if we keep the common base, 4, and subtract the exponents on 4^5 and 4^2.

$$\frac{4^5}{4^2} = 4^{5-2} = 4^3$$

This example suggests another rule for exponents.

Quotient Rule for Exponents

To divide exponential expressions that have the same base, keep the common base and subtract the exponents.

For any nonzero number x and any natural numbers m and n, where $m > n$,

$$\frac{x^m}{x^n} = x^{m-n}$$ Read as "x to the mth power divided by x to the nth power equals x to the m minus nth power."

EXAMPLE 5 Simplify each expression:

a. $\frac{20^{16}}{20^9}$ **b.** $\frac{x^9}{x^3}$ **c.** $\frac{(7.5n)^{12}}{(7.5n)^{11}}$ **d.** $\frac{a^3b^8}{ab^5}$

Strategy In each case, we want to write an equivalent expression using one base and one exponent. We will use the quotient rule for exponents to do this.

Why The quotient rule for exponents is used to divide exponential expressions that have the same base.

Solution

a. $\frac{20^{16}}{20^9} = 20^{16-9}$ Keep the common base, 20, and subtract the exponents.

$= 20^7$ Since 20^7 is a very large number, we will leave the answer in this form. We won't evaluate it.

b. $\frac{x^9}{x^3} = x^{9-3}$ Keep the common base, x, and subtract the exponents.

$= x^6$

c. $\frac{(7.5n)^{12}}{(7.5n)^{11}} = (7.5n)^{12-11}$ Keep the common base, 7.5n, and subtract the exponents.

$= (7.5n)^1$

$= 7.5n$ Any number raised to the first power is simply that number.

Caution

Don't make the mistake of dividing the bases when using the quotient rule. Keep the *same* base.

$$\frac{20^{16}}{20^9} \neq 1^7$$

The Language of Algebra
In this chapter, variables often appear in a denominator. In such cases, we will assume that the variables do not equal 0. That is, we will assume that *there are no divisions by 0.*

d. $\dfrac{a^3b^8}{ab^5} = \dfrac{a^3}{a^1} \cdot \dfrac{b^8}{b^5}$ Group the common bases together. Write a as a^1.

$= a^{3-1}b^{8-5}$ Keep the common base a and subtract the exponents. Keep the common base b and subtract the exponents.

$= a^2b^3$

Self Check 5 Simplify: **a.** $\dfrac{55^{30}}{55^5}$ **b.** $\dfrac{a^5}{a^3}$ **c.** $\dfrac{(8.9t)^8}{(8.9t)^7}$ **d.** $\dfrac{b^{15}c^4}{b^4c}$

Now Try **Problems 51 and 57**

EXAMPLE 6 Simplify: $\dfrac{a^3a^5a^7}{a^4a}$

Strategy We want to write an equivalent expression using one base and one exponent. First, we will use the product rule to simplify the numerator and the denominator. Then, we will use the quotient rule to simplify that result.

Why The expression involves multiplication and division of exponential expressions that have the same base.

Solution We simplify the numerator and denominator separately and proceed as follows.

$\dfrac{a^3a^5a^7}{a^4a} = \dfrac{a^{15}}{a^5}$ In the numerator, keep the common base, a, and add the exponents. In the denominator, keep the common base, a, and add the exponents.

$= a^{15-5}$ Keep the common base, a, and subtract the exponents.

$= a^{10}$

Success Tip
Sometimes, more than one rule for exponents is needed to simplify an expression.

Self Check 6 Simplify: $\dfrac{b^2b^6b}{b^4b^4}$

Now Try **Problem 63**

Recall that like terms are terms with exactly the same variables raised to exactly the same powers. To add or subtract exponential expressions, they must be like terms. To multiply or divide exponential expressions, only the bases need to be the same.

$x^5 + x^2$ These are not like terms; the exponents are different. We cannot add.

$x^2 + x^2 = 2x^2$ These are like terms; we can add. Recall that $x^2 = 1x^2$.

$x^5 \cdot x^2 = x^7$ The bases are the same; we can multiply.

$\dfrac{x^5}{x^2} = x^3$ The bases are the same; we can divide.

4 Raise Exponential Expressions to a Power.

To develop another rule for exponents, we consider $(5^3)^4$. Here, an exponential expression, 5^3, is raised to a power. Since 5^3 is the base and 4 is the exponent, $(5^3)^4$ can be written as $5^3 \cdot 5^3 \cdot 5^3 \cdot 5^3$. Because each of the four factors of 5^3 contains three factors of 5, there are $4 \cdot 3$ or 12 factors of 5.

The Language of Algebra
An exponential expression raised to a power, such as $(5^3)^4$, is also called a *power of a power.*

$$(5^3)^4 = 5^3 \cdot 5^3 \cdot 5^3 \cdot 5^3 = \overbrace{\underbrace{5 \cdot 5 \cdot 5}_{5^3} \cdot \underbrace{5 \cdot 5 \cdot 5}_{5^3} \cdot \underbrace{5 \cdot 5 \cdot 5}_{5^3} \cdot \underbrace{5 \cdot 5 \cdot 5}_{5^3}}^{\text{12 factors of } x} = 5^{12}$$

We can quickly find this result if we keep the common base of 5 and multiply the exponents.

$$(5^3)^4 = 5^{3 \cdot 4} = 5^{12}$$

This example suggests the following rule for exponents.

Power Rule for Exponents

To raise an exponential expression to a power, keep the base and multiply the exponents. For any number x and any natural numbers m and n,

$$(x^m)^n = x^{m \cdot n} = x^{mn}$$ Read as "the quantity of x to the mth power raised to the nth power equals x to the mnth power."

EXAMPLE 7

Simplify: **a.** $(2^3)^7$ **b.** $[(-6)^2]^5$ **c.** $(z^8)^8$

Strategy In each case, we want to write an equivalent expression using one base and one exponent. We will use the power rule for exponents to do this.

Why Each expression is a power of a power.

Solution

a. $(2^3)^7 = 2^{3 \cdot 7} = 2^{21}$ Keep the base, 2, and multiply the exponents.

b. $[(-6)^2]^5 = (-6)^{2 \cdot 5} = (-6)^{10}$ Keep the base, −6, and multiply the exponents. Since $(-6)^{10}$ is a very large number, we will leave the answer in this form.

c. $(z^8)^8 = z^{8 \cdot 8} = z^{64}$ Keep the base, z, and multiply the exponents.

Self Check 7 Simplify: **a.** $(4^6)^5$ **b.** $(y^5)^2$

Now Try Problems 71 and 73

EXAMPLE 8

Simplify: **a.** $(x^2x^5)^2$ **b.** $(z^2)^4(z^3)^3$

Strategy In each case, we want to write an equivalent expression using one base and one exponent. We will use the product and power rules for exponents to do this.

Why The expressions involve multiplication of exponential expressions that have the same base and they involve powers of powers.

Solution

a. $(x^2x^5)^2 = (x^7)^2$ Within the parentheses, keep the common base, x, and add the exponents.

$= x^{14}$ Keep the base, x, and multiply the exponents.

b. $(z^2)^4(z^3)^3 = z^8z^9$ For each power of z raised to a power, keep the base and multiply the exponents.

$= z^{17}$ Keep the common base, z, and add the exponents.

Self Check 8 Simplify: **a.** $(a^4a^3)^3$ **b.** $(a^3)^3(a^4)^2$

Now Try **Problems 77 and 81**

5 Find Powers of Products and Quotients.

To develop more rules for exponents, we consider the expression $(2x)^3$, which is a *power of the product* of 2 and x, and the expression $\left(\frac{2}{x}\right)^3$, which is a *power of the quotient* of 2 and x.

$$\begin{aligned}(2x)^3 &= 2x \cdot 2x \cdot 2x\\ &= (2 \cdot 2 \cdot 2)(x \cdot x \cdot x)\\ &= 2^3x^3\\ &= 8x^3\end{aligned}$$

$$\begin{aligned}\left(\frac{2}{x}\right)^3 &= \frac{2}{x} \cdot \frac{2}{x} \cdot \frac{2}{x} && \text{Assume } x \neq 0.\\ &= \frac{2 \cdot 2 \cdot 2}{x \cdot x \cdot x} && \text{Multiply the numerators. Multiply the denominators.}\\ &= \frac{2^3}{x^3}\\ &= \frac{8}{x^3} && \text{Evaluate: } 2^3 = 8.\end{aligned}$$

These examples suggest the following rules for exponents.

Powers of a Product and a Quotient

To raise a product to a power, raise each factor of the product to that power. To raise a quotient to a power, raise the numerator and the denominator to that power.

For any numbers x and y, and any natural number n,

$$(xy)^n = x^ny^n \quad \text{and} \quad \left(\frac{x}{y}\right)^n = \frac{x^n}{y^n}, \quad \text{where } y \neq 0$$

EXAMPLE 9 Simplify: **a.** $(3c)^4$ **b.** $(x^2y^3)^5$ **c.** $\left(-\frac{1}{4}a^3b\right)^2$

Strategy In each case, we want to write the expression in an equivalent form in which each base is raised to a single power. We will use the power of a product rule for exponents to do this.

Why Within each set of parentheses is a product, and each of those products is raised to a power.

Solution

a. $(3c)^4 = 3^4c^4$ Raise each factor of the product 3c to the 4th power.

$= 81c^4$ Evaluate: $3^4 = 81$.

b. $(x^2y^3)^5 = (x^2)^5(y^3)^5$ Raise each factor of the product x^2y^3 to the 5th power.

$= x^{10}y^{15}$ For each power of a power, keep each base, x and y, and multiply the exponents.

c. $\left(-\frac{1}{4}a^3b\right)^2 = \left(-\frac{1}{4}\right)^2(a^3)^2b^2$ Raise each factor of the product $-\frac{1}{4}a^3b$ to the 2nd power.

$= \frac{1}{16}a^6b^2$ Evaluate: $\left(-\frac{1}{4}\right)^2 = \frac{1}{16}$. Keep the base a and multiply the exponents.

Self Check 9 Simplify: **a.** $(2t)^4$ **b.** $(c^3d^4)^6$ **c.** $\left(-\frac{1}{3}ab^5\right)^3$

Now Try **Problems 85 and 91**

EXAMPLE 10

Simplify: $\dfrac{(a^3b^4)^2}{ab^5}$

Strategy We want to write the expression in an equivalent form in which each base is raised to a single power. We will use the power of a product rule and the quotient rule for exponents to do this.

Why The expression involves a power of a product and it is the quotient of exponential expressions that have the same base.

Solution

$$\frac{(a^3b^4)^2}{ab^5} = \frac{(a^3)^2(b^4)^2}{a^1b^5}$$ In the numerator, raise each factor within the parentheses to the second power. In the denominator, write a as a^1.

$$= \frac{a^6b^8}{a^1b^5}$$ In the numerator, for each power of a power, keep each base, a and b, and multiply the exponents.

$$= a^{6-1}b^{8-5}$$ Keep each of the bases, a and b, and subtract the exponents.

$$= a^5b^3$$

Self Check 10 Simplify: $\dfrac{(c^4d^5)^3}{c^2d^3}$

Now Try **Problem 93**

EXAMPLE 11

Simplify: $\dfrac{(5b)^9}{(5b)^6}$

Strategy We want to write the expression in an equivalent form in which the base is raised to a single power. We will begin by using the quotient rule for exponents to write an equivalent expression.

Why The expression involves division of exponential expressions that have the same base, $5b$.

Solution

$$\frac{(5b)^9}{(5b)^6} = (5b)^{9-6}$$ Keep the common base, $5b$, and subtract the exponents.

$$= (5b)^3$$

$$= 5^3b^3$$ Use the power of a product rule: Raise each factor within the parentheses to the 3rd power.

$$= 125b^3$$ Evaluate 5^3.

Self Check 11 Simplify: $\dfrac{(-2h)^{20}}{(-2h)^{14}}$

Now Try **Problem 97**

EXAMPLE 12

Simplify: **a.** $\left(\dfrac{4}{k}\right)^3$ **b.** $\left(\dfrac{3x^2}{2y^3}\right)^5$

Strategy We want to write each expression in an equivalent form using each base raised to a single power. We will use the power of a quotient rule for exponents to do this.

Why Within each set of parentheses is a quotient, and each of those quotients is raised to a power.

Solution

a. Since $\frac{4}{k}$ is the quotient of 4 and k, the expression $\left(\frac{4}{k}\right)^3$ is a power of a quotient.

$$\left(\frac{4}{k}\right)^3 = \frac{4^3}{k^3} = \frac{64}{k^3}$$ Raise the numerator and denominator to the 3rd power. Then evaluate 4^3.

b. $$\left(\frac{3x^2}{2y^3}\right)^5 = \frac{(3x^2)^5}{(2y^3)^5}$$ Raise the numerator and the denominator to the 5th power.

$$= \frac{3^5(x^2)^5}{2^5(y^3)^5}$$ In the numerator and denominator, raise each factor within the parentheses to the 5th power.

$$= \frac{243x^{10}}{32y^{15}}$$ Evaluate 3^5 and 2^5. For each power of a power, keep the base and multiply the exponents.

Self Check 12 Simplify: **a.** $\left(\frac{x}{7}\right)^3$ **b.** $\left(\frac{2x^3}{3y^2}\right)^4$

Now Try **Problems 101 and 105**

The rules for natural-number exponents are summarized as follows.

Rules for Exponents

If m and n represent natural numbers and there are no divisions by zero, then

Exponent of 1	Product rule	Power rule
$x^1 = x$	$x^m x^n = x^{m+n}$	$(x^m)^n = x^{mn}$
Quotient rule	**Power of a product**	**Power of a quotient**
$\dfrac{x^m}{x^n} = x^{m-n}$	$(xy)^n = x^n y^n$	$\left(\dfrac{x}{y}\right)^n = \dfrac{x^n}{y^n}$

ANSWERS TO SELF CHECKS **1. a.** Base: y, exponent: 4 **b.** Base: $3y$, exponent: 4 **2.** $(x + y)^5$ **3. a.** 7^{15} **b.** x^6 **c.** $(y - 1)^{10}$ **d.** s^8t^7 **5. a.** 55^{25} **b.** a^2 **c.** $8.9t$ **d.** $b^{11}c^3$ **6.** b **7. a.** 4^{30} **b.** y^{10} **8. a.** a^{21} **b.** a^{17} **9. a.** $16t^4$ **b.** $c^{18}d^{24}$ **c.** $-\frac{1}{27}a^3b^{15}$ **10.** $c^{10}d^{12}$ **11.** $64h^6$ **12. a.** $\frac{x^3}{343}$ **b.** $\frac{16x^{12}}{81y^8}$

STUDY SET 5.1

VOCABULARY

Fill in the blank.

1. Expressions such as x^4, 10^3, and $(5t)^2$ are called __________ expressions.
2. Match each expression with the proper description.

$\frac{a^8}{a^2}$ $\quad (a^4b^2)^5 \quad \left(\frac{a^6}{a}\right)^3 \quad (a^8)^4 \quad a^5 \cdot a^3$

a. Product of exponential expressions with the same base
b. Quotient of exponential expressions with the same base
c. Power of an exponential expression
d. Power of a product
e. Power of a quotient

CONCEPTS

Fill in the blanks.

3. a. $(3x)^4 = \square \cdot \square \cdot \square \cdot \square$
   b. $(-5y)(-5y)(-5y) = \square$
4. a. $x = x^{\square}$ b. $x^m x^n = \square$
   c. $(xy)^n = \square$ d. $(a^b)^c = \square$
   e. $\frac{x^m}{x^n} = \square$ f. $\left(\frac{a}{b}\right)^n = \square$
5. To simplify each expression, determine whether you add, subtract, multiply, or divide the exponents.
   a. $\frac{x^8}{x^2}$ b. $b^6 \cdot b^9$
   c. $(n^8)^4$ d. $(a^4b^2)^5$
6. a. To simplify $(2y^3z^2)^4$, what factors within the parentheses must be raised to the fourth power?
   b. To simplify $\left(\frac{y^3}{z^2}\right)^4$, what two expressions must be raised to the fourth power?

Simplify each expression, if possible.

7. a. $x^2 + x^2$ b. $x^2 \cdot x^2$
8. a. $x^2 + x$ b. $x^2 \cdot x$
9. a. $x^3 - x^2$ b. $\frac{x^3}{x^2}$
10. a. $4^2 \cdot 2^4$ b. $\frac{x^3}{y^2}$

NOTATION

Complete each solution to simplify each expression.

11. $(x^4x^2)^3 = (\square)^3 = x^{\square}$
12. $\frac{a^3a^4}{a^2} = \frac{\square}{a^2} = a^{\square - 2} = a^{\square}$

GUIDED PRACTICE

Identify the base and the exponent in each expression. See Example 1.

13. 4^3
14. $(-8)^2$
15. x^5
16. $\left(\frac{5}{x}\right)^3$
17. $(-3x)^2$
18. $(2xy)^{10}$
19. $-\frac{1}{3}y^6$
20. $-x^4$
21. $9m^{12}$
22. $3.14r^4$
23. $(y + 9)^4$
24. $(z - 2)^3$

Write each expression in an equivalent form using an exponent. See Example 2.

25. $4t \cdot 4t \cdot 4t \cdot 4t$
26. $-5u(-5u)(-5u)(-5u)(-5u)$
27. $-4 \cdot t \cdot t \cdot t \cdot t \cdot t$
28. $-5 \cdot u \cdot u \cdot u$
29. $\frac{t}{2} \cdot \frac{t}{2} \cdot \frac{t}{2}$
30. $\frac{x}{c} \cdot \frac{x}{c} \cdot \frac{x}{c} \cdot \frac{x}{c}$
31. $(x - y)(x - y)$
32. $(m + 4)(m + 4)$

Use the product rule for exponents to simplify each expression. Write the results using exponents. See Example 3.

33. $5^3 \cdot 5^4$
34. $3^4 \cdot 3^6$
35. $a^3 \cdot a^3$
36. $m^7 \cdot m^7$
37. bb^2b^3
38. aa^3a^5
39. $(y - 2)^5(y - 2)^2$
40. $(t + 1)^5(t + 1)^3$
41. $(a^2b^3)(a^3b^3)$
42. $(u^3v^5)(u^4v^5)$
43. $cd^4 \cdot cd$
44. $ab^3 \cdot ab^4$

Find an expression that represents the area or volume of each figure. Recall that the formula for the volume of a rectangular solid is V = length · width · height. **See Example 4.**

45.

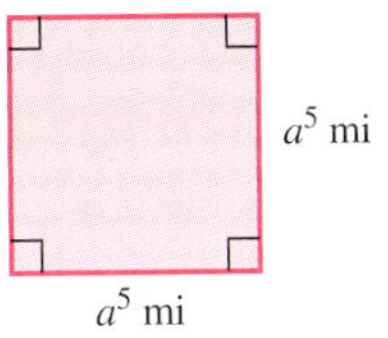

46.

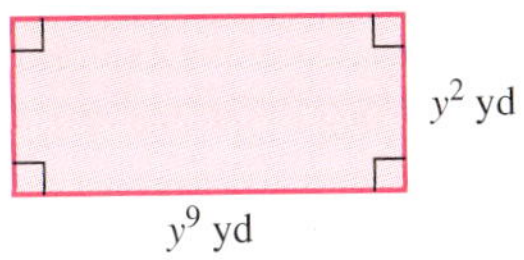

47.

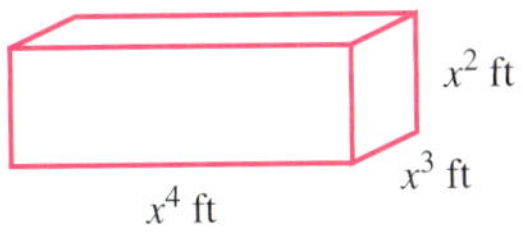

48.

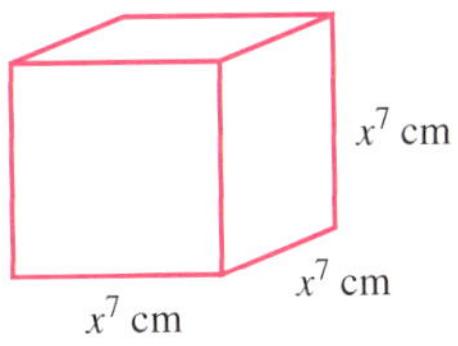

Use the quotient rule for exponents to simplify each expression. Write the results using exponents. **See Example 5.**

49. $\dfrac{8^{12}}{8^4}$ **50.** $\dfrac{10^4}{10^2}$

51. $\dfrac{x^{15}}{x^3}$ **52.** $\dfrac{y^6}{y^3}$

53. $\dfrac{(3.7p)^7}{(3.7p)^2}$ **54.** $\dfrac{(0.25y)^9}{(0.25y)^3}$

55. $\dfrac{(k-2)^{15}}{(k-2)}$ **56.** $\dfrac{(m+8)^{20}}{(m+8)}$

57. $\dfrac{c^3d^7}{cd}$ **58.** $\dfrac{r^8s^9}{rs}$

59. $\dfrac{x^4y^7}{xy^3}$ **60.** $\dfrac{p^7q^{10}}{p^2q^7}$

Use the product and quotient rules for exponents to simplify each expression. **See Example 6.**

61. $\dfrac{y^3y^4}{yy^2}$ **62.** $\dfrac{b^4b^5}{b^2b^3}$

63. $\dfrac{a^2a^3a^4}{a^8}$ **64.** $\dfrac{h^3h^6h}{h^9}$

65. $\dfrac{t^5t^6t}{t^2t^3}$ **66.** $\dfrac{m^5m^{12}m}{m^7m^4}$

67. $\dfrac{s^2s^2s^2}{s^3s}$ **68.** $\dfrac{w^4w^4w^4}{w^2w}$

Use the power rule for exponents to simplify each expression. Write the results using exponents. **See Example 7.**

69. $(3^2)^4$ **70.** $(4^3)^3$

71. $[(-4.3)^3]^8$ **72.** $[(-1.7)^9]^8$

73. $(m^{50})^{10}$ **74.** $(n^{25})^4$

75. $(y^5)^3$ **76.** $(b^3)^6$

Use the product and power rules for exponents to simplify each expression. **See Example 8.**

77. $(x^2x^3)^5$ **78.** $(y^3y^4)^4$

79. $(p^2p^3)^5$ **80.** $(r^3r^4)^2$

81. $(t^3)^4(t^2)^3$ **82.** $(b^2)^5(b^3)^2$

83. $(u^4)^2(u^3)^2$ **84.** $(v^5)^2(v^3)^4$

Use the power of a product rule for exponents to simplify each expression. **See Example 9.**

85. $(6a)^2$ **86.** $(3b)^3$

87. $(5y)^4$ **88.** $(4t)^4$

89. $(-2r^2s^3)^3$ **90.** $(-2x^2y^4)^5$

91. $\left(-\dfrac{1}{3}y^2z^4\right)^5$ **92.** $\left(-\dfrac{1}{4}t^3u^8\right)^2$

Use rules for exponents to simplify each expression. **See Example 10.**

93. $\dfrac{(ab^2)^3}{a^2b^2}$ **94.** $\dfrac{(m^3n^4)^3}{m^3n^6}$

95. $\dfrac{(r^4s^3)^4}{r^3s^9}$ **96.** $\dfrac{(x^2y^5)^5}{x^6y^2}$

Use rules for exponents to simplify each expression. **See Example 11.**

97. $\dfrac{(6k)^7}{(6k)^4}$ **98.** $\dfrac{(-3a)^{12}}{(-3a)^{10}}$

99. $\dfrac{(3q)^5}{(3q)^3}$ **100.** $\dfrac{(ab)^8}{(ab)^4}$

Use the power of a quotient rule for exponents to simplify each expression. **See Example 12.**

101. $\left(\dfrac{a}{b}\right)^3$ **102.** $\left(\dfrac{r}{s}\right)^4$

103. $\left(\dfrac{m}{3}\right)^4$ **104.** $\left(\dfrac{n}{5}\right)^3$

105. $\left(\dfrac{8a^2}{11b^5}\right)^2$ **106.** $\left(\dfrac{7g^4}{6h^3}\right)^2$

107. $\left(\dfrac{3m^4}{2n^5}\right)^5$ **108.** $\left(\dfrac{2s^2}{3t^5}\right)^5$

TRY IT YOURSELF

Simplify each expression.

109. $\left(\dfrac{x^2}{y^3}\right)^5$ **110.** $\left(\dfrac{u^4}{v^2}\right)^6$

111. $y^3y^2y^4$ **112.** y^4yy^6

113. $\dfrac{15^9}{15^6}$ **114.** $\dfrac{25^{13}}{25^7}$

115. $\left(\dfrac{y^3y^5}{yy^2}\right)^3$ **116.** $\left(\dfrac{s^5s^6}{s^2s^2}\right)^4$

117. $(-6a^3b^2)^3$ **118.** $(-10r^3s^2)^2$

119. $\dfrac{(a^2b^2)^{15}}{(ab)^9}$ **120.** $\dfrac{(s^3t^3)^4}{(st)^2}$

121. $(n^4n)^3(n^3)^6$ **122.** $(y^3y)^2(y^2)^2$

123. $\dfrac{(6h)^8}{(6h)^6}$ **124.** $\dfrac{(-7r)^{10}}{(-7r)^8}$

APPLICATIONS

125. ART HISTORY Leonardo da Vinci's drawing relating a human figure to a square and a circle is shown. Find an expression for the following:

a. The area of the square if the man's height is $5x$ feet.

b. The area of the circle if the waist-to-feet distance is $3a$ feet. Leave π in your answer.

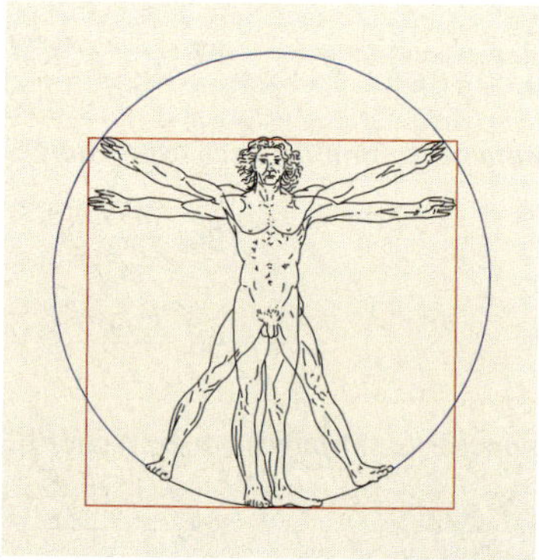

126. PACKAGING A bowling ball fits tightly against all sides of a cardboard box that it is packaged in. Find expressions for the volume of the ball and box. Leave π in your answer.

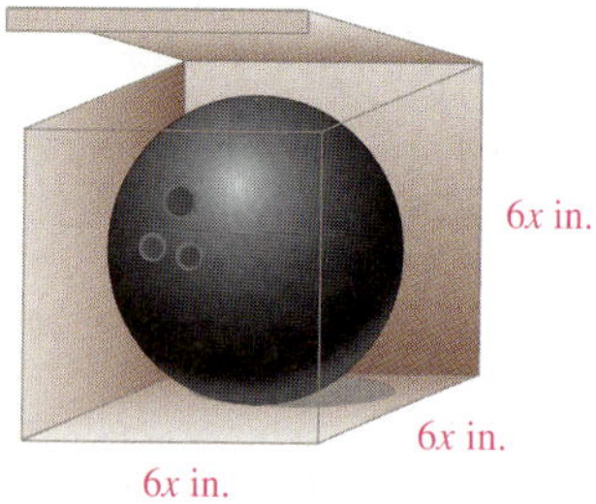

127. CHILDBIRTH Mr. and Mrs. Emory Harrison, of Johnson City, Tennessee, had 13 sons in a row during the 1940s and 1950s. The **probability** of a family of 13 children all being male is $\left(\frac{1}{2}\right)^{13}$. Evaluate this expression.

128. TOYS A Super Ball is dropped from a height of 1 foot and always rebounds to four-fifths of its previous height. The rebound height of the ball after the third bounce is $\left(\frac{4}{5}\right)^3$ feet. Evaluate this expression. Is the third bounce more or less than $\frac{1}{2}$ foot high?

WRITING

129. Explain the mistake in the following work.

$$2^3 \cdot 2^2 = 4^5 = 1{,}024$$

130. Explain why we can simplify $x^4 \cdot x^5$, but cannot simplify $x^4 + x^5$.

REVIEW

Match each equation with its graph below.

131. $y = 2x - 1$

132. $y = 3x - 1$

133. $y = 3$

134. $x = 3$

a.

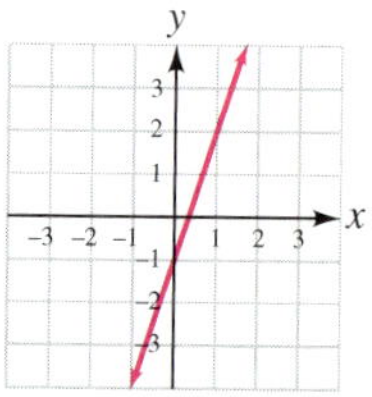

b.

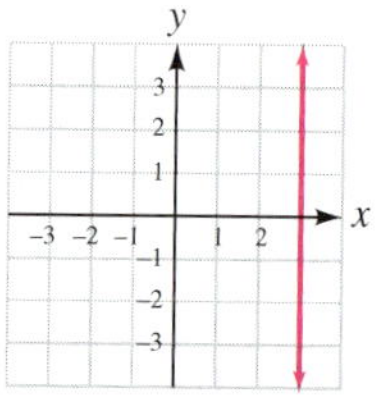

c.

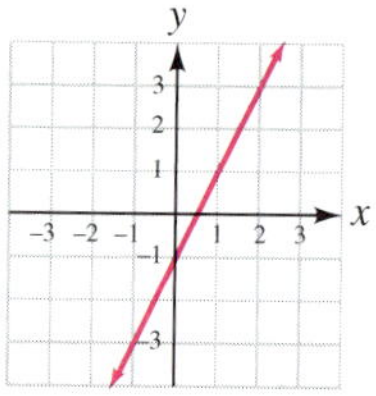

d.

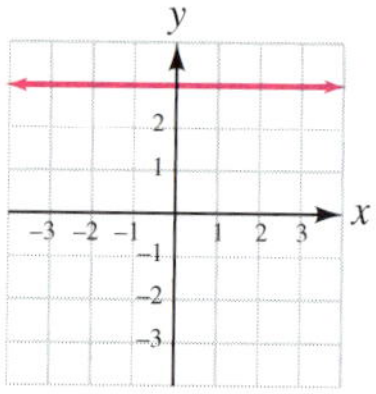

CHALLENGE PROBLEMS

135. Simplify each expression. The variables represent natural numbers.

a. $x^{2m}x^{3m}$

b. $(y^{5c})^4$

c. $\dfrac{m^{8x}}{m^{4x}}$

d. $(2a^{6y})^4$

136. Is the operation of raising to a power commutative? That is, is $a^b = b^a$? Explain.

SECTION 5.2
Zero and Negative Exponents

Objectives

1. Use the zero exponent rule.
2. Use the negative integer exponent rule.
3. Use exponent rules to change negative exponents in fractions to positive exponents.
4. Use all exponent rules to simplify expressions.

We now extend the discussion of natural-number exponents to include exponents that are zero and exponents that are negative integers.

1 Use the Zero Exponent Rule.

To develop the definition of a zero exponent, we will simplify the expression $\frac{5^3}{5^3}$ in two ways and compare the results.

First, we apply the quotient rule for exponents, where we subtract the equal exponents in the numerator and denominator. The result is 5^0. In the second approach, we write 5^3 as $5 \cdot 5 \cdot 5$ and remove the common factors of 5 in the numerator and denominator. The result is 1.

$$\frac{5^3}{5^3} = 5^{3-3} = 5^0 \qquad \frac{5^3}{5^3} = \frac{\overset{1}{\cancel{5}} \cdot \overset{1}{\cancel{5}} \cdot \overset{1}{\cancel{5}}}{\underset{1}{\cancel{5}} \cdot \underset{1}{\cancel{5}} \cdot \underset{1}{\cancel{5}}} = 1$$

These results must be equal.

Since $\frac{5^3}{5^3} = 5^0$ and $\frac{5^3}{5^3} = 1$, we conclude that $5^0 = 1$. This observation suggests the following definition.

Zero Exponents

Any nonzero base raised to the 0 power is 1.

For any nonzero real number x,

$$x^0 = 1$$

EXAMPLE 1 Simplify. Assume $a \neq 0$: **a.** $(-8)^0$ **b.** $\left(\frac{14}{15}\right)^0$ **c.** $(3a)^0$ **d.** $3a^0$

Strategy We note that each exponent is 0. To simplify the expressions, we will identify the base and use the zero-exponent rule.

Why If an expression contains a nonzero base raised to the 0 power, we can replace it with 1.

Solution

a. $(-8)^0 = 1$ Because the base is -8 and the exponent is 0.

b. $\left(\frac{14}{15}\right)^0 = 1$ Because the base is $\frac{14}{15}$ and the exponent is 0.

c. $(3a)^0 = 1$ Because of the parentheses, the base is $3a$. The exponent is 0.

The Language of Algebra
The zero exponent definition does not define 0^0. This expression is called an *indeterminate form*, which is beyond the scope of this book.

d. $3a^0 = 3 \cdot a^0$ Since there are no parentheses, the base is a, not 3a. The exponent is 0.

$= 3 \cdot 1$

$= 3$

Self Check 1 Simplify each expression: **a.** $(0.75)^0$ **b.** $-5c^0d$ **c.** $(5c)^0$

Now Try Problems 13 and 17

2 Use the Negative Integer Exponent Rule.

The Language of Algebra
The *negative integers* are: $-1, -2, -3, -4, -5 \ldots$

To develop the definition of a negative exponent, we will simplify $\frac{6^2}{6^5}$ in two ways and compare the results.

If we apply the quotient rule for exponents, where we subtract the greater exponent in the denominator from the lesser exponent in the numerator, we get 6^{-3}. In the second approach, we remove the two common factors of 6 to get $\frac{1}{6^3}$.

$$\frac{6^2}{6^5} = 6^{2-5} = 6^{-3} \qquad \frac{6^2}{6^5} = \frac{\overset{1}{\cancel{6}} \cdot \overset{1}{\cancel{6}}}{\underset{1}{\cancel{6}} \cdot \underset{1}{\cancel{6}} \cdot 6 \cdot 6 \cdot 6} = \frac{1}{6^3}$$

These must be equal.

Since $\frac{6^2}{6^5} = 6^{-3}$ and $\frac{6^2}{6^5} = \frac{1}{6^3}$, we conclude that $6^{-3} = \frac{1}{6^3}$. Note that 6^{-3} is equal to the reciprocal of 6^3. This observation suggests the following definition.

Negative Exponents

For any nonzero real number x and any integer n,

$$x^{-n} = \frac{1}{x^n}$$

In words, x^{-n} is the reciprocal of x^n.

From the definition, we see that another way to write x^{-n} is to write its reciprocal and change the sign of the exponent. For example,

$$5^{-4} = \frac{1}{5^4}$$ Think of the reciprocal of 5^{-4}, which is $\frac{1}{5^{-4}}$. Then change the sign of the exponent.

EXAMPLE 2 Express using positive exponents and simplify, if possible:
a. 3^{-2} **b.** y^{-1} **c.** $(-2)^{-3}$ **d.** $5^{-2} - 10^{-2}$

Strategy Since each exponent is a negative number, we will use the negative exponent rule.

Why This rule enables us to write an exponential expression that has a negative exponent in an equivalent form using a positive exponent.

Caution
A negative exponent does not indicate a negative number. It indicates a reciprocal. For example,

$$4^{-2} = \frac{1}{4^2} = \frac{1}{16}$$

Solution

a. $3^{-2} = \frac{1}{3^2} = \frac{1}{9}$ Write the reciprocal of 3^{-2} and change the sign of the exponent.

b. $y^{-1} = \frac{1}{y^1} = \frac{1}{y}$ Write the reciprocal of y^{-1} and change the sign of the exponent.

c. $(-2)^{-3} = \frac{1}{(-2)^3}$ Because of the parentheses, the base is −2. Write the reciprocal of $(-2)^{-3}$ and change the exponent.

$= -\frac{1}{8}$ Evaluate: $(-2)^3 = -8$.

d. $5^{-2} - 10^{-2} = \frac{1}{5^2} - \frac{1}{10^2}$ Write the reciprocal of 5^{-2} and 10^{-2} and change the sign of each exponent.

$= \frac{1}{25} - \frac{1}{100}$ Evaluate: $5^2 = 25$ and $10^2 = 100$.

$= \frac{4}{100} - \frac{1}{100}$ Build $\frac{1}{25}$ to have a denominator of 100 so that the fractions can be subtracted: $\frac{1}{25} = \frac{1}{25} \cdot \frac{4}{4} = \frac{4}{100}$.

$= \frac{3}{100}$

Self Check 2 Express using positive exponents and simplify, if possible:
a. 8^{-2} **b.** x^{-5} **c.** $(-3)^{-3}$ **d.** $2^{-3} - 4^{-2}$

Now Try **Problems 25, 29, and 37**

EXAMPLE 3

Simplify. Do not use negative exponents in the answer.
a. $9m^{-3}$ **b.** -5^{-2}

Strategy We note that each exponent is a negative number. We will identify the base for each negative exponent and then use the negative exponent rule.

Why This rule enables us to write an exponential expression that has a negative exponent in an equivalent form using a positive exponent.

Caution
Don't confuse *negative numbers* with *negative exponents*. For example, −2 and 2^{-1} are not the same.

$$2^{-1} = \frac{1}{2^1} = \frac{1}{2}$$

Solution

a. $9m^{-3} = 9 \cdot m^{-3}$ The base is *m*. The exponent is −3.

$= 9 \cdot \frac{1}{m^3}$ Write the reciprocal of m^{-3} and change the sign of the exponent. Since 9 is not part of the base, it is not part of the reciprocal.

$= \frac{9}{m^3}$ Multiply.

b. $-5^{-2} = -1 \cdot 5^{-2}$ The base is 5. The exponent is −2.

$= -1 \cdot \frac{1}{5^2}$ Write the reciprocal of 5^{-2} and change the sign of the exponent. Since −1 is not part of the base, it is not part of the reciprocal.

$= -\frac{1}{25}$ Evaluate 5^2 and multiply. The result is negative because of the − sign in front of -5^{-2}.

Self Check 3 Simplify. Do not use negative exponents in the answer.

a. $12h^{-9}$ **b.** -2^{-4}

Now Try Problems 41 and 45

3 Use Exponent Rules to Change Negative Exponents in Fractions to Positive Exponents.

Negative exponents can appear in the numerator and/or the denominator of a fraction. To develop rules for such situations, we consider the following example.

$$\frac{a^{-4}}{b^{-3}} = \frac{\frac{1}{a^4}}{\frac{1}{b^3}} = \frac{1}{a^4} \div \frac{1}{b^3} = \frac{1}{a^4} \cdot \frac{b^3}{1} = \frac{b^3}{a^4}$$

We can obtain this result in a simpler way. In $\frac{a^{-4}}{b^{-3}}$, we can move a^{-4} from the numerator to the denominator and change the sign of the exponent, and we can move b^{-3} from the denominator to the numerator and change the sign of the exponent.

The Language of Algebra
Factors of a numerator or denominator may be moved *across the fraction bar* if we change the sign of their exponent.

$$\frac{a^{-4}}{b^{-3}} = \frac{b^3}{a^4}$$

This example suggests the following rules.

Changing from Negative to Positive Exponents

A factor can be moved from the denominator to the numerator or from the numerator to the denominator of a fraction if the sign of its exponent is changed.

For any nonzero real numbers x and y, and any integers m and n,

$$\frac{1}{x^{-n}} = x^n \quad \text{and} \quad \frac{x^{-m}}{y^{-n}} = \frac{y^n}{x^m}$$

These rules streamline the process when simplifying fractions involving negative exponents.

EXAMPLE 4 Simplify. Do not use negative exponents in the answer.

a. $\frac{1}{d^{-10}}$ **b.** $\frac{2^{-3}}{3^{-4}}$ **c.** $\frac{-5s^{-2}}{t^{-9}}$

Strategy We will move any factors in the numerator that have a negative exponent to the denominator. Then we will move any factors in the denominator that have a negative exponent to the numerator.

Why In this process, the sign of a negative exponent changes to positive.

Solution

a. $\frac{1}{d^{-10}} = d^{10}$ Move d^{-10} to the numerator and change the sign of the exponent.

Caution

A common error is to mistake the − sign in −5 for a negative exponent. *It is not an exponent.* The −5 should not be moved to the denominator and have its sign changed.

$$\frac{-5s^{-2}}{t^{-9}} \neq \frac{t^9}{5s^2}$$

b. $\frac{2^{-3}}{3^{-4}} = \frac{3^4}{2^3}$ Move 2^{-3} to the denominator and change the sign of the exponent. Move 3^{-4} to the numerator and change the sign of the exponent.

$= \frac{81}{8}$ Evaluate: $3^4 = 81$ and $2^3 = 8$.

c. $\frac{-5s^{-2}}{t^{-9}} = \frac{-5t^9}{s^2}$ Since $-5s^{-2}$ has no parentheses, s is the base. Move only s^{-2} to the denominator and change the sign of the exponent. Do not move −5. Move t^{-9} to the numerator and change the sign of the exponent.

Self Check 4 Simplify. Do not use negative exponents in the answer.

a. $\frac{1}{w^{-5}}$ **b.** $\frac{5^{-2}}{4^{-3}}$ **c.** $\frac{-8h^{-6}}{a^{-7}}$

Now Try **Problems 51, 55, and 59**

When a fraction is raised to a negative power, we can use rules for exponents to change the sign of the exponent. For example, we see that

$$\left(\frac{a}{2}\right)^{-3} = \frac{a^{-3}}{2^{-3}} = \frac{2^3}{a^3} = \left(\frac{2}{a}\right)^3$$

The exponent is the opposite of −3.

The base is the reciprocal of $\frac{a}{2}$.

This process can be streamlined using the following rule.

Negative Exponents and Reciprocals

A fraction raised to a power is equal to the reciprocal of the fraction raised to the opposite power.

For any nonzero real numbers x and y, and any integer n,

$$\left(\frac{x}{y}\right)^{-n} = \left(\frac{y}{x}\right)^n$$

EXAMPLE 5

Simplify: $\left(\frac{4}{m}\right)^{-2}$

Strategy We want to write this fraction that is raised to a negative power in an equivalent form that involves a positive power. We will use the negative exponent and reciprocal rules to do this.

Why It is usually easier to simplify exponential expressions if the exponents are positive.

Solution

$\left(\frac{4}{m}\right)^{-2} = \left(\frac{m}{4}\right)^2$ The base is the fraction $\frac{4}{m}$ and the exponent is −2. Write the reciprocal of the base and change the sign of the exponent.

$= \frac{m^2}{4^2}$ Use the power of a quotient rule: Raise the numerator and denominator to the second power.

$= \frac{m^2}{16}$ Evaluate: $4^2 = 16$.

Self Check 5 Simplify: $\left(\frac{c}{9}\right)^{-2}$

Now Try Problem 65

4 Use All Exponent Rules to Simplify Expressions.

The rules for exponents involving products, powers, and quotients are also true for zero and negative exponents.

Summary of Exponent Rules

If m and n represent integers and there are no divisions by zero, then

Product rule
$$x^m \cdot x^n = x^{m+n}$$

Power rule
$$(x^m)^n = x^{mn}$$

Power of a product
$$(xy)^n = x^n y^n$$

Quotient rule
$$\frac{x^m}{x^n} = x^{m-n}$$

Power of a quotient
$$\left(\frac{x}{y}\right)^n = \frac{x^n}{y^n}$$

Exponents of 0 and 1
$$x^0 = 1 \text{ and } x^1 = x$$

Negative exponent
$$x^{-n} = \frac{1}{x^n}$$

Negative exponents appearing in fractions
$$\frac{1}{x^{-n}} = x^n \qquad \frac{x^{-m}}{y^{-n}} = \frac{y^n}{x^m} \qquad \left(\frac{x}{y}\right)^{-n} = \left(\frac{y}{x}\right)^n$$

The rules for exponents are used to simplify expressions involving products, quotients, and powers. In general, an expression involving exponents is simplified when

- Each base occurs only once
- There are no parentheses
- There are no negative or zero exponents

EXAMPLE 6 Simplify. Do not use negative exponents in the answer.

a. $x^5 \cdot x^{-3}$ **b.** $\frac{x^3}{x^7}$ **c.** $(x^3)^{-2}$ **d.** $(2a^3b^{-5})^3$ **e.** $\left(\frac{3}{b^5}\right)^{-4}$

Strategy In each case, we want to write an equivalent expression using one base and one positive exponent. We will use rules for exponents to do this.

Why These expressions are not in simplest form. In parts a and b, the base x occurs more than once. In parts c, d, and e, there is a negative exponent.

Solution

a. $x^5 \cdot x^{-3} = x^{5+(-3)} = x^2$ Use the product rule: Keep the base, x, and add exponents.

b. $\frac{x^3}{x^7} = x^{3-7}$ Use the quotient rule: Keep the base, x, and subtract the exponents.

$= x^{-4}$ Do the subtraction: $3 - 7 = -4$.

$= \frac{1}{x^4}$ Write the reciprocal of x^{-4} and change the sign of the exponent.

Success Tip

We can use the negative exponent rule to simplify $(x^3)^{-2}$ in an alternate way:

$$(x^3)^{-2} = \frac{1}{(x^3)^2} = \frac{1}{x^6}$$

c. $(x^3)^{-2} = x^{-6}$ Use the power rule: Keep the base, x, and multiply exponents.

$= \frac{1}{x^6}$ Write the reciprocal of x^{-6} and change the sign of the exponent.

d. $(2a^3b^{-5})^3 = 2^3(a^3)^3(b^{-5})^3$ Raise each factor of the product $2a^3b^{-5}$ to the 3rd power.

$= 8a^9b^{-15}$ Use the power rule: Multiply exponents.

$= \frac{8a^9}{b^{15}}$ Move b^{-15} to the denominator and change the sign of the exponent.

e. $\left(\frac{3}{b^5}\right)^{-4} = \left(\frac{b^5}{3}\right)^4$ The base is the fraction $\frac{3}{b^5}$ and its exponent is −4. Write the reciprocal of the base and change the sign of the exponent.

$= \frac{(b^5)^4}{(3)^4}$ Use the power of a quotient rule: Raise the numerator and denominator to the 4th power.

$= \frac{b^{20}}{81}$ Use the power rule: Keep the base, b, and multiply the exponents 5 and 4. Evaluate: $3^4 = 81$.

Self Check 6 Simplify. Do not use negative exponents in the answer.

a. $t^8 \cdot t^{-4}$ **b.** $\frac{a^3}{a^8}$ **c.** $(n^4)^{-5}$

d. $(4c^2d^{-1})^3$ **e.** $\left(\frac{c^4}{2}\right)^{-3}$

Now Try **Problems 69, 73, 77, 81, and 85**

EXAMPLE 7

Simplify. Do not use negative exponents in the answer.

a. $\frac{y^{-4}y^{-3}}{y^{-20}}$ **b.** $\frac{7^{-1}a^3b^4}{6^{-2}a^5b^2}$ **c.** $\left(\frac{x^{-3}y^2}{xy^{-3}}\right)^2$

Strategy In each case, we want to write an equivalent expression that uses each base with a positive exponent only once.

Why These expressions are not in simplest form. The bases occur more than once and the expressions contain negative exponents.

Solution

a. $\frac{y^{-4}y^{-3}}{y^{-20}} = \frac{y^{-7}}{y^{-20}}$ In the numerator, use the product rule: Keep the common base, y, and add exponents: $-4 + (-3) = -7$.

$= y^{-7-(-20)}$ Use the quotient rule: Keep the common base, y, and subtract exponents.

$= y^{13}$ Do the subtraction: $-7 - (-20) = -7 + 20 = 13$.

Alternate solution: To avoid working with negative numbers, we could move each factor across the fraction bar and change the sign of its exponent.

$$\frac{y^{-4}y^{-3}}{y^{-20}} = \frac{y^{20}}{y^4y^3} = \frac{y^{20}}{y^7} = y^{13}$$

Caution

We cannot use this approach if the negative exponents occur in a sum or difference of terms. For example:

$$\frac{y^{-4} + y^{-3}}{y^{-20}} \neq \frac{y^{20}}{y^4 + y^3}$$

You will study this situation in more detail in your next algebra course.

Success Tip

Since the rules for exponents can be applied in different orders, there are several equally valid ways to simplify these expressions.

b. $\dfrac{7^{-1}a^3b^4}{6^{-2}a^5b^2} = \dfrac{6^2a^3b^4}{7^1a^5b^2}$ Move 7^{-1} to the denominator. Change the sign of the exponent. Move 6^{-2} to the numerator. Change the sign of the exponent.

$= \dfrac{36a^{3-5}b^{4-2}}{7}$ Use the quotient rule twice: Keep each base, *a* and *b*, and subtract exponents.

$= \dfrac{36a^{-2}b^2}{7}$

$= \dfrac{36b^2}{7a^2}$ Move a^{-2} to the denominator and change the sign of the exponent.

c. $\left(\dfrac{x^{-3}y^2}{xy^{-3}}\right)^2 = [x^{-3-1}y^{2-(-3)}]^2$ Within the parentheses, use the quotient rule twice: Keep each base, x and y, and subtract exponents.

$= (x^{-4}y^5)^2$

$= x^{-8}y^{10}$ Raise each factor within the parentheses to the second power.

$= \dfrac{y^{10}}{x^8}$ Move x^{-8} to the denominator and change the sign of its exponent. y^{10} is not affected.

Alternate solution: To simplify the expression, we can begin on the "outside" by using the power of a quotient rule first.

$$\left(\frac{x^{-3}y^2}{xy^{-3}}\right)^2 = \frac{(x^{-3})^2(y^2)^2}{x^2(y^{-3})^2} = \frac{x^{-6}y^4}{x^2y^{-6}} = \frac{y^4y^6}{x^6x^2} = \frac{y^{10}}{x^8}$$

Self Check 7 Simplify. Do not use negative exponents in the answer.

a. $\dfrac{a^{-4}a^{-5}}{a^{-3}}$ **b.** $\dfrac{1^{-4}x^5y^3}{9^{-2}x^3y^6}$ **c.** $\left(\dfrac{c^{-2}d^2}{c^4d^{-3}}\right)^3$

Now Try **Problems 89, 93, and 97**

ANSWERS TO SELF CHECKS **1. a.** 1 **b.** $-5d$ **c.** 1 **2. a.** $\frac{1}{64}$ **b.** $\frac{1}{x^5}$ **c.** $-\frac{1}{27}$ **d.** $\frac{1}{16}$ **3. a.** $\frac{12}{h^9}$ **b.** $-\frac{1}{16}$ **4. a.** w^5 **b.** $\frac{64}{25}$ **c.** $-\frac{8a^7}{h^6}$ **5.** $\frac{81}{c^2}$ **6. a.** t^4 **b.** $\frac{1}{a^5}$ **c.** $\frac{1}{n^{20}}$ **d.** $\frac{64c^6}{d^3}$ **e.** $\frac{8}{c^{12}}$ **7. a.** $\frac{1}{a^6}$ **b.** $\frac{81x^2}{y^3}$ **c.** $\frac{d^{15}}{c^{18}}$

STUDY SET 5.2

VOCABULARY

Fill in the blanks.

1. In the expression 5^{-1}, the exponent is a ________ integer.

2. x^{-n} is the __________ of x^n.

CONCEPTS

3. Complete the table.

Expression	Base	Exponent
4^{-2}		
$6x^{-5}$		
$\left(\frac{3}{y}\right)^{-8}$		
-7^{-1}		
$(-2)^{-3}$		
$10a^0$		

4. Complete each rule for exponents.

a. $x^m \cdot x^n =$ 　 **b.** $x^0 =$

c. $(x^m)^n =$ 　 **d.** $(xy)^n =$

e. $\left(\frac{x}{y}\right)^n =$ 　 **f.** $x^{-n} =$

g. $\frac{1}{x^{-n}} =$ 　 **h.** $\frac{x^{-m}}{y^{-n}} =$

i. $\frac{x^m}{x^n} =$ 　 **j.** $\left(\frac{x}{y}\right)^{-n} =$

Complete each table.

5.

x	3^x
2	
1	
0	
−1	
−2	

6.

x	$(-9)^x$
2	
1	
0	
−1	
−2	

7. Fill in the blanks.

a. $2^{-3} = \frac{1}{2}$ 　 **b.** $\frac{1}{t^{-6}} = t$

8. A factor can be moved from the denominator to the numerator or from the numerator to the denominator of a fraction if the ______ of its exponent is changed.

$$\frac{5^{-2}}{6^{-3}} = \frac{6}{5}$$

9. A fraction raised to a power is equal to the __________ of the fraction raise to the opposite power.

$$\left(\frac{3}{d}\right)^{-2} = \left(\frac{d}{3}\right)$$

10. Determine whether each statement is true or false.

a. $6^{-2} = -36$ 　 **b.** $6^{-2} = \frac{1}{36}$

c. $\frac{x^3}{y^{-2}} = \frac{y^2}{x^3}$ 　 **d.** $\frac{-6x^{-5}}{y^{-6}} = \frac{y^6}{6x^5}$

NOTATION

Complete each solution to simplify each expression.

11. $(y^5y^3)^{-5} = (\quad)^{-5} = y \quad = \frac{1}{y}$

12. $\left(\frac{a^2b}{a^{-3}b^3}\right)^3 = (a^{2-} \quad b^{1-})^3$

$= (a \quad b \quad)^3$

$= a \quad b$

$= \frac{a^{15}}{b}$

GUIDED PRACTICE

Simplify each expression. **See Example 1.**

13. 7^0 　 **14.** 9^0

15. $\left(\frac{1}{4}\right)^0$ 　 **16.** $\left(\frac{3}{8}\right)^0$

17. $2x^0$ 　 **18.** $8t^0$

19. $15(-6x)^0$ 　 **20.** $4(-12y)^0$

21. $\left(\frac{a^2b^3}{ab^4}\right)^0$ 　 **22.** $\left(\frac{xyz}{x^2y}\right)^0$

23. $\frac{5}{2x^0}$ 　 **24.** $\frac{4}{3a^0}$

Express using positive exponents and simplify, if possible. **See Example 2.**

25. 2^{-2} 　 **26.** 7^{-2}

27. 6^{-1} 　 **28.** 5^{-1}

29. x^{-9} 　 **30.** y^{-3}

31. b^{-5} 　 **32.** c^{-4}

33. $(-5)^{-1}$ 　 **34.** $(-8)^{-1}$

35. $(-10)^{-3}$ 　 **36.** $(-9)^{-2}$

37. $2^{-2} + 4^{-1}$ 　 **38.** $-9^{-1} + 9^{-2}$

39. $9^0 - 9^{-1}$ 　 **40.** $7^{-1} - 7^0$

Simplify. Do not use negative exponents in the answer. **See Example 3.**

41. $15g^{-6}$ 　 **42.** $16t^{-3}$

43. $5x^{-3}$ 　 **44.** $27m^{-3}$

45. -3^{-3} 　 **46.** -6^{-3}

47. -8^{-2} 　 **48.** -4^{-2}

Simplify. Do not use negative exponents in the answer. **See Example 4.**

49. $\frac{1}{5^{-3}}$ 　 **50.** $\frac{1}{3^{-3}}$

51. $\frac{1}{r^{-20}}$ 　 **52.** $\frac{1}{s^{-30}}$

53. $\frac{8}{s^{-1}}$ 　 **54.** $\frac{6}{k^{-2}}$

55. $\frac{2^{-4}}{3^{-1}}$ 　 **56.** $\frac{7^{-2}}{2^{-3}}$

57. $\frac{a^{-5}}{b^{-2}}$ 　 **58.** $\frac{r^{-6}}{s^{-1}}$

59. $\frac{-4d^{-1}}{p^{-10}}$ 　 **60.** $\frac{-9m^{-1}}{n^{-30}}$

***Simplify.* See Example 5.**

61. $\left(\frac{1}{6}\right)^{-2}$
62. $\left(\frac{1}{7}\right)^{-2}$
63. $\left(\frac{1}{2}\right)^{-3}$
64. $\left(\frac{1}{5}\right)^{-3}$
65. $\left(\frac{c}{d}\right)^{-8}$
66. $\left(\frac{a}{x}\right)^{-10}$
67. $\left(\frac{3}{m}\right)^{-4}$
68. $\left(\frac{2}{t}\right)^{-4}$

***Simplify. Do not use negative exponents in the answer.* See Example 6.**

69. $y^8 \cdot y^{-2}$
70. $m^{10} \cdot m^{-6}$
71. $b^{-7} \cdot b^{14}$
72. $c^{-9} \cdot c^{14}$
73. $\frac{y^4}{y^5}$
74. $\frac{t^7}{t^{10}}$
75. $\frac{h^{-5}}{h^2}$
76. $\frac{y^{-3}}{y^4}$
77. $(x^4)^{-3}$
78. $(y^{-3})^2$
79. $(b^2)^{-4}$
80. $(n^3)^{-5}$
81. $(6s^4t^{-7})^2$
82. $(11r^{10}s^{-3})^2$
83. $(2u^{-2}v^5)^5$
84. $(3w^{-8}x^3)^4$
85. $\left(\frac{4}{x^3}\right)^{-3}$
86. $\left(\frac{2}{b^5}\right)^{-2}$
87. $\left(\frac{y^4}{3}\right)^{-2}$
88. $\left(\frac{p^3}{2}\right)^{-2}$

***Simplify. Do not use negative exponents in the answer.* See Example 7.**

89. $\frac{y^{-3}}{y^{-4}y^{-2}}$
90. $\frac{x^{-12}}{x^{-3}x^{-4}}$
91. $\frac{a^{-5}a^{-9}}{a^{-8}}$
92. $\frac{b^{-2}b^{-3}}{b^{-9}}$
93. $\frac{2^{-1}a^4b^2}{3^{-2}a^2b^4}$
94. $\frac{6^{-2}b^9c^3}{5^{-3}b^4c^8}$
95. $\frac{9^{-2}s^6t}{4^{-3}s^4t^5}$
96. $\frac{2^{-5}m^{10}n^6}{5^{-2}m^6n^{10}}$
97. $\left(\frac{x^2y^{-2}}{x^{-5}y^3}\right)^4$
98. $\left(\frac{r^4s^{-3}}{r^{-3}s^7}\right)^3$
99. $\left(\frac{y^3z^{-2}}{y^{-4}z^3}\right)^2$
100. $\left(\frac{xy^3}{x^{-1}y^{-1}}\right)^3$

TRY IT YOURSELF

Simplify. Do not use negative exponents in the answer. Assume that no variables are 0.

101. $\left(\frac{a^4}{2b}\right)^{-3}$
102. $\left(\frac{n^8}{9m}\right)^{-2}$
103. $\frac{r^{-50}}{r^{-70}}$
104. $\frac{m^{-30}}{m^{-40}}$
105. $(5d^{-2})^3$
106. $(9s^{-6})^2$
107. $-15x^0y$
108. $24g^0h^2$
109. $\left(\frac{4}{h^{10}}\right)^{-2}$
110. $\left(\frac{x^4}{3}\right)^{-4}$
111. $x^{-3} \cdot x^{-3}$
112. $y^{-2} \cdot y^{-2}$
113. $\left(\frac{c^3d^{-4}}{c^{-1}d^5}\right)^3$
114. $\left(\frac{s^2t^{-8}}{s^{-9}t^2}\right)^4$
115. $\frac{2^{-2}g^{-2}h^{-3}}{9^{-1}h^{-3}}$
116. $\frac{5^{-1}x^{-2}y^{-3}}{8^{-2}x^{-11}}$
117. $(2x^3y^{-2})^5$
118. $(3u^{-2}v^3)^3$
119. $\frac{t(t^{-2})^{-2}}{t^{-5}}$
120. $\frac{d(d^{-3})^{-3}}{d^{-7}}$
121. $\frac{-4s^{-5}}{t^{-2}}$
122. $\frac{-9k^{-8}}{m^{-2}}$
123. $(x^{-4}x^3)^3$
124. $(y^{-2}y)^3$

APPLICATIONS

125. UNIT COMPARISONS Consider the relative sizes of the items below. In the measurement column, write the most appropriate power of 10 from the following list. Each power is used only once. (A meter is slightly longer than one yard.)

10^{-5} 10^{-4} 10^{-3} 10^{-2} 10^{-1} 10^0

smallest — largest

Item	Measurement (meter)
Thickness of a dime	
Height of a bathroom sink	
Length of a pencil eraser	
Thickness of soap bubble film	
Length of a cell phone	
Thickness of a piece of paper	

126. ELECTRONICS The total resistance R of a certain circuit is given by

$$R = \left(\frac{1}{R_1} + \frac{1}{R_2}\right)^{-1} + R_3$$

Find R if $R_1 = 4$, $R_2 = 2$, and $R_3 = 1$.

WRITING

127. Explain how you would help a friend understand that 2^{-3} is not equal to -8.

128. Explain the error:

$$\frac{-5x^{-2}}{y^{-2}} = \frac{y^2}{5x^2}$$

REVIEW

Find the slope of the line that passes through the given points.

129. $(1, -4)$ and $(3, -7)$

130. $(1, 3)$ and $(3, -1)$

131. Write an equation of the line having slope $\frac{3}{4}$ and y-intercept -5.

132. Find an equation of the line that passes through $(4, 4)$ and $(-6, -6)$. Write the answer in slope–intercept form.

CHALLENGE PROBLEMS

133. Simplify each expression. Do not use negative exponents in the answer. The variable m represents a positive integer.

a. $r^{5m}r^{-6m}$

b. $\dfrac{x^{3m}}{x^{6m}}$

134. Write an expression equivalent to $\left(\dfrac{2x^3y^7}{3z^5}\right)^9$ that involves only *negative* exponents.

SECTION 5.3
Scientific Notation

Objectives

1. Convert from scientific to standard notation.
2. Write numbers in scientific notation.
3. Perform computations with scientific notation.

Scientists often deal with extremely large and small numbers. For example, the distance from the Earth to the sun is approximately 150,000,000 kilometers. The influenza virus, which causes flu symptoms of cough, sore throat, and headache, has a diameter of 0.00000256 inch.

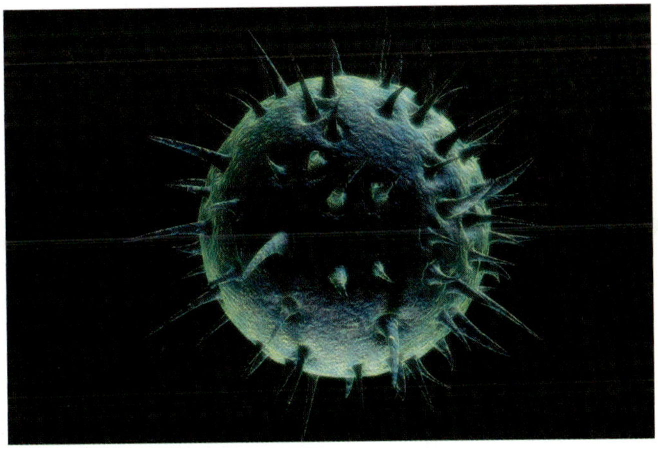

The numbers 150,000,000 and 0.00000256 are written in **standard notation,** which is also called **decimal notation.** Because they contain many zeros, they are difficult to read and cumbersome to work with in calculations. In this section, we will discuss a more convenient form in which we can write such numbers.

1 Convert from Scientific to Standard Notation.

Scientific notation provides a compact way of writing very large or very small numbers.

Scientific Notation

A positive number is written in **scientific notation** when it is written in the form $N \times 10^n$, where $1 \leq N < 10$ and n is an integer.

To write numbers in scientific notation, you need to be familiar with **powers of 10,** like those listed in the table below.

Power of 10	10^{-3}	10^{-2}	10^{-1}	10^0	10^1	10^2	10^3
Value	$\frac{1}{1{,}000} = 0.001$	$\frac{1}{100} = 0.01$	$\frac{1}{10} = 0.1$	1	10	100	1,000

Two examples of numbers written in scientific notation are shown below. Note that each of them is the product of a decimal number (between 1 and 10) and a power of 10.

An integer exponent

$$3.67 \times 10^2 \qquad 2.158 \times 10^{-3}$$

A decimal that is at least 1, but less than 10

Notation
The × sign for multiplication is usually used to write scientific notation. However, some books use a raised dot · instead.

A number written in scientific notation can be converted to standard notation by performing the indicated multiplication. For example, to convert 3.67×10^2, we recall that multiplying a decimal by 100 moves the decimal point 2 places to the right.

$$3.67 \times 10^2 = 3.67 \times 100 = 367.$$

To convert 2.158×10^{-3} to standard notation, we recall that dividing a decimal by 1,000 moves the decimal point 3 places to the left.

$$2.158 \times 10^{-3} = 2.158 \times \frac{1}{10^3} = 2.158 \times \frac{1}{1{,}000} = \frac{2.158}{1{,}000} = 0.002158$$

In 3.67×10^2 and 2.158×10^{-3}, the exponent gives the number of decimal places that the decimal point moves, and the sign of the exponent indicates the direction in which it moves. Applying this observation to several other examples, we have

$5.32 \times 10^6 = 5320000.$ Move the decimal point 6 places to the right.

$1.95 \times 10^{-5} = 0.0000195$ Move the decimal point $|-5| = 5$ places to the left.

$9.7 \times 10^0 = 9.7$ There is no movement of the decimal point.

Success Tip
Since $10^0 = 1$, scientific notation involving 10^0 is easily simplified. For example,

$$9.7 \times 10^0 = 9.7 \times 1 = 9.7$$

The following procedure summarizes our observations.

Converting from Scientific to Standard Notation

1. If the exponent is positive, move the decimal point the same number of places to the right as the exponent.
2. If the exponent is negative, move the decimal point the same number of places to the left as the absolute value of the exponent.

EXAMPLE 1 Convert to standard notation: **a.** 3.467×10^5 **b.** 8.9×10^{-4}

Strategy In each case, we need to identify the exponent on the power of 10 and consider its sign.

Why The exponent gives the number of decimal places that we should move the decimal point. The sign of the exponent indicates whether it should be moved to the right or the left.

The Language of Algebra
Standard notation is also called *decimal notation.*

Solution

a. Since the exponent in 10^5 is 5, the decimal point moves 5 places to the right.

3 4 6 7 0 0. *To move 5 places to the right, two placeholder zeros must be written.*

Thus, $3.467 \times 10^5 = 346{,}700$.

b. Since the exponent in 10^{-4} is -4, the decimal point moves 4 places to the left.

0.0 0 0 8 9 *To move 4 places to the left, three placeholder zeros must be written.*

Thus, $8.9 \times 10^{-4} = 0.00089$

Self Check 1 Convert to standard notation: **a.** 4.88×10^6 **b.** 9.8×10^{-3}

Now Try **Problems 13 and 17**

2 Write Numbers in Scientific Notation.

To write a number in scientific notation ($N \times 10^n$) we first determine N and then n.

EXAMPLE 2 Write each number in scientific notation: **a.** 150,000,000 **b.** 0.00000256 **c.** 432×10^5

Strategy We will write each number as the product of a number between 1 and 10 and a power of 10.

Why Numbers written in scientific notation have the form $N \times 10^n$.

Solution

a. We must write 150,000,000 (the distance from the Earth to the sun) as the product of a number between 1 and 10 and a power of 10. We note that 1.5 lies between 1 and 10. To obtain 150,000,000, we must move the decimal point in 1.5 exactly 8 places to the right.

1.5 0 0 0 0 0 0 0

This will happen if we multiply 1.5 by 10^8. Therefore,

$150{,}000{,}000 = 1.5 \times 10^8$ *This is the distance (in kilometers) from the Earth to the sun.*

Notation

Don't apply the negative exponent rule when writing numbers in scientific notation.

2.56×10^{-6} ~~$2.56 \times \frac{1}{10^6}$~~

Calculators

Scientific notation

Calculators automatically change into scientific notation when a result is too large or too small to fit the answer display.

$(453.46)^5 = 1.917321395 \times 10^{13}$

1.917321395 13

$(0.0005)^{12} = 2.44140625 \times 10^{-40}$

2.44140625 -40

b. We must write 0.00000256 (the diameter in inches of a flu virus) as the product of a number between 1 and 10 and a power of 10. We note that 2.56 lies between 1 and 10. To obtain 0.00000256, the decimal point in 2.56 must be moved 6 places to the left.

0 0 0 0 0 0 2.56

This will happen if we multiply 2.56 by 10^{-6}. Therefore,

$0.00000256 = 2.56 \times 10^{-6}$ This is the diameter (in inches) of a flu virus.

c. The number 432×10^5 is not written in scientific notation because 432 is not a number between 1 and 10. To write this number in scientific notation, we proceed as follows:

$432 \times 10^5 = 4.32 \times 10^2 \times 10^5$ Write 432 in scientific notation.

$= 4.32 \times 10^7$ Use the product rule to find $10^2 \times 10^5$. Keep the base, 10, and add the exponents.

Written in scientific notation, 432×10^5 is 4.32×10^7.

Self Check 2 Write each number in scientific notation:

a. 93,000,000 **b.** 0.00009055

c. 85×10^{-3}

Now Try **Problems 31, 35, and 55**

The results from Example 2 illustrate the following forms to use when converting numbers from standard to scientific notation.

For real numbers between 0 and 1: $\square \times 10^{\text{negative integer}}$

For real numbers at least 1, but less than 10: $\square \times 10^0$

For real numbers greater than or equal to 10: $\square \times 10^{\text{positive integer}}$

3 Perform Computations with Scientific Notation.

Another advantage of scientific notation becomes apparent when we evaluate products or quotients that involve very large or small numbers. If we express those numbers in scientific notation, we can use rules for exponents to make the calculations easier.

NASA

EXAMPLE 3 ***Astronomy.*** Except for the sun, the nearest star visible to the naked eye from most parts of the United States is Sirius. Light from Sirius reaches Earth in about 70,000 hours. If light travels at approximately 670,000,000 mph, how far from Earth is Sirius?

Strategy We can use the formula $d = rt$ to find the distance from Sirius to Earth.

Why We know the *rate* at which light travels and the *time* it takes to travel from Sirius to the Earth.

Solution The rate at which light travels is 670,000,000 mph and the time it takes the light to travel from Sirius to Earth is 70,000 hr. To find the distance from Sirius to Earth, we proceed as follows:

$d = rt$	This is the formula for distance traveled.
$d = 670{,}000{,}000(70{,}000)$	Substitute 670,000,000 for r and 70,000 for t.
$= (6.7 \times 10^8)(7.0 \times 10^4)$	Write each number in scientific notation.
$= (6.7 \cdot 7.0) \times (10^8 \times 10^4)$	Group the decimals together and the powers of 10 together.
$= (6.7 \cdot 7.0) \times 10^{8+4}$	Use the product rule to find $10^8 \times 10^4$. Keep the base, 10, and add exponents.
$= 46.9 \times 10^{12}$	Do the multiplication and the addition.

We note that 46.9 is not between 0 and 1, so 46.9×10^{12} is not written in scientific notation. To answer in scientific notation, we proceed as follows.

$= 4.69 \times 10^1 \times 10^{12}$	Write 46.9 in scientific notation as 4.69×10^1.
$= 4.69 \times 10^{13}$	Use the product rule to find $10^1 \times 10^{12}$. Keep the base, 10, and add the exponents.

Sirius is approximately 4.69×10^{13} or 46,900,000,000,000 miles from Earth.

Now Try Problem 61

EXAMPLE 4

Atoms. As an example of how scientific notation is used in chemistry, we can approximate the weight (in grams) of one atom of the element uranium by evaluating the following expression.

$$\frac{2.4 \times 10^2}{6 \times 10^{23}}$$

Strategy To simplify, we will divide the numbers and powers of 10 separately.

Why We can then use the quotient rule for exponents to simplify the calculations.

Solution

$\frac{2.4 \times 10^2}{6 \times 10^{23}} = \frac{2.4}{6} \times \frac{10^2}{10^{23}}$	Divide the decimals and the powers of 10 separately.
$= \frac{2.4}{6} \times 10^{2-23}$	For the powers of 10, use the quotient rule. Keep the base, 10, and subtract the exponents.
$= 0.4 \times 10^{-21}$	Divide the decimals. Subtract the exponents. The result is not in scientific notation form.
$= 4 \times 10^{-1} \times 10^{-21}$	Write 0.4 in scientific notation as 4×10^{-1}.
$= 4 \times 10^{-22}$	Use the product rule to find $10^{-1} \times 10^{-21}$. Keep the base, 10, and add the exponents.

One atom of uranium weighs 4×10^{-22} gram or 0.0000000000000000000004 g.

Self Check 4 Find the approximate weight (in grams) of one atom of gold by evaluating: $\frac{1.98 \times 10^2}{6 \times 10^{23}}$

Now Try Problem 65

Calculators

Entering scientific notation

To evaluate the expression in Example 4 on a scientific calculator, we enter the numbers using the EE key:

2.4 EE 2 ÷ 6 EE 23 =

92. OIL As of 2006, Saudi Arabia was believed to have crude oil reserves of about 2.643×10^{11} barrels. A barrel contains 42 gallons of oil. Use scientific notation to express Saudi Arabia's oil reserves in gallons. (Source: *infoplease*)

93. INSURED DEPOSITS As of June 2006, the total insured deposits in U.S. banks and savings and loans was approximately 6.4×10^{12} dollars. If this money was invested at a rate of 4% simple annual interest, how much would it earn in 1 year? Use scientific notation to express the answer. (Source: Federal Deposit Insurance Corporation.)

94. CURRENCY As of December 2006, the number of \$20 bills in circulation was approximately 5.96×10^{9}. What was the total value of the currency? Express the answer in scientific notation and standard notation. (Source: The Federal Reserve.)

95. POWERS OF 10 In the United States, we use Latin prefixes in front of "illion" to name extremely large numbers. Write each number in scientific notation.

One million: 1,000,000

One billion: 1,000,000,000

One trillion: 1,000,000,000,000

One quadrillion: 1,000,000,000,000,000

One quintillion: 1,000,000,000,000,000,000

96. SUPERCOMPUTERS As of June 2006, the world's fastest computer was IBM's BlueGene/L System. If it could make 2.81×10^{14} calculations in one second, how many could it make in one minute? Answer in scientific notation.

WRITING

97. In what situations would scientific notation be more convenient than standard notation?

98. To multiply a number by a power of 10, we move the decimal point. Which way, and how far? Explain.

99. 2.3×10^{-3} contains a negative sign but represents a positive number. Explain.

100. Explain why 237.8×10^{8} is not written in scientific notation.

REVIEW

101. If $y = -1$, find the value of $-5y^{55}$.

102. What is the y-intercept of the graph of $y = -3x - 5$?

103. COUNSELING At the end of her first year of practice, a family counselor had 75 clients. At the end of her second year, she had 105 clients. If a linear trend continues, write an equation that gives the number of clients c the counselor will have at the end of t years.

104. Is $(0, -5)$ a solution of $2x + 3y \geq -14$?

CHALLENGE PROBLEMS

105. Consider 2.5×10^{-4}. Answer the following questions in scientific notation form.

a. What is its opposite?

b. What is its reciprocal?

106. a. Write the numbers one million and one millionth in scientific notation.

b. By what number must we multiply one millionth to get one million?

SECTION 5.4
Polynomials

Objectives

1. Know the vocabulary for polynomials.
2. Evaluate polynomials.
3. Graph equations defined by polynomials.

In this section, we will discuss a special type of algebraic expression called a *polynomial.*

1 Know the Vocabulary for Polynomials.

Recall from Chapter 1 that a *term* is a product or quotient of numbers and/or variables. A single number or variable is also a term. Some examples of terms are:

$$14, \quad x, \quad -6y^3, \quad 9cd^2, \quad \text{and} \quad \frac{5}{y}$$

Polynomials

A **polynomial** is a single term or a sum of terms in which all variables have whole-number exponents and no variable appears in a denominator.

The Language of Algebra
The prefix *poly* means many. Some other words that begin with this prefix are *poly*gon, *poly*ester, and *poly*unsaturated.

Caution
The expression $6x^3 + 4x^{-2}$ is not a polynomial, because of the negative exponent on the variable in $4x^{-2}$. Similarly, $y^2 + \frac{5}{y} + 1$ is not a polynomial, because of the variable y appears in the denominator.

Here are some examples of polynomials:

$$3x + 2, \qquad 4y^2 - 2y - 3, \qquad a^3 + 3a^2b + 3ab^2 + b^3, \qquad \text{and} \qquad -8xy^2z$$

The polynomial $3x + 2$ is the sum of two terms, $3x$ and 2, and we say it is a **polynomial in one variable, *x*.** A single number is called a **constant,** and so its last term, 2, is called the **constant term.**

A polynomial is defined as a single term or the sum of several terms. Since $4y^2 - 2y - 3$ can be written as the sum $4y^2 + (-2y) + (-3)$, it has three terms, $4y^2$, $-2y$, and -3. It is written in **descending powers** of y, because the exponents on y decrease from left to right. When a polynomial is written in descending powers, the first term, in this case $4y^2$, is called the **leading term.** The coefficient of the leading term, in this case 4, is called the **leading coefficient.**

A polynomial can have more than one variable. For example, $a^3 + 3a^2b + 3ab^3 + b^3$ is a **polynomial in two variables,** a and b. It has four terms and is written in descending powers of a and **ascending powers** of b. The polynomial $-8xy^2z$ is a polynomial in three variables, x, y, and z, and has only one term.

Polynomials are classified according to the number of terms they have. A polynomial with exactly one term is called a **monomial;** exactly two terms, a **binomial;** and exactly three terms, a **trinomial.** Polynomials with four or more terms have no special names.

The Language of Algebra
The prefix *mono* means one; Jay Leno begins the *Tonight Show* with a monologue. The prefix *bi* means two, as in bicycle or binoculars. The prefix *tri* means three, as in triangle or the *Lord of the Rings Trilogy*.

Polynomials

Monomials	Binomials	Trinomials
$-6x$	$9u - 4$	$5t^2 + 4t + 3$
$5.5x^3y^2$	$-29z^4 - z^2$	$27x^3 - 6x^2 - 2x$
11	$18a^2b + 4ab$	$\frac{1}{2}a^2 + 2ab + b^2$

Polynomials and their terms can be described according to the exponents on their variables.

Degree of a Term of a Polynomial

The **degree of a term** of a polynomial in one variable is the value of the exponent on the variable. If a polynomial is in more than one variable, the **degree of a term** is the sum of the exponents on the variables in that term. The **degree of a nonzero constant** is 0.

Success Tip
The *degree of a term* is the number of variable factors in that term. So, $9x^6$ has degree 6 because it has 6 variable factors:

$$9x^6 = 9 \cdot x \cdot x \cdot x \cdot x \cdot x \cdot x$$

Here are some examples:

$9x^6$ has degree 6.

$-2a^4$ has degree 4.

$47x^2y^{11}$ has degree 13 because $2 + 11 = 13$.

8 has degree 0 since it can be written as $8x^0$.

We determine the *degree of a polynomial* by considering the degrees of each of its terms.

Degree of a Polynomial

The **degree of a polynomial** is the same as the highest degree of any term of the polynomial.

EXAMPLE 1 Use the vocabulary of this section to describe each polynomial:

a. $d^4 + 9d^2 - 16$ **b.** $\frac{1}{2}x^2 - x$

c. $-6y^{14} - 1.5y^9z^9 + 2.5y^8z^{10} + yz^{11}$

Strategy First, we will identify the variable(s) in the polynomial and determine whether it is written in ascending or descending powers. Then we will count the number of terms in the polynomial and determine the degree of each term.

Why The number of terms determines the type of polynomial. The highest degree of any term of the polynomial determines its degree.

The Language of Algebra
To *descend* means to move from higher to lower. When we write $d^4 + 9d^2 - 16$ as $d^4 + 0d^3 + 9d^2 + 0d^1 - 16d^0$, we more clearly see the *descending* powers of d.

Solution

a. $d^4 + 9d^2 - 16$ is a polynomial in one variable that is written in descending powers of d. If we write the subtraction as addition of the opposite, we see that it has 3 terms, d^4, $9d^2$, and -16, and is therefore a trinomial. The highest degree of any of its terms is 4, so it is of degree 4.

$$d^4 + 9d^2 - 16 = \underset{\text{1st Term}}{d^4} + \underset{\text{2nd Term}}{9d^2} + \underset{\text{3rd Term}}{(-16)}$$

Term	Coefficient	Degree
d^4	1	4
$9d^2$	9	2
-16	-16	0

Degree of the polynomial: 4

Success Tip
Recall that *terms* are separated by + symbols and that the *coefficient* of a term is the numerical factor of the term.

b. $\frac{1}{2}x^2 - x$ is a polynomial in one variable. It is written in descending powers of x. Since it has two terms, it is a binomial. The highest degree of any of its terms is 2, so it is of degree 2.

Term	Coefficient	Degree
$\frac{1}{2}x^2$	$\frac{1}{2}$	2
$-x$	-1	1

Degree of the polynomial: 2

c. $-6y^{14} - 1.5y^9z^9 + 2.5y^8z^{10} + yz^{11}$ is a polynomial in two variables, y and z. It is written in descending powers of y and ascending powers of z. It has 4 terms, and therefore has no special name. The highest degree of any term is 18, so it is of degree 18.

Term	Coefficient	Degree
$-6y^{14}$	-6	14
$-1.5y^9z^9$	-1.5	18
$2.5y^8z^{10}$	2.5	18
yz^{11}	1	12

Degree of the polynomial: 18

Self Check 1 Describe each polynomial: **a.** $x^2 + 4x - 16$
b. $-14s^5t + s^4t^3$

Now Try Problems 17 and 35

2 Evaluate Polynomials.

A polynomial can have different values depending on the number that is substituted for its variable (or variables).

EXAMPLE 2 Evaluate $3x^2 + 4x - 5$ for $x = 0$ and $x = -2$.

Strategy We will substitute the given value for each x in the polynomial and follow the rules for the order of operations.

Why To *evaluate a polynomial* means to find its numerical value, once we know the value of its variable.

Caution

Recall that to evaluate $3(-2)^2$, the rules for the order of operations require that we find $(-2)^2$ first, and then multiply that result by 3.

Solution

For x = 0:

$$\begin{aligned} 3x^2 + 4x - 5 &= 3(0)^2 + 4(0) - 5 \\ &= 3(0) + 4(0) - 5 \\ &= 0 + 0 - 5 \\ &= -5 \end{aligned}$$

For x = −2:

$$\begin{aligned} 3x^2 + 4x - 5 &= 3(-2)^2 + 4(-2) - 5 \\ &= 3(4) + 4(-2) - 5 \\ &= 12 + (-8) - 5 \\ &= -1 \end{aligned}$$

Self Check 2 Evaluate $-x^3 + x - 2x + 3$ for $x = -3$.

Now Try **Problem 54**

EXAMPLE 3 ***Supermarket Displays.*** The polynomial $\frac{1}{3}c^3 + \frac{1}{2}c^2 + \frac{1}{6}c$ gives the number of cans used in a display shaped like a square pyramid, having a square base formed by c cans per side. Find the number of cans used in the display.

Strategy We will evaluate the polynomial for $c = 4$.

Why From the illustration, we see that each side of the square base is formed by 4 cans.

Solution

$$\begin{aligned} \frac{1}{3}c^3 + \frac{1}{2}c^2 + \frac{1}{6}c &= \frac{1}{3}(4)^3 + \frac{1}{2}(4)^2 + \frac{1}{6}(4) && \text{Substitute 4 for } c. \\ &= \frac{1}{3}(64) + \frac{1}{2}(16) + \frac{1}{6}(4) && \text{Evaluate the exponential expressions first.} \\ &= \frac{64}{3} + 8 + \frac{2}{3} && \text{Do the multiplication, and then simplify: } \tfrac{4}{6} = \tfrac{2}{3}. \\ &= 30 && \text{Add the fractions: } \tfrac{64}{3} + \tfrac{2}{3} = \tfrac{66}{3} = 22. \end{aligned}$$

There are 30 cans of soup in the display.

Now Try **Problem 81**

In the following example, we evaluate a polynomial in two variables.

EXAMPLE 4 Evaluate $3p^2q - 4pq^2$ for $p = 2$ and $q = -3$.

Strategy We will substitute the given values for each p and q in the polynomial and follow the rules for the order of operations.

Why To evaluate a polynomial means to find its numerical value, once we know the value of its variables.

Solution

$$\begin{aligned} 3p^2q - 4pq^2 &= 3(2)^2(-3) - 4(2)(-3)^2 && \text{Substitute 2 for } p \text{ and } -3 \text{ for } q. \\ &= 3(4)(-3) - 4(2)(9) && \text{Find the powers.} \\ &= -36 - 72 && \text{Do the multiplication.} \\ &= -108 && \text{Do the subtraction.} \end{aligned}$$

Self Check 4 Evaluate $3a^3b^2 + 2a^2b$ for $a = 2$ and $b = -1$.

Now Try **Problem 61**

3 Graph Equations Defined by Polynomials.

In Chapter 3, we graphed equations such as $y = x$ and $y = 2x - 3$. Recall that these equations are called *linear equations* and that their graphs are straight lines. Note that the right side of the first two equations is a polynomial of degree 1.

$y = x$ $y = 2x - 3$ — The degree of each polynomial is 1.

$y = x^2$ — The degree of this polynomial is 2.

$y = x^3 + 1$ — The degree of this polynomial is 3.

We can also graph equations defined by polynomials with degrees greater than 1.

EXAMPLE 5 Graph: $y = x^2$

Strategy We will find several solutions of the equation, plot them on a rectangular coordinate system, and then draw a smooth curve passing through the points.

Why To *graph* an equation in two variables means to make a drawing that represents all of its solutions.

Solution To find some solutions of this equation, we select several values of x that will make the computations easy. Then we find each corresponding value of y. If $x = -3$, we substitute -3 for x in $y = x^2$ and find y.

$$y = x^2 = (-3)^2 = 9$$

Thus, $(-3, 9)$ is a solution. In a similar manner, we find the corresponding y-values for x-values of $-2, -1, 0, 1, 2$, and 3. If we plot the ordered pairs listed in the table and join the points with a smooth curve, we get the graph shown on the next page, which is called a **parabola.**

Success Tip
When constructing a table of solutions, it is wise to select some positive and negative integer-values for x, as well as 0.

The Language of Algebra
The cup-like shape of a *parabola* has many real-life applications. The word *parabolic* (pronounced par • a • bol • ic) means having the form of a parabola. Did you know that a satellite TV dish is more formally known in the electronics industry as a *parabolic* dish?

$y = x^2$

x	y	(x, y)
−3	9	(−3, 9)
−2	4	(−2, 4)
−1	1	(−1, 1)
0	0	(0, 0)
1	1	(1, 1)
2	4	(2, 4)
3	9	(3, 9)

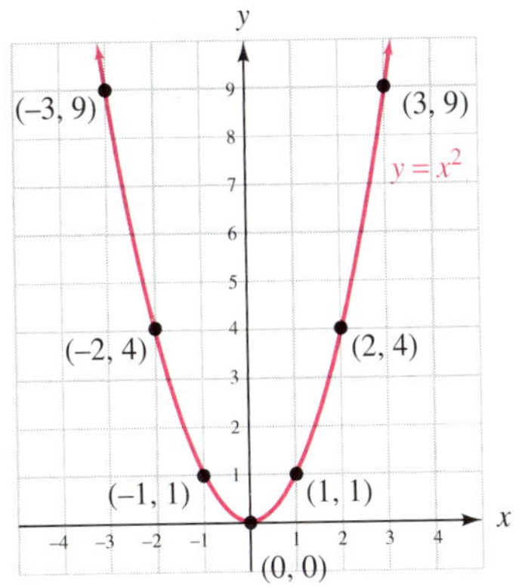

Self Check 5 Graph: $y = x^2 - 2$

Now Try **Problem 69**

EXAMPLE 6 Graph: $y = -x^2 + 2$

Strategy We will find several solutions of the equation, plot them on a rectangular coordinate system, and then draw a smooth curve passing through the points.

Why To *graph* an equation in two variables means to make a drawing that represents all of its solutions.

Solution To make a table of solutions, we select x-values of −3, −2, −1, 0, 1, 2, and 3 and find each corresponding y-value. For example, if $x = -3$, we have

$y = -x^2 + 2$

$y = -(-3)^2 + 2$ Substitute −3 for x.

$y = -(9) + 2$ Evaluate the exponential expression first: $(-3)^2 = 9$.

$y = -7$ Do the addition: $-9 + 2 = -7$.

The ordered pair (−3, −7) is a solution. Six other solutions appear in the table. After plotting each pair, we join the points with a smooth curve to obtain the graph, a parabola opening downward.

$y = -x^2 + 2$

x	y	(x, y)
−3	−7	(−3, −7)
−2	−2	(−2, −2)
−1	1	(−1, 1)
0	2	(0, 2)
1	1	(1, 1)
2	−2	(2, −2)
3	−7	(3, −7)

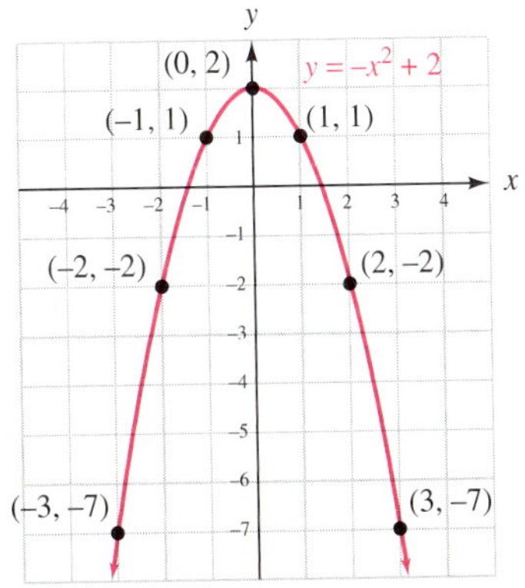

Self Check 6 Graph: $y = -x^2$

Now Try Problem 71

EXAMPLE 7 Graph: $y = x^3 + 1$

Strategy We will find several solutions of the equation, plot them on a rectangular coordinate system, and then draw a smooth curve passing through the points.

Why To *graph* an equation in two variables means to make a drawing that represents all of its solutions.

Solution If we let $x = -2$, we have

$y = x^3 + 1$

$y = (-2)^3 + 1$ Substitute −2 for x.

$y = -8 + 1$ Evaluate the exponential expression first: $(-2)^3 = -8$.

$y = -7$ Do the addition.

The ordered pair $(-2, -7)$ is a solution. This pair and others that satisfy the equation are listed in the table. Plotting the ordered pairs and joining the points with a smooth curve gives us the graph.

$y = x^3 + 1$

x	y	(x, y)
−2	−7	(−2, −7)
−1	0	(−1, 0)
0	1	(0, 1)
1	2	(1, 2)
2	9	(2, 9)

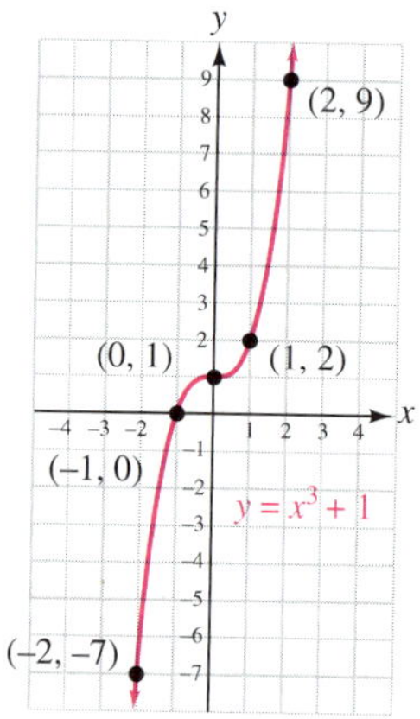

Self Check 7 Graph: $y = x^3 - 1$

Now Try Problem 75

ANSWERS TO SELF CHECKS **1. a.** A trinomial in one variable of degree 2 written in descending powers of x; terms: x^2, $4x$, -16; coefficients: 1, 4, -16; degree: 2, 1, 0 **b.** A binomial in two variables of degree 7 written in descending powers of s and ascending powers of t; terms: $-14s^5t$, s^4t^3; coefficients: -14, 1; degree: 6, 7 **2.** 33 **4.** 16

5.

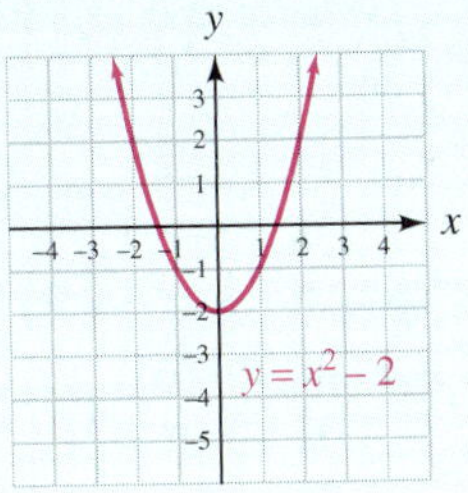

6.

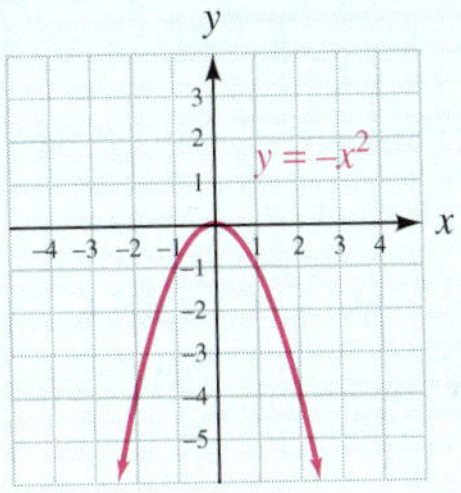

7.

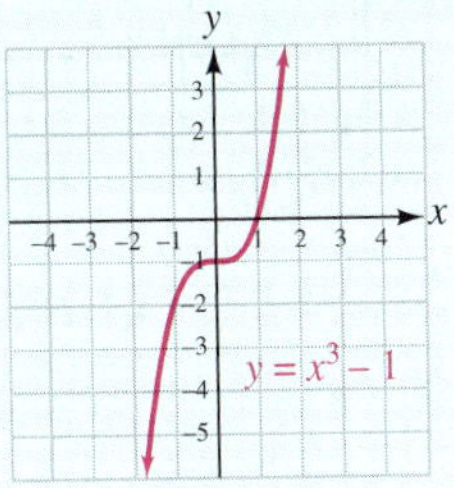

STUDY SET 5.4

VOCABULARY

Fill in the blanks.

1. A __________ is a term or a sum of terms in which all variables have whole-number exponents and no variable appears in a denominator.
2. The ______ of a polynomial are separated by + symbols.
3. $x^3 - 6x^2 + 9x - 2$ is a polynomial in _____ variable, and is written in __________ powers of x and $c^3 + 2c^2d - d^2$ is a polynomial in _____ variables and is written in __________ powers of d.
4. For the polynomial $6x^2 + 3x - 1$, the ________ term is $6x^2$, and the leading __________ is 6. The _________ term is -1.
5. A __________ is a polynomial with exactly one term. A __________ is a polynomial with exactly two terms. A __________ is a polynomial with exactly three terms.
6. The _______ of the term $3x^7$ is 7 because x appears as a factor 7 times: $3 \cdot x \cdot x \cdot x \cdot x \cdot x \cdot x \cdot x$.
7. To _________ the polynomial $x^2 - 2x + 1$ for $x = 6$, we substitute 6 for x and follow the rules for the order of operations.
8. The graph of $y = x^2$ is a cup-shaped curve called a _________.

CONCEPTS

Determine whether each expression is a polynomial.

9. **a.** $x^3 - 5x^2 - 2$ **b.** $x^{-4} - 5x$
 c. $x^2 - \frac{1}{2x} + 3$ **d.** $x^3 - 1$
 e. $x^2 - y^2$ **f.** $a^4 + a^3 + a^2 + a$

10. Fill in the blank so that the term has degree 5.
 a. $9x$ **b.** $-\frac{2}{3}xy$

Make a term-coefficient-degree table like that shown in Example 1 for each polynomial.

11. $8x^2 + x - 7$

Term	Coefficient	Degree

Degree of the polynomial:

12. $y^4 - y^3 + 16y^2 + 3y$

Term	Coefficient	Degree

Degree of the polynomial:

13. $8a^6b^3 - 27ab$

Term	Coefficient	Degree

Degree of the polynomial:

14. $-1.2c^4 + 2.4c^2d^2 - 3.6d^4$

Term	Coefficient	Degree

Degree of the polynomial: ____

NOTATION

15. a. Write $x - 9 + 3x^2 + 5x^3$ in descending powers of x.

b. Write $-2xy + y^2 + x^2$ in ascending powers of y.

16. Complete the solution. Evaluate $-2x^2 + 3x - 1$ for $x = -2$.

$$-2x^2 + 3x - 1 = -2(\quad)^2 + 3(\quad) - 1$$
$$= -2(\quad) + 3(\quad) - 1$$
$$= \quad + (-6) - 1$$
$$= \quad$$

GUIDED PRACTICE

Classify each polynomial as a monomial, a binomial, a trinomial, or none of these. **See Example 1.**

17. $3x + 7$

18. $3y - 5$

19. $y^2 + 4y + 3$

20. $9xy$

21. $\frac{3}{2}z^2$

22. $\frac{3}{5}x^4 - \frac{2}{5}x^3 + \frac{3}{5}x - 1$

23. $t - 32$

24. $12z^4$

25. $s^2 - 23s + 31$

26. $2x^3 - 5x^2 + 6x - 3$

27. $6x^5 - x^4 - 3x^3 + 7$

28. x^3

29. $3m^3n - 4m^2n^2 + mn - 1$

30. $4p^3q^2 + 7p^2q^3 + pq^4 - q^5$

31. $2a^2 - 3ab + b^2$

32. $a^3b - ab^3$

Find the degree of each polynomial. **See Example 1.**

33. $3x^4$

34. $3x^5$

35. $-2x^2 + 3x + 1$

36. $-5x^4 + 3x^2 - 3x$

37. $\frac{1}{3}x - 5$

38. $\frac{1}{2}y^3 + 4y^2$

39. $-5r^2s^2 - r^3s + 3$

40. $4r^2s^3 - 5r^2s^8$

41. $x^{12} + 3x^2y^3$

42. $17ab^5 - 12a^3b$

43. 38

44. -24

45. $\frac{3}{2}m^7 - \frac{3}{4}m^{18}$

46. $\frac{7}{8}t^{10} - \frac{1}{8}t^{16}$

47. $5.5tw - 6.5t^2w - 7.5t^3$

48. $0.4h + 0.6h^4c + 0.6h^5$

Evaluate each expression. **See Examples 2 and 3.**

49. $x^2 - x + 1$ for
a. $x = 2$
b. $x = -3$

50. $x^2 - x + 7$ for
a. $x = 6$
b. $x = -2$

51. $4t^2 + 2t - 8$ for
a. $t = -1$
b. $t = 0$

52. $3s^2 - 2s + 8$ for
a. $s = 1$
b. $s = 0$

53. $\frac{1}{2}a^2 - \frac{1}{4}a$ for
a. $a = 4$
b. $a = -8$

54. $\frac{1}{3}b^2 - \frac{1}{9}b$ for
a. $b = 9$
b. $b = -9$

55. $-9.2x^2 + x - 1.4$ for
a. $x = -1$
b. $x = -2$

56. $-10.3x^2 - x + 6.5$ for
a. $x = -1$
b. $x = -2$

57. $x^3 + 3x^2 + 2x + 4$ for
a. $x = 2$
b. $x = -2$

58. $x^3 - 3x^2 - x + 9$ for
a. $x = 3$
b. $x = -3$

59. $y^4 - y^3 + y^2 + 2y - 1$ for
a. $y = 1$
b. $y = -1$

60. $-y^4 + y^3 + y^2 + y + 1$ for
a. $y = 1$
b. $y = -1$

Evaluate each polynomial for a = −2 and b = 3. **See Example 4.**

61. $6a^2b$

62. $4ab^2$

63. $a^3 + b^3$

64. $a^3 - b^3$

65. $a^2 + 5ab - b^2$

66. $a^3 - 2ab + b^3$

67. $5ab^3 - ab - b + 10$

68. $-a^3b + ab - a - 21$

Construct a table of solutions and then graph the equation. **See Examples 5–7.**

69. $y = x^2 + 1$

70. $y = x^2 - 4$

71. $y = -x^2 - 2$

72. $y = -x^2 + 1$

73. $y = 2x^2 - 3$

74. $y = -2x^2 + 2$

75. $y = x^3 + 2$

76. $y = x^3 + 4$

77. $y = x^3 - 3$

78. $y = x^3 - 2$

79. $y = -x^3 - 1$

80. $y = -x^3$

APPLICATIONS

81. SUPERMARKETS A grocer plans to set up a pyramid-shaped display of cantaloupes like that shown in Example 3. If each side of the square base of the display is made of six cantaloupes, how many will be used in the display?

82. PACKAGING The polynomial $4x^3 - 44x^2 + 120x$ gives the volume (in cubic inches) of the resulting box when a square with sides x inches long is cut from each corner of a box 10 in. × 12 in. piece of cardboard. Find the volume of a box if 3-inch squares are cut out.

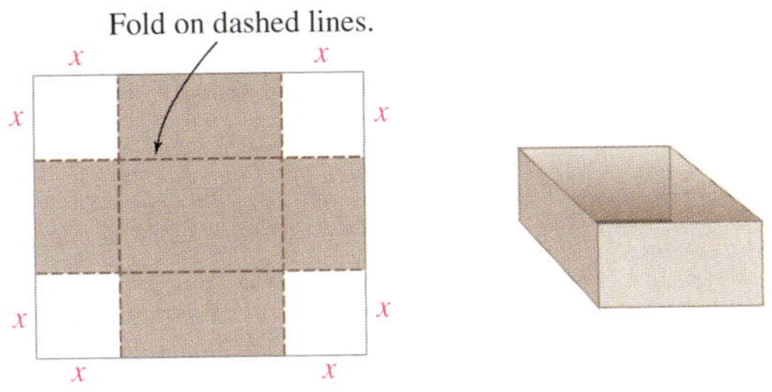

83. STOPPING DISTANCE The number of feet that a car travels before stopping depends on the driver's reaction time and the braking distance, as shown in the illustration. For one driver, the stopping distance is given by the polynomial $0.04v^2 + 0.9v$ where v is the velocity of the car. Find the stopping distance when the driver is traveling at 30 mph.

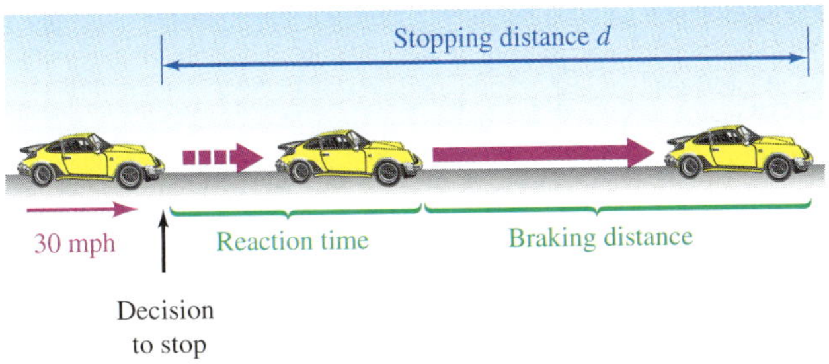

84. SUSPENSION BRIDGES The polynomial $-0.0000001s^4 + 0.0066667s^2 + 400$ approximates the length of the cable between the two vertical towers of a bridge, where s is the sag in the cable (in feet). Estimate the length of the cable if the sag is 24.6 feet.

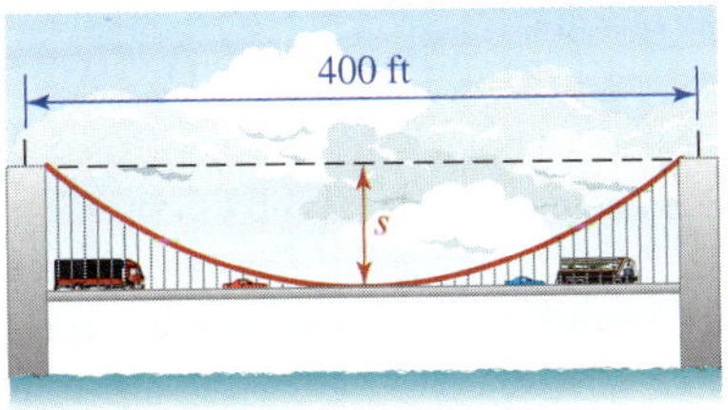

85. ***from Campus to Careers***
Sound Engineering Technician

Many people involved in the recording industry have been impressed by the success of Apple's iTunes Music Store. The polynomial $1.144x^2 + 5.771x + 0.452$ approximates the number of users (in millions) of iTunes. When $x = 0$, the polynomial estimates the number of users (in millions) as of January 2004. When $x = 1$, it estimates the number of users as of January 2005, and so on. Use the polynomial to estimate the number of iTunes users as of January 2008. (Source: *WebSiteOptimization.com*)

86. ONLINE When $x = 0$, the polynomial $-0.619x^2 + 11.778x + 12.171$ approximates the percent of the U.S. population that had gone online by 1995. When $x = 1$, it approximates the percent of the U.S. population that had gone online by 1996, and so on. Use the polynomial to approximate the percent of the U.S. population that had gone online by 2006. (Source: *The State of the News Media 2007*)

87. SCIENCE HISTORY The Italian scientist Galileo Galilei (1564–1642) built an incline plane like that shown to study falling objects. As the ball rolled down, he measured the time it took the ball to travel different distances. Graph the data and then connect the points with a smooth curve.

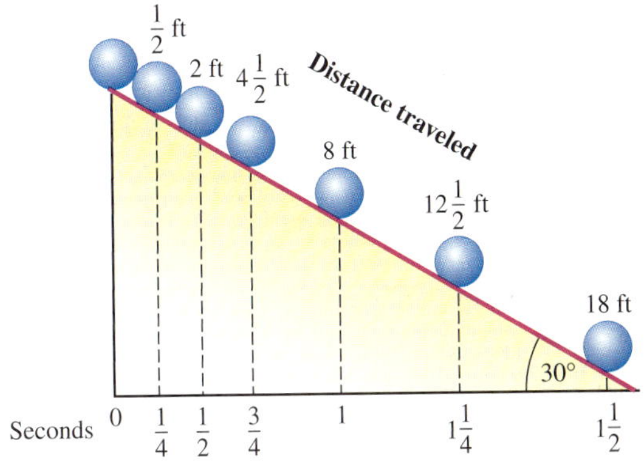

88. DOLPHINS At a marine park, three trained dolphins jump in unison over an arching stream of water whose path can be described by the equation $y = -0.05x^2 + 2x$. Given the takeoff points for each dolphin, how high must each jump to clear the stream of water?

WRITING

89. Describe how to determine the degree of a polynomial.

90. List some words that contain the prefixes *mono, bi,* or *tri.*

91. To graph $y = x^2 - 4$, a table of solutions is constructed and a graph is drawn. Explain the error.

$y = x^2 - 4$

x	y	(x, y)
0	−4	(0, −4)
2	0	(2, 0)

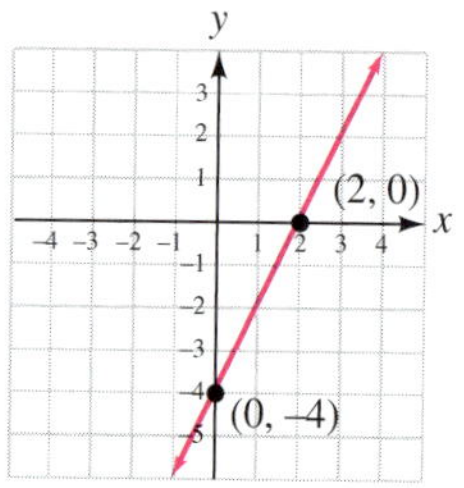

92. The expression $x + y$ is a binomial. Is xy also a binomial? Explain.

REVIEW

Solve each inequality. Write the solution set in interval notation and graph it.

93. $-4(3y + 2) \le 28$

94. $-5 < 3t + 4 \le 13$

Simplify each expression. Do not use negative exponents in the answer.

95. $(x^2x^4)^3$

96. $(a^2)^3(a^3)^2$

97. $\left(\dfrac{y^2y^5}{y^4}\right)^3$

98. $\left(\dfrac{2t^3}{t}\right)^{-4}$

CHALLENGE PROBLEMS

99. Find a three-term polynomial of degree 2 whose value will be 1 when it is evaluated for $x = 2$.

100. Graph: $y = 2x^3 - 3x^2 - 11x + 6$

SECTION 5.5
Adding and Subtracting Polynomials

Objectives

1. Simplify polynomials by combining like terms.
2. Add polynomials.
3. Subtract polynomials.

If we are to add (or subtract) objects, they must be similar. For example, we can add dollars to dollars and inches to inches, but we can't add dollars to inches. If you keep this concept in mind, then adding and subtracting polynomials will be easy. It simply involves combining like terms.

1 Simplify Polynomials by Combining Like Terms.

Recall that **like terms** have the same variables with the same exponents. Only the coefficients may differ.

Like terms	***Unlike terms***	
$-7x$ and $15x$	$-7x$ and $15a$	Different variables
$9.4y^3$ and $1.6y^3$	$9.4y^3$ and $1.6y^2$	Different exponents on the same variable
$\frac{1}{2}x^5y^2$ and $-\frac{1}{3}x^5y^2$	$\frac{1}{2}x^5y^2$ and $-\frac{1}{3}x^2y^5$	Different exponents on different variables

Also recall that to **combine like terms,** we combine their coefficients and keep the same variables with the same exponents. For example,

The Language of Algebra
Simplifying the sum or difference of like terms is called *combining like terms.*

$$4y + 5y = (4 + 5)y \qquad 8x^2 - x^2 = (8 - 1)x^2$$
$$= 9y \qquad\qquad = 7x^2$$

Polynomials with like terms can be simplified by combining like terms.

EXAMPLE 1 Simplify each polynomial by combining like terms:

a. $4x^4 + 81x^4$ **b.** $-0.3r - 0.4r + 0.6r$

c. $17x^2y^2 + 2x^2y - 6x^2y^2$ **d.** $\frac{3}{4}p^2 + \frac{1}{2}q^2 - 7 + \frac{1}{3}p^2 - \frac{5}{4}q^2 + 4$

Strategy We will use the distributive property in reverse to add (or subtract) the coefficients of the like terms. We will keep the same variables raised to the same powers.

Why To *combine like terms* means to add or subtract the like terms in an expression.

Caution
When combining like terms, the exponents on the variables *stay the same.* Don't incorrectly add the exponents.

Solution

a. $4x^4 + 81x^4 = 85x^4$ Think: $(4 + 81)x^4 = 85x^4$.

b. $-0.3r - 0.4r + 0.6r = -0.1r$ Think: $(-0.3 - 0.4 + 0.6)r = -0.1r$.

c. The first and third terms are like terms.

$$17x^2y^2 + 2x^2y - 6x^2y^2 = 11x^2y^2 + 2x^2y$$ Think: $(17 - 6)x^2y^2 = 11x^2y^2$.

d. $\frac{3}{4}p^2 + \frac{1}{2}q^2 - 7 + \frac{1}{3}p^2 - \frac{5}{4}q^2 + 4$

$$= \left(\frac{3}{4} + \frac{1}{3}\right)p^2 + \left(\frac{1}{2} - \frac{5}{4}\right)q^2 - 7 + 4$$ Combine like terms.

$$= \left(\frac{9}{12} + \frac{4}{12}\right)p^2 + \left(\frac{2}{4} - \frac{5}{4}\right)q^2 - 7 + 4$$ Build equivalent fractions: $\frac{3}{4}\cdot\frac{3}{3} = \frac{9}{12}$, $\frac{1}{3}\cdot\frac{4}{4} = \frac{4}{12}$, and $\frac{1}{2}\cdot\frac{2}{2} = \frac{2}{4}$.

$$= \frac{13}{12}p^2 - \frac{3}{4}q^2 - 3$$ Do the additions and the subtraction.

Caution Do not try to clear this expression of fractions by multiplying it by the LCD 12. That strategy works only when we multiply *both sides of an equation* by the LCD.

$$12\left(\frac{3}{4}p^2 + \frac{1}{2}q^2 - 7 + \frac{1}{3}p^2 - \frac{5}{4}q^2 + 4\right)$$

Self Check 1 Simplify each polynomial: **a.** $6m^4 + 3m^4$

b. $-19x + 21x - x$

c. $1.7s^3t + 0.3s^2t - 0.6s^3t$

d. $\frac{1}{8}c^5 + \frac{1}{3}d^5 - 9 + \frac{5}{4}c^5 - \frac{3}{5}d^5 + 1$

Now Try **Problems 13, 27, and 33**

2 Add Polynomials.

When adding polynomials horizontally, each polynomial is usually enclosed within parentheses. For example,

$$(3x^2 + 6x + 7) + (2x - 5)$$

is the sum of a trinomial and a binomial. To find the sum, we reorder and regroup the terms using the commutative and associative properties of addition so that like terms are together.

$$(3x^2 + 6x + 7) + (2x - 5) = 3x^2 + (6x + 2x) + (7 - 5)$$ The x-terms are together. The constant terms are together.

$$= 3x^2 + 8x + 2$$ Combine like terms.

This example suggests the following rule.

Adding Polynomials To add polynomials, combine their like terms.

EXAMPLE 2

Add the polynomials:

a. $(-6a^3 + 5a^2 - 7a + 9) + (4a^3 - 5a^2 - a - 8)$

b. $\left(\frac{1}{2}m^2 + \frac{2}{3}m + 1\right) + \left(\frac{3}{4}m^2 - \frac{7}{9}m - 4\right)$ **c.** $(16g^2 - h^2) + (4g^2 + 2gh + 10h^2)$

Strategy We will reorder and regroup to get the like terms together. Then we will combine like terms.

Why To add polynomials means to combine their like terms.

Success Tip
Combine the like terms in order: a^3-terms first, a^2-terms second, a-terms third, and constants last. Then the answer will be in descending powers of a.

Solution

a. $(-6a^3 + 5a^2 - 7a + 9) + (4a^3 - 5a^2 - a - 8)$

$$= (-6a^3 + 4a^3) + (5a^2 - 5a^2) + (-7a - a) + (9 - 8)$$ Group like terms together.

$$= -2a^3 + 0a^2 + (-8a) + 1$$ Combine like terms.

$$= -2a^3 - 8a + 1$$

b. $\left(\frac{1}{2}m^2 + \frac{2}{3}m + 1\right) + \left(\frac{3}{4}m^2 - \frac{7}{9}m - 4\right)$

$$= \left(\frac{1}{2}m^2 + \frac{3}{4}m^2\right) + \left(\frac{2}{3}m - \frac{7}{9}m\right) + (1 - 4)$$ Group like terms together.

$$= \left(\frac{2}{4}m^2 + \frac{3}{4}m^2\right) + \left(\frac{6}{9}m - \frac{7}{9}m\right) + (1 - 4)$$ Build equivalent fractions: $\frac{1}{2} \cdot \frac{2}{2} = \frac{2}{4}$ and $\frac{2}{3} \cdot \frac{3}{3} = \frac{6}{9}$.

$$= \frac{5}{4}m^2 - \frac{1}{9}m - 3$$ Do the addition and subtraction.

c. $(16g^2 - h^2) + (4g^2 + 2gh + 10h^2)$

$$= (16g^2 + 4g^2) + 2gh + (-h^2 + 10h^2)$$ Group like terms together.

$$= 20g^2 + 2gh + 9h^2$$ Combine like terms.

Self Check 2

Add the polynomials:

a. $(2a^2 - a + 4) + (5a^2 + 6a - 5)$

b. $\left(\frac{3}{2}b^3 + \frac{4}{5}b + 7\right) + \left(\frac{3}{4}b^3 - \frac{11}{10}b - 10\right)$

c. $(7x^2 - 2xy - y^2) + (4x^2 - y^2)$

Now Try Problems 41, 43, and 47

EXAMPLE 3

Trapezoids. Find a polynomial that represents the perimeter of the trapezoid.

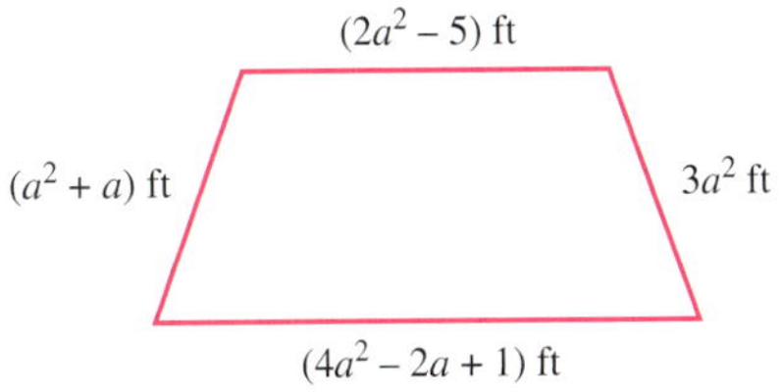

Strategy We will add the polynomials that represent the lengths of the sides of the trapezoid.

Why To find the perimeter of a figure, we find the distance around the figure by finding the sum of the lengths of its sides.

Solution To add the four polynomials that represent the lengths of the sides of the trapezoid, we combine their like terms.

$$(2a^2 - 5) + (a^2 + a) + (4a^2 - 2a + 1) + 3a^2$$
$$= (2a^2 + a^2 + 4a^2 + 3a^2) + (a - 2a) + (-5 + 1) \quad \text{Reorder and regroup terms.}$$
$$= 10a^2 - a - 4 \quad \text{Combine like terms.}$$

The perimeter of the trapezoid is $(10a^2 - a - 4)$ ft.

Now Try Problem 51

The Language of Algebra
A *trapezoid* is a four-sided figure with exactly two sides parallel.

Polynomials can also be added vertically by aligning like terms in columns.

EXAMPLE 4

Add $4x^2 - 3$ and $3x^2 - 8x + 8$ using vertical form.

Strategy First, we will write one polynomial underneath the other and draw a horizontal line beneath them. Then we will add the like terms, column by column, and write each result under the line.

Why *Vertical form* means to use an approach similar to that used in arithmetic to add two numbers.

Solution When performing vertical addition, any missing term may be written with a coefficient of 0. Since the first polynomial does not have an x-term, we insert a placeholder term $0x$ in the second column so that the constant terms line up in the third column.

x^2-terms	x-terms	Constants	
$4x^2$	$+ 0x$	$- 3$	In the x^2-column, find $4x^2 + 3x^2$.
$3x^2$	$- 8x$	$+ 8$	In the x-column, find $0x + (-8x)$.
$7x^2$	$-8x$	$+5$	In the constant column, find $-3 + 8$.

The sum is $7x^2 - 8x + 5$.

Success Tip
In arithmetic, we use vertical form so that we add digits in like place-value columns. In polynomial addition, we combine the like terms in each column.

Hundreds, Tens, Ones

$$\begin{array}{r} 403 \\ +388 \\ \hline \end{array}$$

Self Check 4 Add $4q^2 - 7$ and $2q^2 - 8q + 9$ using vertical form.

Now Try Problem 53

3 Subtract Polynomials.

Recall from Chapter 1 that we can use the distributive property to find the opposite of several terms enclosed within parentheses. For example, we consider $-(2a^2 - a + 9)$.

$$-(2a^2 - a + 9) = -1(2a^2 - a + 9)$$ Replace the − symbol in front of the parentheses with −1.

$$= -2a^2 + a - 9$$ Use the distributive property to remove parentheses.

This example illustrates the following method of subtracting polynomials.

Subtracting Polynomials

To subtract two polynomials, change the signs of the terms of the polynomial being subtracted, drop the parentheses, and combine like terms.

EXAMPLE 5 Subtract the polynomials: **a.** $(3a^2 - 4a - 6) - (2a^2 - a + 9)$ **b.** $(-t^3u + 2t^2u - u + 1) - (-3t^2u - u + 8)$

Strategy In each case, we will change the signs of the terms of the polynomial being subtracted, drop the parentheses, and combine like terms.

Why This is the method for subtracting two polynomials.

Solution

a. $(3a^2 - 4a - 6) - (2a^2 - a + 9)$

$= 3a^2 - 4a - 6 - 2a^2 + a - 9$ Change the sign of each term of $2a^2 - a + 9$ and drop the parentheses.

$= a^2 - 3a - 15$ Combine like terms.

b. $(-t^3u + 2t^2u - u + 1) - (-3t^2u - u + 8)$

$= -t^3u + 2t^2u - u + 1 + 3t^2u + u - 8$ Change the sign of each term of $-3t^2u - u + 8$ and drop the parentheses.

$= -t^3u + 5t^2u - 7$ Combine like terms.

Success Tip
After some practice, you will be able to reorder and regroup the terms of the polynomials in your head to combine them.

Self Check 5 Subtract the polynomials:

a. $(8a^3 - 5a^2 + 5) - (a^3 - a^2 - 7)$

b. $(x^2y - 2x + y - 2) - (6x + 9y - 2)$

Now Try Problems 61 and 71

Polynomials can also be subtracted vertically by aligning like terms in columns.

EXAMPLE 6 Subtract $3x^2 - 2x + 3$ from $2x^2 + 4x - 1$ using vertical form.

Strategy Since $3x^2 - 2x + 3$ is to be subtracted from $2x^2 + 4x - 1$, we will write $3x^2 - 2x + 3$ underneath $2x^2 + 4x - 1$, change the sign of each term of $3x^2 - 2x + 3$ and add, column-by-column.

Why *Vertical form* means to arrange the like terms in columns.

Solution

$$\begin{array}{r} 2x^2 + 4x - 1 \\ \underline{-(3x^2 - 2x + 3)} \end{array} \xrightarrow[\text{and add}]{\text{Change signs}} \begin{array}{r} 2x^2 + 4x - 1 \\ \underline{-3x^2 + 2x - 3} \\ -x^2 + 6x - 4 \end{array}$$

In the x^2-column, find $2x^2 + (-3x^2)$.
In the x-column, find $4x + 2x$.
In the constant column, find $-1 + (-3)$.

The difference is $-x^2 + 6x - 4$.

Self Check 6 Subtract $2p^2 + 2p - 8$ from $5p^2 - 6p + 7$ using vertical form.

Now Try **Problem 73**

EXAMPLE 7 Subtract $1.2a^4 - 0.7a$ from the sum of $0.6a^4 + 1.5a$ and $0.4a^4 - 1.1a$.

Strategy First, we will translate the words of the problem into mathematical symbols. Then we will perform the indicated operations.

Why The words of the problem contain the key phrases *subtract from* and *sum.*

Solution Since $1.2a^4 - 0.7a$, is to be subtracted from the sum, the order must be reversed when we translate to mathematical symbols.

Subtract $1.2a^4$ $0.7a$ from | the sum of $0.6a^4 + 1.5a$ and $0.4a^4 - 1.1a$.

$[(0.6a^4 + 1.5a) + (0.4a^4 - 1.1a)] - (1.2a^4 - 0.7a)$ Use brackets [ ] to enclose the sum.

Next, we remove the grouping symbols to obtain

$= 0.6a^4 + 1.5a + 0.4a^4 - 1.1a - 1.2a^4 + 0.7a$ Change the sign of each term within $(1.2a^4 - 0.7a)$ and drop the parentheses.

$= -0.2a^4 + 1.1a$ Combine like terms.

Self Check 7 Subtract $-0.2q^2 - 0.2q$ from the sum of $0.1q^2 - 0.6q$ and $0.3q^2 + 0.1q$.

Now Try **Problem 87**

EXAMPLE 8 ***Fireworks.*** Two firework shells are fired upward at the same time from different platforms. The height, after t seconds, of the first shell is $(-16t^2 + 160t + 3)$ feet. The height, after t seconds, of a higher-flying second shell is $(-16t^2 + 200t + 1)$ feet.

a. Find a polynomial that represents the difference in the heights of the shells.

b. In 5 seconds, the first shell reaches its peak and explodes. How much higher is the second shell at that time?

Strategy To find the difference in their heights, we will subtract the height of the first shell from the height of the higher-flying second shell.

Why The key word *difference* indicates that we should subtract the polynomials.

Solution

a. Since the height of the higher flying second shell is represented by $-16t^2 + 200t + 1$ and the height of the lower flying shell is represented by $-16t^2 + 160t + 3$, we can find their difference by performing the following subtraction.

$$(-16t^2 + 200t + 1) - (-16t^2 + 160t + 3)$$

$$= -16t^2 + 200t + 1 + 16t^2 - 160t - 3$$ Change the sign of each term of $-16t^2 + 160t + 3$ and remove parentheses.

$$= 40t - 2$$ Combine like terms.

The difference in the heights of the shells t seconds after being fired is $(40t - 2)$ feet.

b. To find the difference in their heights after 5 seconds, we will evaluate the polynomial found in part (a) at a value of 5 seconds. If we substitute 5 for t, we have

$$40t - 2 = 40(5) - 2 = 200 - 2 = 198$$

When the first shell explodes, the second shell will be 198 feet higher than the first shell.

Now Try **Problem 109**

ANSWERS TO SELF CHECKS **1. a.** $9m^4$ **b.** x **c.** $1.1s^3t + 0.3s^2t$ **d.** $\frac{11}{8}c^5 - \frac{4}{15}d^5 - 8$ **2. a.** $7a^2 + 5a - 1$ **b.** $\frac{9}{4}b^3 - \frac{3}{10}b - 3$ **c.** $11x^2 - 2xy - 2y^2$ **4.** $6q^2 - 8q + 2$ **5. a.** $7a^3 - 4a^2 + 12$ **b.** $x^2y - 8x - 8y$ **6.** $3p^2 - 8p + 15$ **7.** $0.6q^2 - 0.3q$

STUDY SET 5.5

VOCABULARY

Fill in the blanks.

1. $(b^3 - b^2 - 9b + 1) + (b^3 - b^2 - 9b + 1)$ is the sum of two ________.
2. $(b^2 - 9b + 11) - (4b^2 - 14b)$ is the ________ of a trinomial and a binomial.
3. ______ terms have the same variables with the same exponents.
4. The polynomial $2t^4 + 3t^3 - 4t^2 + 5t - 6$ is written in ________ powers of t.

CONCEPTS

Fill in the blanks.

5. To add polynomials, ________ their like terms.
6. To subtract polynomials, ________ the signs of the terms of the polynomial being subtracted, drop parentheses, and combine like terms.
7. Simplify each polynomial, if possible.
   a. $2x^2 + 3x^2$ b. $15m^3 - m^3$
   c. $8a^3b - a^3b$ d. $6cd + 4c^2d$
8. What is the result when the addition is done in the x-column?

$$\begin{array}{r} 4x^2 + x - 12 \\ \underline{5x^2 - 8x + 23} \end{array}$$

9. Write without parentheses.
   a. $-(5x^2 - 8x + 23)$ b. $-(-5y^4 + 3y^2 - 7)$
10. What is the result when the subtraction is done in the x-column?

$$\begin{array}{r} 8x^2 - 7x - 1 \\ \underline{-(4x^2 + 6x - 9)} \end{array} \longrightarrow \begin{array}{r} 8x^2 - 7x - 1 \\ \underline{-4x^2 - 6x + 9} \end{array}$$

NOTATION

Fill in the blanks to add (subtract) the polynomials.

11. $(6x^2 + 2x + 3) + (4x^2 - 7x + 1)$
    $= (6x^2 + \underline{\qquad}) + (\underline{\qquad} - 7x) + (3 + \underline{\qquad})$
    $= \underline{\qquad} - 5x + \underline{\qquad}$
12. $(6x^2 + 2x + 3) - (4x^2 - 7x + 1)$
    $= 6x^2 + 2x + 3 \;\underline{\quad}\; 4x^2 \;\underline{\quad}\; 7x - 1$
    $= \underline{\qquad} + 9x + \underline{\qquad}$

GUIDED PRACTICE

***Simplify each polynomial and write it in descending powers of one variable.* See Example 1.**

13. $8t^2 + 4t^2$
14. $15x^2 + 10x^2$
15. $-32u^3 - 16u^3$
16. $-25x^3 - 7x^3$
17. $18x^2 - 19x + 2x^2$
18. $17y^2 - 22y - y^2$
19. $3r^4 - 4r + 7r^4$
20. $-2b^4 + 7b - 3b^4$
21. $10x^2 - 8x + 9x - 9x^2$
22. $-3y^2 - y - 6y^2 + 7y$
23. $\frac{1}{5}x^2 - \frac{3}{8}x + \frac{2}{3}x^2 + \frac{1}{4}x$
24. $\frac{6}{7}y^2 + \frac{1}{2}y - \frac{2}{3}y^2 + \frac{1}{5}y$
25. $0.6x^3 + 0.8x^4 + 0.7x^3 + (-0.8x^4)$
26. $1.9m^4 - 2.4m^6 - 3.7m^4 + 2.8m^6$
27. $\frac{1}{2}st + \frac{3}{2}st$
28. $\frac{2}{5}at + \frac{1}{5}at$
29. $\frac{2}{3}d^2 - \frac{1}{4}c^2 + \frac{5}{6}c^2 - \frac{1}{2}cd + \frac{1}{3}d^2$
30. $\frac{3}{5}s^2 - \frac{2}{5}t^2 - \frac{1}{2}s^2 - \frac{7}{10}st - \frac{3}{10}st$
31. $-4ab + 4ab - ab$
32. $xy - 4xy - 2xy$
33. $4x^2y + 5 - 6x^3y - 3x^2y + 2x^3y$
34. $5b - 9ab^2 + 10a^3b - 8ab^2 - 9a^3b$
35. $-7cd - 8d^2 - 5cd + 8d^2 - 4c^2$
36. $-3rt - 7t^2 - 6rt + 7t^2 - 6r^2$

***Add the polynomials.* See Example 2.**

37. $(3x + 7) + (4x - 3)$
38. $(2y - 3) + (4y + 7)$
39. $(9d^2 + 6d) + (8d - 4d^2)$
40. $(2c^2 - 4c) + (8c - c^2)$
41. $(3q^2 - 5q + 7) + (2q^2 + q - 12)$
42. $(2t^2 + 11t - 15) + (-5t^2 - 13t + 10)$
43. $\left(\frac{2}{3}y^3 + \frac{3}{4}y^2 + \frac{1}{2}\right) + \left(\frac{1}{3}y^3 + \frac{1}{5}y^2 - \frac{1}{6}\right)$
44. $\left(\frac{1}{16}r^6 + \frac{1}{2}r^3 - \frac{11}{12}\right) + \left(\frac{9}{16}r^6 + \frac{9}{4}r^3 + \frac{1}{12}\right)$
45. $(0.3p + 2.1q) + (0.4p - 3q)$
46. $(-0.3r - 5.2s) + (0.8r - 5.2s)$
47. $(2x^2 + xy + 3y^2) + (5x^2 - y^2)$
48. $(-4a^2 - ab + 15b^2) + (5a^2 - b^2)$

Find a polynomial that represents the perimeter of the figure. **See Example 3.**

49.

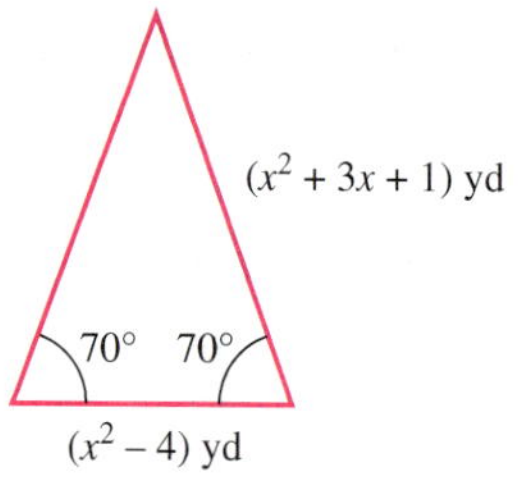

50.

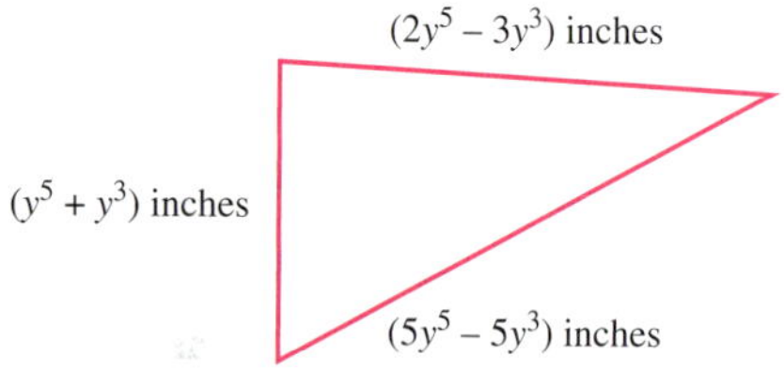

51.

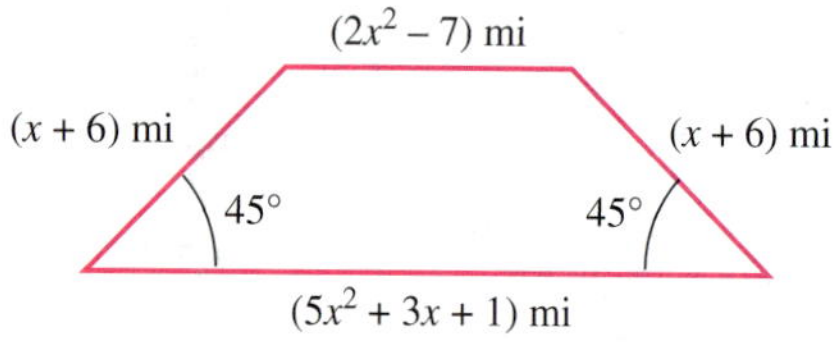

52.

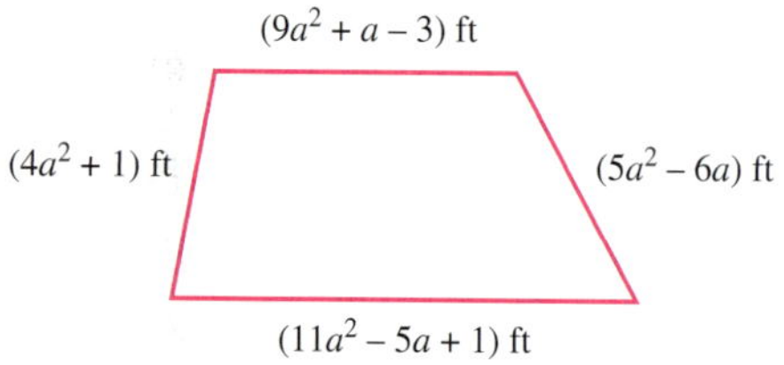

Use vertical form to add the polynomials. **See Example 4.**

53. $\begin{array}{r} 3x^2 + 4x + 5 \\ \underline{2x^2 - 3x + 6} \end{array}$

54. $\begin{array}{r} 6x^3 - 4x^2 + 7 \\ \underline{7x^3 + 9x^2 + 12} \end{array}$

55. $\begin{array}{r} 6a^2 + 7a + 9 \\ \underline{-9a^2 \qquad\ \ - 2} \end{array}$

56. $\begin{array}{r} -2c^2 - 3c - 5 \\ \underline{14c^2 \qquad\ \ - 1} \end{array}$

57. $\begin{array}{r} z^3 + 6z^2 - 7z + 16 \\ \underline{9z^3 - 6z^2 + 8z - 18} \end{array}$

58. $\begin{array}{r} 3x^3 + 4x^2 - 3x + 5 \\ \underline{3x^3 - 4x^2 - \ \ x - 7} \end{array}$

59. $\begin{array}{r} -3x^3y^2 + 4x^2y - 4x + 9 \\ \underline{2x^2y^2 \qquad\qquad + 9x - 3} \end{array}$

60. $\begin{array}{r} 3x^2y^2 + 4xy + 25 \\ \underline{5x^2y^2 \qquad\quad - 12} \end{array}$

Subtract the polynomials. **See Example 5.**

61. $(3a^2 - 2a + 4) - (a^2 - 3a + 7)$

62. $(2b^2 + 3b - 5) - (2b^2 - 4b - 9)$

63. $(9a^2 + 3a) - (2a - 4a^2)$

64. $(4b^2 + 3b) - (7b - b^2)$

65. $(-4h^3 + 5h^2 + 15) - (h^3 - 15)$

66. $(-c^5 + 5c^4 - 12) - (2c^5 - c^4)$

67. $\left(\frac{3}{8}s^8 - \frac{3}{4}s^7\right) - \left(\frac{1}{3}s^8 + \frac{1}{5}s^7\right)$

68. $\left(\frac{5}{6}q^9 - \frac{4}{5}q^8\right) - \left(\frac{1}{4}q^9 + \frac{3}{8}q^8\right)$

69. $(0.03f^2 + 0.25f + 0.91) - (0.17f^2 - 1.18)$

70. $(0.05r^2 - 0.33r) - (0.48\,r^2 + 0.15r + 2.14)$

71. $(5ab + 2b^2) - (2 + ab + b^2)$

72. $(mn + 8n^2) - (6 - 5mn + n^2)$

Use vertical form to subtract the polynomials. **See Example 6.**

73. $\begin{array}{r} 3x^2 + 4x + 5 \\ \underline{-(2x^2 - 2x + 3)} \end{array}$

74. $\begin{array}{r} 6y^2 + 4y + 13 \\ \underline{-(3y^2 - 6y + \ \ 7)} \end{array}$

75. $\begin{array}{r} 4x^3 + 4x^2 - 3x + 10 \\ \underline{-(5x^3 - 2x^2 - 4x - \ \ 4)} \end{array}$

76. $\begin{array}{r} 7m^5 + \ \ m^3 + 9m^2 - m \\ \underline{-(8m^5 - 2m^3 + \ \ m^2 + m)} \end{array}$

77. $\begin{array}{r} 0.8x^3 \qquad\qquad - 2.3x + 0.6 \\ \underline{-(0.2x^3 - 1.2x^2 - 3.6x + 0.9)} \end{array}$

78. $\begin{array}{r} 9.7y^3 \qquad\qquad + \quad y + 1.1 \\ \underline{-(6.3y^3 - 4.4y^2 + 2.7y + 8.8)} \end{array}$

79. $\begin{array}{r} 3x^3y^2 + 4x^2y + 7x + 12 \\ \underline{-(-4x^3y^2 + 6x^2y + 9x - 3)} \end{array}$

80. $\begin{array}{r} -2x^2y^2 \qquad\qquad + 12y^2 \\ \underline{-(10x^2y^2 + 9xy - 24y^2)} \end{array}$

Perform the operations. **See Example 7.**

81. Subtract $(s^2 + 4s + 2)$ from $(5s^2 - s + 9)$.

82. Subtract $(4p^2 - 4p - 40)$ from $(10p^2 - p - 30)$.

83. Subtract $(-y^5 + 5y^4 - 1.2)$ from $(2y^5 - y^4)$.

84. Subtract $(-4w^3 + 5w^2 + 7.6)$ from $(w^3 - 15w^2)$.

85. Find the sum when $(3x^2 + 4x - 7)$ is added to the sum of $(-2x^2 - 7x + 1)$ and $(-4x^2 + 8x - 1)$.

86. Find the difference when $(32x^2 - 17x + 45)$ is subtracted from the sum of $(23x^2 - 12x - 7)$ and $(-11x^2 + 12x + 7)$.

87. Find the difference when $(t^3 - 2t^2 + 2)$ is subtracted from the sum of $(3t^3 + t^2)$ and $(-t^3 + 6t - 3)$.

88. Find the difference when $(-3z^3 - 4z + 7)$ is subtracted from the sum of $(2z^2 + 3z - 7)$ and $(-4z^3 - 2z - 3)$.

89. $(2x^2 - 3x + 1) - (4x^2 - 3x + 2) + (2x^2 + 3x + 2)$

90. $(-3z^2 - 4z + 7) + (2z^2 + 2z - 1) - (2z^2 - 3z + 7)$

91. $(-2.7t^2 + 2.1t - 1.7) + (3.1t^2 - 2.5t + 2.3) - (1.7t^2 - 1.1t)$

92. $(1.04x^2 - 5.01) + (1.33x - 1.91x^2 + 5.02) - (1.07x^2 - 2.07x)$

TRY IT YOURSELF

Perform the operations.

93. $(-8x^2 - 3x) - (-11x^2 + 6x + 10)$

94. $(5m^2 - 8m + 8) - (-20m^2 + m)$

95. $(3x^2 - 3x - 2) + (3x^2 + 4x - 3)$

96. $(4c^2 + 3c - 2) + (3c^2 + 4c + 2)$

97. $\left(\frac{7}{8}r^4 + \frac{5}{9}r^2 - \frac{9}{4}\right) - \left(-\frac{3}{8}r^4 - \frac{2}{3}r^2 - \frac{1}{4}\right)$

98. $\left(\frac{4}{5}t^4 - \frac{1}{3}t^2 + \frac{1}{2}\right) - \left(-\frac{1}{2}t^4 + \frac{3}{8}t^2 - \frac{1}{16}\right)$

99.
$$\begin{array}{r} 8c^2 - 4c - 5 \\ -(-c^2 + 2c + 9) \\ \hline \end{array}$$

100.
$$\begin{array}{r} 3t^3 - 4t^2 - 3t + 5 \\ +11t^3 \qquad\quad - 8t - 2 \\ \hline \end{array}$$

101. $(10 - 2st - 3s^2t) + (4 - 6st)$

102. $(20 - 4rt - 5r^2t) + (10 - 5rt)$

103. $(12.1h^3 + 9.9h^2) - (7.3h^3 + 1.1h^2)$

104. $(5.7n^3 - 2.1n) - (-6.2n^3 - 3.9n)$

APPLICATIONS

105. GREEK ARCHITECTURE

a. Find a polynomial that represents the difference in the heights of the columns.

b. If the columns were stacked one atop the other, to what height would they reach?

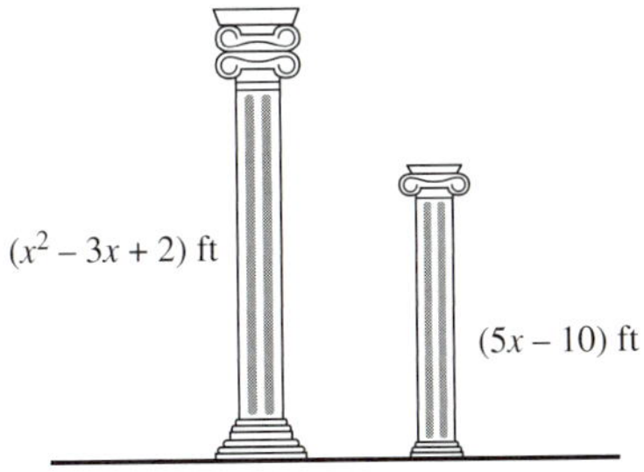

106. JETS Find a polynomial that represents the length of the passenger jet.

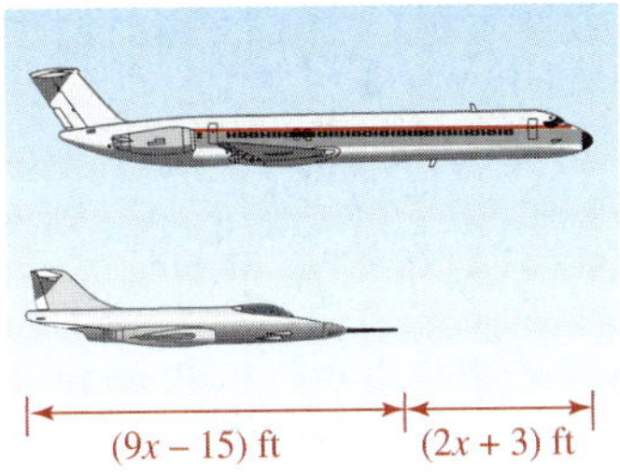

107. PIÑATAS Find a polynomial that represents the length of the rope used to hold up the piñata.

108. READING BLUEPRINTS Find a polynomial that represents

a. the difference in the length and width of the one-bedroom apartment.

b. the perimeter of the apartment.

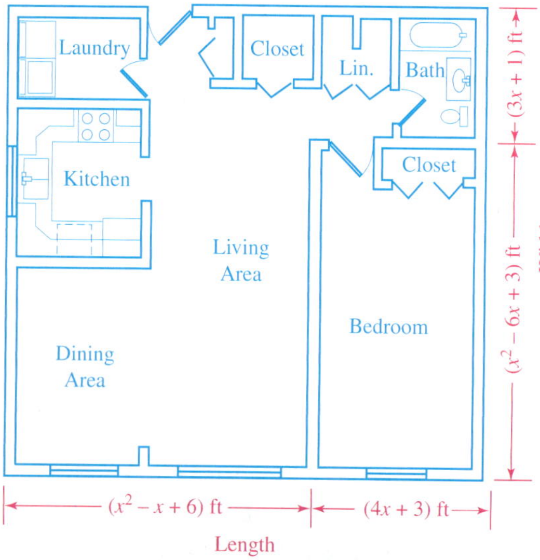

109. NAVAL OPERATIONS Two warning flares are fired upward at the same time from different parts of a ship. The height of the first flare is $(-16t^2 + 128t + 20)$ feet and the height of the higher-traveling second flare is $(-16t^2 + 150t + 40)$ feet, after t seconds.

a. Find a polynomial that represents the difference in the heights of the flares.

b. In 4 seconds, the first flare reaches its peak, explodes, and lights up the sky. How much higher is the second flare at that time?

110. AUTO MECHANICS Find a polynomial that represents the length of the fan belt shown in the diagram in the next column. The dimensions are in inches. Leave π in your answer.

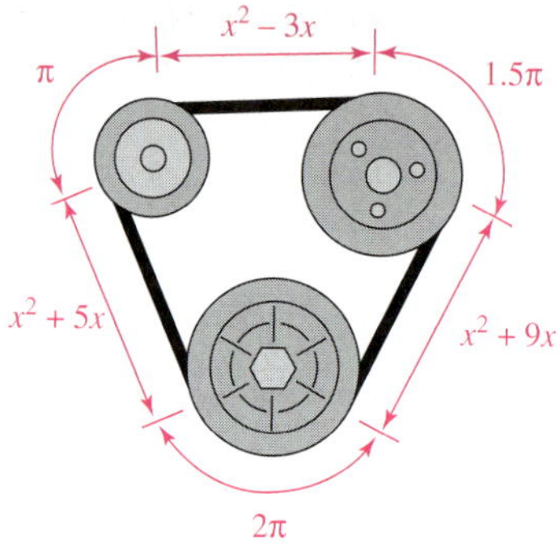

WRITING

111. How do you recognize like terms?

112. Explain why the vertical form used in algebra to add $2x^2 + 4x + 3$ and $5x^2 + 3x + 6$ is similar to the vertical form used in arithmetic to add 243 and 536.

113. Explain the error below.

$$7x^2y + 6x^2y = 13x^4y^2$$

114. Explain the error below.

$$\begin{aligned}(12x^2 - 4) - (3x^2 - 1) &= 12x^2 - 4 - 3x^2 - 1\\ &= 9x^2 - 5\end{aligned}$$

115. A student was asked to simplify $\frac{1}{6}x^2 - 3 + \frac{2}{3}x^2$. Explain the error below:

$$\begin{aligned}6\left(\frac{1}{6}x^2 - 3 + \frac{2}{3}x^2\right) &= x^2 - 18 + 4x^2\\ &= 5x^2 - 18\end{aligned}$$

116. Explain the error below.

Subtract $(2d^2 - d - 3)$ from $(d^2 - 9)$:

$$(2d^2 - d - 3) - (d^2 - 9) = d^2 - d + 6$$

REVIEW

117. What is the sum of the measures of the angles of a triangle?

118. What is the formula for

a. the area of a circle?

b. the area of a triangle?

119. Graph: $y = -\frac{1}{2}x + 2$

120. Graph: $2x + 3y = 9$

CHALLENGE PROBLEMS

121. What polynomial must be added to $2x^2 - x + 3$ so that the sum is $6x^2 - 7x - 8$?

122. Is the sum of two trinomials always a trinomial? Explain why or why not.

SECTION 5.6 Multiplying Polynomials

Objectives

1. Multiply monomials.
2. Multiply a polynomial by a monomial.
3. Multiply binomials.
4. Multiply polynomials.

We now discuss multiplying polynomials. We will begin with the simplest case—finding the product of two monomials.

1 Multiply Monomials.

To find the product of two monomials, such as $8x^2$ and $3x^4$, we use the commutative and associative properties of multiplication to reorder and regroup the factors.

$(8x^2)(3x^4) = (8 \cdot 3)(x^2 \cdot x^4)$ Group the coefficients together and the variables together.

$= 24x^6$ Simplify: $x^2 \cdot x^4 = x^{2+4} = x^6$.

This example suggests the following rule.

Success Tip
In this section, you will see that every polynomial multiplication is a series of monomial multiplications.

Multiplying Monomials

To multiply two monomials, multiply the numerical factors (the coefficients) and then multiply the variable factors.

EXAMPLE 1 Multiply: **a.** $(6r)(r)$ **b.** $3t^4(-2t^5)$ **c.** $\left(\frac{1}{3}a^2b^3\right)(21ab^2)$ **d.** $-4y^5z^2(2y^3z^3)(3yz)$

Strategy We will multiply the numerical factors and then multiply the variable factors.

Why The commutative and associative properties of multiplication enable us to reorder and regroup the factors.

Solution

a. $(6r)(r) = 6r^2$ Recall that $r = 1r$. Think: $6 \cdot 1 = 6$ and $r \cdot r = r^2$.

b. $(3t^4)(-2t^5) = -6t^9$ Think: $3(-2) = -6$ and $t^4 \cdot t^5 = t^{4+5} = t^9$.

c. $\left(\frac{1}{3}a^2b^3\right)(21ab^2) = 7a^3b^5$ Think: $\frac{1}{3} \cdot 21 = \frac{21}{3} = 7$, $a^2 \cdot a = a^3$, and $b^3 \cdot b^2 = b^5$.

d. $-4y^5z^2(2y^3z^3)(3yz) = -24y^9z^6$ Think: $-4(2)(3) = -24$, $y^5 \cdot y^3 \cdot y = y^9$, and $z^2 \cdot z^3 \cdot z = z^6$.

Success Tip
Notice that we *multiply* the coefficients and we *add* the exponents of the like bases.

Self Check 1 Multiply: **a.** $18t(t)$ **b.** $-10d^8(-6d^3)$ **c.** $(16y^{12})(\frac{1}{4}y^2)$ **d.** $(5a^3b^3)(-6a^3b^4)(ab)$

Now Try **Problems 13, 19 and 23**

EXAMPLE 4 Multiply: $(5x - 8)(x + 1)$

Strategy To find the product, we will multiply $x + 1$ by $5x$ and by -8.

Why To multiply two binomials, each term of one binomial must be multiplied by each term of the other binomial.

Solution

$$(5x - 8)(x + 1) = 5x(x + 1) - 8(x + 1)$$ Multiply $x + 1$ by $5x$ and multiply $x + 1$ by -8.

$$= 5x^2 + 5x - 8x - 8$$ Distribute the multiplication by $5x$. Distribute the multiplication by -8.

$$= 5x^2 - 3x - 8$$ Combine like terms.

Self Check 4 Multiply: $(9y + 3)(y - 4)$

Now Try **Problem 49**

> ***The Language of Algebra***
> An *acronym* is an abbreviation of several words in such a way that the abbreviation itself forms a word. The *acronym* FOIL helps us remember the order to follow when multiplying two binomials: First, Outer, Inner, Last.

We can use a shortcut method, called the **FOIL method,** to multiply binomials. FOIL is an acronym for **F**irst terms, **O**uter terms, **I**nner terms, **L**ast terms. To use the FOIL method to multiply $2a + 4$ by $3a + 5$, we

1. multiply the **F**irst terms $2a$ and $3a$ to obtain $6a^2$,
2. multiply the **O**uter terms $2a$ and 5 to obtain $10a$,
3. multiply the **I**nner terms 4 and $3a$ to obtain $12a$, and
4. multiply the **L**ast terms 4 and 5 to obtain 20.

Then we simplify the resulting polynomial, if possible.

Outer, First, Inner, Last

$$(2a + 4)(3a + 5) = \overset{F}{2a(3a)} + \overset{O}{2a(5)} + \overset{I}{4(3a)} + \overset{L}{4(5)}$$

$$= 6a^2 + 10a + 12a + 20$$ Multiply the monomials.

$$= 6a^2 + 22a + 20$$ Combine like terms.

EXAMPLE 5 Multiply: **a.** $(x + 5)(x + 7)$ **b.** $(3x + 4)(2x - 3)$

c. $\left(2r - \frac{1}{2}\right)\left(2r + \frac{5}{2}\right)$ **d.** $(3a^2 - 7b)(a^2 - b)$

Strategy We will use the FOIL method.

Why In each case we are to find the product of two binomials, and the FOIL method is a shortcut for multiplying two binomials.

Success Tip

The area of the large rectangle is given by $(x + 5)(x + 7)$. The sum of the areas of the smaller rectangles is $x^2 + 7x + 5x + 35$ or $x^2 + 12x + 35$. Thus,

$$(x + 5)(x + 7) = x^2 + 12x + 35$$

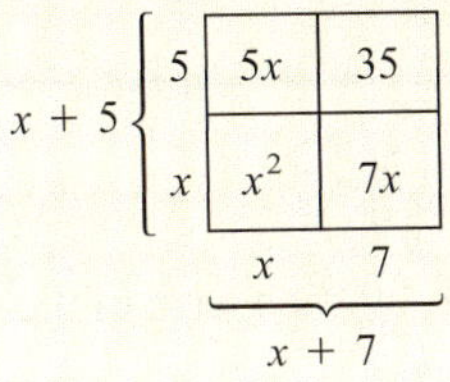

Solution

a.

$$\begin{aligned}(x + 5)(x + 7) &= \overset{F}{x(x)} + \overset{O}{x(7)} + \overset{I}{5(x)} + \overset{L}{5(7)} \\ &= x^2 + 7x + 5x + 35 &&\text{Multiply the monomials.} \\ &= x^2 + 12x + 35 &&\text{Combine like terms.}\end{aligned}$$

b.

$$\begin{aligned}(3x + 4)(2x - 3) &= \overset{F}{3x(2x)} + \overset{O}{3x(-3)} + \overset{I}{4(2x)} + \overset{L}{4(-3)} \\ &= 6x^2 - 9x + 8x - 12 &&\text{Multiply the monomials.} \\ &= 6x^2 - x - 12 &&\text{Combine like terms.}\end{aligned}$$

c.

$$\begin{aligned}\left(2r - \frac{1}{2}\right)\left(2r + \frac{5}{2}\right) &= \overset{F}{2r(2r)} + \overset{O}{2r\left(\frac{5}{2}\right)} - \overset{I}{\frac{1}{2}(2r)} - \overset{L}{\frac{1}{2}\left(\frac{5}{2}\right)} \\ &= 4r^2 + 5r - r - \frac{5}{4} &&\text{Multiply the monomials.} \\ &= 4r^2 + 4r - \frac{5}{4} &&\text{Combine like terms.}\end{aligned}$$

d.

$$\begin{aligned}(3a^2 - 7b)(a^2 - b) &= \overset{F}{3a^2(a^2)} + \overset{O}{3a^2(-b)} - \overset{I}{7b(a^2)} - \overset{L}{7b(-b)} \\ &= 3a^4 - 3a^2b - 7a^2b + 7b^2 &&\text{Multiply the monomials.} \\ &= 3a^4 - 10a^2b + 7b^2 &&\text{Combine like terms.}\end{aligned}$$

Self Check 5 Multiply: **a.** $(y + 3)(y + 1)$
b. $(2a - 1)(3a + 2)$
c. $\left(4x - \frac{1}{2}\right)\left(4x + \frac{3}{4}\right)$
d. $(5y^3 - 2)(2y^3 - 7)$

Now Try **Problems 51, 57, and 59**

4 Multiply Polynomials.

To develop a general rule for multiplying any two polynomials, we will find the product of $2x + 3$ and $3x^2 + 3x + 5$. In the solution, the distributive property is used four times.

$$\begin{aligned}(2x + 3)(3x^2 + 3x + 5) &= (2x + 3)3x^2 + (2x + 3)3x + (2x + 3)5 &&\text{Distribute.} \\ &= (2x + 3)3x^2 + (2x + 3)3x + (2x + 3)5 \\ &= (2x)3x^2 + (3)3x^2 + (2x)3x + (3)3x + (2x)5 + (3)5 &&\text{Distribute.} \\ &= 6x^3 + 9x^2 + 6x^2 + 9x + 10x + 15 &&\text{Multiply the monomials.} \\ &= 6x^3 + 15x^2 + 19x + 15 &&\text{Combine like terms.}\end{aligned}$$

In the third line of the solution, note that each term of $3x^2 + 3x + 5$ has been multiplied by each term of $2x + 3$. This example suggests the following rule.

Multiplying Polynomials

To multiply two polynomials, multiply each term of one polynomial by each term of the other polynomial, and then combine like terms.

EXAMPLE 6 Multiply: $(7y + 3)(6y^2 - 8y + 1)$

Strategy We will multiply each term of the trinomial, $6y^2 - 8y + 1$, by each term of the binomial, $7y + 3$.

Why To multiply two polynomials, we must multiply each term of one polynomial by each term of the other polynomial.

Solution

$$(7y + 3)(6y^2 - 8y + 1)$$

$$\begin{aligned}
&= 7y(6y^2) + 7y(-8y) + 7y(1) + 3(6y^2) + 3(-8y) + 3(1) \\
&= 42y^3 - 56y^2 + 7y + 18y^2 - 24y + 3 && \text{Multiply the monomials.} \\
&= 42y^3 - 38y^2 - 17y + 3 && \text{Combine like terms.}
\end{aligned}$$

Self Check 6 Multiply: $(3a^2 - 1)(2a^4 - a^2 - a)$

Now Try Problem 67

> **Success Tip**
> The FOIL method cannot be applied here—only to products of two binomials.

It is often convenient to multiply polynomials using a vertical form similar to that used to multiply whole numbers.

EXAMPLE 7 Multiply using vertical form: **a.** $(3a^2 - 4a + 7)(2a + 5)$ **b.** $(6y^3 - 5y + 4)(-4y^2 - 3)$

Strategy First, we will write one polynomial underneath the other and draw a horizontal line beneath them. Then, we will multiply each term of the upper polynomial by each term of the lower polynomial.

Why *Vertical form* means to use an approach similar to that used in arithmetic to multiply two numbers.

Solution

a. Multiply:

$$\begin{array}{rrrrl}
 & 3a^2 & -\ 4a & +\ 7 & \\
 & & 2a & +\ 5 & \\
\hline
 & 15a^2 & -\ 20a & +\ 35 & \text{Multiply } 3a^2 - 4a + 7 \text{ by } 5. \\
6a^3 & -\ 8a^2 & +\ 14a & & \text{Multiply } 3a^2 - 4a + 7 \text{ by } 2a. \\
\hline
6a^3 & +\ 7a^2 & -\ 6a & +\ 35 & \text{In each column, combine like terms.}
\end{array}$$

> **Success Tip**
> Multiplying two polynomials in vertical form is much like multiplying two numbers in arithmetic.
>
> $$\begin{array}{r} 347 \\ \times\ 25 \\ \hline 1735 \\ +694\ \ \\ \hline 8675 \end{array}$$

The Language of Algebra The two polynomials written below the horizontal line, $-18y^3 + 15y - 12$ and $-24y^5 + 20y^3 - 16y^2$, are called *partial products*.

b. With this method, it is often necessary to leave a space for a missing term to vertically align like terms.

Multiply:

$$\begin{array}{rrrrrl}
 & 6y^3 & & -\ 5y & +\ 4 & \\
 & & & -4y^2 & -\ 3 & \\
\hline
 & -18y^3 & & +\ 15y & -\ 12 & \text{Multiply } 6y^3 - 5y + 4 \text{ by } -3. \\
-24y^5 & +\ 20y^3 & -\ 16y^2 & & & \text{Multiply } 6y^3 - 5y + 4 \text{ by } -4y^2. \\
\hline
-24y^5 & +\ 2y^3 & -\ 16y^2 & +\ 15y & -\ 12 & \text{Leave a space for any missing powers of } y. \text{ In each column, combine like terms.}
\end{array}$$

Self Check 7 Multiply using vertical form:

a. $(3x + 2)(2x^2 - 4x + 5)$

b. $(-2x^2 + 3)(2x^2 - 4x - 1)$

Now Try **Problem 77**

When finding the product of three polynomials, we begin by multiplying any two of them, and then we multiply that result by the third polynomial.

EXAMPLE 8

Find the product: $-3a(4a + 1)(a - 7)$

Strategy We will find the product of $4a + 1$ and $a - 7$ and then multiply that result by $-3a$.

Why It is wise to perform the most difficult multiplication first. (In this case, that would be the product of the two binomials). Save the simpler multiplication by $-3a$ for last.

Solution

$$\begin{aligned}
-3a(4a + 1)(a - 7) &= -3a(4a^2 - 28a + a - 7) && \text{Multiply the two binomials.} \\
&= -3a(4a^2 - 27a - 7) && \text{Combine like terms within the parentheses.} \\
&= -12a^3 + 81a^2 + 21a && \text{Distribute the multiplication by } -3a.
\end{aligned}$$

Self Check 8 Find the product: $-2y(y + 3)(3y - 2)$

Now Try **Problem 81**

ANSWERS TO SELF CHECKS **1. a.** $18t^2$ **b.** $60d^{11}$ **c.** $4y^{14}$ **d.** $-30a^7b^8$ **2. a.** $20c^5 - 45c^3 - 40c^2$ **b.** $s^6t^4 - s^5t^5 + s^4t^6 - 7s^3t^2$ **c.** $6w^{12} - 12w^6$ **4.** $9y^2 - 33y - 12$ **5. a.** $y^2 + 4y + 3$ **b.** $6a^2 + a - 2$ **c.** $16x^7 + x - \frac{3}{8}$ **d.** $10y^6 - 39y^3 + 14$ **6.** $6a^6 - 5a^4 - 3a^3 + a^2 + a$ **7. a.** $6x^3 - 8x^2 + 7x + 10$ **b.** $-4x^4 + 8x^3 + 8x^2 - 12x - 3$ **8.** $-6y^3 - 14y^2 + 12y$

STUDY SET 5.6

VOCABULARY

Fill in the blanks.

1. $(2x^3)(3x^4)$ is the product of two __________.
2. $(2a - 4)(3a + 5)$ is the product of two __________.
3. In the acronym FOIL, F stands for _____ terms, O for _____ terms, I for _____ terms, and L for _____ terms.
4. $(2a - 4)(3a^2 + 5a - 1)$ is the product of a __________ and a __________.

CONCEPTS

Fill in the blanks.

5. **a.** To multiply two polynomials, multiply _____ term of one polynomial by _____ term of the other polynomial, and then combine like terms.
   b. When multiplying three polynomials, we begin by multiplying _____ two of them, and then we multiply that result by the _____ polynomial.
6. Label each arrow using one of the letters F, O, I, or L. Then fill in the blanks.

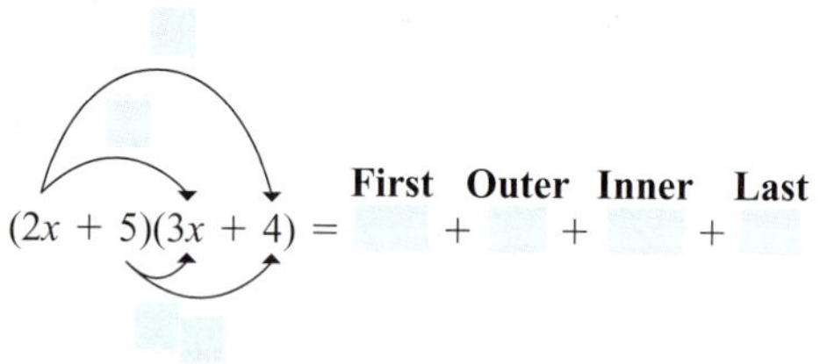

7. Simplify each polynomial by combining like terms.
   a. $6x^2 - 8x + 9x - 12$
   b. $5x^4 + 3ax^2 + 5ax^2 + 3a^2$
8. **a.** Add: $(x - 4) + (x + 8)$
   b. Subtract: $(x - 4) - (x + 8)$
   c. Multiply: $(x - 4)(x + 8)$

NOTATION

Complete each solution.

9. $(9n^3)(8n^2) = (9 \cdot \ \ )(\ \ \cdot n^2) = $ ____
10. $7x(3x^2 - 2x + 5) = \ \ (3x^2) - \ \ (2x) + \ \ (5)$
    $= \ \ \ - 14x^2 + 35x$
11. $(2x + 5)(3x - 2) = 2x(3x) - \ \ (2) + \ \ (3x) - \ \ (2)$
    $= 6x^2 - \ \ \ + \ \ \ - 10$
    $= 6x^2 + \ \ \ - 10$
12.
$$\begin{array}{r} 3x^2 + 4x - 2 \\ 2x + 3 \\ \hline \quad + 12x - 6 \\ 6x^3 + 8x^2 - 4x \quad\quad \\ \hline \quad + 17x^2 + \quad - 6 \end{array}$$

GUIDED PRACTICE

***Multiply.* See Example 1.**

13. $5m(m)$
14. $4s(s)$
15. $(3x^2)(4x^3)$
16. $(-2a^3)(11a^2)$
17. $(1.2c^3)(5c^3)$
18. $(2.5h^4)(2h^4)$
19. $(3b^2)(-2b)(4b^3)$
20. $(3y)(7y^2)(-y^4)$
21. $(2x^2y^3)(4x^3y^2)$
22. $(-5x^3y^6)(2x^2y^2)$
23. $(8a^5)\left(-\frac{1}{4}a^6\right)$
24. $\left(-\frac{2}{3}x^6\right)(9x^3)$

***Multiply.* See Example 2.**

25. $3x(x + 4)$
26. $3a(a + 2)$
27. $-4t(t^2 - 7)$
28. $-6s(s^2 - 3)$
29. $9x^2(x^2 - 2x + 6)$
30. $4y^2(y^2 + 5y - 10)$
31. $-2x^3(3x^2 - x + 1)$
32. $-4b^3(2b^2 - 2b + 2)$
33. $0.3p^5(0.4p^4 - 6p^2)$
34. $0.5u^5(0.4u^6 - 0.5u^3)$
35. $\frac{5}{8}t^2(t^6 + 8t^2)$
36. $\frac{4}{9}a^2(9a^3 + a^2)$
37. $-4x^2z(3x^2 + z^2 + xz - 1)$
38. $-3x^2y(x^2 + y^2 + xy - 1)$
39. $(x^2 - 12x)(6x^{12})$
40. $(w^9 - 11w)(2w^7)$

***Find a polynomial that represents the area of the parallelogram or rectangle.* See Example 3.**

41.

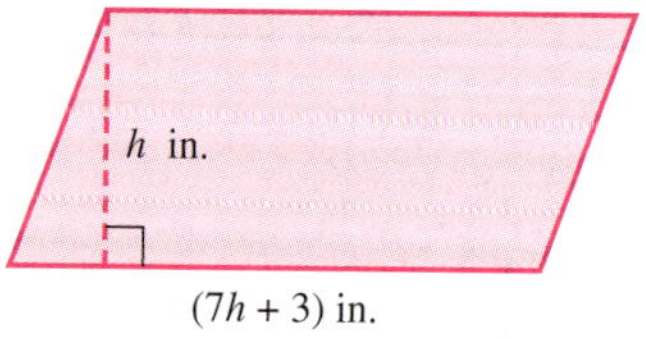

42.

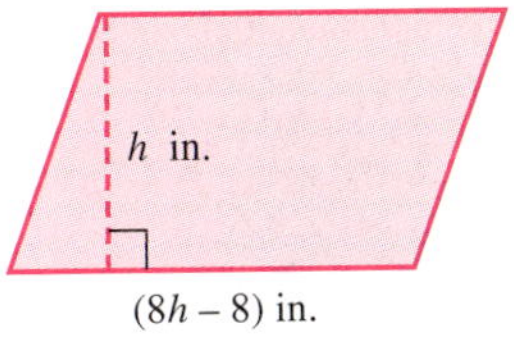

43.

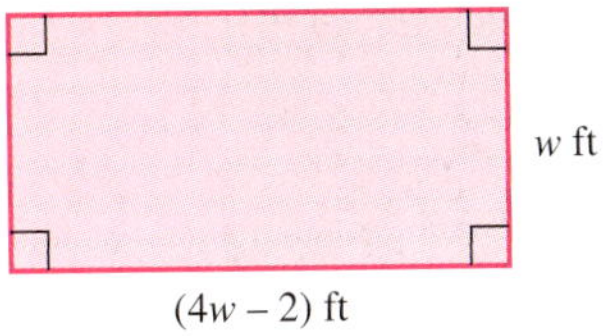

44.

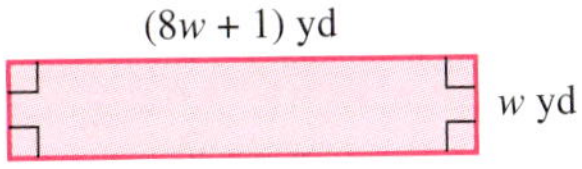

Multiply. **See Examples 4 and 5.**

45. $(y + 3)(y + 5)$

46. $(a + 4)(a + 5)$

47. $(t + 4)(t - 3)$

48. $(x + 7)(x - 6)$

49. $(m + 6)(m - 9)$

50. $(n + 8)(n - 10)$

51. $(4y - 5)(y + 7)$

52. $(3x - 4)(x + 5)$

53. $(2x - 3)(6x - 5)$

54. $(5x - 3)(2x - 3)$

55. $(3.8y - 1)(2y - 1)$

56. $(2.6x - 3)(2x - 1)$

57. $\left(6m - \frac{2}{3}\right)\left(3m - \frac{4}{3}\right)$

58. $\left(8t - \frac{1}{2}\right)\left(4t - \frac{5}{2}\right)$

59. $(t^2 - 3)(t^2 - 4)$

60. $(s^3 - 6)(s^3 - 8)$

61. $(a + b)(a + b)$

62. $(m + n)(m + n)$

63. $(3a - 2b)(4a + b)$

64. $(2t + 3s)(3t - s)$

Multiply. **See Example 6.**

65. $(x + 2)(x^2 - 2x + 3)$

66. $(x - 5)(x^2 + 2x - 3)$

67. $(4t + 3)(t^2 + 2t + 3)$

68. $(3x + 1)(2x^2 - 3x + 1)$

69. $(x - 2)(x^2 + 2x + 4)$

70. $(a + 3)(a^2 - 3a + 9)$

71. $(x^2 + 6x + 7)(2x - 5)$

72. $(y^2 - 2y + 1)(4y + 8)$

73. $(-3x + y)(x^2 - 8xy + 16y^2)$

74. $(3x - y)(x^2 + 3xy - y^2)$

75. $(r^2 - r + 3)(r^2 - 4r - 5)$

76. $(w^2 + w - 9)(w^2 - w + 3)$

Multiply using vertical form. **See Example 7.**

77. $\begin{array}{r} x^2 - 2x + 1 \\ \underline{x + 2} \end{array}$

78. $\begin{array}{r} 5r^2 + r + 6 \\ \underline{2r - 1} \end{array}$

79. $\begin{array}{r} 4x^2 + 3x - 4 \\ \underline{3x + 2} \end{array}$

80. $\begin{array}{r} x^2 - x + 1 \\ \underline{x + 1} \end{array}$

Multiply. **See Example 8.**

81. $4x(2x + 1)(x - 2)$

82. $5a(3a - 2)(2a + 3)$

83. $-3a(a + b)(a - b)$

84. $-2r(r + s)(r + s)$

85. $(-2a^2)(-3a^3)(3a - 2)$

86. $(3x)(-2x^2)(x + 4)$

87. $(x - 4)(x + 1)(x - 3)$

88. $(x + 6)(x - 2)(x - 4)$

TRY IT YOURSELF

Multiply.

89. $(5x - 2)(6x - 1)$

90. $(8x - 1)(3x - 7)$

91. $(3x^2 + 4x - 7)(2x^2)$

92. $(2y^2 - 7y - 8)(3y^3)$

93. $(6x^2z^5)(-3xz^3)$

94. $(-5r^4t^2)(2r^2t)$

95. $\begin{array}{r} 2a^2 + 3a + 1 \\ \underline{3a^2 - 2a + 4} \end{array}$

96. $\begin{array}{r} 3y^2 + 2y - 4 \\ \underline{2y^2 - 4y + 3} \end{array}$

97. $(t + 2s)(9t - 3s)$

98. $(4t - u)(3t + u)$

99. $\left(\frac{1}{2}a\right)(4a^4)(a^5)$

100. $(12b)\left(\frac{7}{6}b\right)(b^4)$

101. $4y(y + 3)(y + 7)$

102. $2t(t + 8)(t + 10)$

103. $8.2pq(2pq - 3p + 5q)$

104. $5.3ab(2ab + 6a - 3b)$

105. $(x + 6)(x^3 + 5x^2 - 4x - 4)$

106. $(x - 8)(x^3 - 4x^2 - 2x - 2)$

107. $\left(4a - \frac{5}{4}r\right)\left(4a + \frac{3}{4}r\right)$

108. $\left(5c - \frac{2}{5}t\right)\left(10c + \frac{1}{5}t\right)$

APPLICATIONS

109. STAMPS Find a polynomial that represents the area of the stamp.

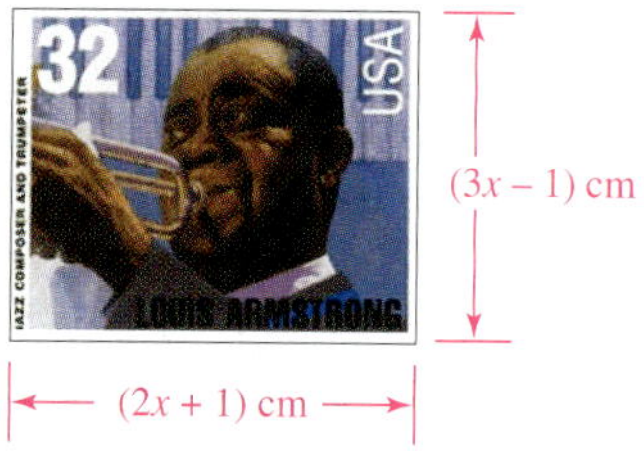

110. PARKING Find a polynomial that represents the total area of the van-accessible parking space and its access aisle.

111. SUNGLASSES An ellipse is an oval-shaped curve. The area of an ellipse is approximately $3.14ab$, where a is its length and b is its width. Find a polynomial that represents the approximate area of one of the elliptical-shaped lenses of the sunglasses.

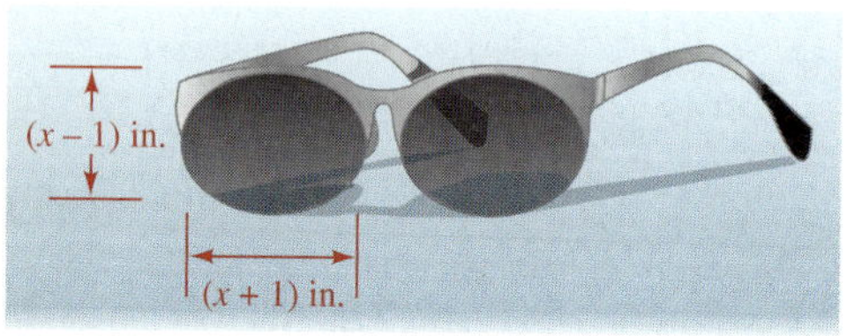

112. GARDENING Refer to the illustration in the next column.

a. Find the area of the region planted with corn, tomatoes, beans, and carrots. Add your answers to find the total area of the garden.

b. Find the length and width of the garden. Multiply your answers to find its area.

c. How do the answers from parts (a) and (b) for the area of the garden compare?

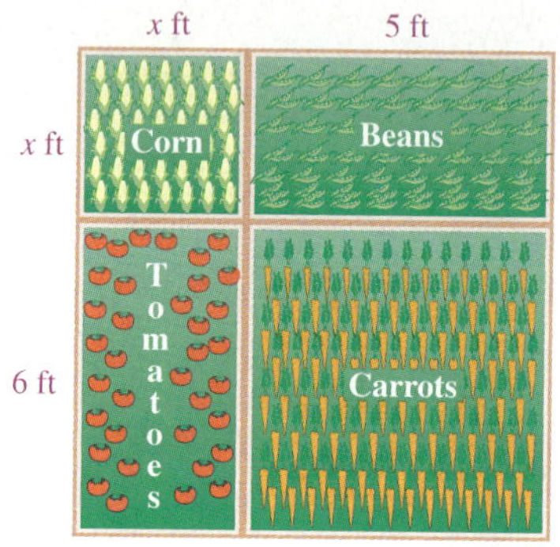

113. LUGGAGE Find a polynomial that represents the volume of the garment bag. (Recall that the formula for the volume of a rectangular solid is $V = lwh$.)

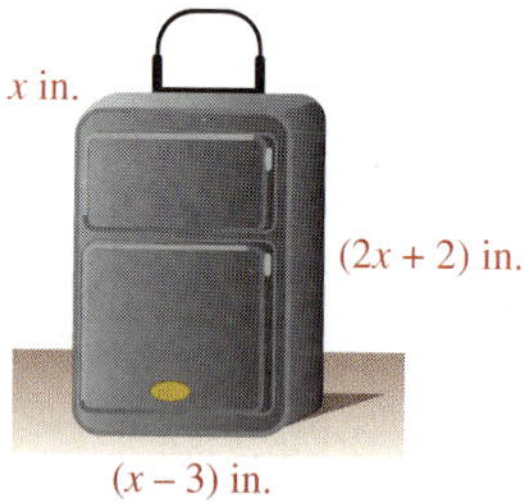

114. BASEBALL Find a polynomial that represents the volume within the batting cage.

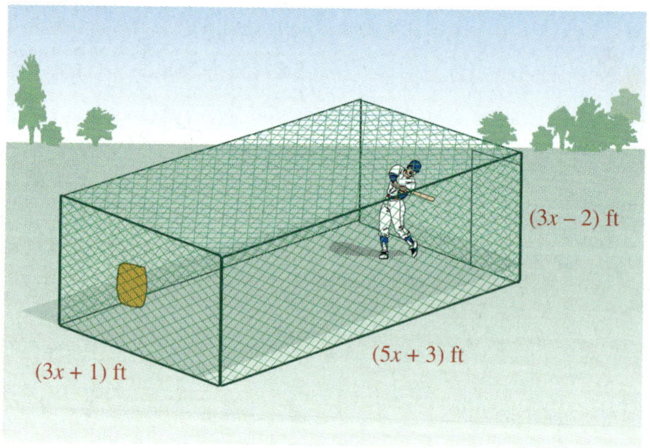

WRITING

115. Is the product of a monomial and a monomial always a monomial? Explain.

116. Explain this diagram.

$$(5x + 6)(7x - 1)$$

117. Explain why the FOIL method cannot be used to find $(3x + 2)(4x^2 - x + 10)$.

118. Explain the error: $(x + 3)(x - 2) = x^2 - 6$

119. Explain why the vertical form used in algebra to multiply $2x^2 + 3x + 1$ and $3x + 2$ is similar to the vertical form used in arithmetic to multiply 231 and 32.

120. Would the OLIF method give the same result as the FOIL method when multiplying two binomials? Explain why or why not.

REVIEW

121. What is the slope of

a. Line 1?

b. Line 2?

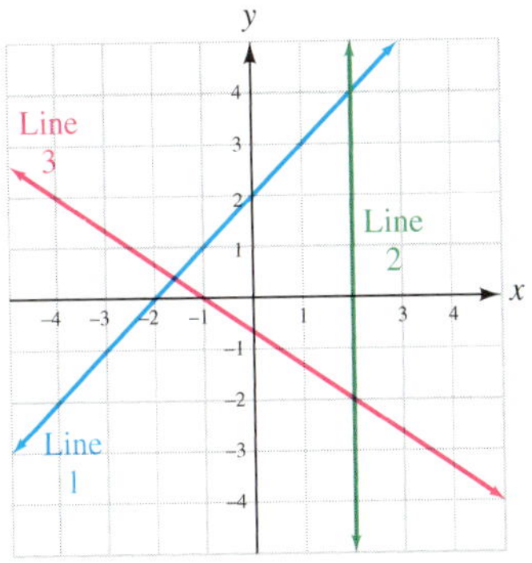

122. What is the slope of

a. Line 3?

b. the x-axis?

123. What is the y-intercept of Line 1?

124. What is the x-intercept of Line 1?

CHALLENGE PROBLEMS

125. a. Find each of the following products.

i. $(x - 1)(x + 1)$

ii. $(x - 1)(x^2 + x + 1)$

iii. $(x - 1)(x^3 + x^2 + x + 1)$

b. Write a product of two polynomials such that the result is $x^5 - 1$.

126. Solve: $(y - 1)(y + 6) = (y - 3)(y - 2) + 8$

SECTION 5.7 Special Products

Objectives

1. Square a binomial.
2. Multiply the sum and difference of the same two terms.
3. Find higher powers of binomials.
4. Simplify expressions containing polynomial multiplication.

Certain products of binomials, called **special products,** occur so often that it is worthwhile to learn their forms.

1 Square a Binomial.

To develop a rule to find the *square of a binomial sum,* we consider $(x + y)^2$. We can use the definition of exponent and the procedure for multiplying two binomials to find the product.

$$(x + y)^2 = (x + y)(x + y) \quad \text{In } (x + y)^2\text{, the base is } (x + y) \text{ and the exponent is 2.}$$
$$= x^2 + xy + xy + y^2 \quad \text{Multiply the binomials.}$$
$$= x^2 + 2xy + y^2 \quad \text{Combine like terms: } xy + xy = 1xy + 1xy = 2xy.$$

Note that the terms of the resulting trinomial are related to the terms of the binomial that was squared.

$$(x + y)^2 = x^2 + 2xy + y^2$$

- y^2: The square of the second term, y.
- $2xy$: Twice the product of the first and second terms, x and y.
- x^2: The square of the first term, x.

To develop a rule to find the *square of a binomial difference,* we consider $(x - y)^2$.

Success Tip

The illustration can be used to visualize a special product. The area of the large square is $(x + y)(x + y) = (x + y)^2$. The sum of the four smaller areas is $x^2 + xy + xy + y^2$ or $x^2 + 2xy + y^2$. Thus,

$$(x + y)^2 = x^2 + 2xy + y^2$$

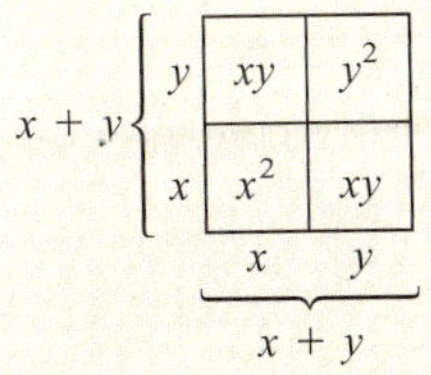

$$(x - y)^2 = (x - y)(x - y)$$ In $(x - y)^2$, the base is $(x - y)$ and the exponent is 2.

$$= x^2 - 2xy + y^2$$ Multiply the binomials.

Combine like terms: $-xy - xy = -2xy$.

Again, the terms of the resulting trinomial are related to the terms of the binomial that was squared.

$$(x - y)^2 = x^2 - 2xy + y^2$$

- y^2: The square of the second term, $-y$.
- $-2xy$: Twice the product of the first and second terms, x and $-y$.
- x^2: The square of the first term, x.

The observations from these two examples suggest the following **special-product rules.**

Squaring a Binomial

The **square of a binomial** is a trinomial, such that:

- Its first term is the square of the first term of the binomial.
- Its last term is the square of the second term of the binomial.
- Its middle term is twice the product of both terms of the binomial.

$$(A + B)^2 = A^2 + 2AB + B^2 \qquad (A - B)^2 = A^2 - 2AB + B^2$$

The Language of Algebra

When squaring a binomial, the result is called a *perfect-square trinomial.*

$$(t + 9)^2 = t^2 + 18t + 81$$

$t^2 + 18t + 81$: *Perfect-square trinomial*

EXAMPLE 1

Find each square: **a.** $(t + 9)^2$ **b.** $(8a - 5)^2$ **c.** $(d + 0.5)^2$ **d.** $\left(c^3 - \frac{7}{2}d\right)^2$

Strategy To find each square of a binomial, we will use one of the special-product rules.

Why This approach is faster than using the FOIL method.

Solution

a. $(t + 9)^2$ is the square of a binomial sum. The first term is t and the second term is 9.

$$(t + 9)^2 = \underbrace{t^2}_{\text{The square of the first term, } t.} + \underbrace{2(t)(9)}_{\text{Twice the product of both terms.}} + \underbrace{9^2}_{\text{The square of the second term, 9.}}$$

$$= t^2 + 18t + 81$$

Caution

The square of a binomial is a *trinomial.* A common error when squaring a binomial is to forget the middle term of the product.

$$(8a - 5)^2 \neq 64a^2 + 25$$

Missing $-80a$

b. $(8a - 5)^2$ is the square of a binomial difference. The first term is $8a$ and the second term is -5.

$$(8a - 5)^2 = \underbrace{(8a)^2}_{\text{The square of the first term, } 8a.} + \underbrace{2(8a)(-5)}_{\text{Twice the product of both terms.}} + \underbrace{(-5)^2}_{\text{The square of the second term, } -5.}$$

$$= 64a^2 - 80a + 25$$ Use the power of a product rule: $(8a)^2 = 8^2a^2 = 64a^2$.

c. $(d + 0.5)^2$ is the square of a binomial sum. The first term is d and the second term is 0.5.

$$(d + 0.5)^2 = \underbrace{(d)^2}_{\text{The square of the first term, } d.} + \underbrace{2(d)(0.5)}_{\text{Twice the product of both terms.}} + \underbrace{(0.5)^2}_{\text{The square of the second term, 0.5.}}$$

$$= d^2 + d + 0.25$$

d. $\left(c^3 - \frac{7}{2}d\right)^2$ is the square of a binomial difference. The first term is c^3 and the second term is $-\frac{7}{2}d$.

$$\left(c^3 - \frac{7}{2}d\right)^2 = \underbrace{(c^3)^2}_{\text{The square of the first term, } c^3.} + \underbrace{2(c^3)\left(-\frac{7}{2}d\right)}_{\text{Twice the product of both terms.}} + \underbrace{\left(-\frac{7}{2}d\right)^2}_{\text{The square of the second term, } -\frac{7}{2}d.}$$

$$= c^6 - 7c^3d + \frac{49}{4}d^2$$ Use rules for exponents to find $(c^3)^2$ and $\left(-\frac{7}{2}d\right)^2$.

Self Check 1 Find each square: **a.** $(r + 6)^2$ **b.** $(7g - 2)^2$ **c.** $(v + 0.8)^2$ **d.** $\left(w^4 - \frac{3}{2}y\right)^2$

Now Try **Problems 9, 19, and 29**

2 Multiply the Sum and Difference of the Same Two Terms.

A final special product that occurs often has the form $(A + B)(A - B)$. In these products, one binomial is the sum of two terms and the other binomial is the difference of the same two terms. To develop a rule to find such products, consider the following multiplication:

$$(x + y)(x - y) = x^2 - xy + xy - y^2$$ Multiply the binomials.

$$= x^2 - y^2$$ Combine like terms: $-xy + xy = 0$.

Note that when we combined like terms, we added opposites. This will always be the case for products of this type; the sum of the outer and inner products will be 0. The first and last products will be squares.

$$(x + y)(x - y) = x^2 - y^2$$

The square of the second term, y.

The square of the first term, x.

These observations suggest a third **special-product rule.**

Success Tip

We can use the FOIL method to find each of the special products discussed in this section. However, these forms occur so often, it is worthwhile to learn the special-product rules.

Multiplying the Sum and Difference of Two Terms

The product of the sum of two terms and difference of the same two terms is the square of the first term minus the square of the second term.

$$(A + B)(A - B) = A^2 - B^2$$

EXAMPLE 2 Multiply: **a.** $(m + 2)(m - 2)$ **b.** $(3y + 4)(3y - 4)$ **c.** $\left(b - \frac{2}{3}\right)\left(b + \frac{2}{3}\right)$ **d.** $(t^4 - 6u)(t^4 + 6u)$

Strategy To find the product of each pair of binomials, we will use the special-product rule for the sum and difference of the same two terms.

Why This approach is faster than using the FOIL method.

The Language of Algebra
When multiplying the sum and difference of two terms, the result is called a *difference of two squares.*

$$(m + 2)(m - 2) = m^2 - 4$$

Difference of two squares

Solution

a. $(m + 2)$ and $(m - 2)$ are the sum and difference of the same two terms, m and 2.

$$(m + 2)(m - 2) = \underbrace{m^2}_{\text{The square of the first term, } m.} - \underbrace{2^2}_{\text{The square of the second term, 2.}}$$

$$= m^2 - 4$$

b. $(3y + 4)$ and $(3y - 4)$ are the sum and difference of the same two terms, $3y$ and 4.

$$(3y + 4)(3y - 4) = \underbrace{(3y)^2}_{\text{The square of the first term, } 3y.} - \underbrace{4^2}_{\text{The square of the second term, 4.}}$$

$$= 9y^2 - 16$$

c. By the commutative property of multiplication, the special-product rule can be written with the factor containing the $-$ symbol first: $(A - B)(A + B) = A^2 - B^2$. Since $\left(b - \frac{2}{3}\right)$ and $\left(b + \frac{2}{3}\right)$ are the difference and sum of the same two terms, b and $\frac{2}{3}$, we have

$$\left(b - \frac{2}{3}\right)\left(b + \frac{2}{3}\right) = \underbrace{b^2}_{\text{The square of the first term, } b.} - \underbrace{\left(\frac{2}{3}\right)^2}_{\text{The square of the second term, } \frac{2}{3}.}$$

$$= b^2 - \frac{4}{9}$$

d. $(t^4 - 6u)$ and $(t^4 + 6u)$ are the difference and sum of the same two terms, t^4 and $6u$.

$$(t^4 - 6u)(t^4 + 6u) = \underbrace{(t^4)^2}_{\text{The square of the first term, } t^4.} - \underbrace{(6u)^2}_{\text{The square of the second term, } 6u.}$$

$$= t^8 - 36u^2$$

Self Check 2 Multiply: **a.** $(b + 4)(b - 4)$
b. $(5m + 9)(5m - 9)$
c. $\left(s - \frac{3}{4}\right)\left(s + \frac{3}{4}\right)$
d. $(c^3 + 2d)(c^3 - 2d)$

Now Try **Problems 35, 37, and 41**

3 Find Higher Powers of Binomials.

When we find the third, fourth, or even higher powers of a binomial, we say that we are **expanding the binomial.** The special-product rules can be used in such cases. The result is an expression that has more terms than the original binomial.

EXAMPLE 3 Expand: $(x + 1)^3$

Strategy We will use a special-product rule to find the third power of $x + 1$.

Why Since $(x + 1)^3$ can be written as $(x + 1)^2(x + 1)$, we can use a special-product rule to quickly find $(x + 1)^2$.

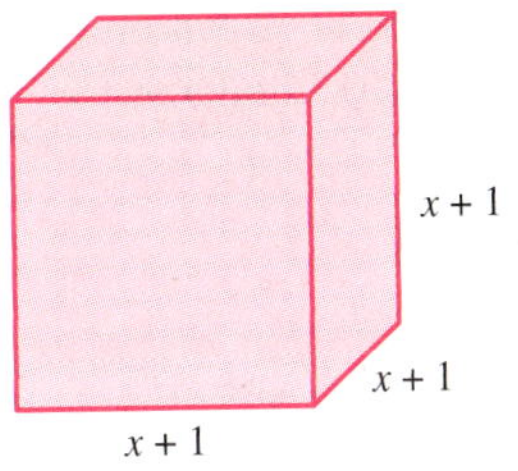

Notation

$(x + 1)^3$ represents the volume of a cube with sides of length $x + 1$ units.

Solution

$$
\begin{aligned}
(x + 1)^3 &= (x + 1)^2(x + 1) \\
&= (x^2 + 2x + 1)(x + 1) && \text{Find } (x + 1)^2 \text{ using the rule for the square of a sum.} \\
&= (x^2 + 2x + 1)(x + 1) && \text{Multiply the binomial and the trinomial.} \\
&= x^2(x) + x^2(1) + 2x(x) + 2x(1) + 1(x) + 1(1) && \text{Multiply each term of } x + 1 \text{ by each term of } x^2 + 2x + 1. \\
&= x^3 + x^2 + 2x^2 + 2x + x + 1 && \text{Multiply the monomials.} \\
&= x^3 + 3x^2 + 3x + 1 && \text{Combine like terms.}
\end{aligned}
$$

Self Check 3 Expand: $(n - 3)^3$

Now Try Problem 49

4 Simplify Expressions Containing Polynomial Multiplication.

We can use a modified version of the order of operations rule to simplify expressions that involve polynomial addition, subtraction, multiplication, and raising to a power.

Order of Operations with Polynomials

1. If possible, simplify any polynomials within parentheses by combining like terms.
2. Square (or expand) all polynomials raised to powers using the FOIL method or a special-product rule.
3. Perform all polynomial multiplications using the distributive property, the FOIL method, or a special-product rule.
4. Perform all polynomial additions and subtractions by combining like terms.

EXAMPLE 4 Simplify each expression:

a. $-8(y^2 - 2y + 3) - 4(2y^2 + y - 6)$
b. $(x + 1)(x - 2) + 3x(x + 3)$
c. $(3y - 2)^2 - (y - 5)(y + 5)$

Strategy We will follow the rules for the order of operations to simplify each expression.

Why If we don't follow the correct order of operations, we can obtain different results that are not equivalent.

Solution

a. The two polynomials within the parentheses do not simplify further and no polynomials are raised to a power. To perform the multiplication, we will use the distributive property twice. Then we will combine like terms.

$$
\begin{aligned}
-8(y^2 - 2y + 3) - 4(2y^2 + y - 6) &= -8y^2 + 16y - 24 - 8y^2 - 4y + 24 && \text{Distribute.} \\
&= -16y^2 + 12y && \text{Add and subtract to combine like terms.}
\end{aligned}
$$

b. The three polynomials within parentheses do not simplify further and no polynomials are raised to a power. To perform the multiplications, we use the FOIL method and the distributive property. Then we will combine like terms.

$(x + 1)(x - 2) + 3x(x + 3)$

$= x^2 - x - 2 + 3x^2 + 9x$ Use the FOIL method to find $(x + 1)(x - 2)$. Distribute the multiplication by $3x$.

$= 4x^2 + 8x - 2$ Combine like terms.

c. The three polynomials within parentheses do not simplify further. To square $3y - 2$, we use a special-product rule. To find the product of $(y - 5)(y + 5)$, we will use the special-product rule for the sum and difference of the same two terms. Then we will combine like terms.

$(3y - 2)^2 - (y - 5)(y + 5)$

$= 9y^2 - 12y + 4 - (y^2 - 25)$ Write $y^2 - 25$ within parentheses so that both terms are subtracted.

$= 9y^2 - 12y + 4 - y^2 + 25$ Change the sign of each term within $(y^2 - 25)$ and drop the parentheses.

$= 8y^2 - 12y + 29$ Combine like terms.

Self Check 4 Simplify each expression:

a. $2(a^2 - 3a) + 5(a^2 + 2a)$

b. $(x - 4)(x + 6) + 5x(2x - 1)$

c. $(a + 9)(a - 9) - (2a - 4)^2$

Now Try **Problems 61, 63, and 67**

EXAMPLE 5 ***Triangles.*** Find a polynomial that represents the area of the triangle.

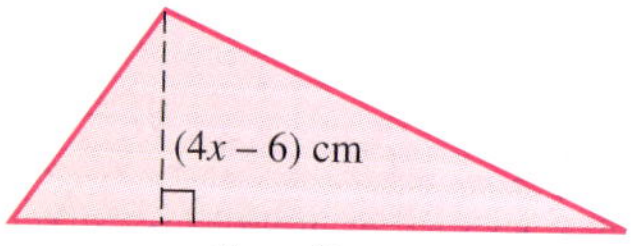

Strategy We will multiply one-half, the length of the base, and the height of the triangle.

Why The area of a triangle is equal to one-half the product of the length of its base and its height.

Solution We begin by substituting $4x + 6$ for the length of the base and $4x - 6$ for the height in the formula for the area of a triangle. It is wise to find $(4x + 6)(4x - 6)$ first, using a special-product rule, and then to multiply that result by $\frac{1}{2}$.

$\text{Area} = \frac{1}{2} \cdot \textbf{base} \cdot \textbf{height}$

$= \frac{1}{2}(4x + 6)(4x - 6)$

$= \frac{1}{2}(16x^2 - 36)$ Use the special product rule for a sum and difference of two terms.

$= 8x^2 - 18$ Distribute the multiplication by $\frac{1}{2}$.

Success Tip
Recall that to multiply three polynomials, begin by multiplying any two of them. Then multiply that result by the third polynomial.

The area of the triangle is $(8x^2 - 18)$ square centimeters, which can be written as $(8x^2 - 18)\text{ cm}^2$.

Now Try Problem 73

ANSWERS TO SELF CHECKS 1. a. $r^2 + 12r + 36$ b. $49g^2 - 28g + 4$ c. $v^2 + 1.6v + 0.64$ d. $w^8 - 3w^4y + \frac{9}{4}y^2$ 2. a. $b^2 - 16$ b. $25m^2 - 81$ c. $s^2 - \frac{9}{16}$ d. $c^6 - 4d^2$ 3. $n^3 - 9n^2 + 27n - 27$ 4. a. $7a^2 + 4a$ b. $11x^2 - 3x - 24$ c. $-3a^2 + 16a - 97$

STUDY SET 5.7

VOCABULARY

Fill in the blanks.

1. Expressions of the form $(x + y)^2$, $(x - y)^2$, and $(x + y)(x - y)$ occur so frequently in algebra that they are called special ________.
2. $(2x + 3)^2$ is the ______ of a binomial and $(a + 6)(a - 6)$ is the product of the sum and difference of the ______ two terms.

CONCEPTS

3. Complete each special product.
   a. $(x + y)^2 = x^2 + 2xy + y^2$
      - x^2: The square of the ______ term
      - $2xy$: ______ the product of the first and second terms
      - y^2: The ______ of the second term
   b. $(x + y)(x - y) = x^2 - y^2$
      - x^2: The ______ of the first term
      - y^2: The square of the ______ term
4. Consider the binomial $5x + 4$.
   a. What is the square of its first term?
   b. What is twice the product of its two terms?
   c. What is the square of its second term?

NOTATION

Complete each solution to find the product.

5. $(x + 4)^2 = \square^2 + 2(x)(\square) + \square^2$
   $= x^2 + \square + 16$
6. $(6r - 1)^2 = (\square)^2 \square 2(6r)(1) + (-1)^2$
   $= \square - \square + 1$
7. $(s + 5)(s - 5) = \square^2 - \square^2$
   $= s^2 - \square$
8. True or false: $(t + 7)(t - 7) = (t - 7)(t + 7)$?

GUIDED PRACTICE

Find each product. See Example 1.

9. $(x + 1)^2$
10. $(y + 7)^2$
11. $(r + 2)^2$
12. $(n + 10)^2$
13. $(m - 6)^2$
14. $(b - 1)^2$
15. $(f - 8)^2$
16. $(w - 9)^2$
17. $(4x + 5)^2$
18. $(6y + 3)^2$
19. $(7m - 2)^2$
20. $(9b - 2)^2$
21. $(1 - 3y)^2$
22. $(1 - 4a)^2$
23. $(y + 0.9)^2$
24. $(d + 0.2)^2$
25. $(a^2 + b^2)^2$
26. $(c^2 + d^2)^2$
27. $(r^2 - s^2)^2$
28. $(t^2 - u^2)^2$
29. $\left(s + \frac{3}{4}\right)^2$
30. $\left(y - \frac{5}{3}\right)^2$
31. $\left(d^4 + \frac{1}{4}\right)^2$
32. $\left(q^6 + \frac{1}{3}\right)^2$

SECTION 5.8
Dividing Polynomials

Objectives

1. Divide a monomial by a monomial.
2. Divide a polynomial by a monomial.
3. Divide a polynomial by a polynomial.

In this section, we will conclude our study of operations with polynomials by discussing division of polynomials. To begin, we consider the simplest case, the quotient of two monomials.

1 Divide a Monomial by a Monomial.

To divide monomials, we can use the method for simplifying fractions or a method that involves a rule for exponents.

Success Tip
In this section, you will see that regardless of the number of terms involved, every polynomial division is a series of monomial divisions.

EXAMPLE 1 Divide the monomials: **a.** $\frac{21x^5}{7x^2}$ **b.** $\frac{10r^6s}{6rs^3}$

Strategy We can use the rules for simplifying fractions and/or the quotient rule for exponents.

Why We need to make sure that the numerator and denominator have no common factors other than 1. If that is the case, then the fraction is in *simplest form.*

Solution

By simplifying fractions

a. $\frac{21x^5}{7x^2} = \frac{3 \cdot \overset{1}{\cancel{7}} \cdot \overset{1}{\cancel{x}} \cdot \overset{1}{\cancel{x}} \cdot x \cdot x \cdot x}{\underset{1}{\cancel{7}} \cdot \underset{1}{\cancel{x}} \cdot \underset{1}{\cancel{x}}}$

$= 3x^3$

b. $\frac{10r^6s}{6rs^3} = \frac{\overset{1}{\cancel{2}} \cdot 5 \cdot \overset{1}{\cancel{r}} \cdot r \cdot r \cdot r \cdot r \cdot r \cdot \overset{1}{\cancel{s}}}{\underset{1}{\cancel{2}} \cdot 3 \cdot \underset{1}{\cancel{r}} \cdot \underset{1}{\cancel{s}} \cdot s \cdot s}$

$= \frac{5r^5}{3s^2}$

Using the rules for exponents

a. $\frac{21x^5}{7x^2} = 3x^{5-2}$ Divide the coefficients.

$= 3x^3$ Subtract the exponents.

b. $\frac{10r^6s}{6rs^3} = \frac{5}{3}r^{6-1}s^{1-3}$ Simplify $\frac{10}{6}$.

$= \frac{5}{3}r^5s^{-2}$ Subtract exponents.

$= \frac{5r^5}{3s^2}$ Move s^{-2} to the denominator and change the sign of the exponent.

Self Check 1 Divide the monomials: **a.** $\frac{30y^4}{5y^2}$ **b.** $\frac{8c^2d^6}{32c^5d^2}$

Now Try **Problems 15 and 23**

2 Divide a Polynomial by a Monomial.

Recall that to add two fractions with the same denominator, we add their numerators and keep their common denominator.

$$\frac{a}{d} + \frac{b}{d} = \frac{a+b}{d}$$

We can use this rule in reverse to divide polynomials by monomials.

Dividing a Polynomial by a Monomial

To divide a polynomial by a monomial, divide each term of the polynomial by the monomial. If A, B, and D represent monomials, where $D \neq 0$, then

$$\frac{A+B}{D} = \frac{A}{D} + \frac{B}{D}$$

EXAMPLE 2 Divide: **a.** $\dfrac{9x^2 + 6x}{3x}$ **b.** $\dfrac{12a^4b^3 - 18a^3b^2 + 2a^2}{6a^2b^2}$

Strategy We will divide each term of the polynomial in the numerator by the monomial in the denominator.

Why A fraction bar indicates division of the numerator by the denominator.

Solution

a. Here, we have a binomial divided by a monomial.

$$\frac{9x^2 + 6x}{3x} = \frac{9x^2}{3x} + \frac{6x}{3x}$$ Divide each term of the numerator, $9x^2 + 6x$, by the denominator, $3x$.

$$= 3x^{2-1} + 2x^{1-1}$$ Do each monomial division. Divide the coefficients. Keep each base and subtract the exponents.

$$= 3x + 2$$ Recall that $x^0 = 1$.

Check: We multiply the divisor, $3x$, and the quotient, $3x + 2$. The result should be the dividend, $9x^2 + 6x$.

$$3x(3x + 2) = 9x^2 + 6x$$ The answer checks.

b. Here, we have a trinomial divided by a monomial.

$$\frac{12a^4b^3 - 18a^3b^2 + 2a^2}{6a^2b^2} = \frac{12a^4b^3}{6a^2b^2} - \frac{18a^3b^2}{6a^2b^2} + \frac{2a^2}{6a^2b^2}$$ Divide each term of the numerator by the denominator, $6a^2b^2$.

$$= 2a^{4-2}b^{3-2} - 3a^{3-2}b^{2-2} + \frac{a^{2-2}}{3b^2}$$ Do each monomial division. Simplify: $\frac{2}{6} = \frac{1}{3}$.

$$= 2a^2b - 3a + \frac{1}{3b^2}$$ $b^{2-2} = b^0 = 1$ and $a^{2-2} = a^0 = 1$.

Recall that the variables in a polynomial must have whole-number exponents. Therefore, the result, $2a^2b - 3a + \frac{1}{3b^2}$, is not a polynomial because the last term has a variable in the denominator.

The Language of Algebra

The names of the parts of a division statement are

Dividend: $9x^2 + 6x$; Divisor: $3x$; Quotient: $3x + 2$

$$\frac{9x^2 + 6x}{3x} = 3x + 2$$

Success Tip

The sum, difference, and product of two polynomials are always polynomials. However, as seen in Example 2b, the quotient of two polynomials is not always a polynomial.

Check:

$$6a^2b^2\left(2a^2b - 3a + \frac{1}{3b^2}\right) = 12a^4b^3 - 18a^3b^2 + 2a^2$$ The answer checks.

Self Check 2 Divide: **a.** $\frac{50h^3 + 15h^2}{5h^2}$ **b.** $\frac{22s^5t^2 - s^4t^3 + 44s^2t^4}{11s^2t^2}$

Now Try **Problems 29, 37, and 45**

3 Divide a Polynomial by a Polynomial.

To divide a polynomial by a polynomial (other than a monomial), we use a method similar to long division in arithmetic.

EXAMPLE 3 Divide $x^2 + 5x + 6$ by $x + 2$.

Strategy We will use the long division method. The dividend is $x^2 + 5x + 6$ and the divisor is $x + 2$.

Why Since the divisor has more than one term, we must use the long division method to divide the polynomials.

Solution We write the division using a long division symbol $\overline{)\quad\quad}$ and proceed as follows:

Step 1: $x + 2\overline{)x^2 + 5x + 6}$ with quotient x

Divide the first term of the dividend by the first term of the divisor: $\frac{x^2}{x} = x$. Write the result, x, above the long division symbol.

Step 2:
$$\begin{array}{r} x \\ x + 2\overline{)x^2 + 5x + 6} \\ \underline{x^2 + 2x} \end{array}$$

Multiply each term of the divisor by x. Write the result, $x^2 + 2x$, under $x^2 + 5x$, and draw a line. Be sure to align the like terms.

Step 3:
$$\begin{array}{r} x \\ x + 2\overline{)x^2 + 5x + 6} \\ \underline{-(x^2 + 2x)} \downarrow \\ 3x + 6 \end{array}$$

Subtract $x^2 + 2x$ from $x^2 + 5x$. Work column by column: $x^2 - x^2 = 0$ and $5x - 2x = 3x$.

Bring down the next term, 6.

Step 4:
$$\begin{array}{r} x + 3 \\ x + 2\overline{)x^2 + 5x + 6} \\ \underline{-(x^2 + 2x)} \\ 3x + 6 \end{array}$$

Divide the first term of $3x + 6$ by the first term of the divisor: $\frac{3x}{x} = 3$. Write $+ 3$ above the long division symbol to form the second term of the quotient.

Step 5:
$$\begin{array}{r} x + 3 \\ x + 2\overline{)x^2 + 5x + 6} \\ \underline{-(x^2 + 2x)} \\ 3x + 6 \\ \underline{3x + 6} \end{array}$$

Multiply each term of the divisor by 3. Write the result, $3x + 6$, under $3x + 6$ and draw a line. Be sure to align the like terms.

The Language of Algebra
Notice how the instruction *to divide* a polynomial by a binomial translates:

Divide $x^2 + 5x + 6$ by $x + 2$

$x + 2\overline{)x^2 + 5x + 6}$

Success Tip
Notice that this method is much like that used for division of whole numbers.

$$\begin{array}{r} 13 \\ 12\overline{)156} \\ \underline{-12}\downarrow \\ 036 \\ \underline{-36} \\ 0 \end{array}$$

Hundreds
Tens
Ones

Success Tip
The long division method aligns like terms vertically.

$$\begin{array}{r} x+3 \\ x+2\overline{)x^2+5x+6} \\ -(x^2+2x) \\ \hline 3x+6 \\ -(3x+6) \\ \hline 0 \end{array}$$

x^2-terms, x-terms, constants

Step 6:
$$\begin{array}{r} x+3 \\ x+2\overline{)x^2+5x+6} \\ -(x^2+2x) \\ \hline 3x+6 \\ -(3x+6) \\ \hline 0 \end{array}$$

Subtract $3x + 6$ from $3x + 6$. Work vertically: $3x - 3x = 0$ and $6 - 6 = 0$.

This is the remainder.

The quotient is $x + 3$ and the remainder is 0.

Step 7: Check the result by verifying that $(x + 2)(x + 3)$ is $x^2 + 5x + 6$.

$$(x+2)(x+3) = x^2 + 3x + 2x + 6$$
$$= x^2 + 5x + 6$$

The answer checks.

Self Check 3 Divide $x^2 + 7x + 12$ by $x + 3$.

Now Try Problem 49

The long division method used in algebra can have a remainder just as long division in arithmetic often does.

EXAMPLE 4 Divide: $(6x^2 - 7x - 2) \div (2x - 1)$

Strategy We will use the long division method. The dividend is $6x^2 - 7x - 2$ and the divisor is $2x - 1$.

Why Since the divisor has more than one term, we must use the long division method to divide the polynomials.

Success Tip
The long-division method is a series of four steps that are repeated:
- Divide
- Multiply
- Subtract
- Bring down the next term

Solution

Step 1:
$$\begin{array}{r} 3x \\ 2x-1\overline{)6x^2-7x-2} \end{array}$$

Divide the first term of the dividend by the first term of the divisor: $\frac{6x^2}{2x} = 3x$. Write the result, $3x$, above the long division symbol.

Step 2:
$$\begin{array}{r} 3x \\ 2x-1\overline{)6x^2-7x-2} \\ 6x^2-3x \\ \hline \end{array}$$

Multiply each term of the divisor by $3x$. Write the result, $6x^2 - 3x$, under $6x^2 - 7x$, and draw a line.

Step 3:
$$\begin{array}{r} 3x \\ 2x-1\overline{)6x^2-7x-2} \\ -(6x^2-3x) \downarrow \\ \hline -4x-2 \end{array}$$

Subtract $6x^2 - 3x$ from $6x^2 - 7x$. Work vertically: $6x^2 - 6x^2 = 0$ and $-7x - (-3x) = -7x + 3x = -4x$. Bring down the next term, -2.

Step 4:
$$\begin{array}{r} 3x-2 \\ 2x-1\overline{)6x^2-7x-2} \\ -(6x^2-3x) \\ \hline -4x-2 \end{array}$$

Divide the first term of $-4x - 2$ by the first term of the divisor: $\frac{-4x}{2x} = -2$. Write -2 above the long division symbol to form the second term of the quotient.

Step 5:

$$\begin{array}{r} 3x - 2 \\ 2x - 1\overline{)6x^2 - 7x - 2} \\ -(\underline{6x^2 - 3x)} \\ -4x - 2 \\ \underline{-4x + 2} \end{array}$$

Multiply each term of the divisor by −2. Write the result, −4x + 2, under −4x − 2, and draw a line.

Success Tip

The long division method for polynomials continues until the degree of the remainder is less than the degree of the divisor. Here, the remainder, −4, has degree 0. The divisor, $2x - 1$, has degree 1. Therefore, the division ends.

Step 6:

$$\begin{array}{r} 3x - 2 \\ 2x - 1\overline{)6x^2 - 7x - 2} \\ -(\underline{6x^2 - 3x)} \\ -4x - 2 \\ -(\underline{-4x + 2)} \\ -4 \end{array}$$

Subtract −4x + 2 from −4x − 2. Work vertically: $-4x - (-4x) = -4x + 4x = 0$ and $-2 - 2 = -4$.

The quotient is $3x - 2$ and the remainder is −4. It is common to write the answer in $Quotient + \frac{remainder}{divisor}$ form as either

$$3x - 2 + \frac{-4}{2x - 1} \quad \text{or} \quad 3x - 2 - \frac{4}{2x - 1}$$

Step 7: We can check the result using the fact that for any division:

$$\underset{\downarrow}{\text{Divisor}} \cdot \underset{\downarrow}{\text{quotient}} + \underset{\downarrow}{\text{remainder}} = \underset{\downarrow}{\text{dividend}}$$

$$\begin{aligned} (2x - 1)(3x - 2) + (-4) &= 6x^2 - 4x - 3x + 2 + (-4) \\ &= 6x^2 - 7x - 2 \end{aligned}$$

The answer checks.

Self Check 4 Divide $(8x^2 + 6x - 3) \div (2x + 3)$. Check the result.

Now Try Problem 63

The division method works best when the terms of the divisor and the dividend are written in descending powers of the variable. If the powers in the dividend or divisor are not in descending order, we use the commutative property of addition to write them that way.

EXAMPLE 5

Divide $4x^2 + 2x^3 + 12 - 2x$ by $x + 3$.

Strategy We will write the dividend in descending powers of x and use the long division method.

Why It is easier to align like terms in columns when the powers of the variable are written in descending order.

Solution

$$
\begin{array}{r}
2x^2 - 2x + 4 \\
x + 3\overline{)2x^3 + 4x^2 - 2x + 12} \\
-(2x^3 + 6x^2) \qquad\qquad \\
-2x^2 - 2x \qquad \\
-(-2x^2 - 6x) \qquad \\
4x + 12 \\
-(4x + 12) \\
0
\end{array}
$$

The first division: $\frac{2x^3}{x} = 2x^2$.

The second division: $\frac{-2x^2}{x} = -2x$.

The third division: $\frac{4x}{x} = 4$.

Check: $(x + 3)(2x^2 - 2x + 4) = 2x^3 - 2x^2 + 4x + 6x^2 - 6x + 12$

$= 2x^3 + 4x^2 - 2x + 12$ The answer checks.

Self Check 5 Divide $x^2 - 10x + 6x^3 + 4$ by $2x - 1$.

Now Try **Problem 65**

When we write the terms of a dividend in descending powers, we must determine whether some powers of the variable are missing. If any are missing, insert such terms with a coefficient of 0 or leave blank spaces for them. This keeps like terms in the same column, which is necessary when performing the subtraction in vertical form.

EXAMPLE 6

Divide: $\frac{27x^3 + 1}{3x + 1}$

Strategy The divisor is $3x + 1$. The dividend, $27x^3 + 1$, does not have an x^2-term or an x-term. We will insert a $0x^2$ term and a $0x$ term as placeholders, and use the long division method.

Why We insert placeholder terms so that like terms will be aligned in the same column when we subtract.

Solution

$$
\begin{array}{r}
9x^2 - 3x + 1 \\
3x + 1\overline{)27x^3 + 0x^2 + 0x + 1} \\
-(27x^3 + 9x^2) \qquad\qquad \\
-9x^2 + 0x \qquad \\
-(-9x^2 - 3x) \qquad \\
3x + 1 \\
-(3x + 1) \\
0
\end{array}
$$

The first division: $\frac{27x^3}{3x} = 9x^2$.

The second division: $\frac{-9x^2}{3x} = -3x$.

The third division: $\frac{3x}{3x} = 1$.

Check: $(3x + 1)(9x^2 - 3x + 1) = 27x^3 - 9x^2 + 3x + 9x^2 - 3x + 1$

$= 27x^3 + 1$ The answer checks.

Self Check 6 Divide $\frac{x^2 - 9}{x - 3}$. Check the result.

Now Try **Problem 73**

EXAMPLE 7 ***Toys.*** The area of an Etch A Sketch screen is represented by the polynomial $(35x^2 + 43x + 12)$ in.^{2}. If the width of the screen is $(5x + 4)$ inches, what expression represents its length?

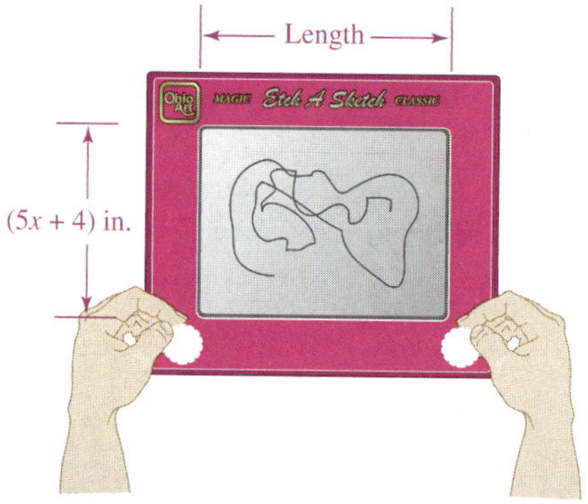

Strategy We will find the length of the screen by dividing its area, $(35x^2 + 43x + 12)$ in.^{2}, by its width, $(5x + 4)$ inches.

Why Recall that the area of a rectangle is given by the formula $A = lw$. If we divide both sides of the formula by w, we see that $l = \frac{A}{w}$.

Solution

$$\text{Length} = \frac{\text{area}}{\text{width}}$$ This is the formula for the length of a rectangle.

$$= \frac{35x^2 + 43x + 12}{5x + 4}$$ Substitute $35x^2 + 43x + 12$ for the area and $5x + 4$ for the width.

To divide $35x^2 + 43x + 12$ by $5x + 4$, we use long division.

$$\begin{array}{r} 7x + 3 \\ 5x + 4\overline{)35x^2 + 43x + 12} \\ -(35x^2 + 28x) \\ \hline 15x + 12 \\ -(15x + 12) \\ \hline 0 \end{array}$$

The first division: $\frac{35x^2}{5x} = 7x$.

The second division: $\frac{15x}{5x} = 3$.

The length of the Etch A Sketch screen is $(7x + 3)$ inches.

Now Try Problem 101

ANSWERS TO SELF CHECKS **1. a.** $6y^2$ **b.** $\frac{d^4}{4c^3}$ **2. a.** $10h + 3$ **b.** $2s^3 - \frac{s^2t}{11} + 4t^2$ **3.** $x + 4$ **4.** $4x - 3 + \frac{6}{2x + 3}$ **5.** $3x^2 + 2x - 4$ **6.** $x + 3$

STUDY SET 5.8

VOCABULARY

Fill in the blanks.

1. The expression $\frac{18x^7}{9x^4}$ is a monomial divided by a ________.

2. The expression $\frac{6x^3y - 4x^2y^2 + 8xy^3 - 2y^4}{2x^4}$ is a ________ divided by a monomial.

3. The expression $\frac{x^2 - 8x + 12}{x - 6}$ is a trinomial divided by a ________.

4.

$$\begin{array}{r} x - 2 \leftarrow \\ x - 6\overline{)x^2 - 8x - 4} \leftarrow \\ -(\underline{x^2 - 6x}) \phantom{- 4} \\ -2x - 4 \\ -(\underline{-2x + 12}) \\ -16 \leftarrow \end{array}$$

CONCEPTS

5. The long division method is a series of four steps that are repeated. Put them in the correct order:

   subtract multiply bring down divide

6. In the following long division, find the answer to the subtraction that must be performed at this stage.

$$\begin{array}{r} x \\ x - 7\overline{)x^2 - 9x - 6} \\ -(\underline{x^2 - 7x}) \end{array}$$

7. Fill in the blanks: To check an answer of a long division, we use the fact that

   Divisor · ________ + remainder = ________

8. Check to see whether the following result of a long division is correct.

$$\frac{x^2 + 4x - 20}{x - 3} = x + 7 + \frac{1}{x - 3}$$

NOTATION

Complete each solution.

9. $\frac{28x^5 - x^3 + 5x^2}{7x^2} = \frac{28x^5}{} - \frac{}{7x^2} + \frac{5x^2}{}$

   $= 4x^{ - } - \frac{x^{3-2}}{} + \frac{5x^{ - }}{7}$

   $= - \frac{x}{7} + \frac{}{}$

10.

$$\begin{array}{r} + 2 \\ x + 2\overline{)x^2 + 4x + 5} \\ -(\underline{x^2 + }) \\ + 5 \\ -(\underline{2x + 4}) \end{array}$$

11. Write the polynomial $2x^2 - 1 + 5x^4$ in descending powers of x and insert placeholders for each missing term.

12. True or false: $6x + 4 + \frac{-3}{x + 2} = 6x + 4 - \frac{3}{x + 2}$

GUIDED PRACTICE

Divide the monomials. See Example 1.

13. $\frac{x^5}{x^2}$

14. $\frac{a^{12}}{a^8}$

15. $\frac{45m^{10}}{9m^5}$

16. $\frac{24n^{12}}{8n^4}$

17. $\frac{12h^8}{9h^6}$

18. $\frac{22b^9}{6b^6}$

19. $\frac{-3d^4}{15d^8}$

20. $\frac{-4x^3}{16x^5}$

21. $\frac{10s^2}{s^3}$

22. $\frac{16y^3}{y^4}$

23. $\frac{8x^3y^2}{40xy^6}$

24. $\frac{3y^3z}{18yz^6}$

25. $\frac{-16r^3y^2}{-4r^2y^7}$

26. $\frac{-35xz^6}{-7x^8z^2}$

27. $\frac{-65rs^2}{15r^2s^5}$

28. $\frac{112uz^4}{-42u^3z^8}$

Divide the polynomial by the monomial. See Example 2.

29. $\frac{6x + 9}{3}$

30. $\frac{8x + 12}{4}$

31. $\frac{9m - 6}{m}$

32. $\frac{10n - 6}{n}$

33. $\frac{a - a^3 + a^4}{a^4}$

34. $\frac{b^2 + b^3 - b^4}{b^4}$

35. $\dfrac{8x^9 - 32x^6}{4x^4}$

36. $\dfrac{30y^8 + 40y^7}{10y^6}$

37. $\dfrac{6h^{12} + 48h^9}{24h^{10}}$

38. $\dfrac{4x^{14} - 36x^8}{36x^{12}}$

39. $\dfrac{-18w^6 - 9}{9w^4}$

40. $\dfrac{-40f^4 + 16}{8f^3}$

41. $\dfrac{9s^8 - 18s^5 + 12s^4}{3s^3}$

42. $\dfrac{16b^{10} + 4b^6 - 20b^4}{4b^2}$

43. $\dfrac{7c^5 + 21c^4 - 14c^3 - 35c}{7c^2}$

44. $\dfrac{12r^{15} - 48r^{12} + r^{10} - 18r^8}{6r^{10}}$

45. $\dfrac{12x^3y^2 - 8x^2y - 4x}{4xy}$

46. $\dfrac{12a^2b^2 - 8a^2b - 4ab}{4ab}$

47. $\dfrac{-25x^2y^3 + 30xy^2 - 5xy}{-5x^2y^2}$

48. $\dfrac{-30a^4b^4 - 15a^3b - 10a^2b^2}{-10a^2b^3}$

***Perform each division.* See Examples 3 and 4.**

49. Divide $x^2 + 8x + 12$ by $x + 2$.

50. Divide $x^2 + 5x + 6$ by $x + 2$.

51. Divide $x^2 - 5x + 6$ by $x - 3$.

52. Divide $x^2 - 12x + 32$ by $x - 4$.

53. $\dfrac{6a^2 + 5a - 6}{2a + 3}$

54. $\dfrac{3b^2 - 5b + 2}{3b - 2}$

55. $\dfrac{3b^2 + 11b + 6}{3b + 2}$

56. $\dfrac{8a^2 + 2a - 3}{2a - 1}$

57. $(x^2 + 6x + 15) \div (x + 5)$

58. $(x^2 + 10x + 30) \div (x + 6)$

59. $a - 5\overline{)a^2 - 17a + 64}$

60. $b - 2\overline{)b^2 - 4b + 6}$

61. $\dfrac{2x^2 + 5x + 2}{2x + 3}$

62. $\dfrac{3x^2 - 8x + 3}{3x - 2}$

63. $\dfrac{6x^2 - 11x + 2}{3x - 1}$

64. $\dfrac{4x^2 + 6x - 1}{2x + 1}$

***Perform each division.* See Example 5.**

65. $x + 2\overline{)3x + 2x^2 - 2}$

66. $x + 3\overline{)-x + 2x^2 - 21}$

67. $(3 + 11x + 10x^2) \div (5x + 3)$

68. $(6x + 1 + 9x^2) \div (3x + 1)$

69. $2x - 7\overline{)-x - 21 + 2x^2}$

70. $2x - 1\overline{)x - 2 + 6x^2}$

71. $3 + 4x\overline{)3 - 5x^2 - 2x + 4x^3}$

72. $2x + 3\overline{)7x^2 - 3 + 4x + 2x^3}$

***Perform each division.* See Example 6.**

73. $(a^2 - 25) \div (a + 5)$

74. $(b^2 - 36) \div (b + 6)$

75. $(x^2 - 1) \div (x - 1)$

76. $(x^2 - 9) \div (x + 3)$

77. $\dfrac{4x^2 - 9}{2x + 3}$

78. $\dfrac{25x^2 - 16}{5x - 4}$

79. $\dfrac{81b^2 - 49}{9b - 7}$

80. $\dfrac{16t^2 - 121}{4t + 11}$

81. $\dfrac{x^3 + 1}{x + 1}$

82. $\dfrac{x^3 - 8}{x - 2}$

83. $\dfrac{y^3 + y}{y - 2}$

84. $\dfrac{a^3 + a}{a + 3}$

TRY IT YOURSELF

Perform each division.

85. Divide $y^2 + 13y + 13$ by $y + 1$.

86. Divide $z^2 + 7z + 14$ by $z + 3$.

87. $\dfrac{15a^8b^2 - 10a^2b^5}{5a^3b^2}$

88. $\dfrac{9a^4b^3 - 16a^3b^4}{12a^2b}$

89. $3x + 2\overline{)2 + 7x + 6x^3 + 10x^2}$

90. $3x - 2\overline{)4x - 4 + 6x^3 - x^2}$

91. $\dfrac{5x^4 - 10x}{25x^3}$

92. $\dfrac{24x^7 - 32x^2}{16x^3}$

93. $\dfrac{a^3 - 1}{a - 1}$

94. $\dfrac{y^3 + 8}{y + 2}$

95. $\dfrac{6x^3 + x^2 + 2x + 1}{3x - 1}$

96. $\dfrac{3y^3 - 4y^2 + 2y + 3}{y + 3}$

97. $\dfrac{8x^{17}y^{20}}{16x^{15}y^{30}}$

98. $\dfrac{21a^{30}b^{15}}{14a^{40}b^{12}}$

99. $(6m^2 - m \quad 40) \div (2m + 5)$

100. $(12d^2 - 20d + 3) \div (6d - 1)$

APPLICATIONS

101. FURNACE FILTERS The area of the furnace filter is $(x^2 - 2x - 24)$ square inches. What expression represents its length?

102. MINI-BLINDS The area covered by the mini-blinds is $(3x^3 - 6x)$ square feet. What expression represents the length of the blinds?

103. POOL The rack shown in the illustration is used to set up the balls for a game of pool. If the perimeter of the rack, in inches, is given by the polynomial $6x^2 - 3x + 9$, what expression represents the approximate length of one side?

104. COMMUNICATIONS Telephone poles were installed every $(2x - 3)$ feet along a stretch of railroad track $(8x^3 - 6x^2 + 5x - 21)$ feet long. What expression represents the number of poles that were used?

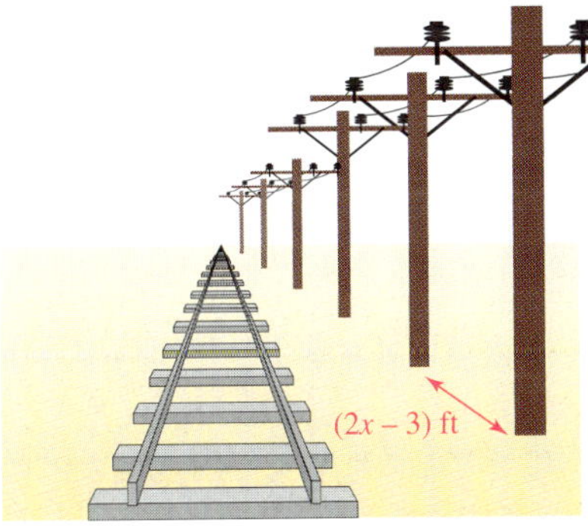

WRITING

105. Explain how to check the following long division.

$$\begin{array}{r} x + 5 \\ 3x + 5\overline{)3x^2 + 20x - 5} \\ -(3x^2 + 5x) \\ 15x - 5 \\ -(15x + 25) \\ -30 \end{array}$$

106. Explain the difference in the methods used to divide $\frac{x^2 - 3x + 2}{x}$ as compared to $\frac{x^2 - 3x + 2}{x - 2}$.

107. How do you know when to stop the long division method when dividing polynomials?

108. When dividing $x^3 + 1$ by $x + 1$, why is it helpful to write $x^3 + 1$ as $x^3 + 0x^2 + 0x + 1$?

REVIEW

109. Write an equation of the line with slope $-\frac{11}{6}$ that passes through $(2, -6)$. Write the answer in slope–intercept form.

110. Solve $S = 2\pi rh + 2\pi r^2$ for h.

111. Evaluate: $-10(18 - 4^2)^3$

112. Evaluate: -5^2

CHALLENGE PROBLEMS

Perform each division.

113. $\dfrac{6a^3 - 17a^2b + 14ab^2 - 3b^3}{2a - 3b}$

114. $(2x^4 + 3x^3 + 3x^2 - 5x - 3) \div (2x^2 - x - 1)$

115. $(x^6 + 2x^4 - 6x^2 - 9) \div (x^2 + 3)$

116. $\dfrac{6x^{6m}y^{6n} + 15x^{4m}y^{7n} - 24x^{2m}y^{8n}}{3x^{2m}y^n}$

CHAPTER 5 Summary & Review

SECTION 5.1 Rules for Exponents

DEFINITIONS AND CONCEPTS	EXAMPLES
An **exponent** indicates repeated multiplication. It tells how many times the **base** is to be used as a factor. Exponent (n), Base (x), n factors of x: $x^n = \underbrace{x \cdot x \cdot x \cdot \cdots \cdot x}_{n \text{ factors of } x}$	Identify the base and the exponent in each expression. $2^6 = 2 \cdot 2 \cdot 2 \cdot 2 \cdot 2 \cdot 2$ 2 is the base and 6 is the exponent. $(-xy)^3 = (-xy)(-xy)(-xy)$ $-xy$ is the base and 3 is the exponent. $5t^4 = 5 \cdot t \cdot t \cdot t \cdot t$ t is the base and 4 is the exponent. $8^1 = 8$ 8 is the base and 1 is the exponent.
Rules for Exponents: If m and n represent integers and there are no divisions by 0, then **Product rule:** $x^m x^n = x^{m+n}$ **Quotient rule:** $\frac{x^m}{x^n} = x^{m-n}$ **Power rule:** $(x^m)^n = x^{m \cdot n} = x^{mn}$ **Power of a product rule:** $(xy)^m = x^m y^m$ **Power of a quotient rule:** $\left(\frac{x}{y}\right)^n = \frac{x^n}{y^n}$	Simplify each expression: $5^2 5^7 = 5^{2+7} = 5^9$ $\frac{t^7}{t^3} = t^{7-3} = t^4$ $(6^3)^7 = 6^{3 \cdot 7} = 6^{21}$ $(2p)^5 = 2^5 p^5 = 32p^5$ $\left(\frac{s}{4}\right)^4 = \frac{s^4}{4^4} = \frac{s^4}{256}$

REVIEW EXERCISES

1. Identify the base and the exponent in each expression.
   a. n^{12}
   b. $(2x)^6$
   c. $3r^4$
   d. $(y - 7)^3$
2. Write each expression in an equivalent form using an exponent.
   a. $m \cdot m \cdot m \cdot m \cdot m$
   b. $-3 \cdot x \cdot x \cdot x \cdot x$
   c. $(x + 8)(x + 8)$
   d. $\left(\frac{1}{2}pq\right)\left(\frac{1}{2}pq\right)\left(\frac{1}{2}pq\right)$

Simplify each expression. Assume there are no divisions by 0.

3. $7^4 \cdot 7^8$
4. $mmnn$
5. $(y^7)^3$
6. $(3x)^4$
7. $\frac{b^{12}}{b^3}$
8. $-b^3 b^4 b^5$
9. $(-16s^3)^2 s^4$
10. $(2.1x^2y)^2$
11. $[(-9)^3]^5$
12. $(a^5)^3(a^2)^4$
13. $\left(\frac{1}{2}x^2x^3\right)^3$
14. $\left(\frac{x^7}{3xy}\right)^2$
15. $\frac{(m - 25)^{16}}{(m - 25)^4}$
16. $\frac{(5y^2z^3)^3}{(yz)^5}$
17. $\frac{a^5a^4a^5}{a^2a}$
18. $\frac{(cd)^9}{(cd)^4}$

Find an expression that represents the area or the volume of each figure, whichever is appropriate.

19.

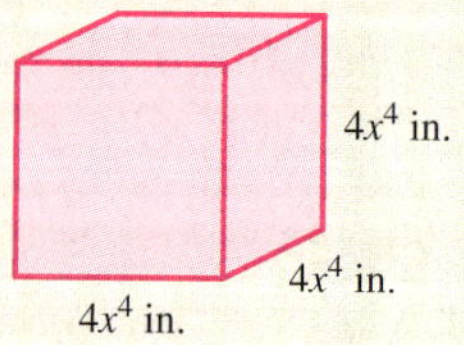

20.

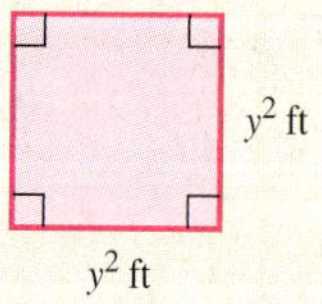

SECTION 5.2 Zero and Negative Exponents

DEFINITIONS AND CONCEPTS	EXAMPLES
Rules for exponents: For any nonzero real numbers x and y and any integers m and n,	Simplify each expression. Do not use negative exponents in the answer.
Zero exponent: $x^0 = 1$	$5^0 = 1$
Negative exponents: $x^{-n} = \frac{1}{x^n}$	$4^{-2} = \frac{1}{4^2} = \frac{1}{16}$ and $7c^{-6} = \frac{7}{c^6}$
Negative to positive rules: $\frac{1}{x^{-n}} = x^n$ $\quad \frac{x^{-m}}{y^{-n}} = \frac{y^n}{x^m}$	$\frac{1}{t^{-8}} = t^8$ and $\frac{2^{-4}}{x^{-6}} = \frac{x^6}{2^4} = \frac{x^6}{16}$
Negative exponents and reciprocals: $\left(\frac{x}{y}\right)^{-n} = \left(\frac{y}{x}\right)^n$	$\left(\frac{x}{10}\right)^{-3} = \left(\frac{10}{x}\right)^3 = \frac{10^3}{x^3} = \frac{1{,}000}{x^3}$

REVIEW EXERCISES

Simplify each expression. Do not use negative exponents in the answer.

21. x^0

22. $(3x^2y^2)^0$

23. $3x^0$

24. 10^{-3}

25. -5^{-2}

26. $\frac{t^4}{t^{10}}$

27. $\frac{8}{x^{-5}}$

28. $-6y^4y^{-5}$

29. $\frac{7^{-2}}{2^{-3}}$

30. $(x^{-3}x^{-4})^{-2}$

31. $\left(\frac{-3r^4r^{-3}}{r^{-3}r^7}\right)^3$

32. $\left(\frac{4z^4}{z^3}\right)^{-2}$

33. $\frac{3^{-2}c^3d^3}{2^{-3}c^2d^8}$

34. $\frac{t^{-30}}{t^{-60}}$

35. $\frac{w(w^{-3})^{-4}}{w^{-9}}$

36. $\left(\frac{4}{f^4}\right)^{-10}$

SECTION 5.3 Scientific Notation

DEFINITIONS AND CONCEPTS	EXAMPLES
A positive number is written in **scientific notation** when it is written in the form $N \times 10^n$, where $1 \le N < 10$ and n is an integer.	Write each number in scientific notation. $32{,}500 = 3.25 \times 10^4$ (4 decimal places) and $0.0025 = 2.5 \times 10^{-3}$ (3 decimal places) Write each number in standard notation. $1.91 \times 10^5 = 191{,}000$ (5 decimal places) and $4.7 \times 10^{-6} = 0.0000047$ (6 decimal places)
Scientific notation provides an easier way to perform computations involving very large or very small numbers.	Use scientific notation to perform the calculation: $\frac{684{,}000{,}000}{456{,}000} = \frac{6.84 \times 10^8}{4.56 \times 10^5} = \frac{6.84}{4.56} \times \frac{10^8}{10^5} = 1.5 \times 10^3$ or 1,500

REVIEW EXERCISES

Write each number in scientific notation.

37. 720,000,000

38. 9,370,000,000,000,000

39. 0.00000000942

40. 0.00013

41. 0.018×10^{-2}

42. 853×10^{3}

Write each number in standard notation.

43. 1.26×10^{5}

44. 3.919×10^{-8}

45. 2.68×10^{0}

46. 5.76×10^{1}

Evaluate each expression by first writing each number in scientific notation. Express the result in scientific notation and standard notation.

47. $\dfrac{(0.000012)(0.000004)}{0.00000016}$

48. $\dfrac{(4,800,000)(20,000,000)}{600,000}$

49. WORLD POPULATION As of January 2007, the world's population was estimated to be 6.57 billion. Write this number in standard notation and in scientific notation.

50. ATOMS The illustration shows a cross section of an atom. How many nuclei (plural for nucleus), placed end to end, would it take to stretch across the atom?

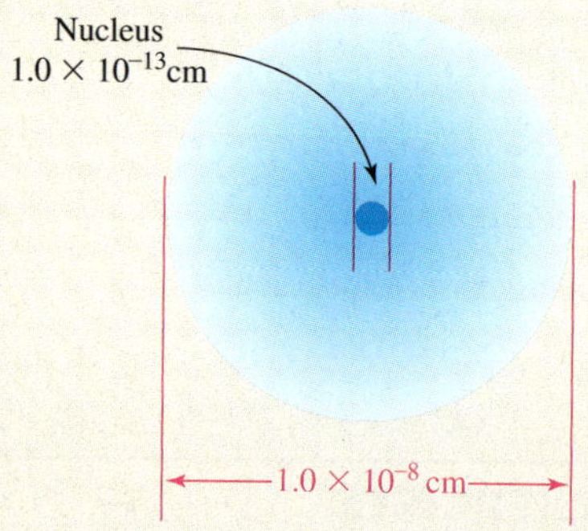

SECTION 5.4 Polynomials

DEFINITIONS AND CONCEPTS	EXAMPLES
A **polynomial** is a single term or a sum of terms in which all variables have whole-number exponents and no variable appears in a denominator.	Polynomials: 32, $-5x^2y^3$, $7p^3 - 14q^3$, $4m^2 + 5m - 12$ Not Polynomials: $y^2 - y^{-5}$, $4x^3 - \frac{7}{x} + 3x$
A polynomial with exactly one term is called a **monomial.** A polynomial with exactly two terms is called a **binomial.** A polynomial with exactly three terms is called a **trinomial.**	***Monomials***: $3x^2$, $-12m^3n^2$ ***Binomials***: $2y^3 + 3y$, $87t - 25$ ***Trinomials***: $3p^2 - 7p + 12$, $4p^2q^3 - 8p^2q^2 + 12p^2q$
The **coefficient of a term** is its numerical factor. The **degree of a term** of a polynomial in one variable is the value of the exponent on the variable. If a polynomial has more than one variable, the **degree of a term** is the sum of the exponents on the variables. The **degree of a nonzero constant** is 0.	***Term*** / ***Coefficient*** / ***Degree of the term*** $6a^7$ / 6 / 7 $-7.3x^5y^4$ / -7.3 / $5 + 4 = 9$ 32 / 32 / 0
The **degree of a polynomial** is equal to the highest degree of any term of the polynomial.	***Polynomial*** / ***Degree of the polynomial*** $7m^3 - 4m^2 + 5m - 12$ / 3 $\frac{1}{2}a^4b + \frac{3}{4}a^3b^2 - \frac{2}{3}a^2b^4$ / $2 + 4 = 6$
To **evaluate a polynomial** for a given value, substitute the value for the variable and follow the rules for the order of operations	Evaluate $3x^2 - 4x + 2$ for $x = 2$. $3x^2 - 4x + 2 = 3(2)^2 - 4(2) + 2$ Substitute 2 for each x. $= 3(4) - 8 + 2$ Evaluate $(2)^2$ first. $= 6$

SECTION 5.4 Polynomials—*continued*

DEFINITIONS AND CONCEPTS	EXAMPLES
We can **graph equations defined by polynomials** such as $y = x^2 - 2$, $y = -x^2$, and $y = x^3 + 1$. The graph of $y = x^2 - 2$ is a cup-shaped curve called a **parabola.**	Graph: $y = x^2 - 2$ Find several solutions of the equation, plot them on a rectangular coordinate system, and then draw a smooth curve passing through the points.

$y = x^2 - 2$

x	y	(x, y)
−2	2	(−2, 2)
−1	−1	(−1, −1)
0	−2	(0, −2)
1	−1	(1, −1)
2	2	(2, 2)

Select x-values. Find y-values. Plot points.

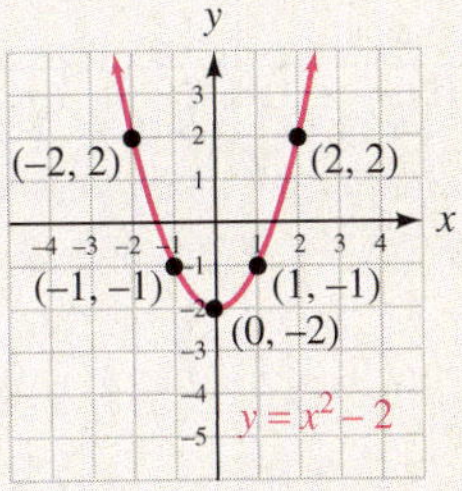

REVIEW EXERCISES

51. Consider the polynomial $3x^3 - x^2 + x + 10$.

a. How many terms does the polynomial have?

b. What is the lead term?

c. What is the coefficient of each term?

d. What is the constant term?

52. Find the degree of each polynomial and classify it as a monomial, binomial, trinomial, or none of these.

a. $13x^7$ **b.** $-16a^2b$

c. $5^3x + x^2$ **d.** $-3x^5 + x - 1$

e. $9xy^2 + 21x^3y^3$ **f.** $4s^4 - 3s^2 + 5s + 4$

53. Evaluate $-x^5 - 3x^4 + 3$ for $x = 0$ and $x = -2$.

54. DIVING The number of inches that the woman deflects the diving board is given by the polynomial $0.1875x^2 - 0.0078125x^3$ where x is the number of feet that she stands from the front anchor point of the board. Find the amount of deflection if she stands on the end of the diving board, 8 feet from the anchor point.

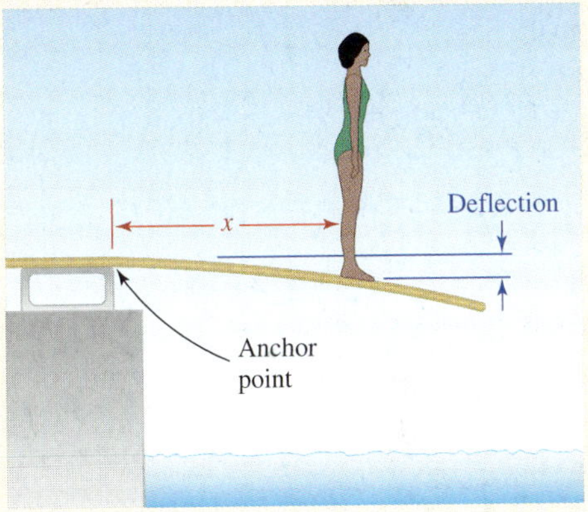

Construct a table of solutions like the one shown here and then graph the equation.

x	−2	−1	0	1	2
y					

55. $y = x^2$

56. $y = x^3 + 1$

SECTION 5.5 Adding and Subtracting Polynomials

DEFINITIONS AND CONCEPTS	EXAMPLES
To **simplify a polynomial,** combine like terms.	Simplify: $3r^4 - 4r^3 + 7r^4 + 8r^2$ $= 10r^4 - 4r^3 + 8r^2$ Combine like terms. Think: $(3 + 7)r^4 = 10r^4$.
To **add polynomials,** combine their like terms.	Add: $(4x^2 + 9x + 4) + (3x^2 - 5x - 1)$ $= (4x^2 + 3x^2) + (9x - 5x) + (4 - 1)$ Group like terms. $= 7x^2 + 4x + 3$ Combine like terms.

SECTION 5.5 Adding and Subtracting Polynomials—*continued*

DEFINITIONS AND CONCEPTS	EXAMPLES
To **subtract two polynomials,** change the signs of the terms of the polynomial being subtracted, drop the parentheses, and combine like terms.	Subtract: $(8a^3b - 4ab^2) - (-3a^3b + 9ab^2)$ $= 8a^3b - 4ab^2 + 3a^3b - 9ab^2$ Change the sign of each term of $-3a^3b + 9ab^2$ and drop the parentheses. $= 11a^3b - 13ab^2$ Combine like terms.

REVIEW EXERCISES

Simplify each polynomial and write the result in descending powers of one variable.

57. $6y^3 + 8y^4 + 7y^3 + (-8y^4)$

58. $4a^2b + 5 - 6a^3b - 3a^2b + 2a^3b + 1$

59. $\frac{5}{6}x^2 + \frac{1}{3}y^2 - \frac{1}{4}x^2 - \frac{3}{4}xy + \frac{2}{3}y^2$

60. $7.6c^5 - 2.1c^3 - 0.9c^5 + 8.1c^4$

Perform the operations.

61. $(2r^6 + 14r^3) + (23r^6 - 5r^3 + 5r)$

62. $(7.1a^2 + 2.2a - 5.8) - (3.4a^2 - 3.9a + 11.8)$

63. $(3r^3s + r^2s^2 - 3rs^3 - 3s^4) + (r^3s - 8r^2s^2 - 4rs^3 + s^4)$

64. $\left(\frac{7}{8}m^4 - \frac{1}{5}m^3\right) - \left(\frac{1}{4}m^4 + \frac{1}{5}m^3\right) - \frac{3}{5}m^3$

65. Find the difference when $(-3z^3 - 4z + 7)$ is subtracted from the sum of $(2z^2 + 3z - 7)$ and $(-4z^3 - 2z - 3)$.

66. GARDENING Find a polynomial that represents the length of the wooden handle of the shovel.

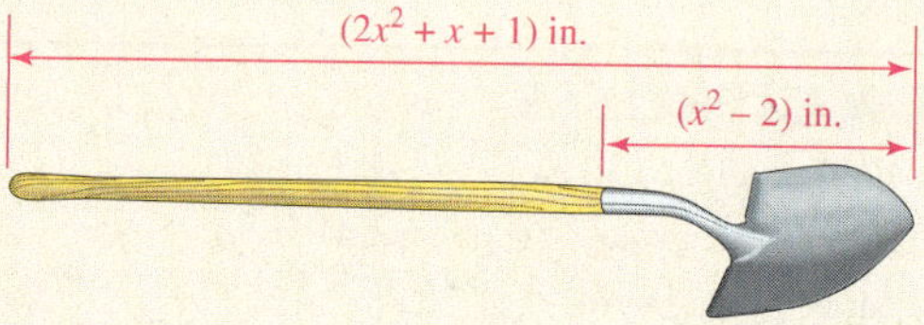

67. Add:

$$\begin{array}{r} 3x^2 + 5x + 2 \\ \underline{x^2 - 3x + 6} \end{array}$$

68. Subtract:

$$\begin{array}{r} 20x^3 + 12x \\ \underline{-(12x^3 + 7x^2 - 7x)} \end{array}$$

SECTION 5.6 Multiplying Polynomials

DEFINITIONS AND CONCEPTS	EXAMPLES
To **multiply two monomials,** multiply the numerical factors (the coefficients) and then multiply the variable factors.	Multiply: $(5p^6)(2p^5) = (5 \cdot 2)(p^6 \cdot p^5)$ Group the coefficients together and the variables together. $= 10p^{11}$ Think: $5 \cdot 2 = 10$ and $p^6 \cdot p^5 = p^{6+5} = p^{11}$.
To **multiply a monomial and a polynomial,** multiply each term of the polynomial by the monomial.	Multiply: $3r^2(2r^4 + 7r^2 - 4)$ $= 3r^2(2r^4) + 3r^2(7r^2) + 3r^2(-4)$ Distribute the multiplication by $3r^2$. $= 6r^6 + 21r^4 - 12r^2$ Multiply the monomials.

SECTION 5.6 Multiplying Polynomials–*continued*

DEFINITIONS AND CONCEPTS	EXAMPLES
To **multiply two binomials,** use the *FOIL method:* F: First O: Outer I: Inner L: Last	Multiply: $(3m + 4)(2m - 5) = 3m(2m) + 3m(-5) + 4(2m) + 4(-5)$ (F O I L) $= 6m^2 - 15m + 8m - 20$ Multiply the monomials. $= 6m^2 - 7m - 20$ Combine like terms.
To **multiply two polynomials,** multiply each term of one polynomial by each term of the other polynomial and then combine like terms.	Multiply: $(a - b)(6a^2 - 4ab + b^2)$ $= a(6a^2) + a(-4ab) + a(b^2) - b(6a^2) - b(-4ab) - b(b^2)$ $= 6a^3 - 4a^2b + ab^2 - 6a^2b + 4ab^2 - b^3$ Multiply the monomials. $= 6a^3 - 10a^2b + 5ab^2 - b^3$ Combine like terms.
When finding the **product of three polynomials,** begin by multiplying any two of them, and then multiply that result by the third polynomial.	Multiply: $-9x^4(x - 1)(x - 7) = -9x^4(x^2 - 7x - x + 7)$ Multiply the two binomials. $= -9x^4(x^2 - 8x + 7)$ Combine like terms within the parentheses. $= -9x^6 + 72x^5 - 63x^4$ Distribute the multiplication by $-9x^4$.

REVIEW EXERCISES

Multiply.

69. $(2x^2)(5x)$

70. $(-6x^4z^3)(x^6z^2)$

71. $5b^3 \cdot 6b^2 \cdot 4b^6$

72. $\frac{2}{3}h^5(3h^9 + 12h^6)$

73. $3n^2(3n^2 - 5n + 2)$

74. $x^2y(y^2 - xy)$

75. $2x(3x^4)(x + 2)$

76. $-a^2b^2(-a^4b^2 + a^3b^3 - ab^4 + 7a)$

77. $(x + 3)(x + 2)$

78. $(2x + 1)(x - 1)$

79. $(3t - 3)(2t + 2)$

80. $(3n^4 - 5n^2)(2n^4 - n^2)$

81. $-a^5(a^2 - b)(5a^2 + b)$

82. $6.6(a - 1)(a + 1)$

83. $\left(3t - \frac{1}{3}\right)\left(6t + \frac{5}{3}\right)$

84. $(5.5 - 6b)(2 - 4b)$

85. $(2a - 3)(4a^2 + 6a + 9)$

86. $(8x^2 + x - 2)(7x^2 + x - 1)$

87. Multiply using vertical form:
$$\begin{array}{r} 4x^2 - 2x + 1 \\ 2x + 1 \\ \hline \end{array}$$

88. Refer to the illustration. Find a polynomial that represents

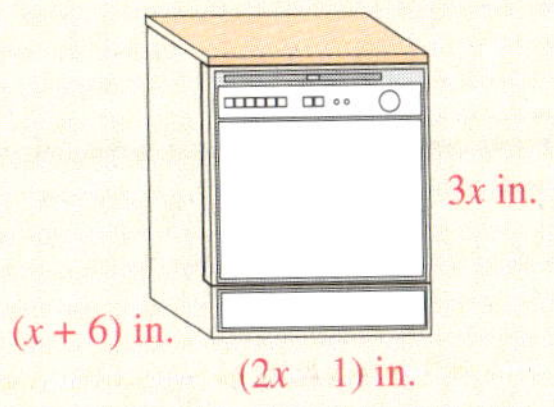

a. the perimeter of the base of the dishwasher.

b. the area of the base of the dishwasher.

c. the volume occupied by the dishwasher.

SECTION 5.7 Special Products

DEFINITIONS AND CONCEPTS	EXAMPLES
The following **special products** occur so often that it is worthwhile to learn their forms. ***Square of a binomial:*** $(A + B)^2 = A^2 + 2AB + B^2$ This is the square of a binomial sum. $(A - B)^2 = A^2 - 2AB + B^2$ This is the square of a binomial difference.	Multiply: $(n + 4)^2 = \underbrace{n^2}_{\text{The square of the first term, } n.} + \underbrace{2(n)(4)}_{\text{Twice the product of both terms.}} + \underbrace{4^2}_{\text{The square of the second term, 4.}}$ $= n^2 + 8n + 16$ Multiply: $(5a - 1)^2 = \underbrace{(5a)^2}_{\text{The square of the first term, } 5a.} + \underbrace{2(5a)(-1)}_{\text{Twice the product of both terms.}} + \underbrace{(-1)^2}_{\text{The square of the second term, } -1.}$ $= 25a^2 - 10a + 1$
Multiplying the Sum and Difference of the Same Two Terms: $(A + B)(A - B) = A^2 - B^2$	Multiply: $(x + 8)(x - 8) = \underbrace{x^2}_{\text{The square of the first term, } x.} - \underbrace{8^2}_{\text{The square of the second term, 8.}}$ $= x^2 - 64$

REVIEW EXERCISES

Find each product.

89. $(a - 3)^2$

90. $(m + 2)^3$

91. $(x + 7)(x - 7)$

92. $(2x - 0.9)(2x + 0.9)$

93. $(2y + 1)^2$

94. $(y^2 + 1)(y^2 - 1)$

95. $(6r^2 + 10s)^2$

96. $-(8a - 3c)^2$

97. $80s(r^2 + s^2)(r^2 - s^2)$

98. $4b(3b - 4)^2$

99. $\left(t - \frac{3}{4}\right)^2$

100. $\left(x + \frac{4}{3}\right)^2$

Perform the operations.

101. $3(9x^2 + 3x + 7) - 2(11x^2 - 5x + 9)$

102. $(5c - 1)^2 - (c + 6)(c - 6)$

103. GRAPHIC ARTS A Dr. Martin Luther King poster has his picture with a $\frac{1}{2}$-inch wide blue border around it. The length of the poster is $(x + 3)$ inches and the width is $(x - 1)$ inches. Find a polynomial that represents the area of the *picture* of Dr. King.

National Archives

104. Find a polynomial that represents the area of the triangle.

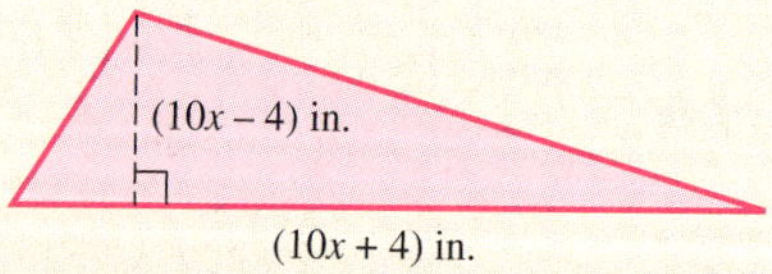

SECTION 5.8 Dividing Polynomials

DEFINITIONS AND CONCEPTS	EXAMPLES
To **divide monomials,** use the method for simplifying fractions and/or the quotient rule for exponents.	Divide the monomials: $\frac{8p^2q}{4pq^3} = \frac{2 \cdot \overset{1}{\cancel{4}} \cdot \overset{1}{\cancel{p}} \cdot p \cdot \overset{1}{\cancel{q}}}{\underset{1}{\cancel{4}} \cdot \underset{1}{\cancel{p}} \cdot \underset{1}{\cancel{q}} \cdot q \cdot q} = \frac{2p}{q^2}$ or $\frac{8p^2q}{4pq^3} = \frac{8}{4}p^{2-1}q^{1-3}$ Keep each base and subtract the exponents. $= 2p^1q^{-2}$ $= \frac{2p}{q^2}$ Move q^{-2} to the denominator and change the sign of the exponent.
To **divide a polynomial by a monomial,** divide each term of the numerator by the denominator	Divide: $\frac{9a^2b^4 - 12a^3b^6}{18ab^5} = \frac{9a^2b^4}{18ab^5} - \frac{12a^3b^6}{18ab^5}$ $= \frac{a}{2b} - \frac{2a^2b}{3}$ Perform each monomial division.
Long division can be used to **divide a polynomial by a polynomial** (other than a monomial). The long-division method is a series of four steps that are repeated: Divide, multiply, subtract, and bring down the next term. When the division has a remainder, write the answer in the form: Quotient + $\frac{\text{remainder}}{\text{divisor}}$	Divide $4x^2 - 4x + 5$ by $2x + 1$. $$\begin{array}{r} 2x - 3 \\ 2x + 1\overline{)4x^2 - 4x + 5} \\ -(4x^2 + 2x) \\ \hline -6x + 5 \\ -(-6x - 3) \\ \hline 8 \end{array}$$ The first division: $\frac{4x^2}{2x} = 2x$. The second division: $\frac{-6x}{2x} = -3$. The remainder is 8. The result is: $2x - 3 + \frac{8}{2x + 1}$
The long division method works best when the terms of the divisor and the dividend are written in **descending powers of the variable.** When the dividend has **missing terms,** insert such terms with a coefficient of 0, or leave a blank space.	Set up each long division. $\frac{5x + x^3 + 3 + 3x^2}{x + 1}$ The terms of the dividend are not in descending powers of x. $x + 1\overline{)x^3 + 3x^2 + 5x + 3}$ $\frac{x^2 - 9}{x - 3}$ The dividend is missing a term. $x - 3\overline{)x^2 + 0x - 9}$

REVIEW EXERCISES

Divide. Do not use negative exponents in the answer.

105. $\frac{16n^8}{8n^5}$

106. $\frac{-14x^2y}{21xy^3}$

107. $\frac{a^{15} - 24a^8}{6a^{12}}$

108. $\frac{15a^5b + ab^2 - 25b}{5a^2b}$

109. $x - 1\overline{)x^2 - 6x + 5}$

110. $\frac{2x^2 + 3 + 7x}{x + 3}$

111. $(15x^2 - 8x - 8) \div (3x + 2)$

112. Divide $25y^2 - 9$ by $5y + 3$.

113. $3x + 1\overline{)-13x - 4 + 9x^3}$

114. $2x - 1\overline{)6x^3 + x^2 + 1}$

115. Use multiplication to show that $(3y^2 + 11y + 6) \div (y + 3)$ is $3y + 2$.

116. BEDDING The area of a rectangular-shaped bed sheet is represented by the polynomial $(4x^3 + 12x^2 + x - 12)$ in.^{2}. If the width of the sheet is $(2x + 3)$ inches, find a polynomial that represents its length.

CHAPTER 5 Test

1. Fill in the blanks.

a. In the expression y^{10}, the _____ is y and the _________ is 10.

b. We call a polynomial with exactly one term a __________, with exactly two terms a _________, and with exactly three terms a _________.

c. The _______ of a term of a polynomial in one variable is the value of the exponent on the variable.

d. $(x + y)^2$, $(x - y)^2$, and $(x + y)(x - y)$ are called _______ products.

2. Use exponents to rewrite $2xxxyyyy$.

Simplify each expression. Do not use negative exponents in the answer.

3. $y^2(yy^3)$

4. $\left(\frac{1}{2}x^3\right)^5(x^2)^3$

5. $3.5x^0$

6. $2y^{-5}y^2$

7. 5^{-3}

8. $\frac{(x + 1)^{15}}{(x + 1)^6}$

9. $\frac{(y^{-5})^{-4}}{yy^{-2}}$

10. $\left(\frac{a^2b^{-1}}{4a^3b^{-2}}\right)^3$

11. $\left(\frac{8}{m^6}\right)^{-2}$

12. $\frac{-6a}{b^{-9}}$

13. Find an expression that represents the volume of a cube that has sides of length $10y^4$ inches.

14. ELECTRICITY One ampere (amp) corresponds to the flow of 6,250,000,000,000,000,000 electrons per second past any point in a direct current (DC) circuit. Write this number in scientific notation.

15. Write 9.3×10^{-5} in standard notation.

16. Evaluate $(2.3 \times 10^{18})(4.0 \times 10^{-15})$. Write the answer in scientific notation and standard notation.

17. Identify $x^4 + 8x^2 - 12$ as a monomial, binomial, or trinomial. Then complete the table.

Term	Coefficient	Degree

Degree of the polynomial

18. Find the degree of the polynomial $3x^3y + 2x^2y^3 - 5xy^2 - 6y$.

19. Complete the table of solutions for $y = x^2 + 2$ and then graph the equation.

x	-2	-1	0	1	2
y					

20. FREE FALL A visitor standing on the rim of the Grand Canyon drops a rock over the side. The distance (in feet) that the rock is from the canyon floor t seconds after being dropped is given by the polynomial $-16t^2 + 5{,}184$. Find the position of the rock 18 seconds after being dropped. Explain your answer.

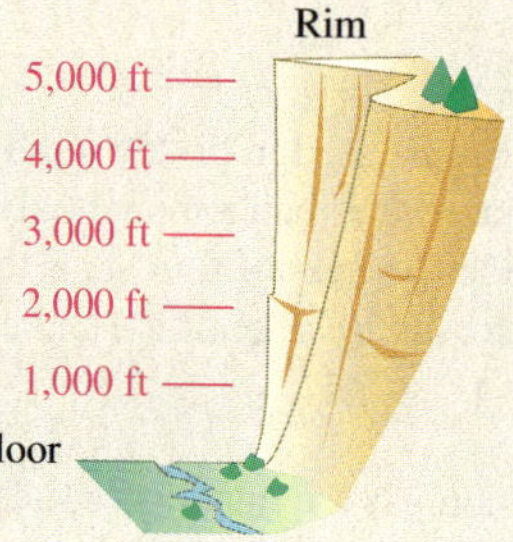

Simplify each polynomial.

21. $\frac{3}{5}x^2 + 6 + \frac{1}{4}x - 8 - \frac{1}{2}x^2 + \frac{1}{3}x$

22. $4a^2b + 5 - 6a^3b - 3a^2b + 2a^3b$

Perform the operations.

23. $(12.1h^3 - 9.9h^2 + 9.5) + (7.3h^3 - 1.2h^2 - 10.1)$

24. Subtract $b^3c - 3bc + 12$ from the sum of $6b^3c - 3bc$ and $b^3c - 2bc$.

25. Subtract:

$$\begin{array}{r} -5y^3 + 4y^2 \qquad + 3 \\ \underline{-(-2y^3 - 14y^2 + 17y - 32)} \end{array}$$

26. Find a polynomial that represents the perimeter of the rectangle.

$(5a^2 + 3a - 1)$ in.

$(a - 9)$ in.

Multiply.

27. $(2x^3y^3)(5x^2y^8)$

28. $9b^3(8b^4)(-b)$

29. $3y^2(y^2 - 2y + 3)$

30. $(x - 5)(3x + 4)$

31. $\left(6t + \frac{1}{2}\right)\left(2t - \frac{3}{2}\right)$

32. $(2x - 3)(x^2 - 2x + 4)$

33. $(1 + 10c)(1 - 10c)$

34. $(7b^3 - 3t)^2$

35. $(2.2a)(a + 5)(a - 3)$

36. Perform the operations: $(x + y)(x - y) + (x + y)^2$

Divide.

37. $\dfrac{6a^2 - 12b^2}{24ab}$

38. $\dfrac{x^2 + x - 6}{x + 3}$

39. $2x - 1\overline{)1 + x^2 + 6x^3}$

40. Find a polynomial that represents the width of a rectangle if its area is represented by the polynomial $(x^2 - 6x + 5)$ ft^2 and the length is $(x - 1)$ ft.

41. Use a check to determine whether $(5m^2 - 29m - 6) \div (5m + 1) = m - 6$.

42. Is $(a + b)^2 = a^2 + b^2$? Show why or why not.

GROUP PROJECT

BINOMIAL MULTIPLICATION AND THE AREA OF RECTANGLES

Overview: In this activity, rectangles are used to visualize binomial multiplication.

Instructions: Form groups of 2 or 3 students. Study the figure. The area of the large rectangle is given by $(x + 2)(x + 3)$. The area of the large rectangle is also the sum of the areas of the four smaller rectangles: $x^2 + 3x + 2x + 6$. Thus,

$$(x + 2)(x + 3) = x^2 + 3x + 2x + 6 = x^2 + 5x + 6$$

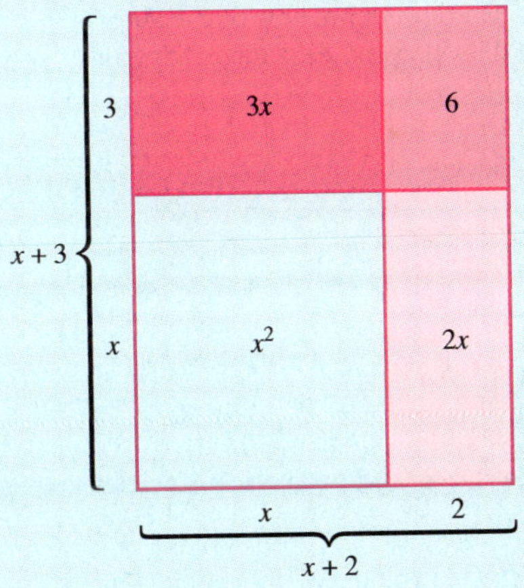

Draw three similar models to represent the following products.

1. $(x + 4)(x + 5)$ 2. $(x + 8)^2$ 3. $x(x + 6)$ 4. $(2x + 1)^2$

Determine the missing number so that the rectangle has the given area.

5. Area: $x^2 + 9x + 14$ 6. Area: $x^2 + 16x + 55$

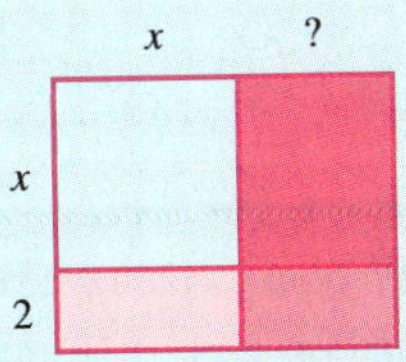

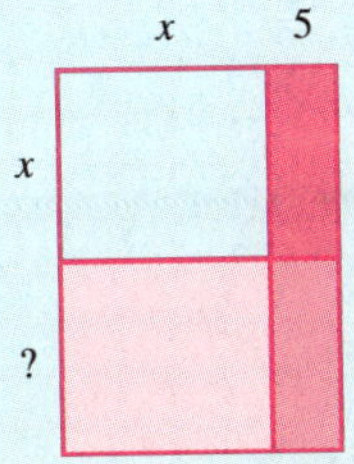

CUMULATIVE REVIEW
Chapters 1–5

1. Use exponents to write the prime factorization of 270. [Section 1.2]

2. a. Use the variables a and b to state the commutative property of addition. [Section 1.4]

   b. Use the variables x, y, and z to state the associative property of multiplication. [Section 1.6]

Evaluate each expression.

3. $3 - 4[-10 - 4(-5)]$ [Section 1.7]

4. $\dfrac{|-45| - 2(-5) + 1^5}{2 \cdot 9 - 2^4}$ [Section 1.7]

Simplify each expression.

5. $27\left(\dfrac{2}{3}x\right)$ [Section 1.9]

6. $3x^2 + 2x^2 - 5x^2$ [Section 1.9]

Solve each equation.

7. $2 - (4x + 7) = 3 + 2(x + 2)$ [Section 2.2]

8. $\dfrac{2}{5}y + 3 = 9$ [Section 2.2]

9. CANDY SALES The circle graph shows how \$6.3 billion in seasonal candy sales for 2005 was spent. Find the candy sales for Halloween. [Section 2.3]

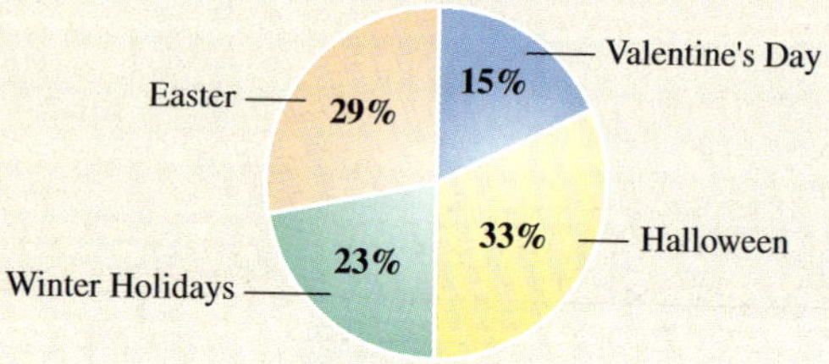

Source: National Confectioners Association

10. AIR CONDITIONING Find the volume of air contained in the duct. Round to the nearest tenth of a cubic foot. [Section 2.4]

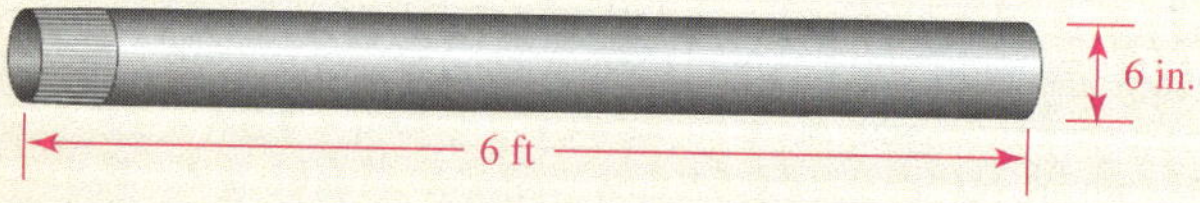

11. ANGLE OF ELEVATION Find x. [Section 2.5]

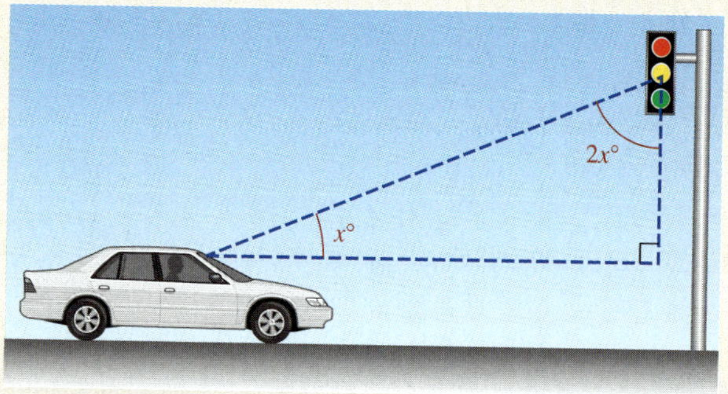

12. LIVESTOCK AUCTION A farmer is going to sell one of her prize hogs at an auction and would like to make \$6,000 after paying a 4% commission to the auctioneer. For what selling price will the farmer make this amount of money? [Section 2.5]

13. STOCK MARKET An investment club invested part of \$45,000 in a high-yield mutual fund that earned 12% annual simple interest. The remainder of the money was invested in Treasury bonds that earned 6.5% simple annual interest. The two investments earned \$4,300 in one year. How much was invested in each account? [Section 2.6]

14. Solve $-4x + 6 > 17$ and graph the solution set. Then describe the graph using interval notation. [Section 2.7]

Graph each equation.

15. $y = 3x$ [Section 3.2]

16. $x = -2$ [Section 3.3]

17. Find the slope of the line passing through $(6, -2)$ and $(-3, 2)$. [Section 3.4]

18. Find the slope and y-intercept of the line. Then write the equation of the line. [Section 3.5]

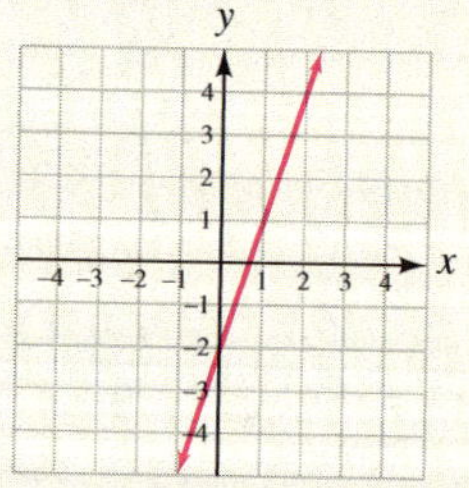

19. Without graphing, determine whether the graphs of $y = \frac{3}{2}x - 1$ and $2x + 3y = 10$ are parallel, perpendicular, or neither. [Section 3.5]

20. Write the equation of the line that passes through $(-2, 10)$ with slope -4. Write the result in slope–intercept form. [Section 3.6]

21. Is $(-2, 1)$ a solution of $2x - 3y \geq -6$? [Section 3.7]

22. If $f(x) = 2x^2 + 3x - 9$, find $f(-5)$. [Section 3.8]

23. Is $\left(\frac{2}{3}, -1\right)$ a solution of the system $\begin{cases} y = -3x + 1 \\ 3x + 3y = -2 \end{cases}$? [Section 4.1]

24. Solve the system $\begin{cases} 3x + 2y = 14 \\ y = \frac{1}{4}x \end{cases}$ by graphing. [Section 4.1]

25. Solve the system $\begin{cases} 2b - 3a = 18 \\ a + 3b = 5 \end{cases}$ by substitution. [Section 4.2]

26. Solve the system $\begin{cases} 8s + 10t = 24 \\ 11s - 3t = -34 \end{cases}$ by elimination (addition). [Section 4.3]

27. VACATIONS One-day passes to Universal Studios Hollywood cost a family of 5 (2 adults and 3 children) \$275. A family of 6 (3 adults and 3 children) paid \$336 for their one-day passes. Find the cost of an adult one-day pass and a child's one-day pass to Universal Studios. [Section 4.4]

28. Graph: $\begin{cases} y \leq 2x - 1 \\ x + 3y > 6 \end{cases}$ [Section 4.5]

Simplify. Do not use negative exponents in the answer.

29. $(-3x^2y^4)^2$ [Section 5.1]

30. $(v^5)^2(v^3)^4$ [Section 5.1]

31. $ab^3c^4 \cdot ab^4c^2$ [Section 5.1]

32. $\left(\frac{4t^3t^4t^5}{3t^2t^6}\right)^3$ [Section 5.1]

33. $(2y)^{-4}$ [Section 5.2]

34. $\frac{a^4b^0}{a^{-3}}$ [Section 5.2]

35. -5^{-2} [Section 5.2]

36. $\left(\frac{a}{x}\right)^{-10}$ [Section 5.2]

Write each number in scientific notation.

37. 615,000 [Section 5.3]

38. 0.0000013 [Section 5.3]

39. Graph: $y = x^2$ [Section 5.4]

40. MUSICAL INSTRUMENTS The amount of deflection of the horizontal beam (in inches) is given by the polynomial $0.01875x^4 - 0.15x^3 + 1.2x$, where x is the distance (in feet) that the gong is hung from one end of the beam. Find the deflection if the gong is hung in the middle of the support. [Section 5.4]

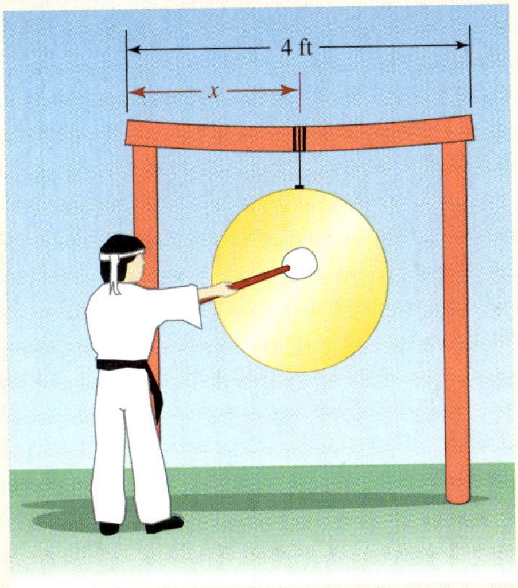

Perform the operations.

41. $(4c^2 + 3c - 2) + (3c^2 + 4c + 2)$ [Section 5.5]

42. Subtract: $\begin{array}{r} 17x^4 - 3x^2 - 65x - 12 \\ \underline{-(23x^4 + 14x^2 + 3x - 23)} \end{array}$ [Section 5.5]

43. $(2t + 3s)(3t - s)$ [Section 5.6]

44. $3x(2x + 3)^2$ [Section 5.7]

45. $5x + 3\overline{)11x + 10x^2 + 3}$ [Section 5.8]

46. $\frac{2x - 32}{16x}$ [Section 5.8]

CHAPTER 6

Factoring and Quadratic Equations

© BananaStock/Jupiterimages

from Campus to Careers

Elementary School Teacher

It has been said that a teacher takes a hand, opens a mind, and touches a heart. That is certainly true for the thousands of dedicated elementary school teachers across the country. Elementary school teachers use their training in mathematics in many ways. Besides teaching math on a daily basis, they calculate student grades, analyze test results, and order instructional materials and supplies. They use measurement and geometry for designing bulletin board displays and they construct detailed schedules so that the classroom time is used wisely.

In **Problem 17** of **Study Set 6.8,** you will determine the maximum dimensions of a bulletin board so that it meets the fire code requirements of an elementary school classroom.

JOB TITLE:
Elementary School Teacher

EDUCATION:
A bachelor's degree and completion of an approved teacher training program

JOB OUTLOOK:
Varies from good to excellent, in some locations

ANNUAL EARNINGS:
U.S. median \$47,897*
*Can vary greatly by region and experience.

FOR MORE INFORMATION:
www.bls.gov/oco/ocos069.htm

Study Skills Workshop

Reading the Textbook

Reading an algebra textbook is different from reading a newspaper or a novel. Here are two ways that you should be reading this textbook.

SKIMMING FOR AN OVERVIEW: This is a quick way to look at material just *before* it is covered in class. It helps you become familiar with the new vocabulary and notation that will be used by your instructor in the lecture. It lays a foundation.

READING FOR UNDERSTANDING: This in-depth type of reading is done more slowly, with a pencil and paper at hand. Don't skip anything—every word counts! You should do just as much writing as you do reading. Highlight the important points and work each example. If you become confused, stop and reread the material until you understand it.

Now Try This

Choose a section from this chapter and . . .

1. . . . quickly skim it. Write down any terms in bold face type and the titles of any properties, definitions, or strategies that are given in the colored boxes.
2. . . . work each *Self Check* problem. Your solutions should look like those in the *Examples*. Be sure to include your own "author notes" (the sentences in red to the right of each step of a solution).

SECTION 6.1
The Greatest Common Factor; Factoring by Grouping

Objectives

1. Find the greatest common factor of a list of terms.
2. Factor out the greatest common factor.
3. Factor by grouping.

> ***Success Tip***
> On the game show *Jeopardy!*, answers are revealed and contestants respond with the appropriate questions. Factoring is similar. Answers to multiplications are given. You are to respond by telling what factors were multiplied.

In Chapter 5, we learned how to multiply polynomials. For example, to multiply $3x + 5$ by $4x$, we use the distributive property, as shown below.

$$\begin{aligned} 4x(3x + 5) &= 4x \cdot 3x + 4x \cdot 5 \\ &= 12x^2 + 20x \end{aligned}$$

In this section, we reverse the previous steps and determine what factors were multiplied to obtain $12x^2 + 20x$. We call that process *factoring the polynomial.*

Multiplication: Given the factors, we find a polynomial. ⟶

$$4x(3x + 5) = 12x^2 + 20x$$

⟵ Factoring: Given a polynomial, we find the factors.

To **factor a polynomial** means to express it as a product of two (or more) polynomials. The first step when factoring a polynomial is to determine whether its terms have any common factors.

1 Find the Greatest Common Factor of a List of Terms.

To determine whether two or more integers have common factors, it is helpful to write them as products of prime numbers. For example, the prime factorizations of 42 and 90 are given below.

$$42 = 2 \cdot 3 \cdot 7 \qquad 90 = 2 \cdot 3 \cdot 3 \cdot 5$$

The highlighting shows that 42 and 90 have one factor of 2 and one factor of 3 in common. To find their *greatest common factor* (*GCF*), we multiply the common factors: $2 \cdot 3 = 6$. Thus, the GCF of 42 and 90 is 6.

The Greatest Common Factor (GCF)

The **greatest common factor (GCF)** of a list of integers is the largest common factor of those integers.

Recall from arithmetic that the factors of a number divide the number exactly, leaving no remainder. Therefore, the greatest common factor of two or more integers is the largest natural number that divides each of the integers exactly.

EXAMPLE 1 Find the GCF of each list of numbers: **a.** 21 and 140 **b.** 24, 60, and 96 **c.** 9, 10, and 30

Strategy We will prime factor each number in the list. Then we will identify the common prime factors and find their product.

Why The product of the common prime factors is the GCF of the numbers in the list.

Solution

a. The prime factorization of each number is shown:

$$21 = 3 \cdot 7$$

$$140 = 2 \cdot 2 \cdot 5 \cdot 7 \quad \text{This can be written as } 2^2 \cdot 5 \cdot 7.$$

Since the only prime factor common to 21 and 140 is 7, the GCF of 21 and 140 is 7.

Success Tip
Note that the GCF, 7, divides 21 and 140 exactly:

$$\frac{21}{7} = 3 \qquad \frac{140}{7} = 20$$

b. To find the GCF of three numbers, we proceed in a similar way by first finding the prime factorization of each number in the list.

$$24 = 2 \cdot 2 \cdot 2 \cdot 3 \quad \text{This can be written as } 2^3 \cdot 3.$$

$$60 = 2 \cdot 2 \cdot 3 \cdot 5 \quad \text{This can be written as } 2^2 \cdot 3 \cdot 5.$$

$$96 = 2 \cdot 2 \cdot 2 \cdot 2 \cdot 2 \cdot 3 \quad \text{This can be written as } 2^5 \cdot 3.$$

The highlighting shows that 24, 60, and 96 have two factors of 2 and one factor of 3 in common. The GCF of 24, 60, and 96 is the product of their common prime factors.

$$\text{GCF} = 2 \cdot 2 \cdot 3 = 2^2 \cdot 3^1 = 12$$

Success Tip
The exponent on any factor in a GCF is the *smallest* exponent that appears on that factor in all of the numbers under consideration.

c. Since there are no prime factors common to 9, 10, and 30, their GCF is 1.

$$9 = 3 \cdot 3$$

$$10 = 2 \cdot 5$$

$$30 = 2 \cdot 3 \cdot 5$$

Self Check 1 Find the GCF of each list of numbers: **a.** 24 and 70 **b.** 22, 25, and 98 **c.** 45, 60, and 75

Now Try **Problems 17 and 21**

To find the greatest common factor of a list of terms, we can use the following approach.

Strategy for Finding the GCF

1. Write each coefficient as a product of prime factors.
2. Identify the numerical and variable factors common to each term.
3. Multiply the common numerical and variable factors identified in Step 2 to obtain the GCF. If there are no common factors, the GCF is 1.

EXAMPLE 2 Find the GCF of each list of terms: **a.** $12x^2$ and $20x$ **b.** $9a^5b^2$, $15a^4b^2$, and $90a^3b^3$

Strategy We will prime factor each coefficient of each term in the list. Then we will identify the numerical and variable factors common to each term and find their product.

Why The product of the common factors is the GCF of the terms in the list.

Solution

a. ***Step 1:*** We write each coefficient, 12 and 20, as a product of prime factors. Recall that an exponent, as in x^2, indicates repeated multiplication.

$12x^2 = 2 \cdot 2 \cdot 3 \cdot x \cdot x$ This can be written as $2^2 \cdot 3 \cdot x^2$.

$20x = 2 \cdot 2 \cdot 5 \cdot x$ This can be written as $2^2 \cdot 5 \cdot x$.

Step 2: There are two common factors of 2 and one common factor of x.

Step 3: We multiply the common factors, 2, 2, and x, to obtain the GCF.

$$\text{GCF} = 2 \cdot 2 \cdot x = 2^2 \cdot x = 4x$$

b. ***Step 1:*** We write the coefficients, 9, 15, and 90, as products of primes. The exponents on the variables represent repeated multiplication.

$9a^5b^2 = 3 \cdot 3 \cdot a \cdot a \cdot a \cdot a \cdot a \cdot b \cdot b$ This can be written as $3^2 \cdot a^5 \cdot b^2$.

$15a^4b^2 = 3 \cdot 5 \cdot a \cdot a \cdot a \cdot a \cdot b \cdot b$ This can be written as $3 \cdot 5 \cdot a^4 \cdot b^2$.

$90a^3b^3 = 2 \cdot 3 \cdot 3 \cdot 5 \cdot a \cdot a \cdot a \cdot b \cdot b \cdot b$ This can be written as $2 \cdot 3^2 \cdot 5 \cdot a^3 \cdot b^3$.

Step 2: The highlighting shows one common factor of 3, three common factors of a, and two common factors of b.

Step 3: $\text{GCF} = 3 \cdot a \cdot a \cdot a \cdot b \cdot b = 3a^3b^2$

Success Tip

One way to identify common factors is to circle them:

$12x^2 = 2 \cdot 2 \cdot 3 \cdot x \cdot x$

$20x = 2 \cdot 2 \cdot 5 \cdot x$

$\text{GCF} = 2 \cdot 2 \cdot x = 4x$

Success Tip

The exponent on any variable in a GCF is the *smallest* exponent that appears on that variable in all of the terms under consideration.

Self Check 2 Find the GCF of each list of terms: **a.** $33c$ and $22c^4$ **b.** $42s^3t^2$, $63s^2t^4$, and $21s^3t^3$

Now Try **Problems 29 and 35**

2 Factor Out the Greatest Common Factor.

The concept of greatest common factor is used to factor polynomials. For example, to factor $12x^2 + 20x$, we note that there are two terms, $12x^2$ and $20x$. We previously determined that the GCF of $12x^2$ and $20x$ is $4x$. With this in mind, we write each term of $12x^2 + 20x$ as a product of the GCF and one other factor. Then we apply the distributive property in reverse: $ab + ac = a(b + c)$.

$$12x^2 + 20x = 4x \cdot 3x + 4x \cdot 5$$ Write $12x^2$ and $20x$ as the product of the GCF, $4x$, and one other factor.

$$= 4x(3x + 5)$$ Write an expression so that the multiplication by $4x$ distributes over the terms $3x$ and 5.

We have found that the factored form of $12x^2 + 20x$ is $4x(3x + 5)$. This process is called **factoring out the greatest common factor.**

The Language of Algebra
We have expressed a *sum of terms* as a *product of factors*.

$12x^2 + 20x = 4x(3x + 5)$

Sum of terms — Factor — Factor

EXAMPLE 3 Factor: **a.** $8m + 24$ **b.** $35a^3b^2 - 14a^2b^3$ **c.** $3x^4 - 5x^3 + x^2$

Strategy First, we will determine the GCF of the terms of the polynomial. Then we will write each term of the polynomial as the product of the GCF and one other factor.

Why We can then use the distributive property to factor out the GCF.

Solution

a. Since the GCF of $8m$ and 24 is 8, we write $8m$ and 24 as the product of 8 and one other factor.

$$8m + 24 = 8 \cdot m + 8 \cdot 3$$

$$= 8(m + 3)$$ Factor out the GCF, 8.

To check, we multiply: $8(m + 3) = 8 \cdot m + 8 \cdot 3 = 8m + 24$. Since we obtain the original polynomial, $8m + 24$, the factorization is correct.

Success Tip
Always verify a factorization by performing the indicated multiplication. The result should be the original polynomial.

Caution Remember to factor out the greatest common factor, not just a common factor. If we factored out 4 in the previous example, we would get

$$8m + 24 = 4(2m + 6)$$

However, the terms in red within parentheses have a common factor of 2, indicating that the factoring is not complete.

b. First, find the GCF of $35a^3b^2$ and $14a^2b^3$.

$$35a^3b^2 = 5 \cdot 7 \cdot a \cdot a \cdot a \cdot b \cdot b$$
$$14a^2b^3 = 2 \cdot 7 \cdot a \cdot a \cdot b \cdot b \cdot b$$
The GCF is $7a^2b^2$.

Now, we write $35a^3b^2$ and $14a^2b^3$ as the product of the GCF, $7a^2b^2$, and one other factor.

$$35a^3b^2 - 14a^2b^3 = 7a^2b^2 \cdot 5a - 7a^2b^2 \cdot 2b$$

$$= 7a^2b^2(5a - 2b)$$ Factor out the GCF, $7a^2b^2$.

We check by multiplying: $7a^2b^2(5a - 2b) = 35a^3b^2 - 14a^2b^3$.

Success Tip
With practice, you will learn how to factor out the GCF and write the correct terms within the parentheses in one step.

Caution
If the GCF of the terms of a polynomial is the same as one of the terms, leave a 1 in place of that term when factoring out the GCF.

c. We factor out the GCF of the three terms, which is x^2.

$$3x^4 - 5x^3 + x^2 = x^2(3x^2) - x^2(5x) + x^2(1) \quad \text{Write the last term, } x^2\text{, as } x^2(1).$$
$$= x^2(3x^2 - 5x + 1) \quad \text{Factor out the GCF, } x^2.$$

We check by multiplying: $x^2(3x^2 - 5x + 1) = 3x^4 - 5x^3 + x^2$.

Self Check 3 Factor: **a.** $6f + 36$ **b.** $48s^2t^2 - 84s^3t$ **c.** $y^6 - 10y^4 - y^3$

Now Try Problems 41, 51, and 53

EXAMPLE 4

Factor -1 from each polynomial: **a.** $-a^3 + 2a^2 - 4$ **b.** $6 - x$

Strategy We will write each term of the polynomial as the product of -1 and one other factor.

Why We can then use the distributive property to factor out the -1.

Solution

Success Tip
This result suggests a quick way to factor out -1. Simply change the sign of each term of $-a^3 + 2a^2 - 4$ and write a $-$ symbol in front of the parentheses.

a.
$$-a^3 + 2a^2 - 4 = (-1)a^3 + (-1)(-2a^2) + (-1)4$$
$$= -1(a^3 - 2a^2 + 4) \quad \text{Factor out } -1.$$
$$= -(a^3 - 2a^2 + 4) \quad \text{The 1 need not be written.}$$

We check by multiplying: $-(a^3 - 2a^2 + 4) = -a^3 + 2a^2 - 4$.

b.
$$6 - x = (-1)(-6) + (-1)x$$
$$= -1(-6 + x) \quad \text{Factor out } -1.$$
$$= -(x - 6) \quad \text{The 1 need not be written. Within the parentheses, write the binomial with the } x\text{-term first.}$$

Self Check 4 Factor -1 from each polynomial: **a.** $-b^4 - 3b^2 + 2$ **b.** $9 - t$

Now Try Problems 63 and 69

EXAMPLE 5

Factor out the opposite of the GCF in $-20m + 30$.

Strategy First, we will determine the GCF of the terms of the polynomial. Then we will write each term of the polynomial as the product of the opposite of the GCF and one other factor.

Why We can then use the distributive property to factor out the opposite of the GCF.

Solution Since the GCF is 10, the opposite of the GCF is -10. We write each term of the polynomial as the product of -10 and another factor. Then we factor out -10.

Success Tip
It is standard practice to factor in such a way that the lead coefficient of the polynomial within the parentheses is positive.

$$-20m + 30 = (-10)(2m) + (-10)(-3)$$
$$= -10(2m - 3)$$

Note that the leading coefficient of the polynomial within the parentheses is positive.

We check by multiplying: $-10(2m - 3) = -20m + 30$.

Self Check 5 Factor out the opposite of the GCF in $-44c + 55$.

Now Try Problem 75

EXAMPLE 6 Factor: $x(x + 4) + 3(x + 4)$

Strategy We will identify the terms of the expression and find their GCF.

Why We can then use the distributive property to factor out the GCF.

The Language of Algebra
We say that the terms have the *common binomial factor* $x + 4$.

Solution The expression has two terms: $\underbrace{x(x + 4)}_{\text{The first term}} + \underbrace{3(x + 4)}_{\text{The second term}}$

The GCF of the terms is the binomial $x + 4$, which can be factored out.

$$x(x + 4) + 3(x + 4) = (x + 4)x + (x + 4)3$$ Write each term as the product of $(x + 4)$ and one other factor.

$$= (x + 4)(x + 3)$$ Factor out the common factor, $(x + 4)$.

Self Check 6 Factor: $2y(y - 1) + 7(y - 1)$

Now Try Problem 81

3 Factor by Grouping.

Although the terms of many polynomials don't have a common factor, other than 1, it is possible to factor some of them by arranging their terms in convenient groups. This method is called **factoring by grouping.**

EXAMPLE 7 Factor by grouping: **a.** $2x^3 + x^2 + 12x + 6$ **b.** $5c - 5d + cd - d^2$

Strategy We will factor out a common factor from the first two terms and a common factor from the last two terms.

Why This will produce a common binomial factor that can then be factored out.

Solution

a. Except for 1, there is no factor that is common to all four terms. However, the first two terms, $2x^3$ and x^2, have a common factor, x^2, and the last two terms, $12x$ and 6, have a common factor, 6.

$$2x^3 + x^2 \quad + \quad 12x + 6$$
$$\downarrow \qquad\qquad \downarrow$$
$$x^2(2x + 1) \quad + \quad 6(2x + 1)$$

We now see that $2x^3 + x^2$ and $12x + 6$ have a common binomial factor, $2x + 1$, which can be factored out.

Caution
Factoring by grouping can be attempted on any polynomial with four or more terms. However, not every such polynomial can be factored in this way.

$$2x^3 + x^2 + 12x + 6 = x^2(2x + 1) + 6(2x + 1)$$ Factor $2x^3 + x^2$ and $12x + 6$.

$$= (2x + 1)(x^2 + 6)$$ Factor out $2x + 1$.

We can check the factorization by multiplying. The result should be the original polynomial.

$$(2x + 1)(x^2 + 6) = 2x^3 + 12x + x^2 + 6$$
$$= 2x^3 + x^2 + 12x + 6$$ Rearrange the terms to get the original polynomial.

Caution
Don't think that $5(c - d) + d(c - d)$ is in factored form. It is a sum of two terms. To be in factored form, the result must be a product.

b. The first two terms have a common factor, 5, and the last two terms have a common factor, d.

$$5c - 5d \quad + \quad cd - d^2 = 5(c - d) + d(c - d)$$ Factor out 5 from $5c - 5d$ and d from $cd - d^2$.

$$= (c - d)(5 + d)$$ Factor out the common binomial factor, $c - d$.

We can check by multiplying:

$$(c - d)(5 + d) = 5c + cd - 5d - d^2$$
$$= 5c - 5d + cd - d^2$$

Self Check 7 Factor by grouping: **a.** $3n^3 + 2n^2 + 9n + 6$
b. $7x - 7y + xy - y^2$

Now Try **Problems 85 and 87**

Factoring a Four-termed Polynomial by Grouping

1. Group the terms of the polynomial so that the first two terms have a common factor and the last two terms have a common factor.
2. Factor out the common factor from each group.
3. Factor out the resulting common binomial factor. If there is no common binomial factor, regroup the terms of the polynomial and repeat steps 2 and 3.

By the multiplication property of 1, we know that 1 is a factor of every term. We can use this observation to factor certain polynomials by grouping.

EXAMPLE 8 Factor: **a.** $x^3 + 6x^2 + x + 6$ **b.** $x^2 - ax - x + a$

Strategy We will follow the steps for factoring a four-termed polynomial.

Why Since the terms of the polynomials do not have a common factor (other than 1), the only option is to attempt to factor these polynomials by grouping.

Solution

a. The first two terms, x^3 and $6x^2$, have a common factor of x^2. The only common factor of the last two terms, x and 6, is 1.

$$x^3 + 6x^2 + x + 6 = x^2(x + 6) + 1(x + 6)$$ Factor out x^2 from $x^3 + 6x^2$. Factor out 1 from $x + 6$.

$$= (x + 6)(x^2 + 1)$$ Factor out the common binomial factor, $x + 6$.

Check the factorization by multiplying.

b. Since x is a common factor of the first two terms, we can factor it out and proceed as follows.

$$x^2 - ax - x + a = x(x - a) - x + a$$ Factor out x from $x^2 - ax$.

When factoring four terms by grouping, if the coefficient of the 3rd term is negative, we often factor out a negative coefficient from the last two terms. If we factor -1 from $-x + a$, a common binomial factor $x - a$ appears within the second set of parentheses, which we can factor out.

$$x^2 - ax - x + a = x(x - a) - 1(x - a)$$ To factor out -1, change the sign of $-x$ and a, and write -1 in front of the parentheses.

$$= (x - a)(x - 1)$$ Factor out the common factor, $x - a$.

Check by multiplying.

Success Tip

When we factor out -1 from the last two terms,

$$x^2 - ax - x + a$$
$$= x(x - a) - 1(x - a)$$

the signs of those terms change within the parentheses. The binomials within both sets of parentheses are then identical.

Self Check 8 Factor: **a.** $a^5 + 11a^4 + a + 11$
b. $b^2 - bc - b + c$

Now Try **Problems 93 and 95**

The next example illustrates that when factoring a polynomial, we should always look for a common factor first.

EXAMPLE 9 Factor: $10k + 10m - 2km - 2m^2$

Strategy Since all four terms have a common factor of 2, we factor it out first. Then we will factor the resulting polynomial by grouping.

Why Factoring out the GCF first makes factoring by any method easier.

Solution After factoring out 2 from all four terms, notice that within the parentheses, the first two terms have a common factor of 5, and the last two terms have a common factor of $-m$.

Success Tip
Since the quantities within parentheses in the second step must be the same, we factor out $-m$ instead of m from $-km - m^2$.

$$
\begin{aligned}
10k + 10m - 2km - 2m^2 &= 2(5k + 5m - km - m^2) && \text{Factor out the GCF, 2.}\\
&= 2[5(k + m) - m(k + m)] && \text{Factor out 5 from } 5k + 5m. \text{ Factor out } -m \text{ from } -km - m^2. \text{ This causes the signs of } -km \text{ and } -m^2 \text{ to change within the second set of parentheses.}\\
&= 2[(k + m)(5 - m)] && \text{Factor out the common binomial factor, } k + m.\\
&= 2(k + m)(5 - m) && \text{Drop the unnecessary brackets.}
\end{aligned}
$$

Check by multiplying.

Self Check 9 Factor: $4t + 4s + 4tz + 4sz$

Now Try **Problem 101**

ANSWERS TO SELF CHECKS **1. a.** 2 **b.** 1 **c.** 15 **2. a.** $11c$ **b.** $21s^2t^2$ **3. a.** $6(f + 6)$ **b.** $12s^2t(4t - 7s)$ **c.** $y^3(y^3 - 10y - 1)$ **4. a.** $-(b^4 + 3b^2 - 2)$ **b.** $-(t - 9)$ **5.** $-11(4c - 5)$ **6.** $(y - 1)(2y + 7)$ **7. a.** $(3n + 2)(n^2 + 3)$ **b.** $(x - y)(7 + y)$ **8. a.** $(a + 11)(a^4 + 1)$ **b.** $(b - c)(b - 1)$ **9.** $4(t + s)(1 + z)$

STUDY SET 6.1

VOCABULARY

Fill in the blanks.

1. To _______ a polynomial means to express it as a product of two (or more) polynomials.

2. GCF stand for ________ ________ _______. When we write $2x + 4$ as $2(x + 2)$, we say that we have ________ out the GCF, 2.

3. To factor $m^3 + 3m^2 + 4m + 12$ by _________, we begin by writing $m^2(m + 3) + 4(m + 3)$.

4. The terms $x(x - 1)$ and $4(x - 1)$ have the common _________ factor $x - 1$.

CONCEPTS

5. Complete each factorization.

a. $6x = 2 \cdot \square \cdot x$ **b.** $35h^2 = 5 \cdot \square \cdot h \cdot \square$

c. $18y^3z = 2 \cdot \square \cdot 3 \cdot \square \cdot y \cdot \square \cdot z$

6. a. Find the GCF of $30x^2$ and $105x^3$.

$$30x^2 = 2 \cdot 3 \cdot 5 \cdot x \cdot x$$
$$105x^3 = 3 \cdot 5 \cdot 7 \cdot x \cdot x \cdot x$$

b. Find the GCF of $12a^2b^2$, $15a^3b$, and $75a^4b^2$.

$$12a^2b^2 = 2 \cdot 2 \cdot 3 \cdot a \cdot a \cdot b \cdot b$$
$$15a^3b = 3 \cdot 5 \cdot a \cdot a \cdot a \cdot b$$
$$75a^4b^2 = 3 \cdot 5 \cdot 5 \cdot a \cdot a \cdot a \cdot a \cdot b \cdot b$$

7. a. Write a binomial such that the GCF of its terms is 2.

b. Write a trinomial such that the GCF of its terms is x.

8. Check to determine whether each factorization is correct.

a. $9y^3 + 5y^2 - 15y = 3y(3y^2 + 2y - 5)$

b. $3s^3 + 2s^2 + 6s + 4 = (3s + 2)(s^2 + 2)$

Fill in the blanks to complete each factorization.

9. $2x^2 + 6x = 2x \cdot x + 2x \cdot 3$

$= \underline{\quad}(\underline{\quad})$

10. $3t^3 - t^2 + 15t - 5 = t^2(3t - 1) + 5(3t - 1)$

$= (\underline{\quad})(\underline{\quad})$

11. Consider the polynomial $2k - 8 + hk - 4h$.

a. How many terms does the polynomial have?

b. Is there a common factor of all the terms, other than 1?

c. What is the GCF of the first two terms and what is the GCF of the last two terms?

12. What is the first step in factoring $8y^2 - 16yz - 6y + 12z$?

NOTATION

Complete each factorization.

13. $8m^2 - 32m + 16 = \underline{\quad}(m^2 - 4m + 2)$

14. $10a^4 - 15a^3 = 5a^{\underline{\quad}}(2a - 3)$

15. $b^3 - 6b^2 + 2b - 12 = \underline{\quad}(b - 6) + \underline{\quad}(b - 6)$

$= (\underline{\quad})(b^2 + 2)$

16. $12 + 8n - 3m - 2mn = 4(3 + 2n) \underline{\quad} m(3 + 2n)$

$= (3 + 2n)(4 \underline{\quad} m)$

GUIDED PRACTICE

Find the GCF of each list of numbers. **See Example 1.**

17. 6, 10
18. 10, 15
19. 18, 24
20. 60, 72
21. 14, 21, 42
22. 16, 24, 48
23. 40, 32, 24
24. 28, 35, 21

Find the GCF of each list of terms. **See Example 2.**

25. m^4, m^3
26. c^2, c^7
27. $15x, 25$
28. $9a, 21$
29. $20c^2, 12c$
30. $18r, 27r^3$
31. $18a^4, 9a^3, 27a^3$
32. $33m^5, 22m^6, 11m^5$
33. $24a^2, 16a^3b, 40ab$
34. $12r^2, 15rs, 9r^2s^2$
35. $6m^4n, 12m^3n^2, 9m^3n^3$
36. $15c^2d^4, 10c^2d, 40c^3d^3$
37. $4(x + 7), 9(x + 7)$
38. $2(y - 1), 5(y - 1)$
39. $4(p - t), p(p - t)$
40. $a(b + c), 3(b + c)$

Factor out the GCF. **See Example 3.**

41. $3x + 6$
42. $2y - 10$
43. $18x + 24$
44. $15s - 35$
45. $18m - 9$
46. $24s + 8$
47. $d^2 - 7d$
48. $a^2 + 9a$
49. $15c^3 + 25$
50. $33h^4 - 22$
51. $24a - 16a^2$
52. $18r - 30r^2$
53. $14x^2 - 7x - 7$
54. $27a^2 - 9a + 9$
55. $t^4 + t^3 + 2t^2$
56. $b^4 - b^3 - 3b^2$
57. $ab + ac - ad$
58. $rs - rt + ru$
59. $21x^2y^3 + 3xy^2$
60. $3x^2y^3 - 9x^4y^3$

Factor out −1 from each polynomial. **See Example 4.**

61. $-a - b$
62. $-x - 2y$
63. $-2x + 5$
64. $-3x + 8$
65. $-3r + 2s - 3$
66. $-6yz + 12xz + 5xy$
67. $-x^2 - x + 16$
68. $-t^2 - 9t + 1$
69. $5 - x$
70. $10 - m$
71. $9 - 4a$
72. $7 - 8b$

Factor each polynomial by factoring out the opposite of the GCF. **See Example 5.**

73. $-3x^2 - 6x$
74. $-4a^2 - 6a$
75. $-4a^2b + 12a^3$
76. $-25x^4 + 30x^2$
77. $-24x^4 - 48x^3 + 36x^2$
78. $-28a^5 - 42a^4 + 14a^3$
79. $-4a^3b^2 + 14a^2b^2 - 10ab^2$
80. $-30x^4y^3 + 24x^3y^2 - 60x^2y$

Factor each expression. **See Example 6.**

81. $y(x + 2) + 3(x + 2)$
82. $r(t + v) + 3(t + v)$
83. $m(p - q) - 5(p - q)$
84. $ab(c - 7) - 12(c - 7)$

Factor by grouping. **See Example 7.**

85. $2x + 2y + ax + ay$
86. $bx + bz + 5x + 5z$
87. $rs - ru + 8sw - 8uw$
88. $12ab - 4ac + 3db - dc$
89. $7m^3 - 2m^2 + 14m - 4$
90. $9s^3 - 2s^2 + 36s - 8$
91. $5x^3 - x^2 + 10x - 2$
92. $6a^3 - a^2 + 18a - 3$

Factor by grouping. **See Example 8.**

93. $ab + ac + b + c$
94. $xy + 3y^2 + x + 3y$

95. $rs + 4s^2 - r - 4s$

96. $tx + tz - x - z$

97. $2ax + 2bx - 3a - 3b$

98. $rx + sx - ry - sy$

99. $mp - np - mq + nq$

100. $9p - 9q - mp + mq$

***Factor by grouping. Remember to factor out the GCF first.* See Example 9.**

101. $ax^3 - 2ax^2 + 5ax - 10a$
102. $x^3y^2 - 2x^2y^2 + 3xy^2 - 6y^2$
103. $6x^3 - 6x^2 + 12x - 12$
104. $3x^3 - 6x^2 + 15x - 30$

TRY IT YOURSELF

Factor.

105. $h^2(14 + r) + 5(14 + r)$
106. $x(y + 9) - 21(y + 9)$
107. $22a^3 - 33a^2$
108. $39r^3 + 26r^2$
109. $ax + bx - a - b$
110. $2xy + y^2 - 2x - y$
111. $15r^8 - 18r^6 - 30r^5$
112. $24cm - 12cn + 16c$
113. $27mp + 9mq - 9np - 3nq$
114. $4abc + 4ac^2 - 2bc - 2c^2$
115. $-60p^2t^2 - 80pt^3$
116. $-25x^5y^7 + 75x^3y^2$
117. $6x^2 - 2xy - 15x + 5y$
118. $m^3 + 5m^2 + m + 5$
119. $2x^3z - 4x^2z + 32xz - 64z$
120. $4a^2b + 12a^2 - 8ab - 24a$
121. $12uvw^3 - 54uv^2w^2$
122. $14xyz - 16x^2y^2z$
123. $x^3 + x^2 + x + 1$
124. $m^4 + m^3 + 2m + 2$

APPLICATIONS

125. GEOMETRY The dimensions of the rectangle shown below can be found by factoring the polynomial that represents its area. Find the polynomials that represent the length and the width of the rectangle.

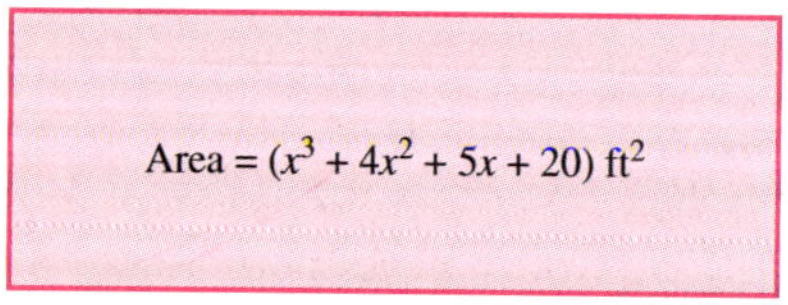

126. INTERIOR DECORATING The expression $\pi rs + \pi Rs$ can be used to find the amount of material needed to make the lamp shade shown. Factor the expression.

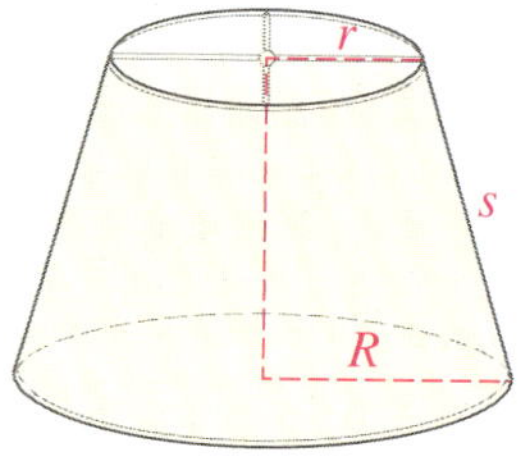

WRITING

127. Explain how to find the GCF of $32a^3$ and $16a^2$.
128. Explain this diagram.

Multiplication ⟶

$$3x^2(5x^2 - 6x + 4) = 15x^4 - 18x^3 + 12x^2$$

⟵ Factoring

129. Explain how factorizations of polynomials are checked. Give an example.
130. Explain the error.

Factor out the GCF: ~~$30a^3 - 12a^2 = 6a(5a^2 - 2a)$~~

REVIEW

131. INSURANCE COSTS A college student's good grades earned her a student discount on her car insurance premium. What was the percent of decrease, to the nearest percent, if her annual premium was lowered from \$1,050 to \$925?
132. CALCULATING GRADES A student has test scores of 68%, 75%, and 79% in a government class. What must she score on the last exam to earn a B (80% or better) in the course?

CHALLENGE PROBLEMS

133. Factor: $6x^{4m}y^n + 21x^{3m}y^{2n} - 15x^{2m}y^{3n}$

134. Factor $ab - b^2 - bc + ac - bc - c^2$ by grouping.

SECTION 6.2
Factoring Trinomials of the Form $x^2 + bx + c$

Objectives

1. Factor trinomials of the form $x^2 + bx + c$.
2. Factor trinomials of the form $x^2 + bx + c$ after factoring out the GCF.
3. Factor trinomials of the form $x^2 + bx + c$ using the grouping method.

In Chapter 5, we learned how to multiply binomials. For example, to multiply $x + 1$ and $x + 2$, we proceed as follows.

$$(x + 1)(x + 2) = x^2 + 2x + x + 2$$
$$= x^2 + 3x + 2$$

To *factor the trinomial* $x^2 + 3x + 2$, we will reverse the multiplication process and determine what factors were multiplied to obtain this result. Since the product of two binomials is often a trinomial, many trinomials factor into two binomials.

Multiplication: Given the binomial factors, we find a trinomial. →

$$(x + 1)(x + 2) = x^2 + 3x + 2$$

← Factoring: Given a trinomial, we find the binomial factors.

The Language of Algebra
Recall that when a polynomial in one variable is written in descending powers of that variable, the coefficient of the first term is called the *leading coefficient.*

To begin the discussion of trinomial factoring, we consider trinomials of the form $x^2 + bx + c$, such as

$$x^2 + 8x + 15, \quad y^2 - 13y + 12, \quad a^2 + a - 20, \quad \text{and} \quad z^2 - 20z - 21$$

In each case, the **leading coefficient**—the coefficient of the squared variable—is 1.

1 Factor Trinomials of the Form $x^2 + bx + c$.

To develop a method for factoring trinomials, we will find the product of $x + 6$ and $x + 4$ and make some observations about the result.

$$(x + 6)(x + 4) = x \cdot x + x \cdot 4 + 6 \cdot x + 6 \cdot 4 \quad \text{Use the FOIL method.}$$
$$= x^2 + 4x + 6x + 24$$
$$= x^2 + 10x + 24$$

First term — Middle term — Last term

The Language of Algebra
If a term of a trinomial is a number only, it is called a *constant term.*

$$x^2 + 10x + 24$$

Constant term (24)

The result is a trinomial, where

- the first term, x^2, is the product of x and x
- the last term, 24, is the product of 6 and 4
- the coefficient of the middle term, 10, is the sum of 6 and 4

These observations suggest a strategy to use to factor trinomials that have 1 as the leading coefficient.

EXAMPLE 1 Factor: $x^2 + 8x + 15$

Strategy We will assume that $x^2 + 8x + 15$ is the product of two binomials and we will use a systematic method to find their terms.

Why Since the terms of $x^2 + 8x + 15$ do not have a common factor (other than 1), the only option available is to try to factor it as the product of two binomials.

Solution We represent the binomials using two sets of parentheses. Since the first term of the trinomial is x^2, we enter x and x as the first terms of its binomial factors.

$$x^2 + 8x + 15 = (x \square)(x \square)$$ Because $x \cdot x$ will give x^2.

The second terms of the binomials must be two integers whose product is 15 and whose sum is 8. Since the integers must have a positive product and a positive sum, we consider only pairs of positive integer factors of 15. The only such pairs, $1 \cdot 15$ and $3 \cdot 5$, are listed in the table. Then we find the sum of each pair and enter each result in the table.

Positive factors of 15	Sum of the positive factors of 15
$1 \cdot 15 = 15$	$1 + 15 = 16$
$3 \cdot 5 = 15$	$3 + 5 = 8$

List all of the pairs of positive integers that multiply to give 15.

Add each pair of factors.

The second row of the table contains the correct pair of integers 3 and 5, whose product is 15 and whose sum is 8. To complete the factorization, we enter 3 and 5 as the second terms of the binomial factors.

$$x^2 + 8x + 15 = (x + 3)(x + 5)$$

We can check the factorization by multiplying:

$$(x + 3)(x + 5) = x^2 + 5x + 3x + 15$$
$$= x^2 + 8x + 15$$ This is the original trinomial.

Notation
By the commutative property of multiplication, the order of the binomial factors in a factorization does not matter. Thus, we can also write:

$x^2 + 8x + 15 = (x + 5)(x + 3)$

Self Check 1 Factor: $y^2 + 7y + 10$

Now Try **Problem 15**

EXAMPLE 2 Factor: $y^2 - 13y + 12$

Strategy We will assume that $y^2 - 13y + 12$ is the product of two binomials and we will use a systematic method to find their terms.

Why Since the terms of $y^2 - 13y + 12$ do not have a common factor (other than 1), the only option available is to try to factor it as the product of two binomials.

Solution We represent the binomials using two sets of parentheses. Since the first term of the trinomial is y^2, the first term of each binomial factor must be y.

$$y^2 - 13y + 12 = (y \square)(y \square)$$ Because $y \cdot y$ will give y^2.

The second terms of the binomials must be two integers whose product is 12 and whose sum is -13. Since the integers must have a positive product and a negative sum, we only consider pairs of negative integer factors of 12. The possible pairs are listed in the table.

Negative factors of 12	Sum of the negative factors of 12
$-1(-12) = 12$	$-1 + (-12) = -13$
$-2(-6) = 12$	$-2 + (-6) = -8$
$-3(-4) = 12$	$-3 + (-4) = -7$

You can stop listing the factors after finding the correct combination.

The Language of Algebra
Make sure you understand this vocabulary: *Many trinomials factor as the product of two binomials.*

Trinomial — Product of two binomials

$$\underbrace{y^2 - 13y + 12}_{\text{Trinomial}} = \underbrace{(y - 1)(y - 12)}_{\text{Product of two binomials}}$$

The first row of the table contains the correct pair of integers -1 and -12, whose product is 12 and whose sum is -13. To complete the factorization, we enter -1 and -12 as the second terms of the binomial factors.

$$y^2 - 13y + 12 = (y - 1)(y - 12)$$

We check the factorization by multiplying:

$$(y - 1)(y - 12) = y^2 - 12y - y + 12$$
$$= y^2 - 13y + 12 \quad \text{This is the original trinomial.}$$

Self Check 2 Factor: $p^2 - 6p + 8$

Now Try **Problem 19**

EXAMPLE 3

Factor: $a^2 + a - 20$

Strategy We will assume that $a^2 + a - 20$ is the product of two binomials and we will use a systematic method to find their terms.

Why Since the terms of $a^2 + a - 20$ do not have a common factor (other than 1), the only option available is to try to factor it as the product of two binomials.

Solution We represent the binomials using two sets of parentheses. Since the first term of the trinomial is a^2, the first term of each binomial factor must be a.

$$a^2 + a - 20 = (a \;\square)(a \;\square) \quad \text{Because } a \cdot a \text{ will give } a^2.$$

To determine the second terms of the binomials, we must find two integers whose product is -20 and whose sum is 1. Because the integers must have a negative product, their signs must be different. The possible pairs are listed in the table.

Factors of -20	Sum of the factors of -20
$1(-20) = -20$	$1 + (-20) = -19$
$2(-10) = -20$	$2 + (-10) = -8$
$4(-5) = -20$	$4 + (-5) = -1$
$5(-4) = -20$	$5 + (-4) = 1$
$10(-2) = -20$	$10 + (-2) = 8$
$20(-1) = -20$	$20 + (-1) = 19$

It is wise to follow an order when listing the factors in the table so that you don't skip the correct combination. Here, the first factors 1, 2, 4, 5, 10, and 20, are listed from least to greatest.

The fourth row of the table contains the correct pair of integers 5 and −4, whose product is −20 and whose sum is 1. To complete the factorization, we enter 5 and −4 as the second terms of the binomial factors.

$$a^2 + a - 20 = (a + 5)(a - 4)$$

Check the factorization by multiplying.

Self Check 3 Factor: $m^2 + m - 42$

Now Try Problem 27

EXAMPLE 4 Factor: $z^2 - 4z - 21$

Strategy We will assume that $z^2 - 4z - 21$ is the product of two binomials and we will use a systematic method to find their terms.

Why Since the terms of $z^2 - 4z - 21$ do not have a common factor (other than 1), the only option available is to try to factor it as the product of two binomials.

Solution We represent the binomials using two sets of parentheses. Since the first term of the trinomial is z^2, the first term of each binomial factor must be z.

$$z^2 - 4z - 21 = (z \;\square)(z \;\square)$$ *Because $z \cdot z$ will give z^2.*

To determine the second terms of the binomials, we must find two integers whose product is −21 and whose sum is −4. Because the integers must have a negative product, their signs must be different. The possible pairs are listed in the table.

Factors of −21	Sum of the factors of −21
$1(-21) = -21$	$1 + (-21) = -20$
$3(-7) = -21$	$3 + (-7) = -4$
$7(-3) = -21$	$7 + (-3) = 4$
$21(-1) = -21$	$21 + (-1) = 20$

The second row of the table contains the correct pair of integers 3 and −7, whose product is −21 and whose sum is −4. To complete the factorization, we enter 3 and −7 as the second terms of the binomial factors.

$$z^2 - 4z - 21 = (z + 3)(z - 7)$$

Check by multiplying.

Self Check 4 Factor: $q^2 - 2q - 24$

Now Try Problem 33

The following guidelines are helpful when factoring trinomials.

Factoring Trinomials Whose Leading Coefficient is 1

To factor a trinomial of the form $x^2 + bx + c$, find two numbers whose product is c and whose sum is b.

1. If c is positive, the numbers have the same sign.
2. If c is negative, the numbers have different signs.

Then write the trinomial as a product of two binomials. You can check by multiplying.

$$x^2 + bx + c = (x \; \square)(x \; \square)$$

The product of these numbers must be c and their sum must be b.

EXAMPLE 5 Factor: $-h^2 + 2h + 63$

Strategy We will factor out -1 and then factor the resulting trinomial.

Why It is easier to factor trinomials that have a positive leading coefficient.

Solution After factoring out -1, we factor the trinomial within the parentheses.

$$\begin{aligned} -h^2 + 2h + 63 &= \mathbf{-1}(h^2 - 2h - 63) && \text{Factor out } -1. \\ &= -(h^2 - 2h - 63) && \text{The 1 need not be written.} \\ &= -(h + 7)(h - 9) && \text{Factor } h^2 - 2h - 63. \end{aligned}$$

Check:

$$\begin{aligned} -(h + 7)(h - 9) &= -(h^2 - 9h + 7h - 63) && \text{Multiply the binomials first.} \\ &= -(h^2 - 2h - 63) && \text{Combine like terms.} \\ &= -h^2 + 2h + 63 && \text{Drop the } - \text{ sign and change the sign of every term within the parentheses.} \end{aligned}$$

The result is the original trinomial.

Self Check 5 Factor: $-x^2 + 11x - 28$

Now Try **Problem 37**

We can factor trinomials in two variables in a similar way.

EXAMPLE 6 Factor: $x^2 - 4xy - 5y^2$

Strategy We will assume that $x^2 - 4xy - 5y^2$ is the product of two binomials and we will use a systematic method to find their terms.

Why Since the terms of $x^2 - 4xy - 5y^2$ do not have a common factor (other than 1), the only option available is to try to factor it as the product of two binomials.

The Language of Algebra
The trinomial $x^2 - 4xy - 5y^2$ is in two variables, x and y. It is written in *descending* powers of x and *ascending* powers of y.

Solution We represent the binomials using two sets of parentheses. Since the first term of the trinomial is x^2, the first term of each binomial factor must be x. Since the third term contains y^2, the last term of each binomial factor must contain y. To complete the factorization, we need to determine the coefficient of each y-term.

$$x^2 - 4xy - 5y^2 = \left(x \boxed{}y\right)\left(x \boxed{}y\right)$$ Because $x \cdot x$ will give x^2 and $y \cdot y$ will give y^2.

The coefficients of y must be two integers whose product is -5 and whose sum is -4. Such a pair is 1 and -5. Instead of writing the first factor as $(x + 1y)$, we write it as $(x + y)$, because $1y = y$.

$$x^2 - 4xy - 5y^2 = (x + y)(x - 5y)$$

Check: $(x + y)(x - 5y) = x^2 - 5xy + xy - 5y^2$
$= x^2 - 4xy - 5y^2$ This is the original trinomial.

Self Check 6 Factor: $s^2 + 6st - 7t^2$

Now Try **Problem 45**

2 Factor Trinomials of the Form $x^2 + bx + c$ After Factoring Out the GCF.

If the terms of a trinomial have a common factor, it should be factored out first. A trinomial is **factored completely** when no factor can be factored further.

EXAMPLE 7

Factor completely: $2x^4 + 26x^3 + 80x^2$

Strategy We will factor out the GCF, $2x^2$, first. Then we will factor the resulting trinomial.

Why The first step in factoring any polynomial is to factor out the GCF. Factoring out the GCF first makes factoring by any method easier.

Caution
When prime factoring 30, for example, you wouldn't stop here because 6 can be factored:

30
/ \
5 6

When factoring polynomials, make sure that no factor can be factored further.

Solution We begin by factoring out the GCF, $2x^2$, from $2x^4 + 26x^3 + 80x^2$.

$$2x^4 + 26x^3 + 80x^2 = 2x^2(x^2 + 13x + 40)$$

Next, we factor $x^2 + 13x + 40$. The integers 8 and 5 have a product of 40 and a sum of 13, so the completely factored form of the given trinomial is

$$2x^4 + 26x^3 + 80x^2 = 2x^2(x + 8)(x + 5)$$ The complete factorization must include $2x^2$.

Check: $2x^2(x + 8)(x + 5) = 2x^2(x^2 + 13x + 40)$
$= 2x^4 + 26x^3 + 80x^2$ This is the original trinomial.

Self Check 7 Factor completely: $4m^5 + 8m^4 - 32m^3$

Now Try **Problem 51**

EXAMPLE 8 Factor completely: $-13g^2 + 36g + g^3$

Strategy We will write the terms of the trinomial in descending powers of g.

Why It is easier to factor a trinomial if its terms are written in descending powers of one variable.

Solution

$$\begin{aligned} -13g^2 + 36g + g^3 &= g^3 - 13g^2 + 36g && \text{Rearrange the terms.} \\ &= g(g^2 - 13g + 36) && \text{Factor out the GCF, } g. \\ &= g(g - 9)(g - 4) && \text{Factor the trinomial.} \end{aligned}$$

Check the factorization by multiplying.

Caution
For multistep factorizations, don't forget to write the GCF in the final factored form.

Self Check 8 Factor completely: $-12t + t^3 + 4t^2$

Now Try **Problem 63**

If a trinomial with integer coefficients cannot be factored using only integers, it is called a **prime trinomial.**

EXAMPLE 9 Factor $x^2 + 2x + 3$, if possible.

Strategy We will assume that $x^2 + 2x + 3$ is the product of two binomials and we will use a systematic method to find their terms.

Why Since the terms of $x^2 + 2x + 3$ do not have a common factor (other than 1), the only option available is to try to factor it as the product of two binomials.

Solution To factor the trinomial, we must find two integers whose product is 3 and whose sum is 2. The possible factorizations are shown in the table.

Factors of 3	Sum of the factors of 3
$1(3) = 3$	$1 + 3 = 4$
$-1(-3) = 3$	$-1 + (-3) = -4$

Since there are no two integers whose product is 3 and whose sum is 2, the trinomial $x^2 + 2x + 3$ cannot be factored and is a *prime trinomial.*

The Language of Algebra
When a trinomial is not factorable using only integers, we say it is *prime* and that it does not factor *over* the integers.

Self Check 9 Factor $x^2 - 4x + 6$, if possible.

Now Try **Problem 71**

3 Factor Trinomials of the Form $x^2 + bx + c$ Using the Grouping Method.

Another way to factor trinomials of the form $x^2 + bx + c$ is to write them as equivalent four-termed polynomials and factor by grouping. To factor $x^2 + 8x + 15$ using this method, we proceed as follows.

1. First, identify b as the coefficient of the x-term, and c as the last term. For trinomials of the form $x^2 + bx + c$, we call c the **key number.**

$$\left.\begin{array}{c} x^2 + bx + c \\ \downarrow \quad \downarrow \\ x^2 + 8x + 15 \end{array}\right\} b = 8 \text{ and } c = 15$$

2. Now find two integers whose product is the key number, 15, and whose sum is $b = 8$. Since the integers must have a positive product and a positive sum, we consider only positive factors of 15.

Key number = 15 $\qquad$ $b = 8$

Positive factors of 15	Sum of the positive factors of 15
$1 \cdot 15 = 15$	$1 + 15 = 16$
$3 \cdot 5 = 15$	$3 + 5 = 8$

The second row of the table contains the correct pair of integers 3 and 5, whose product is the key number 15 and whose sum is $b = 8$.

3. Express the middle term, $8x$, of the trinomial as the *sum of two terms,* using the integers 3 and 5 found in step 2 as coefficients of the two terms.

$$x^2 + 8x + 15 = x^2 + 3x + 5x + 15 \qquad \text{Express } 8x \text{ as } 3x + 5x.$$

4. Factor the equivalent four-term polynomial by grouping:

$$x^2 + 3x + 5x + 15 = x(x + 3) + 5(x + 3) \qquad \text{Factor } x \text{ out of } x^2 + 3x \text{ and 5 out of } 5x + 15.$$

$$= (x + 3)(x + 5) \qquad \text{Factor out } x + 3.$$

Check the factorization by multiplying.

The grouping method is an alternative to the method for factoring trinomials discussed earlier in this section. It is especially useful when the constant term, c, has many factors.

Factoring Trinomials of the Form $x^2 + bx + c$ Using Grouping

To factor a trinomial that has a leading coefficient of 1:

1. Identify b and the key number, c.
2. Find two integers whose product is the key number and whose sum is b.
3. Express the middle term, bx, as the sum (or difference) of two terms. Enter the two numbers found in step 2 as coefficients of x in the form shown below. Then factor the equivalent four-term polynomial by grouping.

$$x^2 + \square x + \square x + c$$

The product of these numbers must be c, and their sum must be b.

4. Check the factorization using multiplication.

EXAMPLE 10 Factor by grouping: $a^2 + a - 20$

Strategy We will express the middle term, a, of the trinomial as the difference of two carefully chosen terms.

Why We want to produce an equivalent four-term polynomial that can be factored by grouping.

Solution Since $a^2 + a - 20 = a^2 + 1a - 20$, we identify b as 1 and the key number c as -20. We must find two integers whose product is -20 and whose sum is 1. Since the integers must have a negative product, their signs must be different.

Key number $= -20$ $b = 1$

Factors of −20	Sum of the factors of −20
$1(-20) = -20$	$1 + (-20) = -19$
$2(-10) = -20$	$2 + (-10) = -8$
$4(-5) = -20$	$4 + (-5) = -1$
$5(-4) = -20$	$5 + (-4) = 1$
$10(-2) = -20$	$10 + (-2) = 8$
$20(-1) = -20$	$20 + (-1) = 19$

The fourth row of the table contains the correct pair of integers 5 and -4, whose product is -20 and whose sum is 1. They serve as the coefficients of $5a$ and $-4a$, the two terms that we use to represent the middle term, a, of the trinomial.

$$a^2 + a - 20 = a^2 + 5a - 4a - 20$$ Express the middle term, a, as $5a - 4a$.

$$= a(a + 5) - 4(a + 5)$$ Factor a out of $a^2 + 5a$ and -4 out of $-4a - 20$.

$$= (a + 5)(a - 4)$$ Factor out $a + 5$.

Check the factorization by multiplying.

Success Tip

We could also express the middle term as $-4a + 5a$. We obtain the same binomial factors, but in reverse order.

$$a^2 - 4a + 5a - 20$$
$$= a(a - 4) + 5(a - 4)$$
$$= (a - 4)(a + 5)$$

Self Check 10 Factor by grouping: $m^2 + m - 42$

Now Try **Problem 27**

EXAMPLE 11 Factor by grouping: $x^2 - 4xy - 5y^2$

Strategy We will express the middle term, $-4xy$, of the trinomial as the sum of two carefully chosen terms.

Why We want to produce an equivalent four-term polynomial that can be factored by grouping.

Solution In $x^2 - 4xy - 5y^2$, we identify b as -4 and the key number c as -5. We must find two integers whose product is -5 and whose sum is -4. Such a pair is -5 and 1. They

Key number $= -5$ $b = -4$

Factors	Sum
$-5(1) = -5$	$-5 + 1 = -4$

serve as the coefficients of $-5xy$ and $1xy$, the two terms that we use to represent the middle term, $-4xy$, of the trinomial.

$$x^2 - 4xy - 5y^2 = x^2 - 5xy + 1xy - 5y^2$$ Express the middle term, $-4xy$, as $-5xy + 1xy$. ($1xy - 5xy$ could also be used.)

$$= x(x - 5y) + y(x - 5y)$$ Factor x out of $x^2 - 5xy$ and y out of $1xy - 5y^2$.

$$= (x - 5y)(x + y)$$ Factor out $x - 5y$.

Check the factorization by multiplying.

Self Check 11 Factor by grouping: $q^2 - 2qt - 24t^2$

Now Try **Problem 45**

EXAMPLE 12

Factor completely: $2x^3 - 20x^2 + 18x$

Strategy We will factor out the GCF, $2x$, first. Then we will factor the resulting trinomial using the grouping method.

Why The first step in factoring any polynomial is to factor out the GCF.

Solution We begin by factoring out the GCF, $2x$, from $2x^3 - 20x^2 + 18x$.

$$2x^3 - 20x^2 + 18x = 2x(x^2 - 10x + 9)$$

Key number = 9 $b = -10$

Factors	Sum
$-9(-1) = 9$	$-9 + (-1) = -10$

To factor $x^2 - 10x + 9$ by grouping, we must find two integers whose product is the key number 9 and whose sum is $b = -10$. Such a pair is -9 and -1.

$$x^2 - 10x + 9 = x^2 - 9x - 1x + 9$$ Express $-10x$ as $-9x - 1x$. ($-1x - 9x$ could also be used.)

$$= x(x - 9) - 1(x - 9)$$ Factor x out of $x^2 - 9x$ and -1 out of $-1x + 9$.

$$= (x - 9)(x - 1)$$ Factor out $x - 9$.

The complete factorization of the original trinomial is

$$2x^3 - 20x^2 + 18x = 2x(x - 9)(x - 1)$$ Don't forget to write the GCF, $2x$.

Check the factorization by multiplying.

Self Check 12 Factor completely: $3m^3 - 27m^2 + 24m$

Now Try **Problem 51**

ANSWERS TO SELF CHECK **1.** $(y + 2)(y + 5)$ **2.** $(p - 2)(p - 4)$ **3.** $(m + 7)(m - 6)$ **4.** $(q + 4)(q - 6)$ **5.** $-(x - 4)(x - 7)$ **6.** $(s + 7t)(s - t)$ **7.** $4m^3(m + 4)(m - 2)$ **8.** $t(t - 2)(t + 6)$ **9.** Prime trinomial **10.** $(m + 7)(m - 6)$ **11.** $(q + 4t)(q - 6t)$ **12.** $3m(m - 8)(m - 1)$

STUDY SET 6.2

VOCABULARY

Fill in the blanks.

1. The trinomial $x^2 - x - 12$ ________ as the product of two binomials: $(x - 4)(x + 3)$.
2. A ________ trinomial cannot be factored by using only integers.
3. The ________ coefficient of $x^2 - 3x + 2$ is 1.
4. A trinomial is factored ____________ when no factor can be factored further.

CONCEPTS

Fill in the blanks.

5. **a.** Before attempting to factor a trinomial, be sure that it is written in ____________ powers of a variable.
   b. Before attempting to factor a trinomial into two binomials, always factor out any __________ factors first.
6. $x^2 + x - 56 = (x \;\square)(x \;\square)$

   The product of these numbers must be ___ ,
   and their sum must be ___ .
7. $x^2 + 5x + 3$ cannot be factored because we cannot find two integers whose product is ___ and whose sum is ___ .
8. Complete the following table.

Factors of 8	Sum of the factors of 8
1(8)	
2( )	
1(8)	
(−4)	

9. Check to determine whether each factorization is correct.
   a. $x^2 - x - 20 = (x + 5)(x - 4)$
   b. $4a^2 + 12a - 16 = 4(a - 1)(a + 4)$
10. Find two integers whose
   a. product is 10 and whose sum is 7.
   b. product is 8 and whose sum is −6.
   c. product is −6 and whose sum is 1.
   d. product is −9 and whose sum is −8.
11. Consider a trinomial of the form $x^2 + bx + c$.
   a. If c is positive, what can be said about the two integers that should be chosen for the factorization?
   b. If c is negative, what can be said about the two integers that should be chosen for the factorization?
12. Fill in each blank to explain how to factor $x^2 + 7x + 10$ by grouping.

   We express the middle term, $7x$, as the sum of _____ terms:

   $$x^2 + 7x + 10 = x^2 + \square x + \square x + 10$$

   The product of these numbers must be ___ ,
   and their sum must be ___ .

NOTATION

13. To factor a trinomial, a student made a table and circled the correct pair of integers, as shown. Complete the factorization of the trinomial.

   $(x \quad)(x \quad)$

Factors	Sum
1(−6)	−5
2(−3)	−1
(3(−2)	1)
6(−1)	5

14. To factor a trinomial by grouping, a student made a table and circled the correct pair of integers, as shown. Enter the correct coefficients.

   $x^2 + \quad x + \quad x + 16$

Key number = 16

Factors	Sum
1 · 16	17
(2 · 8	10)
4 · 4	8

GUIDED PRACTICE

Factor each trinomial. **See Examples 1 and 2 or Example 10.**

15. $x^2 + 3x + 2$
16. $y^2 + 4y + 3$
17. $z^2 + 7z + 12$
18. $x^2 + 7x + 10$
19. $m^2 - 5m + 6$
20. $n^2 - 7n + 10$
21. $t^2 - 11t + 28$
22. $c^2 - 9c + 8$
23. $r^2 - 9r + 18$
24. $y^2 - 17y + 72$
25. $a^2 - 46a + 45$
26. $r^2 - 37r + 36$

Factor each trinomial. **See Examples 3 and 4 or Example 10.**

27. $x^2 + 5x - 24$
28. $u^2 + u - 42$
29. $t^2 + 13t - 48$
30. $m^2 + 2m - 48$
31. $a^2 - 6a - 16$
32. $a^2 - 10a - 39$
33. $b^2 - 9b - 36$
34. $x^2 - 3x - 40$

Factor each trinomial. **See Example 5.**

35. $-x^2 - 7x - 10$
36. $-x^2 + 9x - 20$
37. $-t^2 - t + 30$
38. $-t^2 - 15t + 34$
39. $-r^2 - 3r + 54$
40. $-d^2 - 2d + 63$
41. $-m^2 + 18m - 77$
42. $-n^2 + 14n - 33$

Factor each trinomial. **See Example 6 or Example 11.**

43. $a^2 + 4ab + 3b^2$
44. $a^2 + 6ab + 5b^2$
45. $x^2 - 6xy - 7y^2$
46. $x^2 + 10xy - 11y^2$
47. $r^2 + sr - 2s^2$
48. $m^2 + mn - 6n^2$
49. $a^2 - 5ab + 6b^2$
50. $p^2 - 7pq + 10q^2$

Factor completely. **See Example 7 or Example 12.**

51. $2x^2 + 10x + 12$
52. $3y^2 - 21y + 18$
53. $6a^2 - 30a + 24$
54. $4b^2 + 12b - 16$
55. $5a^2 - 25a + 30$
56. $2b^2 - 20b + 18$
57. $-z^3 + 29z^2 - 100z$
58. $-m^3 + m^2 + 56m$
59. $-n^4 + 28n^3 + 60n^2$
60. $-c^5 + 16c^4 + 80c^3$
61. $4x^4 + 16x^3 + 16x^2$
62. $3a^4 + 30a^3 + 75a^2$

Write each trinomial in descending powers of one variable and factor. **See Example 8.**

63. $80 - 24x + x^2$
64. $y^2 + 100 + 25y$
65. $10y + 9 + y^2$
66. $x^2 - 13 - 12x$
67. $r^2 - 16 + 6r$
68. $u^2 - 12 - u$
69. $4rx + r^2 + 3x^2$
70. $a^2 + 5b^2 + 6ab$

Factor each trinomial, if possible. **See Example 9.**

71. $u^2 + 10u + 15$
72. $v^2 + 9v + 15$
73. $r^2 + 2r - 4$
74. $r^2 - 9r - 12$

TRY IT YOURSELF

Choose the correct method from Section 6.1 or Section 6.2 to factor completely each of the following.

75. $5x + 15 + xy + 3y$
76. $ab + b + 2a + 2$
77. $26n^2 - 8n$
78. $40c^2 - 12c$
79. $a^2 - 4a - 5$
80. $t^2 - 5t - 50$
81. $-x^2 + 21x + 22$
82. $-r^2 + 14r - 45$
83. $4xy - 4x + 28y - 28$
84. $3xy - 3x + 15y - 15$
85. $24b^4 - 48b^3 + 36b^2$
86. $28n^5 - 42n^4 + 28n^3$
87. $x^2 + 4xy + 4y^2$
88. $m^2 - 8mn + 16n^2$
89. $a^2 - 4ab - 12b^2$
90. $p^2 + pq - 6q^2$
91. $r^2 - 2r + 4$
92. $m^2 + 3m - 20$
93. $t(x + 2) + 7(x + 2)$
94. $r(t - v) + 10(t - v)$
95. $s^4 + 11s^3 - 26s^2$
96. $x^4 + 14x^3 + 45x^2$
97. $15s^3 + 75$
98. $33g^4 - 99$
99. $-13y + y^2 - 14$
100. $-3a + a^2 + 2$
101. $2x^2 - 12x + 16$
102. $6t^2 - 18t - 24$

APPLICATIONS

103. PETS The cage shown on the next page is used for transporting dogs. Its volume is $(x^3 + 12x^2 + 27x)$ in.^{3}. The dimensions of the cage can be found by factoring. If the cage is longer than it is tall and taller than it is wide, find the polynomials that represent its length, width, and height.

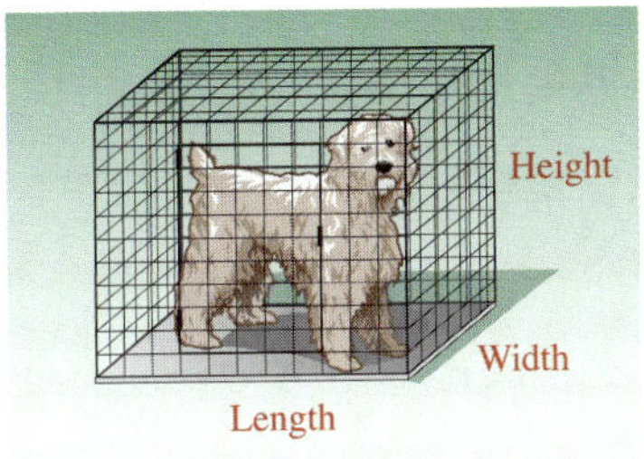

104. PHOTOGRAPHY A picture cube is a clever way to display 6 photographs in a small amount of space. Suppose the surface area of the entire cube is given by the polynomial $(6s^2 + 12s + 6)$ in.^{2}. Find the polynomial that represents the length of an edge of the cube.

WRITING

105. Explain what it means when we say that a trinomial is the product of two binomials. Give an example.

106. Are $2x^2 - 12x + 16$ and $x^2 - 6x + 8$ factored in the same way? Explain.

107. When factoring $x^2 - 2x - 3$, one student got $(x - 3)(x + 1)$, and another got $(x + 1)(x - 3)$. Are both answers acceptable? Explain.

108. In the partial solution shown below, a student began to factor the trinomial. Write a note to the student explaining his mistake.

Factor: $x^2 - 2x - 63$

$(x - \quad)(x - \quad)$

109. Explain the error in the following factorization.

$$x^3 + 8x^2 + 15x = x(x^2 + 8x + 15)$$
$$= (x + 3)(x + 5)$$

110. Explain why the factorization is not complete.

$$2y^2 - 12y + 16 = 2(y^2 - 6y + 8)$$

REVIEW

Simplify each expression. Write each answer without negative exponents.

111. $\dfrac{x^{12}x^{-7}}{x^3x^4}$

112. $\dfrac{a^4a^{-2}}{a^2a^0}$

113. $(x^{-3}x^{-2})^2$

114. $\left(\dfrac{18a^2b^3c^{-4}}{3a^{-1}b^2c}\right)^{-3}$

CHALLENGE PROBLEMS

Factor completely.

115. $x^2 - \dfrac{6}{5}x + \dfrac{9}{25}$

116. $x^2 - 0.5x + 0.06$

117. $x^{2m} - 12x^m - 45$

118. $x^2(y + 1) - 3x(y + 1) - 70(y + 1)$

119. Find all positive integer values of c that make $n^2 + 6n + c$ factorable.

120. Find all integer values of b that make $x^2 + bx - 44$ factorable.

SECTION 6.3
Factoring Trinomials of the Form $ax^2 + bx + c$

Objectives

1. Factor trinomials using the trial-and-check method.
2. Factor trinomials after factoring out the GCF.
3. Factor trinomials using the grouping method.

In this section we will factor trinomials with leading coefficients other than 1, such as

$$2x^2 + 5x + 3, \qquad 6a^2 - 17a + 5, \qquad \text{and} \qquad 4b^2 + 8bc - 45c^2$$

We can use two methods to factor these trinomials. With the first method, we make educated guesses and then check them using multiplication. The correct factorization is determined through a process of elimination. The second method is an extension of factoring by grouping.

1 Factor Trinomials Using the Trial-and-Check Method.

EXAMPLE 1

Factor: $2x^2 + 5x + 3$

Strategy We will assume that $2x^2 + 5x + 3$ is the product of two binomials and we will use a systematic method to find their terms.

Why Since the terms of $2x^2 + 5x + 3$ do not have a common factor (other than 1), the only option available is to try to factor it as the product of two binomials.

Solution We represent the binomials using two sets of parentheses. Since the first term of the trinomial is $2x^2$, we enter $2x$ and x as the first terms of the binomial factors.

$$(2x \square)(x \square)$$ Because $2x \cdot x$ will give $2x^2$.

The second terms of the binomials must be two integers whose product is 3. Since the coefficients of the terms of $2x^2 + 5x + 3$ are positive, we only consider pairs of positive integer factors of 3. Since there is just one such pair, $1 \cdot 3$, we can enter 1 and 3 as the second terms of the binomials, or we can reverse the order and enter 3 and 1.

$$(2x + 1)(x + 3) \quad \text{or} \quad (2x + 3)(x + 1)$$

The first possibility is incorrect, because when we find the outer and inner products and combine like terms, we obtain an incorrect middle term of $7x$.

Outer: $6x$

$$(2x + 1)(x + 3)$$ Multiply and add to find the middle term: $6x + x = 7x$.

Inner: x

The second possibility is correct, because it gives a middle term of $5x$.

Outer: $2x$

$$(2x + 3)(x + 1)$$ Multiply and add to find the middle term: $2x + 3x = 5x$.

Inner: $3x$

Thus,

$$2x^2 + 5x + 3 = (2x + 3)(x + 1)$$

Check the factorization by multiplying:

$$\begin{aligned}(2x + 3)(x + 1) &= 2x^2 + 2x + 3x + 3 \\ &= 2x^2 + 5x + 3\end{aligned}$$ This is the original trinomial.

Self Check 1 Factor: $2x^2 + 5x + 2$

Now Try Problem 19

The Language of Algebra To *interchange* means to put each in the place of the other. We create all of the possible factorizations by *interchanging* the second terms of the binomials.

$$(2x + 1)(x + 3)$$

$$(2x + 3)(x + 1)$$

EXAMPLE 2

Factor: $6a^2 - 17a + 5$

Strategy We will assume that $6a^2 - 17a + 5$ is the product of two binomials and we will use a systematic method to find their terms.

Why Since the terms of $6a^2 - 17a + 5$ do not have a common factor (other than 1), the only option available is to try to factor it as the product of two binomials.

Solution We represent the binomials using two sets of parentheses. Since the first term is $6a^2$, the first terms of the factors must be $6a$ and a or $3a$ and $2a$.

$$(6a \ \square)(a \ \square) \quad \text{or} \quad (3a \ \square)(2a \ \square)$$ Because $6a \cdot a$ or $3a \cdot 2a$ will give $6a^2$.

The second terms of the binomials must be two integers whose product is 5. Since the last term of $6a^2 - 17a + 5$ is positive and the coefficient of the middle term is negative, we only consider negative integer factors of the last term. Since there is just one such pair, $-1(-5)$, we can enter -1 and -5, or we can reverse the order and enter -5 and -1 as second terms of the binomials.

$(6a - 1)(a - 5)$ (outer: $-30a$; inner: $-a$) $\quad -30a - a = -31a$

$(6a - 5)(a - 1)$ (outer: $-6a$; inner: $-5a$) $\quad -6a - 5a = -11a$

$(3a - 1)(2a - 5)$ (outer: $-15a$; inner: $-2a$) $\quad -15a - 2a = -17a$

$(3a - 5)(2a - 1)$ (outer: $-3a$; inner: $-10a$) $\quad -3a - 10a = -13a$

Only the possibility shown in blue gives the correct middle term of $-17a$. Thus,

$$6a^2 - 17a + 5 = (3a - 1)(2a - 5)$$

We check by multiplying: $(3a - 1)(2a - 5) = 6a^2 - 17a + 5$.

Self Check 2 Factor: $6b^2 - 19b + 3$

Now Try Problem 27

EXAMPLE 3 Factor: $3y^2 - 7y - 6$

Strategy We will assume that $3y^2 - 7y - 6$ is the product of two binomials and we will use a systematic method to find their terms.

Why Since the terms of $3y^2 - 7y - 6$ do not have a common factor (other than 1), the only option available is to try to factor it as the product of two binomials.

Solution Since the first term is $3y^2$, the first terms of the binomial factors must be $3y$ and y.

$$(3y \ \square)(y \ \square)$$ Because $3y \cdot y$ will give $3y^2$.

The second terms of the binomials must be two integers whose product is -6. There are four such pairs: $1(-6)$, $-1(6)$, $2(-3)$, and $-2(3)$. When these pairs are entered, and then reversed as second terms of the binomials, there are eight possibilities to consider. Four of them can be discarded because they include a binomial whose terms have a common factor. If the terms of $3y^2 - 7y - 6$ do not have a common factor (other than 1), neither can any of its binomial factors.

Success Tip
If the terms of a trinomial do not have a common factor other than 1, the terms of each of its binomial factors will not have a common factor other than 1.

For 1 and −6: $(3y + 1)(y - 6)$ or $(3y - 6)(y + 1)$

$-18y$, y: $-18y + y = -17y$ — $3y - 6$ has a common factor of 3.

For −1 and 6: $(3y - 1)(y + 6)$ or $(3y + 6)(y - 1)$

$18y$, $-y$: $18y - y = 17y$ — $3y + 6$ has a common factor of 3.

Success Tip
Reversing the signs within the binomial factors reverses the sign of the middle term. For example, the factors 2 and −3 give the middle term $-7y$, while −2 and 3 give the middle term $7y$.

For 2 and −3: $(3y + 2)(y - 3)$ or $(3y - 3)(y + 2)$

$-9y$, $2y$: $-9y + 2y = -7y$ — $3y - 3$ has a common factor of 3.

For −2 and 3: $(3y - 2)(y + 3)$ or $(3y + 3)(y - 2)$

$9y$, $-2y$: $9y - 2y = 7y$ — $3y + 3$ has a common factor of 3.

Success Tip
All eight possible factorizations are listed. In practice, you will often find the correct factorization without having to examine the entire list of possibilities.

Only the possibility shown in green gives the correct middle term of $-7y$. Thus,

$$3y^2 - 7y - 6 = (3y + 2)(y - 3)$$

Check the factorization by multiplying.

Self Check 3 Factor: $5t^2 - 23t - 10$

Now Try Problem 35

EXAMPLE 4

Factor: $4b^2 + 8bc - 45c^2$

Strategy We will assume that $4b^2 + 8bc - 45c^2$ is the product of two binomials and we will use a systematic method to find their terms.

Why Since the terms of $4b^2 + 8bc - 45c^2$ do not have a common factor (other than 1), the only option available is to try to factor it as the product of two binomials.

Notation
The trinomial is in two variables, b and c. It is written in descending powers of b and ascending powers of c.

Solution Since the first term is $4b^2$, the first terms of the binomial factors must be $4b$ and b or $2b$ and $2b$. Since the last term contains c^2, the second terms of the binomial factors must contain c.

$$(4b \,\square\, c)(b \,\square\, c) \text{ or } (2b \,\square\, c)(2b \,\square\, c)$$

Because $4b \cdot b$ or $2b \cdot 2b$ gives $4b^2$, and because $c \cdot c$ gives c^2.

The coefficients of c must be two integers whose product is −45. Since the coefficient of the last term is negative, the signs of the integers must be different. If we pick factors of $4b$ and b for the first terms, and −1 and 45 for the coefficients of c, the multiplication gives an incorrect middle term of $179bc$.

180*bc*

$$(4b - c)(b + 45c) \quad 180bc - bc = 179bc$$

$-bc$

If we pick factors of $4b$ and b for the first terms, and 15 and -3 for the coefficients of c, the multiplication gives an incorrect middle term of $3bc$.

$-12bc$

$$(4b + 15c)(b - 3c) \quad -12bc + 15bc = 3bc$$

15*bc*

If we pick factors of $2b$ and $2b$ for the first terms, and -5 and 9 for the coefficients of c, we have

18*bc*

$$(2b - 5c)(2b + 9c) \quad 18bc - 10bc = 8bc$$

$-10bc$

which gives the correct middle term of $8bc$. Thus,

$$4b^2 + 8bc - 45c^2 = (2b - 5c)(2b + 9c)$$

Check the factorization by multiplying.

Success Tip

When using the trial-and-check method to factor $ax^2 + bx + c$, here is one suggestion: if b is relatively small, test the factorizations that use the smaller factors of a and c first.

Self Check 4 Factor: $4x^2 + 4xy - 3y^2$

Now Try **Problem 43**

Because guesswork is often necessary, it is difficult to give specific rules for factoring trinomials with leading coefficients other than 1. However, the following hints are helpful when using the **trial-and-check method.**

Factoring Trinomials with Leading Coefficients Other Than 1

To factor trinomials with leading coefficients other than 1:

1. Factor out any GCF (including -1 if that is necessary to make a positive in a trinomial of the form $ax^2 + bx + c$).
2. Write the trinomial as a product of two binomials. The coefficients of the first terms of each binomial factor must be factors of a, and the last terms must be factors of c.

   The product of these numbers must be *a*.

   $$ax^2 + bx + c = (\Box x \;\; \Box)(\Box x \;\; \Box)$$

   The product of these numbers must be *c*.
3. If c is positive, the signs within the binomial factors match the sign of b. If c is negative, the signs within the binomial factors are opposites.
4. Try combinations of first terms and second terms until you find the one that gives the proper middle term. If no combination works, the trinomial is prime.
5. Check by multiplying.

2 Factor Trinomials After Factoring Out the GCF.

If the terms of a trinomial have a common factor, the GCF (or the opposite of the GCF) should always be factored out first.

EXAMPLE 5 Factor: $2x^2 - 8x^3 + 3x$

Strategy We will write the trinomial in descending powers of x and factor out the common factor , $-x$.

Why It is easier to factor trinomials that have a positive leading coefficient.

Solution Write the trinomial in descending powers of x: $-8x^3 + 2x^2 + 3x$.

$$-8x^3 + 2x^2 + 3x = -x(8x^2 - 2x - 3)$$ Factor out the opposite of the GCF, $-x$.

We now factor $8x^2 - 2x - 3$. Its factorization has the form

$$(8x \;\square)(x \;\square) \text{ or } (2x \;\square)(4x \;\square)$$ Because $8x \cdot x$ or $4x \cdot 2x$ gives $8x^2$.

The second terms of the binomials must be two integers whose product is -3. There are two such pairs: $1(-3)$ and $-1(3)$. Since the coefficient of middle term $-2x$ is small, we pick the smaller factors of $8x^2$, which are $2x$ and $4x$, for the first terms and 1 and -3 for the second terms.

$$(2x + 1)(4x - 3)$$ (outer: $-6x$; inner: $4x$) $-6x + 4x = -2x$

This factorization gives the correct middle term of $-2x$. Thus,

$$8x^2 - 2x - 3 = (2x + 1)(4x - 3)$$

We can now give the complete factorization of the original trinomial.

$$-8x^3 + 2x^2 + 3x = -x(8x^2 - 2x - 3)$$
$$= -x(2x + 1)(4x - 3)$$

Check the factorization by multiplying.

Self Check 5 Factor: $12y - 2y^3 - 2y^2$

Now Try Problem 53

The Language of Algebra
When asked to *factor* a polynomial, that means we should *factor completely*.

Caution
For multistep factorization, don't forget to write the GCF (or its opposite) in the final factored form.

3 Factoring Trinomials Using the Grouping Method.

Another way to factor a trinomial of the form $ax^2 + bx + c$ is to write it as an equivalent four-termed polynomial and factor it by grouping. For example, to factor $2x^2 + 5x + 3$, we proceed as follows.

1. Identify the values of a, b, and c.

$$\left.\begin{matrix} ax^2 + bx + c \\ \downarrow \quad\;\; \downarrow \quad \downarrow \\ 2x^2 + 5x + 3 \end{matrix}\right\} a = 2,\ b = 5, \text{ and } c = 3$$

Then, find the product ac, called the **key number:** $ac = 2(3) = 6$.

2. Next, find two integers whose product is $ac = 6$ and whose sum is $b = 5$. Since the integers must have a positive product and a positive sum, we consider only positive factors of 6.

Key number = 6 $\qquad$ $b = 5$

Positive factors of 6	Sum of the positive factors of 6
$1 \cdot 6 = 6$	$1 + 6 = 7$
$2 \cdot 3 = 6$	$2 + 3 = 5$

The second row of the table contains the correct pair of integers 2 and 3, whose product is 6 and whose sum is 5.

3. Express the middle term, $5x$, of the trinomial as the *sum of two terms,* using the integers 2 and 3 found in step 2 as coefficients of the two terms.

$$2x^2 + 5x + 3 = 2x^2 + 2x + 3x + 3$$ Express 5x as 2x + 3x.

4. Factor the equivalent four-term polynomial by grouping:

$$2x^2 + 2x + 3x + 3 = 2x(x + 1) + 3(x + 1)$$ Factor 2x out of $2x^2 + 2x$ and 3 out of $3x + 3$.

$$= (x + 1)(2x + 3)$$ Factor out x + 1.

Check by multiplying.

Factoring by grouping is especially useful when the leading coefficient, a, and the constant term, c, have many factors.

Factoring Trinomials by Grouping

To factor a trinomial by grouping:

1. Factor out any GCF (including -1 if that is necessary to make a positive in a trinomial of the form $ax^2 + bx + c$).
2. Identify a, b, and c, and find the key number ac.
3. Find two integers whose product is the key number and whose sum is b.
4. Express the middle term, bx, as the sum (or difference) of two terms. Enter the two numbers found in step 3 as coefficients of x in the form shown below. Then factor the equivalent four-term polynomial by grouping.

$$ax^2 + \square x + \square x + c$$

The product of these numbers must be ac and their sum must be b.

5. Check the factorization by multiplying.

EXAMPLE 6 Factor by grouping: $10x^2 + 13x - 3$

Strategy We will express the middle term, $13x$, of the trinomial as the sum of two carefully chosen terms.

Why We want to produce an equivalent four-term polynomial that can be factored by grouping.

13. Complete the key number table.

Negative factors of 12	Sum of the negative factors of 12
$-1(-12)$	
$-2(\quad)$	
(-4)	

14. Complete the sentence to explain how to factor $3x^2 + 16x + 5$ by grouping.

$$3x^2 + 16x + 5 = 3x^2 + \square\, x + \square\, x + 5$$

The product of these numbers must be and their sum must be .

NOTATION

15. a. Suppose we wish to factor $12b^2 + 20b - 9$ by grouping. Identify a, b, and c.
    b. What is the key number, ac?

16. To factor $6x^2 + 13x + 6$ by grouping, a student made a table and circled the correct pair of integers, as shown. Enter the correct coefficients.

$$6x^2 + \quad x + \quad x + 6$$

$ac = 36 \quad b = 13$

Factors	Sum
$1 \cdot 36$	37
$2 \cdot 18$	20
$3 \cdot 12$	15
$4 \cdot 9$ (circled)	13 (circled)
$6 \cdot 6$	12

Complete each step of the factorization of the trinomial by grouping.

17. $12t^2 + 17t + 6 = 12t^2 + 9t + 8t + 6$
$= \quad (4t + 3) + \quad (4t + 3)$
$= (\quad)(3t + 2)$

18. $35t^2 - 11t - 6 = 35t^2 + 10t - 21t - 6$
$= 5t(7t + 2) \quad 3(7t + 2)$
$= (\quad)(5t - 3)$

GUIDED PRACTICE

Factor. See Example 1 or Example 6.

19. $2x^2 + 3x + 1$
20. $3x^2 + 4x + 1$
21. $3a^2 + 10a + 3$
22. $2b^2 + 7b + 3$
23. $5x^2 + 7x + 2$
24. $7t^2 + 12t + 5$
25. $7x^2 + 18x + 11$
26. $5n^2 + 12n + 7$

Factor. See Example 2 or Example 6.

27. $4x^2 - 8x + 3$
28. $4z^2 - 13z + 3$
29. $8x^2 - 22x + 5$
30. $15a^2 - 28a + 5$
31. $15t^2 - 26t + 7$
32. $10x^2 - 9x + 2$
33. $6y^2 - 13y + 2$
34. $6y^2 - 43y + 7$

Factor. See Example 3 or Example 6.

35. $3x^2 - 2x - 21$
36. $3u^2 - 44u - 15$
37. $5m^2 - 7m - 6$
38. $5y^2 - 18y - 8$
39. $7y^2 + 55y - 8$
40. $7x^2 + 33x - 10$
41. $11y^2 + 7y - 4$
42. $13y^2 + 9y - 4$

Factor. See Example 4.

43. $6r^2 + rs - 2s^2$
44. $3m^2 + 5mn + 2n^2$
45. $4x^2 + 8xy + 3y^2$
46. $4b^2 + 15bc - 4c^2$
47. $8m^2 + 91mn + 33n^2$
48. $2m^2 + 17mn - 9n^2$
49. $15x^2 - xy - 6y^2$
50. $4a^2 - 15ab + 9b^2$

Factor. See Example 5 or Example 7.

51. $-26x + 6x^2 - 20$
52. $-28 + 6a^2 - 2a$
53. $15a + 8a^3 - 26a^2$
54. $16r - 40r^2 + 25r^3$
55. $2u^2 - 6v^2 - uv$
56. $6a^2 + 6b^2 - 13ab$
57. $36y^2 - 88y + 32$
58. $70a^2 - 95a + 30$
59. $130r^2 + 20r - 110$
60. $170h^2 - 210h - 260$
61. $-y^3 - 13y^2 - 12y$
62. $-2xy^2 - 8xy + 24x$
63. $-6x^4 + 15x^3 + 9x^2$
64. $-9y^4 - 3y^3 + 6y^2$
65. $16m^3n + 20m^2n^2 + 6mn^3$
66. $-28u^3v^3 + 26u^2v^4 - 6uv^5$

TRY IT YOURSELF

Factor each polynomial completely. If an expression is prime, so indicate.

67. $6t^2 - 7t - 20$
68. $6w^2 + 13w + 5$
69. $15p^2 - 2pq - q^2$
70. $8c^2 - 10cd + 3d^2$

71. $4t^2 - 16t + 7$

72. $9x^2 - 32x + 15$

73. $8y^2 - 2y - 1$

74. $14y^2 + 11y + 2$

75. $18x^2 + 31x - 10$

76. $20y^2 - 93y - 35$

77. $10u^2 - 13u - 6$

78. $8m^2 + 5m - 10$

79. $3x^2 + x + 6$

80. $2u^2 + 3u + 25$

81. $30r^5 + 63r^4 - 30r^3$

82. $6s^5 - 26s^4 - 20s^3$

83. $6p^2 + pq - q^2$

84. $12m^2 - 11mn + 2n^2$

85. $-12y^2 - 12 + 25y$

86. $-10t^2 + 1 + 3t$

Choose the correct method from Sections 6.1, 6.2, or 6.3 to factor completely each of the following.

87. $m^2 + 3m - 28$

88. $-b^2 - 5b + 24$

89. $6a^3 + 15a^2$

90. $9x^4 + 27x^6$

91. $x^3 - 2x^2 + 5x - 10$

92. $x^3 - x^2 + 2x - 2$

93. $5y^2 + 3 - 8y$

94. $3t^2 + 7 - 10t$

95. $-2x^2 - 10x - 12$

96. $4y^2 + 36y + 72$

97. $12x^3y^3 - 18x^2y^3 + 15x^2y^2$

98. $15c^2d^3 - 25c^3d^2 - 10c^4d^4$

99. $a^2 - 7ab + 10b^2$

100. $x^2 - 13xy + 12y^2$

101. $9u^6 - 71u^5 - 8u^4$

102. $25n^8 - 49n^7 - 2n^6$

APPLICATIONS

103. FURNITURE The area of a desktop is represented by the trinomial $(4x^2 + 20x - 11)$ in.^{2}. Factor it to find the polynomials that represent its length and width.

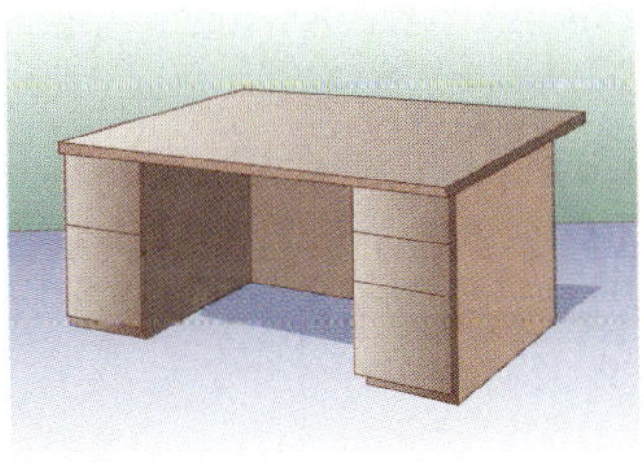

104. STORAGE The volume of an 8-foot-wide portable storage container is represented by the trinomial $(72x^2 + 120x - 400)$ ft^3. Its dimensions can be determined by factoring the trinomial. Find the polynomials that represent height and the length of the container.

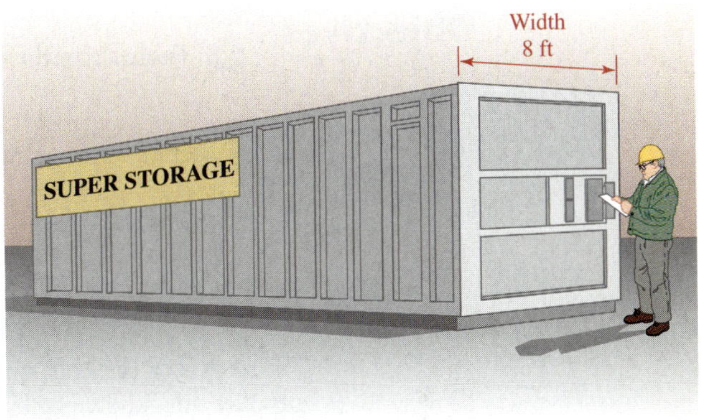

WRITING

105. Two students factor $2x^2 + 20x + 42$ and get two different answers: $(2x + 6)(x + 7)$ and $(x + 3)(2x + 14)$. Do both answers check? Why don't they agree? Is either answer completely correct? Explain.

106. Why is the process of factoring $6x^2 - 5x - 6$ more complicated than the process of factoring $x^2 - 5x - 6$?

107. Suppose a factorization check of $(3x - 9)(5x + 7)$ gives a middle term $-24x$, but a middle term of $24x$ is actually needed. Explain how to quickly obtain the correct factorization.

108. Suppose we want to factor $2x^2 + 7x - 72$. Explain why $(2x - 1)(x + 72)$ is not a wise choice to try first.

REVIEW

Evaluate each expression.

109. -7^2

110. $(-7)^2$

111. 7^0

112. 7^{-2}

113. $\frac{1}{7^{-2}}$

114. $2 \cdot 7^2$

CHALLENGE PROBLEMS

Factor completely.

115. $6a^{10} + 5a^5 - 21$

116. $3x^4y^2 - 29x^2y + 56$

117. $8x^2(c^2 + c - 2) - 2x(c^2 + c - 2) - (c^2 + c - 2)$

118. Find all integer values of b that make $2x^2 + bx - 5$ factorable.

SECTION 6.4
Factoring Perfect-Square Trinomials and the Difference of Two Squares

Objectives

1. Recognize perfect-square trinomials.
2. Factor perfect-square trinomials.
3. Factor the difference of two squares.

In this section, we will discuss a method that can be used to factor two types of trinomials, called *perfect-square trinomials.* We also develop techniques for factoring a type of binomial called the *difference of two squares.*

1 Recognize Perfect-Square Trinomials.

We have seen that the square of a binomial is a trinomial. We have also seen that the special-product rules shown below can be used to quickly find the square of a sum and the square of a difference. The terms of the resulting trinomial are related to the terms of the binomial that was squared.

Success Tip
To prepare for this section, it would be helpful to review Section 5.7 Special Products.

$$(A + B)^2 = A^2 + 2AB + B^2$$

A^2: This is the square of the first term of the binomial.
$2AB$: This is twice the product of the terms of the binomial.
B^2: This is the square of the last term of the binomial.

$$(A - B)^2 = A^2 - 2AB + B^2$$

Trinomials that are squares of a binomial are called **perfect-square trinomials.** Some examples are

$y^2 + 6y + 9$ — Because it is the square of $(y + 3)$: $(y + 3)^2 = y^2 + 6y + 9$

$t^2 - 14t + 49$ — Because it is the square of $(t - 7)$: $(t - 7)^2 = t^2 - 14t + 49$

$4m^2 - 20m + 25$ — Because it is the square of $(2m - 5)$: $(2m - 5)^2 = 4m^2 - 20m + 25$

EXAMPLE 1 Determine whether the following are perfect-square trinomials:

a. $x^2 + 10x + 25$ **b.** $c^2 - 12c - 36$
c. $25y^2 - 30y + 9$ **d.** $4t^2 + 18t + 81$

Strategy We will compare each trinomial, term-by-term, to one of the special-product forms discussed in Section 5.7.

Why If a trinomial matches one of these forms, it is a perfect-square trinomial.

Solution

a. To determine whether this is a perfect-square trinomial, we note that

$$x^2 + 10x + 25$$

The first term is the square of x.
The middle term is twice the product of x and 5: $2 \cdot x \cdot 5 = 10x$.
The last term is the square of 5.

Thus, $x^2 + 10x + 25$ is a perfect-square trinomial.

b. To determine whether this is a perfect-square trinomial, we note that

$$c^2 - 12c - 36$$

The last term, −36, is not the square of a real number.

Since the last term is negative, $c^2 - 12c - 36$ is not a perfect-square trinomial.

The Language of Algebra
The expressions $25y^2$ and 9 are called *perfect squares* because $25y^2 = (5y)^2$ and $9 = 3^2$.

c. To determine whether this is a perfect-square trinomial, we note that

$$25y^2 - 30y + 9$$

The first term is the square of 5y.

The middle term is twice the product of 5y and −3: $2(5y)(-3) = -30y$.

The last term is the square of −3.

Thus, $25y^2 - 30y + 9$ is a perfect-square trinomial.

d. To determine whether this is a perfect-square trinomial, we note that

$$4t^2 + 18t + 81$$

The first term is the square of 2t.

The middle term is not twice the product of 2t and 9, because $2(2t)(9) = 36t$.

The last term is the square of 9.

Thus, $4t^2 + 18t + 81$ is not a perfect-square trinomial.

Self Check 1 Determine whether the following are perfect-square trinomials:

a. $y^2 + 4y + 4$ **b.** $b^2 - 6b - 9$

c. $4z^2 + 4z + 4$ **d.** $49x^2 - 28x + 16$

Now Try **Problems 13 and 17**

2 Factor Perfect-Square Trinomials.

We can factor perfect-square trinomials using the methods previously discussed in Sections 6.2 and 6.3. However, in many cases, we can factor them more quickly by inspecting their terms and applying the special-product rules in reverse.

Factoring Perfect-Square Trinomials

$$A^2 + 2AB + B^2 = (A + B)^2$$
$$A^2 - 2AB + B^2 = (A - B)^2$$

Each of these trinomials factors as the square of a binomial.

When factoring perfect-square trinomials, it is helpful to know the integers that are perfect squares. The number 400, for example, is a perfect-integer square, because $400 = 20^2$.

$1 = 1^2$	$25 = 5^2$	$81 = 9^2$	$169 = 13^2$	$289 = 17^2$
$4 = 2^2$	$36 = 6^2$	$100 = 10^2$	$196 = 14^2$	$324 = 18^2$
$9 = 3^2$	$49 = 7^7$	$121 = 11^2$	$225 = 15^2$	$361 = 19^2$
$16 = 4^2$	$64 = 8^2$	$144 = 12^2$	$256 = 16^2$	$400 = 20^2$

EXAMPLE 2 Factor: **a.** $x^2 + 20x + 100$ **b.** $9x^2 - 30xy + 25y^2$

Strategy The terms of each trinomial do not have a common factor (other than 1). We will determine whether each is a perfect-square trinomial.

Why If it is, we can factor it using a special-product rule in reverse.

Solution

a. $x^2 + 20x + 100$ is a perfect-square trinomial, because:

- The first term x^2 is the square of x.
- The last term 100 is the square of 10.
- The middle term is twice the product of x and 10: $2(x)(10) = 20x$.

To find the factorization, we match $x^2 + 20x + 100$ to the proper rule for factoring a perfect-square trinomial.

$$\begin{array}{c} A^2 + 2\,A\,B + B^2 = (A + B)^2 \\ \downarrow \\ x^2 + 20x + 10^2 = x^2 + 2 \cdot x \cdot 10 + 10^2 = (x + 10)^2 \end{array}$$

Therefore, $x^2 + 20x + 10^2 = (x + 10)^2$. Check by finding $(x + 10)^2$.

b. $9x^2 - 30xy + 25y^2$ is a perfect-square trinomial, because:

- The first term $9x^2$ is the square of $3x$: $(3x)^2 = 9x^2$.
- The last term $25y^2$ is the square of $-5y$: $(-5y)^2 = 25y^2$.
- The middle term is twice the product of $3x$ and $-5y$: $2(3x)(-5y) = -30xy$.

We can use these observations to write the trinomial in one of the special-product forms that then leads to its factorization.

$$9x^2 - 30xy + 25y^2 = (3x)^2 - 2(3x)(5y) + (-5y)^2 \quad -2(3x)(5y) = 2(3x)(-5y).$$
$$= (3x - 5y)^2$$

Therefore, $9x^2 - 30xy + 25y^2 = (3x - 5y)^2$. Check by finding $(3x - 5y)^2$.

Self Check 2 Factor: **a.** $x^2 + 18x + 81$ **b.** $16x^2 - 8xy + y^2$

Now Try **Problems 21 and 31**

> ***Success Tip***
> The sign of the middle term of a perfect-square trinomial is the same as the sign of the second term of the squared binomial.
>
> $A^2 + 2AB + B^2 = (A + B)^2$
> $A^2 - 2AB + B^2 = (A - B)^2$

EXAMPLE 3 Factor completely: $4a^3 - 4a^2 + a$

Strategy We will factor out the GCF, a, first. Then we will factor the resulting perfect-square trinomial using a special-product rule in reverse.

Why The first step in factoring any polynomial is to factor out the GCF.

Solution The terms of $4a^3 - 4a^2 + a$ have the common factor a, which should be factored out first. Within the parentheses, we recognize $4a^2 - 4a + 1$ as a perfect square trinomial of the form $A^2 - 2AB + B^2$, and factor it as such.

$$4a^3 - 4a^2 + a = a(4a^2 - 4a + 1) \quad \text{Factor out } a.$$
$$= a(2a - 1)^2 \quad 4a^2 = (2a)^2, 1 = (-1)^2, \text{ and } -4a = 2(2a)(-1).$$

Self Check 3 Factor completely: $49x^3 - 14x^2 + x$

Now Try **Problem 37**

3 Factor the Difference of Two Squares.

The Language of Algebra
The expression $A^2 - B^2$ is a *difference of two squares*, whereas $(A - B)^2$ is the *square of a difference*. They are not equivalent because $(A - B)^2 \neq A^2 - B^2$.

Recall the special-product rule for multiplying the sum and difference of the same two terms:

$$(A + B)(A - B) = A^2 - B^2$$

The binomial $A^2 - B^2$ is called a **difference of two squares,** because A^2 is the square of A and B^2 is the square of B. If we reverse this rule, we obtain a method for factoring a difference of two squares.

Factoring →

$$A^2 - B^2 = (A + B)(A - B)$$

This pattern is easy to remember if we think of a difference of two squares as the square of a **F**irst quantity minus the square of a **L**ast quantity.

Factoring a Difference of Two Squares

To factor the square of a First quantity minus the square of a Last quantity, multiply the First plus the Last by the First minus the Last.

$$F^2 - L^2 = (F + L)(F - L)$$

EXAMPLE 4 Factor, if possible: **a.** $x^2 - 9$ **b.** $16 - b^2$ **c.** $n^2 - 45$ **d.** $a^2 + 81$

Strategy The terms of each binomial do not have a common factor (other than 1). The only option available is to attempt to factor each as a difference of two squares.

Why If a binomial is a difference of two squares, we can factor it using a special-product rule in reverse.

Notation
By the commutative property of multiplication, the factors of a difference of two squares can be written in either order. For example, we can write:

$$x^2 - 9 = (x - 3)(x + 3)$$

Solution

a. $x^2 - 9$ is the difference of two squares because it can be written as $x^2 - 3^2$. We can match it to the rule for factoring a difference of two squares to find the factorization.

$$\mathbf{F}^2 - \mathbf{L}^2 = (\mathbf{F} + \mathbf{L})(\mathbf{F} - \mathbf{L})$$
$$\downarrow \quad \downarrow \quad \downarrow \quad \downarrow \quad \downarrow \quad \downarrow$$
$$x^2 - 3^2 = (x + 3)(x - 3)$$ 9 is a perfect-integer square: $9 = 3^2$.

Therefore, $x^2 - 9 = (x + 3)(x - 3)$.

Check by multiplying: $(x + 3)(x - 3) = x^2 - 9$.

b. $16 - b^2$ is the difference of two squares because $16 - b^2 = 4^2 - b^2$. Therefore,

$$16 - b^2 = (4 + b)(4 - b)$$ 16 is a perfect-integer square: $16 = 4^2$.

Check by multiplying.

c. Since 45 is not a perfect-integer square, $n^2 - 45$ cannot be factored using integers. It is prime.

d. $a^2 + 81$ can be written $a^2 + 9^2$, and is, therefore, the **sum of two squares.** We might attempt to factor $a^2 + 81$ as $(a + 9)(a + 9)$ or $(a - 9)(a - 9)$. However, the following checks show that neither product is $a^2 + 81$.

$$(a + 9)(a + 9) = a^2 + 18a + 81 \qquad (a - 9)(a - 9) = a^2 - 18a + 81$$

In general, the sum of two squares (with no common factor other than 1) cannot be factored using real numbers. Thus, $a^2 + 81$ is prime.

Self Check 4 Factor, if possible:

a. $c^2 - 4$ **b.** $121 - t^2$

c. $x^2 - 24$ **d.** $s^2 + 36$

Now Try **Problems 45 and 53**

Terms containing variables such as $25x^2$ and $4y^4$ are perfect squares, because they can be written as the square of a quantity. For example:

$$25x^2 = (5x)^2 \qquad \text{and} \qquad 4y^4 = (2y^2)^2$$

EXAMPLE 5

Factor: **a.** $25x^2 - 49$ **b.** $4y^4 - 121z^2$

Strategy In each case, the terms of the binomial do not have a common factor (other than 1). To factor them, we will write each binomial in a form that clearly shows it is a difference of two squares.

Why We can then use a special-product rule in reverse to factor it.

Solution

a. We can write $25x^2 - 49$ in the form $(5x)^2 - 7^2$ and match it to the rule for factoring the difference of two squares:

$$\begin{array}{c} F^2 - L^2 = (F + L)(F - L) \\ \downarrow \quad \downarrow \quad\quad \downarrow \quad \downarrow \quad \downarrow \quad \downarrow \\ (5x)^2 - 7^2 = (5x + 7)(5x - 7) \end{array}$$

Therefore, $25x^2 - 49 = (5x + 7)(5x - 7)$. Check by multiplying.

b. We can write $4y^4 - 121z^2$ in the form $(2y^2)^2 - (11z)^2$ and match it to the rule for factoring the difference of two squares:

$$\begin{array}{c} F^2 - L^2 = (F + L)(F - L) \\ \downarrow \quad \downarrow \quad\quad \downarrow \quad \downarrow \quad \downarrow \quad \downarrow \\ (2y^2)^2 - (11z)^2 = (2y^2 + 11z)(2y^2 - 11z) \end{array}$$

Therefore, $4y^4 - 121z^2 = (2y^2 + 11z)(2y^2 - 11z)$. Check by multiplying.

Success Tip
Remember that a *difference of two squares* is a binomial. Each term is a square and the terms have different signs. The powers of the variables in the terms must be even.

Self Check 5 Factor: **a.** $16y^2 - 9$

b. $9m^2 - 64n^4$

Now Try **Problems 57 and 61**

EXAMPLE 6 Factor completely: $8x^2 - 8$

Strategy We will factor out the GCF, 8, first. Then we will factor the resulting difference of two squares.

Why The first step in factoring any polynomial is to factor out the GCF.

Solution

$$8x^2 - 8 = 8(x^2 - 1) \quad \text{The GCF is 8.}$$
$$= 8(x + 1)(x - 1) \quad \text{Think of } x^2 - 1 \text{ as } x^2 - 1^2 \text{ and factor the difference of two squares.}$$

Check: $8(x + 1)(x - 1) = 8(x^2 - 1)$ Multiply the binomials first.
$= 8x^2 - 8$ Distribute the multiplication by 8.

Self Check 6 Factor completely: $2p^2 - 200$

Now Try **Problem 65**

Sometimes we must factor a difference of two squares more than once to completely factor a polynomial.

EXAMPLE 7 Factor completely: $x^4 - 16$

Strategy The terms of $x^4 - 16$ do not have a common factor (other than 1). To factor this binomial, we will write it in a form that clearly shows it is a difference of two squares.

Why We can then use a special-product rule in reverse to factor it.

Caution
Factoring a polynomial is complete when no factor can be factored further.

Solution

$$x^4 - 16 = (x^2)^2 - 4^2 \quad \text{Write } x^4 \text{ as } (x^2)^2 \text{ and 16 as } 4^2.$$
$$= (x^2 + 4)(x^2 - 4) \quad \text{Factor the difference of two squares.}$$
$$= (x^2 + 4)(x + 2)(x - 2) \quad \text{Factor } x^2 - 4 \text{, which is itself a difference of two squares. The binomial } x^2 + 4 \text{ is a sum of two squares and does not factor further.}$$

Self Check 7 Factor completely: $a^4 - 81$

Now Try **Problem 75**

ANSWERS TO SELF CHECKS **1. a.** Yes **b.** No **c.** No **d.** No **2. a.** $(x + 9)^2$ **b.** $(4x - y)^2$ **3.** $x(7x - 1)^2$ **4. a.** $(c + 2)(c - 2)$ **b.** $(11 + t)(11 - t)$ **c.** Prime **d.** Prime **5. a.** $(4y + 3)(4y - 3)$ **b.** $(3m + 8n^2)(3m - 8n^2)$ **6.** $2(p + 10)(p - 10)$ **7.** $(a^2 + 9)(a + 3)(a - 3)$

STUDY SET 6.4

VOCABULARY

Fill in the blanks.

1. $x^2 + 6x + 9$ is a _______-square trinomial because it is the square of the binomial $x + 3$.
2. The binomial $x^2 - 25$ is called a _________ of two squares and it factors as $(x + 5)(x - 5)$. The binomial $x^2 + 25$ is a _____ of two squares and since it does not factor using integers, it is _______.

CONCEPTS

Fill in the blanks.

3. Consider $25x^2 + 30x + 9$.
   a. The first term is the square of ☐.
   b. The last term is the square of ☐.
   c. The middle term is twice the product of ☐ and ☐.
4. Consider $49x^2 - 28xy + 4y^2$.
   a. The first term is the square of ☐.
   b. The last term is the square of ☐.
   c. The middle term is twice the product of ☐ and ☐.
5. a. $x^2 + 2xy + y^2 = (\square + \square)^2$
   b. $x^2 - 2xy + y^2 = (x \square y)^2$
   c. $x^2 - y^2 = (x \square y)(\square - \square)$
6. a. $36x^2 = (\square)^2$ b. $100x^4 = (\square)^2$
   c. $4x^2 - 9 = (\square)^2 - (\square)^2$
7. List the squares of the integers from 1 through 20.
8. Use multiplication to determine if each factorization is correct.
   a. $9y^2 - 12y + 4 = (3y - 2)^2$
   b. $n^2 - 16 = (n + 8)(n - 8)$

NOTATION

Complete each factorization.

9. $x^2 + 10x + 25 = (x + 5)^{\square}$
10. $9b^2 - 12b + 4 = (3b \square 2)^2$
11. $x^2 - 64 = (x \square 8)(x \square 8)$
12. $16t^2 - 49 = (4t + \square)(4t - \square)$

GUIDED PRACTICE

Determine whether each of the following is a perfect-square trinomial. See Example 1.

13. $x^2 + 18x + 81$
14. $x^2 + 14x + 49$
15. $y^2 + 2y + 4$
16. $y^2 + 4y + 16$
17. $9n^2 - 30n - 25$
18. $9a^2 - 48a - 64$
19. $4y^2 - 12y + 9$
20. $9y^2 - 30y + 25$

Factor. See Example 2.

21. $x^2 + 6x + 9$
22. $x^2 + 10x + 25$
23. $b^2 + 2b + 1$
24. $m^2 + 12m + 36$
25. $c^2 - 12c + 36$
26. $d^2 - 10d + 25$
27. $9y^2 - 24y + 16$
28. $49z^2 - 14z + 1$
29. $9 + 4x^2 + 12x$
30. $121 + 4x^2 + 44x$
31. $36m^2 + 60mn + 25n^2$
32. $25a^2 + 30ab + 9b^2$
33. $81x^2 - 72xy + 16y^2$
34. $9x^2 - 48xy + 64y^2$
35. $49t^2 - 28ts + 4s^2$
36. $81p^2 - 36pq + 4q^2$

Factor completely. See Example 3.

37. $3u^2 - 18u + 27$
38. $3v^2 - 42v + 147$
39. $36x^3 + 12x^2 + x$
40. $4x^4 - 20x^3 + 25x^2$
41. $18a^5 + 84a^4b + 98a^3b^2$
42. $32b^6 + 80b^5c + 50b^4c^2$
43. $-100t^2 + 20t - 1$
44. $-81r^2 - 18r - 1$

Factor completely. If a polynomial can't be factored, write "prime." See Example 4.

45. $x^2 - 4$
46. $x^2 - 9$
47. $x^2 - 16$
48. $x^2 - 25$
49. $36 - y^2$
50. $49 - w^2$
51. $-25 + t^2$
52. $-144 + h^2$
53. $a^2 + b^2$
54. $121a^2 + b^2$
55. $y^2 - 63$
56. $x^2 - 27$

Factor. See Example 5.

57. $25t^2 - 64$
58. $49d^2 - 16$

59. $81y^2 - 1$

60. $400z^2 - 1$

61. $9x^4 - y^2$

62. $4x^2 - z^4$

63. $16c^2 - 49d^4$

64. $36a^2 - 121b^4$

Factor completely. See Example 6.

65. $8x^2 - 32y^2$

66. $2a^2 - 200b^2$

67. $63a^2 - 7$

68. $20x^2 - 5$

69. $x^3 - 144x$

70. $g^3 - 121g$

71. $6x^4 - 6x^2y^2$

72. $4b^2y - 16c^2y$

Factor completely. See Example 7.

73. $81 - s^4$

74. $y^4 - 625$

75. $b^4 - 256$

76. $m^4n^4 - 16$

77. $16t^4 - 16s^4$

78. $2p^4 - 32q^4$

79. $25m^4 - 25$

80. $9 - 9n^4$

TRY IT YOURSELF

Factor completely.

81. $a^4 - 144b^2$

82. $81y^4 - 100z^2$

83. $9x^2y^2 + 30xy + 25$

84. $s^2t^2 - 20st + 100$

85. $t^2 - 20t + 100$

86. $r^2 + 24r + 144$

87. $z^2 - 64$

88. $25 + B^2$

89. $3m^4 - 3n^4$

90. $5a^4 - 80b^4$

91. $25m^2 + 70m + 49$

92. $25x^2 + 20x + 4$

Choose the correct method from Section 6.1, Section 6.2, Section 6.3, or Section 6.4 to factor completely each of the following:

93. $x^2 + x - 42$

94. $rx - sx + r - s$

95. $x^2 - 9$

96. $3a^2 - 4a - 4$

97. $24a^3b - 16a^2b$

98. $20ns^2 - 60nu + 100n$

99. $-2r^2 + 28r - 80$

100. $10s - 39 + s^2$

101. $x^3 + 3x^2 + 4x + 12$

102. $2y^2 - 128z^2$

103. $4b^2 - 20b + 25$

104. $a^2 - 4ab - 12b^2$

APPLICATIONS

105. GENETICS The Hardy–Weinberg equation, one of the fundamental concepts in population genetics, is $p^2 + 2pq + q^2 = 1$, where p represents the frequency of a certain dominant gene and q represents the frequency of a certain recessive gene. Factor the left side of the equation.

106. SIGNAL FLAGS The maritime signal flag for the letter X is shown. Find the polynomial that represents the area of the shaded region and express it in factored form.

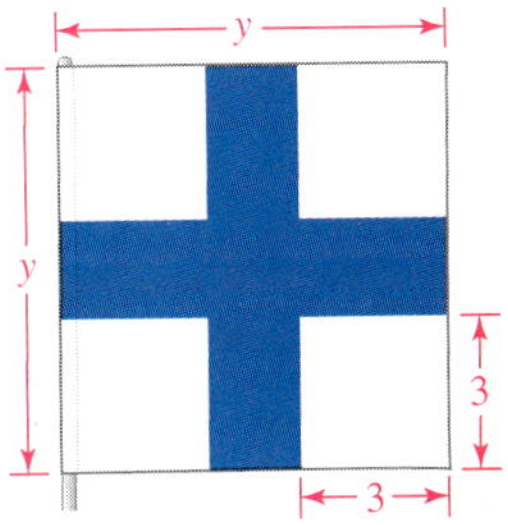

107. PHYSICS The illustration shows a time-sequence picture of a falling apple. Factor the expression, which gives the difference in the distance fallen by the apple during the time interval from t_1 to t_2 seconds.

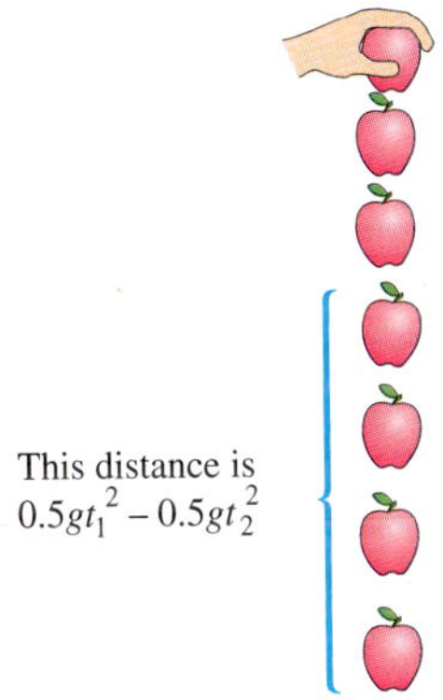

108. DARTS A circular dart board has a series of rings around a solid center, called the bullseye. To find the area of the outer grey ring, we can use the formula $A = \pi R^2 - \pi r^2$. Factor the expression on the right side of the equation.

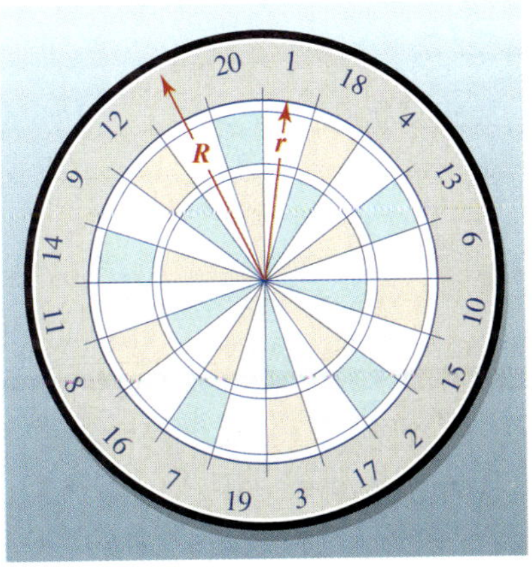

WRITING

109. When asked to factor $x^2 - 25$, one student wrote $(x + 5)(x - 5)$, and another student wrote $(x - 5)(x + 5)$. Are both answers correct? Explain.

110. Explain the error that was made in the following factorization:

$$x^2 - 100 = (x + 50)(x - 50)$$

111. Explain why the following factorization isn't complete.

$$x^4 - 625 = (x^2 + 25)(x^2 - 25)$$

112. Explain why $a^2 + 2a + 1$ is a perfect square trinomial and why $a^2 + 4a + 1$ isn't a perfect square trinomial.

REVIEW

Perform each division.

113. $\dfrac{5x^2 + 10y^2 - 15xy}{5xy}$

114. $\dfrac{-30c^2d^2 - 15c^2d - 10cd^2}{-10cd}$

115. $2a - 1\overline{)a - 2 + 6a^2}$

116. $4b + 3\overline{)4b^3 - 5b^2 - 2b + 3}$

CHALLENGE PROBLEMS

117. For what value of c does $80x^2 - c$ factor as $5(4x + 3)(4x - 3)$?

118. Find all values of b so that $0.16x^2 + bxy + 0.25y^2$ is a perfect square trinomial.

Factor completely.

119. $81x^6 + 36x^3y^2 + 4y^4$

120. $p^2 + p + \dfrac{1}{4}$

121. $c^2 + 1.6c + 0.64$

122. $x^{2n} - y^{4n}$

123. $(x + 5)^2 - y^2$

124. $\dfrac{1}{2} - 2a^2$

125. $c^2 - \dfrac{1}{16}$

126. $t^2 - \dfrac{9}{25}$

SECTION 6.5
Factoring the Sum and Difference of Two Cubes

Objective 1 Factor the sum and difference of two cubes.

In this section we will discuss how to factor two types of binomials, called the *sum* and the *difference of two cubes*.

1 Factor the Sum and Difference of Two Cubes.

We have seen that the sum of two squares, such as $x^2 + 4$ or $25a^2 + 9b^2$, cannot be factored. However, the sum of two cubes and the difference of two cubes can be factored.

The sum of two cubes

$x^3 + 8$

This is x cubed. This is 2 cubed: $2^3 = 8$.

The difference of two cubes

$a^3 - 64b^3$

This is a cubed. This is $4b$ cubed: $(4b)^3 = 64b^3$.

To find rules for factoring the sum of two cubes and the difference of two cubes, we need to find the products shown below. Note that each term of the trinomial is multiplied by each term of the binomial.

$$(x + y)(x^2 - xy + y^2) = x^3 - x^2y + xy^2 + x^2y - xy^2 + y^3$$
$$= x^3 + y^3 \quad \text{Combine like terms: } -x^2y + x^2y = 0 \text{ and } xy^2 - xy^2 = 0.$$

The Language of Algebra
The expression $x^3 + y^3$ is a *sum of two cubes*, whereas $(x + y)^3$ is the *cube of a sum*. If you expand $(x + y)^3$, you will see that $(x + y)^3 \neq x^3 + y^3$.

$$(x - y)(x^2 + xy + y^2) = x^3 + x^2y + xy^2 - x^2y - xy^2 - y^3$$
$$= x^3 - y^3 \qquad \text{Combine like terms.}$$

These results justify the rules for factoring the **sum and difference of two cubes.** They are easier to remember if we think of a sum (or a difference) of two cubes as the cube of a **F**irst quantity plus (or minus) the cube of the **L**ast quantity.

Factoring the Sum and Difference of Two Cubes

To factor the cube of a First quantity plus the cube of a Last quantity, multiply the First plus the Last by the First squared, minus the First times the Last, plus the Last squared.

$$F^3 + L^3 = (F + L)(F^2 - FL + L^2)$$

To factor the cube of a First quantity minus the cube of a Last quantity, multiply the First minus the Last by the First squared, plus the First times the Last, plus the Last squared.

$$F^3 - L^3 = (F - L)(F^2 + FL + L^2)$$

To factor the sum or difference of two cubes, it's helpful to know the cubes of integers from 1 to 10. The number 216, for example, is a **perfect-integer cube,** because $216 = 6^3$.

$1 = 1^3$	$27 = 3^3$	$125 = 5^3$	$343 = 7^3$	$729 = 9^3$
$8 = 2^3$	$64 = 4^3$	$216 = 6^3$	$512 = 8^3$	$1{,}000 = 10^3$

EXAMPLE 1

Factor: $x^3 + 8$

Strategy We will write $x^3 + 8$ in a form that clearly shows it is the sum of two cubes.

Why We can then use the rule for factoring the sum of two cubes.

Solution $x^3 + 8$ is the sum of two cubes because it can be written as $x^3 + 2^3$. We can match it to the rule for factoring the sum of two cubes to find its factorization.

$$F^3 + L^3 = (F + L)(F^2 - F\ L + L^2)$$
$$\downarrow \quad \downarrow \quad \downarrow \quad \downarrow \quad \downarrow \quad \downarrow \quad \downarrow \quad \downarrow$$
$$x^3 + 2^3 = (x + 2)(x^2 - x \cdot 2 + 2^2)$$
$$= (x + 2)(x^2 - 2x + 4)$$

To write the trinomial factor:
- Square the first term of the binomial factor.
- Multiply the terms of the binomial factor.
- Square the last term of the binomial factor.

$x^2 - 2x + 4$ does not factor.

Caution
A common error is to try to factor $x^2 - 2x + 4$. It is not a perfect square trinomial, because the middle term needs to be $-4x$. Furthermore, it cannot be factored by the methods of Section 6.2. It is prime.

Therefore, $x^3 + 8 = (x + 2)(x^2 - 2x + 4)$. We can check by multiplying.

$$(x + 2)(x^2 - 2x + 4) = x^3 + 2x^2 - 2x^2 - 4x + 4x + 8$$
$$= x^3 + 8 \qquad \text{This is the original binomial.}$$

Self Check 1 Factor: $h^3 + 27$

Now Try **Problem 17**

Terms containing variables such as $64b^3$ and m^6 are also perfect cubes, because they can be written as the cube of a quantity:

$$64b^3 = (4b)^3 \quad \text{and} \quad m^6 = (m^2)^3$$

EXAMPLE 2 Factor: $a^3 - 64b^3$

Strategy We will write $a^3 - 64b^3$ in a form that clearly shows it is the difference of two cubes.

Why We can then use the rule for factoring the difference of two cubes.

Solution $a^3 - 64b^3$ is the difference of two cubes because it can be written as $a^3 - (4b)^3$. We can match it to the rule for factoring the difference of two cubes to find its factorization.

$$F^3 - L^3 = (F - L)(F^2 + F\ L + L^2)$$

$$\downarrow \quad \downarrow \quad \downarrow \quad \downarrow \quad \downarrow \quad \downarrow \quad \downarrow \quad \downarrow$$

$$a^3 - (4b)^3 = (a - 4b)[a^2 + a \cdot 4b + (4b)^2]$$

$$= (a - 4b)(a^2 + 4ab + 16b^2) \quad a^2 + 4ab + 16b^2 \text{ does not factor.}$$

Therefore, $a^3 - 64b^3 = (a - 4b)(a^2 + 4ab + 16b^2)$. Check by multiplying.

Self Check 2 Factor: $8c^3 - 1$

Now Try **Problem 37**

You should memorize the rules for factoring the sum and the difference of two cubes. Note that the right side of each rule has the form

(a binomial)(a trinomial)

and that there is a relationship between the signs that appear in these forms.

The Sum of Two Cubes

The same sign

$$F^3 + L^3 = (F + L)(F^2 - FL + L^2)$$

Opposite signs Always plus

The Difference of Two Cubes

The same sign

$$F^3 - L^3 = (F - L)(F^2 + FL + L^2)$$

Opposite signs Always plus

If the terms of a binomial have a common factor, the GCF (or the opposite of the GCF) should always be factored out first.

EXAMPLE 3 Factor: $-2t^5 + 250t^2$

Strategy We will factor out the common factor, $-2t^2$. We can then factor the resulting binomial as a difference of two cubes.

Why The first step in factoring any polynomial is to factor out the GCF, or its opposite.

Solution

$$-2t^5 + 250t^2 = -2t^2(t^3 - 125) \quad \text{Factor out the opposite of the GCF, } -2t^2.$$

$$= -2t^2(t - 5)(t^2 + 5t + 25) \quad \text{Factor } t^3 - 125.$$

Therefore, $-2t^5 + 250t^2 = -2t^2(t - 5)(t^2 + 5t + 25)$. Check by multiplying.

Self Check 3 Factor: $4c^3 + 4d^3$

Now Try **Problem 43**

ANSWERS TO SELF CHECKS **1.** $(h + 3)(h^2 - 3h + 9)$ **2.** $(2c - 1)(4c^2 + 2c + 1)$ **3.** $4(c + d)(c^2 - cd + d^2)$

STUDY SET 6.5

VOCABULARY

Fill in the blanks.

1. $x^3 + 27$ is the _____ of two cubes and $a^3 - 125$ is the difference of two ______.

2. The factorization of $x^3 + 8$ is $(x + 2)(x^2 - 2x + 4)$. The first factor is a binomial and the second is a __________.

CONCEPTS

Fill in the blanks.

3. a. $F^3 + L^3 = (\square + \square)(F^2 - FL + L^2)$

b. $F^3 - L^3 = (F \square L)(\square + FL + \square)$

4. $m^3 + 64$

This is $\square$ cubed. This is $\square$ cubed.

5. $216n^3 - 125$

This is $\square$ cubed. This is $\square$ cubed.

6. a. $x^3 + 64y^3 = (\square)^3 + (\square)^3$

b. $8x^3 - 27 = (\square)^3 - (\square)^3$

7. List the first ten positive integer cubes.

8. $(x - 2)(x^2 + 2x + 4)$ is the factorization of what binomial?

9. Use multiplication to determine if the factorization is correct.

$$b^3 + 27 = (b + 3)(b^2 + 3b + 9)$$

10. The factorization of $y^3 + 27$ is $(y + 3)(y^2 - 3y + 9)$. Is this factored completely, or does $y^2 - 3y + 9$ factor further?

NOTATION

Complete each factorization.

11. $a^3 + 8 = (a + 2)(a^2 - \square + 4)$

12. $x^3 - 1 = (x - 1)(x^2 + \square + 1)$

13. $b^3 + 27 = (\square)(b^2 - 3b + 9)$

14. $z^3 - 125 = (z - 5)(\square + 5z + \square)$

Give an example of each type of expression.

15. a. the sum of two cubes

b. the cube of a sum

16. a. the difference of two cubes

b. the cube of a difference

GUIDED PRACTICE

Factor. **See Example 1.**

17. $y^3 + 125$

18. $b^3 + 216$

19. $a^3 + 64$

20. $n^3 + 1$

21. $n^3 + 512$

22. $t^3 + 729$

23. $8 + t^3$

24. $27 + y^3$

25. $a^3 + 1{,}000b^3$

26. $8u^3 + w^3$

27. $125c^3 + 27d^3$

28. $64m^3 + 343n^3$

Factor. **See Example 2.**

29. $a^3 - 27$

30. $r^3 - 8$

31. $m^3 - 343$

32. $y^3 - 216$

33. $216 - v^3$

34. $125 - t^3$

35. $8s^3 - t^3$

36. $27a^3 - b^3$

37. $1{,}000a^3 - w^3$

38. $s^3 - 64t^3$

39. $64x^3 - 27y^3$

40. $27x^3 - 1{,}000y^3$

***Factor completely.* See Example 3.**

41. $2x^3 + 2$

42. $8y^3 + 8$

43. $3d^3 + 81$

44. $2x^3 + 54$

45. $x^4 - 216x$

46. $x^5 - 125x^2$

47. $64m^3x - 8n^3x$

48. $16r^4 - 128rs^3$

TRY IT YOURSELF

Choose the correct method from Section 6.1 through Section 6.5 and factor completely.

49. $x^2 + 8x + 16$

50. $64p^3 - 27$

51. $9r^2 - 16s^2$

52. $-63 - 13x + 6x^2$

53. $xy - ty + sx - st$

54. $12p^2 + 14p - 6$

55. $4p^3 + 32q^3$

56. $56a^4 - 15a^3 + a^2$

57. $16c^3t^2 + 20c^2t^3 + 6ct^4$

58. $-t^2 - 9t + 1$

59. $36e^4 - 36$

60. $3(z + 4) - a(z + 4)$

61. $35a^3b^2 - 14a^2b^3 + 14a^3b^3$

62. $-y^2 - 15y + 34$

63. $36r^2 + 60rs + 25s^2$

64. $16u^2 - 16$

APPLICATIONS

65. MAILING BREAKABLES Write a polynomial that describes the amount of space in the larger box that must be filled with styrofoam chips if the smaller box containing a glass tea cup is to be placed within the larger box for mailing. Then factor the polynomial.

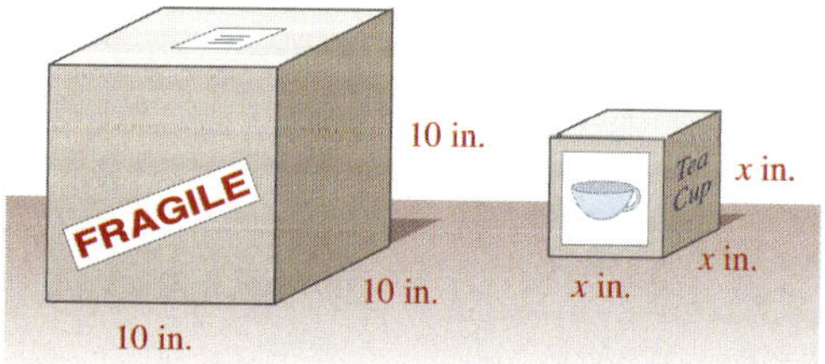

66. MELTING ICE In one hour, the block of ice shown below had melted to the size shown on the right. Write a polynomial that describes the volume of ice that melted away. Then factor the polynomial.

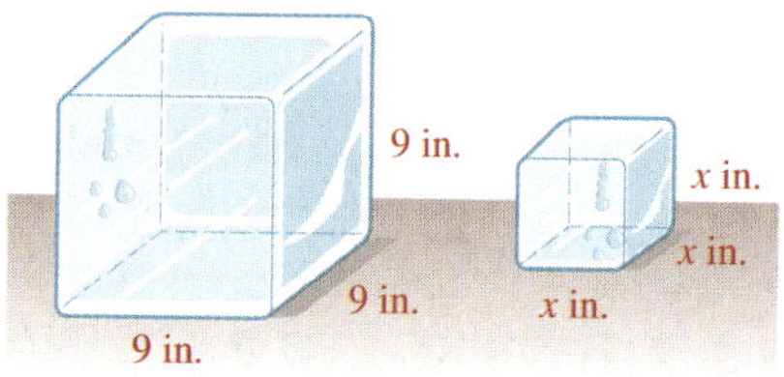

WRITING

67. Explain why $x^3 - 25$ is not a difference of two cubes.

68. Explain this diagram. Then draw a similar diagram for the difference of two cubes.

The same

$$F^3 + L^3 = (F + L)(F^2 - FL + L^2)$$

Opposite Always plus

REVIEW

69. When expressed as a decimal, is $\frac{7}{9}$ a terminating or a repeating decimal?

70. Solve: $x + 20 = 4x - 1 + 2x$

71. Write the set of integers.

72. Solve: $2x + 2 = \frac{2}{3}x - 2$

73. Evaluate $2x^2 + 5x - 3$ for $x = -3$.

74. Check to determine whether 4 is a solution of $3(m - 8) + 2m = 4 - (m + 2)$.

CHALLENGE PROBLEMS

75. Consider: $x^6 - 1$

a. Write the binomial as a difference of two squares. Then factor.

b. Write the binomial as a difference of two cubes. Then factor.

76. What binomial multiplied by $(a^2b^2 + 7ab + 49)$ gives a difference of two cubes?

Factor completely.

77. $x^6 - y^9$

78. $\frac{125}{8}s^3 + \frac{1}{27}t^3$

79. $64x^{12} + y^{15}z^{18}$

80. $x^{3m} - y^{3n}$

SECTION 6.6
A Factoring Strategy

The factoring methods discussed so far will be used in the remaining chapters to simplify expressions and solve equations. In such cases, we must determine the factoring method—it will not be specified. This section will give you practice in selecting the appropriate factoring method to use given a randomly chosen polynomial.

The following strategy is helpful when factoring polynomials.

Steps for Factoring a Polynomial

1. Is there a common factor? If so, factor out the GCF, or the opposite of the GCF so that the leading coefficient is positive.
2. How many terms does the polynomial have?

   If it has *two terms,* look for the following problem types:

   a. The difference of two squares
   b. The sum of two cubes
   c. The difference of two cubes

   If it has *three terms,* look for the following problem types:

   a. A perfect-square trinomial
   b. If the trinomial is not a perfect square, use the trial-and-check-method or the grouping method.

   If it has *four or more terms,* try to factor by grouping.
3. Can any factors be factored further? If so, factor them completely.
4. Does the factorization check? Check by multiplying.

EXAMPLE 1 Factor: $2x^4 - 162$

Strategy We will answer the four questions listed in the *Steps for Factoring a Polynomial.*

Why The answers to these questions help us determine which factoring techniques to use.

Solution ***Is there a common factor?*** Yes. Factor out the GCF, which is 2.

$$2x^4 - 162 = 2(x^4 - 81)$$

How many terms does it have? The polynomial within the parentheses, $x^4 - 81$, has two terms. It is a difference of two squares.

$$2x^4 - 162 = 2(x^4 - 81) \qquad \text{Think of } x^4 - 81 \text{ as } (x^2)^2 - 9^2.$$
$$= 2(x^2 + 9)(x^2 - 9)$$

Is it factored completely? No. $x^2 - 9$ is also the difference of two squares and can be factored.

$$2x^4 - 162 = 2(x^4 - 81)$$
$$= 2(x^2 + 9)(x^2 - 9) \qquad \text{Think of } x^2 - 9 \text{ as } x^2 - 3^2.$$
$$= 2(x^2 + 9)(x + 3)(x - 3) \qquad x^2 + 9 \text{ is a sum of two squares and does not factor.}$$

Therefore, $2x^4 - 162 = 2(x^2 + 9)(x + 3)(x - 3)$.

The Language of Algebra
Recall that *to factor a polynomial* means to express it as a product of two (or more) polynomials.

The Language of Algebra
Remember that the instruction to *factor* means to *factor completely.* A polynomial is *factored completely* when no factor can be factored further.

Does it check? Yes.

$$\begin{aligned} 2(x^2+9)(x+3)(x-3) &= 2(x^2+9)(x^2-9) && \text{Multiply } (x+3)(x-3) \text{ first.} \\ &= 2(x^4-81) && \text{Multiply } (x^2+9)(x^2-9). \\ &= 2x^4-162 && \text{This is the original polynomial.} \end{aligned}$$

Self Check 1 Factor: $11a^6 - 11a^2$

Now Try Problem 21

EXAMPLE 2

Factor: $-4c^5d^2 - 12c^4d^3 - 9c^3d^4$

Strategy We will answer the four questions listed in the *Steps for Factoring a Polynomial.*

Why The answers to these questions help us determine which factoring techniques to use.

Solution ***Is there a common factor?*** Yes. Factor out the opposite of the GCF, $-c^3d^2$, so that the leading coefficient is positive.

$$-4c^5d^2 - 12c^4d^3 - 9c^3d^4 = -c^3d^2(4c^2 + 12cd + 9d^2)$$

How many terms does it have? The polynomial within the parentheses has three terms. It is a perfect-square trinomial because $4c^2 = (2c)^2$, $9d^2 = (3d)^2$, and $12cd = 2 \cdot 2c \cdot 3d$.

$$\begin{aligned} -4c^5d^2 - 12c^4d^3 - 9c^3d^4 &= -c^3d^2(4c^2 + 12cd + 9d^2) \\ &= -c^3d^2(2c+3d)^2 \end{aligned}$$

Is it factored completely? Yes. The binomial $2c + 3d$ does not factor further.

Therefore, $-4c^5d^2 - 12c^4d^3 - 9c^3d^4 = -c^3d^2(2c + 3d)^2$.

Does it check? Yes.

$$\begin{aligned} -c^3d^2(2c+3d)^2 &= -c^3d^2(4c^2 + 12cd + 9d^2) && \text{Use a special-product rule.} \\ &= -4c^5d^2 - 12c^4d^3 - 9c^3d^4 && \text{This is the original polynomial.} \end{aligned}$$

Self Check 2 Factor: $-32h^4 - 80h^3 - 50h^2$

Now Try Problem 33

EXAMPLE 3

Factor: $y^4 - 3y^3 + y - 3$

Strategy We will answer the four questions listed in the *Steps for Factoring a Polynomial.*

Why The answers to these questions help us determine which factoring techniques to use.

Solution ***Is there a common factor?*** No. There is no common factor (other than 1).

How many terms does it have? Since the polynomial has four terms, we will try factoring by grouping.

$$y^4 - 3y^3 + y - 3 = y^3(y - 3) + 1(y - 3)$$ Factor out y^3 from $y^4 - 3y^3$. Factor out 1 from $y - 3$.

$$= (y - 3)(y^3 + 1)$$

Is it factored completely? No. We can factor $y^3 + 1$ as a sum of two cubes.

$$y^4 - 3y^3 + y - 3 = y^3(y - 3) + 1(y - 3)$$
$$= (y - 3)(y^3 + 1)$$ Think of $y^3 + 1$ as $y^3 + 1^3$.
$$= (y - 3)(y + 1)(y^2 - y + 1)$$ $y^2 - y + 1$ does not factor further.

Therefore, $y^4 - 3y^3 + y - 3 = (y - 3)(y + 1)(y^2 - y + 1)$.

Does it check? Yes.

$$(y - 3)(y + 1)(y^2 - y + 1) = (y - 3)(y^3 + 1)$$ Multiply the last two factors.
$$= y^4 + y - 3y^3 - 3$$ Use the FOIL method.
$$= y^4 - 3y^3 + y - 3$$ This is the original polynomial.

Self Check 3 Factor: $b^4 + b^3 + 8b + 8$

Now Try **Problem 37**

EXAMPLE 4 Factor: $32n - 4n^2 + 4n^3$

Strategy We will answer the four questions listed in the *Steps for Factoring a Polynomial.*

Why The answers to these questions help us determine which factoring techniques to use.

Solution ***Is there a common factor?*** Yes. When we write the terms in descending powers of n, we see that the GCF is $4n$.

$$4n^3 - 4n^2 + 32n = 4n(n^2 - n + 8)$$

How many terms does it have? The polynomial within the parentheses has three terms. It is not a perfect-square trinomial because the last term, 8, is not a perfect-integer square.

Negative factors of 8	Sum of the negative factors of 8
$-1(-8) = 8$	$-1 + (-8) = -9$
$-2(-4) = 8$	$-2 + (-8) = -10$

To factor the trinomial $n^2 - n + 8$, we must find two integers whose product is 8 and whose sum is -1. As we see in the table, there are no such integers. Thus, $n^2 - n + 8$ is prime.

Is it factored completely? Yes.

Therefore, $4n^3 - 4n^2 + 32n = 4n(n^2 - n + 8)$. Remember to write the GCF, $4n$, from the first step.

Does it check? Yes.

$$4n(n^2 - n + 8) = 4n^3 - 4n^2 + 32n$$ This is the original polynomial.

Self Check 4 Factor: $6m^2 - 54m + 6m^3$

Now Try **Problem 45**

EXAMPLE 5 Factor: $3y^3 - 4y^2 - 4y$

Strategy We will answer the four questions listed in the *Steps for Factoring a Polynomial.*

Why The answers to these questions help us determine which factoring techniques to use.

Solution ***Is there a common factor?*** Yes. The GCF is y.

$$3y^3 - 4y^2 - 4y = y(3y^2 - 4y - 4)$$

How many terms does it have? The polynomial within the parentheses has three terms. It is not a perfect-square trinomial because the first term, $3y^2$, is not a perfect square.

If we use grouping to factor $3y^2 - 4y - 4$, the key number is $ac = 3(-4) = -12$. We must find two integers whose product is -12 and whose sum is $b = -4$.

Key number $= -12$ $\qquad b = -4$

Factors of -12	Sum of the factors of -12
$2(-6) = -12$	$2 + (-6) = -4$

From the table, the correct pair is 2 and -6. They serve as the coefficients of $2y$ and $-6y$, the two terms that we use to represent the middle term, $-4y$, of the trinomial.

$$3y^2 - 4y - 4 = 3y^2 + 2y - 6y - 4$$ Express $-4y$ as $2y - 6y$.

$$= y(3y + 2) - 2(3y + 2)$$ Factor y from the first two terms and factor -2 the last two terms.

$$= (3y + 2)(y - 2)$$ Factor out $3y + 2$.

The trinomial $3y^2 - 4y - 4$ factors as $(3y + 2)(y - 2)$.

Is it factored completely? Yes. Because $3y + 2$ and $y - 2$ do not factor.

Therefore, $3y^3 - 4y^2 - 4y = y(3y + 2)(y - 2)$. Remember to write the GCF, y, from the first step.

Does it check? Yes.

$$y(3y + 2)(y - 2) = y(3y^2 - 4y - 4)$$ Multiply the binomials.

$$= 3y^3 - 4y^2 - 4y$$ This is the original polynomial.

Self Check 5 Factor: $6y^3 + 21y^2 - 12y$

Now Try **Problem 67**

ANSWERS TO SELF CHECKS **1.** $11a^2(a^2 + 1)(a + 1)(a - 1)$ **2.** $-2h^2(4h + 5)^2$ **3.** $(b + 1)(b + 2)(b^2 - 2b + 4)$ **4.** $6m(m^2 + m - 9)$ **5.** $3y(2y - 1)(y + 4)$

STUDY SET 6.6

VOCABULARY

Fill in the blanks.

1. To factor a polynomial means to express it as a ________ of two (or more) polynomials.
2. A polynomial is factored ________ when each factor is prime.

CONCEPTS

For each of the following polynomials, which factoring method would you use first?

3. $2x^5y - 4x^3y$
4. $9b^2 + 12y - 5$
5. $x^2 + 18x + 81$
6. $ax + ay - x - y$
7. $x^3 + 27$
8. $y^3 - 64$
9. $m^2 + 3mn + 2n^2$
10. $16 - 25z^2$
11. What is the first question that should be asked when using the strategy of this section to factor a polynomial?
12. Use multiplication to determine whether the factorization is correct.

$$5c^3d^2 - 40c^2d^3 + 35cd^4 = 5cd^2(c - 7d)(c - d)$$

NOTATION

Complete each factorization.

13. $6m^3 - 28m^2 + 16m = 2m(3m^2 - \square + 8)$
 $= 2m(3m - 2)(\square - 4)$
14. $2a^3 + 3a^2 - 2a - 3$
 $= \square(2a + 3) - 1(\square + 3)$
 $= (\square)(a^2 - 1)$
 $= (2a + 3)(a + 1)(\square)$

TRY IT YOURSELF

The following is a list of random factoring problems. Factor each expression completely. If a expression is not factorable, write "prime." **See Examples 1–5.**

15. $2b^2 + 8b - 24$
16. $32 - 2t^4$
17. $8p^3q^7 + 4p^2q^3$
18. $8m^2n^3 - 24mn^4$
19. $2 + 24y + 40y^2$
20. $6r^2 + 3rs - 18s^2$
21. $8x^4 - 8$
22. $t - 90 + t^2$
23. $14c - 147 + c^2$
24. $ab^2 - 4a + 3b^2 - 12$
25. $x^2 + 7x + 1$
26. $3a^3 + 24b^3$
27. $-2x^5 + 128x^2$
28. $16 - 40z + 25z^2$
29. $a^2c + a^2d^2 + bc + bd^2$
30. $6t^4 + 14t^3 - 40t^2$
31. $-9x^2y^2 + 6xy - 1$
32. $x^2y^2 - 2x^2 - y^2 + 2$
33. $-20m^3 - 100m^2 - 125m$
34. $5x^3y^3z^4 + 25x^2y^4z^2 - 35x^3y^2z^5$
35. $2c^2 - 5cd - 3d^2$
36. $125p^3 - 64y^3$
37. $p^4 - 2p^3 - 8p + 16$
38. $a^2 + 8a + 3$
39. $a^2(x - a) - b^2(x - a)$
40. $70p^4q^3 - 35p^4q^2 + 49p^5q^2$
41. $a^2b^2 - 144$
42. $-16x^4y^2z + 24x^5y^3z^4 - 15x^2y^3z^7$
43. $2x^3 + 10x^2 + x + 5$
44. $u^2 - 18u + 81$
45. $8v^2 - 14v^3 + v^4$
46. $28 - 3m - m^2$
47. $x^4 - 13x^2 + 36$
48. $81r^4 - 256$
49. $8a^2x^3 - 2b^2x$
50. $12x^2 + 14x - 6$
51. $6x^2 - 14x + 8$
52. $12x^2 - 12$
53. $4x^2y^2 + 4xy^2 + y^2$
54. $81r^4s^2 - 24rs^5$
55. $4m^5 + 500m^2$
56. $ae + bf + af + be$
57. $x^4 - 2x^2 - 8$
58. $6x^2 - x - 16$
59. $4x^2 + 9y^2$
60. $x^4y + 216xy^4$
61. $16a^5 - 54a^2$
62. $25x^2 - 16y^2$

63. $27x - 27y - 27z$

64. $12x^2 + 52x + 35$

65. $xy - ty + xs - ts$

66. $bc + b + cd + d$

67. $35x^8 - 2x^7 - x^6$

68. $x^3 - 25$

69. $5(x - 2) + 10y(x - 2)$

70. $16x^2 - 40x^3 + 25x^4$

71. $49p^2 + 28pq + 4q^2$

72. $x^2y^2 - 6xy - 16$

73. $4t^2 + 36$

74. $r^5 + 3r^3 + 2r^2 + 6$

75. $m^2n^2 - 9m^2 + 3n^2 - 27$

76. $z^2 + 6yz^2 + 9y^2z^2$

WRITING

77. Which factoring method do you find the most difficult? Why?

78. What four questions make up the factoring strategy for polynomials discussed in this section?

79. What does it mean to factor a polynomial?

80. How is a factorization checked?

REVIEW

81. Graph the real numbers $-3, 0, 2$, and $-\frac{3}{2}$ on a number line.

82. Graph the interval $(-2, 3]$ on a number line.

83. Graph: $y = \frac{1}{2}x + 1$

84. Graph: $y < 2 - 3x$

CHALLENGE PROBLEMS

Factor completely using rational numbers.

85. $x^6 - 4x^3 - 12$

86. $x(x - y) - y(y - x)$

87. $24 - x^3 + 8x^2 - 3x$

88. $25b^2 + 14b + \frac{49}{25}$

89. $x^9 + y^6$

90. $\frac{1}{4} - \frac{u^2}{81}$

SECTION 6.7
Solving Quadratic Equations by Factoring

Objectives

1. Define quadratic equations.
2. Solve quadratic equations using the zero-factor property.
3. Solve third-degree equations by factoring.

The factoring methods that we have discussed have many applications in algebra. In this section, we will use factoring to solve *quadratic equations*. These equations are different from those that we solved in Chapter 2. They contain a term where the variable is raised to the second power, such as x^2 or t^2.

1 Define Quadratic Equations.

In a linear, or first degree equation, such as $2x + 3 = 8$, the exponent on the variable is an unwritten 1. A quadratic, or second degree equation, has a term in which the exponent on the variable is 2, and has no other terms of higher degree.

Quadratic Equations

A **quadratic equation** is an equation that can be written in the **standard form**

$$ax^2 + bx + c = 0$$

where a, b, and c represent real numbers, and $a \neq 0$.

The Language of Algebra
Quadratic equations involve the square of a variable, not the 4th power as *quad* might suggest. Why is this? Because the origin of the word *quadratic* is the Latin word *quadratus,* meaning square.

Some examples of quadratic equations are

$$x^2 - 2x - 63 = 0, \qquad x^2 - 25 = 0, \qquad \text{and} \qquad 2x^2 + 3x = 2$$

The first two equations are in standard form. To write the third equation in standard form, we subtract 2 from both sides to get $2x^2 + 3x - 2 = 0$.

2 Solve Quadratic Equations Using the Zero-Factor Property.

To **solve a quadratic equation,** we find all values of the variable that make the equation true. The methods that we used to solve linear equations in Chapter 2 cannot be used to solve a quadratic equation, because we cannot isolate the variable on one side of the equation. However, we can often solve quadratic equations using factoring and the following property of real numbers.

The Zero-Factor Property

When the product of two real numbers is 0, at least one of them is 0.

If a and b represent real numbers, and

$$\text{if } ab = 0, \text{ then} \quad a = 0 \quad \text{or} \quad b = 0$$

EXAMPLE 1 Solve: $(4x - 1)(x + 6) = 0$

Strategy We will set $4x - 1$ equal to 0 and $x + 6$ equal to 0 and solve each equation.

Why If the product of $4x - 1$ and $x + 6$ is 0, then, by the zero-factor property, $4x - 1$ must equal 0, or $x + 6$ must equal 0.

Caution
It would not be helpful to multiply $(4x - 1)$ and $(x + 6)$. We want the left side of the equation to be in factored form so that we can use the zero-factor property.

Solution If $(4x - 1)(x + 6) = 0$ is to be a true statement, then either

$$4x - 1 = 0 \quad \text{or} \quad x + 6 = 0$$

Now we solve each of these linear equations.

$$4x - 1 = 0$$
$$4x = 1 \quad \text{Add 1 to both sides.}$$
$$x = \frac{1}{4} \quad \text{Divide both sides by 4.}$$

or

$$x + 6 = 0$$
$$x = -6 \quad \text{Subtract 6 from both sides.}$$

The Language of Algebra
In the zero-factor property, the word *or* means one or the other or both. If the product of two numbers is 0, then one factor is 0, or the other factor is 0, or both factors can be 0.

The results must be checked separately to see whether each of them produces a true statement. We substitute $\frac{1}{4}$ and then -6 for x in the original equation and evaluate the left side.

Check $\frac{1}{4}$**:**

$$(4x - 1)(x + 6) = 0$$
$$\left[4\left(\frac{1}{4}\right) - 1\right]\left(\frac{1}{4} + 6\right) \stackrel{?}{=} 0$$
$$(1 - 1)\left(\frac{25}{4}\right) \stackrel{?}{=} 0$$
$$0\left(\frac{25}{4}\right) \stackrel{?}{=} 0$$
$$0 = 0 \quad \text{True}$$

Check **−6:**

$$(4x - 1)(x + 6) = 0$$
$$[4(-6) - 1](-6 + 6) \stackrel{?}{=} 0$$
$$(-24 - 1)(0) \stackrel{?}{=} 0$$
$$-25(0) \stackrel{?}{=} 0$$
$$0 = 0 \quad \text{True}$$

The resulting true statements indicate that $(4x - 1)(x + 6) = 0$ has two solutions: $\frac{1}{4}$ and -6. Recall from Chapter 2 that the *solution set* of an equation is the set of all numbers that make the equation true. Thus, the solution set is $\{-6, \frac{1}{4}\}$.

Self Check 1 Solve: $(x - 12)(5x + 6) = 0$

Now Try **Problem 15**

In Example 1, the left side of $(4x - 1)(x + 6) = 0$ is in factored form and the right side is 0, so we can immediately use the zero-factor property. However, to solve many quadratic equations, we must factor before using the zero-factor property.

EXAMPLE 2 Solve: $x^2 - 2x - 63 = 0$

Strategy We will factor the trinomial on the left side of the equation and use the zero-factor property.

Why To use the zero-factor property, we need one side of the equation to be factored completely and the other side to be 0.

Solution

$x^2 - 2x - 63 = 0$		This is the equation to solve.
$(x + 7)(x - 9) = 0$		Factor the trinomial, $x^2 - 2x - 63$.
$x + 7 = 0$ or	$x - 9 = 0$	Set each factor equal to 0.
$x = -7$	$x = 9$	Solve each linear equation.

To check the results, we substitute -7 and then 9 for x in the original equation and evaluate the left side.

Check -7:

$$x^2 - 2x - 63 = 0$$
$$(-7)^2 - 2(-7) - 63 \stackrel{?}{=} 0$$
$$49 - (-14) - 63 \stackrel{?}{=} 0$$
$$63 - 63 \stackrel{?}{=} 0$$
$$0 = 0 \quad \text{True}$$

Check 9:

$$x^2 - 2x - 63 = 0$$
$$(9)^2 - 2(9) - 63 \stackrel{?}{=} 0$$
$$81 - 18 - 63 \stackrel{?}{=} 0$$
$$63 - 63 \stackrel{?}{=} 0$$
$$0 = 0 \quad \text{True}$$

The solutions of $x^2 - 2x - 63 = 0$ are -7 and 9, and the solution set is $\{-7, 9\}$.

Self Check 2 Solve: $x^2 + 5x + 6 = 0$

Now Try **Problem 27**

Success Tip

When you see the word *solve* in this example, you probably think of steps from Chapter 2 such as combining like terms, distributing, or doing something to both sides. However, to solve this quadratic equation, we begin by factoring $x^2 - 2x - 63$.

The previous examples suggest the following strategy to solve quadratic equations by factoring.

The Factoring Method for Solving a Quadratic Equation

1. Write the equation in standard form: $ax^2 + bx + c = 0$ or $0 = ax^2 + bx + c$.
2. Factor completely.
3. Use the zero-factor property to set each factor equal to 0.
4. Solve each resulting linear equation.
5. Check the results in the original equation.

With this method, we factor *expressions* to solve *equations.*

Notation
Although $x^2 - 25 = 0$ is missing a term involving x, it is a quadratic equation in standard form $ax^2 + bx + c = 0$, where $a = 1$, $b = 0$, and $c = -25$.

EXAMPLE 3 Solve: $x^2 - 25 = 0$

Strategy We will factor the binomial on the left side of the equation and use the zero-factor property.

Why To use the zero-factor property, we need one side of the equation to be factored completely and the other side to be 0.

Solution We factor the difference of two squares on the left side of the equation and proceed as follows.

$x^2 - 25 = 0$ — This is the equation to solve.

$(x + 5)(x - 5) = 0$ — Factor the difference of two squares, $x^2 - 25$.

$x + 5 = 0$ or $x - 5 = 0$ — Set each factor equal to 0.

$x = -5$ | $x = 5$ — Solve each linear equation.

Check each result by substituting it into the original equation.

Check -5:	Check 5:
$x^2 - 25 = 0$	$x^2 - 25 = 0$
$(-5)^2 - 25 \stackrel{?}{=} 0$	$5^2 - 25 \stackrel{?}{=} 0$
$25 - 25 \stackrel{?}{=} 0$	$25 - 25 \stackrel{?}{=} 0$
$0 = 0$ True	$0 = 0$ True

The solutions of $x^2 - 25 = 0$ are -5 and 5, and the solution set is $\{-5, 5\}$.

Self Check 3 Solve: $x^2 - 49 = 0$

Now Try Problem 35

EXAMPLE 4 Solve: $6x^2 = 12x$

Strategy We will subtract $12x$ from both sides of the equation to get 0 on the right side. Then we will factor the resulting binomial and use the zero-factor property.

Why To use the zero-factor property, we need one side of the equation to be factored completely and the other side to be 0.

Solution The equation is not in standard form, $ax^2 + bx + c = 0$. To get 0 on the right side, we proceed as follows.

> **Caution**
>
> A creative, but incorrect, approach to solve $6x^2 = 12x$ is to divide both sides by $6x$.
>
> $$\frac{6x^2}{6x} = \frac{12x}{6x}$$
>
> You will obtain $x = 2$; however, you will lose the second solution, 0.

$6x^2 = 12x$	This is the equation to solve.
$6x^2 - 12x = 12x - 12x$	Use the subtraction property of equality to get 0 on the right side: Subtract 12x from both sides.
$6x^2 - 12x = 0$	Combine like terms: 12x − 12x = 0. This equation is in standard form.

To solve this equation, we factor the left side and proceed as follows.

$6x(x - 2) = 0$	Factor out the GCF, 6x.
$6x = 0$ or $x - 2 = 0$	Set each factor equal to 0.
$x = \frac{0}{6}$, $x = 0$ | $x = 2$	Solve each equation.

The solutions are 0 and 2 and the solution set is $\{0, 2\}$. Check each solution in the original equation, $6x^2 = 12x$.

Self Check 4 Solve: $5x^2 = 25x$

Now Try **Problem 47**

EXAMPLE 5 Solve: $2x^2 - 2 = -3x$

Strategy We will add $3x$ to both sides of the equation to get 0 on the right side. Then we will factor the resulting trinomial and use the zero-factor property.

Why To use the zero-factor property, we need one side of the equation to be factored completely and the other side to be 0.

Solution The equation is not in standard form, $ax^2 + bx + c = 0$. To get 0 on the right side, we proceed as follows.

$2x^2 - 2 = -3x$	This is the equation to solve.
$2x^2 + 3x - 2 = -3x + 3x$	Use the addition property of equality to get 0 on the right side: Add 3x to both sides.
$2x^2 + 3x - 2 = 0$	Combine like terms: −3x + 3x = 0. This equation is in standard form.
$(2x - 1)(x + 2) = 0$	Factor the trinomial.
$2x - 1 = 0$ or $x + 2 = 0$	Set each factor equal to 0.
$2x = 1$, $x = \frac{1}{2}$ | $x = -2$	Solve each equation.

The solutions are $\frac{1}{2}$ and -2 and the solution set is $\left\{-2, \frac{1}{2}\right\}$. Check each solution in the original equation, $2x^2 - 2 = -3x$.

Self Check 5 Solve: $3x^2 - 8 = -10x$

Now Try **Problem 51**

Unlike linear equations, quadratic equations have two solutions. In some cases, however, the two solutions are the same number.

EXAMPLE 6 Solve: $x(9x - 12) = -4$

Strategy To write the equation in standard form, we will distribute the multiplication by x and add 4 to both sides. Then we will factor the resulting trinomial and use the zero-factor property.

Why To use the zero-factor property, we need one side of the equation to be factored completely and the other side to be 0.

Caution
To use the zero-factor property, one side of the equation must be 0. In this example, it would be incorrect to set each factor equal to -4.

~~$x = -4$ or $9x - 12 = -4$~~

If the product of two numbers is -4, one of them does not have to be -4. For example, $2(-2) = -4$.

Solution

$$x(9x - 12) = -4 \quad \text{This is the equation to solve.}$$
$$9x^2 - 12x = -4 \quad \text{Distribute the multiplication by } x.$$
$$9x^2 - 12x + 4 = -4 + 4 \quad \text{To get 0 on the right side, add 4 to both sides.}$$
$$9x^2 - 12x + 4 = 0 \quad \text{Combine like terms: } -4 + 4 = 0. \text{ This equation is in standard form.}$$
$$(3x - 2)(3x - 2) = 0 \quad \text{Factor the trinomial, } 9x^2 - 12x + 4.$$
$$3x - 2 = 0 \quad \text{or} \quad 3x - 2 = 0 \quad \text{Set each factor equal to 0.}$$
$$3x = 2 \quad\quad 3x = 2 \quad \text{Solve each equation.}$$
$$x = \frac{2}{3} \quad\quad x = \frac{2}{3}$$

After solving both equations, we see that $\frac{2}{3}$ is a *repeated solution*. Thus, the solution set is $\left\{\frac{2}{3}\right\}$. Check this result by substituting it into the original equation.

Self Check 6 Solve: $x(4x + 12) = -9$

Now Try **Problem 59**

3 Solve Third-Degree Equations by Factoring.

Some equations involving polynomials with degrees higher than 2 can also be solved by using the factoring method. In such cases, we use an extension of the zero-factor property: When the product of two *or more* real numbers is 0, at least one of them is 0.

EXAMPLE 7 Solve: $6x^3 + 12x = 17x^2$

Strategy This equation is not quadratic, because it contains a term involving x^3. However, we can solve it by using factoring. First we get 0 on the right side by subtracting $17x^2$ from both sides. Then we factor the polynomial on the left side and use an extension of the zero-factor property.

Why To use the zero-factor property, we need one side of the equation to be factored completely and the other side to be 0.

The Language of Algebra
Since the highest degree of any term in $6x^3 + 12x = 17x^2$ is 3, it is called a *third-degree* equation. Note that it has three solutions.

Solution

$$6x^3 + 12x = 17x^2$$ This is the equation to solve.

$$6x^3 - 17x^2 + 12x = 17x^2 - 17x^2$$ To get 0 on the right side, subtract $17x^2$ from both sides.

$$6x^3 - 17x^2 + 12x = 0$$ Combine like terms: $17x^2 - 17x^2 = 0$.

$$x(6x^2 - 17x + 12) = 0$$ Factor out the GCF, x.

$$x(2x - 3)(3x - 4) = 0$$ Factor the trinomial, $6x^2 - 17x + 12$.

If $x(2x - 3)(3x - 4) = 0$, then at least one of the factors is equal to 0.

$x = 0$ or $2x - 3 = 0$ or $3x - 4 = 0$ Set each factor equal to 0.

$2x = 3$ $\quad$ $3x = 4$ Solve each equation.

$x = \frac{3}{2}$ $\quad$ $x = \frac{4}{3}$

The solutions are $0, \frac{3}{2}$, and $\frac{4}{3}$ and the solution set is $\{0, \frac{4}{3}, \frac{3}{2}\}$. Check each solution in the original equation, $6x^3 + 12x = 17x^2$.

Self Check 7 Solve: $10x^3 + x^2 = 2x$

Now Try Problem 67

ANSWERS TO SELF CHECKS **1.** $12, -\frac{6}{5}$ **2.** $-2, -3$ **3.** $-7, 7$ **4.** $0, 5$ **5.** $\frac{2}{3}, -4$ **6.** $-\frac{3}{2}$ **7.** $0, \frac{2}{5}, -\frac{1}{2}$

STUDY SET 6.7

VOCABULARY

Fill in the blanks.

1. $2x^2 + 3x - 1 = 0$ and $x^2 - 36 = 0$ are examples of __________ equations.

2. $ax^2 + bx + c = 0$ is called the __________ form of a quadratic equation.

3. The __________ property states that if the product of two numbers is 0, at least one of them is 0: If $ab = 0$, then $a =$ ___ or $b =$ ___.

4. Since the highest degree of any term in $x^3 - 5x^2 - 6x = 0$ is 3, it is called a ______-degree equation.

CONCEPTS

5. Which of the following are quadratic equations?

a. $x^2 + 2x - 10 = 0$ **b.** $2x - 10 = 0$

c. $x^2 = 15x$ **d.** $x^3 + x^2 + 2x = 0$

6. Write each equation in the standard form $ax^2 + bx + c = 0$.

a. $x^2 + 2x = 6$ **b.** $x^2 = 5x$

c. $3x(x - 8) = -9$ **d.** $4x^2 = 25$

7. Set $5x + 4$ equal to 0 and solve for x.

8. What step should be performed first to solve $x^2 - 6x - 16 = 0$?

9. What step (or steps) should be performed first before factoring is used to solve each equation?

a. $x^2 + 7x = -6$

b. $x(x + 7) = 3$

10. Check to determine whether the given number is a solution of the given quadratic equation.

a. $x^2 - 4x = 0;\ 4$

b. $x^2 - 2x - 7 = 0;\ -2$

NOTATION

Complete each solution to solve the equation.

11. $(x - 1)(x + 7) = 0$

$x - 1 = \_\_\_$ or $\_\_\_ = 0$

$x = 1$ | $x = \_\_\_$

12. $7y^2 + 14y = 0$

$\_\_\_(y + 2) = 0$

$7y = 0$ $\_\_\_$ $y + 2 = 0$

$y = \_\_\_$ | $y = -2$

13. $p^2 - p - 6 = 0$

$(\_\_\_ - 3)(p + 2) = 0$

$\_\_\_ = 0$ or $p + 2 = \_\_\_$

$p = \_\_\_$ | $p = \_\_\_$

14. $4y^2 - 25 = 0$

$(2y + \_\_\_)(2y - \_\_\_) = 0$

$2y + 5 = \_\_\_$ or $\_\_\_ = 0$

$2y = \_\_\_$ | $2y = 5$

$y = -\frac{5}{2}$ | $y = \_\_\_$

GUIDED PRACTICE

***Solve each equation.* See Example 1.**

15. $(x - 3)(x - 2) = 0$

16. $(x + 2)(x + 3) = 0$

17. $(x + 7)(x - 7) = 0$

18. $(x - 8)(x + 8) = 0$

19. $6x(2x - 5) = 0$

20. $5x(5x + 7) = 0$

21. $-7a(3a + 10) = 0$

22. $-6t(2t - 9) = 0$

23. $t(t - 6)(t + 8) = 0$

24. $n(n + 1)(n - 6) = 0$

25. $(x - 1)(x + 2)(x - 3) = 0$

26. $(x + 2)(x + 3)(x - 4) = 0$

***Solve each equation.* See Example 2.**

27. $x^2 - 13x + 12 = 0$

28. $x^2 + 7x + 6 = 0$

29. $x^2 - 4x - 21 = 0$

30. $x^2 + 2x - 15 = 0$

31. $x^2 - 9x + 8 = 0$

32. $x^2 - 14x + 45 = 0$

33. $a^2 + 8a + 15 = 0$

34. $a^2 - 17a + 60 = 0$

***Solve each equation.* See Example 3.**

35. $x^2 - 81 = 0$

36. $x^2 - 36 = 0$

37. $t^2 - 25 = 0$

38. $m^2 - 49 = 0$

39. $4x^2 - 1 = 0$

40. $9y^2 - 1 = 0$

41. $9y^2 - 49 = 0$

42. $16z^2 - 25 = 0$

***Solve each equation.* See Example 4.**

43. $w^2 = 7w$

44. $x^2 = 5x$

45. $s^2 = 16s$

46. $p^2 = 20p$

47. $4y^2 = 12y$

48. $5m^2 = 15m$

49. $3x^2 = -8x$

50. $3s^2 = -4s$

***Solve each equation.* See Example 5.**

51. $3x^2 + 5x = 2$

52. $3x^2 + 14x = -8$

53. $2x^2 + x = 3$

54. $2x^2 - 5x = -2$

55. $5x^2 + 1 = 6x$

56. $6x^2 + 1 = 5x$

57. $2x^2 - 3x = 20$

58. $2x^2 - 3x = 14$

***Solve each equation.* See Example 6.**

59. $4r(r + 7) = -49$

60. $5m(5m + 8) = -16$

61. $9a(a - 3) = 3a - 25$

62. $3x(3x + 10) = 6x - 16$

63. $z(z - 7) = -12$

64. $p(p + 1) = 6$

65. $(n + 8)(n - 3) = -30$

66. $(2s + 5)(s + 1) = -1$

***Solve each equation.* See Example 7.**

67. $x^3 + 3x^2 + 2x = 0$

68. $x^3 - 7x^2 + 10x = 0$

69. $k^3 - 27k - 6k^2 = 0$

70. $j^3 - 22j - 9j^2 = 0$

71. $x^3 - 6x^2 = -9x$

72. $m^3 - 8m^2 + 16m = 0$

73. $2x^3 = 2x(x + 2)$

74. $x^3 + 7x^2 = x^2 - 9x$

TRY IT YOURSELF

Solve each equation.

75. $4x^2 = 81$

76. $9y^2 = 64$

77. $x^2 - 16x + 64 = 0$

78. $h^2 + 2h + 1 = 0$

79. $(2s - 5)(s + 6) = 0$

80. $h(3h - 4)(h + 1) = 0$

81. $3b^2 - 30b = 6b - 60$

82. $2m^2 - 8m = 2m - 12$

83. $k^3 + k^2 - 20k = 0$

84. $n^3 - 6n^2 + 8n = 0$

85. $x^2 - 100 = 0$

86. $z^2 - 25 = 0$

87. $3y^2 - 14y - 5 = 0$

88. $4y^2 - 11y - 3 = 0$

89. $(x - 2)(x^2 - 8x + 7) = 0$

90. $(x - 1)(x^2 + 5x + 6) = 0$

91. $4a^2 + 1 = 8a + 1$

92. $3b^2 - 6 = 12b - 6$

93. $2b(6b + 13) = -12$

94. $5f(5f - 16) = -15$

95. $3a^3 + 4a^2 + a = 0$

96. $10b^3 - 15b^2 - 25b = 0$

97. $-15x^2 + 2 + 7x = 0$

98. $-8x^2 + 3 - 10x = 0$

99. $4p^2 - 121 = 0$

100. $q^2 - \frac{1}{4} = 0$

101. $d(8d - 9) = -1$

102. $6n^3 - 6n = 0$

WRITING

103. Explain the zero-factor property.

104. Find the error in the following solution.

$$x(x + 1) = 6$$

$x = 6$ or $x + 1 = 6$

$x = 5$

The solutions are 6 and 5.

105. A student solved $x^2 - 5x + 6 = 0$ and obtained two solutions: 2 and 3. Explain the error in his check.

Check:

$$x^2 - 5x + 6 = 0$$

$$2^2 - 5(3) + 6 \stackrel{?}{=} 0$$

$$4 - 15 + 6 \stackrel{?}{=} 0$$

$$-5 = 0 \quad \text{False}$$

2 is not a solution. 3 is not a solution.

106. In this section, we solved quadratic equations by factoring. Did we always obtain two different solutions? Explain.

107. What is wrong with the step used to solve $x^2 = 2x$ shown below?

$$x^2 = 2x$$

$$\frac{x^2}{x} = \frac{2x}{x}$$

$$x = 2$$

The solution is 2.

108. Explain the error in the following solution.

Factor: $x^2 - 5x + 6$

$$(x - 2)(x - 3) = 0$$

$x - 2 = 0$ or $x - 3 = 0$

$x = 2$ | $x = 3$

The solutions are 2 and 3.

REVIEW

109. EXERCISE A doctor advises a patient to exercise at least 15 minutes but less than 30 minutes per day. Use a compound inequality to express the range of these times t in minutes.

110. SNACKS A bag of peanuts is worth \$0.30 less than the same size bag of cashews. Equal amounts of peanuts and cashews are used to make 40 bags of a mixture that is worth \$1.05 per bag. How much is a bag of cashews worth?

CHALLENGE PROBLEMS

Solve each equation.

111. $x^4 - 625 = 0$

112. $2a^3 + a^2 - 32a - 16 = 0$

113. $(x - 3)^2 = 2x + 9$

114. $(x + 3)^2 = (2x - 1)^2$

SECTION 6.8
Applications of Quadratic Equations

Objectives

1. Solve problems involving geometric figures.
2. Solve problems involving consecutive integers.
3. Solve problems using the Pythagorean theorem.
4. Solve problems given the quadratic equation model.

In Chapter 2, we solved mixture, investment, and uniform motion problems. To model those situations, we used *linear equations* in one variable. We will now consider situations that are modeled by *quadratic equations*.

1 Solve Problems Involving Geometric Figures.

We can use the five-step problem solving strategy and the factoring method for solving quadratic equations to find the dimensions of certain figures, given their area.

EXAMPLE 1 ***Painting.*** In 2002, the pop art painting *American Sweetheart,* by artist Robert Indiana, sold for $614,500. The area of the rectangular painting is 32 square feet. Find the dimensions of the painting if it is twice as long as it is wide.

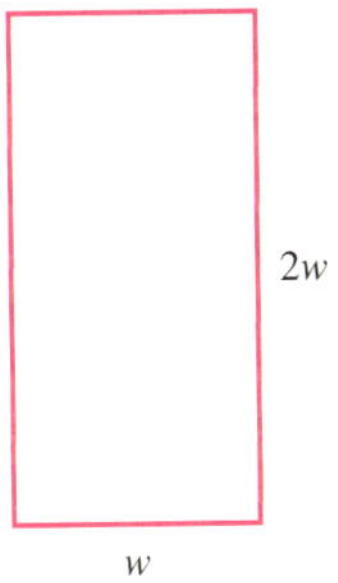

Analyze the Problem

- The area of the painting is 32 ft^2.
- The length is twice as long as the width.
- Find the length and width.

Form an Equation

Since the length is related to the width, let w = the width of the painting in feet. Then $2w$ = the length of the painting. To form an equation, we use the formula for the area of a rectangle, $A = lw$, where $A = 32$.

The area of the rectangle	equals	the length	times	the width.
32	=	$2w$	$\cdot$	w

Solve the Equation

$$32 = 2w \cdot w$$

$$32 = 2w^2$$ Multiply $2w$ and w. Note that this is a quadratic equation.

$$0 = 2w^2 - 32$$ To get 0 on the left side, subtract 32 from both sides.

$$0 = 2(w^2 - 16)$$ Factor out the GCF, 2.

$$0 = 2(w + 4)(w - 4)$$ Factor the difference of two squares, $w^2 - 16$.

$$w + 4 = 0 \quad \text{or} \quad w - 4 = 0$$ Since 2 cannot equal 0, discard that possibility. Set each factor that contains a variable equal to 0.

$$\cancel{w = -4} \quad\quad w = 4$$ Solve each equation.

The Language of Algebra
When solving real-world application problems, we *discard* or *reject* any solutions of equations that do not make sense, such as a negative width of a painting.

State the Conclusion The solutions of the equation are −4 and 4. Since w represents the width of the picture, and the width cannot be negative, we discard −4. Thus, the width of the picture is 4 feet and the length is $2 \cdot 4 = 8$ feet.

Check the Result A rectangle with dimensions 4 feet by 8 feet has an area of 32 ft^2, and the length is twice the width. The answers check.

Now Try **Problem 13**

EXAMPLE 2 ***Windmills.*** The height of a triangular canvas sail of a windmill is 1 foot less than twice the length of its base. If the sail has an area of 22.5 ft^2, find the length of the base and the height.

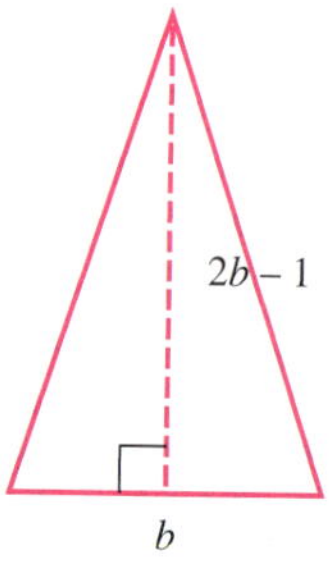

Analyze the Problem

- The height is 1 ft less than twice the length of the base.
- The area is 22.5 ft^2.
- Find the length of the base and the height.

Form an Equation Since the height is related to the length of the base, we let b = the length of the base of the sail in feet. Then $2b - 1$ = the height of the sail. To form an equation, we use the formula for the area of a triangle: $A = \frac{1}{2}bh$, where $A = 22.5$.

The area of the triangle	equals	one-half	times	the length of the base	times	the height.
22.5	=	$\frac{1}{2}$	$\cdot$	b	$\cdot$	$(2b - 1)$

Solve the Equation

$$22.5 = \frac{1}{2}b(2b - 1)$$

$$2 \cdot 22.5 = 2 \cdot \frac{1}{2}b(2b - 1)$$ To clear the equation of the fraction, multiply both sides by 2.

$$45 = b(2b - 1)$$ Multiply: $2 \cdot 22.5 = 45$ and $2 \cdot \frac{1}{2} = 1$.

$$45 = 2b^2 - b$$ Distribute the multiplication by b. Note that this is a quadratic equation.

$$0 = 2b^2 - b - 45$$ To get 0 on the left side, subtract 45 from both sides.

$$0 = (2b + 9)(b - 5)$$ Factor the trinomial.

$$2b + 9 = 0 \quad \text{or} \quad b - 5 = 0$$ Set each factor equal to 0.

$$2b = -9 \qquad b = 5$$ Solve each equation.

$$b = -\frac{9}{2} \text{ (crossed out)}$$

> **Caution**
> A common error is to incorrectly "distribute" the 2 on the right side of the equation.
>
> $2 \cdot 22.5 = 2 \cdot \frac{1}{2}b(2b - 1)$ (crossed out)

State the Conclusion The solutions of the equation are $-\frac{9}{2}$ and 5. Since b represents the length of the base of the sail, and it cannot be negative, we discard $-\frac{9}{2}$. The length of the base is then 5 feet, and the height is $2(5) - 1 = 9$ feet.

Check the Result A triangle with height 9 feet and base 5 feet has area $\frac{1}{2}(9)(5) = 22.5 \text{ ft}^2$, and the height is 1 foot less than twice the base. The answers check.

Now Try Problem 19

2 Solve Problems Involving Consecutive Integers.

Consecutive integers are integers that follow one another, such as 15 and 16. When solving consecutive integer problems, if we let x = the first integer, then:

- two consecutive integers are x and $x + 1$
- two consecutive even integers are x and $x + 2$
- two consecutive odd integers are x and $x + 2$

EXAMPLE 3 ***Women's Tennis.*** In the 1998 Australian Open, sisters Venus and Serena Williams played against each other for the first time as professionals. Venus was victorious over her younger sister. At that time, their ages were consecutive integers whose product was 272. How old were Venus and Serena when they met in this match?

Analyze the Problem

- Venus is older than Serena.
- Their ages were consecutive integers.
- The product of their ages was 272.
- Find Venus' and Serena's age when they played this match.

Form an Equation Let x = Serena's age when she played in the 1998 Australian Open. Since their ages were consecutive integers, and since Venus is older, we let $x + 1$ = Venus' age. The word *product* indicates multiplication.

Serena's age	times	Venus' age	was	272.
x	$\cdot$	$(x + 1)$	$=$	272

Solve the Equation

$$x(x + 1) = 272$$

$$x^2 + x = 272$$ Distribute the multiplication by x. Note that this is a quadratic equation.

$$x^2 + x - 272 = 0$$ Subtract 272 from both sides to make the right side 0.

$$(x + 17)(x - 16) = 0$$ Factor $x^2 + x - 272$. Two numbers whose product is -272 and whose sum is 1 are 17 and -16.

$$x + 17 = 0 \quad \text{or} \quad x - 16 = 0$$ Set each factor equal to 0.

$$x = -17 \quad | \quad x = 16$$ Solve each equation.

State the Conclusion The solutions of the equation are -17 and 16. Since x represents Serena's age, and it cannot be negative, we discard -17. Thus, Serena Williams was 16 years old and Venus Williams was $16 + 1 = 17$ years old when they played against each other for the first time as professionals.

Success Tip
The prime factorization of 272 is helpful in determining that $272 = 17 \cdot 16$.

$$\begin{array}{r|l} 2 & 272 \\ 2 & 136 \\ 2 & 68 \\ 2 & 34 \\ & 17 \end{array}$$

16 (the product of the four factors of 2)

Check the Result Since 16 and 17 are consecutive integers, and since $16 \cdot 17 = 272$, the answers check.

Now Try Problem 23

3 Solve Problems Using the Pythagorean Theorem.

Caution
It doesn't matter which leg of the right triangle is labeled a and which is labeled b. However, the hypotenuse must be labeled c.

A **right triangle** is a triangle that has a 90° (right) angle. The longest side of a right triangle is the **hypotenuse,** which is the side opposite the right angle. The remaining two sides are the **legs** of the triangle. The **Pythagorean theorem** provides a formula relating the lengths of the three sides of a right triangle.

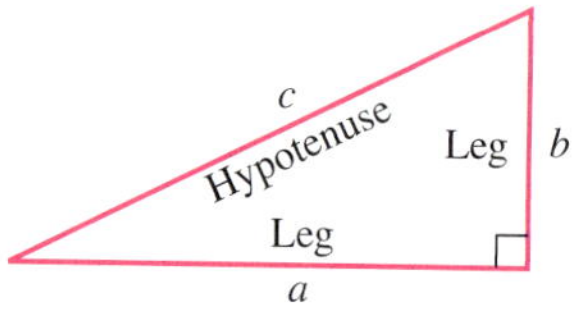

The Pythagorean Theorem

If a and b are the lengths of the legs of a right triangle and c is the length of the hypotenuse, then

$$a^2 + b^2 = c^2$$

In a right triangle, the sum of the squares of the lengths of the two legs is equal to the square of the length of the hypotenuse.

Pythagoras

The Language of Algebra
A *theorem* is a mathematical statement that can be proved. The *Pythagorean theorem* is named after *Pythagoras,* a Greek mathematician who lived about 2,500 years ago. He is thought to have been the first to prove the theorem.

EXAMPLE 4 ***Right Triangles.*** The longer leg of a right triangle is 3 units longer than the shorter leg. If the hypotenuse is 6 units longer than the shorter leg, find the lengths of the sides of the triangle.

Analyze the Problem We begin by drawing a right triangle and labeling the legs and the hypotenuse.

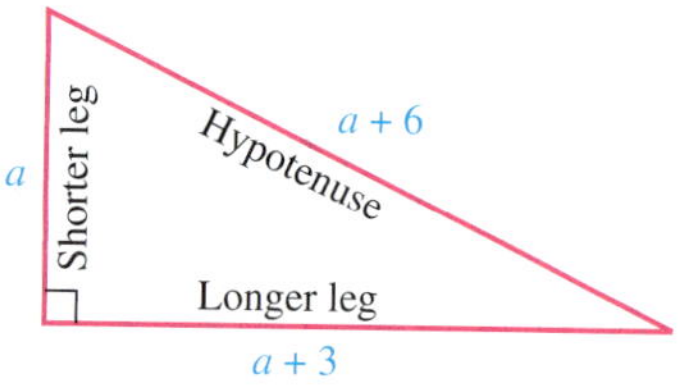

Form an Equation Let a = length of the shorter leg. Then the length of the hypotenuse is $a + 6$ and the length of the longer leg is $a + 3$. By the Pythagorean theorem, we have

(The length of the shorter leg)^2	plus	(the length of the longer leg)^2	equals	(the length of the hypotenuse)^2
a^2	$+$	$(a + 3)^2$	$=$	$(a + 6)^2$

Solve the Equation

$$a^2 + (a + 3)^2 = (a + 6)^2$$

$$a^2 + a^2 + 6a + 9 = a^2 + 12a + 36 \quad \text{Find } (a + 3)^2 \text{ and } (a + 6)^2.$$

$$2a^2 + 6a + 9 = a^2 + 12a + 36 \quad \text{On the left side: } a^2 + a^2 = 2a^2.$$

$$a^2 - 6a - 27 = 0 \quad \text{To get 0 on the right side, subtract } a^2, 12a, \text{ and 36 from both sides. This is a quadratic equation.}$$

$$(a - 9)(a + 3) = 0 \quad \text{Factor the trinomial.}$$

$$a - 9 = 0 \quad \text{or} \quad a + 3 = 0 \quad \text{Set each factor equal to 0.}$$

$$a = 9 \quad \Big| \quad a = -3 \quad \text{Solve each equation.}$$

State the Conclusion Since a side cannot have a negative length, we discard the solution −3. Thus, the shorter leg is 9 units long, the hypotenuse is 9 + 6 = 15 units long, and the longer leg is 9 + 3 = 12 units long.

Check the Result The longer leg, 12, is 3 units longer than the shorter leg, 9. The hypotenuse, 15, is 6 units longer than the shorter leg, 9, and the side lengths satisfy the Pythagorean theorem. So the results check.

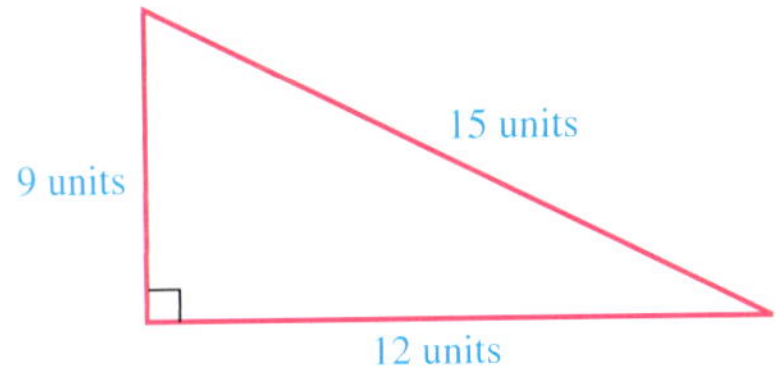

$$9^2 + 12^2 \stackrel{?}{=} 15^2$$
$$81 + 144 \stackrel{?}{=} 225$$
$$225 = 225$$

Now Try Problem 31

4 Solve Problems Given the Quadratic Equation Model.

A quadratic equation can be used to describe the height of an object that is projected upward, such as a ball thrown into the air or an arrow shot into the sky.

EXAMPLE 5 *College Pranks.* A student uses rubber tubing to launch a water balloon from the roof of his dormitory. The height h (in feet) of the balloon, t seconds after being launched, is given by the formula $h = -16t^2 + 48t + 64$. After how many seconds will the balloon hit the ground?

Analyze the Problem When the water balloon hits the ground, its height will be 0 feet. To find the time that it takes for the balloon to hit the ground, we set h equal to 0, and solve the quadratic equation for t.

Form an Equation

$$h = -16t^2 + 48t + 64$$
$$0 = -16t^2 + 48t + 64$$ Substitute 0 for the height, h. This is a quadratic equation.

Solve the Equation

$$0 = -16t^2 + 48t + 64$$
$$0 = -16(t^2 - 3t - 4)$$ Factor out the opposite of the GCF, −16.
$$0 = -16(t + 1)(t - 4)$$ Factor the trinomial.
$$t + 1 = 0 \quad \text{or} \quad t - 4 = 0$$ Since −16 cannot equal 0, discard that possibility. Set each factor that contains a variable equal to 0.
$$t = -1 \qquad t = 4$$ Solve each equation.

Success Tip
Note that the common factor, −16, divides −16, 48, and 64 exactly:

$$\frac{-16}{-16} = 1 \qquad \frac{48}{-16} = -3 \qquad \frac{64}{-16} = -4$$

State the Conclusion The equation has two solutions, -1 and 4. Since t represents time, and, in this case, time cannot be negative, we discard -1. The second solution, 4, indicates that the balloon hits the ground 4 seconds after being launched.

Check the Result Check this result by substituting 4 for t in $h = -16t^2 + 48t + 64$. You should get $h = 0$.

Now Try Problem 37

STUDY SET 6.8

VOCABULARY

Fill in the blanks.

1. Integers that follow one another, such as 6 and 7, are called ____________ integers.
2. A ______ triangle is a triangle that has a 90° angle.
3. The longest side of a right triangle is the ____________. The remaining two sides are the _____ of the triangle.
4. The ____________ theorem is a formula that relates the lengths of the three sides of a right triangle.

CONCEPTS

5. A rectangle has an area of 40 in.^{2}. The length is 3 inches longer than the width. Which rectangle below meets these conditions?

i.
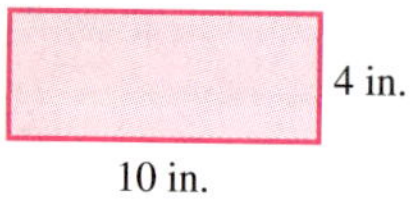

ii.
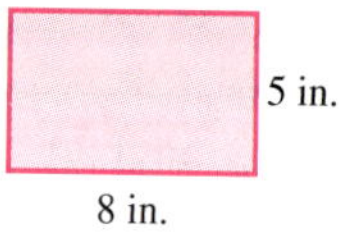

6. A triangle has an area of 15 ft^2. The height is 7 feet less than twice the length of the base. Which triangle below meets these conditions?

i.
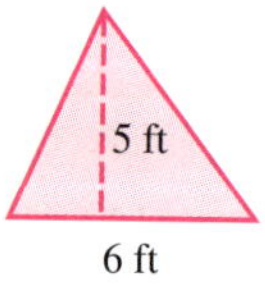

ii.
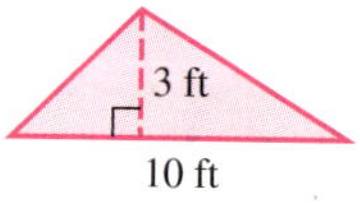

7. Multiply both sides of the equation by 2. ***Do not solve.***

$$10 = \frac{1}{2}b(b + 5)$$

8. Fill in the blanks.
   a. If the length of the hypotenuse of a right triangle is c and the lengths of the other two legs are a and b, then ______ $= c^2$.
   b. In a right triangle, the sum of the ________ of the lengths of the two legs is equal to the square of the length of the ____________.

9. **a.** What kind of triangle is shown?
   b. What are the lengths of the legs of the triangle?
   c. How much longer is the hypotenuse than the shorter leg?

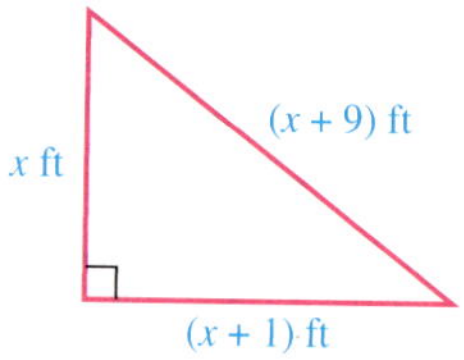

10. A ball is thrown into the air. Its height h in feet, t seconds after being released, is given by the formula $h = -16t^2 + 24t + 6$. When the ball hits the ground, what is the value of h?

NOTATION

Complete the solution to solve the equation.

11. $0 = -16t^2 + 32t + 48$

$0 = \_\_\_\_(t^2 - 2t - 3)$

$0 = -16(t - 3)(t + \_\_\_\_)$

$t - 3 = \_\_\_\_$ or $t + 1 = \_\_\_\_$

$t = \_\_\_\_$ | $t = \_\_\_\_$

12. Fill in the blanks.
   a. Consecutive integers can be represented by x and ______.
   b. Consecutive odd integers can be represented by x and ______.
   c. Consecutive even integers can be represented by x and ______.

APPLICATIONS

Geometry Problems

13. FLAGS The length of the flag of Australia is twice as long as the width. Find the dimensions of an Australian flag if its area is 18 ft^2.

14. BILLIARDS Pool tables are rectangular, and their length is twice the width. Find the dimensions of a pool table if it occupies 50 ft^2 of floor space.

15. *X*-RAYS. A rectangular-shaped x-ray film has an area of 80 square inches. The length is 2 inches longer than the width. Find its width and length.

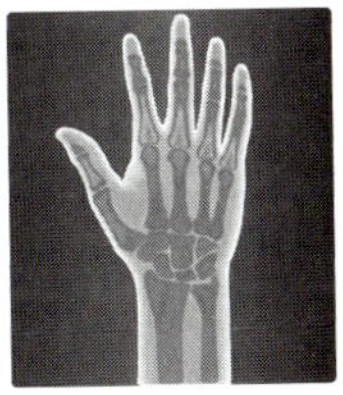

16. INSULATION The area of the rectangular slab of foam insulation in the illustration is 36 square meters. Find the dimensions of the slab.

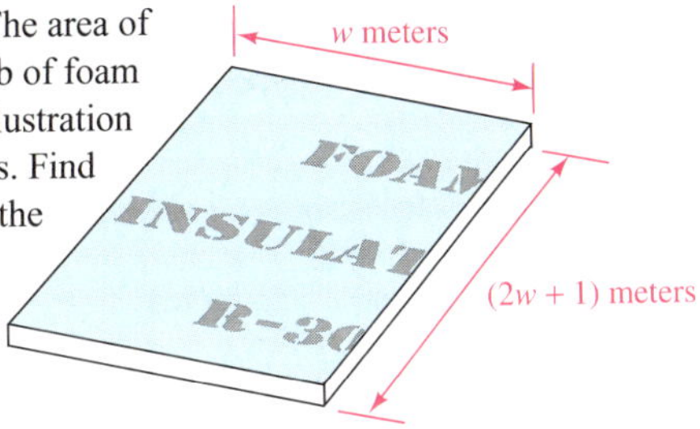

17. ***from Campus to Careers***
Bulletin Boards

Suppose you are an elementary school teacher. You want to order a rectangular bulletin board to mount on a classroom wall that has an area of 90 square feet. Fire code requirements allow for no more than 30% of a classroom wall to be covered by a bulletin board. If the length of the board to be three times as long as the width, what are the dimensions of the largest bulletin board that meets fire code?

© BananaStock/Jupiterimages

18. TUBING Refer to the diagram in the next column. A piece of cardboard in the shape of a parallelogram is twisted to form the tube. The parallelogram has an area of 60 square inches. If its height h is 7 inches more than the length of the base b, what is the length of the base? (*Hint:* The formula for the area of a parallelogram is $A = bh$.)

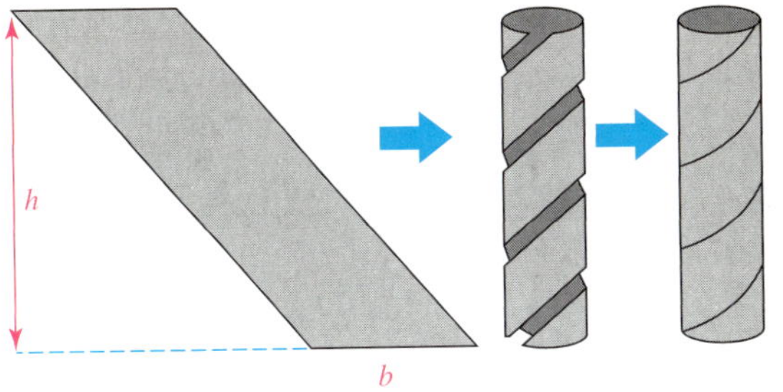

19. JEANS The height of the triangular-shaped logo on a pair of jeans is 1 centimeter less than the length of its base. If the area of the logo is 15 square centimeters, find the length of the base and the height.

20. SHUFFLEBOARD The area of the numbered triangle on a shuffle board court is 27 ft^2. Its height is 3 feet more than the length of the base. Find the length of the base and the height.

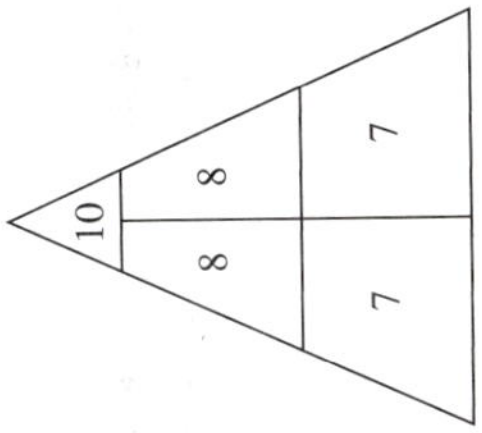

21. SAILBOATS Refer to the diagram of a sail shown here. The length of the *luff* is 3 times longer than the length of the *foot* of the sail. Find the length of the foot and the length of the luff.

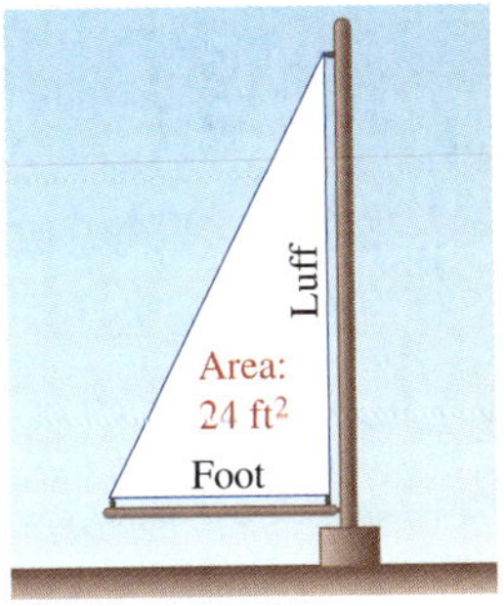

22. DESIGNING TENTS The length of the base of the triangular sheet of canvas above the door of a tent is 2 feet more than twice its height. The area is 30 square feet. Find the height and the length of the base of the triangle.

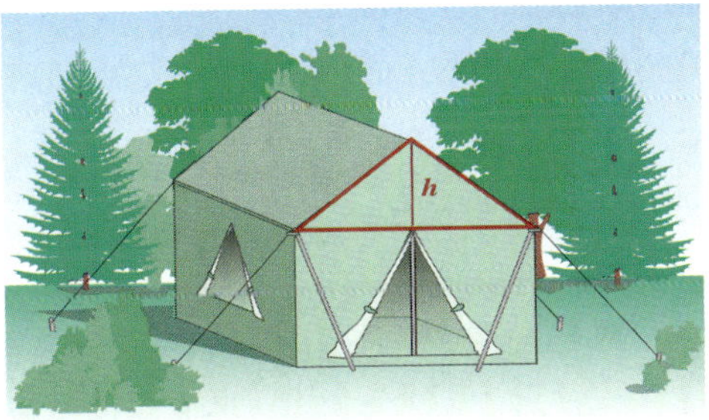

Consecutive Integer Problems

23. NASCAR The car numbers of drivers Kasey Kahne and Scott Riggs are consecutive positive integers whose product is 90. If Kahne's car number is the smaller, what is the number of each car.

24. BASEBALL Catcher Thurman Munson and pitcher Whitey Ford are two of the sixteen New York Yankees who have had their uniform numbers retired. Their uniform numbers are consecutive integers whose product is 240. If Munson's was the smaller number, determine the uniform number of each player.

25. CUSTOMER SERVICE At a pharmacy, customers take a ticket to reserve their turn for service. If the product of the ticket number now being served and the next ticket number to be served is 156, what number is now being served?

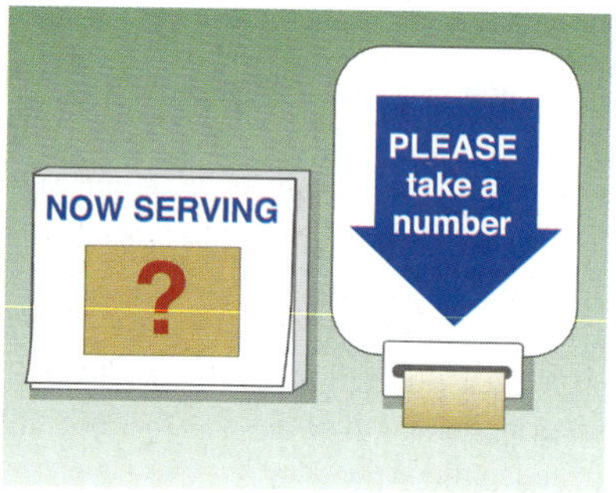

26. HISTORY Delaware was the first state to enter the Union and Hawaii was the 50th. If we order the positions of entry for the rest of the states, we find that Kentucky entered the Union right after Vermont, and the product of their order-of-entry numbers is 210. Use the given information to complete these statements:

Kentucky was the ____ th state to enter the Union.

Vermont was the ____ th state to enter the Union.

27. PLOTTING POINTS The x-coordinate and y-coordinate of a point in quadrant I are consecutive odd integers whose product is 143. Find the coordinates of the point.

28. PRESIDENTS George Washington was born on 2-22-1732 (February 22, 1732). He died in 1799 at the age of 67. The month in which he died and the day of the month on which he died are consecutive even integers whose product is 168. When did Washington die?

Pythagorean Theorem Problems

29. HIGH-ROPES ADVENTURES COURSES A builder of a high-ropes adventure course wants to secure a pole by attaching a support cable from the anchor stake 8 yards from its base to a point 6 yards up the pole. How long should the cable be?

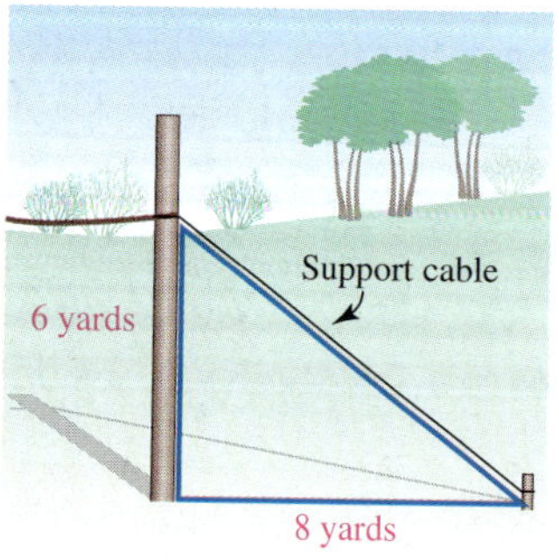

30. WIND DAMAGE A tree was blown over in a wind storm. Find x. Then find the height of the tree when it was standing upright.

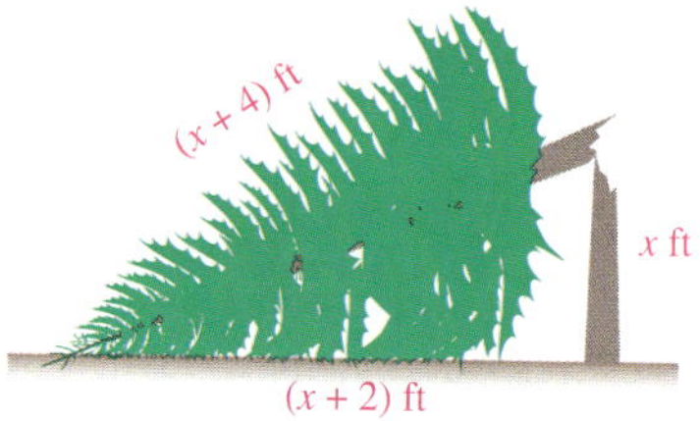

31. MOTO X Find x, the height of the landing ramp.

32. GARDENING TOOLS The dimensions (in millimeters) of the teeth of a pruning saw blade are given in the illustration. Find each length.

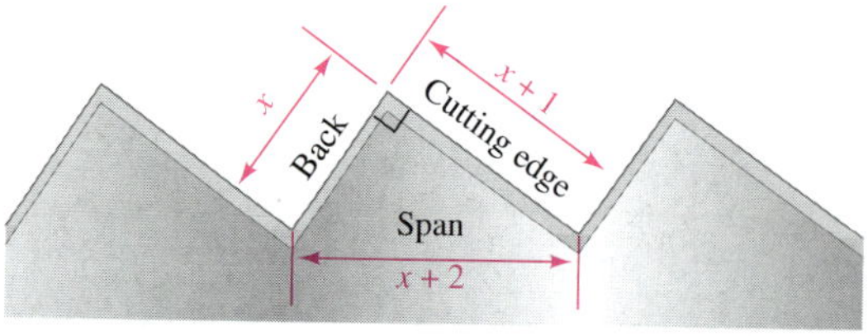

33. BOATING The inclined ramp of the boat launch is 8 meters longer than the rise of the ramp. The run is 7 meters longer than the rise. How long are the three sides of the ramp?

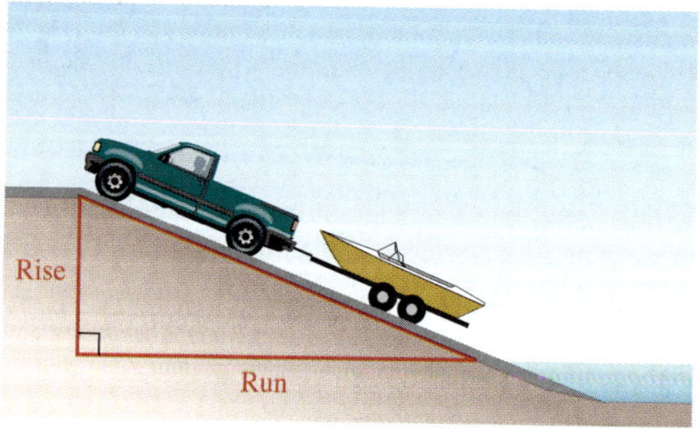

34. CAR REPAIRS To create some space to work under the front end of a car, a mechanic drives it up steel ramps. A ramp is 1 foot longer than the back, and the base is 2 feet longer than the back of the ramp. Find the length of each side of the ramp.

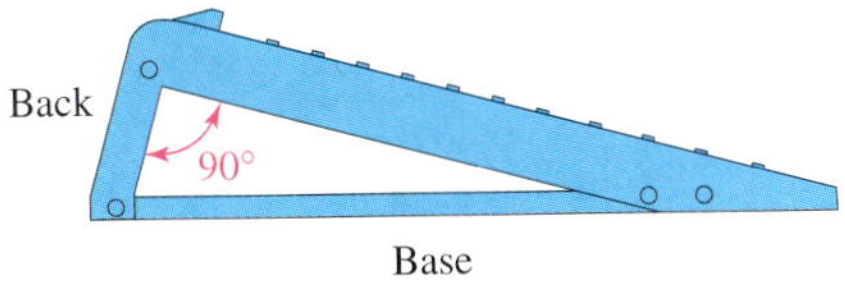

Quadratic Equation Model Problems

35. THRILL RIDES At the peak of a roller coaster ride, a rider's sunglasses fly off his head. The height h (in feet) of the glasses, t seconds after he loses them, is given by $h = -16t^2 + 64t + 80$. After how many seconds will the glasses hit the ground? (*Hint:* Factor out -16.)

36. PARADES A celebrity on the top of a parade float is tossing pieces of candy to the people on the street below. The height h (in feet) of a piece of candy, t seconds after being thrown, is given by $h = -16t^2 + 16t + 32$. After how many seconds will the candy hit the ground? (*Hint:* Factor out -16.)

37. SOFTBALL A pitcher can throw a fastball underhand at 63 feet per second (about 45 mph). If she throws a ball into the air with that velocity, its height h in feet, t seconds after being released, is given by $h = -16t^2 + 63t + 4$. After the ball is thrown, in how many seconds will it hit the ground? (*Hint:* Factor out -16.)

38. OFFICIATING Before a football game, a coin toss is used to determine which team will kick off. The height h (in feet) of a coin above the ground t seconds after being flipped up into the air is given by $h = -16t^2 + 22t + 3$. How long does a team captain have to call heads or tails if it must be done while the coin is in the air? (*Hint:* Factor out -1.)

39. DOLPHINS Refer to the illustration. The height h in feet reached by a dolphin t seconds after breaking the surface of the water is given by $h = -16t^2 + 32t$. How long will it take the dolphin to jump out of the water and touch the trainer's hand?

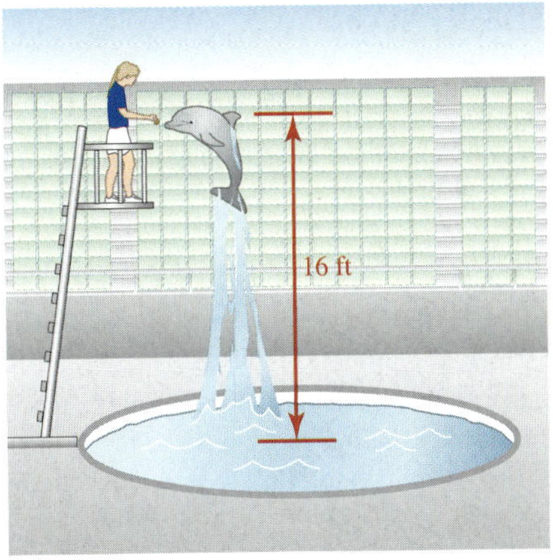

40. EXHIBITION DIVING In Acapulco, Mexico, men diving from a cliff to the water 64 feet below are quite a tourist attraction. A diver's height h above the water (in feet), t seconds after diving, is given by $h = -16t^2 + 64$. How long does a dive last?

41. CHOREOGRAPHY For the finale of a musical, 36 dancers are to assemble in a triangular-shaped series of rows, where each row has one more dancer than the previous row. The illustration shows the beginning of such a formation. The relationship between the number of rows r and the number of dancers d is given by $d = \frac{1}{2}r(r + 1)$. Determine the number of rows in the formation.

42. CRAFTS The illustration shows how a wall hanging can be created by stretching yarn from peg to peg across a wooden ring. The relationship between the number of pegs p placed evenly around the ring and the number of yarn segments s that criss-cross the ring is given by the formula $s = \frac{p(p - 3)}{2}$. How many pegs are needed if the designer wants 27 segments to criss-cross the ring? (*Hint:* Multiply both sides of the equation by 2.)

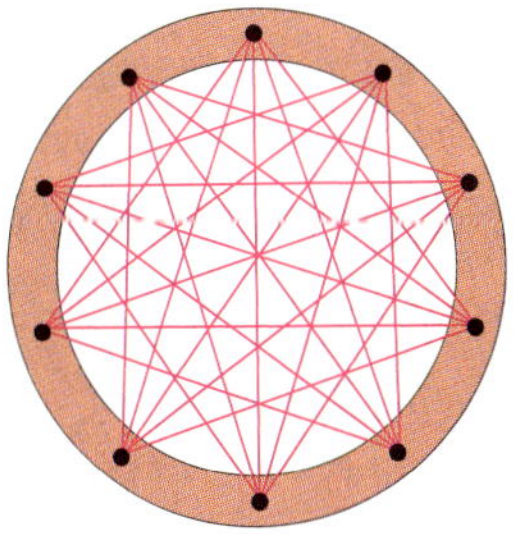

WRITING

43. A student was asked to solve the following problem: *The length of a rectangular room is 2 feet more than twice the width. If the area of the room is 60 square feet, find its dimensions.* Here is the student's solution:

Since $10 \cdot 6 = 60$, the length of the room is 10 feet and the width is 6 feet.

Explain why his solution is incorrect.

44. Suppose that to find the length of the base of a triangle, you write a quadratic equation and solve it to find $b = 6$ or $b = -8$. Explain why one solution should be discarded.

45. What error is apparent in the following illustration?

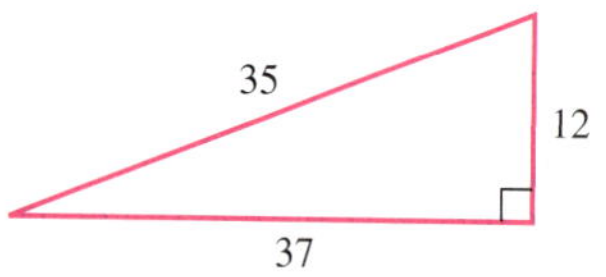

46. When naming the legs of a right triangle, explain why it doesn't matter which leg you label a and which leg you label b.

REVIEW

Find each special product.

47. $(5b - 2)^2$

48. $(2a + 3)^2$

49. $(s^2 + 4)^2$

50. $(m^2 - 1)^2$

51. $(9x + 6)(9x - 6)$

52. $(5b + 2)(5b - 2)$

CHALLENGE PROBLEMS

53. POOL BORDERS The owners of a 10-meter wide by 25-meter long rectangular swimming pool want to surround the pool with a crushed-stone border of uniform width. They have enough stone to cover 74 square meters. How wide should they make the border?

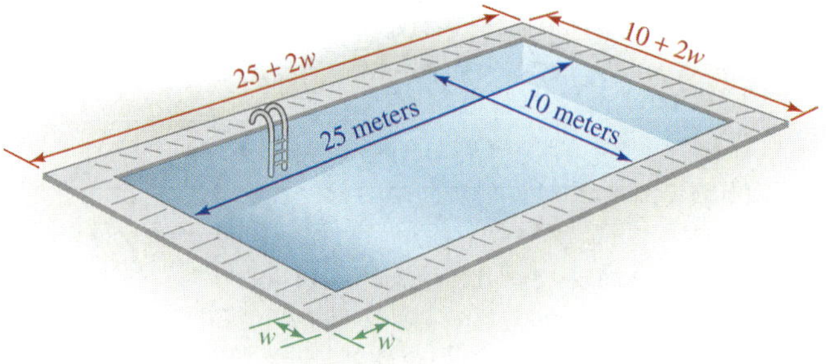

54. Find h.

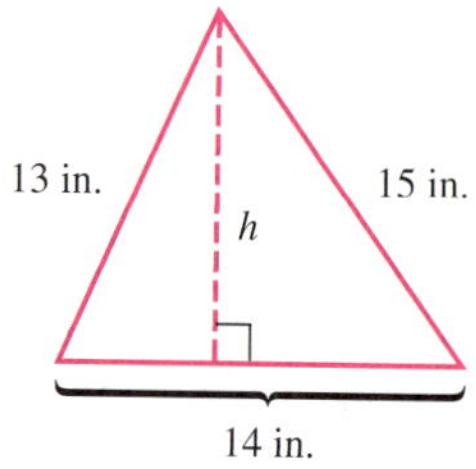

CHAPTER 6
Summary & Review

SECTION 6.1 The Greatest Common Factor; Factoring by Grouping

DEFINITIONS AND CONCEPTS	EXAMPLES
Factoring is multiplication reversed. To **factor a polynomial** means to express it as a product of two (or more) polynomials.	Multiplication: Given the factors, we find a polynomial. ⟶ $2x(5x + 3) = 10x^2 + 6x$ ⟵ Factoring: Given a polynomial, we find the factors.
A natural number is in **prime-factored form** when it is written as the product of prime numbers.	The prime-factored form of 28 is $2 \cdot 2 \cdot 7 = 2^2 \cdot 7$.

SECTION 6.1 The Greatest Common Factor; Factoring by Grouping—*continued*

DEFINITIONS AND CONCEPTS	EXAMPLES
To find the **greatest common factor, GCF,** of a list of terms **1.** Write each coefficient as a product of prime factors. **2.** Identify the numerical and variable factors common to each term. **3.** Multiply the common numerical and variable factors identified in Step 2 to obtain the GCF. If there are no common factors, the GCF is 1.	Find the GCF of $35x^4$, $63x^3$, and $42x^2$. $35x^4 = 5 \cdot 7 \cdot x \cdot x \cdot x \cdot x$ $63x^3 = 3 \cdot 3 \cdot 7 \cdot x \cdot x \cdot x$ $42x^2 = 2 \cdot 3 \cdot 7 \cdot x \cdot x$ GCF $= 7 \cdot x \cdot x = 7x^2$
The first step of factoring a polynomial is to see whether the terms of the polynomial have a common factor. If they do, **factor out the GCF.**	Factor: $35x^4 + 63x^3 - 42x^2$ $= 7x^2(5x^2 + 9x - 6)$ Factor out the GCF, $7x^2$. Use multiplication to check the factorization: $7x^2(5x^2 + 9x - 6) = 35x^4 + 63x^3 - 42x^2$ This is the original polynomial.
If a polynomial has four terms, try **factoring by grouping.** **1.** Group the terms of the polynomial so that the first two terms have a common factor and the last two terms have a common factor. **2.** Factor out the common factor from each group. **3.** Factor out the resulting common binomial factor. If there is no common binomial factor, regroup the terms of the polynomial and repeat steps 2 and 3.	Factor: $ax - bx + ay - by$ $= x(a - b) + y(a - b)$ Factor out x from $ax - bx$ and y from $ay - by$. $= (a - b)(x + y)$ Factor out the common binomial factor, $(a - b)$.

REVIEW EXERCISES

Find the prime-factorization of each number.

1. 35

2. 96

Find the GCF of each list.

3. 28 and 35

4. $36a^4$, $54a^3$, and $126a^6$

Factor.

5. $3x + 9y$

6. $5ax^2 + 15a$

7. $7s^5 + 14s^3$

8. $\pi ab - \pi ac$

9. $24x^3 + 60x^2 - 48x$

10. $x^5y^3z^2 + xy^5z^3 - xy^3z^2$

11. $-5ab^2 + 10a^2b - 15ab$

12. $4(x - 2) - x(x - 2)$

Factor out −1.

13. $-a - 7$

14. $-4t^2 + 3t - 1$

Factor.

15. $2c + 2d + ac + ad$

16. $3xy + 18x - 5y - 30$

17. $2a^3 + 2a^2 - a - 1$

18. $4m^2n + 12m^2 - 8mn - 24m$

SECTION 6.2 Factoring Trinomials of the Form $x^2 + bx + c$

DEFINITIONS AND CONCEPTS	EXAMPLES
Many trinomials factor as the product of two binomials. To **factor a trinomial** of the form $x^2 + bx + c$, whose **leading coefficient is 1,** find two integers whose product is c and whose sum is b. $x^2 + bx + c = (x \square)(x \square)$ The product of these numbers must be c and their sum must be b. Use the FOIL method to check the factorization.	Factor: $p^2 + 7p + 12$ $= (p \square)(p \square)$ $= (p + 3)(p + 4)$ **Check:** $(p + 3)(p + 4) = p^2 + 4p + 3p + 12$ $= p^2 + 7p + 12$
Before factoring a trinomial, write it in **descending powers** of one variable. Also, factor out -1 if that is necessary to make the **leading coefficient positive.**	Factor: $7q - q^2 - 6$ $= -q^2 + 7q - 6$ Write the terms in descending powers of q. $= -(q^2 - 7q + 6)$ Factor out -1. $= -(q - 1)(q - 6)$ Factor the trinomial.
If a trinomial cannot be factored using only integers, it is called a **prime trinomial.**	$t^2 + 2t - 5$ is a prime trinomial because there are no two integers whose product is -5 and whose sum is 2.
The GCF should always be factored out first. A trinomial is **factored completely** when no factor can be factored further. Use multiplication to check the factorization.	Factor completely: $3m^3 - 6m^2 - 24m$ $= 3m(m^2 - 2m - 8)$ Factor out the GCF, $3m$, first. $= 3m(m - 4)(m + 2)$ Factor the trinomial.
To factor a trinomial of the form $x^2 + bx + c$ by **grouping,** write it as a equivalent four-term polynomial: $x^2 + \square x + \square x + c$ The product of these numbers must be c, and their sum must be b. Then factor the four-term polynomial by grouping. Use the FOIL method to check the factorization.	Factor by grouping: $p^2 + 7p + 12$ We must find two numbers whose product is $c = 12$ and whose sum is $b = 7$. Two such numbers are 4 and 3. They serve as the coefficients of $4p$ and $3p$, the two terms that we use to represent the middle term, $7p$, of the trinomial. $p^2 + 7p + 12 = p^2 + 4p + 3p + 12$ Express $7p$ as $4p + 3p$. $= p(p + 4) + 3(p + 4)$ Factor p out of $p^2 + 4p$ and 3 out of $3p + 12$. $= (p + 4)(p + 3)$ Factor out $(p + 4)$.

Positive factors of 12	Sum of positive factors of 12
$1 \cdot 12 = 12$	$1 + 12 = 13$
$2 \cdot 6 = 12$	$2 + 6 = 8$
$3 \cdot 4 = 12$	$3 + 4 = 7$

REVIEW EXERCISES

19. What is the leading coefficient of $x^2 + 8x - 9$?

20. Complete the table.

Factors of 6	Sum of the factors of 6
1(6)	
2(3)	
−1(−6)	
−2(−3)	

Factor each trinomial, if possible.

21. $x^2 + 2x - 24$

22. $x^2 - 18x - 40$

23. $x^2 - 14x + 45$

24. $t^2 + 10t + 15$

25. $-y^2 + 15y - 56$

26. $10y + 9 + y^2$

27. $c^2 + 3cd - 10d^2$

28. $-3mn + m^2 + 2n^2$

29. Explain how we can check to determine whether $(x - 4)(x + 5)$ is the factorization of $x^2 + x - 20$.

30. Explain why $x^2 + 7x + 11$ is prime.

Completely factor each trinomial.

31. $5a^5 + 45a^4 - 50a^3$

32. $-4x^2y - 4x^3 + 24xy^2$

SECTION 6.3 Factoring Trinomials of the Form $ax^2 + bx + c$

DEFINITIONS AND CONCEPTS	EXAMPLES
We can use the **trial-and-check method** to factor trinomials with **leading coefficients other than 1.** Write the trinomial as the product of two binomials and determine four integers. The product of these numbers must be *a*. $ax^2 + bx + c = (\square x \;\square)(\square x \;\square)$ The product of these numbers must be *c*. Use the FOIL method to check the factorization.	Factor: $2x^2 - 5x - 12$ Since the first term is $2x^2$, the first terms of the binomial factors must be $2x$ and x. $(2x \;\square)(x \;\square)$ Because $2x \cdot x$ will give $2x^2$ The second terms of the binomials must be two integers whose product is -12. There are six such pairs: $12(-1)$, $6(-2)$, $3(-4)$, $4(-3)$, $2(-6)$, and $1(-12)$ The pair in blue gives the correct middle term when we use the FOIL method to check: Outer: $-8x$ $(2x + 3)(x - 4)$ Combine like terms: $-8x + 3x = -5x$. Inner: $3x$ Thus, $2x^2 - 5x - 12 = (2x + 3)(x - 4)$.
To factor $ax^2 + bx + c$ by **grouping,** write it as an equivalent four-term polynomial: $ax^2 + \square x + \square x + c$ The product of these numbers must be *ac*, and their sum must be *b*. Then factor the four-term polynomial by grouping. Use the FOIL method to check your work.	Factor by grouping: $2x^2 - 5x - 12$ We must find two numbers whose product is $ac = 2(-12) = -24$ and whose sum is $b = -5$. Two such numbers are -8 and 3. They serve as the coefficients of $-8x$ and $3x$, the two terms that we use to represent the middle term, $-5x$, of the trinomial. $2x^2 - 5x - 12 = 2x^2 - 8x + 3x - 12$ Express $-5x$ as $-8x + 3x$. $= 2x(x - 4) + 3(x - 4)$ $= (x - 4)(2x + 3)$ Factor out $(x - 4)$.

REVIEW EXERCISES

Factor each trinomial completely, if possible.

33. $2x^2 - 5x - 3$

34. $35y^2 + 11y - 10$

35. $-3x^2 + 13x + 30$

36. $-33p^2 - 6p + 18p^3$

37. $4b^2 - 17bc + 4c^2$

38. $7y^2 + 7y - 18$

39. ENTERTAINING The rectangular-shaped area occupied by a table setting is $(12x^2 - x - 1)$ square inches. Factor the polynomial to find the binomials that represent the length and width of the table setting.

40. In the following work, a student began to factor $5x^2 - 8x + 3$. Explain his mistake.

$(5x - \quad)(x + \quad)$

SECTION 6.4 Factoring Perfect-Square Trinomials and the Difference of Two Squares

DEFINITIONS AND CONCEPTS	EXAMPLES
Trinomials that are squares of a binomial are called **perfect-square trinomials.** We can factor perfect-square trinomials by applying the special-product rules in reverse. $A^2 + 2AB + B^2 = (A + B)^2$ $A^2 - 2AB + B^2 = (A - B)^2$	Factor: $g^2 + 8g + 16$ and $m^2 - 18mn + 81n^2$ We match each trinomial to a special-product form shown in the left column. $g^2 + 8g + 16 = g^2 + 2 \cdot 4 \cdot g + 4^2 = (g + 4)^2$ $m^2 - 18mn + 81n^2 = m^2 - 2 \cdot m \cdot 9n + (-9n)^2 = (m - 9n)^2$
To factor the **difference of two squares,** use the rule $F^2 - L^2 = (F + L)(F - L)$ It will be helpful to review the table of **squares of integers** shown on page 507.	Factor: $25b^2 - 36$ $= (5b)^2 - 6^2$ This is a difference of two squares. $= (5b + 6)(5b - 6)$
In general, the **sum of two squares** (with no common factor other than 1) cannot be factored using real numbers.	$x^2 + 100$ and $36y^2 + 49$ are prime polynomials.

REVIEW EXERCISES

Factor completely, if possible.

41. $x^2 + 10x + 25$

42. $9y^2 + 16 - 24y$

43. $-z^2 + 2z - 1$

44. $25a^2 + 20ab + 4b^2$

45. $x^2 - 9$

46. $49t^2 - 121y^2$

47. $x^2y^2 - 400$

48. $8at^2 - 32a$

49. $c^4 - 256$

50. $h^2 + 36$

SECTION 6.5 Factoring the Sum and Difference of Two Cubes

DEFINITIONS AND CONCEPTS	EXAMPLES
To factor the **sum** and **difference of two cubes,** use the following rules. $F^3 + L^3 = (F + L)(F^2 - FL + L^2)$ $F^3 - L^3 = (F - L)(F^2 + FL + L^2)$ It will be helpful to review the table of **cubes of integers** shown on page 515.	Factor: $p^3 + 64$ and $125a^3 - 27b^3$ We match each binomial to a factoring rule shown in the left column. $p^3 + 64 = p^3 + 4^3$ This is a sum of two cubes. $= (p + 4)(p^2 - p \cdot 4 + 4^2)$ $= (p + 4)(p^2 - 4p + 16)$ $125a^3 - 27b^3 = (5a)^3 - (3b)^3$ This is a difference of two cubes. $= (5a - 3b)[(5a)^2 + 5a \cdot 3b + (3b)^2]$ $= (5a - 3b)(25a^2 + 15ab + 27b^2)$

REVIEW EXERCISES

Factor each polynomial completely.

51. $b^3 + 1$

52. $x^3 - 216$

53. $p^3 + 125q^3$

54. $16x^5 - 54x^2y^3$

SECTION 6.6 A Factoring Strategy

DEFINITIONS AND CONCEPTS	EXAMPLES
To factor a random polynomial, use the **factoring strategy** discussed in Section 6.6 on page 519. Remember that the instruction to factor means to **factor completely.** A polynomial is factored completely when no factor can be factored further.	Factor: $a^5 + 8a^2 + 4a^3 + 32$ ***Is there a common factor?*** No. There is no common factor (other than 1). ***How many terms does it have?*** Since the polynomial has four terms, try factoring by grouping. $a^5 + 8a^2 + 4a^3 + 32 = a^2(a^3 + 8) + 4(a^3 + 8)$ Factor a^2 from $a^5 + 8a^2$ and 4 from $4a^3 + 32$. $= (a^3 + 8)(a^2 + 4)$ Factor out $a^3 + 8$. ***Is it factored completely?*** No. We can factor $a^3 + 8$ as a sum of two cubes. $a^5 + 8a^2 + 4a^3 + 32$ $= (a^3 + 8)(a^2 + 4)$ $a^2 + 4$ is prime. $= (a + 2)(a^2 - 2a + 4)(a^2 + 4)$ $a^2 - 2a + 4$ is prime. ***Does it check?*** Use multiplication to check.

REVIEW EXERCISES

Factor each polynomial completely, if possible.

55. $14y^3 + 6y^4 - 40y^2$
56. $5s^2t + 5s^2u^2 + 5tv + 5u^2v$
57. $j^4 - 16$
58. $-3j^3 - 24$
59. $400x + 400 - m^2x - m^2$
60. $12w^4 - 36w^3 + 27w^2$
61. $2t^3 + 10$
62. $121p^2 + 36q^2$
63. $x^2z + 64y^2z + 16xyz$
64. $18c^3d^2 - 12c^3d - 24c^2d$

SECTION 6.7 Solving Quadratic Equations by Factoring

DEFINITIONS AND CONCEPTS	EXAMPLES
A **quadratic equation** is an equation that can be written in the **standard form** $ax^2 + bx + c = 0$, where a, b, and c are real numbers and $a \neq 0$.	Examples of quadratic equations are: $5x^2 + 25x = 0$, $4a^2 - 9 = 0$, and $y^2 - 13y = 6$
The Zero-Factor Property If the product of two or more numbers is 0, then at least one of the numbers is 0.	If $(x + 2)(x - 3) = 0$ then, $x + 2 = 0$ or $x - 3 = 0$.
To use the **factoring method to solve a quadratic equation:** **1.** Write the equation in standard form: $ax^2 + bx + c = 0$ or $0 = ax^2 + bx + c$ **2.** Factor completely. **3.** Use the *zero-factor property* to set each factor equal to 0. **4.** Solve each resulting linear equation. **5.** Check each result in the original equation.	Solve: $5x^2 + 25x = 0$ $5x(x + 5) = 0$ $5x = 0$ or $x + 5 = 0$ $x = 0$ | $x = -5$ The solutions are 0 and −5. Solve: $4a^2 - 9 = 0$ $(2a + 3)(2a - 3) = 0$ $2a + 3 = 0$ or $2a - 3 = 0$ $2a = -3$ | $2a = 3$ $a = -\frac{3}{2}$ | $a = \frac{3}{2}$ The solutions are $-\frac{3}{2}$ and $\frac{3}{2}$. Check each result in the original equation.

SECTION 6.7 Solving Quadratic Equations by Factoring—*continued*

DEFINITIONS AND CONCEPTS	EXAMPLES
To use the zero-factor property to solve a quadratic equation, we need one side of the equation to be factored completely and the other side to be 0.	Solve: $5y^2 - 13y = 6$ $5y^2 - 13y - 6 = 6 - 6$ To get 0 on the right side, subtract 6 from both sides. $5y^2 - 13y - 6 = 0$ $(5y + 2)(y - 3) = 0$ $5y + 2 = 0$ or $y - 3 = 0$ $5y = -2$ | $y = 3$ $y = -\frac{2}{5}$ The solutions are $-\frac{2}{5}$ and 3. Check each result in the original equation.

REVIEW EXERCISES

Solve each equation by factoring.

65. $8x(x - 6) = 0$

66. $(4x - 7)(x + 1) = 0$

67. $x^2 + 2x = 0$

68. $x^2 - 9 = 0$

69. $144x^2 - 25 = 0$

70. $a^2 - 7a + 12 = 0$

71. $2t^2 + 28t + 98 = 0$

72. $2x - x^2 = -24$

73. $5a^2 - 6a + 1 = 0$

74. $2p^3 = 2p(p + 2)$

SECTION 6.8 Applications of Quadratic Equations

DEFINITIONS AND CONCEPTS	EXAMPLES
To solve application problems, use the **five-step problem solving strategy:** **1.** Analyze the problem **2.** Form an equation **3.** Solve the equation **4.** State the conclusion **5.** Check the result	Find two consecutive positive integers whose product is 72. **Analyze the Problem** *Consecutive integers* are integers that follow each other. The word *product* indicates multiplication. **Form an Equation** Let $x =$ the smaller positive integer. Then $x + 1 =$ the larger integer. The smaller integer times the larger integer equals 72. $x \cdot (x + 1) = 72$ **Solve the Equation** $x(x + 1) = 72$ $x^2 + x - 72 = 0$ Remove parentheses. To get 0 on the right side, subtract 72 from both sides. $(x + 9)(x - 8) = 0$ Factor the trinomial. $x + 9 = 0$ or $x - 8 = 0$ Set each factor equal to 0. $x = -9$ | $x = 8$ Solve each linear equation. **State the Conclusion** Since we are looking for positive integers, the solution -9 must be discarded. Thus, the smaller integer is 8 and the larger integer is $x + 1 = 9$. **Check the Result** The integers 8 and 9 are consecutive positive integers and their product is 72.

SECTION 6.8 Applications of Quadratic Equations—*continued*

DEFINITIONS AND CONCEPTS	EXAMPLES
The Pythagorean Theorem: If a and b are the lengths of the legs of a right triangle and c is the length of the hypotenuse, then $a^2 + b^2 = c^2$	To show that a triangle with sides of 5, 12, and 13 units is a right triangle, we verify that $5^2 + 12^2 = 13^2$. $5^2 + 12^2 \stackrel{?}{=} 13^2$ $25 + 144 \stackrel{?}{=} 169$ $169 = 169$ True 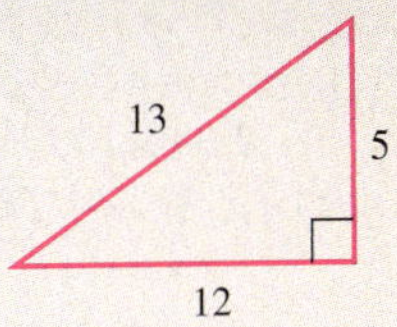

REVIEW EXERCISES

75. CONSTRUCTION The face of the triangular concrete panel has an area of 45 square meters, and its base is 3 meters longer than twice its height. Find the length of its base.

76. ACADEMY AWARDS Meryl Streep and Katherine Hepburn are the two most-nominated actresses for Oscars. The number of times Streep has been nominated and the number of times Hepburn has been nominated are consecutive even integers whose product is 168. How many times was each actress nominated?

77. TIGHTROPE WALKERS A circus performer intends to walk up a taut cable shown in the illustration to a platform at the top of a pole. How high above the ground is the platform?

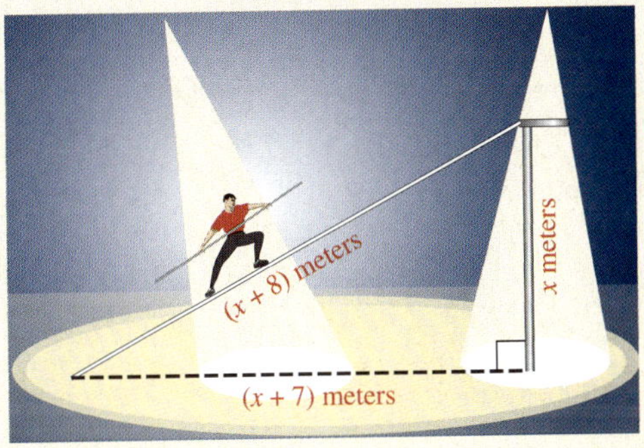

78. BALLOONING A hot-air balloonist accidentally dropped his camera overboard while traveling at a height of 1,600 ft. The height h in feet of the camera t seconds after being dropped is given by $h = -16t^2 + 1{,}600$. In how many seconds will the camera hit the ground?

CHAPTER 6 Test

1. Fill in the blanks.

a. The letters GCF stand for ________ ________ ______.

b. To factor a polynomial means to express it as a _______ of two (or more) polynomials.

c. The ___________ theorem provides a formula relating the lengths of the three sides of a right triangle.

d. $y^2 - 25$ is a _________ of two squares.

e. The trinomial $x^2 + x - 6$ factors as the product of two _________: $(x + 3)(x - 2)$.

2. a. Find the prime factorizations of 45 and 30.

b. Find the greatest common factor of $45x^4$ and $30x^3$.

Factor completely. If an expression cannot be factored, write "prime."

3. $4x + 16$

4. $q^2 - 81$

5. $30a^2b^3 - 20a^3b^2 + 5ab$

6. $x^2 + 9$

7. $2x(x + 1) + 3(x + 1)$

8. $x^2 + 4x + 3$

9. $-x^2 + 9x + 22$

10. $60x^2 - 32x^3 + x^4$

11. $9a - 9b + ax - bx$

12. $2a^2 + 5a - 12$

13. $18x^2 + 60xy + 50y^2$

14. $x^3 + 8$

15. $20m^8 - 15m^6$

16. $3a^3 - 81$

17. $16x^4 - 81$

18. $a^3 + 5a^2 + a + 5$

19. CHECKERS The area of a square checkerboard is represented by $(25x^2 - 40x + 16)$ in.^{2}. Find the polynomial that represents the length of a side of the checkerboard.

20. Factor $x^2 - 3x - 54$. Show a check of your answer.

Solve each equation.

21. $(x + 3)(x - 2) = 0$

22. $x^2 - 25 = 0$

23. $36x^2 - 6x = 0$

24. $x^2 + 6x = -9$

25. $6x^2 + x - 1 = 0$

26. $a(a - 7) = 18$

27. $x^3 + 7x^2 = -6x$

28. DRIVING SAFETY All cars have a blind spot where it is difficult for the driver to see a car behind and to the right. The area of the rectangular blind spot shown is 54 ft^2. Its length is 3 feet longer than its width. Find its dimensions.

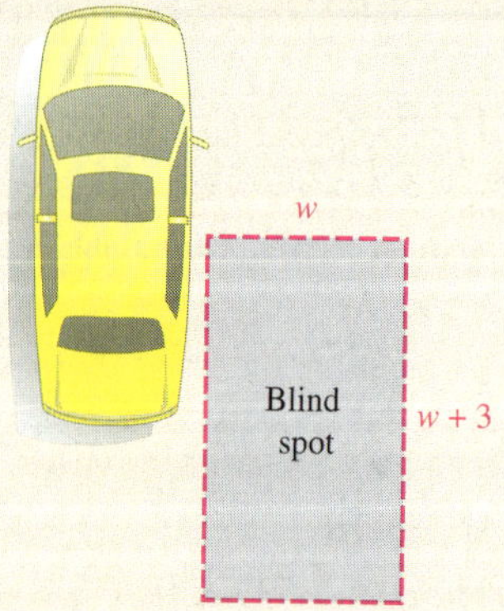

29. ROCKETRY The height h, in feet, of a toy rocket t seconds after being launched is given by $h = -16t^2 + 80t$. After how many seconds will the rocket hit the ground?

30. ATV'S The area of a triangular-shaped safety flag on an all-terrain vehicle is 33 in.^{2}. Its height is 1 inch less than twice the length of the base. Find the length of the base and the height of the flag.

31. Find two consecutive positive integers whose product is 156.

32. Find the length of the hypotenuse of the right triangle shown.

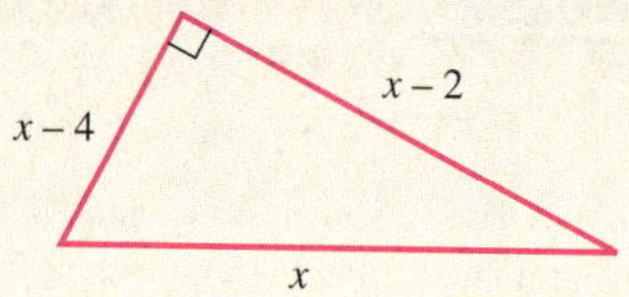

33. What is a quadratic equation? Give an example.

34. If the product of two numbers is 0, what conclusion can be drawn about the numbers?

GROUP PROJECT

FACTORING MODELS

Overview: In this activity, you will construct geometric models to find factorizations of several trinomials.

Instructions: Form groups of 2 or 3 students.

1. Copy and cut out each of the following figures. On each figure, write its area.

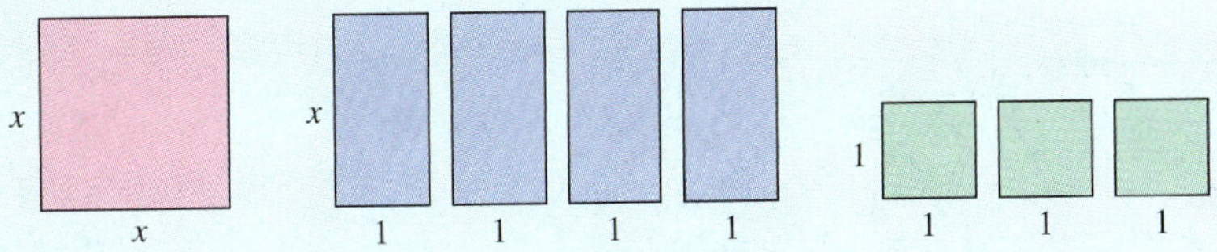

Write a trinomial that represents the *sum* of the areas of the eight figures by combining any like terms: _____ + _____ + _____

2. Now assemble the eight figures to form the large rectangle shown below.

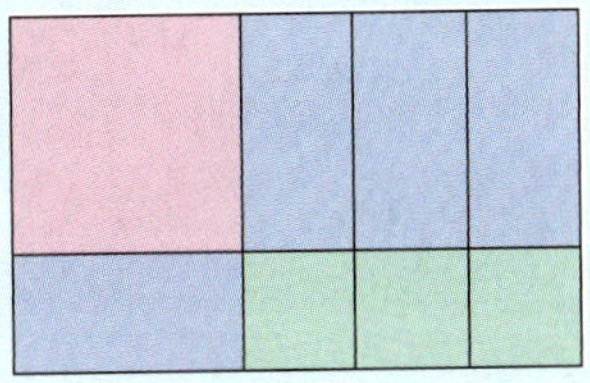

Write an expression that represents the *length* of the rectangle: _____ + _____

Write an expression that represents the *width* of the rectangle: _____ + _____

Express the area of the rectangle as the product of its length and width:

(_________)(_________)

3. The set of figures used in step 1 and the set of figures used in step 2 are the same. Therefore, the expressions for the areas must be equal. Set your answers from steps 1 and 2 equal to find the factorization of the trinomial $x^2 + 4x + 3$.

_______________ = _______________
Answer from step 1 Answer from step 2

4. Make a new model to find the factorization of $x^2 + 5x + 4$. (*Hint:* You will need to make one more 1-by-x figure and one more 1-by-1 figure.)

_______________ = _______________

5. Make a new model to find the factorization of $2x^2 + 5x + 2$. (*Hint:* You will need to make one more x-by-x figure.)

_______________ = _______________

CUMULATIVE REVIEW
Chapters 1–6

1. HEART RATES Refer to the graph. Determine the difference in the maximum heart beat rate for a 70-year-old as compared to someone half that age. [Section 1.1]

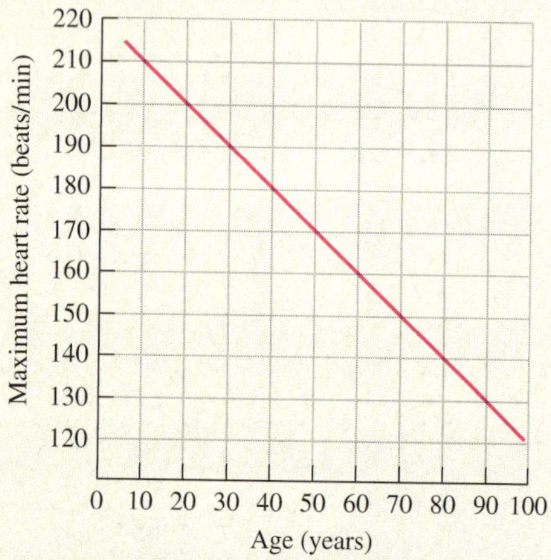

Based on data from *Cardiopulmonary Anatomy and Physiology: Essentials for Respiratory Care*, 2nd ed.

2. Find the prime factorization of 250. [Section 1.2]

3. Find the quotient: $\frac{16}{5} \div \frac{10}{3}$ [Section 1.2]

4. Write $\frac{124}{125}$ as a decimal. [Section 1.3]

5. Determine whether each statement is true or false. [Section 1.2]

a. Every integer is a whole number.

b. Every integer is a rational number.

c. π is a real number.

6. Which division is undefined, $\frac{0}{5}$ or $\frac{5}{0}$? [Section 1.6]

Evaluate each expression.

7. $3 + 2[-1 - 4(5)]$ [Section 1.7]

8. $\dfrac{|-25| - 2(-5)}{9 - 2^4}$ [Section 1.7]

9. What is -3 cubed? [Section 1.7]

10. What is the value of x twenty-dollar bills? [Section 1.8]

11. Evaluate $\frac{-x - a}{y - b}$ for $x = -2$, $y = 1$, $a = 5$, and $b = 2$. [Section 1.8]

12. Identify the coefficient of each term in the expression $8x^2 - x + 9$. [Section 1.8]

Simplify each expression.

13. $-8y^2 - 5y^2 + 6$ [Section 1.9]

14. $3z + 2(y - z) + y$ [Section 1.9]

Solve each equation.

15. $-(3a + 1) + a = 2$ [Section 2.2]

16. $2 - (4x + 7) = 3 + 2(x + 2)$ [Section 2.2]

17. $\dfrac{3t - 21}{2} = t - 6$ [Section 2.2]

18. $-\dfrac{1}{3} - \dfrac{x}{5} = \dfrac{3}{2}$ [Section 2.2]

19. WATERMELONS The heaviest watermelon on record weighed 270 pounds. If watermelon is 92% water by weight, what was its water weight? (Round to the nearest pound.) [Section 2.3]

20. Find the distance traveled by a truck traveling for $5\frac{1}{2}$ hours at a rate of 60 miles per hour. [Section 2.4]

21. What is the formula for simple interest? [Section 2.4]

22. GEOMETRY TOOLS A compass is adjusted so that the distance between the pointed ends is 2 inches. Then a circle is drawn. What will the area of the circle be? Round to the nearest tenth of a square inch. [Section 2.4]

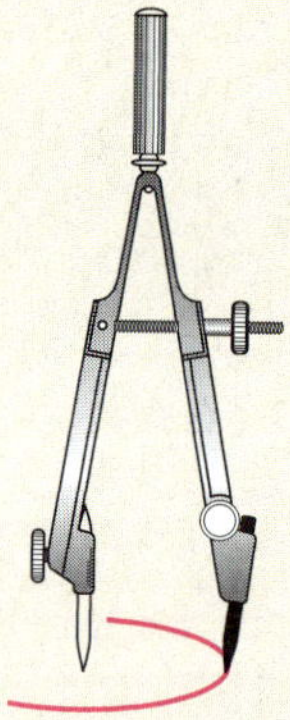

23. Solve $A = P + Prt$ for t. [Section 2.4]

24. HISTORY George Washington was the first president of the United States. John Adams was the second, and Thomas Jefferson was the third, and so on. Grover Cleveland was president two *different* times, as shown in the illustration. The sum of the numbers of Cleveland's presidencies is 46. Find these two numbers. [Section 2.5]

25. PHOTOGRAPHIC CHEMICALS A photographer wishes to mix 6 liters of a 5% acetic acid solution with a 10% solution to get a 7% solution. How many liters of 10% solution must be added? [Section 2.6]

26. Solve: $-\frac{x}{2} + 4 > 5$. Write the solution set in interval notation and graph it. [Section 2.8]

27. Is $(-2, 5)$ a solution of $3x + 2y = 4$? [Section 3.2]

28. Graph: $y = 2x - 3$ [Section 3.2]

29. Is the graph of $x = 3$ a vertical or horizontal line? [Section 3.3]

30. If two lines are parallel, what can be said about their slopes? [Section 3.4]

31. BOTTLED WATER Refer to the graph below. Determine the rate of change in the number of gallons of bottled water that the average American drank in a year for 1995 through 2005. [Section 3.4]

U.S. Consumption of Bottled Water 1995–2005

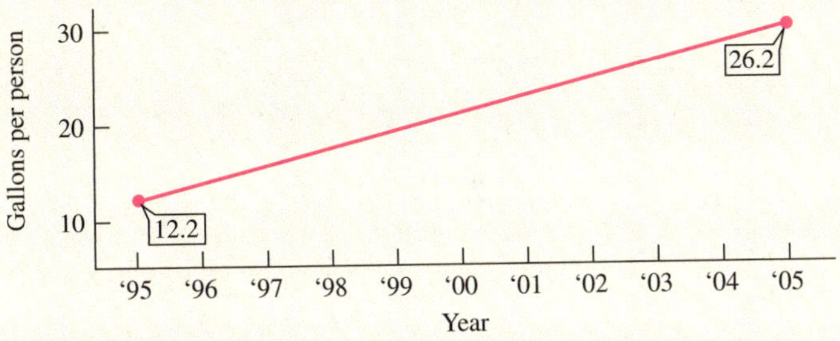

Source: Beverage Marketing Coproration, 2005

32. Find the slope and the y-intercept of the graph of $3x - 3y = 6$. [Section 3.5]

33. Find an equation of the line passing through $(-2, 5)$ and $(-3, -2)$. Write the equation in slope–intercept form. [Section 3.5]

34. Graph the line passing through $(-4, 1)$ that has slope -3. [Section 3.6]

35. Graph: $8x + 4y \geq -24$ [Section 3.7]

36. If $f(x) = 3x^2 - 2x + 1$, find $f(-2)$. [Section 3.8]

37. Is $\left(\frac{1}{2}, 1\right)$ a solution of the system $\begin{cases} 4x - y = 1 \\ 2x + y = 2 \end{cases}$? [Section 4.1]

38. Solve the system $\begin{cases} 3x - 2y = 6 \\ x - y = 1 \end{cases}$ by graphing. [Section 4.1]

39. Solve the system $\begin{cases} y = -4x + 1 \\ 4x - y = 5 \end{cases}$ by substitution. [Section 4.2]

40. Solve the system $\begin{cases} 5a + 3b = -8 \\ 2a + 9b = 2 \end{cases}$ by elimination (addition). [Section 4.3]

41. FUNDRAISING A Rotary Club held a city-wide recycling drive. They collected a total of 14 tons of newspaper and cardboard that earned them \$356. They were paid \$31 per ton for the newspaper and \$18 per ton for the cardboard. How many tons of each did they collect? [Section 4.4]

42. Graph: $\begin{cases} 4x + 3y \geq 12 \\ y < 4 \end{cases}$ [Section 4.5]

Simplify each expression. Write each answer without negative exponents.

43. $-y^2(4y^3)$ [Section 5.1]

44. $\dfrac{(x^2y^5)^5}{(x^3y)^2}$ [Section 5.1]

45. $\left(\dfrac{b^5}{b^{-2}}\right)$ [Section 5.2]

46. $2x^0$ [Section 5.1]

47. Write 0.00009011 in scientific notation. [Section 5.3]

48. Write 1,700,000 in scientific notation. [Section 5.3]

49. Find the degree of $7y^3 + 4y^2 + y + 3$. [Section 5.4]

50. Graph: $y = x^3 + 2$ [Section 5.4]

Perform the operations.

51. $(x^2 - 3x + 8) - (3x^2 + x + 3)$ [Section 5.5]

52. $4b^3(2b^2 - 2b)$ [Section 5.6]

53. $(3x - 2)(x + 4)$ [Section 5.6]

54. $(y - 6)^2$ [Section 5.7]

55. $\dfrac{12a^2b^2 - 8a^2b - 4ab}{4ab}$ [Section 5.8]

56. $x - 3\overline{)2x^2 - 5x - 3}$ [Section 5.8]

57. PLAYPENS Find an expression that represents the

a. perimeter of the playpen. [Section 5.5]

b. area of the floor of the playpen. [Section 5.6]

c. volume of the playpen. [Section 5.6]

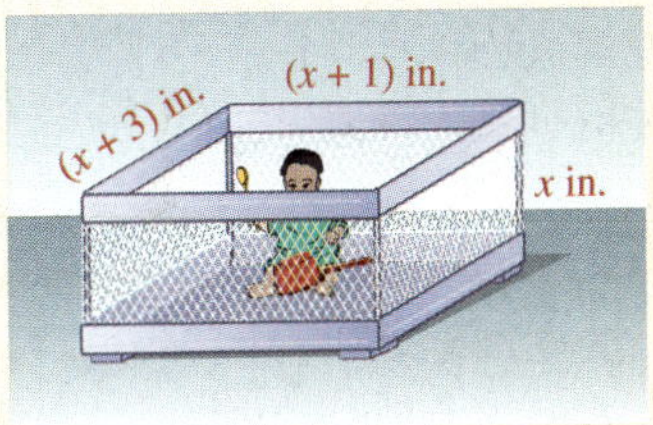

58. Find the GCF of $24x^5y^8$ and $54x^6y$. [Section 6.1]

Factor completely.

59. $9b^3 - 27b^2$ [Section 6.1]

60. $ax + bx + ay + by$ [Section 6.1]

61. $u^2 - 3 + 2u$ [Section 6.2]

62. $10x^2 + x - 2$ [Section 6.3]

63. $4a^2 - 12a + 9$ [Section 6.4]

64. $9z^2 - 1$ [Section 6.4]

65. $t^3 - 8$ [Section 6.5]

66. $3a^2b^2 - 6a^2 - 3b^2 + 6$ [Section 6.6]

Solve each equation.

67. $15s^2 - 20s = 0$ [Section 6.7]

68. $2x^2 - 5x = -2$ [Section 6.7]

CHAPTER 7

Rational Expressions and Equations

© BananaStock/SuperStock

from Campus to Careers
Recreation Director

People of all ages enjoy participating in activities, such as arts and crafts, camping, sports, and the performing arts. Recreation directors plan, organize, and oversee these activities in local playgrounds, camps, community centers, religious organizations, theme parks, and tourist attractions. The job of recreation director requires mathematical skills such as budgeting, scheduling, and forecasting trends.

Problem 33 of **Study Set 7.7** involves an area of responsibility for many recreation directors—swimming pools.

JOB TITLE:
Recreation Director

EDUCATION:
An associate's or bachelor's degree in parks and recreation is preferred.

JOB OUTLOOK:
Employment is expected to increase between 9% to 17% through the year 2014.

ANNUAL EARNINGS:
From $36,000 to an average of $55,000, up to $75,000 or more

FOR MORE INFORMATION:
http://stats.bls.gov/oco/ocos058.htm

Study Skills Workshop
Study Groups

Study groups give students an opportunity to ask their classmates questions, share ideas, compare lecture notes, and review for tests. If something like this interests you, here are some suggestions.

GROUP SIZE: A study group should be small—from 3 to 6 people is best.

TIME AND PLACE: You should meet regularly in a place where you can spread out and talk without disturbing others.

GROUND RULES: The study group will be more effective if, early on, you agree on some rules to follow.

Now Try This

Would you like to begin a study group? If so, you need to answer the following questions.

Who will be in your group? Where will your group meet? How often will it meet? For how long will each session last? Will you have a group leader? What will be the leader's responsibilities? What will you try to accomplish each session? How will the members prepare for each meeting? Will you follow a set agenda each session? How will the members share contact information? When will you discuss ways to improve the study sessions?

SECTION 7.1
Simplifying Rational Expressions

Objectives

1. Evaluate rational expressions.
2. Find numbers that cause a rational expression to be undefined.
3. Simplify rational expressions.
4. Simplify rational expressions that have factors that are opposites.

Fractions that are the quotient of two integers are *rational numbers.* Examples are $\frac{1}{2}$ and $\frac{9}{5}$. Fractions such as

$$\frac{3}{2y}, \qquad \frac{x}{x+2}, \qquad \text{and} \qquad \frac{2a^2-8a}{a^2-6a+8}$$

that are the quotient of two polynomials are called **rational expressions.**

Rational Expressions

A **rational expression** is an expression of the form $\frac{A}{B}$, where A and B are polynomials and B does not equal 0.

1 Evaluate Rational Expressions.

Rational expressions can have different values depending on the number that is substituted for the variable.

EXAMPLE 1 Evaluate $\frac{2x - 1}{x^2 + 1}$ for $x = -3$ and for $x = 0$.

Strategy We will replace each x in the rational expression with the given value of the variable. Then we will evaluate the numerator and denominator separately, and simplify, if possible.

Why Recall from Chapter 1 that to *evaluate an expression* means to find its numerical value, once we know the value of its variable.

Solution

For $x = -3$:

$$\frac{2x - 1}{x^2 + 1} = \frac{2(-3) - 1}{(-3)^2 + 1} = \frac{-6 - 1}{9 + 1} = -\frac{7}{10}$$

For $x = 0$:

$$\frac{2x - 1}{x^2 + 1} = \frac{2(0) - 1}{(0)^2 + 1} = \frac{0 - 1}{0 + 1} = -1$$

Self Check 1 Evaluate $\frac{2x - 1}{x^2 + 1}$ for $x = 7$.

Now Try **Problem 13**

2 Find Numbers that Cause a Rational Expression to be Undefined.

The fraction bar in a rational expression indicates division. Since division by 0 is undefined, we must make sure that the denominator of a rational expression is not equal to 0.

EXAMPLE 2 Find all real numbers for which the rational expression is undefined: **a.** $\frac{7x}{x - 5}$ **b.** $\frac{3x - 2}{x^2 - x - 6}$ **c.** $\frac{8}{x^2 + 1}$

Strategy To find the real numbers for which each rational expression is undefined, we will find the values of the variable that make the *denominator* 0.

Why We don't need to examine the numerator of the rational expression; it can be any value, including 0. It's a denominator of 0 that makes a rational expression undefined, because a denominator of 0 indicates division by 0.

Solution

a. The denominator of $\frac{7x}{x - 5}$ will be 0 if we replace x with 5.

$$\frac{7x}{x - 5} = \frac{7(5)}{5 - 5} = \frac{35}{0}$$

Since $\frac{35}{0}$ is undefined, the rational expression $\frac{7x}{x - 5}$ is undefined for $x = 5$.

b. $\frac{3x - 2}{x^2 - x - 6}$ will be undefined for values of x that make the denominator 0. To find these values, we set $x^2 - x - 6$ equal to 0, and solve for x.

$$x^2 - x - 6 = 0 \quad \text{Set the denominator of } \frac{3x-2}{x^2-x-6} \text{ equal to 0.}$$

$$(x - 3)(x + 2) = 0 \quad \text{To solve the quadratic equation, factor the trinomial.}$$

$$x - 3 = 0 \quad \text{or} \quad x + 2 = 0 \quad \text{Set each factor equal to 0.}$$

$$x = 3 \quad | \quad x = -2 \quad \text{Solve each equation.}$$

> **The Language of Algebra**
> Another way that Example 2 could be phrased is: State the *restrictions* on the variable. For $\frac{3x - 2}{x^2 - x - 6}$, we can state the *restrictions* by writing $x \neq 3$ and $x \neq -2$.

Since 3 and -2 make the denominator 0, the rational expression $\frac{3x - 2}{x^2 - x - 6}$ is undefined for $x = 3$ and $x = -2$.

***For* $x = 3$:**

$$\frac{3x - 2}{x^2 - x - 6} = \frac{3(3) - 2}{3^2 - 3 - 6} = \frac{9 - 2}{9 - 3 - 6} = \frac{7}{0} \quad \text{This expression is undefined.}$$

***For* $x = -2$:**

$$\frac{3x - 2}{x^2 - x - 6} = \frac{3(-2) - 2}{(-2)^2 - (-2) - 6} = \frac{-6 - 2}{4 + 2 - 6} = \frac{-8}{0} \quad \text{This expression is undefined.}$$

c. No matter what real number is substituted for x, the denominator, $x^2 + 1$, will not be equal to 0. (A number squared plus 1 cannot equal 0.) Thus, no real numbers make $\frac{8}{x^2 + 1}$ undefined.

Self Check 2 Find all real numbers for which the rational expression is undefined: **a.** $\frac{x}{x + 9}$ **b.** $\frac{9x + 7}{x^2 - 25}$ **c.** $\frac{4 - x}{x^2 + 64}$

Now Try **Problems 25 and 33**

3 Simplify Rational Expressions.

> **Success Tip**
> Unless otherwise stated, we will now assume that the variables in rational expressions represent real numbers for which the denominator is not zero.

In Section 1.2, we simplified fractions by removing a factor equal to 1. For example, to simplify $\frac{6}{15}$, we factor 6 and 15, and then remove the factor $\frac{3}{3}$.

$$\frac{6}{15} = \frac{2 \cdot 3}{5 \cdot 3} = \frac{2}{5} \cdot \frac{3}{3} = \frac{2}{5} \cdot 1 = \frac{2}{5}$$

To streamline this process, we can replace $\frac{3}{3}$ in $\frac{2 \cdot 3}{5 \cdot 3}$ with the equivalent fraction $\frac{1}{1}$.

$$\frac{6}{15} = \frac{2 \cdot 3}{5 \cdot 3} = \frac{2 \cdot \overset{1}{\cancel{3}}}{5 \cdot \underset{1}{\cancel{3}}} = \frac{2}{5} \quad \text{We are removing } \frac{3}{3} = 1.$$

We can simplify rational expressions in a similar manner using a procedure that is based on the following property.

The Fundamental Property of Rational Expressions

If A, B, and C are polynomials, and B and C are not 0,

$$\frac{AC}{BC} = \frac{A}{B}$$

A rational expression is **simplified** if its numerator and denominator have no common factors other than 1. To simplify a rational expression, follow these steps.

Simplifying Rational Expressions

1. Factor the numerator and denominator completely to determine their common factors.
2. Remove factors equal to 1 by replacing each pair of factors common to the numerator and denominator with the equivalent fraction $\frac{1}{1}$.
3. Multiply the remaining factors in the numerator and in the denominator.

EXAMPLE 3 Simplify: $\dfrac{21x^3y}{14x^2y^2}$

Strategy We will write the numerator and denominator in factored form and then remove pairs of factors that are equal to 1.

Why The rational expression is simplified when the numerator and denominator have no common factor other than 1.

Solution

$$\frac{21x^3y}{14x^2y^2} = \frac{3 \cdot 7 \cdot x \cdot x \cdot x \cdot y}{2 \cdot 7 \cdot x \cdot x \cdot y \cdot y}$$ Factor the numerator and the denominator.

$$= \frac{3 \cdot \overset{1}{\cancel{7}} \cdot \overset{1}{\cancel{x}} \cdot \overset{1}{\cancel{x}} \cdot x \cdot \overset{1}{\cancel{y}}}{2 \cdot \underset{1}{\cancel{7}} \cdot \underset{1}{\cancel{x}} \cdot \underset{1}{\cancel{x}} \cdot \underset{1}{\cancel{y}} \cdot y}$$ Simplify by replacing $\frac{7}{7}$, $\frac{x}{x}$, and $\frac{y}{y}$ with the equivalent fraction $\frac{1}{1}$. This removes the factor $\frac{7 \cdot x \cdot x \cdot y}{7 \cdot x \cdot x \cdot y}$, which is equal to 1.

$$= \frac{3x}{2y}$$ Multiply the remaining factors in the numerator: $3 \cdot 1 \cdot 1 \cdot 1 \cdot x \cdot 1 = 3x$. Multiply the remaining factors in the denominator: $2 \cdot 1 \cdot 1 \cdot 1 \cdot 1 \cdot y = 2y$.

We say that $\frac{21x^3y}{14x^2y^2}$ simplifies to $\frac{3x}{2y}$.

An alternate approach is to use rules for exponents to simplify the rational expression.

$$\frac{21x^3y}{14x^2y^2} = \frac{3 \cdot \overset{1}{\cancel{7}} \cdot x^{3-2}y^{1-2}}{2 \cdot \underset{1}{\cancel{7}}} = \frac{3x^1y^{-1}}{2} = \frac{3x}{2y}$$ To divide exponential expressions with the same base, keep the base and subtract the exponents.

Self Check 3 Simplify: $\frac{32a^3b^2}{24ab^4}$

Now Try **Problem 39**

To simplify rational expressions, we often make use of the factoring methods discussed in Chapter 6.

EXAMPLE 4 Simplify: **a.** $\dfrac{30t - 6}{36}$ **b.** $\dfrac{x^2 + 13x + 12}{x^2 + 12x}$

Strategy We will begin by factoring the numerator and denominator. Then we will remove any factors common to the numerator and denominator.

Why We need to make sure that the numerator and denominator have no common factor other than 1. When this is the case, the rational expression is simplified.

Solution

a. $\frac{30t - 6}{36} = \frac{6(5t - 1)}{6 \cdot 6}$ Factor the numerator: The GCF is 6. Factor the denominator.

$= \frac{\overset{1}{\cancel{6}}(5t - 1)}{\underset{1}{\cancel{6}} \cdot 6}$ Simplify by removing a factor equal to 1. Replace $\frac{6}{6}$ with $\frac{1}{1}$.

$= \frac{5t - 1}{6}$ Multiply the remaining factors in the numerator: $1 \cdot (5t - 1) = 5t - 1$. Multiply the remaining factors in the denominator: $1 \cdot 6 = 6$.

b. $\frac{x^2 + 13x + 12}{x^2 + 12x} = \frac{(x + 1)(x + 12)}{x(x + 12)}$ Factor the numerator. Factor the denominator: The GCF is x.

$= \frac{(x + 1)\overset{1}{\cancel{(x + 12)}}}{x\underset{1}{\cancel{(x + 12)}}}$ Simplify by replacing $\frac{x + 12}{x + 12}$ with the equivalent fraction $\frac{1}{1}$. This removes the factor $\frac{x + 12}{x + 12} = 1$.

$= \frac{x + 1}{x}$ This rational expression cannot be simplified further.

The Language of Algebra Sometimes, a common factor of the numerator and denominator has two or more terms. Here, we remove the common *binomial factor*, $x + 12$.

Self Check 4 Simplify: **a.** $\frac{4t - 20}{12}$ **b.** $\frac{x^2 - x - 6}{x^2 - 3x}$

Now Try Problems 43 and 47

Caution When simplifying rational expressions, we can only remove factors common to the entire numerator and denominator. *It is incorrect to remove any terms common to the numerator and denominator.*

$\frac{\overset{1}{\cancel{x}} + 1}{\underset{1}{\cancel{x}}}$ x is a term of $x + 1$.

$\frac{a^2 - 3a + \overset{1}{\cancel{2}}}{a + \underset{1}{\cancel{2}}}$ 2 is a term of $a^2 - 3a + 2$ and a term of $a + 2$.

$\frac{\overset{1}{\cancel{y^2}} - 36}{\underset{1}{\cancel{y^2}} - y - 7}$ y^2 is a term of $y^2 - 36$ and a term of $y^2 - y - 7$.

EXAMPLE 5

Simplify: **a.** $\frac{3x^2 - 8x - 3}{2x^5 - 18x^3}$ **b.** $\frac{(x - y)^4}{x^2 - 2xy + y^2}$

Strategy We will begin by factoring the numerator and denominator using the methods discussed in Chapter 5. Then we will remove any factors common to the numerator and denominator.

Why We need to make sure that the numerator and denominator have no common factor other than 1. When this is the case, then the rational expression is simplified.

Solution

a. $\frac{3x^2 - 8x - 3}{2x^5 - 18x^3} = \frac{(3x + 1)(x - 3)}{2x^3(x^2 - 9)}$ Factor the trinomial in the numerator. Factor the denominator: The GCF is $2x^3$.

$= \frac{(3x + 1)(x - 3)}{2x^3(x + 3)(x - 3)}$ In the denominator, factor the difference of two squares, $x^2 - 9$.

The Language of Algebra When a rational expression is simplified, the result is an *equivalent expression*. This means that $\frac{3x^2 - 8x - 3}{2x^5 - 18x^3}$ has the same value as $\frac{3x + 1}{2x^3(x + 3)}$ for all values of x, except those that make either denominator 0.

$$= \frac{(3x + 1)\overset{1}{\cancel{(x - 3)}}}{2x^3(x + 3)\underset{1}{\cancel{(x - 3)}}}$$

Simplify by replacing $\frac{x-3}{x-3}$ with the equivalent fraction $\frac{1}{1}$. This removes the factor $\frac{x-3}{x-3} = 1$.

$$= \frac{3x + 1}{2x^3(x + 3)}$$

It is not necessary to perform the multiplication $2x^3(x + 3)$ in the result. It is usually more convenient to leave the denominator in factored form.

b. $\frac{(x - y)^4}{x^2 - 2xy + y^2} = \frac{(x - y)^4}{(x - y)^2}$

In the denominator, factor the perfect square trinomial $x^2 - 2xy + y^2$.

$$= \frac{(x - y)(x - y)(x - y)(x - y)}{(x - y)(x - y)}$$

Write the repeated multiplication indicated by each exponent.

$$= \frac{\overset{1}{\cancel{(x - y)}}\overset{1}{\cancel{(x - y)}}(x - y)(x - y)}{\underset{1}{\cancel{(x - y)}}\underset{1}{\cancel{(x - y)}}}$$

Simplify by replacing each $\frac{x-y}{x-y}$ with $\frac{1}{1}$.

$$= (x - y)^2$$

Use an exponent to write the repeated multiplication in the numerator.

Notation

It is not necessary to find $(x - y)^2$. The result can be presented in factored form.

Self Check 5 Simplify: **a.** $\frac{4x^2 - 4x - 15}{8x^3 - 50x}$ **b.** $\frac{(a + 3b)^5}{a^2 + 6ab + 9b^2}$

Now Try **Problems 51 and 53**

EXAMPLE 6

Simplify: $\frac{5(x + 3) - 5}{7(x + 3) - 7}$

Strategy We will begin by simplifying the numerator, $5(x + 3) - 5$, and the denominator, $7(x + 3) - 7$, separately. Then we will factor each result and remove any common factors.

Why We cannot immediately remove $x + 3$ because it is not a factor of the *entire* numerator and the *entire* denominator.

Solution

$$\frac{5(x + 3) - 5}{7(x + 3) - 7} = \frac{5x + 15 - 5}{7x + 21 - 7}$$

Use the distributive property in the numerator and in the denominator.

$$= \frac{5x + 10}{7x + 14}$$

Combine like terms: $15 - 5 = 10$ and $21 - 7 = 14$.

$$= \frac{5(x + 2)}{7(x + 2)}$$

Factor the numerator: The GCF is 5.
Factor the denominator: The GCF is 7.

$$= \frac{5\overset{1}{\cancel{(x + 2)}}}{7\underset{1}{\cancel{(x + 2)}}}$$

Simplify by replacing $\frac{x+2}{x+2}$ with the equivalent fraction $\frac{1}{1}$. This removes the factor $\frac{x+2}{x+2} = 1$.

$$= \frac{5}{7}$$

The Language of Algebra

Some rational expressions cannot be simplified. For example, to attempt to simplify the following rational expression, we factor its numerator and denominator. Since there are no common factors, we say it *does not simplify* or that it is in *simplest form.*

$$\frac{x^2 + x - 2}{x^2 + x} = \frac{(x + 2)(x - 1)}{x(x + 1)}$$

Self Check 6 Simplify: $\frac{4(x-2)+4}{3(x-2)+3}$

Now Try Problem 55

4 Simplify Rational Expressions That Have Factors That Are Opposites.

If the terms of two polynomials are the same, except that they are opposite in sign, the polynomials are **opposites.** For example, the following pairs of polynomials are opposites.

$2a-1$ and $1-2a$
Compare terms: $2a$ and $-2a$; -1 and 1.

$-3x^2-x+5$ and $3x^2+x-5$
Compare terms: $-3x^2$ and $3x^2$; $-x$, and x; 5 and -5.

Success Tip
When a difference is reversed, the original binomial and the resulting binomial are opposites. Here are some pairs of opposites:

$b-11$ and $11-b$
x^2-4 and $4-x^2$

We have seen that the quotient of two real numbers that are opposites is always -1:

$$\frac{2}{-2}=-1 \qquad \frac{-78}{78}=-1 \qquad \frac{3.5}{-3.5}=-1$$

Likewise, the quotient of two binomials that are opposites is always -1.

EXAMPLE 7 Simplify: $\frac{2a-1}{1-2a}$

Strategy We will rearrange the terms of the numerator, $2a-1$, and factor out -1.

Why This step is useful when the numerator and denominator contain factors that are opposites, such as $2a-1$ and $1-2a$. It produces a common factor that can be removed.

Solution

$$\frac{2a-1}{1-2a}=\frac{-1+2a}{1-2a}$$
Think of the numerator, $2a-1$, as $2a+(-1)$. Then change the order of the terms: $2a+(-1)=-1+2a$.

$$=\frac{-1(1-2a)}{1-2a}$$
Factor out -1 from the two terms of the numerator.

$$=\frac{-1\overset{1}{\cancel{(1-2a)}}}{\underset{1}{\cancel{1-2a}}}$$
Simplify by replacing $\frac{1-2a}{1-2a}$ with the equivalent fraction $\frac{1}{1}$. This removes the factor $\frac{1-2a}{1-2a}=1$.

$$=\frac{-1}{1}$$
Multiply the remaining factors in the numerator.

$$=-1$$
Any number divided by 1 is itself.

Success Tip
The result would have been the same if, instead, we had factored out -1 from the denominator, $1-2a$.

Self Check 7 Simplify: $\frac{3p-2}{2-3p}$

Now Try Problem 59

In general, we have this fact.

The Quotient of Opposites

The quotient of any nonzero polynomial and its opposite is -1.

Caution

Don't incorrectly apply the rule for opposites to a rational expression such as $\frac{x+1}{1+x}$. This is the quotient of a number and itself. The result is 1, not -1.

$$\frac{x+1}{1+x} = \frac{\overset{1}{\cancel{x+1}}}{\underset{1}{\cancel{x+1}}} = 1$$

For each of the following rational expressions, the numerator and denominator are opposites. Thus, each expression is equal to -1.

$$\frac{x-6}{6-x} = -1 \qquad \frac{2a-9b}{9b-2a} = -1 \qquad \frac{-3x^2-x+5}{3x^2+x-5} = -1$$

This fact can be used to simplify certain rational expressions by removing a factor equal to -1. If a factor of the numerator is the opposite of a factor of the denominator, we can replace them with the equivalent fraction $\frac{-1}{1}$, as shown in the following example.

EXAMPLE 8 Simplify, if possible: **a.** $\frac{y^2-1}{3-3y}$ **b.** $\frac{t+8}{t-8}$

Strategy We will begin by factoring the numerator and denominator. Then we look for common factors, or factors that are opposites, and remove them.

Why We need to make sure that the numerator and denominator have no common factor (or opposite factors) other than 1. When this is the case, then the rational expression is simplified.

Solution

a.
$$\frac{y^2-1}{3-3y} = \frac{(y+1)(y-1)}{3(1-y)}$$
Factor the numerator. Factor the denominator.

$$= \frac{(y+1)\overset{-1}{\cancel{(y-1)}}}{3\underset{1}{\cancel{(1-y)}}}$$
Since $y-1$ and $1-y$ are opposites, simplify by replacing $\frac{y-1}{1-y}$ with the equivalent fraction $\frac{-1}{1}$. This removes the factor $\frac{y-1}{1-y} = -1$.

$$= \frac{-(y+1)}{3}$$

This result may be written in several other equivalent forms.

$$\frac{-(y+1)}{3} = -\frac{y+1}{3}$$
The $-$ symbol in $-(y+1)$ can be written in the front of the fraction, and the parentheses can be dropped.

$$\frac{-(y+1)}{3} = \frac{-y-1}{3}$$
The $-$ symbol in $-(y+1)$ represents a factor of -1. Distribute the multiplication by -1 in the numerator.

$$\frac{-(y+1)}{3} = \frac{y+1}{-3}$$
The $-$ symbol in $-(y+1)$ can be applied to the denominator. However, we don't usually use this form.

Caution

A $-$ symbol in front of a fraction may be applied to the numerator or to the denominator, but not to both:

$$-\frac{y+1}{3} \neq \frac{-(y+1)}{-3}$$

b. The binomials $t+8$ and $t-8$ are not opposites because their first terms do not have opposite signs. Thus, $\frac{t+8}{t-8}$ does not simplify.

Self Check 8 Simplify, if possible: **a.** $\frac{m^2-100}{10m-m^2}$
b. $\frac{2x-3}{2x+3}$

Now Try **Problem 63**

ANSWERS TO SELF CHECKS **1.** $\frac{13}{50}$ **2. a.** -9 **b.** $-5, 5$ **c.** None **3.** $\frac{4a^2}{3b^2}$ **4. a.** $\frac{t-5}{3}$ **b.** $\frac{x+2}{x}$ **5. a.** $\frac{2x+3}{2x(2x+5)}$ **b.** $(a+3b)^3$ **6.** $\frac{4}{3}$ **7.** -1 **8. a.** $-\frac{m+10}{m}$ **b.** Does not simplify

STUDY SET 7.1

VOCABULARY

Fill in the blanks.

1. A quotient of two polynomials, such as $\frac{x^2 + x}{x^2 - 3x}$, is called a ________ expression.
2. To simplify a rational expression, we remove common ________ of the numerator and denominator.
3. Because of the division by 0, the expression $\frac{8}{0}$ is ________.
4. The binomials $x - 15$ and $15 - x$ are called ________, because their terms are the same, except that they are opposite in sign.

CONCEPTS

5. When we simplify $\frac{x^2 + 5x}{4x + 20}$, the result is $\frac{x}{4}$. These equivalent expressions have the same value for all real numbers, except $x = -5$. Show that they have the same value for $x = 1$.
6. Determine whether each pair of polynomials are opposites. Write *yes* or *no*.
   a. $y + 7$ and $y - 7$
   b. $b - 20$ and $20 - b$
   c. $x^2 + 2x - 1$ and $-x^2 - 2x - 1$
7. Simplify each expression, if possible.
   a. $\frac{x - 8}{x - 8}$
   b. $\frac{x - 8}{8 - x}$
   c. $\frac{x + 8}{8 + x}$
   d. $\frac{x + 8}{x}$
8. Simplify each expression.
   a. $\frac{(x + 2)(x - 2)}{(x + 1)(x + 2)}$
   b. $\frac{y(y - 2)}{9(2 - y)}$
   c. $\frac{(2m + 7)(m - 5)}{(2m + 7)}$
   d. $\frac{x \cdot x}{x \cdot x(x - 30)}$

NOTATION

Complete the solution to simplify the rational expression.

9. $$\frac{x^2 + 2x + 1}{x^2 + 4x + 3} = \frac{(x + 1)(\quad + 1)}{(x + 3)(x + \quad)}$$
$$= \frac{\overset{1}{(x + 1)(x + 1)}}{(x + 3)\underset{1}{}}$$
$$= \frac{x + 1}{}$$

10. In the following table, a student's answers to three homework problems are compared with the answers in the back of the book. Are the answers equivalent?

Answer	Book's answer	Equivalent?
$\frac{-3}{x + 3}$	$-\frac{3}{x + 3}$	
$\frac{-x + 4}{6x + 1}$	$\frac{-(x - 4)}{6x + 1}$	
$\frac{x + 7}{(x - 4)(x + 2)}$	$\frac{x + 7}{(x + 2)(x - 4)}$	

GUIDED PRACTICE

Evaluate each expression for $x = 6$. See Example 1.

11. $\frac{x - 2}{x - 5}$
12. $\frac{3x - 2}{x - 2}$
13. $\frac{x^2 - 4x - 12}{x^2 + x - 2}$
14. $\frac{x^2 - 1}{x^3 - 1}$
15. $\frac{-x + 1}{x^2 - 5x - 6}$
16. $\frac{-2x^2 - 3}{x - 6}$

Evaluate each expression for $y = -3$. See Example 1.

17. $\frac{y + 5}{3y - 2}$
18. $\frac{2y + 9}{y^2 + 25}$
19. $\frac{y^2 + 9}{9 - y^2}$
20. $\frac{-y - 11}{y^2 + 2y - 3}$
21. $\frac{-y}{y^2 - y + 6}$
22. $\frac{y^3}{3y^2 + 1}$

Find all real numbers for which the rational expression is undefined. See Example 2.

23. $\frac{x + 5}{8x}$
24. $\frac{4x - 1}{6x}$
25. $\frac{15}{x - 2}$
26. $\frac{5x}{x + 5}$
27. $\frac{15x + 2}{x^2 + 6}$
28. $\frac{x^2 - 4x}{x^2 + 4}$
29. $\frac{x + 1}{2x - 1}$
30. $\frac{-6x}{3x - 1}$
31. $\frac{30x}{x^2 - 36}$
32. $\frac{2x - 15}{x^2 - 49}$
33. $\frac{15}{x^2 + x - 2}$
34. $\frac{x - 20}{x^2 + 2x - 8}$

Simplify each expression. See Example 3.

35. $\frac{45}{9a}$
36. $\frac{48}{16y}$
37. $\frac{6x^2}{4x^2}$
38. $\frac{9x}{6x}$

$$= \frac{9 \cdot \overset{1}{\cancel{7}} \cdot \overset{1}{\cancel{x}} \cdot 1}{1 \cdot \underset{1}{\cancel{7}} \cdot \underset{1}{\cancel{x}}}$$ Write 63x in factored form as 9 · 7 · x. Then simplify by removing a factor equal to 1: $\frac{7x}{7x}$.

$$= 9$$ Because $\frac{9}{1} = 9$.

b. $5a\left(\frac{3a - 1}{a}\right) = \frac{5a}{1}\left(\frac{3a - 1}{a}\right)$ Write 5a as a fraction: $5a = \frac{5a}{1}$.

$$= \frac{5\overset{1}{\cancel{a}}(3a - 1)}{1 \cdot \underset{1}{\cancel{a}}}$$ Multiply the numerators and multiply the denominators. Then simplify by removing a factor equal to 1: $\frac{a}{a}$.

$$= 5(3a - 1)$$

Note that $5(3a - 1)$ can be written as $15a - 5$.

Self Check 3 Multiply and simplify the result, if possible: **a.** $36b\left(\frac{1}{6b}\right)$ **b.** $4x\left(\frac{x + 3}{x}\right)$

Now Try **Problems 37 and 41**

2 Divide Rational Expressions.

Recall that one number is the **reciprocal** of another if their product is 1. To find the reciprocal of a fraction, we invert its numerator and denominator. We have seen that to divide fractions, we multiply the first fraction by the reciprocal of the second fraction.

$$\frac{4}{7} \div \frac{3}{5} = \frac{4}{7} \cdot \frac{5}{3}$$ Invert $\frac{3}{5}$ and change the division to a multiplication.

$$= \frac{20}{21}$$ Multiply the numerators and multiply the denominators.

We use the same procedure to divide rational expressions.

Dividing Rational Expressions

To divide two rational expressions, multiply the first by the reciprocal of the second. Then, if possible, we factor and simplify.

For any two rational expressions, $\frac{A}{B}$ and $\frac{C}{D}$, where $\frac{C}{D} \neq 0$,

$$\frac{A}{B} \div \frac{C}{D} = \frac{A}{B} \cdot \frac{D}{C} = \frac{AD}{BC}$$

EXAMPLE 4 Divide: **a.** $\frac{a}{13} \div \frac{17}{26}$ **b.** $\frac{9x}{35y} \div \frac{15x^2}{14}$

Strategy We will use the rule for dividing rational expressions. After multiplying by the reciprocal, we will factor the monomials that are not prime, and remove any common factors of the numerator and denominator.

Why We want to give the result in simplified form, which requires that the numerator and denominator have no common factor other than 1.

Solution

a. $\frac{a}{13} \div \frac{17}{26} = \frac{a}{13} \cdot \frac{26}{17}$ Multiply by the reciprocal of $\frac{17}{26}$.

$= \frac{a \cdot 2 \cdot 13}{13 \cdot 17}$ Multiply the numerators and denominators. Then factor 26 as $2 \cdot 13$.

$= \frac{a \cdot 2 \cdot \overset{1}{\cancel{13}}}{\underset{1}{\cancel{13}} \cdot 17}$ Simplify by removing common factors of the numerator and denominator.

$= \frac{2a}{17}$ Multiply the remaining factors in the numerator. Multiply the remaining factors in the denominator.

b. $\frac{9x}{35y} \div \frac{15x^2}{14} = \frac{9x}{35y} \cdot \frac{14}{15x^2}$ Multiply by the reciprocal of $\frac{15x^2}{14}$.

$= \frac{3 \cdot 3 \cdot x \cdot 2 \cdot 7}{5 \cdot 7 \cdot y \cdot 3 \cdot 5 \cdot x \cdot x}$ Multiply the numerators and denominators. Then factor 9, 35, 14, and $15x^2$.

$= \frac{3 \cdot \overset{1}{\cancel{3}} \cdot \overset{1}{\cancel{x}} \cdot 2 \cdot \overset{1}{\cancel{7}}}{5 \cdot \underset{1}{\cancel{7}} \cdot y \cdot \underset{1}{\cancel{3}} \cdot 5 \cdot \underset{1}{\cancel{x}} \cdot x}$ Simplify by removing factors equal to 1.

$= \frac{6}{25xy}$ Multiply the remaining factors in the numerator. Multiply the remaining factors in the denominator.

Caution
When dividing rational expressions, always write the result in simplest form, by removing any factors common to the numerator and denominator.

Self Check 4 Divide and simplify the result, if possible: $\frac{8a}{3b} \div \frac{16a^2}{9b^2}$

Now Try **Problems 45 and 49**

EXAMPLE 5

Divide: $\frac{x^2 + x}{3x - 15} \div \frac{(x + 1)^2}{6x - 30}$

Strategy To find the quotient, we will use the rule for dividing rational expressions. After multiplying by the reciprocal, we will factor the binomials that are not prime, and remove any common factors of the numerator and denominator.

Why We want to give the result in simplified form, which requires that the numerator and denominator have no common factor other than 1.

Solution

$\frac{x^2 + x}{3x - 15} \div \frac{(x + 1)^2}{6x - 30}$

$= \frac{x^2 + x}{3x - 15} \cdot \frac{6x - 30}{(x + 1)^2}$ Multiply by the reciprocal of $\frac{(x + 1)^2}{6x - 30}$.

$= \frac{x(x + 1) \cdot 2 \cdot 3(x - 5)}{3(x - 5)(x + 1)(x + 1)}$ Multiply the numerators and multiply the denominators. Then factor the binomials. Write $(x + 1)^2$ as repeated multiplication.

$= \frac{x\overset{1}{\cancel{(x + 1)}} \cdot 2 \cdot \overset{1}{\cancel{3}}\overset{1}{\cancel{(x - 5)}}}{\underset{1}{\cancel{3}}\underset{1}{\cancel{(x - 5)}}\underset{1}{\cancel{(x + 1)}}(x + 1)}$ Simplify by removing common factors of the numerator and denominator.

The Language of Algebra
To find the reciprocal of $\frac{(x + 1)^2}{6x - 30}$, we invert it. To *invert* means to turn upside down: $\frac{6x - 30}{(x + 1)^2}$. Some amusement park thrill rides have giant loops where the riders become *inverted*.

$$= \frac{2x}{x + 1}$$ Multiply the remaining factors in the numerator. Multiply the remaining factors in the denominator.

Self Check 5 Divide and simplify the result, if possible:
$$\frac{z^2 - 9}{z^2 + 4z + 3} \div \frac{z^2 - 3z}{(z + 1)^2}$$

Now Try **Problem 59**

EXAMPLE 6

Divide: $\dfrac{2x^2 - 3xy - 2y^2}{2x + y} \div (4y^2 - x^2)$

Strategy We begin by writing $4y^2 - x^2$ as a rational expression by inserting a denominator 1. Then we will use the rule for dividing rational expressions.

Why Writing $4y^2 - x^2$ over 1 is helpful when we invert its numerator and denominator to find its reciprocal.

Solution

$$\frac{2x^2 - 3xy - 2y^2}{2x + y} \div (4y^2 - x^2)$$

$$= \frac{2x^2 - 3xy - 2y^2}{2x + y} \div \frac{4y^2 - x^2}{1}$$ Write $4y^2 - x^2$ as a fraction with a denominator of 1.

$$= \frac{2x^2 - 3xy - 2y^2}{2x + y} \cdot \frac{1}{4y^2 - x^2}$$ Multiply by the reciprocal of $\frac{4y^2 - x^2}{1}$.

$$= \frac{(2x + y)(x - 2y) \cdot 1}{(2x + y)(2y + x)(2y - x)}$$ Multiply the numerators and denominators. Then factor $2x^2 - 3xy - 2y^2$ and $4y^2 - x^2$.

$$= \frac{\overset{1}{\cancel{(2x + y)}}\overset{-1}{\cancel{(x - 2y)}} \cdot 1}{\underset{1}{\cancel{(2x + y)}}(2y + x)\underset{1}{\cancel{(2y - x)}}}$$ Since $x - 2y$ and $2y - x$ are opposites, simplify by replacing $\frac{x - 2y}{2y - x}$ with $\frac{-1}{1}$.

$$= \frac{-1}{2y + x}$$

Note that $\frac{-1}{2y + x}$ can be written as $-\frac{1}{2y + x}$.

Self Check 6 Divide and simplify the result, if possible: $(b - a) \div \dfrac{a^2 - b^2}{a^2 + ab}$

Now Try **Problem 67**

3 Convert Units of Measurement.

We can use the concepts discussed in this section to make conversions from one unit of measure to another. *Unit conversion factors* play an important role in this process. A **unit conversion factor** is a fraction that has a value of 1. For example, we can use the fact that 1 square yard = 9 square feet to form two unit conversion factors:

$$\frac{1 \text{ yd}^2}{9 \text{ ft}^2} = 1$$ Read as "1 square yard per 9 square feet."

$$\frac{9 \text{ ft}^2}{1 \text{ yd}^2} = 1$$ Read as "9 square feet per 1 square yard."

Success Tip

Remember that unit conversion factors are equal to 1. Some examples are:

$$\frac{12 \text{ in.}}{1 \text{ ft}} = 1 \qquad \frac{60 \text{ min}}{1 \text{ hr}} = 1$$

Since a unit conversion factor is equal to 1, multiplying a measurement by a unit conversion factor does not change the measurement, it only changes the units of measure.

EXAMPLE 7 ***Carpeting.*** A roll of carpeting is 12 feet wide and 150 feet long. Find the number of square yards of carpeting on the roll.

Strategy We will begin by determining the number of square feet of carpeting on the roll. Then we will multiply that result by a unit conversion factor.

Why A properly chosen unit conversion factor can convert the number of square feet of carpeting on the roll to the number of square yards on the roll.

Solution When unrolled, the carpeting forms a rectangular shape with an area of $12 \cdot 150 = 1{,}800$ square feet. We will multiply 1,800 ft^2 by a unit conversion factor such that the units of ft^2 are removed and the units of yd^2 are introduced. Since 1 yd^2 = 9 ft^2, we will use $\frac{1\text{yd}^2}{9\text{ft}^2}$.

$$\frac{1{,}800\text{ ft}^2}{1\text{ roll}} = \frac{1{,}800\text{ ft}^2}{1\text{ roll}} \cdot \frac{\mathbf{1\text{ yd}^2}}{\mathbf{9\text{ ft}^2}}$$ Multiply by a unit conversion factor that relates yd^2 to ft^2.

$$= \frac{1{,}800\,\cancel{\text{ft}^2}}{1\text{ roll}} \cdot \frac{1\text{ yd}^2}{9\,\cancel{\text{ft}^2}}$$ Remove the units of ft^2 that are common to the numerator and denominator.

$$= \frac{200\text{ yd}^2}{1\text{ roll}}$$ Divide 1,800 by 9 to get 200.

There are 200 yd^2 of carpeting on the roll.

Self Check 7 Convert 5,400 ft^2 to square yards.

Now Try **Problem 71**

EXAMPLE 8 ***The Speed of Light.*** The speed with which light moves through space is about 186,000 miles per second. Express this speed in miles per minute.

Strategy The speed of light can be expressed as $\frac{186{,}000\text{ mi}}{1\text{ sec}}$. We will multiply that fraction by a unit conversion factor.

Why A properly chosen unit conversion factor can convert the number of miles traveled per second to the number of miles traveled per minute.

Solution We will multiply $\frac{186{,}000\text{ mi}}{1\text{ sec}}$ by a unit conversion factor such that the units of seconds are removed and the units of minutes are introduced. Since 60 seconds = 1 minute, we will use $\frac{60\text{ sec}}{1\text{ min}}$.

$$\frac{186{,}000\text{ mi}}{1\text{ sec}} = \frac{186{,}000\text{ mi}}{1\text{ sec}} \cdot \frac{\mathbf{60\text{ sec}}}{\mathbf{1\text{ min}}}$$ Multiply by a unit conversion factor that relates seconds to minutes.

$$= \frac{186{,}000 \text{ mi}}{1 \text{ sec}} \cdot \frac{60 \text{ sec}}{1 \text{ min}}$$ Remove the units of seconds that are common to the numerator and denominator.

$$= \frac{11{,}160{,}000 \text{ mi}}{1 \text{ min}}$$ Multiply 186,000 and 60 to get 11,160,000.

The speed of light is about 11,160,000 miles per minute.

Self Check 8 A mosquito flaps it wings about 600 times per second. How many times is that per minute?

Now Try **Problem 75**

ANSWERS TO SELF CHECKS **1. a.** $\frac{12y}{(y + 6)(y - 4)}$ **b.** $\frac{3a}{11}$ **2. a.** $\frac{(n - 3)(3n - 2)}{2}$ **b.** $-\frac{2(m + 1)}{3m + 1}$ **3. a.** 6 **b.** $4x + 12$ **4.** $\frac{3b}{2a}$ **5.** $\frac{z + 1}{z}$ **6.** $-a$ **7.** 600 yd^2 **8.** 36,000 flaps per minute

STUDY SET 7.2

VOCABULARY

Fill in the blanks.

1. The ________ of $\frac{x^2 + 6x + 1}{10x}$ is $\frac{10x}{x^2 + 6x + 1}$.

2. A _____ conversion factor is a fraction that is equal to 1, such as $\frac{3 \text{ ft}}{1 \text{ yd}}$.

CONCEPTS

Fill in the blanks.

3. a. To multiply rational expressions, multiply their ________ and multiply their ________. To divide two rational expressions, multiply the first by the ________ of the second. In symbols,

b. $\frac{A}{B} \cdot \frac{C}{D} = \frac{\square}{\square}$ and $\frac{A}{B} \div \frac{C}{D} = \frac{A}{B} \cdot \frac{\square}{\square}$

Simplify each expression.

4. $\frac{(x + 7) \cdot 2 \cdot 5}{5(x + 1)(x + 7)(x - 9)}$

5. $\frac{y \cdot y \cdot y(15 - y)}{y(y - 15)(y + 1)}$

6. a. Write $3x + 5$ in fractional form.

b. What is the reciprocal of $18x$?

7. Find the product of the rational expression and its reciprocal.

$$\frac{3}{x + 2} \cdot \frac{x + 2}{3}$$

8. Use the fact that 1 tablespoon = 3 teaspoons to write two unit conversion factors.

NOTATION

9. What units are common to the numerator and denominator?

$$\frac{45 \text{ ft}}{1} \cdot \frac{1 \text{ yd}}{3 \text{ ft}}$$

10. a. What fact is indicated by the unit conversion factor $\frac{1 \text{ day}}{24 \text{ hours}}$?

b. Fill in the blank: $\frac{1 \text{ day}}{24 \text{ hours}} = \square$.

GUIDED PRACTICE

***Multiply, and then simplify, if possible.* See Example 1.**

11. $\frac{3}{7} \cdot \frac{y}{2}$

12. $\frac{2}{7} \cdot \frac{z}{3}$

13. $\frac{y + 2}{y} \cdot \frac{3}{y^2}$

14. $\frac{4}{a + 1} \cdot \frac{a}{7}$

15. $\frac{35n}{12} \cdot \frac{16}{7n^2}$

16. $\frac{11m}{21} \cdot \frac{14}{55m^3}$

17. $\frac{2x^2y}{3xy} \cdot \frac{3xy^2}{2}$

18. $\frac{2x^2z}{z} \cdot \frac{5x}{z}$

***Multiply, and then simplify, if possible.* See Example 2.**

19. $\frac{x + 5}{5} \cdot \frac{x}{x + 5}$

20. $\frac{a - 9}{9} \cdot \frac{8a}{a - 9}$

21. $\frac{x-2}{x} \cdot \frac{2x}{2-x}$

22. $\frac{y-3}{y} \cdot \frac{3y}{3-y}$

23. $\frac{2x+6}{x+3} \cdot \frac{3}{4x}$

24. $\frac{3y-9}{y-3} \cdot \frac{y}{3y^2}$

25. $\frac{(x+1)^2}{x+2} \cdot \frac{x+2}{x+1}$

26. $\frac{(y-3)^2}{y-5} \cdot \frac{y-5}{y-3}$

27. $\frac{x^2-x}{x} \cdot \frac{3x-6}{3-3x}$

28. $\frac{5z-10}{z+2} \cdot \frac{3}{6-3z}$

29. $\frac{x^2+x-6}{5x} \cdot \frac{5x-10}{x+3}$

30. $\frac{z^2+4z-5}{5z-5} \cdot \frac{5z}{z+5}$

31. $\frac{m^2-2m-3}{2m+4} \cdot \frac{m^2-4}{m^2+3m+2}$

32. $\frac{p^2-p-6}{3p-9} \cdot \frac{2p^2-5p-3}{p^2-3p}$

33. $\frac{6a^2}{a^2+6a+9} \cdot \frac{(a+3)^4}{4a^5}$

34. $\frac{9b^3}{b^2-8b+16} \cdot \frac{(b-4)^4}{15b^8}$

Multiply, and then simplify, if possible. **See Example 3.**

35. $7m\left(\frac{5}{m}\right)$

36. $9p\left(\frac{10}{p}\right)$

37. $15x\left(\frac{x+1}{5x}\right)$

38. $30t\left(\frac{t-7}{10t}\right)$

39. $12y\left(\frac{5y-8}{6y}\right)$

40. $16x\left(\frac{3x+8}{4x}\right)$

41. $24\left(\frac{3a-5}{2a}\right)$

42. $28\left(\frac{8-3t}{4t}\right)$

Divide, and then simplify, if possible. **See Example 4.**

43. $\frac{2}{y} \div \frac{4}{3}$

44. $\frac{3}{a} \div \frac{9}{5}$

45. $\frac{3a}{25} \div \frac{1}{5}$

46. $\frac{3y}{8} \div \frac{3}{2}$

47. $\frac{x^3}{18y} \div \frac{x}{6y}$

48. $\frac{21x}{z^2} \div \frac{7x^3}{z^5}$

49. $\frac{27p^4}{35q} \div \frac{9p}{21q}$

50. $\frac{12}{25s^5} \div \frac{10}{15s^2}$

Divide, and then simplify, if possible. **See Example 5.**

51. $\frac{9a-18}{28} \div \frac{9a^3}{35}$

52. $\frac{3x+6}{40} \div \frac{3x^2}{24}$

53. $\frac{x^2-4}{3x+6} \div \frac{2-x}{x+2}$

54. $\frac{x^2-9}{5x+15} \div \frac{3-x}{x+3}$

55. $\frac{m^2+m-20}{m} \div \frac{4-m}{m}$

56. $\frac{n^2+4n-21}{n} \div \frac{3-n}{n}$

57. $\frac{t^2+5t-14}{t} \div \frac{t-2}{t}$

58. $\frac{r^2+12r+11}{r} \div \frac{r+11}{r}$

59. $\frac{x^2-2x-35}{3x^2+27x} \div \frac{3x^2+17x+10}{18x^2+12x}$

60. $\frac{x^2-x-6}{2x^2+9x+10} \div \frac{x^2-25}{2x^2+15x+25}$

61. $\frac{36c^2-49d^2}{3d^3} \div \frac{12c+14d}{d^4}$

62. $\frac{25y^2-16z^2}{2yz} \div \frac{10y-8z}{y^2}$

Divide, and then simplify, if possible. **See Example 6.**

63. $\frac{x^2-1}{3x-3} \div (x+1)$

64. $\frac{x^2-16}{x-4} \div (3x+12)$

65. $\frac{n^2-10n+9}{n-9} \div (n-1)$

66. $\frac{r^2-11r+18}{r-9} \div (r-2)$

67. $\frac{2r-3s}{12} \div (4r^2-12rs+9s^2)$

68. $\frac{3m+n}{18} \div (9m^2+6mn+n^2)$

69. $24n^2 \div \frac{18n^3}{n-1}$

70. $12m \div \frac{16m^2}{m+4}$

Complete each unit conversion. **See Examples 7 and 8.**

71. $\frac{150 \text{ yards}}{1} \cdot \frac{3 \text{ feet}}{1 \text{ yard}} = ?$

72. $\frac{60 \text{ inches}}{1} \cdot \frac{1 \text{ feet}}{12 \text{ inches}} = ?$

73. $\frac{6 \text{ pints}}{1} \cdot \frac{1 \text{gallon}}{8 \text{ pints}} = ?$

74. $\frac{4 \text{ cups}}{1} \cdot \frac{1 \text{gallon}}{16 \text{ cups}} = ?$

75. $\frac{30 \text{ miles}}{1 \text{ hour}} \cdot \frac{1 \text{ hour}}{60 \text{ minute}} = ?$

76. $\frac{300 \text{ meters}}{3 \text{ months}} \cdot \frac{12 \text{ months}}{1 \text{ year}} = ?$

77. $\frac{30 \text{ meters}}{1 \text{ seconds}} \cdot \frac{60 \text{ seconds}}{1 \text{ minutes}} = ?$

78. $\frac{288 \text{ inches}^2}{1 \text{ year}} \cdot \frac{1 \text{ feet}^2}{144 \text{ inches}^2} = ?$

TRY IT YOURSELF

Perform the operations and simplify, if possible.

79. $\frac{b^2-5b+6}{b^2-10b+16} \div \frac{b^2+2b}{b^2-6b-16}$

80. $\frac{m^2+m-6}{m^2-6m+9} \div \frac{m^2-4}{m^2-9}$

81. $\frac{5x+5}{25} \cdot \frac{5}{(x+1)^3}$

82. $\frac{7t-7}{28} \cdot \frac{4}{(t-1)^4}$

83. $10h\left(\frac{5h-3}{2h}\right)$

84. $33r\left(\frac{5r+4}{11r}\right)$

85. $\dfrac{n^2-9}{n^2-3n} \div \dfrac{n+3}{n^2-n}$

86. $\dfrac{b^2-b}{b+2} \div \dfrac{b^2-2b}{b^2-4}$

87. $\dfrac{10r^2s}{6rs^2} \cdot \dfrac{3r^3}{2rs}$

88. $\dfrac{3a^3b}{25cd^3} \cdot \dfrac{5cd^2}{6ab}$

89. $\dfrac{7}{3p^3} \cdot \dfrac{p+2}{p}$

90. $\dfrac{5t^2}{11} \cdot \dfrac{2t}{t-5}$

91. $\dfrac{5x^2+13x-6}{x+3} \div \dfrac{5x^2-17x+6}{x-2}$

92. $\dfrac{3p^2+5p-2}{p^3+2p^2} \div \dfrac{6p^2+13p-5}{2p^3+5p^2}$

93. $\dfrac{4x^2-12xy+9y^2}{x^3y^2} \cdot \dfrac{x^3y}{4x^2-9y^2}$

94. $\dfrac{ab^4}{25a^2-16b^2} \cdot \dfrac{25a^2-40ab+16b^2}{a^2b^4}$

APPLICATIONS

95. GEOMETRY Find the area of the rectangle.

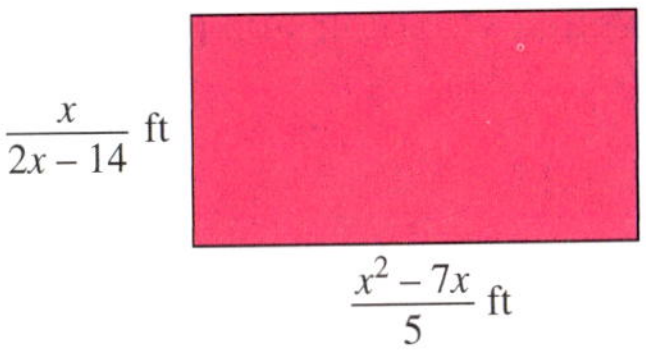

96. MOTION The table contains algebraic expressions for the rate an object travels and the time traveled at that rate. Complete the table.

Rate (mph)	Time (hr)	Distance (mi)
$\dfrac{k^2+k-6}{k-3}$	$\dfrac{k^2-9}{k^2-4}$	

97. TALKING According to the *Sacramento Bee* newspaper, the number of words an average man speaks a day is about 12,000. How many words does an average man speak in 1 year? (*Hint:* 365 days = 1 year.)

98. CLASSROOM SPACE The recommended size of an elementary school classroom in the United States is approximately 900 square feet. Convert this to square yards.

99. NATURAL LIGHT According to the University of Georgia School Design and Planning Laboratory, the basic classroom should have at least 72 square feet of windows for natural light. Convert this to square yards.

100. TRUCKING A cement truck holds 9 cubic yards of concrete. How many cubic feet of concrete does it hold? (*Hint:* 27 cubic feet = 1 cubic yard.)

101. BEARS The maximum speed a grizzly bear can run is about 30 miles per hour. What is its maximum speed in miles per minute?

102. FUEL ECONOMY Use the information that follows to determine the miles per fluid ounce of gasoline for city and for highway driving for the Dodge Dakota Pickup. (*Hint:* 1 gallon = 128 fluid ounces.)

2007 Dodge Dakota Pickup

Fuel Economy

Fuel Type	Regular
MPG (city)	**16**
MPG (highway)	**20**

103. TV TRIVIA On the comedy television series *Green Acres* (1965–1971), New York socialites Oliver Wendell Douglas (played by Eddie Albert) and his wife, Lisa Douglas (played by Eva Gabor), move from New York to purchase a 160-acre farm in Hooterville. Convert this to square miles. (*Hint:* 1 square mile = 640 acres.)

104. CAMPING The capacity of backpacks is usually given in cubic inches. Convert a backpack capacity of 5,400 cubic inches to cubic feet. (*Hint:* 1 cubic foot = 1,728 cubic inches.)

WRITING

105. Explain how to multiply rational expressions.

106. To divide rational expressions, you must first know how to multiply rational expressions. Explain why.

107. Explain why 60 miles per hour and 1 mile per minute are the same speed.

108. Explain why the unit conversion factor $\frac{1 \text{ ft}}{12 \text{ in.}}$ is equal to 1.

REVIEW

109. HARDWARE A brace has a length that is 2 inches less than twice the width of the shelf that it supports. The brace is anchored to the wall 8 inches below the shelf. Find the width of the shelf and the length of the brace.

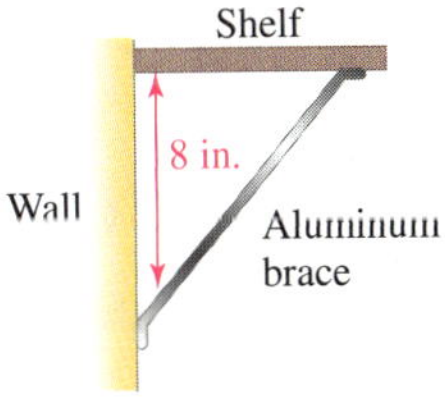

110. Solve $A = \frac{1}{2}h(b+d)$ for b.

CHALLENGE PROBLEMS

Perform the operations. Simplify, if possible.

111. $\dfrac{c^3-2c^2+5c-10}{c^2-c-2} \cdot \dfrac{c^3+c^2-5c-5}{c^4-25}$

112. $\dfrac{x^3-y^3}{x^3+y^3} \div \dfrac{x^3+x^2y+xy^2}{x^2y-xy^2+y^3}$

113. $\dfrac{-x^3+x^2+6x}{3x^3+21x^2} \div \left(\dfrac{2x+4}{3x^2} \div \dfrac{2x+14}{x^2-3x}\right)$

114. $\dfrac{x^2-y^2}{2x^4-2x^3} \div \left(\dfrac{x-y}{2x^2} \div \dfrac{x+y}{x^2+2xy+y^2}\right)$

SECTION 7.3
Adding and Subtracting with Like Denominators; Least Common Denominators

Objectives

1. Add and subtract rational expressions that have the same denominator.
2. Find the least common denominator.
3. Build rational expressions into equivalent expressions.

In this section, we extend the rules for adding and subtracting fractions to problems involving addition and subtraction of rational expressions.

1 Add and Subtract Rational Expressions That Have the Same Denominator.

Recall from Chapter 1 that to add (or subtract) fractions that have the same denominator, we add (or subtract) their numerators and write the sum (or difference) over the common denominator. For example,

$$\frac{3}{7} + \frac{2}{7} = \frac{3+2}{7} = \frac{5}{7} \quad \text{and} \quad \frac{18}{25} - \frac{9}{25} = \frac{18-9}{25} = \frac{9}{25}$$

The Language of Algebra
We can describe $\frac{3}{7}$ and $\frac{2}{7}$ as having the *same* denominator, *common* denominators, or *like* denominators.

We use the same procedure to add and subtract rational expressions with like denominators.

Adding and Subtracting Rational Expressions That Have the Same Denominator

To add (or subtract) rational expressions that have same denominator, add (or subtract) their numerators and write the sum (or difference) over the common denominator. Then, if possible, factor and simplify.

If $\frac{A}{D}$ and $\frac{B}{D}$ are rational expressions,

$$\frac{A}{D} + \frac{B}{D} = \frac{A+B}{D} \quad \text{and} \quad \frac{A}{D} - \frac{B}{D} = \frac{A-B}{D}$$

EXAMPLE 1 Add: **a.** $\frac{x}{8} + \frac{3x}{8}$ **b.** $\frac{4s-9}{9t} + \frac{7}{9t}$

Strategy We will add the numerators and write the sum over the common denominator. Then, if possible, we will factor and simplify.

Why This is the rule for adding rational expressions, such as these, that have the same denominator.

Solution

a. The given rational expressions have the same denominator, 8.

$$\frac{x}{8} + \frac{3x}{8} = \frac{x+3x}{8}$$

$$= \frac{4x}{8}$$ Combine like terms in the numerator: $x + 3x = 4x$. This result can be simplified.

The Language of Algebra
Caution
We *do not* add rational expressions by adding numerators and adding denominators!

$$\frac{x}{8} + \frac{3x}{8} = \frac{4x}{16}$$ (crossed out)

The same caution applies when subtracting rational expressions.

$$= \frac{\overset{1}{\cancel{4}} \cdot x}{2 \cdot \underset{1}{\cancel{4}}}$$ Factor 8 as 4 · 2. Then simplify by removing a factor equal to 1.

$$= \frac{x}{2}$$

b. The given rational expressions have the same denominator, $9t$.

$$\frac{4s - 9}{9t} + \frac{7}{9t} = \frac{4s - 9 + 7}{9t}$$ Add the numerators. Write the sum over the common denominator, $9t$.

$$= \frac{4s - 2}{9t}$$ Combine like terms in the numerator: $-9 + 7 = -2$.

To attempt to simplify the result, we factor the numerator to get $\frac{2(2s - 1)}{9t}$. Since the numerator and denominator do not have any common factors, $\frac{4s - 2}{9t}$ cannot be simplified. Thus,

$$\frac{4s - 9}{9t} + \frac{7}{9t} = \frac{4s - 2}{9t}$$

Notation

The numerator of the result may be written two ways:

Not factored	Factored
$\frac{4s - 2}{9t}$	$\frac{2(2s - 1)}{9t}$

Check with your instructor to see which form he or she prefers.

Self Check 1 Add and simplify the result, if possible: **a.** $\frac{2x}{15} + \frac{4x}{15}$ **b.** $\frac{3m - 8}{23n} + \frac{2}{23n}$

Now Try Problems 17 and 21

EXAMPLE 2

Add: **a.** $\frac{3x + 21}{5x + 10} + \frac{8x + 1}{5x + 10}$ **b.** $\frac{x^2 + 9x - 7}{2x(x - 6)} + \frac{x^2 - 9x}{(x - 6)2x}$

Strategy We will add the numerators and write the sum over the common denominator. Then, if possible, we will factor and simplify.

Why This is the rule for adding rational expressions that have the same denominator.

Solution

a. $$\frac{3x + 21}{5x + 10} + \frac{8x + 1}{5x + 10} = \frac{3x + 21 + 8x + 1}{5x + 10}$$ Add the numerators. Write the sum over the common denominator, $5x + 10$.

$$= \frac{11x + 22}{5x + 10}$$ Combine like terms in the numerator: $3x + 8x = 11x$ and $21 + 1 = 22$.

$$= \frac{11\overset{1}{\cancel{(x + 2)}}}{5\underset{1}{\cancel{(x + 2)}}}$$ Factor the numerator: The GCF is 11. Factor the denominator: The GCF is 5. Then simplify by removing a factor equal to 1.

$$= \frac{11}{5}$$

Caution

When adding or subtracting rational expressions, always write the result in simplest form by removing any factors common to the numerator and denominator.

b. By the commutative property of multiplication, $2x(x - 6) = (x - 6)2x$. Therefore, the denominators are the same. We add the numerators and write the sum over the common denominator.

Caution

Don't make the mistake of removing the 2's in the result. This is incorrect because 2 is not a factor of the *entire* numerator.

$$\frac{\overset{1}{\cancel{2}}x^2 - 7}{\underset{1}{\cancel{2}}x(x - 6)}$$

$$\frac{x^2 + 9x - 7}{2x(x - 6)} + \frac{x^2 - 9x}{(x - 6)2x} = \frac{x^2 + 9x - 7 + x^2 - 9x}{2x(x - 6)}$$

$$= \frac{2x^2 - 7}{2x(x - 6)}$$

Combine like terms in the numerator: $x^2 + x^2 = 2x^2$ and $9x - 9x = 0$.

Since the numerator, $2x^2 - 7$, does not factor, $\frac{2x^2 - 7}{2x(x - 6)}$ is in simplest form.

Self Check 2 Add and simplify the result, if possible: **a.** $\frac{m + 3}{3m - 9} + \frac{m - 9}{3m - 9}$ **b.** $\frac{c^2 - c}{(c - 1)(c + 2)} + \frac{c^2 - 10c}{(c + 2)(c - 1)}$

Now Try **Problems 25 and 27**

EXAMPLE 3

Subtract: $\frac{x + 6}{x^2 + 4x - 5} - \frac{1}{x^2 + 4x - 5}$

Strategy We will subtract the numerators and write the sum over the common denominator. Then, if possible, we will factor and simplify.

Why This is the rule for subtracting rational expressions that have the same denominator.

Solution

$$\frac{x + 6}{x^2 + 4x - 5} - \frac{1}{x^2 + 4x - 5} = \frac{x + 6 - 1}{x^2 + 4x - 5}$$

Subtract the numerators. Write the difference over the common denominator, $x^2 + 4x - 5$.

$$= \frac{x + 5}{x^2 + 4x - 5}$$

Combine like terms in the numerator: $6 - 1 = 5$.

$$= \frac{\overset{1}{\cancel{x + 5}}}{\underset{1}{\cancel{(x + 5)}}(x - 1)}$$

Factor the denominator. Then simplify by removing a factor equal to 1.

$$= \frac{1}{x - 1}$$

Self Check 3 Subtract and simplify the result, if possible: $\frac{n - 3}{n^2 - 16} - \frac{1}{n^2 - 16}$

Now Try **Problem 37**

EXAMPLE 4

Subtract: **a.** $\frac{x^2 + 10x}{x + 3} - \frac{4x - 9}{x + 3}$

b. $\frac{x^2}{(x + 7)(x - 8)} - \frac{-x^2 + 14x}{(x + 7)(x - 8)}$

Strategy We will use the rule for subtracting rational expressions that have the same denominators. In both cases, it is important to note that the numerator of the second fraction has *two* terms.

Why We must make sure that entire numerator (not just the first term) of the second fraction is subtracted.

> **Caution**
>
> Don't make this common error by forgetting to write $4x - 9$ within parentheses.
>
> $$\frac{x^2 + 10x - \cancel{4x - 9}}{x + 3}$$

Solution

a. To subtract the numerators, each term of $4x - 9$ must be subtracted from $x^2 + 10x$.

This − symbol applies to the entire numerator 4x − 9.

This numerator is written within parentheses to make sure that we subtract both of its terms.

$$\frac{x^2 + 10x}{x + 3} - \frac{4x - 9}{x + 3} = \frac{x^2 + 10x - (4x - 9)}{x + 3}$$ *Subtract the numerators. Write the difference over the common denominator.*

$$= \frac{x^2 + 10x - 4x + 9}{x + 3}$$ *In the numerator, use the distributive property: $-(4x - 9) = -1(4x - 9) = -4x + 9$.*

$$= \frac{x^2 + 6x + 9}{x + 3}$$ *Combine like terms in the numerator: $10x - 4x = 6x$.*

$$= \frac{(x + 3)(x + 3)}{x + 3}$$ *To see if the result simplifies, factor the numerator.*

$$= \frac{\overset{1}{\cancel{(x + 3)}}(x + 3)}{\underset{1}{\cancel{x + 3}}}$$ *Simplify by removing a factor equal to 1.*

$$= x + 3$$

> **Notation**
>
> A fraction bar is a grouping symbol. Think of parentheses around the terms of the numerator and the denominator.
>
> $$\frac{4x - 9}{x + 3} = \frac{(4x - 9)}{(x + 3)}$$

b. We subtract the numerators and write the difference over the common denominator.

$$\frac{x^2}{(x + 7)(x - 8)} - \frac{-x^2 + 14x}{(x + 7)(x - 8)} = \frac{x^2 - (-x^2 + 14x)}{(x + 7)(x - 8)}$$ *Write the second numerator within parentheses.*

$$= \frac{x^2 + x^2 - 14x}{(x + 7)(x - 8)}$$ *Use the distributive property: $-(-x^2 + 14x) = x^2 - 14x$.*

$$= \frac{2x^2 - 14x}{(x + 7)(x - 8)}$$ *In the numerator, combine like terms: $x^2 + x^2 = 2x^2$.*

> **Success Tip**
>
> $-(-x^2 + 14x)$ means $-1(-x^2 + 14x)$. The multiplication by -1 changes the signs of each term within the parentheses:
>
> $$-(-x^2 + 14x) = x^2 - 14x$$

In an attempt to simplify, we can factor $2x^2 - 14x$ as $2x(x - 7)$. However, the numerator and denominator have no common factors. The result is in simplest form.

Self Check 4 Subtract and simplify the result, if possible:

a. $\frac{x^2 + 3x}{x - 1} - \frac{5x - 1}{x - 1}$

b. $\frac{3y^2}{(y + 3)(y - 3)} - \frac{-3y^2 + y}{(y + 3)(y - 3)}$

Now Try **Problems 41 and 47**

2 Find the Least Common Denominator.

We will now discuss two skills that are needed for adding and subtracting rational expressions that have unlike denominators. To begin, let's consider

$$\frac{11}{8x} + \frac{7}{18x^2}$$

To add these expressions, we must express them as equivalent expressions with a common denominator. The **least common denominator (LCD)** is usually the easiest one to use. The least common denominator of several rational expressions can be found as follows.

Finding the LCD

1. Factor each denominator completely.
2. The LCD is a product that uses each different factor obtained in step 1 the greatest number of times it appears in any one factorization.

EXAMPLE 5 Find the LCD of each pair of rational expressions:

a. $\frac{11}{8x}$ and $\frac{7}{18x^2}$ **b.** $\frac{20}{x}$ and $\frac{4x}{x-9}$

Strategy We will begin by factoring completely the denominator of each rational expression. Then we will form a product using each factor the greatest number of times it appears in any one factorization.

Why Since the LCD must contain the factors of each denominator, we need to write each denominator in factored form.

Solution

a. $8x = 2 \cdot 2 \cdot 2 \cdot x$ Prime factor 8.

$18x^2 = 2 \cdot 3 \cdot 3 \cdot x \cdot x$ Prime factor 18. Factor x^2.

The factorizations of $8x$ and $18x^2$ contain the factors 2, 3, and x. The LCD of $\frac{11}{8x}$ and $\frac{7}{18x^2}$ should contain each factor of $8x$ and $18x^2$ the greatest number of times it appears in any one factorization.

The greatest number of times the factor 2 appears is three times.
The greatest number of times the factor 3 appears is twice.
The greatest number of times the factor x appears is twice.

$$\text{LCD} = 2 \cdot 2 \cdot 2 \cdot 3 \cdot 3 \cdot x \cdot x$$
$$= 72x^2$$

The LCD for $\frac{11}{8x}$ and $\frac{7}{18x^2}$ is $72x^2$.

b. Since the denominators of $\frac{20}{x}$ and $\frac{4x}{x-9}$ are completely factored, the factor x appears once and the factor $x - 9$ appears once. Thus, the LCD is $x(x - 9)$.

Success Tip

The factorizations can be written:

$$8x = 2^3 \cdot x$$
$$18x^2 = 2 \cdot 3^2 \cdot x^2$$

Note that the highest power of each factor is used to form the LCD.

$$\text{LCD} = 2^3 \cdot 3^2 \cdot x^2 = 72x^2$$

Self Check 5 Find the LCD of each pair of rational expressions:

a. $\frac{y+7}{6y^3}$ and $\frac{7}{75y}$ **b.** $\frac{a-3}{a+3}$ and $\frac{21}{a}$

Now Try Problems 53 and 59

EXAMPLE 6 Find the LCD of each pair of rational expressions:

a. $\frac{x}{7x+7}$ and $\frac{x-2}{5x+5}$ **b.** $\frac{6-x}{x^2+8x+16}$ and $\frac{15x}{x^2-16}$

Strategy We will begin by factoring completely each binomial and trinomial in the denominators of the rational expressions. Then we will form a product using each factor the greatest number of times it appears in any one factorization.

Why Since the LCD must contain the factors of each denominator, we need to write each denominator in factored form.

Solution

a. Factor each denominator completely.

$$7x + 7 = 7(x + 1) \quad \text{The GCF is 7.}$$
$$5x + 5 = 5(x + 1) \quad \text{The GCF is 5.}$$

The factorizations of $7x + 7$ and $5x + 5$ contain the factors 7, 5, and $x + 1$. The LCD of $\frac{x}{7x + 7}$ and $\frac{x - 2}{5x + 5}$ should contain each factor of $7x + 7$ and $5x + 5$ the greatest number of times it appears in any one factorization.

The greatest number of times the factor 7 appears is once.
The greatest number of times the factor 5 appears is once.
The greatest number of times the factor x + 1 appears is once.

$$\text{LCD} = 7 \cdot 5 \cdot (x + 1) = 35(x + 1)$$

Notation

Rather than performing the multiplication, it is often better to leave an LCD in factored form:

$$\text{LCD} = 35(x + 1)$$

b. Factor each denominator completely.

$$x^2 + 8x + 16 = (x + 4)(x + 4) \quad \text{Factor the trinomial.}$$
$$x^2 - 16 = (x + 4)(x - 4) \quad \text{Factor the difference of two squares.}$$

The factorizations of $x^2 + 8x + 16$ and $x^2 - 16$ contain the factors $x + 4$ and $x - 4$.

The greatest number of times the factor x + 4 appears is twice.
The greatest number of times the factor x − 4 appears is once.

$$\text{LCD} = (x + 4)(x + 4)(x - 4) = (x + 4)^2(x - 4)$$

Self Check 6 Find the LCD of each pair of rational expressions:

a. $\frac{x^3}{x^2 - 6x}$ and $\frac{25x}{2x - 12}$

b. $\frac{m + 1}{m^2 - 9}$ and $\frac{6m^2}{m^2 - 6m + 9}$

Now Try **Problems 63 and 69**

3 Build Rational Expressions into Equivalent Expressions.

Recall from Chapter 1 that writing a fraction as an equivalent fraction with a larger denominator is called **building the fraction.** For example, to write $\frac{3}{5}$ as an equivalent fraction with a denominator of 35, we multiply it by 1 in the form of $\frac{7}{7}$:

$$\frac{3}{5} = \frac{3}{5} \cdot \frac{7}{7} = \frac{21}{35} \quad \begin{array}{l}\text{Multiply the numerators.}\\ \text{Multiply the denominators.}\end{array}$$

To add and subtract rational expressions with different denominators, we must write them as equivalent expressions having a common denominator. To do so, we build rational expressions.

Building Rational Expressions

To build a rational expression, multiply it by 1 in the form of $\frac{c}{c}$, where c is any nonzero number or expression.

EXAMPLE 7 Write each rational expression as an equivalent expression with the indicated denominator: **a.** $\frac{7}{15n}$, denominator $30n^3$

b. $\frac{6x}{x + 4}$, denominator $(x + 4)(x - 4)$

Strategy We will begin by asking, "By what must we multiply the given denominator to get the required denominator?"

Why The answer to that question helps us determine the form of 1 to be used to build an equivalent rational expression.

Solution

a. We need to multiply the denominator of $\frac{7}{15n}$ by $2n^2$ to obtain a denominator of $30n^3$. It follows that $\frac{2n^2}{2n^2}$ is the form of 1 that should be used to build an equivalent expression.

$$\frac{7}{15n} = \frac{7}{15n} \cdot \frac{2n^2}{2n^2}$$ Multiply the given rational expression by 1, in the form of $\frac{2n^2}{2n^2}$.

$$= \frac{14n^2}{30n^3}$$ Multiply the numerators. Multiply the denominators.

The Language of Algebra

We say that $\frac{7}{15n}$ and $\frac{14n^2}{30n^3}$ are *equivalent expressions* because they have the same value for all values of n, except those that make either denominator 0.

b. We need to multiply the denominator of $\frac{6x}{x + 4}$ by $x - 4$ to obtain a denominator of $(x + 4)(x - 4)$. It follows that $\frac{x - 4}{x - 4}$ is the form of 1 that should be used to build an equivalent expression.

$$\frac{6x}{x + 4} = \frac{6x}{x + 4} \cdot \frac{x - 4}{x - 4}$$ Multiply the given rational expression by 1, in the form of $\frac{x-4}{x-4}$.

$$= \frac{6x(x - 4)}{(x + 4)(x - 4)}$$ Multiply the numerators. Multiply the denominators.

$$= \frac{6x^2 - 24x}{(x + 4)(x - 4)}$$ In the numerator, distribute the multiplication by 6x. Leave the denominator in factored form.

Notation To get this answer, we multiplied the factors in the numerator to obtain a polynomial in unfactored form: $6x^2 - 24x$. However, we left the denominator in factored form. This approach is beneficial in the next section when we add and subtract rational expressions with unlike denominators.

Self Check 7 Write each rational expression as an equivalent expression with the indicated denominator: **a.** $\frac{7}{20m^2}$, denominator $60m^3$

b. $\frac{2c}{c + 1}$, denominator $(c + 1)(c + 3)$

Now Try **Problem 77**

EXAMPLE 8 Write $\frac{x + 1}{x^2 + 6x}$ as an equivalent expression with a denominator of $x(x + 6)(x + 2)$.

Strategy We will begin by factoring the denominator of $\frac{x + 1}{x^2 + 6x}$. Then we will compare the factors of $x^2 + 6x$ to those of $x(x + 6)(x + 2)$.

Why This comparison will enable us to answer the question, "By what must we multiply $x^2 + 6x$ to obtain $x(x + 6)(x + 2)$?"

Solution We factor the denominator to determine what factors are missing.

$$\frac{x+1}{x^2+6x} = \frac{x+1}{x(x+6)}$$ Factor out the GCF, x, from $x^2 + 6x$.

It is now apparent that we need to multiply the denominator by $x + 2$ to obtain a denominator of $x(x + 6)(x + 2)$. It follows that $\frac{x+2}{x+2}$ is the form of 1 that should be used to build an equivalent expression.

$$\frac{x+1}{x^2+6x} = \frac{x+1}{x(x+6)} \cdot \frac{x+2}{x+2}$$ Multiply the given rational expression by 1, in the form of $\frac{x+2}{x+2}$.

$$= \frac{(x+1)(x+2)}{x(x+6)(x+2)}$$ Multiply the numerators. Multiply the denominators.

$$= \frac{x^2+3x+2}{x(x+6)(x+2)}$$ In the numerator, use the FOIL method to multiply $(x + 1)(x + 2)$. Leave the denominator in factored form.

Notation

When building rational expressions, write the numerator of the result as a polynomial in unfactored form. Write the denominator in factored form.

Self Check 8 Write $\frac{x-3}{x^2-4x}$ as an equivalent expression with a denominator of $x(x - 4)(x + 8)$.

Now Try Problem 83

ANSWERS TO SELF CHECKS 1. a. $\frac{2x}{5}$ b. $\frac{3m-6}{23n}$ or $\frac{3(m-2)}{23n}$ 2. a. $\frac{2}{3}$ b. $\frac{2c^2-11c}{(c-1)(c+2)}$ or $\frac{c(2c-11)}{(c-1)(c+2)}$ 3. $\frac{1}{n+4}$ 4. a. $x - 1$ b. $\frac{6y^2-y}{(y+3)(y-3)}$ 5. a. $150y^3$ b. $a(a + 3)$ 6. a. $2x(x - 6)$ b. $(m + 3)(m - 3)^2$ 7. a. $\frac{21m}{60m^3}$ b. $\frac{2c^2+6c}{(c+1)(c+3)}$ 8. $\frac{(x-3)(x+8)}{x(x-4)(x+8)}$

STUDY SET 7.3

VOCABULARY

Fill in the blanks.

1. The rational expressions $\frac{7}{6n}$ and $\frac{n+1}{6n}$ have the common ____________ $6n$.
2. The _____ _________ denominator of $\frac{x-8}{x+6}$ and $\frac{6-5x}{x}$ is $x(x + 6)$.
3. To ______ a rational expression, we multiply it by a form of 1. For example: $\frac{2}{n^2} \cdot \frac{8}{8} = \frac{16}{8n^2}$
4. $\frac{2}{n^2}$ and $\frac{16}{8n^2}$ are __________ expressions. They have the same value for all values of n, except for $n = 0$.

CONCEPTS

Fill in the blanks.

5. To add or subtract rational expressions that have the same denominator, add or subtract the ___________, and write the sum or difference over the common ___________. In symbols, $\frac{A}{D} + \frac{B}{D} =$ ______ and $\frac{A}{D} - \frac{B}{D} =$ ______
6. To find the least common denominator of several rational expressions, ______ each denominator completely. The LCD is a product that uses each different factor the ________ number of times it appears in any one factorization.
7. The sum of two rational expressions is $\frac{4x+4}{5(x+1)}$. Factor the numerator and then simplify the result.

8. Factor each denominator completely.

a. $\dfrac{17}{40x^2}$ **b.** $\dfrac{x+25}{2x^2-6x}$

9. Consider the following factorizations.

$$18x - 36 = 2 \cdot 3 \cdot 3 \cdot (x-2)$$
$$3x - 6 = 3(x-2)$$

a. What is the greatest number of times the factor 3 appears in any one factorization?

b. What is the greatest number of times the factor $x - 2$ appears in any one factorization?

10. Fill in the blanks. To write $\frac{x}{x-9}$ as an equivalent rational expression with a denominator of $3x(x-9)$, we need to multiply the denominator by ____. It follows that ____ is the form of 1 that should be used to build $\frac{x}{x-9}$.

NOTATION

Complete the solution.

11. $\dfrac{5}{x} - \dfrac{x-1}{x} = \dfrac{5 - (\quad)}{x}$

$= \dfrac{5 \;\square\; x \;\square\; 1}{x}$

$= \dfrac{\square - x}{x}$

12. The type of multiplication that is used to build rational expressions is shown below. Fill in the blanks.

a. $\dfrac{4x}{5} \cdot \dfrac{2}{2} = \dfrac{\square}{10}$ **b.** $\dfrac{3}{t} \cdot \dfrac{t-2}{t-2} = \dfrac{\square}{t(t-2)}$

c. $\dfrac{m+1}{m-3} \cdot \dfrac{m-5}{m-5} = \dfrac{\square}{(m-3)(m-5)}$

GUIDED PRACTICE

Add and simplify the result, if possible. **See Examples 1 and 2.**

13. $\dfrac{9}{x} + \dfrac{2}{x}$

14. $\dfrac{4}{s} + \dfrac{4}{s}$

15. $\dfrac{x}{18} + \dfrac{5}{18}$

16. $\dfrac{7}{10} + \dfrac{3y}{10}$

17. $\dfrac{x}{9} + \dfrac{2x}{9}$

18. $\dfrac{5x}{7} + \dfrac{9x}{7}$

19. $\dfrac{a-5}{3a^3} + \dfrac{5}{3a^3}$

20. $\dfrac{b^3-8}{10b^4} + \dfrac{8}{10b^4}$

21. $\dfrac{x+3}{2y} + \dfrac{x+5}{2y}$

22. $\dfrac{y+2}{10z} + \dfrac{y+4}{10z}$

23. $\dfrac{2}{r^2-3r-10} + \dfrac{r}{r^2-3r-10}$

24. $\dfrac{1}{h^2-4h-5} + \dfrac{h}{h^2-4h-5}$

25. $\dfrac{3x-5}{x-2} + \dfrac{6x-13}{x-2}$

26. $\dfrac{8x-7}{x+3} + \dfrac{2x+37}{x+3}$

27. $\dfrac{a^2+a}{4a^2-8a} + \dfrac{2a^2-7a}{4a^2-8a}$

28. $\dfrac{3b^2+16b}{6b^2+9b} + \dfrac{7b^2-b}{6b^2+9b}$

Subtract and simplify the result, if possible. **See Example 3.**

29. $\dfrac{2x}{25} - \dfrac{x}{25}$

30. $\dfrac{16c}{11} - \dfrac{4c}{11}$

31. $\dfrac{35t}{99} - \dfrac{13t}{99}$

32. $\dfrac{44y}{72} - \dfrac{35y}{72}$

33. $\dfrac{m-1}{6m^2} - \dfrac{5}{6m^2}$

34. $\dfrac{c+7}{4c^4} - \dfrac{3}{4c^4}$

35. $\dfrac{17a}{2a+4} - \dfrac{7a}{2a+4}$

36. $\dfrac{10b}{3b-18} - \dfrac{4b}{3b-18}$

37. $\dfrac{t}{t^2+t-2} - \dfrac{1}{t^2+t-2}$

38. $\dfrac{r}{r^2-2r-3} - \dfrac{3}{r^2-2r-3}$

39. $\dfrac{11w+6}{3w(w-9)} - \dfrac{11w}{3w(w-9)}$

40. $\dfrac{y+8}{2y(y-14)} - \dfrac{y}{2y(y-14)}$

Subtract and simplify the result, if possible. **See Example 4.**

41. $\dfrac{3y-2}{2y+6} - \dfrac{2y-5}{2y+6}$

42. $\dfrac{5x+8}{3x+15} - \dfrac{3x-2}{3x+15}$

43. $\dfrac{6x^2}{3x+2} - \dfrac{11x+10}{3x+2}$

44. $\dfrac{8a^2}{2a+5} - \dfrac{4a^2+25}{2a+5}$

45. $\dfrac{6x-5}{3xy} - \dfrac{3x-5}{3xy}$

46. $\dfrac{7x+7}{5y} - \dfrac{2x+7}{5y}$

47. $\dfrac{2-p}{p^2-p} - \dfrac{-p+2}{p^2-p}$

48. $\dfrac{2-7n}{n^2+5} - \dfrac{-7n+2}{n^2+5}$

49. $\dfrac{8}{9-3x^2} - \dfrac{-6x+8}{9-3x^2}$

50. $\dfrac{5}{10-5t^2} - \dfrac{-15t+5}{10-5t^2}$

51. $\dfrac{-4x}{3x^2-7x+2} - \dfrac{-3x-2}{3x^2-7x+2}$

52. $\dfrac{-3c}{5c^2-16c+3} - \dfrac{-2c-3}{5c^2-16c+3}$

Find the LCD of each pair of rational expressions. **See Examples 5 and 6.**

53. $\dfrac{1}{2x}, \dfrac{9}{6x}$

54. $\dfrac{4}{9y}, \dfrac{11}{3y}$

55. $\dfrac{33}{15a^3}, \dfrac{9}{10a}$

56. $\dfrac{m-21}{12m^4}, \dfrac{m+1}{18m}$

57. $\dfrac{35}{3a^2b}, \dfrac{23}{a^2b^3}$

58. $\dfrac{27}{c^3d}, \dfrac{17}{2c^2d^3}$

59. $\dfrac{8}{c}, \dfrac{8-c}{c+2}$

60. $\dfrac{d^2-5}{d+9}, \dfrac{d-3}{d}$

61. $\frac{3x+1}{3x-1}, \frac{3x}{3x+1}$

62. $\frac{b+1}{b-1}, \frac{b}{b+1}$

63. $\frac{b-9}{4b+8}, \frac{b}{6}$

64. $\frac{b^2-b}{10b-15}, \frac{11b}{10}$

65. $\frac{6-k}{2k+4}, \frac{11}{8k}$

66. $\frac{5m+6}{4m+12}, \frac{7}{6m}$

67. $\frac{-2x}{x^2-1}, \frac{5x}{x+1}$

68. $\frac{7-y^2}{y^2-4}, \frac{y-49}{y+2}$

69. $\frac{4x-5}{x^2-4x-5}, \frac{3x+1}{x^2-25}$

70. $\frac{44}{s^2-9}, \frac{s+9}{s^2-s-6}$

71. $\frac{5n^2-16}{2n^2+13n+20}, \frac{3n^2}{n^2+8n+16}$

72. $\frac{4y+25}{y^2+10y+25}, \frac{y^2-7}{2y^2+17y+35}$

***Build each rational expression into an equivalent expression with the given denominator.* See Examples 7 and 8.**

73. $\frac{5}{r}$; $10r$

74. $\frac{4}{y}$; $7y$

75. $\frac{8}{x}$; x^2y

76. $\frac{7}{y}$; xy^2

77. $\frac{9}{4b}$; $12b^2$

78. $\frac{7}{6c}$; $30c^2$

79. $\frac{3x}{x+1}$; $(x+1)^2$

80. $\frac{5y}{y-2}$; $(y-2)^2$

81. $\frac{x-2}{x}$; $x(x+3)$

82. $\frac{y-4}{y}$; $y(y-9)$

83. $\frac{t+5}{4t+8}$; $20(t+2)$

84. $\frac{x+7}{3x-15}$; $6(x-5)$

85. $\frac{y+3}{y^2-5y+6}$; $4y(y-2)(y-3)$

86. $\frac{3x-4}{x^2+3x+2}$; $8x(x+1)(x+2)$

87. $\frac{12-h}{h^2-81}$; $3(h+9)(h-9)$

88. $\frac{m^2}{m^2-100}$; $9(m+10)(m-10)$

TRY IT YOURSELF

Perform the operations. Then simplify, if possible.

89. $\frac{3t}{t^2-8t+7} - \frac{3}{t^2-8t+7}$

90. $\frac{10x}{x^2-2x+1} - \frac{10}{x^2-2x+1}$

91. $\frac{c}{c^2-d^2} - \frac{d}{c^2-d^2}$

92. $\frac{b}{b^2-4} - \frac{2}{b^2-4}$

93. $\frac{11n}{(n+4)(n-2)} - \frac{4n-1}{(n-2)(n+4)}$

94. $\frac{1}{(t-1)(t+1)} - \frac{6-t}{(t+1)(t-1)}$

95. $\frac{11}{36y} + \frac{9}{36y}$

96. $\frac{13}{24w} + \frac{17}{24w}$

97. $\frac{3x^2}{x+1} - \frac{-x+2}{x+1}$

98. $\frac{8b^2}{3b-2} - \frac{-b^2+4}{3b-2}$

99. $\frac{5r-27}{3r^2-9r} + \frac{4r}{3r^2-9r}$

100. $\frac{9a}{5a^2+25a} + \frac{a+50}{5a^2+25a}$

APPLICATIONS

101. GEOMETRY What is the difference of the length and width of the rectangle?

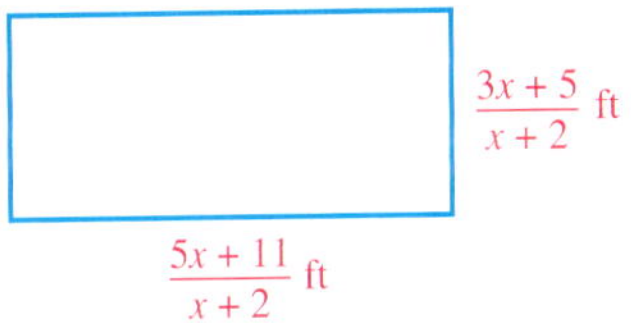

102. GEOMETRY What is the perimeter of the rectangle in Problem 101?

WRITING

103. Explain how to add fractions with the same denominator.

104. Explain how to find a least common denominator.

105. Explain the error in the following solution:

$$\frac{2x+3}{x+5} - \frac{x+2}{x+5} = \frac{2x+3-x+2}{x+5}$$
$$= \frac{x+5}{x+5}$$
$$= 1$$

106. Explain the error in the following solution:

$$\frac{y+4}{y} - \frac{1}{y} = \frac{y+4-1}{y+y}$$
$$= \frac{y+3}{2y}$$

107. Explain why the LCD of $\frac{5}{h^2}$ and $\frac{3}{h}$ is h^2 and not h^3.

108. Explain how multiplication by 1 is used to build a rational expression.

REVIEW

Give the formula for . . .

109. **a.** simple interest

b. the area of a triangle

c. the perimeter of a rectangle

110. **a.** the slope of a line

b. distance traveled

c. the area of a circle

CHALLENGE PROBLEMS

Perform the operations. Simplify the result, if possible.

111. $\frac{3xy}{x-y} - \frac{x(3y-x)}{x-y} - \frac{x(x-y)}{x-y}$

112. $\frac{9t^3 - 12t^2}{27t^3 - 64} - \frac{-3t + 4}{27t^3 - 64}$

113. $\frac{2a^2 + 2}{a^3 + 8} + \frac{a^3 + a}{a^3 + 8}$

114. Find the LCD of these rational expressions.

$$\frac{2}{a^3 + 8}, \frac{a}{a^2 - 4}, \frac{2a + 5}{a^3 - 8}$$

SECTION 7.4
Adding and Subtracting with Unlike Denominators

Objectives

1. Add and subtract rational expressions that have unlike denominators.
2. Add and subtract rational expressions that have denominators that are opposites.

We have discussed a method for finding the least common denominator (LCD) of two rational expressions. We have also built rational expressions into equivalent expressions having a given denominator. We will now use these skills to add and subtract rational expressions with unlike denominators.

1 Add and Subtract Rational Expressions That Have Unlike Denominators.

The following steps summarize how to add or subtract rational expressions that have different denominators.

Adding and Subtracting Rational Expressions That Have Unlike Denominators

1. Find the LCD.
2. Rewrite each rational expression as an equivalent expression with the LCD as the denominator. To do so, build each fraction using a form of 1 that involves any factor(s) needed to obtain the LCD.
3. Add or subtract the numerators and write the sum or difference over the LCD.
4. Simplify the result, if possible.

EXAMPLE 1 Add: $\frac{9x}{7} + \frac{3x}{5}$

Strategy We will use the procedure for adding rational expressions that have unlike denominators. The first step is to determine thc LCD.

Why If we are to add (or subtract) fractions, their denominators must be the same. Since the denominators of these rational expressions are different, we cannot add them in their present form.

$$\underset{\text{sevenths}}{\frac{9x}{7}} + \underset{\text{fifths}}{\frac{3x}{5}}$$

Not the same number

Solution

Step 1: The denominators are 7 and 5. The LCD is $7 \cdot 5 = 35$.

Step 2: We need to multiply the denominator of $\frac{9x}{7}$ by 5 and we need to multiply the denominator of $\frac{3x}{5}$ by 7 to obtain the LCD, 35. It follows that $\frac{5}{5}$ and $\frac{7}{7}$ are the forms of 1 that should be used to write the equivalent rational expressions.

$$\frac{9x}{7} + \frac{3x}{5} = \frac{9x}{7} \cdot \frac{5}{5} + \frac{3x}{5} \cdot \frac{7}{7}$$ Build the rational expressions so that each has a denominator of 35.

$$= \frac{45x}{35} + \frac{21x}{35}$$ Multiply the numerators. Multiply the denominators.

Step 3: $$= \frac{45x + 21x}{35}$$ Add the numerators. Write the sum over the common denominator.

$$= \frac{66x}{35}$$ Combine like terms in the numerator: $45x + 21x = 66x$.

Step 4: Since 66 and 35 have no common factor other than 1, the result cannot be simplified.

> **Caution**
>
> In Step 2, *don't simplify* $\frac{45x}{35}$ and $\frac{21x}{35}$, because that will take you back to the original problem.

Self Check 1 Add: $\frac{y}{2} + \frac{6y}{7}$

Now Try **Problem 13**

EXAMPLE 2 Subtract: $\frac{13}{18b^2} - \frac{1}{24b}$

Strategy We will use the procedure for subtracting rational expressions that have unlike denominators. The first step is to determine the LCD.

Why If we are to subtract fractions, their denominators must be the same. Since the denominators of these rational expressions are different, we cannot subtract them in their present form.

Solution

Step 1: To find the LCD, we form a product that uses each different factor of $18b^2$ and $24b$ the greatest number of times it appears in any one factorization.

$$\left.\begin{aligned} 18b^2 &= 2 \cdot 3 \cdot 3 \cdot b \cdot b \\ 24b &= 2 \cdot 2 \cdot 2 \cdot 3 \cdot b \end{aligned}\right\} \text{LCD} = 2 \cdot 2 \cdot 2 \cdot 3 \cdot 3 \cdot b \cdot b = 72b^2$$

Step 2: We need to multiply $18b^2$ by 4 to obtain $72b^2$, and $24b$ by $3b$ to obtain $72b^2$. It follows that we should use $\frac{4}{4}$ and $\frac{3b}{3b}$ to build the equivalent rational expressions.

Notation
When checking your answers with those in the back of the book, remember that the results can often be presented in several equivalent forms. For example, this result can also be expressed as

$$\frac{-3b + 52}{72b^2}$$

$$\frac{13}{18b^2} - \frac{1}{24b} = \frac{13}{18b^2} \cdot \frac{4}{4} - \frac{1}{24b} \cdot \frac{3b}{3b}$$ Build the rational expressions so that each has a denominator of $72b^2$.

$$= \frac{52}{72b^2} - \frac{3b}{72b^2}$$ Multiply the numerators. Multiply the denominators.

Step 3: $$= \frac{52 - 3b}{72b^2}$$ Subtract the numerators. Write the difference over the common denominator.

Step 4: Since $52 - 3b$ does not factor, the result cannot be simplified.

Self Check 2 Subtract: $\frac{5}{21z^2} - \frac{3}{28z}$

Now Try **Problem 29**

EXAMPLE 3 Add: $\frac{3}{2x + 18} + \frac{27}{x^2 - 81}$

Strategy We use the procedure for adding rational expressions when the denominators are binomials. The first step is to find the LCD.

Why Since the denominators are different, we cannot add these rational expressions in their present form.

Solution After factoring the denominators, we see that the greatest number of times each of the factors 2, $x + 9$, and $x - 9$ appear in any one of the factorizations is once.

$$\left.\begin{aligned} 2x + 18 &= 2(x + 9) \\ x^2 - 81 &= (x + 9)(x - 9) \end{aligned}\right\} \text{LCD} = 2(x + 9)(x - 9)$$

Since we need to multiply $2(x + 9)$ by $x - 9$ to obtain the LCD and $(x + 9)(x - 9)$ by 2 to obtain the LCD, $\frac{x - 9}{x - 9}$ and $\frac{2}{2}$ are the forms of 1 to use to build the equivalent rational expressions.

$$\frac{3}{2x + 18} + \frac{27}{x^2 - 81} = \frac{3}{2(x + 9)} + \frac{27}{(x + 9)(x - 9)}$$ Write each denominator in factored form.

$$= \frac{3}{2(x + 9)} \cdot \frac{x - 9}{x - 9} + \frac{27}{(x + 9)(x - 9)} \cdot \frac{2}{2}$$ Build the expressions so that each has a denominator of $2(x + 9)(x - 9)$.

$$= \frac{3x - 27}{2(x + 9)(x - 9)} + \frac{54}{2(x + 9)(x - 9)}$$ Multiply numerators. Multiply denominators.

Although it is not required, the factors of each denominator are written in the same order.

$$= \frac{3x - 27 + 54}{2(x + 9)(x - 9)}$$ Add the numerators. Write the sum over the common denominator.

$$= \frac{3x + 27}{2(x + 9)(x - 9)}$$ Combine like terms in the numerator: $-27 + 54 = 27$.

Success Tip
To build the first rational expression, we use the distributive property to multiply the numerators: Note that we don't multiply out the denominators.

$$\frac{3}{2(x + 9)} \cdot \frac{x - 9}{x - 9}$$

The result is $\frac{3x - 27}{2(x + 9)(x - 9)}$.

Caution
Always write the result in simplest form by removing any factors common to the numerator and denominator.

$$= \frac{\overset{1}{3(\cancel{x+9})}}{\underset{1}{2(\cancel{x+9})(x-9)}}$$ Factor the numerator. Then simplify the expression by removing a factor equal to 1.

$$= \frac{3}{2(x-9)}$$

Self Check 3 Add: $\frac{2}{5x+25} + \frac{4}{x^2-25}$

Now Try **Problem 37**

EXAMPLE 4 Subtract: $\frac{x}{x-1} - \frac{x-6}{x-4}$

Strategy We use the same procedure for subtracting rational expressions when the denominators are binomials. The first step is to find the LCD.

Why Since the denominators are different, we cannot subtract these rational expressions in their present form.

Solution The denominators of $\frac{x}{x-1}$ and $\frac{x-6}{x-4}$ are completely factored. The factor $x - 1$ appears once and the factor $x - 4$ appears once. Thus, the LCD $= (x-1)(x-4)$.

We need to multiply the first denominator by $x - 4$ to obtain the LCD and the second denominator by $x - 1$ to obtain the LCD. It follows that $\frac{x-4}{x-4}$ and $\frac{x-1}{x-1}$ are the forms of 1 to use to build the equivalent rational expressions.

Success Tip
To build the second rational expression, we use the FOIL method to multiply the numerators: Note that we don't multiply out the denominators.

$$\frac{x-6}{x-4} \cdot \frac{x-1}{x-1}$$

The result is $\frac{x^2-7x+6}{(x-4)(x-1)}$.

$$\frac{x}{x-1} - \frac{x-6}{x-4} = \frac{x}{x-1} \cdot \frac{x-4}{x-4} - \frac{x-6}{x-4} \cdot \frac{x-1}{x-1}$$ Build the rational expressions so that each has a denominator of $(x-1)(x-4)$.

$$= \frac{x^2-4x}{(x-1)(x-4)} - \frac{x^2-7x+6}{(x-4)(x-1)}$$ Multiply the numerators. Multiply the denominators.

By the commutative property of multiplication, these are like denominators.

$$= \frac{x^2-4x-(x^2-7x+6)}{(x-1)(x-4)}$$ Subtract the numerators. Remember the parentheses. Write the difference over the common denominator.

$$= \frac{x^2-4x-x^2+7x-6}{(x-1)(x-4)}$$ In the numerator, use the distributive property: $-(x^2-7x+6) = -1(x^2-7x+6) = -x^2+7x-6$.

$$= \frac{3x-6}{(x-1)(x-4)}$$ Combine like terms in the numerator: $x^2 - x^2 = 0$ and $-4x + 7x = 3x$.

The numerator factors as $3(x-2)$. Since the numerator and denominator have no common factor, the result is in simplest form.

Self Check 4 Subtract: $\frac{x}{x+9} - \frac{x-7}{x+8}$

Now Try **Problem 45**

EXAMPLE 5 Subtract: $\dfrac{m}{m^2 + 5m + 6} - \dfrac{2}{m^2 + 3m + 2}$

Strategy We use the same procedure for subtracting rational expressions when the denominators are trinomials. The first step is to find the LCD.

Why Since the denominators are different, we cannot subtract these rational expressions in their present form.

Solution Factor each denominator and form the LCD.

$$\left.\begin{aligned} m^2 + 5m + 6 &= (m + 2)(m + 3) \\ m^2 + 3m + 2 &= (m + 2)(m + 1) \end{aligned}\right\} \text{LCD} = (m + 2)(m + 3)(m + 1)$$

Examining the factored forms, we see that the first denominator must be multiplied by $m + 1$, and the second must be multiplied by $m + 3$ to obtain the LCD. To build the expressions, we will use $\frac{m+1}{m+1}$ and $\frac{m+3}{m+3}$.

$$\frac{m}{m^2 + 5m + 6} - \frac{2}{m^2 + 3m + 2}$$

$$= \frac{m}{(m + 2)(m + 3)} - \frac{2}{(m + 2)(m + 1)}$$ Write each denominator in factored form.

$$= \frac{m}{(m + 2)(m + 3)} \cdot \frac{m + 1}{m + 1} - \frac{2}{(m + 2)(m + 1)} \cdot \frac{m + 3}{m + 3}$$ Build each expression, so that it has a denominator of $(m + 2)(m + 3)(m + 1)$.

$$= \frac{m^2 + m}{(m + 2)(m + 3)(m + 1)} - \frac{2m + 6}{(m + 2)(m + 1)(m + 3)}$$ Multiply numerators. Multiply denominators.

By the commutative property of multiplication, these are like denominators.

$$= \frac{m^2 + m - (2m + 6)}{(m + 2)(m + 3)(m + 1)}$$ Subtract the numerators. Remember the parentheses. Write the difference over the common denominator.

$$= \frac{m^2 + m - 2m - 6}{(m + 2)(m + 3)(m + 1)}$$ Use the distributive property: $-(2m + 6) = -1(2m + 6) = -2m - 6$.

$$= \frac{m^2 - m - 6}{(m + 2)(m + 3)(m + 1)}$$ Combine like terms in the numerator.

$$= \frac{(m - 3)\overset{1}{\cancel{(m + 2)}}}{\underset{1}{\cancel{(m + 2)}}(m + 3)(m + 1)}$$ Factor the numerator and simplify the expression by removing a factor equal to 1.

$$= \frac{m - 3}{(m + 3)(m + 1)}$$

Self Check 5 Subtract: $\dfrac{b}{b^2 - 2b - 8} - \dfrac{6}{b^2 + b - 20}$

Now Try **Problem 57**

EXAMPLE 6 Add: $\dfrac{4b}{a - 5} + b$

Strategy We will begin by writing the second addend, b, as $\frac{b}{1}$ and then find the LCD.

Why To add b to the rational expression, $\frac{4b}{a-5}$, we must rewrite b as a rational expression.

Solution The LCD of $\frac{4b}{a-5}$ and $\frac{b}{1}$ is $1(a-5)$, or simply $a-5$. Since we must multiply the denominator of $\frac{b}{1}$ by $a-5$ to obtain the LCD, we will use $\frac{a-5}{a-5}$ to write an equivalent rational expression.

$$\frac{4b}{a-5}+b=\frac{4b}{a-5}+\frac{b}{1}\cdot\frac{a-5}{a-5}$$
Build $\frac{b}{1}$ so that it has a denominator of $a-5$.

$$=\frac{4b}{a-5}+\frac{ab-5b}{a-5}$$
Multiply numerators: $b(a-5)=ab-5b$.
Multiply denominators: $1(a-5)=a-5$.

$$=\frac{4b+ab-5b}{a-5}$$
Add the numerators. Write the sum over the common denominator.

$$=\frac{ab-b}{a-5}$$
Combine like terms in the numerator: $4b-5b=-b$.

Although the numerator factors as $b(a-1)$, the numerator and denominator do not have a common factor. Therefore, the result is in simplest form.

Success Tip

Since the denominator of $\frac{4b}{a-5}$ is the LCD, we *do not* need to build it by multiplying by it a form of 1. The step below is unnecessary.

$$\frac{4b}{a-5}\cdot\frac{1}{1}$$

Self Check 6 Add: $\frac{10y}{n+4}+y$

Now Try **Problem 65**

2 Add and Subtract Rational Expressions That Have Denominators That Are Opposites.

Recall that two polynomials are **opposites** if their terms are the same but they are opposite in sign. For example, $x-4$ and $4-x$ are opposites. If we multiply one of these binomials by -1, the subtraction is reversed, and the result is the other binomial.

$$-1(x-4)=-x+4$$
$$=4-x$$
Write the expression with 4 first.

$$-1(4-x)=-4+x$$
$$=x-4$$
Write the expression with x first.

These results suggest a general fact.

Multiplying by −1

When a polynomial is multiplied by -1, the result is its opposite.

This fact can be used when adding or subtracting rational expressions whose denominators are opposites.

EXAMPLE 7 Subtract: $\frac{x}{x-7}-\frac{1}{7-x}$

Strategy We note that the denominators are opposites. Either can serve as the LCD; we will choose $x-7$. To obtain a common denominator, we will multiply $\frac{1}{7-x}$ by $\frac{-1}{-1}$.

Why When $7-x$ is multiplied by -1, the subtraction is reversed, and the result is $x-7$.

Solution We must multiply the denominator of $\frac{1}{7-x}$ by -1 to obtain the LCD. It follows that $\frac{-1}{-1}$ should be the form of 1 that is used to write an equivalent rational expression.

$$\frac{x}{x-7} - \frac{1}{7-x} = \frac{x}{x-7} - \frac{1}{7-x} \cdot \frac{-1}{-1}$$ Build $\frac{1}{7-x}$ so that it has a denominator of $x - 7$.

$$= \frac{x}{x-7} - \frac{-1}{-7+x}$$ Multiply the numerators. Multiply the denominators.

$$= \frac{x}{x-7} - \frac{-1}{x-7}$$ Rewrite the second denominator: $-7 + x = x - 7$.

$$= \frac{x-(-1)}{x-7}$$ Subtract the numerators. Remember the parentheses. Write the difference over the common denominator.

$$= \frac{x+1}{x-7}$$ Simplify the numerator.

The result does not simplify.

Success Tip

Either denominator can serve as the LCD. However, it is common to have a result whose denominator is written in descending powers of the variable. Therefore, we chose $x - 7$, as opposed to $7 - x$, as the LCD.

Self Check 7 Add: $\frac{n}{n-8} + \frac{12}{8-n}$

Now Try Problem 71

ANSWERS TO SELF CHECK 1. $\frac{19y}{14}$ 2. $\frac{20-9z}{84z^2}$ 3. $\frac{2}{5(x-5)}$ 4. $\frac{6x+63}{(x+9)(x+8)}$ 5. $\frac{b+3}{(b+2)(b+5)}$ 6. $\frac{ny+14y}{n+4}$ 7. $\frac{n-12}{n-8}$

STUDY SET 7.4

VOCABULARY

Fill in the blanks.

1. $\frac{x}{x-7}$ and $\frac{1}{x-7}$ have like denominators. $\frac{x+5}{x-7}$ and $\frac{4x}{x+7}$ have ________ denominators.
2. Two polynomials are ________ if their terms are the same, but are opposite in sign.

CONCEPTS

3. Write each denominator in factored form.
   a. $\frac{x+1}{20x^2}$
   b. $\frac{3x^2-4}{x^2+4x-12}$
4. The factorizations of the denominators of two rational expressions are given. Find the LCD.
   a. $12a = 2 \cdot 2 \cdot 3 \cdot a$
      $18a^2 = 2 \cdot 3 \cdot 3 \cdot a \cdot a$
   b. $x^2 - 36 = (x+6)(x-6)$
      $3x - 18 = 3(x-6)$
5. What is the LCD for $\frac{x-1}{x+6}$ and $\frac{1}{x+3}$?
6. The LCD for $\frac{1}{9n^2}$ and $\frac{37}{15n^3}$ is $3 \cdot 3 \cdot 5 \cdot n \cdot n \cdot n = 45n^3$. If we want to add these rational expressions, what form of 1 should be used
   a. to build $\frac{1}{9n^2}$?
   b. to build $\frac{37}{15n^3}$?

Fill in the blanks.

7. To build $\frac{x}{x+2}$ so that it has a denominator of $5(x+2)$, we multiply it by 1 in the form of ___.
8. To build $\frac{8x}{2-x}$ so that it has a denominator of $x - 2$, we multiply it by 1 in the form of ___.

NOTATION

Complete the solution.

9. $\frac{2}{5} + \frac{7}{3x} = \frac{2}{5} \cdot \frac{\square}{3x} + \frac{7}{3x} \cdot \frac{\square}{5}$

$= \frac{6x}{\square} + \frac{35}{\square}$

$= \frac{6x + \square}{15x}$

10. Are the student's answers and the book's answers equivalent?

Student's answer	Book's answer	Equivalent?
$\frac{m^2 + 2m}{(m-1)(m-4)}$	$\frac{m^2 + 2m}{(m-4)(m-1)}$	
$\frac{-5x^2 - 7}{4x(x+3)}$	$-\frac{5x^2 - 7}{4x(x+3)}$	
$\frac{-2x}{x - y}$	$-\frac{2x}{x - y}$	

GUIDED PRACTICE

Perform the operations. Simplify, if possible. **See Example 1.**

11. $\frac{x}{3} + \frac{2x}{7}$

12. $\frac{y}{4} + \frac{3y}{5}$

13. $\frac{5y}{6} + \frac{5y}{3}$

14. $\frac{4x}{3} + \frac{x}{6}$

15. $\frac{7}{8} - \frac{4}{t}$

16. $\frac{5}{3} - \frac{2}{m}$

17. $\frac{4b}{3} - \frac{5b}{12}$

18. $\frac{21y}{12} - \frac{7y}{6}$

Perform the operations. Simplify, if possible. **See Example 2.**

19. $\frac{7}{m^2} - \frac{2}{m}$

20. $\frac{6}{n^2} - \frac{2}{n}$

21. $\frac{3}{x^2} + \frac{17}{x}$

22. $\frac{7}{c} + \frac{14}{c^2}$

23. $\frac{3}{5p} - \frac{5}{10p}$

24. $\frac{15}{16a} - \frac{3}{4a}$

25. $\frac{1}{6t} - \frac{11}{8t}$

26. $\frac{3}{10a} - \frac{13}{15a}$

27. $\frac{11}{5x} - \frac{5}{6x}$

28. $\frac{5}{9y} - \frac{1}{4y}$

29. $\frac{1}{6c^4} + \frac{8}{9c^2}$

30. $\frac{7}{8b^2} + \frac{5}{6b^3}$

Perform the operations. Simplify, if possible. **See Example 3.**

31. $\frac{1}{2a + 4} + \frac{5}{a^2 - 4}$

32. $\frac{5}{p^2 - 9} + \frac{2}{3p + 9}$

33. $\frac{2}{3a - 2} + \frac{5}{9a^2 - 4}$

34. $\frac{2}{5b - 3} + \frac{5}{25b^2 - 9}$

35. $\frac{4}{a + 2} - \frac{7}{a^2 + 4a + 4}$

36. $\frac{9}{b^2 - 2b + 1} - \frac{2}{b - 1}$

37. $\frac{6}{5m^2 - 5m} - \frac{3}{5m - 5}$

38. $\frac{9}{2c^2 - 2c} - \frac{5}{2c - 2}$

Perform the operations. Simplify, if possible. **See Example 4.**

39. $\frac{9}{t + 3} + \frac{8}{t + 2}$

40. $\frac{2}{m - 3} + \frac{7}{m - 4}$

41. $\frac{3x}{2x - 1} - \frac{2x}{2x + 3}$

42. $\frac{2y}{5y - 1} - \frac{2y}{3y + 2}$

43. $\frac{1}{5x} + \frac{7x}{x + 5}$

44. $\frac{10h}{h - 3} + \frac{7}{9h}$

45. $\frac{x}{x + 1} + \frac{x - 1}{x}$

46. $\frac{t - 2}{t} + \frac{t}{t + 3}$

47. $\frac{s + 7}{s + 3} - \frac{s - 3}{s + 7}$

48. $\frac{t + 5}{t - 5} - \frac{t - 5}{t + 5}$

49. $\frac{3m}{m - 2} - \frac{m - 3}{m + 5}$

50. $\frac{2x}{x + 2} - \frac{x + 1}{x - 3}$

Perform the operations. Simplify, if possible. **See Example 5.**

51. $\frac{4}{s^2 + 5s + 4} + \frac{s}{s^2 + 2s + 1}$

52. $\frac{d}{d^2 + 6d + 5} - \frac{3}{d^2 + 5d + 4}$

53. $\frac{5}{x^2 - 9x + 8} - \frac{3}{x^2 - 6x - 16}$

54. $\frac{3}{t^2 + t - 6} + \frac{1}{t^2 + 3t - 10}$

55. $\frac{2}{a^2 + 4a + 3} + \frac{1}{a + 3}$

56. $\frac{1}{c + 6} + \frac{4}{c^2 + 8c + 12}$

57. $\frac{8}{y^2 - 16} - \frac{7}{y^2 - y - 12}$

58. $\frac{6}{s^2 - 9} - \frac{5}{s^2 - s - 6}$

Perform the operations. Simplify, if possible. **See Example 6.**

59. $\frac{8}{x} + 6$

60. $\frac{2}{y} + 7$

61. $\frac{9}{x - 4} + x$

62. $\frac{9}{m + 4} + 9$

63. $b - \frac{3}{a^2}$

64. $c - \frac{5}{3b}$

65. $\frac{x + 2}{x + 1} - 5$

66. $\frac{y + 8}{y - 8} - 4$

***Perform the operations. Simplify, if possible.* See Example 7.**

67. $\frac{7}{a - 4} + \frac{5}{4 - a}$

68. $\frac{4}{b - 6} + \frac{b}{6 - b}$

69. $\frac{c}{7c - d} - \frac{d}{d - 7c}$

70. $\frac{a}{5a - 3b} - \frac{b}{3b - 5a}$

71. $\frac{3d - 3}{d - 9} - \frac{3d}{9 - d}$

72. $\frac{2x + 2}{x - 2} - \frac{2x}{2 - x}$

73. $\frac{g}{g^2 - 4} + \frac{2}{4 - g^2}$

74. $\frac{h}{h^2 - 49} + \frac{7}{49 - h^2}$

TRY IT YOURSELF

Perform the operations and simplify, if possible.

75. $\frac{j}{j^2 + 9j + 20} - \frac{4}{j^2 + 7j + 12}$

76. $\frac{r}{r^2 + 5r + 6} - \frac{2}{r^2 + 3r + 2}$

77. $\frac{10}{x - 1} + y$

78. $\frac{3}{s - 8} + t$

79. $\frac{b}{b + 1} - \frac{b - 1}{b + 2}$

80. $\frac{x}{x - 2} - \frac{x + 2}{x + 3}$

81. $\frac{y}{y - 1} - \frac{4}{1 - y}$

82. $\frac{1}{t - 7} - \frac{t}{7 - t}$

83. $\frac{n}{5} - \frac{n - 2}{15}$

84. $\frac{m}{9} - \frac{m + 1}{27}$

85. $\frac{y + 2}{5y^2} + \frac{y + 4}{15y}$

86. $\frac{x + 3}{x^2} + \frac{x + 5}{2x}$

87. $\frac{x}{x - 2} + \frac{4 + 2x}{x^2 - 4}$

88. $\frac{y}{y + 3} - \frac{2y - 6}{y^2 - 9}$

89. $\frac{7}{3a} + \frac{1}{a - 2}$

90. $\frac{5}{9x} + \frac{4}{x + 6}$

Perform the operations and simplify, if possible. Be careful to apply the correct method because these problems involve addition, subtraction, multiplication, and division.

91. **a.** $\frac{5}{2x} + \frac{4x}{15}$ **b.** $\frac{5}{2x} \cdot \frac{4x}{15}$

92. **a.** $\frac{2a + 4}{3} - \frac{9}{a + 2}$ **b.** $\frac{2a + 4}{3} \cdot \frac{9}{a + 2}$

93. **a.** $\frac{t}{t - 5} - \frac{t}{t^2 - 25}$ **b.** $\frac{t}{t - 5} \div \frac{t}{t^2 - 25}$

94. **a.** $\frac{1}{m + 2} - \frac{2}{m^2 + 4m + 4}$ **b.** $\frac{1}{m + 2} \div \frac{2}{m^2 + 4m + 4}$

APPLICATIONS

95. Find the total height of the funnel.

96. What is the difference between the diameter of the opening at the top of the funnel and the diameter of its spout?

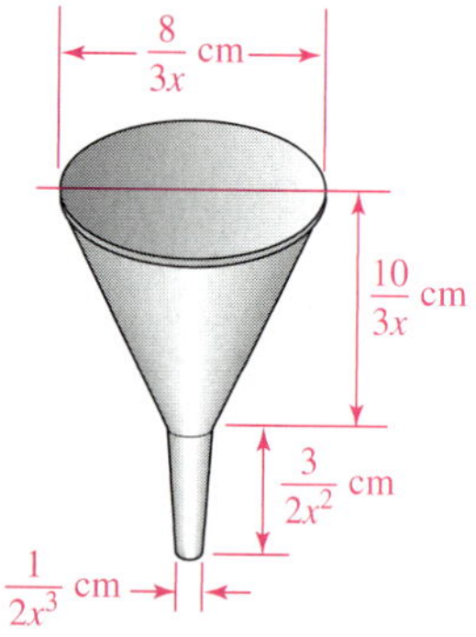

WRITING

97. Explain the error:

$$\frac{3}{x} + \frac{8}{y} = \frac{3 + 8}{x + y} = \frac{11}{x + y}$$

98. Explain how to add two rational expressions with unlike denominators.

99. When will the LCD of two rational expressions be the product of the denominators of those rational expressions? Give an example.

100. Explain how multiplication by $\frac{-1}{-1}$ is used in this section.

REVIEW

101. Find the slope and y-intercept of the graph of $y = 8x + 2$.

102. Find the slope and y-intercept of the graph of $3x + 4y = -36$.

103. What is the slope of the graph of $y = 2$?

104. Is the graph of the equation $x = 0$ the x-axis or the y-axis?

CHALLENGE PROBLEMS

Perform the operations and simplify the result, if possible.

105. $\frac{a}{a - 1} - \frac{2}{a + 2} + \frac{3(a - 2)}{a^2 + a - 2}$

106. $\frac{2x}{x^2 - 3x + 2} + \frac{2x}{x - 1} - \frac{x}{x - 2}$

107. $\frac{1}{a + 1} + \frac{a^2 - 7a + 10}{2a^2 - 2a - 4} \cdot \frac{2a^2 - 50}{a^2 + 10a + 25}$

108. $1 - \frac{(x - 2)^2}{(x + 2)^2}$

SECTION 7.5
Simplifying Complex Fractions

Objectives

1. Simplify complex fractions using division.
2. Simplifying complex fractions using the LCD.

A rational expression whose numerator and/or denominator contain fractions is called a **complex rational expression** or a **complex fraction.** The expression above the main fraction bar of a complex fraction is the numerator, and the expression below the main fraction bar is the denominator. Two examples of complex fractions are:

$$\frac{\frac{5x}{3}}{\frac{2x}{9}} \qquad \frac{\frac{1}{2}-\frac{1}{x}}{\frac{x}{3}+\frac{1}{5}}$$

← Numerator of complex fraction →

← Main fraction bar →

← Denominator of complex fraction →

In this section, we will discuss two methods for simplifying complex fractions. To **simplify a complex fraction** means to write it in the form $\frac{A}{B}$, where A and B are polynomials that have no common factors.

1 Simplify Complex Fractions Using Division.

One method for simplifying complex fractions uses the fact that the main fraction bar indicates division.

Simplifying Complex Fractions Method 1: Using Division

1. Add or subtract in the numerator and/or denominator so that the numerator is a single fraction and the denominator is a single fraction.
2. Perform the indicated division by multiplying the numerator of the complex fraction by the reciprocal of the denominator.
3. Simplify the result, if possible.

EXAMPLE 1 Simplify: $\frac{\frac{5x^2}{3}}{\frac{2x^3}{9}}$

Strategy We will perform the division indicated by the main fraction bar using the procedure for dividing rational expressions from Section 7.2.

Why We can skip the first step of method 1 and immediately divide because the numerator and the denominator of the complex fraction are already single fractions.

Solution

$$\frac{\frac{5x^2}{3}}{\frac{2x^3}{9}} = \frac{5x^2}{3} \div \frac{2x^3}{9}$$

Write the division indicated by the main fraction bar using a ÷ symbol.

The Language of Algebra
The second step of this method could also be phrased: Perform the division by *inverting the denominator of the complex fraction and multiplying.*

$$= \frac{5x^2}{3} \cdot \frac{9}{2x^3}$$ To divide rational expressions, multiply the first by the reciprocal of the second.

$$= \frac{5x^2 \cdot 9}{3 \cdot 2x^3}$$ Multiply the numerators. Multiply the denominators.

$$= \frac{5 \cdot \overset{1}{\cancel{x}} \cdot \overset{1}{\cancel{x}} \cdot \overset{1}{\cancel{3}} \cdot 3}{\underset{1}{\cancel{3}} \cdot 2 \cdot \underset{1}{\cancel{x}} \cdot \underset{1}{\cancel{x}} \cdot x}$$ Factor 9 as 3 · 3. Then simplify by removing factors equal to 1.

$$= \frac{15}{2x}$$ Multiply the remaining factors in the numerator. Multiply the remaining factors in the denominator.

Self Check 1 Simplify: $\frac{\frac{7y^3}{8}}{\frac{21y^2}{20}}$

Now Try Problem 19

EXAMPLE 2 Simplify: $\frac{\frac{1}{2} - \frac{1}{x}}{\frac{x}{3} + \frac{1}{5}}$

Strategy We will simplify the expressions above and below the main fraction bar separately to write $\frac{1}{2} - \frac{1}{x}$ and $\frac{x}{3} + \frac{1}{5}$ as single fractions. Then we will perform the indicated division.

Why The numerator and the denominator of the complex fraction must be written as single fractions before dividing.

Solution To write the numerator as a single fraction, we build $\frac{1}{2}$ and $\frac{1}{x}$ to have an LCD of $2x$, and then subtract. To write the denominator as a single fraction, we build $\frac{x}{3}$ and $\frac{1}{5}$ to have an LCD of 15, and then add.

$$\frac{\frac{1}{2} - \frac{1}{x}}{\frac{x}{3} + \frac{1}{5}} = \frac{\frac{1}{2} \cdot \frac{x}{x} - \frac{1}{x} \cdot \frac{2}{2}}{\frac{x}{3} \cdot \frac{5}{5} + \frac{1}{5} \cdot \frac{3}{3}}$$
← The LCD for the numerator is 2x. Build each fraction so that each has a denominator of 2x.
← The LCD for the denominator is 15. Build each fraction so that each has a denominator of 15.

$$= \frac{\frac{x}{2x} - \frac{2}{2x}}{\frac{5x}{15} + \frac{3}{15}}$$ Multiply the numerators and multiply the denominators.

$$= \frac{\frac{x - 2}{2x}}{\frac{5x + 3}{15}}$$ Subtract in the numerator and add in the denominator of the complex fraction.

Now that the numerator and the denominator of the complex fraction are single fractions, we perform the indicated division.

$$\frac{\frac{x-2}{2x}}{\frac{5x+3}{15}} = \frac{x-2}{2x} \div \frac{5x+3}{15}$$ Write the division indicated by the main fraction bar using a ÷ symbol.

$$= \frac{x-2}{2x} \cdot \frac{15}{5x+3}$$ Multiply by the reciprocal of $\frac{5x+3}{15}$.

$$= \frac{15(x-2)}{2x(5x+3)}$$ Multiply the numerators. Multiply the denominators.

Since the numerator and denominator have no common factor, the result does not simplify.

Notation

The result after simplifying a complex fraction can often have several equivalent forms. The result for Example 2 could be written:

$$\frac{15x-30}{2x(5x+3)}$$

Self Check 2 Simplify: $\frac{\frac{1}{3}+\frac{1}{x}}{\frac{x}{5}-\frac{1}{2}}$

Now Try **Problem 27**

EXAMPLE 3

Simplify: $\frac{\frac{6}{x}+y}{\frac{6}{y}+x}$

Strategy We will simplify the expressions above and below the main fraction bar separately to write $\frac{6}{x}+y$ and $\frac{6}{y}+x$ as single fractions. Then we will perform the indicated division.

Why The numerator and the denominator of the complex fraction must be written as single fractions before dividing.

Solution To write $\frac{6}{x}+y$ as a single fraction, we build y into a fraction with a denominator of x and add. To write $\frac{6}{y}+x$ as a single fraction, we build x into a fraction with a denominator of y and add.

$$\frac{\frac{6}{x}+y}{\frac{6}{y}+x} = \frac{\frac{6}{x}+\frac{y}{1}\cdot\frac{x}{x}}{\frac{6}{y}+\frac{x}{1}\cdot\frac{y}{y}}$$

← Write y as $\frac{y}{1}$. The LCD for the numerator is x. Build $\frac{y}{1}$ so that it has a denominator of x.

← Write x as $\frac{x}{1}$. The LCD for the denominator is y. Build $\frac{x}{1}$ so that it has a denominator of y.

$$= \frac{\frac{6}{x}+\frac{xy}{x}}{\frac{6}{y}+\frac{xy}{y}}$$ Multiply in the numerator and multiply in the denominators.

$$= \frac{\frac{6+xy}{x}}{\frac{6+xy}{y}}$$ Add in the numerator and in the denominator of the complex fraction.

Now that the numerator and the denominator of the complex fraction are single fractions, we can perform the division.

The Language of Algebra
After multiplying a complex fraction by $\frac{LCD}{LCD}$ and performing the multiplications, the numerator and denominator of the complex fraction will be *cleared of fractions*.

$$= \frac{\left(\frac{1}{8} - \frac{1}{y}\right)8y^2}{\left(\frac{8-y}{4y^2}\right)8y^2}$$ ← Multiply the numerators. ← Multiply the denominators.

$$= \frac{\frac{1}{8}(8y^2) - \frac{1}{y}(8y^2)}{\left(\frac{8-y}{4y^2}\right)(8y^2)}$$ Distribute the multiplication by $8y^2$.

$$= \frac{y^2 - 8y}{(8-y)2}$$ Perform each of the three multiplications by $8y^2$.

$$= \frac{y\overset{-1}{\cancel{(y-8)}}}{\underset{1}{\cancel{(8-y)}}2}$$ In the numerator, factor out the GCF, y. Since $y - 8$ and $8 - y$ are opposites, simplify by replacing $\frac{y-8}{8-y}$ with $\frac{-1}{1}$.

$$= -\frac{y}{2}$$

Self Check 5 Simplify: $\frac{\frac{10-n}{5n^2}}{\frac{1}{10} - \frac{1}{n}}$

Now Try Problem 55

EXAMPLE 6

Simplify: $\frac{1}{1 + \frac{1}{x+1}}$

Strategy Although either method can be used, we will use method 2 to simplify this complex fraction.

Why Method 2 is often easier when the complex fraction contains a sum or difference.

Solution The only fraction within the complex fraction has the denominator $x + 1$. Therefore, the LCD is $x + 1$. We multiply the complex fraction by a factor equal to 1, using the LCD: $\frac{x+1}{x+1} = 1$.

Success Tip
Simplifying using the LCD (method 2) works well when the complex fraction has sums and/or differences in the numerator and/or denominator.

$$\frac{1}{1 + \frac{1}{x+1}} = \frac{1}{1 + \frac{1}{x+1}} \cdot \frac{x+1}{x+1}$$

$$= \frac{1(x+1)}{\left(1 + \frac{1}{x+1}\right)(x+1)}$$ Multiply the numerators. Multiply the denominators.

$$= \frac{1(x+1)}{1(x+1) + \frac{1}{x+1}(x+1)}$$ In the denominator, distribute the multiplication by $x + 1$.

$$= \frac{x+1}{x+1+1}$$ Perform each of the three multiplications by $x + 1$.

$$= \frac{x+1}{x+2}$$ Combine like terms in the denominator.

The result does not simplify.

Self Check 6 Simplify: $\dfrac{2}{\dfrac{1}{x+2}+2}$

Now Try **Problem 63**

ANSWERS TO SELF CHECK 1. $\frac{5y}{6}$ 2. $\frac{10(x+3)}{3x(2x-5)}$ 3. $\frac{b}{a}$ 4. $\frac{15(x-4)}{4x(3x+5)}$ 5. $-\frac{2}{n}$ 6. $\frac{2(x+2)}{2x+5}$

STUDY SET 7.5

VOCABULARY

Fill in the blanks.

1. The expression $\dfrac{\frac{2}{3}-\frac{1}{x}}{\frac{x-3}{4}}$ is called a ________ rational expression or a ________ fraction.
2. In a complex fraction, the numerator is above the ______ fraction bar and the ____________ is below it.

CONCEPTS

Fill in the blanks.

3. Method 1: To ________ a complex fraction, write its numerator and denominator as single fractions. Then perform the indicated ________ by multiplying the numerator of the complex fraction by the _________ of the denominator.
4. Method 2: To ________ a complex fraction, find the LCD of ____ fractions within the complex fraction. Multiply the complex fraction by 1 in the form ____. Then perform the operations.
5. Consider: $\dfrac{\frac{x-3}{4}}{\frac{1}{12}-\frac{x}{6}}$
   a. What is the numerator of the complex fraction? Is it a single fraction?
   b. What is the denominator of the complex fraction? Is it a single fraction?
6. Consider the complex fraction: $\dfrac{\frac{1}{y}-\frac{1}{3}}{\frac{5}{6}+\frac{1}{y}}$
   a. What is the LCD of all fractions in the complex fraction?
   b. To simplify the complex fraction using method 2, it should be multiplied by what form of 1?

NOTATION

Fill in the blanks to simplify each complex fraction.

7. $\dfrac{\frac{12}{y^2}}{\frac{4}{y^3}} = \dfrac{12}{y^2} \quad \square \quad \dfrac{4}{y^3}$

8. $\dfrac{\left(\frac{1}{5}-\frac{1}{a}\right)}{\left(\frac{a}{4}+\frac{2}{a}\right)} \cdot \dfrac{20a}{20a} = \dfrac{\frac{1}{5}(\quad)-\frac{1}{a}(\quad)}{\frac{a}{4}(\quad)+\frac{2}{a}(\quad)}$

$$= \frac{4a - \square}{\square + \square}$$

GUIDED PRACTICE

Simplify each complex fraction. **See Example 1.**

9. $\dfrac{\frac{2}{3}}{\frac{3}{4}}$

10. $\dfrac{\frac{3}{5}}{\frac{2}{7}}$

11. $\dfrac{\frac{x}{2}}{\frac{6}{5}}$

12. $\dfrac{\frac{9}{4}}{\frac{7}{x}}$

13. $\dfrac{\frac{x}{y}}{\frac{1}{x}}$

14. $\dfrac{\frac{y}{x}}{\frac{x}{xy}}$

15. $\dfrac{\frac{n}{8}}{\frac{1}{n^2}}$

16. $\dfrac{\frac{1}{m}}{\frac{m^3}{15}}$

17. $\dfrac{\frac{4a}{11}}{\frac{6a}{55}}$

18. $\dfrac{\frac{14}{15m}}{\frac{21}{25m}}$

19. $\dfrac{-\frac{x^4}{30}}{\frac{7x}{15}}$

20. $\dfrac{-\frac{5x^2}{24}}{\frac{x^5}{56}}$

21. $\dfrac{\frac{10x}{x-3}}{\frac{6}{x-3}}$

22. $\dfrac{\frac{18a}{a-4}}{\frac{12}{a-4}}$

23. $\dfrac{\frac{4t-8}{t^2}}{\frac{8t-16}{t^5}}$

24. $\dfrac{\frac{9m-27}{m^6}}{\frac{2m-6}{m^8}}$

Simplify each complex fraction. **See Examples 2 or 4.**

25. $\dfrac{\frac{1}{2}+\frac{3}{4}}{\frac{3}{2}+\frac{1}{4}}$

26. $\dfrac{\frac{2}{3}-\frac{5}{2}}{\frac{2}{3}-\frac{3}{2}}$

27. $\dfrac{\frac{1}{4}+\frac{1}{y}}{\frac{y}{3}-\frac{1}{2}}$

28. $\dfrac{\frac{2}{x}-\frac{1}{3}}{\frac{2}{3}+\frac{x}{5}}$

29. $\dfrac{\frac{1}{y}-\frac{5}{2}}{\frac{3}{y}}$

30. $\dfrac{\frac{1}{6}-\frac{5}{s}}{\frac{2}{s}}$

31. $\dfrac{\frac{4}{c}-\frac{c}{6}}{\frac{2}{c}}$

32. $\dfrac{\frac{10}{n}-\frac{n}{4}}{\frac{8}{n}}$

33. $\dfrac{\frac{2}{s}-\frac{2}{s^2}}{\frac{4}{s^3}+\frac{4}{s^2}}$

34. $\dfrac{\frac{2}{x^3}-\frac{2}{x}}{\frac{4}{x}+\frac{8}{x^2}}$

35. $\dfrac{\frac{1}{a^2b}-\frac{5}{ab}}{\frac{3}{ab}-\frac{7}{ab^2}}$

36. $\dfrac{\frac{3}{ab^2}+\frac{6}{a^2b}}{\frac{6}{a}-\frac{9}{b^2}}$

Simplify each complex fraction. **See Examples 3 or 5.**

37. $\dfrac{\frac{2}{3}+1}{\frac{1}{3}+1}$

38. $\dfrac{\frac{3}{5}-2}{\frac{2}{5}-2}$

39. $\dfrac{\frac{1}{x}-3}{\frac{5}{x}+2}$

40. $\dfrac{\frac{1}{y}+3}{\frac{3}{y}-2}$

41. $\dfrac{\frac{2}{x}+2}{\frac{4}{x}+2}$

42. $\dfrac{\frac{3}{x}-3}{\frac{9}{x}-3}$

43. $\dfrac{\frac{3y}{x}-y}{y-\frac{y}{x}}$

44. $\dfrac{\frac{y}{x}+3y}{y+\frac{2y}{x}}$

45. $\dfrac{4-\frac{1}{8h}}{12+\frac{3}{4h}}$

46. $\dfrac{12+\frac{1}{3b}}{12-\frac{1}{b^2}}$

47. $\dfrac{1-\frac{9}{d^2}}{2+\frac{6}{d}}$

48. $\dfrac{1-\frac{16}{a^2}}{\frac{12}{a}+3}$

Simplify each complex fraction. **See Examples 4 and 5.**

49. $\dfrac{\frac{1}{6}-\frac{2}{x}}{\frac{1}{6}+\frac{1}{x}}$

50. $\dfrac{\frac{3}{4}+\frac{1}{y}}{\frac{5}{6}-\frac{1}{y}}$

51. $\dfrac{\frac{a}{7}-\frac{7}{a}}{\frac{1}{a}+\frac{1}{7}}$

52. $\dfrac{\frac{t}{9}-\frac{9}{t}}{\frac{1}{t}+\frac{1}{9}}$

53. $\dfrac{\frac{m}{n}+\frac{n}{m}}{\frac{m}{n}-\frac{n}{m}}$

54. $\dfrac{\frac{2a}{b}-\frac{b}{a}}{\frac{2a}{b}+\frac{b}{a}}$

55. $\dfrac{\frac{d+2}{2}}{\frac{d}{3}-\frac{d}{4}}$

56. $\dfrac{\frac{d^2}{4}+\frac{4d}{5}}{\frac{d+1}{2}}$

57. $\dfrac{\frac{2}{c^2}}{\frac{1}{c}+\frac{5}{4}}$

58. $\dfrac{\frac{7}{s^2}}{\frac{1}{s}+\frac{10}{3}}$

59. $\dfrac{\frac{2}{x}}{\frac{2}{y}-\frac{4}{x}}$

60. $\dfrac{\frac{2y}{3}}{\frac{2y}{3}-\frac{8}{y}}$

Simplify each complex fraction. **See Example 6.**

61. $\dfrac{\frac{1}{x+1}}{1+\frac{1}{x+1}}$

62. $\dfrac{\frac{1}{x-1}}{1-\frac{1}{x-1}}$

63. $\dfrac{\frac{x}{x+2}}{\frac{x}{x+2}+x}$

64. $\dfrac{\frac{2}{x-2}}{\frac{2}{x-2}-1}$

65. $\dfrac{3+\frac{3}{x-1}}{3-\frac{3}{x-1}}$

66. $\dfrac{2-\frac{2}{x+1}}{2+\frac{2}{x+1}}$

67. $\dfrac{m-\frac{1}{2m+1}}{1-\frac{m}{2m+1}}$

68. $\dfrac{1-\frac{r}{2r+1}}{r-\frac{1}{2r+1}}$

TRY IT YOURSELF

Simplify each complex fraction.

69. $\dfrac{\frac{1}{p}+\frac{1}{q}}{\frac{1}{p}}$

70. $\dfrac{\frac{m}{n}+1}{1-\frac{m}{n}}$

71. $\dfrac{\frac{40x^2}{20x}}{9}$

72. $\dfrac{\frac{18n^2}{6n}}{13}$

73. $\dfrac{\frac{1}{c}+\frac{1}{2}}{\frac{1}{c^2}-\frac{1}{4}}$

74. $\dfrac{\frac{1}{m}-\frac{1}{n}}{\frac{m}{n}-\frac{n}{m}}$

75. $\dfrac{\frac{1}{r+1}+1}{\frac{3}{r-1}+1}$

76. $\dfrac{5+\frac{1}{n+7}}{4-\frac{2}{n+7}}$

77. $\dfrac{\frac{b^2-81}{18a^2}}{\frac{4b-36}{9a}}$

78. $\dfrac{\frac{8x-64}{y}}{\frac{x^2-64}{y^2}}$

79. $\dfrac{1+\frac{6}{t}+\frac{8}{t^2}}{1+\frac{1}{t}-\frac{12}{t^2}}$

80. $\dfrac{1-p+\frac{2}{p}}{\frac{6}{p^2}+\frac{1}{p}-1}$

81. $\dfrac{1}{\frac{1}{x}+\frac{1}{y}}$

82. $\dfrac{1}{\frac{b}{a}-\frac{a}{b}}$

83. $\dfrac{-\frac{25}{16x^2}}{\frac{15}{32x^5}}$

84. $\dfrac{\frac{21}{8g^3}}{-\frac{35}{16g^8}}$

APPLICATIONS

85. SLOPE We can use the slope formula to find the slope of a line that passes through $\left(\frac{1}{2}, \frac{1}{3}\right)$ and $\left(\frac{3}{4}, \frac{5}{8}\right)$:

$$m = \dfrac{\frac{5}{8}-\frac{1}{3}}{\frac{3}{4}-\frac{1}{2}}$$

Simplify the complex fraction to find m.

86. PITCHING The earned run average (ERA) is a statistic that gives the average number of earned runs a pitcher allows. For a softball pitcher, this is based on a six-inning game. The formula for ERA is

$$\text{ERA} = \dfrac{\text{earned runs}}{\frac{\text{innings pitched}}{6}}$$

Simplify the complex fraction on the right side of the formula.

87. ELECTRONICS In electronic circuits, resistors are tiny components that limit the flow of an electric current. An important formula about two resistors in a circuit is

$$\text{Total resistance} = \dfrac{1}{\frac{1}{R_1}+\frac{1}{R_2}}$$

(Recall that R_1 is read as R sub one.)

Simplify the complex fraction on the right side of the formula.

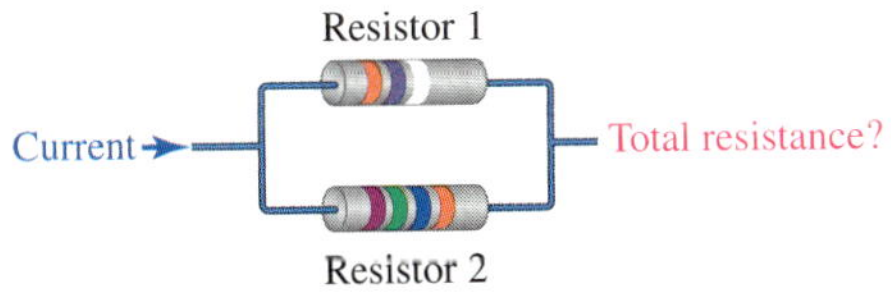

88. DATA ANALYSIS Use the data in the table to find the average measurement for the three-trial experiment.

	Trial 1	Trial 2	Trial 3
Measurement	$\frac{k}{2}$	$\frac{k}{3}$	$\frac{k}{2}$

WRITING

89. What is a complex fraction? Give several examples.

90. Explain how to use method 1 to simplify: $\dfrac{1+\frac{1}{x}}{3-\frac{1}{x}}$

91. Explain how to use method 2 to simplify the expression in Problem 90.

92. a. List an advantage and a disadvantage of using method 1 to simplify a complex fraction.

b. List an advantage and a disadvantage of using method 2 to simplify a complex fraction.

REVIEW

Simplify each expression. Write each answer without using parentheses or negative exponents.

93. $(8x)^0$

94. $\left(-\frac{3r}{4r^3}\right)^4$

95. $\left(\frac{4x^3}{5x^{-3}}\right)^{-2}$

96. $\left(\frac{12xy^{-3}}{3x^{-2}y^2}\right)^{-2}$

CHALLENGE PROBLEMS

Simplify.

97. $\dfrac{\dfrac{h}{h^2+3h+2}}{\dfrac{4}{h+2}-\dfrac{4}{h+1}}$

98. $\dfrac{\dfrac{2}{b^2-1}-\dfrac{3}{ab-a}}{\dfrac{3}{ab-a}-\dfrac{2}{b^2-1}}$

99. $a+\dfrac{a}{1+\dfrac{a}{a+1}}$

100. $\dfrac{y^{-2}+1}{y^{-2}-1}$

SECTION 7.6
Solving Rational Equations

Objectives

1. Solve rational equations.
2. Solve for a specified variable in a formula.

In Chapter 2, we solved equations such as $\frac{1}{6}x+\frac{5}{2}=\frac{1}{3}$ by multiplying both sides by the LCD. With this approach, the equation that results is equivalent to the original equation, but easier to solve because it is cleared of fractions.

In this section, we will extend the fraction-clearing strategy to solve another type of equation, called a *rational equation.*

Rational Equations

A **rational equation** is an equation that contains one or more rational expressions.

Rational equations often have a variable in a denominator. Some examples are:

$$\frac{2x}{3}=\frac{x}{6}+\frac{3}{2} \qquad \frac{2}{x}+\frac{1}{4}=\frac{5}{2x} \qquad \frac{11x}{x-5}=6+\frac{55}{x-5}$$

1 Solve Rational Equations.

To **solve a rational equation,** we find all the values of the variable that make the equation true. Any value of the variable that makes a denominator in a rational equation equal to 0 cannot be a solution of the equation. Such a number must be rejected, because division by 0 is undefined.

The following steps can be used to solve rational equations.

Strategy for Solving Rational Equations

1. Determine which numbers cannot be solutions of the equation.
2. Multiply both sides of the equation by the LCD of all rational expressions in the equation. This clears the equation of fractions.
3. Solve the resulting equation.
4. Check all possible solutions in the original equation.

EXAMPLE 1 Solve: $\frac{2x}{3} = \frac{x}{6} + \frac{3}{2}$

Strategy We will use the multiplication property of equality to clear this rational equation of fractions by multiplying both sides by the LCD.

Why Equations that contain only integers are usually easier to solve than equations that contain fractions.

Solution There are no restrictions on x, because no value of x ever makes a denominator 0. Since the denominators are 3, 6, and 2, we multiply both sides of the equation by the LCD, 6.

$$\frac{2x}{3} = \frac{x}{6} + \frac{3}{2}$$

$$6\left(\frac{2x}{3}\right) = 6\left(\frac{x}{6} + \frac{3}{2}\right)$$ Multiply both sides of the equation by the LCD of $\frac{2x}{3}$, $\frac{x}{6}$, and $\frac{3}{2}$, which is 6.

$$6\left(\frac{2x}{3}\right) = 6\left(\frac{x}{6}\right) + 6\left(\frac{3}{2}\right)$$ Distribute the multiplication by 6.

$$2 \cdot \overset{1}{\cancel{3}}\left(\frac{2x}{\underset{1}{\cancel{3}}}\right) = \overset{1}{\cancel{6}}\left(\frac{x}{\underset{1}{\cancel{6}}}\right) + \overset{1}{\cancel{2}} \cdot 3\left(\frac{3}{\underset{1}{\cancel{2}}}\right)$$ Perform the three multiplications by 6 by first removing common factors of the numerator and denominator. Try to do this step in your head.

$$4x = x + 9$$ Simplify. Note that the fractions have been cleared.

$$3x = 9$$ To eliminate x on the right side, subtract x from both sides.

$$x = 3$$ To undo the multiplication by 3, divide both sides by 3.

Caution

Always enclose the left and right sides of an equation within parentheses when multiplying both sides by the LCD.

$$\left(\frac{2x}{3}\right) = \left(\frac{x}{6} + \frac{3}{2}\right)$$

Notation

Here is an alternate way to remove common factors when multiplying by the LCD:

$$\overset{2}{\cancel{6}}\left(\frac{2x}{\underset{1}{\cancel{3}}}\right) \text{ and } \overset{3}{\cancel{6}}\left(\frac{3}{\underset{1}{\cancel{2}}}\right)$$

To check, we replace each x with 3 in the original equation.

$$\frac{2x}{3} = \frac{x}{6} + \frac{3}{2}$$ This is the original equation.

$$\frac{2(3)}{3} \stackrel{?}{=} \frac{3}{6} + \frac{3}{2}$$ Substitute 3 for x.

$$2 \stackrel{?}{=} \frac{1}{2} + \frac{3}{2}$$ Simplify: $\frac{2(\overset{1}{\cancel{3}})}{\underset{1}{\cancel{3}}} = 2$ and $\frac{3}{6} = \frac{1}{2}$.

$$2 = 2$$ $\frac{1}{2} + \frac{3}{2} = \frac{4}{2} = 2.$

Since we obtain a true statement, 3 is the solution of $\frac{2x}{3} = \frac{x}{6} + \frac{3}{2}$. The solution set is $\{3\}$.

Self Check 1 Solve: $\frac{3x}{5} = \frac{x}{2} + \frac{1}{10}$

Now Try Problem 15

EXAMPLE 2 Solve: $\frac{2}{x} + \frac{1}{4} = \frac{5}{2x}$

Strategy This equation contains two rational expressions that have a variable in their denominator. We begin by asking, "What value(s) of x make either denominator 0?" Then we will clear the equation of fractions by multiplying both sides by the LCD.

Why If a number makes the denominator of a rational expression 0, that number cannot be a solution of the equation because division by 0 is undefined.

Solution If x is 0, the denominators of $\frac{2}{x}$ and $\frac{5}{2x}$ are 0 and the expressions would be undefined. Therefore, 0 cannot be a solution.

Since the denominators are x, 4, and $2x$, we multiply both sides of the equation by the LCD, $4x$, to clear the equation of fractions.

$$\frac{2}{x} + \frac{1}{4} = \frac{5}{2x}$$

$$4x\left(\frac{2}{x} + \frac{1}{4}\right) = 4x\left(\frac{5}{2x}\right)$$

Write each side of the equation within parentheses, and then multiply both sides by $4x$.

$$4x\left(\frac{2}{x}\right) + 4x\left(\frac{1}{4}\right) = 4x\left(\frac{5}{2x}\right)$$

Distribute the multiplication by $4x$.

$$\overset{1}{4\cancel{x}}\left(\frac{2}{\underset{1}{\cancel{x}}}\right) + \overset{1}{\cancel{4}x}\left(\frac{1}{\underset{1}{\cancel{4}}}\right) = 2 \cdot \overset{1}{\cancel{2}} \cdot \overset{1}{\cancel{x}}\left(\frac{5}{\underset{1}{\cancel{2}} \cdot \underset{1}{\cancel{x}}}\right)$$

On the right side, factor $4x$ as $2 \cdot 2 \cdot x$. Perform the three multiplications by $4x$ by first removing common factors of each numerator and denominator. Try to do this step in your head.

$$8 + x = 10$$

Simplify. Note that the fractions have been cleared.

$$x = 2$$

To solve the resulting equation, subtract 8 from both sides.

The solution of $\frac{2}{x} + \frac{1}{4} = \frac{5}{2x}$ is 2. The solution set is $\{2\}$. Check by substituting 2 for each x in the original equation.

> ### *Caution*
> After multiplying both sides by the LCD and simplifying, the equation should not contain any fractions. If it does, check for an algebraic error, or perhaps your LCD is incorrect.

> ### *Success Tip*
> Don't confuse procedures. To *simplify the expression* $\frac{2}{x} + \frac{1}{4}$, we build each fraction to have the LCD $4x$, add the numerators, and write the sum over the LCD. To *solve the equation* $\frac{2}{x} + \frac{1}{4} = \frac{5}{2x}$, we multiply both sides by the LCD $4x$ to eliminate the denominators.

Self Check 2 Solve: $\frac{1}{6} + \frac{4}{3x} = \frac{5}{x}$

Now Try **Problem 23**

EXAMPLE 3 Solve: $y - \frac{12}{y} = 4$

Strategy Since the only denominator is y, we will multiply both sides of the equation by y.

Why Multiplying both sides by y will clear the equation of the fraction, $\frac{12}{y}$.

Solution If y is 0, the denominator of $\frac{12}{y}$ is 0 and the fraction would be undefined. Therefore, 0 cannot be a solution.

$$y - \frac{12}{y} = 4$$

$$y\left(y - \frac{12}{y}\right) = y(4)$$

Write each side of the equation within parentheses and then multiply both sides by the LCD, y.

$$y(y) - y\left(\frac{12}{y}\right) = y(4)$$

Distribute the multiplication by y.

$$y^2 - 12 = 4y$$

Simplify: $\overset{1}{\cancel{y}}\left(\frac{12}{\underset{1}{\cancel{y}}}\right) = 12$. Note that the fraction has been cleared.

> ### *Caution*
> By the multiplication property of equality, each term on both sides of the equation must be multiplied by the LCD. Here, it would be incorrect to only multiply the second term, $\frac{12}{y}$, by the LCD.
>
> $\cancel{y - y \cdot \frac{12}{y} = 4}$

We can solve the resulting quadratic equation using the factoring method.

$$y^2 - 4y - 12 = 0$$ Subtract 4y from both sides to get 0 on the right side.

$$(y - 6)(y + 2) = 0$$ Factor the trinomial.

$$y - 6 = 0 \quad \text{or} \quad y + 2 = 0$$ Set each factor equal to 0.

$$y = 6 \quad \Big| \quad y = -2$$ Solve each equation.

There are two possible solutions, 6 and −2, to check.

Check y = 6:

$$y - \frac{12}{y} = 4$$

$$6 - \frac{12}{6} \stackrel{?}{=} 4$$

$$6 - 2 \stackrel{?}{=} 4$$

$$4 = 4$$ True

Check y = −2:

$$y - \frac{12}{y} = 4$$ This is the original equation.

$$-2 - \frac{12}{-2} \stackrel{?}{=} 4$$

$$-2 - (-6) \stackrel{?}{=} 4$$

$$4 = 4$$ True

The solutions of $y - \frac{12}{y} = 4$ are 6 and −2.

Self Check 3 Solve: $x - \frac{24}{x} = -5$

Now Try **Problem 29**

EXAMPLE 4 Solve: $\frac{11x}{x - 5} = 6 + \frac{55}{x - 5}$

Strategy Since both denominators are $x - 5$, we multiply both sides by the LCD, $x - 5$.

Why This will clear the equation of fractions.

Solution If x is 5, the denominators of $\frac{11x}{x-5}$ and $\frac{55}{x-5}$ are 0, and the expressions are undefined. Therefore, 5 cannot be a solution of the equation.

$$\frac{11x}{x - 5} = 6 + \frac{55}{x - 5}$$

$$(x - 5)\left(\frac{11x}{x - 5}\right) = (x - 5)\left(6 + \frac{55}{x - 5}\right)$$ Write each side of the equation within parentheses and then multiply both sides by x − 5.

$$\frac{\overset{1}{\cancel{(x - 5)}}}{}\left(\frac{11x}{\underset{1}{\cancel{x - 5}}}\right) = (x - 5)6 + \overset{1}{\cancel{(x - 5)}}\left(\frac{55}{\underset{1}{\cancel{x - 5}}}\right)$$ Distribute the multiplication by x − 5. Remove the common binomial factor (x − 5) of the numerator and denominator.

$$11x = (x - 5)6 + 55$$ Simplify. Note that the fractions have been cleared.

$$11x = 6x - 30 + 55$$ To solve the resulting equation, distribute the multiplication by 6.

$$11x = 6x + 25$$ Combine like terms: −30 + 55 = 25.

$$5x = 25$$ To eliminate 6x on the right side, subtract 6x from both sides.

$$x = 5$$ To undo the multiplication by 5, divide both sides by 5.

Caution
Even if you do not make an arithmetic or algebraic error when solving a rational equation, a possible solution may not check.

The Language of Algebra
Extraneous means not a vital part. Mathematicians speak of *extraneous* solutions. Rock groups don't want *extraneous* sounds (like feedback) coming from their amplifiers. Artists erase *extraneous* marks on their sketches.

We have determined that 5 makes both denominators in the original equation 0. Therefore, 5 cannot be a solution. Since 5 is the only possible solution, and it must be rejected, it follows that $\frac{11x}{x-5} = 6 + \frac{55}{x-5}$ has no solution. The solution set is written as { } or ∅.

When solving an equation, a possible solution that does not satisfy the original equation is called an **extraneous solution.** In this example, 5 is an extraneous solution.

Self Check 4 Solve $\frac{9x}{x-6} = 3 + \frac{54}{x-6}$, if possible.

Now Try Problem 39

EXAMPLE 5 Solve: $\frac{x+5}{x+3} + \frac{1}{x^2+2x-3} = 1$

Strategy We will multiply both sides by the LCD of the two rational expressions in the equation. But first, we must factor the second denominator.

Why To determine the restrictions on the variable and to find the LCD, we need to write $x^2 + 2x - 3$ in factored form.

Solution Since the trinomial $x^2 + 2x - 3$ factors as $(x+3)(x-1)$, we can write the given equation as:

$$\frac{x+5}{x+3} + \frac{1}{(x+3)(x-1)} = 1$$ If x is −3, the first denominator is 0. If x is −3 or 1, the second denominator is 0.

We see that −3 and 1 cannot be solutions of the equation, because they make rational expressions in the equation undefined.

Since the denominators are $x + 3$ and $(x+3)(x-1)$, we multiply both sides of the equation by the LCD, $(x+3)(x-1)$, to clear the fractions.

$$(x+3)(x-1)\left[\frac{x+5}{x+3} + \frac{1}{(x+3)(x-1)}\right] = (x+3)(x-1)[1]$$ Write each side within brackets [ ].

$$\overset{1}{\cancel{(x+3)}}(x-1)\frac{x+5}{\underset{1}{\cancel{x+3}}} + \overset{1}{\cancel{(x+3)}}\overset{1}{\cancel{(x-1)}}\frac{1}{\underset{1}{\cancel{(x+3)}}\underset{1}{\cancel{(x-1)}}} = (x+3)(x-1)1$$ Distribute the multiplication by (x + 3)(x − 1) and remove common factors.

$$(x-1)(x+5) + 1 = (x+3)(x-1)$$ Simplify.

To solve the resulting equation, we multiply the binomials on the left side and the right side, and proceed as follows.

$$x^2 + 4x - 5 + 1 = x^2 + 2x - 3$$ Find (x − 1)(x + 5) and (x + 3)(x − 1).

$$x^2 + 4x - 4 = x^2 + 2x - 3$$ Combine like terms: −5 + 1 = −4.

$$4x - 4 = 2x - 3$$ Subtract x^2 from both sides.

$$2x - 4 = -3$$ To eliminate 2x on the right side, subtract 2x from both sides.

$$2x = 1$$ To undo the subtraction of 4, add 4 to both sides.

$$x = \frac{1}{2}$$ To undo the multiplication by 2, divide both sides by 2.

A check will show that $\frac{1}{2}$ is the solution of the original equation.

The Language of Algebra
We say that the LCD is a *multiplier* that clears a rational equation of fractions.

Self Check 5 Solve: $\frac{1}{x+3} + \frac{1}{x-3} = \frac{5}{x^2-9}$

Now Try Problem 47

2 Solve for a Specified Variable in a Formula.

Many formulas are expressed as rational equations. To solve such formulas for a specified variable, we use the same steps, in the same order, as we do when solving rational equations having only one variable.

EXAMPLE 6 ***Determining a Child's Dosage.*** The formula $C = \frac{AD}{A+12}$ is called **Young's rule.** It is a way to find the approximate child's dose C of a medication, where A is the age of the child in years and D is the recommended dosage for an adult. Solve the formula for D.

Strategy As we have done in the previous examples, we will begin by multiplying both sides of the equation by the LCD to clear it of the fraction.

Why To isolate D on the right side of the equation, we must first isolate the term AD on that side. That calls for clearing the right side of the denominator $A + 12$.

Solution

$$C = \frac{AD}{A+12}$$

$$(A+12)(C) = (A+12)\left(\frac{AD}{A+12}\right)$$ Write each side of the formula within parentheses, and then multiply both sides by the LCD, $A + 12$.

$$(A+12)C = AD$$ Simplify the right side: $\overset{1}{(\cancel{A+12})}\left(\frac{AD}{\underset{1}{\cancel{A+12}}}\right)$.

$$AC + 12C = AD$$ Distribute the multiplication by C.

$$\frac{AC+12C}{A} = D$$ To undo the multiplication by A on the right side and isolate D, divide both sides by A.

Solving Young's rule for D, we have $D = \frac{AC+12C}{A}$.

Self Check 6 Solve $R = \frac{eS}{T-10}$ for S.

Now Try Problem 61

© Natalee Hazelwcod/Alamy

EXAMPLE 7 ***Photography.*** The design of a camera lens uses the formula $\frac{1}{f} = \frac{1}{p} + \frac{1}{q}$, where f is the focal length of the lens, p is the distance from the lens to the object, and q is the distance from the lens to the image. Solve the formula for q.

Strategy We will begin by multiplying both sides of the equation by the LCD.

Why It will be easier to isolate q if there are no fractions.

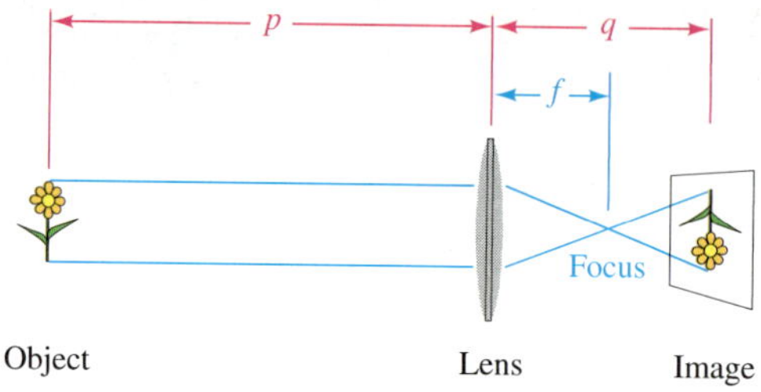

Solution

$$\frac{1}{f} = \frac{1}{p} + \frac{1}{q}$$

$$fpq\left(\frac{1}{f}\right) = fpq\left(\frac{1}{p} + \frac{1}{q}\right)$$ Write each side of the formula within parentheses and then multiply both sides by the LCD, fpq.

$$\overset{1}{\cancel{f}}pq\left(\frac{1}{\underset{1}{\cancel{f}}}\right) = f\overset{1}{\cancel{p}}q\left(\frac{1}{\underset{1}{\cancel{p}}}\right) + fp\overset{1}{\cancel{q}}\left(\frac{1}{\underset{1}{\cancel{q}}}\right)$$ Distribute the multiplication by fpq and then remove the common factors of each numerator and denominator.

$$pq = fq + fp$$ Simplify.

> **Caution**
> A common error, is to divide both sides of $pq = fq + fp$ by p to solve for q:
>
> $$\cancel{q = \frac{fq + fp}{p}}$$
>
> The formula is not solved for q because q appears on *both sides* of the equation.

If we subtract fq from both sides, all terms that contain q will be on the left side.

$$pq - fq = fp$$ Subtract fq from both sides.

$$q(p - f) = fp$$ Factor out the GCF, q, from the two terms on the left side.

$$\frac{q(p - f)}{p - f} = \frac{fp}{p - f}$$ To undo the multiplication by $(p - f)$ and isolate q, divide both sides by $p - f$.

$$q = \frac{fp}{p - f}$$ Simplify the left side: $\frac{q\overset{1}{\cancel{(p - f)}}}{\underset{1}{\cancel{p - f}}} = q$.

Solving the formula for q, we have $q = \frac{fp}{p - f}$.

Self Check 7 Solve the formula in Example 7 for p.

Now Try **Problem 63**

ANSWERS TO SELF CHECK **1.** 1 **2.** 22 **3.** 3, −8 **4.** No solution **5.** $\frac{5}{2}$ **6.** $S = \frac{RT - 10R}{e}$ **7.** $p = \frac{fq}{q - f}$

STUDY SET 7.6

VOCABULARY

Fill in the blanks.

1. Equations that contain one or more rational expressions, such as $\frac{x}{x + 2} = 4 + \frac{10}{x + 2}$, are called ________ equations.

2. To *solve* a rational equation we find all the values of the variable that make the equation _____.

3. To *clear* a rational equation of fractions, ________ both sides by the LCD of all rational expressions in the equation.

4. When solving a rational equation, if we obtain a number that does not satisfy the original equation, the number is called an __________ solution.

CONCEPTS

5. Is 5 a solution of the given rational equation?

a. $\frac{1}{x-1} = 1 - \frac{3}{x-1}$

b. $\frac{x}{x-5} = 3 + \frac{5}{x-5}$

6. A student was asked to solve a rational equation. The first step of his solution is as follows:

$$12x\left(\frac{5}{x} + \frac{2}{3}\right) = 12x\left(\frac{7}{4x}\right)$$

a. What equation was he asked to solve?

b. What LCD is used to clear the equation of fractions?

7. Consider the rational equation $\frac{x}{x-3} = \frac{1}{x} + \frac{2}{x-3}$.

a. What values of x make a denominator 0?

b. What values of x make a rational expression undefined?

c. What numbers can't be solutions of the equation?

8. A student solved a rational equation and found 8 to be a possible solution. When she checked 8, she obtained $\frac{3}{0} = \frac{1}{0} + \frac{2}{3}$. What conclusion can be drawn?

By what should both sides of the equation be multiplied to clear it of fractions?

9. a. $\frac{1}{y} = 20 - \frac{5}{y}$ b. $\frac{x}{x^2-4} = \frac{4}{x-2}$

10. a. $\frac{x}{5} = \frac{3x}{10} + \frac{7}{2x}$ b. $\frac{2x}{x-6} = 4 + \frac{1}{x-6}$

11. Perform each multiplication.

a. $4x\left(\frac{3}{4x}\right)$ b. $(x+6)(x-2)\left(\frac{3}{x-2}\right)$

12. Fill in the blanks.

$$\underbrace{8x\left(\frac{3}{4x}\right)}_{\square} = \underbrace{8x\left(\frac{1}{8x}\right)}_{\square} + \underbrace{8x\left(\frac{5}{4}\right)}_{\square}$$

NOTATION

Complete the solution to solve the equation.

13. $\frac{2}{a} + \frac{1}{2} = \frac{7}{2a}$

$\square\left(\frac{2}{a} + \frac{1}{2}\right) = \square\left(\frac{7}{2a}\right)$

$\square\left(\frac{2}{a}\right) + \square\left(\frac{1}{2}\right) = \square\left(\frac{7}{2a}\right)$

$\square + a = \square$

$4 + a - 4 = 7 - \square$

$a = \square$

14. Can $5x\left(\frac{2}{x} + \frac{4}{5}\right)$ be written as $5x \cdot \frac{2}{x} + \frac{4}{5}$? Explain.

GUIDED PRACTICE

Solve each equation and check the result. If an equation has no solution, so indicate. **See Examples 1 and 2.**

15. $\frac{2}{3} = \frac{1}{2} + \frac{x}{6}$

16. $\frac{7}{4} = \frac{x}{8} + \frac{5}{2}$

17. $\frac{s}{12} - \frac{s}{2} = \frac{5s}{4}$

18. $\frac{n}{18} - \frac{n}{6} = \frac{4n}{3}$

19. $\frac{x}{18} = \frac{1}{3} - \frac{x}{2}$

20. $\frac{x}{4} = \frac{1}{2} - \frac{3x}{20}$

21. $\frac{5}{3k} + \frac{1}{k} = -2$

22. $\frac{3}{4h} + \frac{2}{h} = 1$

23. $\frac{1}{4} - \frac{5}{6} = \frac{1}{a}$

24. $\frac{5}{9} - \frac{1}{3} = \frac{1}{b}$

25. $\frac{1}{8} + \frac{2}{b} - \frac{1}{12} = 0$

26. $\frac{1}{14} + \frac{2}{n} - \frac{2}{21} = 0$

Solve each equation and check the result. If an equation has no solution, so indicate. **See Example 3.**

27. $x + \frac{8}{x} = 6$

28. $z - \frac{16}{z} = 6$

29. $\frac{10}{t} - t = 3$

30. $\frac{7}{p} - p = -6$

31. $\frac{20}{c} + c = -9$

32. $d = 4 + \frac{21}{d}$

33. $4 + \frac{15}{p} = 3p$

34. $2x = 6 + \frac{8}{x}$

Solve each equation and check the result. If an equation has no solution, so indicate. **See Example 4.**

35. $\frac{2}{y+1} + 5 = \frac{12}{y+1}$

36. $\frac{3}{p+6} - 2 = \frac{7}{p+6}$

37. $\frac{x}{x-5} - \frac{5}{x-5} = 3$

38. $\frac{3}{y-2} + 1 = \frac{3}{y-2}$

39. $\frac{a^2}{a+2} - a = \frac{4}{a+2}$

40. $\frac{z^2}{z+1} + 2 = \frac{1}{z+1}$

41. $\frac{5a}{a+1} - 4 = \frac{3}{a+1}$

42. $\frac{4}{b-3} = \frac{b+5}{b-3} - 5$

43. $\frac{z-4}{z-3} = \frac{z+2}{z+1}$

44. $\frac{a+2}{a+8} = \frac{a-3}{a-2}$

45. $\frac{2}{3-t} = \frac{-t}{t+3}$

46. $\frac{n}{n+1} = \frac{6}{n+7}$

Solve each equation and check the result. If an equation has no solution, so indicate. **See Example 5.**

47. $\frac{2x}{x^2 + x - 2} + \frac{2}{x + 2} = 1$

48. $\frac{4x}{x^2 + 2x - 3} + \frac{3}{x + 3} = 1$

49. $\frac{4}{y^2 - 4} = \frac{1}{y - 2} + \frac{1}{y + 2}$

50. $\frac{2w}{w^2 - 9} = \frac{1}{w + 3} - \frac{4}{w - 3}$

51. $\frac{3}{x - 2} + \frac{1}{x} = \frac{6x + 4}{x^2 - 2x}$

52. $\frac{x}{x - 1} - \frac{12}{x^2 - x} = \frac{-1}{x - 1}$

53. $\frac{2}{4m + 12} - \frac{m + 1}{3m + 9} = \frac{m}{2m + 6}$

54. $\frac{5}{4y - 4} + \frac{y - 2}{2y - 2} = \frac{y}{5y - 5}$

Solve each formula for the indicated variable. **See Examples 6 and 7.**

55. $\frac{P}{n} = rt$ for P

56. $\frac{F}{m} = a$ for F

57. $\frac{a}{b} = \frac{c}{d}$ for d

58. $\frac{pc}{s} = \frac{t}{r}$ for c

59. $h = \frac{2A}{b + d}$ for A

60. $T = \frac{3R}{M - n}$ for R

61. $I = \frac{E}{R + r}$ for r

62. $\frac{S}{k + h} = E$ for k

63. $\frac{1}{a} + \frac{1}{b} = 1$ for a

64. $\frac{1}{a} - \frac{1}{b} = 1$ for b

65. $\frac{5}{x} - \frac{4}{y} = \frac{5}{z}$ for x

66. $\frac{2}{c} + \frac{2}{d} = \frac{1}{h}$ for c

67. $\frac{1}{r} + \frac{1}{s} = \frac{1}{t}$ for r

68. $\frac{1}{x} - \frac{1}{y} = \frac{1}{z}$ for x

69. $F = \frac{L^2}{6d} + \frac{d}{2}$ for L^2

70. $H = \frac{J^3}{cd} - \frac{K^3}{d}$ for J^3

TRY IT YOURSELF

Solve each equation and check the result. If an equation has no solution, so indicate.

71. $\frac{1}{3} + \frac{2}{x - 3} = 1$

72. $\frac{3}{5} + \frac{7}{x + 2} = 2$

73. $\frac{7}{q^2 - q - 2} + \frac{1}{q + 1} = \frac{3}{q - 2}$

74. $\frac{3}{x - 1} - \frac{1}{x + 9} = \frac{18}{x^2 + 8x - 9}$

75. $\frac{1}{8} + \frac{2}{y} = \frac{1}{y} + \frac{1}{10}$

76. $\frac{7}{10} + \frac{4}{c} = \frac{1}{c} + \frac{11}{15}$

77. $4 - \frac{8}{x + 1} = \frac{8x}{x + 1}$

78. $\frac{x}{x - 2} = \frac{2}{x - 2} + 2$

79. $\frac{3}{x + 1} = \frac{x - 2}{x + 1} + \frac{x - 2}{2}$

80. $\frac{2}{x - 1} + \frac{x - 2}{3} = \frac{4}{x - 1}$

81. $\frac{3}{x} + 2 = 3$

82. $\frac{2}{x} + 9 = 11$

83. $\frac{3}{5d} + \frac{4}{3} = \frac{9}{10d}$

84. $\frac{2}{3d} + \frac{1}{4} = \frac{11}{6d}$

85. $\frac{n}{n^2 - 9} + \frac{n + 8}{n + 3} = \frac{n - 8}{n - 3}$

86. $\frac{7}{x - 5} = \frac{40}{x^2 - 25} + \frac{3}{x + 5}$

87. $y + \frac{2}{3} = \frac{2y - 12}{3y - 9}$

88. $1 - \frac{3}{b} = \frac{-8b}{b^2 + 3b}$

89. $\frac{a - 1}{7} - \frac{a - 2}{14} = \frac{1}{2}$

90. $\frac{3x - 1}{6} - \frac{x + 3}{2} = \frac{3x + 4}{3}$

For each expression, perform the indicated operations and then simplify, if possible. Solve each equation and check the result.

91. a. $\frac{a}{3} + \frac{3}{5} + \frac{a}{15}$ **b.** $\frac{a}{3} + \frac{3}{5} = \frac{a}{15}$

92. a. $\frac{1}{6x} - \frac{2}{x - 6}$ **b.** $\frac{1}{6x} = \frac{2}{x - 6}$

93. a. $\frac{x}{x - 2} - \frac{1}{x - 3}$ **b.** $\frac{x}{x - 2} - \frac{1}{x - 3} = 1$

94. a. $\frac{u^2 + 1}{u^2 - u} - \frac{u}{u - 1}$ **b.** $\frac{u^2 + 1}{u^2 - u} - \frac{u}{u - 1} = \frac{1}{u}$

APPLICATIONS

95. MEDICINE Radioactive tracers are used for diagnostic work in nuclear medicine. The ***effective half-life H*** of a radioactive material in an organism is given by the formula $H = \frac{RB}{R + B}$ where R is the radioactive half-life and B is the biological half-life of the tracer. Solve the formula for R.

96. CHEMISTRY Charles's law describes the relationship between the volume and temperature of a gas that is kept at a constant pressure. It can be expressed as $\frac{V_1}{V_2} = \frac{T_1}{T_2}$ where V_1 and V_2 are variables representing two different volumes, and T_1 and T_2 are variables representing two different temperatures. (Recall that the notation V_1 is read as *V sub one.*) Solve for V_2.

97. ELECTRONICS Most electronic circuits require resistors to make them work properly. Resistors are components that limit current. An important formula about resistors in a circuit is $\frac{1}{r} = \frac{1}{r_1} + \frac{1}{r_2}$. Solve for r.

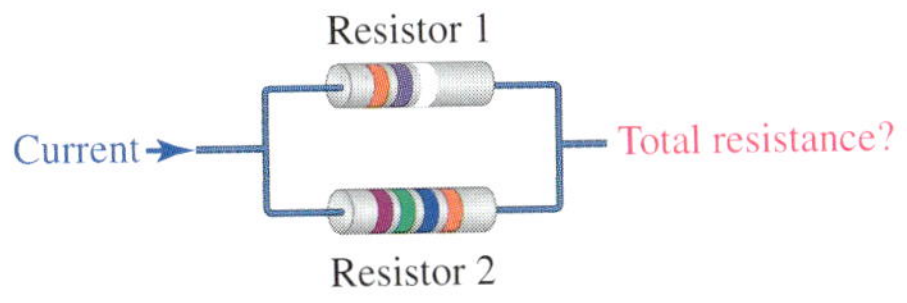

98. MATHEMATICAL FORMULAS To quickly find the sum $\frac{1}{2} + \frac{1}{4} + \frac{1}{8} + \frac{1}{16} + \frac{1}{32} + \frac{1}{64} + \frac{1}{128}$, mathematicians use the formula $S = \frac{a(1 - r^n)}{1 - r}$. Solve the formula for a.

WRITING

99. Explain how the multiplication property of equality is used to solve rational equations. Give an example.

100. When solving rational equations, how do you know whether a solution is extraneous?

101. What is meant by clearing a rational equation of fractions? Give an example.

102. Explain the difference between the procedure used to simplify $\frac{1}{x} + \frac{1}{3}$ and the procedure used to solve $\frac{1}{x} + \frac{1}{3} = \frac{1}{2}$.

REVIEW

103. UNIFORMS A cheerleading squad had their school mascot embroidered on the front of their uniform sweaters. They were charged \$18.50 per sweater plus a one time set up fee of \$75. If the project cost \$445, how many sweaters were embroidered?

104. GEOMETRY The vertex angle of an isosceles triangle is 46°. Find the measure of each base angle.

CHALLENGE PROBLEMS

105. Solve: $x^{-2} + 2x^{-1} + 1 = 0$

106. ENGINES A formula that is used in the design and testing of diesel engines is $E = 1 - \frac{T_4 - T_1}{a(T_3 - T_2)}$. Solve the formula for T_1.

SECTION 7.7
Problem Solving Using Rational Equations

Objectives

1. Solve number problems.
2. Solve uniform motion problems.
3. Solve shared-work problems.
4. Solve investment problems.

We will now use the five-step problem-solving strategy to solve application problems from a variety of areas, including banking, petroleum engineering, sports, and travel. In each case, we will use a rational equation to model the situation. We begin with an example in which we find an unknown number.

1 Solve Number Problems.

EXAMPLE 1 ***Number Problem.*** If the same number is added to both the numerator and the denominator of the fraction $\frac{3}{5}$, the result is $\frac{4}{5}$. Find the number.

Analyze the Problem

- Begin with the fraction $\frac{3}{5}$.
- Add the same number to the numerator and to the denominator.
- The result is $\frac{4}{5}$.
- Find the number.

Form an Equation Let n = the unknown number. To form an equation, add the unknown number to the numerator and to the denominator of $\frac{3}{5}$. Then set the result equal to $\frac{4}{5}$.

$$\frac{3+n}{5+n} = \frac{4}{5}$$

Solve the Equation To solve this rational equation, we begin by clearing it of fractions.

$$\frac{3+n}{5+n} = \frac{4}{5}$$

$$5(5+n)\left(\frac{3+n}{5+n}\right) = 5(5+n)\left(\frac{4}{5}\right)$$ Multiply both sides by the LCD, $5(5+n)$. Then remove common factors of the numerator and denominator.

$$5(3+n) = (5+n)4$$ Simplify.

$$15 + 5n = 20 + 4n$$ Distribute the multiplication by 5 and by 4.

$$15 + n = 20$$ To isolate the variable term of the left side, subtract $4n$ from both sides.

$$n = 5$$ To undo the addition of 15, subtract 15 from both sides.

State the Conclusion The number is 5.

Check the Result When we add 5 to both the numerator and denominator of $\frac{3}{5}$, we get

$$\frac{3+5}{5+5} = \frac{8}{10} = \frac{4}{5}$$

The result checks.

Now Try Problem 13

2 Solve Uniform Motion Problems.

Recall that we use the distance formula $d = rt$ to solve motion problems. The relationship between distance, rate, and time can be expressed in another way by solving for t.

$$d = rt$$ Distance = rate · time.

$$\frac{d}{r} = \frac{rt}{r}$$ To undo the multiplication by r and isolate t, divide both sides by r.

$$\frac{d}{r} = t$$ Simplify the right side: $\frac{r \cdot t}{r} = t$.

$$t = \frac{d}{r}$$

The Language of Algebra In uniform motion problems, the word *speed* is often used in place of the word *rate*. For example, we can say a car travels at a *rate* of 50 mph or its *speed* is 50 mph.

This result suggests an alternate form of the distance formula, Time $= \frac{\text{distance}}{\text{rate}}$, that is used to solve the next example.

EXAMPLE 2 ***Runners.*** A coach can run 10 miles in the same amount of time as his best student-athlete can run 12 miles. If the student runs 1 mile per hour (mph) faster than the coach, find the running speeds of the coach and the student.

Analyze the Problem

- The coach runs 10 miles in the same time that the student runs 12 miles.
- The student runs 1 mph faster than the coach.
- Find the speed that each runs.

Form an Equation

Since the student's speed is 1 mph faster than the coach's, let r = the speed that the coach can run. Then, $r + 1$ = the speed that the student can run. The expressions for the rates are entered in the Rate column of the table. The distances run by the coach and by the student are entered in the Distance column of the table.

Using $t = \frac{d}{r}$, we find that the time it takes the coach to run 10 miles, at a rate of r mph, is $\frac{10}{r}$ hours. Similarly, we find that the time it takes the student to run 12 miles, at a rate of $(r + 1)$ mph, is $\frac{12}{r + 1}$ hours. These expressions are entered in the Time column of the table.

	Rate	· Time	= Distance
Coach	r	$\frac{10}{r}$	10
Student	$r + 1$	$\frac{12}{r+1}$	12

Enter this information first.

Divide the distance by the rate to obtain an expression for the time: $t = \frac{d}{r}$.

The time it takes the coach to run 10 miles	is the same as	the time it takes the student to run 12 miles.
$\frac{10}{r}$	$=$	$\frac{12}{r + 1}$

Solve the Equation

To solve this rational equation, we begin by clearing it of fractions.

$$\frac{10}{r} = \frac{12}{r + 1}$$

$$\overset{1}{\cancel{r}}(r + 1)\left(\frac{10}{\underset{1}{\cancel{r}}}\right) = r\overset{1}{\cancel{(r + 1)}}\left(\frac{12}{\underset{1}{\cancel{r + 1}}}\right)$$ Multiply both sides by the LCD, $r(r + 1)$. Then remove common factors of the numerator and denominator.

$$(r + 1)10 = 12r$$ Simplify.

$$10r + 10 = 12r$$ Distribute the multiplication by 10.

$$10 = 2r$$ To isolate the variable term on the right, subtract $10r$ from both sides.

$$5 = r$$ To undo the multiplication by 2, divide both sides by 2.

If $r = 5$, then $r + 1 = 6$.

Caution

Check the result *in the words of the problem*, not by substituting it into the equation. Why? The equation may have been solved correctly, but the danger is you may have formed it incorrectly.

State the Conclusion The coach's running speed is 5 mph and the student's running speed is 6 mph.

Check the Result The coach will run 10 miles in $\frac{10 \text{ miles}}{5 \text{ mph}} = 2$ hours. The student will run 12 miles in $\frac{12 \text{ miles}}{6 \text{ mph}} = 2$ hours. The times are the same; the results check.

Now Try **Problem 23**

3 Solve Shared-Work Problems.

Problems in which two or more people (or machines) work together to complete a job are called *shared-work problems.* To solve such problems, we must determine the **rate of work** for each person (or machine) involved. For example, suppose it takes you 4 hours to clean your house. Your rate of work can be expressed as $\frac{1}{4}$ of the job is completed per hour. If someone else takes 5 hours to clean the same house, they complete $\frac{1}{5}$ of the job per hour. In general, a rate of work can be determined in the following way.

Rate of Work

If a job can be completed in t units of time, the rate of work can be expressed as:

$\frac{1}{t}$ of the job is completed per unit of time.

To solve shared-work problems, we must also determine what fractional part of a job is completed. To do this, we use the formula

Work completed = rate of work · time worked or $W = rt$

EXAMPLE 3 ***Payroll.*** At the end of a pay period, it takes the president of a company 15 minutes to sign all of her employees' payroll checks. What fractional part of the job is completed if the president signs checks for 10 minutes?

Strategy We will begin by finding the president's check-signing rate. Then we can use the formula $W = rt$ to find the part of the job that is completed.

Why We know the time worked is 10 minutes. To use the work formula to find what part of the job is completed, we also need to know the president's work rate.

Solution If all of the checks can be signed in 15 minutes, the president's work rate is $\frac{1}{15}$ job per minute. Substituting into the work formula, we have

$W = rt$

$= \frac{1}{15} \cdot 10$ Substitute $\frac{1}{15}$ for r, the work rate, and 10 for t, the time worked.

$= \frac{10}{15}$ Multiply.

$= \frac{2}{3}$ Simplify by removing a common factor of 5.

In 10 minutes, the president will complete $\frac{2}{3}$ of the job of signing the payroll checks.

Self Check 3 It takes a farmer 8 days to harvest a wheat crop. What part of the job is completed in 6 days?

Now Try **Problem 5**

EXAMPLE 4 ***Filling a Tank.*** An inlet pipe can fill an oil storage tank in 7 days, and a second inlet pipe can fill the same tank in 9 days. If both pipes are used, how long will it take to fill the tank?

Analyze the Problem

- The first pipe can fill the tank in 7 days.
- The second pipe can fill the tank in 9 days.
- How long will it take the two pipes, working together, to fill the tank?

Success Tip
The tables and the formulas used to solve uniform motion problems and shared-work problems are similar:

Motion: $d = rt$ Work: $W = rt$

Form an Equation

Let x = the number of days it will take to fill the tank if both pipes are used. It is helpful to organize the facts of the problem in a table. Since the pipes will be open for the same amount of time as they fill the tank, enter x as the time worked for each pipe.

The first pipe can fill the tank in 7 days; its rate working alone is $\frac{1}{7}$ of the job per day. The second pipe can fill the tank in 9 days; its rate working alone is $\frac{1}{9}$ of the job per day. To determine the work completed by each pipe, multiply the rate by the time.

	Rate ·	Time =	Work completed
1st pipe	$\frac{1}{7}$	x	$\frac{x}{7}$
2nd pipe	$\frac{1}{9}$	x	$\frac{x}{9}$

Enter this information first. Multiply to get each of these entries: $W = rt$.

In shared-work problems, the number 1 represents one whole job completed. So we have

The part of job done by 1st pipe	plus	part of job done by 2nd pipe	equals	1 job completed.
$\frac{x}{7}$	$+$	$\frac{x}{9}$	$=$	1

Solve the Equation

$$\frac{x}{7} + \frac{x}{9} = 1$$ This is a rational equation.

$$63\left(\frac{x}{7} + \frac{x}{9}\right) = 63(1)$$ Clear the equation of fractions by multiplying both sides by the LCD, 63.

$$63\left(\frac{x}{7}\right) + 63\left(\frac{x}{9}\right) = 63$$ Distribute the multiplication by 63.

$$9x + 7x = 63$$ Simplify the left side: $\overset{1}{\cancel{7}} \cdot 9\left(\frac{x}{\underset{1}{\cancel{7}}}\right) = 9x$ and $7 \cdot \overset{1}{\cancel{9}}\left(\frac{x}{\underset{1}{\cancel{9}}}\right) = 7x$.

$$16x = 63 \quad \text{Combine like terms.}$$

$$x = \frac{63}{16} \quad \text{To undo the multiplication by 16 and isolate } x \text{, divide both sides by 16.}$$

Notation
For this application problem, the mixed-number form of the result, $3\frac{15}{16}$ days, is preferable to the improper fraction form, $\frac{63}{16}$ days.

State the Conclusion If both pipes are used, it will take $\frac{63}{16}$ or $3\frac{15}{16}$ days to fill the tank.

Check the Result To check, we use the work formula and multiply each rate by the time. In $\frac{63}{16}$ days, the first pipe fills $\frac{1}{7} \cdot \frac{63}{16} = \frac{9}{16}$ of the tank and the second pipe fills $\frac{1}{9} \cdot \frac{63}{16} = \frac{7}{16}$ of the tank. The sum of these efforts, $\frac{9}{16} + \frac{7}{16}$, is $\frac{16}{16}$ or 1 full tank. The result checks.

Now Try **Problem 31**

Strategy for Solving Work Problems

Equations that model shared-work problems involving two people (or machines) have the form

$$\frac{x}{a} + \frac{x}{b} = 1$$

where x represents the time they work together on the job, and a and b represent the respective times each worker needs to complete the job alone.

Example 4 can be solved in a different way by considering the amount of work done by each pipe in 1 day. As before, if we let x = the number of days it will take to fill the tank if both inlet pipes are used, then together, in 1 day, they will complete $\frac{1}{x}$ of the job. If we add what the first pipe can do in 1 day to what the second pipe can do in 1 day, the sum is what they can do together in 1 day.

What the first inlet pipe can do in 1 day	plus	what the second inlet pipe can do in 1 day	equals	what they can do together in 1 day.
$\frac{1}{7}$	$+$	$\frac{1}{9}$	$=$	$\frac{1}{x}$

To solve the equation, begin by clearing it of fractions.

$$\frac{1}{7} + \frac{1}{9} = \frac{1}{x}$$

$$63x\left(\frac{1}{7} + \frac{1}{9}\right) = 63x\left(\frac{1}{x}\right) \quad \text{Multiply both sides by the LCD, } 63x.$$

$$9x + 7x = 63 \quad \text{Distribute the multiplication by } 63x \text{ and simplify.}$$

$$16x = 63 \quad \text{Combine like terms.}$$

$$x = \frac{63}{16} \quad \text{To isolate } x \text{, divide both sides by 16.}$$

This is the same answer as the one obtained in Example 4.

4 Solve Investment Problems.

We have used the interest formula $I = Prt$ to solve investment problems. The relationship between interest, principal, rate, and time can be expressed in another way, by solving for P.

$I = Prt$ Interest = principal · rate · time.

$\frac{I}{rt} = \frac{Prt}{rt}$ To undo the multiplication by rt and isolate P, divide both sides by rt.

$\frac{I}{rt} = P$ Simplify the right side: $\frac{P \cdot \overset{1}{\cancel{r}} \cdot \overset{1}{\cancel{t}}}{\underset{1}{\cancel{r}} \cdot \underset{1}{\cancel{t}}} = P$.

$P = \frac{I}{rt}$ Reverse the sides of the equation so that P is on the left.

This alternate form of the interest formula, Principal $= \frac{\text{Interest}}{\text{rate} \cdot \text{time}}$, is used to solve the next example.

EXAMPLE 5 ***Comparing Investments.*** An amount of money invested for one year in bonds will earn \$120. At a bank, that same amount of money will only earn \$75 interest, because the interest rate paid by the bank is 3% less than that paid by the bonds. Find the rate of interest paid by each investment.

Analyze the Problem

- The investment in bonds earns \$120 in one year.
- The same amount of money, invested in a bank, earns \$75 in one year.
- The interest rate paid by the bank is 3% less than that paid by the bonds.
- Find the bond's rate of interest and the bank's rate of interest.

Form an Equation Since the interest rate paid by the bank is 3% less than that paid by the bonds, let r = the bond's rate of interest, and $r - 0.03$ = the bank's interest rate. (Recall that $3\% = 0.03$.)

If an investment earns \$120 interest in 1 year at some rate r, we can use $P = \frac{I}{rt}$ to find that the principal invested was $\frac{120}{r}$ dollars. Similarly, if another investment earns \$75 interest in 1 year at some rate $r - 0.03$, the principal invested was $\frac{75}{r - 0.03}$ dollars. We can organize the facts of the problem in a table.

	Principal ·	Rate	· Time	= Interest
Bonds	$\frac{120}{r}$	r	1	120
Bank	$\frac{75}{r - 0.03}$	$r - 0.03$	1	75

Divide to get each of these entries: $P = \frac{I}{rt}$. Enter this information first.

The amount invested in the bonds	equals	the amount invested in the bank.
$\frac{120}{r}$	=	$\frac{75}{r - 0.03}$

Solve the Equation

$$\frac{120}{r} = \frac{75}{r - 0.03}$$ This is a rational equation.

$$\overset{1}{\cancel{r}}(r - 0.03)\left(\frac{120}{\underset{1}{\cancel{r}}}\right) = \left(\frac{75}{\underset{1}{\cancel{r - 0.03}}}\right)r\overset{1}{\cancel{(r - 0.03)}}$$ Multiply both sides by the LCD, r(r − 0.03). Then remove common factors of the numerator and denominator.

$$(r - 0.03)120 = 75r$$ Simplify.

$$120r - 3.6 = 75r$$ Distribute the multiplication by 120.

$$45r - 3.6 = 0$$ To isolate the variable term on the left side, subtract 75r from both sides.

$$45r = 3.6$$ Add 3.6 to both sides.

$$r = 0.08$$ To undo the multiplication by 45 and isolate r, divide both sides by 45.

If $r = 0.08$, then $r - 0.03 = 0.05$.

State the Conclusion The bonds pay 0.08, or 8%, interest. The bank's interest rate is 5%.

Check the Result The amount invested at 8% that will earn \$120 interest in 1 year is $\frac{120}{(0.08)1} = \$1{,}500$. The amount invested at 5% that will earn \$75 interest in 1 year is $\frac{75}{(0.05)1} = \$1{,}500$. The amounts invested in the bonds and the bank are the same. The results check.

Now Try **Problem 41**

ANSWER TO SELF CHECK 3. $\frac{3}{4}$

STUDY SET 7.7

VOCABULARY

Fill in the blanks.

1. In this section, problems that involve:
   - moving vehicles are called uniform ________ problems.
   - depositing money are called ___________ problems.
   - people completing jobs are called shared-_______ problems.
2. In the formula $W = rt$, the variable W stands for the ______ completed, r is the _____, and t is the _____.

CONCEPTS

3. Choose the equation that can be used to solve the following problem: *If the same number is added to the numerator and the denominator of the fraction $\frac{5}{8}$, the result is $\frac{2}{3}$. Find the number.*

   (i) $\frac{5}{8} + x = \frac{2}{3}$ **(ii)** $\frac{5 + x}{8} = \frac{2}{3}$

   (iii) $\frac{5 + x}{8 + x} = \frac{2}{3}$ **(iv)** $\frac{5}{8} = \frac{2 + x}{3 + x}$

4. Fill in the blank: If a job can be completed in t hours, then the rate of work can be expressed as $\frac{1}{\;}$ of the job is completed per hour.

5. It takes a night security officer 45 minutes to check each of the doors in an office building to make sure they are locked. What is the officer's rate of work?

6. It takes an elementary school teacher 4 hours to make out the semester report cards. What part of the job does she complete in x hours?

7. **a.** Solve $d = rt$ for t.

b. Solve $I = Prt$ for P.

8. Complete the table.

	r	$\cdot\ t$	$= d$
Snowmobile	r		4
4×4 truck	$r - 5$		3

9. Complete the table.

	Rate	· Time	= Work completed
1st printer	$\frac{1}{15}$	x	
2nd printer	$\frac{1}{8}$	x	

10. Complete the table.

	P	$\cdot\ r$	$\cdot\ t$	$= I$
City savings bank		r	1	50
Credit union		$r - 0.02$	1	75

NOTATION

11. Write $\frac{55}{9}$ days using a mixed number.

12. **a.** Write 9% as a decimal.

b. Write 0.035 as a percent.

GUIDED PRACTICE

Solve each of these number problems. **See Example 1.**

13. If the same number is added to both the numerator and the denominator of $\frac{2}{5}$, the result is $\frac{2}{3}$. Find the number.

14. If the same number is subtracted from both the numerator and the denominator of $\frac{11}{13}$, the result is $\frac{3}{4}$. Find the number.

15. If the denominator of $\frac{3}{4}$ is increased by a number, and the numerator is doubled, the result is 1. Find the number.

16. If a number is added to the numerator of $\frac{7}{8}$, and the same number is subtracted from the denominator, the result is 2. Find the number.

17. If a number is added to the numerator of $\frac{3}{4}$, and twice as much is added to the denominator, the result is $\frac{4}{7}$. Find the number.

18. If a number is added to the numerator of $\frac{5}{7}$, and twice as much is subtracted from the denominator, the result is 8. Find the number.

19. The sum of a number and its reciprocal is $\frac{13}{6}$. Find each number.

20. The sum of the reciprocals of two consecutive even integers is $\frac{7}{24}$. Find each integer.

APPLICATIONS

21. COOKING If the same number is added to both the numerator and the denominator of the amount of butter used in the following recipe for toffee, the result is the amount of brown sugar to be used. Find the number.

Toffee

$\frac{2}{3}$ cup pecans, chopped

$\frac{3}{4}$ cup brown sugar (packed)

$\frac{1}{2}$ package (6-oz size) semisweet chocolate

$\frac{1}{4}$ cup butter

22. TAPE MEASURES If the same number is added to both the numerator and the denominator of the first measurement, the result is the second measurement. Find the number.

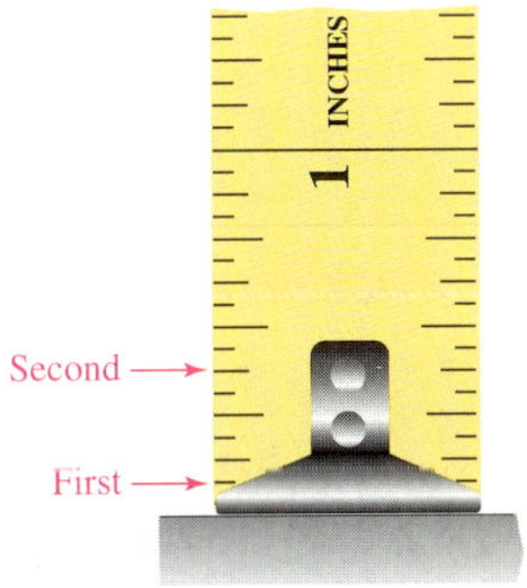

23. TOUR DE FRANCE Maurice Garin of France won the first Tour de France bicycle road race in 1903. In 2005, American Lance Armstrong won his seventh consecutive Tour de France. Armstrong's average speed in 2005 was 10 mph faster than Garin's in 1903. In the time it took Garin to ride 80 miles, Armstrong could have ridden 130 miles. Find each cyclist average speed.

24. PHYSICAL FITNESS A woman can bicycle 28 miles in the same time as it takes her to walk 8 miles. She can ride 10 mph faster than she can walk. How fast can she walk?

25. PACKAGING FRUIT The diagram on the next page shows how apples are processed for market. Although the second conveyor belt is shorter, an apple spends the same amount of time on each belt because the second conveyor moves 1 foot per second slower than the first. Determine the speed of each conveyor belt.

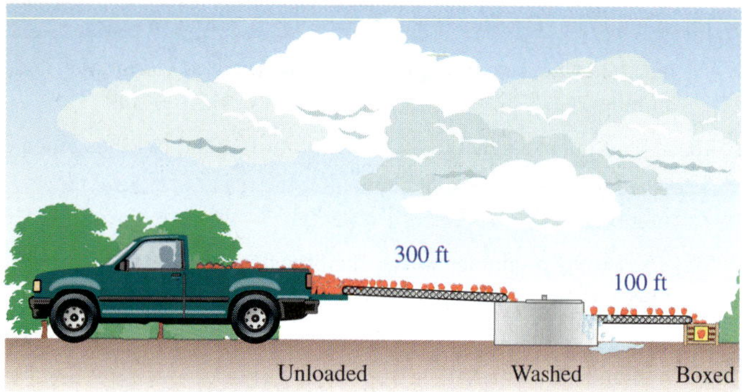

26. COMPARING TRAVEL A plane can fly 300 miles in the same time as it takes a car to go 120 miles. If the car travels 90 mph slower than the plane, find the speed of the plane.

27. BIRDS IN FLIGHT Although flight speed is dependent upon the weather and the wind, in general, a Canada goose can fly about 10 mph faster than a great blue heron. In the same time that a Canada goose travels 120 miles, a great blue heron travels 80 miles. Find their flying speeds.

28. FAST CARS The top speed of a Dodge Charger SRT8 is 33 mph less than the top speed of a Chevrolet Corvette Z06. At their top speeds, a Corvette can travel 6 miles in the same time that a Charger can travel 5 miles. Find the top speed of each car.

© Jim West/Alamy

29. WIND SPEED When a plane flies downwind, the wind pushes the plane so that its speed is the *sum* of the speed of the plane in still air and the speed of the wind. Traveling upwind, the wind pushes against the plane so that its speed is the *difference* of the speed of the plane in still air and the speed of the wind. Suppose a plane that travels 255 mph in still air can travel 300 miles downwind in the same time as it takes to travel 210 miles upwind. Complete the following table and find the speed of the wind, represented by x.

	Rate ·	Time =	Distance
Downwind	$255 + x$		300
Upwind	$255 - x$		210

30. BOATING A boat that travels 18 mph in still water can travel 22 miles downstream in the same time as it takes to travel 14 miles upstream. Find the speed of the current in the river. (See problem 29.)

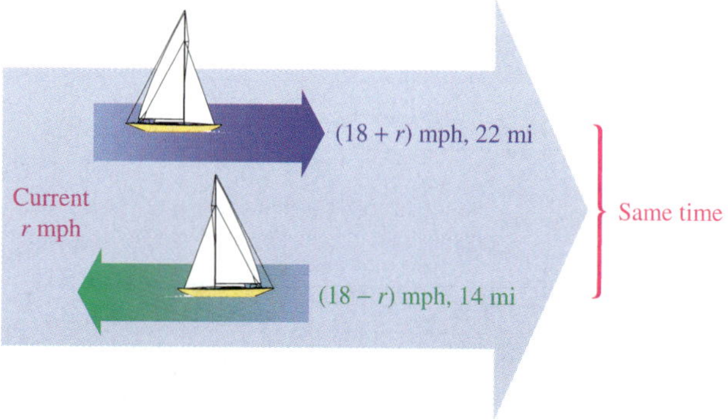

31. ROOFING HOUSES A homeowner estimates that it will take her 7 days to roof her house. A professional roofer estimates that he could roof the house in 4 days. How long will it take if the homeowner helps the roofer?

32. HOLIDAY DECORATING One crew can put up holiday decorations in the mall in 8 hours. A second crew can put up the decorations in 10 hours. How long will it take if both crews work together to decorate the mall?

33. ***from Campus to Careers***
Recreation Director

Suppose you are a recreation director at a summer camp. The water in the camp swimming pool was drained out for the winter and it is now time to refill the pool. One pipe can fill the empty pool in 12 hours and another can fill the empty pool in 18 hours. Suppose both pipes are opened at 8:00 A.M. and you have scheduled a swimming activity for 2:00 P.M. that day. Will the pool be filled by then?

© BananaStock/SuperStock

34. GROUNDSKEEPING It takes a groundskeeper 45 minutes to prepare a softball field for a game. It takes his assistant 55 minutes to prepare the same field. How long will it take if they work together to prepare the field?

35. FILLING A POOL One inlet pipe can fill an empty pool in 4 hours, and a drain can empty the pool in 8 hours. How long will it take the pipe to fill the pool if the drain is left open?

36. SEWAGE TREATMENT A sludge pool is filled by two inlet pipes. One pipe can fill the pool in 15 days, and the other can fill it in 21 days. However, if no sewage is added, continuous waste removal will empty the pool in 36 days. How long will it take the two inlet pipes to fill an empty sludge pool?

37. GRADING PAPERS On average, it takes a teacher 30 minutes to grade a set of quizzes. It takes her teacher's aide twice as long to do the same grading. How long will it take if they work together to grade a set of quizzes?

38. DOG KENNELS It takes the owner/operator of a dog kennel 6 hours to clean all of the cages. It takes his assistant 2 hours more than that to clean the same cages. How long will it take if they work together?

39. PRINTERS It takes a printer 6 hours to print the class schedules for all of the students enrolled in a community college. A faster printer can print the schedules in 4 hours. How long will it take the two printers working together to print $\frac{3}{4}$ of the class schedules?

40. OFFICE WORK In 5 hours, a secretary can address 100 envelopes. Another secretary can address 100 envelopes in 6 hours. How long would it take the secretaries, working together, to address 300 envelopes. (*Hint:* Think of addressing 300 envelopes as three 100-envelope jobs.)

41. COMPARING INVESTMENTS An amount of money invested for 1 year in tax-free bonds will earn \$300. In a certain credit union account, that same amount of money will only earn \$200 interest in a year, because the interest paid is 2% less than that paid by the bonds. Find the rate of interest paid by each investment.

42. COMPARING INVESTMENTS An amount of money invested for 1 year in a savings account will earn \$1,500. That same amount of money, invested in a mini-mall development will earn \$6,500 interest in a year, because the interest paid is 10% more than that paid by the savings account. Find the rate of interest paid by each investment.

43. COMPARING INVESTMENTS Two certificates of deposit (CDs) pay interest at rates that differ by 1%. Money invested for 1 year in the first CD earns \$175 interest. The same principal invested in the second CD earns \$200. Find the two rates of interest.

44. COMPARING INTEREST RATES Two bond funds pay interest at rates that differ by 2%. Money invested for 1 year in the first fund earns \$315 interest. The same amount invested in the second fund earns \$385. Find the lower rate of interest.

WRITING

45. In Example 4, one inlet pipe could fill an oil tank in 7 days, and another could fill the same tank in 9 days. We were asked to find how long it would take if both pipes were used. Explain why each of the following approaches is incorrect.

The time it would take to fill the tank

- is the *sum* of the lengths of time it takes each pipe to fill the tank: 7 days + 9 days = 16 days.
- is the *difference* in the lengths of time it takes each pipe to fill the tank: 9 days − 7 days = 2 days.
- is the *average* of the lengths of time it takes each pipe to fill the tank:

$$\frac{7 \text{ days} + 9 \text{ days}}{2} = \frac{16 \text{ days}}{2} = 8 \text{ days}.$$

46. Write a shared-work problem that can be modeled by the equation:

$$\frac{x}{3} + \frac{x}{4} = 1$$

REVIEW

47. Solve using substitution: $\begin{cases} x + y = 4 \\ y = 3x \end{cases}$

48. Solve using elimination (addition): $\begin{cases} 5x - 4y = 19 \\ 3x + 2y = 7 \end{cases}$

49. Use a check to determine whether $\frac{21}{5}$ is a solution of: $x + 20 = 4x - 1 + 2x$

50. Solve: $4x^2 + 8x = 0$

51. Evaluate $2x^2 + 5x - 3$ for $x = -3$.

52. Solve $T - R = ma$ for R.

CHALLENGE PROBLEMS

53. RIVER TOURS A river boat tour begins by going 60 miles upstream against a 5-mph current. There, the boat turns around and returns with the current. What still-water speed should the captain use to complete the tour in 5 hours?

54. TRAVEL TIME A company president flew 680 miles one way in the corporate jet, but returned in a smaller plane that could fly only half as fast. If the total travel time was 6 hours, find the speeds of the planes.

55. SALES A dealer bought some radios for a total of \$1,200. She gave away 6 radios as gifts, sold the rest for \$10 more than she paid for each radio, and broke even. How many radios did she buy?

56. FURNACE REPAIRS A repairman purchased several furnace-blower motors for a total cost of \$210. If his cost per motor had been \$5 less, he could have purchased one additional motor. How many motors did he buy at the regular rate?

SECTION 7.8
Proportions and Similar Triangles

Objectives

1. Write ratios and rates in simplest form.
2. Solve proportions.
3. Use proportions to solve problems.
4. Use proportions to solve problems involving similar triangles.

In this section, we will discuss a problem-solving tool called a *proportion.* A proportion is a type of rational equation that involves two *ratios* or two *rates.*

1 Write Ratios and Rates in Simplest Form.

Ratios enable us to compare numerical quantities. Here are some examples.

- To prepare fuel for a lawnmower, gasoline is mixed with oil in a 50-to-1 ratio.
- In the stock market, winning stocks might outnumber losers by a ratio of 7 to 4.
- Gold is combined with other metals in the ratio of 14 to 10 to make 14-karat jewelry.

Ratios A **ratio** is the quotient of two numbers or the quotient of two quantities that have the same units.

There are three ways to write a ratio: as a fraction, using the word *to,* or with a colon. For example, the comparison of the number of winning stocks to the number of losing stocks mentioned earlier can be written as

$$\frac{7}{4}, \qquad 7 \text{ to } 4, \qquad \text{or} \qquad 7:4$$

Each of these forms can be read as "the ratio of 7 to 4."

EXAMPLE 1 Translate each phrase into a ratio written in fractional form: **a.** The ratio of 5 to 9 **b.** 12 ounces to 2 pounds

Strategy To translate. we need to identify the number (or quantity) before the word *to* and the number (or quantity) after it.

Why The number before the word *to* is the numerator of the ratio and the number after it is the denominator.

Solution

a. The ratio of 5 to 9 is written $\frac{5}{9}$.

b. To write a ratio of two quantities with the same units, we must express 2 pounds in terms of ounces. Since 1 pound = 16 ounces, 2 pounds = 32 ounces. The ratio of 12 ounces to 32 ounces can be simplified so that no units appear in the final form.

Notation
A ratio that is the quotient of two quantities having the same units should be simplified so that no units appear in the final answer.

$$\frac{12 \text{ ounces}}{32 \text{ ounces}} = \frac{3 \cdot \overset{1}{\cancel{4}} \; \overset{1}{\cancel{\text{ounces}}}}{\underset{1}{\cancel{4}} \cdot 8 \; \underset{1}{\cancel{\text{ounces}}}} = \frac{3}{8}$$

Self Check 1 Translate each phrase into a ratio written in fractional form:
a. The ratio of 15 to 2 **b.** 12 hours to 2 days

Now Try **Problem 25**

A quotient that compares quantities with different units is called a **rate.** For example, if the 495-mile drive from New Orleans to Dallas takes 9 hours, the average rate of speed is the quotient of the miles driven and the length of time the trip takes.

$$\text{Average rate of speed} = \frac{495 \text{ miles}}{9 \text{ hours}} = \frac{\overset{1}{\cancel{9}} \cdot 55 \text{ miles}}{\underset{1}{\cancel{9}} \cdot 1 \text{ hours}} = \frac{55 \text{ miles}}{1 \text{ hour}}$$

Rates A **rate** is a quotient of two quantities that have different units.

2 Solve Proportions.

If two ratios or two rates are equal, we say that they are *in proportion.*

Proportion A **proportion** is a mathematical statement that two ratios or two rates are equal.

The Language of Algebra
The word *proportion* implies a comparative relationship in size. For a picture to appear realistic, the artist must draw the shapes in the proper *proportion.* Remember the Y2K scare? The massive computer failures predicted by some experts were blown way out of *proportion.*

Some examples of proportions are

$$\frac{1}{2} = \frac{3}{6} \qquad \frac{3 \text{ waiters}}{7 \text{ tables}} = \frac{9 \text{ waiters}}{21 \text{ tables}} \qquad \frac{a}{b} = \frac{c}{d}$$

- The proportion $\frac{1}{2} = \frac{3}{6}$ can be read as "1 is to 2 as 3 is to 6."
- The proportion $\frac{3 \text{ waiters}}{7 \text{ tables}} = \frac{9 \text{ waiters}}{21 \text{ tables}}$ can be read as "3 waiters is to 7 tables as 9 waiters is to 21 tables."
- The proportion $\frac{a}{b} = \frac{c}{d}$ can be read as "a is to b as c is to d."

Each of the four numbers in a proportion is called a **term.** The first and fourth terms are called the **extremes,** and the second and third terms are called the **means.**

$$\begin{matrix} \text{First term} \rightarrow a \\ \text{Second term} \rightarrow b \end{matrix} \; \frac{a}{b} = \frac{c}{d} \; \begin{matrix} c \leftarrow \text{Third term} \\ d \leftarrow \text{Fourth term} \end{matrix}$$

a and d are the extremes. b and c are the means.

For the proportion $\frac{a}{b} = \frac{c}{d}$, we can show that the product of the extremes, ad, is equal to the product of the means, bc, by multiplying both sides of the proportion by bd, and observing that $ad = bc$.

$$\frac{a}{b} = \frac{c}{d}$$

$$\overset{1}{\cancel{b}}d \cdot \frac{a}{\underset{1}{\cancel{b}}} = b\overset{1}{\cancel{d}} \cdot \frac{c}{\underset{1}{\cancel{d}}}$$ To clear the fractions, multiply both sides by the LCD, *bd*. Remove common factors of the numerator and denominator.

$$ad = bc$$ Simplify: $\frac{b}{b} = 1$ and $\frac{d}{d} = 1$.

Since $ad = bc$, the product of the extremes equals the product of the means.

The same products ad and bc can be found by multiplying diagonally in the proportion $\frac{a}{b} = \frac{c}{d}$. We call ad and bc **cross products.**

ad bc

$$\frac{a}{b} = \frac{c}{d}$$

The Fundamental Property of Proportions

In a proportion, the product of the extremes is equal to the product of the means.

If $\frac{a}{b} = \frac{c}{d}$, then $ad = bc$ and if $ad = bc$, then $\frac{a}{b} = \frac{c}{d}$.

EXAMPLE 2 Determine whether each equation is a proportion:

a. $\frac{3}{7} = \frac{9}{21}$ **b.** $\frac{8}{3} = \frac{13}{5}$

Strategy We will check to see whether the product of the extremes is equal to the product of the means.

Why If the product of the extremes equals the product of the means, the equation is a proportion. If the cross products are not equal, the equation is not a proportion.

Solution

a. The product of the extremes is $3 \cdot 21 = 63$. The product of the means is $7 \cdot 9 = 63$. Since the cross products are equal, $\frac{3}{7} = \frac{9}{21}$ is a proportion.

$3 \cdot 21 = 63$ $7 \cdot 9 = 63$

$$\frac{3}{7} = \frac{9}{21}$$ Each cross product is 63.

b. The product of the extremes is $8 \cdot 5 = 40$. The product of the means is $3 \cdot 13 = 39$. Since the cross products are not equal, the equation is not a proportion: $\frac{8}{3} \neq \frac{13}{5}$.

$8 \cdot 5 = 40$ $3 \cdot 13 = 39$

$$\frac{8}{3} = \frac{13}{5}$$ One cross product is 40 and the other is 39.

Caution

We cannot remove common factors "across" an = symbol.

$$\frac{\overset{1}{\cancel{3}}}{7} = \frac{9}{\underset{7}{\cancel{21}}}$$

When this is done, the original proportion, $\frac{3}{7} = \frac{9}{21}$, which we found to be true, is made false: $\frac{1}{7} = \frac{9}{7}$.

Self Check 2 Determine whether the equation $\frac{6}{13} = \frac{24}{53}$ is a proportion.

Now Try Problems 29

The fundamental property of proportions provides us with a way to solve proportions.

EXAMPLE 3 Solve: $\frac{3}{2} = \frac{9}{x}$

Strategy To solve for x, we will set the cross products equal.

Why This equation is a proportion, and in a proportion the product of the means equals the product of the extremes.

Solution

$$\frac{3}{2} = \frac{9}{x}$$ This is the given proportion.

$$3 \cdot x = 2 \cdot 9$$ Find each cross product and set them equal.

$$3x = 18$$ Do the multiplications.

$$\frac{3x}{3} = \frac{18}{3}$$ To isolate x, divide both sides by 3.

$$x = 6$$

Check: To check the result, we substitute 6 for x in $\frac{3}{2} = \frac{9}{x}$ and find the cross products.

$3 \cdot 6 = 18$ $\quad 2 \cdot 9 = 18$

$$\frac{3}{2} \stackrel{?}{=} \frac{9}{6}$$ Each cross product is 18.

Since the cross products are equal, the solution of $\frac{3}{2} = \frac{9}{x}$ is 6. The solution set is $\{6\}$.

Self Check 3 Solve: $\frac{15}{x} = \frac{25}{40}$

Now Try **Problem 35**

Caution

Remember that a cross product is the product of the means or extremes of a proportion. It would be incorrect to try to compute cross products to solve $\frac{12}{18} = \frac{4}{x} + \frac{1}{2}$ because there is more than one term on the right side of the equation. It is *not* a proportion.

EXAMPLE 4 Solve: $\frac{a}{2} = \frac{4}{a-2}$

Strategy To solve for a, we will set the cross products equal.

Why Since this equation is a proportion, the product of the means equals the product of the extremes.

Solution

$$\frac{a}{2} = \frac{4}{a-2}$$ This is the given proportion.

$$a(a-2) = 2 \cdot 4$$ Find each cross product and set them equal. Don't forget to write the parentheses.

$$a^2 - 2a = 8$$ On the left hand side, distribute the multiplication by a. This is a quadratic equation.

$$a^2 - 2a - 8 = 0$$ To get 0 on the right side of the equation, subtract 8 from both sides.

$$(a+2)(a-4) = 0$$ Factor $a^2 - 2a - 8$.

Success Tip

Since proportions are rational equations, they can also be solved by multiplying both sides by the LCD. Here, an alternate approach is to multiply both sides by $2(a-2)$:

$$2(a-2)\left(\frac{a}{2}\right) = 2(a-2)\left(\frac{4}{a-2}\right)$$

$$a + 2 = 0 \quad \text{or} \quad a - 4 = 0 \qquad \text{Set each factor equal to 0.}$$
$$a = -2 \qquad\qquad a = 4 \qquad \text{Solve each equation.}$$

The solutions are -2 and 4. Verify this using a check.

Self Check 4 Solve: $\frac{6}{c} = \frac{c-1}{5}$

Now Try Problem 47

3 Use Proportions to Solve Problems.

We can use proportions to solve many problems. If we are given a ratio (or rate) comparing two quantities, the words of the problem can be translated into a proportion, and we can solve it to find the unknown.

EXAMPLE 5 ***Grocery Shopping.*** If 6 apples cost \$1.38, how much will 16 apples cost?

Analyze the Problem We know the cost of 6 apples; we are to find the cost of 16 apples.

Form a Proportion Let $c =$ the cost of 16 apples. If we compare the number of apples to their cost, the two ratios must be equal.

6 apples is to \$1.38 as 16 apples is to \$c.

$$\frac{6}{1.38} = \frac{16}{c} \qquad \begin{array}{l}\text{Number of apples} \rightarrow 6,\ 16 \leftarrow \text{Number of apples}\\ \text{Cost} \rightarrow 1.38,\ c \leftarrow \text{Cost}\end{array}$$

> ***The Language of Algebra***
> Remember that the word *to* separates the numerator and denominator of a ratio. If the units are written outside the ratio, we can write 6 apples is *to* \$1.38 as
> $$\frac{6}{1.38}$$

Solve the Proportion We drop the units, find each cross product, set them equal, and then solve the resulting equation for c.

$$6 \cdot c = 1.38(16) \qquad \text{In a proportion, the product of the extremes equals the product of the means.}$$
$$6c = 22.08 \qquad \text{Multiply: } 1.38(16) = 22.08.$$
$$\frac{6c}{6} = \frac{22.08}{6} \qquad \text{To undo the multiplication by 6 and isolate } c\text{, divide both sides by 6.}$$
$$c = 3.68 \qquad \text{Recall that } c \text{ represents the cost of 16 apples.}$$

State the Conclusion Sixteen apples will cost \$3.68.

Check the Result We can use estimation to check the result. 16 apples are about 3 times as many as 6 apples, which cost \$1.38. If we multiply \$1.38 by 3, we get an estimate of the cost of 16 apples: $\$1.38 \cdot 3 = \$4.14$. The result, \$3.68, seems reasonable.

Self Check 5 If 9 tickets to a concert cost \$112.50, how much will 15 tickets cost?

Now Try Problem 71

CAUTION When solving problems using proportions, we must make sure that the units of both numerators are the same and the units of both denominators are the same. In Example 5, it would be incorrect to write

Cost of 6 apples → $\frac{1.38}{6} = \frac{16}{c}$ ← 16 apples

6 apples → ← Cost of 16 apples

EXAMPLE 6

Miniatures. A **scale** is a ratio (or rate) that compares the size of a model, drawing, or map with the size of an actual object. The scale indicates that 1 inch on the model carousel is equivalent to 160 inches on the actual carousel. How wide should the model be if the actual carousel is 35 feet wide?

Analyze the Problem We are asked to determine the width of the miniature carousel if a ratio of 1 inch to 160 inches is used. We would like the width of the model to be given in inches, not feet, so we will express the 35-foot width of the actual carousel as $35 \cdot 12 = 420$ inches.

Form a Proportion Let w = the width of the model. The ratios of the dimensions of the model to the corresponding dimensions of the actual carousel are equal.

1 inch is to 160 inches as w inches is to 420 inches.

Model → $\frac{1}{160} = \frac{w}{420}$ ← Model

Actual size → ← Actual size

Solve the Proportion We drop the units, find each cross product, set them equal, and then solve the resulting equation for w.

$420 = 160w$ In a proportion, the product of the extremes is equal to the product of the means.

$\frac{420}{\mathbf{160}} = \frac{160w}{\mathbf{160}}$ To undo the multiplication by 160 and isolate w, divide both sides by 160.

$2.625 = w$ Recall that w represents the width of the model.

State the Conclusion The width of the miniature carousel should be 2.625 in., or $2\frac{5}{8}$ in.

Check the Result A width of $2\frac{5}{8}$ in. is approximately 3 in. When we write the ratio of the model's approximate width to the width of the actual carousel, we get $\frac{3}{420} = \frac{1}{140}$, which is about $\frac{1}{160}$. The answer seems reasonable.

Now Try **Problem 85**

The Language of Algebra

Architects, interior decorators, landscapers, and automotive engineers are a few of the professionals who construct *scale* drawings or *scale* models of the projects they are designing.

When shopping, *unit prices* can be used to compare costs of different sizes of the same brand to determine the best buy. The **unit price** gives the cost per unit, such as cost per ounce, cost per pound, or cost per sheet. We can find the unit price of an item using a proportion.

EXAMPLE 7 ***Comparison Shopping.*** Which size of toothpaste is the better buy?

\$2.19

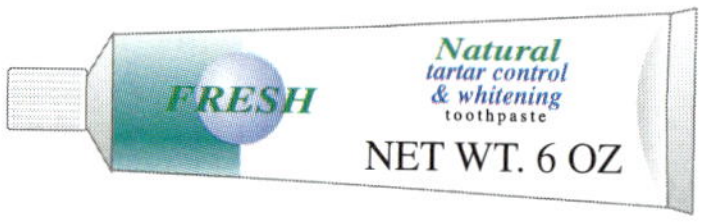

\$2.79

Solution To find the unit price for each tube, we let x = the price of 1 ounce of toothpaste. Then we set up and solve the following proportions.

For the 4-ounce tube:

Ounces → 4, Price → 2.19; ← Ounce 1, ← Price x

$$\frac{4}{2.19} = \frac{1}{x}$$
$$4x = 2.19$$
$$x = \frac{2.19}{4}$$
$$x \approx 0.55$$

The unit price is approximately \$0.55.

For the 6-ounce tube:

Ounces → 6, Price → 2.79; ← Ounce 1, ← Price x

$$\frac{6}{2.79} = \frac{1}{x}$$
$$6x = 2.79$$
$$x = \frac{2.79}{6}$$
$$x \approx 0.47$$

The unit price is approximately \$0.47.

The price of 1 ounce of toothpaste from the 4-ounce tube is about 55¢. The price for 1 ounce of toothpaste from the 6-ounce tube is about 47¢. Since the 6-ounce tube has the lower unit price, it is the better buy.

Self Check 7 Which is the better buy: 3 pounds of hamburger for \$6.89 or 5 pounds for \$12.49?

Now Try **Problem 89**

The Language of Algebra
A unit price indicates the cost of 1 unit of an item, such as 1 ounce of bottled water or 1 pound of hamburger. In advanced mathematics, we study unit circles—circles that have a radius of 1 unit.

4 Use Proportions to Solve Problems Involving Similar Triangles.

If two angles of one triangle have the same measures as two angles of a second triangle, the triangles have the same shape. Triangles with the same shape, but not necessarily the same size, are called **similar triangles.** In the following figure, $\Delta ABC \sim \Delta DEF$. (Read the symbol $\sim$ as "is similar to.")

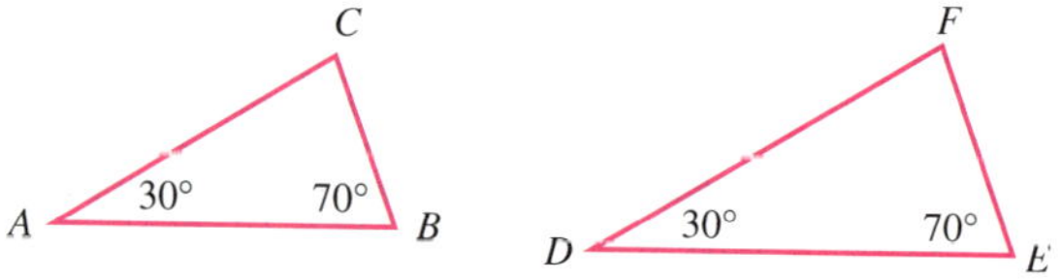

Property of Similar Triangles

If two triangles are **similar,** all pairs of corresponding sides are in proportion.

For the similar triangles previously shown, the following proportions are true.

$$\frac{AB}{DE} = \frac{BC}{EF}, \quad \frac{BC}{EF} = \frac{CA}{FD}, \quad \text{and} \quad \frac{CA}{FD} = \frac{AB}{DE}$$

Read *AB* as "the length of segment *AB*."

EXAMPLE 8 ***Finding the Height of a Tree.*** A tree casts a shadow 18 feet long at the same time as a woman 5 feet tall casts a shadow 1.5 feet long. Find the height of the tree.

Analyze the Problem The figure shows the similar triangles determined by the tree and its shadow and the woman and her shadow. Since the triangles are similar, the lengths of their corresponding sides are in proportion. We can use this fact to find the height of the tree.

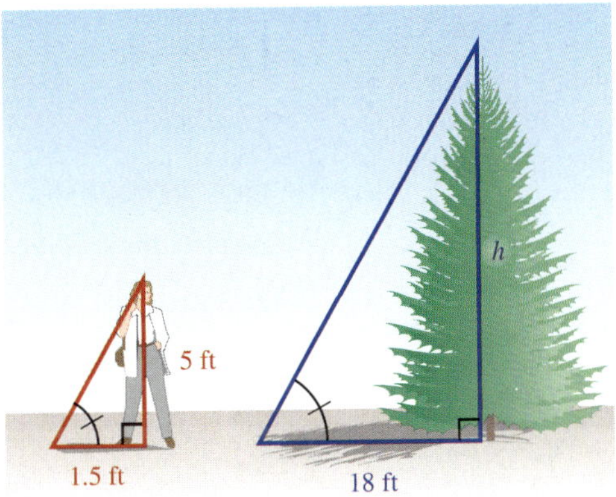

Each triangle has a right angle. Since the sun's rays strike the ground at the same angle, the angles highlighted with a tick mark have the same measure. Therefore, two angles of the smaller triangle have the same measures as two angles of the larger triangle; the triangles are similar.

Success Tip

Similar triangles do not have to be positioned the same. When they are placed differently, be careful to match their corresponding letters correctly. Here, $\Delta RST \sim \Delta MNO$.

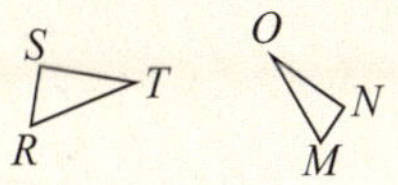

Form a Proportion If we let h = the height of the tree, we can find h by solving the following proportion.

$$\frac{h}{5} = \frac{18}{1.5} \qquad \frac{\text{Height of the tree}}{\text{Height of the woman}} = \frac{\text{Length of shadow of the tree}}{\text{Length of shadow of the woman}}$$

Solve the Proportion

$1.5h = 5(18)$ In a proportion, the product of the extremes equals the product of the means.

$1.5h = 90$ Multiply.

$\frac{1.5h}{1.5} = \frac{90}{1.5}$ To undo the multiplication by 1.5 and isolate h, divide both sides by 1.5.

$h = 60$ Do the decimal division, $1.5\overline{)90}$, to get 60.

State the Conclusion The tree is 60 feet tall.

Check the Result $\frac{18}{1.5} = 12$ and $\frac{60}{5} = 12$. Since the ratios are the same, the result checks.

Now Try **Problems 55 and 97**

ANSWERS TO SELF CHECKS **1. a.** $\frac{15}{2}$ **b.** $\frac{1}{4}$ **2.** No **3.** 24 **4.** $-5, 6$ **5.** \$187.50 **7.** 3 lb for \$6.89

STUDY SET 7.8

VOCABULARY

Fill in the blanks.

1. A ______ is the quotient of two numbers or the quotient of two quantities with the same units. A ______ is a quotient of two quantities that have different units.
2. A ___________ is a mathematical statement that two ratios or two rates are equal.
3. In $\frac{50}{3} = \frac{x}{9}$, the terms 50 and 9 are called the _________ and the terms 3 and x are called the ________ of the proportion.
4. The ______ products for the proportion $\frac{5}{2} = \frac{6}{x}$ are $5x$ and 12.
5. Examples of _____ prices are \$1.65 per gallon, 17¢ per day, and \$50 per foot.
6. Two triangles with the same shape, but not necessarily the same size, are called ________ triangles.

CONCEPTS

7. Fill in the blanks: In a proportion, the product of the extremes is ______ to the product of the means. In symbols,

$$\text{If } \frac{a}{b} = \frac{c}{d}, \text{ then } \_\_\_\_ = \_\_\_\_.$$

8. Is 45 a solution of $\frac{5}{3} = \frac{75}{x}$?
9. SNACKS In a sample of 25 bags of potato chips, 2 were found to be underweight. Complete the following proportion that could be used to find the number of underweight bags that would be expected in a shipment of 1,000 bags of potato chips.

$$\frac{\text{Number of bags} \rightarrow \_\_\_\_}{\text{Number underweight} \rightarrow \_\_\_\_} = \frac{\_\_\_\_ \leftarrow \text{Number of bags}}{\_\_\_\_ \leftarrow \text{Number underweight}}$$

10. MINIATURES A model of the Seattle Space Needle is to be made using a scale of 2 inches to 35 feet. Complete the following proportion to determine the height h of the model.

$$\frac{2}{35} = \frac{\_\_\_\_}{\_\_\_\_}$$

11. KLEENEX Complete the following proportion that can be used to find the unit price of facial tissue if a box of 85 tissues sells for \$2.19.

$$\frac{\text{Price} \rightarrow \_\_\_\_}{\text{Number of sheets} \rightarrow 85} = \frac{x \leftarrow \text{Price}}{\_\_\_\_ \leftarrow \text{Number of sheets}}$$

12. The two triangles shown in the following illustration are similar. Complete the proportion.

$$\frac{x}{\_\_\_\_} = \frac{\_\_\_\_}{10}$$

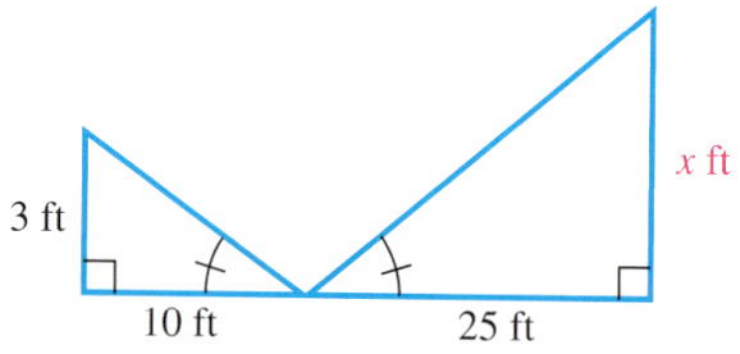

NOTATION

Complete the solution.

13. Solve for x: $\frac{12}{18} = \frac{x}{24}$

$$12 \cdot 24 = 18 \cdot \_\_\_\_$$
$$\_\_\_\_ = 18x$$
$$\frac{288}{\_\_\_\_} = \frac{18x}{\_\_\_\_}$$
$$16 = x$$

14. Write the ratio of 25 to 4 in two other forms.
15. Fill in the blanks: The proportion $\frac{20}{1.6} = \frac{100}{8}$ can be read: 20 is to 1.6 ____ 100 is ____ 8.
16. Fill in the blank: We read $\Delta XYZ \sim \Delta MNO$ as: triangle XYZ is ________ to triangle MNO.

GUIDED PRACTICE

***Translate each ratio into a fraction in simplest form.* See Example 1.**

17. 4 boxes to 15 boxes
18. 2 miles to 9 miles
19. 18 watts to 24 watts
20. 11 cans to 121 cans
21. 30 days to 24 days
22. 45 people to 30 people
23. 90 minutes to 3 hours
24. 20 inches to 2 feet
25. 8 quarts to 4 gallons
26. 6 feet to 12 yards
27. 6,000 feet to 1 mile (*Hint:* 1 mi = 5,280 ft)
28. 5 tons to 4,000 pounds (*Hint:* 1 ton = 2,000 lb)

***Determine whether each equation is a true proportion.* See Example 2.**

29. $\frac{7}{3} = \frac{14}{6}$
30. $\frac{7}{16} = \frac{3}{7}$
31. $\frac{5}{8} = \frac{12}{19.4}$
32. $\frac{9}{32} = \frac{4.5}{16}$

Solve each proportion. See Example 3.

33. $\frac{2}{3} = \frac{x}{6}$

34. $\frac{3}{6} = \frac{x}{8}$

35. $\frac{63}{g} = \frac{9}{2}$

36. $\frac{27}{x} = \frac{9}{4}$

37. $\frac{x+1}{5} = \frac{3}{15}$

38. $\frac{x-1}{7} = \frac{2}{21}$

39. $\frac{5-x}{17} = \frac{13}{34}$

40. $\frac{4-x}{13} = \frac{11}{26}$

41. $\frac{15}{7b+5} = \frac{5}{2b+1}$

42. $\frac{8}{3n+6} = \frac{16}{3n-3}$

43. $\frac{8x}{3} = \frac{11x+9}{4}$

44. $\frac{3x}{16} = \frac{x+2}{5}$

Solve each proportion. See Example 4.

45. $\frac{2}{3x} = \frac{x}{6}$

46. $\frac{y}{4} = \frac{4}{y}$

47. $\frac{b-5}{3} = \frac{2}{b}$

48. $\frac{2}{q} = \frac{q-3}{2}$

49. $\frac{a-4}{a} = \frac{15}{a+4}$

50. $\frac{s}{s-5} = \frac{s+5}{24}$

51. $\frac{t+3}{t+5} = \frac{-1}{2t}$

52. $\frac{5h}{14h+3} = \frac{1}{h}$

Each pair of triangles is similar. Find the missing side length. See Example 8.

53.

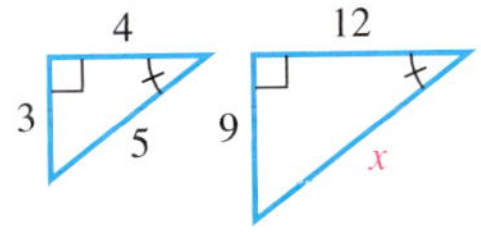

54.

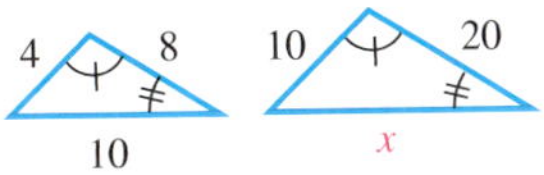

55.

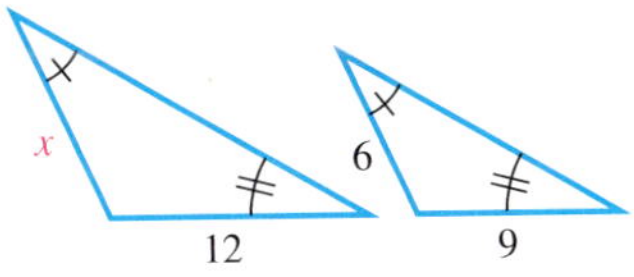

56.

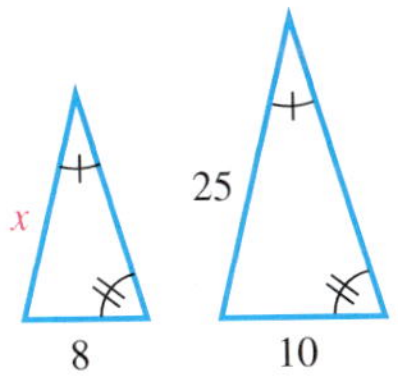

TRY IT YOURSELF

Solve each proportion.

57. $\frac{x-1}{x+1} = \frac{2}{3x}$

58. $\frac{2}{x+6} = \frac{-2x}{5}$

59. $\frac{x+1}{4} = \frac{3x}{8}$

60. $\frac{x-1}{9} = \frac{2x}{3}$

61. $\frac{y-4}{y+1} = \frac{y+3}{y+6}$

62. $\frac{r-6}{r-8} = \frac{r+1}{r-4}$

63. $\frac{c}{10} = \frac{10}{c}$

64. $\frac{-6}{r} = \frac{r}{-6}$

65. $\frac{m}{3} = \frac{4}{m+1}$

66. $\frac{n}{2} = \frac{5}{n+3}$

67. $\frac{3}{3b+4} = \frac{2}{5b-6}$

68. $\frac{2}{4d-1} = \frac{3}{2d+1}$

APPLICATIONS

69. GEAR RATIOS Write each ratio in two ways: as a fraction in simplest form and using a colon.

a. The number of teeth of the larger gear to the number of teeth of the smaller gear

b. The number of teeth of the smaller gear to the number of teeth of the larger gear

70. FACULTY–STUDENT RATIOS At a college, there are 300 faculty members and 2,850 students. Find the rate of faculty to students. (This is often referred to as the faculty-to-student ratio, even though the units are different.)

71. SHOPPING FOR CLOTHES If shirts are on sale at two for $25, how much do five shirts cost?

72. COMPUTING A PAYCHECK Billie earns $412 for a 40-hour week. If she missed 10 hours of work last week, how much did she get paid?

73. COOKING A recipe for spaghetti sauce requires four 16-ounce bottles of ketchup to make 2 gallons of sauce. How many bottles of ketchup are needed to make 10 gallons of sauce?

74. MIXING PERFUME A perfume is to be mixed in the ratio of 3 drops of pure essence to 7 drops of alcohol. How many drops of pure essence should be mixed with 56 drops of alcohol?

75. CPR A first aid handbook states that when performing cardiopulmonary resuscitation on an adult, the ratio of chest compressions to breaths should be 5:2. If 210 compressions were administered to an adult patient, how many breaths should have been given?

76. COOKING A recipe for wild rice soup follows. Find the amounts of chicken broth, rice, and flour needed to make 15 servings.

Wild Rice Soup

A sumptuous side dish with a nutty flavor

3 cups chicken broth	1 cup light cream
$\frac{2}{3}$ cup uncooked rice	2 tablespoons flour
$\frac{1}{4}$ cup sliced onions	$\frac{1}{8}$ teaspoon pepper
$\frac{1}{2}$ cup shredded carrots	Serves: 6

77. NUTRITION The table shows the nutritional facts about a 10-oz chocolate milkshake sold by a fast-food restaurant. Use the information to complete the table for the 16-oz shake. Round to the nearest unit when an answer is not exact.

	Calories	Fat (gm)	Protein (gm)
10-oz chocolate milkshake	355	8	9
16-oz chocolate milkshake			

78. ENGINEERING A portion of a bridge is shown. Use the fact that $\frac{AB}{BC}$ is in proportion to $\frac{FE}{ED}$ to find FE.

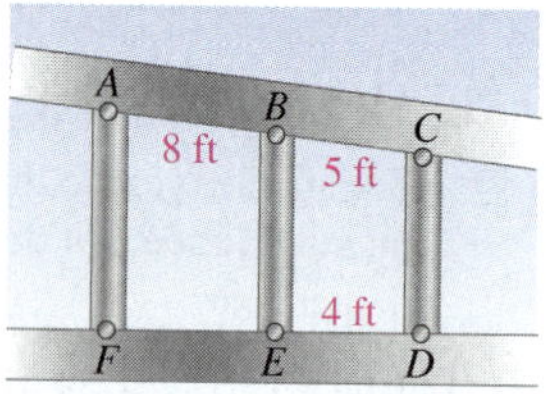

79. QUALITY CONTROL Out of a sample of 500 men's shirts, 17 were rejected because of crooked collars. How many crooked collars would you expect to find in a run of 15,000 shirts?

80. PHOTO ENLARGEMENTS The 3-by-5 photo is to be blown up to the larger size. Find x.

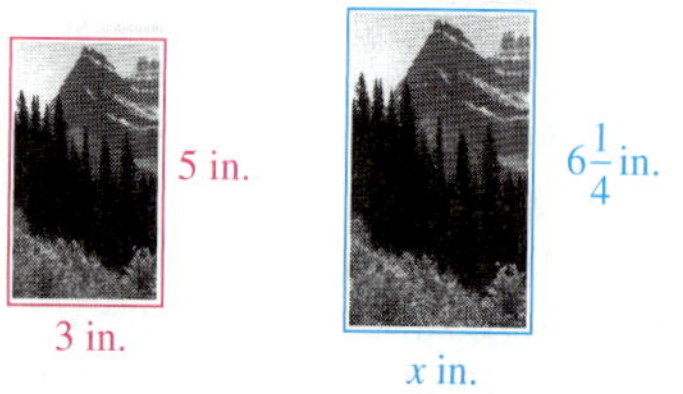

81. MIXING FUEL The instructions on a can of oil intended to be added to lawnmower gasoline are shown below. Are these instructions correct? (*Hint:* There are 128 ounces in 1 gallon.)

Recommended	Gasoline	Oil
50 to 1	6 gal	16 oz

82. DRIVER'S LICENSES Of the 50 states, Alabama has one of the highest ratios of licensed drivers to residents. If the ratio is 399:500 and the population of Alabama is about 4,500,000, how many residents of that state have a driver's license?

83. CAPTURE–RELEASE METHOD To estimate the ground squirrel population on his acreage, a farmer trapped, tagged, and then released a dozen squirrels. Two weeks later, the farmer trapped 35 squirrels and noted that 3 were tagged. Use this information to estimate the number of ground squirrels on his acreage.

84. CONCRETE A 2:3 concrete mix means that for every two parts of sand, three parts of gravel are used. How much sand should be used in a mix composed of 25 cubic feet of gravel?

85. MODEL RAILROADS An HO scale model railroad engine is 6 inches long. If the HO scale is 1 to 87, how long is a real engine, in inches? In feet?

86. MODEL RAILROADS An N scale model railroad caboose is 4.5 inches long. If the N scale is 1 to 160, how long is a real caboose, in inches? In feet?

87. BLUEPRINTS The scale for the drawing shown means that a $\frac{1}{4}$-inch length $\left(\frac{1}{4}''\right)$ on the drawing corresponds to an actual size of 1 foot (1′-0″). Suppose the length of the kitchen is $2\frac{1}{2}$ inches on the drawing. How long is the actual kitchen?

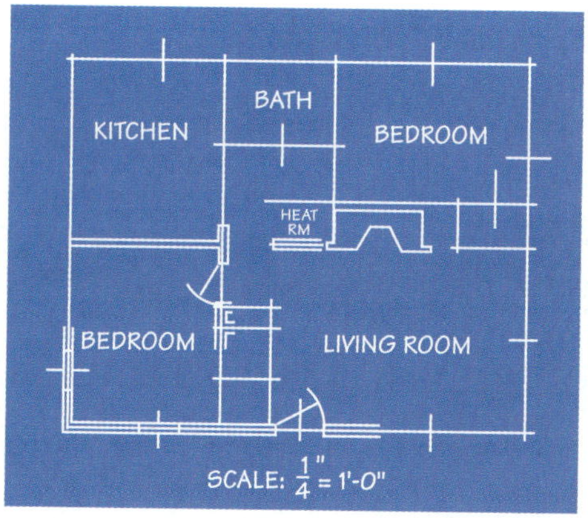

88. THE TITANIC A 1:144 scale model of the *Titanic* is to be built. If the ship was 882 feet long, find the length of the model.

***For each of the following purchases, determine the better buy.* See Example 7.**

89. Trumpet lessons: 45 minutes for \$25 or 60 minutes for \$35

90. Memory for a computer: 128 megabytes for \$26 or 512 megabytes for \$110

91. Business cards: 100 for \$9.99 or 150 for \$12.99

92. Dog food: 20 pounds for \$7.49 or 44 pounds for \$14.99

93. Soft drinks: 6-pack for \$1.50 or a case (24 cans) for \$6.25

94. Donuts: A dozen for \$6.24 or a baker's dozen (13) for \$6.65

95.

96. 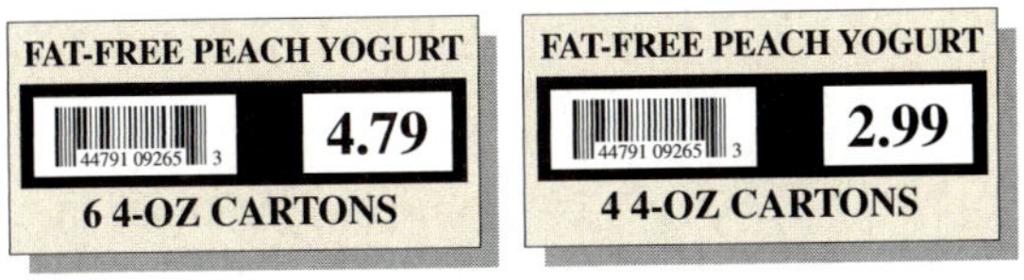

97. HEIGHT OF A TREE A tree casts a shadow of 26 feet at the same time as a 6-foot man casts a shadow of 4 feet. Find the height of the tree.

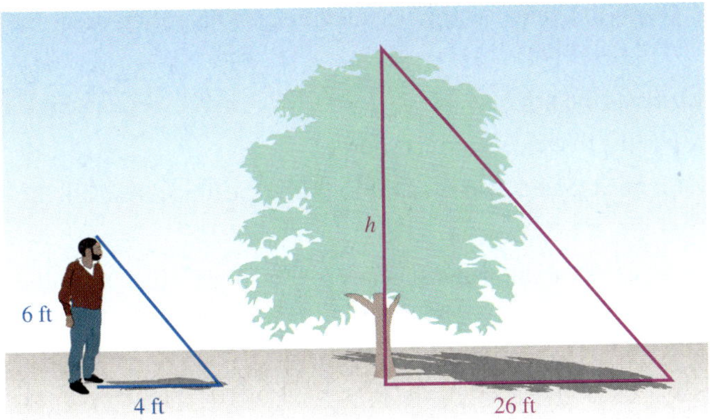

98. HEIGHT OF A BUILDING A man places a mirror on the ground and sees the reflection of the top of a building, as shown. The two triangles in the illustration are similar. Find the height, h, of the building.

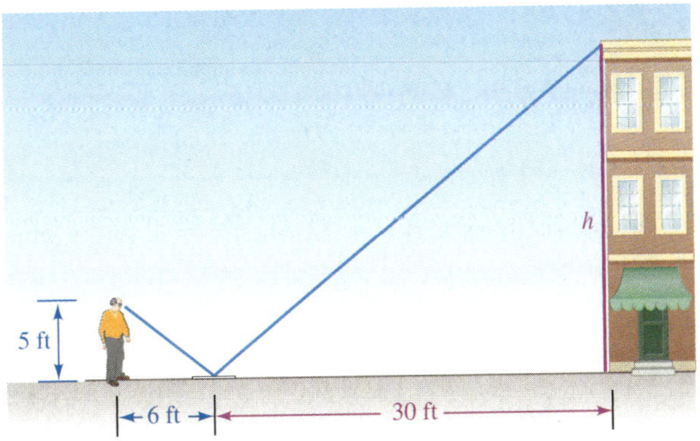

99. SURVEYING To find the width of a river, a surveyor laid out the following similar triangles. Find w.

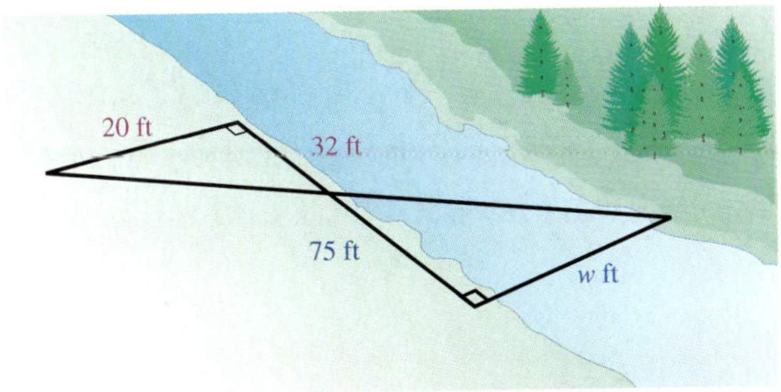

100. FLIGHT PATHS An airplane ascends 100 feet as it flies a horizontal distance of 1,000 feet. How much altitude will it gain as it flies a horizontal distance of 1 mile? (*Hint:* 5,280 feet=1 mile.)

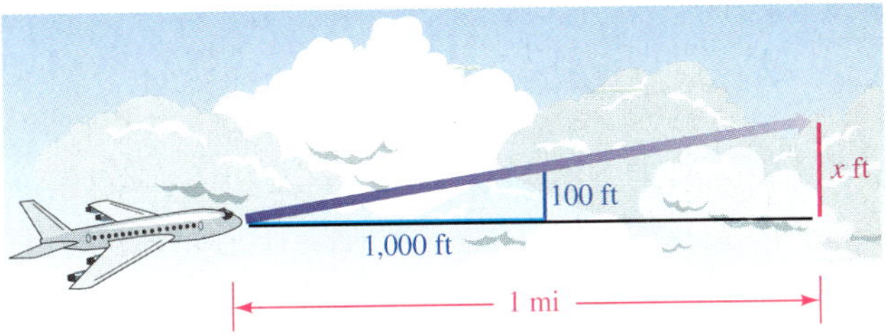

WRITING

101. Explain how to solve the equation $\frac{7}{6} = \frac{2}{x}$ and how to simplify the expression $\frac{7}{6} \cdot \frac{2}{x}$.

102. Explain why the concept of cross products cannot immediately be used to solve the equation:

$$\frac{x}{3} - \frac{3x}{4} = \frac{1}{12}$$

103. What are similar triangles?

104. What is a unit price? Give an example.

REVIEW

105. Change $\frac{9}{10}$ to a percent.

106. Change $33\frac{1}{3}\%$ to a fraction.

107. Find 30% of 1,600.

108. SHOPPING Maria bought a dress for 25% off the original price of $98. How much did the dress cost?

CHALLENGE PROBLEMS

109. Suppose $\frac{a}{b} = \frac{c}{d}$. Write three other proportions using a, b, c, and d.

110. Verify that $\frac{3}{5} = \frac{12}{20} = \frac{3 + 12}{5 + 20}$. Is the following rule always true? Explain.

$$\frac{a}{b} = \frac{c}{d} = \frac{a + c}{b + d}$$

SECTION 7.9
Variation

Objectives

1. Solve direct variation problems.
2. Solve inverse variation problems.

If the value of one quantity depends on the value of another quantity, we can often describe that relationship using the language of variation:

- The sales tax on an item *varies* as the price.
- The intensity of light *varies* as the distance from its source.
- The pressure exerted by water on an object *varies* as the depth of the object beneath the surface.

In this section, we will discuss two types of variation and see how to model them algebraically.

1 Solve Direct Variation Problems.

One type of variation, called **direct variation,** is represented by an equation of the form $y = kx$, where k is a nonzero constant. Two variables are said to *vary directly* if one is a constant multiple of the other.

Direct Variation

The words *y varies directly as x* or *y is directly proportional to x* mean that

$$y = kx$$

for some nonzero constant k, called the **constant of variation.**

EXAMPLE 1 Suppose y varies directly as x. If $y = 12$ when $x = 4$, find y when $x = 6$.

Strategy We will use the equation $y = kx$ to solve this problem.

Why The words *varies directly* indicate that we should use the direct variation equation $y = kx$.

Solution We can use the given pair of values of x and y to determine the constant of variation k.

$y = kx$ This is the equation that models direct variation.

$12 = k(4)$ Substitute 4 for x and 12 for y.

$3 = k$ To isolate k on the right side, divide both sides by 4. This is the constant of variation.

Since $y = kx$ and $k = 3$, we have $y = 3x$.

We can use $y = 3x$ to find other pairs of values of x and y. When $x = 6$, we see that

$y = 3(6) = 18$ Substitute 6 for x.

Thus, when $x = 6$, the value of y is 18.

Success Tip

If we divide both sides of $y = kx$ by x, we obtain $\frac{y}{x} = k$. Thus, for the direct variation model, k is simply the quotient of one pair of values of x and y.

Self Check 1 Suppose y varies directly as x. If $y = 24$ when $x = 3$, find y when $x = 5$.

Now Try **Problem 17**

The Language of Algebra
The phrase *"is directly proportional to"* is also used to indicate direct variation.

Scientists have found that the distance a spring will stretch is directly proportional to the force applied to it. The more force that is applied to the spring, the more it will stretch.

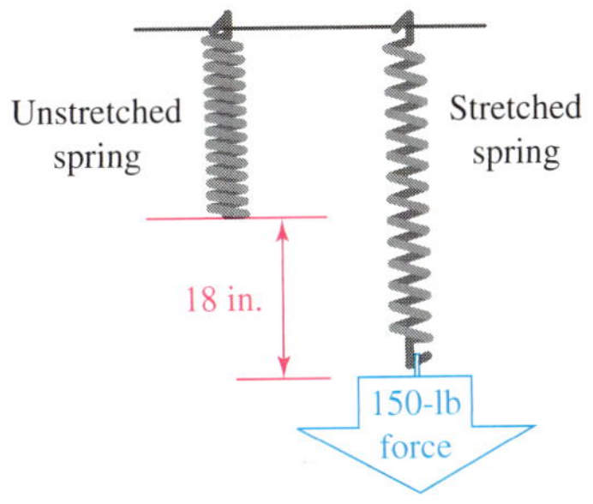

We could use the equation $y = kx$ to model this direct variation. However, in application problems, the variables x and y are often replaced with letters that better describe the quantities involved. If we let d represent the distance stretched and f represent the force applied, this relationship can be represented by the equation

$d = kf$ where k is the constant of variation

Suppose that a 150-pound weight stretches a spring 18 inches. We can find the constant of variation for the spring by substituting 150 for f and 18 for d in the equation $d = kf$ and solving for k:

Success Tip
The value that we found for k is for this specific example. Another spring made out of a different type of steel will more than likely have a different value of k.

$d = kf$	This equation models direct variation.
$18 = k(150)$	Substitute 18 for d and 150 for f.
$\frac{18}{150} = k$	To isolate k on the right side, divide both sides by 150.
$\frac{3}{25} = k$	Simplify the fraction: $\frac{18}{150} = \frac{\overset{1}{\cancel{6}} \cdot 3}{\underset{1}{\cancel{6}} \cdot 25} = \frac{3}{25}$. This is the constant of variation.

Thus, the equation that describes the relationship between the distance the spring will stretch and the amount of force applied to it is $d = \frac{3}{25}f$.

Once the value of k is known, other pairs of values can be found. To find the distance that the same spring will stretch when a lighter, 50-pound weight is used, we proceed as follows:

$d = \frac{3}{25}f$	This equation models direct variation for the specific spring used in this example.
$d = \frac{3}{25}(50)$	Substitute 50 for f.
$d = 6$	Multiply: $\frac{3}{25}\left(\frac{50}{1}\right) = \frac{3 \cdot 2 \cdot \overset{1}{\cancel{25}}}{\underset{1}{\cancel{25}} \cdot 1} = 6$.

The spring will stretch 6 inches when a 50-pound weight is used.

Success Tip
For any direct variation equation of the form $y = kx$, where $k > 0$:

as x increases, y increases

The table on the next page shows some other possible values for f and d as determined by the equation $d = \frac{3}{25}f$. When these ordered pairs are graphed and a straight line is drawn through them, it is apparent that as the force f applied to a spring increases, the distance d that it stretches increases in a predictable way. Furthermore, the slope of the graph is $\frac{3}{25}$, which is the constant of variation.

$d = \frac{3}{25}f$

f	d
0	0
25	3
50	6
75	9
100	12

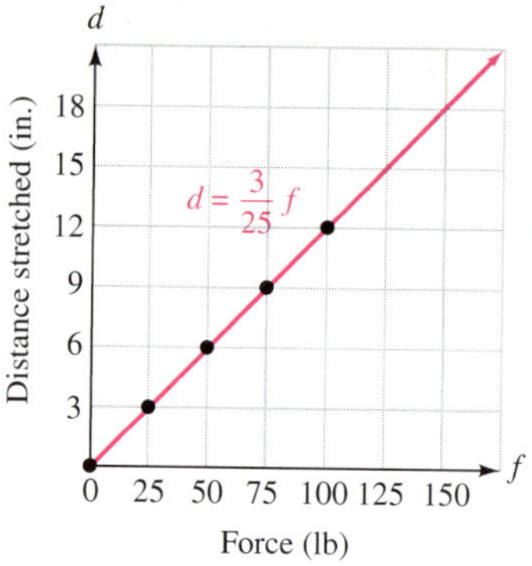

This straight-line graph shows that d varies directly as f.

We can use the following steps to solve variation problems.

Strategy for Solving Variation Problems

1. Translate the verbal model into an equation.
2. Substitute the values of a pair of variables into the equation from step 1 to determine the value of k.
3. Substitute the value of k into the equation from step 1.
4. Substitute the remaining given value into the equation from step 3 and solve for the unknown variable.

© Ablestock/Alamy

EXAMPLE 2 ***Geology.*** The weight of an object on Earth varies directly as its weight on the moon. If a rock weighed 5 pounds on the moon and 30 pounds on Earth, what would be the weight on Earth of a larger rock weighing 26 pounds on the moon?

Strategy We will follow the strategy for solving a direct variation problem.

Why In the words of the problem, the phrase *varies directly* indicates that a direct variation model should be used.

Solution

Step 1: We let e represent the weight (in pounds) of an object on Earth and m the weight (in pounds) of the object on the moon. Translating the words *weight on Earth varies directly as its weight on the moon,* we get the equation

$e = km$ This equation models direct variation.

Step 2: To find the constant of variation, k, we substitute 30 for e and 5 for m.

$e = km$

$30 = k(5)$ Substitute 30 for e and 5 for m.

$6 = k$ To isolate k, divide both sides by 5. This is the constant of variation.

Step 3: We now substitute the value of k into the equation from step 1. The equation describing the relationship between the weight of an object on Earth and on the moon is

$e = 6m$

Step 4: We can find the weight of the larger rock on Earth by substituting 26 for m in the equation from step 3.

$$e = 6m$$

$$e = 6(26) \quad \text{Substitute 26 for } m.$$

$$e = 156$$

The rock would weigh 156 pounds on Earth.

Self Check 2 The cost of a bus ticket varies directly with the number of miles traveled. If a ticket for a 180-mile trip cost \$45, what would a ticket for a 1,500-mile trip cost?

Now Try Problem 35

2 Solve Inverse Variation Problems.

Another type of variation, called **inverse variation,** is represented by an equation of the form $y = \frac{k}{x}$, where k is a constant. Two variables are said to *vary inversely* if one is a constant multiple of the reciprocal of the other.

Inverse Variation

The words *y varies inversely as x* or *y is inversely proportional to x* mean that

$$y = \frac{k}{x}$$

for some nonzero constant k, called the **constant of variation.**

EXAMPLE 3 Suppose y varies inversely as x. If $y = 5$ when $x = 20$, find y when $x = 50$.

Strategy We will use the equation $y = \frac{k}{x}$ to solve this problem.

Why The words *varies inversely* indicate that we should use the inverse variation equation $y = \frac{k}{x}$.

Solution We can use the given pair of values of x and y to determine the constant of variation, k, in $y = \frac{k}{x}$.

$$y = \frac{k}{x} \quad \text{This is the equation that models inverse variation.}$$

$$5 = \frac{k}{20} \quad \text{Substitute 20 for x and 5 for y.}$$

$$20 \cdot 5 = k \quad \text{To isolate } k \text{ on the right side, multiply both sides by 20.}$$

$$100 = k \quad \text{This is the constant of variation.}$$

Success Tip

If we multiply both sides of $y = \frac{k}{x}$ by x, we get $xy = k$. Thus, for the inverse variation model, k is simply the product of one pair of values of x and y. (Assume $x \neq 0$.)

Since $y = \frac{k}{x}$ and $k = 100$, we have

$$y = \frac{100}{x}$$

We can use $y = \frac{100}{x}$ to find other pairs of values of x and y. When $x = 50$, we see that

$$y = \frac{100}{50} = 2 \quad \text{Substitute 50 for } x.$$

Thus, when $x = 50$, the value of y is 2.

Self Check 3 Suppose y varies inversely as x. If $y = 25$ when $x = 3$, find y when $x = 15$.

Now Try **Problem 25**

The Language of Algebra
The phrase *"is inversely proportional to"* is also used to indicate inverse variation.

Suppose that the time (in hours) that it takes to paint a house is inversely proportional to the size of the painting crew. As the number of painters increases, the time that it takes to paint the house decreases. If n represents the number of painters and t represents the time it takes to paint the house, this relationship can be expressed by the equation

$$t = \frac{k}{n} \quad \text{where } k \text{ is the constant of variation}$$

If we know that a crew of 8 can paint the house in 12 hours, we can find the constant of variation by substituting 8 for n and 12 for t in the equation $t = \frac{k}{n}$ and solving for k:

$$t = \frac{k}{n} \quad \text{This equation models inverse variation.}$$

$$12 = \frac{k}{8} \quad \text{Substitute 8 for } n \text{ and 12 for } t.$$

$$8 \cdot 12 = k \quad \text{To isolate } k \text{ on the right side, multiply both sides by 8.}$$

$$96 = k \quad \text{This is the constant of variation.}$$

The equation describing the relationship between the size of the painting crew and the time it takes to paint the house is $t = \frac{96}{n}$. We can use this equation to find the time it will take a crew of any size to paint the house. For example, to find the time it would take a four-person crew, we substitute 4 for n in the equation $t = \frac{96}{n}$.

$$t = \frac{96}{n} \quad \text{This equation models inverse variation for the specific house used in this example.}$$

$$t = \frac{96}{4} \quad \text{Substitute 4 for } n.$$

$$t = 24$$

It would take a four-person crew 24 hours to paint the house.

The following table shows some possible values for n and t as determined by the equation $t = \frac{96}{n}$. When these ordered pairs are graphed and a smooth curve is drawn through them, it is clear that as the number of painters n increases, the time t decreases, in a predictable way.

Success Tip

For any Inverse variation equation of the form $y = \frac{k}{x}$, where $k > 0$:

as x increases, y decreases

$$t = \frac{96}{n}$$

n	t
2	48
3	32
4	24
6	16
8	12
12	8
16	6
24	4

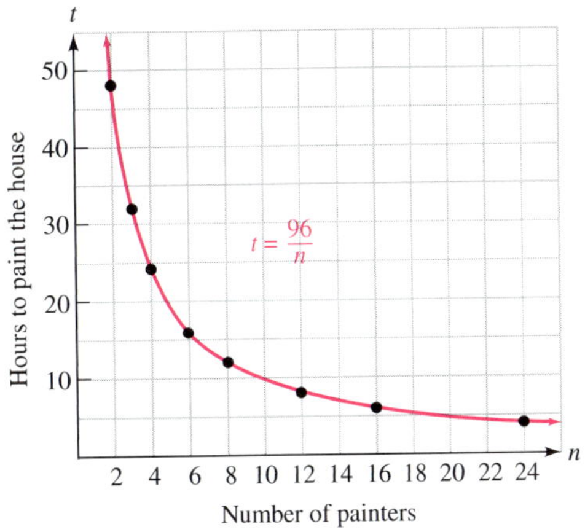

This curved graph shows that t varies inversely as n.

EXAMPLE 4 ***Chemistry.*** The volume occupied by a fixed weight of gas (held at a constant temperature) varies inversely as the pressure placed on it. If hydrogen gas occupies a volume of 22.5 cubic inches when placed under 3 pounds per square inch (psi) of pressure, find the volume occupied by the hydrogen gas when the pressure is 7.5 psi.

Strategy We will follow the strategy for solving an inverse variation problem.

Why In the words of the problem, the phrase *varies inversely* indicates that an inverse variation model should be used.

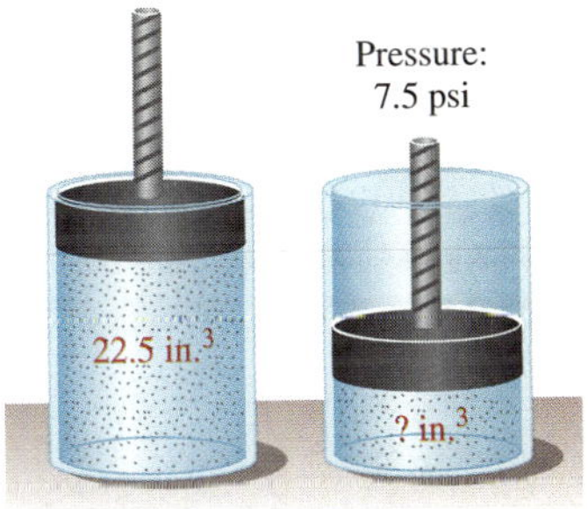

The volume occupied by gas decreases as pressure increases.

Solution

Step 1: We let V represent the volume occupied by the gas and p represent the pressure. Translating the words *volume occupied by a gas varies inversely as the pressure,* wc gct thc cquation

$$V = \frac{k}{p} \quad \text{This equation models inverse variation.}$$

Step 2: To find the constant of variation, k, we substitute 22.5 for V and 3 for p.

$$V = \frac{k}{p}$$

$$22.5 = \frac{k}{3} \quad \text{Substitute 3 for } p \text{ and 22.5 for } V.$$

$$67.5 = k \quad \text{To isolate } k \text{, multiply both sides by 3. This is the constant of variation.}$$

Step 3: The equation describing the relationship between the volume occupied by the gas and the pressure placed on it is

$$V = \frac{67.5}{p}$$

Step 4: We can find the volume occupied by the gas when a pressure of 7.5 psi is placed on it by substituting 7.5 for p in the equation and evaluating the right side.

$$V = \frac{67.5}{p}$$

$$V = \frac{67.5}{7.5} \quad \text{Substitute 7.5 for } p.$$

$$V = 9$$

The hydrogen gas will occupy a volume of 9 cubic inches when the pressure placed on it is 7.5 psi.

Self Check 4 Find the volume occupied by the hydrogen gas in Example 4 when the pressure placed on it is 5 psi.

Now Try **Problem 37**

ANSWERS TO SELF CHECKS **1.** 40 **2.** \$375 **3.** 5 **4.** 13.5 in.^{3}

STUDY SET 7.9

VOCABULARY

Fill in the blanks. Assume that k is a constant.

1. The equation $y = kx$ defines ______ variation and the equation $y = \frac{k}{x}$ defines ______ variation.

2. In $y = kx$ and in $y = \frac{k}{x}$, we call k the ______ of variation.

CONCEPTS

Determine whether each graph represents direct variation or inverse variation.

3.

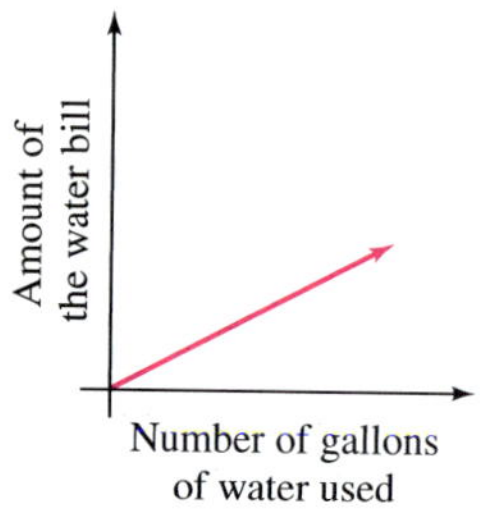

4.

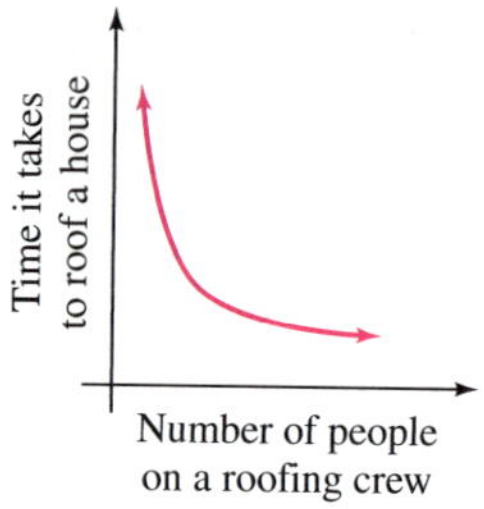

5.

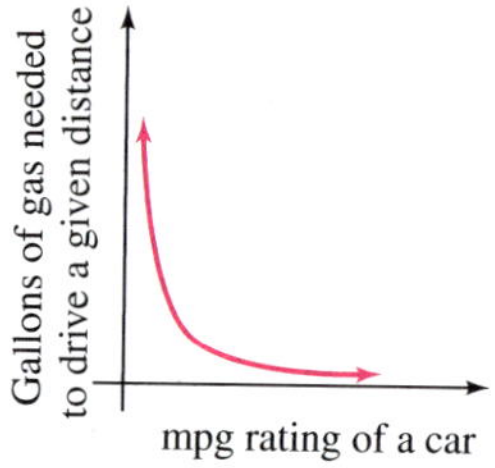

6.

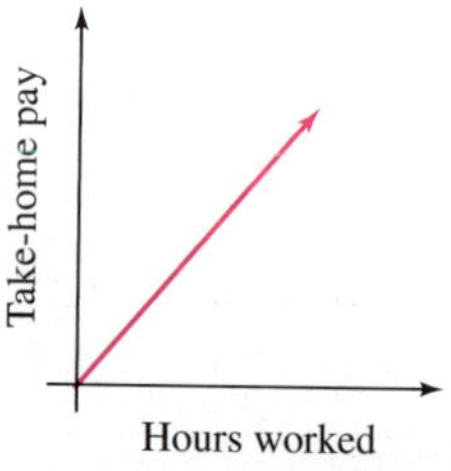

Complete each graph by sketching either a direct variation or an inverse variation.

7.

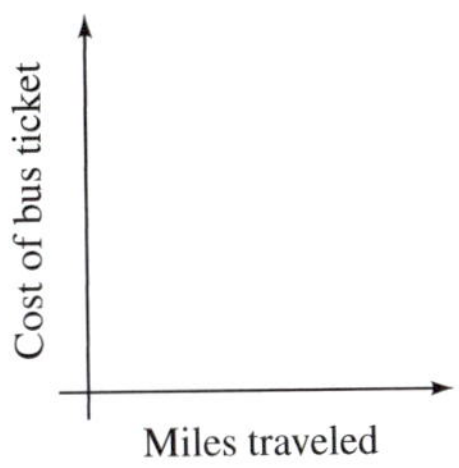

8.

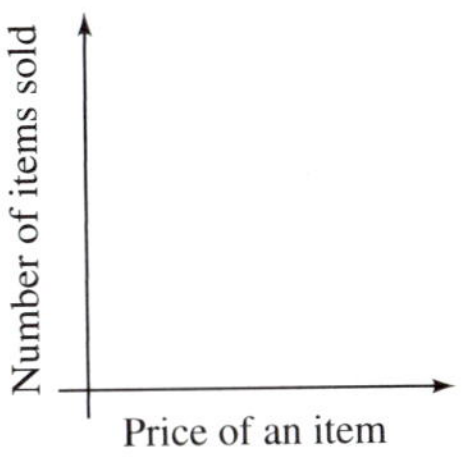

9.

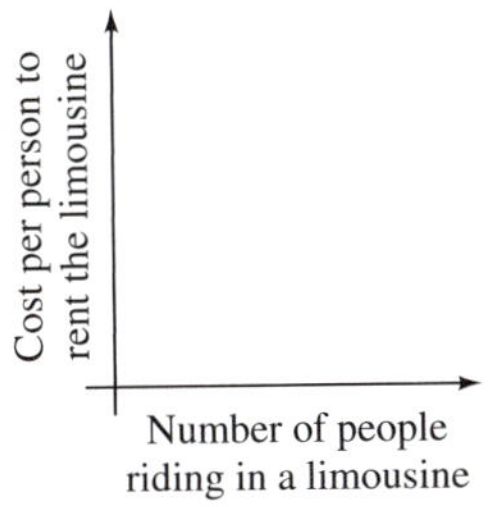

10.

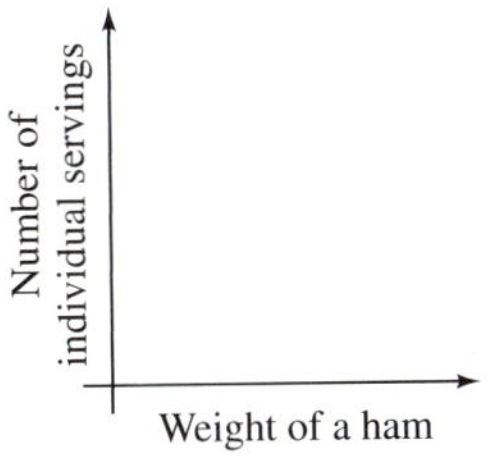

Write an equation to describe each variation. Use k for the constant of variation.

11. SWIMMING When you swim underwater, the pressure p in your ears varies directly as the depth d at which you swim.

12. FARMING The number of bushels b of corn that a farmer harvests varies directly as a, the number of acres he plants.

13. GEOLOGY The amount of dust d in a desert region varies inversely as the amount of rainfall r.

14. WORD PROCESSING The number of words w that can be printed on an 8 in. by 11 in. piece of paper varies inversely as the size s of the font used.

NOTATION

15. Determine whether the equation describes direct variation.

a. $y = kx$ **b.** $y = k + x$

c. $y = \frac{k}{x}$ **d.** $m = kc$

16. Determine whether each equation describes inverse variation.

a. $y = kx$ **b.** $y = \frac{k}{x}$

c. $y = \frac{x}{k}$ **d.** $d = \frac{k}{g}$

GUIDED PRACTICE

***Solve each direct variation problem.* See Examples 1 and 2.**

17. y varies directly as x. If $y = 10$ when $x = 2$, find y when $x = 7$.

18. A varies directly as z. If $A = 30$ when $z = 5$, find A when $z = 9$.

19. r varies directly as s. If $r = 21$ when $s = 7$, find r when $s = 12$.

20. h varies directly as m. If $h = 22$ when $m = 11$, find h when $m = 3$.

21. s varies directly as t. If $s = 1.2$ when $t = 4$, find s when $t = 30$.

22. y varies directly as x. If $y = 1.6$ when $x = 2$, find y when $x = 20$.

23. d is directly proportional to t. If $d = 21$ when $t = 6$, find d when $t = 4$.

24. b is directly proportional to c. If $b = 16$ when $c = 18$, find b when $c = 27$.

***Solve each inverse variation problem.* See Examples 3 and 4.**

25. y varies inversely as x. If $y = 8$ when $x = 2$, find y when $x = 4$.

26. V varies inversely as p. If $V = 30$ when $p = 5$, find V when $p = 6$.

27. r varies inversely as t. If $r = 40$ when $t = 10$, find r when $t = 200$.

28. J varies inversely as v. If $J = 90$ when $v = 5$, find J when $v = 45$.

29. p varies inversely as x. If $p = 6$ when $x = 4$, find p when $x = 1.5$.

30. a varies inversely as d. If $a = 6$ when $d = 3$, find a when $d = 1.2$.

31. q is inversely proportional to s. If $q = 6$ when $s = 9$, find q when $s = 24$.

32. w is inversely proportional to s. If $w = 8$ when $s = 6$, find w when $s = 32$.

APPLICATIONS

***Solve each direct variation problem.* See Example 2.**

33. DRIVING The distance that a car can travel without refueling varies directly as the number of gallons of gasoline in the tank. If a car can go 360 miles on a full tank of gas (15 gallons), how far can it go on 7 gallons?

34. GRAVITY The force of gravity acting on an object varies directly as the mass of the object. The force on a mass of 5 kilograms is 49 newtons. What is the force acting on a mass of 12 kilograms?

35. MEDICATIONS To fight ear infections in children, doctors often prescribe Ceclor. The recommended dose in milligrams is directly proportional to the child's body weight in pounds. The correct dosage for a 20-pound child is 124 milligrams. What would be the correct dosage for a 28-pound child?

36. DOSAGES The recommended dose (in milligrams) of Demerol, a preoperative medication given to children, varies directly as the child's weight in pounds. The proper dosage for a child weighing 30 pounds is 18 milligrams. What would be the correct dosage for a child weighing 45 pounds?

***Solve each inverse variation problem.* See Example 4.**

37. TRAVELING The time it takes a car to travel a certain distance varies inversely as its rate of speed. If a certain trip takes 3 hours at 50 miles per hour, how long will the trip take at 60 miles per hour?

38. GEOMETRY For a fixed area, the length of a rectangle is inversely proportional to its width. A rectangle has a width of 12 feet and a length of 20 feet. If its width is increased to 12.5 feet, find the length that will maintain the same area.

39. ELECTRICITY The current in an electric circuit varies inversely as the resistance. If the current in a circuit is 30 amps when the resistance is 4 ohms, what will the current be for a resistance of 15 ohms?

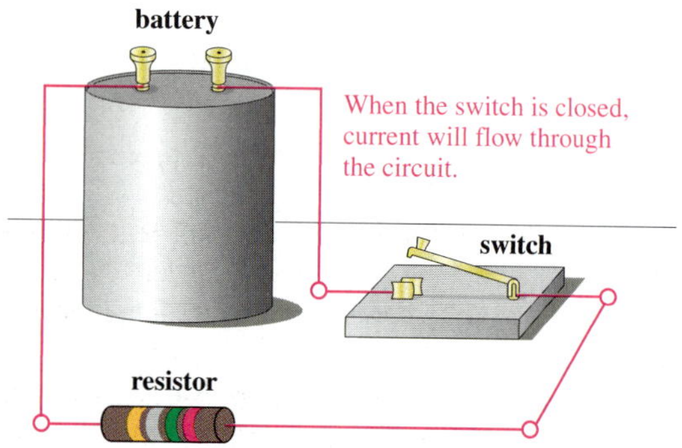

40. FARMING The length of time a given number of bushels of corn will last when feeding cattle varies inversely as the number of animals. If a certain number of bushels will feed 25 cows for 10 days, how long will the feed last for 10 cows?

TRY IT YOURSELF

41. PULLEYS The speeds (in revolutions per minute) of two pulleys connected by a belt are inversely proportional to their diameters. If a pulley 24 inches in diameter, making 120 revolutions per minute, is belted to a second pulley 16 inches in diameter, how many revolutions per minute does the smaller pulley make?

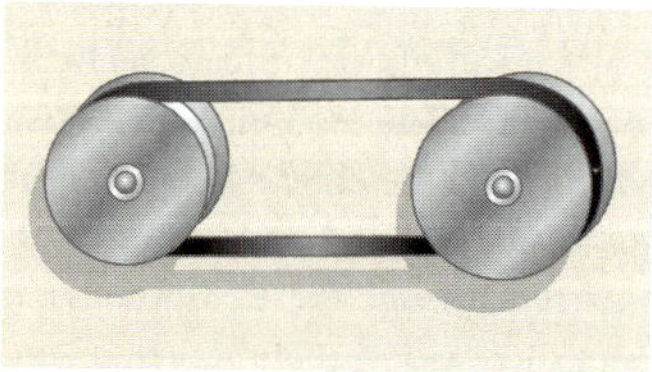

42. CIDER For the following recipe, the number of inches of stick cinnamon to use varies directly as the number of servings of spiced cider to be made. How many inches of stick cinnamon are needed to make 36 servings?

Hot Spiced Cider

8 cups apple cider or apple juice
$\frac{1}{4}$ to $\frac{1}{2}$ cup packed brown sugar
6 inches stick cinnamon
1 teaspoon whole allspice
1 teaspoon whole cloves
8 thin orange wedges or slices (optional)
8 whole cloves (optional) Makes 8 servings

43. GRAVITY The weight of an object on the moon is directly proportional to its weight on earth; six pounds on Earth weighs 1 pound on the moon. What would the scale register if the astronaut in the illustration were weighed on the moon?

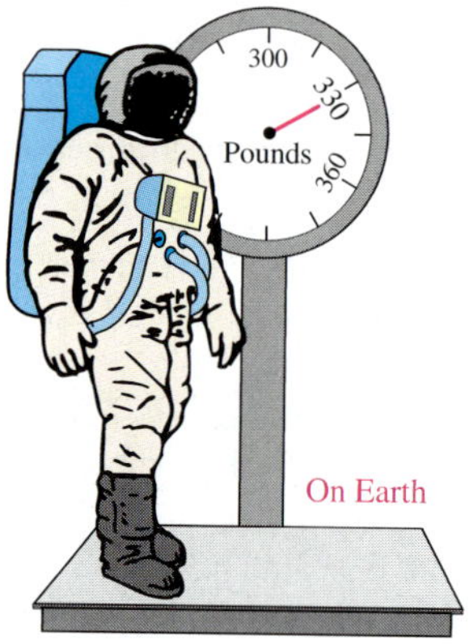

44. SEESAWS When a seesaw is balanced, the distance (in feet) each person is from the fulcrum is inversely proportional to that person's weight. Use the information in the illustration to determine how far away from the fulcrum that Brandon is sitting.

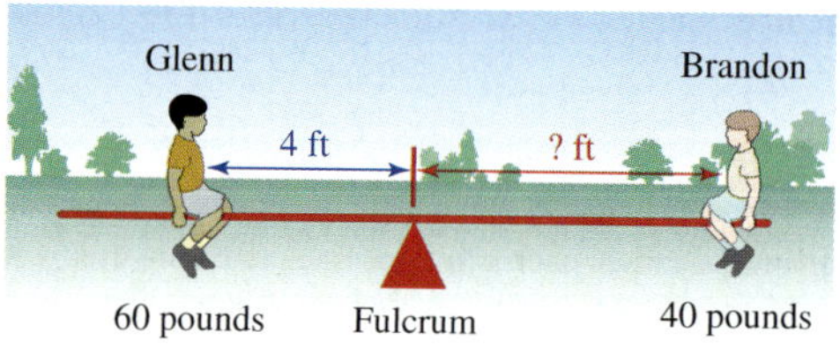

45. HOOKE'S LAW The distance a spring will stretch varies directly as the force applied to it. Suppose that a 15-kilogram weight stretches a spring 24 centimeters. Find the distance that the same spring will stretch when a heavier, 25-kilogram, weight is used.

46. ARCHITECTURE The total number of windows needed in the construction of an apartment building varies directly as the number of floors. If a 4-story building requires 176 windows, how many windows does an 11-story building require?

47. CHEMISTRY The volume occupied by a fixed weight of gas (held at a constant temperature) varies inversely as the pressure placed on it. If a nitrogen gas occupies a volume of 40 cubic meters under a pressure of 8 atmospheres, find the volume that the gas occupies when the pressure is changed to 6 atmospheres.

48. DEPRECIATION Assume that the value of a machine varies inversely as its age. If a drill press is worth $300 when it is 2 years old, find its value when it is 6 years old. How much has the machine depreciated over that 4-year period?

WRITING

49. Give examples of two quantities that vary directly and two quantities that do not.

50. What is the difference between direct variation and inverse variation?

51. What is a constant of variation?

52. COMPUTER PRINTERS Is there a direct variation or an inverse variation between each pair of quantities? Explain. Draw a graph to support your answer.
   a. The time it takes to print a term paper and the speed of the printer.
   b. The time it takes to print a term paper and the length of the term paper.

REVIEW

Solve each equation.

53. $(a - 1)(a^2 + 5a + 6) = 0$

54. $(b - 2)(b^2 - 8b + 7) = 0$

55. $x^3 - 6x^2 - 27x = 0$

56. $6t^3 + 35t^2 - 6t = 0$

CHALLENGE PROBLEMS

57. WIND ENERGY The power produced by a propeller wind turbine varies directly as the cube of the wind speed. If a wind speed of 15 mph produces 5.4 kilowatts of power, how much power would be produced by a 30-mph wind?

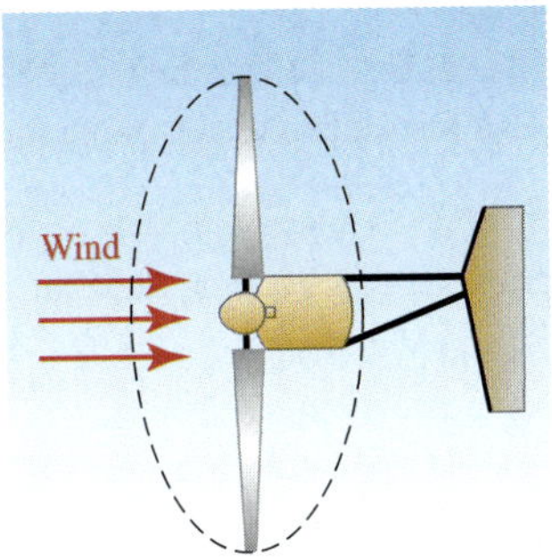

58. GRAVITY The force with which the Earth attracts an object above the Earth's surface varies inversely as the square of the distance of the object from the center of the Earth. An object 4,000 miles from the center of the Earth is attracted with a force of 80 pounds. Find the force of attraction if the object were 5,000 miles from the center of the Earth.

59. LANDING AIRCRAFT The runway distance required to land a single-engine airplane varies directly as the square of its touchdown speed. If a touchdown speed of 70 mph requires a landing distance of 1,470 feet, what landing distance is needed for a touchdown speed of 80 mph?

60. CONSTRUCTION The time it takes to build a highway varies directly as the length of the road and inversely as the number of workers. It takes 100 workers 4 weeks to build 2 miles of highway. How long will it take 80 workers to build 10 miles of highway?

CHAPTER 7
Summary & Review

SECTION 7.1 Simplifying Rational Expressions

DEFINITIONS AND CONCEPTS	EXAMPLES
A **rational expression** is an expression of the form $\frac{A}{B}$, where A and B are polynomials and B does not equal 0.	Rational expressions: $\frac{8}{7t}$, $\frac{a}{a-3}$, and $\frac{4x^2-16x}{x^2-6x+8}$
To **evaluate a rational expression,** we substitute the values of its variables and simplify.	Evaluate $\frac{3x+1}{x-2}$ for $x = 5$. $\frac{3x+1}{x-2} = \frac{3(5)+1}{5-2} = \frac{16}{3}$ Substitute 5 for x.
To find the real numbers for which a **rational expression is undefined,** find the values of the variable that make the denominator 0.	For which real numbers is $\frac{11}{2x-3}$ undefined? $2x - 3 = 0$ Set the denominator equal to 0 and solve for x. $2x = 3$ $x = \frac{3}{2}$ The expression is undefined for $x = \frac{3}{2}$.
To **simplify a rational expression:** **1.** Factor the numerator and the denominator completely. **2.** Remove factors equal to 1. **3.** Multiply the remaining factors in the numerator and denominator and simplify, if possible.	Simplify: $\frac{x^2-4}{x^2-7x+10} = \frac{(x+2)\overset{1}{\cancel{(x-2)}}}{(x-5)\underset{1}{\cancel{(x-2)}}}$ Factor and simplify. $= \frac{x+2}{x-5}$
The quotient of any nonzero expression and its **opposite** is -1.	$\frac{2t-3}{3-2t} = -1$ Because $2t - 3$ and $3 - 2t$ are opposites.

REVIEW EXERCISES

1. Find the values of x for which the rational expression $\frac{x-1}{x^2-16}$ is undefined.

2. Evaluate $\frac{x^2-1}{x-5}$ for $x = -2$.

Simplify each rational expression, if possible. Assume that no denominators are zero.

3. $\frac{3x^2}{6x^3}$

4. $\frac{5xy^2}{2x^2y^2}$

5. $\frac{x^2}{x^2+x}$

6. $\frac{a^2-4}{a+2}$

7. $\frac{3p-2}{2-3p}$

8. $\frac{8-x}{x^2-5x-24}$

9. $\frac{2x^2-16x}{2x^2-18x+16}$

10. $\frac{x^2+x-2}{x^2-x-2}$

11. $\frac{x^2-2xy+y^2}{(x-y)^3}$

12. $\frac{4(t+3)+8}{3(t+3)+6}$

13. Explain the error in the following work: $\frac{x+1}{x} = \frac{\overset{1}{\cancel{x}}+1}{\underset{1}{\cancel{x}}} = \frac{2}{1} = 2$.

14. DOSAGES Cowling's rule is a formula that can be used to determine the dosage of a prescription medication for children. If C is the proper child's dosage, D is an adult dosage, and A is the child's age in years, then $C = \frac{D(A+1)}{24}$. Find the daily dosage of an antibiotic for an 11-year-old child if the adult daily dosage is 300 milligrams.

SECTION 7.2 Multiplying and Dividing Rational Expressions

DEFINITIONS AND CONCEPTS	EXAMPLES
To **multiply rational expressions,** multiply their numerators and multiply their denominators. $\frac{A}{B} \cdot \frac{C}{D} = \frac{AC}{BD}$ Then simplify, if possible.	Multiply: $\frac{4b}{b+2} \cdot \frac{7}{b} = \frac{4b \cdot 7}{(b+2)b}$ $= \frac{4\overset{1}{\cancel{b}} \cdot 7}{(b+2)\underset{1}{\cancel{b}}}$ Simplify. $= \frac{28}{b+2}$
To find the **reciprocal** of a rational expression, invert its numerator and denominator.	The reciprocal of $\frac{c}{c+7}$ is $\frac{c+7}{c}$.
To **divide rational expressions,** multiply the first expression by the reciprocal of the second. $\frac{A}{B} \div \frac{C}{D} = \frac{A}{B} \cdot \frac{D}{C} = \frac{AD}{BC}$ Then simplify, if possible.	Divide: $\frac{t}{t+1} \div \frac{8}{t^2+t} = \frac{t}{t+1} \cdot \frac{t^2+t}{8}$ $= \frac{t \cdot t\overset{1}{\cancel{(t+1)}}}{\underset{1}{\cancel{(t+1)}}8}$ Factor and simplify. $= \frac{t^2}{8}$
A **unit conversion factor** is a fraction that has a value of 1.	$\frac{1 \text{ yd}^2}{9 \text{ ft}^2} = 1$ and $\frac{1 \text{ mi}}{5{,}280 \text{ ft}} = 1$

REVIEW EXERCISES

Multiply and simplify, if possible.

15. $\frac{3xy}{2x} \cdot \frac{4x}{2y^2}$

16. $56x\left(\frac{12}{7x}\right)$

17. $\frac{x^2-1}{x^2+2x} \cdot \frac{x}{x+1}$

18. $\frac{x^2+x}{3x-15} \cdot \frac{6x-30}{x^2+2x+1}$

Divide and simplify, if possible.

19. $\frac{3x^2}{5x^2y} \div \frac{6x}{15xy^2}$

20. $\frac{x^2-x-6}{1-2x} \div \frac{x^2-2x-3}{2x^2+x-1}$

21. Determine whether the given fraction is a unit conversion factor.

a. $\frac{1 \text{ ft}}{12 \text{ in.}}$ **b.** $\frac{60 \text{ min}}{1 \text{ day}}$

c. $\frac{2{,}000 \text{ lb}}{1 \text{ ton}}$ **d.** $\frac{1 \text{ gal}}{4 \text{ qt}}$

22. TRAFFIC SIGNS Convert the speed limit on the sign from miles per hour to miles per minute.

SECTION 7.3 Adding and Subtracting with Like Denominators; Least Common Denominators

DEFINITIONS AND CONCEPTS	EXAMPLES
To **add (or subtract) rational expressions** that have the same denominator, add (or subtract) their numerators and write the sum (or difference) over their common denominator. $\frac{A}{D} + \frac{B}{D} = \frac{A+B}{D}$ $\frac{A}{D} - \frac{B}{D} = \frac{A-B}{D}$	Add: $\frac{2b}{3b-9} + \frac{b}{3b-9} = \frac{2b+b}{3b-9}$ $= \frac{\overset{1}{\cancel{3}}b}{\underset{1}{\cancel{3}}(b-3)}$ Factor and simplify. $= \frac{b}{b-3}$

SECTION 7.3 –continued

DEFINITIONS AND CONCEPTS	EXAMPLES
Then simplify, if possible.	Subtract: $\frac{x+1}{x} - \frac{x-1}{x} = \frac{x+1-(x-1)}{x}$ Don't forget the parentheses. $= \frac{x+1-x+1}{x}$ $= \frac{2}{x}$ Combine like terms.
To find the **LCD** of several fractions, factor each denominator completely. Form a product using each different factor the greatest number of times it appears in any one factorization.	Find the LCD of $\frac{3}{x^3-x^2}$ and $\frac{x}{x^2-1}$. $x^3 - x^2 = x \cdot x \cdot (x-1)$ $x^2 - 1 = (x+1)(x-1)$ LCD $= x \cdot x \cdot (x-1)(x+1)$
To **build an equivalent rational expression,** multiply the given expression by 1 in the form of $\frac{c}{c}$ where $c \neq 0$.	$\frac{7}{4t} = \frac{7}{4t} \cdot \frac{3t}{3t} = \frac{21t}{12t^2}$ and $\frac{x+1}{x-7} = \frac{x+1}{x-7} \cdot \frac{x-1}{x-1} = \frac{(x+1)(x-1)}{(x-7)(x-1)} = \frac{x^2-1}{x^2-8x+7}$

REVIEW EXERCISES

Add or subtract and simplify, if possible.

23. $\frac{13}{15d} - \frac{8}{15d}$

24. $\frac{x}{x+y} + \frac{y}{x+y}$

25. $\frac{3x}{x-7} - \frac{x-2}{x-7}$

26. $\frac{a}{a^2-2a-8} + \frac{2}{a^2-2a-8}$

Find the LCD of each pair of rational expressions.

27. $\frac{12}{x}, \frac{1}{9}$

28. $\frac{1}{2x^3}, \frac{5}{8x}$

29. $\frac{7}{m}, \frac{m+2}{m-8}$

30. $\frac{x}{5x+1}, \frac{5x}{5x-1}$

31. $\frac{6-a}{a^2-25}, \frac{a^2}{a-5}$

32. $\frac{4t+25}{t^2+10t+25}, \frac{t^2-7}{2t^2+17t+35}$

Build each rational expression into an equivalent fraction having the denominator shown in red.

33. $\frac{9}{a}$, $7a$

34. $\frac{2y+1}{x-9}$, $x(x-9)$

35. $\frac{b+7}{3b-15}$, $6(b-5)$

36. $\frac{9r}{r^2+6r+5}$, $(r+1)(r-4)(r+5)$

SECTION 7.4 Adding and Subtracting with Unlike Denominators

DEFINITIONS AND CONCEPTS	EXAMPLES
To **add (or subtract) rational expressions** with unlike denominators: 1. Find the LCD. 2. Write each rational expression as an equivalent expression whose denominator is the LCD. 3. Add (or subtract) the numerators and write the sum (or difference) over the LCD. 4. Simplify the resulting rational expression if possible.	Add: $\frac{4x}{x} + \frac{2}{x-1} = \frac{4x}{x} \cdot \frac{x-1}{x-1} + \frac{2}{x-1} \cdot \frac{x}{x}$ The LCD is $x(x-1)$. $= \frac{4x(x-1)}{x(x-1)} + \frac{2x}{x(x-1)}$ $= \frac{4x^2 - 4x + 2x}{x(x-1)}$ Distribute the multiplication by 4x. $= \frac{4x^2 - 2x}{x(x-1)}$ Combine like terms. $= \frac{2\overset{1}{\cancel{x}}(2x-1)}{\underset{1}{\cancel{x}}(x-1)}$ Factor and simplify. $= \frac{2(2x-1)}{x-1}$
When a polynomial is multiplied by -1, the result is its opposite. This fact is used when adding or subtracting rational expressions whose **denominators are opposites.**	Add: $\frac{c}{c-4} + \frac{1}{4-c} = \frac{c}{c-4} + \frac{1}{4-c} \cdot \frac{-1}{-1}$ $= \frac{c}{c-4} + \frac{-1}{c-4}$ $-1(4-c) = c-4$ $= \frac{c-1}{c-4}$

REVIEW EXERCISES

Add or subtract and simplify, if possible.

37. $\frac{1}{7} - \frac{1}{a}$

38. $\frac{x}{x-1} + \frac{1}{x}$

39. $\frac{2t+2}{t^2+2t+1} - \frac{1}{t+1}$

40. $\frac{x+2}{2x} - \frac{2-x}{x^2}$

41. $\frac{6}{b-1} - \frac{b}{1-b}$

42. $\frac{8}{c} + 6$

43. $\frac{n+7}{n+3} - \frac{n-3}{n+7}$

44. $\frac{4}{t+2} - \frac{7}{(t+2)^2}$

45. $\frac{6}{a^2-9} - \frac{5}{a^2-a-6}$

46. $\frac{2}{3y-6} + \frac{3}{4y+8}$

47. Working on a homework assignment, a student added two rational expressions and obtained $\frac{-5n^3-7}{3n(n+6)}$. The answer given in the back of the book was $-\frac{5n^3+7}{3n(n+6)}$. Are the answers equivalent?

48. DIGITAL VIDEO CAMERAS Find the perimeter and the area of the LED screen of the camera.

SECTION 7.5 Simplifying Complex Fractions

DEFINITIONS AND CONCEPTS	EXAMPLES
Complex fractions contain fractions in their numerators and/or their denominators.	Complex fractions: $\dfrac{\frac{2}{t}}{\frac{5}{4t}}$ and $\dfrac{\frac{3}{m}+\frac{m}{4}}{\frac{m}{2}}$
To **simplify a complex fraction:** **Method 1** Write the numerator and the denominator as single rational expressions and perform the indicated division.	Simplify: $\dfrac{\frac{3}{m}+\frac{m}{2}}{\frac{m}{4}} = \dfrac{\frac{3}{m}\cdot\frac{2}{2}+\frac{m}{2}\cdot\frac{m}{m}}{\frac{m}{4}}$ In the numerator, build to have an LCD of 2*m*. $= \dfrac{\frac{6}{2m}+\frac{m^2}{2m}}{\frac{m}{4}}$ The main fraction bar indicates division. $= \dfrac{\frac{6+m^2}{2m}}{\frac{m}{4}}$ Add the fractions in the numerator. $= \dfrac{(6+m^2)\cdot \overset{1}{\cancel{2}}\cdot 2}{\underset{1}{\cancel{2}}m\cdot m}$ Multiply by the reciprocal of $\frac{m}{4}$. Factor and simplify. $= \dfrac{12+2m^2}{m^2}$ Distribute the multiplication by 2.
Method 2 Determine the LCD of all the rational expressions in the complex fraction and multiply the complex fraction by 1, written in the form $\frac{\text{LCD}}{\text{LCD}}$.	Simplify: $\dfrac{\frac{3}{m}+\frac{m}{2}}{\frac{m}{4}} = \dfrac{\frac{3}{m}+\frac{m}{2}}{\frac{m}{4}}\cdot\dfrac{4m}{4m}$ The LCD for all the rational expressions is 4*m*. $= \dfrac{\frac{3}{m}\cdot 4m+\frac{m}{2}\cdot 4m}{\frac{m}{4}\cdot 4m}$ In the numerator, distribute the multiplication by 4*m*. $= \dfrac{12+2m^2}{m^2}$ Perform each multiplication by 4*m*.

REVIEW EXERCISES

Simplify each complex fraction.

49. $\dfrac{\frac{n^4}{30}}{\frac{7n}{15}}$

50. $\dfrac{\frac{r^2-81}{18s^2}}{\frac{4r-36}{9s}}$

51. $\dfrac{\frac{1}{y}+1}{\frac{1}{y}-1}$

52. $\dfrac{\frac{7}{a^2}}{\frac{1}{a}+\frac{10}{3}}$

53. $\dfrac{\frac{2}{x-1}+\frac{x-1}{x+1}}{\frac{1}{x^2-1}}$

54. $\dfrac{\frac{1}{x^2y}-\frac{5}{xy}}{\frac{3}{xy}-\frac{7}{xy^2}}$

SECTION 7.6 Solving Rational Equations

DEFINITIONS AND CONCEPTS

To **solve a rational equation** we use the multiplication property of equality to clear the equation of fractions. Use these steps:

1. Determine which numbers cannot be solutions.
2. Multiply both sides of the equation by the LCD of the rational expressions contained in the equation.
3. Solve the resulting equation.
4. Check all possible solutions in the *original* equation. A possible solution that does not satisfy the original equation is called an **extraneous solution.**

EXAMPLES

Solve: $\frac{y}{y-2} - 1 = \frac{1}{y}$ Since no denominators can be 0, $y \neq 2$ and $y \neq 0$.

$$y(y-2)\left(\frac{y}{y-2} - 1\right) = y(y-2)\left(\frac{1}{y}\right)$$ The LCD is $y(y-2)$.

$$\overset{1}{y(\cancel{y-2})}\left(\frac{y}{\underset{1}{\cancel{y-2}}}\right) - y(y-2)1 = \overset{1}{\cancel{y}}(y-2)\left(\frac{1}{\underset{1}{\cancel{y}}}\right)$$ Simplify.

$$y \cdot y - y(y-2) = (y-2) \cdot 1$$

$$y^2 - y^2 + 2y = y - 2$$

$$2y = y - 2$$ Combine like terms.

$$y = -2$$

REVIEW EXERCISES

Solve each equation and check the result. If an equation has no solution, so indicate.

55. $\frac{3}{x} = \frac{2}{x-1}$

56. $\frac{a}{a-5} = 3 + \frac{5}{a-5}$

57. $\frac{2}{3t} + \frac{1}{t} = \frac{5}{9}$

58. $a = \frac{3a-50}{4a-24} - \frac{3}{4}$

59. $\frac{4}{x+2} - \frac{3}{x+3} = \frac{6}{x^2+5x+6}$

60. $\frac{3}{x+1} - \frac{x-2}{2} = \frac{x-2}{x+1}$

61. ENGINEERING The efficiency E of a Carnot engine is given by the following formula. Solve it for T_1.

$$E = 1 - \frac{T_2}{T_1}$$

62. Solve for y: $\frac{1}{x} = \frac{1}{y} + \frac{1}{z}$

SECTION 7.7 Problem Solving Using Rational Equations

DEFINITIONS AND CONCEPTS

To solve application problems, follow these steps:

1. Analyze the problem.
2. Form an equation.
3. Solve the equation.
4. State the conclusion.
5. Check the result.

Rate of Work: If a job can be completed in t units of time, the rate of work can be expressed as $\frac{1}{t}$ of the job is completed per unit of time.

Shared work problems:
Work completed = rate of work · time worked

EXAMPLES

WASHING CARS Working alone, Carlos can wash the family SUV in 30 minutes. Victor, his brother, can wash the same SUV in 20 minutes working alone. How long will it take them if they wash the SUV together?

Analyze the Problem Let x = the number of minutes it will take Carlos and Victor, working together, to wash the SUV. Enter the data in a table.

	Rate	· Time	= Work Completed
Carlos	$\frac{1}{30}$	x	$\frac{x}{30}$
Victor	$\frac{1}{20}$	x	$\frac{x}{20}$

Form an equation The part of the job done by Carlos plus the part of the job done by Victor equals 1 job completed.

$$\frac{x}{30} + \frac{x}{20} = 1$$

SECTION 7.7 Problem Solving Using Rational Equations–*continued*

DEFINITIONS AND CONCEPTS | **EXAMPLES**

Solve the equation

$$60\left(\frac{x}{30}+\frac{x}{20}\right)=60(1) \quad \text{Multiply both sides by the LCD, 60.}$$

$$60\left(\frac{x}{30}\right)+60\left(\frac{x}{20}\right)=60(1) \quad \text{On the left side, distribute the multiplication by 60.}$$

$$2x+3x=60 \quad \text{Perform each multiplication by 60.}$$

$$5x=60 \quad \text{Combine like terms.}$$

$$x=\frac{60}{5} \quad \text{Divide both sides by 5.}$$

$$x=12$$

State the Conclusion Working together, it will take Carlos and Victor 12 minutes to wash the family SUV.

Check the Result In 12 minutes, Carlos will do $\frac{12}{30}=\frac{24}{60}$ of the job and Victor will do $\frac{12}{20}=\frac{36}{60}$ of the job. Together they will do $\frac{24}{60}+\frac{36}{60}=\frac{60}{60}$ or 1 whole job. The result checks.

Definitions and Concepts	Examples
Uniform motion problems: $\text{Time}=\frac{\text{distance}}{\text{rate}}$	See Example 2 in Section 7.7.
Investment problems: $\text{Principal}=\frac{\text{Interest}}{\text{rate}\cdot\text{time}}$	See Example 5 in Section 7.7.

REVIEW EXERCISES

63. NUMBER PROBLEM If a number is subtracted from the denominator of $\frac{4}{5}$ and twice as much is added to the numerator, the result is 5. Find the number.

64. EXERCISE A woman can bicycle 30 miles in the same time that it takes her to jog 10 miles. If she can ride 10 mph faster than she can jog, how fast can she jog?

65. HOUSE CLEANING A maid can clean a house in 4 hours. What is her rate of work?

66. HOUSE PAINTING If a homeowner can paint a house in 14 days and a professional painter can paint it in 10 days, how long will it take if they work together?

67. INVESTMENTS In 1 year, a student earned $100 interest on money she deposited at a savings and loan. She later learned that the money would have earned $120 if she had deposited it at a credit union, because the credit union paid 1% more interest at the time. Find the rate she received from the savings and loan.

68. WIND SPEED A plane flies 400 miles downwind in the same amount of time as it takes to travel 320 miles upwind. If the plane can fly at 360 mph in still air, find the velocity of the wind.

SECTION 7.8 Proportions and Similar Triangles

DEFINITIONS AND CONCEPTS	EXAMPLES
A **ratio** is the quotient of two numbers with the same units.	Ratios: $\frac{2}{3}$, $\frac{1}{50}$, and 2:3
A **rate** is the quotient of two quantities with different units.	Rates: $\frac{4\text{ oz}}{6\text{ lb}}$, $\frac{525\text{ mi}}{15\text{ hr}}$, and $\frac{\$1.95}{2\text{ lb}}$

SECTION 7.8 Proportions and Similar Triangles–*continued*

DEFINITIONS AND CONCEPTS	EXAMPLES
A **proportion** is a statement that two ratios or two rates are equal. In the proportion $\frac{a}{b} = \frac{c}{d}$, a and d are the **extremes** and b and c are the **means.**	A proportion: $\frac{4}{9} = \frac{28}{63}$ Extremes: 4 and 63 Means: 9 and 28
In any proportion, the product of the extremes is equal to the product of the means. (The **cross products** are equal.)	A proportion: $\frac{4}{9} = \frac{28}{63}$ Cross product: $4 \cdot 63 = 252$ Cross product: $9 \cdot 28 = 252$
To **solve a proportion,** set the product of the extremes equal to the product of the means and solve the resulting equation.	Solve the proportion: $\frac{3}{2} = \frac{x}{10}$ $3 \cdot 10 = 2 \cdot x$ Set the cross products equal. $30 = 2x$ $15 = x$ Solve for x.
Triangles with the same shape but not necessarily the same size are called **similar triangles.** The lengths of the corresponding sides of two similar triangles are in proportion.	In these similar triangles: $\frac{a}{d} = \frac{b}{e} = \frac{c}{f}$ 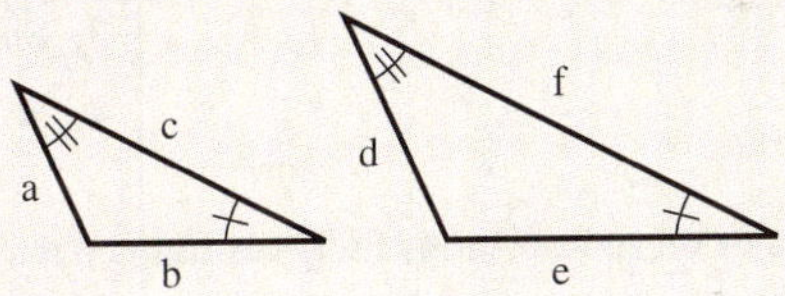
A **scale** is a ratio (or rate) that compares the size of a model to the size of an actual object.	See Example 6 in Section 7.8.
Unit prices can be used to compare costs of different sizes of the same brand to determine the best buy.	For the same item, a cost of $\frac{\$1.95}{1 \text{ lb}}$ is a better buy than a cost of $\frac{\$1.99}{1 \text{ lb}}$. See Example 7 in Section 7.8.

REVIEW EXERCISES

Determine whether each equation is a true proportion.

69. $\frac{4}{7} = \frac{20}{34}$

70. $\frac{5}{7} = \frac{30}{42}$

Solve each proportion.

71. $\frac{3}{x} = \frac{6}{9}$

72. $\frac{x}{3} = \frac{x}{5}$

73. $\frac{x-2}{5} = \frac{x}{7}$

74. $\frac{2x}{x+4} = \frac{3}{x-1}$

75. DENTISTRY The diagram below was displayed in a dentist's office. According to the diagram, if the dentist has 340 adult patients, how many will develop gum disease?

3 out of 4 adults will develop gum disease.

76. UTILITY POLES A telephone pole casts a shadow 12 feet long at the same time that a man 6 feet tall casts a shadow of 3.6 feet. How tall is the pole?

77. PORCELAIN FIGURINES A model of a flutist, standing and playing at a music stand, was made using a 1/12th scale. If the scale model is 5.5 inches tall, how tall is the flutist?

78. COMPARISON SHOPPING Which is the better buy for recordable compact discs: 150 for \$60 or 250 for \$98?

32. CURRENCY EXCHANGE RATES Preparing for a visit to London, a New York resident exchanged 3,500 U.S. dollars for British pounds. (A pound is the basic monetary unit of Great Britain.) If the exchange rate was 100 U.S. dollars for 51 British pounds, how many British pounds did the traveler receive?

33. TV TOWERS A television tower casts a shadow 114 feet long at the same time that a 6-foot-tall television reporter casts a shadow of 4 feet. Find the height of the tower.

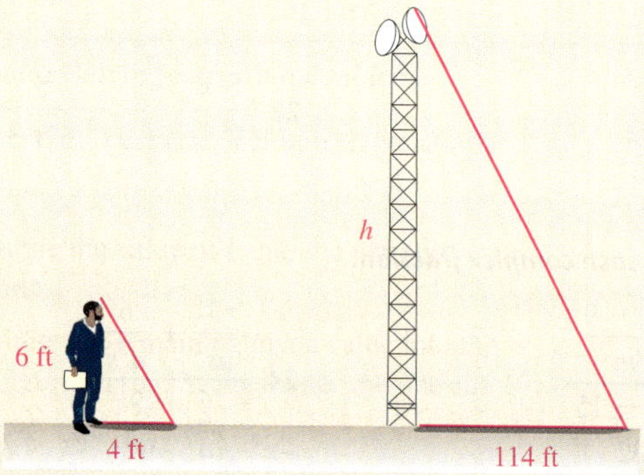

34. COMPARISON SHOPPING Which is the better buy for fabric softener: 80 sheets for \$3.89 or 120 sheets for \$6.19?

35. CLEANING HIGHWAYS One highway worker can pick up all the trash on a strip of highway in 7 hours, and his helper can pick up the trash in 9 hours. How long will it take them if they work together?

36. PHYSICAL FITNESS A man roller-blades at a rate 6 miles per hour faster than he jogs. In the same time it takes him to roller-blade 5 miles he can jog 2 miles. How fast does he jog?

37. NUMBER PROBLEM If a number is subtracted from the numerator of $\frac{5}{8}$ and twice as much is added to the denominator, the result is $\frac{1}{4}$. Find the number.

38. Explain what it means to clear the following equation of fractions.

$$\frac{u}{u-1} + \frac{1}{u} = \frac{u^2+1}{u^2-u}$$

Why is this a helpful first step in solving the equation?

39. POGO STICKS The force required to compress a spring varies directly with the change in the length of the spring. If a force of 130 pounds compresses the spring on the pogo stick 6.5 inches, how much force is required to compress the spring 5 inches?

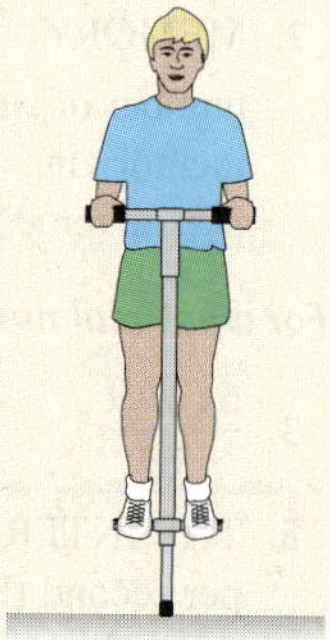

40. Assume that r varies inversely with s. If $r = 40$ when $s = 10$, find r when $s = 15$.

GROUP PROJECT

WHAT IS π?

Overview: In this activity, you will discover an important fact about the ratio of the circumference to the diameter of a circle.

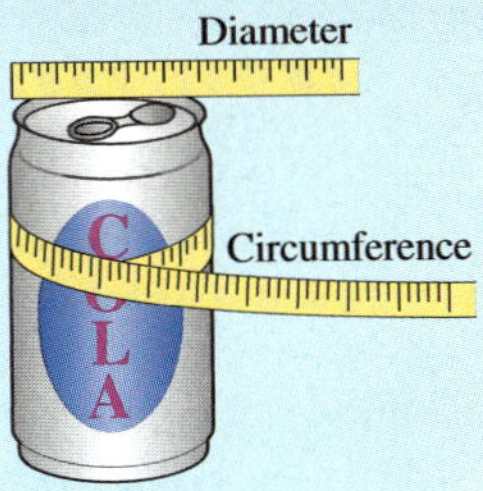

Instructions: Form groups of 2 or 3 students. With a piece of string or a cloth tape measure, find the circumference and the diameter of objects that are circular in shape. You can measure anything that is round: for example, a coin, the top of a can, a tire, or a wastepaper basket. Enter your results in a table, as shown below. Convert each measurement to a decimal, and then use a calculator to determine a decimal approximation of the ratio of the circumference C to diameter d.

Object	Circumference	Diameter	$\frac{C}{d}$ (approx.)
A quarter	$2\frac{15}{16}$ in. = 2.9375 in.	$\frac{15}{16}$ in. = 0.9375 in.	3.13333

Since early history, mathematicians have known that the ratio of the circumference to the diameter of a circle is the same for any size circle, approximately 3. Today, following centuries of study, we know that this ratio is exactly 3.141592653589. . . .

$$\frac{C}{d} = 3.141592653589\ldots$$

The Greek letter π (pi) is used to represent the ratio of circumference to diameter:

$$\pi = \frac{C}{d}, \quad \text{where } \pi = 3.141592653589\ldots$$

Are the ratios in your table numerically close to π? Give some reasons why they aren't exactly 3.141592653589 . . . in each case.

CUMULATIVE REVIEW
Chapters 1–7

1. Determine whether each statement is true or false. [Section 1.3]
   a. Every integer is a whole number.
   b. 0 is not a rational number.
   c. π is an irrational number.
   d. The set of integers is the set of whole numbers and their opposites.
2. Insert the proper symbol, $<$ or $>$, in the blank to make a true statement.

   $|2 - 4| \quad -(-6)$ [Section 1.5]
3. Evaluate: $9^2 - 3[45 - 3(6 + 4)]$ [Section 1.7]
4. Find the average (mean) test score of a student in a history class with scores of 80, 73, 61, 73, and 98. [Section 1.7]
5. Simplify: $8(c + 7) - 2(c - 3)$ [Section 1.9]
6. Solve: $\frac{4}{5}d = -4$ [Section 2.1]
7. Solve: $2 - 3(x - 5) = 4(x - 1)$ [Section 2.2]
8. GRAND KING SIZE BEDS Because Americans are taller compared to 100 years ago, bed manufacturers are making larger models. Find the percent of increase in sleeping area of the new grand king size bed compared to the standard king size. [Section 2.3]

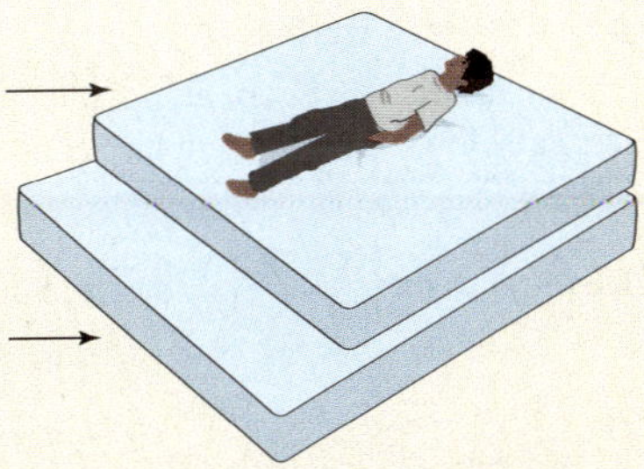

9. Solve $A - c = 2B + r$ for B. [Section 2.4]

10. Change 40°C to degrees Fahrenheit. [Section 2.4]

11. Find the volume of a pyramid that has a square base, measuring 6 feet on a side, and whose height is 20 feet. [Section 2.4]

12. BLENDING TEA One grade of tea (worth \$6.40 per pound) is to be mixed with another grade (worth \$4 per pound) to make 20 pounds of a mixture that will be worth \$5.44 per pound. How much of each grade of tea must be used? [Section 2.6]

13. SPEED OF A PLANE Two planes are 6,000 miles apart and their speeds differ by 200 mph. If they travel toward each other and meet in 5 hours, find the speed of the slower plane. [Section 2.6]

14. Solve $7x + 2 \geq 4x - 1$. Write the solution set in interval notation and graph it. [Section 2.7]

15. Graph: $y = 2x - 3$ [Section 3.2]

16. Find the slope of the line passing through $(-1, 3)$ and $(3, -1)$. [Section 3.4]

17. CUTTING STEEL The graph shows the amount of wear (in millimeters) on a cutting blade for a given length of a cut (in meters). Find the rate of change in the length of the cutting blade. [Section 3.4]

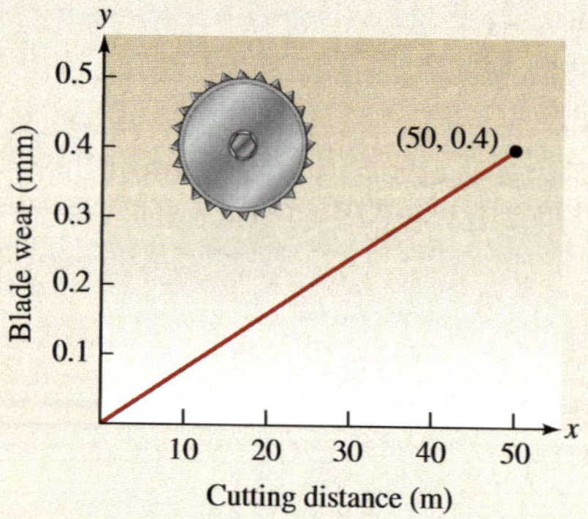

18. What is the slope of a line perpendicular to the line $y = -\frac{7}{8}x - 6$? [Section 3.5]

19. Write an equation of the line that has slope 3 and passes through the point (1, 5). Write the answer in slope-intercept form. [Section 3.6]

20. Graph: $3x - 2y \leq 6$ [Section 3.7]

21. If $f(x) = -3x^2 - 6x$, find $f(-2)$. [Section 3.8]

22. Fill in the blanks. The set of all possible input values for a function is called the ________ and the set of all output values is called the ________. [Section 3.8]

23. Solve the system $\begin{cases} x + y = 1 \\ y = x + 5 \end{cases}$ by graphing. [Section 4.1]

24. Solve the system: $\begin{cases} x = 3y - 1 \\ 2x - 3y = 4 \end{cases}$ [Section 4.2]

25. Solve the system: $\begin{cases} 2x + 3y = -1 \\ 3x + 5y = -2 \end{cases}$ [Section 4.3]

26. POKER After a night of cards, a poker player finished with some red chips (worth \$5 ech) and some blue chips (worth \$10 each). He received \$190 when he cashed in the 23 chips. How many of each colored chip did he finish with? [Section 4.4]

Simplify each expression. Write each answer without using negative exponents.

27. x^4x^3 [Section 5.1]

28. $(x^2x^3)^5$ [Section 5.1]

29. $\left(\dfrac{y^3y}{2yy^2}\right)^3$ [Section 5.1]

30. $\left(\dfrac{-2a}{b}\right)^5$ [Section 5.1]

31. $(a^{-2}b^3)^{-4}$ [Section 5.2]

32. $\dfrac{9b^0b^3}{3b^{-3}b^4}$ [Section 5.2]

33. Write 290,000 in scientific notation. [Section 5.3]

34. What is the degree of the polynomial $5x^3 - 4x + 16$ [Section 5.4]

35. Graph: $y = -x^3$ [Section 5.4]

36. CONCENTRIC CIRCLES The area of the ring between the two concentric circles of radius r and R is given by the formula

$$A = \pi(R + r)(R - r)$$

Do the multiplication on the right-hand side of the equation. [Section 5.7]

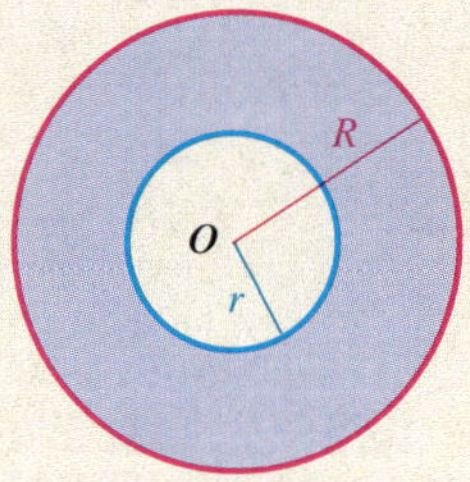

Perform the operations.

37. $(3x^2 - 3x - 2) + (3x^2 + 4x - 3)$ [Section 5.5]

38. $\left(\frac{1}{16}t^3 + \frac{1}{2}t^2 - \frac{1}{6}t\right) - \left(\frac{9}{16}t^3 + \frac{9}{4}t^2 - \frac{1}{12}t\right)$ [Section 5.5]

39. $(2x^2y^3)(3x^2y^2)$ [Section 5.6]

40. $(2y - 5)(3y + 7)$ [Section 5.6]

41. $-4x^2z(3x^2 - z)$ [Section 5.6]

42. $(3a - 4)^2$ [Section 5.7]

43. $\dfrac{6x + 9}{3}$ [Section 5.8]

44. $2x + 3\overline{)2x^3 + 7x^2 + 4x - 3}$ [Section 5.8]

Factor each polynomial completely, if possible.

45. $k^3t - 3k^2t$ [Section 6.1]

46. $2ab + 2ac + 3b + 3c$ [Section 6.1]

47. $u^2 - 18u + 81$ [Section 6.2]

48. $-r^2 + 2 + r$ [Section 6.2]

49. $u^2 + 10u + 15$ [Section 6.2]

50. $6x^2 - 63 - 13x$ [Section 6.3]

51. $2a^2 - 200b^2$ [Section 6.4]

52. $b^3 + 125$ [Section 6.5]

Solve each equation by factoring.

53. $5x^2 + x = 0$ [Section 6.7]

54. $6x^2 - 5x = -1$ [Section 6.7]

55. COOKING The electric griddle shown has a cooking surface of 160 square inches. Find the length and the width of the griddle. [Section 6.7]

56. For what values of x is the rational expression $\frac{3x^2}{x^2 - 25}$ undefined? [Section 7.1]

Perform the operations. Simplify, if possible.

57. $\dfrac{2x^2 - 8x}{x^2 - 6x + 8}$ [Section 7.1]

58. $\dfrac{x^2 - 16}{4 - x} \div \dfrac{3x + 12}{x^3}$ [Section 7.2]

59. $\dfrac{8m^2}{2m + 5} - \dfrac{4m^2 + 25}{2m + 5}$ [Section 7.3]

60. $\dfrac{4}{x - 3} + \dfrac{5}{3 - x}$ [Section 7.4]

61. $\dfrac{m}{m^2 + 5m + 6} - \dfrac{2}{m^2 + 3m + 2}$ [Section 7.4]

62. Simplify: $\dfrac{2 - \dfrac{2}{x + 1}}{2 + \dfrac{2}{x}}$ [Section 7.5]

Solve each equation.

63. $\dfrac{7}{5x} - \dfrac{1}{2} = \dfrac{5}{6x} + \dfrac{1}{3}$ [Section 7.6]

64. $\dfrac{u}{u - 1} + \dfrac{1}{u} = \dfrac{u^2 + 1}{u^2 - u}$ [Section 7.6]

65. DRAINING A TANK If one outlet pipe can drain a tank in 24 hours, and another pipe can drain the tank in 36 hours, how long will it take for both pipes to drain the tank? [Section 7.7]

66. HEIGHT OF A TREE A tree casts a shadow of 29 feet at the same time as a vertical yardstick casts a shadow of 2.5 feet. Find the height of the tree. [Section 7.8]

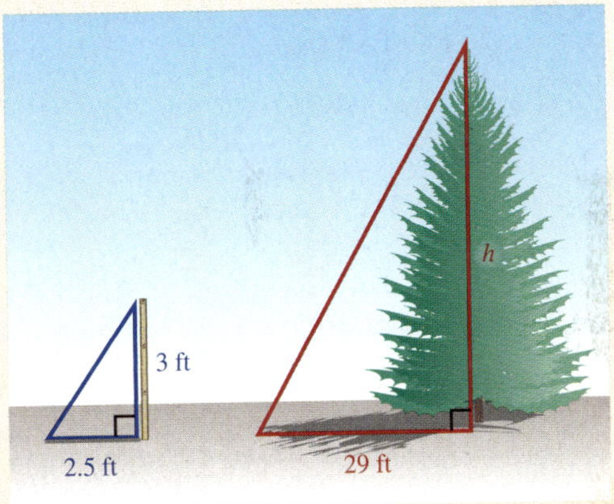

67. FORESTRY For certain types of hardwood trees, the diameter of the tree trunk varies directly as the age of the tree A 48-year old hardwood tree has a diameter of 16 inches. Find the diameter of a hardwood tree that is 84 years old. [Section 7.9]

68. Suppose y varies inversely as x. If $y = 8$ when x is 6, find y when x is 64. [Section 7.9]

CHAPTER 8

Radical Expressions and Equations

from Campus to Careers
Crime Scene Investigator

Crime scene investigators are often the first people on the scene after a crime has been committed. They are responsible for collecting evidence that can lead to criminal convictions. Most investigators spend some time each day in their office and outside in the field. The job description calls for strong interest in the sciences and strong technical skills.

Problem 91 of **Study Set 8.1** involves a skill required of many crime scene investigators—calculating the speed of a car prior to an accident from the length of its skid marks.

JOB TITLE:
Crime Scene Investigator

EDUCATION:
Some positions require a 4-year degree while others only high school graduation or a GED.

JOB OUTLOOK:
Employment is expected to increase between 9% and 17% through the year 2014.

ANNUAL EARNINGS:
Mean annual salary $58,750, mean hourly wage $28.24

FOR MORE INFORMATION:
www.worldwidelearn.com/criminal-justice

Study Skills Workshop
Preparing for a Final Exam

Final exams can be stressful for many students because the number of topics to study can seem overwhelming. Here are some suggestions to help reduce the stress and prepare you for the test.

GET ORGANIZED: Gather all of your notes, study sheets, homework assignments, and especially all of your returned tests to review.

TALK WITH YOUR INSTRUCTOR: Ask your instructor to list the topics that may appear on the final and those that won't be covered.

MANAGE YOUR TIME: Adjust your daily schedule one week before the final so that it includes extended periods of study time.

Now Try This

1. Review your old tests. Make a list of the test problems that you are still unsure about and see a tutor or your instructor to get help.
2. Make a practice final exam that includes one or more of each type of problem that may appear on the test.
3. Make a detailed study plan. Determine when, where, and what you will study each day for one week before the final.

SECTION 8.1
An Introduction to Square Roots

Objectives

1. Find square roots of perfect squares.
2. Approximate irrational square roots.
3. Find square roots of variable expressions.
4. Use the Pythagorean theorem to solve problems.
5. Use the distance formula.

Addition and subtraction are reverse operations, and so are multiplication and division. In this section, we will discuss another pair of reverse operations: raising a number to a power and finding the root of a number.

1 Find Square Roots of Perfect Squares.

When we raise a number to the second power, we are squaring it, or finding its **square.**

- The square of 5 is 25 because $5^2 = 25$.
- The square of -5 is 25, because $(-5)^2 = 25$.

We can reverse the squaring process to find **square roots** of numbers. For example, to find the square roots of 25, we ask ourselves "What number, when squared, is equal to 25?" There are two possible answers.

- 5 is a square root of 25, because $5^2 = 25$.
- -5 is a square root of 25, because $(-5)^2 = 25$.

In general, we have the following definition.

The Definition of Square Root

The number b is a **square root** of the number a if $b^2 = a$.

Every positive number has two square roots, one positive and one negative. For example, the two square roots of 9 are 3 and -3, and the two square roots of 144 are 12 and -12. The number 0 is the only real number with exactly one square root. In fact, it is its own square root, because $0^2 = 0$.

A **radical symbol** $\sqrt{}$ represents the **positive** or **principal square root** of a positive number. When reading this symbol, we usually drop the word *positive* (or *principal*) and simply say *square root.* Since 3 is the positive square root of 9, we can write

The Language of Algebra
The word *radical* comes from the Latin word *radix,* meaning root. The radical symbol has evolved over the years from ℞ in the 1300s, to √ in the 1500s, to the familiar $\sqrt{}$ in the 1600s.

$\sqrt{9} = 3$ $\sqrt{9}$ represents the positive number whose square is 9. Read as "the square root of 9 is 3."

The symbol $-\sqrt{}$ is used to represent the **negative square root** of a positive number. It is the opposite of the principal square root. Since -12 is the negative square root of 144, we can write

$-\sqrt{144} = -12$ Read as "the negative square root of 144 is -12" or "the opposite of the square root of 144 is -12." The notation $-\sqrt{144}$ represents the negative number whose square is 144.

If the number under the radical symbol is 0, we have $\sqrt{0} = 0$.

Square Root Notation

If a is a positive real number,

1. $\sqrt{a}$ represents the **positive** or **principal square root** of a. It is the positive number we square to get a.
2. $-\sqrt{a}$ represents the **negative square root** of a. It is the opposite of the principal square root of a: $-\sqrt{a} = -1 \cdot \sqrt{a}$.
3. The principal square root of 0 is 0: $\sqrt{0} = 0$.

The number or variable expression within (under) a radical symbol is called the **radicand,** and the radical symbol and radicand are called a **radical.**

Radical symbol → $\sqrt{16}$ ← Radicand
Radical

An algebraic expression containing a radical is called a **radical expression.** Some examples of radical expressions are

$$\sqrt{100}, \quad \sqrt{2} + 3, \quad \sqrt{x^2}, \quad \text{and} \quad \sqrt{\frac{a-1}{49}}$$

EXAMPLE 1

Evaluate each square root:

a. $\sqrt{16}$ **b.** $\sqrt{1}$ **c.** $\sqrt{0.36}$ **d.** $\sqrt{\frac{4}{9}}$ **e.** $-\sqrt{225}$

Strategy In each case, we will determine what positive number, when squared, produces the radicand.

Why The radical symbol $\sqrt{}$ indicates that the positive square root (principal square root) of the number written under it should be found.

Solution

a. $\sqrt{16} = 4$ Ask: What positive number, when squared, is 16? The answer is 4 because $4^2 = 16$.

b. $\sqrt{1} = 1$ Ask: What positive number, when squared, is 1? The answer is 1 because $1^2 = 1$.

c. $\sqrt{0.36} = 0.6$ Ask: What positive number, when squared, is 0.36? The answer is 0.6 because $(0.6)^2 = 0.36$.

d. $\sqrt{\frac{4}{9}} = \frac{2}{3}$ Ask: What positive number, when squared, is $\frac{4}{9}$? The answer is $\frac{2}{3}$ because $\left(\frac{2}{3}\right)^2 = \frac{4}{9}$.

e. $-\sqrt{225}$ is the opposite of the square root of 225. Since $\sqrt{225} = 15$, we have

$-\sqrt{225} = -15$ Because $-\sqrt{225} = -1 \cdot \sqrt{225} = -1 \cdot 15 = -15$.

The Language of Algebra
The symbol $\sqrt{}$ is read as "square root." It can also be read as "radical." For example, $\sqrt{16}$ can be read as "*radical* 16."

Self Check 1 Evaluate each square root: **a.** $\sqrt{49}$ **b.** $\sqrt{121}$ **c.** $-\sqrt{0.09}$ **d.** $\sqrt{\frac{1}{25}}$

Now Try Problem 17

Caution
Although $-\sqrt{4}$ and $\sqrt{-4}$ look similar, these radical expressions have very different meanings:

$-\sqrt{4} = -2$

$\sqrt{-4}$ is not a real number.

Not every number has a real-number square root. For example, to find $\sqrt{-4}$, we ask "What number, when squared, is equal to -4?" Since the square of a real number is always positive or 0, no real number squared is -4. Therefore, $\sqrt{-4}$ is not a real number. If we attempt to evaluate $\sqrt{-4}$ using a calculator, an error message is displayed.

Scientific calculator

Graphing calculator

Since square roots of negative numbers like $\sqrt{-4}$, $\sqrt{-1}$, and $\sqrt{-39}$ are not real numbers, they do not correspond to any point on the real number line. They are examples of **imaginary numbers,** which are discussed later in the text.

We summarize three important facts about square roots as follows.

Square roots

1. If a is a perfect square, then $\sqrt{a}$ is rational.
2. If a is a positive number that is not a perfect square, then $\sqrt{a}$ is irrational.
3. If a is a negative number, then $\sqrt{a}$ is not a real number.

2 Approximate Irrational Square Roots.

A table of square roots

n	$\sqrt{n}$
1	1.000
2	1.414
3	1.732
4	2.000
5	2.236

A number such as 16, 1, 0.36, $\frac{4}{9}$, or 225 that is the square of some rational number is called a **perfect square.** In Example 1, we saw that the square root of a perfect square is a rational number.

If a positive number is not a perfect square, its square root is irrational. For example, $\sqrt{3}$, $\sqrt{29}$, and $\sqrt{83}$ are irrational numbers because 3, 29, and 83 are not perfect squares. Since these square roots are irrational, they cannot be written as the quotient of two integers. Furthermore, their decimal representations are nonterminating and nonrepeating. We can find decimal approximations for them using the square root key $\sqrt{\ }$ on a calculator or from the table of square roots found in Appendix II.

$\sqrt{3} \approx 1.732$ Use a table. Read ≈ as "is approximately equal to."

$\sqrt{29} \approx 5.385164807$ Use a calculator. The number of decimal places displayed can vary slightly by model.

$\sqrt{83} \approx 9.11$ Use a calculator and round to two decimal places.

Success Tip
Estimation is helpful when approximating square roots. For example, $\sqrt{83}$ must be a number between 9 and 10, because $\sqrt{81} < \sqrt{83} < \sqrt{100}$.

Remember that the values found for irrational square roots are only approximations. If we square 9.11, which is an approximation of $\sqrt{83}$, the result is not quite 83.

$(9.11)^2 = 82.9921$

Square roots appear in many applied problems.

EXAMPLE 2 ***Pendulums.*** The **period of a pendulum** is the time required for the pendulum to swing back and forth to complete one cycle. To find the period P (in seconds) of a pendulum L feet long, we can use the formula

$P = 1.11\sqrt{L}$ Read $1.11\sqrt{L}$ as "1.11 times the square root of L."

Find the period of a clock pendulum that is 5 feet long.

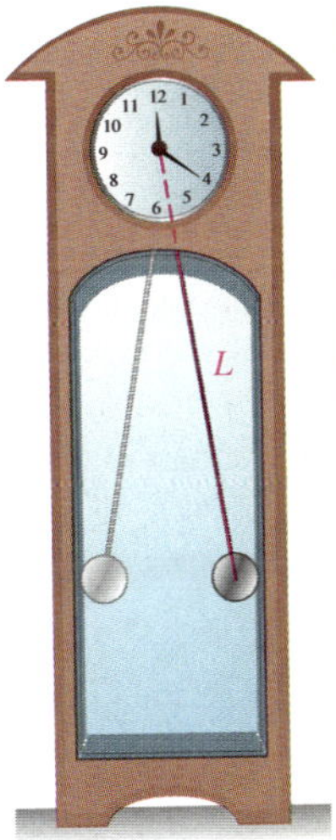

Strategy We will substitute 5 for L in the formula, use a calculator to approximate $\sqrt{5}$, and multiply that value by 1.11.

Why $1.11\sqrt{L}$ means $1.11 \cdot \sqrt{L}$.

Solution

$P = 1.11\sqrt{L}$

$P = 1.11\sqrt{5}$

$P \approx 2.482035455$ On a scientific calculator, enter 1.11 × 5 $\sqrt{\ }$ = .

The period is approximately 2.5 seconds.

Self Check 2 Find the period of a pendulum that is 3 feet long. Round to the nearest tenth of a second.

Now Try **Problem 93**

EXAMPLE 3 Classify each square root as rational, irrational, or not a real number: **a.** $\sqrt{55}$ **b.** $\sqrt{-81}$ **c.** $-\sqrt{400}$

Strategy We need to determine whether the radicand is positive or negative and whether it is a perfect square.

Why If a positive number is a perfect square, its square root is rational. If a positive number is not a perfect square, its square root is irrational. The square root of a negative number is not a real number.

Solution

a. Since 55 is positive, but not a perfect square, $\sqrt{55}$ is an irrational number. If we use a calculator and round to two decimal places, we find that $\sqrt{55} \approx 7.42$.

b. $\sqrt{-81}$ is not a real number because it is the square root of a negative number.

c. Since $400 = (20)^2$, it is a perfect square and $-\sqrt{400}$ is rational: $-\sqrt{400} = -20$.

Self Check 3 Classify each square root as rational, irrational, or not a real number: **a.** $\sqrt{-6}$ **b.** $-\sqrt{37}$ **c.** $\sqrt{\frac{16}{9}}$

Now Try **Problem 51**

3 Find Square Roots of Variable Expressions.

The Language of Algebra
We can state the restriction on the variables used in this chapter in another way: All variables represent *nonnegative* real numbers.

We have seen that the square root of a negative number is not a real number. To avoid negative radicands in this chapter, we will assume that *any variables within square root symbols do not represent negative real numbers.*

Variable expressions can be perfect squares. For example, x^2, x^4, x^6, and x^8 are perfect squares because

$$x^2 = (x)^2, \quad x^4 = (x^2)^2, \quad x^6 = (x^3)^2, \quad \text{and} \quad x^8 = (x^4)^2$$

These examples illustrate an important fact: *even powers of a variable expression are perfect squares.* We can use this fact to simplify certain radical expressions.

EXAMPLE 4 Find each square root. All variables represent nonnegative real numbers. **a.** $\sqrt{x^2}$ **b.** $\sqrt{a^4}$ **c.** $\sqrt{y^{16}}$ **d.** $\sqrt{25b^6}$ **e.** $\sqrt{144t^{14}}$

Strategy In each case, we will determine what expression, when squared, produces the radicand.

Why The radical symbol $\sqrt{}$ indicates that the positive square root of the expression written under it should be found.

Solution

a. $\sqrt{x^2} = x$ Ask: What expression, when squared, is x^2? The answer is x because $(x)^2 = x^2$.

b. $\sqrt{a^4} = a^2$ Ask: What expression, when squared, is a^4? The answer is a^2 because $(a^2)^2 = a^4$.

Caution

$\sqrt{y^{16}} \neq y^4$ because $(y^4)^2 \neq y^{16}$.

Instead, $\sqrt{y^8} = y^4$ because $(y^4)^2 = y^8$.

c. $\sqrt{y^{16}} = y^8$ Ask: What expression, when squared, is y^{16}? The answer is y^8 because $(y^8)^2 = y^{16}$.

d. $\sqrt{25b^6} = 5b^3$ Ask: What expression, when squared, is $25b^6$? The answer is $5b^3$ because $(5b^3)^2 = 25b^6$.

e. $\sqrt{144t^{14}} = 12t^7$ Ask: What expression, when squared, is $144t^{14}$? The answer is $12t^7$ because $(12t^7)^2 = 144t^{14}$.

Self Check 4 Find each square root: **a.** $\sqrt{m^4}$ **b.** $\sqrt{b^2}$ **c.** $\sqrt{t^8}$ **d.** $\sqrt{81c^2}$

Now Try **Problem 67**

The results from Example 4 suggest that an easy way to find the square root of a variable raised to an even power is to *divide the exponent by 2.*

Divide 2 by 2: $\sqrt{x^2} = x^1 = x$ Divide 4 by 2: $\sqrt{a^4} = a^2$ Divide 16 by 2: $\sqrt{y^{16}} = y^8$

4 Use the Pythagorean Theorem to Solve Problems.

The Pythagorean theorem is a formula that relates the lengths of the sides of a right triangle.

The Pythagorean Theorem

If a and b are the lengths of the legs of a right triangle and c is the length of its hypotenuse,

$$a^2 + b^2 = c^2$$

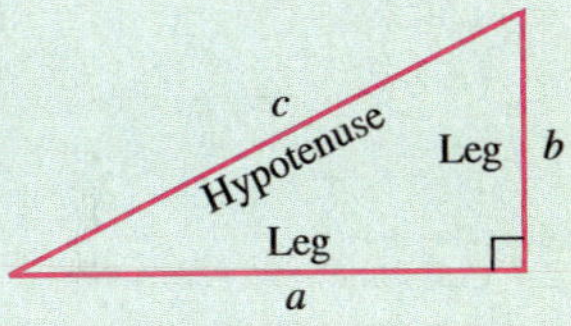

In words, the Pythagorean theorem can be stated as:

In a right triangle, the square of the hypotenuse is equal to the sum of the squares of the other two sides.

When the lengths of two sides of a right triangle are given, we can use the Pythagorean theorem and the concept of square root to find the length of the third side of the triangle.

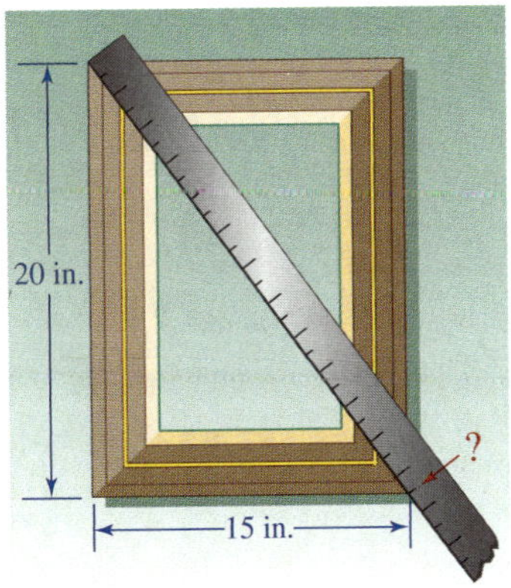

EXAMPLE 5 ***Picture Frames.*** After nailing two pieces of molding together, a frame maker checks her work by making a diagonal measurement. What should she read on the yardstick if the sides of the frame form a right angle?

Analyze the Problem The 15- and 20-inch sides of the frame in the figure are the legs and the diagonal is the hypotenuse of a right triangle. We need to find the length of the hypotenuse.

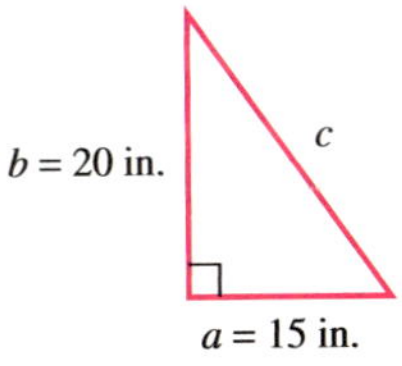

Form an Equation We can use the Pythagorean theorem to form an equation. We substitute 15 for a, 20 for b, and let c represent the length of the hypotenuse.

Solve the Equation

$a^2 + b^2 = c^2$ This is the Pythagorean theorem.

$15^2 + 20^2 = c^2$ Substitute 15 for a and 20 for b.

$225 + 400 = c^2$ $15^2 = 225$ and $20^2 = 400$.

$625 = c^2$

To find c, we ask "What number, when squared, is equal to 625?" There are two such numbers: the positive square root of 625 and the negative square root of 625. Since c represents the *length* of the hypotenuse, and it cannot be negative, it follows that c is the positive square root of 625.

$625 = c^2$

$\sqrt{625} = c$

$25 = c$ $\sqrt{625} = 25$ because $(25)^2 = 625$.

The Language of Algebra
If the square of one side of a triangle is equal to the sum of the squares of the other two sides, the triangle is a right triangle. This is called the *converse* of the Pythagorean theorem.

State the Conclusion
The diagonal measurement should be 25 inches.

Check the Result
Sides of length 15, 20, and 25 inches satisfy the Pythagorean theorem.

$15^2 + 20^2 \stackrel{?}{=} 25^2$

$225 + 400 \stackrel{?}{=} 625$

$625 = 625$ True

EXAMPLE 6

Find the missing side length of the right triangle. Give the exact length and an approximation to two decimal places.

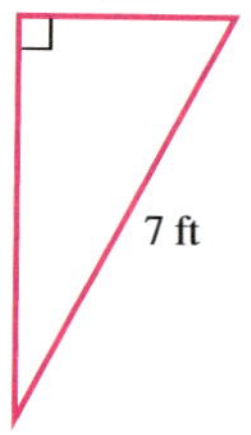

Strategy We will use the Pythagorean theorem to find the missing length.

Why The triangle is a right triangle, and we know the lengths of two of its sides.

Solution If we substitute 4 for a, 7 for c, and let b represent the unknown length of the other leg in the Pythagorean theorem formula, we have

Caution
When using the Pythagorean theorem, we can let a represent the length of either leg of the right triangle. We then let b represent the length of the other leg. But c must always represent the length of the hypotenuse.

$a^2 + b^2 = c^2$

$4^2 + b^2 = 7^2$ The length of one leg is 4 ft. The length of the hypotenuse is 7 ft.

$16 + b^2 = 49$

$b^2 = 33$ To isolate b^2, subtract 16 from both sides.

$b = \sqrt{33}$ If $b^2 = 33$, then b must be a square root of 33. Because b represents a length, it must be the positive square root of 33.

$b \approx 5.744562647$ Use a calculator to find an approximation.

The length of the leg is exactly $\sqrt{33}$ feet, which is approximately 5.74 feet.

Self Check 6
Find the length of the leg of the right triangle. Give the exact length and an approximation to two decimal places.

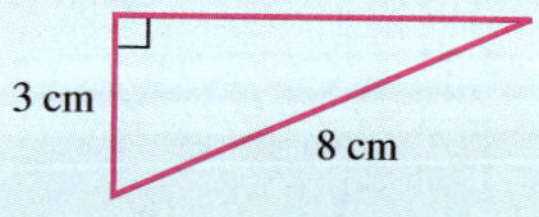

Now Try Problem 75

Recall from geometry that an **isosceles triangle** has at least two sides of equal length. An **isosceles right triangle** is a right triangle with two legs of equal length and angles that measure 45°, 45°, and 90°.

EXAMPLE 7 ***Roof Design.*** The gable end of a roof is an isosceles right triangle with a span of 48 feet. Find the distance from the eaves to the peak of the roof.

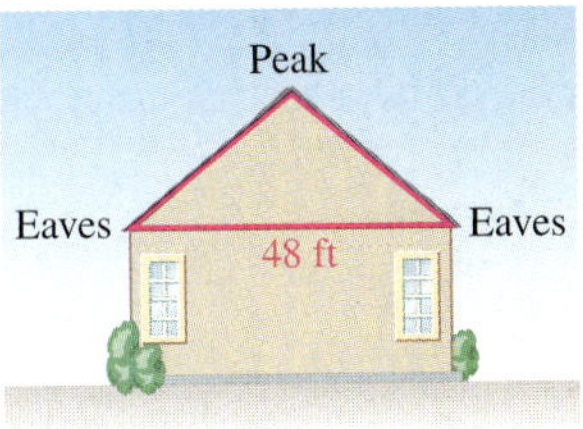

Analyze the Problem The two equal sides of the isosceles triangle form the two legs of a right triangle, and the span of 48 feet is the length of the hypotenuse.

The Language of Algebra The word *isosceles* comes from the Greek word *isokelēs*, meaning equally legged.

Form an Equation Let x represent the length (in feet) of each leg of the isosceles right triangle, which is the eaves-to-peak distance. We can use the Pythagorean theorem to form an equation.

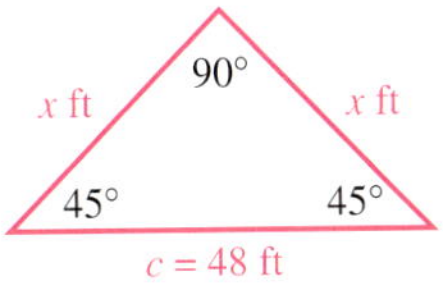

Solve the Equation

$a^2 + b^2 = c^2$ — This is the Pythagorean theorem.

$x^2 + x^2 = 48^2$ — Substitute x for a and b, and 48 for c.

$2x^2 = 2{,}304$ — Combine like terms: $x^2 + x^2 = 2x^2$.

$x^2 = 1{,}152$ — To isolate x^2, divide both sides by 2.

$x = \sqrt{1{,}152}$ — If $x^2 = 1{,}152$, then x must be a square root of 1,152. Because x represents a length, it must be the positive square root of 1,152.

$x \approx 33.9411255$ — Use a calculator to find an approximation.

State the Conclusion The eaves-to-peak distance is exactly $\sqrt{1{,}152}$ feet. Rounded to the nearest foot, the distance is approximately 34 feet.

Check the Result For an eaves-to-peak distance of 34 feet, we have $(34)^2 + (34)^2 = 1{,}156 + 1{,}156 = 2{,}312$, which is approximately $(48)^2 = 2{,}304$. The result seems reasonable.

Now Try **Problem 101**

5 Use the Distance Formula.

We can use the Pythagorean theorem to develop a formula for finding the distance between two points with coordinates (x_1, y_1) and (x_2, y_2) on a rectangular coordinate system. The distance d between the points is the length of the hypotenuse of the triangle in the figure. The two legs have lengths $x_2 - x_1$ and $y_2 - y_1$.

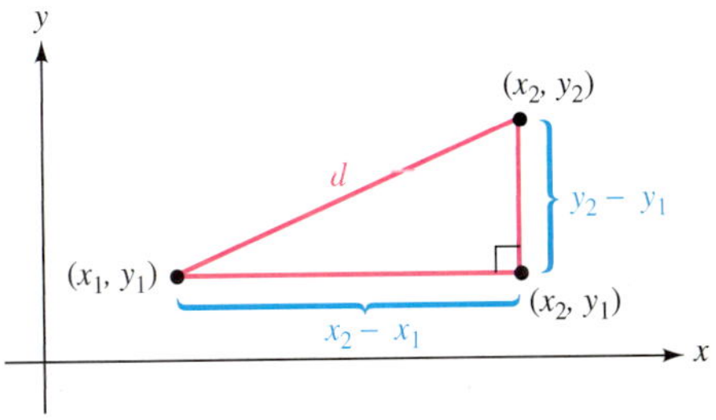

By the Pythagorean theorem, we have

$$d^2 = (x_2 - x_1)^2 + (y_2 - y_1)^2$$

Because d represents the distance between two points, it must be equal to the positive square root of $(x_2 - x_1)^2 + (y_2 - y_1)^2$. Thus,

$$d = \sqrt{(x_2 - x_1)^2 + (y_2 - y_1)^2}$$

We call this result the **distance formula.**

The Distance Formula

The distance between the points with coordinates (x_1, y_1) and (x_2, y_2) is given by

$$d = \sqrt{(x_2 - x_1)^2 + (y_2 - y_1)^2}$$

EXAMPLE 8 Find the distance between $(-4, 5)$ and $(3, -1)$.

Strategy We will substitute the coordinates into the distance formula and solve for d.

Why In the formula, d represents the distance between the points.

Solution To use the distance formula, it doesn't matter which point we call (x_1, y_1) and which point we call (x_2, y_2). If we choose $(x_1, y_1) = (-4, 5)$ and $(x_2, y_2) = (3, -1)$, we have

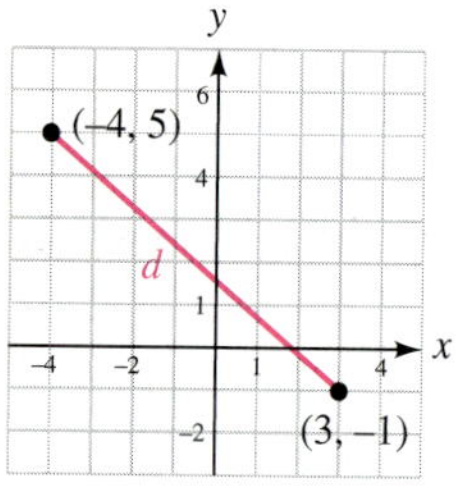

$$d = \sqrt{(x_2 - x_1)^2 + (y_2 - y_1)^2}$$ This is the distance formula.

$$d = \sqrt{[3 - (-4)]^2 + (-1 - 5)^2}$$ Substitute 3 for x_2, -4 for x_1, -1 for y_2, and 5 for y_1.

We will evaluate the entire expression under the radical symbol first, and then find the square root of that result.

$$d = \sqrt{7^2 + (-6)^2}$$ Perform each subtraction within the parentheses.

$$d = \sqrt{49 + 36}$$ Evaluate each exponential expression.

$$= \sqrt{85}$$ Do the addition.

$$\approx 9.219544457$$ Use a calculator to find an approximation.

The distance between the two points is exactly $\sqrt{85}$ units, or approximately 9.22 units.

Self Check 8 Find the distance between $(-6, 4)$ and $(-1, 2)$.

Now Try **Problem 83**

ANSWERS TO SELF CHECKS **1. a.** 7 **b.** 11 **c.** -0.3 **d.** $\frac{1}{5}$ **2.** 1.9 sec **3. a.** Not a real number **b.** Irrational **c.** Rational **4. a.** m^2 **b.** b **c.** t^4 **d.** $9c$ **6.** $\sqrt{55}$ cm $\approx$ 7.42 cm **8.** $\sqrt{29} \approx 5.39$

STUDY SET 8.1

VOCABULARY

Fill in the blanks.

1. The number b is a ______ root of the number a if $b^2 = a$.

2. The symbol $\sqrt{}$ is called a ______ symbol. It represents the ______ or principal square root of a number. The symbol $-\sqrt{}$ is used to represent the ______ square root of a number.

3. The number or variable expression under a radical symbol is called the ______.

4. A number such as 25, 49, or $\frac{4}{81}$ that is the square of some rational number, is called a ______ square.

5. The ______ theorem is a formula that relates the lengths of the sides of a right triangle:

$$\boxed{} + \boxed{} = \boxed{}$$

6. A(n) ______ right triangle is shown here.

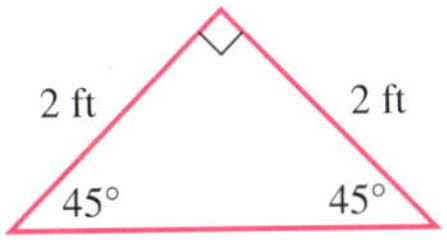

CONCEPTS

Fill in the blanks.

7. Every positive number has ______ square roots, one positive and one negative. The positive square root of 25 is ______ and the negative square root of 25 is ______. In symbols, we write

$$\sqrt{25} = \boxed{} \quad \text{and} \quad -\sqrt{25} = \boxed{}$$

8. To find the positive square root of 36, we ask "What positive number, when ______, is equal to 36?"

9. The number 0 is the only real number with exactly one square root: $\sqrt{0} = \boxed{}$

10. $\sqrt{4} = 2$ because $(\boxed{})^2 = 4$ and $\sqrt{x^4} = x^2$ because $(\boxed{})^2 = x^4$.

11. Identify the radicand of each radical expression.

a. $\sqrt{64}$ **b.** $-\sqrt{36y^4}$

c. $2\sqrt{d}$ **d.** $\sqrt{\frac{16}{25}}$

12. Determine whether each statement is true or false.

a. $-\sqrt{36} = \sqrt{-36}$ **b.** $\sqrt{m^{16}} = m^4$

13. Fill in the blanks: $\sqrt{29}$ must be a number between 5 and 6, because $\sqrt{\boxed{}} < \sqrt{29} < \sqrt{\boxed{}}$.

14. Evaluate: $\sqrt{(5-2)^2 + (8-4)^2}$

NOTATION

Complete the solution.

15. The legs of a right triangle measure 5 and 12 centimeters. Find the length of the hypotenuse.

$$\begin{aligned} a^2 + b^2 &= c^2 \\ \boxed{}^2 + 12^2 &= c^2 \\ 25 + \boxed{} &= c^2 \\ \boxed{} &= c^2 \\ \sqrt{169} &= \boxed{} \\ \boxed{} &= c \end{aligned}$$

The length of the hypotenuse is ______ cm.

16. Fill in the blanks. The distance formula is:

$$d = \sqrt{(x_2 - \boxed{})^2 + (\boxed{} - y_1)^2}$$

GUIDED PRACTICE

Evaluate each square root without using a calculator. **See Example 1.**

17. $\sqrt{64}$ **18.** $\sqrt{25}$

19. $\sqrt{36}$ **20.** $\sqrt{1}$

21. $\sqrt{100}$ **22.** $\sqrt{144}$

23. $\sqrt{400}$ **24.** $\sqrt{900}$

25. $-\sqrt{81}$ **26.** $-\sqrt{36}$

27. $\sqrt{1.21}$ **28.** $\sqrt{1.69}$

29. $\sqrt{169}$ **30.** $\sqrt{196}$

31. $\sqrt{\frac{9}{25}}$ **32.** $\sqrt{\frac{9}{49}}$

33. $-\sqrt{\frac{1}{64}}$ **34.** $-\sqrt{\frac{1}{16}}$

35. $-\sqrt{0.04}$ **36.** $-\sqrt{0.64}$

37. $-\sqrt{289}$ **38.** $-\sqrt{324}$

39. $\sqrt{2,500}$ **40.** $\sqrt{625}$

Use a calculator to approximate each square root to three decimal places. **See Example 2.**

41. $\sqrt{3}$ **42.** $\sqrt{2}$

43. $\sqrt{95}$ **44.** $\sqrt{99}$

45. $\sqrt{428}$ **46.** $\sqrt{844}$

47. $2\sqrt{3}$ **48.** $3\sqrt{2}$

Classify each square root as rational, irrational, or not a real number. **See Example 3.**

49. $\sqrt{9}$ **50.** $\sqrt{17}$

51. $\sqrt{-21}$ **52.** $\sqrt{99}$

53. $\sqrt{33}$ **54.** $-\sqrt{\frac{81}{121}}$

55. $-\sqrt{0.25}$ **56.** $\sqrt{-47}$

Find each square root. All variables represent nonnegative real numbers. **See Example 4.**

57. $\sqrt{m^2}$ **58.** $\sqrt{b^2}$

59. $\sqrt{t^4}$ **60.** $\sqrt{a^6}$

61. $\sqrt{c^{10}}$ **62.** $\sqrt{r^8}$

63. $\sqrt{n^{12}}$ **64.** $\sqrt{t^{18}}$

65. $\sqrt{x^{36}}$ **66.** $\sqrt{y^{64}}$

67. $\sqrt{4y^2}$ **68.** $\sqrt{9b^2}$

69. $\sqrt{64b^4}$ **70.** $\sqrt{100s^{16}}$

71. $-\sqrt{49s^{14}}$ **72.** $-\sqrt{4a^4}$

Refer to the right triangle. **See Examples 5 and 6.**

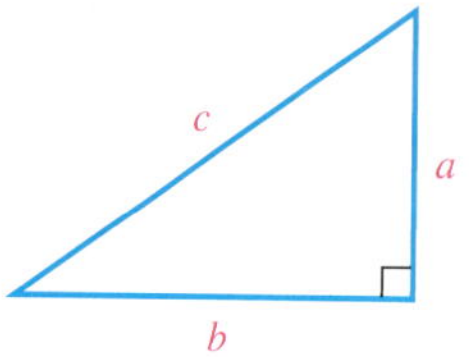

73. Find c if $a = 4$ and $b = 3$.

74. Find c if $a = 5$ and $b = 12$.

75. Find a if $b = 15$ and $c = 17$.

76. Find a if $b = 16$ and $c = 34$.

77. Find b if $a = 45$ and $c = 53$.

78. Find a if $b = 7$ and $c = 25$.

79. Find b if $c = 125$ and $a = 44$.

80. Find c if $a = 176$ and $b = 57$.

For each isosceles right triangle, find the length of the legs. Give an exact answer and an approximation to two decimal places. **See Example 7.**

81.

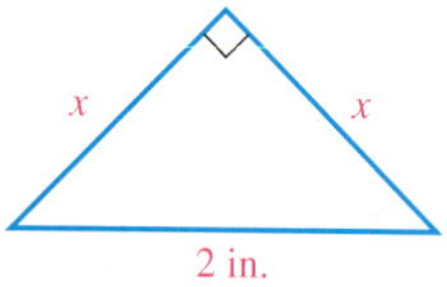

82.

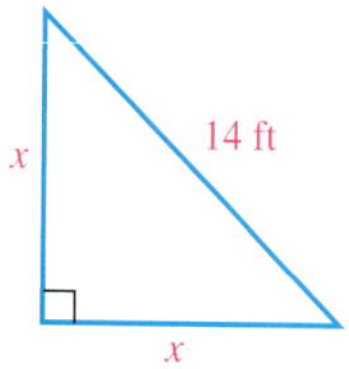

Find the distance between the two points. If an answer contains a radical, give an exact answer and an approximation to two decimal places. **See Example 8.**

83. (4, 6) and (1, 2)

84. (9, 8) and (3, 0)

85. (−2, −8) and (3, 4)

86. (−5, −2) and (7, 3)

87. (6, −5) and (0, −4)

88. (−2, −5) and (6, −6)

89. (9, −1) and (7, −6)

90. (5, −6) and (−1, 1)

APPLICATIONS

If an answer contains a radical, give an exact answer and an approximation to two decimal places. **See Examples 2, 5, and 7.**

91. ***from Campus to Careers***

Crime Scene Investigator

© "CSI: Crime Scene Investigation," CBS/Landov

Suppose you are a crime scene investigator and you need to determine whether the driver of a crashed car was speeding prior to an accident. You measure the car's skid marks on the pavement and find them to be 144 feet long. The formula $s = 4.5\sqrt{d}$ can be used to determine how fast the car was traveling. The variable s represents the speed (in mph) of the car and d the distance (in feet) of the skid when the driver hit the brakes. If the speed limit for the street was 35 mph, was the driver speeding? If so, by how many mph?

92. FREE FALL The time t (in seconds) that it takes an object dropped from a distance d (in feet) to fall to the ground is given by the formula $t = 0.25\sqrt{d}$. How long will it take a pair of sunglasses to fall to the ground if a sightseer drops them from the top of a 400-foot-tall observation tower?

93. BOATING To approximate the maximum speed (in knots) for certain types of boats, navigators use the formula $s = 1.34\sqrt{L}$, where s represents the maximum speed of the boat and L represents its length (in feet) at the waterline. Find the approximate maximum speed of the boat shown below.

94. CROSSWORD PUZZLES If the area of a square is A square units, then the length of a side of the square is $\sqrt{A}$ units. The world's largest crossword puzzle can be purchased from the *Hammacher Schlemmer Gift Catalog* for $29.95. The crossword puzzle has a square shape and can be hung on a wall. If it covers 7,056 square inches of wall space, find the length of one of its sides. Answer in inches and also in feet.

95. DRAFTING Among the tools used in drafting are the 30-60-90 and the 45-45-90 triangles shown.

a. Find the length of the hypotenuse of the 45-45-90 triangle if it is $\sqrt{2}$ times as long as a leg.

b. Find the length of the side opposite the 60° angle of the 30-60-90 triangle if it is $\sqrt{3}$ times as long as one-half of the hypotenuse.

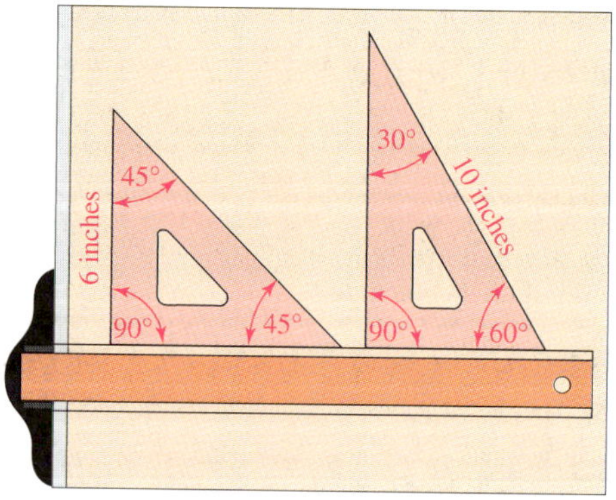

96. GARDENING A rectangular garden has sides of 28 and 45 feet. Find the length of a path that extends from one corner to the opposite corner.

97. BASEBALL A baseball diamond is a square, with each side 90 feet long. How far is it from home plate to second base?

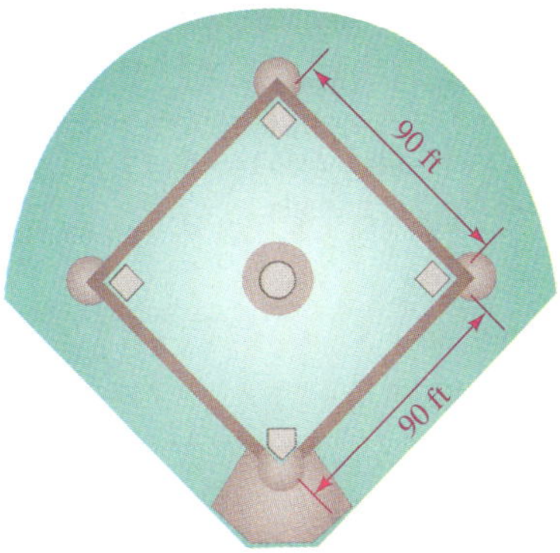

98. LCD FLAT PANEL HDTV The size of a television screen is the diagonal distance from the upper left to the lower right corner. What is the size of the screen shown in the illustration? Round to the nearest inch.

99. CARPENTRY A carpenter's square is made of flat steel or aluminum and is shaped like an L. It is used for framing, roofing, and stairway work.

a. How long is each side of the carpenter's square shown below? (The units are inches.)

b. What should be the diagonal measurement (shown in red) that will guarantee that the sides of the square form a 90° angle?

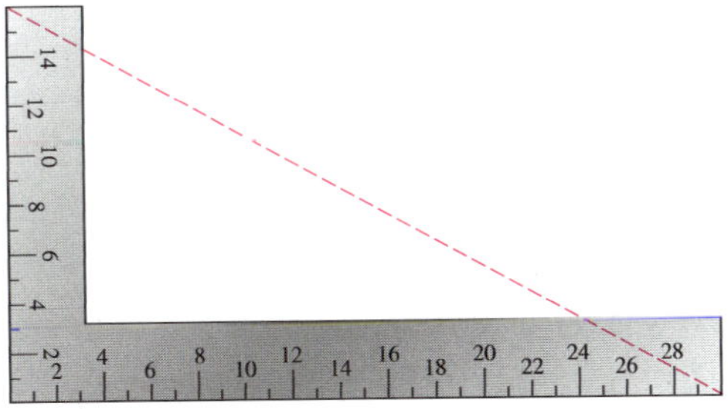

100. BOXING The diagonal distance from corner to corner of a square boxing ring is 24 feet. Find the length of a side of the ring.

101. FOOTBALL On first down and ten, a quarterback tells his tight end to go out 6 yards, cut 45° to the right, and run 6 yards, as shown in the illustration. The tight end follows instructions, catches a pass, and is tackled immediately.

a. Find x.

b. Does he gain the necessary 10 yards for a first down?

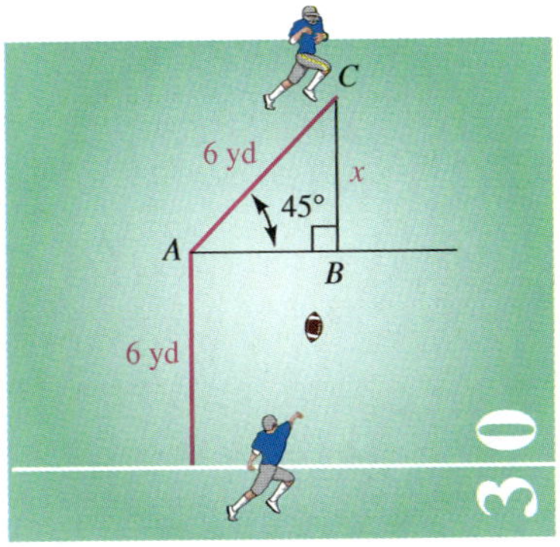

102. THE WIZARD OF OZ In the 1939 classic movie, the Scarecrow was in search of a brain. Once he received an honorary degree in "Thinkology" from the Wizard, he tried to impress his friends by stating: "*The sum of the square roots of any two sides of an isosceles triangle is equal to the square root of the remaining side.*" What well-known mathematical fact was the Scarecrow attempting to recite? Explain the errors that he made. (Over 50 years later, in an episode of *The Simpsons,* Homer quotes the Scarecrow word-for-word after finding a pair of eyeglasses in a public restroom.)

103. DECK DESIGNS The plans for a patio deck shown in the illustration call for three redwood braces directly under the hot tub. Use the distance formula to find the length of each brace. Round to the nearest tenth of a meter.

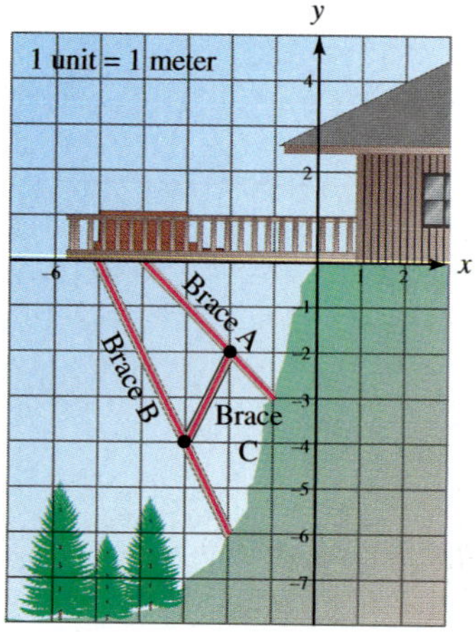

104. NAVIGATION A ship is sailing from Italy to Tunisia. The captain wants to travel a course that is always the same distance from a point on the coast of Sardinia as it is from a point on the coast of Sicily. (See the illustration below.) Determine if the ship is on the correct course if its current position is (9, 6).

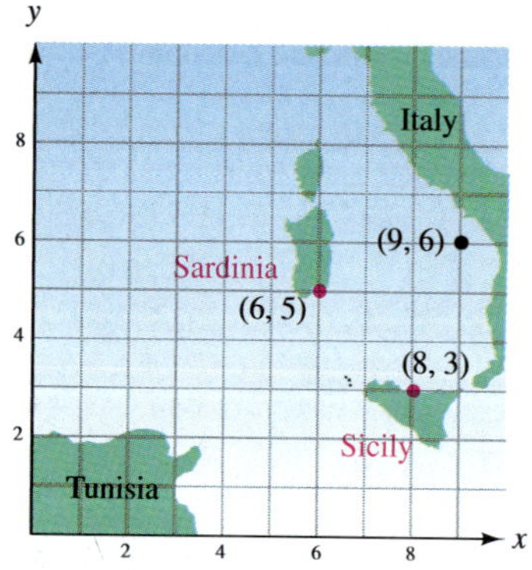

WRITING

105. Consider the statement $\sqrt{26} \approx 5.1$. Explain why an $\approx$ symbol is used instead of an $=$ symbol.

106. Suppose you are told that $\sqrt{10} \approx 3.16$. Explain how another key on your calculator (besides the square root key $\sqrt{}$) could be used to see whether this is a reasonable approximation.

107. A calculator was used to find $\sqrt{-16}$. Explain the meaning of the message on the display.

108. Explain why the Pythagorean theorem does not hold for this triangle.

$$3^2 + 4^2 = 6^2$$
$$9 + 16 \stackrel{?}{=} 36$$
$$25 \neq 36$$

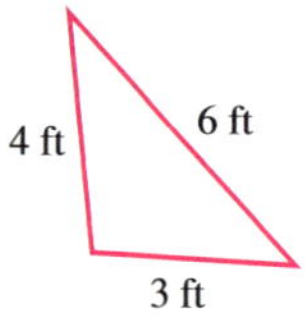

REVIEW

109. Add: $(3s^2 - 3s - 2) + (3s^2 + 4s - 3)$

110. Subtract: $(3c^2 - 2c + 4) - (c^2 - 3c + 7)$

111. Multiply: $(3x - 2)(x + 4)$

112. Divide: $x^2 + 13x + 12$ by $x + 1$

CHALLENGE PROBLEMS

113. SHORTCUTS Instead of walking on the sidewalk, students take a diagonal shortcut across a rectangular vacant lot, as shown in the illustration. How much distance do they save?

114. Evaluate: $\sqrt{\sqrt{\sqrt{256}}}$

115. Simplify $\sqrt{16x^{16n}}$, where n is a positive integer.

116. Graph $y = \sqrt{x}$ by first completing the table of solutions. Then plot the points and draw a smooth line through them.

x	0	1	3	4	5	9	12	16
y								

SECTION 8.2
Simplifying Square Roots

Objectives

1. Use the product rule to simplify square roots.
2. Use prime factorization to simplify square roots.
3. Simplify square roots of variable expressions.
4. Use the quotient rule to simplify square roots.

Square roots such as $\sqrt{9}$ and $\sqrt{16}$ are easy to find because their radicands are perfect squares. Square roots with radicands that are not perfect squares, such as $\sqrt{12}$ and $\sqrt{200}$, can be approximated using a calculator, or we can write them in a simpler exact form using the *product rule for square roots.*

1 Use the Product Rule to Simplify Square Roots.

To introduce the product rule for square roots, we will find $\sqrt{4 \cdot 25}$ and $\sqrt{4}\sqrt{25}$, and compare the results.

Notation
The product $\sqrt{4}\sqrt{25}$ can also be written as $\sqrt{4} \cdot \sqrt{25}$.

Square root of a product	***Product of square roots***
$\sqrt{4 \cdot 25} = \sqrt{100}$	$\sqrt{4}\sqrt{25} = 2 \cdot 5$
$= 10$	$= 10$

Read as "the square root of 4 times the square root of 25."

In each case, the answer is 10. Thus, $\sqrt{4 \cdot 25} = \sqrt{4}\sqrt{25}$. This result illustrates the *product rule for square roots.*

The Product Rule for Square Roots

The square root of the product of two nonnegative numbers is equal to the product of their square roots.

For any nonnegative real numbers a and b,

$$\sqrt{a \cdot b} = \sqrt{a}\sqrt{b}$$

A square root is in **simplified form** when each of the following is true.

Simplified Form of a Square Root

1. Except for 1, the radicand has no perfect-square factors.
2. No fraction appears in the radicand.
3. No radical appears in the denominator of a fraction.

To simplify square roots, we follow these steps.

Simplifying Square Roots

1. Write the radicand as a product of the greatest perfect-square factor and one other factor.
2. Use the product rule for square roots to write the expression as a product of square roots.
3. Find the square root of the perfect-square factor.

Recall th
squares i
factor dif

The La
When ref
the instru
remove a
from the

The La
The instr
approxim
same thi

Simplify:
Approxim

Self Check 8 Simplify: $\sqrt{\frac{99b^4}{16b^2}}$

Now Try Problem 81

ANSWERS TO SELF CHECKS 1. $3\sqrt{5}$ 2. $6\sqrt{2}$ 3. a. $2\sqrt{35}$ b. Cannot be simplified 4. a. $x^2\sqrt{x}$ b. $b^5\sqrt{b}$ 5. $36\sqrt{2y}$ 6. a. $7p^4\sqrt{p}$ b. $5y\sqrt{3y}$ 7. a. $\frac{6}{11}$ b. $\frac{\sqrt{29}}{8}$ c. $\frac{6\sqrt{2}}{13}$ 8. $\frac{3b\sqrt{11}}{4}$

STUDY SET 8.2

VOCABULARY

Fill in the blanks.

1. "To ________ $\sqrt{20}$" means to write it as $2\sqrt{5}$.
2. The perfect-square factors of 200 are 4, 25, and 100. Of these, 100 is the ________ perfect-square factor of 200.
3. The expression $\sqrt{4 \cdot 3}$ is a square root of a ________ and $\sqrt{4}\sqrt{3}$ is a product of two ________ roots.
4. The expression $\sqrt{\frac{40}{9}}$ is a square root of a ________ and $\frac{\sqrt{40}}{\sqrt{9}}$ is a quotient of two ________ roots.

CONCEPTS

5. Fill in the blanks.
   a. The square root of the product of two positive numbers is equal to the ________ of their square roots. In symbols,

   $\sqrt{a \cdot b} =$ ________

   b. The square root of the quotient of two positive numbers is equal to the ________ of their square roots. In symbols,

   $\sqrt{\frac{a}{b}} =$ ________

6. Complete each list of perfect squares.
   a. 1, 4, 9, ____, 25, 36, ____, 64, 81, ____, 121, ____, . . .
   b. x^2, ____, x^6, ____, ____, x^{12}, ____, . . .
7. Which perfect square, 1, 4, 9, 16, 25, 36, 49, 64, 81, or 100, is the greatest perfect-square factor of the given number?
   a. 20
   b. 98
   c. 54
   d. 48
8. Complete each factorization using a perfect square.
   a. $24 = \_\_ \cdot 6$
   b. $63 = \_\_ \cdot 7$
   c. $r^5 = \_\_ \cdot r$
   d. $t^9 = \_\_ \cdot t$
9. To simplify $\sqrt{40}$, which one of the following factorizations should be used?

   $\sqrt{20 \cdot 2}$ $\sqrt{40 \cdot 1}$ $\sqrt{8 \cdot 5}$ $\sqrt{4 \cdot 10}$

10. To simplify $\sqrt{n^5}$, which one of the following factorizations should be used?

    $\sqrt{n^2 \cdot n^3}$ $\sqrt{n^3 \cdot n^2}$ $\sqrt{n^4 \cdot n}$

11. Use the square root of a product property to simplify each expression.
    a. $\sqrt{81 \cdot 2} = \sqrt{\_\_}\sqrt{\_\_} = \_\_\sqrt{\_\_}$
    b. $\sqrt{m^4 \cdot m} = \sqrt{\_\_}\sqrt{\_\_} = \_\_\sqrt{\_\_}$
12. Use the square root of a quotient property to simplify the expression.

    $\sqrt{\frac{17}{25}} = \frac{\sqrt{\_\_}}{\sqrt{\_\_}} = \frac{\sqrt{\_\_}}{\_\_}$

NOTATION

13. We can read $2\sqrt{6}$ as "2 ________ the square root of 6" or as "2 ________ 6."
14. Write each radical expression in better form.
    a. $3 \cdot 2\sqrt{15}$
    b. $\sqrt{7} \cdot 9$

GUIDED PRACTICE

Simplify, if possible. **See Examples 1–2.**

15. $\sqrt{20}$
16. $\sqrt{18}$
17. $\sqrt{27}$
18. $\sqrt{28}$
19. $\sqrt{50}$
20. $\sqrt{54}$
21. $\sqrt{24}$
22. $\sqrt{88}$
23. $\sqrt{48}$
24. $\sqrt{32}$
25. $\sqrt{44}$
26. $\sqrt{60}$
27. $\sqrt{15}$
28. $\sqrt{21}$
29. $\sqrt{63}$
30. $\sqrt{75}$

Simplify. **See Example 3.**

31. $\sqrt{98}$
32. $\sqrt{128}$
33. $\sqrt{180}$
34. $\sqrt{147}$

35. $\sqrt{192}$
36. $\sqrt{216}$
37. $\sqrt{375}$
38. $\sqrt{405}$

***Simplify.* All variables represent nonnegative real numbers. See Examples 4–5.**

39. $\sqrt{x^7}$
40. $\sqrt{n^3}$
41. $\sqrt{n^9}$
42. $\sqrt{t^7}$
43. $\sqrt{2d^{11}}$
44. $\sqrt{3x^9}$
45. $\sqrt{4k}$
46. $\sqrt{9p}$
47. $\sqrt{9y}$
48. $\sqrt{16x}$
49. $\sqrt{10b}$
50. $\sqrt{30s}$
51. $4\sqrt{12x}$
52. $9\sqrt{20y}$
53. $5\sqrt{54q}$
54. $7\sqrt{18k}$

***Simplify.* All variables represent nonnegative real numbers. See Example 6.**

55. $\sqrt{24t^3}$
56. $\sqrt{27p^3}$
57. $\sqrt{32x^5}$
58. $\sqrt{52y^5}$
59. $2\sqrt{75x^7}$
60. $3\sqrt{98y^7}$
61. $16\sqrt{18t^{11}}$
62. $4\sqrt{50m^{11}}$

***Simplify.* See Example 7.**

63. $\sqrt{\frac{25}{9}}$
64. $\sqrt{\frac{36}{49}}$
65. $\sqrt{\frac{81}{64}}$
66. $\sqrt{\frac{121}{144}}$
67. $\sqrt{\frac{6}{121}}$
68. $\sqrt{\frac{15}{64}}$
69. $\sqrt{\frac{23}{64}}$
70. $\sqrt{\frac{35}{144}}$
71. $\sqrt{\frac{20}{49}}$
72. $\sqrt{\frac{50}{9}}$
73. $\sqrt{\frac{75}{16}}$
74. $\sqrt{\frac{32}{25}}$

***Simplify.* All variables represent positive real numbers. See Example 8.**

75. $\sqrt{\frac{a^4}{4a}}$
76. $\sqrt{\frac{n^4}{81n}}$
77. $\sqrt{\frac{r^{10}}{225r}}$
78. $\sqrt{\frac{m^4}{225m}}$
79. $\sqrt{\frac{72x^3}{x}}$
80. $\sqrt{\frac{108b^3}{b}}$
81. $\sqrt{\frac{125n^5}{64n}}$
82. $\sqrt{\frac{72q^7}{25q^3}}$

TRY IT YOURSELF

***Simplify.* All variables represent positive real numbers.**

83. $\sqrt{75t}$
84. $\sqrt{24s}$
85. $\sqrt{\frac{48}{81}}$
86. $\sqrt{\frac{27}{64}}$
87. $\sqrt{h^7}$
88. $\sqrt{s^9}$
89. $3\sqrt{72}$
90. $2\sqrt{28}$
91. $\sqrt{\frac{75q^2}{16q^4}}$
92. $\sqrt{\frac{128n^5}{81n^7}}$
93. $\sqrt{162}$
94. $\sqrt{486}$
95. $12\sqrt{16y^5}$
96. $4\sqrt{36y^5}$
97. $\sqrt{50x^4}$
98. $\sqrt{75y^4}$

APPLICATIONS

99. STUDYING PAST CULTURES A coordinate system of 1 ft × 1 ft squares was constructed from string to record the location of objects found during an excavation at a dig site. Use the distance formula (or the Pythagorean theorem) to determine the exact distance between a piece of pottery found at point A and an arrowhead found at point B. Express the answer in simplified radical form. Then approximate the distance to the nearest hundredth of a foot.

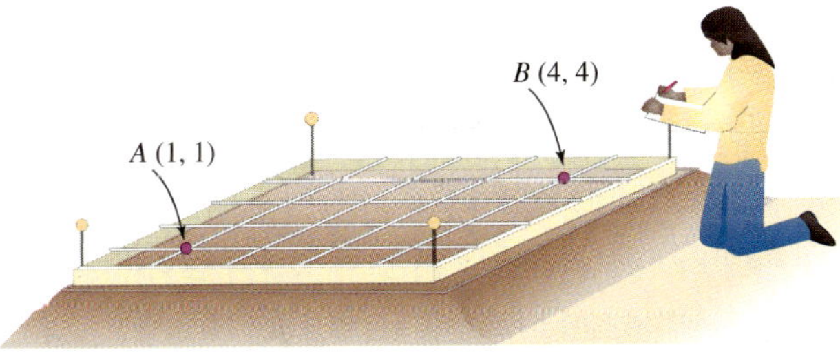

100. SQUARE DANCING A dance floor is a square with sides of length 12 yards. Use the Pythagorean theorem to find the exact length of the diagonal shown in the illustration. Express the answer in simplified radical form. Then approximate the length to the nearest yard.

101. SQUARE ROOT SPIRAL The pattern shown below is formed by a collection of right triangles, where the hypotenuse of one becomes a leg of the next. The length of each segment in red is 1 unit. Simplify any square roots that are not in simplified form.

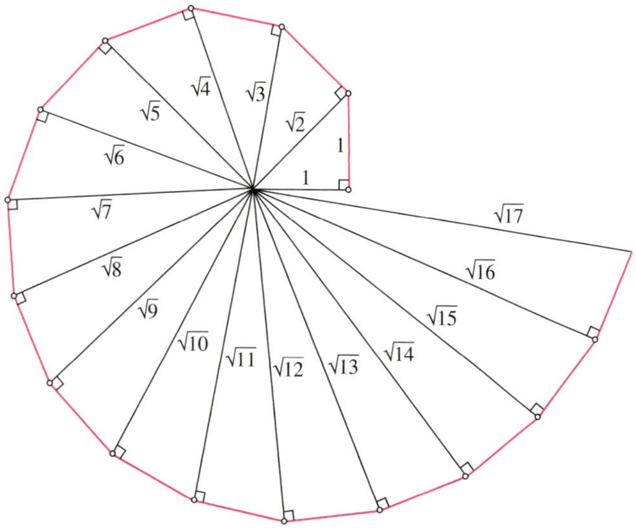

102. AMUSEMENT PARK RIDES The time t (in seconds) that it takes the "Swashbuckler" pirate ship ride to swing from one extreme to the other is given by the formula $t = \pi\sqrt{\frac{L}{32}}$. Use the value of L from the illustration to find t and express it in simplified radical form. Leave π in the answer. Then approximate the time to the nearest tenth of a second.

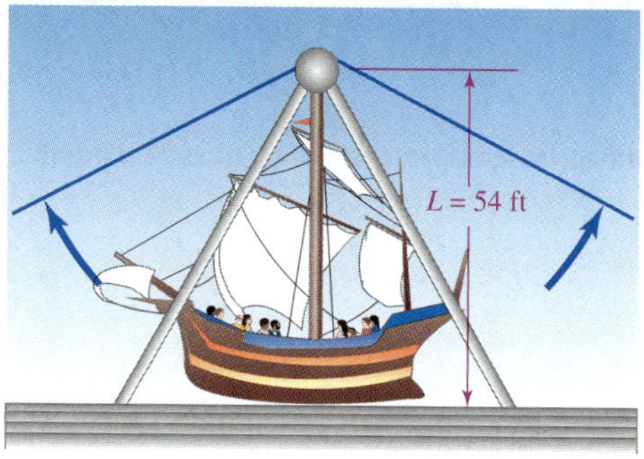

WRITING

103. Explain why each square root is not in simplified form.

a. $\sqrt{8}$ **b.** $\sqrt{\frac{7}{4}}$

104. Explain each property in words.

a. $\sqrt{a \cdot b} = \sqrt{a}\sqrt{b}$ **b.** $\sqrt{\frac{a}{b}} = \frac{\sqrt{a}}{\sqrt{b}}$

105. Explain why we cannot simplify $\sqrt{42}$.

106. Does $\sqrt{9x^9} = 3x^3$? Explain why or why not.

REVIEW

Solve each system.

107. $\begin{cases} y = 2x - 6 \\ 2x + y = 6 \end{cases}$

108. $\begin{cases} 2x + y = -2 \\ -2x - 3y = -6 \end{cases}$

109. $\begin{cases} 3x + 4y = -7 \\ 2x - y = -1 \end{cases}$

110. $\begin{cases} 2x + 3y = 8 \\ 3x - 2y = -1 \end{cases}$

CHALLENGE PROBLEMS

111. Simplify. All variables represent positive numbers.

$$\sqrt{\frac{196a^8b^5c^9}{50abc^8}}$$

112. Find the radicand: $\sqrt{} = 2a^4b\sqrt{7a}$

SECTION 8.3
Adding and Subtracting Radical Expressions

Objectives

1. Add and subtract square roots.
2. Simplify square roots in a sum or difference.

We have learned how to combine like terms. In this section, we will use a similar procedure to combine *like radicals*.

1 Add and Subtract Square Roots.

Recall that we use the distributive property to simplify a sum or difference of like terms. For example,

$$3x + 5x = (3 + 5)x \qquad \text{and} \qquad 9y - 6y = (9 - 6)y$$
$$= 8x \qquad\qquad\qquad\qquad = 3y$$

In each case, we say that we have *combined like terms.*

The distributive property can also be used to simplify certain sums and differences of square roots. For example,

$$3\sqrt{2} + 5\sqrt{2} = (3 + 5)\sqrt{2} \qquad \text{and} \qquad 9\sqrt{15} - 6\sqrt{15} = (9 - 6)\sqrt{15}$$
$$= 8\sqrt{2} \qquad\qquad\qquad\qquad = 3\sqrt{15}$$

In the above examples, the terms $3\sqrt{2}$ and $5\sqrt{2}$ and the terms $9\sqrt{15}$ and $6\sqrt{15}$ are called **like radical terms.** In these examples, we say that we have *combined like radical terms,* or more simply, *combined like radicals.*

Like Radicals Square root radical terms are said to be **like radicals** when they have the same radicand.

Like radicals	*Unlike radicals*
$9\sqrt{6}$ and $4\sqrt{6}$	$8\sqrt{5}$ and $5\sqrt{3}$
The same radicand	Different radicands

EXAMPLE 1 Simplify, if possible, by combining like radicals:

a. $4\sqrt{7} + 6\sqrt{7}$ **b.** $2\sqrt{m} - 3\sqrt{m}$

c. $5\sqrt{14} + \sqrt{14} - 9\sqrt{14}$ **d.** $\sqrt{10} + \sqrt{6}$

Strategy First, we will see if the radicands of the square roots are the same. If they are, we can use the distributive property in reverse to add (or subtract) like radicals.

Why We must check the radicands first because only square roots with the same radicand can be combined.

Solution

a. Since $4\sqrt{7}$ and $6\sqrt{7}$ are like radicals, we can add them.

$$4\sqrt{7} + 6\sqrt{7} = 10\sqrt{7}$$ To combine like radicals, think: $(4 + 6)\sqrt{7} = 10\sqrt{7}$.

b. Since $2\sqrt{m}$ and $3\sqrt{m}$ are like radicals, we can subtract them.

$$2\sqrt{m} - 3\sqrt{m} = -1\sqrt{m}$$ Think: $(2 - 3)\sqrt{m} = -1\sqrt{m}$.

$$= -\sqrt{m}$$ Because $-1\sqrt{m} = -\sqrt{m}$.

c. Since each radicand is 14, the three radical terms are like radicals.

$$5\sqrt{14} + \sqrt{14} - 9\sqrt{14} = 5\sqrt{14} + 1\sqrt{14} - 9\sqrt{14}$$ $\sqrt{14} = 1\sqrt{14}$

$$= -3\sqrt{14}$$ Think: $(5 + 1 - 9)\sqrt{14} = -3\sqrt{14}$.

d. Since $\sqrt{10}$ and $\sqrt{6}$ have different radicands, they are unlike radicals and we cannot add them. The expression $\sqrt{10} + \sqrt{6}$ does not simplify further.

Success Tip

Just as 4 apples plus 6 apples is 10 apples, we see that 4 square roots of 7 plus 6 square roots of 7 is 10 square roots of 7.

$$4\sqrt{7} + 6\sqrt{7} = 10\sqrt{7}$$

Caution
We have discussed the product and quotient rules for square roots. There is no sum nor difference rule for square roots. In general,

$$\sqrt{a} + \sqrt{b} \neq \sqrt{a+b}$$
$$\sqrt{a} - \sqrt{b} \neq \sqrt{a-b}$$

CAUTION A common error is to add the radicands to simplify $\sqrt{10} + \sqrt{6}$ and get $\sqrt{16}$. However, the sum of two square roots does not equal the square root of the sum. To see why, we can approximate $\sqrt{10}$ with 3.16 and $\sqrt{6}$ with 2.45 and add. The sum is not equal to $\sqrt{16}$, which is 4.

$$\sqrt{10} + \sqrt{6} \neq \sqrt{16} \quad \text{since} \quad 3.16 + 2.45 \neq 4$$

Self Check 1 Simplify, if possible: **a.** $20\sqrt{5} + 30\sqrt{5}$ **b.** $6\sqrt{b} - 4\sqrt{b}$ **c.** $\sqrt{21} + 4\sqrt{21} - 12\sqrt{21}$ **d.** $\sqrt{29} - \sqrt{2}$

Now Try **Problems 9, 17, 25, and 31**

2 Simplify Square Roots in a Sum or Difference.

If a sum or difference involves unlike radicals, make sure that each one is written in simplified form. After doing so, like radicals may result that can be combined.

EXAMPLE 2 Perform the indicated operations:
a. $\sqrt{18} + \sqrt{8}$ **b.** $\sqrt{108} - 8\sqrt{12} + 2\sqrt{27}$

Strategy Since the radicals in each part are unlike radicals, we cannot add or subtract them in their current form. However, we will simplify the radicals and hope that like radicals result.

Why Only like radicals can be combined.

Solution

a. Recall that to write a square root in simplified form we express the radicand as a product of the greatest perfect-square factor and one other factor. Then we use the product rule for square roots.

$\sqrt{18} + \sqrt{8} = \sqrt{9 \cdot 2} + \sqrt{4 \cdot 2}$ — Factor 18 and 8 using their greatest perfect square factors.

$= \sqrt{9}\sqrt{2} + \sqrt{4}\sqrt{2}$ — The square root of a product is equal to the product of the square roots.

$= 3\sqrt{2} + 2\sqrt{2}$ — Evaluate: $\sqrt{9} = 3$ and $\sqrt{4} = 2$.

$= 5\sqrt{2}$ — To combine like radicals, think: $(3 + 2)\sqrt{2} = 5\sqrt{2}$.

The Language of Algebra
The phrase *combining like radicals* refers to the operations of addition and subtraction.

b. When we write each radical in simplified form, the results are three like radicals.

$\sqrt{108} - 8\sqrt{12} + 2\sqrt{27} = \sqrt{36 \cdot 3} - 8\sqrt{4 \cdot 3} + 2\sqrt{9 \cdot 3}$ — Factor using the greatest perfect squares.

$= \sqrt{36}\sqrt{3} - 8\sqrt{4}\sqrt{3} + 2\sqrt{9}\sqrt{3}$ — The square root of a product is equal to the product of the square roots.

$= 6\sqrt{3} - 8 \cdot 2\sqrt{3} + 2 \cdot 3\sqrt{3}$ — Evaluate: $\sqrt{36} = 6$, $\sqrt{4} = 2$, and $\sqrt{9} = 3$.

$= 6\sqrt{3} - 16\sqrt{3} + 6\sqrt{3}$ — Do the multiplications.

$= -4\sqrt{3}$ — Think: $(6 - 16 + 6)\sqrt{3} = -4\sqrt{3}$.

Self Check 2 Perform the indicated operations: **a.** $\sqrt{75} + \sqrt{12}$
b. $3\sqrt{50} - 6\sqrt{98} - \sqrt{128}$

Now Try **Problems 33 and 45**

EXAMPLE 3

Perform the indicated operations, if possible:

a. $\sqrt{44x^2} + x\sqrt{99}$ **b.** $\sqrt{27x} - \sqrt{20x}$

Strategy Since the radicals in each part are unlike radicals, we cannot add or subtract them in their current form. However, we will simplify the radicals and hope that like radicals result.

Why Only like radicals can be combined.

Solution

a. $\sqrt{44x^2} + x\sqrt{99}$

$= \sqrt{4x^2 \cdot 11} + x\sqrt{9 \cdot 11}$ Factor $44x^2$ and 99.

$= \sqrt{4x^2}\sqrt{11} + x\sqrt{9}\sqrt{11}$ The square root of a product is equal to the product of the square roots.

$= 2x\sqrt{11} + 3x\sqrt{11}$ $\sqrt{4x^2} = 2x$ and $\sqrt{9} = 3$.

$= 5x\sqrt{11}$ To combine like radicals, think: $(2x + 3x)\sqrt{11} = 5x\sqrt{11}$.

b. $\sqrt{27x} - \sqrt{20x} = \sqrt{9 \cdot 3x} - \sqrt{4 \cdot 5x}$ Factor $27x$ and $20x$.

$= \sqrt{9}\sqrt{3x} - \sqrt{4}\sqrt{5x}$ The square root of a product is equal to the product of the square roots.

$= 3\sqrt{3x} - 2\sqrt{5x}$ Evaluate: $\sqrt{9} = 3$ and $\sqrt{4} = 2$.

Since the remaining radicands are different, the radicals are unlike and the expression cannot be simplified further.

Self Check 3 Perform the indicated operations, if possible:
a. $\sqrt{24y^2} + y\sqrt{54}$ **b.** $\sqrt{75a} - \sqrt{72a}$

Now Try **Problems 49 and 51**

When we discuss multiplying radicals in the next section, we will need to know how to simplify four-termed expressions like those in the following example.

EXAMPLE 4

Perform the indicated operations, if possible:

a. $\sqrt{121} + 3\sqrt{11} + 5\sqrt{11} + 15$

b. $4\sqrt{y^2} - 6\sqrt{y} + 6\sqrt{y} - 9$

Strategy We will examine each term of the expression looking for radicals that can be simplified. If there are any, we will simplify them.

Why If there are any like radicals, we can combine them.

Solution

a. $\sqrt{121} + 3\sqrt{11} + 5\sqrt{11} + 15$

$= 11 + 3\sqrt{11} + 5\sqrt{11} + 15$ — Evaluate: $\sqrt{121} = 11$.

$= 8\sqrt{11} + 26$ — Combine like radicals: $(3 + 5)\sqrt{11} = 8\sqrt{11}$. Combine like terms: $11 + 15 = 26$.

b. $4\sqrt{y^2} - 6\sqrt{y} + 6\sqrt{y} - 9$

$= 4y - 6\sqrt{y} + 6\sqrt{y} - 9$ — Find the square root: $\sqrt{y^2} = y$.

$= 4y - 9$ — Combine like radicals: $-6\sqrt{y} + 6\sqrt{y} = 0$.

Self Check 4

Perform the indicated operations, if possible:

a. $\sqrt{36} + 9\sqrt{6} + 3\sqrt{6} + 27$

b. $16\sqrt{c^2} + 40\sqrt{c} - 40\sqrt{c} - 100$

Now Try **Problems 61 and 65**

ANSWERS TO SELF CHECKS **1. a.** $50\sqrt{5}$ **b.** $2\sqrt{b}$ **c.** $-7\sqrt{21}$ **d.** Cannot be simplified **2. a.** $7\sqrt{3}$ **b.** $-35\sqrt{2}$ **3. a.** $5y\sqrt{6}$ **b.** $5\sqrt{3a} + 6\sqrt{2a}$ **4. a.** $12\sqrt{6} + 33$ **b.** $16c - 100$

STUDY SET 8.3

VOCABULARY

Fill in the blanks.

1. Square roots such as $\sqrt{2}$ and $5\sqrt{2}$, that have the same radicand, are called like ________.
2. To simplify $2\sqrt{2} + 3\sqrt{2}$, we ________ like radicals.

CONCEPTS

Determine whether each pair of radicals are like radicals.

3. **a.** $3\sqrt{3}, 4\sqrt{3}$ **b.** $9\sqrt{a}, 9\sqrt{7a}$
4. **a.** $6\sqrt{10}, 10\sqrt{6}$ **b.** $-\sqrt{5y}, \sqrt{5y}$
5. Fill in the blanks.
   a. $5\sqrt{6} + 3\sqrt{6} = (\ \ + \ \ )\sqrt{6} = \ \ \sqrt{6}$
   b. $2\sqrt{n} - 9\sqrt{n} = (\ \ - \ \ )\sqrt{n} = \ \ \sqrt{n}$
6. Write each radical in simplified form.
   a. $\sqrt{18}$ **b.** $\sqrt{4x^3}$

NOTATION

7. Complete the solution to simplify the expression.

$$9\sqrt{5} - 3\sqrt{20} = 9\sqrt{5} - 3\sqrt{\quad \cdot 5}$$
$$= 9\sqrt{5} - 3\sqrt{\quad}\sqrt{5}$$
$$= 9\sqrt{5} - 3 \cdot \quad \sqrt{5}$$
$$= 9\sqrt{5} - \quad$$
$$= \quad$$

8. Write each expression in simpler form.
   a. $-1\sqrt{x}$ **b.** $1\sqrt{x}$

GUIDED PRACTICE

Perform the indicated operations, if possible, by combining like radicals. All variables represent nonnegative real numbers. **See Example 1.**

9. $5\sqrt{7} + 4\sqrt{7}$
10. $3\sqrt{10} + 4\sqrt{10}$
11. $14\sqrt{21} - 4\sqrt{21}$
12. $7\sqrt{3} - 2\sqrt{3}$
13. $-9\sqrt{21} + 6\sqrt{21}$
14. $-2\sqrt{5} + 8\sqrt{5}$
15. $\sqrt{3} + \sqrt{15}$
16. $\sqrt{6} + 8\sqrt{13}$

17. $8\sqrt{n} + \sqrt{n}$
18. $6\sqrt{m} + \sqrt{m}$
19. $-\sqrt{y} + \sqrt{y}$
20. $3\sqrt{t} - 3\sqrt{t}$
21. $\sqrt{x} - 4\sqrt{x}$
22. $\sqrt{t} - 9\sqrt{t}$
23. $\sqrt{2} + \sqrt{x}$
24. $\sqrt{3} - 19\sqrt{y}$
25. $2\sqrt{11} + 3\sqrt{11} + 5\sqrt{11}$
26. $2\sqrt{5} + 6\sqrt{5} + 9\sqrt{5}$
27. $4\sqrt{2} + 4\sqrt{2} - 4\sqrt{2}$
28. $9\sqrt{3} - 9\sqrt{3} + 9\sqrt{3}$
29. $5 + 3\sqrt{3} + 3\sqrt{3}$
30. $\sqrt{5} + 2 + 3\sqrt{5}$
31. $\sqrt{2} + \sqrt{3} + \sqrt{5}$
32. $\sqrt{3} + \sqrt{7} + \sqrt{11}$

Perform the indicated operations, if possible, by combining like radicals. **See Example 2.**

33. $\sqrt{12} + \sqrt{27}$
34. $\sqrt{20} + \sqrt{45}$
35. $\sqrt{18} - \sqrt{8}$
36. $\sqrt{32} - \sqrt{18}$
37. $\sqrt{12} - \sqrt{48}$
38. $\sqrt{48} - \sqrt{75}$
39. $\sqrt{288} - 3\sqrt{200}$
40. $\sqrt{80} - \sqrt{245}$
41. $2\sqrt{45} + 2\sqrt{80}$
42. $3\sqrt{80} + 3\sqrt{125}$
43. $2\sqrt{80} - 3\sqrt{125}$
44. $3\sqrt{245} - 2\sqrt{180}$
45. $\sqrt{20} + \sqrt{45} + \sqrt{80}$
46. $\sqrt{48} + \sqrt{27} + \sqrt{75}$
47. $3\sqrt{200} - \sqrt{75} + \sqrt{48}$
48. $6\sqrt{20} + \sqrt{80} - \sqrt{125}$

Perform the indicated operations, if possible, by combining like radicals. All variables represent nonnegative real numbers. **See Example 3.**

49. $\sqrt{72a} - \sqrt{98a}$
50. $\sqrt{25b} - \sqrt{49b}$
51. $\sqrt{18y} - \sqrt{27y}$
52. $\sqrt{49x} + \sqrt{8x}$
53. $\sqrt{2x^2} + \sqrt{8x^2}$
54. $\sqrt{3y^2} - \sqrt{12y^2}$
55. $3\sqrt{54b^2} + 5\sqrt{24b^2}$
56. $3\sqrt{24x^4} + 2\sqrt{54x^4}$
57. $\sqrt{2d^3} + \sqrt{8d^3}$
58. $\sqrt{3a^3} - \sqrt{12a^3}$
59. $\sqrt{32x^5} - \sqrt{18x^5}$
60. $\sqrt{27y^3} - \sqrt{48y^3}$

Perform the indicated operations, if possible, by combining like radicals. All variables represent nonnegative real numbers. **See Example 4.**

61. $\sqrt{4} + \sqrt{2} + 3\sqrt{2} + 3$
62. $\sqrt{25} + 2\sqrt{5} - 3\sqrt{5} - 6$
63. $\sqrt{y^2} + 3\sqrt{y} - 5\sqrt{y} - 15$
64. $\sqrt{z^2} - 2\sqrt{z} + 7\sqrt{z} + 12$
65. $2\sqrt{t^2} + 5\sqrt{t} - 2\sqrt{t} - 5$
66. $4\sqrt{y^2} + 6\sqrt{y} - 6\sqrt{y} - 9$
67. $3\sqrt{a^2} + 3\sqrt{a} - 3\sqrt{a} - 1$
68. $6\sqrt{p^2} + 21\sqrt{p} - 4\sqrt{p} - 14$

TRY IT YOURSELF

Perform the indicated operations, if possible. All variables represent nonnegative real numbers.

69. $8\sqrt{6} - 5\sqrt{2} - 3\sqrt{6}$
70. $3\sqrt{2} - 3\sqrt{15} - 4\sqrt{15}$
71. $\sqrt{24} + \sqrt{150} + \sqrt{240}$
72. $\sqrt{28} + \sqrt{63} + \sqrt{18}$
73. $2\sqrt{5} + 2\sqrt{5}$
74. $\sqrt{15} + \sqrt{15}$
75. $-1 + 2\sqrt{r} - 3\sqrt{r}$
76. $-8 - 5\sqrt{c} + 4\sqrt{c}$
77. $2\sqrt{28} + 2\sqrt{112}$
78. $4\sqrt{63} + 6\sqrt{112}$
79. $15\sqrt{b^2} + 20\sqrt{b} - 3\sqrt{b} - 4$
80. $25\sqrt{m^2} + 10\sqrt{m} - 10\sqrt{m} - 4$
81. $\sqrt{48} - \sqrt{8} + \sqrt{27} - \sqrt{32}$
82. $\sqrt{162} + \sqrt{50} - \sqrt{75} - \sqrt{108}$
83. $6\sqrt{40y} - 2\sqrt{360z}$
84. $3\sqrt{20x} + 2\sqrt{45y}$

APPLICATIONS

85. PLAYGROUND EQUIPMENT Find the exact total length of pipe necessary to construct the frame of the swing set. Then approximate the length to the nearest foot.

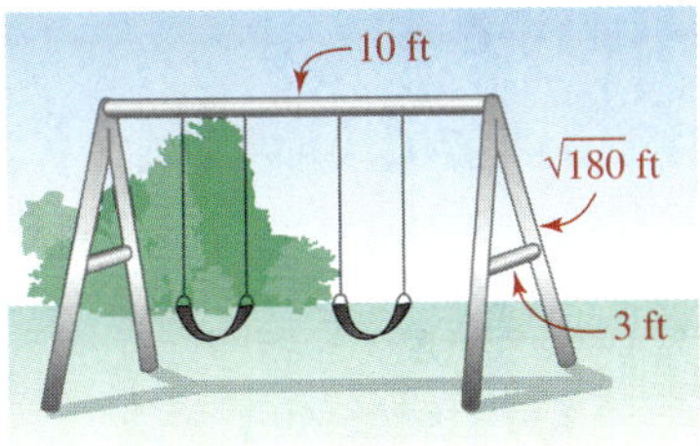

86. HARDWARE Find the exact difference in the lengths of the "arms" of the door-closing device. Then approximate the difference to the nearest inch.

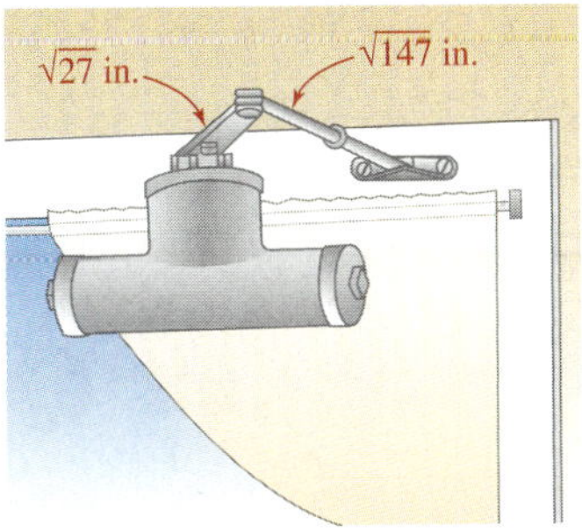

87. CAMPING The length l of a center support pole for a tent is given by the formula $l = 0.5s\sqrt{3}$, where s is the length of the side of the tent. Find the exact total length of the four poles needed for the parents' and children's tents. Then approximate the total length to the nearest tenth of a foot.

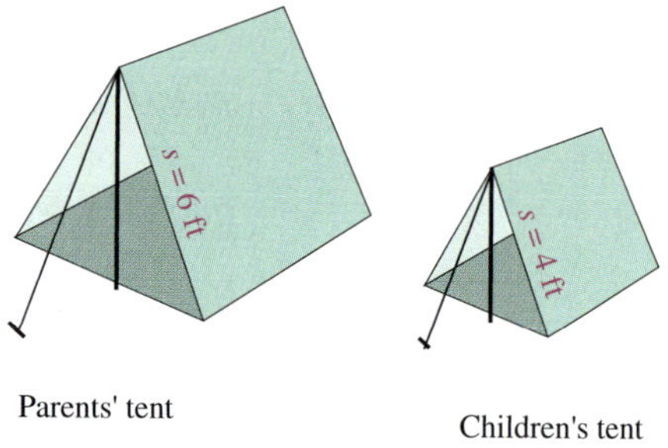

88. SURVEYING The length of each lot line of a swimming pool complex is shown. Find the exact perimeter of the irregular-shaped lot. Then approximate the perimeter to the nearest foot.

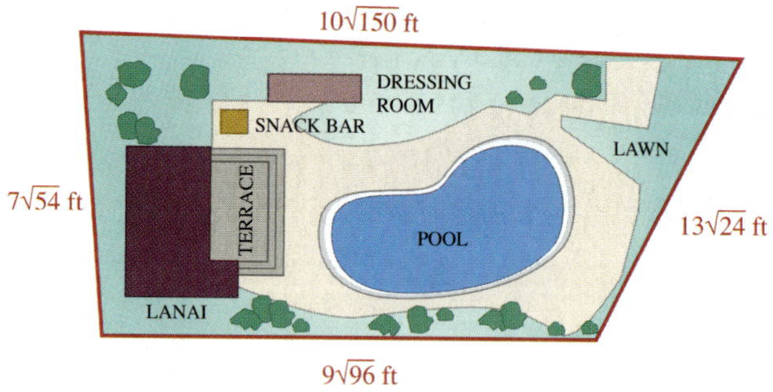

WRITING

89. Are $2\sqrt{3}$ and $3\sqrt{2}$ like radicals? Explain.

90. Does $\sqrt{9} + \sqrt{16} = \sqrt{25}$? Explain.

91. Explain why $\sqrt{6} + \sqrt{5}$ cannot be simplified further.

92. Is $2\sqrt{5} - \sqrt{5} = 2$? Explain.

93. Some of the instructions in this Study Set contain the sentence "All variables represent nonnegative real numbers." Why is this sentence necessary?

94. $\sqrt{2}$ and $\sqrt{50}$ are not like radicals. Can they be added? Explain.

REVIEW

Simplify each expression. Write each answer without using negative exponents.

95. 3^{-2}

96. $\dfrac{1}{3^{-2}}$

97. -3^2

98. -3^{-2}

99. x^{-3}

100. $\dfrac{1}{x^{-3}}$

101. 3^0

102. x^0

CHALLENGE PROBLEMS

103. Fill in the blank: $2y\sqrt{175y} - \sqrt{} = 0$

104. Add: $0.1\sqrt{50x^3} + \dfrac{1}{8}\sqrt{32x^3}$

SECTION 8.4
Multiplying and Dividing Radical Expressions

Objectives

1. Multiply square roots.
2. Find powers of square roots.
3. Multiply radical expressions.
4. Divide radical expressions.
5. Rationalize denominators.

We will now discuss methods used to multiply and divide square roots.

1 Multiply Square Roots.

We have used the product rule for square roots to write square roots in simplified form.

$$\sqrt{a \cdot b} = \sqrt{a} \cdot \sqrt{b}$$

We can use this rule in reverse to multiply square roots.

The Product Rule for Square Roots

The product of the square roots of two nonnegative numbers is equal to the square root of the product of those numbers.

For any nonnegative real numbers a and b,

$$\sqrt{a} \cdot \sqrt{b} = \sqrt{a \cdot b}$$

EXAMPLE 1 Multiply and then simplify, if possible:

a. $\sqrt{3} \cdot \sqrt{2}$ **b.** $\sqrt{6x}(\sqrt{8x})$ **c.** $\sqrt{a^3}\sqrt{5a^8}$

Strategy To multiply the square roots, we will multiply their radicands and write the product within a square root symbol. Then, we will simplify the result, if possible.

Why The product of the square roots of two nonnegative numbers is the square root of the product of those numbers.

Notation

Multiplication of radicals can be shown in several ways:

$\sqrt{3} \cdot \sqrt{2} = \sqrt{3}(\sqrt{2}) = \sqrt{3}\sqrt{2}$

Solution

a. $\sqrt{3} \cdot \sqrt{2} = \sqrt{3 \cdot 2} = \sqrt{6}$ Multiply the radicands, 3 and 2, to get 6, and write the square root of that product.

Since 6 does not have a perfect square factor, $\sqrt{6}$ does not simplify.

b. $\sqrt{6x}(\sqrt{8x}) = \sqrt{6x \cdot 8x} = \sqrt{48x^2}$ Multiply the radicands, 6x and 8x, to get $48x^2$, and write the square root of that product.

Since $48x^2$ has a perfect square factor of $16x^2$, we can simplify $\sqrt{48x^2}$.

$\sqrt{48x^2} = \sqrt{16x^2 \cdot 3}$ Factor $48x^2$ as $16x^2 \cdot 3$.

$= \sqrt{16x^2}\sqrt{3}$ The square root of a product is the product of the square roots.

$= 4x\sqrt{3}$ Find the square root of the perfect square: $\sqrt{16x^2} = 4x$.

Success Tip

We use the product rule for square roots

To multiply:

$\sqrt{a^3}(\sqrt{5a^8}) = \sqrt{a^3 \cdot 5a^8}$

and to simplify:

$\sqrt{a^{10} \cdot 5a} = \sqrt{a^{10}}\sqrt{5a}$

c. $\sqrt{a^3}\sqrt{5a^8} = \sqrt{a^3 \cdot 5a^8} = \sqrt{5a^{11}}$ Multiply the radicands: $a^3 \cdot 5a^8 = 5a^{8+3} = 5a^{11}$. Then write the square root of that product.

Since a^{11} has a perfect square factor of a^{10}, we can simplify $\sqrt{5a^{11}}$.

$\sqrt{5a^{11}} = \sqrt{a^{10} \cdot 5a}$ Factor $5a^{11}$ as $a^{10} \cdot 5a$.

$= \sqrt{a^{10}}\sqrt{5a}$ The square root of a product is the product of the square roots.

$= a^5\sqrt{5a}$ Find the square root of the perfect square: $\sqrt{a^{10}} = a^5$.

Self Check 1 Multiply and simplify, if possible: **a.** $\sqrt{5} \cdot \sqrt{3}$

b. $\sqrt{10y}(\sqrt{2y})$ **c.** $\sqrt{5b^3}\sqrt{2b^6}$

Now Try **Problems 13, 21, and 23**

EXAMPLE 2 Multiply and simplify, if possible:

a. $6\sqrt{5} \cdot 7$ **b.** $2\sqrt{3} \cdot \sqrt{12}$ **c.** $3\sqrt{6} \cdot 4\sqrt{5}$

Strategy We will use the commutative and associative properties of multiplication to reorder and regroup the factors in each expression. Then we will multiply and simplify the result, if possible.

Why We want to write the factors that are integers together and the factors that are square roots together so that we can find their products separately.

Caution
Note the difference.

If 7 is *not* a radicand:
$6\sqrt{5} \cdot 7 = 42\sqrt{5}$

If 7 is a radicand:
$6\sqrt{5} \cdot \sqrt{7} = 6\sqrt{35}$

Solution

a. $6\sqrt{5} \cdot 7 = 6 \cdot 7 \cdot \sqrt{5} = 42\sqrt{5}$ Multiply the integer factors, 6 and 7, to get 42. The result does not simplify.

b. $2\sqrt{3} \cdot \sqrt{12} = 2\sqrt{36}$ Multiply the radicands, 3 and 12, to get 36, and write the square root of that product.

$= 2 \cdot 6$ Simplify the result: $\sqrt{36} = 6$.

$= 12$

c. $3\sqrt{6} \cdot 4\sqrt{5} = 3 \cdot 4 \cdot \sqrt{6} \cdot \sqrt{5}$ Write the integer factors together and the square root factors together.

$= 12\sqrt{6 \cdot 5}$ Multiply 3 and 4 and multiply $\sqrt{6}$ and $\sqrt{5}$.

$= 12\sqrt{30}$

Since 30 does not have a perfect-square factor, the result does not simplify.

Self Check 2 Multiply and simplify, if possible: **a.** $2\sqrt{2} \cdot 6$ **b.** $9\sqrt{5} \cdot \sqrt{20}$

Now Try **Problems 25 and 27**

2 Find Powers of Square Roots.

By definition, when the square root of a positive number is squared, the result is that positive number. In symbols, we have the following fact.

The Square of a Square Root

For any nonnegative real number a,

$$\left(\sqrt{a}\right)^2 = a$$

EXAMPLE 3 Find: **a.** $\left(\sqrt{5}\right)^2$ **b.** $\left(\sqrt{x+1}\right)^2$ **c.** $\left(2\sqrt{7}\right)^2$

Strategy We will use the rule $\left(\sqrt{a}\right)^2 = a$ to find each power.

Why Each expression involves the square of the square root of a positive number.

Success Tip
Since $\left(\sqrt{a}\right)^2 = \sqrt{a} \cdot \sqrt{a}$, it also follows that $\sqrt{a} \cdot \sqrt{a} = a$, if a is a positive real number.

Solution

a. $\left(\sqrt{5}\right)^2 = 5$ The square of the square root of a positive number is that number.

b. $\left(\sqrt{x+1}\right)^2 = x + 1$ If $x > 0$, then x represents a positive number, and $x + 1$ is positive. The square of the square root of a positive number is that number.

c. We can use the power of a product rule for exponents to evaluate $\left(2\sqrt{7}\right)^2$.

$\left(2\sqrt{7}\right)^2 = 2^2\left(\sqrt{7}\right)^2$ Raise each factor of the product $2\sqrt{7}$ to the second power.

$= 4 \cdot 7$ Simplify: $2^2 = 4$ and $\left(\sqrt{7}\right)^2 = 7$.

$= 28$

Self Check 3 Find: **a.** $(\sqrt{11})^2$ **b.** $(\sqrt{3y}+5)^2$ **c.** $(5\sqrt{2})^2$

Now Try **Problems 29, 33, and 35**

3 Multiply Radical Expressions.

To multiply radical expressions that contain more than one term, we use the same methods that were used to multiply polynomials with more than one term.

EXAMPLE 4 Multiply and simplify, if possible:

a. $\sqrt{7}(\sqrt{7}-3)$ **b.** $2\sqrt{2x}(\sqrt{x}+3\sqrt{2})$

Strategy As with polynomials, we will multiply each term within the parentheses by the term outside the parentheses.

Why This is an application of the distributive property.

Solution

Caution
A common error is to incorrectly subtract 7 and 3:
~~$7-3\sqrt{7}=4\sqrt{7}$~~

a.
$$\sqrt{7}(\sqrt{7}-3)=\sqrt{7}\sqrt{7}-\sqrt{7}\cdot 3 \quad \text{Distribute the multiplication by } \sqrt{7}.$$
$$=7-3\sqrt{7} \quad \text{Multiply: } \sqrt{7}\sqrt{7}=(\sqrt{7})^2=7.$$

b.
$$2\sqrt{2x}(\sqrt{x}+3\sqrt{2})=2\sqrt{2x}\sqrt{x}+2\sqrt{2x}\cdot 3\sqrt{2} \quad \text{Distribute the multiplication by } 2\sqrt{2x}.$$
$$=2\sqrt{2x^2}+6\sqrt{4x} \quad \text{Multiply radicands and write the square root of each product.}$$
$$=2\sqrt{x^2\cdot 2}+6\sqrt{4\cdot x} \quad \text{To simplify } \sqrt{2x^2}\text{, factor } 2x^2\text{. To simplify } \sqrt{4x}\text{, factor } 4x.$$
$$=2\sqrt{x^2}\sqrt{2}+6\sqrt{4}\sqrt{x} \quad \text{The square root of a product is the product of square roots.}$$
$$=2x\sqrt{2}+6\cdot 2\sqrt{x} \quad \text{Find the square roots of the perfect squares: } \sqrt{x^2}=x \text{ and } \sqrt{4}=2.$$
$$=2x\sqrt{2}+12\sqrt{x} \quad \text{To simplify the last term, multiply: } 6\cdot 2=12.$$

Since $2x\sqrt{2}$ and $12\sqrt{x}$ are unlike radicals, we cannot add them.

Self Check 4 Multiply and simplify, if possible:

a. $\sqrt{3}(\sqrt{3}+16)$

b. $4\sqrt{5n}(\sqrt{n}-2\sqrt{5})$

Now Try **Problems 37 and 41**

$= \sqrt{4}\sqrt{2}$

$= 2\sqrt{2}$

c. $\frac{\sqrt{18x^5}}{\sqrt{2x^3}} = \sqrt{\frac{18x^5}{2x^3}}$ The quotient of the square roots is equal to the square root of the quotient.

$= \sqrt{9x^2}$ Simplify the radicand: $\frac{18x^5}{2x^3} = \frac{\overset{1}{\cancel{2}} \cdot 9 \cdot \overset{1}{\cancel{x}} \cdot \overset{1}{\cancel{x}} \cdot \overset{1}{\cancel{x}} \cdot x \cdot x}{\underset{1}{\cancel{2}} \cdot \underset{1}{\cancel{x}} \cdot \underset{1}{\cancel{x}} \cdot \underset{1}{\cancel{x}}} = 9x^2$.

$= 3x$ Find the square root of $9x^2$.

Self Check 7 Divide and simplify, if possible:

a. $\frac{\sqrt{42}}{\sqrt{7}}$ **b.** $\frac{\sqrt{90}}{\sqrt{2}}$ **c.** $\frac{\sqrt{125d^5}}{\sqrt{5d}}$

Now Try Problems 65 and 71

5 Rationalize Denominators.

The radical expression $\frac{1}{\sqrt{2}}$ is not in simplified form because a radical appears in the denominator of the fraction. There is a way to change the denominator of the fraction from a radical representing an irrational number to an expression representing a rational number. In this process, called **rationalizing the denominator,** we multiply the fraction by a form of 1 and use the fact that $\sqrt{a} \cdot \sqrt{a} = a$.

EXAMPLE 8 Rationalize the denominator: $\frac{1}{\sqrt{2}}$

Strategy We look at the denominator of the fraction and ask, "By what must we multiply $\sqrt{2}$ to obtain a rational number?" The answer is $\sqrt{2}$, because $\sqrt{2} \cdot \sqrt{2} = 2$. Then we multiply the fraction by 1 in the form $\frac{\sqrt{2}}{\sqrt{2}}$.

Why This will result in a new fraction that has the rational number 2 for its denominator.

Solution

$\frac{1}{\sqrt{2}} = \frac{1}{\sqrt{2}} \cdot \frac{\sqrt{2}}{\sqrt{2}}$ To build an equivalent fraction, multiply by 1 in the form $\frac{\sqrt{2}}{\sqrt{2}} = 1$.

$= \frac{\sqrt{2}}{2}$ Multiply the numerators: $1 \cdot \sqrt{2} = \sqrt{2}$. Multiply the denominators: $\sqrt{2} \cdot \sqrt{2} = \sqrt{4} = 2$.

Thus, $\frac{1}{\sqrt{2}} = \frac{\sqrt{2}}{2}$.

Self Check 8 Rationalize the denominator: $\frac{11}{\sqrt{6}}$

Now Try Problem 73

The Language of Algebra

Since $\sqrt{2}$ is an irrational number, $\frac{1}{\sqrt{2}}$ has an *irrational* denominator. Since 2 is a rational number, $\frac{\sqrt{2}}{2}$ has a *rational* denominator.

One reason for rationalizing denominators is historical. The concept of rationalizing the denominator remains with us from the days when all arithmetic was done by hand. It makes evaluation by long division easier. For example, to estimate the decimal value of $\frac{1}{\sqrt{2}}$, we can divide 1 by 1.414213, or we can estimate the decimal value of $\frac{\sqrt{2}}{2}$, by dividing 1.414213 by 2. The results are approximately the same, but the second calculation is easier.

$$\frac{1}{\sqrt{2}} \approx \frac{1}{1.414213} \approx 0.707107 \qquad \frac{\sqrt{2}}{2} \approx \frac{1.414213}{2} \approx 0.707107$$

With the calculators and computers that are now available, either of these divisions are rather simple.

Rationalizing a denominator is important today because it gives us a standard form in which we can write radical expressions. This helps when we compare such radicals and enables us to combine many radical expressions.

The procedure for rationalizing square root denominators is as follows.

Rationalizing Denominators

To **rationalize a square root denominator,** multiply the numerator and denominator of the given fraction by the square root that appears in its denominator, or by a square root that makes a perfect-square radicand in the denominator.

Caution

Do not attempt to remove a common factor of 3 from the numerator and denominator of $\frac{\sqrt{15}}{3}$. The numerator, $\sqrt{15}$, does not have a factor of 3.

$$\frac{\sqrt{15}}{3} = \frac{\sqrt{3} \cdot \sqrt{5}}{3}$$

Success Tip

We usually simplify a radical expression before rationalizing the denominator.

EXAMPLE 9 Simplify each expression: **a.** $\sqrt{\frac{5}{3}}$ **b.** $\frac{4}{\sqrt{24x}}$

Strategy In each case, we will rationalize a denominator.

Why The expression will then be written in simplest form.

Solution

a. The expression $\sqrt{\frac{5}{3}}$ is not in simplified form, because the radicand is a fraction. To write it in simplified form, we use the quotient rule for square roots and then rationalize the denominator.

$$\sqrt{\frac{5}{3}} = \frac{\sqrt{5}}{\sqrt{3}}$$ The square root of a quotient is equal to the quotient of the square roots.

$$= \frac{\sqrt{5}}{\sqrt{3}} \cdot \frac{\sqrt{3}}{\sqrt{3}}$$ To build an equivalent fraction, multiply by $\frac{\sqrt{3}}{\sqrt{3}} = 1$.

$$= \frac{\sqrt{15}}{3}$$ Multiply the numerators and multiply the denominators: $\sqrt{3} \cdot \sqrt{3} = \sqrt{9} = 3$.

b. We could begin by multiplying $\frac{4}{\sqrt{24x}}$ by $\frac{\sqrt{24x}}{\sqrt{24x}}$. However, to work with smaller numbers, we simplify $\sqrt{24x}$ first and then rationalize the denominator.

$$\frac{4}{\sqrt{24x}} = \frac{4}{2\sqrt{6x}}$$ $\sqrt{24x} = \sqrt{4 \cdot 6x} = \sqrt{4}\sqrt{6x} = 2\sqrt{6x}$.

$$= \frac{4}{2\sqrt{6x}} \cdot \frac{\sqrt{6x}}{\sqrt{6x}}$$ To build an equivalent fraction, multiply by $\frac{\sqrt{6x}}{\sqrt{6x}} = 1$.

$$= \frac{4\sqrt{6x}}{2 \cdot 6x}$$ Multiply the numerators and multiply the denominators.

$$= \frac{\overset{1}{\cancel{2}} \cdot \overset{1}{\cancel{2}}\sqrt{6x}}{\underset{1}{\cancel{2}} \cdot \underset{1}{\cancel{2}} \cdot 3x}$$ Factor 4 as 2 · 2 and 6x as 2 · 3x. Simplify by removing the common factors.

$$= \frac{\sqrt{6x}}{3x}$$

Self Check 9 Rationalize each denominator: **a.** $\sqrt{\frac{6}{7}}$ **b.** $\frac{15}{\sqrt{50y}}$

Now Try **Problems 77 and 83**

We will now discuss a method to rationalize denominators of fractions that have two terms.

One-term denominators	***Two-term denominators***
$\frac{1}{\sqrt{2}}, \frac{\sqrt{5}}{\sqrt{3}}, \frac{4}{\sqrt{24x}}$	$\frac{5}{\sqrt{6}-1}, \frac{3}{\sqrt{7}+2}$

To rationalize the denominator of $\frac{5}{\sqrt{6}-1}$, we multiply the numerator and denominator by $\sqrt{6}+1$, because the product $(\sqrt{6}-1)(\sqrt{6}+1)$ contains no radicals. Radical expressions such as $\sqrt{6}-1$ and $\sqrt{6}+1$ are said to be **conjugates** of each other. Notice that they differ only in the sign separating the terms.

EXAMPLE 10

Rationalize each denominator: **a.** $\frac{2}{\sqrt{6}-1}$ **b.** $\frac{\sqrt{2}-\sqrt{y}}{\sqrt{3}+\sqrt{y}}$

Strategy To rationalize each two-term denominator, we will multiply the expression by a form of 1 that uses the conjugate of the denominator.

Why Since the product of a radical expression and its conjugate contains no radicals, this step clears the denominator of radicals.

Solution

a. The conjugate of $\sqrt{6}-1$ is $\sqrt{6}+1$.

$$\frac{2}{\sqrt{6}-1} = \frac{2}{\sqrt{6}-1} \cdot \frac{\sqrt{6}+1}{\sqrt{6}+1}$$ To build an equivalent fraction, multiply by $\frac{\sqrt{6}+1}{\sqrt{6}+1} = 1$.

$$= \frac{2(\sqrt{6}+1)}{(\sqrt{6}-1)(\sqrt{6}+1)}$$ Multiply the numerators and multiply the denominators.

$$= \frac{2(\sqrt{6}+1)}{(\sqrt{6})^2-1^2}$$ In the denominator, use a special product formula: $(A-B)(A+B) = A^2 - B^2$.

$$= \frac{2(\sqrt{6}+1)}{6-1}$$ After evaluating $(\sqrt{6})^2$, the denominator no longer contains a radical.

$$= \frac{2(\sqrt{6}+1)}{5}$$ In the denominator, subtract: $6-1=5$.

$$= \frac{2\sqrt{6}+2}{5}$$ In the numerator, distribute the multiplication by 2.

Success Tip
It is best to leave the numerator, $2(\sqrt{6}+1)$, in factored form until we are sure that the denominator does not contain a factor of 2 or $\sqrt{6}+1$.

b. The conjugate of the denominator, $\sqrt{3} + \sqrt{y}$, is $\sqrt{3} - \sqrt{y}$.

$$\frac{\sqrt{2} - \sqrt{y}}{\sqrt{3} + \sqrt{y}} = \frac{\sqrt{2} - \sqrt{y}}{\sqrt{3} + \sqrt{y}} \cdot \frac{\sqrt{3} - \sqrt{y}}{\sqrt{3} - \sqrt{y}}$$

To build an equivalent fraction, multiply by $\frac{\sqrt{3} - \sqrt{y}}{\sqrt{3} - \sqrt{y}} = 1$.

$$= \frac{(\sqrt{2} - \sqrt{y})(\sqrt{3} - \sqrt{y})}{(\sqrt{3} + \sqrt{y})(\sqrt{3} - \sqrt{y})}$$

Multiply the numerators and multiply the denominators.

$$= \frac{\sqrt{6} - \sqrt{2y} - \sqrt{3y} + y}{(\sqrt{3})^2 - (\sqrt{y})^2}$$

In the numerator, use the FOIL method. In the denominator, use a special product formula: $(A - B)(A + B) = A^2 - B^2$.

$$= \frac{\sqrt{6} - \sqrt{2y} - \sqrt{3y} + y}{3 - y}$$

After simplifying the denominator, it no longer contains radicals.

Self Check 10 Rationalize the denominator: **a.** $\frac{5}{\sqrt{7} + 2}$ **b.** $\frac{\sqrt{x} - \sqrt{7}}{\sqrt{x} - \sqrt{2}}$

Now Try **Problems 85 and 95**

ANSWERS TO SELF CHECKS **1. a.** $\sqrt{15}$ **b.** $2y\sqrt{5}$ **c.** $b^4\sqrt{10b}$ **2. a.** $12\sqrt{2}$ **b.** 90 **3. a.** 11 **b.** $3y + 5$ **c.** 50 **4. a.** $3 + 16\sqrt{3}$ **b.** $4n\sqrt{5} - 40\sqrt{n}$ **5. a.** $\sqrt{66} - \sqrt{42} + \sqrt{55} - \sqrt{35}$ **b.** $5a + \sqrt{5a} - 6$ **6. a.** $15 - 2\sqrt{15n} + n$ **b.** -2 **7. a.** $\sqrt{6}$ **b.** $3\sqrt{5}$ **c.** $5d^2$ **8.** $\frac{11\sqrt{6}}{6}$ **9. a.** $\frac{\sqrt{42}}{7}$ **b.** $\frac{3\sqrt{2y}}{2y}$ **10. a.** $\frac{5\sqrt{7} - 10}{3}$ **b.** $\frac{x + \sqrt{2x} - \sqrt{7x} - \sqrt{14}}{x - 2}$

STUDY SET 8.4

VOCABULARY

Fill in the blanks.

1. The __________ of $\frac{1}{\sqrt{3}}$ is an irrational number.
2. Two fractions that represent the same number, but have different forms, such as $\frac{1}{\sqrt{2}}$ and $\frac{\sqrt{2}}{2}$, are called __________ fractions.
3. To __________ the denominator of $\frac{4}{\sqrt{5}}$, we multiply the fraction by $\frac{\sqrt{5}}{\sqrt{5}}$.
4. The __________ of $1 + \sqrt{2}$ is $1 - \sqrt{2}$.

CONCEPTS

Fill in the blanks. All variables represent nonnegative numbers.

5. **a.** $\sqrt{a} \cdot \sqrt{b} = \sqrt{}$ **b.** $\frac{\sqrt{a}}{\sqrt{b}} = \sqrt{}$
   c. $(\sqrt{a})^2 = \_\_$ **d.** $\sqrt{b} \cdot \sqrt{b} = \_\_$
6. **a.** To rationalize the denominator of $\frac{4}{\sqrt{3}}$, we multiply the fraction by $\frac{\sqrt{3}}{\sqrt{3}}$, which is a form of __.
   b. To rationalize the denominator of $\frac{5}{\sqrt{6} + 3}$, we multiply the fraction by __.
7. Explain why each expression is not in simplified radical form.
   a. $\sqrt{\frac{3}{4}}$ **b.** $\frac{1}{\sqrt{10}}$
8. Perform each operation, if possible.
   a. $\sqrt{2} + \sqrt{3}$ **b.** $\sqrt{2} \cdot \sqrt{3}$
   c. $\sqrt{2} - \sqrt{3}$ **d.** $\frac{\sqrt{2}}{\sqrt{3}}$

NOTATION

Complete each solution.

9. $7\sqrt{2} \cdot 4\sqrt{3} = 7 \cdot \_\_ \cdot \sqrt{2} \cdot \_\_$
   $= 28\sqrt{}$

10. $(\sqrt{x}+\sqrt{2})(\sqrt{x}-\sqrt{2})$

$= \sqrt{x}\,\square - \sqrt{x}\cdot\sqrt{2} + \sqrt{2}\cdot\square - \sqrt{2}\cdot\sqrt{2}$

$= \square - \sqrt{\square} + \sqrt{2x} - \square$

$= x - \square$

11. $\frac{5}{\sqrt{7}} = \frac{5}{\sqrt{7}}\cdot\frac{\sqrt{\square}}{\sqrt{\square}}$

$= \frac{5\sqrt{7}}{\square}$

12. $\frac{4}{\sqrt{x}+1} = \frac{4}{\sqrt{x}+1}\cdot\frac{\sqrt{x}\;\square\;1}{\sqrt{x}\;\square\;1}$

$= \frac{4(\sqrt{x}-1)}{(\sqrt{x})^{\square}-1^{\square}}$

$= \frac{4\sqrt{x}-\square}{\square-\square}$

GUIDED PRACTICE

Multiply and simplify, if possible. All variables represent nonnegative real numbers. **See Examples 1 and 2.**

13. $\sqrt{3}\cdot\sqrt{5}$
14. $\sqrt{2}\cdot\sqrt{7}$
15. $\sqrt{7}\cdot\sqrt{7}$
16. $\sqrt{10}\cdot\sqrt{10}$
17. $\sqrt{5}\cdot\sqrt{10}$
18. $\sqrt{2}\cdot\sqrt{6}$
19. $3\sqrt{2}(\sqrt{x})$
20. $4\sqrt{3x}(\sqrt{5})$
21. $\sqrt{5d}\cdot\sqrt{8d}$
22. $\sqrt{3b}\cdot\sqrt{15b}$
23. $\sqrt{5x^3}\sqrt{x^5}$
24. $\sqrt{a^7}\sqrt{2a^3}$
25. $(3\sqrt{5})5$
26. $(4\sqrt{3})7$
27. $5\sqrt{3}\cdot 2\sqrt{5}$
28. $2\sqrt{5}\cdot 5\sqrt{2}$

Find each power. All variables represent nonnegative real numbers. **See Example 3.**

29. $(\sqrt{6})^2$
30. $(\sqrt{11})^2$
31. $(\sqrt{y})^2$
32. $(\sqrt{m})^2$
33. $(\sqrt{2b+7})^2$
34. $(\sqrt{3t+8})^2$
35. $(2\sqrt{3})^2$
36. $(3\sqrt{5})^2$

Multiply and simplify, if possible. All variables represent nonnegative real numbers. **See Example 4.**

37. $\sqrt{2}(\sqrt{2}+1)$
38. $\sqrt{5}(\sqrt{5}+2)$
39. $3\sqrt{3}(\sqrt{27}-\sqrt{2})$
40. $2\sqrt{2}(\sqrt{8}-\sqrt{3})$
41. $\sqrt{x}(\sqrt{3x}-2)$
42. $\sqrt{y}(\sqrt{y}+5)$
43. $\sqrt{m}(\sqrt{8m}+9)$
44. $\sqrt{n}(\sqrt{50n}-4)$

Multiply and simplify, if possible. All variables represent nonnegative real numbers. **See Example 5.**

45. $(\sqrt{2}+1)(\sqrt{2}-1)$
46. $(\sqrt{3}-1)(\sqrt{3}+1)$
47. $(\sqrt{2}-\sqrt{3})(\sqrt{3}+\sqrt{5})$
48. $(\sqrt{3}+\sqrt{5})(\sqrt{5}-\sqrt{2})$
49. $(\sqrt{2x}+3)(\sqrt{8x}-6)$
50. $(\sqrt{5y}-3)(\sqrt{20y}+6)$
51. $(2\sqrt{7}-x)(3\sqrt{2}+x)$
52. $(4\sqrt{2}-\sqrt{x})(\sqrt{x}+2\sqrt{3})$

Find each product and simplify, if possible. All variables represent nonnegative real numbers. **See Example 6.**

53. $(5+\sqrt{3})^2$
54. $(4+\sqrt{2})^2$
55. $(a+\sqrt{7})^2$
56. $(t+\sqrt{6})^2$
57. $(\sqrt{5}-\sqrt{m})^2$
58. $(\sqrt{3}-\sqrt{v})^2$
59. $(\sqrt{6}+\sqrt{3})(\sqrt{6}-\sqrt{3})$
60. $(\sqrt{8}+\sqrt{7})(\sqrt{8}-\sqrt{7})$
61. $(\sqrt{11}-y)(\sqrt{11}+y)$
62. $(\sqrt{2}-s)(\sqrt{2}+s)$
63. $(\sqrt{7c}-3)(\sqrt{7c}+3)$
64. $(\sqrt{5r}+4)(\sqrt{5r}-4)$

Simplify. All variables represent positive real numbers. **See Example 7.**

65. $\frac{\sqrt{60}}{\sqrt{6}}$
66. $\frac{\sqrt{115}}{\sqrt{5}}$
67. $\frac{\sqrt{a^7}}{\sqrt{a^3}}$
68. $\frac{\sqrt{b^{11}}}{\sqrt{b^5}}$
69. $\frac{\sqrt{18x}}{\sqrt{25x}}$
70. $\frac{\sqrt{27y}}{\sqrt{16y}}$
71. $\frac{\sqrt{27x}}{\sqrt{75x^3}}$
72. $\frac{\sqrt{50z}}{\sqrt{98z^3}}$

Rationalize each denominator and simplify. All variables represent positive real numbers. **See Example 8 and 9.**

73. $\frac{1}{\sqrt{3}}$
74. $\frac{1}{\sqrt{5}}$
75. $\frac{4}{\sqrt{19}}$
76. $\frac{9}{\sqrt{2}}$
77. $\sqrt{\frac{12}{5}}$
78. $\sqrt{\frac{24}{7}}$
79. $\frac{6}{\sqrt{27}}$
80. $\frac{4}{\sqrt{20}}$

81. $\dfrac{10}{\sqrt{x}}$

82. $\dfrac{12}{\sqrt{y}}$

83. $\dfrac{\sqrt{9}}{\sqrt{2x}}$

84. $\dfrac{\sqrt{4}}{\sqrt{3z}}$

Rationalize each denominator and simplify. All variables represent positive real numbers. See Example 10.

85. $\dfrac{3}{\sqrt{3}-1}$

86. $\dfrac{5}{\sqrt{7}-1}$

87. $\dfrac{8}{\sqrt{7}+2}$

88. $\dfrac{9}{\sqrt{11}+3}$

89. $\dfrac{5}{\sqrt{3}+\sqrt{2}}$

90. $\dfrac{3}{\sqrt{3}-\sqrt{2}}$

91. $\dfrac{\sqrt{3}}{5+\sqrt{x}}$

92. $\dfrac{\sqrt{2}}{3-\sqrt{b}}$

93. $\dfrac{\sqrt{5}+\sqrt{7}}{\sqrt{2}-\sqrt{5}}$

94. $\dfrac{\sqrt{7}-\sqrt{2}}{\sqrt{5}+\sqrt{2}}$

95. $\dfrac{\sqrt{3}-\sqrt{a}}{\sqrt{5}+\sqrt{a}}$

96. $\dfrac{\sqrt{y}-\sqrt{2}}{\sqrt{y}+\sqrt{2}}$

TRY IT YOURSELF

Perform the operations and simplify, if possible.

97. $\sqrt{x}\left(\sqrt{14x}+\sqrt{2}\right)$

98. $3\sqrt{z}\left(\sqrt{z}-\sqrt{5}\right)$

99. $\left(10\sqrt{x}\right)^2$

100. $\left(\sqrt{5y}+2\right)^2$

101. $\sqrt{\dfrac{13}{7}}$

102. $\sqrt{\dfrac{3}{11}}$

103. $\left(3p+\sqrt{5}\right)^2$

104. $\left(6t-\sqrt{3}\right)^2$

105. $\sqrt{6}\sqrt{11}$

106. $\sqrt{5}\sqrt{13}$

107. $\dfrac{\sqrt{12x^3}}{\sqrt{27x}}$

108. $\dfrac{\sqrt{32}}{\sqrt{98x^2}}$

109. $\left(\sqrt{15}+\sqrt{2}\right)\left(\sqrt{15}-\sqrt{2}\right)$

110. $\left(\sqrt{14x}+\sqrt{3}\right)\left(\sqrt{14x}-\sqrt{3}\right)$

111. $7\sqrt{3}\left(2\sqrt{5}\right)$

112. $3\sqrt{5}\left(4\sqrt{7}\right)$

113. $\dfrac{\sqrt{3}}{7+\sqrt{3}}$

114. $\dfrac{\sqrt{r}-3}{\sqrt{r}+3}$

115. $\sqrt{6n^4}\cdot\sqrt{10n^5}$

116. $\sqrt{5m^2}\cdot\sqrt{10m^5}$

117. $\left(2\sqrt{x}+2\right)\left(\sqrt{x}-6\right)$

118. $\left(\sqrt{s}-1\right)\left(3\sqrt{s}+2\right)$

119. $\dfrac{3}{\sqrt{32}}$

120. $\dfrac{5}{\sqrt{18}}$

APPLICATIONS

121. SET LIGHTING The director of a stage play requested bright downlighting over the portion of the set shown. Find the area of the rectangular region of light. (*Hint:* Use the grid marks and the Pythagorean theorem to find the length and the width of the rectangle first.)

1 ft x 1 ft grid

122. GEOMETRY Find the perimeter and the area of the rectangle.

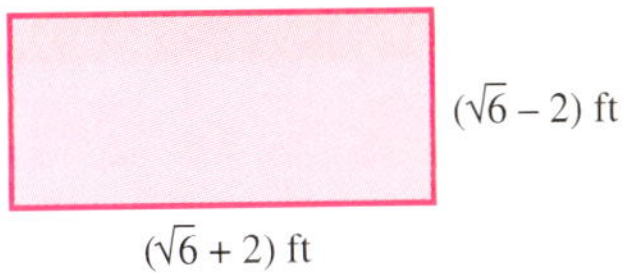

123. TUNING GUITARS The formula $f = 0.772\sqrt{\frac{T}{u}}$ can be used to calculate the pitch of a string on a Fender Stratocaster electric guitar. In the formula, f is related to the vibration of the string, T represents the tension in the string, and u is related to the thickness of the string. Rewrite the right side of the formula by rationalizing the denominator.

124. THE PERFECT SMILE Throughout history, **golden rectangles** have been considered the most pleasing to the human eye. They can be seen in art, architecture, and nature. The ratio of the length to the width of any golden rectangle is $\frac{2}{\sqrt{5}-1}$. Some cosmetic dentists feel the front two teeth should form a golden rectangle. Rationalize the denominator of this ratio.

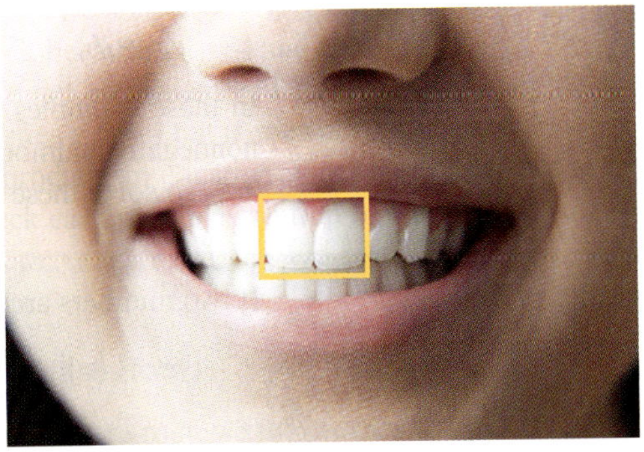

Strategy for Solving Radical Equations Containing Square Roots

1. Isolate a radical term on one side of the equation.
2. Square both sides of the equation.
3. Solve the resulting equation.
4. Check the possible solutions in the original equation. This step is required.

EXAMPLE 2

Solve: $\sqrt{2x - 1} = 9$

Strategy Since the radical is already isolated on one side, we can move to Step 2 of the strategy for solving radical equations and square both sides of the equation.

Why Squaring both sides will produce, on the left side, the expression $\left(\sqrt{2x - 1}\right)^2$ that simplifies to $2x - 1$. This step clears the equation of the radical.

Solution

$$\sqrt{2x - 1} = 9$$

$$\left(\sqrt{2x - 1}\right)^2 = (9)^2$$ Use the squaring property of equality and square both sides.

$$2x - 1 = 81$$ Simplify: $\left(\sqrt{2x - 1}\right)^2 = 2x - 1$ and $(9)^2 = 81$. The radical has been removed.

$$2x = 82$$ To solve the resulting equation, add 1 to both sides.

$$x = 41$$ To isolate x, divide both sides by 2.

The Language of Algebra
Possible solutions are also called *potential* or *proposed* solutions.

Check the possible solution, 41, in the original equation.

Evaluate the left side. Do not square both sides when checking!

$$\sqrt{2x - 1} = 9$$ This is the original equation.

$$\sqrt{2(41) - 1} \stackrel{?}{=} 9$$ Substitute 41 for x.

$$\sqrt{82 - 1} \stackrel{?}{=} 9$$ Do the multiplication within the radical.

$$\sqrt{81} \stackrel{?}{=} 9$$ Do the subtraction within the radical.

$$9 = 9$$ True

The solution is 41. The solution set is written $\{41\}$.

Self Check 2 Solve: $\sqrt{3x - 5} = 2$

Now Try Problem 19

EXAMPLE 3

Solve: $a = \sqrt{a^2 + 3a - 3}$

Strategy Since the radical is isolated on the right side, we will square both sides.

Why Squaring both sides will produce, on the right side, the expression $\left(\sqrt{a^2 + 3a - 3}\right)^2$ that simplifies to $a^2 + 3a - 3$. This step clears the equation of the radical.

$$a = \sqrt{a^2 + 3a - 3}$$

$$(a)^2 = \left(\sqrt{a^2 + 3a - 3}\right)^2$$ Square both sides.

$$a^2 = a^2 + 3a - 3$$ Simplify: $\left(\sqrt{a^2 + 3a - 3}\right)^2 = a^2 + 3a - 3$. The radical has been removed.

$$a^2 - a^2 = a^2 + 3a - 3 - a^2$$ To solve the resulting equation, eliminate a^2 from the left side by subtracting a^2 from both sides.

$$0 = 3a - 3$$ Combine like terms: $a^2 - a^2 = 0$. Notice that both terms involving a^2 drop out.

$$3 = 3a$$ To isolate the variable term, $3a$, add 3 to both sides.

$$1 = a$$ To isolate a, divide both sides by 3.

Check:

$$a = \sqrt{a^2 + 3a - 3}$$ This is the original equation.

$$1 \stackrel{?}{=} \sqrt{(1)^2 + 3(1) - 3}$$ Substitute 1 for a.

$$1 \stackrel{?}{=} \sqrt{1 + 3 - 3}$$ Evaluate the expression within the radical.

$$1 \stackrel{?}{=} \sqrt{1}$$

$$1 = 1$$ True

The solution is 1. The solution set is $\{1\}$.

Self Check 3 Solve: $b = \sqrt{b^2 - 2b + 24}$

Now Try Problem 27

3 Identify Radical Equations That Have No Solutions.

EXAMPLE 4

Solve: $\sqrt{x + 2} + 5 = 3$

Strategy Since 5 is outside the square root symbol, there are two terms on the left side of the equation. To isolate the radical, we will subtract 5 from both sides.

Why This will put the equation in a form where we can square both sides to clear the radical.

Solution

$$\sqrt{x + 2} + 5 = 3$$

$$\sqrt{x + 2} + 5 - 5 = 3 - 5$$ To isolate the radical, subtract 5 from both sides.

$$\sqrt{x + 2} = -2$$ The square root term is by itself on one side of the equation.

We now square both sides to clear the equation of the radical.

$$\left(\sqrt{x + 2}\right)^2 = (-2)^2$$ Square both sides.

$$x + 2 = 4$$ Simplify: $\left(\sqrt{x + 2}\right)^2 = x + 2$ and $(-2)^2 = 4$.

$$x = 2$$ To solve the resulting equation, subtract 2 from both sides.

Success Tip

It is apparent that $\sqrt{x + 2} = -2$ has no solution. Since $\sqrt{x + 2}$ is the principal square root, and cannot be negative, there is no real number x that could make $\sqrt{x + 2}$ equal to a negative number.

Check:

$\sqrt{x + 2} + 5 = 3$ This is the original equation.

$\sqrt{2 + 2} + 5 \stackrel{?}{=} 3$ Substitute 2 for x.

$\sqrt{4} + 5 \stackrel{?}{=} 3$ Do the addition within the radical.

$2 + 5 \stackrel{?}{=} 3$ Find the square root.

$7 = 3$ False

Since the result is a false statement, 2 is not a solution of the original equation and must be discarded. The original equation has no solution. The solution set is the empty set, written as $\varnothing$.

Self Check 4 Solve: $\sqrt{x - 2} + 8 = 1$

Now Try Problem 39

Example 4 shows that squaring both sides of an equation can lead to possible solutions that do not satisfy the original equation. We call such numbers **extraneous solutions.** In that example, we found that 2 is an extraneous solution of $\sqrt{x + 2} + 5 = 3$.

4 Solve Radical Equations by Squaring a Binomial.

Sometimes, after clearing an equation of a radical, the result is a quadratic equation.

EXAMPLE 5 Solve: $\sqrt{3 - x} - x = -3$

Strategy Since $-x$ is outside the radical, there are two terms on the left side of the equation. To isolate the radical, we will add x to both sides.

Why This will put the equation in a form where we can square both sides to clear the radical.

Solution

$\sqrt{3 - x} - x = -3$

$\sqrt{3 - x} - x + x = x - 3$ Add x to both sides. On the right side, it is best to add the x-term in front of −3.

$\sqrt{3 - x} = x - 3$ The radical is now isolated.

We then square both sides to clear the equation of the radical.

$\left(\sqrt{3 - x}\right)^2 = (x - 3)^2$ Square both sides.

$3 - x = x^2 - 6x + 9$ Simplify: $\left(\sqrt{3 - x}\right)^2 = 3 - x$. Use a special product formula to find $(x - 3)^2$.

To solve the resulting quadratic equation, we write it in standard form so that the left side is 0.

$3 = x^2 - 5x + 9$ To eliminate x on the left side, add x to both sides.

$0 = x^2 - 5x + 6$ To get 0 on the left side, subtract 3 from both sides.

$0 = (x - 3)(x - 2)$ Factor $x^2 - 5x + 6$.

$x - 3 = 0$ or $x - 2 = 0$ Set each factor equal to 0.

$x = 3$ | $x = 2$ Solve each equation.

Success Tip
Even if you are certain that no algebraic mistakes were made when solving a radical equation, you must still check your solutions. Squaring both sides can introduce extraneous solutions that must be discarded.

Since there are two possible solutions, we check each one in the original equation.

***The check for* 3:**

$$\sqrt{3 - x} - x = -3$$
$$\sqrt{3 - 3} - 3 \stackrel{?}{=} -3$$
$$\sqrt{0} - 3 \stackrel{?}{=} -3$$
$$0 - 3 \stackrel{?}{=} -3$$
$$-3 = -3 \quad \text{True}$$

***The check for* 2:**

$$\sqrt{3 - x} - x = -3$$
$$\sqrt{3 - 2} - 2 \stackrel{?}{=} -3$$
$$\sqrt{1} - 2 \stackrel{?}{=} -3$$
$$1 - 2 \stackrel{?}{=} -3$$
$$-1 = -3 \quad \text{False}$$

Since a true statement results when 3 is substituted for x, 3 is a solution. Since a false statement results when 2 is substituted for x, 2 is an extraneous solution. The solution set is $\{3\}$.

Self Check 5 Solve: $\sqrt{x + 4} - x = -2$

Now Try **Problem 43**

5 Solve Equations Containing Two Square Roots.

To solve an equation containing two square roots, we want to have one radical on the left side and one radical on the right side.

EXAMPLE 6 Solve: $\sqrt{x + 12} = 3\sqrt{x + 4}$

Strategy We will square both sides to clear the equation of both radicals.

Why We can immediately square both sides since each square root term is by itself on one side of the equation. This step will clear the equation of both radicals.

Caution
Recall that to raise a product to a power, we raise each factor of the product to that power:

$$(xy)^n = x^n y^n$$

Use this rule for exponents to simplify $\left(3\sqrt{x + 4}\right)^2$. A common error is to forget to square 3.

Solution

$$\sqrt{x + 12} = 3\sqrt{x + 4}$$
$$\left(\sqrt{x + 12}\right)^2 = \left(3\sqrt{x + 4}\right)^2 \quad \text{Square both sides.}$$
$$x + 12 = 3^2\left(\sqrt{x + 4}\right)^2$$

On the left, simplify: $\left(\sqrt{x + 12}\right)^2 = x + 12$. On the right, raise each factor of the product $3\sqrt{x + 4}$ to the second power.

$$x + 12 = 9(x + 4)$$

Simplify: $3^2 = 9$ and $\left(\sqrt{x + 4}\right)^2 = x + 4$. The square roots have been removed.

$$x + 12 = 9x + 36$$

To solve the resulting equation, distribute the multiplication by 9.

$$12 = 8x + 36 \quad \text{To eliminate } x \text{ on the left side, subtract } x \text{ from both sides.}$$
$$-24 = 8x \quad \text{To isolate the variable term, } 8x\text{, subtract 36 from both sides.}$$
$$-3 = x \quad \text{To isolate } x\text{, divide both sides by 8.}$$

Check:

$$\sqrt{x + 12} = 3\sqrt{x + 4} \quad \text{This is the original equation.}$$
$$\sqrt{-3 + 12} \stackrel{?}{=} 3\sqrt{-3 + 4} \quad \text{Substitute } -3 \text{ for } x.$$
$$\sqrt{9} \stackrel{?}{=} 3\sqrt{1} \quad \text{Evaluate the expressions within the radicals.}$$
$$3 \stackrel{?}{=} 3 \cdot 1 \quad \text{Find each square root.}$$
$$3 = 3 \quad \text{True}$$

The solution is -3. The solution set is $\{-3\}$.

Self Check 6 Solve: $\sqrt{x-4} = 2\sqrt{x-16}$

Now Try **Problem 59**

6 Solve Problems Modeled by a Radical Equation.

Radical equations can be used to model many real-world situations.

© Walter Bibikow/Getty Images

EXAMPLE 7 ***Bridges.*** The time t in seconds that it takes an object to fall d feet is given by the formula $t = \frac{\sqrt{d}}{4}$. To find the height of the Benjamin Franklin Bridge in Philadelphia, a man stands in the center section and drops a coin into the water. If it takes the coin 3 seconds to hit the water, how high above the water is the bridge?

Strategy We will substitute 3 for t in the formula and solve for d.

Why Since the coin fell for 3 seconds, $t = 3$. The height of the bridge above the water is the same as the distance d that the coin fell.

Solution

$$t = \frac{\sqrt{d}}{4} \quad \text{This is the equation that models the situation.}$$

$$3 = \frac{\sqrt{d}}{4} \quad \text{Substitute 3 for } t.$$

$$12 = \sqrt{d} \quad \text{To clear the equation of the fraction, multiply both sides by 4.}$$

$$(12)^2 = \left(\sqrt{d}\right)^2 \quad \text{To clear the equation of the radical, square both sides.}$$

$$144 = d \quad \text{Simplify each side.}$$

The bridge is 144 feet above the water. Check this result in the original equation.

Self Check 7 To find the depth of a well, a farmer dropped a rock down it. If the rock fell straight downward for 2 seconds before he heard a splash, how deep was the water level?

Now Try **Problem 87**

ANSWERS TO SELF CHECKS **1.** 100 **2.** 3 **3.** 12 **4.** No solution **5.** 5 **6.** 20 **7.** 64 ft

STUDY SET 8.5

VOCABULARY

Fill in the blanks.

1. A ________ equation is an equation that contains a variable in a radicand.
2. To ________ the radical in $\sqrt{x} + 1 = 6$ means to get $\sqrt{x}$ by itself on one side of the equation.
3. Squaring both sides of an equation can lead to possible solutions that do not satisfy the original equation. We call such numbers ________ solutions.
4. The squaring property of equality states that if two numbers are equal, their ________ are equal.

CONCEPTS

Fill in the blanks.

5. The squaring property of equality states that if $a = b$, then $a^2 =$ ___.
6. To clear $\sqrt{x+1} = 5$ of the radical, we ________ both sides of the equation.
7. Use properties of algebra to isolate each radical term on one side of the equation. DO NOT SOLVE THE EQUATION.
   a. $\sqrt{x-4} - 1 = 2$
   b. $8 = \sqrt{x} - x$
8. Simplify each expression.
   a. $\left(\sqrt{x}\right)^2$
   b. $\left(\sqrt{2x+3}\right)^2$
   c. $\left(4\sqrt{x}\right)^2$
   d. $\left(\sqrt{n^2 - 2n + 6}\right)^2$

NOTATION

Complete each solution.

9. Solve: $\sqrt{x-3} = 5$

$$\left(\sqrt{x-3}\right)^{\square} = 5^{\square}$$
$$\square = 25$$
$$x = \square$$

10. Determine whether -1 is a solution of the equation.

$$2\sqrt{3x+4} = \sqrt{5x+9}$$
$$2\sqrt{3(\quad)+4} \stackrel{?}{=} \sqrt{5(\quad)+9}$$
$$2\sqrt{\quad + 4} \stackrel{?}{=} \sqrt{\quad + 9}$$
$$2\sqrt{1} \stackrel{?}{=} \sqrt{\quad}$$
$$2 \cdot \quad \stackrel{?}{=} 2$$
$$2 = 2$$

Thus, ___ is the solution.

GUIDED PRACTICE

Solve each equation. See Example 1.

11. $\sqrt{x} = 3$
12. $\sqrt{x} = 5$
13. $\sqrt{y} = 12$
14. $\sqrt{y} = 20$
15. $\sqrt{2a} = 4$
16. $\sqrt{3a} = 9$
17. $\sqrt{4n} = 6$
18. $\sqrt{4n} = 8$

Solve each equation. See Example 2.

19. $\sqrt{x+3} = 2$
20. $\sqrt{x+2} = 3$
21. $\sqrt{5-T} = 10$
22. $\sqrt{3-T} = 2$
23. $\sqrt{6x+19} = 7$
24. $\sqrt{2x+6} = 4$
25. $\sqrt{5x-5} = 5$
26. $\sqrt{4x-7} = 7$

Solve each equation. See Example 3.

27. $x = \sqrt{x^2 - 2x + 16}$
28. $x = \sqrt{x^2 + 3x - 21}$
29. $c = \sqrt{c^2 - 3c + 39}$
30. $2s = \sqrt{4s^2 - 2s + 10}$
31. $\sqrt{9q^2 - 5q + 10} = 3q$
32. $\sqrt{m^2 - 7m + 21} = m$
33. $\sqrt{4m^2 + 6m + 6} = -2m$
34. $\sqrt{9t^2 + 4t + 20} = -3t$

Solve each equation. See Example 4.

35. $\sqrt{x} = -6$
36. $\sqrt{x} = -7$
37. $\sqrt{r} + 4 = 0$
38. $\sqrt{r} + 1 = 0$
39. $\sqrt{2x+7} + 4 = 1$
40. $\sqrt{15a+3} + 5 = 4$
41. $\sqrt{6-2b} - 7 = -9$
42. $\sqrt{3-12t} - 2 = -3$

Solve each equation. See Example 5.

43. $\sqrt{3a+7} - a = 3$
44. $-1 = \sqrt{2a-2} - a$
45. $b = \sqrt{5b+1} - 1$
46. $1 = \sqrt{t+1} - t$
47. $x = \sqrt{x^2 - 15} + 3$
48. $v = \sqrt{v^2 - 16} + 2$
49. $1 = \sqrt{x^2 + x + 4} - x$
50. $-1 = \sqrt{x^2 - 4x + 9} - x$
51. $y = 9 + \sqrt{y-3}$
52. $m = \sqrt{m-7} + 9$
53. $\sqrt{15-3t} + 5 = t$
54. $\sqrt{1-8s} - s = 4$

Solve each equation. See Example 6.

55. $\sqrt{10-3x} = \sqrt{2x+20}$
56. $\sqrt{1-2x} = \sqrt{x+10}$
57. $\sqrt{3c-8} = \sqrt{c}$
58. $\sqrt{2x} = \sqrt{x+8}$
59. $5\sqrt{a} = \sqrt{10a+15}$
60. $\sqrt{17m-4} = 4\sqrt{m}$
61. $2\sqrt{3x+4} = \sqrt{5x+9}$
62. $\sqrt{3x+6} = 2\sqrt{2x-11}$

TRY IT YOURSELF

Solve each equation.

63. $5\sqrt{x} - 11 = 9$
64. $5\sqrt{x} - 6 = 4$

We can use this definition to simplify exponential expressions that have rational exponents with a numerator of 1.

EXAMPLE 6 Evaluate: **a.** $64^{1/3}$ **b.** $9^{1/2}$ **c.** $(-8)^{1/3}$ **d.** $-16^{1/4}$

Strategy We will identify the base and the exponent of the exponential expression so that we can write the exponential expression in an equivalent radical form.

Why We know how to evaluate square roots, cube roots, and fourth roots.

Solution

a. In the exponential expression $64^{1/3}$, the base is 64 and the exponent is $\frac{1}{3}$. The base is the same as the radicand of the corresponding radical. The denominator of the rational exponent is the same as the radical's index.

Root

$$64^{1/3} = \sqrt[3]{64} = 4$$ Because the denominator of the exponent is 3, find the cube root of the base, 64.

Base

b. $9^{1/2} = \sqrt{9} = 3$ Because the denominator of the exponent is 2, find the square root of the base, 9.

c. $(-8)^{1/3} = \sqrt[3]{-8} = -2$ Because the denominator of the exponent is 3, find the cube root of the base, −8.

d. $-16^{1/4} = -\sqrt[4]{16} = -2$ Because the denominator of the exponent is 4, find the fourth root of the base, 16.

Caution

Consider $-16^{1/4}$. Because of the lack of parentheses, the exponent $\frac{1}{4}$ only applies to the base 16, not −16.

Self Check 6 Evaluate: **a.** $144^{1/2}$ **b.** $(-27)^{1/3}$ **c.** $-81^{1/4}$

Now Try **Problems 69, 71, and 73**

5 Evaluate Expressions of the Form $a^{m/n}$.

We can extend the definition of $a^{1/n}$ to cover fractional exponents for which the numerator is not 1. For example, because $4^{3/2}$ can be written as $(4^{1/2})^3$, we have

$$4^{3/2} = (4^{1/2})^3 = \left(\sqrt{4}\right)^3 = 2^3 = 8$$

Because $4^{3/2}$ can also be written as $(4^3)^{1/2}$, we have

$$4^{3/2} = (4^3)^{1/2} = 64^{1/2} = \sqrt{64} = 8$$

In general, $a^{m/n}$ can be written as $(a^{1/n})^m$ or as $(a^m)^{1/n}$. Since $(a^{1/n})^m = \left(\sqrt[n]{a}\right)^m$ and $(a^m)^{1/n} = \sqrt[n]{a^m}$, we make the following definition.

Definition of $a^{m/n}$

If m and n represent positive integers ($n \neq 1$) and $\sqrt[n]{a}$ represents a real number,

$$a^{m/n} = \left(\sqrt[n]{a}\right)^m \quad \text{and} \quad a^{m/n} = \sqrt[n]{a^m}$$

We can use this definition to evaluate exponential expressions that have rational exponents with a numerator that is not 1. To avoid large numbers, we usually find the root of the base first and then calculate the power using the relationship $a^{m/n} = \left(\sqrt[n]{a}\right)^m$.

EXAMPLE 7 Evaluate: **a.** $125^{4/3}$ **b.** $81^{3/4}$ **c.** $-25^{3/2}$ **d.** $(-27)^{2/3}$

Strategy We will identify the base and the exponent of the exponential expression so that we can write the exponential expression in an equivalent radical form.

Why We know how to evaluate square roots, cube roots, and fourth roots.

Solution

a. In the exponential expression $125^{4/3}$, the base is 125 and the exponent is $\frac{4}{3}$. The base is the same as the radicand of the corresponding radical. The denominator of the rational exponent is the same as the radical's index. The numerator of the rational exponent indicates the power to which the radical base is raised.

Power

Root

$$125^{4/3} = \left(\sqrt[3]{125}\right)^4 = (5)^4 = 625$$

Base

Because the exponent is 4/3, find the cube root of the base, 125, to get 5. Then find the fourth power of 5.

b. $81^{3/4} = \left(\sqrt[4]{81}\right)^3 = (3)^3 = 27$ Because the exponent is 3/4, find the fourth root of the base, 81, to get 3. Then find the third power of 3.

c. For the exponential expression $-25^{3/2}$, the base is 25, not -25.

$$-25^{3/2} = -\left(\sqrt{25}\right)^3 = -(5)^3 = -125$$

Because the exponent is 3/2, find the square root of the base, 25, to get 5. Then find the third power of 5.

d. Because of the parentheses, the base of the exponential expression $(-27)^{2/3}$ is -27.

$$(-27)^{2/3} = \left(\sqrt[3]{-27}\right)^2 = (-3)^2 = 9$$

Because the exponent is 2/3, find the cube root of the base, -27, to get -3. Then find the second power of -3.

Caution

We can also evaluate $x^{m/n}$ using $\sqrt[n]{x^m}$; however, the resulting radicand is often extremely large. For example,

$$\begin{aligned} 81^{3/4} &= \sqrt[4]{81^3} \\ &= \sqrt[4]{531{,}441} \\ &= 27 \end{aligned}$$

Self Check 7 Evaluate: **a.** $100^{3/2}$ **b.** $(-8)^{2/3}$ **c.** $-32^{4/5}$

Now Try **Problems 81, 85, and 91**

We define negative rational exponents in the following way.

Definition of $a^{-m/n}$

If $a^{-m/n}$ is a nonzero real number,

$$a^{-m/n} = \frac{1}{a^{m/n}}$$

From the definition, we see that another way to write $a^{-m/n}$ is to write its reciprocal and change the sign of its exponent.

EXAMPLE 8 Evaluate: **a.** $27^{-2/3}$ **b.** $36^{-1/2}$ **c.** $-625^{-3/4}$

Strategy We will identify the base and the exponent of the exponential expression so that we can write the given exponential expression in an equivalent radical form.

Why We know how to evaluate square roots, cube roots, and fourth roots.

Solution

a. In the exponential expression $27^{-2/3}$, the base is 27 and the exponent is $-\frac{2}{3}$. Because the exponent is $-2/3$, we write the reciprocal of $27^{-2/3}$ and change the sign of the exponent. Then we find the cube root of the base, 27, to get 3. Finally, we find the second power of 3.

Reciprocal; change sign

$$27^{-2/3} = \frac{1}{27^{2/3}} = \frac{1}{(\sqrt[3]{27})^2} = \frac{1}{(3)^2} = \frac{1}{9}$$

Root

Power

> **Caution**
> A negative exponent does not indicate a negative number. For example,
>
> $$27^{-2/3} = \frac{1}{9}$$
>
> $$36^{-1/2} = \frac{1}{6}$$

b. $36^{-1/2} = \frac{1}{36^{1/2}} = \frac{1}{\sqrt{36}} = \frac{1}{6}$

c. In $-625^{-3/4}$, the base is 625.

$$-625^{-3/4} = -\frac{1}{625^{3/4}} = -\frac{1}{(\sqrt[4]{625})^3} = -\frac{1}{(5)^3} = -\frac{1}{125}$$

Self Check 8 Evaluate: **a.** $49^{-1/2}$ **b.** $8^{-4/3}$ **c.** $-16^{-1/4}$

Now Try **Problems 93, 97, and 101**

ANSWERS TO SELF CHECKS **1. a.** 1 **b.** -2 **c.** -6 **2. a.** 2 **b.** -4 **c.** Not a real number **d.** 3 **3. a.** n^2 **b.** $3x^4$ **c.** m^3 **4. a.** s^5 **b.** $10b$ **c.** x^3 **5. a.** $4\sqrt{2}$ **b.** $2\sqrt[4]{11}$ **c.** $\frac{2\sqrt[3]{5}}{7}$ **6. a.** 12 **b.** -3 **c.** -3 **7. a.** 1,000 **b.** 4 **c.** -16 **8. a.** $\frac{1}{7}$ **b.** $\frac{1}{16}$ **c.** $-\frac{1}{2}$

STUDY SET 8.6

VOCABULARY

Fill in the blanks.

1. We read $\sqrt[3]{8}$ as "the ______ root of 8" and $\sqrt[4]{16}$ as "the ______ root of 16."
2. For the radical $\sqrt[3]{27}$, the ______ is 3 and the ______ is 27.
3. "To ______ $\sqrt[3]{54}$" means to write it as $3\sqrt[3]{2}$.
4. The expression $25^{3/2}$ has a ______ (fractional) exponent.

CONCEPTS

Fill in the blanks.

5. **a.** $\sqrt[3]{64} = 4$ because $(\quad)^3 = 64$.
   b. $\sqrt[4]{16x^4} = 2x$ because $(\quad)^4 = 16x^4$.
   c. $\sqrt[5]{-32} = -2$ because $(-2)^5 = $ ______.

6. a. Write the first seven perfect cubes.

1, ____, ____, ____, ____, ____, 343

b. Write the first five perfect fourth powers.

1, ____, ____, 256, ____

7. a. $\sqrt[n]{a \cdot b} =$ ____ **b.** $\sqrt[n]{\frac{a}{b}} =$ ____

8. a. $x^{1/n} =$ ____ **b.** $x^{m/n} = (\ \_\_\_\_\ )^m$

c. $x^{-m/n} = \frac{1}{\_\_\_\_}$

9. a. To simplify $\sqrt[3]{24}$, which of the following expressions should be used?

$\sqrt[3]{12 \cdot 2}$ $\sqrt[3]{4 \cdot 6}$ $\sqrt[3]{8 \cdot 3}$ $\sqrt[3]{24 \cdot 1}$

b. To simplify $\sqrt[4]{48}$, which of the following expressions should be used?

$\sqrt[4]{24 \cdot 2}$ $\sqrt[4]{16 \cdot 3}$ $\sqrt[4]{8 \cdot 6}$ $\sqrt[4]{48 \cdot 1}$

10. a. Find the cube root of the perfect cube to simplify $\sqrt[3]{8}\sqrt[3]{3}$.

b. Find the fourth root of the perfect fourth power to simplify $\sqrt[4]{16}\sqrt[4]{3}$.

NOTATION

Write each exponential expression in an equivalent form involving a radical. Do not evaluate it.

11. a. $25^{1/2}$ **b.** $(-27)^{2/3}$

12. a. $36^{-1/2}$ **b.** $16^{-5/4}$

GUIDED PRACTICE

***Evaluate each cube root.* See Example 1.**

13. $\sqrt[3]{8}$ **14.** $\sqrt[3]{27}$

15. $\sqrt[3]{0}$ **16.** $\sqrt[3]{1}$

17. $\sqrt[3]{-125}$ **18.** $\sqrt[3]{-1}$

19. $\sqrt[3]{-64}$ **20.** $\sqrt[3]{-27}$

21. $-\sqrt[3]{-1}$ **22.** $-\sqrt[3]{343}$

23. $-\sqrt[3]{-27}$ **24.** $-\sqrt[3]{64}$

25. $\sqrt[3]{729}$ **26.** $\sqrt[3]{512}$

27. $\sqrt[3]{\frac{1}{125}}$ **28.** $\sqrt[3]{\frac{1}{1{,}000}}$

***Find each root, if possible.* See Example 2.**

29. $\sqrt[4]{16}$ **30.** $\sqrt[4]{81}$

31. $\sqrt[4]{256}$ **32.** $-\sqrt[4]{625}$

33. $\sqrt[4]{-625}$ **34.** $\sqrt[4]{-16}$

35. $\sqrt[5]{-243}$ **36.** $\sqrt[5]{-1}$

37. $\sqrt[5]{32}$ **38.** $\sqrt[5]{-32}$

39. $-\sqrt[4]{\frac{1}{81}}$ **40.** $\sqrt[4]{\frac{1}{256}}$

***Find each root.* See Examples 3 and 4.**

41. $\sqrt[4]{16b^{12}}$ **42.** $\sqrt[4]{81h^{16}}$

43. $\sqrt[6]{s^{12}}$ **44.** $\sqrt[6]{d^{18}}$

45. $\sqrt[6]{t^{30}}$ **46.** $\sqrt[6]{p^{36}}$

47. $\sqrt[5]{y^5}$ **48.** $\sqrt[7]{x^7}$

49. $\sqrt[3]{m^3}$ **50.** $\sqrt[3]{n^{12}}$

51. $\sqrt[3]{27a^6}$ **52.** $\sqrt[3]{-64b^9}$

53. $\sqrt[5]{x^{15}}$ **54.** $\sqrt[5]{t^{10}}$

55. $\sqrt[5]{32v^{30}}$ **56.** $\sqrt[5]{32x^5}$

***Simplify.* See Example 5.**

57. $\sqrt[3]{24}$ **58.** $\sqrt[4]{32}$

59. $\sqrt[3]{-128}$ **60.** $\sqrt[3]{-250}$

61. $\sqrt[4]{162}$ **62.** $\sqrt[4]{80}$

63. $\sqrt[3]{72}$ **64.** $\sqrt[3]{189}$

65. $\sqrt[3]{\frac{250}{27}}$ **66.** $\sqrt[3]{\frac{54}{125}}$

67. $\sqrt[5]{64}$ **68.** $\sqrt[5]{160}$

***Evaluate.* See Example 6.**

69. $81^{1/2}$ **70.** $100^{1/2}$

71. $8^{1/3}$ **72.** $27^{1/3}$

73. $\left(\frac{1}{16}\right)^{1/4}$ **74.** $\left(\frac{1}{81}\right)^{1/4}$

75. $-144^{1/2}$ **76.** $-400^{1/2}$

77. $(-125)^{1/3}$ **78.** $(-8)^{1/3}$

79. $\left(\frac{9}{64}\right)^{1/2}$ **80.** $\left(\frac{4}{49}\right)^{1/2}$

***Evaluate.* See Example 7.**

81. $16^{3/2}$ **82.** $25^{3/2}$

83. $27^{2/3}$ **84.** $8^{4/3}$

85. $(-125)^{2/3}$ **86.** $(-64)^{2/3}$

87. $4^{5/2}$ **88.** $9^{5/2}$

89. $\left(\frac{1}{16}\right)^{3/4}$ **90.** $\left(-\frac{243}{32}\right)^{3/5}$

91. $-32^{3/5}$ **92.** $-32^{2/5}$

***Evaluate.* See Example 8.**

93. $8^{-1/3}$ **94.** $81^{-1/2}$

95. $25^{-3/2}$ **96.** $9^{-3/2}$

97. $16^{-5/4}$ **98.** $16^{-3/4}$

99. $(-27)^{-2/3}$ **100.** $8^{-5/3}$

101. $-81^{-3/2}$ **102.** $-125^{-2/3}$

103. $32^{-3/5}$ **104.** $(-32)^{-2/5}$

TRY IT YOURSELF

105. Simplify each root.

a. $\sqrt{32}$ **b.** $\sqrt[3]{32}$

c. $\sqrt[4]{32}$ **d.** $\sqrt[5]{32}$

106. Evaluate.

a. $16^{1/2}$ **b.** $16^{3/4}$

c. $16^{-1/4}$ **d.** $16^{-5/2}$

APPLICATIONS

107. WINDMILLS The power generated by a windmill is related to the speed of the wind by the formula $S = \sqrt[3]{\frac{P}{0.02}}$, where S is the speed of the wind (in mph) and P is the power (in watts). Find the speed of the wind when the windmill is producing 20 watts of power.

108. DEPRECIATION The formula $r = 1 - \sqrt[n]{\frac{S}{C}}$ gives the annual depreciation rate r of an item that had an original cost of C dollars and has a useful life of n years and a salvage value of S dollars. Use the information in the illustration to find (in percent) the annual depreciation rate for the new piece of sound equipment. (In the memo, \$81K = \$81,000.)

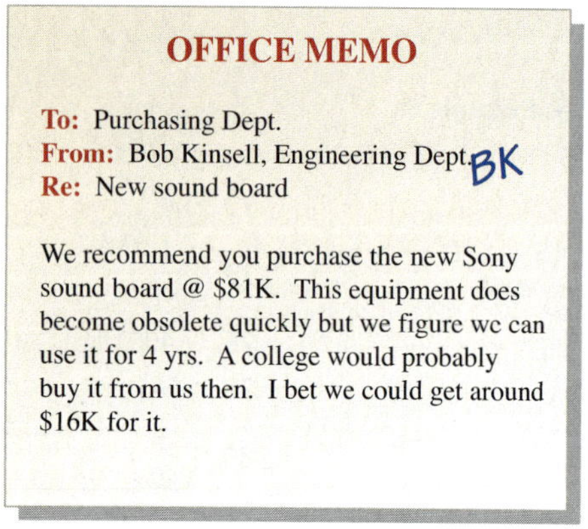
OFFICE MEMO

To: Purchasing Dept.
From: Bob Kinsell, Engineering Dept. BK
Re: New sound board

We recommend you purchase the new Sony sound board @ \$81K. This equipment does become obsolete quickly but we figure we can use it for 4 yrs. A college would probably buy it from us then. I bet we could get around \$16K for it.

109. HOLIDAY DECORATIONS Find the length s of each string of colored lights used to decorate an evergreen tree in the manner shown if $s = (r^2 + h^2)^{1/2}$.

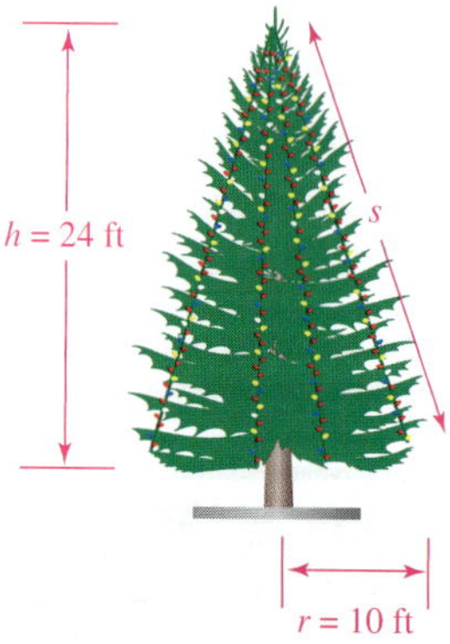

110. VISIBILITY The distance d in miles that a person in an airplane can see to the horizon on a clear day is given by the formula $d = 1.22a^{1/2}$, where a is the altitude of the plane in feet. Find d. Round to the nearest tenth of a mile.

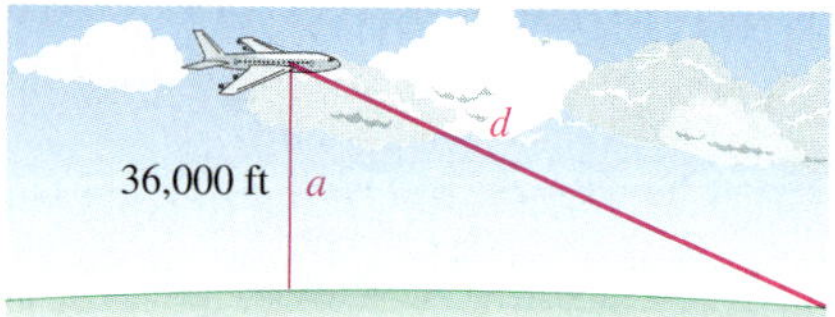

111. SPEAKERS The formula $A = V^{2/3}$ can be used to find the area A of one face of a cube if its volume V is known. Find the amount of floor space occupied by a speaker if it is a cube with a volume of 216 cubic inches.

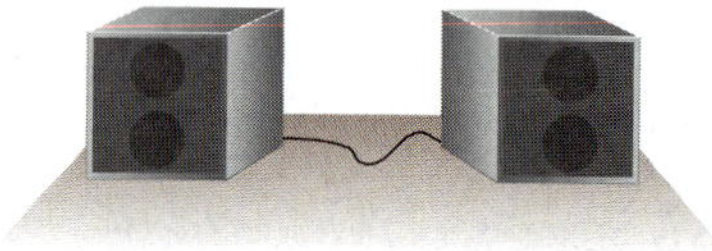

112. MEDICAL TESTS Before a series of X-rays are taken, a patient is injected with a special contrast mixture that highlights obstructions in his blood vessels. The amount of the original dose of contrast material remaining in the patient's bloodstream h hours after it is injected is given by $h^{-3/2}$. How much of the contrast material remains in the patient's bloodstream 4 hours after the injection?

WRITING

113. Explain why a negative number can have a real number for its cube root yet cannot have a real number for its fourth root.

114. What is a rational exponent? Give several examples.

115. Explain this statement: In the expression $16^{3/2}$, the number 3/2 requires that two operations be performed on 16.

116. To evaluate $27^{4/3}$, which expression would be easier to use, $(\sqrt[3]{27})^4$ or $\sqrt[3]{27^4}$? Explain.

REVIEW

Graph each equation.

117. $x = 3$

118. $y = -3$

119. $2x - y = -4$

120. $y = 4x - 4$

CHALLENGE PROBLEMS

121. Graph $y = \sqrt[3]{x}$ by first completing the table of solutions. Then plot the points and draw a smooth curve through them.

x	-27	-8	-1	0	1	8	27
y							

122. Evaluate: $\left(\frac{25}{49}\right)^{-1.5}$

CHAPTER 8 Summary & Review

SECTION 8.1 An Introduction to Square Roots

DEFINITIONS AND CONCEPTS	EXAMPLES
The number b is a **square root** of a if $b^2 = a$. The **positive** or **principal square root** of a positive number a, written as $\sqrt{a}$, is the positive square root of a. The **negative square root** of a is written as $-\sqrt{a}$.	3 is a square root of 9 because $3^2 = 9$. $\sqrt{49} = 7$ because $7^2 = 49$. $\sqrt{25} = 5$ because $5^2 = 25$. $\sqrt{16a^6} = 4a^3$ because $(4a^3)^2 = 16a^6$.
The expression within a **radical symbol** $\sqrt{}$ is called a **radicand.**	In $\sqrt{49}$, the radicand is 49. In $\sqrt{16a^6}$, the radicand is $16a^6$.
A number such as 16, 1, 0.36, $\frac{4}{9}$, or 225, that is the square of some rational number, is called a **perfect square.** The square root of a perfect square is a rational number. If a positive number is not a perfect square, its square root is an irrational number. Square roots of negative numbers are not real numbers.	Since 5 is not a perfect square, $\sqrt{5}$ is irrational. Its decimal value is approximately 2.236. Since no real number squared is -4, $\sqrt{-4}$ is not a real number.

Pythagorean theorem: If a and b are the lengths of the legs of a right triangle and c is the length of the hypotenuse, then $a^2 + b^2 = c^2$.

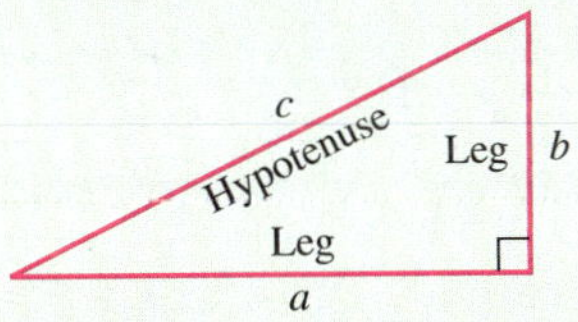

Find the length of the third side of the right triangle.

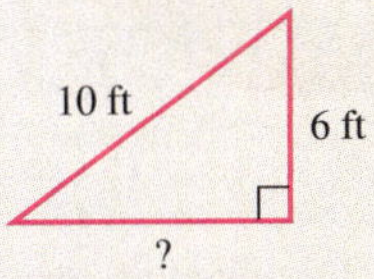

$a^2 + b^2 = c^2$ This is the Pythagorean theorem.

$6^2 + b^2 = 10^2$ Substitute 6 for a and 10 for c.

$36 + b^2 = 100$

$b^2 = 64$ To isolate b^2, subtract 36 from both sides.

$b = \sqrt{64}$ Since b must be positive, find the positive square root of 64.

$b = 8$

The length of the third side of the triangle is 8 ft.

Distance formula: The distance between points with coordinates (x_1, y_1) and (x_2, y_2) is given by

$$d = \sqrt{(x_2 - x_1)^2 + (y_2 - y_1)^2}$$

Find the distance between (2, 0) and (−2, 3).

$$d = \sqrt{(x_2 - x_1)^2 + (y_2 - y_1)^2}$$

$$d = \sqrt{(-2 - 2)^2 + (3 - 0)^2} \quad \text{Substitute.}$$

$$d = \sqrt{(-4)^2 + 3^2}$$

$$d = \sqrt{25}$$

$$d = 5$$

REVIEW EXERCISES

1. Fill in the blanks: $\sqrt{36} = 6$ because $(\quad)^2 =$ ____
2. Determine whether the statement is true or false: $-\sqrt{4} = \sqrt{-4}$

Find each square root. DO NOT USE A CALCULATOR.

3. $\sqrt{25}$
4. $\sqrt{49}$
5. $-\sqrt{144}$
6. $\sqrt{\frac{16}{81}}$
7. $\sqrt{0.64}$
8. $\sqrt{1}$

Use a calculator to approximate each expression to three decimal places.

9. $\sqrt{21}$

10. $-2\sqrt{7}$

11. Determine whether each number is rational, irrational, or not a real number:

$$\sqrt{-2}, \sqrt{68}, \sqrt{81}, \sqrt{3}$$

12. ROAD SIGNS To find the maximum velocity a car can safely travel around a curve without skidding, we can use the formula $v = \sqrt{2.5r}$, where v is the velocity in mph and r is the radius of the curve in feet. How should the road sign be labeled if it is to be posted in front of a curve with a radius of 360 feet?

Simplify. All variables represent positive real numbers.

13. $\sqrt{x^2}$

14. $\sqrt{4b^4}$

15. $-\sqrt{y^{12}}$

16. $\sqrt{9h^{16}}$

Refer to the right triangle.

17. Find c where $a = 6$ and $b = 8$.

18. Find b where $a = 8$ and $c = 17$.

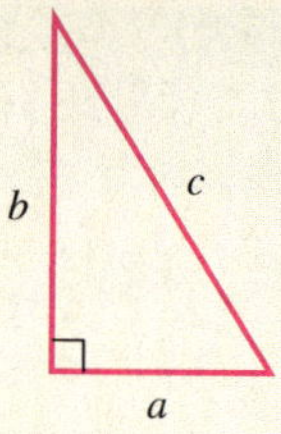

19. THEATER SEATING How much higher is the seat at the top of the incline than the one at the bottom?

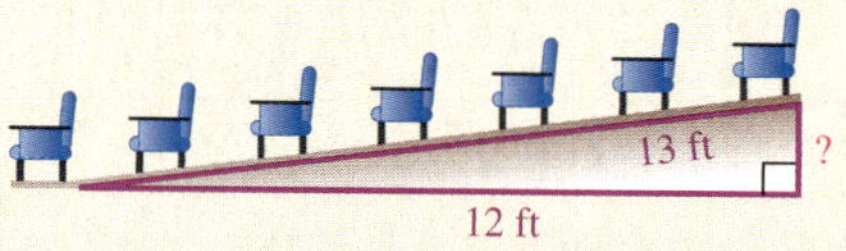

20. TRIANGLES The length of the hypotenuse of an isosceles right triangle is 3 inches. Find the length of a leg of the triangle. Round to the nearest hundredth of an inch.

Find the distance between the points. If an answer is not exact, round to the nearest hundredth.

21. $(-7, 12), (-4, 8)$

22. $(-15, -3), (-10, -16)$

SECTION 8.2 Simplifying Square Roots

DEFINITIONS AND CONCEPTS	EXAMPLES
Simplified form of a radical: 1. Except for 1, the radicand has no perfect-square factors. 2. No fraction appears in the radicand. 3. No radical appears in the denominator.	These radicals are *not* in simplified form: $\sqrt{18m^3}$ — The radicand has a perfect-square factor of $9m^2$. $\sqrt{\frac{5}{64}}$ — The radicand is a fraction. $\frac{1}{\sqrt{3}}$ — A radical appears in the denominator.
The **product and quotient rules for square roots** can be used to simplify radical expressions. $\sqrt{a \cdot b} = \sqrt{a}\sqrt{b}$ $\sqrt{\frac{a}{b}} = \frac{\sqrt{a}}{\sqrt{b}} \quad (b \neq 0)$	Simplify: $\sqrt{18m^3} = \sqrt{9m^2 \cdot 2m}$ — Write $18m^3$ as a product of its greatest perfect-square factor, $9m^2$, and one other factor. $= \sqrt{9m^2}\sqrt{2m}$ — The square root of a product is equal to the product of the square roots. $= 3m\sqrt{2m}$ — Find the square root of the perfect square: $\sqrt{9m^2} = 3m$. Simplify: $\sqrt{\frac{5}{64}} = \frac{\sqrt{5}}{\sqrt{64}}$ — The square root of a quotient is equal to the quotient of the square roots. $= \frac{\sqrt{5}}{8}$ — In the denominator, evaluate: $\sqrt{64} = 8$.

REVIEW EXERCISES

Simplify. All variables represent positive real numbers.

23. $\sqrt{32}$

24. $\sqrt{x^5}$

25. $\sqrt{80x^2}$

26. $-2\sqrt{63}$

27. $\sqrt{250t^3}$

28. $\sqrt{\frac{16}{25}}$

29. $\sqrt{\frac{60}{49}}$

30. $\sqrt{\frac{242x^4}{169x^2}}$

31. FITNESS EQUIPMENT The length of the sit-up board can be found using the Pythagorean theorem. Find its length. Express the answer in simplified radical form. Then express your result as a decimal approximation rounded to the nearest tenth.

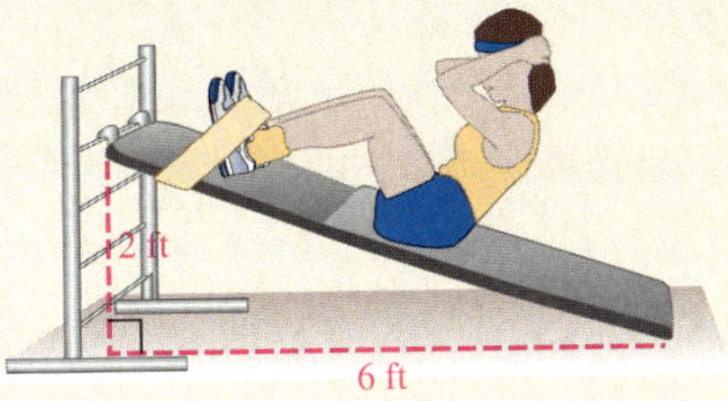

32. Determine whether the statement is true or false: $\sqrt{x^9} = x^3$.

SECTION 8.3 Adding and Subtracting Radical Expressions

DEFINITIONS AND CONCEPTS	EXAMPLES
Square root radicals are called **like radicals** when they have the same radicand.	Like radicals: $4\sqrt{2}$ and $5\sqrt{2}$ *The same radicand* Unlike radicals: $3\sqrt{6}$ and $7\sqrt{3}$ *Different radicands*
Radical expressions can be added or subtracted if they contain like radicals. To **combine like radicals** we use the distributive property in reverse.	Add: $3\sqrt{7} + 5\sqrt{7} = (3 + 5)\sqrt{7}$ $= 8\sqrt{7}$ Subtract: $8\sqrt{2y} - 2\sqrt{2y} = (8 - 2)\sqrt{2y}$ $= 6\sqrt{2y}$
If a sum or difference involves unlike radicals, make sure that each one is written in simplified form. After doing so, like radicals may result that can be combined.	Add: $\sqrt{12} + \sqrt{75} = \sqrt{4}\sqrt{3} + \sqrt{25}\sqrt{3}$ Simplify $\sqrt{12}$ and $\sqrt{75}$. $= 2\sqrt{3} + 5\sqrt{3}$ $= 7\sqrt{3}$ Combine like radicals.

REVIEW EXERCISES

Perform the operations, if possible. All variables represent nonnegative real numbers.

33. $\sqrt{10} + \sqrt{10}$

34. $6\sqrt{x} - \sqrt{x}$

35. $\sqrt{2} + \sqrt{8} - \sqrt{18}$

36. $\sqrt{3} + 4 + \sqrt{27} - 7$

37. $5\sqrt{28} - 3\sqrt{63}$

38. $3\sqrt{5y^3} - 5y\sqrt{20y}$

39. Explain why we cannot add $3\sqrt{5}$ and $5\sqrt{3}$.

40. GARDENING Find the difference in the lengths of the two wires used to secure the tree.

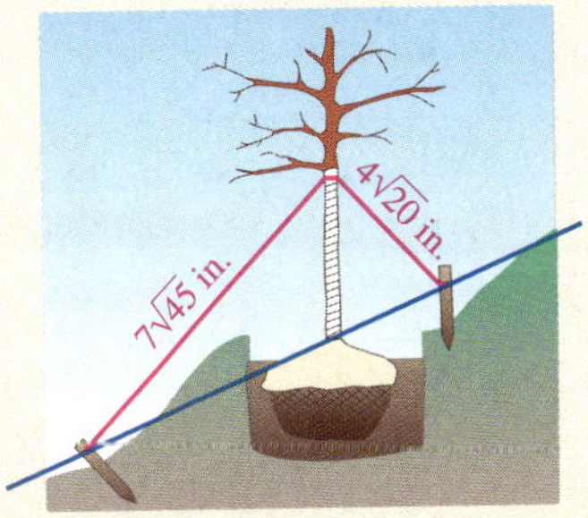

SECTION 8.4 Multiplying and Dividing Radical Expressions

DEFINITIONS AND CONCEPTS	EXAMPLES
To **multiply square roots:** The product of two square roots is equal to the square root of the product. $\sqrt{a} \cdot \sqrt{b} = \sqrt{a \cdot b}$	Multiply: $3\sqrt{5} \cdot 2\sqrt{7} = 6\sqrt{5 \cdot 7} = 6\sqrt{35}$ Multiply the radicands, 5 and 7, to get 35, and write the square root of that product. Multiply and simplify: $\sqrt{2a^5} \cdot \sqrt{6a} = \sqrt{2a^5 \cdot 6a}$ $= \sqrt{12a^6}$ Since $12a^6$ has a perfect-square factor of $4a^6$, we can simplify $\sqrt{12a^6} = \sqrt{4a^6 \cdot 3} = 2a^3\sqrt{3}$
The Square of a Square Root: For any nonnegative real number a, $(\sqrt{a})^2 = a$.	$(\sqrt{5})^2 = 5$ $\quad (\sqrt{z+1})^2 = z + 1$ $\quad (3\sqrt{a})^2 = 9a$
Radical expressions with more than one term are multiplied as if they were polynomials.	Multiply: $2\sqrt{x}(3 - 4\sqrt{x}) = 2\sqrt{x} \cdot 3 - 2\sqrt{x} \cdot 4\sqrt{x}$ Distribute $2\sqrt{x}$. $= 6\sqrt{x} - 8\sqrt{x^2}$ $= 6\sqrt{x} - 8x$ Simplify $\sqrt{x^2}$.
Use the FOIL method to multiply two radical expressions, each having two terms.	Multiply: $(\sqrt{a} - 2)(\sqrt{a} + 6) = \overset{F}{\sqrt{a} \cdot \sqrt{a}} + \overset{O}{6\sqrt{a}} - \overset{I}{2\sqrt{a}} - \overset{L}{2 \cdot 6}$ $= \sqrt{a^2} + 4\sqrt{a} - 12$ Combine like radicals. $= a + 4\sqrt{a} - 12$ Simplify $\sqrt{a^2}$.
To **divide square roots:** The quotient of two square roots is equal to the square root of the quotient. $\dfrac{\sqrt{a}}{\sqrt{b}} = \sqrt{\dfrac{a}{b}}$ $\quad (b \neq 0)$	Divide: $\dfrac{\sqrt{70}}{\sqrt{10}} = \sqrt{\dfrac{70}{10}} = \sqrt{7}$ Divide the radicands, 70 by 10, to get 7, and write the square root of that quotient. The result does not simplify.
To **rationalize a square root denominator,** multiply the numerator and denominator of the given fraction by the square root that appears in its denominator, or by a square root that makes a perfect-square radicand in the denominator.	Rationalize the denominator: $\dfrac{3}{\sqrt{5}} = \dfrac{3}{\sqrt{5}} \cdot \dfrac{\sqrt{5}}{\sqrt{5}}$ Build an equivalent fraction by multiplying by a form of 1. $= \dfrac{3\sqrt{5}}{5}$ Multiply the numerators. Multiply the denominators: $\sqrt{5} \cdot \sqrt{5} = \sqrt{25} = 5$.
To **rationalize a two-term denominator,** multiply the expression by a form of 1 that uses the **conjugate** of the denominator.	Rationalize the denominator: $\dfrac{3 - \sqrt{a}}{2 + \sqrt{a}} = \dfrac{3 - \sqrt{a}}{2 + \sqrt{a}} \cdot \dfrac{2 - \sqrt{a}}{2 - \sqrt{a}}$ $2 - \sqrt{a}$ is the conjugate of $2 + \sqrt{a}$. $= \dfrac{6 - 3\sqrt{a} - 2\sqrt{a} + \sqrt{a^2}}{4 - \sqrt{a^2}}$ $= \dfrac{6 - 5\sqrt{a} + a}{4 - a}$ Simplify.

REVIEW EXERCISES

Perform the operations. All variables represent nonnegative real numbers.

41. $\sqrt{2}\sqrt{3}$

42. $(-5\sqrt{5})^2$

43. $(3\sqrt{3x^3})(4\sqrt{6x^2})$

44. $(\sqrt{15} + 3x)^2$

45. $\sqrt{2}(\sqrt{8} - \sqrt{18})$

46. $(\sqrt{3} + \sqrt{5})(\sqrt{3} - \sqrt{5})$

47. $(\sqrt{x})^2$

48. $(\sqrt{t - 1})^2$

49. Find the area of the rectangle. Express the answer in simplified radical form. Then express your result as a decimal approximation to the nearest tenth.

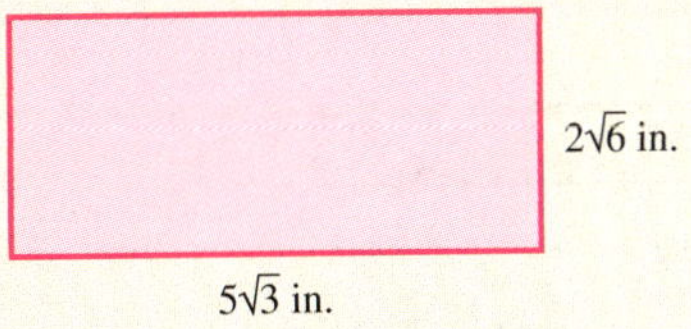

50. PATIO TILES The length of the diagonal of a square tile is 1 foot. It can be shown using the Pythagorean theorem that the length of one side of the tile is $\frac{1}{\sqrt{2}}$ foot. Rationalize the denominator of this expression.

Rationalize the denominator. All variables represent positive real numbers.

51. $\frac{9}{\sqrt{7}}$

52. $\sqrt{\frac{3}{a}}$

53. $\frac{16\sqrt{3}}{\sqrt{5}}$

54. $\frac{11}{\sqrt{75}}$

55. $\frac{1}{\sqrt{8x}}$

56. $\frac{\sqrt{7}}{\sqrt{7} - \sqrt{2}}$

57. $\frac{\sqrt{a}}{\sqrt{a} + 1}$

58. $\frac{2 + \sqrt{b}}{\sqrt{b} - 3}$

SECTION 8.5 Solving Radical Equations

DEFINITIONS AND CONCEPTS

The **squaring property of equality:**

If $a = b$, then $a^2 = b^2$.

The **square of a square root:** For any nonnegative real number a,

$$(\sqrt{a})^2 = a$$

To **solve radical equations** containing square roots:

1. Isolate a radical term on one side of the equation.
2. Square both sides of the equation.
3. Solve the resulting equation.
4. Check the possible solutions in the original equation. Discard any **extraneous solutions.**

EXAMPLES

Solve: $\sqrt{x + 2} - 3 = 5$

$\sqrt{x + 2} - 3 + 3 = 5 + 3$	To isolate the radical, add 3 to both sides.
$\sqrt{x + 2} = 8$	Do the additions.
$(\sqrt{x + 2})^2 = (8)^2$	Use the squaring property of equality and square both sides.
$x + 2 = 64$	Simplify the square of a square root: $(\sqrt{x + 2})^2 = x + 2$.
$x = 62$	To isolate x, subtract 2 from both sides.

Check: $\sqrt{x + 2} - 3 = 5$	The original equation.
$\sqrt{62 + 2} - 3 \stackrel{?}{=} 5$	Substitute 62 for x.
$\sqrt{64} - 3 \stackrel{?}{=} 5$	Evaluate the left side.
$5 = 5$	True

Since we obtain a true statement, 62 is the solution.

Rationalize each denominator. All variables represent positive real numbers.

29. $\dfrac{2}{\sqrt{7}}$

30. $\dfrac{\sqrt{3}}{\sqrt{x} - 8}$

Solve each equation and check the result.

31. $\sqrt{x} = 15$

32. $\sqrt{2 - x} - 2 = 6$

33. $\sqrt{3x + 9} = 2\sqrt{x + 1}$

34. $x = \sqrt{x - 1} + 1$

35. Check to determine whether 0 is a solution of $\sqrt{3x + 1} + 2 = 4$. Explain.

36. CARPENTRY A carpenter uses a tape measure to see whether the wall he just put up is perfectly square with the floor. Explain what mathematical concept he is applying. If the wall is positioned correctly, what should the measurement on the tape read?

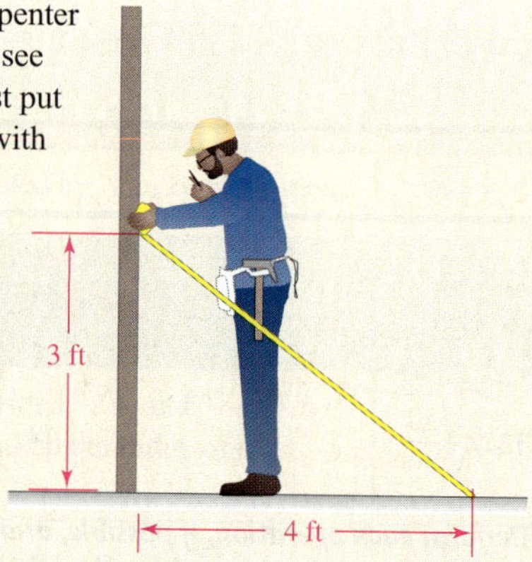

Find each root, if possible. All variables represent nonnegative real numbers.

37. $\sqrt[3]{-125}$

38. $\sqrt[4]{-256}$

39. $\sqrt[3]{\dfrac{1}{64}}$

40. $\sqrt[5]{-32x^{10}}$

Simplify. All variables represent nonnegative real numbers.

41. $\sqrt[4]{81x^4}$

42. $\sqrt[3]{88}$

Evaluate.

43. $144^{1/2}$

44. $8^{2/3}$

45. $(-27)^{-4/3}$

46. $-9^{-1/2}$

GROUP PROJECT

A SPIRAL OF ROOTS

Overview: In this activity, you will gain a better understanding of square roots and get additional practice using the Pythagorean theorem.

Instructions: Form groups of 2 students. You will need an $8\frac{1}{2} \times 11$ inch piece of white, unlined paper, a 3×5 inch index card (to trace a right angle), a ruler, and a pencil.

Draw an isosceles right triangle with legs 1 inch long near the right edge of the paper. Use the Pythagorean theorem to determine the length of the hypotenuse.

Next, construct a second right triangle using the hypotenuse of the first triangle as one leg. Draw a second leg that is 1 inch long. Find the length of the hypotenuse of Triangle 2 using the Pythagorean theorem. Continue creating right triangles, using the previous hypotenuse as one leg and drawing a second leg that is 1 inch long. For each right triangle created, use the Pythagorean theorem to find the length of its hypotenuse. Make a list of the lengths of each hypotenuse, and look for a pattern.

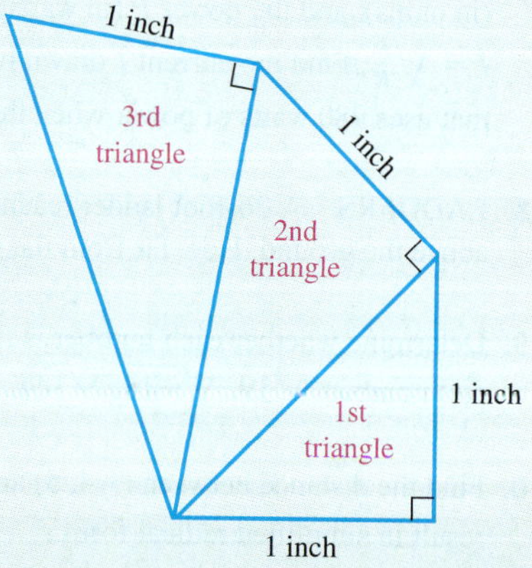

CUMULATIVE REVIEW
Chapters 1–8

1. Determine whether each statement is true or false. [Section 1.3]

a. All whole numbers are integers.

b. π is a rational number.

c. A real number is either rational or irrational.

2. Evaluate: $\dfrac{-3(3+2)^2-(-5)}{17-|-22|}$ [Section 1.7]

3. Simplify: $3p - 6(p + z) + p$ [Section 1.9]

4. Solve: $2 - (4x + 7) = 3 + 2(x + 2)$ [Section 2.2]

5. BACKPACKS Pediatricians advise that children should not carry more than 20% of their own body weight in a backpack. According to this warning, how much weight can a fifth-grade girl who weighs 85 pounds safely carry in her backpack? [Section 2.3]

6. SURFACE AREA The total surface area A of a box with dimensions l, w, and h is given by the formula $A = 2lw + 2wh + 2lh$. If $A = 202$ square inches, $l = 9$ inches, and $w = 5$ inches, find h. [Section 2.4]

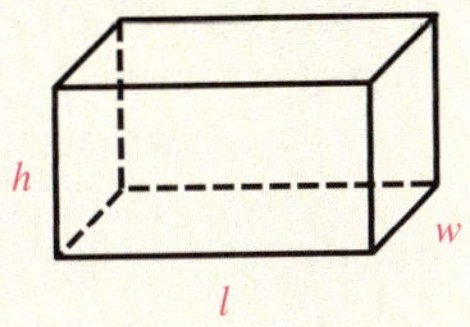

7. SEARCH AND RESCUE Two search and rescue teams leave base at the same time, looking for a lost boy. The first team, on foot, heads north at 2 mph and the other, on horseback, south at 4 mph. How long will it take them to search a distance of 21 miles between them? [Section 2.6]

8. BLENDING COFFEE A store sells regular coffee for \$8 a pound and gourmet coffee for \$14 a pound. Using 40 pounds of the gourmet coffee, the owner makes a blend to put on sale for \$10 a pound. How many pounds of regular coffee should he use? [Section 2.6]

9. Solve: $3 - 3x \geq 6 + x$. Graph the solution set. Then describe the graph using interval notation. [Section 2.7]

10. Is $(-6, -7)$ a solution of $4x - 3y = -4$? [Section 3.1]

Graph each equation.

11. $y = \frac{1}{2}x$ [Section 3.2]

12. $3x - 4y = 12$ [Section 3.3]

13. $x = 5$ [Section 3.3]

14. $y = 2x^2 - 3$ [Section 5.4]

15. SHOPPING On the graph in the next column, the line approximates the growth in retail sales for U.S. shopping centers during the years 1994–2005. Find the rate of increase in sales by finding the slope of the line. [Section 3.4]

Sales at U.S. Shopping Centers

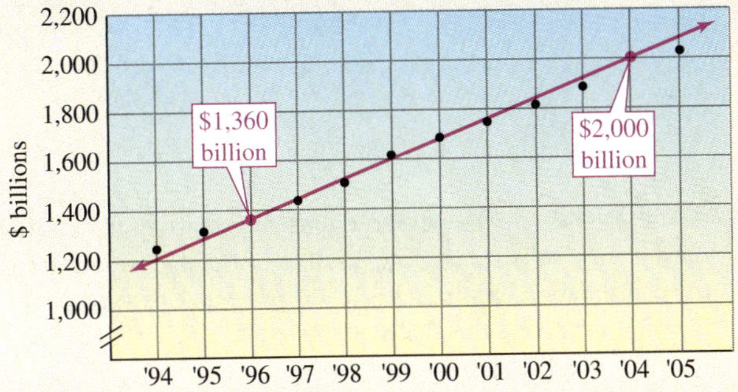

Based on data from International Council of Shopping Centers

16. What is the slope of the line defined by each equation? [Section 3.5]

a. $y = 3x - 7$

b. $2x + 3y = -10$

17. Find the equation of the line passing through $(-2, 5)$ and $(4, 8)$. Write the answer in slope–intercept form. [Section 3.6]

18. If $f(x) = 2x^2 - 3x + 1$, find $f(-3)$. [Section 3.8]

19. BOATING The graph shows the vertical distance from a point on the tip of a propeller to the centerline as the propeller spins. Is this the graph of a function? [Section 3.8]

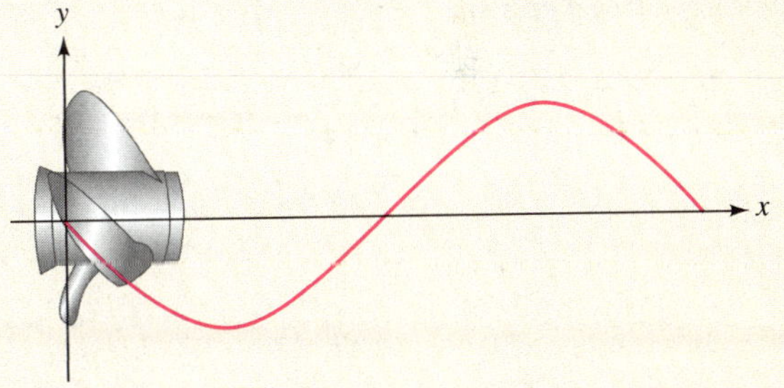

20. Solve the system $\begin{cases} x + y = 4 \\ y = x + 6 \end{cases}$ by graphing. [Section 4.1]

Solve each system of equations.

21. $\begin{cases} x = y + 4 \\ 2x + y = 5 \end{cases}$ [Section 4.2]

22. $\begin{cases} 3s + 4t = 5 \\ 2s - 3t = -8 \end{cases}$ [Section 4.3]

23. FINANCIAL PLANNING In investing \$6,000 of a couple's money, a financial planner put some of it into a savings account paying 6% annual interest. The rest was invested in a riskier mini-mall development plan paying 12% annually. The combined interest earned for the first year was \$540. How much money was invested at each rate? Use two variables to solve this problem. [Section 4.4]

24. Graph: $\begin{cases} 3x + 2y \geq 6 \\ x + 3y \leq 6 \end{cases}$ [Section 4.5]

Simplify. Use only positive exponents in your answers.

25. $(x^5)^2(x^7)^3$ [Section 5.1]

26. $\left(\dfrac{a^3b}{c^4}\right)^5$ [Section 5.1]

27. $4^{-3} \cdot 4^{-2} \cdot 4^5$ [Section 5.2]

28. $(2a^{-2}b^3)^{-4}$ [Section 5.2]

29. ASTRONOMY The **parsec,** a unit of distance used in astronomy, is 3×10^{16} meters. The distance to Betelgeuse, a star in the constellation Orion, is 1.6×10^2 parsecs. Use scientific notation to express this distance in meters. [Section 5.3]

30. Write 0.0000000043 in scientific notation. [Section 5.3]

Perform the operations.

31. $(3a^2 - 2a + 4) - (a^2 - 3a + 7)$ [Section 5.5]

32. $0.3p^5(0.4p^4 - 6p^2)$ [Section 5.6]

33. $(-3t + 2s)(2t - 3s)$ [Section 5.6]

34. $(4b - 8)^2$ [Section 5.7]

35. $\left(6b + \dfrac{1}{2}\right)\left(6b - \dfrac{1}{2}\right)$ [Section 5.7]

36. $x + 2\overline{)2x^2 + 3x - 2}$ [Section 5.8]

Factor completely, if possible.

37. $12x^2y - 6xy^2 + 9xy^3$ [Section 6.1]

38. $2x^2 + 2xy - 3x - 3y$ [Section 6.1]

39. $x^2 + 7x + 10$ [Section 6.2]

40. $6a^2 - 7a - 20$ [Section 6.3]

41. $6 + 3x^2 + x$ [Section 6.3]

42. $25a^2 - 70ab + 49b^2$ [Section 6.4]

43. $a^3 + 8b^3$ [Section 6.5]

44. $2x^5 - 32x$ [Section 6.6]

Solve each equation.

45. $x^2 + 3x + 2 = 0$ [Section 6.7]

46. $5x^2 = 10x$ [Section 6.7]

47. $6x^2 - x = 2$ [Section 6.7]

48. $a^2 - 25 = 0$ [Section 6.7]

49. CHILDREN'S STICKERS A rectangular-shaped sticker has an area of 20 cm^2. The width is 1 cm shorter than the length. Find the length of the sticker. [Section 6.8]

50. Simplify: $\dfrac{x^2 + 2x + 1}{x^2 - 1}$ [Section 7.1]

Perform the operations. Simplify, if possible.

51. $\dfrac{p^2 - p - 6}{3p - 9} \div \dfrac{p^2 + 6p + 9}{p^2 - 9}$ [Section 7.2]

52. $\dfrac{12x^2}{7 - x} \cdot \dfrac{x - 7}{20x^3}$ [Section 7.2]

53. $\dfrac{13}{15a} - \dfrac{8}{15a}$ [Section 7.3]

54. $\dfrac{x + 2}{x + 5} - \dfrac{x - 3}{x + 7}$ [Section 7.4]

55. $\dfrac{1}{6b^4} - \dfrac{8}{9b^2}$ [Section 7.4]

56. $\dfrac{\frac{1}{x} + \frac{1}{y}}{\frac{1}{x} - \frac{1}{y}}$ [Section 7.5]

57. Solve: $\dfrac{7}{a^2 - a - 2} + \dfrac{1}{a + 1} = \dfrac{3}{a - 2}$ [Section 7.6]

58. FILLING A POOL An inlet pipe can fill an empty swimming pool in 5 hours, and another inlet pipe can fill the pool in 4 hours. How long will it take both pipes to fill the pool? [Section 7.7]

59. ONLINE SALES A company found that, on average, it made 9 online sales transactions for every 500 hits on its Internet Web site. If the company's Web site had 360,000 hits in one year, how many sales transactions did it have that year? [Section 7.8]

60. CAFETERIAS The length of time the punch in a dispenser at a children's summer camp will last varies inversely as the number of children at the camp. If the punch in the dispenser is consumed by 45 children in 3 days, how long will the punch last for 30 children? [Section 7.9]

61. CARGO SPACE How wide a piece of plywood can be stored diagonally in the back of the van? [Section 8.1]

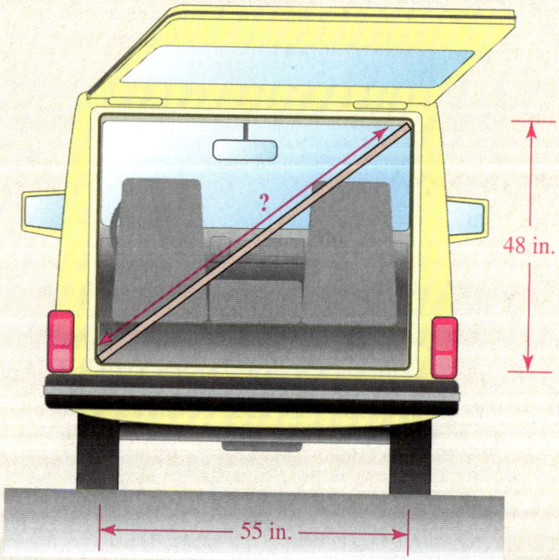

62. Explain why $\sqrt{-4}$ is not a real number. [Section 8.1]

Simplify. All variables represent nonnegative real numbers.

63. $\sqrt{64}$ [Section 8.1]

64. $\sqrt{32x^3}$ [Section 8.2]

65. $-\sqrt[3]{-27m^6}$ [Section 8.6]

66. $\sqrt[4]{81t^4}$ [Section 8.6]

Perform each operation and simplify. All variables represent nonnegative real numbers.

67. $\sqrt{48} - \sqrt{8} + \sqrt{27} - \sqrt{32}$ [Section 8.3]

68. $(\sqrt{y} - 4)(\sqrt{y} - 5)$ [Section 8.4]

Rationalize the denominator. All variables represent positive real numbers.

69. $\dfrac{4}{\sqrt{5}}$ [Section 8.4]

70. $\sqrt{\dfrac{17}{2x}}$ [Section 8.4]

71. Solve: $\sqrt{6x + 19} - 5 = 2$ [Section 8.5]

72. Evaluate: $16^{3/2}$ [Section 8.6]

CHAPTER 9

Quadratic Equations

from Campus to Careers
Graphic Designer

Graphic designers combine their artistic talents with their mathematical skills to create attractive advertisements, photography, packaging, and even Web pages. They perform arithmetic computations with fractions and decimals to determine proper font sizes and page margins. They apply algebraic concepts such as graphing and equation solving to plan page layouts. And they use formulas to create production schedules and calculate costs.

Graphic designers often begin a project with the final product in mind and work backwards to determine the specifics. In **Problem 81** of **Study Set 9.3,** you will do just that. Given the area of a triangular-shaped poster, you are to find the length of its base and its height by writing and then solving a quadratic equation.

JOB TITLE:
Graphic Designer
EDUCATION:
A bachelor's degree is required for most entry-level positions.
JOB OUTLOOK:
Employment is expected to increase between 9% and 17% through the year 2014.
ANNUAL EARNINGS:
Median annual salary for entry-level designers is \$32,000 and for staff level designers is \$42,500.
FOR MORE INFORMATION:
www.bls.gov/oco/home.htm

Study Skills Workshop

Preparing for Your Next Math Course

Before moving on to a new mathematics course, it's worthwhile to take some time to reflect on your effort and performance in this course.

Now Try This

As this course draws to a close, here are some questions to ask yourself.

1. How was my attendance?
2. Was I organized? Did I have the right materials?
3. Did I follow a regular schedule?
4. Did I pay attention in class and take good notes?
5. Did I spend the appropriate amount of time on homework?
6. How did I prepare for tests? Did I have a test-taking strategy?
7. Was I part of a study group? If not, why not? If so, was it worthwhile?
8. Did I ever seek extra help from a tutor or from my instructor?
9. In what topics was I the strongest? In what topics was I the weakest?
10. If I had it to do over, would I do anything differently?

SECTION 9.1

Solving Quadratic Equations: The Square Root Property

Objectives

1. Use the square root property to solve equations of the form $x^2 = c$.
2. Use the square root property to solve equations of the form $(ax + b)^2 = c$.
3. Solve problems modeled by quadratic equations.

Recall that a **quadratic equation** can be written in the form $ax^2 + bx + c = 0$, where a, b, and c represent real numbers and $a \neq 0$. Some examples are

$$x^2 - x - 6 = 0, \qquad 4x^2 + 4x + 1 = 0, \qquad \text{and} \qquad x^2 - 16 = 0$$

We have solved quadratic equations like these using factoring in combination with the zero-factor property. To review this method, let's solve $x^2 - 16 = 0$.

$$x^2 - 16 = 0$$

$$(x + 4)(x - 4) = 0 \qquad \text{Factor the difference of two squares.}$$

$$x + 4 = 0 \quad \text{or} \quad x - 4 = 0 \qquad \text{Set each factor equal to 0.}$$

$$x = -4 \quad \big| \quad x = 4 \qquad \text{Solve each equation.}$$

The solutions are −4 and 4.

1 Use the Square Root Property to Solve Equations of the Form $x^2 = c$.

We will now solve $x^2 - 16 = 0$ in another way. This time, we will ignore the zero-factor property condition that requires 0 on one side of the equation. Instead, we will add 16 to both sides to isolate x^2.

$$x^2 - 16 = 0$$

$$x^2 = 16 \quad \text{Add 16 to both sides.}$$

We see that x must be a number whose square is 16. Therefore, x must be a square root of 16. Since every positive number has two square roots, one positive and one negative, we have

$$x = \sqrt{16} \quad \text{or} \quad x = -\sqrt{16}$$

$$x = 4 \quad | \quad x = -4$$

As with the factor method, we find that the solutions of $x^2 - 16 = 0$ are 4 and -4. This approach illustrates how the following *square root property* can be used to solve certain types of quadratic equations.

The Square Root Property of Equations

For any nonnegative real number c, if $x^2 = c$, then

$$x = \sqrt{c} \quad \text{or} \quad x = -\sqrt{c}$$

We can write the conclusion of the square root property in more compact form, called **double-sign notation:**

$$x = \pm\sqrt{c} \quad \text{Read as "x is equal to the positive or negative square root of c."}$$

EXAMPLE 1 Solve: **a.** $x^2 - 5 = 0$ **b.** $x^2 = 12$ **c.** $3a^2 + 1 = 11$ **d.** $n^2 = -4$

Strategy We will use properties of equality to isolate the *squared term* on one side of the equation. Then we will use the square root property to isolate the variable itself.

Why To solve the original equation, we want to find a simpler equivalent equation of the form $x = \pm$ **a number**, whose solutions are obvious.

Solution

a. We can use the addition property of equality to isolate x^2.

$$x^2 - 5 = 0 \quad \text{This is the equation to solve.}$$

$$x^2 = 5 \quad \text{Add 5 to both sides.}$$

Now we use the square root property to isolate x.

$$x = \sqrt{5} \quad \text{or} \quad x = -\sqrt{5}$$

$$x = \pm\sqrt{5} \quad \text{Use double-sign notation.}$$

Success Tip
The square root property is especially helpful when solving quadratic equations of the form $x^2 - c = 0$, where c is not a perfect square.

The check for $\sqrt{5}$:

$$x^2 - 5 = 0$$

$$(\sqrt{5})^2 - 5 \stackrel{?}{=} 0$$

$$5 - 5 \stackrel{?}{=} 0$$

$$0 = 0 \quad \text{True}$$

The check for $-\sqrt{5}$:

$$x^2 - 5 = 0$$

$$(-\sqrt{5})^2 - 5 \stackrel{?}{=} 0$$

$$5 - 5 \stackrel{?}{=} 0$$

$$0 = 0 \quad \text{True}$$

Since each statement is true, the solutions are $\sqrt{5}$ and $-\sqrt{5}$, or $\pm\sqrt{5}$. The solution set is written as $\{-\sqrt{5}, \sqrt{5}\}$ or $\{\pm\sqrt{5}\}$.

b. Since x^2 is already isolated, we simply apply the square root property.

$x^2 = 12$ — This is the equation to solve.

$x = \sqrt{12}$ or $x = -\sqrt{12}$ — To isolate x, use the square root property.

$x = \pm\sqrt{12}$ — Use double-sign notation.

$x = \pm 2\sqrt{3}$ — Simplify the radical: $\sqrt{12} = \sqrt{4 \cdot 3} = \sqrt{4}\sqrt{3} = 2\sqrt{3}$.

Verify that $2\sqrt{3}$ and $-2\sqrt{3}$ are solutions by checking each in the original equation.

Caution
When using the square root property to solve an equation, always write the ± symbol, or you will lose one of the solutions.

c. $3a^2 + 1 = 11$ — This is the equation to solve.

$3a^2 = 10$ — To isolate the term $3a^2$, subtract 1 from both sides.

$a^2 = \frac{10}{3}$ — To isolate a^2, divide both sides by 3.

$a = \pm\sqrt{\frac{10}{3}}$ — To isolate a, use the square root property: $a = \sqrt{\frac{10}{3}}$ or $a = -\sqrt{\frac{10}{3}}$. Use double-sign notation.

To rationalize the denominator of the result, we proceed as follows:

$$x = \pm\sqrt{\frac{10}{3}} = \pm\frac{\sqrt{10}}{\sqrt{3}} \cdot \frac{\sqrt{3}}{\sqrt{3}} = \pm\frac{\sqrt{30}}{3}$$

Verify that $\frac{\sqrt{30}}{3}$ and $-\frac{\sqrt{30}}{3}$ are solutions by checking each in the original equation.

The Language of Algebra
The *exact* solutions are $\pm\frac{\sqrt{30}}{3}$. To the nearest hundredth, the *approximate* solutions are ± 1.83.

d. $n^2 = -4$ — This is the equation to solve.

$n = \pm\sqrt{-4}$ — Use the square root property.

Since the square root of -4 is not a real number, the equation has no real-number solutions.

Self Check 1
Solve: **a.** $x^2 - 21 = 0$ **b.** $b^2 = 54$
c. $5m^2 - 1 = 6$ **d.** $x^2 = -25$

Now Try Problems 23, 25, 31, and 33

2 Use the Square Root Property to Solve Equations of the Form $(ax + b)^2 = c$.

We can extend the square root property to solve equations that involve the square of a binomial.

EXAMPLE 2
Solve: **a.** $(x - 3)^2 = 36$ **b.** $(x + 1)^2 = 50$
c. $(2s - 4)^2 = 7$

Strategy Instead of a variable squared on the left side of the equation, we have a quantity squared. We still use the square root property to solve each equation.

Why We want to eliminate the square on the binomial, so that we can eventually isolate the variable on one side of the equation.

> **Caution**
> It might be tempting to square the binomial on the left side of $(x - 3)^2 = 36$. However, that causes unnecessary, additional steps to be used to solve the equation in another way.
>
> $(x - 3)^2 = 36$
> ~~$x^2 - 6x + 9 = 36$~~

Solution

a.

$(x - 3)^2 = 36$	This is the equation to solve.
$x - 3 = \sqrt{36}$ or $x - 3 = -\sqrt{36}$	Use the square root property.
$x - 3 = \pm\sqrt{36}$	Use double-sign notation.
$x - 3 = \pm 6$	Evaluate: $\sqrt{36} = 6$.
$x = 3 \pm 6$	To isolate x, add 3 to both sides. It is standard practice to write the 3 in front of the ± symbol.

We read 3 ± 6 as "3 plus or minus 6." To find the solutions, we perform the calculation using a plus symbol + and then using a minus symbol −.

$x = 3 + 6$ or $x = 3 - 6$

$x = 9$ | $x = -3$

***The check for* 9:**	***The check for* −3:**
$(x - 3)^2 = 36$	$(x - 3)^2 = 36$
$(9 - 3)^2 \stackrel{?}{=} 36$	$(-3 - 3)^2 \stackrel{?}{=} 36$
$(6)^2 \stackrel{?}{=} 36$	$(-6)^2 \stackrel{?}{=} 36$
$36 = 36$ True	$36 = 36$ True

The solutions are 9 and −3.

> ***The Language of Algebra***
> The ± symbol is often seen in political polls. A candidate with 48% support (±4%) could be between $48 + 4 = 52\%$ and $48 - 4 = 44\%$.

b.

$(x + 1)^2 = 50$	This is the equation to solve.
$x + 1 = \pm\sqrt{50}$	By the square root property, $x + 1 = \sqrt{50}$ or $x + 1 = -\sqrt{50}$. Use double-sign notation.
$x + 1 = \pm 5\sqrt{2}$	$\sqrt{50} = \sqrt{25 \cdot 2} = \sqrt{25}\sqrt{2} = 5\sqrt{2}$.
$x = -1 \pm 5\sqrt{2}$	To isolate x, subtract 1 from both sides (or add −1 to both sides). It is standard practice to write the −1 in front of the ± symbol.

Use a check to verify that the solutions are $-1 + 5\sqrt{2}$ and $-1 - 5\sqrt{2}$.

> **Caution**
> Since 2 is not a common factor of the entire numerator, it would be incorrect to simplify the solutions as shown:
>
> ~~$s = \dfrac{\overset{2}{\cancel{4}} \pm \sqrt{7}}{\underset{1}{\cancel{2}}}$~~

c.

$(2s - 4)^2 = 7$	This is the equation to solve.
$2s - 4 = \pm\sqrt{7}$	By the square root property, $2s - 4 = \sqrt{7}$ or $2s - 4 = -\sqrt{7}$. Use double-sign notation.
$2s = 4 \pm \sqrt{7}$	To isolate the variable term 2s, add 4 to both sides.
$s = \dfrac{4 \pm \sqrt{7}}{2}$	To isolate s, divide both sides by 2.

Use a check to verify that the solutions are $\frac{4 + \sqrt{7}}{2}$ and $\frac{4 - \sqrt{7}}{2}$.

Self Check 2 Solve each equation: **a.** $(x - 2)^2 = 64$ **b.** $(x + 3)^2 = 98$ **c.** $(3r - 1)^2 = 3$

Now Try **Problems 35, 41, and 45**

Caution Equations like $(x - 3)^2 = -16$ and $(5y + 6)^2 = -4$ have no real-number solutions because no real number squared is negative.

If one side of the equation factors as the square of a binomial, we can use the methods of Example 2 to solve the equation.

EXAMPLE 3 Solve: $x^2 + 16x + 64 = 2$

Strategy We will attempt to factor the trinomial on the left side of the equation. Our hope is that it factors as the square of a binomial.

Why If the equation can be written in the form $(ax + b)^2 = c$, we can use the square root property to solve it.

Solution Since $x^2 + 16x + 64 = x^2 + 2 \cdot x \cdot 8 + 8^2$, it is a perfect-square trinomial and factors as $(x + 8)^2$.

$x^2 + 16x + 64 = 2$ This is the equation to solve.

$(x + 8)^2 = 2$ Factor the perfect-square trinomial $x^2 + 16x + 64$.

$x + 8 = \pm\sqrt{2}$ By the square root property, $x + 8 = \sqrt{2}$ or $x + 8 = -\sqrt{2}$. Use double-sign notation.

$x = -8 \pm \sqrt{2}$ To isolate x, subtract 8 from both sides.

As an informal check, we can use a calculator to approximate $-8 + \sqrt{2}$ and $-8 - \sqrt{2}$. Then we can substitute each approximation into the original equation.

The check for $-8 + \sqrt{2} \approx -6.6$:

$$x^2 + 16x + 64 = 2$$
$$(-6.6)^2 + 16(-6.6) + 64 \stackrel{?}{=} 2$$
$$1.96 \approx 2$$

The check for $-8 - \sqrt{2} \approx -9.4$:

$$x^2 + 16x + 64 = 2$$
$$(-9.4)^2 + 16(-9.4) + 64 \stackrel{?}{=} 2$$
$$1.96 \approx 2$$

In each case, the sides are approximately equal. This suggests that the results, $-8 + \sqrt{2}$ and $-8 - \sqrt{2}$, are reasonable.

Self Check 3 Solve: $x^2 - 14x + 49 = 11$

Now Try **Problem 47**

> ***The Language of Algebra***
> Recall that trinomials that are squares of binomials are called **perfect-square trinomials.**

> ***Success Tip***
> To check, we must substitute $-8 + \sqrt{2}$ and then $-8 - \sqrt{2}$ for x in $x^2 + 16x + 64 = 2$. Those calculations would be difficult by hand. Instead, we can perform an informal check using approximations of the possible solutions.

3 Solve Problems Modeled by Quadratic Equations.

The equation solving methods discussed in this section can be used to solve a variety of real-world applications that are modeled by quadratic equations.

EXAMPLE 4 ***Movie Stunts.*** In a scene for an action movie, a stuntwoman falls from the top of a 95-foot-tall building into a 10-foot-tall airbag directly below her on the ground. The formula $d = 16t^2$ gives the distance d in feet that she falls in t seconds. For how many seconds will she fall before making contact with the airbag?

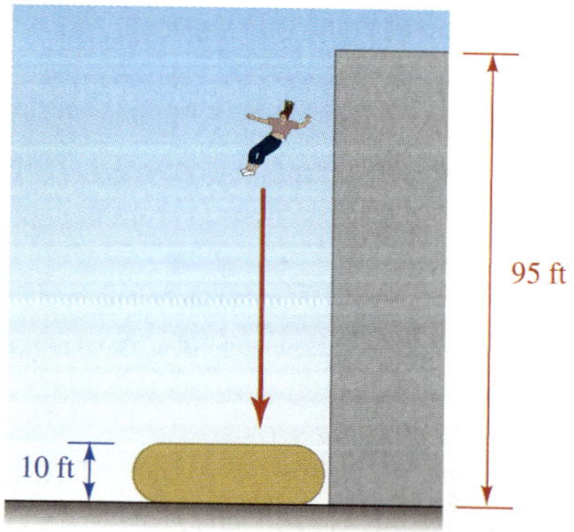

Solution The woman will fall a distance of $95 - 10 = 85$ feet before making contact with the airbag. To find the number of seconds that the fall will last, we substitute 85 for d in the formula and solve for t, the time.

$d = 16t^2$ This is the formula that models the situation.

$85 = 16t^2$ Substitute 85 for d, the distance in feet, that the stuntwoman falls.

The resulting quadratic equation is easily solved by the square root property.

$\frac{85}{16} = t^2$ To isolate t^2, divide both sides by 16.

$\pm\sqrt{\frac{85}{16}} = t$ To isolate t, use the square root property. Use double-sign notation.

$\pm\frac{\sqrt{85}}{\sqrt{16}} = t$ Use the quotient rule for square roots: The square root of a quotient is the quotient of square roots.

$\pm\frac{\sqrt{85}}{4} = t$ Evaluate: $\sqrt{16} = 4$. Since $85 = 5 \cdot 17$, we cannot simplify $\sqrt{85}$.

The stuntwoman will fall for $\frac{\sqrt{85}}{4}$ seconds (approximately 2.3 seconds) before making contact with the airbag. We discard the other solution, $-\frac{\sqrt{85}}{4}$, because a negative time does not make sense in this example.

Now Try **Problem 75**

ANSWERS TO SELF CHECKS **1. a.** $\pm\sqrt{21}$ **b.** $\pm 3\sqrt{6}$ **c.** $\pm\frac{\sqrt{35}}{5}$ **d.** No real-number solutions **2. a.** 10, -6 **b.** $-3 \pm 7\sqrt{2}$ **c.** $\frac{1 \pm \sqrt{3}}{3}$ **3.** $7 \pm \sqrt{11}$

STUDY SET 9.1

VOCABULARY

Fill in the blanks.

1. $x^2 - 15 = 0$ is an example of a ________ equation.

2. $x^2 + 6x + 9$ is a perfect ________ trinomial because $x^2 + 6x + 9 = (x + 3)^2$.

CONCEPTS

Fill in the blanks.

3. The square root property of equations: If $x^2 = c$, then $x =$ ____ or $x =$ ____.

4. a. If $x^2 = 5$, then $x = \pm$ ____.

b. If $(x - 2)^2 = 7$, then $x - 2 = \pm$ ____.

5. Use a property of equality to isolate the variable on the left side of the equation.

a. $x + 9 = \pm\sqrt{2}$

b. $6x = 3 \pm \sqrt{2}$

6. Rationalize the denominator: $x = \pm\sqrt{\frac{7}{2}}$

7. Is $2\sqrt{5}$ a solution of $x^2 - 20$?

8. Is -4 a solution of $(x - 1)^2 = 25$?

NOTATION

9. Write the statement $x = \sqrt{6}$ or $x = -\sqrt{6}$ using a $\pm$ symbol (double-sign notation).

10. Fill in the blanks: $2 \pm \sqrt{3}$ is read as "Two ______ or ______ the square root of three."

GUIDED PRACTICE

Use the square root property to solve each equation, if possible. See Example 1.

11. $x^2 - 36 = 0$
12. $x^2 - 4 = 0$
13. $b^2 - 9 = 0$
14. $a^2 - 25 = 0$
15. $x^2 = \frac{49}{16}$
16. $x^2 = \frac{81}{121}$
17. $4x^2 = 400$
18. $3m^2 = 27$
19. $5x^2 = 125$
20. $4x^2 = 16$
21. $x^2 - 6 = 0$
22. $x^2 - 7 = 0$
23. $b^2 - 17 = 0$
24. $a^2 - 26 = 0$
25. $m^2 = 20$
26. $n^2 = 32$
27. $t^2 = 72$
28. $n^2 = 75$
29. $2x^2 + 8 = 23$
30. $3m^2 + 5 = 18$
31. $6r^2 - 3 = 4$
32. $2w^2 - 9 = 12$
33. $x^2 = -81$
34. $y^2 = -100$

Use the square root property to solve each equation. See Example 2.

35. $(x + 1)^2 = 25$
36. $(x - 1)^2 = 49$
37. $(x + 2)^2 = 81$
38. $(x + 3)^2 = 16$
39. $(x - 2)^2 = 8$
40. $(x + 2)^2 = 50$
41. $(s + 9)^2 = 63$
42. $(t - 11)^2 = 45$
43. $(3x + 1)^2 - 18 = 0$
44. $(6y + 5)^2 - 72 = 0$
45. $(5c - 10)^2 - 6 = 0$
46. $(4n + 8)^2 - 17 = 0$

Use the square root property to solve each equation. See Example 3.

47. $x^2 + 2x + 1 = 10$
48. $x^2 + 8x + 16 = 6$
49. $x^2 - 18x + 81 = 7$
50. $x^2 - 14x + 49 = 19$
51. $a^2 - 6a + 9 = 40$
52. $b^2 - 10b + 25 = 90$
53. $m^2 + 4m + 4 = 75$
54. $m^2 + 16m + 64 = 80$

TRY IT YOURSELF

Use the square root property to solve each equation, if possible.

55. $(x + 12)^2 = 27$
56. $(m + 1)^2 = 32$
57. $m^2 = 98$
58. $n^2 = 99$
59. $b^2 - 12b + 36 = 2$
60. $a^2 - 18a + 81 = 5$
61. $(y - 15)^2 - 8 = 0$
62. $(y - 5)^2 - 12 = 0$
63. $t^2 = \frac{1}{144}$
64. $d^2 = \frac{1}{9}$
65. $4(t - 7)^2 - 12 = 0$
66. $2(t - 6)^2 - 22 = 0$
67. $h^2 + 25 = 0$
68. $4r^2 + 16 = 0$
69. $5x^2 + 1 = 18$
70. $7x^2 + 3 = 6$
71. $x^2 - 14 = 0$
72. $x^2 - 46 = 0$
73. $(8y + 9)^2 = 44$
74. $(6y + 13)^2 = 99$

APPLICATIONS

75. LIGHTHOUSES The 144-foot-tall Tybee Island Lighthouse is located near Savannah, Georgia. If an object is dropped from the top of the lighthouse, how long would it take for it to hit the ground?

76. SKYSCRAPERS A downtown office building on north side of the Chicago River is 784 feet tall. If an object is dropped from the top of the building, how long would it take for it to hit the ground?

77. SCIENCE HISTORY Legend has it that Galileo Galilei (1564–1642) dropped two objects having different weights from the leaning tower of Pisa in order to prove that they fall at the same rate. If a steel ball is dropped from the lowest side of the tower, and falls 183 feet, how long will it take to hit the ground? Round to the nearest tenth of a second.

78. STUDYING MICROGRAVITY NASA's Glenn Research Center in Cleveland, Ohio, has a 435-foot drop tower that begins on the surface and descends into Earth like a mineshaft. How long will it take a sealed container to fall 435 feet? Round to the nearest tenth of a second.

79. DAREDEVILS In 1873, Henry Bellini combined a tightrope walk with a leap into the Niagara River below, where he was picked up by a boat. If the rope was 200 feet above the water, for how many seconds did he fall before hitting the water? Round to the nearest tenth of a second.

80. ROLLER COASTERS Soaring to a height of 420 feet, the *Top Thrill Dragster* at Cedar Point Amusement Park in Sandusky, Ohio, is one of the tallest and fastest roller coasters in the world. If a rider accidentally dropped a camera as the coaster reached its highest point, how long would it take the camera to hit the ground? Round to the nearest tenth.

81. PRO WRESTLING A WWE (World Wrestling Entertainment) ring is square in shape and has an area of 400 ft^2. Find the length of a side of the ring. (*Hint:* Use the formula for the area of a square, $A = s^2$.)

82. CHESS A tournament chessboard is square in shape and has an area of 441 $in.^2$. Find the length of a side of the board. (*Hint:* Use the formula for the area of a square, $A = s^2$.)

83. ESCAPE VELOCITY The speed at which a rocket must be fired for it to leave the earth's gravitational attraction is called the *escape velocity*. If the escape velocity v, in miles per hour, is given by $\frac{v^2}{2g} = R$, where $g = 78{,}545$ and $R = 3{,}960$, find v. Round to the nearest mi/hr.

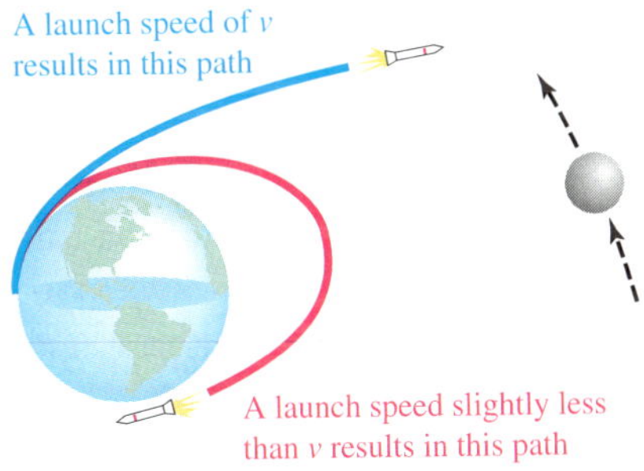

84. CAROUSELS In 1999, the city of Lancaster, Pennsylvania, considered installing a classic Dentzel carousel in an abandoned downtown building. After learning that the circular-shaped carousel (like that shown above) would occupy 2,376 square feet of floor space the proposal was determined to be impractical because of the large remodeling costs. Find the radius of the carousel to the nearest tenth of one foot. (*Hint:* Use the formula for the area of a circle, $A = \pi r^2$.)

WRITING

85. Explain why the equation $x^2 + 16 = 0$ has no real-number solutions.

86. Explain why the notation $6 \pm \sqrt{2}$ represents two real numbers.

87. Explain the error in the following work.

Solve: $x^2 = 7$

$$x = \sqrt{7}$$

88. Explain the error in the following work.

Solve: $x^2 = 28$

$$x = \pm\sqrt{28}$$
$$x = 2 \pm \sqrt{7}$$

REVIEW

Solve each equation.

89. $\sqrt{5x - 6} = 2$

90. $\sqrt{6x + 1} + 2 = 7$

91. $2\sqrt{x} = \sqrt{5x - 16}$

92. $\sqrt{22y + 86} = y + 9$

CHALLENGE PROBLEMS

Solve each equation.

93. $25\left(x + \frac{1}{3}\right)^2 = 144$

94. $100(0.2x - 1.2)^2 = 64$

SECTION 9.2
Solving Quadratic Equations: Completing the Square

Objectives

1. Complete the square to write perfect-square trinomials.
2. Solve quadratic equations with leading coefficients of 1 by completing the square.
3. Solve quadratic equations with leading coefficients other than 1 by completing the square.

In Section 9.1, we used the square root property to solve equations such as $(x - 3)^2 = 36$ and $(x + 1)^2 = 50$, whose left side is a binomial squared and right side is a constant. We also solved equations such as $x^2 + 16x + 64 = 2$ in a similar way by first factoring the perfect-square trinomial on the left side.

In this section, we will discuss a procedure that enables us to solve quadratic equations such as $x^2 + 4x = -3$, whose left side is not a perfect-square trinomial. To make the left side a perfect-square trinomial we will use a procedure called *completing the square.*

1 Complete the Square to Write Perfect-Square Trinomials.

Consider the following perfect-square trinomials and their factored forms.

$$x^2 + 2bx + b^2 = (x + b)^2 \quad \text{and} \quad x^2 - 2bx + b^2 = (x - b)^2$$

In each of these perfect-square trinomials, the third term is the square of one-half of the coefficient of x.

- In $x^2 + 2bx + b^2$, the coefficient of x is $2b$. If we find $\frac{1}{2} \cdot 2b$, which is b, and square it, we get the third term, b^2.
- In $x^2 - 2bx + b^2$, the coefficient of x is $-2b$. If we find $\frac{1}{2}(-2b)$, which is $-b$, and square it, we get the third term: $(-b)^2 = b^2$.

We can use these observations to change certain binomials into perfect-square trinomials. For example, to change $x^2 + 12x$ into a perfect-square trinomial, we find one-half of the coefficient of x, square the result, and add the square to $x^2 + 12x$.

$$x^2 + 12x + \square$$

First, find one-half of the coefficient of x.

$$\frac{1}{2} \cdot 12 = 6 \qquad 6^2 = 36$$

Then square the result.

Finally, add the square to the binomial to create a trinomial.

We obtain the perfect-square trinomial $x^2 + 12x + 36$ that factors as $(x + 6)^2$. By adding 36 to $x^2 + 12x$, we **completed the square** on $x^2 + 12x$.

Completing the Square

To complete the square on $x^2 + bx$, add the square of one-half of the coefficient of x:

$$x^2 + bx + \left(\frac{1}{2}b\right)^2$$

EXAMPLE 1 Complete the square and factor the resulting perfect-square trinomial: **a.** $x^2 + 6x$ **b.** $x^2 - 5x$ **c.** $x^2 + \frac{7}{3}x$

Strategy We will add the square of one-half of the coefficient of x to the given binomial.

Why Adding such a term will change the binomial into a perfect-square trinomial that will factor.

Solution

a. In $x^2 + 6x$, the coefficient of x is 6. One-half of 6 is 3, and $3^2 = 9$. If we add 9 to $x^2 + 6x$, it becomes a perfect-square trinomial.

$x^2 + 6x + 9$ To complete the square: $\frac{1}{2} \cdot 6 = 3$ and $3^2 = 9$. Add 9 to the binomial.

This trinomial factors as $(x + 3)^2$. We can check using multiplication.

Check: $(x + 3)^2 = (x + 3)(x + 3) = x^2 + 3x + 3x + 9 = x^2 + 6x + 9$

b. In $x^2 - 5x$, the coefficient of x is -5. One-half of -5 is $-\frac{5}{2}$, and $\left(-\frac{5}{2}\right)^2 = \frac{25}{4}$. If we add $\frac{25}{4}$ to $x^2 - 5x$, it becomes a perfect-square trinomial.

$x^2 - 5x + \frac{25}{4}$ To complete the square: $\frac{1}{2}(-5) = -\frac{5}{2}$ and $\left(-\frac{5}{2}\right)^2 = \frac{25}{4}$. Add $\frac{25}{4}$ to the binomial.

This trinomial factors as $\left(x - \frac{5}{2}\right)^2$. We can check using multiplication.

Check: $\left(x - \frac{5}{2}\right)^2 = \left(x - \frac{5}{2}\right)\left(x - \frac{5}{2}\right) = x^2 - \frac{5}{2}x - \frac{5}{2}x + \frac{25}{4} = x^2 - 5x + \frac{25}{4}$

c. In $x^2 + \frac{7}{3}x$, the coefficient of x is $\frac{7}{3}$. One-half of $\frac{7}{3}$ is $\frac{7}{6}$ and $\left(\frac{7}{6}\right)^2 = \frac{49}{36}$. If we add $\frac{49}{36}$ to $x^2 + \frac{7}{3}x$, it becomes a perfect-square trinomial.

$x^2 + \frac{7}{3}x + \frac{49}{36}$ To complete the square: $\frac{1}{2}\left(\frac{7}{3}\right) = \frac{7}{6}$ and $\left(\frac{7}{6}\right)^2 = \frac{49}{36}$. Add $\frac{49}{36}$ to the binomial.

This trinomial factors as $\left(x + \frac{7}{6}\right)^2$. Check this result using multiplication.

The Language of Algebra
When we add 9 to $x^2 + 6x$, we say that we have *completed the square* on $x^2 + 6x$.

Caution
When we complete the square on a binomial, we are not writing an equivalent trinomial expression. It would be incorrect to use an = symbol between the two.

~~$x^2 - 5x = x^2 - 5x + \frac{25}{4}$~~

Self Check 1 Complete the square for each expression and factor the resulting perfect-square trinomial:
a. $y^2 - 8y$
b. $y^2 + 3y$
c. $x^2 + \frac{7}{5}x$

Now Try **Problems 13, 15, and 23**

2 Solve Quadratic Equations with Leading Coefficients of 1 by Completing the Square.

We can use completing the square to solve quadratic equations.

EXAMPLE 2 Solve by completing the square: $x^2 + 4x = 12$

Strategy We will use the addition property of equality and add the square of one-half of the coefficient of x to both sides of the equation.

Why This will create a perfect-square trinomial on the left side that will factor as the square of a binomial. Then we can use the square root property to solve for x.

Solution In $x^2 + 4x$, the coefficient of x is 4. One-half of 4 is 2, and $2^2 = 4$. If we add 4 to $x^2 + 4x$, it becomes a perfect-square trinomial.

$x^2 + 4x = 12$ This is the equation to solve.

$x^2 + 4x + 4 = 12 + 4$ To complete the square on the left side, add 4 to both sides.

$(x + 2)^2 = 16$ Factor the perfect-square trinomial. Add on the right side.

$x + 2 = \pm\sqrt{16}$ By the square root property, $x + 2 = \sqrt{16}$ or $x + 2 = -\sqrt{16}$. Use double-sign notation.

$x + 2 = \pm 4$ Evaluate: $\sqrt{16} = 4$.

$x = -2 \pm 4$ To isolate x, subtract 2 from both sides.

$x = -2 + 4$ or $x = -2 - 4$ To find the solutions, perform the calculation using a + symbol and then using a − symbol.

$x = 2$ | $x = -6$

The solutions are 2 and −6 and the solution set is $\{-6, 2\}$. Verify this by substituting each one in the original equation.

Success Tip
We could also solve $x^2 + 4x = 12$ by writing it in the form $x^2 + 4x - 12 = 0$ and using factoring:

$$(x + 6)(x - 2) = 0$$
$$x = -6 \quad \text{or} \quad x = 2$$

Self Check 2 Solve: $x^2 + 4x = 32$

Now Try **Problem 25**

EXAMPLE 3 Solve by completing the square: $x^2 + 8x - 5 = 0$

Strategy We will use the addition property of equality and add 5 to both sides. Then we will proceed as in Example 2.

Why To prepare to complete the square, we need to isolate the variable terms, x^2 and $8x$, on the left side of the equation and the constant term on the right side.

Solution

$x^2 + 8x - 5 = 0$ This is the equation to solve.

$x^2 + 8x = 5$ Add 5 to both sides so that the constant term is on the right side.

In $x^2 + 8x$, the coefficient of x is 8. One-half of 8 is 4, and $4^2 = 16$. If we add 16 to $x^2 + 8x$, it becomes a perfect-square trinomial.

$x^2 + 8x + 16 = 5 + 16$ To complete the square on the left side, add 16 to both sides.

$(x + 4)^2 = 21$ Factor the perfect-square trinomial. Add on the right side.

$x + 4 = \pm\sqrt{21}$ By the square root property, $x + 4 = \sqrt{21}$ or $x + 4 = -\sqrt{21}$. Use double-sign notation.

$x = -4 \pm \sqrt{21}$ To isolate x, subtract 4 from both sides.

The Language of Algebra
In $x^2 + 8x - 5 = 0$, x^2 and $8x$ are called *variable terms* and −5 is called the *constant term*.

> **Success Tip**
> Example 2 but not Example 3 can be solved by factoring. These observations illustrate that the method of completing the square can be used to solve any quadratic equation.

The solutions are $-4 + \sqrt{21}$ and $-4 - \sqrt{21}$.

We can also approximate each solution by substituting the decimal approximation of $\sqrt{21}$, which is 4.582575695:

$$-4 + \sqrt{21} \approx -4 + 4.582575695 \approx 0.58 \qquad -4 - \sqrt{21} \approx -4 - 4.582575695 \approx -8.58$$

To the nearest hundredth, the solutions are 0.58 and -8.58.

Self Check 3 Solve: $x^2 + 10x - 4 = 0$

Now Try **Problem 33**

EXAMPLE 4

Solve by completing the square: $a^2 - 7a = 2$

Strategy We will use the addition property of equality and add the square of one-half of the coefficient of a to both sides of the equation.

Why This will create a perfect-square trinomial on the left side that will factor as the square of a binomial. Then we can use the square root property to solve for a.

Solution In $a^2 - 7a$, the coefficient of a is -7. One-half of -7 is $-\frac{7}{2}$, and $\left(-\frac{7}{2}\right)^2 = \frac{49}{4}$. If we add $\frac{49}{4}$ to $a^2 - 7a$, it becomes a perfect-square trinomial.

> **The Language of Algebra**
> The phrase "one-half of" translates to multiplication by $\frac{1}{2}$. So, one-half of -7 is written as:
> $$\frac{1}{2}(-7) = -\frac{7}{2}$$

$$a^2 - 7a = 2$$ This is the equation to solve.

$$a^2 - 7a + \frac{49}{4} = 2 + \frac{49}{4}$$ To complete the square on the left side, add $\frac{49}{4}$ to both sides.

$$\left(a - \frac{7}{2}\right)^2 = \frac{2}{1} \cdot \frac{4}{4} + \frac{49}{4}$$ On the left side, factor the perfect-square trinomial. On the right side, prepare to add by writing 2 as $\frac{2}{1}$ and building it so that its denominator is 4.

$$\left(a - \frac{7}{2}\right)^2 = \frac{8}{4} + \frac{49}{4}$$ On the right side, multiply $\frac{2}{1} \cdot \frac{4}{4} = \frac{8}{4}$.

$$\left(a - \frac{7}{2}\right)^2 = \frac{57}{4}$$ On the right side, add the fractions.

$$a - \frac{7}{2} = \pm\sqrt{\frac{57}{4}}$$ Use the square root property.

$$a - \frac{7}{2} = \pm\frac{\sqrt{57}}{2}$$ Use the quotient rule to simplify: $\sqrt{\frac{57}{4}} = \frac{\sqrt{57}}{\sqrt{4}} = \frac{\sqrt{57}}{2}$.

$$a = \frac{7}{2} \pm \frac{\sqrt{57}}{2}$$ To isolate a, add $\frac{7}{2}$ to both sides.

$$a = \frac{7 \pm \sqrt{57}}{2}$$ Write the sum (and difference) over the common denominator 2.

The solutions are $\frac{7 + \sqrt{57}}{2}$ and $\frac{7 - \sqrt{57}}{2}$. If we approximate them to the nearest hundredth, we have

$$\frac{7 + \sqrt{57}}{2} \approx 7.27 \quad \text{and} \quad \frac{7 - \sqrt{57}}{2} \approx -0.27$$

> **Success Tip**
> As an informal check, we can substitute each approximation into the original equation to see whether the sides are approximately equal. For $\frac{7 + \sqrt{57}}{2} \approx 7.27$ we have:
> $$a^2 - 7a = 2$$
> $$(7.27)^2 - 7(7.27) \stackrel{?}{=} 2$$
> $$1.9629 \approx 2$$

Self Check 4 Solve: $b^2 - 3b = 3$

Now Try Problem 41

Comment Completing the square sometimes leads to equations such as $(x - 1)^2 = -9$ and $(y + 6)^2 = -25$. Recall from Section 9.1 that equations like these have no real-number solutions because no real number squared is negative.

3 Solve Quadratic Equations with Leading Coefficients Other Than 1 by Completing the Square.

The method of completing the square can be used to solve any quadratic equation. However, when the coefficient of the squared variable (called the **leading coefficient**) is not 1, we must make it 1 before we can complete the square.

EXAMPLE 5 Solve by completing the square: $2x^2 + 5x + 1 = 0$

Strategy We will use the division property of equality and divide both sides by 2 so that the coefficient of x^2 is 1.

Why We create a leading coefficient that is 1 so that we can complete the square to solve the equation.

Solution

$2x^2 + 5x + 1 = 0$ — This is the equation to solve.

$x^2 + \frac{5}{2}x + \frac{1}{2} = 0$ — To make the leading coefficient 1, divide both sides by 2: $\frac{2x^2}{2} + \frac{5x}{2} + \frac{1}{2} = \frac{0}{2}$.

$x^2 + \frac{5}{2}x = -\frac{1}{2}$ — Subtract $\frac{1}{2}$ both sides so that the constant term is on the right side.

$x^2 + \frac{5}{2}x + \frac{25}{16} = -\frac{1}{2} + \frac{25}{16}$ — Complete the square: $\frac{1}{2}\left(\frac{5}{2}\right) = \frac{5}{4}$ and $\left(\frac{5}{4}\right)^2 = \frac{25}{16}$. Add $\frac{25}{16}$ to both sides.

$\left(x + \frac{5}{4}\right)^2 = -\frac{1}{2} \cdot \frac{8}{8} + \frac{25}{16}$ — On the left side, factor the perfect-square trinomial. On the right, prepare to add by building $-\frac{1}{2}$ so that its denominator is 16.

$\left(x + \frac{5}{4}\right)^2 = -\frac{8}{16} + \frac{25}{16}$ — On the right side, multiply $-\frac{1}{2} \cdot \frac{8}{8} = -\frac{8}{16}$.

$\left(x + \frac{5}{4}\right)^2 = \frac{17}{16}$ — On the right side, add the fractions.

$x + \frac{5}{4} = \pm\sqrt{\frac{17}{16}}$ — Use the square root property.

$x + \frac{5}{4} = \pm\frac{\sqrt{17}}{4}$ — Use the quotient rule to simplify: $\sqrt{\frac{17}{16}} = \frac{\sqrt{17}}{\sqrt{16}} = \frac{\sqrt{17}}{4}$.

$x = -\frac{5}{4} \pm \frac{\sqrt{17}}{4}$ — To isolate x, subtract $\frac{5}{4}$ from both sides.

$x = \frac{-5 \pm \sqrt{17}}{4}$ — Write the sum (and difference) over the common denominator 4.

The solutions are $\frac{-5 \pm \sqrt{17}}{4}$. Check each one in the original equation.

Success Tip
We could also make the leading coefficient 1 by multiplying both sides by $\frac{1}{2}$:

$$\frac{1}{2}(2x^2 + 5x + 1) = \frac{1}{2} \cdot 0$$

Self Check 5 Solve: $3x^2 + x - 1 = 0$

Now Try **Problem 53**

To solve a quadratic equation in x by completing the square, we follow these steps.

Completing the Square to Solve a Quadratic Equation in *x*

1. If the coefficient of x^2 is 1, go to step 2. If it is not 1, make it 1 by dividing both sides of the equation by the coefficient of x^2.
2. Get all variable terms on one side of the equation and constants on the other side.
3. Complete the square by finding one-half of the coefficient of x, squaring the result, and adding the square to both sides of the equation.
4. Factor the perfect-square trinomial as the square of a binomial.
5. Solve the resulting equation using the square root property.
6. Check your answers in the original equation.

EXAMPLE 6 Solve by completing the square: $4x^2 - 3 = 24x$

Strategy We will use the addition and subtraction properties of equality to get the variable terms on one side of the equation and the constant term on the other. Then we will use the division property of equality and divide both sides by 4 so that the coefficient of x^2 is 1.

Why This will create a leading coefficient of 1 so that we can complete the square to solve the equation.

Solution

$4x^2 - 3 = 24x$ This is the equation to solve.

$4x^2 - 24x = 3$ To have both variable terms on the left side, subtract 24x from both sides. To have the constant term on the right, add 3 to both sides.

$x^2 - 6x = \frac{3}{4}$ To make the coefficient of the x^2 term 1, divide both sides by 4: $\frac{4x^2}{4} - \frac{24x}{4} = \frac{3}{4}$.

$x^2 - 6x + 9 = \frac{3}{4} + 9$ Complete the square: $\frac{1}{2}(-6) = -3$ and $(-3)^2 = 9$. Add 9 to both sides.

$(x - 3)^2 = \frac{39}{4}$ On the left side, factor. On the right side, express 9 as $\frac{36}{4}$ and add to $\frac{3}{4}$ to get $\frac{39}{4}$.

$x - 3 = \pm\sqrt{\frac{39}{4}}$ Use the square root property.

$x - 3 = \pm\frac{\sqrt{39}}{2}$ Use the quotient rule to simplify: $\sqrt{\frac{39}{4}} = \frac{\sqrt{39}}{\sqrt{4}} = \frac{\sqrt{39}}{2}$.

$x = 3 \pm \frac{\sqrt{39}}{2}$ To isolate x, add 3 to both sides.

$x = \frac{6}{2} \pm \frac{\sqrt{39}}{2}$ To write the solutions in compact form, express 3 as a fraction with denominator 2: $3 = \frac{3}{1} \cdot \frac{2}{2} = \frac{6}{2}$.

$x = \frac{6 \pm \sqrt{39}}{2}$ Write the sum (and difference) over the common denominator 2.

Caution
A common error is to add a constant to one side of an equation to complete the square and forget to add it to the other side.

The exact solutions are $\frac{6 \pm \sqrt{39}}{2}$. Check each one in the original equation. We can approximate the solutions using a calculator. To the nearest hundredth, we have

$$\frac{6 + \sqrt{39}}{2} \approx 6.12 \qquad \frac{6 - \sqrt{39}}{2} \approx -0.12$$

Self Check 6 Solve: $4d^2 - 1 = 32d$

Now Try **Problem 63**

ANSWERS TO SELF CHECKS **1. a.** $y^2 - 8y + 16 = (y - 4)^2$ **b.** $y^2 + 3y + \frac{9}{4} = \left(y + \frac{3}{2}\right)^2$ **c.** $x^2 + \frac{7}{5}x + \frac{49}{100} = \left(x + \frac{7}{10}\right)^2$ **2.** $-8, 4$ **3.** $-5 \pm \sqrt{29}$ **4.** $\frac{3 \pm \sqrt{21}}{2}$ **5.** $\frac{-1 \pm \sqrt{13}}{6}$ **6.** $\frac{8 \pm \sqrt{65}}{2}$

STUDY SET 9.2

VOCABULARY

Fill in the blanks.

1. When we add 9 to $x^2 + 6x$, we say that we have completed the ______ on $x^2 + 6x$.

2. The ______ coefficient of $5x^2 - 2x + 7$ is 5 and the ______ term is 7.

CONCEPTS

3. Find one-half of the given number and square the result.

a. 6 **b.** -5

4. Factor each perfect-square trinomial.

a. $x^2 + 14x + 49$ **b.** $x^2 - 2x + 1$

5. Fill in the blank: To complete the square on $x^2 + 8x$, add the square of ______ of the coefficient of x.

6. Fill in the blanks to complete the square. Then factor the resulting perfect-square trinomial.

a. $x^2 + 8x + \square = (x + \square)^2$

b. $x^2 - 9x + \square = \left(x - \square\right)^2$

7. What is the first step to solve the equation by completing the square? **Do not solve.**

a. $x^2 + 9x + 7 = 0$

b. $4x^2 + 5x - 16 = 0$

8. Check to determine whether $3 + \sqrt{7}$ is a solution of $x^2 - 6x + 2 = 0$.

9. Determine whether each statement is true or false.

a. Any quadratic equation can be solved by the factoring method.

b. Any quadratic equation can be solved by the completing the square.

10. Simplify: $\sqrt{\frac{19}{4}}$

NOTATION

11. Translate to mathematical symbols: *the square of one-half of nine.* Then evaluate the expression.

12. Fill in the blanks: $\frac{5}{3} \pm \frac{\sqrt{17}}{3} = \frac{\square}{\square}$

GUIDED PRACTICE

***Complete the square and factor the resulting perfect-square trinomial.* See Example 1.**

13. $x^2 + 2x$

14. $x^2 + 12x$

15. $x^2 - 4x$

16. $x^2 - 14x$

17. $a^2 - 7a$

18. $b^2 + 11b$

19. $x^2 + x$

20. $x^2 - x$

21. $b^2 - \frac{2}{3}b$

22. $t^2 - \frac{6}{5}t$

23. $x^2 - \frac{5}{2}x$

24. $x^2 - \frac{7}{6}x$

***Solve each equation by completing the square.* See Example 2.**

25. $x^2 + 4x = 5$

26. $x^2 + 6x = 7$

27. $g^2 - 2g = 15$

28. $s^2 - 5s = 14$

29. $x^2 + 6x = -8$

30. $x^2 + 8x = -12$

31. $k^2 - 8k = -12$

32. $p^2 - 4p = -3$

***Solve each equation by completing the square.* See Example 3.**

33. $x^2 + 8x - 6 = 0$

34. $x^2 + 6x - 2 = 0$

35. $x^2 - 2x - 17 = 0$

36. $x^2 + 10x - 7 = 0$

37. $x^2 + 6x + 4 = 0$

38. $x^2 + 8x + 6 = 0$

39. $s^2 - 4s - 3 = 0$

40. $t^2 - 6t + 3 = 0$

***Solve each equation by completing the square.* See Example 4.**

41. $x^2 - 7x = 5$

42. $x^2 + 5x = 7$

43. $b^2 - 5b = 10$

44. $b^2 - 3b = 5$

45. $t^2 + 3t = 20$

46. $a^2 + a = 3$

47. $x^2 + x = 9$

48. $x^2 + 7x = 6$

***Solve each equation by completing the square.* See Example 5.**

49. $3x^2 - 18x - 21 = 0$

50. $5x^2 - 20x - 25 = 0$

51. $2x^2 + 24x + 44 = 0$

52. $2x^2 + 32x + 6 = 0$

53. $2x^2 - 7x - 3 = 0$

54. $3x^2 - 5x - 1 = 0$

55. $4a^2 - 9a + 1 = 0$

56. $4a^2 - 11a + 1 = 0$

57. $5b^2 + 3b - 4 = 0$

58. $2x^2 + 9x + 3 = 0$

59. $3x^2 + 5x - 5 = 0$

60. $5b^2 + 7b - 2 = 0$

***Solve each equation by completing the square.* See Example 6.**

61. $4x^2 - 13 = 24x$

62. $4d^2 - 9 = 16d$

63. $9a^2 - 5 = -18a$

64. $9r^2 - 7 = 36r$

65. $4t^2 + 11 = 48t$

66. $4x^2 + 13 = 40x$

67. $16x^2 + 3 = -64x$

68. $16b^2 + 1 = -32b$

***Solve each equation by completing the square. Approximate the solutions to the nearest hundredth.* See Example 6.**

69. $x^2 - 2x - 4 = 0$

70. $x^2 - 4x - 2 = 0$

71. $n^2 - 9n = 5$

72. $a^2 + 7a = 4$

73. $2x^2 = -6x - 1$

74. $3x^2 = -4x + 3$

75. $3x^2 - 4 = -2x$

76. $2x^2 + 3 = 10x$

TRY IT YOURSELF

Solve each equation by completing the square.

77. $x^2 = -7x - 2$

78. $y^2 = -3y + 6$

79. $9x^2 - 36x - 1 = 0$

80. $16y^2 - 32y + 1 = 0$

81. $2t^2 - 4 = -5t$

82. $2s^2 - 16 = 3s$

83. $3n^2 - 8n = -4$

84. $3a^2 - 11a = 4$

85. $x^2 - 2x - 5 = 0$

86. $x^2 + 4x + 1 = 0$

87. $x^2 - 12x = -35$

88. $x^2 + 10x = -24$

APPLICATIONS

89. GEOMETRY Completing the square can be applied to geometric figures. For example, the area of the figure in part (a) is $x^2 + x + x + x + x$, or simply $x^2 + 4x$. Note that the figure is not a square because it is missing a corner. Determine the number of small squares that were added in part (b) to "complete the square." Then fill in the blanks.

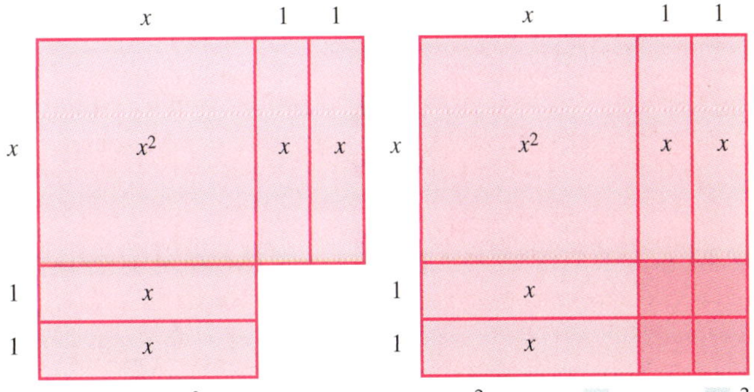

Area: $x^2 + 4x$
(a)

Area: $x^2 + 4x + \_\_ = (x + \_\_)^2$
(b)

90. GEOMETRY Draw a figure similar to that in part (a) of Problem 89 that has area $x^2 + 8x$. Then draw part (b) to determine the number of small squares that must be added to "complete the square." Fill in the blanks:

$$\text{Area: } x^2 + 8x + \square = (x + \square)^2$$

WRITING

91. Give an example of a perfect-square trinomial. Why do you think the word "perfect" is used to describe it?

92. Explain why completing the square on $x^2 + 5x$ is more difficult than completing the square on $x^2 + 4x$.

93. Find the error in the following solution.

Solve:
$$x^2 + 8x = 5$$
$$x^2 + 8x + 16 = 5$$
$$(x + 4)^2 = 5$$
$$x + 4 = \pm\sqrt{5}$$
$$x = -4 \pm \sqrt{5}$$

94. Try to solve $x^2 - 4x = -40$ and then explain why it has no real-number solutions.

REVIEW

Perform the indicated operation and simplify when possible.

95. $\dfrac{x+3}{x-3} \cdot \dfrac{x+1}{x^2-9}$

96. $\dfrac{x+3}{x-3} \div \dfrac{x+1}{x^2-9}$

97. $\dfrac{x+3}{x-3} + \dfrac{x+1}{x^2-9}$

98. $\dfrac{x+3}{x-3} - \dfrac{x+1}{x^2-9}$

CHALLENGE PROBLEMS

Solve each equation by completing the square.

99. $0.2x^2 + 0.4x + 0.1 = 0$

100. $0.03x^2 + 0.06x = 0.02$

101. $x(x + 3) - \dfrac{1}{2} = -2$

102. $x[(x - 2) + 3] = 3\left(x - \dfrac{2}{9}\right)$

SECTION 9.3
Solving Quadratic Equations: The Quadratic Formula

Objectives

1. Use the quadratic formula to solve quadratic equations.
2. Identify quadratic equations with no real-number solutions.
3. Determine the most efficient method to use to solve a quadratic equation.
4. Solve problems modeled by quadratic equations.

We can solve any quadratic equation by completing the square, but the work is often lengthy and involved. In this section, we will develop a formula that will enable us to solve quadratic equations with much less effort.

1 Use the Quadratic Formula to Solve Quadratic Equations.

To develop a formula that will produce the solutions of any given quadratic equation, we start with the **general quadratic equation** $ax^2 + bx + c = 0$, with $a > 0$, and solve it for x by completing the square.

$$ax^2 + bx + c = 0$$

$$\frac{ax^2}{a} + \frac{bx}{a} + \frac{c}{a} = \frac{0}{a}$$ Divide both sides by a so that the coefficient of x^2 is 1.

$$x^2 + \frac{b}{a}x + \frac{c}{a} = 0$$ Simplify: $\frac{ax^2}{a} = x^2$. Write $\frac{bx}{a}$ as $\frac{b}{a}x$.

$$x^2 + \frac{b}{a}x = -\frac{c}{a}$$ Subtract $\frac{c}{a}$ from both sides so that only the variable terms are on the left side of the equation and the constant is on the right side.

The Language of Algebra
Your instructor may ask you to *derive* the quadratic formula. That means to solve $ax^2 + bx + c = 0$ for x, using the series of steps shown here.

We can complete the square on $x^2 + \frac{b}{a}x$ by adding the square of one-half of the coefficient of x. Since the coefficient of x is $\frac{b}{a}$, we have $\frac{1}{2} \cdot \frac{b}{a} = \frac{b}{2a}$ and $\left(\frac{b}{2a}\right)^2 = \frac{b^2}{4a^2}$.

$$x^2 + \frac{b}{a}x + \frac{b^2}{4a^2} = -\frac{c}{a} + \frac{b^2}{4a^2}$$ To complete the square, add $\frac{b^2}{4a^2}$ to both sides.

$$x^2 + \frac{b}{a}x + \frac{b^2}{4a^2} = -\frac{4ac}{4aa} + \frac{b^2}{4a^2}$$ On the right side, build $-\frac{c}{a}$ by multiplying it by $\frac{4a}{4a}$. Now the fractions on that side have the common denominator $4a^2$.

$$\left(x + \frac{b}{2a}\right)^2 = \frac{b^2 - 4ac}{4a^2}$$ On the left side, factor the perfect-square trinomial. On the right side, add the fractions. In the numerator, write $-4ac + b^2$ as $b^2 - 4ac$.

$$x + \frac{b}{2a} = \pm\sqrt{\frac{b^2 - 4ac}{4a^2}}$$ Use the square root property.

$$x + \frac{b}{2a} = \pm\frac{\sqrt{b^2 - 4ac}}{\sqrt{4a^2}}$$ On the right side, the square root of a quotient is the quotient of square roots.

$$x + \frac{b}{2a} = \pm\frac{\sqrt{b^2 - 4ac}}{2a}$$ On the right side, simplify the denominator. Since $a > 0$, $\sqrt{4a^2} = 2a$.

$$x = -\frac{b}{2a} \pm \frac{\sqrt{b^2 - 4ac}}{2a}$$ To isolate x, subtract $\frac{b}{2a}$ from both sides.

$$x = \frac{-b \pm \sqrt{b^2 - 4ac}}{2a}$$ Combine the fractions. Write the sum (and difference) over the common denominator 2a.

This result is called the **quadratic formula.** To develop this formula, we assumed that a was positive. If a is negative, similar steps are used, and we obtain the same result. This formula is very useful and should be memorized.

The Quadratic Formula

The solutions of $ax^2 + bx + c = 0$, with $a \neq 0$, are given by

$$x = \frac{-b \pm \sqrt{b^2 - 4ac}}{2a}$$

Read as "x equals the opposite of b plus or minus the square root of b squared minus 4ac, all over 2a."

EXAMPLE 1 Use the quadratic formula to solve: $4x^2 + x - 3 = 0$

Strategy We will compare the given equation to the general quadratic equation $ax^2 + bx + c = 0$ to identify a, b, and c.

Why To use the quadratic formula, we need to know what numbers to substitute for a, b, and c in $x = \frac{-b \pm \sqrt{b^2 - 4ac}}{2a}$.

Solution

$4x^2 + x - 3 = 0$ This is the equation to solve.

$4x^2 + 1x - 3 = 0$ Write x as 1x.

$\uparrow \quad \uparrow \quad \uparrow$

$ax^2 + bx + c = 0$

We see that $a = 4$, $b = 1$, and $c = -3$. To find the solutions of $4x^2 + x - 3 = 0$, we substitute these numbers into the formula and evaluate the right side.

Caution
When writing the quadratic formula, be careful to draw the fraction bar so that it includes the entire numerator. Do not write

~~$x = -b \pm \frac{\sqrt{b^2 - 4ac}}{2a}$~~

$$x = \frac{-b \pm \sqrt{b^2 - 4ac}}{2a}$$ This is the quadratic formula.

$$x = \frac{-1 \pm \sqrt{1^2 - 4(4)(-3)}}{2(4)}$$ Substitute 4 for a, 1 for b, and -3 for c.

$$x = \frac{-1 \pm \sqrt{1 - (-48)}}{8}$$ Evaluate the power and multiply within the radical. Multiply in the denominator.

$$x = \frac{-1 \pm \sqrt{49}}{8}$$ Evaluate the expression within the radical: $1 - (-48) = 1 + 48 = 49$.

$$x = \frac{-1 \pm 7}{8}$$ Evaluate: $\sqrt{49} = 7$.

To find the first solution, evaluate the expression using the + symbol. To find the second solution, evaluate the expression using the − symbol.

$$x = \frac{-1 + 7}{8} \quad \text{or} \quad x = \frac{-1 - 7}{8}$$

$$x = \frac{6}{8} \qquad\qquad x = \frac{-8}{8}$$

$$x = \frac{3}{4} \qquad\qquad x = -1$$

Verify that $\frac{3}{4}$ and -1 are the solutions by substituting them into the original equation. The solution set is $\left\{-1, \frac{3}{4}\right\}$.

Self Check 1 Solve: $3x^2 + x - 2 = 0$

Now Try **Problem 17**

EXAMPLE 2

Use the quadratic formula to solve $5x^2 + 1 = 5x$. Approximate the solutions to the nearest hundredth.

Strategy We will use the subtraction property of equality to get 0 on the right side of the equation. Then we will compare the resulting equation to $ax^2 + bx + c = 0$ to identify a, b, and c.

Why We need to know a, b, and c to use the quadratic formula.

Solution

$$5x^2 + 1 = 5x$$ This is the equation to solve.

$$5x^2 - 5x + 1 = 0$$ To get 0 on the right side, subtract $5x$ from both sides.

$$\uparrow \quad \uparrow \quad \uparrow$$

$$ax^2 + bx + c = 0$$

We see that $a = 5$, $b = -5$, and $c = 1$. To find the solutions, we substitute these values into the quadratic formula and evaluate the right side.

$$x = \frac{-b \pm \sqrt{b^2 - 4ac}}{2a}$$ This is the quadratic formula.

> ***Success Tip***
> Example 1 but not Example 2 can be solved by factoring. These observations illustrate that the quadratic formula can be used to solve any quadratic equation.

$$x = \frac{-(-5) \pm \sqrt{(-5)^2 - 4(5)(1)}}{2(5)}$$ Substitute 5 for *a*, −5 for *b*, and 1 for *c*.

$$x = \frac{5 \pm \sqrt{25 - 20}}{10}$$ Simplify: −(−5) = 5. Evaluate the power and multiply within the radical. Multiply in the denominator.

$$x = \frac{5 \pm \sqrt{5}}{10}$$ Subtract within the radical.

This fraction does not simplify further. Thus, the solutions are $\frac{5 \pm \sqrt{5}}{10}$. We can use a calculator to approximate the solutions. To the nearest hundredth,

$$\frac{5 + \sqrt{5}}{10} \approx 0.72 \qquad \frac{5 - \sqrt{5}}{10} \approx 0.28$$

When each approximation is substituted into the original equation, and the calculations are made, notice that the sides are approximately equal. This suggests that the results, $\frac{5 + \sqrt{5}}{10}$ and $\frac{5 - \sqrt{5}}{10}$, are reasonable.

$$\begin{aligned} 5x^2 + 1 &= 5x \\ 5(0.72)^2 + 1 &\stackrel{?}{=} 5(0.72) \\ 3.592 &\approx 3.6 \end{aligned} \qquad \begin{aligned} 5x^2 + 1 &= 5x \\ 5(0.28)^2 + 1 &\stackrel{?}{=} 5(0.28) \\ 1.392 &\approx 1.4 \end{aligned}$$

Self Check 2 Solve $4x^2 + 2 = 7x$. Approximate the solutions to the nearest hundredth.

Now Try **Problem 31**

EXAMPLE 3

Use the quadratic formula to solve: $3t^2 - 4 = -2t$

Strategy We will use the addition property of equality to get 0 on the right side of the equation. Then we will compare the resulting equation to $at^2 + bt + c = 0$ to identify *a*, *b*, and *c*.

Why We need to know *a*, *b*, and *c* to use the quadratic formula.

Solution

$$3t^2 - 4 = -2t$$ This is the equation to solve.

$$3t^2 + 2t - 4 = 0$$ To get 0 on the right side, add 2t to both sides.

$$\uparrow \quad \uparrow \quad \uparrow$$

$$at^2 + bt + c = 0$$

We see that $a = 3$, $b = 2$, and $c =$ 4. To find the solutions, we substitute these values into the quadratic formula and evaluate the right side.

> ***Caution***
> Since the variable in the equation is *t*, not *x*, we must change the variable in the quadratic formula to reflect this:
> $$t = \frac{-b \pm \sqrt{b^2 - 4ac}}{2a}$$

$$t = \frac{-b \pm \sqrt{b^2 - 4ac}}{2a}$$ This is the quadratic formula, where *x* is replaced by *t*.

$$t = \frac{-2 \pm \sqrt{2^2 - 4(3)(-4)}}{2(3)}$$ Substitute 3 for *a*, 2 for *b*, and −4 for *c*.

$$t = \frac{-2 \pm \sqrt{4 - (-48)}}{6}$$ Evaluate the power and multiply within the radical. Multiply in the denominator.

$$t = \frac{-2 \pm \sqrt{52}}{6}$$ Subtract within the radical.

$$t = \frac{-2 \pm 2\sqrt{13}}{6}$$ Simplify: $\sqrt{52} = \sqrt{4}\sqrt{13} = 2\sqrt{13}$.

This fraction can be simplified. We will factor the numerator and denominator and remove a common factor.

$$t = \frac{2(-1 \pm \sqrt{13})}{2 \cdot 3}$$ Factor out the GCF, 2, from the two terms in the numerator. In the denominator, factor 6 as $2 \cdot 3$.

$$t = \frac{\overset{1}{\cancel{2}}(-1 \pm \sqrt{13})}{\underset{1}{\cancel{2}} \cdot 3}$$ Simplify the fraction by removing the common factor 2 in the numerator and denominator: $\frac{2}{2} = 1$.

$$t = \frac{-1 \pm \sqrt{13}}{3}$$

Thus, the solutions are $\frac{-1 \pm \sqrt{13}}{3}$. Check each in the original equation.

Caution

To simplify expressions such as $\frac{-2 \pm 2\sqrt{13}}{6}$, factor the numerator and denominator first. Then remove a common factor of the *entire numerator*. Some common errors are shown:

$$\frac{-2 \pm \overset{1}{\cancel{2}}\sqrt{13}}{\underset{1}{\cancel{2}} \cdot 3} \qquad \frac{\overset{1}{\cancel{-2}} \pm 2\sqrt{13}}{\underset{1}{\cancel{2}} \cdot 3}$$

Self Check 3 Solve: $2s^2 - 1 = -2s$

Now Try **Problem 37**

To solve a quadratic equation in x using the quadratic formula, we follow these steps.

Solving a Quadratic Equation in *x* Using the Quadratic Formula

1. Write the equation in standard form: $ax^2 + bx + c = 0$.
2. Identify a, b, and c.
3. Substitute the values for a, b, and c in the quadratic formula

$$x = \frac{-b \pm \sqrt{b^2 - 4ac}}{2a}$$

and evaluate the right side to obtain the solutions.

2 Identify Quadratic Equations with No Real-Number Solutions.

The next example shows that some quadratic equations have no real-number solutions.

EXAMPLE 4 Use the quadratic formula to solve $x^2 - x + 6 = 0$, if possible.

Strategy We will compare the given equation to the general quadratic equation $ax^2 + bx + c = 0$ to identify a, b, and c.

Why We need to know a, b, and c to use the quadratic formula.

Success Tip
When identifying a and b, recall that the term x^2 has an understood coefficient 1 and the term $-x$ has an understood coefficient -1.

$x^2 = 1x^2$ and $-x = -1x$

Caution
When you obtain a result such as $x = \frac{1 \pm \sqrt{-19}}{2}$, don't forget to write, *"No real-number solutions."*

Solution

$$x^2 - x + 6 = 0 \quad \text{This is the equation to solve.}$$

$$1x^2 - 1x + 6 = 0 \quad \text{Write } x^2 \text{ as } 1x^2 \text{ and write } -x \text{ as } -1x.$$

$$\uparrow \quad \uparrow \quad \uparrow$$

$$ax^2 + bx + c = 0$$

We see that $a = 1$, $b = -1$, and $c = 6$ and substitute these values into the quadratic formula.

$$x = \frac{-b \pm \sqrt{b^2 - 4ac}}{2a}$$

$$x = \frac{-(-1) \pm \sqrt{(-1)^2 - 4(1)(6)}}{2(1)} \quad \text{Substitute 1 for } a, -1 \text{ for } b, \text{ and 6 for } c.$$

$$x = \frac{1 \pm \sqrt{1 - 24}}{2} \quad \text{Evaluate the power and multiply within the radical. Multiply in the denominator.}$$

$$x = \frac{1 \pm \sqrt{-23}}{2} \quad \text{Subtract within the radical.}$$

Since $\sqrt{-23}$ is not a real number, $x^2 - x + 6 = 0$ has no real-number solutions.

Self Check 4 Does $x^2 - x + 1 = 0$ have any real-number solutions?

Now Try **Problem 45**

3 Determine the Most Efficient Method to Solve a Quadratic Equation.

We have discussed four methods that are used to solve quadratic equations. The following table shows some advantages and disadvantages of each method.

Method	Advantages	Disadvantages	Examples
Factoring and the zero-factor property	It can be very fast. When each factor is set equal to 0, the resulting equations are usually easy to solve.	Some polynomials may be difficult to factor and others impossible.	$x^2 - 2x - 24 = 0$ $4a^2 + a = 0$
Square root property	It is the fastest way to solve equations of the form $ax^2 = n$ or $(ax + b)^2 = n$, where n is a number.	It only applies to equations that are in these forms.	$x^2 = 27$ $(2y + 3)^2 = 25$
Completing the square*	It can be used to solve any quadratic equation. It works well with equations of the form $x^2 + bx = n$, where b is even.	It involves more steps than the other methods. The algebra can be cumbersome if the leading coefficient is not 1.	$t^2 - 14t = 9$ $x^2 + 4x + 1 = 0$
Quadratic formula	It can be used to solve any quadratic equation.	It involves several computations where sign errors can be made. Often the result must be simplified.	$x^2 + 3x - 33 = 0$ $4s^2 - 10s + 5 = 0$

*The quadratic formula is just a condensed version of completing the square and is usually easier to use. However, you need to know how to complete the square because it is used in more advanced mathematics courses.

To determine the most efficient method for a given equation, we can use the following strategy.

Strategy for Solving Quadratic Equations

1. See whether the equation is in a form such that the **square root method** is easily applied.
2. See whether the equation is in a form such that the **completing the square method** is easily applied.
3. If neither Step 1 nor Step 2 is reasonable, write the equation in $ax^2 + bx + c = 0$ form.
4. See whether the equation can be solved using the **factoring method.**
5. If you can't factor, solve the equation by the **quadratic formula.**

4 Solve Problems Modeled by Quadratic Equations.

The equation solving methods discussed in this section can be used to solve a variety of real-world applications that are modeled by quadratic equations.

EXAMPLE 5 ***Sailing.*** The height of a triangular sail is 4 feet more than the length of the base. If the sail has an area of 30 square feet, find the length of its base and the height.

Analyze the Problem

- The height of the sail is 4 feet more than the length of the base.
- The area of the sail is 30 ft^2.
- Find the length of the base and height of the sail.

Form an Equation

If we let $b =$ the length of the base in feet of the triangular sail, then $b + 4 =$ the height in feet. We can use the formula for the area of a triangle, $A = \frac{1}{2}bh$, to form an equation.

$\frac{1}{2}$	times	the length of the base	times	the height	equals	the area of the triangle.
$\frac{1}{2}$	$\cdot$	b	$\cdot$	$(b + 4)$	$=$	30

Success Tip
It is usually easier to clear quadratic equations of fractions before attempting to solve them.

Solve the Equation

$$\frac{1}{2}b(b + 4) = 30$$

$b(b + 4) = 60$ To clear the equation of the fraction, multiply both sides by 2.

$b^2 + 4b = 60$ Distribute the multiplication by b.

Since the coefficient of the b-term is the even number 4, this equation can be solved quickly by completing the square.

Success Tip
This equation could also be solved by writing it in the form $b^2 + 4b - 60 = 0$ and factoring: $(b + 10)(b - 6) = 0$. Or, the quadratic formula could be used, where $a = 1$, $b = 4$, and $c = -60$.

$$b^2 + 4b = 60$$

$b^2 + 4b + 4 = 60 + 4$ Complete the square: $\frac{1}{2}(4) = 2$ and $(2)^2 = 4$. Add 4 to both sides.

$(b + 2)^2 = 64$ On the left side, factor the perfect-square trinomial. On the right, add.

$b + 2 = \pm\sqrt{64}$ Use the square root property.

$b = -2 \pm 8$ To isolate b, subtract 2 from both sides. Evaluate: $\sqrt{64} = 8$.

$b = -2 + 8$	or	$b = -2 - 8$	To find the solutions, perform the calculation using a + symbol and then using a − symbol.
$b = 6$		~~$b = -10$~~	Discard the solution −10. The length of the base cannot be negative.

State the Conclusion The length of the base of the sail is 6 feet. Since the height is given by $b + 4$, the height of the sail is $6 + 4 = 10$ feet.

Check the Result A height of 10 feet is 4 feet more than the length of the base, which is 6 feet. Also, the area of the triangle is $\frac{1}{2}(6)(10) = 30 \text{ ft}^2$. The results check.

Now Try **Problem 79**

EXAMPLE 6

Televisions. A television's screen size is measured diagonally. For the 42-inch plasma television shown in the illustration, the screen's height is 16 inches less than its length. What are the height and length of the screen?

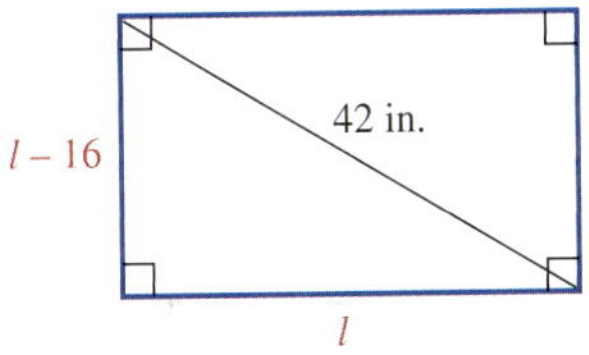

Analyze the Problem A sketch of the screen shows that two adjacent sides and the diagonal form a right triangle. The length of the hypotenuse is 42 inches.

Form an Equation If we let l = the length of the screen in inches, then $l - 16$ represents the height of the screen in inches. We can use the Pythagorean theorem to form an equation.

$a^2 + b^2 = c^2$	This is the Pythagorean theorem.
$l^2 + (l - 16)^2 = 42^2$	Substitute l for a, $l - 16$ for b, and 42 for c.
$l^2 + l^2 - 32l + 256 = 1{,}764$	Find $(l - 16)^2$ and 42^2.
$2l^2 - 32l - 1{,}508 = 0$	To get 0 on the right side of the equation, subtract 1,764 from both sides.
$l^2 - 16l - 754 = 0$	Divide both sides of the equation by 2: $\frac{2l^2}{2} - \frac{32l}{2} - \frac{1{,}508}{2} = \frac{0}{2}$.

Success Tip

To solve $2l^2 - 32l - 1{,}508 = 0$, we can substitute $a = 2$, $b = -32$, and $c = -1{,}508$ into the quadratic formula. Or, we can divide both sides by 2 and solve the equivalent equation $l^2 - 16l - 754 = 0$, and use smaller values: $a = 1$, $b = -16$, and $c = -754$.

Solve the Equation Because of the large constant term, −754, we will not attempt to solve this quadratic equation by factoring. Instead, we will use the quadratic formula, with $a = 1$, $b = -16$, and $c = -754$.

$$l = \frac{-b \pm \sqrt{b^2 - 4ac}}{2a}$$ In the quadratic formula, replace x with l.

$$l = \frac{-(-16) \pm \sqrt{(-16)^2 - 4(1)(-754)}}{2(1)}$$ Substitute 1 for a, −16 for b, and −754 for c.

$$l = \frac{16 \pm \sqrt{256 - (-3{,}016)}}{2}$$ Evaluate the power and multiply within the radical. Multiply in the denominator.

$$l = \frac{16 \pm \sqrt{3{,}272}}{2}$$ Subtract within the radical.

We can use a calculator to approximate each one to the nearest tenth. The negative solution is discarded because the length of the screen cannot be negative.

$$\frac{16 + \sqrt{3{,}272}}{2} \approx 36.6 \quad \text{or} \quad \frac{16 - \sqrt{3{,}272}}{2} \approx -20.6$$

State the Conclusion The length of the television screen is approximately 36.6 inches. Since the height is $l - 16$, the height is approximately $36.6 - 16$ or 20.6 inches.

Check the Result The sum of the squares of the lengths of the sides is $(36.6)^2 + (20.6)^2 = 1{,}763.92$. The square of the length of the hypotenuse is $42^2 = 1{,}764$. Since these are approximately equal, the results seem reasonable.

Now Try Problem 83

ANSWERS TO SELF CHECKS 1. $\frac{2}{3}, -1$ 2. $\frac{7 + \sqrt{17}}{8} \approx 1.39, \frac{7 - \sqrt{17}}{8} \approx 0.36$ 3. $\frac{-1 \pm \sqrt{3}}{2}$ 4. No

STUDY SET 9.3

VOCABULARY

Fill in the blanks.

1. The general ________ equation is $ax^2 + bx + c = 0$, where $a \neq 0$.
2. Complete the quadratic formula.

$$x = \frac{\qquad}{\qquad}$$

CONCEPTS

3. Write each equation in $ax^2 + bx + c = 0$ form.
   a. $x^2 + 2x = -5$
   b. $3x^2 = -2x + 1$
4. For each quadratic equation, find a, b, and c.
   a. $x^2 + 5x + 6 = 0$
   b. $8x^2 - x = 10$
5. Divide both sides of $2x^2 - 4x + 8 = 0$ by 2, and then find a, b, and c.
6. Evaluate each expression.
   a. $\frac{-2 \pm \sqrt{2^2 - 4(1)(-8)}}{2(1)}$
   b. $\frac{-(-1) \pm \sqrt{(-1)^2 - 4(2)(-4)}}{2(2)}$
7. a. How many terms does the expression $10 \pm 15\sqrt{2}$ have?
   b. What common factor do the terms have?
8. Simplify each expression.
   a. $\frac{-1 \pm \sqrt{45}}{2(7)}$
   b. $\frac{-(-4) \pm \sqrt{(-4)^2 - 4(2)(-9)}}{2(2)}$
9. A student used the quadratic formula to solve an equation and obtained

$$x = \frac{-3 \pm \sqrt{15}}{2}$$

   a. How many solutions does the equation have?
   b. What are they?
   c. Approximate each to the nearest hundredth.

10. Match each quadratic equation with the best method for solving it. Each answer can be used only once.

a. $x^2 + 7x - 8 = 0$	**i.** Square root method
b. $x^2 + 19x - 2 = 0$	**ii.** Factoring method
c. $x^2 = 21$	**iii.** Complete the square
d. $x^2 + 10x = 5$	**iv.** Quadratic formula

NOTATION

Complete the solution.

11. Solve: $x^2 - 5x - 6 = 0$

$$x = \frac{-b \pm \sqrt{b^2 - 4ac}}{2a}$$

$$x = \frac{-(\quad) \pm \sqrt{(-5)^2 - 4(1)(\quad)}}{2(\quad)}$$

$$x = \frac{5 \pm \sqrt{25 + \quad}}{2}$$

$$x = \frac{5 \pm \sqrt{\quad}}{2}$$

$$x = \frac{\quad \pm 7}{2}$$

$$x = \frac{5 + \quad}{2} = \quad \text{ or } \quad x = \frac{5 - \quad}{2} = \quad$$

12. Fill in the blanks: In reading $\frac{-b \pm \sqrt{b^2 - 4ac}}{2a}$ we say, "the ________ of b, plus or ______ the square root of b ________ minus $4ac$, all ______ $2a$."

GUIDED PRACTICE

Use the quadratic formula to solve each equation. See Example 1.

13. $x^2 + 5x + 6 = 0$

14. $x^2 + 5x + 4 = 0$

15. $x^2 + 7x + 12 = 0$

16. $x^2 + 8x + 15 = 0$

17. $4x^2 + 3x - 1 = 0$

18. $2x^2 + 3x - 2 = 0$

19. $6x^2 + 5x - 6 = 0$

20. $4x^2 + 4x - 3 = 0$

21. $3x^2 - 5x - 2 = 0$

22. $3x^2 - 8x - 3 = 0$

23. $5x^2 - 13x - 6 = 0$

24. $2x^2 - x - 1 = 0$

Use the quadratic formula to solve each equation. Approximate the solutions to nearest hundredth. See Example 2.

25. $x^2 + 1 = -3x$

26. $x^2 + 2 = -5x$

27. $x^2 - 4 = -7x$

28. $x^2 - 2 = -3x$

29. $3x^2 - x = 3$

30. $5x^2 - 3x = 1$

31. $7x^2 - 3x = 1$

32. $4x^2 - x = 2$

33. $4x^2 = 7x - 2$

34. $3x^2 = 5x + 4$

35. $2x^2 = 7 - 9x$

36. $6x^2 = 3 - 11x$

Use the quadratic formula to solve each equation, if possible. See Examples 3 and 4.

37. $2n^2 + 10n + 11 = 0$

38. $2m^2 + 8m + 5 = 0$

39. $2d^2 - 6d + 1 = 0$

40. $2s^2 + 2s - 5 = 0$

41. $3x^2 - 8x + 2 = 0$

42. $3x^2 - 6x - 2 = 0$

43. $4x^2 - 12x = -1$

44. $4x^2 + 12x = 3$

45. $3m^2 + 5 = 2m$

46. $2n^2 + 3n = -3$

47. $x^2 = 1 - 2x$

48. $x^2 = 4 + 2x$

49. $a^2 + 4a = 3$

50. $d^2 + 6d = 5$

51. $7x^2 - x = -8$

52. $9x^2 - 2x = -4$

TRY IT YOURSELF

Use the most efficient method to solve each equation. If a solution contains a radical, give the exact solution and then approximate it to the nearest hundredth.

53. $(2y - 1)^2 = 25$

54. $(y + 1)^2 = 64$

55. $2x^2 + x = 5$

56. $2x^2 - x + 2 = 0$

57. $x^2 - 12x + 35 = 0$

58. $m^2 + 14m + 49 = 0$

59. $r^2 + 9r + 1 = 0$

60. $b^2 = 18$

61. $5x^2 - 2x = 8$

62. $6x^2 - 6x - 1 = 0$

63. $x^2 + 3x = 9$

64. $x^2 + 7x = -1$

65. $7x^2 + 6x + 4 = 0$

66. $3x^2 - x = 1$

67. $4c^2 + 16c = 0$

68. $2t^2 - 162 = 0$

69. $3y^2 = 18$

70. $25x - 50x^2 = 0$

71. $x^2 - 4x = 29$

72. $a^2 + 20a = 3$

73. $t^2 - 24t + 144 = 0$

74. $3a^2 = 2a + 3$

75. $x^2 - 63 = 0$

76. $y^2 - 80 = 0$

Solve each equation and approximate each solution to the nearest tenth.

77. $2.4x^2 - 9.5x + 6.2 = 0$

78. $-1.7x^2 + 0.5x + 0.9 = 0$

APPLICATIONS

79. KITCHEN FLOOR PLANS To minimize the number of steps that a cook must take when preparing meals, designers carefully plan the *kitchen work triangle* (the area between the sink, refrigerator, and range). The one leg of the work triangle shown is 2 feet longer than the other leg, and the area covered is 24 ft^2. Find the length of each leg of the triangle.

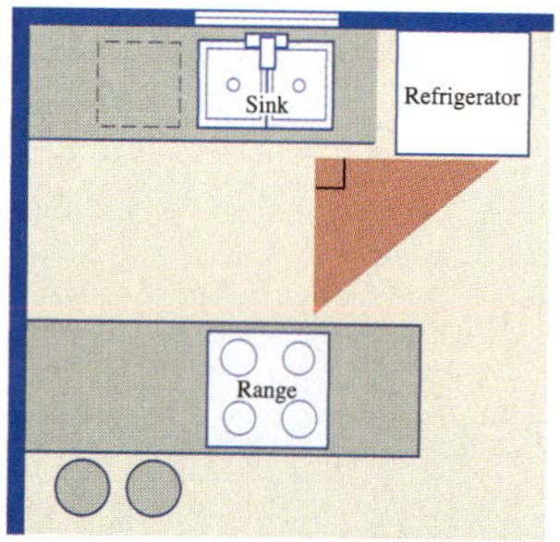

80. EARTHQUAKES After a powerful earthquake, a store owner nailed 50-inch-long boards across a broken display window. Find the height and length of the window, if the height is 10 inches less than the length.

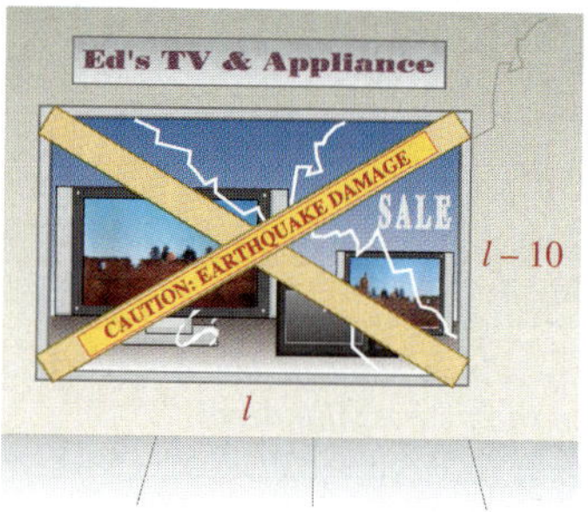

81. ***from Campus to Careers***
Graphic Designer

A poster that shows former UCLA basketball coach John Wooden's *Pyramid of Success* has an area of 80 square inches. The base of the triangular-shaped poster is 6 inches longer than the height. Find the length of the base and the height of the poster.

© Mark Richards/PhotoEdit

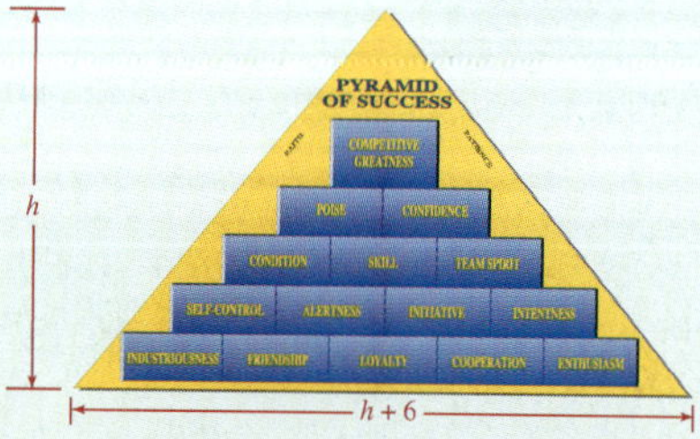

82. COMICS A comic strip occupies 96 square centimeters of space in a newspaper. The length of the rectangular space is 4 centimeters more than twice its width. Find its dimensions.

83. THE ABACUS The Chinese abacus shown consists of a frame, parallel wires, and beads that are moved to perform arithmetic computations. The frame is 21 centimeters wider than it is high. Find its height and width.

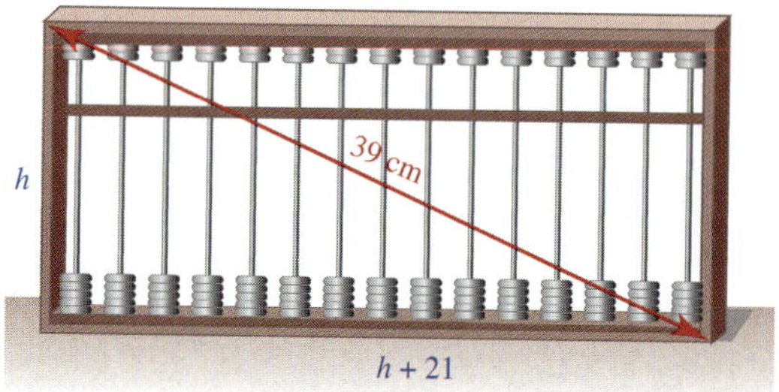

84. BADMINTON A badminton court occupies 880 square feet of floor space of a high school gym. If the length of the court is 16 feet less than three times its width, find its dimensions.

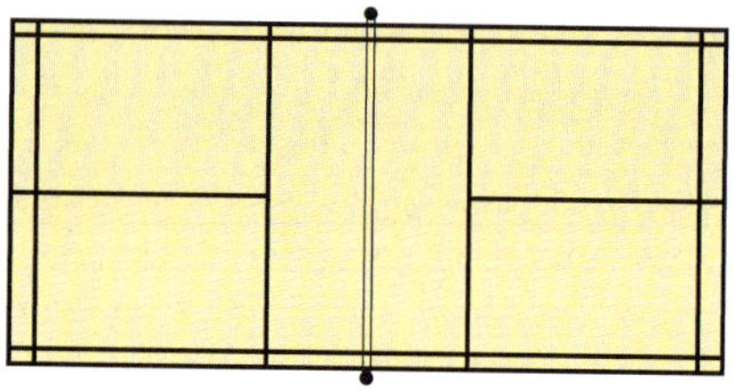

85. INVESTING We can use the formula $A = P(1 + r)^2$ to find the amount \$$A$ that \$$P$ will become when invested at an annual rate of r% for 2 years. What interest rate is needed to make \$5,000 grow to \$5,724.50 in 2 years?

86. RETAILING When a wholesaler sells n compact disc players, his revenue R is given by the formula $R = 150n - \frac{1}{2}n^2$. How many players would he have to sell to receive \$11,250? (*Hint:* Multiply both sides of the equation by -2.)

WRITING

87. Do you agree with the following statement? Explain. *The quadratic formula is the easiest method to use to solve quadratic equations.*

88. In this section, we used diagrams like the following when solving quadratic equations. What is its purpose?

$$5x^2 + 6x + 7 = 0$$
$$\uparrow \quad \uparrow \quad \uparrow$$
$$ax^2 + bx + c = 0$$

89. Write the quadratic formula in words.

90. At times, certain types of solutions to applied problems are discarded. Explain. Give an example of such a situation.

91. Solve $x^2 - 8x + 16 = 0$ using the quadratic formula. Explain why the equation has only one solution.

92. Find the error in the following work.

Solve: $x^2 + 4x - 5 = 0$

$$x = -4 \pm \frac{\sqrt{16 - 4(1)(-5)}}{2}$$

REVIEW

Solve each equation for the specified variable.

93. $A = p + prt$; for r

94. $F = \dfrac{GMm}{d^2}$; for M

95. $\dfrac{1}{r} = \dfrac{1}{r_1} + \dfrac{1}{r_2}$; for r

96. $2E = \dfrac{T - t}{9}$; for t

CHALLENGE PROBLEMS

Use the quadratic formula to solve each equation.

97. $\dfrac{2}{3}x^2 - \dfrac{1}{3} = \dfrac{5}{9}x$

98. $\dfrac{1}{x - 1} = \dfrac{1}{4} - \dfrac{2}{x}$

99. $\dfrac{(n + 5)(2n + 1)}{2} = 3.5$

100. $2x^2 + \sqrt{5}x - 3 = 0$

101. DECKING The owner of a pool wants to surround it with a concrete deck of uniform width (shown in gray). If he can afford 368 square feet of decking, how wide can he make the deck?

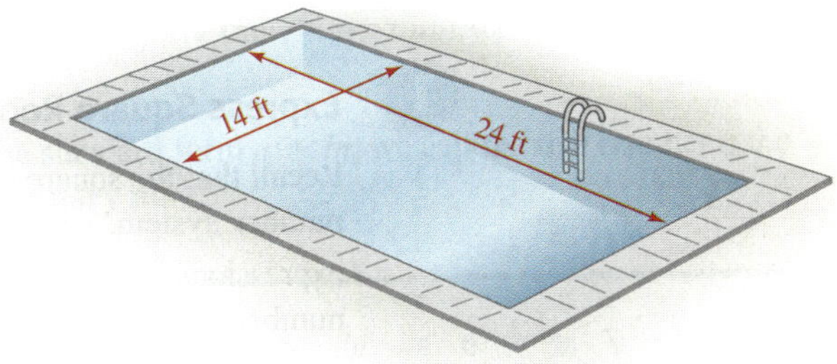

102. METAL FABRICATION A square piece of tin, 12 inches on a side, is to have four equal squares cut from its corners, as shown. If the edges are then to be folded up to make a box with a floor area of 64 square inches, find the depth of the box.

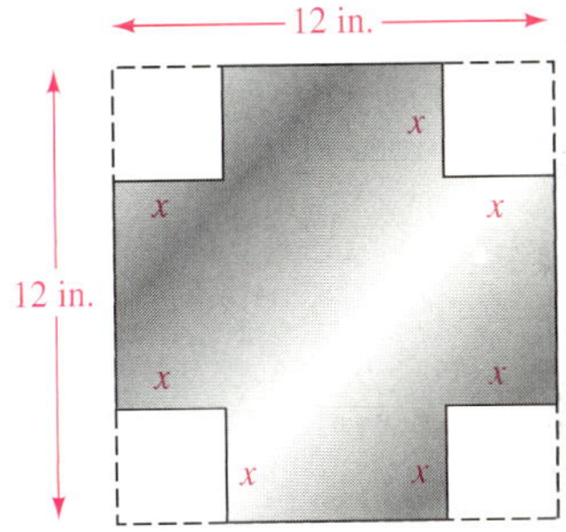

SECTION 9.4
Complex Numbers

Objectives

1. Express square roots of negative numbers in terms of i.
2. Write complex numbers in $a + bi$ form.
3. Add and subtract complex numbers.
4. Multiply and divide complex numbers.
5. Solve quadratic equations with complex-number solutions.

Earlier in this chapter, we saw that some quadratic equations do not have real-number solutions. When we used the square root property or the quadratic formula to solve them, we encountered the square root of a negative number.

3 Add and Subtract Complex Numbers.

Adding and subtracting complex numbers is similar to adding and subtracting polynomials.

Addition and Subtraction of Complex Numbers

1. To add complex numbers, add their real parts and add their imaginary parts.
2. To subtract complex numbers, add the opposite of the complex number being subtracted.

EXAMPLE 3 Find each sum or difference: **a.** $(5 + 2i) + (1 + 8i)$ **b.** $(-6 - 5i) - (3 - 4i)$ **c.** $11i + (-2 + 6i)$

Strategy To add the complex numbers, we will add their real parts and add their imaginary parts. To subtract the complex numbers, we will add the opposite of the complex number to be subtracted.

Why We perform the indicated operations as if the complex numbers were polynomials with i as a variable.

Solution

a. $(5 + 2i) + (1 + 8i) = (5 + 1) + (2 + 8)i$ — (5 + 1): The sum of the real parts. (2 + 8): The sum of the imaginary parts.

$= 6 + 10i$

b. $(-6 - 5i) - (3 - 4i) = (-6 - 5i) + (-3 + 4i)$ — Add; the opposite. To find the opposite, change the sign of each term of $3 - 4i$.

$= [-6 + (-3)] + (-5 + 4)i$ — Add the real parts. Add the imaginary parts.

$= -9 - i$

c. $11i + (-2 + 6i) = -2 + (11 + 6)i$ — Add the imaginary parts.

$= -2 + 17i$

Success Tip

i is not a variable, but it is helpful to think of it as one when adding, subtracting, and multiplying. For example:

$$4i + 3i = 7i$$
$$8i - 6i = 2i$$
$$i \cdot i = i^2$$

Self Check 3 Find the sum or difference. Write each result in the form $a + bi$:
a. $(3 - 5i) + (-2 + 6i)$
b. $(-4 - i) - (-1 - 6i)$ **c.** $9 + (16 - 4i)$

Now Try Problems 33 and 35

4 Multiply and Divide Complex Numbers.

Multiplying complex numbers is similar to multiplying polynomials.

EXAMPLE 4 Find each product: **a.** $5i(4 - i)$ **b.** $(2 + 5i)(6 + 4i)$
c. $(3 + 5i)(3 - 5i)$

Strategy We will use the distributive property or the FOIL method to find the products.

Why We perform the indicated operations as if the complex numbers were polynomials with i as a variable.

Solution

a. $5i(4 - i) = 5i \cdot 4 - 5i \cdot i$ Distribute the multiplication by $5i$.

$= 20i - 5i^2$

$= 20i - 5(-1)$ Replace i^2 with -1.

$= 20i + 5$

$= 5 + 20i$ Write the result in the form $a + bi$.

b. $(2 + 5i)(6 + 4i) = 12 + 8i + 30i + 20i^2$ (F O I L) Use the FOIL method.

$= 12 + 38i + 20(-1)$ $8i + 30i = 38i$. Replace i^2 with -1.

$= 12 + 38i - 20$

$= -8 + 38i$ Subtract: $12 - 20 = -8$.

c. $(3 + 5i)(3 - 5i) = 9 - 15i + 15i - 25i^2$ (F O I L) Use the FOIL method.

$= 9 - 25(-1)$ Add: $-15i + 15i = 0$. Replace i^2 with -1.

$= 9 + 25$

$= 34$

Written in the form $a + bi$, the product is $34 + 0i$.

Self Check 4 Find the product. Write each result in the form $a + bi$:

a. $4i(3 - 9i)$ **b.** $(3 - 2i)(5 - 4i)$

c. $(2 + 6i)(2 - 6i)$

Now Try Problems 43 and 45

In Example 4c, we saw that the product of the imaginary numbers $3 + 5i$ and $3 - 5i$ is the real number 34. We call $3 + 5i$ and $3 - 5i$ *complex conjugates* of each other.

Complex Conjugates

The complex numbers $a + bi$ and $a - bi$ are called **complex conjugates.**

Success Tip

To find the conjugate of a complex number, write it in standard form and change the sign between the real and imaginary parts from + to − or from − to +.

For example,

- $7 + 4i$ and $7 - 4i$ are complex conjugates.
- $5 - i$ and $5 + i$ are complex conjugates.
- $-6i$ and $6i$ are complex conjugates, because $-6i = 0 - 6i$ and $6i = 0 + 6i$.

In general, the product of the complex number $a + bi$ and its complex conjugate $a - bi$ is the real number $a^2 + b^2$, as the following work shows:

$(a + bi)(a - bi) = a^2 - abi + abi - b^2i^2$ (F O I L) Use the FOIL method.

$= a^2 - b^2(-1)$ Add: $-abi + abi = 0$. Replace i^2 with -1.

$= a^2 + b^2$

We can use this fact when dividing by a complex number. The process that we use is similar to rationalizing denominators.

EXAMPLE 5 Write each quotient in the form $a + bi$: **a.** $\frac{6}{7 - 4i}$ **b.** $\frac{3 - i}{2 + i}$

Strategy We will build each fraction by multiplying it by a form of 1 that uses the conjugate of the denominator.

Why This step produces a real number in the denominator so that the result can then be written in the form $a + bi$.

Solution

a. We want to build a fraction equivalent to $\frac{6}{7 - 4i}$ that does not have i in the denominator. To make the denominator, $7 - 4i$, a real number, we need to multiply it by its complex conjugate, $7 + 4i$. It follows that $\frac{7 + 4i}{7 + 4i}$ should be the form of 1 that is used to build $\frac{6}{7 - 4i}$.

$$\frac{6}{7 - 4i} = \frac{6}{7 - 4i} \cdot \frac{7 + 4i}{7 + 4i}$$ To build an equivalent fraction, multiply by $\frac{7 + 4i}{7 + 4i} = 1$.

$$= \frac{42 + 24i}{49 - 16i^2}$$ To multiply the numerators, distribute the multiplication by 6. To multiply the denominators, find $(7 - 4i)(7 + 4i)$.

$$= \frac{42 + 24i}{49 - 16(-1)}$$ Replace i^2 with -1. The denominator no longer contains i.

$$= \frac{42 + 24i}{49 + 16}$$ Simplify the denominator.

$$= \frac{42 + 24i}{65}$$ This notation represents the sum of two fractions that have the common denominator 65: $\frac{42}{65}$ and $\frac{24i}{65}$.

$$= \frac{42}{65} + \frac{24}{65}i$$ Write the result in the form $a + bi$.

b. We can make the denominator of $\frac{3 - i}{2 + i}$ a real number by multiplying it by the complex conjugate of $2 + i$, which is $2 - i$.

$$\frac{3 - i}{2 + i} = \frac{3 - i}{2 + i} \cdot \frac{2 - i}{2 - i}$$ To build an equivalent fraction, multiply by $\frac{2 - i}{2 - i} = 1$.

$$= \frac{6 - 3i - 2i + i^2}{4 - i^2}$$ To multiply the numerators, find $(3 - i)(2 - i)$. To multiply the denominators, find $(2 + i)(2 - i)$.

$$= \frac{6 - 5i + (-1)}{4 - (-1)}$$ Replace i^2 with -1. The denominator no longer contains i.

$$= \frac{5 - 5i}{5}$$ Simplify the numerator and the denominator.

$$= \frac{5}{5} - \frac{5i}{5}$$ Write each term of the numerator over the denominator, 5.

$$= 1 - i$$ Simplify each fraction.

The Language of Algebra
Recall that the word *conjugate* was used in Chapter 8 when we rationalized the denominators of radical expressions such as $\frac{5}{\sqrt{6} - 1}$.

Caution
A common mistake is to replace i with -1. Remember, $i \neq -1$. By definition, $i = \sqrt{-1}$ and $i^2 = -1$.

Self Check 5 Write each quotient in standard form: **a.** $\frac{5}{4-i}$ **b.** $\frac{5-3i}{4+2i}$

Now Try Problems 49 and 53

As the results of Example 5 show, we use the following rule to divide complex numbers.

Division of Complex Numbers

To divide complex numbers, multiply the numerator and denominator by the complex conjugate of the denominator.

5 Solve Quadratic Equations with Complex-Number Solutions.

We have seen that certain quadratic equations do not have real-number solutions. In the complex number system, all quadratic equations have solutions. We can write their solutions in the form $a + bi$.

EXAMPLE 6 Solve each equation. Express the solutions in the form $a + bi$:
a. $x^2 + 25 = 0$ **b.** $(y-3)^2 = -54$

Strategy We will use the square root property to solve each equation.

Why It is the fastest way to solve equations of this form.

Solution

a. $x^2 + 25 = 0$ This is the equation to solve.

$x^2 = -25$ To isolate x^2, subtract 25 from both sides.

$x = \pm\sqrt{-25}$ Use the square root property. Note that the radicand is negative.

$x = \pm 5i$ Write $\sqrt{-25}$ in terms of i: $\sqrt{-25} = \sqrt{-1}\sqrt{25} = 5i$.

To express the solutions in the form $a + bi$, we must write 0 for the real part, a. Thus, the solutions are $0 + 5i$ and $0 - 5i$, or more simply, $0 \pm 5i$.

Notation
It is acceptable to use $a - bi$ as a substitute for the form $a + (-b)i$. In Example 6, we write $0 - 5i$ instead of $0 + (-5)i$.

b. $(y-3)^2 = -54$ This is the equation to solve.

$y - 3 = \pm\sqrt{-54}$ Use the square root property. Note that the radicand is negative.

$y - 3 = \pm 3i\sqrt{6}$ Write $\sqrt{-54}$ in terms of i: $\sqrt{-54} = \sqrt{-1}\sqrt{9}\sqrt{6} = 3i\sqrt{6}$.

$y = 3 \pm 3i\sqrt{6}$ To isolate y, add 3 to both sides.

The solutions are $3 + 3i\sqrt{6}$ and $3 - 3i\sqrt{6}$, or more simply, $3 \pm 3i\sqrt{6}$.

Self Check 6 Solve each equation. Express the solutions in the form $a + bi$:
a. $x^2 + 100 = 0$ **b.** $(d-3)^2 = -48$

Now Try Problems 57 and 63

EXAMPLE 7 Solve $4t^2 - 6t + 3 = 0$. Express the solutions in the form $a + bi$.

Strategy We use the quadratic formula to solve the equation.

Why $4t^2 - 6t + 3$ does not factor.

Solution

$$t = \frac{-b \pm \sqrt{b^2 - 4ac}}{2a}$$ In the quadratic formula, replace *x* with *t*.

$$t = \frac{-(-6) \pm \sqrt{(-6)^2 - 4(4)(3)}}{2(4)}$$ Substitute 4 for *a*, −6 for *b*, and 3 for *c*.

$$t = \frac{6 \pm \sqrt{36 - 48}}{8}$$ Evaluate the power and multiply within the radical. Multiply in the denominator.

$$t = \frac{6 \pm \sqrt{-12}}{8}$$ Subtract within the radical: 36 − 48. Note that the radicand is negative.

$$t = \frac{6 \pm 2i\sqrt{3}}{8}$$ Write $\sqrt{-12}$ in terms of *i*: $\sqrt{-12} = \sqrt{-1}\sqrt{12} = \sqrt{-1}\sqrt{4}\sqrt{3} = 2i\sqrt{3}$

$$t = \frac{\overset{1}{\cancel{2}}(3 \pm i\sqrt{3})}{\underset{1}{\cancel{2}} \cdot 4}$$ Factor out the GCF 2 from $6 \pm 2i\sqrt{3}$ and factor 8 as $2 \cdot 4$. Then remove the common factor 2: $\frac{2}{2} = 1$.

$$t = \frac{3 \pm i\sqrt{3}}{4}$$ This notation represents the sum and difference of two fractions that have the common denominator 4: $\frac{3}{4}$ and $\frac{i\sqrt{3}}{4}$.

Writing each result as a complex number in the form $a + bi$, the solutions are

$$\frac{3}{4} + \frac{\sqrt{3}}{4}i \quad \text{and} \quad \frac{3}{4} - \frac{\sqrt{3}}{4}i \quad \text{or more simply} \quad \frac{3}{4} \pm \frac{\sqrt{3}}{4}i$$

Self Check 7 Solve $a^2 + 2a + 3 = 0$. Express the solutions in the form $a + bi$.

Now Try **Problem 65**

ANSWERS TO SELF CHECKS **1. a.** $9i$ **b.** $-i\sqrt{11}$ **c.** $2i\sqrt{7}$ **d.** $\frac{3i\sqrt{3}}{10}$ **2. a.** $-18 + 0i$ **b.** $0 + 6i$ **c.** $1 + 2i\sqrt{6}$ **3. a.** $1 + i$ **b.** $-3 + 5i$ **c.** $25 - 4i$ **4. a.** $36 + 12i$ **b.** $7 - 22i$ **c.** $40 + 0i$ **5. a.** $\frac{20}{17} + \frac{5}{17}i$ **b.** $\frac{7}{10} - \frac{11}{10}i$ **6. a.** $0 \pm 10i$ **b.** $3 \pm 4i\sqrt{3}$ **7.** $-1 \pm i\sqrt{2}$

STUDY SET 9.4

VOCABULARY

Fill in the blanks.

1. $9 + 2i$ is an example of a ________ number. The _____ part is 9 and the ________ part is 2.
2. The ________ number i is used to define complex numbers.

CONCEPTS

Fill in the blanks.

3. **a.** $i =$ ______ **b.** $i \cdot i = i$ ___ $=$ ______
4. $\sqrt{-25} = \sqrt{\quad \cdot 25} =$ ______ $\sqrt{25} = 5$ ___
5. **a.** $5i + 3i =$ ______ **b.** $5i - 3i =$ ______
6. The product of any complex number and its complex conjugate is a _____ number.
7. To write the quotient $\frac{2-3i}{6-i}$ as a complex number in standard form, we multiply it by ______.
8. $\frac{3 \pm \sqrt{-4}}{5} = \frac{3 \pm \quad}{5}$
9. Determine whether each statement is true or false.
   a. Every real number is a complex number.
   b. $2 + 7i$ is an imaginary number.
   c. $\sqrt{-16}$ is a real number.
   d. In the complex number system, all quadratic equations have solutions.
10. Give the complex conjugate of each number.
   a. $2 - 9i$ **b.** $-8 + i$
   c. $4i$ **d.** $-11i$

NOTATION

11. Write each expression so it is clear that i is not within the radical symbol.
   a. $\sqrt{7}i$ **b.** $2\sqrt{3}i$
12. Write $\frac{3-4i}{5}$ in the form $a + bi$.

GUIDED PRACTICE

***Write each expression in terms of i.* See Example 1.**

13. $\sqrt{-9}$
14. $\sqrt{-4}$
15. $\sqrt{-7}$
16. $\sqrt{-11}$
17. $\sqrt{-24}$
18. $\sqrt{-28}$
19. $-\sqrt{-32}$
20. $-\sqrt{-72}$
21. $5\sqrt{-81}$
22. $6\sqrt{-49}$
23. $\sqrt{-\frac{25}{9}}$
24. $\sqrt{-\frac{121}{144}}$

***Write each number in the form a + bi.* See Example 2.**

25. 12
26. -27
27. $\sqrt{-100}$
28. $\sqrt{-64}$
29. $6 + \sqrt{-16}$
30. $14 + \sqrt{-25}$
31. $-9 - \sqrt{-49}$
32. $-45 - \sqrt{-36}$

***Perform the operations. Write all answers in the form a + bi.* See Example 3.**

33. $(3 + 4i) + (5 - 6i)$
34. $(5 + 3i) - (6 - 9i)$
35. $(7 - 3i) - (4 + 2i)$
36. $(8 + 3i) + (-7 - 2i)$
37. $(14 - 4i) - 9i$
38. $(20 - 5i) - 17i$
39. $15 + (-3 - 9i)$
40. $-25 + (18 - 9i)$

***Perform the operations. Write all answers in the form a + bi.* See Example 4.**

41. $3(2 - i)$
42. $9(-4 - 4i)$
43. $-5i(5 - 5i)$
44. $2i(7 + 2i)$
45. $(3 - 2i)(2 + 3i)$
46. $(3 - i)(2 + 3i)$
47. $(4 + i)(3 - i)$
48. $(1 - 5i)(1 - 4i)$

***Write each quotient in the form a + bi.* See Example 5.**

49. $\frac{5}{2-i}$
50. $\frac{26}{3-2i}$
51. $\frac{-4i}{7-2i}$
52. $\frac{5i}{6+2i}$
53. $\frac{2+3i}{2-3i}$
54. $\frac{2-5i}{2+5i}$
55. $\frac{4-3i}{7-i}$
56. $\frac{4+i}{4-i}$

***Solve each equation. Write all solutions in the form a + bi.* See Example 6.**

57. $x^2 + 9 = 0$
58. $x^2 + 100 = 0$
59. $d^2 + 8 = 0$
60. $a^2 + 27 = 0$

61. $(x + 3)^2 = -1$

62. $(x + 2)^2 = -25$

63. $(x - 11)^2 = -75$

64. $(x - 22)^2 = -18$

***Solve each equation. Write all solutions in the form a + bi.* See Example 7.**

65. $x^2 - 3x + 4 = 0$

66. $y^2 + y + 3 = 0$

67. $2x^2 + x + 1 = 0$

68. $2x^2 + 3x + 3 = 0$

TRY IT YOURSELF

Perform the operations. Write all answers in the form a + bi.

69. $\dfrac{3}{5 + i}$

70. $\dfrac{-4}{7 - 2i}$

71. $(6 - i) + (9 + 3i)$

72. $(5 - 4i) + (3 + 2i)$

73. $(-3 - 8i) - (-3 - 9i)$

74. $(-1 - 8i) - (-1 - 7i)$

75. $(2 + i)(2 + 3i)$

76. $2i(7 + 2i)$

77. $\dfrac{3 - 2i}{3 + 2i}$

78. $\dfrac{3 + 2i}{3 + i}$

79. $(10 - 9i) + (-1 + i)$

80. $(32 - 3i) + (-44 + 15i)$

81. $-4(3 + 4i)$

82. $-7(5 - 3i)$

83. $(2 + i)^2$

84. $(3 - 2i)^2$

Solve each equation. Write all solutions in the form a + bi.

85. $2x^2 + x = -5$

86. $4x^2 = -7x - 4$

87. $(x - 4)^2 = -45$

88. $(x - 9)^2 = -80$

89. $b^2 + 2b + 2 = 0$

90. $t^2 - 2t + 6 = 0$

91. $x^2 = -36$

92. $x^2 = -49$

93. $x^2 = -\dfrac{16}{9}$

94. $x^2 = -\dfrac{25}{4}$

95. $3x^2 + 2x + 1 = 0$

96. $3x^2 - 4x + 2 = 0$

APPLICATIONS

97. ELECTRICITY In an AC (alternating current) circuit, if two sections are connected in series and have the same current in each section, the voltage is given by $V = V_1 + V_2$. Find the total voltage in a given circuit if the voltages in the individual sections are $V_1 = 10.31 - 5.97i$ and $V_2 = 8.14 + 3.79i$.

98. ELECTRONICS The impedance Z in an AC (alternating current) circuit is a measure of how much the circuit impedes (hinders) the flow of current through it. The impedance is related to the voltage V and the current I by the formula

$$V = IZ$$

If a circuit has a current of $(0.5 + 2.0i)$ amps and an impedance of $(0.4 - 3.0i)$ ohms, find the voltage.

WRITING

99. What unusual situation discussed at the beginning of this section illustrated the need to define the square root of a negative number?

100. Explain the difference between the opposite of a complex number and its conjugate.

101. What is an imaginary number?

102. In this section, we have seen that $i^2 = -1$. From your previous experience in this course, what is unusual about that fact?

REVIEW

Rationalize the denominator.

103. $\dfrac{1}{\sqrt{7}}$

104. $\dfrac{\sqrt{3}}{\sqrt{10}}$

105. $\dfrac{8}{\sqrt{x} - 2}$

106. $\dfrac{\sqrt{3}}{5 + \sqrt{x}}$

CHALLENGE PROBLEMS

Perform the operations.

107. $\left(2\sqrt{2} + i\sqrt{2}\right)\left(3\sqrt{2} - i\sqrt{2}\right)$

108. $\dfrac{\sqrt{5} - i\sqrt{3}}{\sqrt{5} + i\sqrt{3}}$

SECTION 9.5
Graphing Quadratic Equations

Objectives

1. Understand the vocabulary used to describe parabolas.
2. Find the intercepts of a parabola.
3. Determine the vertex of a parabola.
4. Graph equations of the form $y = ax^2 + bx + c$.
5. Solve quadratic equations graphically.

In this section, we will combine our graphing skills with our equation solving skills to graph *quadratic equations in two variables.*

1 Understand the Vocabulary Used to Describe Parabolas.

Equations that can be written in the form $y = ax^2 + bx + c$, where $a \neq 0$, are called **quadratic equations in two variables.** Some examples are

$$y = x^2 - 2x - 3 \qquad y = -2x^2 - 8x - 8 \qquad y = x^2 + x$$

In Section 5.4, we graphed $y = x^2$, a quadratic equation in two variables. To do this, we constructed a table of solutions, plotted points, and joined them with a smooth curve, called a **parabola.** The parabola opens upward, and the lowest point on the graph, called the **vertex,** is the point (0, 0). If we fold the graph paper along the y-axis, the two sides of the parabola match. We say that the graph is *symmetric about the y-axis* and we call the y-axis the **axis of symmetry.**

The Language of Algebra
An *axis of symmetry* (or *line of symmetry*) divides a parabola into two matching sides. The sides are said to be *mirror images* of each other.

$y = x^2$

x	y	(x, y)
−3	9	(−3, 9)
−2	4	(−2, 4)
−1	1	(−1, 1)
0	0	(0, 0)
1	1	(1, 1)
2	4	(2, 4)
3	9	(3, 9)

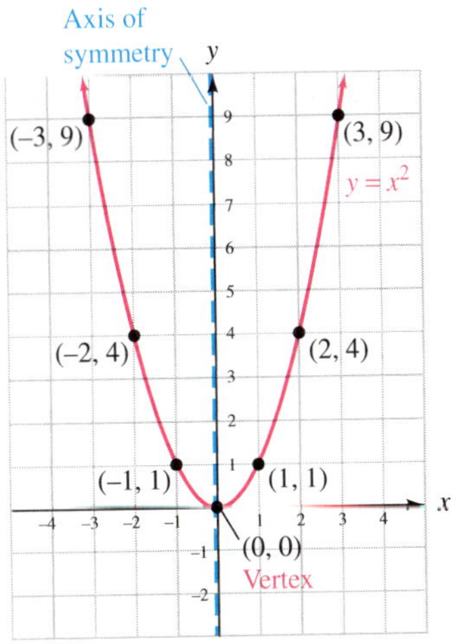

In Section 5.4, we also graphed $y = -x^2 + 2$, another quadratic equation in two variables. The resulting parabola opens downward, and the **vertex** (in this case, the highest point on the graph) is the point (0, 2). The axis of symmetry is the y-axis.

The Language of Algebra
In the equation $y = ax^2 + bx + c$, each value of x determines exactly one value of y. Therefore, the equation defines y to be a function of x and we could write $f(x) = ax^2 + bx + c$. Your instructor may ask you to use the vocabulary and notation of functions throughout this section.

$y = -x^2 + 2$

x	y	(x, y)
-3	-7	$(-3, -7)$
-2	-2	$(-2, -2)$
-1	1	$(-1, 1)$
0	2	$(0, 2)$
1	1	$(1, 1)$
2	-2	$(2, -2)$
3	-7	$(3, -7)$

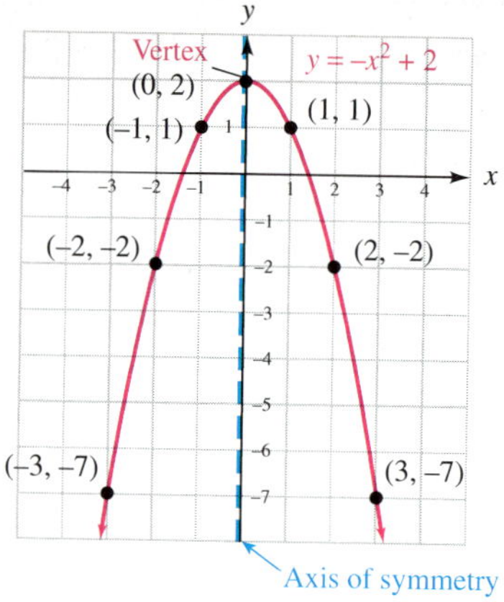

For the equation $y = x^2$, the coefficient of the x^2 term is the positive number 1. For $y = -x^2 + 2$, the coefficient of the x^2 term is the negative number -1. These observations suggest the following fact.

Graphs of Quadratic Equations

The graph of $y = ax^2 + bx + c$, where $a \neq 0$, is a parabola. It opens upward when $a > 0$ and downward when $a < 0$.

The Language of Algebra
The word *parabolic* (pronounced *par • a • BOL • ic*) means having the form of a parabola. For example, the light bulb in most flashlights is surrounded by a *parabolic* reflecting mirror.

Parabolic shapes can be seen in a wide variety of real-world settings. These shapes can be modeled by quadratic equations in two variables.

The path of a stream of water

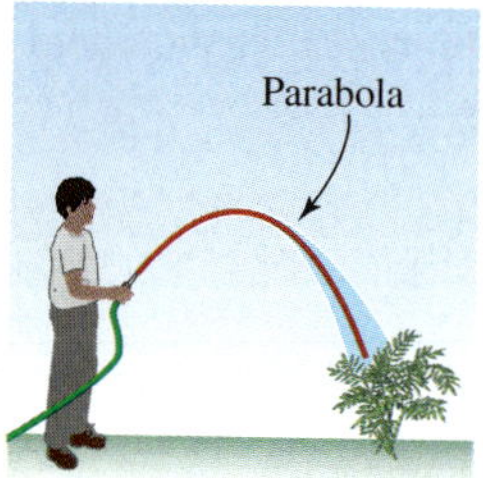

The shape of a satellite antenna dish

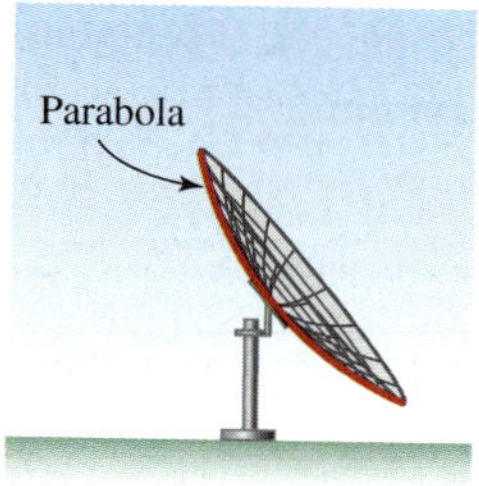

The path of a thrown object

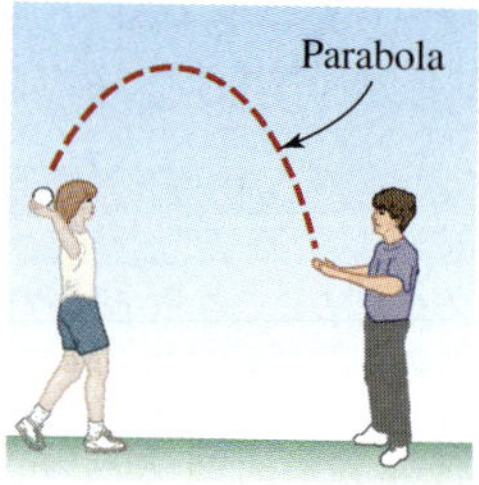

2 Find the Intercepts of a Parabola.

When graphing quadratic equations, it is helpful to know the x- and y-intercepts of the parabola. To find the intercepts of a parabola, we use the same steps that we used to find the intercepts of the graphs of linear equations.

Finding Intercepts

To find the y-intercept, substitute 0 for x in the given equation and solve for y.

To find the x-intercepts, substitute 0 for y in the given equation and solve for x.

EXAMPLE 1 Find the y- and x-intercepts of the graph of $y = x^2 - 2x - 3$.

Strategy To find the y-intercept of the graph, we will let $x = 0$ and find y. To find the x-intercepts of the graph, we will let $y = 0$ and solve the resulting equation for x.

Why A point on the y-axis has an x-coordinate of 0. A point on the x-axis has a y-coordinate of 0.

Solution We let $x = 0$ and evaluate the right side to find the y-intercept.

$$y = x^2 - 2x - 3 \quad \text{This is the given equation.}$$
$$y = 0^2 - 2(0) - 3 \quad \text{Substitute 0 for x.}$$
$$y = -3 \quad \text{Evaluate the right side.}$$

The y-intercept is $(0, -3)$. We note that the y-coordinate of the y-intercept is the same as the value of the constant term c on the right side of $y = x^2 - 2x - 3$.

Next, we let $y = 0$ and solve the resulting quadratic equation to find the x-intercepts.

$$y = x^2 - 2x - 3 \quad \text{This is the given equation.}$$
$$0 = x^2 - 2x - 3 \quad \text{Substitute 0 for y.}$$
$$0 = (x - 3)(x + 1) \quad \text{Factor the trinomial.}$$
$$x - 3 = 0 \quad \text{or} \quad x + 1 = 0 \quad \text{Set each factor equal to 0.}$$
$$x = 3 \quad\quad x = -1$$

Since there are two solutions, the graph has two x-intercepts: $(3, 0)$ and $(-1, 0)$.

Self Check 1 Find the y- and x-intercepts of the graph of $y = x^2 + 6x + 8$.

Now Try Problem 17

3 Determine the Vertex of a Parabola.

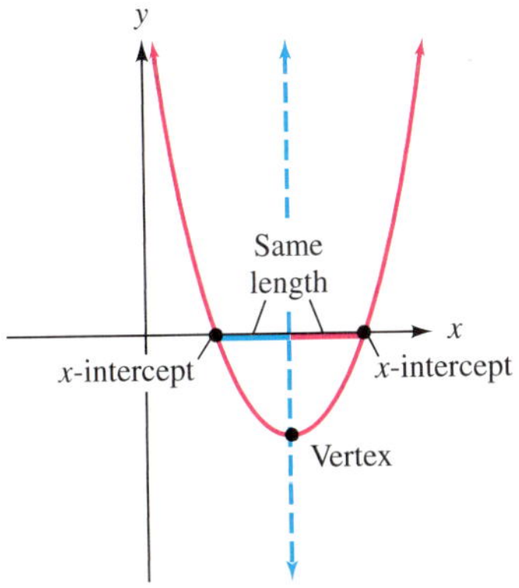

It is usually easier to graph a quadratic equation when we know the coordinates of the vertex of its graph. Because of symmetry, if a parabola has two x-intercepts, the x-coordinate of the vertex is exactly midway between them. We can use this fact to derive a formula to find the vertex of a parabola.

In general, if a parabola has two x-intercepts, they can be found by solving $0 = ax^2 + bx + c$ for x. We can use the quadratic formula to find the solutions. They are

$$x = \frac{-b - \sqrt{b^2 - 4ac}}{2a} \quad \text{and} \quad x = \frac{-b + \sqrt{b^2 - 4ac}}{2a}$$

Thus, the parabola's x-intercepts are $\left(\frac{-b - \sqrt{b^2 - 4ac}}{2a}, 0\right)$ and $\left(\frac{-b + \sqrt{b^2 - 4ac}}{2a}, 0\right)$.

Since the x-value of the vertex of a parabola is halfway between the two x-intercepts, we can find this value by finding the average, or $\frac{1}{2}$ of the sum of the x-coordinates of the x-intercepts.

$$x = \frac{1}{2}\left(\frac{-b - \sqrt{b^2 - 4ac}}{2a} + \frac{-b + \sqrt{b^2 - 4ac}}{2a}\right)$$

$$x = \frac{1}{2}\left(\frac{-b - \sqrt{b^2 - 4ac} + (-b) + \sqrt{b^2 - 4ac}}{2a}\right) \quad \text{Add the numerators and keep the common denominator.}$$

$$x = \frac{1}{2}\left(\frac{-2b}{2a}\right)$$ $-b + (-b) = -2b.$ $-\sqrt{b^2-4ac} + \sqrt{b^2 - 4ac} = 0.$

$$x = \frac{-b}{2a}$$ Simplify.

This result is true even if the graph has no x-intercepts.

Finding the Vertex of a Parabola

The graph of the quadratic equation $y = ax^2 + bx + c$ is a parabola whose vertex has an x-coordinate of $\frac{-b}{2a}$. To find the y-coordinate of the vertex, substitute $\frac{-b}{2a}$ for x into the equation and find y.

EXAMPLE 2 Find the vertex of the graph of $y = x^2 - 2x - 3$.

Strategy We will compare the given equation to the general form $y = ax^2 + bx + c$ to identify a and b.

Why To use the vertex formula, we need to know a and b.

Solution From the following diagram, we see that $a = 1$, $b = -2$, and $c = -3$.

$$y = 1x^2 - 2x - 3$$
$$\uparrow \quad \uparrow \quad \uparrow$$
$$y = ax^2 + bx + c$$

To find the x-coordinate of the vertex, we substitute the values for a and b into the formula $x = \frac{-b}{2a}$.

$$x = \frac{-b}{2a} = \frac{-(-2)}{2(1)} = 1$$

The x-coordinate of the vertex is 1. To find the y-coordinate, we substitute 1 for x in the original equation.

$y = x^2 - 2x - 3$ This is the given equation.

$y = 1^2 - 2(1) - 3$ Substitute 1 for x.

$y = -4$ Evaluate the right side.

The vertex of the parabola is $(1, -4)$.

Self Check 2 Find the vertex of the graph of $y = x^2 + 6x + 8$.

Now Try Problem 21

Success Tip
An easy way to remember the vertex formula is to note that $x = \frac{-b}{2a}$ is part of the quadratic formula:

$$x = \frac{-b \pm \sqrt{b^2 - 4ac}}{2a}$$

4 Graph Equations of the Form $y = ax^2 + bx + c$.

Much can be determined about the graph of $y = ax^2 + bx + c$ from the coefficients a, b, and c. We can use this information to help graph the equation.

Graphing a Quadratic Equation $y = ax^2 + bx + c$

1. ***Test for opening upward/downward*** Determine whether the parabola opens upward or downward. If $a > 0$, the graph opens upward. If $a < 0$, the graph opens downward.
2. ***Find the vertex/axis of symmetry*** The x-coordinate of the vertex of the parabola is $x = \frac{-b}{2a}$. To find the y-coordinate, substitute $\frac{-b}{2a}$ for x into the equation and find y. The axis of symmetry is the vertical line passing through the vertex.
3. ***Find the intercepts*** To find the y-intercept, substitute 0 for x in the given equation and solve for y. The result will be c. Thus, the y-intercept is $(0, c)$.
   To find the x-intercepts (if any), substitute 0 for y and solve the resulting quadratic equation $ax^2 + bx + c = 0$. If no real-number solutions exist, the graph has no x-intercepts.
4. ***Plotting points/using symmetry*** To find two more points on the graph, select a convenient value for x and find the corresponding value of y. Plot that point and its mirror image on the opposite side of the axis of symmetry.
5. ***Draw the parabola*** Draw a smooth curve through the located points.

EXAMPLE 3 Graph: $y = x^2 - 2x - 3$

Strategy We will use the five-step procedure described above to sketch the graph of the equation.

Why This strategy is usually faster than making a table of solutions and plotting points.

Solution ***Upward/downward:*** The equation is in the form $y = ax^2 + bx + c$, with $a = 1, b = -2$, and $c = -3$. Since $a > 0$, the parabola opens upward.

Vertex/axis of symmetry: In Example 2, we found that the vertex of the graph is $(1, -4)$. The axis of symmetry will be the vertical line passing through $(1, -4)$. See part (a) of the figure on the next page.

Intercepts: In Example 1, we found that the y-intercept is $(0, -3)$ and x-intercepts are $(3, 0)$ and $(-1, 0)$.

If the point $(0, -3)$, which is 1 unit to the left of the axis of symmetry, is on the graph, the point $(2, -3)$, which is 1 unit to the right of the axis of symmetry, is also on the graph. See part (a).

Success Tip
When graphing, remember that any point on a parabola to the right of the axis of symmetry yields a second point to the left of the axis of symmetry, and vice versa. Think of it as two-for-one.

Plotting points/using symmetry: It would be helpful to locate two more points on the graph. To find a solution of $y = x^2 - 2x - 3$, we select a convenient value for x, say -2, and find the corresponding value of y.

$y = x^2 - 2x - 3$ This is the equation to graph.

$y = (-2)^2 - 2(-2) - 3$ Substitute -2 for x.

$y = 5$ Evaluate the right side.

Success Tip
To find additional points on the graph, select values of x that are close to the x-coordinate of the vertex.

$y = x^2 - 2x - 3$

x	y	(x, y)
-2	5	$(-2, 5)$

Thus, the point $(-2, 5)$ lies on the parabola. If the point $(-2, 5)$, which is 3 units to the left of the axis of symmetry, is on the graph, the point $(4, 5)$, which is 3 units to the right of the axis of symmetry, is also on the graph. See part (a).

Draw a smooth curve through the points: The completed graph of $y = x^2 - 2x - 3$ is shown in part (b).

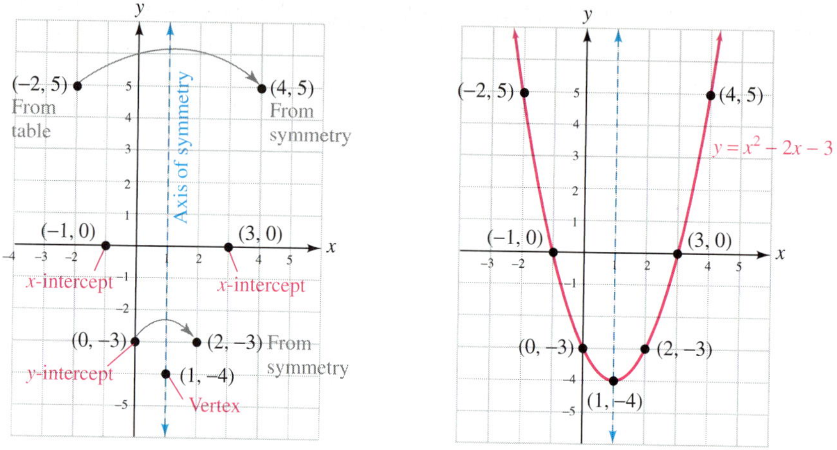

(a) (b)

Self Check 3 Use your results from Self Checks 1 and 2 to help graph $y = x^2 + 6x + 8$.

Now Try Problem 25

EXAMPLE 4

Graph: $y = -2x^2 - 8x - 8$

Strategy We will follow the five-step procedure to sketch the graph of the equation.

Why This strategy is usually faster than making a table of solutions and plotting points.

> **Success Tip**
> The most important point to find when graphing a quadratic equation in two variables is the vertex.

Solution ***Upward/downward:*** The equation is in the form $y = ax^2 + bx + c$, with $a = -2$, $b = -8$, and $c = -8$. Since $a < 0$, the parabola opens downward.

Vertex/axis of symmetry: To find the x-coordinate of the vertex, we substitute -2 for a and -8 for b into the formula $x = \frac{-b}{2a}$.

$$x = \frac{-b}{2a} = \frac{-(-8)}{2(-2)} = -2$$

The x-coordinate of the vertex is -2. To find the y-coordinate, we substitute -2 for x in the original equation and find y.

$$y = -2x^2 - 8x - 8 \quad \text{This is the equation to graph.}$$

$$y = -2(-2)^2 - 8(-2) - 8 \quad \text{Substitute } -2 \text{ for } x.$$

$$y = 0 \quad \text{Evaluate the right side.}$$

The vertex of the parabola is the point $(-2, 0)$. The axis of symmetry is the vertical line passing through $(-2, 0)$. See part (a) of the figure.

Intercepts: Since $c = -8$, the y-intercept of the parabola is $(0, -8)$. The point $(-4, -8)$, which is 2 units to the left of the axis of symmetry, must also be on the graph. See part (a).

To find the x-intercepts, we let $y = 0$ and solve the resulting quadratic equation.

Success Tip

To solve $0 = -2x^2 - 8x - 8$, instead of dividing both sides by -2, we could factor out the common factor -2 and proceed as follows:

$$0 = -2x^2 - 8x - 8$$
$$0 = -2(x^2 + 4x + 4)$$
$$0 = -2(x + 2)(x + 2)$$

$y = -2x^2 - 8x - 8$

x	y	(x, y)
-3	-2	$(-3, -2)$

$y = -2x^2 - 8x - 8$ This is the equation to graph.

$0 = -2x^2 - 8x - 8$ Substitute 0 for y.

$0 = x^2 + 4x + 4$ Divide both sides by -2: $\frac{0}{-2} = \frac{-2x^2}{-2} - \frac{8x}{-2} - \frac{8}{-2}$.

$0 = (x + 2)(x + 2)$ Factor the trinomial.

$x + 2 = 0$ or $x + 2 = 0$ Set each factor equal to 0.

$x = -2$ | $x = -2$

Since the solutions are the same, the graph has only one x-intercept: $(-2, 0)$. This point is the vertex of the parabola and has already been plotted.

Plotting points/using symmetry: It would be helpful to know two more points on the graph. To find a solution of $y = -2x^2 - 8x - 8$, we select a convenient value for x, say -3, and find the corresponding value for y.

$y = -2x^2 - 8x - 8$ This is the equation to graph.

$y = -2(-3)^2 - 8(-3) - 8$ Substitute -3 for x.

$y = -2$ Evaluate the right side.

Thus, the point $(-3, -2)$ lies on the parabola. We plot $(-3, -2)$ and then use symmetry to determine that $(-1, -2)$ is also on the graph. See part (a).

Draw a smooth curve through the points: The completed graph of $y = -2x^2 - 8x - 8$ is shown in part (b).

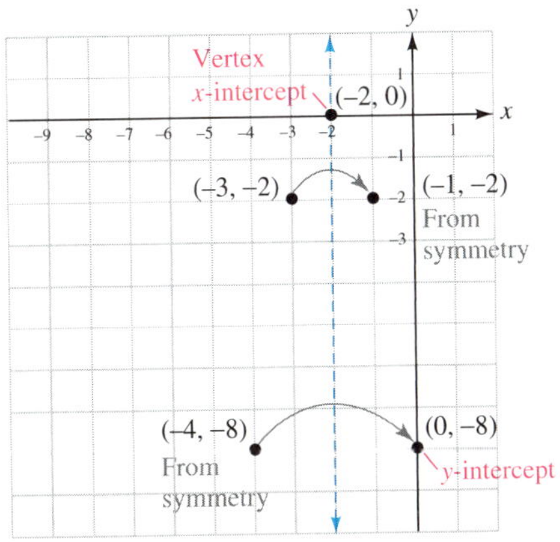

(a)

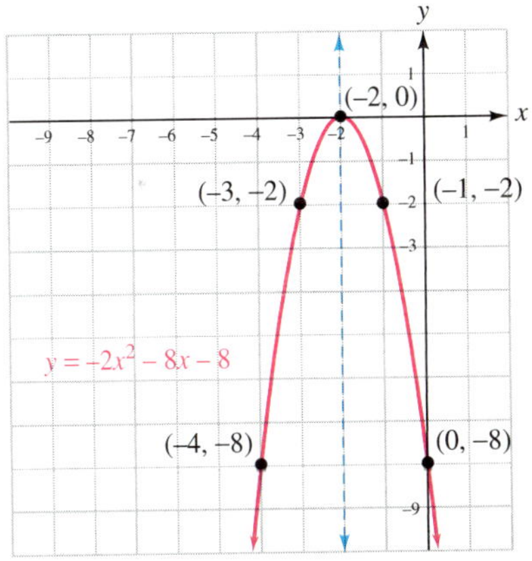

(b)

Self Check 4 Graph: $y = -2x^2 + 4x - 2$

Now Try **Problem 29**

EXAMPLE 5 Graph: $y = x^2 + 2x - 2$

Strategy We will follow the five-step procedure to sketch the graph of the equation.

Why This strategy is usually faster than making a table of solutions and plotting points.

Solution ***Upward/downward:*** The equation is in the form $y = ax^2 + bx + c$, with $a = 1$, $b = 2$, and $c = -2$. Since $a > 0$, the parabola opens upward.

Vertex/axis of symmetry: To find the x-coordinate of the vertex, we substitute 1 for a and 2 for b into the formula $x = \frac{-b}{2a}$.

$$x = \frac{-b}{2a} = \frac{-2}{2(1)} = -1$$

The x-coordinate of the vertex is -1. To find the y-coordinate, we substitute -1 for x in the original equation and find y.

$y = x^2 + 2x - 2$ This is the equation to graph.

$y = (-1)^2 + 2(-1) - 2$ Substitute -1 for x.

$y = -3$ Evaluate the right side.

The vertex of the parabola is the point $(-1, -3)$. The axis of symmetry is the vertical line passing through $(-1, -3)$. See part (a) of the figure.

Intercepts: Since $c = -2$, the y-intercept of the parabola is $(0, -2)$. The point $(-2, -2)$, which is one unit to the left of the axis of symmetry, must also be on the graph. See part (a).

To find the x-intercepts, we let $y = 0$ and solve the resulting quadratic equation.

$y = x^2 + 2x - 2$ This is the equation to graph.

$0 = x^2 + 2x - 2$ Substitute 0 for y.

Since $x^2 + 2x - 2$ does not factor, we will use the quadratic formula to solve for x.

$$x = \frac{-2 \pm \sqrt{2^2 - 4(1)(-2)}}{2(1)}$$ In the quadratic formula, substitute 1 for a, 2 for b, and -2 for c.

$$x = \frac{-2 \pm \sqrt{12}}{2}$$ Evaluate the right side.

$$x = \frac{-2 \pm 2\sqrt{3}}{2}$$ Simplify the radical: $\sqrt{12} = \sqrt{4 \cdot 3} = 2\sqrt{3}$.

$$x = \frac{\overset{1}{\cancel{2}}(-1 \pm \sqrt{3})}{\underset{1}{\cancel{2}}}$$ Factor out the GCF, 2, from the terms in the numerator. Then simplify by removing the common factor of 2 in the numerator and denominator.

$$x = -1 \pm \sqrt{3}$$

The x-intercepts of the graph are $\left(-1 + \sqrt{3}, 0\right)$ and $\left(-1 - \sqrt{3}, 0\right)$. To help to locate their position on the graph, we can use a calculator to approximate the two irrational numbers. See part (a) on the next page.

$$-1 + \sqrt{3} \approx 0.7 \qquad \text{and} \qquad -1 - \sqrt{3} \approx -2.7$$

$y = x^2 + 2x - 2$

x	y	(x, y)
2	6	(2, 6)

Plotting points/using symmetry: To find two more points on the graph, we let $x = 2$, substitute 2 for x in $y = x^2 + 2x - 2$, and find that y is 6. We plot (2, 6) and then use symmetry to determine that $(-4, 6)$ is also on the graph. See part (a).

Draw a smooth curve through the points: The completed graph of $y = x^2 + 2x - 2$ is shown in part (b).

> **Success Tip**
> To plot the x-intercepts, we use the approximations $(0.7, 0)$ and $(-2.7, 0)$ to locate their position on the x-axis. However, we label the points on the graph using the exact values: $\left(-1 + \sqrt{3}, 0\right)$ and $\left(-1 - \sqrt{3}, 0\right)$.

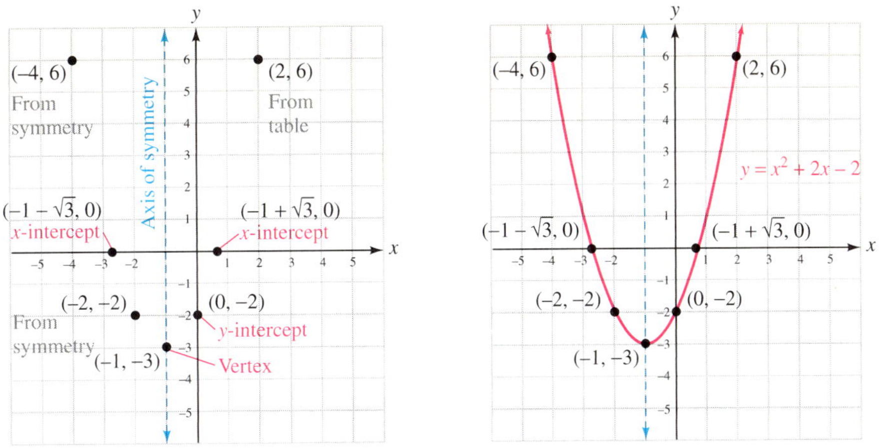

(a) (b)

Self Check 5 Graph: $y = x^2 - 4x - 3$

Now Try Problem 43

5 Solve Quadratic Equations Graphically.

The number of distinct x-intercepts of the graph of $y = ax^2 + bx + c$ is the same as the number of distinct real-number solutions of $ax^2 + bx + c = 0$. For example, the graph of $y = x^2 + x - 2$ in part (a) of the following figure has two x-intercepts, and $x^2 + x - 2 = 0$ has two real-number solutions. In part (b), the graph has one x-intercept, and the corresponding equation has one real-number solution. In part (c), the graph does not have an x-intercept, and the corresponding equation does not have any real-number solutions. Note that the solutions of each equation are given by the x-coordinates of the x-intercepts of each respective graph.

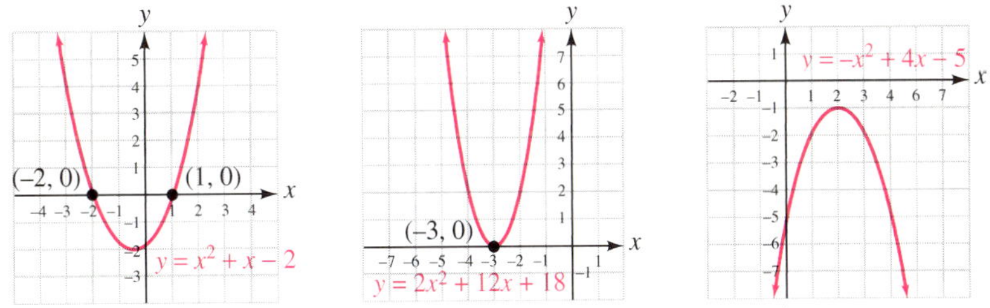

$x^2 + x - 2 = 0$ has two solutions, -2 and 1.
(a)

$2x^2 + 12x + 18 = 0$ has one repeated solution, -3.
(b)

$-x^2 + 4x - 5 = 0$ has no real-number solutions.
(c)

ANSWERS TO SELF CHECKS **1.** y-intercept: $(0, 8)$; x-intercepts: $(-2, 0)$, $(-4, 0)$ **2.** $(-3, -1)$

3. **4.** **5.**

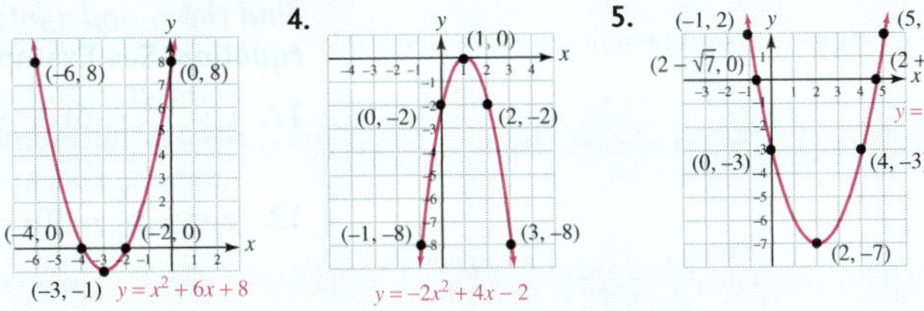

56. TRACK AND FIELD Sketch the parabolic path traveled by the long-jumper's center of gravity from the take-off board to the landing. Let the x-axis represent the ground.

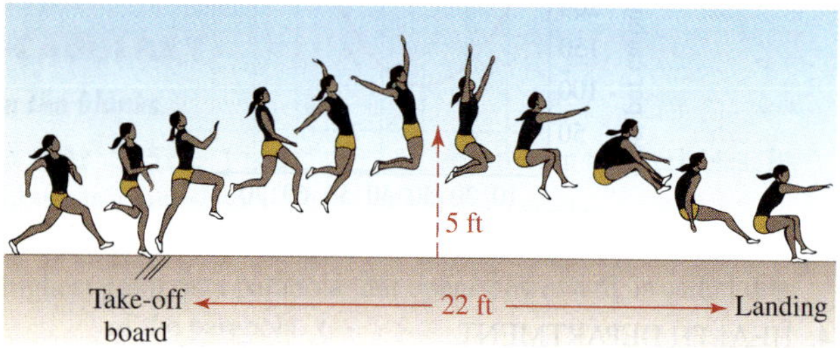

WRITING

57. A mirror is held against the y-axis of the graph of a quadratic equation. What fact about parabolas does this illustrate?

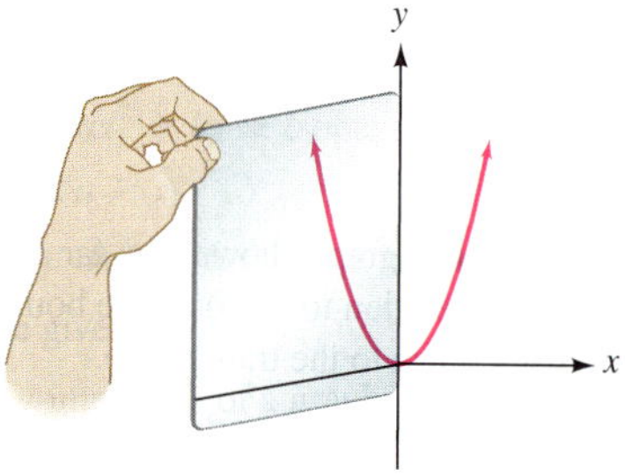

58. Use the example of a stream of water from a drinking fountain to explain the concept of the vertex of a parabola.

59. Explain why the y-intercept of the graph of the quadratic equation $y = ax^2 + bx + c$ is $(0, c)$.

60. Explain how to determine from its equation whether the graph of a parabola opens upward or downward.

61. Is it possible for the graph of a parabola with equation of the form $y = ax^2 + bx + c$ not to have a y-intercept? Explain.

62. Sketch the graphs of parabolas with zero, one, and two x-intercepts. Can the graph of a quadratic equation of the form $y = ax^2 + bx + c$ have more than two x-intercepts? Explain why or why not.

REVIEW

Simplify each expression.

63. $\sqrt{8} - \sqrt{50} + \sqrt{72}$

64. $(4\sqrt{x})(-2\sqrt{x})$

65. $3\sqrt{z}(\sqrt{4z} - \sqrt{z})$

66. $\sqrt[3]{27y^3z^6}$

CHALLENGE PROBLEMS

67. Graph $y = x^2 - x - 2$ and use the graph to solve the quadratic equation $-2 = x^2 - x - 2$.

68. What quadratic equation is graphed here?

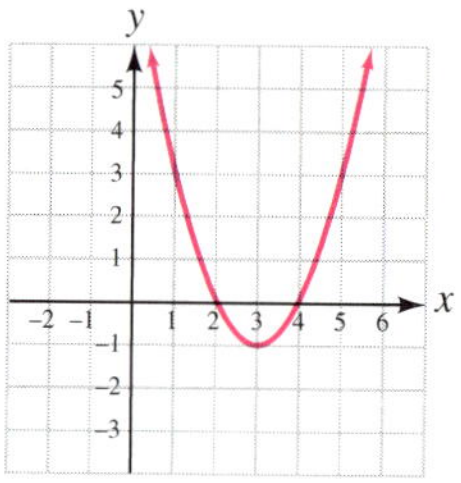

CHAPTER 9
Summary & Review

SECTION 9.1 Solving Quadratic Equations: The Square Root Property

DEFINITIONS AND CONCEPTS	EXAMPLES
We can use the **square root property** to solve equations of the form $x^2 = c$, where $c > 0$. The two solutions are $x = \sqrt{c}$ or $x = -\sqrt{c}$ We can write $x = \sqrt{c}$ or $x = -\sqrt{c}$ in more compact form using **double-sign notation:** $x = \pm\sqrt{c}$	Solve: $x^2 = 27$ $x = \sqrt{27}$ or $x = -\sqrt{27}$ Use the square root property. $x = \pm\sqrt{27}$ Use double-sign notation. $x = \pm 3\sqrt{3}$ Simplify: $\sqrt{27} = \sqrt{9 \cdot 3} = \sqrt{9} \cdot \sqrt{3} = 3\sqrt{3}$. Solve: $(x - 3)^2 = 5$ $x - 3 = \pm\sqrt{5}$ Use the square root property and double-sign notation. $x = 3 \pm \sqrt{5}$ To isolate x, add 3 to both sides. The exact solutions are $3 + \sqrt{5}$ and $3 - \sqrt{5}$. If we approximate the solutions to the nearest hundredth, we have $3 + \sqrt{5} \approx 3 + 2.236067978 \approx 5.24$ Use a calculator. $3 - \sqrt{5} \approx 3 - 2.236067978 \approx 0.76$ Solve: $x^2 - 22x + 121 = 25$ $(x - 11)^2 = 25$ Factor the perfect-square trinomial. $x - 11 = \pm\sqrt{25}$ Use the square root property. $x = 11 \pm 5$ Add 11 to both sides and simplify $\sqrt{25}$. $x = 11 + 5$ or $x = 11 - 5$ $x = 16$ | $x = 6$

REVIEW EXERCISES

Use the square root property to solve each equation.

1. $x^2 = 64$
2. $t^2 - 8 = 0$
3. $2x^2 - 1 = 149$
4. $(x - 1)^2 = 25$
5. $(9x - 8)^2 = 40$
6. $4(x - 2)^2 - 9 = 0$
7. $p^2 - 20p + 100 = 9$
8. $9m^2 + 6m + 1 = 6$

Use the square root property to find all real-number solutions of each equation. Round each solution to the nearest hundredth.

9. $x^2 = 12$
10. $(x - 1)^2 = 55$
11. $m^2 + 36 = 0$
12. $(2x - 3)^2 = -8$

13. CLIFF DIVERS The La Quebrada Cliff Divers of Acapulco, Mexico, perform daily for the public by diving 148 feet from ocean-side cliffs into the sea below. Find the length of time of a dive. Round to the nearest one tenth of one second. (*Hint:* Use the formula $d = 16t^2$.)

14. RUBIK'S CUBE The area of one of the square faces of a Rubik's cube is $\frac{81}{16}$ in.^{2}. Find the length of one side of a Rubik's cube. Express your answer as a mixed number. (*Hint:* Use the formula for the area of a square, $A = s^2$.)

SECTION 9.2 Solving Quadratic Equations: Completing the Square

DEFINITIONS AND CONCEPTS	EXAMPLES
To **complete the square** on $x^2 + bx$, add the square of one-half of the coefficient of x. $x^2 + bx + \left(\frac{1}{2}b\right)^2$	Complete the square on $x^2 + 12x$ and factor the resulting perfect-square trinomial. $x^2 + 12x + 36$ The coefficient of x is 12. To complete the square: $\frac{1}{2} \cdot 12 = 6$ and $6^2 = 36$. Add 36 to the binomial. This trinomial factors as $(x + 6)^2$. We can check using multiplication.
To **solve a quadratic equation in x by completing the square:** 1. If necessary, divide both sides of the equation by the coefficient of x^2 to make its coefficient 1. 2. Get all variable terms on one side of the equation and all constants on the other side. 3. Complete the square. 4. Factor the perfect-square trinomial. 5. Solve the resulting equation by using the square root property. 6. Check your answers in the original equation.	Solve: $3x^2 - 12x + 6 = 0$ $\frac{3x^2}{3} - \frac{12}{3}x + \frac{6}{3} = 0$ To make the leading coefficient 1, divide both sides by 3. $x^2 - 4x + 2 = 0$ Do the divisions. $x^2 - 4x = -2$ Subtract 2 from both sides so that the constant term, −2, is on the right side. $x^2 - 4x + 4 = -2 + 4$ The coefficient of x is −4. To complete the square: $\frac{1}{2}(-4) = -2$ and $(-2)^2 = 4$. Add 4 to both sides. $(x - 2)^2 = 2$ Factor the perfect-square trinomial on the left side. Add on the right side. $x - 2 = \pm\sqrt{2}$ Use the square root property. $x = 2 \pm \sqrt{2}$ To isolate x, add 2 to both sides. The solutions are $2 + \sqrt{2}$ and $2 - \sqrt{2}$. We can approximate each solution. To the nearest one hundredth $2 + \sqrt{2} \approx 3.41$ $2 - \sqrt{2} \approx 0.59$

REVIEW EXERCISES

Complete the square to make each expression a perfect square trinomial. Then factor.

15. $x^2 + 4x$ **16.** $t^2 - 5t$

Solve each quadratic equation by completing the square.

17. $x^2 - 8x + 15 = 0$ **18.** $x^2 = -5x + 14$

19. $x^2 + 2x = 5$ **20.** $4x^2 - 16x = 7$

21. $2x^2 - 2x - 1 = 0$ **22.** $3x^2 + 5x + 2 = 0$

Solve each quadratic equation by completing the square. Round each solution to the nearest hundredth.

23. $x^2 + 4x + 1 = 0$ **24.** $x^2 - 7x = 5$

SECTION 9.3 Solving Quadratic Equations: The Quadratic Formula

DEFINITIONS AND CONCEPTS

To **solve a quadratic equation in x using the quadratic formula:**

1. Write the equation in standard form:

$$ax^2 + bx + c = 0$$

2. Identify a, b, and c.
3. Substitute the values for a, b, and c in the quadratic formula

$$x = \frac{-b \pm \sqrt{b^2 - 4ac}}{2a}$$

and evaluate the right side to obtain the solutions.

EXAMPLES

Use the quadratic formula to solve: $3x^2 - 2x = 2$

$3x^2 - 2x - 2 = 0$ To get 0 on the right side, subtract 2 from both sides.

Here, $a = 3$, $b = -2$, and $c = -2$.

$x = \frac{-b \pm \sqrt{b^2 - 4ac}}{2a}$ This is the quadratic formula.

$x = \frac{-(-2) \pm \sqrt{(-2)^2 - 4(3)(-2)}}{2(3)}$ Substitute 3 for a, −2 for b, and −2 for c.

$x = \frac{2 \pm \sqrt{4 - (-24)}}{6}$ Evaluate within the radical. Multiply in the denominator.

$x = \frac{2 \pm \sqrt{28}}{6}$ Add the opposite: $4 - (-24) = 4 + 24 = 28$.

$x = \frac{2 \pm 2\sqrt{7}}{6}$ Simplify: $\sqrt{28} = \sqrt{4}\sqrt{7} = 2\sqrt{7}$.

$x = \frac{\overset{1}{\cancel{2}}(1 \pm \sqrt{7})}{\underset{1}{\cancel{2}} \cdot 3}$ Factor out the GCF, 2, in the numerator. In the denominator, factor 6. Remove the common factor, 2.

$x = \frac{1 \pm \sqrt{7}}{3}$ The solutions are $\frac{1 \pm \sqrt{7}}{3}$.

A **strategy for solving quadratic equations** is given on page 756.

A suggested method for solving each quadratic equation is given.

Factor method	***Square root property***
$x^2 - 3x - 18 = 0$	$(x - 1)^2 = 18$
Complete the square	***Quadratic formula***
$x^2 + 6x - 11 = 0$ (↑ even coefficient)	$3x^2 - 9x + 1 = 0$

REVIEW EXERCISES

Write each equation in $ax^2 + bx + c = 0$ form and find a, b, and c.

25. $x^2 + 2x = -5$

26. $6x^2 = 2x + 1$

Use the quadratic formula to find all real-number solutions of each equation.

27. $x^2 - 2x - 15 = 0$
28. $6x^2 = 7x + 3$

29. $p^2 - 4 = 2p$
30. $x^2 + 7 = 6x$

31. $3x^2 + 3x = 1$
32. $5x^2 + x = 1$

33. $7x^2 - x + 2 = 0$
34. $2x^2 + 6x = 5$

Use the most efficient method to find all real-number solutions of each equation.

35. $4x^2 + 16x = 0$
36. $(y + 3)^2 = 16$

37. $3g^2 - 81 = 0$
38. $3x^2 - 6x = -1$

39. $2x^2 + 2x - 5 = 0$
40. $a^2 = 4a - 4$

41. $(2x - 5)^2 = 64$
42. $a^2 - 2a + 5 = 0$

43. Use the quadratic formula to solve $3x^2 + 2x - 2 = 0$. Give the solutions in exact form and then rounded to the nearest hundredth.

44. SECURITY GATES The length of the frame for an iron gate is 14 feet longer than the width. A diagonal cross brace is 26 feet long. Find the width and length of the gate frame.

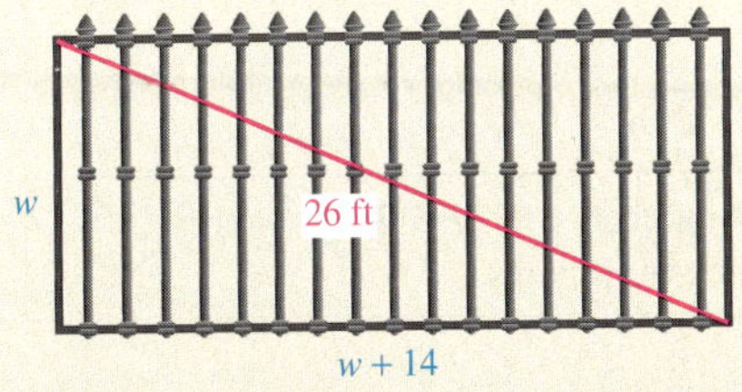

45. THE GRAND CANYON The depth of the Grand Canyon at the South Rim is almost one mile. Suppose a visitor standing on the rim tosses a rock upward over the canyon. The time t (in seconds) that it takes for the rock to hit the bottom of the canyon can be found by solving the quadratic equation $0 = -16t^2 + 8t + 5{,}040$. Find t.

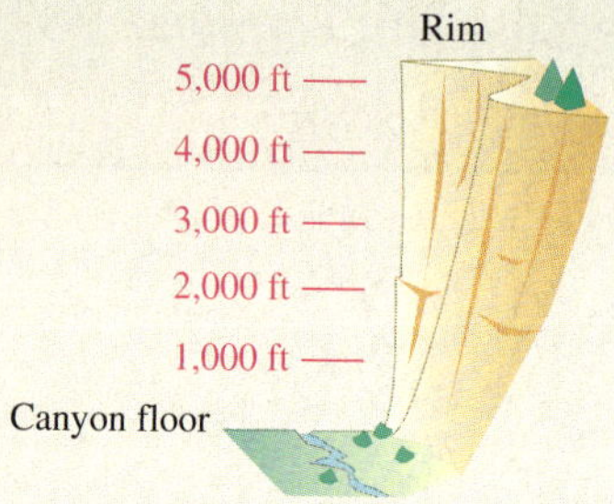

46. GEOMETRY The triangle shown has an area of 24 square inches. Find its height.

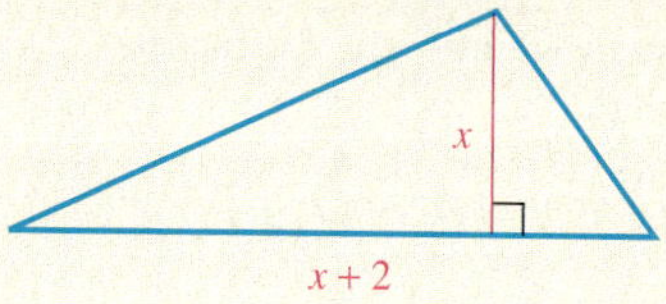

SECTION 9.4 Complex Numbers

DEFINITIONS AND CONCEPTS	EXAMPLES
The **imaginary number** i is defined as $i = \sqrt{-1}$ From the definition, it follows that $i^2 = -1$. A **complex number** is any number that can be written in the form $a + bi$, where a and b are real numbers and $i = \sqrt{-1}$. We call a the **real part** and b the **imaginary part.** If $b \neq 0$, $a + bi$ is also an **imaginary number.**	Write each expression in terms of i: $\sqrt{-100} = \sqrt{-1 \cdot 100} = \sqrt{-1}\sqrt{100} = i \cdot 10 = 10i$ $\sqrt{-75} = \sqrt{-1 \cdot 75} = \sqrt{-1}\sqrt{75} = i\sqrt{25}\sqrt{3} = 5i\sqrt{3}$ or $5\sqrt{3}i$ Show that each number is a complex number by writing it in the form $a + bi$. $10 = 10 + 0i$ 10 is the real part and 0 is the imaginary part. $5i = 0 + 5i$ 0 is the real part and 5 is the imaginary part. $-2 - \sqrt{-16} = -2 - 4i$ −2 is the real part and −4 is the imaginary part.
Adding and subtracting complex numbers is similar to adding and subtracting polynomials. To **add two complex numbers,** add their real parts and add their imaginary parts. To **subtract two complex numbers,** add the opposite of the complex number being subtracted.	Add: $(3 - 4i) + (5 + 7i) = (3 + 5) + (-4 + 7)i$ Add the real parts. Add the imaginary parts. $= 8 + 3i$ Subtract: $(3 - 4i) - (5 + 7i) = (3 - 4i) + (-5 - 7i)$ Add the opposite of $5 + 7i$. $= (3 - 5) + [-4 + (-7)]i$ Add the real parts. Add the imaginary parts. $= -2 - 11i$
Multiplying complex numbers is similar to multiplying polynomials.	Find each product: $7i(4 - 9i) = 28i - 63i^2 = 28i - 63(-1) = 28i + 63 = 63 + 28i$ $(2 + 3i)(4 - 3i) = 8 - 6i + 12i - 9i^2 = 8 + 6i - 9(-1) = 8 + 6i + 9 = 17 + 6i$
The complex numbers $a + bi$ and $a - bi$ are called **complex conjugates.**	The complex numbers $4 - 3i$ and $4 + 3i$ are complex conjugates.

SECTION 9.4 Complex Numbers–*continued*

DEFINITIONS AND CONCEPTS	EXAMPLES
To write the **quotient of two complex numbers** in the form $a + bi$, multiply the numerator and denominator by the complex conjugate of the denominator. The process is similar to rationalizing denominators.	Write each quotient in the form $a + bi$: $\frac{6}{4-3i} = \frac{6}{4-3i} \cdot \frac{4+3i}{4+3i} = \frac{24+18i}{16-9i^2} = \frac{24+18i}{16-9(-1)} = \frac{24+18i}{16+9} = \frac{24+18i}{25} = \frac{24}{25} + \frac{18}{25}i$ $\frac{2+i}{1+i} = \frac{2+i}{1+i} \cdot \frac{1-i}{1-i} = \frac{2-2i+i-i^2}{1-i^2} = \frac{2-i-(-1)}{1-(-1)} = \frac{3-i}{2} = \frac{3}{2} - \frac{1}{2}i$
Some quadratic equations have **complex solutions that are imaginary numbers.**	Use the quadratic formula to solve: $p^2 + p + 3 = 0$ Here, $a = 1$, $b = 1$, and $c = 3$. $p = \frac{-b \pm \sqrt{b^2 - 4ac}}{2a}$ In the quadratic forumla, replace x with p. $p = \frac{-1 \pm \sqrt{1^2 - 4(1)(3)}}{2(1)}$ Substitute 1 for a, 1 for b, and 3 for c. $p = \frac{-1 \pm \sqrt{-11}}{2}$ Evaluate the power and multiply within the radical. Multiply in the denominator. $p = \frac{-1 \pm i\sqrt{11}}{2}$ Write $\sqrt{-11}$ in terms of i. $p = -\frac{1}{2} \pm \frac{\sqrt{11}}{2}i$ Write the solutions in the form $a + bi$.

REVIEW EXERCISES

Write each expression in terms of i.

47. $\sqrt{-25}$

48. $\sqrt{-18}$

49. $-\sqrt{-49}$

50. $\sqrt{-\frac{9}{64}}$

51. Complete the diagram.

Complex numbers

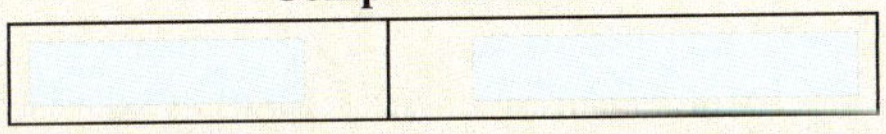

52. Determine whether each statement is true or false.

a. Every real number is a complex number.

b. $3 - 4i$ is an imaginary number.

c. $\sqrt{-4}$ is a real number.

d. i is a real number.

Give the complex conjugate of each number.

53. $3 + 6i$

54. $-1 - 7i$

55. $19i$

56. $-i$

Perform the operations. Write all answers in the form a + bi.

57. $(3 + 4i) + (5 - 6i)$

58. $(7 - 3i) - (4 + 2i)$

59. $3i(2 - i)$

60. $(2 + 3i)(3 - i)$

61. $\frac{2 + 3i}{2 - 3i}$

62. $\frac{3}{5 + i}$

Solve each equation. Write all solutions in the form a + bi.

63. $x^2 + 9 = 0$

64. $3x^2 = -16$

65. $(p - 2)^2 = -24$

66. $(q + 3)^2 = -54$

67. $x^2 + 2x = -2$

68. $2x^2 - 3x + 2 = 0$

SECTION 9.5 Graphing Quadratic Equations

DEFINITIONS AND CONCEPTS

The **vertex** of a parabola is the lowest (or highest) point on the parabola.

A vertical line through the vertex of a parabola that opens upward or downward is called its **axis of symmetry.**

EXAMPLES

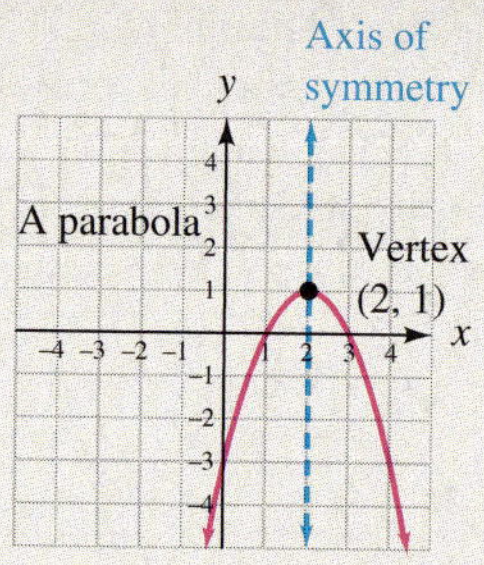

DEFINITIONS AND CONCEPTS

Equations that can be written in the form $y = ax^2 + bx + c$, where $a \neq 0$, are called **quadratic equations in two variables.**

The graph of $y = ax^2 + bx + c$ is a **parabola.** Much can be determined about the graph from the coefficients a, b, and c.

The parabola opens **upward** when $a > 0$ and **downward** when $a < 0$.

The x-coordinate of the **vertex** of the parabola is $x = \frac{-b}{2a}$. To find the y-coordinate of the vertex, substitute $\frac{-b}{2a}$ for x in the equation of the parabola and find y.

To find the ***y*-intercept,** substitute 0 for x in the given equation and solve for y. To find the ***x*-intercepts,** substitute 0 for y in the given equation and solve for x.

The number of distinct x-intercepts of the graph of a quadratic equation $y = ax^2 + bx + c$ is the same as the number of distinct **real-number solutions** of $ax^2 + bx + c = 0$.

EXAMPLES

Graph: $y = x^2 + 6x + 8$

Upward/downward: The equation is in the form $y = ax^2 + bx + c$, with $a = 1$, $b = 6$, and $c = 8$. Since $a > 0$, the parabola opens upward.

Vertex/axis of symmetry: The x-coordinate of the vertex is

$$\frac{-b}{2a} = \frac{-6}{2(1)} = -3$$

The y-coordinate of the vertex is:

$$y = x^2 + 6x + 8 \quad \text{The equation to graph.}$$
$$y = (-3)^2 + 6(-3) + 8$$
$$y = 9 - 18 + 8$$
$$y = -1$$

The vertex of the parabola is $(-3, -1)$.

Intercepts: Since $c = 8$, the y-intercept of the parabola is $(0, 8)$. The point $(-6, 8)$, which is 3 units to the left of the axis of symmetry, must also be on the graph.

To find the x-intercepts of the graph of $y = x^2 + 6x + 8$, we set $y = 0$ and solve for x.

$$0 = x^2 + 6x + 8$$
$$0 = (x + 4)(x + 2)$$
$$x + 4 = 0 \quad \text{or} \quad x + 2 = 0$$
$$x = -4 \quad \Big| \quad x = -2$$

The x-intercepts of the graph are $(-4, 0)$ and $(-2, 0)$.

Plotting points/using symmetry: To locate two more points on the graph, we let $x = -1$ and find the corresponding value of y.

$$y = (-1)^2 + 6(-1) + 8 = 3$$

Thus, the point $(-1, 3)$ lies on the parabola. We use symmetry to determine that $(-5, 3)$ is also on the graph.

SECTION 9.5 Graphing Quadratic Equations–*continued*

DEFINITIONS AND CONCEPTS	EXAMPLES
	Draw a smooth curve through the points: The completed graph of $y = x^2 + 6x + 8$ is shown here.

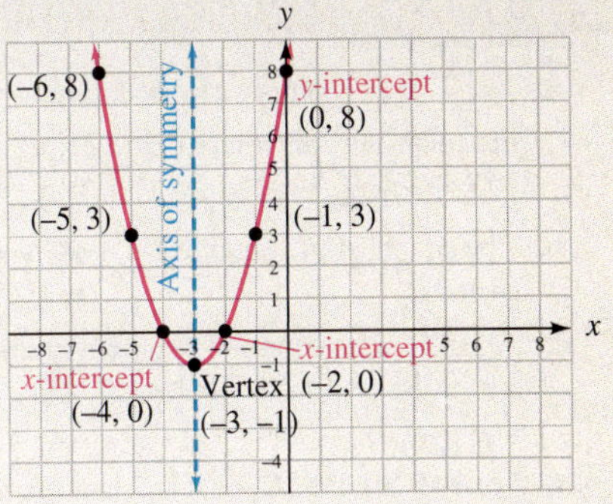

Since the graph of $y = x^2 + 6x + 8$ (shown above) has two x-intercepts, $(-4, 0)$ and $(-2, 0)$, the equation $x^2 + 6x + 8 = 0$ has two distinct real-number solutions, -4 and -2.

REVIEW EXERCISES

69. Refer to the figure.

a. What are the x-intercepts of the parabola?

b. What is the y-intercept of the parabola?

c. What is the vertex of the parabola?

d. Draw the axis of symmetry of the parabola on the graph.

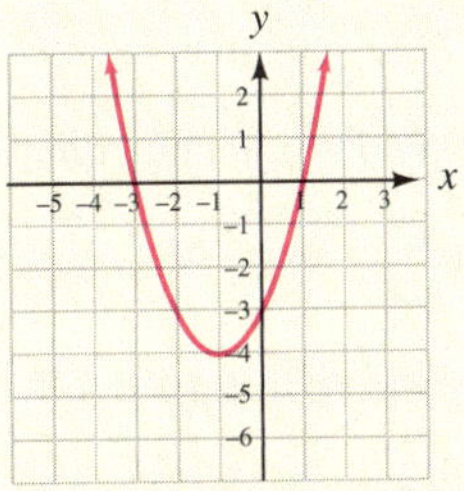

70. The point $(0, -3)$ lies on the parabola graphed above. Use symmetry to determine the coordinates of another point that lies on the parabola.

Find the vertex of the graph of each quadratic equation and tell in which direction the parabola opens. Do not draw the graph.

71. $y = 2x^2 - 4x + 7$ **72.** $y = -3x^2 + 18x - 11$

Find the x- and y-intercepts of the graph of each quadratic equation.

73. $y = x^2 + 6x + 5$ **74.** $y = x^2 + 2x + 3$

Graph each quadratic equation by finding the vertex, the x- and y-intercepts, and the axis of symmetry of its graph.

75. $y = x^2 + 2x - 3$ **76.** $y = -2x^2 + 4x - 2$

77. $y = -x^2 - 2x + 5$ **78.** $y = x^2 + 4x - 1$

79. The graphs of three quadratic equations in two variables are shown. Fill in the blanks.

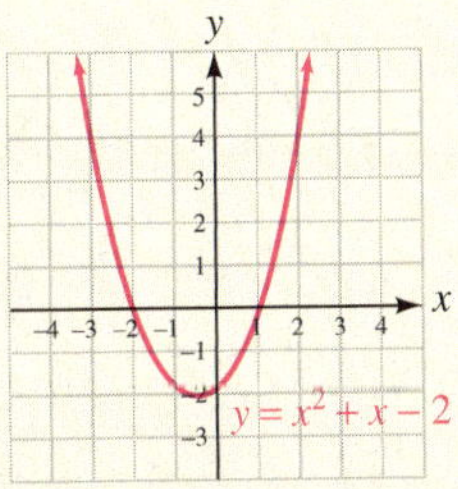

$x^2 + x - 2 = 0$ has ____ real-number solution(s). Give the solution(s): ____

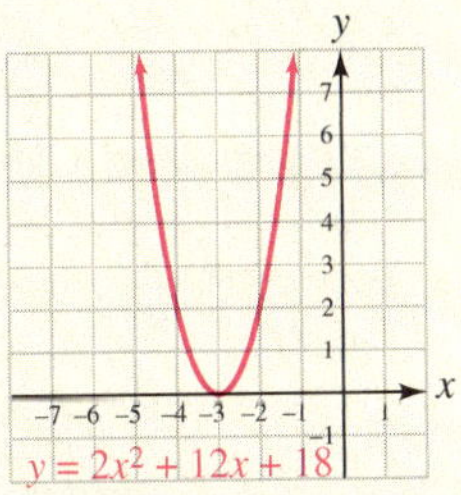

$2x^2 + 12x + 18 = 0$ has ____ repeated real-number solution. Give the solution(s): ____

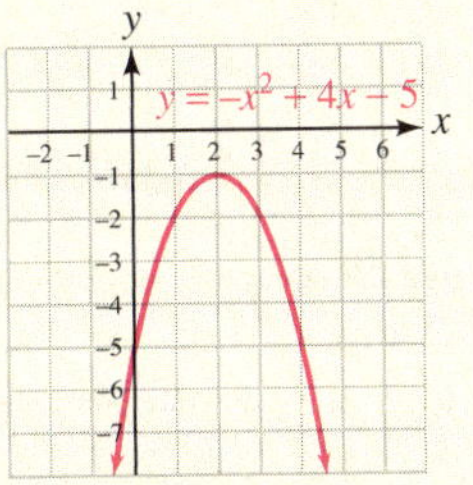

$-x^2 + 4x - 5 = 0$ has ____ real-number solution(s).

80. MANUFACTURING What important information can be obtained from the vertex of the parabola in the graph below?

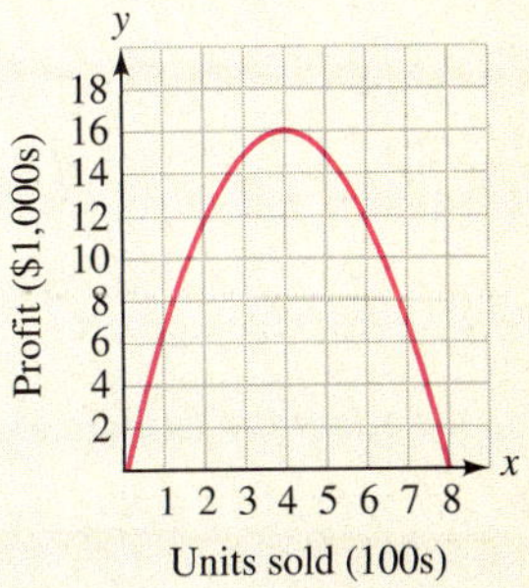

CHAPTER 9 Test

1. Fill in the blanks.

a. A ________ number is any number that can be written in the form $a + bi$, where a and b are real numbers and $i = \sqrt{-1}$.

b. The graph of the equation $y = x^2$ is a cup-shaped figure called a ________.

c. A ________ equation can be written in the form $ax^2 + bx + c = 0$, where a, b, and c represent real numbers and $a \neq 0$.

d. For $7 + 8i$, the real part is 7 and the ________ part is 8.

e. The ________ coefficient of $3x^2 + 8x - 9$ is 3.

2. Write the statement $x = \sqrt{5}$ or $x = -\sqrt{5}$ using double-sign notation.

Solve each equation by the square root method.

3. $x^2 = 17$

4. $(x - 2)^2 = 3$

5. $4y^2 - 25 = 0$

6. $x^2 + 16x + 64 = 24$

7. Explain why the equation $m^2 + 49 = 0$ has no real-number solutions.

8. ARCHERY The area of a circular archery target is 5,026 cm^2. What is the radius of the target? Round to the nearest centimeter.

Complete the square and factor the resulting perfect-square trinomial.

9. $x^2 - 14x$

10. $a^2 - \frac{5}{3}a$

11. Complete the square to solve $a^2 + 2a - 4 = 0$. Give the exact solutions and then round them to the nearest hundredth.

12. Complete the square to solve: $2x^2 = 3x + 2$

Use the quadratic formula to solve each equation.

13. $2x^2 - 5x - 12 = 0$

14. $5x^2 + 11x = -3$

15. Solve $3x^2 - 2x - 2 = 0$ using the quadratic formula. Give the exact solutions, and then approximate them to the nearest hundredth.

16. NEW YORK CITY The rectangular Samsung sign in Times Square is a full color LED screen that has an area of 2,665 ft^2. Its length is 17 feet less than twice its width. Find the width and length of the sign.

© Hugh Sitton/Getty Images

Use the most efficient method to solve each equation.

17. $x^2 - 4x = -2$

18. $(3b + 1)^2 = 16$

19. $u^2 - 24 = 0$

20. $3x^2 + 2x + 1 = 0$

Write each expression in terms of i.

21. $\sqrt{-100}$

22. $-\sqrt{-18}$

Perform the operations. Write all answers in the form a + bi.

23. $(8 + 3i) + (-7 - 2i)$

24. $(5 + 3i) - (6 - 9i)$

25. $(2 - 4i)(3 + 2i)$

26. $\frac{3 - 2i}{3 + 2i}$

27. ADVERTISING When a business runs x advertisements per week on television, the number y of air conditioners it sells is given by the graph. What important information can be obtained from the vertex?

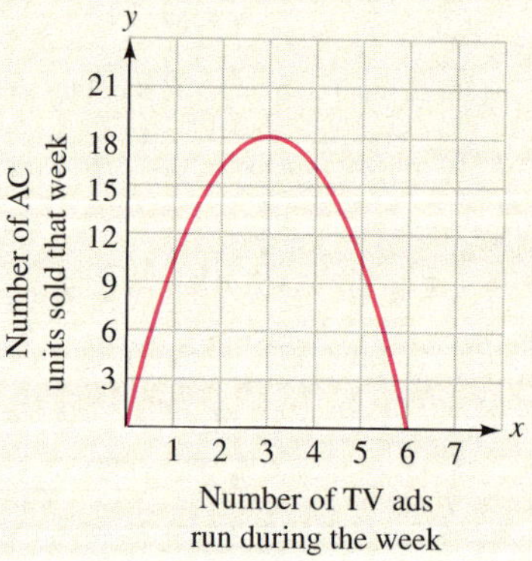

28. Fill in the blanks: The graph of $y = ax^2 + bx + c$ opens downward when a ___ 0 and upward when a ___ 0.

Graph each quadratic equation by finding the vertex, the x- and y-intercepts, and the axis of symmetry of its graph.

29. $y = x^2 + 6x + 5$

30. $y = -x^2 + 6x - 7$

GROUP PROJECT

Overview: In this activity, you will learn how to predict the type of solutions that a quadratic equation has, before solving it.

Instructions: Form groups of 2 or 3 students. The expression $b^2 - 2ac$ is called the **discriminant** of the quadratic equation $ax^2 + bx + c = 0$. It provides us with information about the solutions of a quadratic equation.

- If $b^2 - 4ac > 0$, the equation has two different real-number solutions.
- If $b^2 - 4ac = 0$, the equation has one repeated rational-number solution.
- If $b^2 - 4ac < 0$, the equation has two different imaginary-number solutions.

Find the value of the discriminant for each of the following equations. Then, use the answer to predict the type of solutions that the equation has. Next, solve the equation. Do the solutions match your predictions about them?

1. $x^2 - 2x - 1 = 0$ 2. $x^2 - 4x + 4 = 0$ 3. $x^2 + 2x + 2 = 0$

CUMULATIVE REVIEW
Chapters 1–9

1. Determine whether each statement is true or false. [Section 1.3]

a. Every rational number can be written as a ratio of two integers.

b. The set of real numbers corresponds to all points on the number line.

c. The whole numbers and their opposites form the set of integers.

2. DRIVING SAFETY In cold-weather climates, salt is spread on roads to keep snow and ice from bonding to the pavement. This allows snowplows to remove accumulated snow quickly. According to the graph, when is the accident rate the highest? [Section 1.4]

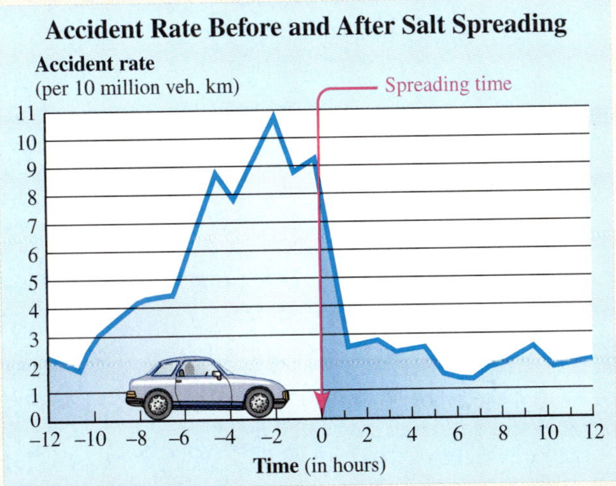

Based on data from the Salt Institute

3. Evaluate: $-4 + 2[-7 - 3(-9)]$ [Section 1.7]

4. Evaluate: $\left|\frac{4}{5} \cdot 10 - 12\right|$ [Section 1.7]

5. Evaluate $(x - a)^2 + (y - b)^2$ for $x = -2, y = 1, a = 5$, and $b = -3$. [Section 1.8]

6. Simplify: $3p - 6(p - 9) + p$ [Section 1.9]

7. Solve $\frac{5}{6}k = 10$ and check the result. [Section 2.2]

8. Solve $-(3a + 1) + a = 2$ and check the result. [Section 2.2]

9. LOOSE CHANGE The Coinstar machines that are in many grocery stores count unsorted coins and print out a voucher that can be exchanged for cash at the checkout stand. However, to use this service, a processing fee is charged. If a boy turned in a jar of coins worth \$50 and received a voucher for \$45.55, what was the processing fee (expressed as a percent) charged by Coinstar? [Section 2.3]

10. Solve $T = 2r + 2t$ for r. [Section 2.4]

11. SELLING A HOME At what price should a home be listed if the owner wants to make \$330,000 on its sale after paying a 4% real estate commission? [Section 2.5]

12. BUSINESS LOANS Last year, a women's professional organization made two small-business loans totaling \$28,000 to young women beginning their own businesses. The money was lent at 7% and 10% simple interest rates. If the annual income the organization received from these loans was \$2,560, what was each loan amount? [Section 2.6]

13. Solve $5x + 7 < 2x + 1$ and graph the solution set. Then use interval notation to describe the solution. [Section 2.7]

14. Is $(-5, -3)$ a solution of $2x - 3y = -1$? [Section 3.1]

Graph each equation or inequality.

15. $y = -x + 2$ [Section 3.2]

16. $2y - 2x = 6$ [Section 3.3]

17. $y = -3$ [Section 3.3]

18. $y < 3x$ [Section 3.7]

19. Find the slope of the line passing through $(-2, -2)$ and $(-12, -8)$. [Section 3.4]

20. TV NEWS The graph in red approximates the evening news viewership on all networks for the years 1995-2005. Find the rate of decrease over this period of time. [Section 3.4]

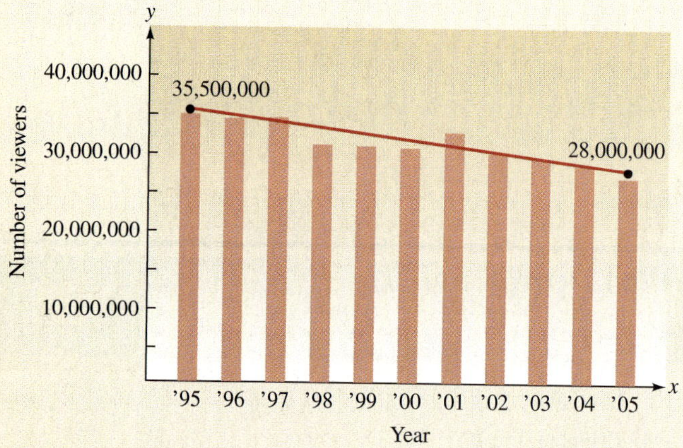

Source: The State of the News Media, 2006

21. What is the slope of the line defined by $4x + 5y = 6$? [Section 3.5]

22. Write the equation of the line whose graph has slope -2 and y-intercept $(0, 1)$. [Section 3.5]

23. Are the graphs of $y = 4x + 9$ and $x + 4y = -10$ parallel, perpendicular, or neither? [Section 3.5]

24. Write the equation of the line whose graph has slope $\frac{1}{4}$ and passes through the point $(8, 1)$. Write the equation in slope–intercept form. [Section 3.6]

25. Graph the line passing through $(-2, -1)$ and having slope $\frac{4}{3}$. [Section 3.6]

26. If $f(x) = 3x^2 + 3x - 8$, find $f(-1)$. [Section 3.8]

27. Find the domain and range of the relation: $\{(1, 8), (4, -3), (-4, 2), (5, 8)\}$ [Section 3.8]

28. Is this the graph of a function? [Section 3.8]

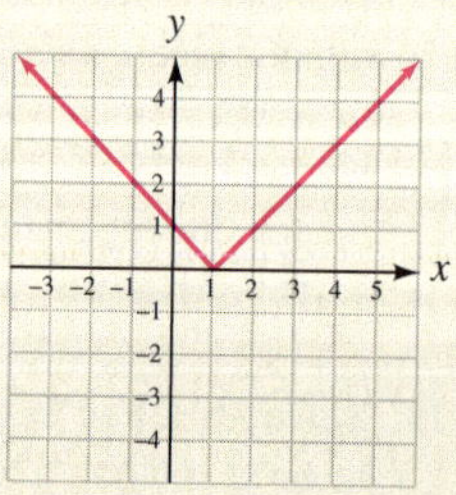

29. Solve using the graphing method. [Section 4.1]

$$\begin{cases} x + y = 1 \\ y = x + 5 \end{cases}$$

30. Solve using the substitution method.

$$\begin{cases} y = 2x + 5 \\ x + 2y = -5 \end{cases}$$ [Section 4.2]

31. Solve using the elimination (addition) method.

$$\begin{cases} \frac{3}{5}s + \frac{4}{5}t = 1 \\ -\frac{1}{4}s + \frac{3}{8}t = 1 \end{cases}$$ [Section 4.3]

32. AVIATION With the wind, a plane can fly 3,000 miles in 5 hours. Against the wind, the trip takes 6 hours. Find the airspeed of the plane (the speed in still air). Use two variables to solve this problem. [Section 4.4]

33. MIXING CANDY How many pounds of each candy must be mixed to obtain 48 pounds of candy that would be worth \$4.50 per pound? Use two variables to solve this problem. [Section 4.4]

34. Solve the system of linear inequalities.

$$\begin{cases} 3x + 4y > -7 \\ 2x - 3y \geq 1 \end{cases}$$ [Section 4.5]

Simplify each expression. Write each answer without using parentheses or negative exponents.

35. $y^3(y^2y^4)$ [Section 5.1]

36. $\left(\frac{b^2}{3a}\right)^3$ [Section 5.1]

37. $\frac{10a^4a^{-2}}{5a^2a^0}$ [Section 5.2]

38. $\left(\frac{21x^{-2}y^2z^{-2}}{7x^3y^{-1}}\right)^{-2}$ [Section 5.2]

39. FIVE-CARD POKER The odds against being dealt the hand shown are about 2.6×10^6 to 1. Express 2.6×10^6 using standard notation. [Section 5.3]

40. Write 0.00073 in scientific notation. [Section 5.3]

41. Graph: $y = x^3 - 2$ [Section 5.4]

42. Write a polynomial that represents the perimeter of the rectangle. [Section 5.5]

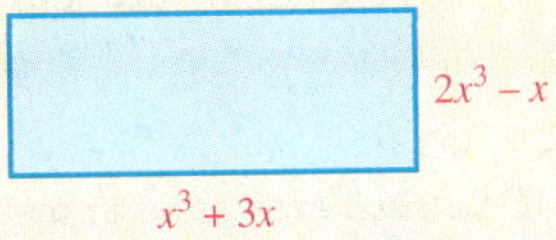

Perform the operations.

43. $4(4x^3 + 2x^2 - 3x - 8) - 5(2x^3 - 3x + 8)$ [Section 5.5]

44. $(-2a^3)(3a^2)$ [Section 5.6]

45. $(2b - 1)(3b + 4)$ [Section 5.6]

46. $(3x + y)(2x^2 - 3xy + y^2)$ [Section 5.6]

47. $(2x + 5y)^2$ [Section 5.7]

48. $(9m^2 - 1)(9m^2 + 1)$ [Section 5.7]

49. $\dfrac{12a^3b - 9a^2b^2 + 3ab}{6a^2b}$ [Section 5.8]

50. $x - 3\overline{)2x^2 - 3 - 5x}$ [Section 5.8]

Factor each expression completely.

51. $6a^2 - 12a^3b + 36ab$ [Section 6.1]

52. $2x + 2y + ax + ay$ [Section 6.1]

53. $x^2 - 6x - 16$ [Section 6.2]

54. $30y^5 + 63y^4 - 30y^3$ [Section 6.3]

55. $t^4 - 16$ [Section 6.4]

56. $b^3 + 125$ [Section 6.5]

Solve each equation by factoring.

57. $3x^2 + 8x = 0$ [Section 6.7]

58. $15x^2 - 2 = 7x$ [Section 6.7]

59. HEIGHT OF A TRIANGLE The triangle shown has an area of 22.5 square inches. Find its height. [Section 6.8]

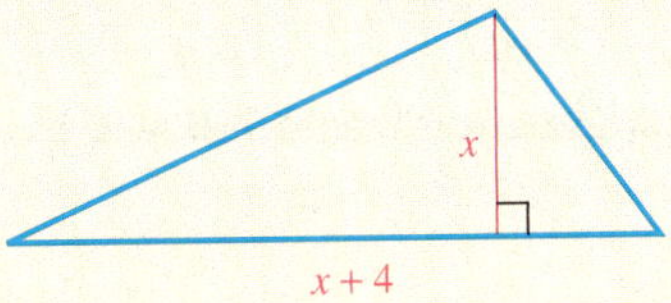

60. For what value is $\frac{x}{x+8}$ undefined? [Section 7.1]

Simplify each expression.

61. $\dfrac{3x^2 - 27}{x^2 + 3x - 18}$ [Section 7.1]

62. $\dfrac{a - 15}{15 - a}$ [Section 7.1]

Perform the operations and simplify when possible.

63. $\dfrac{x^2 - x - 6}{2x^2 + 9x + 10} \div \dfrac{x^2 - 25}{2x^2 + 15x + 25}$ [Section 7.2]

64. $\dfrac{1}{s^2 - 4s - 5} + \dfrac{s}{s^2 - 4s - 5}$ [Section 7.3]

65. $\dfrac{x + 5}{xy} - \dfrac{x - 1}{x^2y}$ [Section 7.4]

66. $\dfrac{x}{x - 2} + \dfrac{3x}{x^2 - 4}$ [Section 7.4]

Simplify each complex fraction.

67. $\dfrac{\dfrac{9m - 27}{m^6}}{\dfrac{2m - 6}{m^8}}$ [Section 7.5]

68. $\dfrac{\dfrac{5}{y} + \dfrac{4}{y + 1}}{\dfrac{4}{y} - \dfrac{5}{y + 1}}$ [Section 7.5]

Solve each equation.

69. $\dfrac{2p}{3} - \dfrac{1}{p} = \dfrac{2p - 1}{3}$ [Section 7.6]

70. $\dfrac{7}{q^2 - q - 2} + \dfrac{1}{q + 1} = \dfrac{3}{q - 2}$ [Section 7.6]

71. Solve the formula $\frac{1}{a} + \frac{1}{b} = 1$ for a. [Section 7.6]

72. ROOFING A homeowner estimates that it will take him 7 days to roof his house. A professional roofer estimates that he can roof the house in 4 days. How long will it take if the homeowner helps the roofer? [Section 7.7]

73. LOSING WEIGHT If a person cuts his or her daily calorie intake by 100, it will take 350 days for that person to lose 10 pounds. How long will it take for the person to lose 25 pounds? [Section 7.8]

74. ΔABC and ΔDEC are similar triangles. Find x. [Section 7.8]

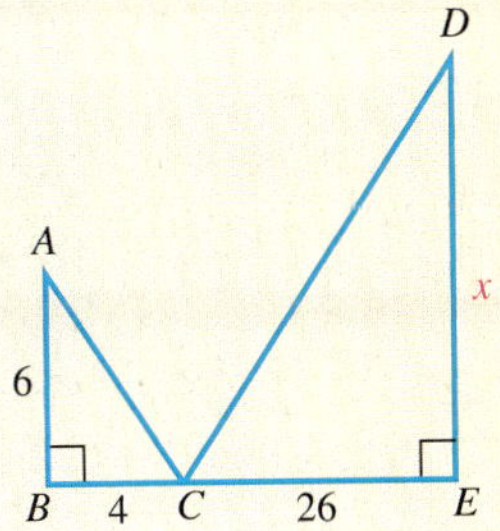

75. Suppose w varies directly as x. If $w = 1.2$ when $x = 4$, find w when $x = 30$. [Section 7.9]

76. GEARS The speed of a gear varies inversely with the number of teeth. If a gear with 10 teeth makes 3 revolutions per second, how many revolutions per second will a gear with 25 teeth make? [Section 7.9]

Simplify each radical expression. All variables represent positive numbers.

77. $\sqrt{100x^2}$ [Section 8.1]

78. $-\sqrt{18b^3}$ [Section 8.2]

Perform the indicated operation.

79. $3\sqrt{24} + \sqrt{54}$ [Section 8.3]

80. $(\sqrt{2} + 1)(\sqrt{2} - 3)$ [Section 8.4]

Rationalize the denominator.

81. $\dfrac{8}{\sqrt{10}}$ [Section 8.4]

82. $\dfrac{\sqrt{2}}{3 - \sqrt{a}}$ [Section 8.4]

Solve each equation.

83. $\sqrt{6x + 1} + 2 = 7$ [Section 8.5]

84. $\sqrt{3t + 7} = t + 3$ [Section 8.5]

Simplify each radical expression. All variables represent positive numbers.

85. $\sqrt[3]{\dfrac{27m^3}{8n^6}}$ [Section 8.6]

86. $\sqrt[4]{16}$ [Section 8.6]

Evaluate each expression.

87. $25^{3/2}$ [Section 8.6]

88. $(-8)^{-4/3}$ [Section 8.6]

Solve each equation.

89. $t^2 = 75$ [Section 9.1]

90. $(6y + 5)^2 - 72 = 0$ [Section 9.1]

91. STORAGE CUBES The diagonal distance across the face of each of the stacking cubes is 15 inches. What is the height of the entire storage arrangement? Round to the nearest tenth of an inch. [Section 9.1]

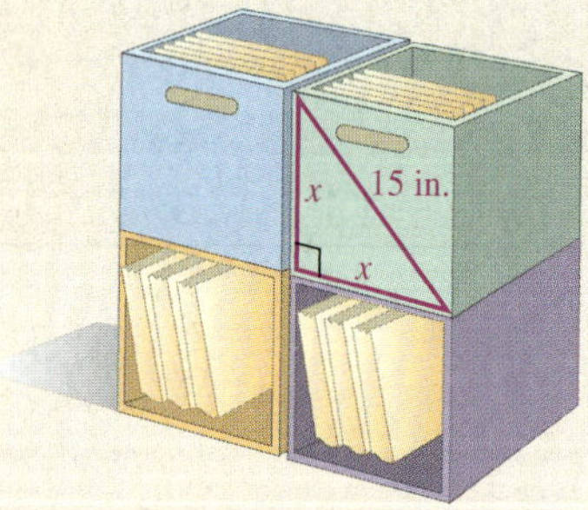

92. Solve $x^2 + 8x + 12 = 0$ by completing the square. [Section 9.2]

93. Solve $4x^2 - x - 2 = 0$ using the quadratic formula. Give the exact solutions, and then approximate each to the nearest hundredth. [Section 9.3]

94. QUILTS According to the *Guinness Book of World Records 1998,* the world's largest quilt was made by the Seniors' Association of Saskatchewan, Canada, in 1994. If the length of the rectangular quilt is 11 feet less than twice its width and it has an area of 12,865 ft^2, find its width and length. [Section 9.3]

Write each expression in terms of i.

95. $\sqrt{-49}$ [Section 9.4]

96. $\sqrt{-54}$ [Section 9.4]

Perform the operations. Express each answer in the form a + bi.

97. $(2 + 3i) - (1 - 2i)$ [Section 9.4]

98. $(7 - 4i) + (9 + 2i)$ [Section 9.4]

99. $(3 - 2i)(4 - 3i)$ [Section 9.4]

100. $\dfrac{3 - i}{2 + i}$ [Section 9.4]

Solve each equation. Express the solutions in the form a + bi.

101. $x^2 + 16 = 0$ [Section 9.4]

102. $x^2 - 4x = -5$ [Section 9.4]

103. Graph the quadratic equation $y = 2x^2 + 8x + 6$. Find the vertex, the x- and y-intercepts, and the axis of symmetry of the graph. [Section 9.5]

104. POWER OUTPUT The graph shows the power output (in horsepower, hp) of a certain engine for various engine speeds (in revolutions per minute, rpm). For what engine speed does the power output reach a maximum? [Section 9.5]

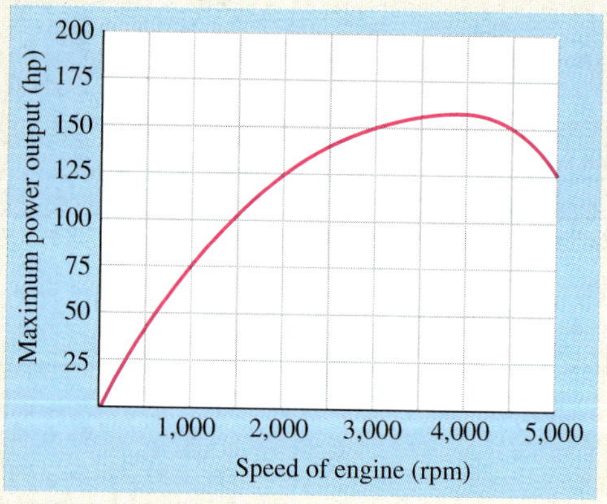

APPENDIX 2
Roots and Powers

n	n^2	$\sqrt{n}$	n^3	$\sqrt[3]{n}$	n	n^2	$\sqrt{n}$	n^3	$\sqrt[3]{n}$
1	1	1.000	1	1.000	51	2,601	7.141	132,651	3.708
2	4	1.414	8	1.260	52	2,704	7.211	140,608	3.733
3	9	1.732	27	1.442	53	2,809	7.280	148,877	3.756
4	16	2.000	64	1.587	54	2,916	7.348	157,464	3.780
5	25	2.236	125	1.710	55	3,025	7.416	166,375	3.803
6	36	2.449	216	1.817	56	3,136	7.483	175,616	3.826
7	49	2.646	343	1.913	57	3,249	7.550	185,193	3.849
8	64	2.828	512	2.000	58	3,364	7.616	195,112	3.871
9	81	3.000	729	2.080	59	3,481	7.681	205,379	3.893
10	100	3.162	1,000	2.154	60	3,600	7.746	216,000	3.915
11	121	3.317	1,331	2.224	61	3,721	7.810	226,981	3.936
12	144	3.464	1,728	2.289	62	3,844	7.874	238,328	3.958
13	169	3.606	2,197	2.351	63	3,969	7.937	250,047	3.979
14	196	3.742	2,744	2.410	64	4,096	8.000	262,144	4.000
15	225	3.873	3,375	2.466	65	4,225	8.062	274,625	4.021
16	256	4.000	4,096	2.520	66	4,356	8.124	287,496	4.041
17	289	4.123	4,913	2.571	67	4,489	8.185	300,763	4.062
18	324	4.243	5,832	2.621	68	4,624	8.246	314,432	4.082
19	361	4.359	6,859	2.668	69	4,761	8.307	328,509	4.102
20	400	4.472	8,000	2.714	70	4,900	8.367	343,000	4.121
21	441	4.583	9,261	2.759	71	5,041	8.426	357,911	4.141
22	484	4.690	10,648	2.802	72	5,184	8.485	373,248	4.160
23	529	4.796	12,167	2.844	73	5,329	8.544	389,017	4.179
24	576	4.899	13,824	2.884	74	5,476	8.602	405,224	4.198
25	625	5.000	15,625	2.924	75	5,625	8.660	421,875	4.217
26	676	5.099	17,576	2.962	76	5,776	8.718	438,976	4.236
27	729	5.196	19,683	3.000	77	5,929	8.775	456,533	4.254
28	784	5.292	21,952	3.037	78	6,084	8.832	474,552	4.273
29	841	5.385	24,389	3.072	79	6,241	8.888	493,039	4.291
30	900	5.477	27,000	3.107	80	6,400	8.944	512,000	4.309
31	961	5.568	29,791	3.141	81	6,561	9.000	531,441	4.327
32	1,024	5.657	32,768	3.175	82	6,724	9.055	551,368	4.344
33	1,089	5.745	35,937	3.208	83	6,889	9.110	571,787	4.362
34	1,156	5.831	39,304	3.240	84	7,056	9.165	592,704	4.380
35	1,225	5.916	42,875	3.271	85	7,225	9.220	614,125	4.397
36	1,296	6.000	46,656	3.302	86	7,396	9.274	636,056	4.414
37	1,369	6.083	50,653	3.332	87	7,569	9.327	658,503	4.431
38	1,444	6.164	54,872	3.362	88	7,744	9.381	681,472	4.448
39	1,521	6.245	59,319	3.391	89	7,921	9.434	704,969	4.465
40	1,600	6.325	64,000	3.420	90	8,100	9.487	729,000	4.481
41	1,681	6.403	68,921	3.448	91	8,281	9.539	753,571	4.498
42	1,764	6.481	74,088	3.476	92	8,464	9.592	778,688	4.514
43	1,849	6.557	79,507	3.503	93	8,649	9.644	804,357	4.531
44	1,936	6.633	85,184	3.530	94	8,836	9.695	830,584	4.547
45	2,025	6.708	91,125	3.557	95	9,025	9.747	857,375	4.563
46	2,116	6.782	97,336	3.583	96	9,216	9.798	884,736	4.579
47	2,209	6.856	103,823	3.609	97	9,409	9.849	912,673	4.595
48	2,304	6.928	110,592	3.634	98	9,604	9.899	941,192	4.610
49	2,401	7.000	117,649	3.659	99	9,801	9.950	970,299	4.626
50	2,500	7.071	125,000	3.684	100	10,000	10.000	1,000,000	4.642

APPENDIX 3
Answers to Selected Exercises

Study Set Section 1.1 (page 8)

1. sum, difference, product, quotient **3.** constant **5.** equation
7. horizontal **9. a.** Equation **b.** Algebraic Expression
11. a. Algebraic Expression **b.** Equation
13. Addition, multiplication, division; t
15.

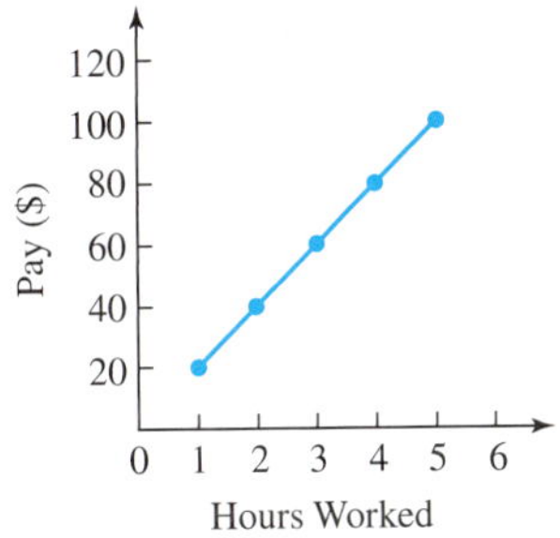

17. is, not, equal, to
19. $5 \cdot 6$, 5(6) **21.** $4x$
23. $2w$ **25.** $\frac{32}{x}$ **27.** $\frac{55}{5}$
29. 300 **31.** 15-year-old machinery is worth $35,000.
33. The product of 8 and 2 equals 16.
35. The difference of 11 and 9 equals 2.
37. The sum of x and 2 equals 10.
39. The quotient of 66 and 11 equals 6. **41.** $p = 100 - d$
43. $7d = h$ **45.** $s = 3c$ **47.** $w = e + 1{,}200$ **49.** $p = r - 600$
51. $\frac{l}{4} = m$ **53.** 390, 400, 405 **55.** 1,300; 1,200; 1,100
57. $\frac{e}{12}$ **59.** $2c$
61. 90, 75, 60, 45, 30, 15, 0

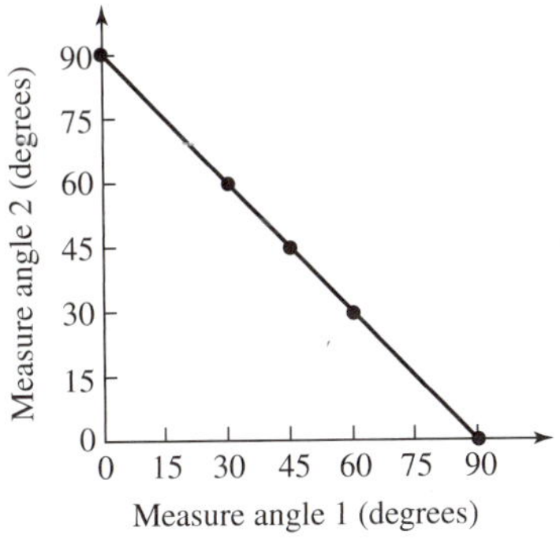

Study Set Section 1.2 (page 22)

1. multiplied **3.** prime-factored **5.** equivalent
7. least or lowest **9. a.** 1 **b.** a **c.** $\frac{a \cdot c}{b \cdot d}$ **d.** $\frac{a \cdot d}{b \cdot c}$ **e.** $\frac{a+b}{d}$
f. $\frac{a-b}{d}$ **11. a.** 1 **b.** 1 **13. a.** $\frac{5}{5}, \frac{25}{30}$ **b.** 2, 7, $\frac{2}{7}$
15. $3 \cdot 5 \cdot 5$ **17.** $2 \cdot 2 \cdot 7$ **19.** $3 \cdot 3 \cdot 3 \cdot 3$ **21.** $3 \cdot 3 \cdot 13$
23. $2 \cdot 2 \cdot 5 \cdot 11$ **25.** $2 \cdot 3 \cdot 11 \cdot 19$ **27.** $\frac{5}{48}$ **29.** $\frac{21}{55}$ **31.** $\frac{15}{8}$
33. $\frac{42}{25}$ **35.** $\frac{3}{9}$ **37.** $\frac{24}{54}$ **39.** $\frac{35}{5}$ **41.** $\frac{35}{7}$ **43.** $\frac{1}{3}$ **45.** $\frac{6}{7}$
47. $\frac{3}{8}$ **49.** Lowest terms **51.** $\frac{2}{3}$ **53.** $\frac{4}{25}$ **55.** $\frac{6}{5}$ **57.** $\frac{4}{7}$
59. $\frac{5}{24}$ **61.** $\frac{22}{35}$ **63.** $\frac{41}{45}$ **65.** $\frac{3}{20}$ **67.** 24 **69.** 4 **71.** $\frac{7}{9}$
73. $\frac{7}{20}$ **75.** $32\frac{2}{3}$ **77.** $2\frac{1}{2}$ **79.** $\frac{5}{9}$ **81.** $5\frac{19}{48}$ **83.** $\frac{19}{15}$
85. 70 **87.** $13\frac{3}{4}$ **89.** $\frac{1}{2}$ **91.** $\frac{14}{5}$ **93.** $\frac{8}{5}$ **95.** $\frac{3}{35}$ **97.** $\frac{9}{4}$
99. $\frac{3}{10}$ **101.** $1\frac{9}{11}$ **103.** $\frac{1}{7}$ **105. a.** $\frac{7}{32}$ in. **b.** $\frac{3}{32}$ in.
107. $5\frac{3}{4}$ lb **109.** $40\frac{1}{2}$ in. **115.** 150, 180

Study Set Section 1.3 (page 32)

1. whole **3.** integers **5.** Negatives; positives **7.** rational
9. irrational **11.** real **13. a.** −$15 million
b. $\frac{5}{16}$ in. or $+\frac{5}{16}$ in. **15. a.** −20 **b.** $\frac{2}{3}$ **17.** −14 and −4
19. square, root **21.** is, approximately, equal, to **23.** Greek
25. −4, −5
27.

	5	0	−3	$\frac{7}{8}$	0.17	$-9\frac{1}{4}$	$\sqrt{2}$	π
Real	✓	✓	✓	✓	✓	✓	✓	✓
Irrational							✓	✓
Rational	✓	✓	✓	✓	✓	✓		
Integer	✓	✓	✓					
Whole	✓	✓						
Natural	✓							

29. True **31.** False **33.** True **35.** True **37.** > **39.** >
41. < **43.** > **45.** < **47.** > **49.** 0.625 **51.** $0.0\overline{3}$
53. $0.0\overline{16}$ **55.** 0.42
57.

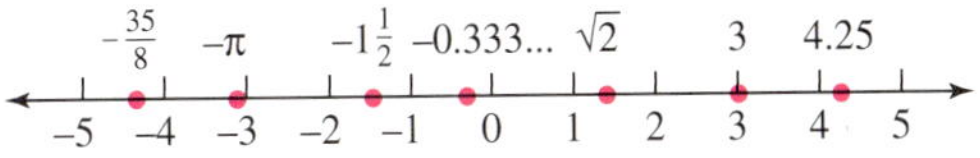

59.

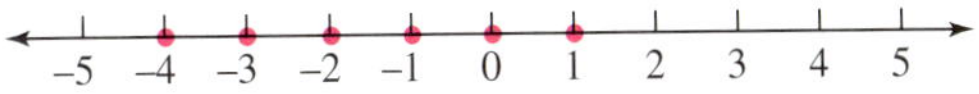

61. 83 **63.** $\frac{4}{3}$ **65.** 11 **67.** 6.1 **69.** >
71. < **73.** = **75.** = **77.** < **79.** >
81. Natural, whole, integers: 9; rational: 9, $\frac{15}{16}$, $3\frac{1}{8}$, 1.765; irrational: 2π, 3π, $\sqrt{89}$; real: all **83.** Arrow 1; $|-6| > |5|$
85. a. 2000; −$81 billion **b.** 1990; −$40 billion **93.** $\frac{4}{9}$
95. $2\frac{5}{23}$

Study Set Section 1.4 (page 41)

1. sum **3.** commutative, associative **5. a.** 6 **b.** −9.2
7. a. Negative **b.** Positive **9. a.** $1 + (-5)$ **b.** $-80.5 + 15$
c. $20 + 4$ **d.** $3 + 2.1$ **11.** Step 1: Commutative Property of Addition; Step 2: Associative Property of Addition

13. a. $x + y = y + x$ **b.** $(x + y) + z = x + (y + z)$ **15.** -9 **17.** -17 **19.** -74 **21.** -10.3 **23.** $-\frac{17}{12}$ **25.** $-\frac{7}{20}$ **27.** -3 **29.** 39 **31.** 0 **33.** 2.25 **35.** $-\frac{4}{15}$ **37.** $\frac{3}{8}$ **39.** 16 **41.** -15 **43.** -21 **45.** -26 **47.** 0.67 **49.** 195 **51.** 215 **53.** -112 **55.** $1\frac{2}{3}$ **57.** 15.4 **59.** 9 **61.** -5 **63.** 1 **65.** -1.7 **67.** 70 **69.** -6.6 **71.** $-\frac{1}{8}$ **73.** -14 **75.** 0 **77.** -22.1 **79.** 2,167 **81.** 68 **83.** $-\frac{15}{28}$ **85.** -0.9 **87.** 2,150 m **89.** Woods: -18, Kite: -6, Tolles: -5, Watson: -4 **91.** 1,242.86 **93.** $-\$99{,}650{,}000$ **95.** $-1, 3$ **97.** 79 feet above sea level **99.** $-\$23.2$ million **103.** True **105.** -9 and 3

Study Set Section 1.5 (page 49)

1. Subtraction **3.** range **5. a.** -12 **b.** $\frac{1}{5}$ **c.** -2.71 **d.** 0 **7. a.** 8 **b.** -8 **9.** No; $7 + (-8) \neq 15$ **11. a.** $1 - (-7) = 8$ **b.** $-(-2) = 2$ **c.** $-|-3| = -3$ **d.** $2 - 6 = -4$ **13.** 55 **15.** x **17.** -25 **19.** $-\frac{3}{16}$ **21.** -3 **23.** -13 **25.** -10 **27.** 11 **29.** -6 **31.** 1 **33.** 2 **35.** 40 **37.** 5 **39.** -69 **41.** -88 **43.** 12 **45.** -1.1 **47.** -3.5 **49.** -2.31 **51.** $-\frac{1}{2}$ **53.** $-\frac{5}{16}$ **55.** $-\frac{5}{12}$ **57.** 22 **59.** -25 **61.** -11 **63.** -50 **65.** -7 **67.** -1 **69.** 256 **71.** 0 **73.** -2.1 **75.** $\frac{47}{56}$ **77.** 3 **79.** -47.5 **81.** 149 **83.** -171 **85.** 4.63 **87.** $-\frac{19}{12}$ **89.** 160°F **91.** 21 **93.** Orlando: $-40{,}000$; LaGuardia: 37,000 **95.** 1,030 ft **97. a.** iii **b.** $-\$116.1$ billion **99.** $-68, -78, -147$ **105.** $2 \cdot 3 \cdot 5$ **107.** True

Study Set Section 1.6 (page 59)

1. product, quotient **3.** associative **5. a.** positive **b.** negative **7. a.** $-3, 3, -9$ **b.** 0, 8, 0 **9. a.** a **b.** 1 **c.** 0 **d.** Undefined **11. a.** $8 \cdot 5$ **b.** $(-2 \cdot 6)9$ **c.** $\frac{1}{5}$ **d.** 1 **13. a.** NEG **b.** Not possible to tell **c.** POS **d.** NEG **15.** $-4(-5) = 20$ **17.** -4 **19.** -16 **21.** -60 **23.** -66 **25.** -0.48 **27.** $-\frac{1}{4}$ **29.** 7 **31.** 54 **33.** 9 **35.** -441 **37.** 2.4 **39.** $\frac{1}{12}$ **41.** 0 **43.** 66 **45.** -720 **47.** -861 **49.** -216 **51.** 16 **53.** $\frac{9}{7}$; 1 **55.** $-\frac{1}{13}$; 1 **57.** 10 **59.** 3 **61.** -4 **63.** -17 **65.** -1 **67.** 1 **69.** -9 **71.** -0.005 **73.** 0 **75.** Undefined **77.** $-\frac{5}{12}$ **79.** $\frac{15}{4}$ **81.** -4.7 **83.** -520 **85.** $\frac{1}{24}$ **87.** -11 **89.** $1\frac{1}{2}$ **91.** 30.24 **93.** $-\frac{3}{8}$ **95.** $-\frac{3}{20}$ **97.** 30.3 **99.** $\frac{15}{16}$ **101.** -67 **103.** 6 **105.** \$8,000 **107.** $-72°$ **109. a.** ii **b.** 36 lb **111.** $-51°$F **113.** $-\$1{,}100, -\$400, -\$1{,}100$ **115. a.** $5, -10$ **b.** $2.5, -5$ **c.** $7.5, -15$ **d.** $10, -20$ **121.** -5 **123.** $1.08\overline{3}$

Study Set Section 1.7 (page 71)

1. base, exponent, power **3.** exponent **5.** order **7. a.** Subtraction **b.** Division **c.** Addition **d.** Power **9. a.** Parentheses, brackets, braces, absolute value symbols, fraction bar **b.** Innermost: parentheses; outermost: brackets **11. a.** -5 **b.** 5 **13.** $3, 9, 27, 54, -73$ **15.** 8^3 **17.** $7^3 12^2$ **19.** x^3 **21.** $r^4 s^2$ **23.** 49 **25.** 216 **27.** 625 **29.** 0.01 **31.** $-\frac{1}{64}$ **33.** $\frac{8}{27}$ **35.** $36, -36$ **37.** $64, -64$ **39.** -17 **41.** 30 **43.** 43 **45.** 8 **47.** -34 **49.** -118 **51.** 86 **53.** -8 **55.** -44 **57.** 0 **59.** -148 **61.** 100 **63.** -32 **65.** 53 **67.** -86 **69.** -392 **71.** 3 **73.** -19 **75.** $\frac{1}{2}$ **77.** 0 **79.** $-\frac{8}{9}$ **81.** 13 **83.** -8 **85.** -31 **87.** 11 **89.** 1 **91.** -500 **93.** -376 **95.** 12 **97.** 39 **99.** Undefined **101.** -110 **103.** -54 **105.** $\frac{1}{8}$ **107.** 10 **109.** -1 **111.** 2^2 square units, 3^2 square units, 4^2 square units **113.** About 6 **115. a.** \$11,875 **b.** \$95 **117.** 81 in. **123.** $-17, -5$

Study Set Section 1.8 (page 82)

1. expressions **3.** terms **5.** coefficient **7.** $7, 14, 21, 7w$ **9.** $(12 - h)$ in. **11.** $(x + 20)$ ounces **13. a.** $b - 15$ **b.** $p + 15$ **15.** 5, 25, 45 **17. a.** $8y$ **b.** $2cd$ **c.** Commutative **19. a.** 4 **b.** $3, 11, -1, 9$ **21.** Term **23.** Factor **25.** $l + 15$ **27.** $50x$ **29.** $\frac{w}{l}$ **31.** $P + \frac{2}{3}p$ **33.** $k^2 - 2{,}005$ **35.** $2a - 1$ **37.** $\frac{1{,}000}{n}$ **39.** $2p + 90$ **41.** $3(35 + h + 300)$ **43.** $p - 680$ **45.** $4d - 15$ **47.** $2(200 + t)$ **49.** $|a - 2|$ **51.** $0.1d$ or $\frac{1}{10}d$ **53.** Three-fourths of r **55.** 50 less than t **57.** The product of x, y, and z **59.** Twice m, increased by 5 **61.** $(x + 2)$ in. **63.** $(36 - x)$ in. **65.** $60h$ **67.** $\frac{i}{12}$ **69.** $\$8x$ **71.** $49x$¢ **73.** $\$2t$ **75.** $\$25(x + 2)$ **77.** 2 **79.** 13 **81.** 20 **83.** -12 **85.** -5 **87.** $-\frac{1}{5}$ **89.** 17 **91.** 36 **93.** 255 **95.** 8 **97.** $-1, -2, -28$ **99.** 41, 11, 2 **101.** $150, -450$ **103.** 0, 0, 5 **105. a.** Let x = weight of the Element, $2x - 340$ = weight of the Hummer **b.** 6,400 lb **107. a.** Let x = age of Apple; $x + 80$ = age of IBM; $x - 9$ = age of Dell **b.** IBM: 112; Dell: 23 **113.** 60 **115.** $\frac{8}{27}$

Study Set Section 1.9 (page 93)

1. simplify **3.** distributive **5.** like **7. a.** 4, 9, 36 **b.** Associative Property of Multiplication **9. a.** + **b.** − **c.** − **d.** + **11. a.** $10x$ **b.** Can't be simplified **c.** $-42x$ **d.** Can't be simplified **e.** $18x$ **f.** $3x + 5$ **13. a.** $6(h - 4)$ **b.** $-(z + 16)$ **15.** $12t$ **17.** $63m$ **19.** $-35q$ **21.** $300t$ **23.** $11.2x$ **25.** $60c$ **27.** $-96m$ **29.** g **31.** $5x$ **33.** $6y$ **35.** $5x + 15$ **37.** $-12x - 27$ **39.** $9x + 10$ **41.** $0.4x - 1.6$ **43.** $36c - 42$ **45.** $-78c + 18$ **47.** $30t + 90$ **49.** $4a - 1$ **51.** $24t + 16$ **53.** $2w - 4$ **55.** $56y + 32$ **57.** $50a - 75b + 25$ **59.** $-x + 7$ **61.** $5.6y - 7$ **63.** $3x, -2x$ **65.** $-3m^3, -m^3$ **67.** $10x$ **69.** 0 **71.** $20b^2$ **73.** r **75.** $28y$ **77.** $-s^3$ **79.** $-3.6c$ **81.** $0.4r$ **83.** $\frac{4}{5}t$ **85.** $-\frac{5}{8}x$ **87.** $-6y - 10$ **89.** $-2x + 5$ **91.** $9m^2 + 6m - 4$ **93.** $4x^2 - 3x + 9$ **95.** $7z - 15$ **97.** $s^2 - 12$ **99.** $-41r + 130$ **101.** $8x - 9$ **103.** $12c + 34$ **105.** $-10r$ **107.** $-20r$ **109.** $3a$ **111.** $9r - 16$ **113.** $-6x$ **115.** $c - 13$ **117.** $a^3 - 8$ **119.** $12x$ **121.** $(4x + 8)$ ft **125.** 2

Chapter 1 Review (page 96)

1. 1 hr; 100 cars **2.** 100 **3.** 7 P.M. **4.** 12 A.M. (midnight) **5.** The difference of 15 and 3 equals 12. **6.** The sum of 15 and 3 equals 18. **7.** The quotient of 15 and 3 equals 5. **8.** The product of 15 and 3 equals 45. **9. a.** $4 \cdot 9$; $4(9)$ **b.** $\frac{9}{3}$ **10. a.** $8b$ **b.** Prt **11. a.** Equation

b. Expression **12.** 10, 15, 25 **13. a.** $2 \cdot 12$, $3 \cdot 8$ (Answers may vary) **b.** $2 \cdot 2 \cdot 6$ (Answers may vary) **c.** 1, 2, 3, 4, 6, 8, 12, 24 **14.** Equivalent **15.** $2 \cdot 3^3$ **16.** $3 \cdot 7^2$ **17.** $5 \cdot 7 \cdot 11$ **18.** Prime **19.** $\frac{4}{7}$ **20.** $\frac{4}{3}$ **21.** $\frac{40}{64}$ **22.** $\frac{36}{3}$ **23.** 90 **24.** 210 **25.** $\frac{7}{64}$ **26.** $\frac{5}{21}$ **27.** $\frac{16}{45}$ **28.** $3\frac{1}{4}$ **29.** $\frac{2}{5}$ **30.** $\frac{5}{22}$ **31.** $\frac{59}{60}$ **32.** $\frac{5}{18}$ **33.** $52\frac{1}{2}$ million **34.** $\frac{17}{96}$ in. **35. a.** 0 **b.** $\{\ldots, -2, -1, 0, 1, 2, \ldots\}$ **36.** −206 ft **37. a.** < **b.** > **38. a.** $\frac{7}{10}$ **b.** $\frac{14}{3}$ **39.** 0.004 **40.** $0.7\overline{72}$

41.

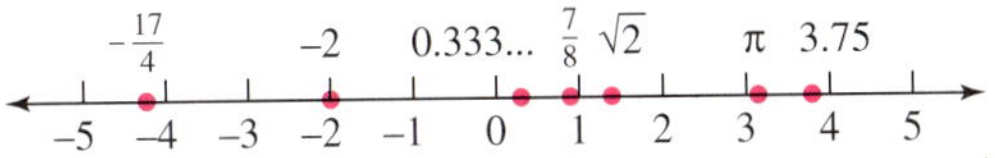

42. Natural: 8; whole: 0, 8; integers: 0, −12, 8; rational: $-\frac{4}{5}$, 99.99, 0, −12, $4\frac{1}{2}$, 0.666 . . . , 8; irrational: $\sqrt{2}$; real: all **43.** False **44.** False **45.** True **46.** True **47.** > **48.** < **49.** −82 **50.** 12 **51.** −7 **52.** 0 **53.** −11 **54.** −12.3 **55.** $-\frac{3}{16}$ **56.** 11 **57. a.** Commutative Property of Addition **b.** Associative Property of Addition **c.** Addition Property of Opposites (Inverse Property of Addition) **d.** Addition Property of 0 (Identity Property of Addition) **58.** 118°F **59. a.** −10 **b.** 3 **60. a.** $\frac{9}{16}$ **b.** −4 **61.** −19 **62.** $-\frac{14}{15}$ **63.** 5 **64.** 5.7 **65.** −10 **66.** −29 **67.** 65,233 ft; 65,233 + (−36,205) = 29,028 **68.** 428 B.C.; (−428); −428 + 81 = −347 **69.** −56 **70.** 1 **71.** 12 **72.** −12 **73.** 6.36 **74.** −2 **75.** $-\frac{2}{15}$ **76.** 0 **77.** High: 3, low: −4.5 **78. a.** Associative Property of Multiplication **b.** Commutative Property of Multiplication **c.** Multiplication Property of 1 (Identity Property of Multiplication) **d.** Inverse Property of Multiplication **79.** −1 **80.** −17 **81.** 3 **82.** $-\frac{6}{5}$ **83.** Undefined **84.** −4.5 **85.** 0, 18, 0 **86.** −$360 **87. a.** 8^5 **b.** $9\pi r^2$ **88. a.** 81 **b.** $-\frac{8}{27}$ **c.** 32 **d.** 50 **89.** 17 **90.** −36 **91.** −169 **92.** 23 **93.** −420 **94.** $-\frac{7}{19}$ **95.** 113 **96.** Undefined **97. a.** $(-9)^2 = 81$ **b.** $-9^2 = -81$ **98.** $20 **99. a.** 3 **b.** 1 **100. a.** 16, −5, 25 **b.** $\frac{1}{2}$, 1 **101.** $h + 25$ **102.** $3s - 15$ **103.** $\frac{1}{2}t - 6$ **104.** $|2 - a^2|$ **105.** $(n + 4)$ in. **106.** $(b - 4)$ in. **107.** $10d$ **108.** $(x - 5)$ years **109.** 30, $10d$ **110.** 0, 19, −16 **111.** 40 **112.** −36 **113.** $-28w$ **114.** $24x$ **115.** $2.08f$ **116.** r **117.** $5x + 15$ **118.** $-2x - 3 + y$ **119.** $3c - 6$ **120.** $12.6c + 29.4$ **121.** $9p$ **122.** $-7m$ **123.** $4n$ **124.** $-p - 18$ **125.** $0.1k^2$ **126.** $8a^3 - 1$ **127.** w **128.** $4h - 15$ **129.** $(4x + 4)$ ft **130. a.** x **b.** $-x$ **c.** $4x + 1$ **d.** $4x - 1$

Chapter 1 Test (page 105)

1. a. equivalent **b.** product **c.** reciprocal **d.** like, terms **e.** undefined **2. a.** $24 **b.** 5 hr **3.** 3, 20, 70 **4.** $2 \cdot 2 \cdot 3 \cdot 3 \cdot 5 = 2^2 \cdot 3^2 \cdot 5$ **5.** $\frac{2}{5}$ **6.** $\frac{3}{2} = 1\frac{1}{2}$ **7.** $\frac{27}{35}$ **8.** $6\frac{11}{15}$ **9.** $3.57 **10.** $0.8\overline{3}$

11.

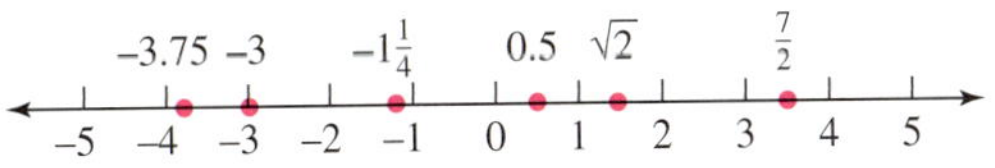

12. a. True **b.** False **c.** True **d.** True **13.** A real number is any number that is either a rational or an irrational number.

14. a. > **b.** < **c.** < **d.** > **15.** A gain of 0.6 of a rating point **16.** −2 **17.** $\frac{3}{8}$ **18. a.** −6 **b.** −6 + (−4) = −10 **19. a.** 14 **b.** 14(−9) = −126 **20.** −30 **21.** −2.44 **22.** 0 **23.** $-\frac{27}{125}$ **24.** 0 **25.** −3 **26.** 50 **27.** 14 **28. a.** Associative Property of Addition **b.** The Distributive Property **c.** Commutative Property of Multiplication **d.** Inverse Property of Multiplication **e.** Identity Property of Addition **29. a.** 9^5 **b.** $3x^2z^3$ **30.** 170 **31.** −12 **32.** −100 **33.** −351 **34.** 36 **35.** 4, 17, −59 **36.** $2w - 7$ **37. a.** $x - 2$ **b.** $25q$¢ **38.** 3; 5 **39.** $-20x$ **40.** $224t$ **41.** $-4a + 4$ **42.** $-5.9d^3$ **43.** $14x + 3$ **44.** $(18x + 6)$ ft

Study Set Section 2.1 (page 117)

1. equation **3.** solve **5.** equivalent **7. a.** $x + 6$ **b.** Neither **c.** No **d.** Yes **9. a.** c, c **b.** c, c **11. a.** x **b.** y **c.** t **d.** h **13.** 5, 5, 50, 50, $\stackrel{?}{=}$, 45, 50 **15. a.** Is possibly equal to **b.** Yes **17.** No **19.** No **21.** No **23.** No **25.** Yes **27.** No **29.** No **31.** Yes **33.** Yes **35.** Yes **37.** 71 **39.** 18 **41.** −0.9 **43.** 3 **45.** $\frac{8}{9}$ **47.** 3 **49.** $-\frac{1}{25}$ **51.** −2.3 **53.** 45 **55.** 0 **57.** 21 **59.** −2.64 **61.** 20 **63.** 15 **65.** −6 **67.** 4 **69.** 4 **71.** 7 **73.** 1 **75.** −6 **77.** 20 **79.** 0.5 **81.** −18 **83.** $-\frac{4}{21}$ **85.** 13 **87.** 2.5 **89.** $-\frac{8}{3}$ **91.** $\frac{13}{20}$ **93.** 4 **95.** −5 **97.** −200 **99.** 95 **101.** 65° **103.** $6,000,000 **109.** 0 **111.** $45 - x$

Study Set Section 2.2 (page 127)

1. equation **3.** identity **5.** subtraction, multiplication **7. a.** $-2x - 8 = -24$ **b.** $-20 = 3x - 16$ **9. a.** $12x$ **b.** $2x$ **11.** 10 **13.** +7, +7, 2, 2, 14, $\stackrel{?}{=}$, 28, 21, 14 **15.** 6 **17.** 5 **19.** −7 **21.** 18 **23.** 16 **25.** 12 **27.** $\frac{10}{3}$ **29.** $-\frac{5}{2}$ **31.** 5 **33.** −0.25 **35.** 2.9 **37.** −4 **39.** $\frac{11}{5}$ **41.** −1 **43.** −6 **45.** 0.04 **47.** −6 **49.** −11 **51.** 7 **53.** −11 **55.** 1 **57.** $\frac{9}{2}$ **59.** 3 **61.** −20 **63.** 6 **65.** $\frac{2}{15}$ **67.** $-\frac{12}{5}$ **69.** $\frac{27}{5}$ **71.** 5 **73.** 200 **75.** 1,000 **77.** 200 **79.** $\frac{5}{4}$ **81.** −1 **83.** 1 **85.** 80 **87.** All real numbers **89.** No solution **91.** No solution **93.** All real numbers **95.** $\frac{1}{4}$ **97.** 30 **99.** −11 **101.** No solution **103.** 1 **105.** $\frac{52}{9}$ **107.** −6 **109.** −5 **115.** Commutative Property of Multiplication **117.** Associative Property of Addition

Study Set Section 2.3 (page 135)

1. Percent **3.** multiplication, is **5.** $\frac{51}{100}$, 0.51, 51% **7. a.** 2,449 **b.** 2,449, what, 14,792 **9. a.** $12 = 0.40 \cdot x$ **b.** $99 = x \cdot 200$ **c.** $x = 0.66 \cdot 3$ **11. a.** 0.35 **b.** 0.085 **c.** 1.5 **d.** 0.0275 **13.** 312 **15.** 26% **17.** 300 **19.** 46.2 **21.** 2.5% **23.** 1,464 **25.** 0.48 oz **27. a.** $925 billion **b.** $600 billion **29.** $10.45 **31.** $24.20 **33.** 60%, 40% **35.** 19% **37.** No (66%) **39.** 120 **41. a.** 5 g; 25% **b.** 20 g **43.** 2000–2001, about 9% **45.** 12% **47.** 3% **49.** $75 **51.** $300 **53.** $95,000 **55.** $25,600 **61.** $\frac{12}{5} = 2\frac{2}{5}$ **63.** No

Study Set Section 2.4 (page 147)

1. formula **3.** volume **5. a.** $d = rt$ **b.** $r = c + m$ **c.** $p = r - c$ **d.** $I = Prt$ **7.** 11,176,920 mi, 65,280 ft **9.** $Ax, -Ax, B, B, B$ **11.** \$240 million **13.** \$931 **15.** 3.5% **17.** \$6,000 **19.** 2.5 mph **21.** 4.5 hours **23.** 185°C **25.** −454°F **27.** 20 in. **29.** 1,885 mm^3 **31.** $c = r - m$ **33.** $b = P - a - c$ **35.** $R = \frac{E}{I}$ **37.** $l = \frac{V}{wh}$ **39.** $r = \frac{C}{2\pi}$ **41.** $h = \frac{3V}{B}$ **43.** $f = \frac{s}{w}$ **45.** $r = \frac{T - 2t}{2}$ **47.** $x = \frac{C - By}{A}$ **49.** $m = \frac{2K}{v^2}$ **51.** $c = 3A - a - b$ **53.** $t = T - 18E$ **55.** $r^2 = \frac{s}{4\pi}$ **57.** $v^2 = \frac{2Kg}{w}$ **59.** $r^3 = \frac{3V}{4\pi}$ **61.** $M = 4.2B + 19.8$ **63.** $h = \frac{S - 2\pi r^2}{2\pi r}$ **65.** $y = -3x + 9$ **67.** $y = \frac{1}{3}x + 3$ **69.** $y = -\frac{3}{4}x - 4$ **71.** $b = \frac{2A}{h} - d$ or $b = \frac{2A - hd}{h}$ **73.** $c = \frac{72 - 8w}{7}$ **75.** 87, 89, 91

77.

Income Statement (dollars amounts in millions)	Quarter ending Sep 04	Quarter ending Sep 05
Revenue	1,806.2	1,886.0
Cost of goods sold	1,543.4	1,638.9
Operating profit	262.8	247.1

79. 14 in. **81.** 50 in. **83.** 25 in., 2.5 in. **85.** 18.1 in.^{2} **87.** 2,463 ft^2 **89.** 3,150 cm^2 **91.** 6 in. **93.** 8 ft **95.** 348 ft^3 **97.** 254 in.^{2} **99.** $D = \frac{L - 3.25r - 3.25R}{2}$ **105.** 137.76 **107.** 15%

Study Set Section 2.5 (page 158)

1. consecutive **3.** vertex, base

5.

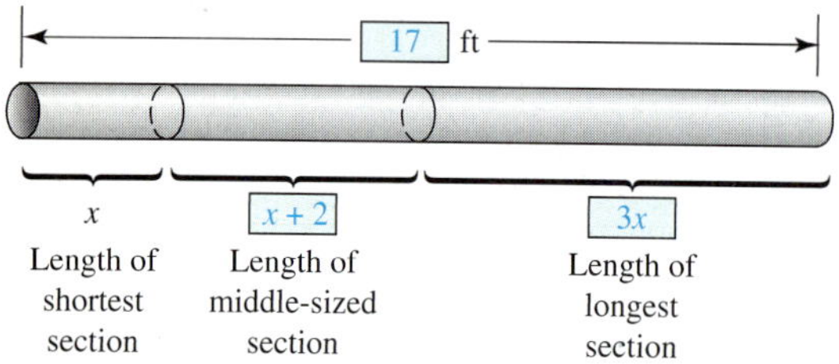

7. \$0.03x **9.** 180° **11. a.** $x + 1$ **b.** $x + 2$ **13.** 4 ft, 8 ft **15.** 102 mi, 108 mi, 114 mi, 120 mi **17.** 7.3 ft, 10.7 ft **19.** 250 calories in ice cream, 600 calories in pie **21.** 7 **23.** 580 **25.** 20 **27.** \$50,000 **29.** \$5,250 **31.** Ronaldo: 15, Mueller: 14 **33.** *Friends:* 236; *Leave It to Beaver:* 234 **35.** Jan. 8, 10, 12 **37.** Width: 27 ft, length: 78 ft **39.** 21 in. by 30.25 in. **41.** 7 ft, 7 ft, 11 ft **43.** 20° **45.** 42.5°, 70°, 67.5° **47.** 22°, 68° **53.** −24 **55.** $-\frac{40}{37}$ **57.** 1

Study Set Section 2.6 (page 169)

1. investment, motion **3.** $30{,}000 - x$ **5.** $r - 150$ **7.** $35t + 45t = 80, 35t, t, 45t, 80$ **9. a.** $0.50(6) + 0.25x = 0.30(6 + x), 0.50(6), 0.25x, 6 + x, 0.30(6 + x)$ **b.** $0.06x + 0.03(10 - x) = 0.05(10), 0.06x, 10 - x, 0.03(10 - x), 0.05(10)$ **11.** 0.06, 0.152 **13.** 4 **15.** 6,000 **17.** \$15,000 at 4%; \$10,000 at 7% **19.** Silver: \$1,500; gold: \$2,000 **21.** \$26,000 **23.** 822: \$9,000; 721: \$6,000 **25.** \$4,900 **27.** 2 hr **29.** $\frac{1}{4}$ hr = 15 min **31.** 1 hr **33.** 4 hr **35.** 55 mph **37.** 50 gal **39.** 4%: 5 gal; 1%: 10 gal **41.** 32 ounces of 8%; 32 ounces of 22% **43.** 6 gal **45.** 50 lb **47.** 20 scoops **49.** 15 **51.** \$4.25 **53.** 17 **55.** 90 **57.** 40 pennies, 20 dimes, 60 nickels **59.** 2-pointers: 50; 3-pointers: 4 **63.** $-50x + 125$ **65.** $-3x + 3$ **67.** $16y - 16$

Study Set Section 2.7 (page 183)

1. inequality **3.** interval **5. a.** same **b.** positive **c.** negative **7.** $x > 32$ **9. a.** ≤ **b.** ∞ **c.** [or] **d.** > **11.** 5, 5, 12, 4, 4, 3 **13. a.** Yes **b.** No **15. a.** No **b.** Yes

17. $(-\infty, 5)$

19. $(-3, 1]$

21. $x < -1, (-\infty, -1)$

23. $-7 < x \le 2, (-7, 2]$

25. $(3, \infty)$
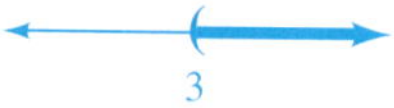

27. $[10, \infty)$

29. $(-\infty, 6)$

31. $(-\infty, 48]$

33. $[2, \infty)$

35. $[3, \infty)$
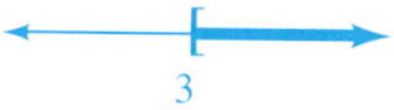

37. $(7, \infty)$

39. $(-\infty, 0.4]$

41. $[16, \infty)$

43. $(-\infty, 0)$

45. $[-10, \infty)$

47. $(-\infty, -2)$

49. $(-5, \infty)$

51. $(-\infty, 1.5]$

53. $(-\infty, 20]$

55. $(0, \infty)$

57. $\left(\frac{5}{4}, \infty\right)$

59. $\left(-\infty, \frac{3}{2}\right]$
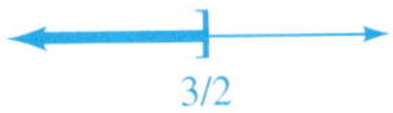

61. $(7, 10)$

63. $[-10, 0]$

65. $[-6, 10]$

67. $[2, 3)$

69. $(-3, 6]$

71. $(-5, -2)$

73. $\left[\frac{9}{4}, \infty\right)$
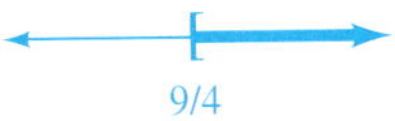

75. $(-\infty, -40]$

77. $(-2, 1]$

79. $(-\infty, 2]$

81. $(-\infty, -27)$

83. $\left(-\infty, \frac{1}{8}\right]$

85. $[-13, \infty)$

87. $(6, \infty)$

89. $[-32, 48]$

91. $\left(-\infty, -\frac{11}{4}\right)$

93. $\left[-\frac{3}{8}, \infty\right)$

95. $(-\infty, -1]$

97. $\left(\frac{6}{7}, \infty\right)$

99. 98% or better **101.** More than 27 mpg **103.** 19 ft or less **105.** More than 5 ft **107.** 40 or less **109.** 12.5 in. or less **111.** **a.** 26 lb $\le w \le$ 31 lb **b.** 12 lb $\le w \le$ 14 lb **c.** 18.5 lb $\le w \le$ 20.5 lb **d.** 11 lb $\le w \le$ 13 lb **115.** 1, -3, 6

Chapter 2 Review (page 187)

1. Yes **2.** No **3.** No **4.** No **5.** Yes **6.** Yes **7.** equation **8.** True **9.** 21 **10.** 32 **11.** -20.6 **12.** 107 **13.** 24 **14.** 2 **15.** -9 **16.** -7.8 **17.** 0 **18.** $-\frac{16}{5}$ **19.** 2 **20.** -30.6 **21.** 30 **22.** -19 **23.** 4 **24.** 1 **25.** $\frac{5}{4}$ **26.** $\frac{47}{13}$ **27.** 6 **28.** $-\frac{22}{75}$ **29.** 5 **30.** 1 **31.** Identity; all real numbers **32.** Contradiction; no solution **33.** **a.** Percent **b.** discount **c.** commission **34.** 192.4 **35.** 142.5 **36.** 12% **37.** Broadband: 139.3 million; dial-up: 45.3 million **38.** \$26.74 **39.** No **40.** \$450 **41.** \$150 **42.** 1,567% **43.** \$176 **44.** \$11,800 **45.** 8 min **46.** 4.5% **47.** 1,949°F **48.** **a.** 168 in. **b.** 1,440 in.^{2} **c.** 4,320 in.^{3} **49.** 76.5 m^2 **50.** 144 in.^{2} **51.** **a.** 50.27 cm **b.** 201 cm^2 **52.** 9.4 ft^3 **53.** 120 ft^3 **54.** 381.70 in.^{3} **55.** $h = \frac{A}{2\pi r}$ **56.** $G = 3A - 3BC + K$ **57.** $t = \frac{4C}{s} + d$ **58.** $y = \frac{3}{4}x + 4$ **59.** 8 ft **60.** 200 **61.** \$2,500,000 **62.** Labonte: 43; Petty: 45 **63.** 24.875 in. $\times$ 29.875 in. $\left(24\frac{7}{8} \text{ in.} \times 29\frac{7}{8} \text{ in.}\right)$ **64.** 76.5°, 76.5° **65.** \$16,000 at 7%, \$11,000 at 9% **66.** 20 **67.** $1\frac{2}{3}$ hr = 1 hr 40 min **68.** 12, 4 **69.** 10 lb of each **70.** 2 gal

71. $(-\infty, 1)$

72. $(-\infty, 12]$

73. $\left(\frac{5}{4}, \infty\right)$

74. $[3, \infty)$

75. $(-\infty, 40]$
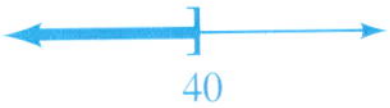

76. $(9, \infty)$

77. $(6, 11)$

78. $\left(-\frac{7}{2}, \frac{3}{2}\right]$

79. 2.40 g $\le w \le$ 2.53 g **80.** 48 inches or less

Chapter 2 Test (page 194)

1. **a.** solve **b.** Percent **c.** circumference **d.** inequality **e.** multiplication, equality **2.** No **3.** 2 **4.** -5 **5.** 22 **6.** $-\frac{1}{4}$ **7.** 1,336 **8.** All real numbers (an identity) **9.** $\frac{7}{4}$ **10.** -4 **11.** No solution (a contradiction) **12.** 0 **13.** 12.16 **14.** \$76,000 **15.** 6% **16.** \$30 **17.** \$295 **18.** -10°C **19.** 393 in.^{3} **20.** $r = \frac{A - P}{Pt}$ **21.** 20 in.^{2} **22.** 22 min, 8 min **23.** \$120,000 **24.** 380 mi, 280 mi **25.** Green: 16 lb; herbal: 4 lb **26.** 412, 413 **27.** $\frac{3}{5}$ hr **28.** 10 liters **29.** 68° **30.** \$5,250

31. $[-3, \infty)$

32. $(-\infty, 6.4)$

33. $[-7, 4)$

34. 180 words

Study Set Section 3.1 (page 204)

1. ordered **3.** axis, axis, origin **5.** rectangular **7.** **a.** origin, left, up **b.** origin, right, down **9.** **a.** I and II

b. II and III **c.** IV **11.** (3, 5) is an ordered pair, $3(5) = 3 \cdot 5$
13. Yes **15.** Horizontal
17.

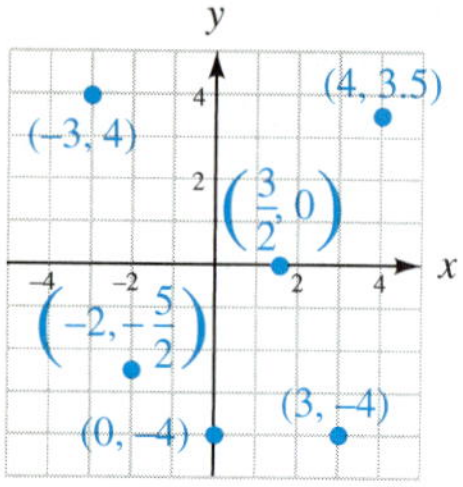

19. (4, 3), (0, 4), (−5, 0), (−4, −5), (3, −3)
21. a. 60 beats/min **b.** 10 min
23. a. 5 min and 50 min after starting **b.** 20 min
25. a. 2 hr **b.** −1,000 ft **27. a.** It ascends (rises) 500 ft
b. −500 ft **29.** Rivets: (−6, 0), (−2, 0), (2, 0), (6, 0); welds: (−4, 3), (0, 3), (4, 3); anchors: (−6, −3), (6, −3)
31. (G, 2), (G, 3), (G, 4) **33. a.** 8
b. It represents the patient's left side. **35.** (2, 4); 12 sq. units
37.

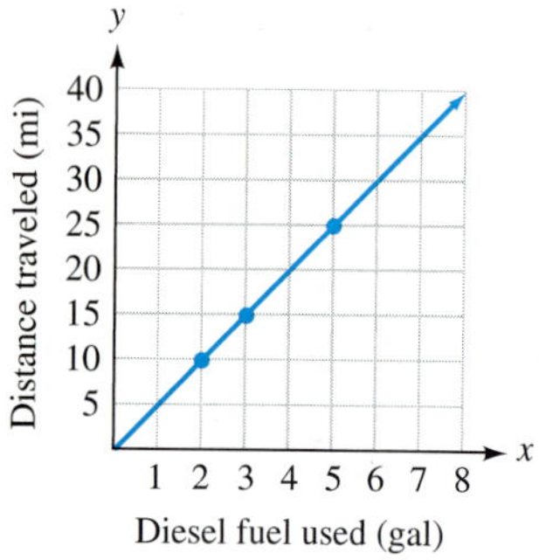

a. 20 mi **b.** 6 gal
c. 35 mi

39.

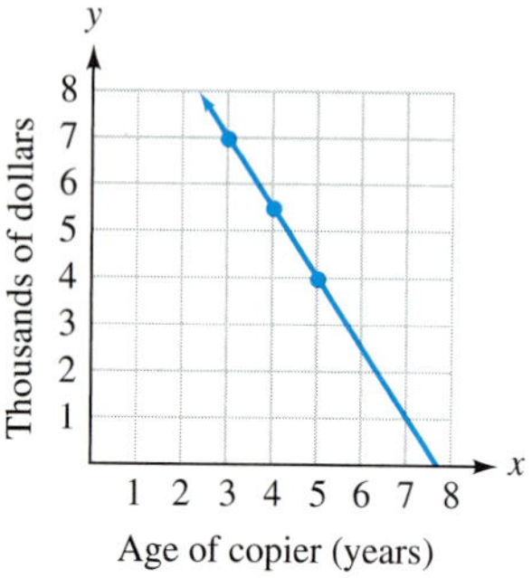

a. A 3-yr-old copier is worth \$7,000. **b.** \$1,000
c. 6 yr **45.** $h = \frac{3(AC + T)}{2}$
47. −1

Study Set Section 3.2 (page 217)

1. two **3.** table **5.** linear **7. a.** 2 **b.** Yes **c.** No
d. Infinitely many **9.** solution, point
11. a. −5, 0, 5 (Answers may vary)
b. −10, 0, 10 (Answers may vary) **13.** 6, −2, 2, 6
15. a. $y^1 = \frac{1}{2}x^1 + 7$ **b.** The exponent on x is not 1. **17.** Yes
19. No **21.** Yes **23.** Yes **25.** No **27.** No **29.** 11
31. 4 **33.** 13 **35.** $-\frac{8}{7}$
37.

x	y	(x, y)
8	12	(8, 12)
6	8	(6, 8)

39.

x	y	(x, y)
−5	−13	(−5, −13)
−1	−1	(−1, −1)

41.

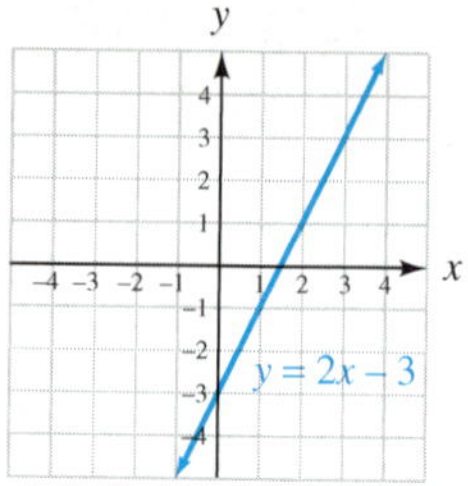

43.

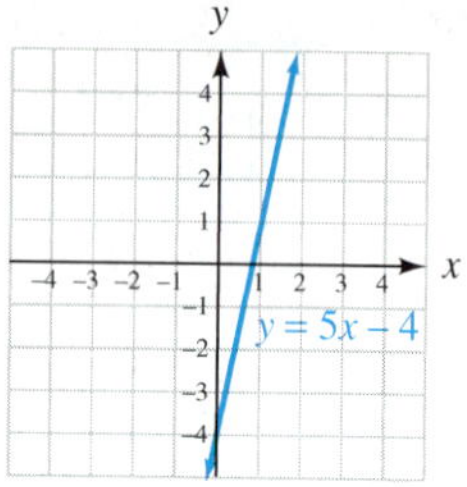

45.

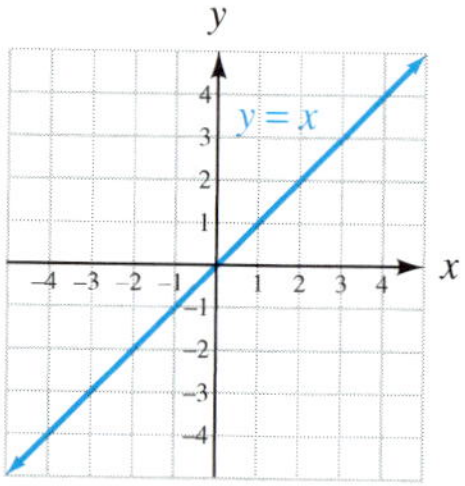

47.

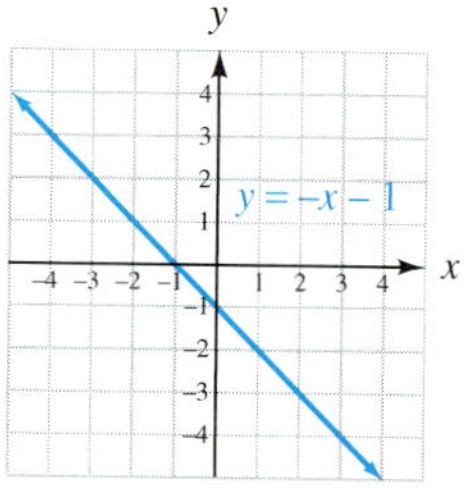

49.

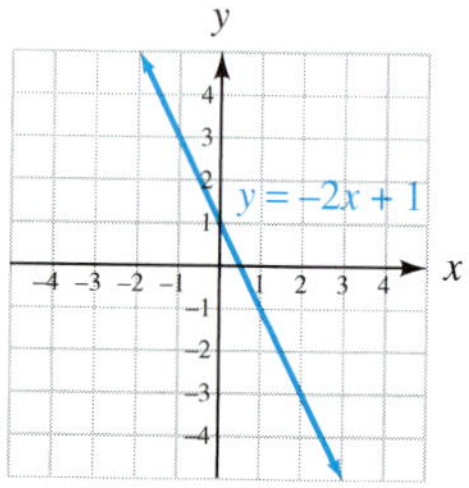

51.

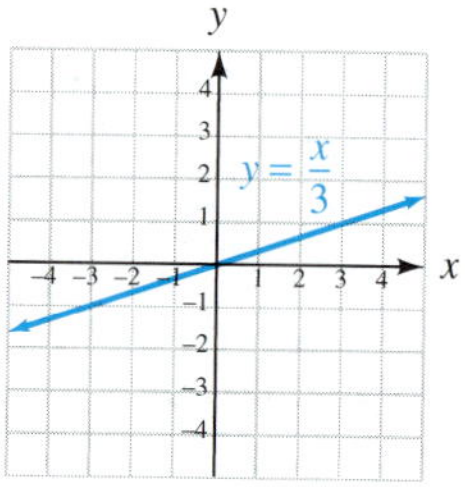

53.

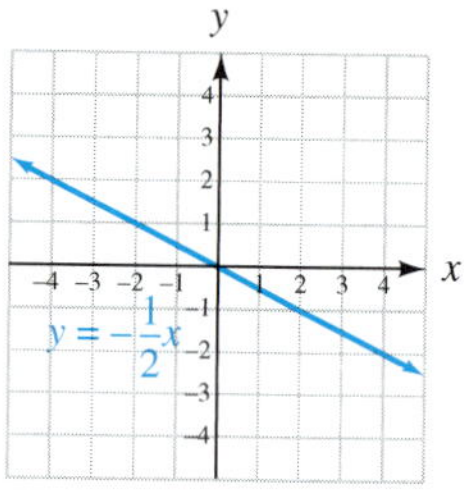

55.

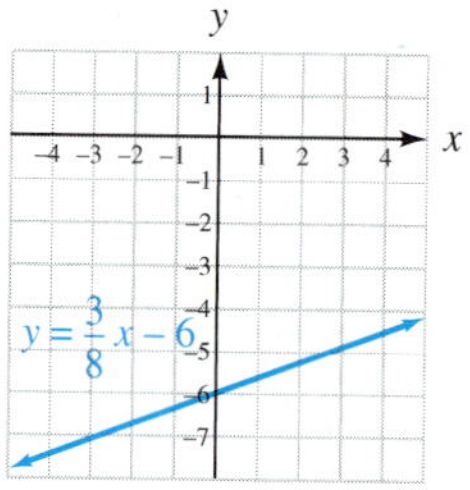

57.

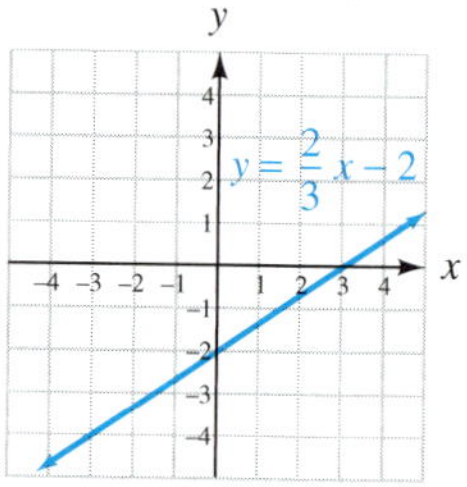

59.

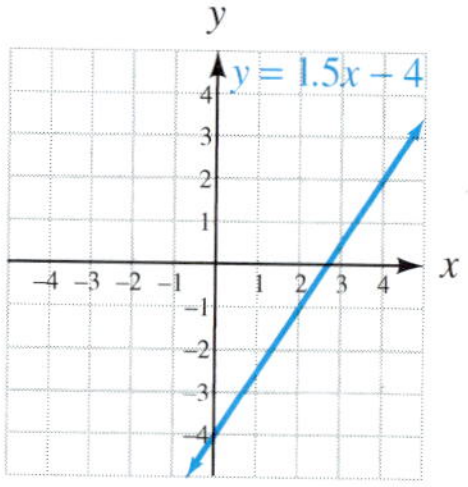

61.

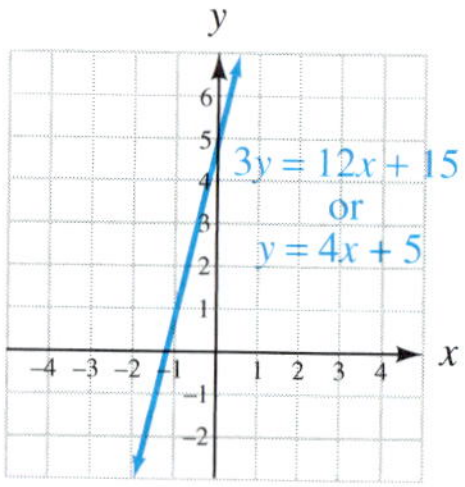

63.

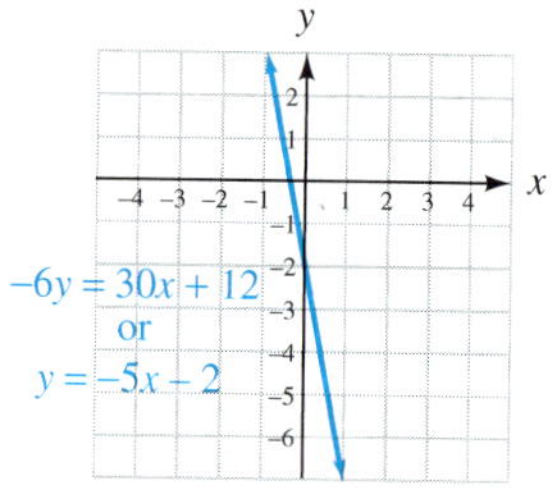

65.

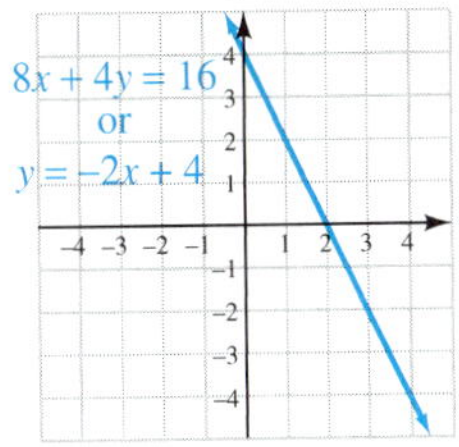

67.

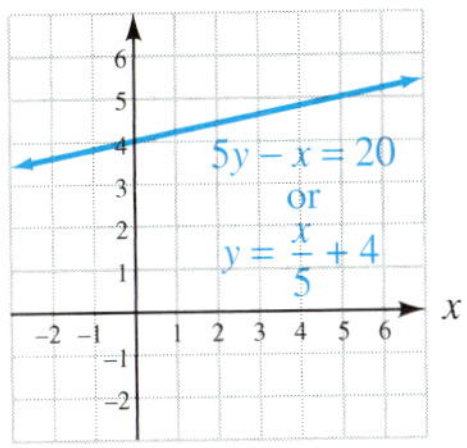

69.

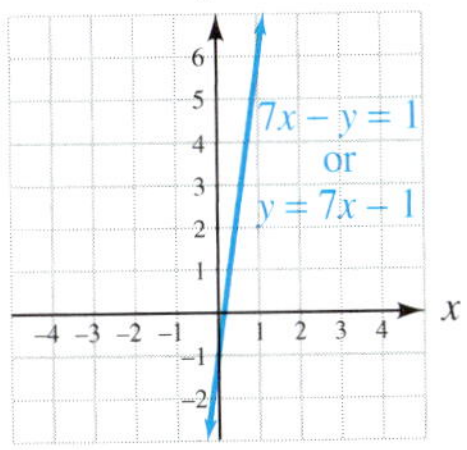

71.

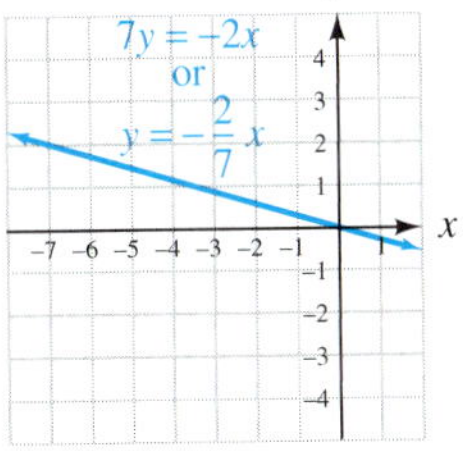

73.

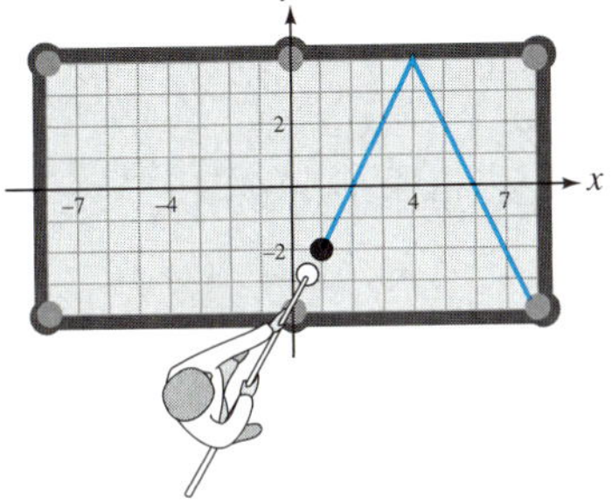

75. 125 hr

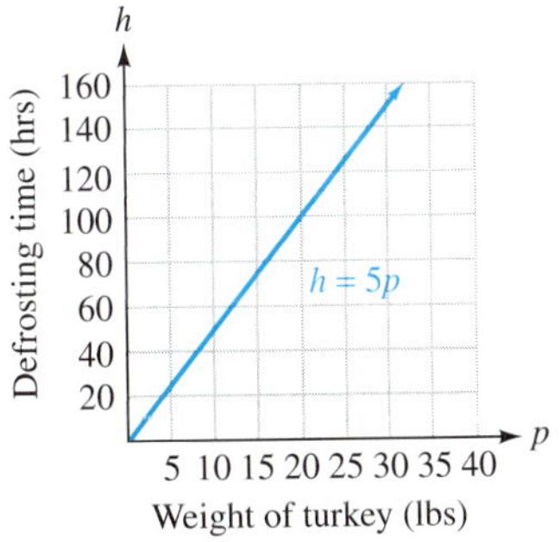

77. 3 oz

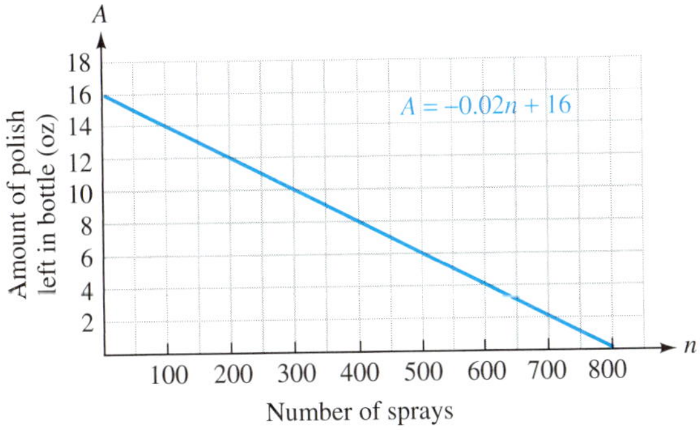

79. About \$95

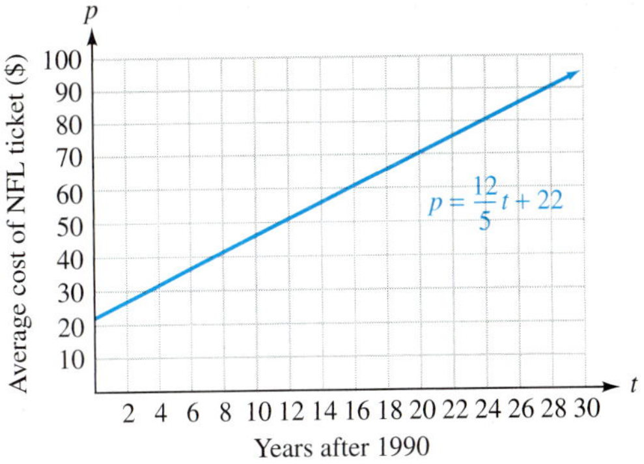

81. About 180

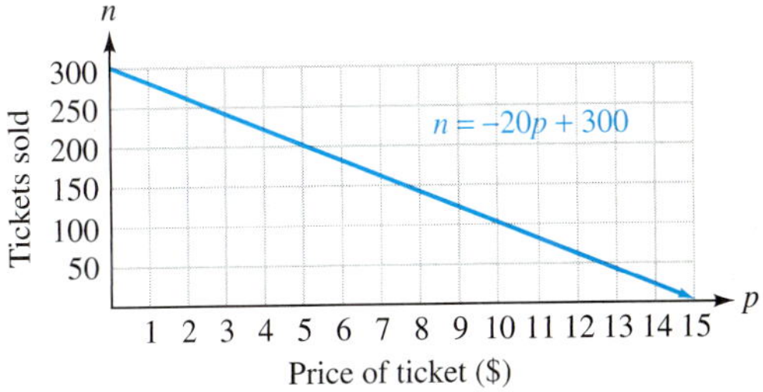

91. $5 + 4c$ **93.** 904.8 ft^3

Study Set Section 3.3 (page 229)

1. x-intercept **3.** horizontal, vertical **5. a.** 0, y **b.** 0, x

7. a. y-intercept: (0, 40); \$40,000

b. x-intercept: (30, 0); 30 years after purchase **9.** $y = 0; x = 0$

11. x-intercept: (4, 0), y-intercept: (0, 3)

13. x-intercept: $(-5, 0)$, y-intercept: $(0, -4)$

15. y-intercept: (0, 2)

17. x-intercept: $\left(-2\frac{1}{2}, 0\right)$; y-intercept: $\left(0, \frac{2}{3}\right)$ (Answers may vary)

19. (3, 0); (0, 8) **21.** $(4, 0); (0, -14)$ **23.** $(-2, 0); \left(0, -\frac{10}{3}\right)$

25. $\left(\frac{3}{2}, 0\right)$; (0, 9)

27.

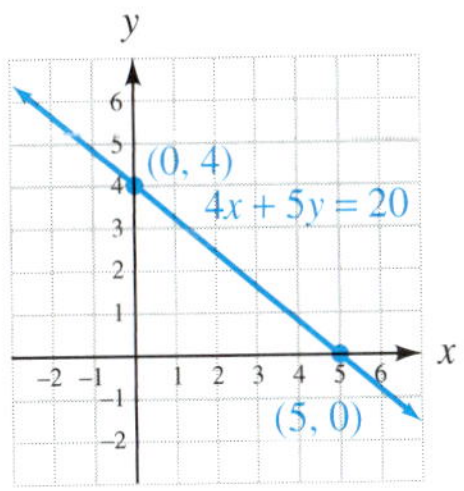

29.

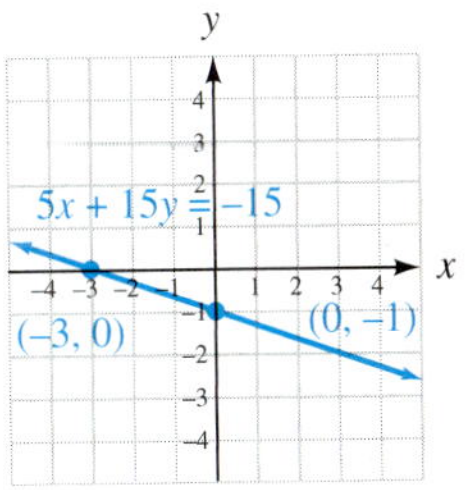

31.

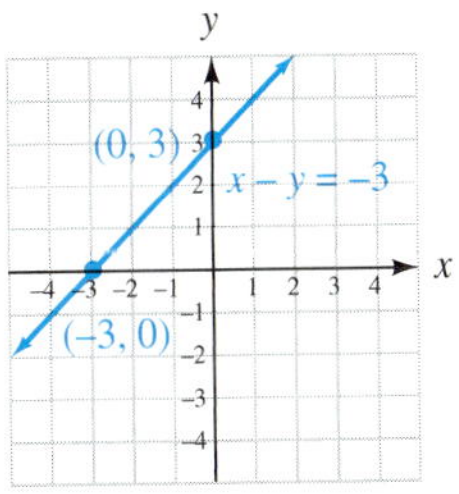

33.

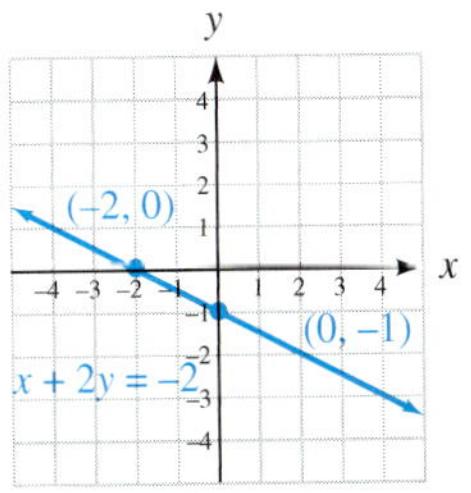

35.

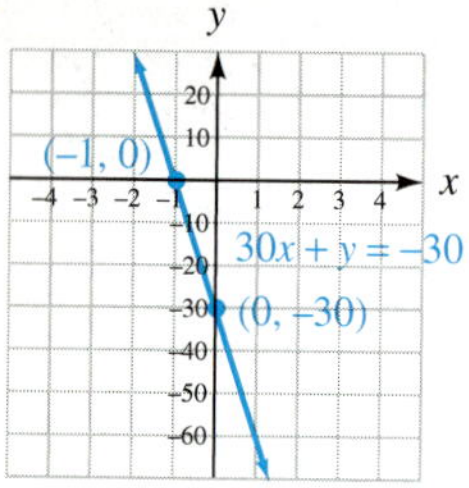

37.

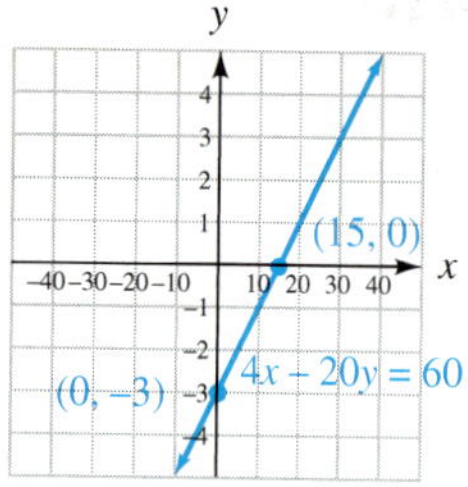

39.

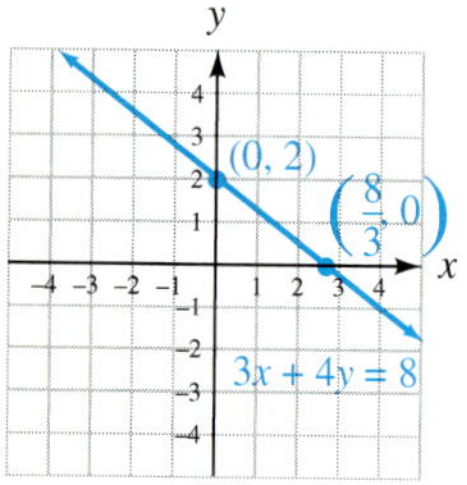

41.

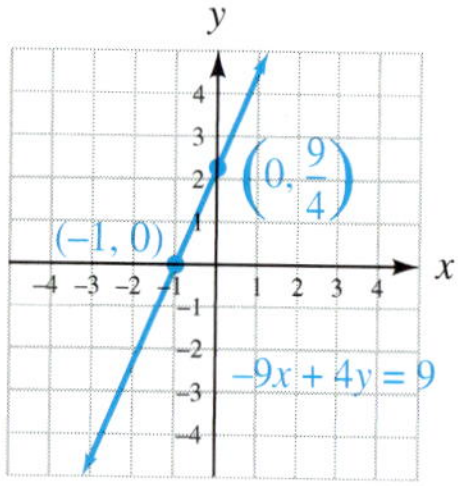

43.

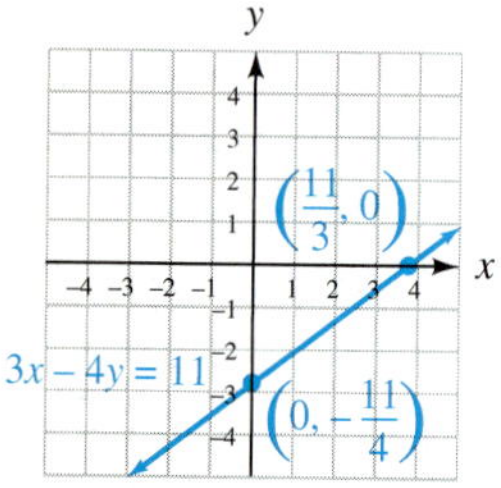

45.

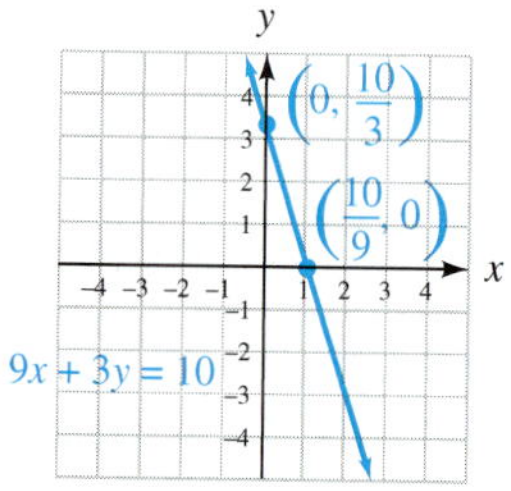

47.

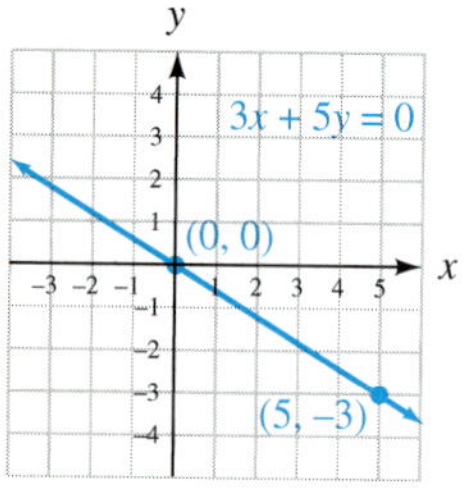

49.

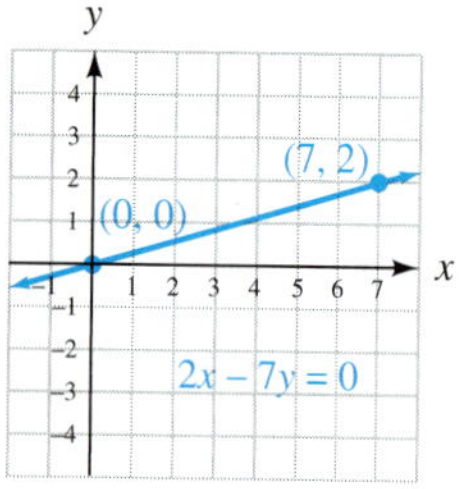

51.

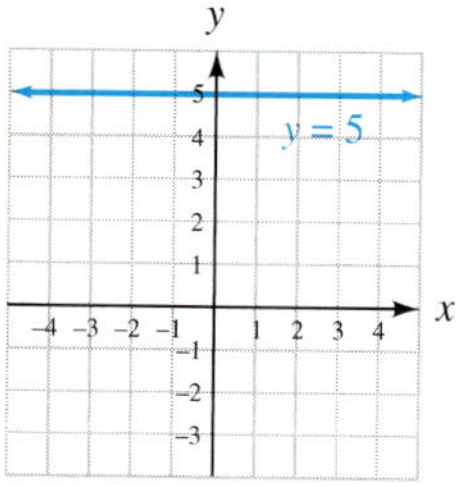

53.

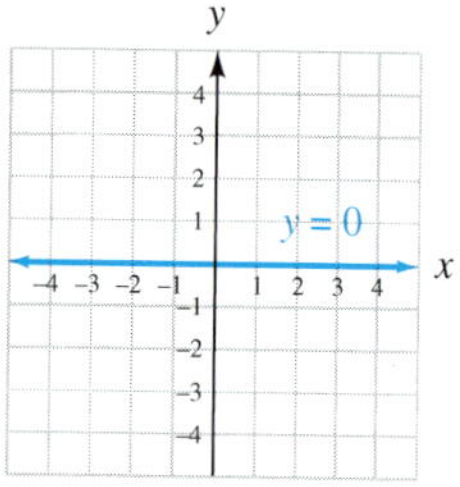

55.

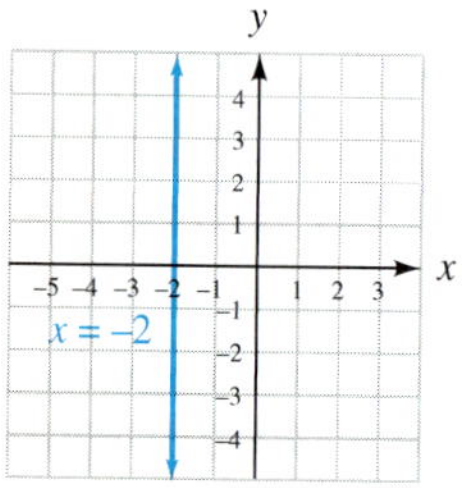

57.

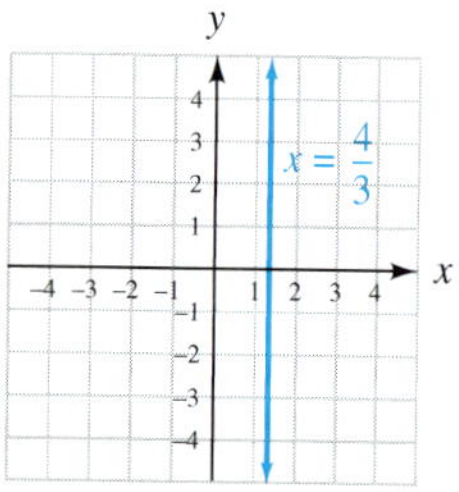

59. $y = 2$

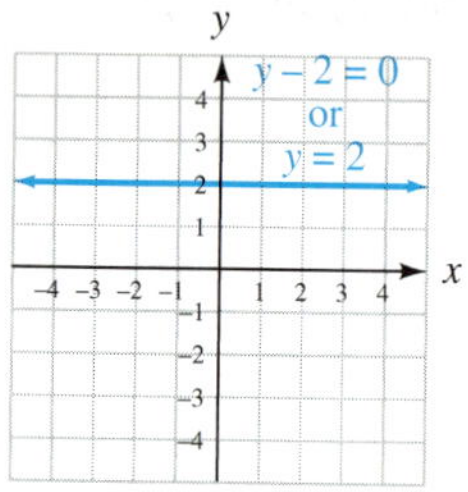

61. $x = 1.5$

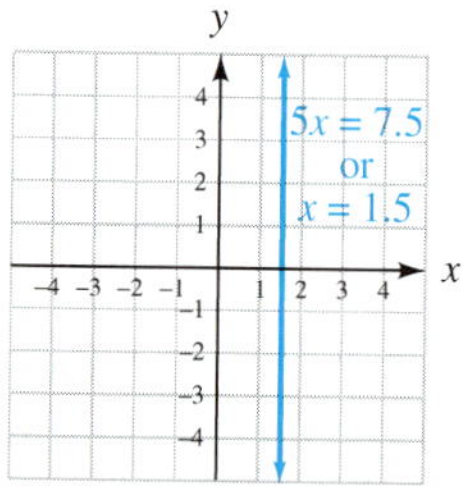

63.

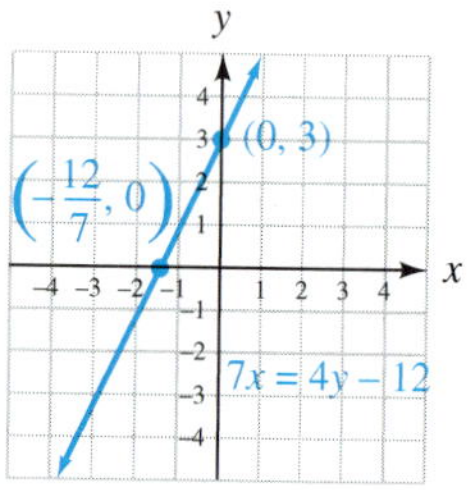

65.

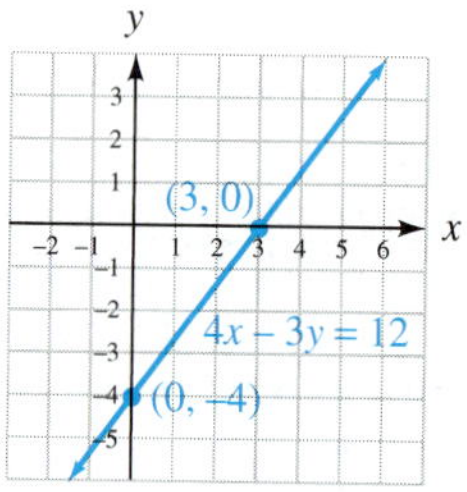

67.

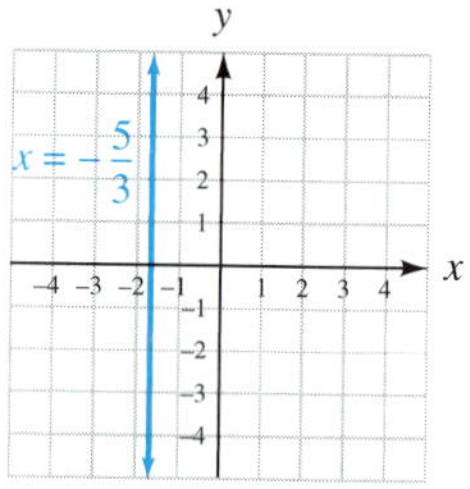

69.

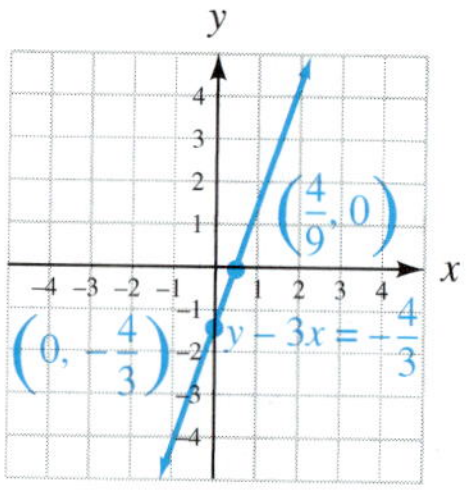

71.

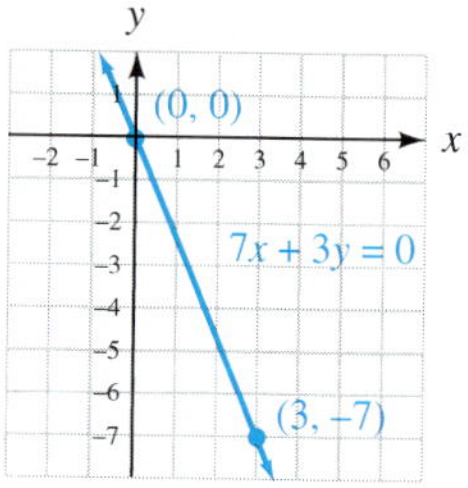

73.

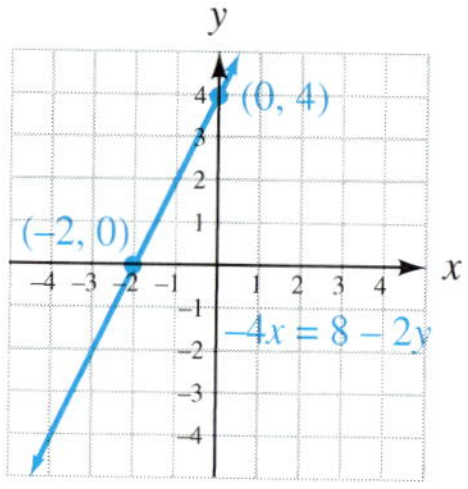

75.

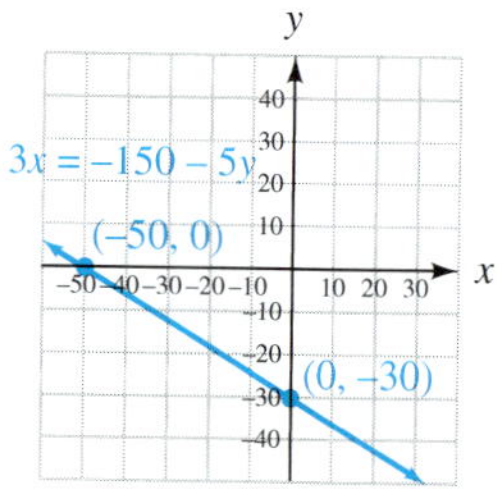

77. $y = -1$

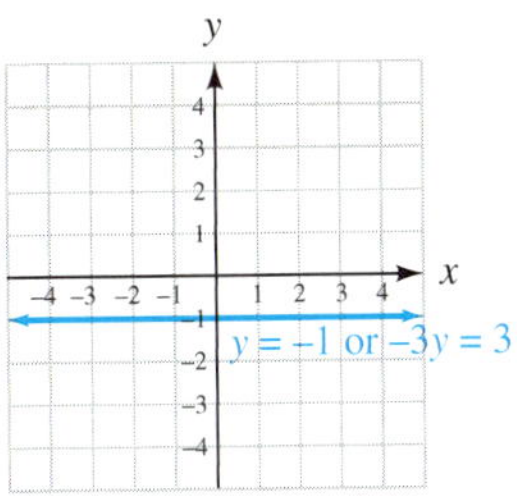

79. a. About $-270°C$ **b.** 0 milliliters

81.

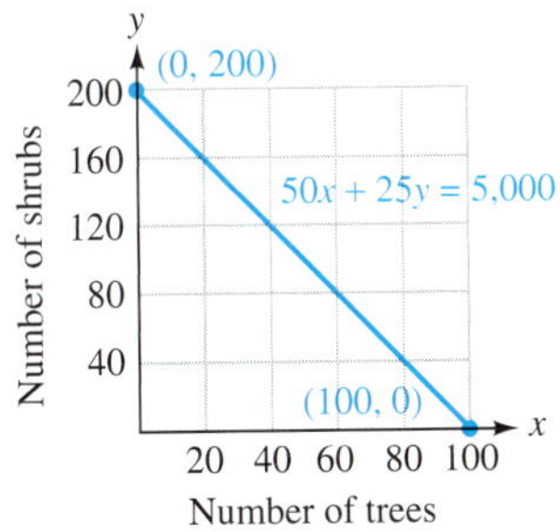

a. If only shrubs are purchased, he can buy 200.
b. If only trees are purchased, he can buy 100. **87.** $\frac{1}{5}$ **89.** $2x - 6$

Study Set Section 3.4 (page 242)

1. slope, ratio **3.** change **5. a.** Line 2 **b.** Line 1 **c.** Line 4 **d.** Line 3 **7. a.** Line 1 **b.** Line 1 **9.** same **11. a.** 0 **b.** Undefined **c.** $\frac{1}{4}$ **d.** 2 **13.** 40% **15. a.** $-\frac{1}{6}$ **b.** $\frac{8}{7}$ **c.** 1 **17. a.** $m = \frac{y_2 - y_1}{x_2 - x_1}$ **b.** sub, sub, over (divided by), two, one **19.** 1 **21.** $\frac{2}{3}$ **23.** $\frac{4}{3}$ **25.** -2 **27.** 0 **29.** $-\frac{1}{5}$ **31.** $\frac{1}{2}$ **33.** 1 **35.** -3 **37.** $\frac{5}{4}$ **39.** $-\frac{1}{2}$ **41.** $\frac{3}{5}$ **43.** 0 **45.** Undefined **47.** $-\frac{2}{3}$ **49.** -4.75 **51.** 0 **53.** $\frac{7}{5}$ **55.** $-\frac{2}{5}$ **57.** $m = \frac{3}{4}$ **59.** $m = 0$ **61.** 0 **63.** 0 **65.** Undefined **67.** Undefined **69.** 0 **71.** Undefined **73.** Parallel **75.** Perpendicular **77.** Neither **79.** Perpendicular **81.** Parallel **83.** Neither **85.** $\frac{5}{9}$ **87.** $-\frac{2}{3}$ **89.** -1 **91.** $\frac{1}{2}$ **93.** $-\frac{2}{5}$ **95.** $\frac{1}{20}$; 5% **97.** 4%, 8%, 12% **99.** Front: $\frac{3}{2}$; side: $\frac{3}{5}$ **101.** -875 gal per hr **103.** 380 lb per yr **109.** 40 lb licorice; 20 lb gumdrops

Study Set Section 3.5 (page 254)

1. slope–intercept **3. a.** No **b.** No **c.** Yes **d.** No **5. a.** $y = 2x + 8$ **b.** $y = -5x - 3$ **c.** $y = \frac{x}{3} - 1$ **d.** $y = \frac{9}{5}x + 4$ **7.** $-2x$, $5y$, 5, 5, 5, $-\frac{2}{5}$, 3, $-\frac{2}{5}$, (0, 3) **9.** -2, -3 **11.** 4, (0, 2) **13.** -5, (0, -8) **15.** 4, (0, -9) **17.** -1, (0, 11) **19.** -20, (0, 1) **21.** $\frac{1}{2}$, (0, 6) **23.** $\frac{1}{4}$, $\left(0, -\frac{1}{2}\right)$ **25.** -5, (0, 0) **27.** $\frac{2}{3}$, (0, 0) **29.** 1, (0, 0) **31.** 0, (0, -2) **33.** 0, $\left(0, -\frac{2}{5}\right)$ **35.** -1, (0, 8) **37.** $\frac{1}{6}$, (0, -1) **39.** -2, (0, 7) **41.** $-\frac{3}{2}$, (0, 1) **43.** $-\frac{2}{3}$, (0, 2) **45.** $\frac{3}{5}$, (0, -3) **47.** $\frac{4}{3}$, (0, -4) **49.** 1, $\left(0, -\frac{11}{6}\right)$

51. $y = 5x - 3$

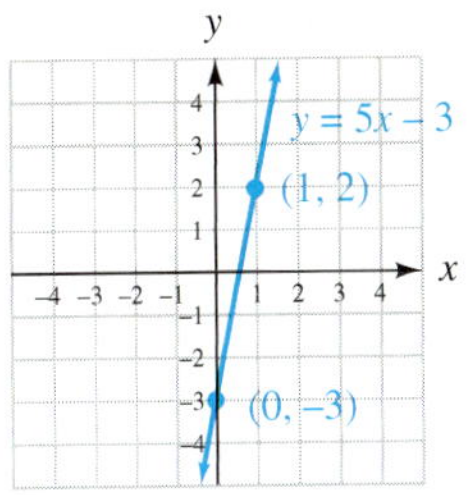

53. $y = -3x + 6$

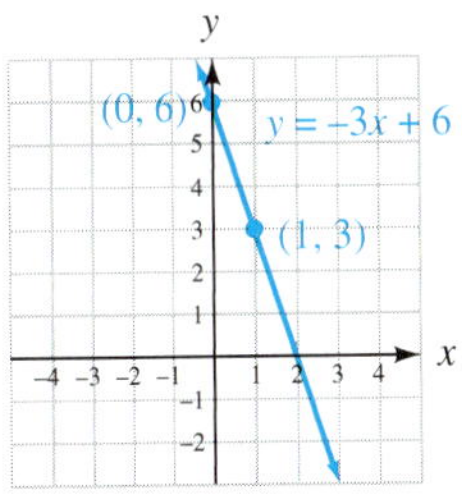

55. $y = \frac{1}{4}x - 2$

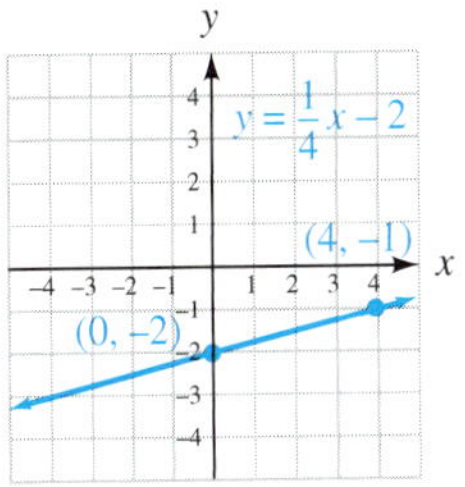

57. $y = -\frac{8}{3}x + 5$

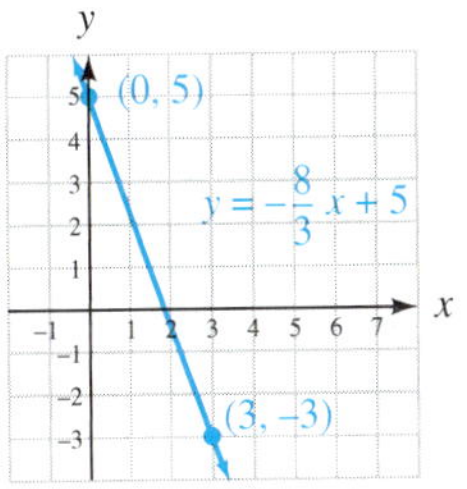

59. $y = 5x - 1$ **61.** $y = -2x + 3$ **63.** $y = \frac{4}{5}x - 2$ **65.** $y = -\frac{5}{3}x + 2$ **67.** 3, (0, 3)

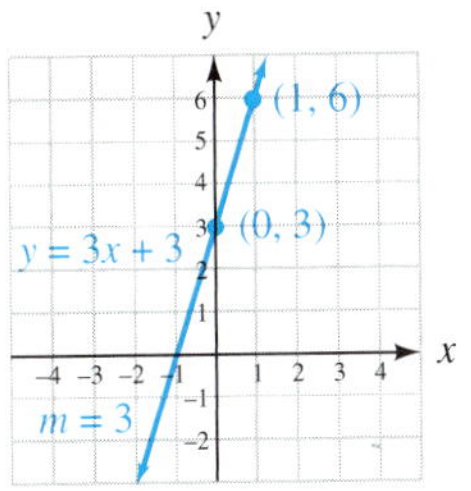

69. $-\frac{1}{2}$, (0, 2)

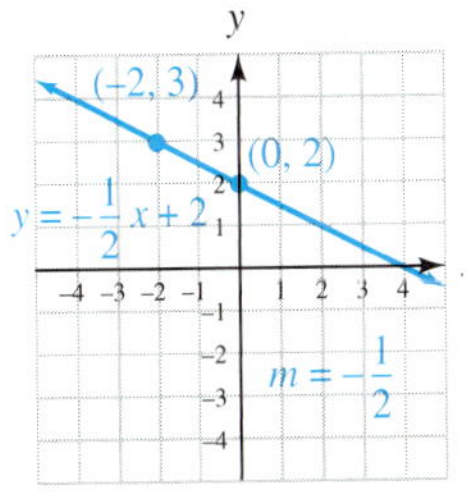

71. $-3, (0, 0)$

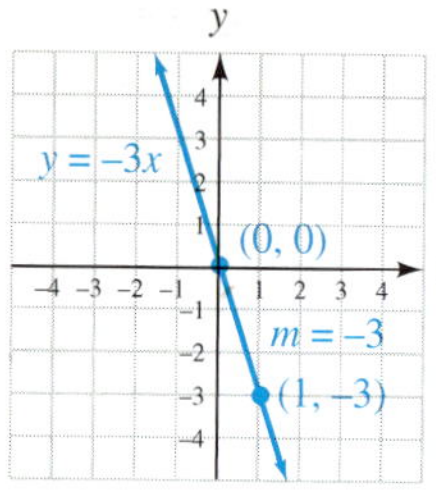

73. $-4, (0, -4)$

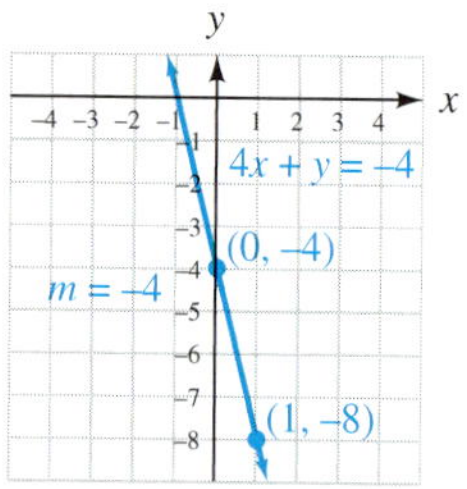

75. $-\frac{3}{4}, (0, 4)$

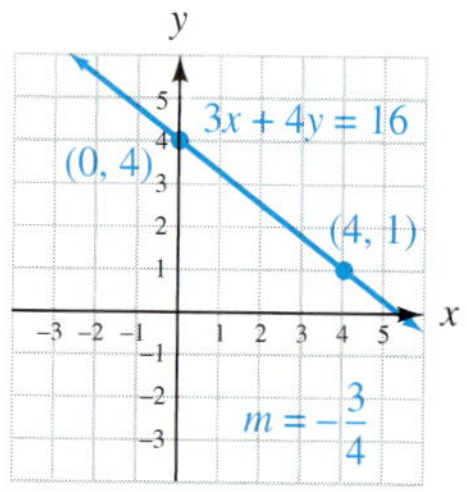

77. $2, (0, -1)$

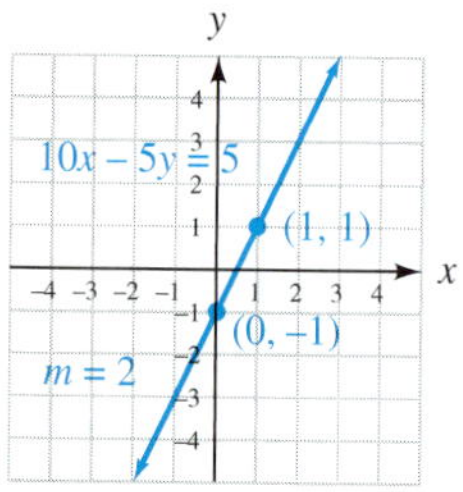

79. Parallel **81.** Perpendicular **83.** Neither **85.** Perpendicular **87.** Parallel **89.** Perpendicular **91. a.** $c = 2{,}000h + 5{,}000$ **b.** \$21,000 **93.** $F = 5t - 10$ **95.** $c = -20m + 500$

97. a. $c = 5x + 20$

b and c.

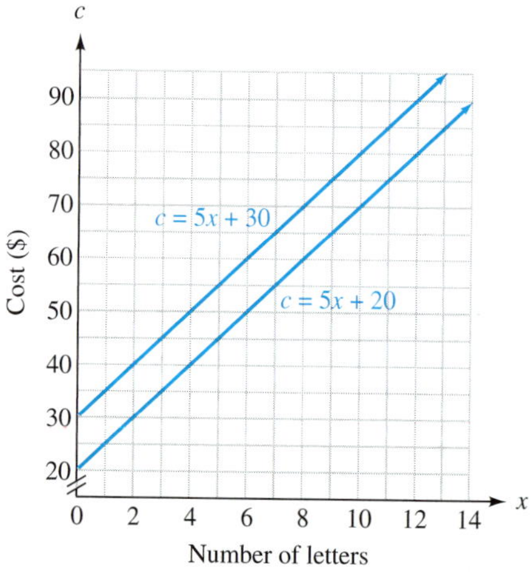

99. $c = 0.89t + 16.63$ **103.** 42 ft, 45 ft, 48 ft, 51 ft

Study Set Section 3.6 (page 264)

1. point–slope, sub, times, minus, one **3. a.** point–slope **b.** slope–intercept **5. a.** $(-2, -3)$ **b.** $\frac{5}{6}$ **c.** $y + 3 = \frac{5}{6}(x + 2)$ **7.** (67, 170), (79, 220) **9.** 5, −1, +, 2, 3 **11.** point–slope, slope–intercept **13.** $y - 1 = 3(x - 2)$ **15.** $y + 1 = \frac{4}{5}(x + 5)$ **17.** $y = 2x - 1$ **19.** $y = -5x - 37$ **21.** $y = -3x$ **23.** $y = \frac{1}{5}x - 1$ **25.** $y = -\frac{4}{3}x + 4$ **27.** $y = -\frac{11}{6}x - \frac{7}{3}$ **29.** $y = 2x + 5$ **31.** $y = -\frac{1}{2}x + 1$ **33.** $y = 5$ **35.** $y = \frac{1}{10}x + \frac{1}{2}$ **37.** $x = -8$ **39.** $y = \frac{1}{2}x$ **41.** $x = 4$ **43.** $y = 5$

45.

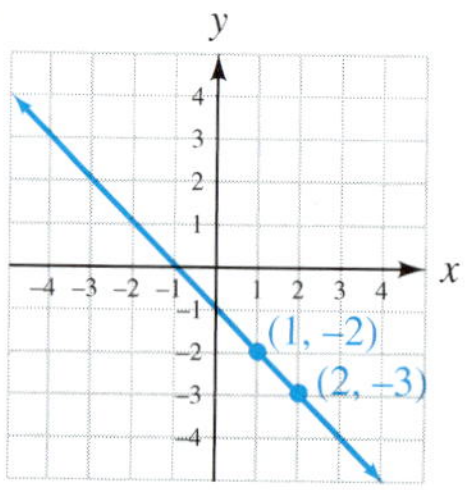

47.

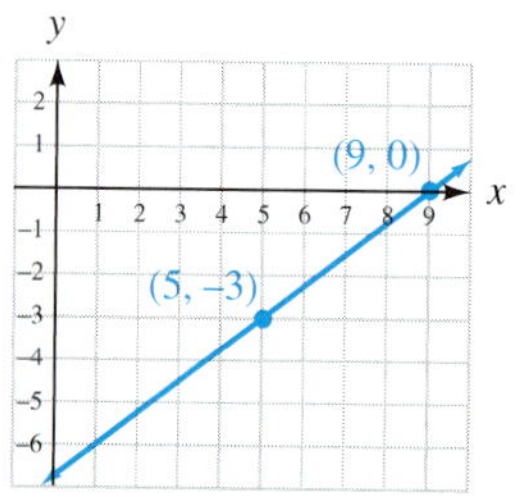

49.

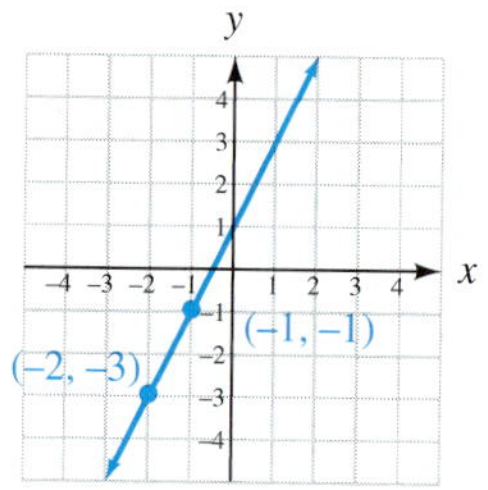

51.

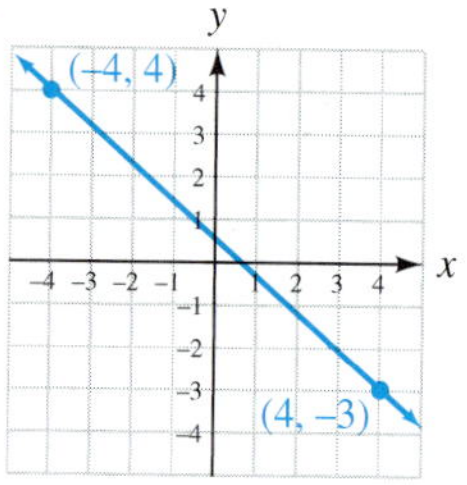

53. $y = \frac{1}{4}x - \frac{5}{4}$ **55.** $y = 12$ **57.** $y = -\frac{2}{3}x + 2$ **59.** $y = 8x + 4$ **61.** $x = -3$ **63.** $y = 7x$ **65.** $y = -4x - 9$ **67.** $y = \frac{2}{7}x - 2$ **69.** $y = \frac{1}{10}x$ **71.** $x = -\frac{1}{8}$ **73.** $h = 3.9r + 28.9$ **75.** $y = -\frac{2}{5}x + 4, y = -7x + 70, x = 10$ **77. a.** $y = -40m + 920$ **b.** 440 yd^3 **79.** $l = \frac{25}{4}r + \frac{1}{4}$ **81. a.** $y = -\frac{3}{10}x + \frac{283}{10}$ or $y = -0.3x + 28.3$ **b.** 16.3 gal **87.** 17 in. by 39 in.

Study Set Section 3.7 (page 275)

1. inequality **3.** satisfies **5.** half-planes **7.** Yes **9.** dashed, solid **11.** The half-plane opposite that in which the test point lies **13. a.** Yes **b.** No **c.** No **d.** Yes **15. a.** Is less than **b.** Is greater than or equal to **c.** Is less than or equal to **d.** Is possibly greater than **17.** $=, <$ **19.** Yes **21.** No **23.** No **25.** Yes

27.

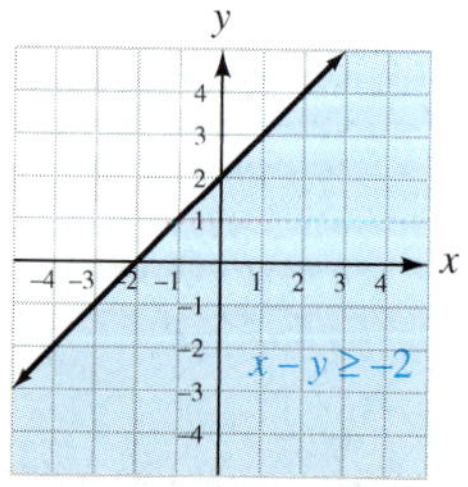

29.

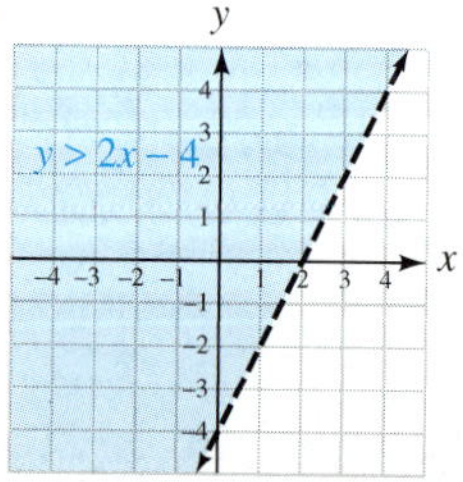

31.
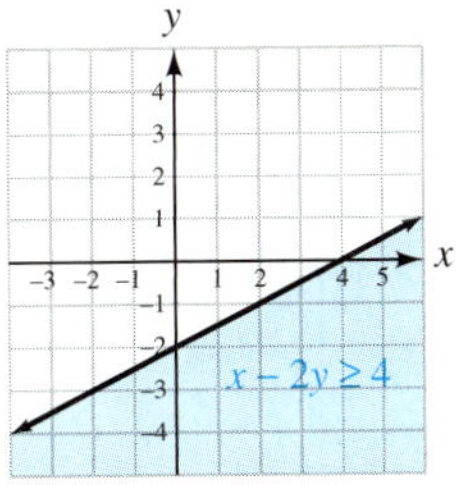

33.
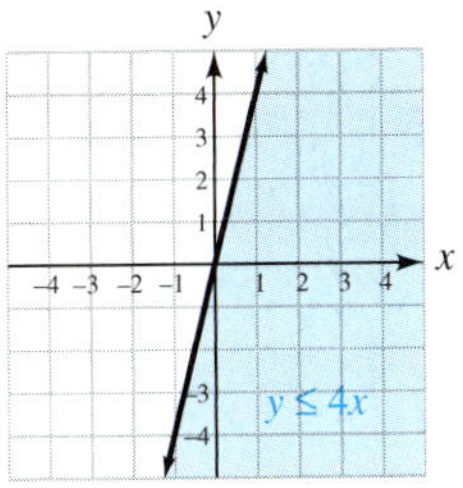

35.
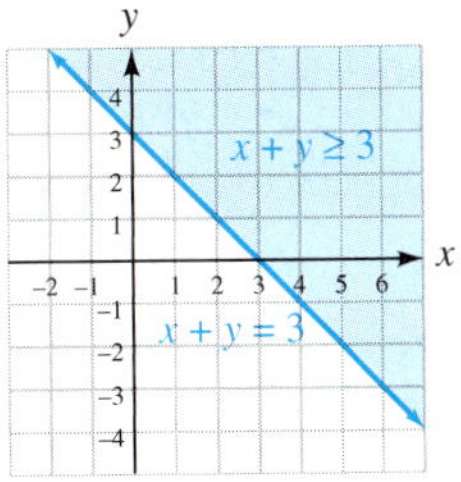

37.
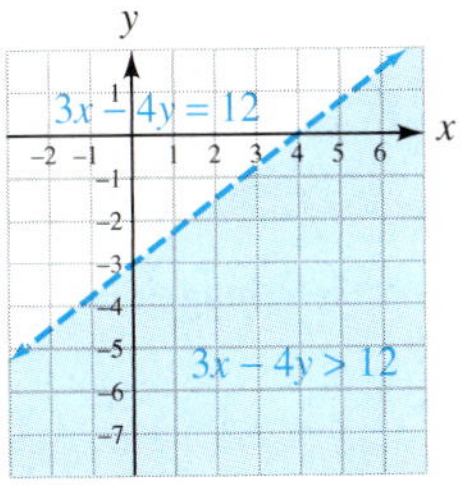

39.
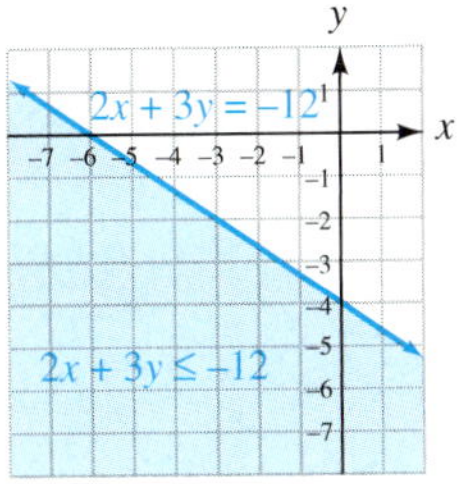

41.
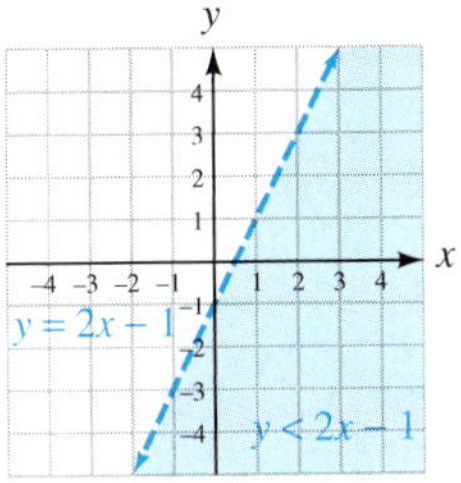

43.
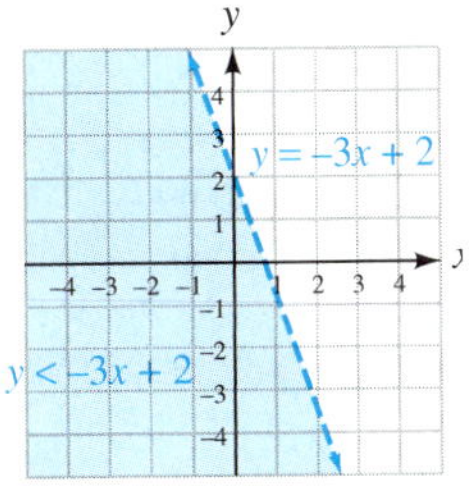

45.
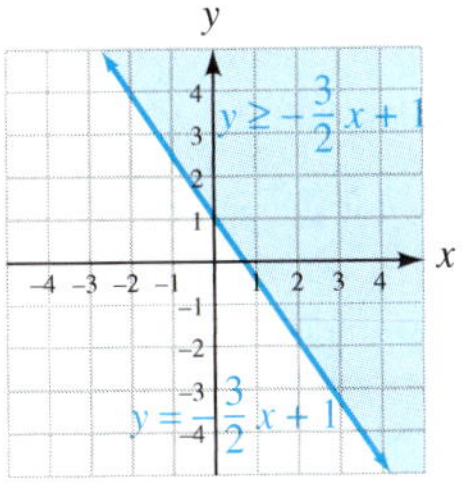

47.
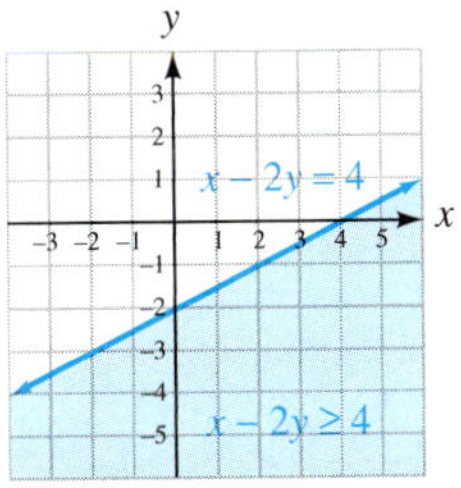

49.
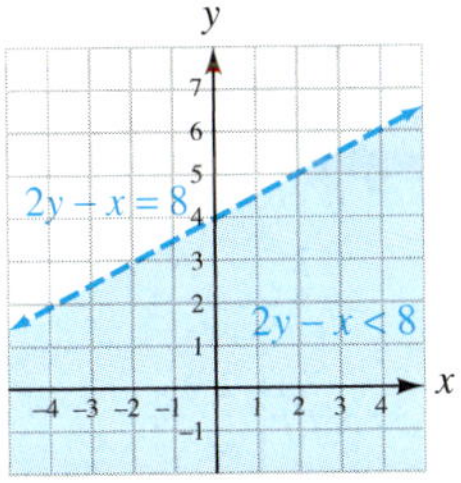

51.
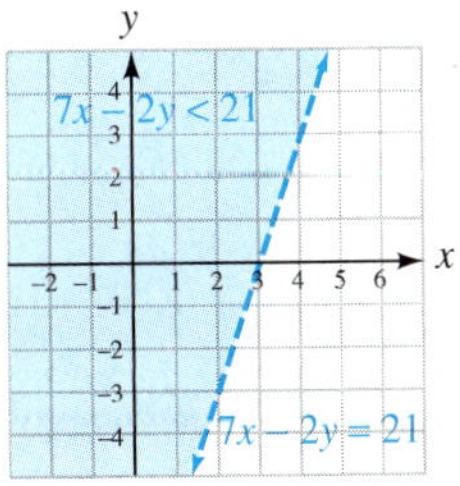

53.
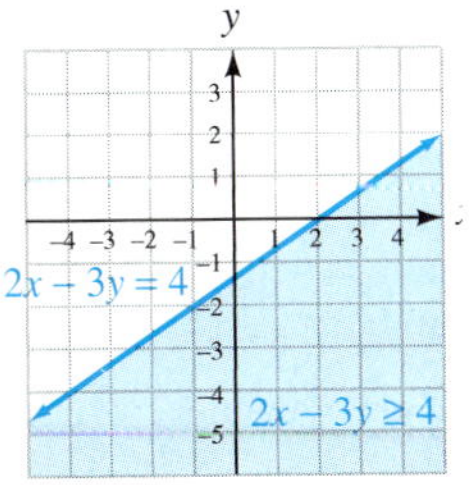

55.
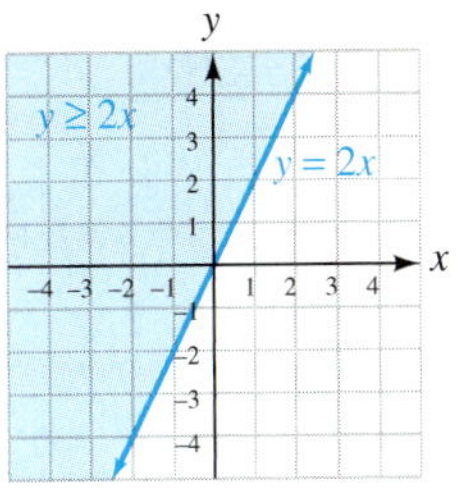

57.
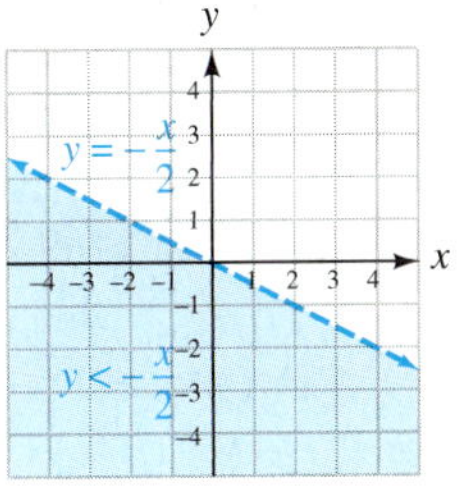

59.
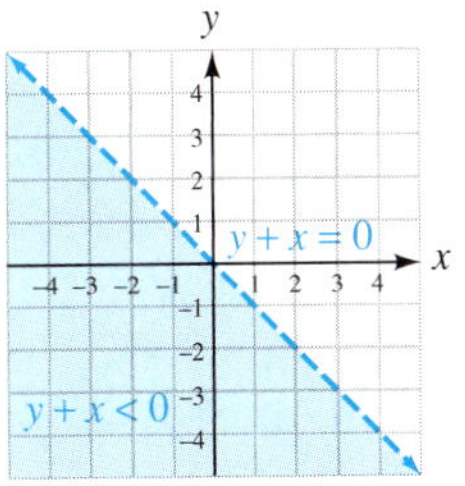

61.
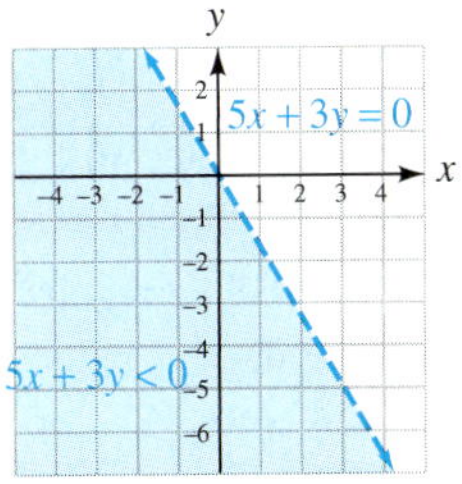

63.
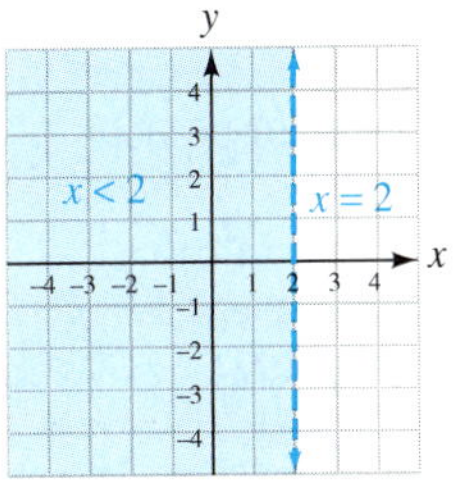

65.
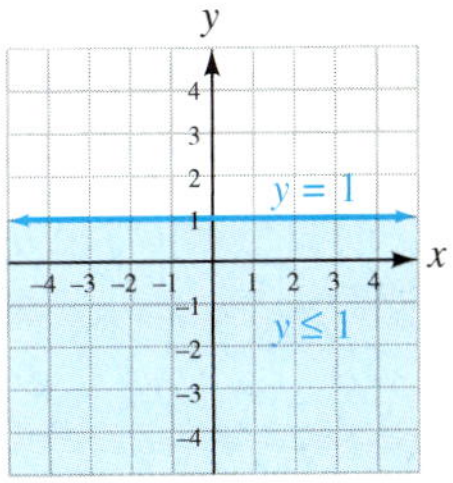

67.
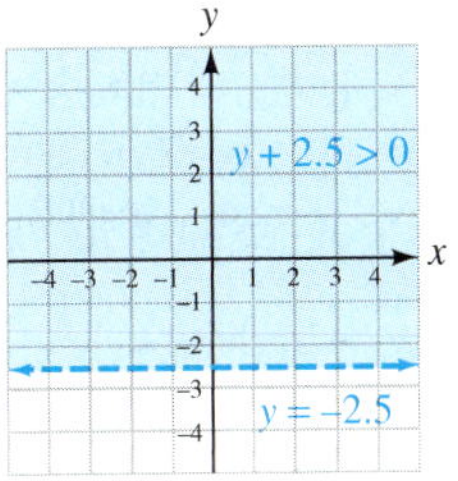

69.
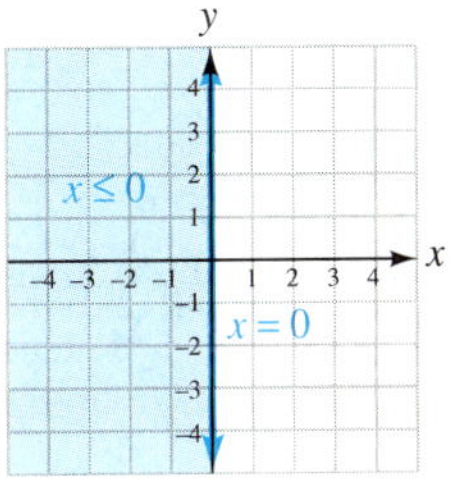

71. No **73.** ii

75. (10, 10), (20, 10), (10, 20); Answers may vary

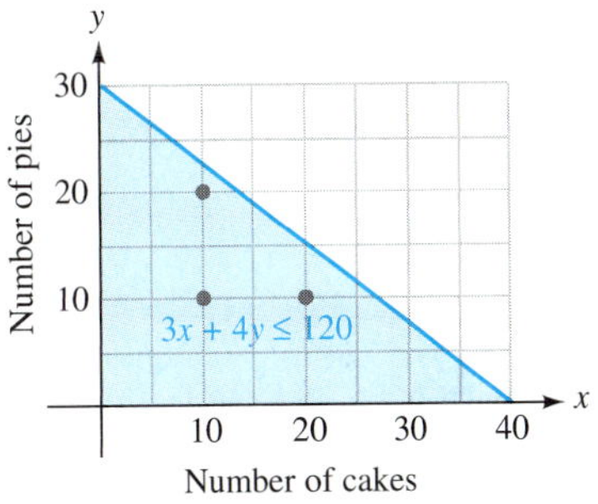

77. (40, 30), (30, 40), (40, 20); Answers may vary

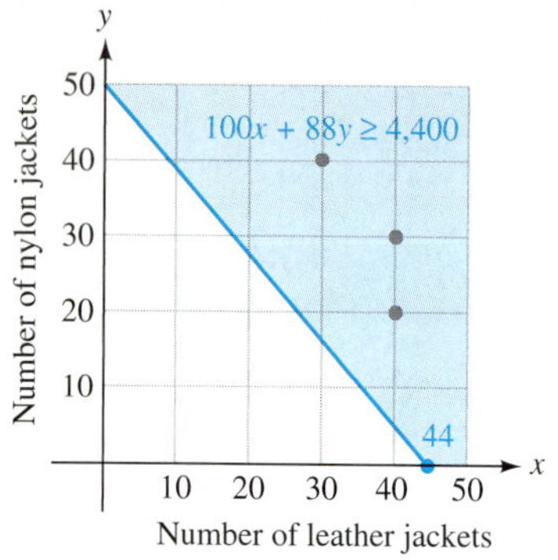

83. $t = \frac{A - P}{Pr}$
85. $15x + 22$

Study Set Section 3.8 (page 288)

1. relation **3.** domain, range **5.** value

7.

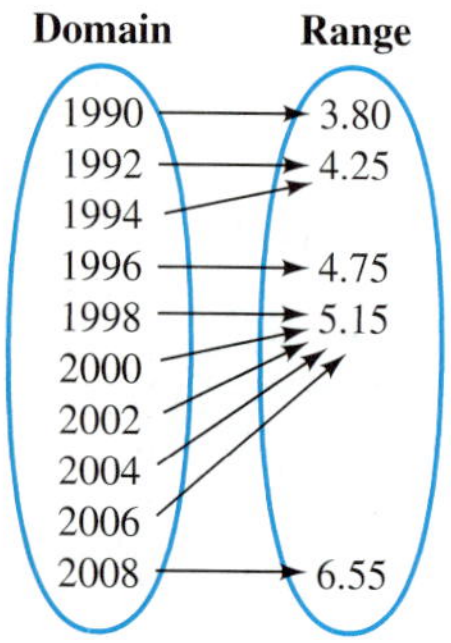

9. 33 **11.** of **13.** 4, 5, (4, 5) **15.** Domain: $\{-6, -1, 6, 8\}$; range: $\{-10, -5, -1, 2\}$ **17.** Domain: $\{-8, 0, 6\}$; range: $\{9, 50\}$ **19.** Yes; domain: $\{10, 20, 30\}$; range: $\{20, 40, 60\}$ **21.** No; (4, 2), (4, 4), (4, 6) **23.** Yes; domain: $\{1, 2, 3, 4, 5\}$; range: $\{7, 8, 15, 16, 23\}$ **25.** No; $(-1, 0)$, $(-1, 2)$ **27.** No; (3, 4), $(3, -4)$ or (4, 3), $(4, -3)$ **29.** Yes; domain: $\{-3, 1, 5, 6\}$; range: $\{-8, 0, 4, 9\}$ **31. a.** 3 **b.** -9 **c.** 0 **d.** 199 **33. a.** 0.32 **b.** 18 **c.** 2,000,000 **d.** $\frac{1}{32}$ **35. a.** 7 **b.** 14 **c.** 0 **d.** 1 **37. a.** 0 **b.** 990 **c.** -24 **d.** 210 **39. a.** 36 **b.** 0 **c.** 9 **d.** 4 **41.** 1.166

43.

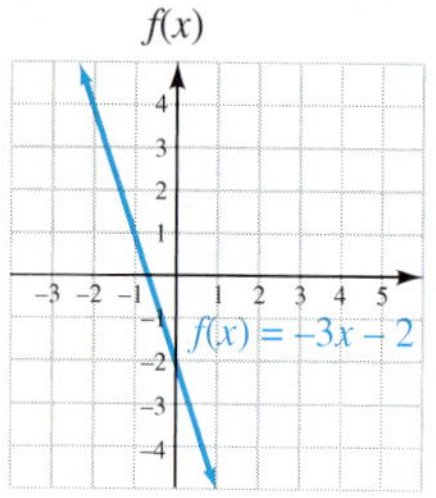

x	f(x)
−2	4
−1	1
0	−2
1	−5

45.

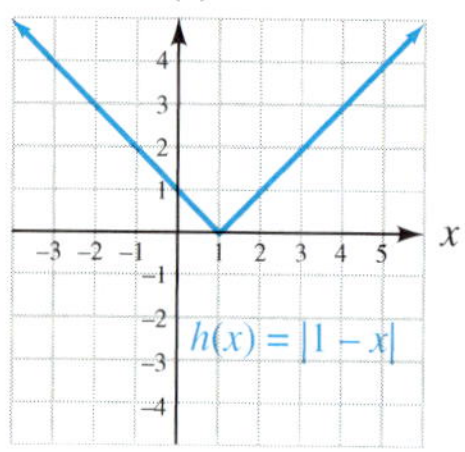

x	h(x)
−2	3
−1	2
0	1
1	0
2	1
3	2
4	3

47.

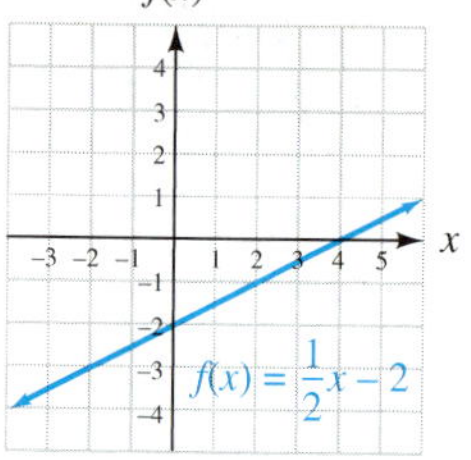

49.

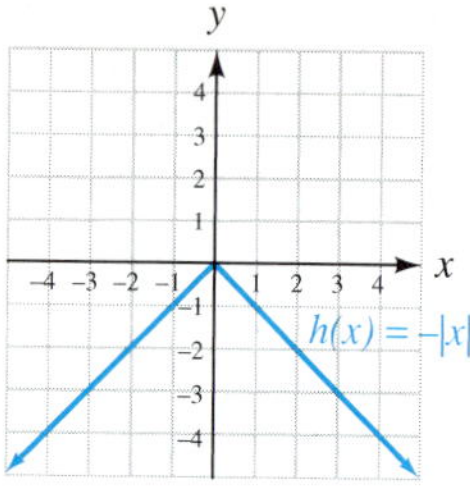

51. Yes **53.** No; (3, 4), $(3, -1)$ (Answers may vary) **55.** No; (0, 2), $(0, -4)$ (Answers may vary) **57.** No; (3, 0), (3, 1) (Answers may vary) **59.** $f(x) = |x|$ **61.** \$900 **63.** 78.5 ft^2, 1,256.6 ft^2 **69.** 80 lb

Chapter 3 Review (page 293)

1.

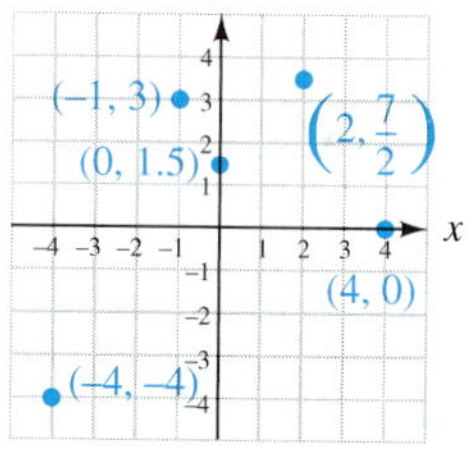

2. (158, 21.5) **3.** Quadrant III **4.** (0, 0) **5.** (1, 4); 36 square units **6. a.** 2,500; week 2 **b.** 1,000 **c.** 1st week and 5th week **7.** Yes

8.

x	y	(x, y)
−2	−6	(−2, −6)
−8	3	(−8, 3)

9. $y = x^2 + 1$ and $y - x^3 = 0$ **10. a.** True **b.** False

11.

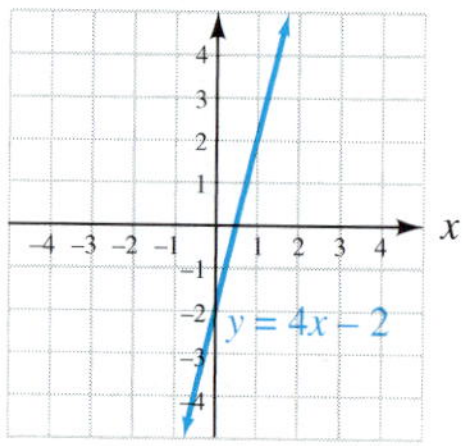

12.

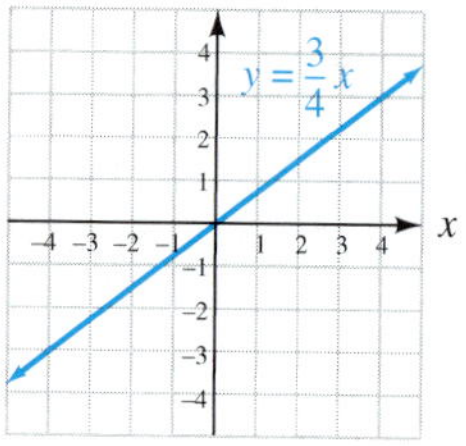

13.

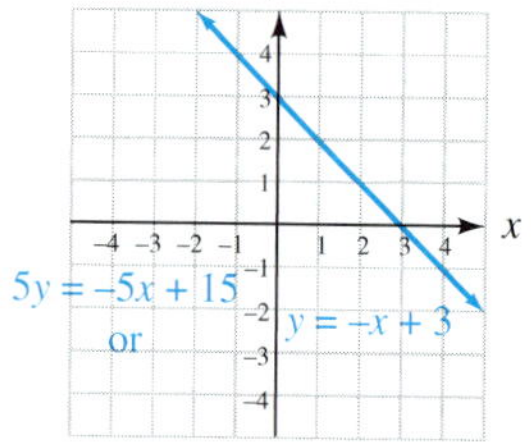

14.

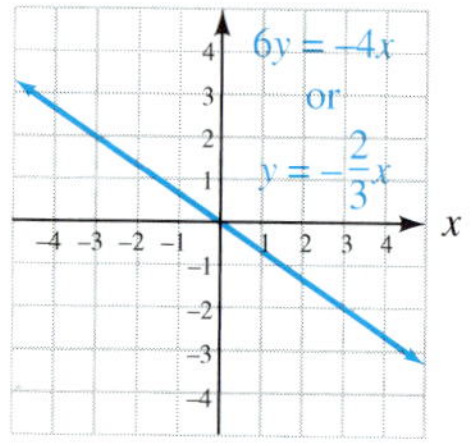

15. About \$190 new

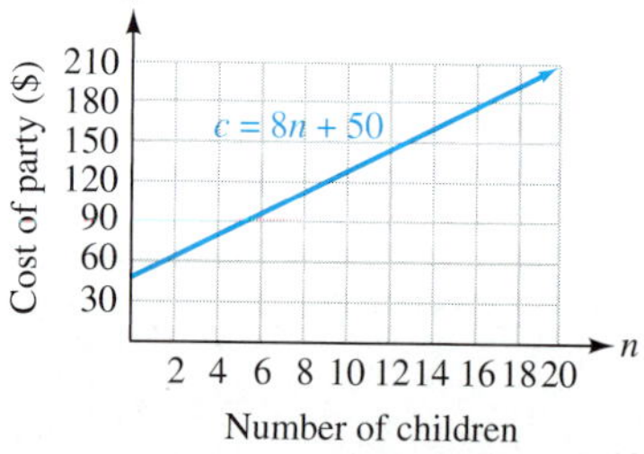

16. a. False **b.** True **17.** $(-3, 0)$, $(0, 2.5)$
18. $(0, 25{,}000)$; the equipment was originally valued at \$25,000. $(10, 0)$; in 10 years, the sound equipment had no value.
19. x-intercept: $(-2, 0)$; y-intercept: $(0, 4)$

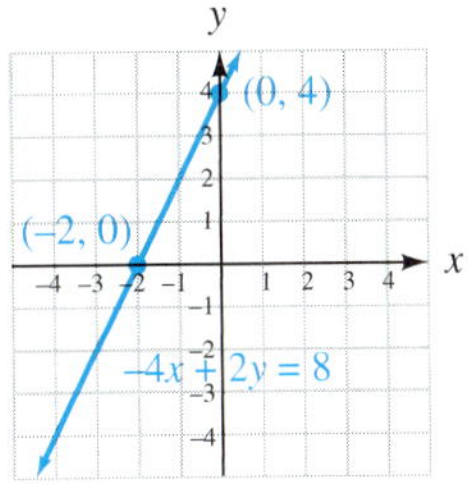

20. x-intercept: $\left(\frac{13}{5}, 0\right)$; y-intercept: $\left(0, -\frac{13}{4}\right)$

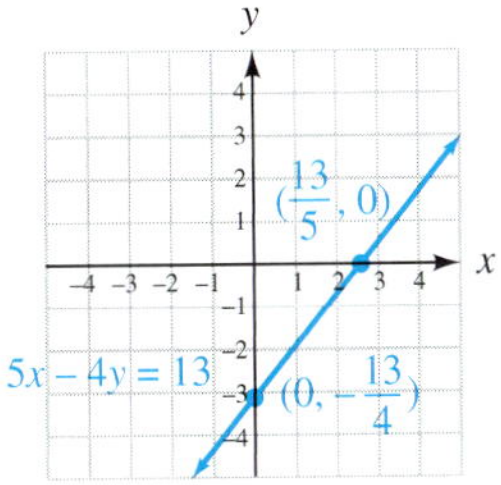

21.

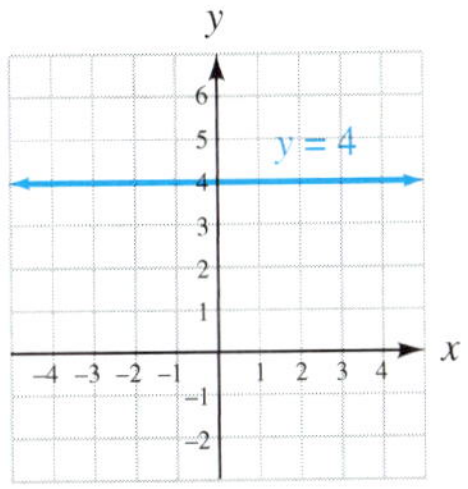

22.

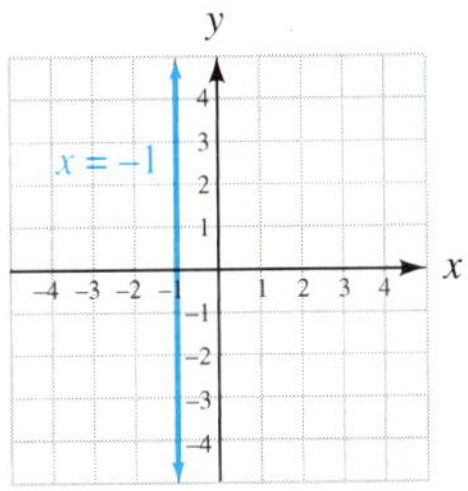

23. $\frac{1}{4}$ **24.** $-\frac{7}{8}$ **25.** -7 **26.** $-\frac{3}{2}$
27.

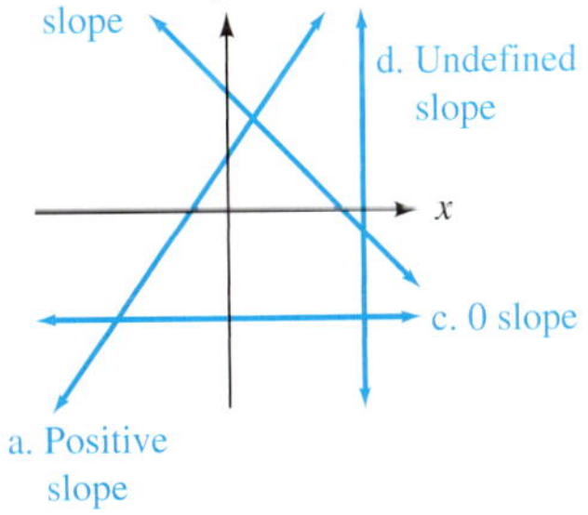

28. $\frac{3}{4}$ **29.** 8.3% **30. a.** -4.5 million people per yr **b.** 4.05 million people per yr **31.** They are neither. **32.** $-\frac{7}{5}$
33. $m = \frac{3}{4}$; y-intercept: $(0, -2)$ **34.** $m = -4$; y-intercept: $(0, 0)$
35. $m = \frac{1}{8}$; y-intercept: $(0, 10)$ **36.** $m = -\frac{7}{5}$; y-intercept: $\left(0, -\frac{21}{5}\right)$

37. $y = 4x - 1$

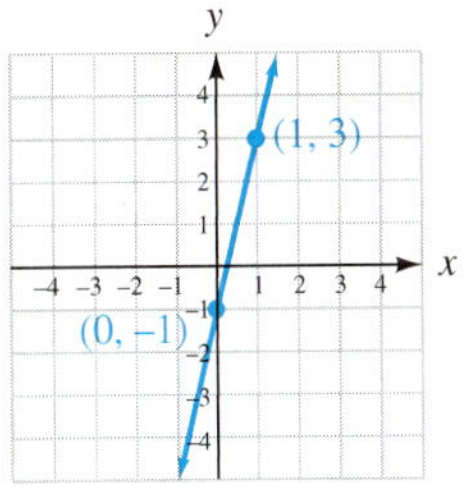

38. $y = \frac{3}{2}x - 3$
39. $m = 3$; y-intercept: $(0, -5)$

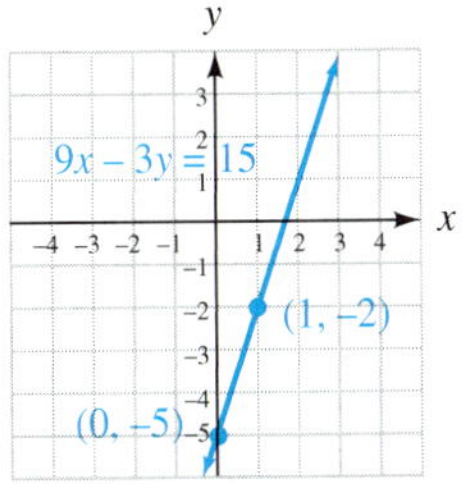

40. a. $c = 300w + 75{,}000$ **b.** 90,600 **41.** Parallel
42. Perpendicular
43. $y = 3x + 2$

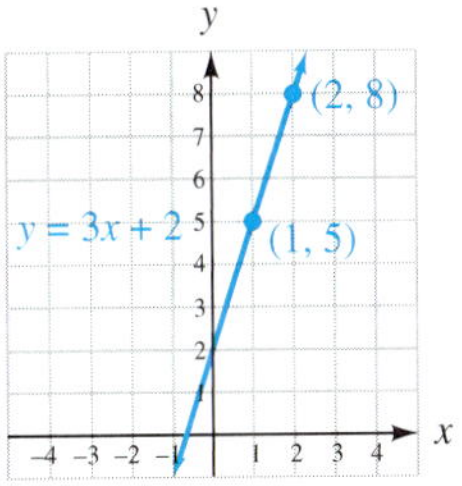

44. $y = -\frac{1}{2}x - 3$

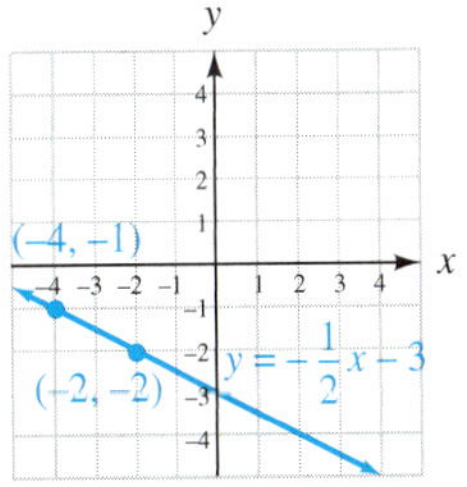

45. $y = \frac{2}{3}x + 5$ **46.** $y = -8$ **47.** $f = -35x + 450$
48. a. $y = 1{,}200x + 19{,}800$ **b.** \$43,800 **49. a.** Yes **b.** Yes **c.** Yes **d.** No **50.** $=, >$
51.

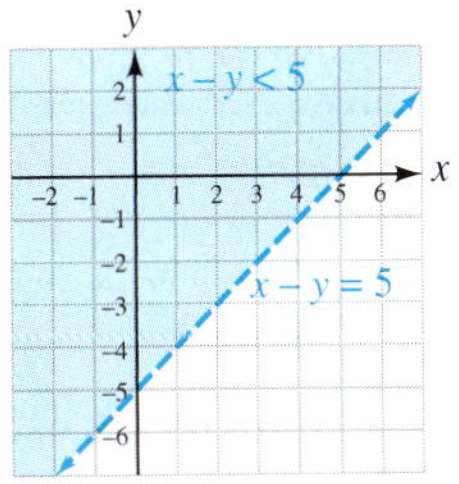

52.

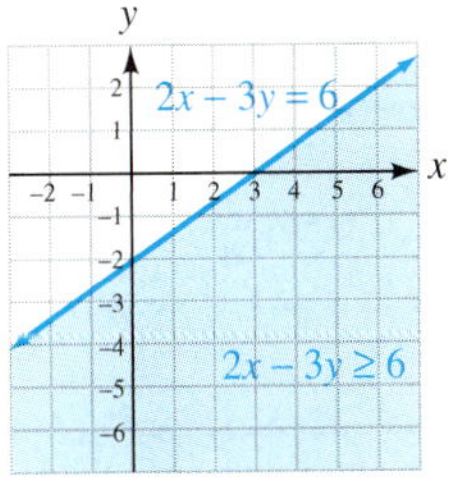

53.

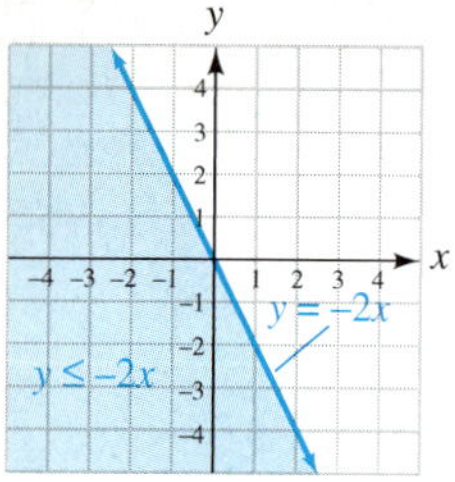

54.

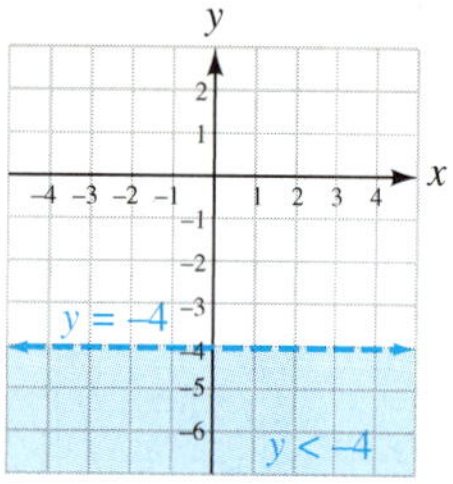

55. **a.** True **b.** False **c.** False

56. (2, 4), (5, 3), (6, 2); Answers may vary

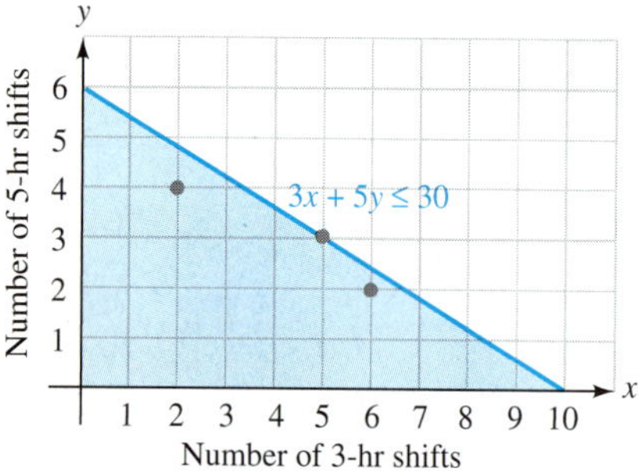

57. Domain: $\{-5, 0, 4, 7\}$; range: $\{-11, -3, 4, 9\}$

58. Domain: $\{-6, 1, 2, 15\}$; range: $\{-8, -2, 9\}$

59. Yes; domain: $\{1, 4, 8\}$; range: $\{0, 6, 9\}$

60. Yes; domain: $\{2, 3, 5, 6\}$; range: $\{1, 4\}$

61. Yes; domain: $\{3, 5, 7, 9\}$; range: $\{9, 25, 49, 81\}$

62. No; $(-1, 2)$, $(-1, 4)$ **63.** domain, range **64.** $f(x)$ **65.** -3

66. 0 **67.** 21 **68.** $-\frac{7}{4}$ **69.** -5 **70.** 37 **71.** -2

72. -8 **73.** No; (1, 0.5), (1, 4), (Answers may vary) **74.** Yes

75.

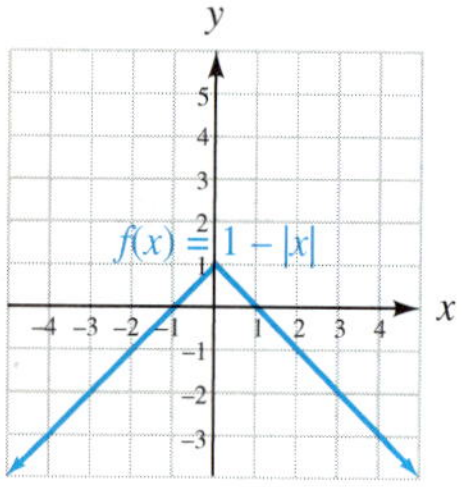

x	$f(x)$
0	1
1	0
2	−1
−1	0
−2	−1
−3	−2

76. 1,004.8 in.^{3}

Chapter 3 Test (page 302)

1. **a.** axis, axis **b.** solution **c.** linear **d.** slope **e.** function

2. 10 **3.** 60 **4.** 1 day before and the 3rd day of the holiday

5. 50 dogs were in the kennel when the holiday began.

6.

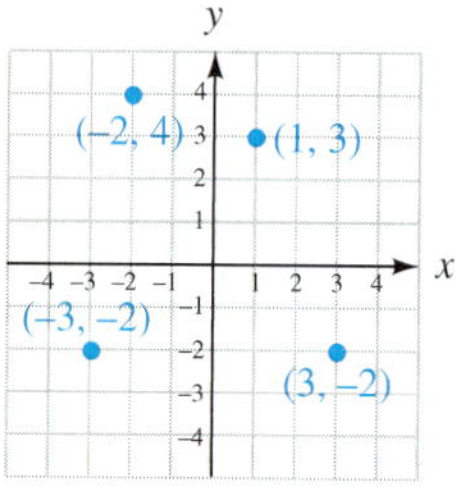

7. $A(2, 4)$, $B(-3, 3)$, $C(-2, -3)$, $D(4, -3)$ **8.** **a.** III **b.** IV

9. Yes **10.**

x	y	(x, y)
2	1	(2, 1)
−6	3	(−6, 3)

11. **a.** False **b.** True **12.**

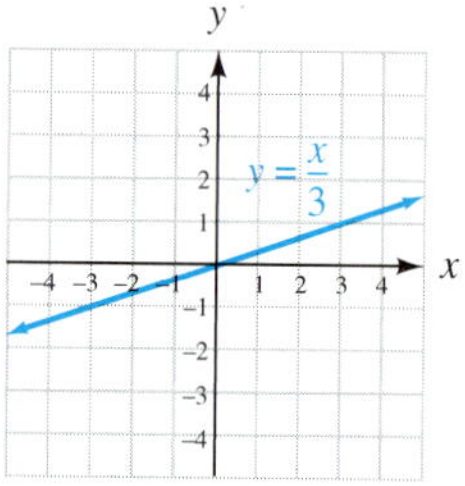

13. x-intercept: (3, 0); y-intercept: $(0, -2)$

14.

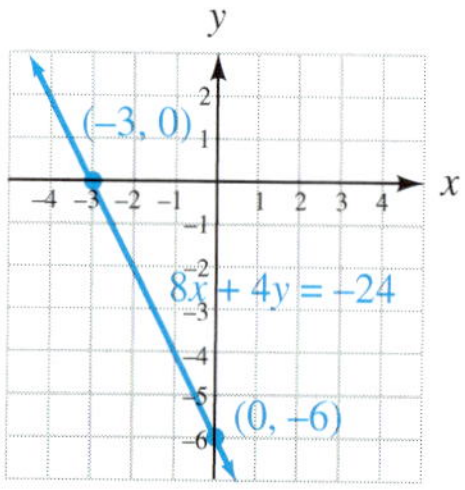

15. $\frac{8}{7}$ **16.** -1 **17.** 0

18. 10% **19.** Perpendicular

20. Parallel **21.** -15 ft per mi

22. 25 ft per mi

23.

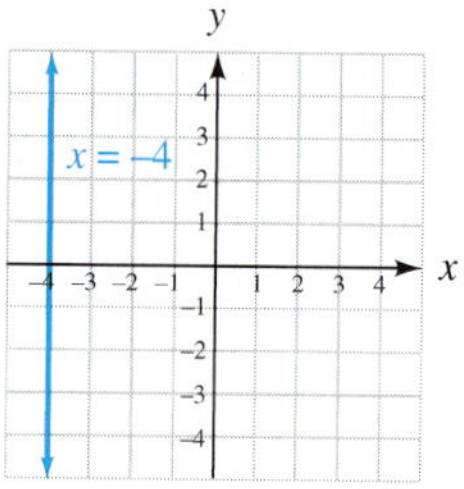

24.

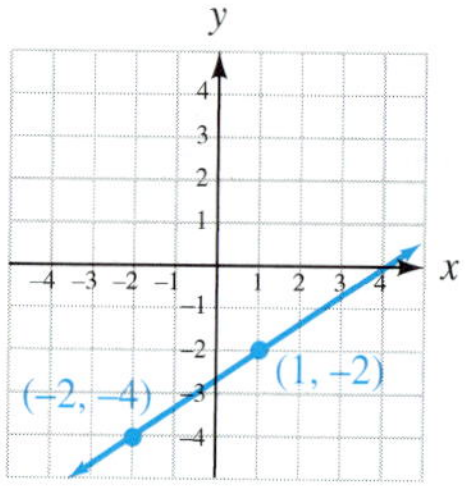

25. $m = -\frac{1}{2}$; (0, 4) **26.** $y = 7x + 19$ **27.** $y = -2x - 5$

28. **a.** $v = -1{,}500x + 15{,}000$ **b.** $3,000 **29.** Yes

30. $y = -\frac{1}{5}T + 41$

31.

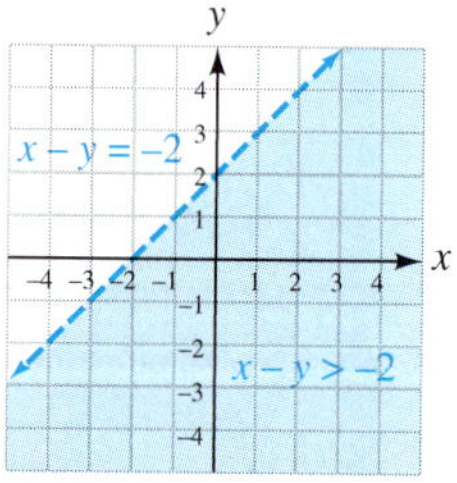

32. Domain: $\{-4, 0, 1, 5\}$; range: $\{-8, 3, 12\}$

33. Yes; domain: $\{1, 2, 3, 4\}$; range: $\{1, 2, 3, 4\}$

34. No; $(-3, 9)$, $(-3, -7)$;

35. No; (2, 3.5), $(2, -3.5)$; (Answers may vary) **36.** Yes; domain: $\{5, 10, 15, 20, 25\}$; range: $\{12\}$

37. -13 **38.** 756

39. $C(45) = 28.50$; it costs $28.50 to make 45 calls.

40.

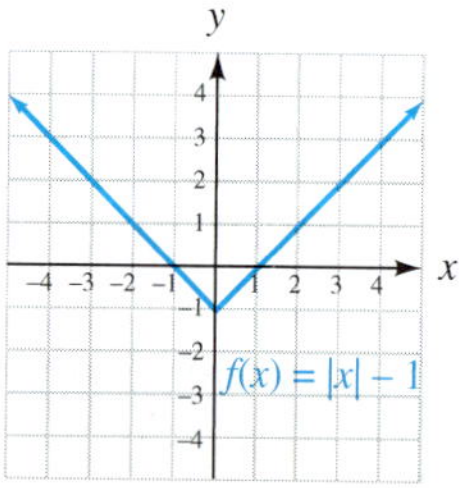

Cumulative Review Chapters 1–3 (page 305)

1. $2^2 \cdot 3^3$ **2.** 0.004 **3. a.** True **b.** True **c.** True **4.** -15
5. -0.77 **6.** -945 **7.** 30 **8.** 2 **9.** 32 **10.** $500 - x$
11. $3, -2$ **12. a.** $2x + 8$ **b.** $-2x + 8$ **13.** $4a + 10$
14. $-63t$ **15.** $4b^2$ **16.** 0 **17.** 4 **18.** $-160a$ **19.** $-3y$
20. $7x - 12$ **21.** 6 **22.** 2.9 **23.** 9 **24.** -19 **25.** $\frac{1}{7}$
26. 1 **27.** $-\frac{55}{6}$ **28.** No solution, contradiction **29.** -99
30. $-\frac{1}{4}$ **31.** 1,100 **32.** $h = \frac{S - 2\pi r^2}{2\pi r}$ **33.** $3\frac{1}{8}$ in., $\frac{39}{64}$ in.^{2}
34. 45°
35.

	% acid	Liters	Amount of acid
50% solution	0.50	x	$0.50x$
25% solution	0.25	$13 - x$	$0.25(13 - x)$
30% mixture	0.30	13	0.30(13)

36. 7.5 hr **37.** 80 lb candy corn, 120 lb gumdrops
38. $(-\infty, 48]$ 

39. $(0, \infty)$ 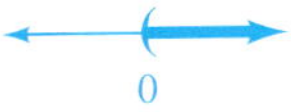

40. I and II **41.** No
42.
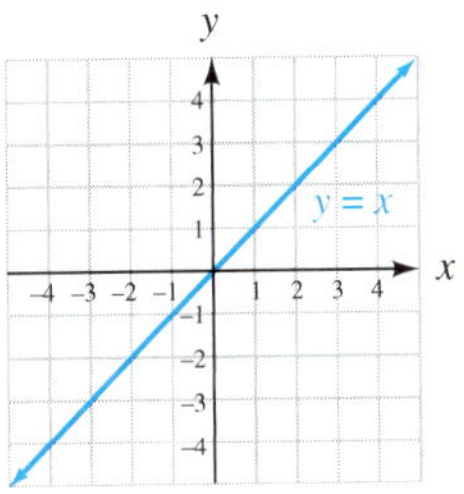

43.
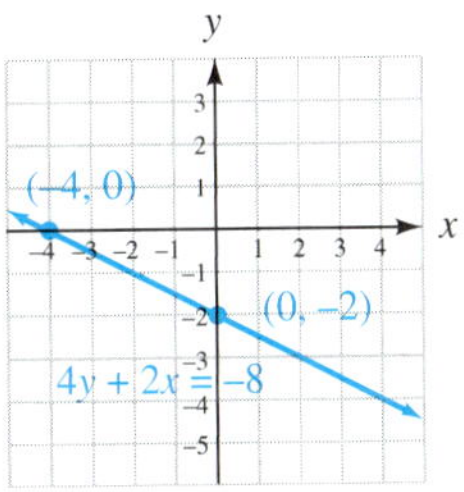

44. 0 **45.** $-\frac{10}{7}$ **46.** $\frac{7}{12}$ **47.** $\frac{2}{3}$, (0, 2) **48.** $y = -2x + 1$
49. $y + 9 = -\frac{7}{8}(x - 2)$; $y = -\frac{7}{8}x - \frac{29}{4}$ **50.** Yes
51.
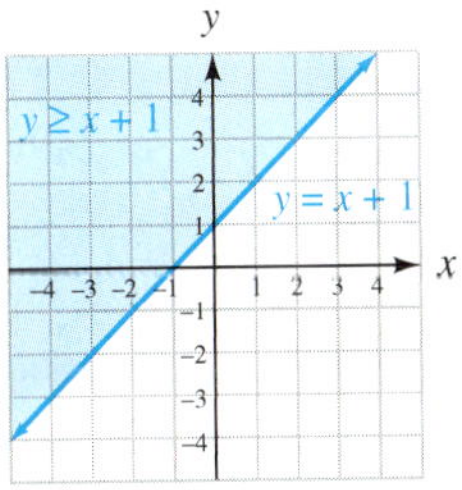

52.
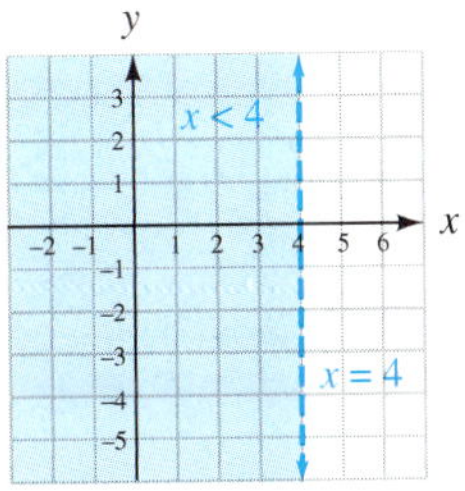

53. 78 **54.** No

Study Set Section 4.1 (page 316)

1. system **3.** intersection **5.** consistent, inconsistent
7. a. True **b.** True **9. a.** $-5, 2$ **b.** 3, 3, (0, −2) **11.** No solution; independent **13.** Yes **15.** Yes **17.** No **19.** No
21. No **23.** Yes **25.** (3, 2)
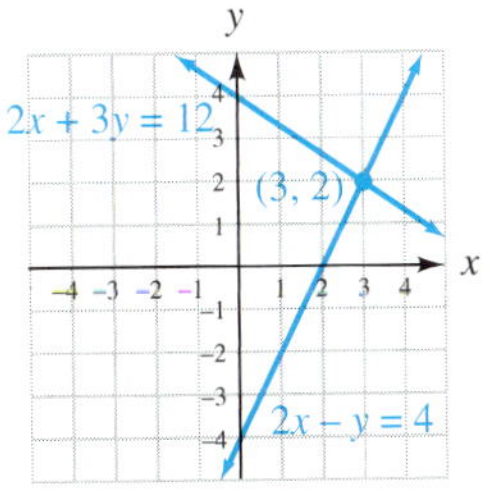

27. (−1, 5)
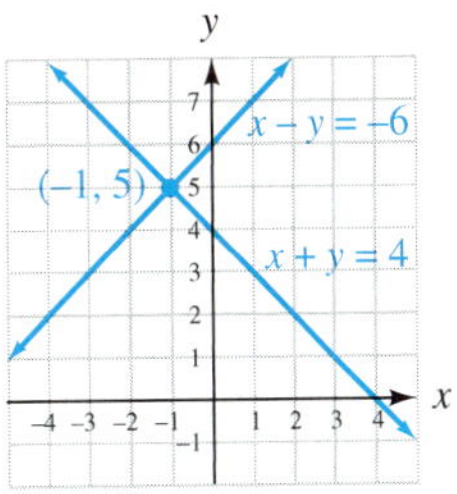

29. (−2, 0)
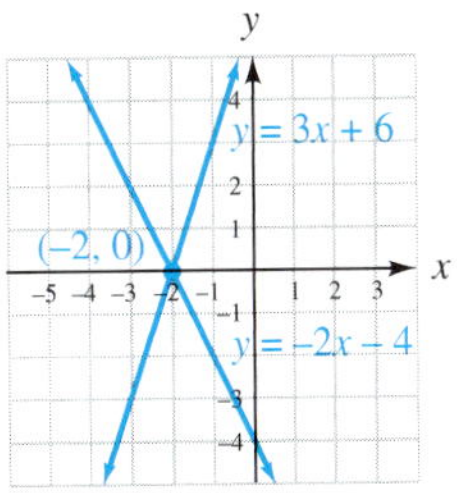

31. Infinitely many solutions
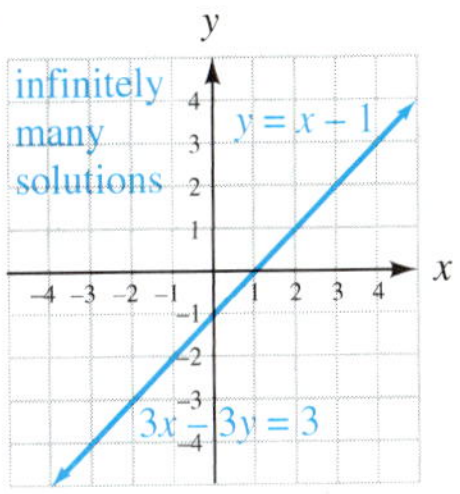

33. No solution
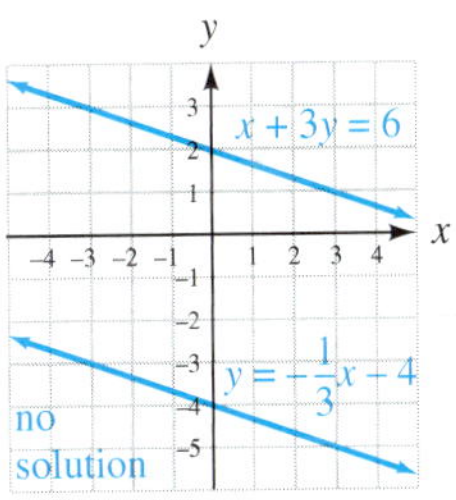

35. (4, −6)
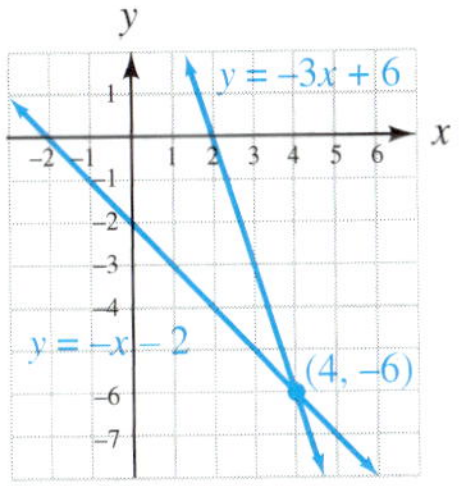

37. (5, −2)
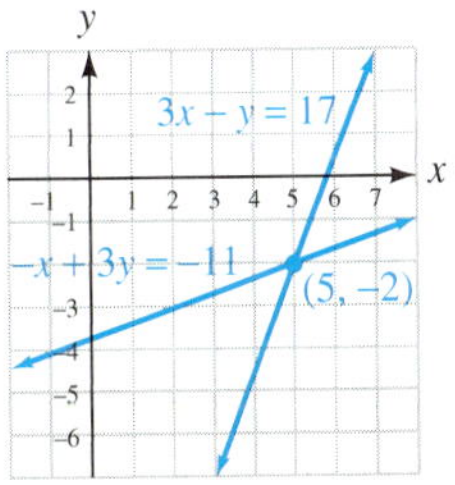

39. (1, 1)

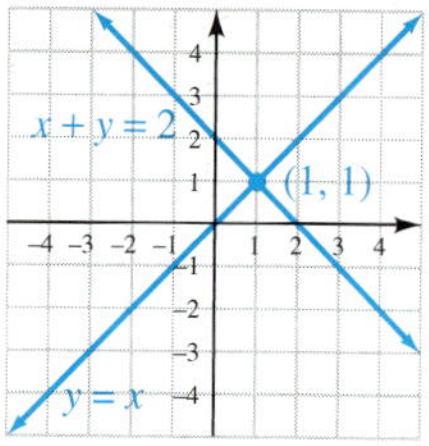

41. (−4, 0)

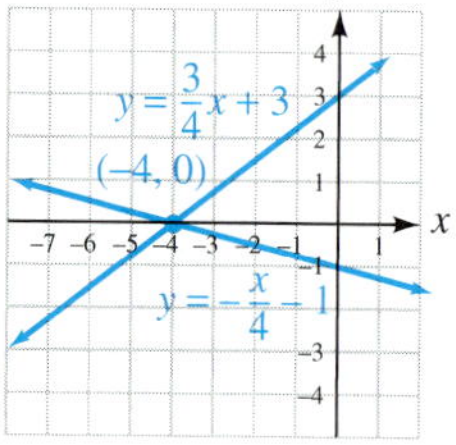

43. No solution

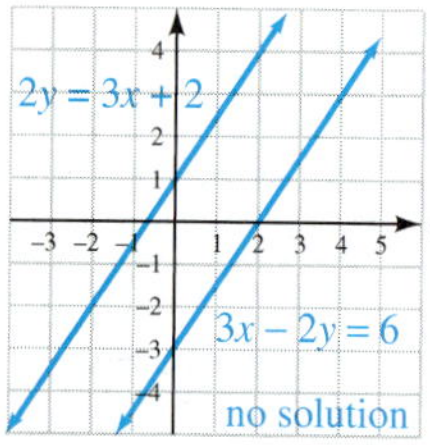

45. Infinitely many solutions

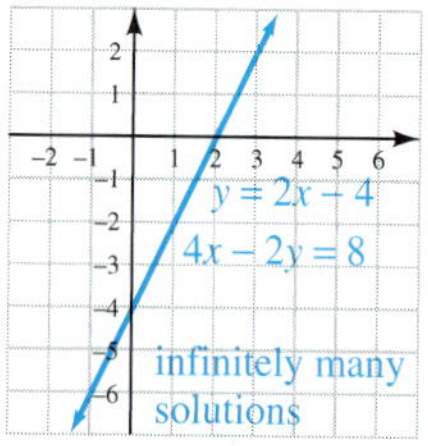

47. (3, −1)

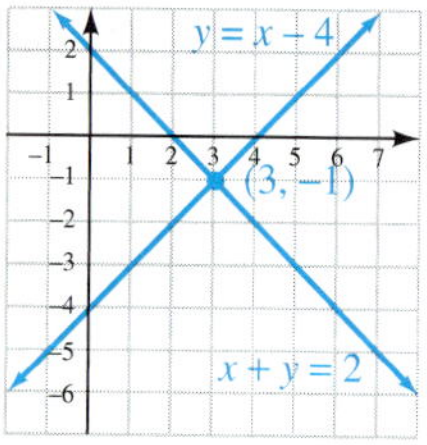

49. (−6, 1)

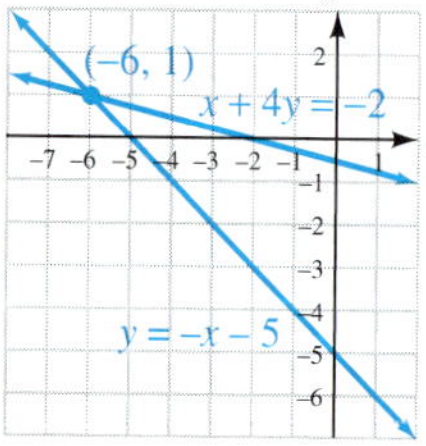

51. (3, 0)

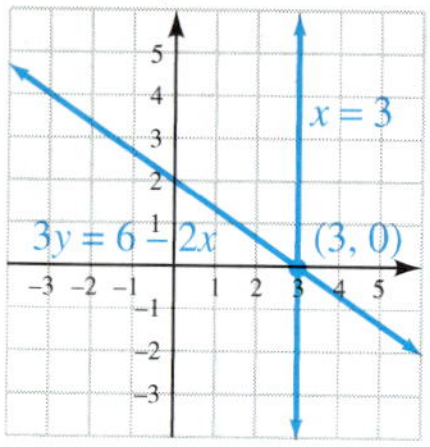

53. (−2, −3)

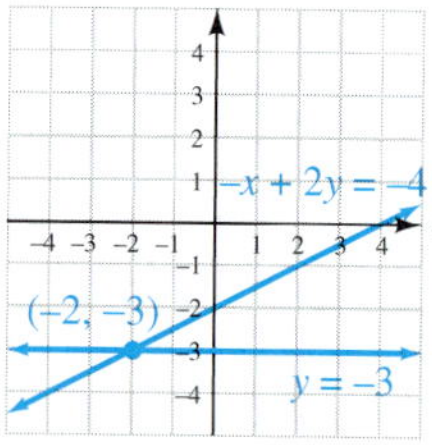

55. (4, −4)

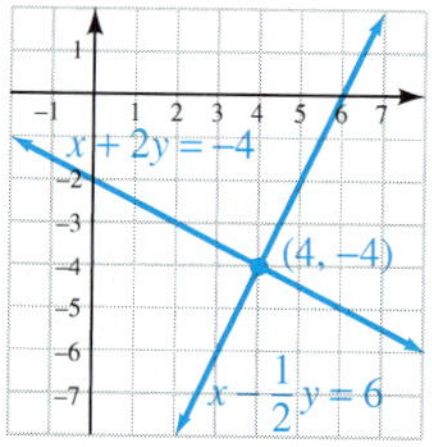

57. 1 solution **59.** Same line; infinitely many solutions **61.** No solution **63.** 1 solution **65.** (1, 3) **67.** No solution **69.** 1994; about 4,100 **71. a.** Houston, New Orleans, St. Augustine **b.** St. Louis, Memphis, New Orleans **c.** New Orleans **73. a.** The incumbent; 7% **b.** November 2 **c.** The challenger; 3 **75.** 10 mi **83.** $[-3, \infty)$

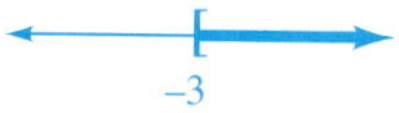

85. $[-8, \infty)$

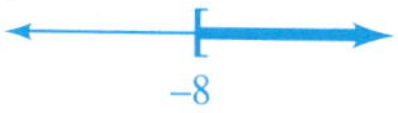

87. $(-\infty, 16)$

Study Set Section 4.2 (page 327)

1. substituting **3.** $y = -3x$ **5.** $x + 3(x - 4) = 8$ **7.** Substitute 3 for a in the second equation. **9. a.** No **b.** ii **11.** $3x$, 4, −2, −2, −6, (−2, −6) **13.** (2, 4) **15.** (3, 0) **17.** (−1, −1) **19.** (−10, 2) **21.** (−3, −1) **23.** $\left(\frac{1}{2}, \frac{1}{3}\right)$ **25.** (4, −2) **27.** (−4, −9) **29.** (3, 2) **31.** (−5, 5) **33.** $\left(\frac{2}{3}, -\frac{1}{3}\right)$ **35.** $\left(-4, \frac{5}{4}\right)$ **37.** (−2, 3) **39.** (−4, −6) **41.** (3, −2) **43.** (−5, −1) **45.** (−6, 4) **47.** $\left(\frac{1}{5}, 4\right)$ **49.** $\left(10, \frac{15}{2}\right)$ **51.** (9, 11) **53.** (−4, −1) **55.** (4, 2) **57.** No solution **59.** Infinitely many solutions **61.** No solution **63.** Infinitely many solutions **65.** (3, −2) **67.** (1, 1) **69.** No solution **71.** $\left(\frac{1}{3}, \frac{2}{3}\right)$ **73.** Infinitely many solutions **75.** (−10, −24) **77.** $\left(\frac{1}{2}, 2\right)$ **79.** $\left(-1, \frac{2}{3}\right)$ **81.** Angle of approach: 40°; angle of departure: 37° **87.** $3^3 \cdot 7$ **89.** 5/6 **91.** 21/40

Study Set Section 4.3 (page 338)

1. opposites **3.** $7y$ and $-7y$ **5. a.** $5a = -4$ **b.** $-4y = 1$ **7. a.** -2 **b.** 3 **9. a.** Multiply both sides by 15. **b.** Multiply both sides by 10. **11.** $2x$, 1, 1, (1, 4) **13.** (3, 2) **15.** (3, −2) **17.** (−2, −3) **19.** (0, 8) **21.** (−3, 4) **23.** (0, −2) **25.** (−12, 1) **27.** (−5, 10) **29.** (3, 11) **31.** (−8, −15) **33.** (−2, 7) **35.** (1, 1) **37.** (12, −9) **39.** $\left(1, -\frac{5}{2}\right)$ **41.** (−4, −1) **43.** (−4, 5) **45.** (−2, 5) **47.** (2, −1) **49.** (6, −2) **51.** (3, 0) **53.** (6, 8) **55.** (−4, −5) **57.** (−1, 2) **59.** (3, 2) **61.** $\left(\frac{7}{25}, -\frac{1}{25}\right)$ **63.** $\left(\frac{13}{75}, \frac{14}{75}\right)$ **65.** No solution **67.** No solution **69.** Infinitely many solutions **71.** Infinitely many solutions **73.** (2, 3) **75.** $\left(\frac{3}{4}, \frac{1}{3}\right)$ **77.** $\left(\frac{1}{3}, 3\right)$ **79.** Infinitely many solutions **81.** $\left(\frac{10}{3}, \frac{10}{3}\right)$ **83.** (10, 9) **85.** (1, −1) **87.** (4, −2) **89.** 1991 **95.** $y = -\frac{11}{6}x - \frac{7}{3}$ **97.** −80

Study Set Section 4.4 (page 351)

1. complementary, supplementary **3.** $x + y = 20, y = 2x - 1$ **5.** $x + y = 180, y = x - 25$ **7.** $5x + 2y = 10$ **9.** $x + c, x - c$ **11. a.** $(x + y)$ mL **b.** 33% **13.** 20°, 70° **15.** 50°, 130° **17.** 22 ft, 29 ft **19.** President: \$400,000; vice president: \$208,100 **21.** 65°, 115° **23.** 96 ft, 70 ft **25.** 15 m, 10 m **27.** Printer: \$2; copier: \$15 **29.** \$29.50 for a 10 × 14; \$21.00 for an 8 × 10 **31.** Elvis: 29¢; Lucy: 34¢ **33.** 85, 63 **35.** Nursing: \$2,000; business: \$3,000 **37.** International fund: \$18,500; offshore bank: \$21,500 **39.** 4% account: \$11,000; biotech: \$11,000 **41.** 25 mph, 5 mph **43.** 180 mph, 20 mph **45.** 8 gal 6% salt water, 24 gal 2% salt water **47.** 4% solution: 48 oz; 12% solution: 80 oz **49.** 52 lb \$8.75, 48 lb \$3.75 **51.** $\frac{20}{3} = 6\frac{2}{3}$ pints of mushrooms; $\frac{40}{3} = 13\frac{1}{3}$ pints of olives **55.** $(-\infty, 4)$

57. $(-1, 2]$

Study Set Section 4.5 (page 362)

1. inequalities **3.** intersection **5. a.** $3x - y = 5$ **b.** Dashed **7.** Slope: $4 = \frac{4}{1}$, y-intercept: (0, −3) **9. a.** No **b.** Above **11. a.** Yes **b.** No **c.** No **13. a.** ii **b.** iii **c.** iv **d.** i

15.

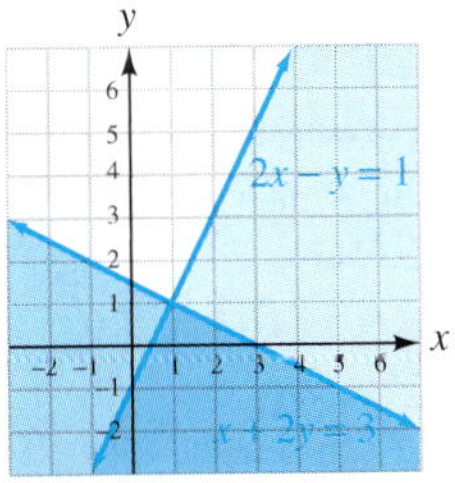

17.

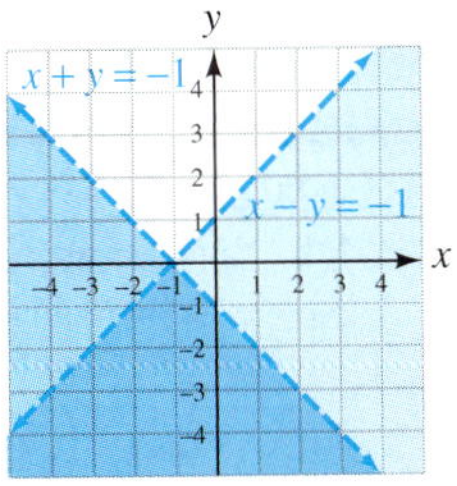

19.

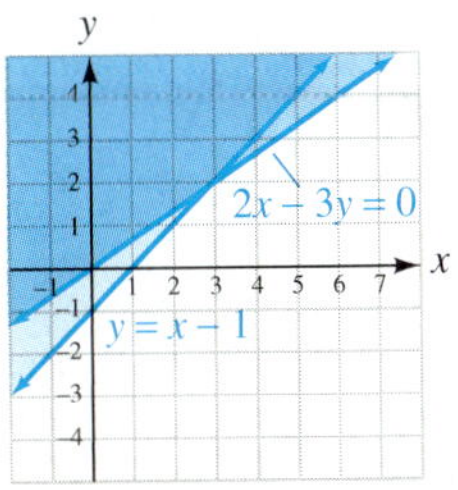

21.

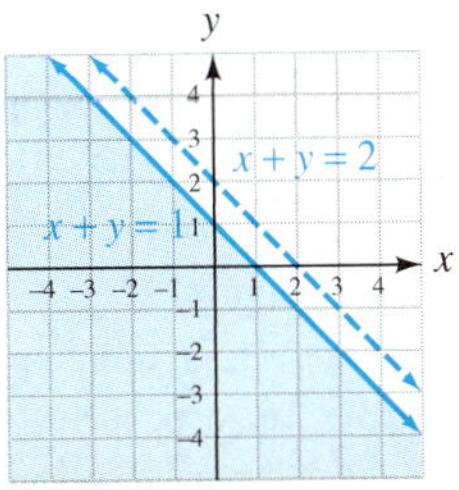

23.

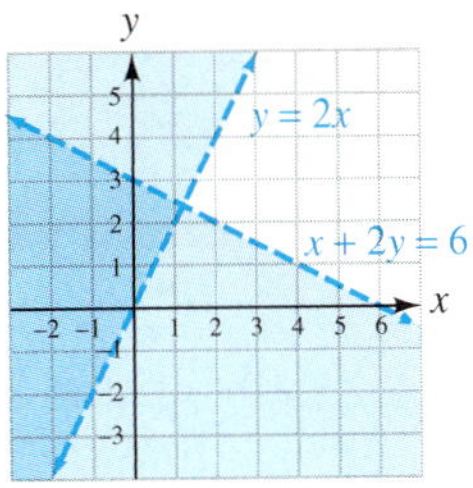

25.

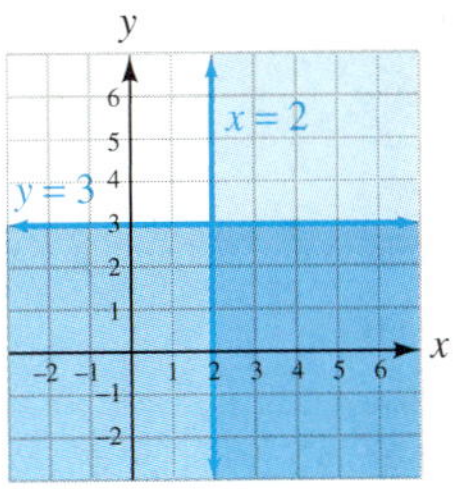

27.

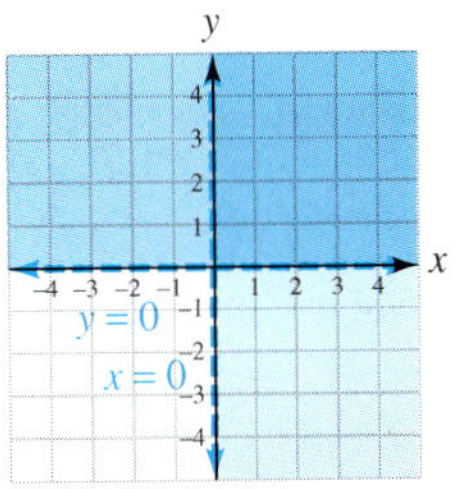

29.

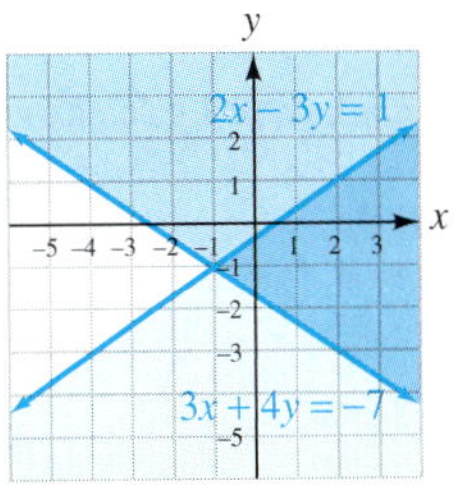

31.

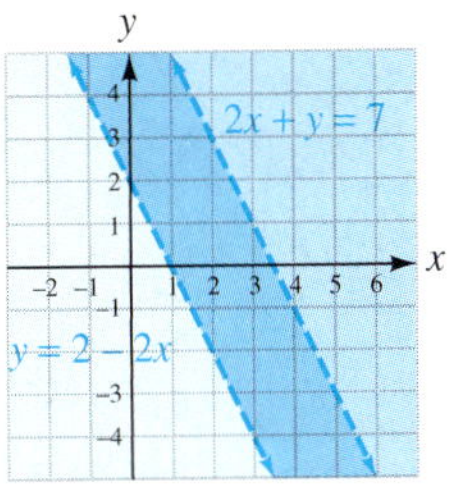

33.

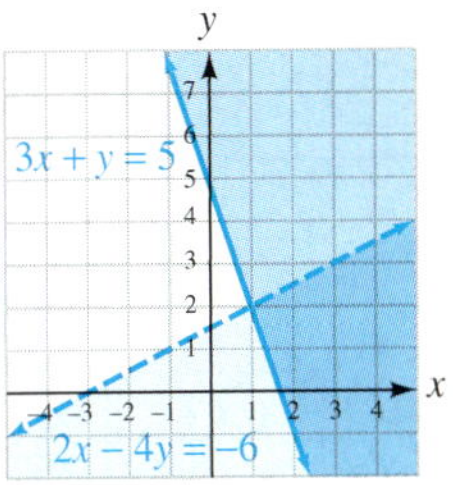

35.

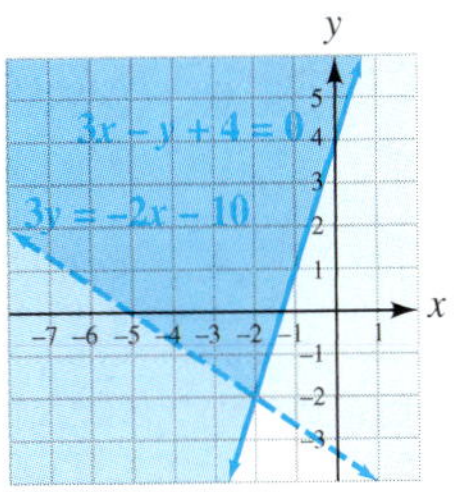

37.

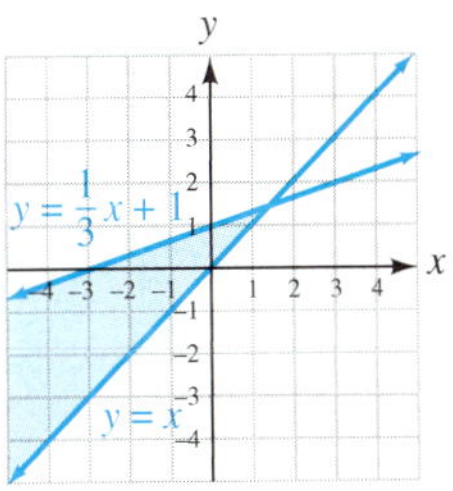

39.

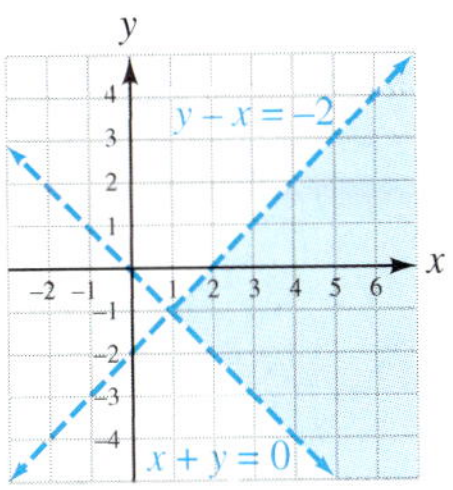

41.

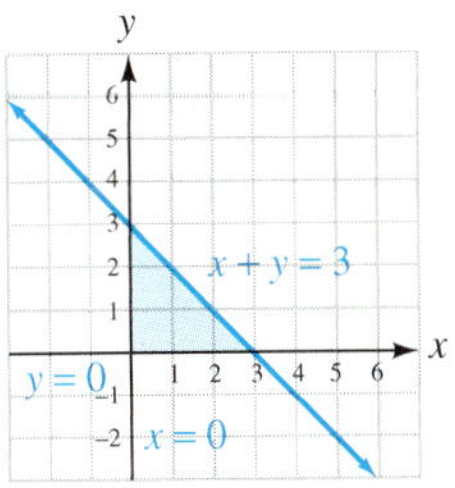

43.

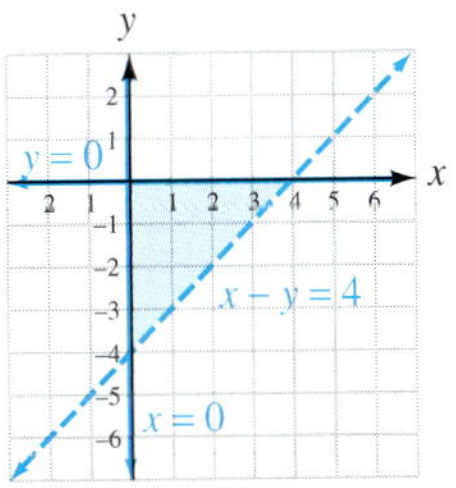

45. (c)

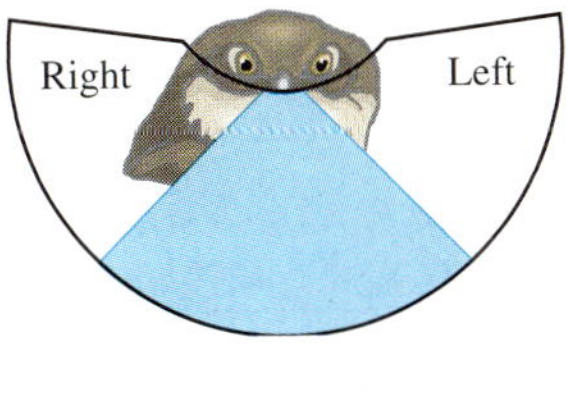

47. 1 \$10 CD and 2 \$15 CDs; 4 \$10 CDs and 1 \$15 CD

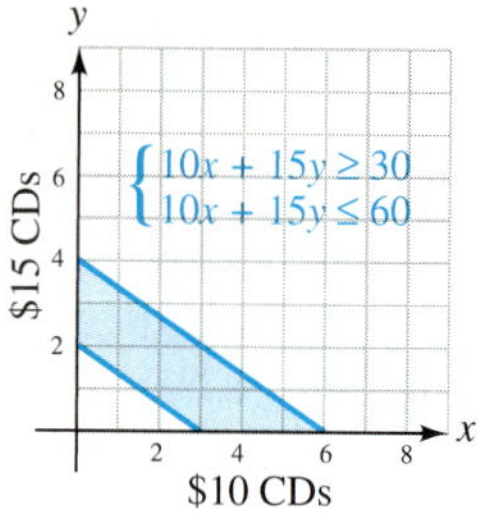

49. 2 desk chairs and 4 side chairs; 1 desk chair and 5 side chairs

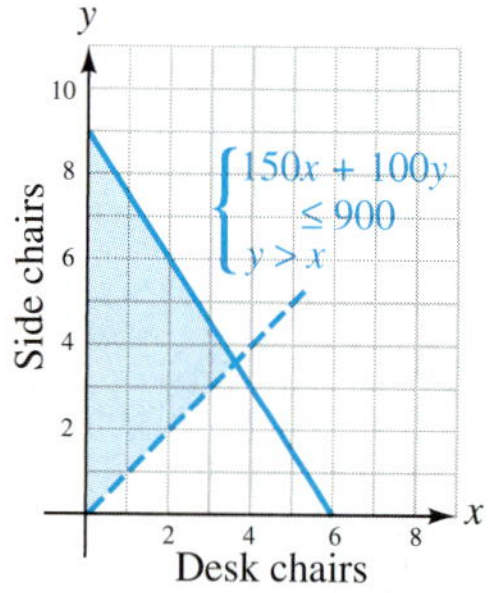

51.

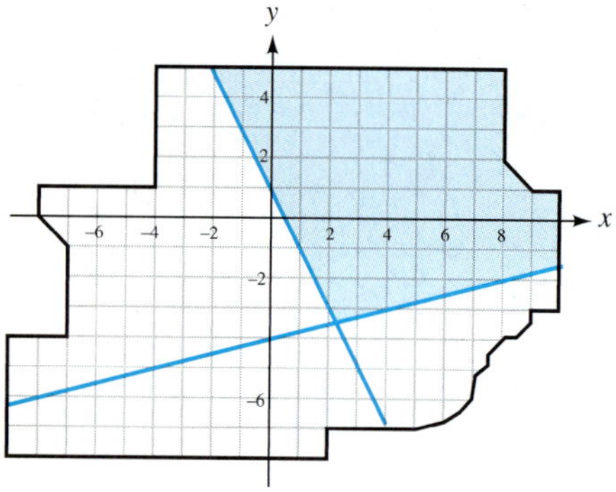

57. $6t$
59. $-2w + 4$
61. $-\frac{5}{8}x$
63. $9r - 16$

Chapter 4 Review (page 366)

1. Yes **2.** No **3.** (4, 3)

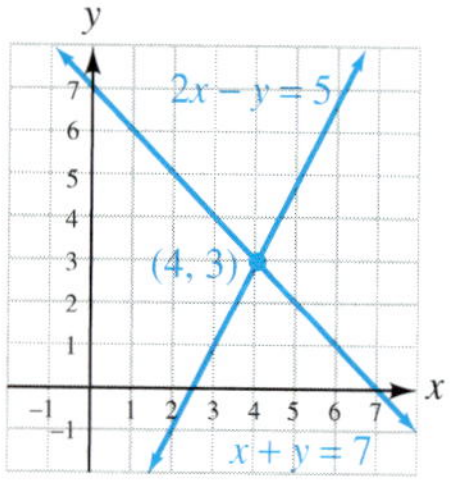

4. (3, −1)

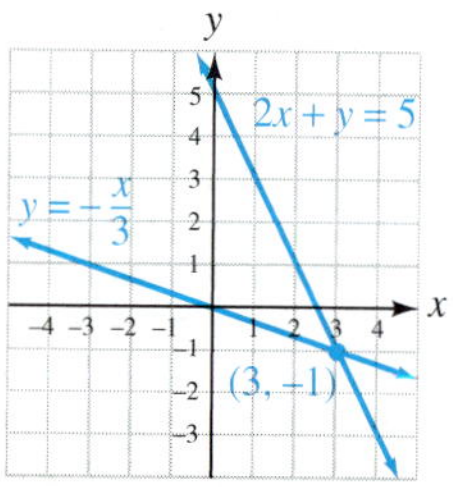

5. Infinitely many solutions

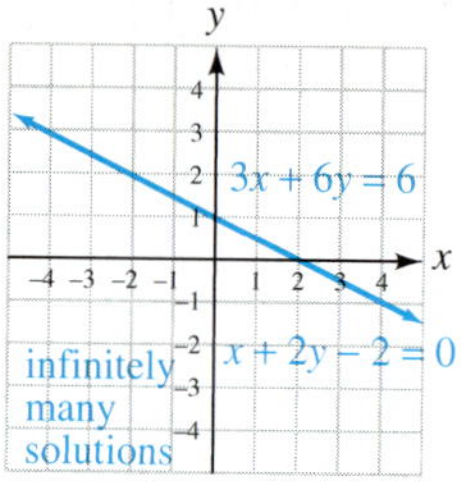

6. No solution

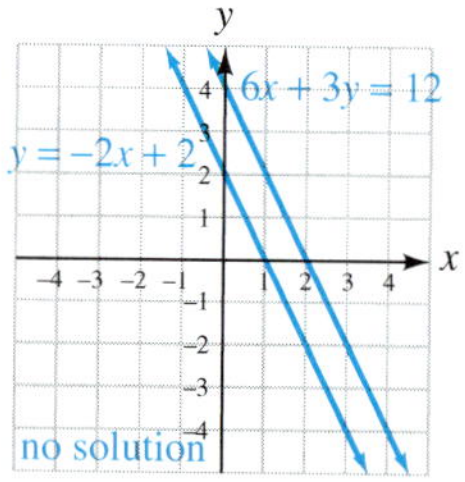

7. No solution **8.** (1978, 5,600,000); In 1978, the same number of males as females were enrolled in college, about 5.6 million of each. **9.** (5, 0) **10.** (3, 3) **11.** $\left(-\frac{1}{2}, \frac{7}{2}\right)$ **12.** (1, −2) **13.** Infinitely many solutions **14.** (12, 10) **15. a.** No solution **b.** Two parallel lines **c.** Inconsistent system **16.** one **17.** $\begin{cases} 4x + 2y = 7 \\ 5x - 3y = -6 \end{cases}$ **18.** one **19.** (3, −5) **20.** $\left(3, \frac{1}{2}\right)$ **21.** (−1, 7) **22.** (0, 9) **23.** Infinitely many solutions **24.** (1, −1) **25.** (−5, 2) **26.** No solution **27.** Elimination; no variables have a coefficient of 1 or −1. **28.** Substitution; equation 1 is solved for x. **29.** Las Vegas; 2,000 ft; Baltimore: 100 ft **30.** Base: 21 ft; extension: 14 ft **31.** 65°, 25° **32.** 10,800 yd^2 **33. a.** $0.02x$, $0.09y$, $0.08(100)$ **b.** $5(s + w)$, $7(s - w)$ **c.** $0.11x$, $0.06y$ **d.** $4x$, $8y$, $10(5)$ **34.** 12 lb worms, 18 lb bears **35.** 3 mph **36.** \$16.40, \$10.20 **37.** \$750 **38.** $13\frac{1}{3}$ gal 40%, $6\frac{2}{3}$ gal 70%

39.

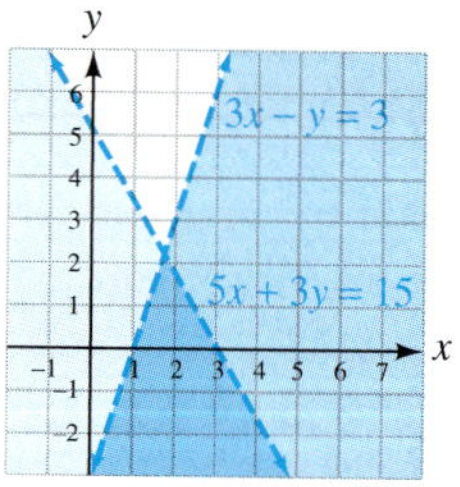

40.

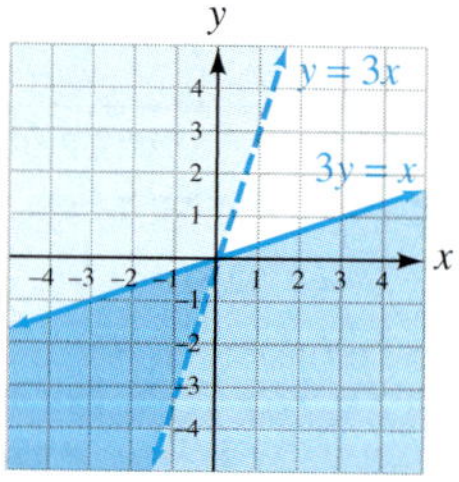

41. $10x + 20y \geq 40$, $10x + 20y \leq 60$; (3, 1): 3 shirts and 1 pair of pants; (1, 2): 1 shirt and 2 pairs of pants (Answers may vary)

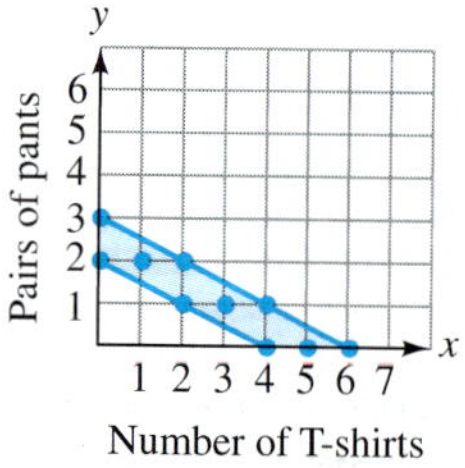

42. a. Yes **b.** No

Chapter 4 Test (page 372)

1. Yes **2.** No **3. a.** solution **b.** consistent **c.** inconsistent **d.** independent **e.** dependent **4.** Since the lines have different slopes, they will intersect at one point. The system has 1 solution.

5. (2, 3)

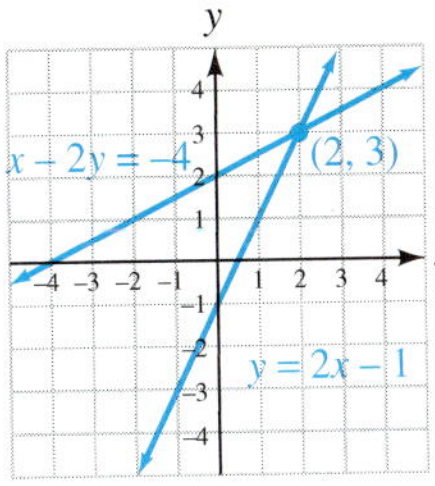

6. No solution

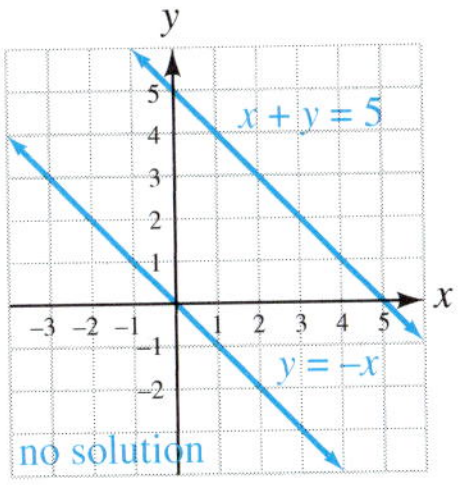

7. (30, 3,000); if 30 items are sold, the salesperson gets paid the same by both plans, \$3,000. **8.** Plan 1 **9.** (−2, −3) **10.** Infinitely many solutions **11.** (2, 4) **12.** (−3, 3) **13.** No solution **14.** (−1, −1) **15.** (5, 14) **16.** (0, 0) **17.** 8 mi, 14 mi **18.** 3 adult tickets; 4 child tickets **19.** \$6,000, \$4,000 **20.** 165 mph, 15 mph **21.** Larger: 70°, smaller: 20° **22.** 5%: 4 pints; 20%: 8 pints **23.** \$1.50 sunscreen: 3 oz; \$0.80 sunscreen: 7 oz

24.

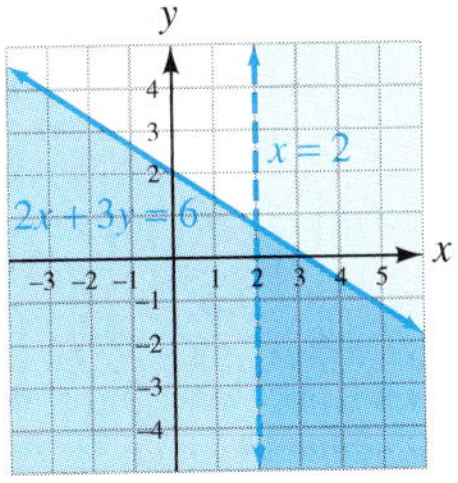

25. (1, 2), (2, 2), (3, 1) (Answers may vary)

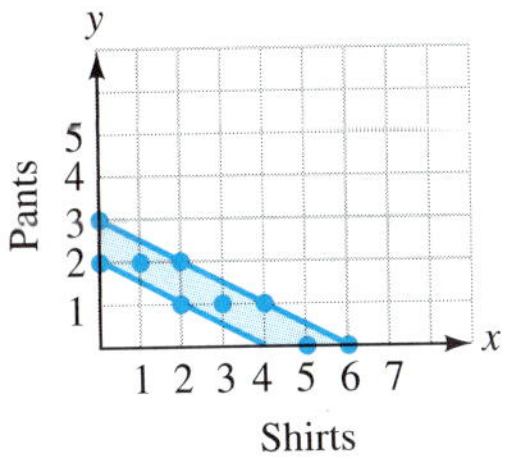

26. No

27. II **28.** No **29.**

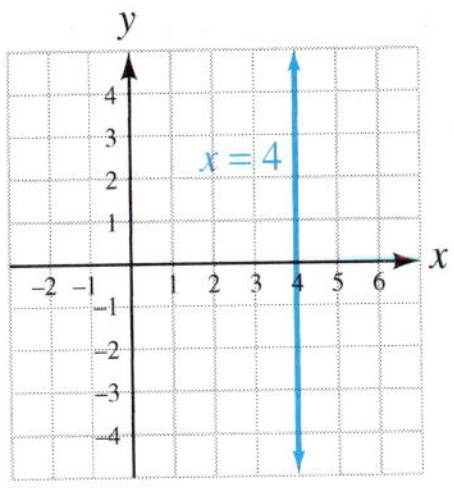

30.

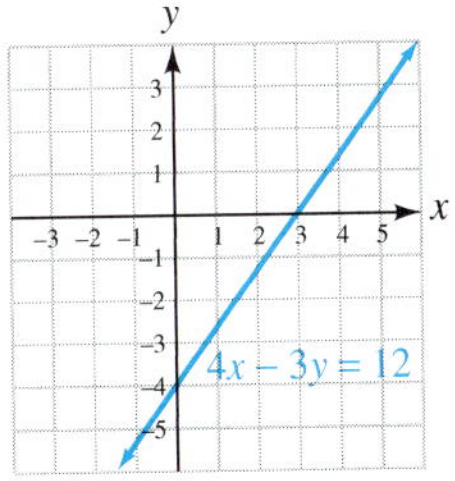

31. $\frac{1}{2}$ **32.** 0 **33.** $\frac{2}{3}$ **34.** Perpendicular **35.** $y = \frac{2}{3}x + 5$ **36.** $y = \frac{3}{4}x + \frac{11}{2}$

37. $y = 4$ **38.**

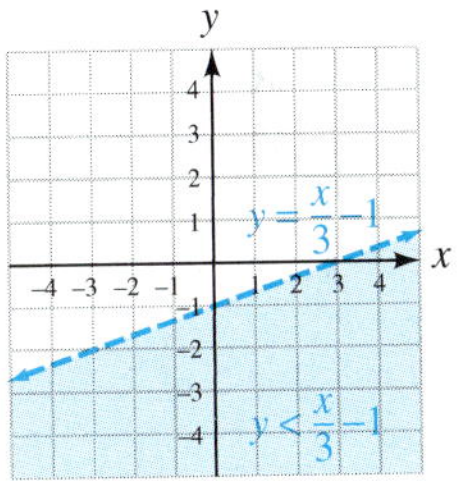

39. 1 **40.** No

41. (−6, 1)

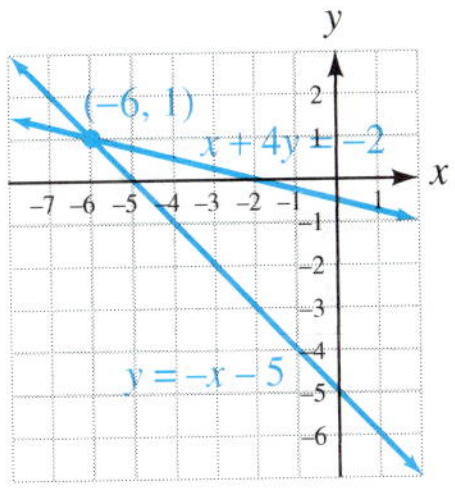

42.

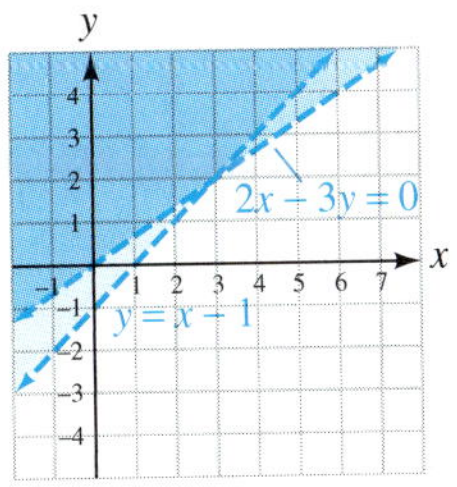

43. (4, 1) **44.** Noodles: 2 servings, rice: 3 servings

Cumulative Review Chapters 1–4 (page 374)

1. a. 1993 **b.** 2005 **2.** $2^2 \cdot 5^2$ **3.** $\frac{5}{8}$ **4.** $\frac{22}{35}$ **5.** Irrational **6.**

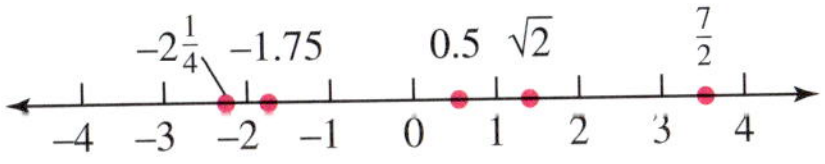

7. $0.\overline{6}$

8. Associative property of multiplication **9.** 0 **10.** 1 **11.** 4 **12.** −2 **13.** $250 - x$ **14.** $10d$ cents **15.** r **16.** $18x$ **17.** $3d - 11$ **18.** $-78c + 18$ **19.** 13 **20.** 41 **21.** $\frac{10}{9}$ **22.** −24 **23.** 140 **24.** $h = \frac{2A}{b + B}$ **25.** 20 lb of \$3.80 candy; 10 lb of \$4.40 candy **26.** $(-\infty, -14)$

Study Set Section 5.1 (page 388)

1. exponential **3. a.** $3x \cdot 3x \cdot 3x \cdot 3x$ **b.** $(-5y)^3$ **5. a.** Subtract **b.** Add **c.** Multiply **d.** Multiply **7. a.** $2x^2$ **b.** x^4 **9. a.** Doesn't simplify **b.** x **11.** x^6, 18 **13.** Base 4, exponent 3 **15.** Base x, exponent 5 **17.** Base $-3x$, exponent 2 **19.** Base y, exponent 6 **21.** Base m, exponent 12 **23.** Base $y + 9$, exponent 4 **25.** $(4t)^4$ **27.** $-4t^5$ **29.** $\left(\frac{t}{2}\right)^3$ **31.** $(x - y)^2$ **33.** 5^7 **35.** a^6 **37.** b^6 **39.** $(y - 2)^7$ **41.** a^5b^6 **43.** c^2d^5 **45.** a^{10} mi^2 **47.** x^9 ft^3 **49.** 8^8 **51.** x^{12} **53.** $(3.7p)^5$ **55.** $(k - 2)^{14}$ **57.** c^2d^6 **59.** x^3y^4 **61.** y^4 **63.** a **65.** t^7 **67.** s^2 **69.** 3^8 **71.** $(-4.3)^{24}$ **73.** m^{500} **75.** y^{15} **77.** x^{25} **79.** p^{25} **81.** t^{18} **83.** u^{14} **85.** $36a^2$

87. $625y^4$ **89.** $-8r^6s^9$ **91.** $-\frac{1}{243}y^{10}z^{20}$ **93.** ab^4 **95.** $r^{13}s^3$ **97.** $216k^3$ **99.** $9q^2$ **101.** $\frac{a^3}{b^3}$ **103.** $\frac{m^4}{81}$ **105.** $\frac{64a^4}{121b^{10}}$ **107.** $\frac{243m^{20}}{32n^{25}}$ **109.** $\frac{x^{10}}{y^{15}}$ **111.** y^9 **113.** 15^3 **115.** y^{15} **117.** $-216a^9b^6$ **119.** $a^{21}b^{21}$ **121.** n^{33} **123.** $36h^2$ **125. a.** $25x^2$ ft^2 **b.** $9a^2\pi$ ft^2 **127.** $\frac{1}{8{,}192}$ **131.** c **133.** d

Study Set Section 5.2 (page 398)

1. negative **3.**

Expression	Base	Exponent
4^{-2}	4	-2
$6x^{-5}$	x	-5
$\left(\frac{3}{y}\right)^{-8}$	$\frac{3}{y}$	-8
-7^{-1}	7	-1
$(-2)^{-3}$	-2	-3
$10a^0$	a	0

5.

x	3^x
2	9
1	3
0	1
-1	$\frac{1}{3}$
-2	$\frac{1}{9}$

7. a. 3 **b.** 6 **9.** reciprocal, 2 **11.** y^8, -40, 40 **13.** 1 **15.** 1 **17.** 2 **19.** 15 **21.** 1 **23.** $\frac{5}{2}$ **25.** $\frac{1}{4}$ **27.** $\frac{1}{6}$ **29.** $\frac{1}{x^9}$ **31.** $\frac{1}{b^5}$ **33.** $-\frac{1}{5}$ **35.** $-\frac{1}{1{,}000}$ **37.** $\frac{1}{2}$ **39.** $\frac{8}{9}$ **41.** $\frac{15}{g^6}$ **43.** $\frac{5}{x^3}$ **45.** $-\frac{1}{27}$ **47.** $\frac{1}{64}$ **49.** 125 **51.** r^{20} **53.** $8s$ **55.** $\frac{3}{16}$ **57.** $\frac{b^2}{a^5}$ **59.** $-\frac{4p^{10}}{d}$ **61.** 36 **63.** 8 **65.** $\frac{d^8}{c^8}$ **67.** $\frac{m^4}{81}$ **69.** y^6 **71.** b^7 **73.** $\frac{1}{y}$ **75.** $\frac{1}{h^7}$ **77.** $\frac{1}{x^{12}}$ **79.** $\frac{1}{b^8}$ **81.** $\frac{36s^8}{t^{14}}$ **83.** $\frac{32v^{25}}{u^{10}}$ **85.** $\frac{x^9}{64}$ **87.** $\frac{9}{y^8}$ **89.** y^3 **91.** $\frac{1}{a^6}$ **93.** $\frac{9a^2}{2b^2}$ **95.** $\frac{64s^2}{81t^4}$ **97.** $\frac{x^{28}}{y^{20}}$ **99.** $\frac{y^{14}}{z^{10}}$ **101.** $\frac{8b^3}{a^{12}}$ **103.** r^{20} **105.** $\frac{125}{d^6}$ **107.** $-15y$ **109.** $\frac{h^{20}}{16}$ **111.** $\frac{1}{x^6}$ **113.** $\frac{c^{12}}{d^{27}}$ **115.** $\frac{9}{4g^2}$ **117.** $\frac{32x^{15}}{y^{10}}$ **119.** t^{10} **121.** $-\frac{4t^2}{s^5}$ **123.** $\frac{1}{x^3}$

125.

Item	Measurement (meter)
Thickness of a dime	10^{-3}
Height of a bathroom sink	10^0
Length of a pencil eraser	10^{-2}
Thickness of soap bubble film	10^{-5}
Length of a cell phone	10^{-1}
Thickness of a piece of paper	10^{-4}

129. $-\frac{3}{2}$ **131.** $y = \frac{3}{4}x - 5$

Study Set Section 5.3 (page 406)

1. scientific, standard **3.** right, left **5. a.** positive **b.** negative **7. a.** 7.7 **b.** 5.0 **c.** 8 **9. a.** $(5.1 \times 1.5)(10^9 \times 10^{22})$ **b.** $\frac{8.8}{2.2} \times \frac{10^{30}}{10^{19}}$ **11.** 1, 10, integer **13.** 230 **15.** 812,000 **17.** 0.00115 **19.** 0.000976 **21.** 6,001,000 **23.** 2.718 **25.** 0.06789 **27.** 0.00002 **29.** 2.3×10^4 **31.** 1.7×10^6 **33.** 6.2×10^{-2} **35.** 5.1×10^{-6} **37.** 5.0×10^9 **39.** 3.0×10^{-7} **41.** 9.09×10^8 **43.** 3.45×10^{-2} **45.** 9.0×10^0 **47.** 1.1×10^1 **49.** 1.718×10^{18} **51.** 1.23×10^{-14} **53.** 7.3×10^5 **55.** 2.018×10^{17} **57.** 7.3×10^{-5} **59.** 3.602×10^{-19} **61.** 7.14×10^5; 714,000 **63.** 4.032×10^{-3}; 0.004032 **65.** 4.0×10^{-4}; 0.0004 **67.** 3.0×10^4; 30,000 **69.** 4.3×10^{-3}; 0.0043 **71.** 3.08×10^{-2}; 0.0308 **73.** 2.0×10^5; 200,000 **75.** 7.5×10^{-11}; 0.000000000075 **77.** $9.038030748 \times 10^{15}$ **79.** $1.734152992 \times 10^{-12}$ **81.** 2.57×10^{13} mi **83.** 197,000,000 mi^2; 109,000,000,000,000,000 mi^2; 14,600,000 mi^2 **85.** 4.5×10^{-10} oz **87.** g, x, u, v, i, m, r **89.** 1.7×10^{-18} g **91.** 3.09936×10^{16} ft **93.** 2.56×10^{11} dollars **95.** 1.0×10^6, 1.0×10^9, 1.0×10^{12}, 1.0×10^{15}, 1.0×10^{18} **101.** 5 **103.** $c = 30t + 45$

Study Set Section 5.4 (page 415)

1. polynomial **3.** one, descending, two, ascending **5.** monomial, binomial, trinomial **7.** evaluate **9. a.** Yes **b.** No **c.** No **d.** Yes **e.** Yes **f.** Yes **11.**

Term	Coefficient	Degree
$8x^2$	8	2
x	1	1
-7	-7	0
Degree of the polynomial 2		

13.

Term	Coefficient	Degree
$8a^6b^3$	8	9
$-27ab$	-27	2
Degree of the polynomial 9		

15. a. $5x^3 + 3x^2 + x - 9$ **b.** $x^2 - 2xy + y^2$ **17.** Binomial **19.** Trinomial **21.** Monomial **23.** Binomial **25.** Trinomial **27.** None of these **29.** None of these **31.** Trinomial **33.** 4th **35.** 2nd **37.** 1st **39.** 4th **41.** 12th **43.** 0th **45.** 18th **47.** 3rd **49. a.** 3 **b.** 13 **51. a.** -6 **b.** -8 **53. a.** 7 **b.** 34 **55. a.** -11.6 **b.** -40.2 **57. a.** 28 **b.** 4 **59. a.** 2 **b.** 0 **61.** 72 **63.** 19 **65.** -35 **67.** -257

69.

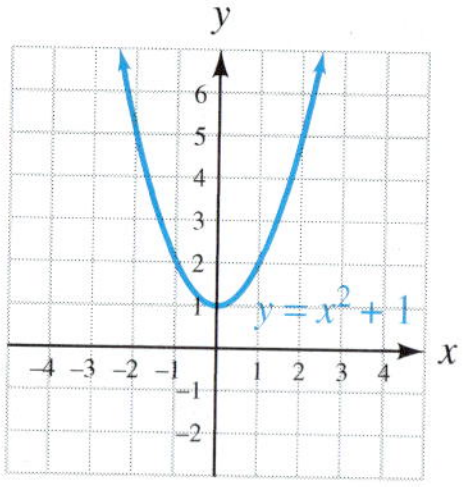

71.

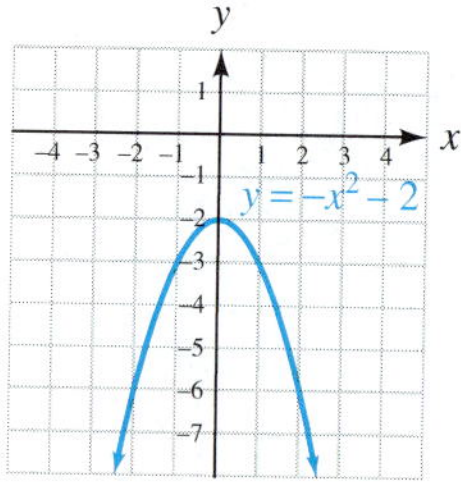

73.

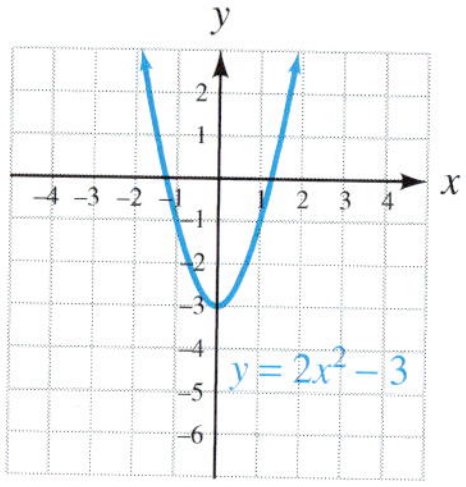

75.

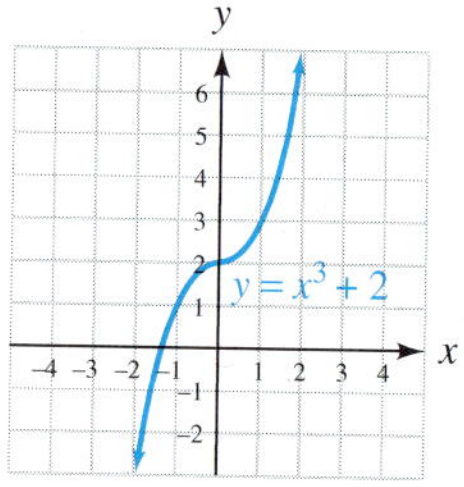

77.

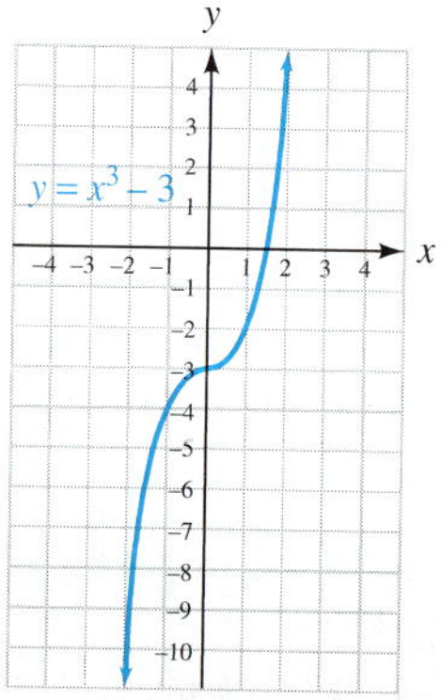

79.

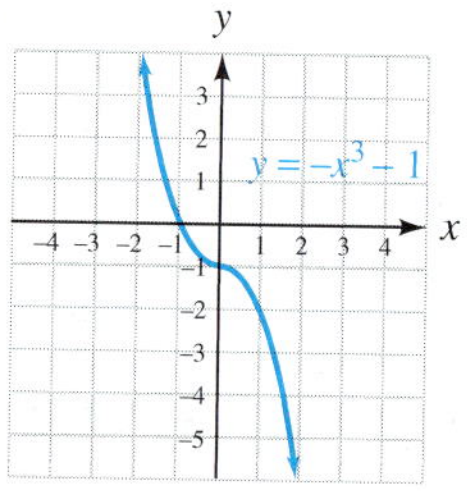

81. 91 **83.** 63 ft **85.** About 42 million

87.
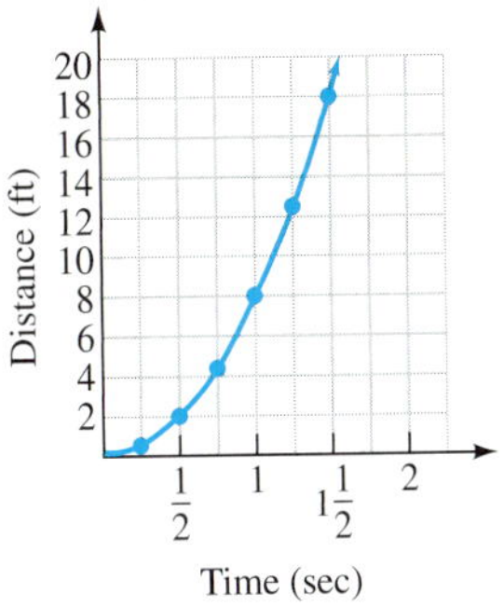

93. $[-3, \infty)$

95. x^{18} 97. y^9

Study Set Section 5.5 (page 425)

1. polynomials 3. Like 5. combine 7. a. $5x^2$ b. $14m^3$ c. $7a^3b$ d. $6cd + 4c^2d$ 9. a. $-5x^2 + 8x - 23$ b. $5y^4 - 3y^2 + 7$ 11. $4x^2, 2x, 1, 10x^2, 4$ 13. $12t^2$ 15. $-48u^3$ 17. $20x^2 - 19x$ 19. $10r^4 - 4r$ 21. $x^2 + x$ 23. $\frac{13}{15}x^2 - \frac{1}{8}x$ 25. $1.3x^3$ 27. $2st$ 29. $\frac{7}{12}c^2 - \frac{1}{2}cd + d^2$ 31. $-ab$ 33. $-4x^3y + x^2y + 5$ 35. $-4c^2 - 12cd$ 37. $7x + 4$ 39. $5d^2 + 14d$ 41. $5q^2 - 4q - 5$ 43. $y^3 + \frac{19}{20}y^2 + \frac{1}{3}$ 45. $0.7p - 0.9q$ 47. $7x^2 + xy + 2y^2$ 49. $(3x^2 + 6x - 2)$ yd 51. $(7x^2 + 5x + 6)$ mi 53. $5x^2 + x + 11$ 55. $-3a^2 + 7a + 7$ 57. $10z^3 + z - 2$ 59. $-x^3y^2 + 4x^2y + 5x + 6$ 61. $2a^2 + a - 3$ 63. $13a^2 + a$ 65. $-5h^3 + 5h^2 + 30$ 67. $\frac{1}{24}s^8 - \frac{19}{20}s^7$ 69. $-0.14f^2 + 0.25f + 2.09$ 71. $b^2 + 4ab - 2$ 73. $x^2 + 6x + 2$ 75. $-x^3 + 6x^2 + x + 14$ 77. $0.6x^3 + 1.2x^2 + 1.3x - 0.3$ 79. $7x^3y^2 - 2x^2y - 2x + 15$ 81. $4s^2 - 5s + 7$ 83. $3y^5 - 6y^4 + 1.2$ 85. $-3x^2 + 5x - 7$ 87. $t^3 + 3t^2 + 6t - 5$ 89. $3x + 1$ 91. $-1.3t^2 + 0.7t + 0.6$ 93. $3x^2 - 9x - 10$ 95. $6x^2 + x - 5$ 97. $\frac{5}{4}r^4 + \frac{11}{9}r^2 - 2$ 99. $9c^2 - 6c - 14$ 101. $-3s^2t - 8st + 14$ 103. $4.8h^3 + 8.8h^2$ 105. a. $(x^2 - 8x + 12)$ ft b. $(x^2 + 2x - 8)$ ft 107. $(2a^2 + 6a + 5)$ in. 109. a. $(22t + 20)$ ft b. 108 ft 117. 180° 119.
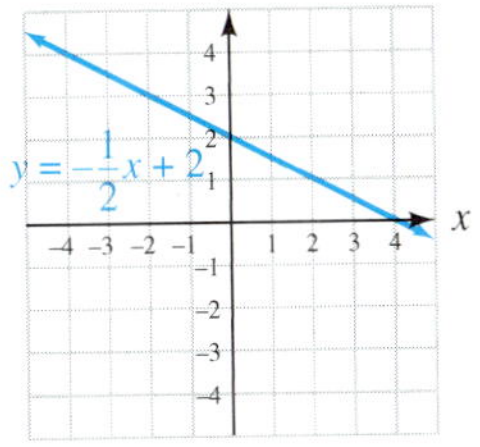

Study Set Section 5.6 (page 436)

1. monomials 3. first, outer, inner, last 5. a. each, each b. any, third 7. a. $6x^2 + x - 12$ b. $5x^4 + 8ax^2 + 3a^2$ 9. $8, n^3, 72n^5$ 11. $2x, 5, 5, 4x, 15x, 11x$ 13. $5m^2$ 15. $12x^5$ 17. $6c^6$ 19. $-24b^6$ 21. $8x^5y^5$ 23. $-2a^{11}$ 25. $3x^2 + 12x$ 27. $-4t^3 + 28t$ 29. $9x^4 - 18x^3 + 54x^2$ 31. $-6x^5 + 2x^4 - 2x^3$ 33. $0.12p^9 - 1.8p^7$ 35. $\frac{5}{8}t^8 + 5t^4$ 37. $-12x^4z - 4x^2z^3 - 4x^3z^2 + 4x^2z$ 39. $6x^{14} - 72x^{13}$ 41. $(7h^2 + 3h)$ in.^{2} 43. $(4w^2 - 2w)$ ft^2 45. $y^2 + 8y + 15$ 47. $t^2 + t - 12$ 49. $m^2 - 3m - 54$ 51. $4y^2 + 23y - 35$ 53. $12x^2 - 28x + 15$ 55. $7.6y^2 - 5.8y + 1$ 57. $18m^2 - 10m + \frac{8}{9}$ 59. $t^4 - 7t^2 + 12$ 61. $a^2 + 2ab + b^2$ 63. $12a^2 - 5ab - 2b^2$ 65. $x^3 - x + 6$ 67. $4t^3 + 11t^2 + 18t + 9$ 69. $x^3 - 8$ 71. $2x^3 + 7x^2 - 16x - 35$ 73. $-3x^3 + 25x^2y - 56xy^2 + 16y^3$ 75. $r^4 - 5r^3 + 2r^2 - 7r - 15$ 77. $x^3 - 3x + 2$ 79. $12x^3 + 17x^2 - 6x - 8$ 81. $8x^3 - 12x^2 - 8x$ 83. $-3a^3 + 3ab^2$ 85. $18a^6 - 12a^5$ 87. $x^3 - 6x^2 + 5x + 12$ 89. $30x^2 - 17x + 2$ 91. $6x^4 + 8x^3 - 14x^2$ 93. $-18x^3z^8$ 95. $6a^4 + 5a^3 + 5a^2 + 10a + 4$ 97. $9t^2 + 15st - 6s^2$ 99. $2a^{10}$ 101. $4y^3 + 40y^2 + 84y$ 103. $16.4p^2q^2 - 24.6p^2q + 41pq^2$ 105. $x^4 + 11x^3 + 26x^2 - 28x - 24$ 107. $16a^2 - 2ar - \frac{15}{16}r^2$ 109. $(6x^2 + x - 1)$ cm^2 111. $(3.14x^2 - 3.14)$ in.^{2} 113. $(2x^3 - 4x^2 - 6x)$ in.^{3} 121. a. 1 b. Undefined 123. (0, 2)

Study Set Section 5.7 (page 445)

1. products 3. a. square, Twice, first b. second, square 5. $x, 4, 4, 8x$ 7. $s, 5, 25$ 9. $x^2 + 2x + 1$ 11. $r^2 + 4r + 4$ 13. $m^2 - 12m + 36$ 15. $f^2 - 16f + 64$ 17. $16x^2 + 40x + 25$ 19. $49m^2 - 28m + 4$ 21. $1 - 6y + 9y^2$ 23. $y^2 + 1.8y + 0.81$ 25. $a^4 + 2a^2b^2 + b^4$ 27. $r^4 - 2r^2s^2 + s^4$ 29. $s^2 + \frac{3}{2}s + \frac{9}{16}$ 31. $d^8 + \frac{1}{2}d^4 + \frac{1}{16}$ 33. $x^2 - 9$ 35. $d^2 - 49$ 37. $4p^2 - 49$ 39. $9n^2 - 1$ 41. $c^2 - \frac{9}{16}$ 43. $36b^2 - \frac{1}{4}$ 45. $0.16 - 81m^4$ 47. $25 - 36g^2$ 49. $x^3 + 12x^2 + 48x + 64$ 51. $n^3 - 18n^2 + 108n - 216$ 53. $8g^3 - 36g^2 + 54g - 27$ 55. $a^3 + 3a^2b + 3ab^2 + b^3$ 57. $8m^3 + 12m^2n + 6mn^2 + n^3$ 59. $n^4 - 8n^3 + 24n^2 - 32n + 16$ 61. $-x^2 + 20x - 8$ 63. $3t^2 + 12t - 9$ 65. $2x^2 + xy - y^2$ 67. $24a^2 - 10a + 65$ 69. $-80d^3 + 40d^2 - 5d$ 71. $4d^5 - 4dg^6$ 73. $(2x^2 - 2)$ yd^2 75. $(9x^2 + 6x + 1)$ ft^2 77. $4v^6 - 32v^3 + 64$ 79. $12x^3 + 36x^2 + 27x$ 81. $16f^2 - 0.16$ 83. $r^4 + 20r^2s + 100s^2$ 85. $6x - 2$ 87. $4a^2 - 12ab + 9b^2$ 89. $n^2 - 36$ 91. $36m + 36$ 93. $25m^2 - 12m + \frac{36}{25}$ 95. $8e^3 + 12e^2 + 6e + 1$ 97. $x^2 - 4x + 4$ 99. $13x^2 - 8x + 5$ 101. $36 - 24d^3 + 4d^6$ 103. $64x^2 + 48x + 9$ 105. $(x^2 + 12x + 36)$ in.^{2} 107. $\pi hR^2 - \pi hr^2$ 113. $3^3 \cdot 7$ 115. $\frac{5}{6}$ 117. $\frac{21}{40}$

Study Set Section 5.8 (page 455)

1. monomial 3. binomial 5. Divide, multiply, subtract, bring down 7. quotient, dividend 9. $7x^2, x^3, 7x^2, 5, 2, 7, 2, 2, 4x^3, \frac{5}{7}$ 11. $5x^4 + 0x^3 + 2x^2 + 0x - 1$ 13. x^3 15. $5m^5$ 17. $\frac{4h^2}{3}$ 19. $-\frac{1}{5d^4}$ 21. $\frac{10}{s}$ 23. $\frac{x^2}{5y^4}$ 25. $\frac{4r}{y^5}$ 27. $-\frac{13}{3rs^3}$ 29. $2x + 3$ 31. $9 - \frac{6}{m}$ 33. $\frac{1}{a^3} - \frac{1}{a} + 1$ 35. $2x^5 - 8x^2$ 37. $\frac{h^2}{4} + \frac{2}{h}$ 39. $-2w^2 - \frac{1}{w^4}$ 41. $3s^5 - 6s^2 + 4s$ 43. $c^3 + 3c^2 - 2c - \frac{5}{c}$ 45. $3x^2y - 2x - \frac{1}{y}$ 47. $5y - \frac{6}{x} + \frac{1}{xy}$ 49. $x + 6$ 51. $x - 2$ 53. $3a - 2$ 55. $b + 3$ 57. $x + 1 + \frac{10}{x + 5}$ 59. $a - 12 + \frac{4}{a - 5}$ 61. $x + 1 + \frac{-1}{2x + 3}$ 63. $2x - 3 + \frac{-1}{3x - 1}$ 65. $2x - 1$ 67. $2x + 1$ 69. $x + 3$ 71. $x^2 - 2x + 1$ 73. $a - 5$ 75. $x + 1$ 77. $2x - 3$ 79. $9b + 7$ 81. $x^2 - x + 1$ 83. $y^2 + 2y + 5 + \frac{10}{y - 2}$ 85. $y + 12 + \frac{1}{y + 1}$ 87. $3a^5 - \frac{2b^3}{a}$ 89. $2x^2 + 2x + 1$ 91. $\frac{x}{5} - \frac{2}{5x^2}$ 93. $a^2 + a + 1$

95. $2x^2 + x + 1 + \frac{2}{3x - 1}$ **97.** $\frac{x^2}{2y^{10}}$ **99.** $3m - 8$
101. $(x - 6)$ in. **103.** $(2x^2 - x + 3)$ in. **109.** $y = -\frac{11}{6}x - \frac{7}{3}$
111. -80

Chapter 5 Review (page 458)

1. a. Base n, exponent 12 **b.** Base $2x$, exponent 6 **c.** Base r, exponent 4 **d.** Base $y - 7$, exponent 3 **2. a.** m^5 **b.** $-3x^4$ **c.** $(x + 8)^2$ **d.** $\left(\frac{1}{2}pq\right)^3$ **3.** 7^{12} **4.** m^2n^2 **5.** y^{21} **6.** $81x^4$ **7.** b^9 **8.** $-b^{12}$ **9.** $256s^{10}$ **10.** $4.41x^4y^2$ **11.** $(-9)^{15}$ **12.** a^{23} **13.** $\frac{1}{8}x^{15}$ **14.** $\frac{x^{12}}{9y^2}$ **15.** $(m - 25)^{12}$ **16.** $125yz^4$ **17.** a^{11} **18.** c^5d^5 **19.** $64x^{12}$ in.^{3} **20.** y^4 ft^2 **21.** 1 **22.** 1 **23.** 3 **24.** $\frac{1}{1{,}000}$ **25.** $-\frac{1}{25}$ **26.** $\frac{1}{t^6}$ **27.** $8x^5$ **28.** $-\frac{6}{y}$ **29.** $\frac{8}{49}$ **30.** x^{14} **31.** $-\frac{27}{r^9}$ **32.** $\frac{1}{16z^2}$ **33.** $\frac{8c}{9d^5}$ **34.** t^{30} **35.** w^{22} **36.** $\frac{f^{40}}{4^{10}}$ **37.** 7.2×10^8 **38.** 9.37×10^{15} **39.** 9.42×10^{-9} **40.** 1.3×10^{-4} **41.** 1.8×10^{-4} **42.** 8.53×10^5 **43.** 126,000 **44.** 0.00000003919 **45.** 2.68 **46.** 57.6 **47.** 3.0×10^{-4}; 0.0003 **48.** 1.6×10^8; 160,000,000 **49.** 6,570,000,000; 6.57×10^9 **50.** $1.0 \times 10^5 = 100{,}000$ **51. a.** 4 **b.** $3x^3$ **c.** 3, -1, 1, 10 **d.** 10 **52. a.** 7th, monomial **b.** 3rd, monomial **c.** 2nd, binomial **d.** 5th, trinomial **e.** 6th, binomial **f.** 4th, none of these **53.** 3, -13 **54.** 8 in.

55.

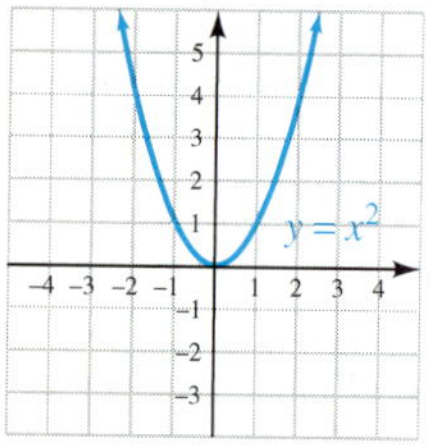

56.

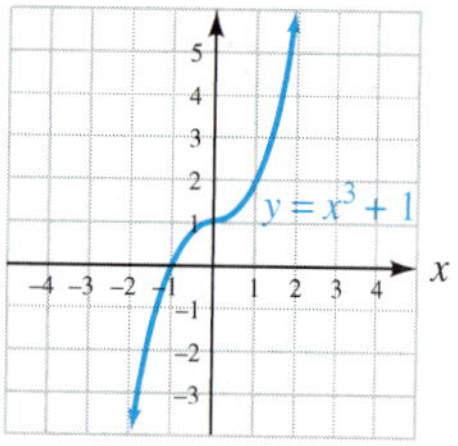

57. $13y^3$ **58.** $-4a^3b + a^2b + 6$ **59.** $\frac{7}{12}x^2 - \frac{3}{4}xy + y^2$ **60.** $6.7c^5 + 8.1c^4 - 2.1c^3$ **61.** $25r^6 + 9r^3 + 5r$ **62.** $3.7a^2 + 6.1a - 17.6$ **63.** $4r^3s - 7r^2s^2 - 7rs^3 - 2s^4$ **64.** $\frac{5}{8}m^4 - m^3$ **65.** $-z^3 + 2z^2 + 5z - 17$ **66.** $(x^2 + x + 3)$ in. **67.** $4x^2 + 2x + 8$ **68.** $8x^3 - 7x^2 + 19x$ **69.** $10x^3$ **70.** $-6x^{10}z^5$ **71.** $120b^{11}$ **72.** $2h^{14} + 8h^{11}$ **73.** $9n^4 - 15n^3 + 6n^2$ **74.** $x^2y^3 - x^3y^2$ **75.** $6x^6 + 12x^5$ **76.** $a^6b^4 - a^5b^5 + a^3b^6 - 7a^3b^2$ **77.** $x^2 + 5x + 6$ **78.** $2x^2 - x - 1$ **79.** $6t^2 - 6$ **80.** $6n^8 - 13n^6 + 5n^4$ **81.** $-5a^9 + 4a^7b + a^5b^2$ **82.** $6.6a^2 - 6.6$ **83.** $18t^2 + 3t - \frac{5}{9}$ **84.** $24b^2 - 34b + 11$ **85.** $8a^3 - 27$ **86.** $56x^4 + 15x^3 - 21x^2 - 3x + 2$ **87.** $8x^3 + 1$ **88. a.** $(6x + 10)$ in. **b.** $(2x^2 + 11x - 6)$ in.^{2} **c.** $(6x^3 + 33x^2 - 18x)$ in.^{3} **89.** $a^2 - 6a + 9$ **90.** $m^3 + 6m^2 + 12m + 8$ **91.** $x^2 - 49$ **92.** $4x^2 - 0.81$ **93.** $4y^2 + 4y + 1$ **94.** $y^4 - 1$ **95.** $36r^4 + 120r^2s + 100s^2$ **96.** $-64a^2 + 48ac - 9c^2$ **97.** $80r^4s - 80s^5$ **98.** $36b^3 - 96b^2 + 64b$ **99.** $t^2 - \frac{3}{2}t + \frac{9}{16}$ **100.** $x^2 + \frac{8}{3}x + \frac{16}{9}$ **101.** $5x^2 + 19x + 3$ **102.** $24c^2 - 10c + 37$ **103.** $(x^2 - 4)$ in.^{2} **104.** $(50x^2 - 8)$ in.^{2} **105.** $2n^3$ **106.** $-\frac{2x}{3y^2}$ **107.** $\frac{a^3}{6} - \frac{4}{a^4}$ **108.** $3a^3 + \frac{b}{5a} - \frac{5}{a^2}$ **109.** $x - 5$ **110.** $2x + 1$ **111.** $5x - 6 + \frac{4}{3x + 2}$ **112.** $5y - 3$ **113.** $3x^2 - x - 4$ **114.** $3x^2 + 2x + 1 + \frac{2}{2x - 1}$ **115.** $(y + 3)(3y + 2) = 3y^2 + 11y + 6$ **116.** $(2x^2 + 3x - 4)$ in.

Chapter 5 Test (page 466)

1. a. base, exponent **b.** monomial, binomial, trinomial **c.** degree **d.** special **2.** $2x^3y^4$ **3.** y^6 **4.** $\frac{1}{32}x^{21}$ **5.** 3.5 **6.** $\frac{2}{y^3}$ **7.** $\frac{1}{125}$ **8.** $(x + 1)^9$ **9.** y^{21} **10.** $\frac{b^3}{64a^3}$ **11.** $\frac{m^{12}}{64}$ **12.** $-6ab^9$ **13.** $1{,}000y^{12}$ in.^{3} **14.** 6.25×10^{18} **15.** 0.000093 **16.** 9.2×10^3; 9,200 **17.** Trinomial

Term	Coefficient	Degree
x^4	1	4
$8x^2$	8	2
-12	-12	0
Degree of the polynomial 4		

18. 5th degree **19.**

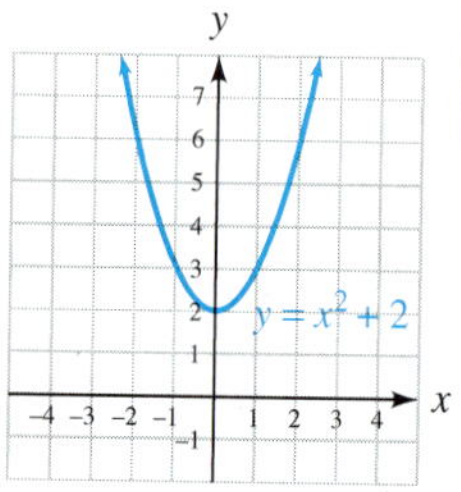

x	-2	-1	0	1	2
y	6	3	2	3	6

20. 0 ft; the rock hits the canyon floor 18 seconds after being dropped. **21.** $\frac{1}{10}x^2 + \frac{7}{12}x - 2$ **22.** $-4a^3b + a^2b + 5$ **23.** $19.4h^3 - 11.1h^2 - 0.6$ **24.** $6b^3c - 2bc - 12$ **25.** $-3y^3 + 18y^2 - 17y + 35$ **26.** $(10a^2 + 8a - 20)$ in. **27.** $10x^5y^{11}$ **28.** $-72b^8$ **29.** $3y^4 - 6y^3 + 9y^2$ **30.** $3x^2 - 11x - 20$ **31.** $12t^2 - 8t - \frac{3}{4}$ **32.** $2x^3 - 7x^2 + 14x - 12$ **33.** $1 - 100c^2$ **34.** $49b^6 - 42b^3t + 9t^2$ **35.** $2.2a^3 + 4.4a^2 - 33a$ **36.** $2x^2 + 2xy$ **37.** $\frac{a}{4b} - \frac{b}{2a}$ **38.** $x - 2$ **39.** $3x^2 + 2x + 1 + \frac{2}{2x - 1}$ **40.** $(x - 5)$ ft **41.** Yes; $(5m + 1)(m - 6) = 5m^2 - 29m - 6$ **42.** No; $(a + b)^2 = a^2 + 2ab + b^2$

Cumulative Review Chapters 1–5 (page 468)

1. $2 \cdot 3^3 \cdot 5$ **2. a.** $a + b = b + a$ **b.** $(xy)z = x(yz)$ **3.** -37 **4.** 28 **5.** $18x$ **6.** 0 **7.** -2 **8.** 15 **9.** \$2.079 billion **10.** 1.2 ft^3 **11.** 30 **12.** \$6,250 **13.** Mutual fund: \$25,000; bonds: \$20,000 **14.**

15.

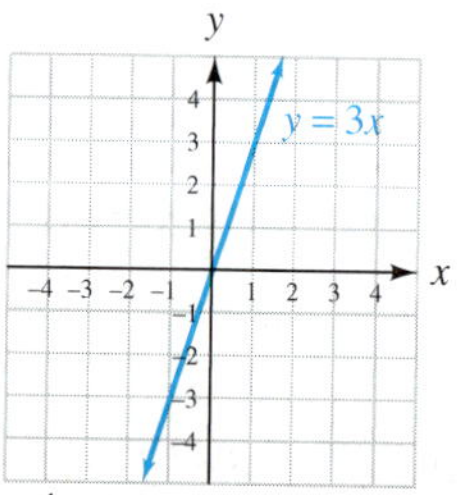

16.

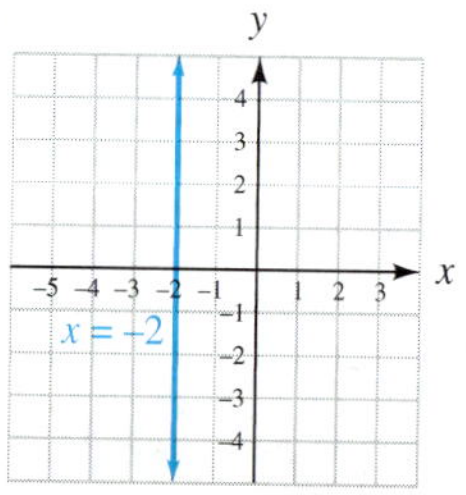

17. $-\frac{4}{9}$ **18.** $m = 3$, $(0, -2)$; $y = 3x - 2$ **19.** Perpendicular **20.** $y = -4x + 2$ **21.** No **22.** 26 **23.** No

24. (4, 1)

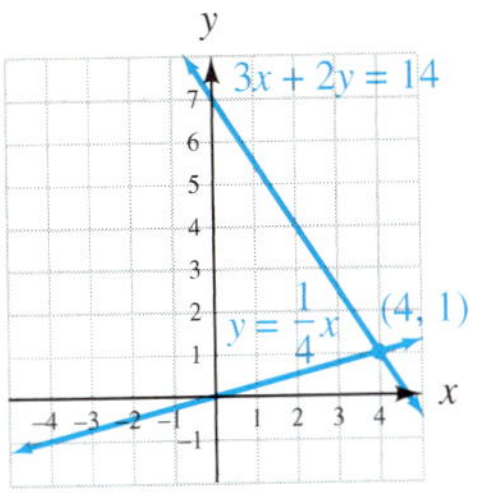

25. (−4, 3) **26.** (−2, 4)
27. Adult: \$61; child: \$51

28.

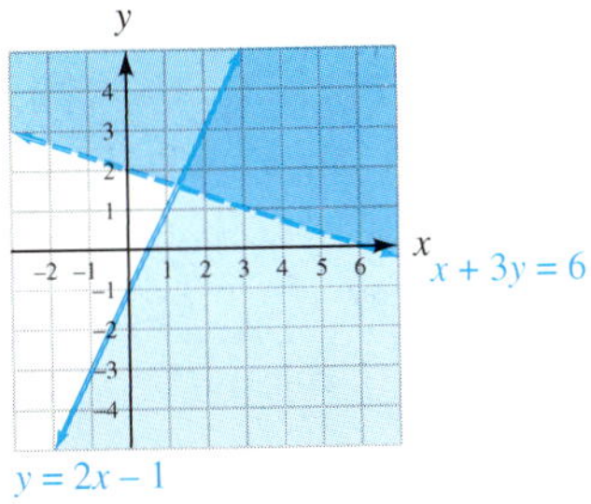

29. $9x^4y^8$ **30.** v^{22}

31. $a^2b^7c^6$ **32.** $\frac{64t^{12}}{27}$ **33.** $\frac{1}{16y^4}$ **34.** a^7 **35.** $-\frac{1}{25}$ **36.** $\frac{x^{10}}{a^{10}}$
37. 6.15×10^5 **38.** 1.3×10^{-6}

39.

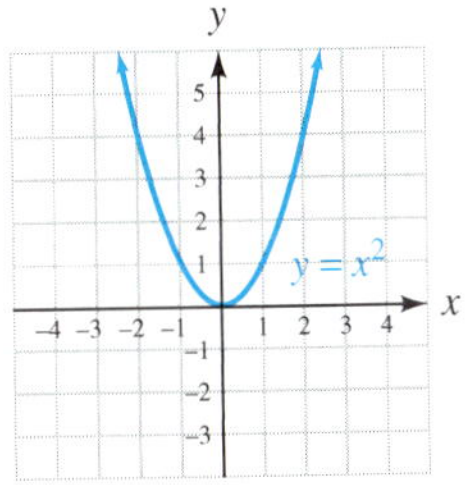

40. 1.5 in. **41.** $7c^2 + 7c$
42. $-6x^4 - 17x^2 - 68x + 11$
43. $6t^2 + 7st - 3s^2$
44. $12x^3 + 36x^2 + 27x$
45. $2x + 1$ **46.** $\frac{1}{8} - \frac{2}{x}$

Study Set Section 6.1 (page 480)

1. factor **3.** grouping **5. a.** 3 **b.** 7, h **c.** 3, y, y
7. a. $2x + 4$ (Answers may vary) **b.** $x^3 + x^2 + x$ (Answers may vary) **9.** $2x$, $x + 3$ **11. a.** 4 **b.** No **c.** 2; h **13.** 8
15. b^2, 2, $b - 6$ **17.** 2 **19.** 6 **21.** 7 **23.** 8 **25.** m^3
27. 5 **29.** $4c$ **31.** $9a^3$ **33.** $8a$ **35.** $3m^3n$ **37.** $x + 7$
39. $p - t$ **41.** $3(x + 2)$ **43.** $6(3x + 4)$ **45.** $9(2m - 1)$
47. $d(d - 7)$ **49.** $5(3c^3 + 5)$ **51.** $8a(3 - 2a)$
53. $7(2x^2 - x - 1)$ **55.** $t^2(t^2 + t + 2)$ **57.** $a(b + c - d)$
59. $3xy^2(7xy + 1)$ **61.** $-(a + b)$ **63.** $-(2x - 5)$
65. $-(3r - 2s + 3)$ **67.** $-(x^2 + x - 16)$ **69.** $-(-5 + x)$ or $-(x - 5)$ **71.** $-(-9 + 4a)$ or $-(4a - 9)$ **73.** $-3x(x + 2)$
75. $-4a^2(b - 3a)$ **77.** $-12x^2(2x^2 + 4x - 3)$
79. $-2ab^2(2a^2 - 7a + 5)$ **81.** $(x + 2)(y + 3)$
83. $(p - q)(m - 5)$ **85.** $(x + y)(2 + a)$ **87.** $(s - u)(r + 8w)$
89. $(7m - 2)(m^2 + 2)$ **91.** $(5x - 1)(x^2 + 2)$ **93.** $(b + c)(a + 1)$
95. $(r + 4s)(s - 1)$ **97.** $(2x - 3)(a + b)$ **99.** $(m - n)(p - q)$
101. $a(x - 2)(x^2 + 5)$ **103.** $6(x^2 + 2)(x - 1)$
105. $(14 + r)(h^2 + 5)$ **107.** $11a^2(2a - 3)$ **109.** $(a + b)(x - 1)$
111. $3r^5(5r^3 - 6r - 10)$ **113.** $3(3p + q)(3m - n)$
115. $-20pt^2(3p + 4t)$ **117.** $(3x - y)(2x - 5)$
119. $2z(x - 2)(x^2 + 16)$ **121.** $6uvw^2(2w - 9v)$
123. $(x + 1)(x^2 + 1)$ **125.** $(x^2 + 5)$ ft; $(x + 4)$ ft **131.** 12%

Study Set Section 6.2 (page 493)

1. factors **3.** leading **5. a.** descending **b.** common **7.** 3, 5
9. a. No **b.** Yes **11. a.** They are both positive or both negative.
b. One will be positive, the other negative. **13.** +3, −2
15. $(x + 2)(x + 1)$ **17.** $(z + 4)(z + 3)$ **19.** $(m - 3)(m - 2)$
21. $(t - 7)(t - 4)$ **23.** $(r - 3)(r - 6)$ **25.** $(a - 45)(a - 1)$
27. $(x + 8)(x - 3)$ **29.** $(t - 3)(t + 16)$ **31.** $(a - 8)(a + 2)$
33. $(b - 12)(b + 3)$ **35.** $-(x + 5)(x + 2)$ **37.** $-(t + 6)(t - 5)$
39. $-(r + 9)(r - 6)$ **41.** $-(m - 7)(m - 11)$
43. $(a + 3b)(a + b)$ **45.** $(x - 7y)(x + y)$ **47.** $(r + 2s)(r - s)$
49. $(a - 3b)(a - 2b)$ **51.** $2(x + 3)(x + 2)$ **53.** $6(a - 4)(a - 1)$
55. $5(a - 3)(a - 2)$ **57.** $-z(z - 4)(z - 25)$
59. $-n^2(n - 30)(n + 2)$ **61.** $4x^2(x + 2)(x + 2) = 4x^2(x + 2)^2$
63. $(x - 4)(x - 20)$ **65.** $(y + 9)(y + 1)$ **67.** $(r - 2)(r + 8)$
69. $(r + 3x)(r + x)$ **71.** Prime **73.** Prime **75.** $(x + 3)(5 + y)$
77. $2n(13n - 4)$ **79.** $(a - 5)(a + 1)$ **81.** $-(x - 22)(x + 1)$
83. $4(y - 1)(x + 7)$ **85.** $12b^2(2b^2 - 4b + 3)$
87. $(x + 2y)(x + 2y) = (x + 2y)^2$ **89.** $(a - 6b)(a + 2b)$
91. Prime **93.** $(x + 2)(t + 7)$ **95.** $s^2(s + 13)(s - 2)$
97. $15(s^3 + 5)$ **99.** $(y - 14)(y + 1)$ **101.** $2(x - 2)(x - 4)$
103. $(x + 9)$ in., x in., $(x + 3)$ in. **111.** $\frac{1}{x^2}$ **113.** $\frac{1}{x^{10}}$

Study Set Section 6.3 (page 503)

1. leading **3.** $5y$, y, 1, 3 **5.** $10x$ and x, $5x$ and $2x$
7. a. descending, GCF, coefficient **b.** $3s^2$ **c.** $-(2d^2 - 19d + 8)$
9. negative **11.** different **13.** −13; −6, −8; −3, −7
15. a. 12, 20, −9 **b.** −108 **17.** $3t$, 2, $4t + 3$
19. $(2x + 1)(x + 1)$ **21.** $(3a + 1)(a + 3)$ **23.** $(5x + 2)(x + 1)$
25. $(7x + 11)(x + 1)$ **27.** $(2x - 3)(2x - 1)$
29. $(4x - 1)(2x - 5)$ **31.** $(5t - 7)(3t - 1)$ **33.** $(6y - 1)(y - 2)$
35. $(3x + 7)(x - 3)$ **37.** $(5m + 3)(m - 2)$ **39.** $(7y - 1)(y + 8)$
41. $(11y - 4)(y + 1)$ **43.** $(3r + 2s)(2r - s)$
45. $(2x + 3y)(2x + y)$ **47.** $(8m + 3n)(m + 11n)$
49. $(5x + 3y)(3x - 2y)$ **51.** $2(3x + 2)(x - 5)$
53. $a(2a - 5)(4a - 3)$ **55.** $(2u + 3v)(u - 2v)$
57. $4(9y - 4)(y - 2)$ **59.** $10(13r - 11)(r + 1)$
61. $-y(y + 12)(y + 1)$ **63.** $-3x^2(2x + 1)(x - 3)$
65. $2mn(4m + 3n)(2m + n)$ **67.** $(2t - 5)(3t + 4)$
69. $(3p - q)(5p + q)$ **71.** $(2t - 1)(2t - 7)$
73. $(4y + 1)(2y - 1)$ **75.** $(18x - 5)(x + 2)$ **77.** Prime
79. Prime **81.** $3r^3(5r - 2)(2r + 5)$ **83.** $(3p - q)(2p + q)$
85. $-(4y - 3)(3y - 4)$ **87.** $(m + 7)(m - 4)$ **89.** $3a^2(2a + 5)$
91. $(x - 2)(x^2 + 5)$ **93.** $(5y - 3)(y - 1)$ **95.** $-2(x + 2)(x + 3)$
97. $3x^2y^2(4xy - 6y + 5)$ **99.** $(a - 5b)(a - 2b)$
101. $u^4(9u + 1)(u - 8)$ **103.** $(2x + 11)$ in., $(2x - 1)$ in.
109. −49 **111.** 1 **113.** 49

Study Set Section 6.4 (page 512)

1. perfect **3. a.** $5x$ **b.** 3 **c.** $5x$, 3 **5. a.** x, y **b.** −
c. +, x, y **7.** 1, 4, 9, 16, 25, 36, 49, 64, 81, 100, 121, 144, 169, 196, 225, 256, 289, 324, 361, 400 **9.** 2 **11.** +, − **13.** Yes
15. No **17.** No **19.** Yes **21.** $(x + 3)^2$ **23.** $(b + 1)^2$
25. $(c - 6)^2$ **27.** $(3y - 4)^2$ **29.** $(2x + 3)^2$ **31.** $(6m + 5n)^2$
33. $(9x - 4y)^2$ **35.** $(7t - 2s)^2$ **37.** $3(u - 3)^2$ **39.** $x(6x + 1)^2$
41. $2a^3(3a + 7b)^2$ **43.** $-(10t - 1)^2$ **45.** $(x + 2)(x - 2)$
47. $(x + 4)(x - 4)$ **49.** $(6 + y)(6 - y)$ **51.** $(t + 5)(t - 5)$
53. Prime **55.** Prime **57.** $(5t + 8)(5t - 8)$
59. $(9y + 1)(9y - 1)$ **61.** $(3x^2 + y)(3x^2 - y)$
63. $(4c + 7d^2)(4c - 7d^2)$ **65.** $8(x + 2y)(x - 2y)$
67. $7(3a + 1)(3a - 1)$ **69.** $x(x + 12)(x - 12)$
71. $6x^2(x + y)(x - y)$ **73.** $(9 + s^2)(3 + s)(3 - s)$

75. $(b^2 + 16)(b + 4)(b - 4)$ **77.** $16(t^2 + s^2)(t + s)(t - s)$
79. $25(m^2 + 1)(m + 1)(m - 1)$ **81.** $(a^2 + 12b)(a^2 - 12b)$
83. $(3xy + 5)^2$ **85.** $(t - 10)^2$ **87.** $(z + 8)(z - 8)$
89. $3(m^2 + n^2)(m + n)(m - n)$ **91.** $(5m + 7)^2$
93. $(x + 7)(x - 6)$ **95.** $(x + 3)(x - 3)$ **97.** $8a^2b(3a - 2)$
99. $-2(r - 10)(r - 4)$ **101.** $(x + 3)(x^2 + 4)$ **103.** $(2b - 5)^2$
105. $(p + q)^2$ **107.** $0.5g(t_1 + t_2)(t_1 - t_2)$ **113.** $\frac{x}{y} + \frac{2y}{x} - 3$
115. $3a + 2$

Study Set Section 6.5 (page 517)

1. sum., cubes **3. a.** F, L **b.** $-, F^2, L^2$ **5.** $6n, 5$ **7.** 1, 8, 27, 64, 125, 216, 343, 512, 729, 1,000 **9.** No **11.** $2a$ **13.** $b + 3$
15. a. $x^3 + 8$ (Answers may vary.) **b.** $(x + 8)^3$
17. $(y + 5)(y^2 - 5y + 25)$ **19.** $(a + 4)(a^2 - 4a + 16)$
21. $(n + 8)(n^2 - 8n + 64)$ **23.** $(2 + t)(4 - 2t + t^2)$
25. $(a + 10b)(a^2 - 10ab + 100b^2)$
27. $(5c + 3d)(25c^2 - 15cd + 9d^2)$ **29.** $(a - 3)(a^2 + 3a + 9)$
31. $(m - 7)(m^2 + 7m + 49)$ **33.** $(6 - v)(36 + 6v + v^2)$
35. $(2s - t)(4s^2 + 2st + t^2)$ **37.** $(10a - w)(100a^2 + 10aw + w^2)$
39. $(4x - 3y)(16x^2 + 12xy + 9y^2)$ **41.** $2(x + 1)(x^2 - x + 1)$
43. $3(d + 3)(d^2 - 3d + 9)$ **45.** $x(x - 6)(x^2 + 6x + 36)$
47. $8x(2m - n)(4m^2 + 2mn + n^2)$ **49.** $(x + 4)^2$
51. $(3r + 4s)(3r - 4s)$ **53.** $(x - t)(y + s)$
55. $4(p + 2q)(p^2 - 2pq + 4q^2)$ **57.** $2ct^2(4c + 3t)(2c + t)$
59. $36(e^2 + 1)(e + 1)(e - 1)$ **61.** $7a^2b^2(5a - 2b + 2ab)$
63. $(6r + 5s)^2$ **65.** $(1{,}000 - x^3)$ in.^{3}; $(10 - x)(100 + 10x + x^2)$
69. Repeating **71.** $\{\ldots, -3, -2, -1, 0, 1, 2, 3, \ldots\}$ **73.** 0

Study Set Section 6.6 (page 523)

1. product **3.** Factor out the GCF **5.** Perfect-square trinomial
7. Sum of two cubes **9.** Trinomial factoring **11.** Is there a common factor? **13.** $14m, m$ **15.** $2(b + 6)(b - 2)$
17. $4p^2q^3(2pq^4 + 1)$ **19.** $2(2y + 1)(10y + 1)$
21. $8(x^2 + 1)(x + 1)(x - 1)$ **23.** $(c + 21)(c - 7)$ **25.** Prime
27. $-2x^2(x - 4)(x^2 + 4x + 16)$ **29.** $(c + d^2)(a^2 + b)$
31. $-(3xy - 1)^2$ **33.** $-5m(2m + 5)^2$ **35.** $(2c + d)(c - 3d)$
37. $(p - 2)^2(p^2 + 2p + 4)$ **39.** $(x - a)(a + b)(a - b)$
41. $(ab + 12)(ab - 12)$ **43.** $(x + 5)(2x^2 + 1)$
45. $v^2(v^2 - 14v + 8)$ **47.** $(x + 2)(x - 2)(x + 3)(x - 3)$
49. $2x(2ax + b)(2ax - b)$ **51.** $2(3x - 4)(x - 1)$
53. $y^2(2x + 1)^2$ **55.** $4m^2(m + 5)(m^2 - 5m + 25)$
57. $(x + 2)(x - 2)(x^2 + 2)$ **59.** Prime
61. $2a^2(2a - 3)(4a^2 + 6a + 9)$ **63.** $27(x - y - z)$
65. $(x - t)(y + s)$ **67.** $x^6(7x + 1)(5x - 1)$
69. $5(x - 2)(1 + 2y)$ **71.** $(7p + 2q)^2$ **73.** $4(t^2 + 9)$
75. $(n + 3)(n - 3)(m^2 + 3)$
81.

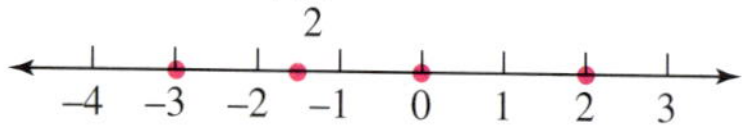

83.

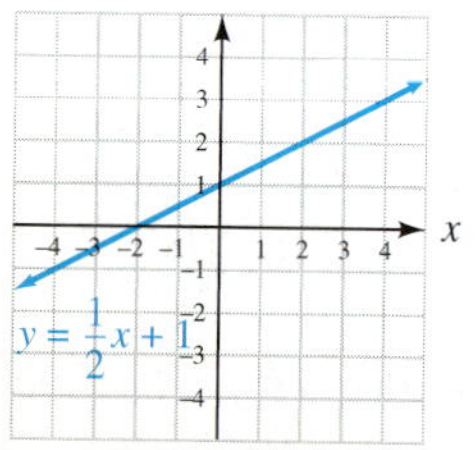

Study Set Section 6.7 (page 530)

1. quadratic **3.** zero-factor, 0, 0 **5. a.** Yes **b.** No **c.** Yes **d.** No **7.** $-\frac{4}{5}$ **9. a.** Add 6 to both sides. **b.** Distribute the multiplication by x and subtract 3 from both sides. **11.** $0, x + 7, -7$
13. $p, p - 3, 0, 3, -2$ **15.** 3, 2 **17.** −7, 7 **19.** $0, \frac{5}{2}$
21. $0, -\frac{10}{3}$ **23.** 0, 6, −8 **25.** 1, −2, 3
27. 12, 1 **29.** −3, 7 **31.** 8, 1 **33.** −3, −5 **35.** −9, 9
37. −5, 5 **39.** $-\frac{1}{2}, \frac{1}{2}$ **41.** $-\frac{7}{3}, \frac{7}{3}$ **43.** 0, 7 **45.** 0, 16
47. 0, 3 **49.** $0, -\frac{8}{3}$ **51.** $-2, \frac{1}{3}$ **53.** $-\frac{3}{2}, 1$ **55.** $\frac{1}{5}, 1$
57. $-\frac{5}{2}, 4$ **59.** $-\frac{7}{2}$ **61.** $\frac{5}{3}$ **63.** 3, 4 **65.** −3, −2
67. 0, −1, −2 **69.** 0, 9, −3 **71.** 0, 3 **73.** 0, −1, 2
75. $-\frac{9}{2}, \frac{9}{2}$ **77.** 8 **79.** $\frac{5}{2}, -6$ **81.** 2, 10 **83.** 0, −5, 4
85. −10, 10 **87.** $-\frac{1}{3}, 5$ **89.** 2, 7, 1 **91.** 0, 2 **93.** $-\frac{2}{3}, -\frac{3}{2}$
95. $0, -1, -\frac{1}{3}$ **97.** $\frac{2}{3}, -\frac{1}{5}$ **99.** $-\frac{11}{2}, \frac{11}{2}$ **101.** $\frac{1}{8}, 1$
109. 15 min $\leq t <$ 30 min

Study Set Section 6.8 (page 538)

1. consecutive **3.** hypotenuse, legs **5.** ii. **7.** $20 = b(b + 5)$
9. a. A right triangle **b.** x ft; $(x + 1)$ ft **c.** 9 ft
11. −16, 1, 0, 0, 3, −1 **13.** Width: 3 ft; length: 6 ft **15.** 8 in., 10 in. **17.** 3 ft by 9 ft **19.** Base: 6 cm; height: 5 cm
21. Foot: 4 ft; luff: 12 feet **23.** Kahne: 9; Riggs: 10 **25.** 12
27. (11, 13) **29.** 10 yd **31.** 8 ft **33.** 5 m, 12 m, 13 m
35. 5 sec **37.** 4 sec **39.** 1 sec **41.** 8 **47.** $25b^2 - 20b + 4$
49. $s^4 + 8s^2 + 16$ **51.** $81x^2 - 36$

Chapter 6 Review (page 542)

1. $5 \cdot 7$ **2.** $2^5 \cdot 3$ **3.** 7 **4.** $18a^3$ **5.** $3(x + 3y)$
6. $5a(x^2 + 3)$ **7.** $7s^3(s^2 + 2)$ **8.** $\pi a(b - c)$
9. $12x(2x^2 + 5x - 4)$ **10.** $xy^3z^2(x^4 + y^2z - 1)$
11. $-5ab(b - 2a + 3)$ **12.** $(x - 2)(4 - x)$ **13.** $-(a + 7)$
14. $-(4t^2 - 3t + 1)$ **15.** $(c + d)(2 + a)$ **16.** $(y + 6)(3x - 5)$
17. $(a + 1)(2a^2 - 1)$ **18.** $4m(n + 3)(m - 2)$ **19.** 1
20.

Factors of 6	Sum of the factors of 6
1(6)	7
2(3)	5
−1(−6)	−7
−2(−3)	−5

21. $(x + 6)(x - 4)$
22. $(x - 20)(x + 2)$
23. $(x - 5)(x - 9)$
24. Prime
25. $-(y - 8)(y - 7)$
26. $(y + 9)(y + 1)$
27. $(c + 5d)(c - 2d)$
28. $(m - 2n)(m - n)$ **29.** Multiply **30.** There are no two integers whose product is 11 and whose sum is 7.
31. $5a^3(a + 10)(a - 1)$ **32.** $-4x(x + 3y)(x - 2y)$
33. $(2x + 1)(x - 3)$ **34.** $(7y + 5)(5y - 2)$
35. $-(3x + 5)(x - 6)$ **36.** $3p(6p + 1)(p - 2)$
37. $(4b - c)(b - 4c)$ **38.** Prime **39.** $(4x + 1)$ in., $(3x - 1)$ in.
40. The signs of the second terms must be negative. **41.** $(x + 5)^2$
42. $(3y - 4)^2$ **43.** $-(z - 1)^2$ **44.** $(5a + 2b)^2$
45. $(x + 3)(x - 3)$ **46.** $(7t + 11y)(7t - 11y)$
47. $(xy + 20)(xy - 20)$ **48.** $8a(t + 2)(t - 2)$
49. $(c^2 + 16)(c + 4)(c - 4)$ **50.** Prime
51. $(b + 1)(b^2 - b + 1)$ **52.** $(x - 6)(x^2 + 6x + 36)$
53. $(p + 5q)(p^2 - 5pq + 25q^2)$
54. $2x^2(2x - 3y)(4x^2 + 6xy + 9y^2)$ **55.** $2y^2(3y - 5)(y + 4)$

56. $5(t + u^2)(s^2 + v)$ **57.** $(j^2 + 4)(j + 2)(j - 2)$
58. $-3(j + 2)(j^2 - 2j + 4)$ **59.** $(x + 1)(20 + m)(20 - m)$
60. $3w^2(2w - 3)^2$ **61.** $2(t^3 + 5)$ **62.** Prime **63.** $z(x + 8y)^2$
64. $6c^2d(3cd - 2c - 4)$ **65.** 0, 6 **66.** $\frac{7}{4}, -1$ **67.** 0, −2
68. −3, 3 **69.** $-\frac{5}{12}, \frac{5}{12}$ **70.** 3, 4 **71.** −7 **72.** 6, −4
73. $1, \frac{1}{5}$ **74.** 0, −1, 2 **75.** 15 m **76.** Streep: 14, Hepburn: 12
77. 5m **78.** 10 sec

Chapter 6 Test (page 549)

1. a. greatest, common, factor **b.** product **c.** Pythagorean **d.** difference **e.** binomials **2. a.** $45 = 3^2 \cdot 5; 30 = 2 \cdot 3 \cdot 5$ **b.** $15x^3$ **3.** $4(x + 4)$ **4.** $(q + 9)(q - 9)$
5. $5ab(6ab^2 - 4a^2b + 1)$ **6.** Prime **7.** $(x + 1)(2x + 3)$
8. $(x + 3)(x + 1)$ **9.** $-(x - 11)(x + 2)$ **10.** $x^2(x - 30)(x - 2)$
11. $(a - b)(9 + x)$ **12.** $(2a - 3)(a + 4)$ **13.** $2(3x + 5y)^2$
14. $(x + 2)(x^2 - 2x + 4)$ **15.** $5m^6(4m^2 - 3)$
16. $3(a - 3)(a^2 + 3a + 9)$ **17.** $(4x^2 + 9)(2x + 3)(2x - 3)$
18. $(a + 5)(a^2 + 1)$ **19.** $(5x - 4)$ in.
20. $(x - 9)(x + 6) = x^2 + 6x - 9x - 54 = x^2 - 3x - 54$
21. −3, 2 **22.** −5, 5 **23.** $0, \frac{1}{6}$ **24.** −3 **25.** $\frac{1}{3}, -\frac{1}{2}$
26. 9, −2 **27.** 0, −1, −6 **28.** 6 ft by 9 ft **29.** 5 sec
30. Base: 6 in.; height: 11 in. **31.** 12, 13 **32.** 10
33. A quadratic equation is an equation that can be written in the form $ax^2 + bx + c = 0$; $x^2 - 2x + 1 = 0$. (Answers may vary.)
34. At least one of them is 0.

Cumulative Review Chapters 1–6 (page 552)

1. About 35 beats/min difference **2.** $2 \cdot 5^3$ **3.** $\frac{24}{25}$ **4.** 0.992
5. a. False **b.** True **c.** True **6.** $\frac{5}{0}$ **7.** −39 **8.** −5
9. −27 **10.** $20x **11.** 3 **12.** 8, −1, 9 **13.** $-13y^2 + 6$
14. $3y + z$ **15.** $-\frac{3}{2}$ **16.** −2 **17.** 9 **18.** $-\frac{55}{6}$ **19.** 248 lb
20. 330 mi **21.** $I = Prt$ **22.** 12.6 in.^{2} **23.** $t = \frac{A - P}{Pr}$
24. 22nd president, 24th president **25.** 4 L
26. $(-\infty, -2)$

27. Yes

28.

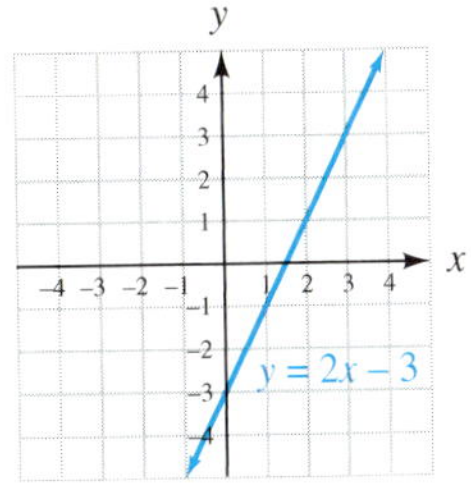

29. Vertical line **30.** They are the same. **31.** 1.4 gal/yr **32.** 1; (0, −2) **33.** $y = 7x + 19$

34.

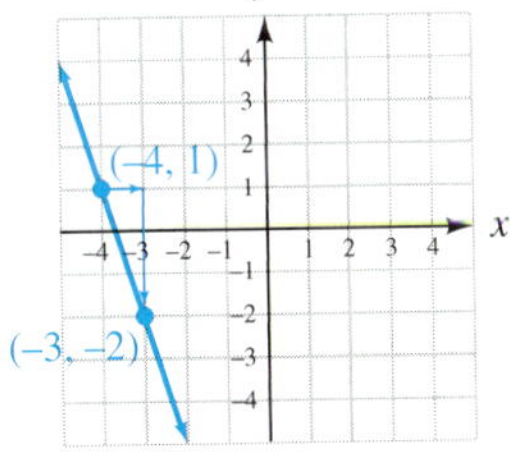

35.

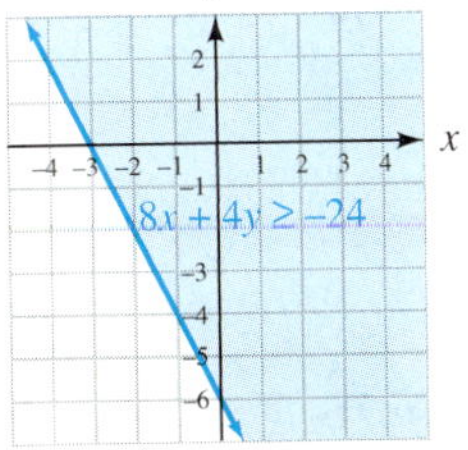

36. 17 **37.** Yes **38.**

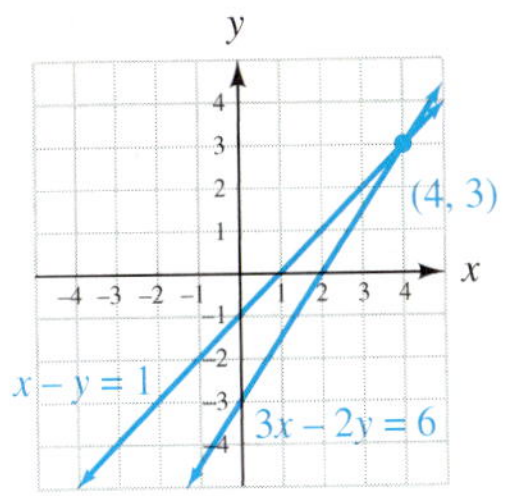

39. $\left(\frac{3}{4}, -2\right)$ **40.** $\left(-2, \frac{2}{3}\right)$ **41.** Newspaper: 8 tons; cardboard: 6 tons **42.**

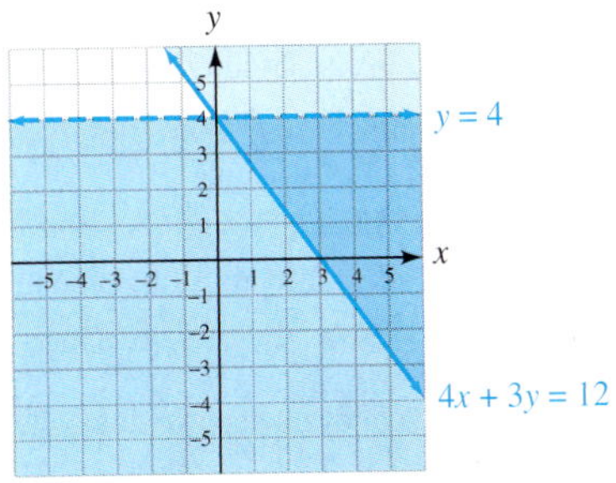

43. $-4y^5$

44. x^4y^{23} **45.** b^7 **46.** 2 **47.** 9.011×10^{-5} **48.** 1.7×10^6
49. 3 **50.**

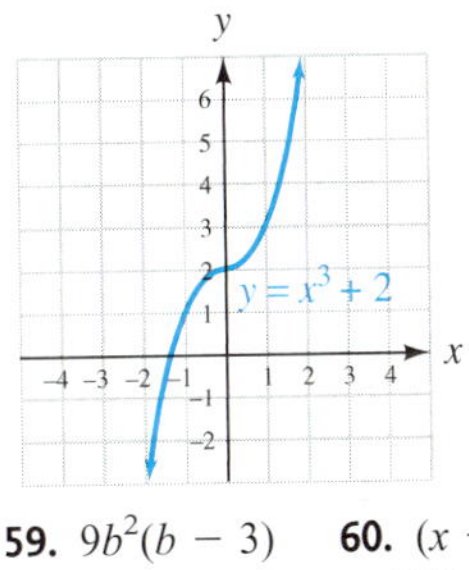

51. $-2x^2 - 4x + 5$
52. $8b^5 - 8b^4$
53. $3x^2 + 10x - 8$
54. $y^2 - 12y + 36$
55. $3ab - 2a - 1$
56. $2x + 1$
57. a. $(4x + 8)$ in. **b.** $(x^2 + 4x + 3)$ in.^{2} **c.** $(x^3 + 4x^2 + 3x)$ in.^{3}
58. $6x^5y$ **59.** $9b^2(b - 3)$ **60.** $(x + y)(a + b)$
61. $(u + 3)(u - 1)$ **62.** $(2x + 1)(5x - 2)$ **63.** $(2a - 3)^2$
64. $(3z + 1)(3z - 1)$ **65.** $(t - 2)(t^2 + 2t + 4)$
66. $3(b^2 - 2)(a + 1)(a - 1)$ **67.** $0, \frac{4}{3}$ **68.** $\frac{1}{2}, 2$

Study Set Section 7.1 (page 564)

1. rational **3.** undefined **5.** $\frac{6}{24} = \frac{1}{4}$ **7. a.** 1 **b.** −1 **c.** 1 **d.** Does not simplify **9.** $x, 1, x + 1, x + 3$ **11.** 4 **13.** 0
15. Undefined **17.** $-\frac{2}{11}$ **19.** Undefined **21.** $\frac{1}{6}$ **23.** 0
25. 2 **27.** None **29.** $\frac{1}{2}$ **31.** −6, 6 **33.** −2, 1 **35.** $\frac{5}{a}$
37. $\frac{3}{2}$ **39.** $\frac{7c^2}{3d^2}$ **41.** $\frac{9a^2}{11b}$ **43.** $\frac{2x + 1}{y}$ **45.** $\frac{1}{3}$ **47.** $\frac{x + 2}{x - 4}$
49. Does not simplify **51.** $\frac{1}{2b + 1}$ **53.** $\frac{m - n}{7(m + n)}$ or $\frac{m - n}{7m + 7n}$
55. $\frac{10}{3}$ **57.** $\frac{2x}{x - 6}$ **59.** −1 **61.** $-\frac{1}{2}$ **63.** $-\frac{1}{a + 1}$
65. $-\frac{5}{m + 5}$ **67.** $\frac{1}{a}$ **69.** $-\frac{x + 2}{x + 1}$ **71.** −6 **73.** $\frac{x + 1}{x - 1}$ **75.** $\frac{3x}{y}$
77. $\frac{4 - x}{4 + x}$ or $-\frac{x - 4}{x + 4}$ **79.** 4 **81.** $\frac{3(x + 3)}{2x + 1}$ or $\frac{3x + 9}{2x + 1}$ **83.** $-\frac{3x + 11}{x + 3}$
85. Does not simplify **87.** $\frac{2u - 3}{u^3}$ **89.** $(2x + 3)^2$ **91.** 9
93. $\frac{3x}{5y}$ **95.** $85\frac{1}{3}$ **97.** 2, 1.6, and 1.2 milligrams per liter
103. a. $(a + b) + c = a + (b + c)$ **b.** $ab = ba$

Study Set Section 7.2 (page 573)

1. reciprocal **3. a.** numerators, denominators, reciprocal **b.** AC, BD, D, C **5.** $-\frac{y^2}{y + 1}$ **7.** 1 **9.** ft **11.** $\frac{3y}{14}$
13. $\frac{3(y + 2)}{y^3}$ or $\frac{3y + 6}{y^3}$ **15.** $\frac{20}{3n}$ **17.** x^2y^2 **19.** $\frac{x}{5}$ **21.** −2

23. $\frac{3}{2x}$ **25.** $x + 1$ **27.** $-(x - 2)$ or $-x + 2$ **29.** $\frac{(x - 2)^2}{x}$
31. $\frac{(m - 2)(m - 3)}{2(m + 2)}$ **33.** $\frac{3(a + 3)^2}{2a^3}$ **35.** 35 **37.** $3x + 3$
39. $10y - 16$ **41.** $\frac{36a - 60}{a}$ **43.** $\frac{3}{2y}$ **45.** $\frac{3a}{5}$ **47.** $\frac{x^2}{3}$ **49.** $\frac{9p^3}{5}$
51. $\frac{5(a - 2)}{4a^3}$ **53.** $-\frac{x + 2}{3}$ **55.** $-(m + 5)$ or $-m - 5$ **57.** $t + 7$
59. $\frac{2(x - 7)}{x + 9}$ **61.** $\frac{d(6c - 7d)}{6}$ **63.** $\frac{1}{3}$ **65.** 1 **67.** $\frac{1}{12(2r - 3s)}$
69. $\frac{4(n - 1)}{3n}$ **71.** 450 ft **73.** $\frac{3}{4}$ gal **75.** $\frac{1}{2}$ mi per min
77. 1,800 m per min **79.** $\frac{b - 3}{b}$ **81.** $\frac{1}{(x + 1)^2}$ **83.** $25h - 15$
85. $n - 1$ **87.** $\frac{5r^3}{2s^2}$ **89.** $\frac{7(p + 2)}{3p^4}$ or $\frac{7p + 14}{3p^4}$ **91.** $\frac{x - 2}{x - 3}$
93. $\frac{2x - 3y}{y(2x + 3y)}$ **95.** $\frac{x^2}{10}$ ft^2 **97.** 4,380,000 **99.** 8 yd^2
101. $\frac{1}{2}$ mi per min **103.** $\frac{1}{4}$ mi^2 **109.** $w = 6$ in., $l = 10$ in.

Study Set Section 7.3 (page 583)

1. denominator **3.** build **5.** numerators, denominator, $A + B, D, A - B, D$ **7.** $\frac{4}{5}$ **9. a.** Twice **b.** Once
11. $x - 1, -, +, 6$ **13.** $\frac{11}{x}$ **15.** $\frac{x + 5}{18}$ **17.** $\frac{x}{3}$ **19.** $\frac{1}{3a^2}$
21. $\frac{x + 4}{y}$ **23.** $\frac{1}{r - 5}$ **25.** 9 **27.** $\frac{3}{4}$ **29.** $\frac{x}{25}$ **31.** $\frac{2t}{9}$
33. $\frac{m - 6}{6m^2}$ **35.** $\frac{5a}{a + 2}$ **37.** $\frac{1}{t + 2}$ **39.** $\frac{2}{w(w - 9)}$ **41.** $\frac{1}{2}$
43. $2x - 5$ **45.** $\frac{1}{y}$ **47.** 0 **49.** $\frac{2x}{3 - x^2}$ **51.** $-\frac{1}{3x - 1}$ **53.** $6x$
55. $30a^3$ **57.** $3a^2b^3$ **59.** $c(c + 2)$ **61.** $(3x + 1)(3x - 1)$
63. $12(b + 2)$ **65.** $8k(k + 2)$ **67.** $(x + 1)(x - 1)$
69. $(x + 1)(x + 5)(x - 5)$ **71.** $(2n + 5)(n + 4)^2$ **73.** $\frac{50}{10r}$
75. $\frac{8xy}{x^2y}$ **77.** $\frac{27b}{12b^2}$ **79.** $\frac{3x^2 + 3x}{(x + 1)^2}$ **81.** $\frac{x^2 + x - 6}{x(x + 3)}$ **83.** $\frac{5t + 25}{20(t + 2)}$
85. $\frac{4y^2 + 12y}{4y(y - 2)(y - 3)}$ **87.** $\frac{36 - 3h}{3(h + 9)(h - 9)}$ **89.** $\frac{3}{t - 7}$ **91.** $\frac{1}{c + d}$
93. $\frac{7n + 1}{(n + 4)(n - 2)}$ **95.** $\frac{5}{9y}$ **97.** $3x - 2$ **99.** $\frac{3}{r}$ **101.** $\frac{2x + 6}{x + 2}$ ft
109. a. $I = Prt$ **b.** $A = \frac{1}{2}bh$ **c.** $P = 2l + 2w$

Study Set Section 7.4 (page 592)

1. unlike **3. a.** $2 \cdot 2 \cdot 5 \cdot x \cdot x$ **b.** $(x - 2)(x + 6)$
5. $(x + 6)(x + 3)$ **7.** $\frac{5}{5}$ **9.** $3x, 5, 15x, 15x, 35$ **11.** $\frac{13x}{21}$
13. $\frac{5y}{2}$ **15.** $\frac{7t - 32}{8t}$ **17.** $\frac{11b}{12}$ **19.** $\frac{7 - 2m}{m^2}$ **21.** $\frac{17x + 3}{x^2}$
23. $\frac{1}{10p}$ **25.** $-\frac{29}{24t}$ **27.** $\frac{41}{30x}$ **29.** $\frac{16c^2 + 3}{18c^4}$ **31.** $\frac{a + 8}{2(a + 2)(a - 2)}$
33. $\frac{6a + 9}{(3a + 2)(3a - 2)}$ **35.** $\frac{4a + 1}{(a + 2)^2}$ **37.** $\frac{6 - 3m}{5m(m - 1)}$
39. $\frac{17t + 42}{(t + 3)(t + 2)}$ **41.** $\frac{2x^2 + 11x}{(2x - 1)(2x + 3)}$ **43.** $\frac{35x^2 + x + 5}{5x(x + 5)}$
45. $\frac{2x^2 - 1}{x(x + 1)}$ **47.** $\frac{14s + 58}{(s + 3)(s + 7)}$ **49.** $\frac{2m^2 + 20m - 6}{(m - 2)(m + 5)}$
51. $\frac{s^2 + 8s + 4}{(s + 4)(s + 1)(s + 1)}$ **53.** $\frac{2x + 13}{(x - 8)(x - 1)(x + 2)}$ **55.** $\frac{1}{a + 1}$
57. $\frac{1}{(y + 3)(y + 4)}$ **59.** $\frac{6x + 8}{x}$ **61.** $\frac{x^2 - 4x + 9}{x - 4}$ **63.** $\frac{a^2b - 3}{a^2}$
65. $\frac{-4x - 3}{x + 1}$ or $-\frac{4x + 3}{x + 1}$ **67.** $\frac{2}{a - 4}$ **69.** $\frac{c + d}{7c - d}$ **71.** $\frac{6d - 3}{d - 9}$
73. $\frac{1}{g + 2}$ **75.** $\frac{j - 5}{(j + 3)(j + 5)}$ **77.** $\frac{xy - y + 10}{x - 1}$ **79.** $\frac{2b + 1}{(b + 1)(b + 2)}$
81. $\frac{y + 4}{y - 1}$ **83.** $\frac{2n + 2}{15}$ **85.** $\frac{y^2 + 7y + 6}{15y^2}$ **87.** $\frac{x + 2}{x - 2}$
89. $\frac{10a - 14}{3a(a - 2)}$ **91. a.** $\frac{75 + 8x^2}{30x}$ **b.** $\frac{2}{3}$ **93. a.** $\frac{t^2 + 4t}{(t - 5)(t + 5)}$
b. $t + 5$ **95.** $\frac{20x + 9}{6x^2}$ cm **101.** 8; (0, 2) **103.** 0

Study Set Section 7.5 (page 601)

1. complex, complex **3.** simplify, division, reciprocal
5. a. $\frac{x - 3}{4}$, yes **b.** $\frac{1}{12} - \frac{x}{6}$; no **7.** $\div$ **9.** $\frac{8}{9}$ **11.** $\frac{5x}{12}$ **13.** $\frac{x^2}{y}$
15. $\frac{n^3}{8}$ **17.** $\frac{10}{3}$ **19.** $-\frac{x^3}{14}$ **21.** $\frac{5x}{3}$ **23.** $\frac{t^3}{2}$ **25.** $\frac{5}{7}$
27. $\frac{3y + 12}{4y^2 - 6y}$ **29.** $\frac{2 - 5y}{6}$ **31.** $\frac{24 - c^2}{12}$ **33.** $\frac{s^2 - s}{2 + 2s}$ **35.** $\frac{b - 5ab}{3ab - 7a}$
37. $\frac{5}{4}$ **39.** $\frac{1 - 3x}{5 + 2x}$ **41.** $\frac{1 + x}{2 + x}$ **43.** $\frac{3 - x}{x - 1}$ **45.** $\frac{32h - 1}{96h + 6}$
47. $\frac{d - 3}{2d}$ **49.** $\frac{x - 12}{x + 6}$ **51.** $a - 7$ **53.** $\frac{m^2 + n^2}{m^2 - n^2}$ **55.** $\frac{6d + 12}{d}$
57. $\frac{8}{4c + 5c^2}$ **59.** $\frac{y}{x - 2y}$ **61.** $\frac{1}{x + 2}$ **63.** $\frac{1}{x + 3}$ **65.** $\frac{x}{x - 2}$
67. $2m - 1$ **69.** $\frac{q + p}{q}$ **71.** $18x$ **73.** $\frac{2c}{2 - c}$ **75.** $\frac{r - 1}{r + 1}$
77. $\frac{b + 9}{8a}$ **79.** $\frac{t + 2}{t - 3}$ **81.** $\frac{xy}{y + x}$ **83.** $-\frac{10x^3}{3}$ **85.** $\frac{7}{6}$
87. $\frac{R_1R_2}{R_2 + R_1}$ **93.** 1 **95.** $\frac{25}{16x^{12}}$

Study Set Section 7.6 (page 610)

1. rational **3.** multiply **5. a.** Yes **b.** No **7. a.** 3, 0 **b.** 3, 0 **c.** 3, 0 **9. a.** y **b.** $(x + 2)(x - 2)$ **11. a.** 3 **b.** $3(x + 6)$ **13.** $2a, 2a, 2a, 2a, 2a, 4, 7, 4, 3$ **15.** 1 **17.** 0
19. $\frac{3}{5}$ **21.** $-\frac{4}{3}$ **23.** $-\frac{12}{7}$ **25.** -48 **27.** 2, 4 **29.** $-5, 2$
31. $-4, -5$ **33.** $3, -\frac{5}{3}$ **35.** 1 **37.** No solution; 5 is extraneous
39. No solution; -2 is extraneous **41.** 7 **43.** 1 **45.** $-1, 6$
47. 0, 3 **49.** No solution; 2 is extraneous **51.** -3 **53.** $\frac{1}{5}$
55. $P = nrt$ **57.** $d = \frac{bc}{a}$ **59.** $A = \frac{h(b + d)}{2}$ **61.** $r = \frac{E - IR}{I}$
63. $a = \frac{b}{b - 1}$ **65.** $x = \frac{5yz}{5y + 4z}$ **67.** $r = \frac{st}{s - t}$
69. $L^2 = 6dF - 3d^2$ **71.** 6 **73.** 1 **75.** -40 **77.** No solution; -1 is extraneous **79.** $-4, 3$ **81.** 3 **83.** $\frac{9}{40}$
85. 0 **87.** 1, 2 **89.** 7 **91. a.** $\frac{2a + 3}{5}$ **b.** $-\frac{9}{4}$
93. a. $\frac{x^2 - 4x + 2}{(x - 2)(x - 3)}$ **b.** 4 **95.** $R = \frac{HB}{B - H}$ **97.** $r = \frac{r_1r_2}{r_2 + r_1}$
103. 20

Study Set Section 7.7 (page 620)

1. motion, investment, work **3.** iii **5.** $\frac{1}{45}$ of the job per minute
7. a. $t = \frac{d}{r}$ **b.** $P = \frac{I}{rt}$ **9.** $\frac{x}{15}, \frac{x}{8}$ **11.** $6\frac{1}{9}$ days **13.** 4 **15.** 2
17. 5 **19.** $\frac{2}{3}, \frac{3}{2}$ **21.** 8 **23.** Garin: 16 mph, Armstrong: 26 mph
25. 1st: $1\frac{1}{2}$ ft per sec, 2nd: $\frac{1}{2}$ ft per sec **27.** Canada goose: 30 mph, great blue heron: 20 mph **29.** $\frac{300}{255 + x}, \frac{210}{255 - x}$; 45 mph
31. $2\frac{6}{11}$ days **33.** No, after the pipes are opened, the swimming is scheduled to take place in 6 hours. It takes 7.2 hr (7 hr 12 min) to fill the pool. **35.** 8 hr **37.** 20 min **39.** $1\frac{4}{5}$ hr $= 1.8$ hr
41. Credit union: 4%, bonds: 6% **43.** 7% and 8% **47.** (1, 3)
49. Yes **51.** 0

Study Set Section 7.8 (page 632)

1. ratio, rate **3.** extremes, means **5.** unit **7.** equal, *ad*, *bc*
9. 25, 2, 1,000, x **11.** 2.19, 1 **13.** x, 288, 18, 18 **15.** as, to
17. $\frac{4}{15}$ **19.** $\frac{3}{4}$ **21.** $\frac{5}{4}$ **23.** $\frac{1}{2}$ **25.** $\frac{1}{2}$ **27.** $\frac{25}{22}$ **29.** Yes
31. No **33.** 4 **35.** 14 **37.** 0 **39.** $-\frac{3}{2}$ **41.** -2 **43.** -27
45. $2, -2$ **47.** $6, -1$ **49.** $-1, 16$ **51.** $-\frac{5}{2}, -1$ **53.** 15
55. 8 **57.** $-\frac{1}{3}, 2$ **59.** 2 **61.** $-\frac{27}{2}$ **63.** $-10, 10$ **65.** $-4, 3$
67. $\frac{26}{9}$ **69. a.** $\frac{3}{2}$, 3:2 **b.** $\frac{2}{3}$, 2:3 **71.** \$62.50 **73.** 20
75. 84 **77.** 568, 13, 14 **79.** 510 **81.** Not exactly, but close
83. 140 **85.** 522 in.; 43.5 ft **87.** 10 ft **89.** 45 min for \$25

91. 150 for \$12.99 **93.** 6-pack for \$1.50 **95.** 24 12-oz bottles **97.** 39 ft **99.** $46\frac{7}{8}$ ft **105.** 90% **107.** 480

Study Set Section 7.9 (page 642)

1. direct, inverse **3.** Direct **5.** Inverse

7.

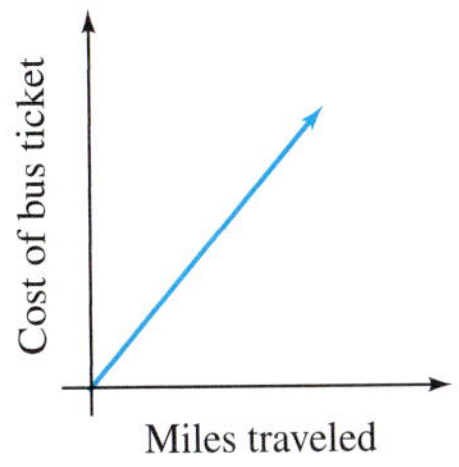

9.

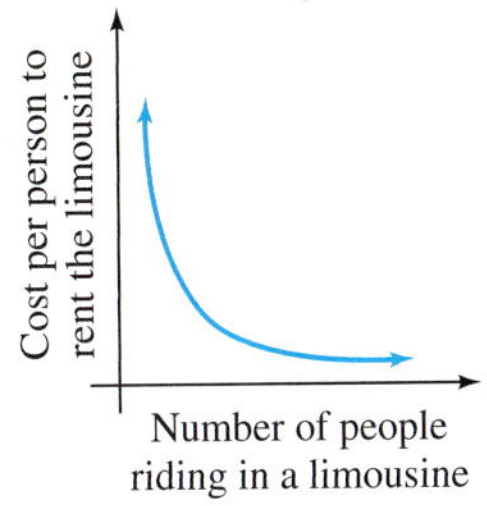

11. $p = kd$ **13.** $d = \frac{k}{r}$ **15. a.** Yes **b.** No **c.** No **d.** Yes **17.** 35 **19.** 36 **21.** 9 **23.** 14 **25.** 4 **27.** 2 **29.** 16 **31.** 2.25 or $\frac{9}{4}$ **33.** 168 mi **35.** 173.6 mg **37.** $2\frac{1}{2}$ hr **39.** 8 amps **41.** 180 **43.** 55 lb **45.** 40 cm **47.** $53\frac{1}{3}$ m^3 **53.** 1, −2, −3 **55.** 0, −3, 9

Chapter 7 Review (page 646)

1. 4, −4 **2.** $-\frac{3}{7}$ **3.** $\frac{1}{2x}$ **4.** $\frac{5}{2x}$ **5.** $\frac{x}{x+1}$ **6.** $a - 2$ **7.** −1 **8.** $-\frac{1}{x+3}$ **9.** $\frac{x}{x-1}$ **10.** Does not simplify **11.** $\frac{1}{x-y}$ **12.** $\frac{4}{3}$ **13.** x is not a common factor of the numerator and the denominator; x is a term of the numerator. **14.** 150 mg **15.** $\frac{3x}{y}$ **16.** 96 **17.** $\frac{x-1}{x+2}$ **18.** $\frac{2x}{x+1}$ **19.** $\frac{3y}{2}$ **20.** $-x - 2$ **21. a.** Yes **b.** No **c.** Yes **d.** Yes **22.** $\frac{1}{3}$ mi per min **23.** $\frac{1}{3d}$ **24.** 1 **25.** $\frac{2x+2}{x-7}$ **26.** $\frac{1}{a-4}$ **27.** $9x$ **28.** $8x^3$ **29.** $m(m - 8)$ **30.** $(5x + 1)(5x - 1)$ **31.** $(a + 5)(a - 5)$ **32.** $(2t + 7)(t + 5)^2$ **33.** $\frac{63}{7a}$ **34.** $\frac{2xy + x}{x(x-9)}$ **35.** $\frac{2b+14}{6(b-5)}$ **36.** $\frac{9r^2 - 36r}{(r+1)(r-4)(r+5)}$ **37.** $\frac{a-7}{7a}$ **38.** $\frac{x^2+x-1}{x(x-1)}$ **39.** $\frac{1}{t+1}$ **40.** $\frac{x^2+4x-4}{2x^2}$ **41.** $\frac{b+6}{b-1}$ **42.** $\frac{6c+8}{c}$ **43.** $\frac{14n+58}{(n+3)(n+7)}$ **44.** $\frac{4t+1}{(t+2)^2}$ **45.** $\frac{1}{(a+3)(a+2)}$ **46.** $\frac{17y-2}{12(y-2)(y+2)}$ **47.** Yes **48.** $\frac{14x+28}{(x+6)(x-1)}$ units, $\frac{12}{(x+6)(x-1)}$ square units **49.** $\frac{n^3}{14}$ **50.** $\frac{r+9}{8s}$ **51.** $\frac{1+y}{1-y}$ **52.** $\frac{21}{3a+10a^2}$ **53.** $x^2 + 3$ **54.** $\frac{y-5xy}{3xy-7x}$ **55.** 3 **56.** No solution; 5 is extraneous **57.** 3 **58.** 2, 4 **59.** 0 **60.** −4, 3 **61.** $T_1 = \frac{T_2}{1-E}$ **62.** $y = \frac{xz}{z-x}$ **63.** 3 **64.** 5 mph **65.** $\frac{1}{4}$ of the job per hr **66.** $5\frac{5}{6}$ days **67.** 5% **68.** 40 mph **69.** No **70.** Yes **71.** $\frac{9}{2}$ **72.** 0 **73.** 7 **74.** 4, $-\frac{3}{2}$ **75.** 255 **76.** 20 ft **77.** 5 ft 6 in. **78.** 250 for \$98 **79.** $c = kt$ **80.** $f = \frac{k}{L}$ **81.** \$2,000 **82.** 600 **83.** 1.25 amps **84.** Inverse variation

Chapter 7 Test (page 655)

1. a. rational **b.** similar **c.** proportion **d.** build **e.** factors **2.** 10 words **3.** 0 **4.** −3, 2 **5.** 3,360,000 or 3,360K bits per minute **6.** 5 is not a common factor of the numerator, and therefore cannot be removed. 5 is a term of the numerator. **7.** $\frac{8x}{9y}$ **8.** −7 **9.** $\frac{x+1}{2x+3}$ **10.** 1 **11.** $3c^2d^3$ **12.** $(n + 1)(n + 5)(n - 5)$ **13.** $\frac{5y^2}{4}$ **14.** $\frac{x+1}{3(x-2)}$ **15.** $-\frac{x^2}{3}$ **16.** $\frac{1}{6}$ **17.** 3 **18.** $\frac{4n - 5mn}{10m}$ **19.** $\frac{2x^2+x+1}{x(x+1)}$ **20.** $\frac{2a+7}{a-1}$ **21.** $\frac{c^2-4c+9}{c-4}$ **22.** $\frac{1}{(t+3)(t+2)}$ **23.** $\frac{12}{5m}$ **24.** $\frac{a+2s}{2as^2-3a^2}$ **25.** 11 **26.** No solution, 6 is extraneous **27.** 1 **28.** 1, 2 **29.** $\frac{2}{3}$ **30.** $B = \frac{HR}{R-H}$ **31.** Yes **32.** 1,785 **33.** 171 ft **34.** 80 sheets for \$3.89 **35.** $3\frac{15}{16}$ hr **36.** 4 mph **37.** 2 **38.** We multiply both sides of the equation by the LCD of the rational expressions appearing in the equation. The resulting equation is easier to solve. **39.** 100 lb **40.** $\frac{80}{3}$ or $26\frac{2}{3}$

Cumulative Review Chapters 1–7 (page 657)

1. a. False **b.** False **c.** True **d.** True **2.** < **3.** 36 **4.** 77 **5.** $6c + 62$ **6.** −5 **7.** 3 **8.** About 26% **9.** $B = \frac{A - c - r}{2}$ **10.** 104°F **11.** 240 ft^3 **12.** 12 lb of the \$6.40 tea and 8 lb of the \$4 tea **13.** 500 mph **14.** $[-1, \infty)$

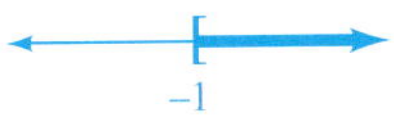

15.

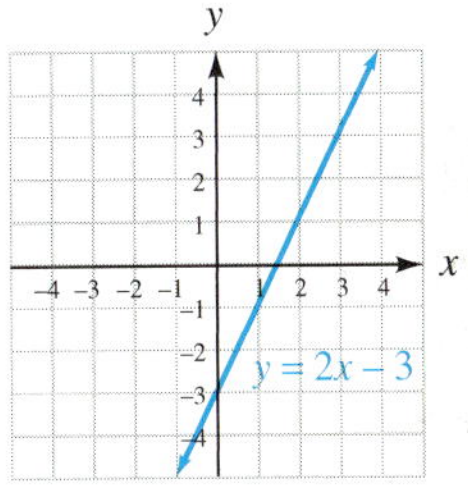

16. −1 **17.** 0.008 mm/m **18.** $\frac{8}{7}$ **19.** $y = 3x + 2$

20.

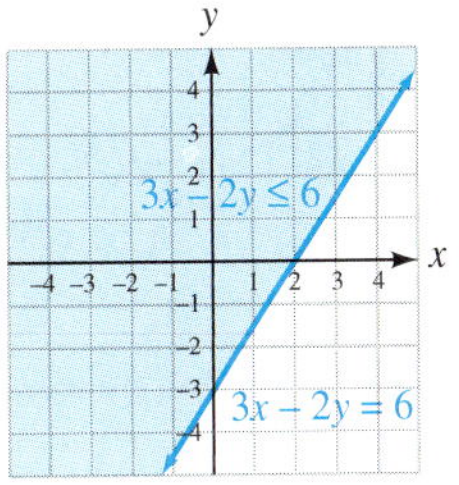

21. 0 **22.** domain, range

23.

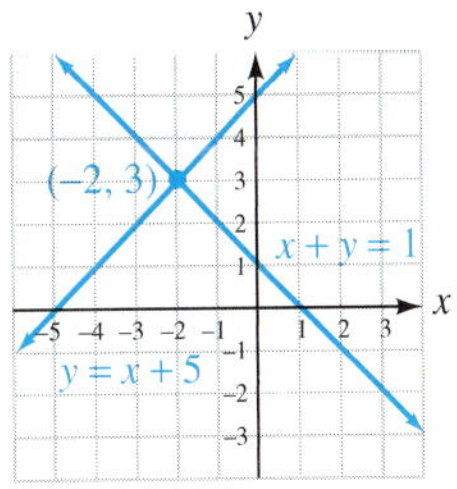

24. (5, 2) **25.** (1, −1) **26.** Red: 8, blue: 15 **27.** x^7 **28.** x^{25} **29.** $\frac{y^3}{8}$ **30.** $-\frac{32a^5}{b^5}$ **31.** $\frac{a^8}{b^{12}}$ **32.** $3b^2$ **33.** 2.9×10^5 **34.** 3

35.

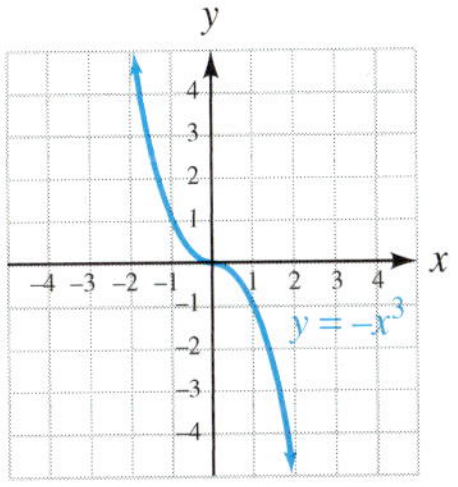

36. $A = \pi R^2 - \pi r^2$ **37.** $6x^2 + x - 5$ **38.** $-\frac{1}{2}t^3 - \frac{7}{4}t^2 - \frac{1}{12}$ **39.** $6x^4y^5$ **40.** $6y^2 - y - 35$ **41.** $-12x^4z + 4x^2z^2$ **42.** $9a^2 - 24a + 16$ **43.** $2x + 3$ **44.** $x^2 + 2x - 1$ **45.** $k^2t(k - 3)$ **46.** $(b + c)(2a + 3)$ **47.** $(u - 9)^2$ **48.** $-(r - 2)(r + 1)$ **49.** Prime **50.** $(2x - 9)(3x + 7)$ **51.** $2(a + 10b)(a - 10b)$ **52.** $(b + 5)(b^2 - 5b + 25)$ **53.** 0, $-\frac{1}{5}$ **54.** $\frac{1}{3}, \frac{1}{2}$ **55.** 16 in., 10 in. **56.** 5, −5 **57.** $\frac{2x}{x-2}$ **58.** $-\frac{x^3}{3}$

59. $2m-5$ **60.** $-\frac{1}{x-3}$ **61.** $\frac{m-3}{(m+3)(m+1)}$ **62.** $\frac{x^2}{(x+1)^2}$
63. $\frac{17}{25}$ **64.** 2 **65.** $14\frac{2}{5}$ hr **66.** 34.8 ft **67.** 28 in.
68. $\frac{3}{4}=0.75$

Study Set Section 8.1 (page 670)

1. square **3.** radicand **5.** Pythagorean, a^2, b^2, c^2 **7.** two, $5, -5, 5, -5$ **9.** 0 **11. a.** 64 **b.** $36y^4$ **c.** d **d.** $\frac{16}{25}$ **13.** 25, 36 **15.** 5, 144, 169, c, 13, 13 **17.** 8 **19.** 6 **21.** 10 **23.** 20 **25.** -9 **27.** 1.1 **29.** 13 **31.** $\frac{3}{5}$ **33.** $-\frac{1}{8}$ **35.** -0.2 **37.** -17 **39.** 50 **41.** 1.732 **43.** 9.747 **45.** 20.688 **47.** 3.464 **49.** Rational **51.** Not a real number **53.** Irrational **55.** Rational **57.** m **59.** t^2 **61.** c^5 **63.** n^6 **65.** x^{18} **67.** $2y$ **69.** $8b^2$ **71.** $-7s^7$ **73.** 5 **75.** 8 **77.** 28 **79.** 117 **81.** $\sqrt{2}$ in. $\approx$ 1.41 in. **83.** 5 **85.** 13 **87.** $\sqrt{37}\approx 6.08$ **89.** $\sqrt{29}\approx 5.39$ **91.** 54 mph; 19 mph **93.** 10.72 knots **95. a.** $6\sqrt{2}$ in. $\approx$ 8.49 in. **b.** $5\sqrt{3}$ in. $\approx$ 8.66 in. **97.** $\sqrt{16{,}200}$ ft $\approx$ 127.28 ft **99. a.** 16 in., 30 in. **b.** 34 in. **101. a.** $\sqrt{18}$ yd $\approx$ 4.24 yd **b.** Yes: 6 yd + 4.24 yd > 10 yd **103.** Brace A: $\sqrt{18}$ m $\approx$ 4.24 m; Brace B: $\sqrt{45}\approx$ 6.71 m; Brace C: $\sqrt{5}$ m $\approx$ 2.24 m **109.** $6s^2+s-5$ **111.** $3x^2+10x-8$

Study Set Section 8.2 (page 682)

1. simplify **3.** product, square **5. a.** product, $\sqrt{a}\sqrt{b}$ **b.** quotient, $\frac{\sqrt{a}}{\sqrt{b}}$ **7. a.** 4 **b.** 49 **c.** 9 **d.** 16 **9.** $\sqrt{4\cdot 10}$ **11. a.** 81, 2, 9, 2 **b.** m^4, m, m^2, m **13.** times, radical **15.** $2\sqrt{5}$ **17.** $3\sqrt{3}$ **19.** $5\sqrt{2}$ **21.** $2\sqrt{6}$ **23.** $4\sqrt{3}$ **25.** $2\sqrt{11}$ **27.** Cannot be simplified **29.** $3\sqrt{7}$ **31.** $7\sqrt{2}$ **33.** $6\sqrt{5}$ **35.** $8\sqrt{3}$ **37.** $5\sqrt{15}$ **39.** $x^3\sqrt{x}$ **41.** $n^4\sqrt{n}$ **43.** $d^5\sqrt{2d}$ **45.** $2\sqrt{k}$ **47.** $3\sqrt{y}$ **49.** Cannot be simplified **51.** $8\sqrt{3x}$ **53.** $15\sqrt{6q}$ **55.** $2t\sqrt{6t}$ **57.** $4x^2\sqrt{2x}$ **59.** $10x^3\sqrt{3x}$ **61.** $48t^5\sqrt{2t}$ **63.** $\frac{5}{3}$ **65.** $\frac{9}{8}$ **67.** $\frac{\sqrt{6}}{11}$ **69.** $\frac{\sqrt{23}}{8}$ **71.** $\frac{2\sqrt{5}}{7}$ **73.** $\frac{5\sqrt{3}}{4}$ **75.** $\frac{a\sqrt{a}}{2}$ **77.** $\frac{r^4\sqrt{r}}{15}$ **79.** $6x\sqrt{2}$ **81.** $\frac{5n^2\sqrt{5}}{8}$ **83.** $5\sqrt{3t}$ **85.** $\frac{4\sqrt{3}}{9}$ **87.** $h^3\sqrt{h}$ **89.** $18\sqrt{2}$ **91.** $\frac{5\sqrt{3}}{4q}$ **93.** $9\sqrt{2}$ **95.** $48y^2\sqrt{y}$ **97.** $5x^2\sqrt{2}$ **99.** $\sqrt{18}$ ft $=3\sqrt{2}$ ft; 4.24 ft **101.** $\sqrt{4}=2, \sqrt{8}=2\sqrt{2}, \sqrt{9}=3, \sqrt{12}=2\sqrt{3}, \sqrt{16}=4$ **107.** (3, 0) **109.** $(-1,-1)$

Study Set Section 8.3 (page 688)

1. radicals **3. a.** Yes **b.** No **5. a.** 5, 3, 8 **b.** 2, 9, -7 **7.** 4, 4, 2, $6\sqrt{5}$, $3\sqrt{5}$ **9.** $9\sqrt{7}$ **11.** $10\sqrt{21}$ **13.** $-3\sqrt{21}$ **15.** Does not simplify **17.** $9\sqrt{n}$ **19.** 0 **21.** $-3\sqrt{x}$ **23.** Does not simplify **25.** $10\sqrt{11}$ **27.** $4\sqrt{2}$ **29.** $5+6\sqrt{3}$ **31.** Does not simplify **33.** $5\sqrt{3}$ **35.** $\sqrt{2}$ **37.** $-2\sqrt{3}$ **39.** $-18\sqrt{2}$ **41.** $14\sqrt{5}$ **43.** $-7\sqrt{5}$ **45.** $9\sqrt{5}$ **47.** $30\sqrt{2}-\sqrt{3}$ **49.** $-\sqrt{2a}$ **51.** $3\sqrt{2y}-3\sqrt{3y}$ **53.** $3x\sqrt{2}$ **55.** $19b\sqrt{6}$ **57.** $3d\sqrt{2d}$ **59.** $x^2\sqrt{2x}$ **61.** $5+4\sqrt{2}$ **63.** $y-2\sqrt{y}-15$ **65.** $2t+3\sqrt{t}-5$ **67.** $3a-1$ **69.** $5\sqrt{6}-5\sqrt{2}$ **71.** $7\sqrt{6}+4\sqrt{15}$ **73.** $4\sqrt{5}$ **75.** $-1-\sqrt{r}$ **77.** $12\sqrt{7}$ **79.** $15b+17\sqrt{b}-4$ **81.** $7\sqrt{3}-6\sqrt{2}$ **83.** $12\sqrt{10y}-12\sqrt{10z}$ **85.** $(16+24\sqrt{5})$ ft $\approx$ 70 ft **87.** $10\sqrt{3}$ft $\approx$ 17.3 ft **95.** $\frac{1}{9}$ **97.** -9 **99.** $\frac{1}{x^3}$ **101.** 1

Study Set Section 8.4 (page 699)

1. denominator **3.** rationalize **5. a.** $a\cdot b$ **b.** $\frac{a}{b}$ **c.** a **d.** b **7. a.** The radicand is a fraction. **b.** There is a radical in the denominator. **9.** 4, $\sqrt{3}$, 6 **11.** 7, 7, 7 **13.** $\sqrt{15}$ **15.** 7 **17.** $5\sqrt{2}$ **19.** $3\sqrt{2x}$ **21.** $2d\sqrt{10}$ **23.** $x^4\sqrt{5}$ **25.** $15\sqrt{5}$ **27.** $10\sqrt{15}$ **29.** 6 **31.** y **33.** $2b+7$ **35.** 12 **37.** $2+\sqrt{2}$ **39.** $27-3\sqrt{6}$ **41.** $x\sqrt{3}-2\sqrt{x}$ **43.** $2m\sqrt{2}+9\sqrt{m}$ **45.** 1 **47.** $\sqrt{6}+\sqrt{10}-3-\sqrt{15}$ **49.** $4x-18$ **51.** $6\sqrt{14}+2x\sqrt{7}-3x\sqrt{2}-x^2$ **53.** $28+10\sqrt{3}$ **55.** $a^2+2a\sqrt{7}+7$ **57.** $5-2\sqrt{5m}+m$ **59.** 3 **61.** $11-y^2$ **63.** $7c-9$ **65.** $\sqrt{10}$ **67.** a^2 **69.** $\frac{3\sqrt{2}}{5}$ **71.** $\frac{3}{5x}$ **73.** $\frac{\sqrt{3}}{3}$ **75.** $\frac{4\sqrt{19}}{19}$ **77.** $\frac{2\sqrt{15}}{5}$ **79.** $\frac{2\sqrt{3}}{3}$ **81.** $\frac{10\sqrt{x}}{x}$ **83.** $\frac{3\sqrt{2x}}{2x}$ **85.** $\frac{3\sqrt{3}+3}{2}$ **87.** $\frac{8\sqrt{7}-16}{3}$ **89.** $5\sqrt{3}-5\sqrt{2}$ **91.** $\frac{5\sqrt{3}-\sqrt{3x}}{25-x}$ **93.** $-\frac{\sqrt{10}+5+\sqrt{14}+\sqrt{35}}{3}$ **95.** $\frac{\sqrt{15}-\sqrt{3a}-\sqrt{5a}+a}{5-a}$ **97.** $x\sqrt{14}+\sqrt{2x}$ **99.** $100x$ **101.** $\frac{\sqrt{91}}{7}$ **103.** $9p^2+6p\sqrt{5}+5$ **105.** $\sqrt{66}$ **107.** $\frac{2x}{3}$ **109.** 13 **111.** $14\sqrt{15}$ **113.** $\frac{7\sqrt{3}-3}{46}$ **115.** $2n^4\sqrt{15n}$ **117.** $2x-10\sqrt{x}-12$ **119.** $\frac{3\sqrt{2}}{8}$ **121.** Length: $\sqrt{8^2+2^2}=\sqrt{68}=2\sqrt{17}$ ft; width: $\sqrt{4^2+1^2}=\sqrt{17}$ ft; area: 34 ft^2 **123.** $f=0.772\frac{\sqrt{Tu}}{u}$ **129.** 4.5

Study Set Section 8.5 (page 709)

1. radical **3.** extraneous **5.** b^2 **7. a.** $\sqrt{x-4}=3$ **b.** $x+8=\sqrt{x}$ **9.** 2, 2, $x-3$, 28 **11.** 9 **13.** 144 **15.** 8 **17.** 9 **19.** 1 **21.** -95 **23.** 5 **25.** 6 **27.** 8 **29.** 13 **31.** 2 **33.** -1 **35.** No solution **37.** No solution **39.** No solution **41.** No solution **43.** $-2, -1$ **45.** 0, 3 **47.** 4 **49.** 3 **51.** 12 **53.** 5 **55.** -2 **57.** 4 **59.** 1 **61.** -1 **63.** 16 **65.** -2, 4 **67.** 5 **69.** 49 **71.** 25 **73.** $\frac{1}{4}$, 1 **75.** 4 **77.** 9 **79.** 7 **81.** 36 **83.** 1, 3 **85.** 9 **87.** 169 ft **89.** 64.4 ft **91.** 288 ft **93.** 2,010 ft **99.** $\frac{5}{4}$ **101.** $\frac{10}{9}$

Study Set Section 8.6 (page 718)

1. cube, fourth **3.** simplify **5. a.** 4 **b.** $2x$ **c.** -32 **7. a.** $\sqrt[n]{a}\sqrt[n]{b}$ **b.** $\frac{\sqrt[n]{a}}{\sqrt[n]{b}}$ **9. a.** $\sqrt[3]{8\cdot 3}$ **b.** $\sqrt[4]{16\cdot 3}$ **11. a.** $\sqrt{25}$ **b.** $(\sqrt[3]{-27})^2$ **13.** 2 **15.** 0 **17.** -5 **19.** -4 **21.** 1 **23.** 3 **25.** 9 **27.** $\frac{1}{5}$ **29.** 2 **31.** 4 **33.** Not a real number **35.** -3 **37.** 2 **39.** $-\frac{1}{3}$ **41.** $2b^3$ **43.** s^2 **45.** t^5 **47.** y **49.** m **51.** $3a^2$ **53.** x^3 **55.** $2v^6$ **57.** $2\sqrt[3]{3}$ **59.** $-4\sqrt[3]{2}$ **61.** $3\sqrt[4]{2}$ **63.** $2\sqrt[3]{9}$ **65.** $\frac{5\sqrt[3]{2}}{3}$ **67.** $2\sqrt[5]{2}$ **69.** 9 **71.** 2 **73.** $\frac{1}{2}$ **75.** -12 **77.** -5 **79.** $\frac{3}{8}$ **81.** 64 **83.** 9 **85.** 25 **87.** 32 **89.** $\frac{1}{8}$ **91.** -8 **93.** $\frac{1}{2}$

95. $\frac{1}{125}$ **97.** $\frac{1}{32}$ **99.** $\frac{1}{9}$ **101.** $-\frac{1}{729}$ **103.** $\frac{1}{8}$ **105. a.** $4\sqrt{2}$ **b.** $2\sqrt[3]{4}$ **c.** $2\sqrt[4]{2}$ **d.** 2 **107.** 10 mph **109.** 26 ft **111.** 36 in.^{2}

117.

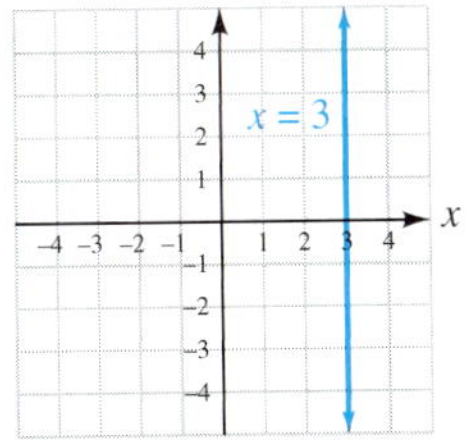

119.

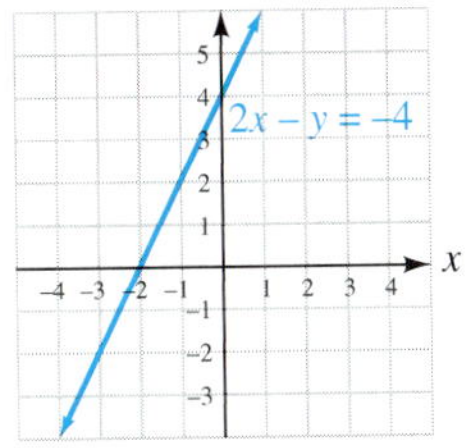

Chapter 8 Review (page 721)

1. 6, 36 **2.** False **3.** 5 **4.** 7 **5.** −12 **6.** $\frac{4}{9}$ **7.** 0.8 **8.** 1 **9.** 4.583 **10.** −5.292 **11.** Not a real number, irrational, rational, irrational **12.** 30 mph **13.** x **14.** $2b^2$ **15.** $-y^6$ **16.** $3h^8$ **17.** 10 **18.** 15 **19.** 5 ft **20.** 2.12 in. **21.** 5 **22.** $\sqrt{194} \approx 13.93$ **23.** $4\sqrt{2}$ **24.** $x^2\sqrt{x}$ **25.** $4x\sqrt{5}$ **26.** $-6\sqrt{7}$ **27.** $5t\sqrt{10t}$ **28.** $\frac{4}{5}$ **29.** $\frac{2\sqrt{15}}{7}$ **30.** $\frac{11x\sqrt{2}}{13}$ **31.** $2\sqrt{10}$ ft; 6.3 ft **32.** False **33.** $2\sqrt{10}$ **34.** $5\sqrt{x}$ **35.** 0 **36.** $-3 + 4\sqrt{3}$ **37.** $\sqrt{7}$ **38.** $-7y\sqrt{5y}$ **39.** The radicands are different. **40.** $13\sqrt{5}$ in. **41.** $\sqrt{6}$ **42.** 125 **43.** $36x^2\sqrt{2x}$ **44.** $15 + 6x\sqrt{15} + 9x^2$ **45.** −2 **46.** −2 **47.** x **48.** $t - 1$ **49.** $18\sqrt{2}$ in.^{2}; 25.5 in.^{2} **50.** $\frac{\sqrt{2}}{2}$ **51.** $\frac{9\sqrt{7}}{7}$ **52.** $\frac{\sqrt{3a}}{a}$ **53.** $\frac{16\sqrt{15}}{5}$ **54.** $\frac{11\sqrt{3}}{15}$ **55.** $\frac{\sqrt{2x}}{4x}$ **56.** $\frac{7 + \sqrt{14}}{5}$ **57.** $\frac{a - \sqrt{a}}{a - 1}$ **58.** $\frac{b + 5\sqrt{b} + 6}{b - 9}$ **59.** 81 **60.** 45 **61.** No solution **62.** −3 **63.** −2 **64.** −2, 4 **65.** 64 ft **66.** 96 ft **67.** −3 **68.** −5 **69.** 3 **70.** 2 **71.** 0 **72.** −1 **73.** $\frac{1}{4}$ **74.** 4 **75.** x **76.** $3y^2$ **77.** $2a^3$ **78.** b^4 **79.** $3\sqrt[3]{2}$ **80.** $2\sqrt[4]{5}$ **81.** $-\frac{2\sqrt[3]{2}}{7}$ **82.** $\frac{2\sqrt[3]{7}}{5}$ **83.** 7 **84.** −10 **85.** 216 **86.** $\frac{4}{9}$ **87.** $\frac{1}{8}$ **88.** −243 **89.** $\frac{1}{64}$ of the original dose **90.** No real number squared is −64.

Chapter 8 Test (page 727)

1. a. radical **b.** radicand **c.** Pythagorean **d.** cube, root **e.** index **2.** 4, 16 **3.** 10 **4.** $-\frac{8}{3}$ **5.** 0.5 **6.** 1 **7.** 7 amps **8.** 10 ft **9.** Irrational, not a real number, rational **10.** $6\sqrt{2}$ **11.** $2x$ **12.** $3x\sqrt{6x}$ **13.** $\frac{5\sqrt{2}}{7}$ **14.** $3a^2\sqrt{a}$ **15.** $9\sqrt{5}$ **16.** $3\sqrt{11}$ **17.** $5b^2\sqrt{3}$ **18.** $10\sqrt{6}$ **19.** $2\sqrt{6} + 3\sqrt{2}$ **20.** $-5x\sqrt{2x}$ **21.** −1 **22.** $-24x^2\sqrt{6x}$ **23.** $5x - 21\sqrt{x} + 4$ **24.** $12t$ **25.** $x + 1$ **26.** $x + 2\sqrt{x} + 1$ **27.** $(6\sqrt{2} + 2\sqrt{10})$ in. **28.** They are not like radicals. **29.** $\frac{2\sqrt{7}}{7}$ **30.** $\frac{\sqrt{3x} + 8\sqrt{3}}{x - 64}$ **31.** 225 **32.** −62 **33.** 5 **34.** 2, 1 **35.** No; when 0 is substituted for x, the result is not true. **36.** The Pythagorean theorem; 5 ft **37.** −5 **38.** Not a real number **39.** $\frac{1}{4}$ **40.** $-2x^2$ **41.** $3x$ **42.** $2\sqrt[3]{11}$ **43.** 12 **44.** 4 **45.** $\frac{1}{81}$ **46.** $-\frac{1}{3}$

Cumulative Review Chapters 1–8 (page 729)

1. a. True **b.** False **c.** True **2.** −14 **3.** $-2p - 6z$ **4.** −2 **5.** 17 lb **6.** 4 in. **7.** 3.5 hr **8.** 80 **9.** $\left(-\infty, -\frac{3}{4}\right]$

10. No

11.

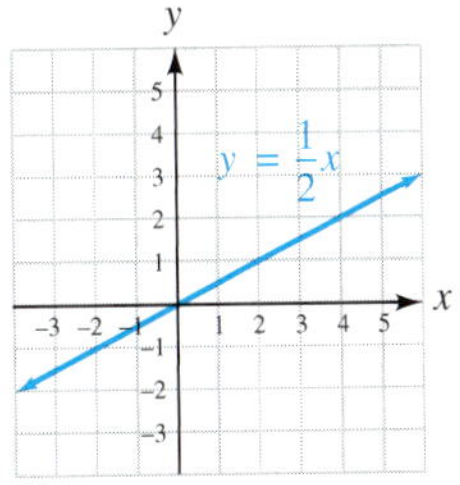

12.

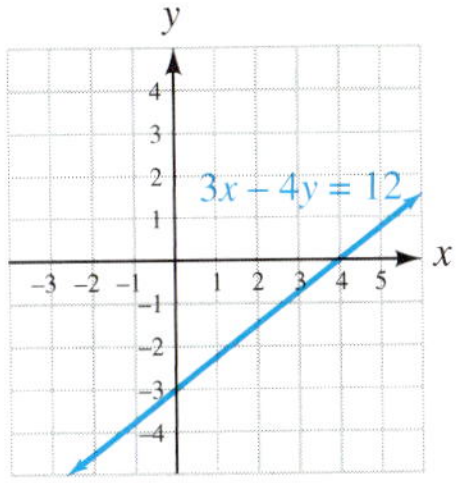

13.

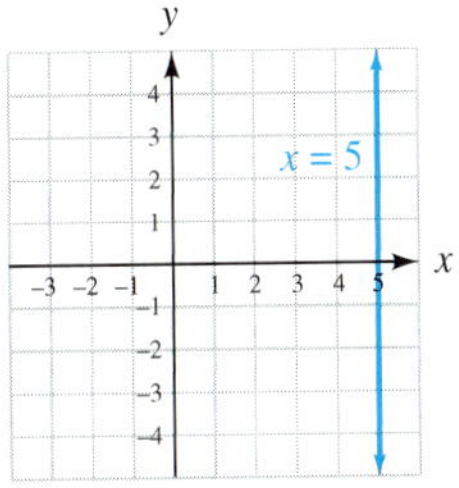

14.

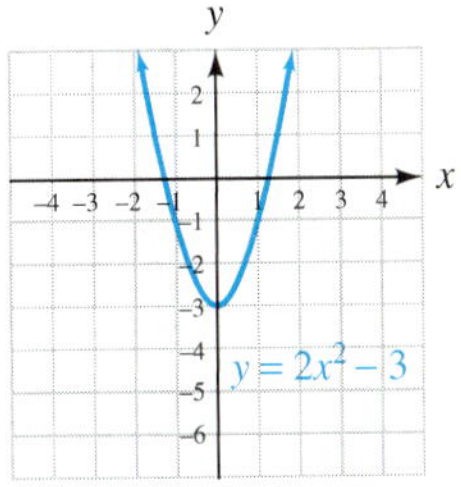

15. \$80 billion per yr **16. a.** 3 **b.** $-\frac{2}{3}$ **17.** $y = \frac{1}{2}x + 6$ **18.** 28 **19.** Yes **20.** $(-1, 5)$

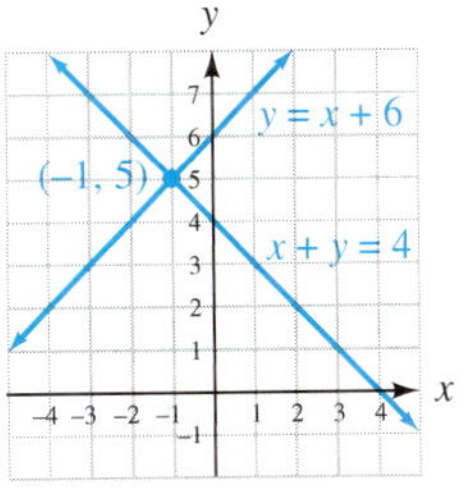

21. $(3, -1)$ **22.** $(-1, 2)$ **23.** 6%: \$3,000; 12%: \$3,000 **24.**

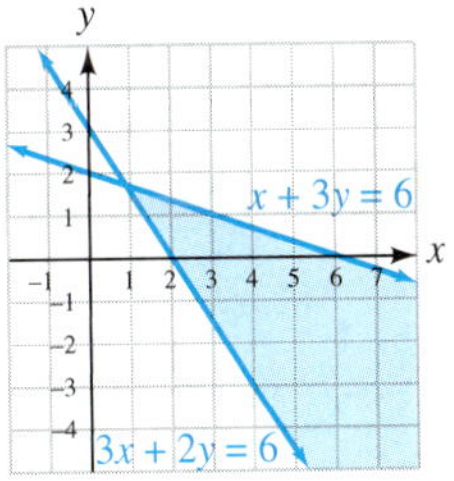

25. x^{31} **26.** $\frac{a^{15}b^5}{c^{20}}$ **27.** 1 **28.** $\frac{a^8}{16b^{17}}$ **29.** 4.8×10^{18} m **30.** 4.3×10^{-9} **31.** $2a^2 + a - 3$ **32.** $0.12p^9 - 1.8p^7$ **33.** $-6t^2 + 13st - 6s^2$ **34.** $16b^2 - 64b + 64$ **35.** $36b^2 - \frac{1}{4}$ **36.** $2x - 1$ **37.** $3xy(4x - 2y + 3y^2)$ **38.** $(x + y)(2x - 3)$ **39.** $(x + 5)(x + 2)$ **40.** $(3a + 4)(2a - 5)$ **41.** Prime **42.** $(5a - 7b)^2$ **43.** $(a + 2b)(a^2 - 2ab + 4b^2)$ **44.** $2x(x^2 + 4)(x + 2)(x - 2)$ **45.** −1, −2 **46.** 0, 2 **47.** $\frac{2}{3}, -\frac{1}{2}$ **48.** −5, 5 **49.** 5 cm **50.** $\frac{x + 1}{x - 1}$ **51.** $\frac{(p - 3)(p + 2)}{3(p + 3)}$ **52.** $-\frac{3}{5x}$ **53.** $\frac{1}{3a}$ **54.** $\frac{7x + 29}{(x + 5)(x + 7)}$ **55.** $\frac{3 - 16b^2}{18b^4}$ **56.** $\frac{y + x}{y - x}$ **57.** 1 **58.** $2\frac{2}{9}$ hr **59.** 6,480 **60.** $4\frac{1}{2}$ days **61.** 73 in. **62.** No real number squared is −4. **63.** 8 **64.** $4x\sqrt{2x}$ **65.** $3m^2$ **66.** $3t$ **67.** $7\sqrt{3} - 6\sqrt{2}$ **68.** $y - 9\sqrt{y} + 20$ **69.** $\frac{4\sqrt{5}}{5}$ **70.** $\frac{\sqrt{34x}}{2x}$ **71.** 5 **72.** 64

Study Set Section 9.1 (page 739)

1. quadratic **3.** $\sqrt{c}, -\sqrt{c}$ **5. a.** $x = -9 \pm \sqrt{2}$ **b.** $x = \frac{3 \pm \sqrt{2}}{6}$ **7.** Yes **9.** $x = \pm\sqrt{6}$ **11.** ± 6 **13.** ± 3 **15.** $\pm\frac{7}{4}$ **17.** ± 10 **19.** ± 5 **21.** $\pm\sqrt{6}$ **23.** $\pm\sqrt{17}$ **25.** $\pm 2\sqrt{5}$ **27.** $\pm 6\sqrt{2}$ **29.** $\pm\frac{\sqrt{30}}{2}$ **31.** $\pm\frac{\sqrt{42}}{6}$ **33.** No real-number solutions **35.** $-6, 4$ **37.** $7, -11$ **39.** $2 \pm 2\sqrt{2}$ **41.** $-9 \pm 3\sqrt{7}$ **43.** $\frac{-1 \pm 3\sqrt{2}}{3}$ **45.** $\frac{10 \pm \sqrt{6}}{5}$ **47.** $-1 \pm \sqrt{10}$ **49.** $9 \pm \sqrt{7}$ **51.** $3 \pm 2\sqrt{10}$ **53.** $-2 \pm 5\sqrt{3}$ **55.** $-12 \pm 3\sqrt{3}$ **57.** $\pm 7\sqrt{2}$ **59.** $6 \pm \sqrt{2}$ **61.** $15 \pm 2\sqrt{2}$ **63.** $\pm\frac{1}{12}$ **65.** $7 \pm \sqrt{3}$ **67.** No real-number solutions **69.** $\pm\frac{\sqrt{85}}{5}$ **71.** $\pm\sqrt{14}$ **73.** $\frac{-9 \pm 2\sqrt{11}}{8}$ **75.** 3 sec **77.** 3.4 sec **79.** 3.5 sec **81.** 20 ft **83.** 24,941 mi/hr **89.** 2 **91.** 16

Study Set Section 9.2 (page 748)

1. square **3. a.** 9 **b.** $\frac{25}{4}$ **5.** one-half **7. a.** Subtract 7 from both sides **b.** Divide both sides by 4 **9. a.** False **b.** True **11.** $\left(\frac{1}{2} \cdot 9\right)^2 = \frac{81}{4}$ **13.** $x^2 + 2x + 1 = (x + 1)^2$ **15.** $x^2 - 4x + 4 = (x - 2)^2$ **17.** $a^2 - 7a + \frac{49}{4} = \left(a - \frac{7}{2}\right)^2$ **19.** $x^2 + x + \frac{1}{4} = \left(x + \frac{1}{2}\right)^2$ **21.** $b^2 - \frac{2}{3}b + \frac{1}{9} = \left(b - \frac{1}{3}\right)^2$ **23.** $x^2 - \frac{5}{2}x + \frac{25}{16} = \left(x - \frac{5}{4}\right)^2$ **25.** $-5, 1$ **27.** $-3, 5$ **29.** $-2, -4$ **31.** $2, 6$ **33.** $-4 \pm \sqrt{22}$ **35.** $1 \pm 3\sqrt{2}$ **37.** $-3 \pm \sqrt{5}$ **39.** $2 \pm \sqrt{7}$ **41.** $\frac{7 \pm \sqrt{69}}{2}$ **43.** $\frac{5 \pm \sqrt{65}}{2}$ **45.** $\frac{-3 \pm \sqrt{89}}{2}$ **47.** $\frac{-1 \pm \sqrt{37}}{2}$ **49.** $-1, 7$ **51.** $-6 \pm \sqrt{14}$ **53.** $\frac{7 \pm \sqrt{73}}{4}$ **55.** $\frac{9 \pm \sqrt{65}}{8}$ **57.** $\frac{-3 \pm \sqrt{89}}{10}$ **59.** $\frac{-5 \pm \sqrt{85}}{6}$ **61.** $-\frac{1}{2}, \frac{13}{2}$ **63.** $\frac{-3 \pm \sqrt{14}}{3}$ **65.** $\frac{12 \pm \sqrt{133}}{2}$ **67.** $\frac{-8 \pm \sqrt{61}}{4}$ **69.** $3.24, -1.24$ **71.** $9.52, -0.52$ **73.** $-0.18, -2.82$ **75.** $0.87, -1.54$ **77.** $\frac{-7 \pm \sqrt{41}}{2}$ **79.** $\frac{6 \pm \sqrt{37}}{3}$ **81.** $\frac{-5 \pm \sqrt{57}}{4}$ **83.** $2, \frac{2}{3}$ **85.** $1 \pm \sqrt{6}$ **87.** $7, 5$ **89.** $4; 4, 2$ **95.** $\frac{x + 1}{(x - 3)^2}$ **97.** $\frac{x^2 + 7x + 10}{x^2 - 9}$

Study Set Section 9.3 (page 758)

1. quadratic **3. a.** $x^2 + 2x + 5 = 0$ **b.** $3x^2 + 2x - 1 = 0$ **5.** $1, -2, 4$ **7. a.** 2 **b.** 5 **9. a.** 2 **b.** $\frac{-3 + \sqrt{15}}{2}, \frac{-3 - \sqrt{15}}{2}$ **c.** $0.44, -3.44$ **11.** $-5, -6, 1, 24, 49, 5, 7, 6, 7, -1$ **13.** $-2, -3$ **15.** $-3, -4$ **17.** $\frac{1}{4}, -1$ **19.** $\frac{2}{3}, -\frac{3}{2}$ **21.** $2, -\frac{1}{3}$ **23.** $3, -\frac{2}{5}$ **25.** $\frac{-3 \pm \sqrt{5}}{2}$; $-0.38, -2.62$ **27.** $\frac{-7 \pm \sqrt{65}}{2}$; $0.53, -7.53$ **29.** $\frac{1 \pm \sqrt{37}}{6}$; $1.18, -0.85$ **31.** $\frac{3 \pm \sqrt{37}}{14}$; $0.65, -0.22$ **33.** $\frac{7 \pm \sqrt{17}}{8}$; $1.39, 0.36$ **35.** $\frac{-9 \pm \sqrt{137}}{4}$; $0.68, -5.18$ **37.** $\frac{-5 \pm \sqrt{3}}{2}$ **39.** $\frac{3 \pm \sqrt{7}}{2}$ **41.** $\frac{4 \pm \sqrt{10}}{3}$ **43.** $\frac{3 \pm 2\sqrt{2}}{2}$ **45.** No real-number solutions **47.** $-1 \pm \sqrt{2}$ **49.** $-2 \pm \sqrt{7}$ **51.** No real-number solutions **53.** $-2, 3$ **55.** $\frac{-1 \pm \sqrt{41}}{4}$; $-1.85, 1.35$ **57.** $5, 7$ **59.** $\frac{-9 \pm \sqrt{77}}{2}$; $-0.11, -8.89$ **61.** $\frac{1 \pm \sqrt{41}}{5}$; $1.48, -1.08$ **63.** $\frac{-3 \pm 3\sqrt{5}}{2}$; $1.85, -4.85$ **65.** No real-number solutions **67.** $-4, 0$ **69.** $\pm\sqrt{6}$; ± 2.45 **71.** $2 \pm \sqrt{33}$; $7.74, -3.74$ **73.** 12 **75.** $\pm 3\sqrt{7}$; ± 7.94 **77.** 0.8, 3.1 **79.** 6 ft, 8 ft **81.** 16 in., 10 in. **83.** 15 cm by 36 cm **85.** 7% **93.** $r = \frac{A - p}{pt}$ **95.** $r = \frac{r_1 r_2}{r_2 + r_1}$

Study Set Section 9.4 (page 769)

1. complex. real, imaginary **3. a.** $\sqrt{-1}$ **b.** $2, -1$ **5. a.** $8i$ **b.** $2i$ **7.** $\frac{6 + i}{6 + i}$ **9. a.** True **b.** True **c.** False **d.** True **11. a.** $i\sqrt{7}$ **b.** $2i\sqrt{3}$ **13.** $3i$ **15.** $i\sqrt{7}$ or $\sqrt{7}i$ **17.** $2i\sqrt{6}$ or $2\sqrt{6}i$ **19.** $-4i\sqrt{2}$ or $-4\sqrt{2}i$ **21.** $45i$ **23.** $\frac{5}{3}i$ **25.** $12 + 0i$ **27.** $0 + 10i$ **29.** $6 + 4i$ **31.** $-9 - 7i$ **33.** $8 - 2i$ **35.** $3 - 5i$ **37.** $14 - 13i$ **39.** $12 - 9i$ **41.** $6 - 3i$ **43.** $-25 - 25i$ **45.** $12 + 5i$ **47.** $13 - i$ **49.** $2 + i$ **51.** $\frac{8}{53} - \frac{28}{53}i$ **53.** $-\frac{5}{13} + \frac{12}{13}i$ **55.** $\frac{31}{50} - \frac{17}{50}i$ **57.** $0 \pm 3i$ **59.** $0 \pm 2i\sqrt{2}$ **61.** $-3 \pm i$ **63.** $11 \pm 5i\sqrt{3}$ **65.** $\frac{3}{2} \pm \frac{\sqrt{7}}{2}i$ **67.** $-\frac{1}{4} \pm \frac{\sqrt{7}}{4}i$ **69.** $\frac{15}{26} - \frac{3}{26}i$ **71.** $15 + 2i$ **73.** $0 + i$ **75.** $1 + 8i$ **77.** $\frac{5}{13} - \frac{12}{13}i$ **79.** $9 - 8i$ **81.** $-12 - 16i$ **83.** $3 + 4i$ **85.** $-\frac{1}{4} \pm \frac{\sqrt{39}}{4}i$ **87.** $4 \pm 3i\sqrt{5}$ **89.** $-1 \pm i$ **91.** $0 \pm 6i$ **93.** $0 \pm \frac{4}{3}i$ **95.** $-\frac{1}{3} \pm \frac{\sqrt{2}}{3}i$ **97.** $18.45 - 2.18i$ **103.** $\frac{\sqrt{7}}{7}$ **105.** $\frac{8\sqrt{x} + 16}{x - 4}$

Study Set Section 9.5 (page 780)

1. quadratic, parabola **3.** intercepts, intercept **5.** $<, 0$ **7. a.** 0 **b.** x **9. a.** Parabola **b.** $(1, 0), (3, 0)$ **c.** $(0, -3)$ **d.** $(2, 1)$ **e.** It is a vertical line through $(2, 1)$. **11.** $(2, -2), (-1, 1)$

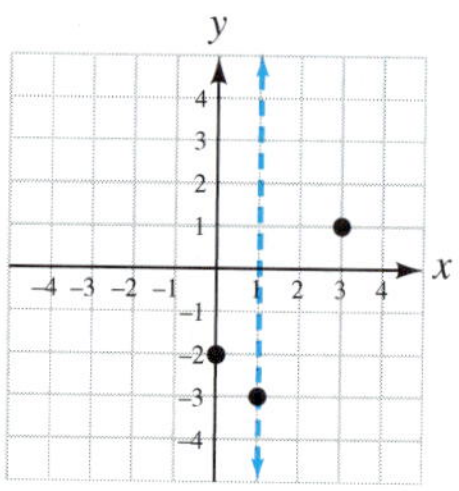

13. Two; $-3, 1$ **15. a.** $2, 4, -8$ **b.** -1 **17.** $(4, 0), (2, 0)$; $(0, 8)$ **19.** $(-3, 0), (-7, 0)$; $(0, -21)$ **21.** $(1, -1)$ **23.** $(3, 1)$

25.

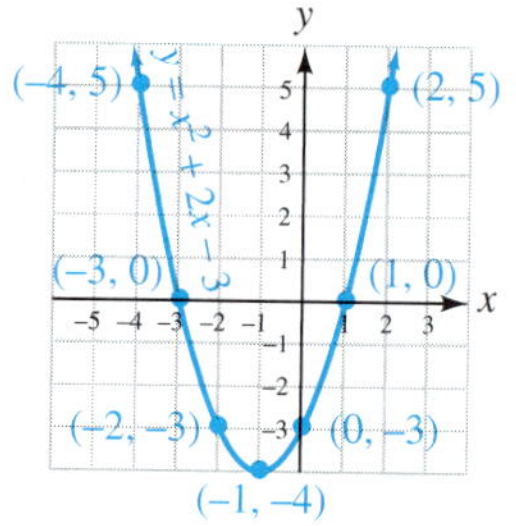

27.

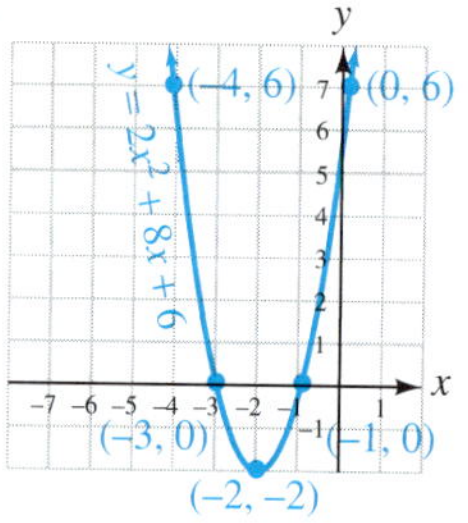

29.

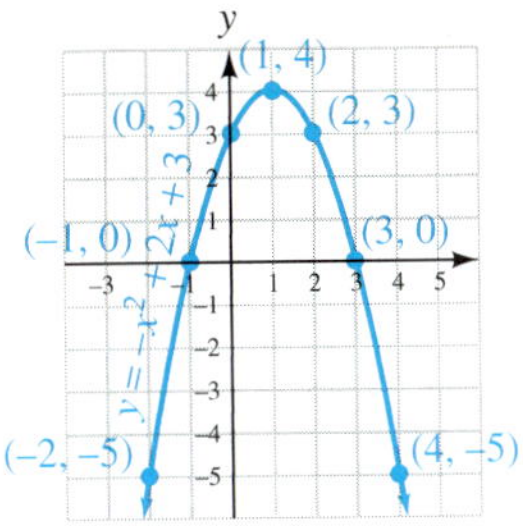

31.

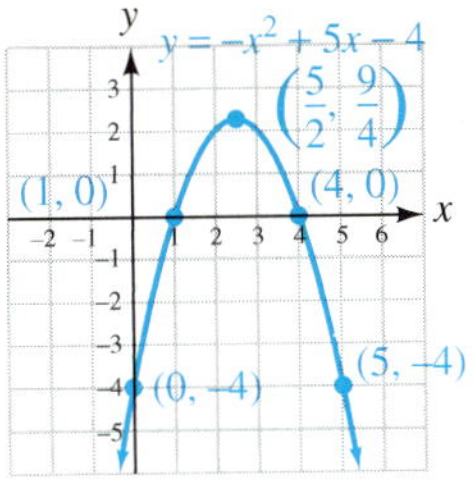

33.

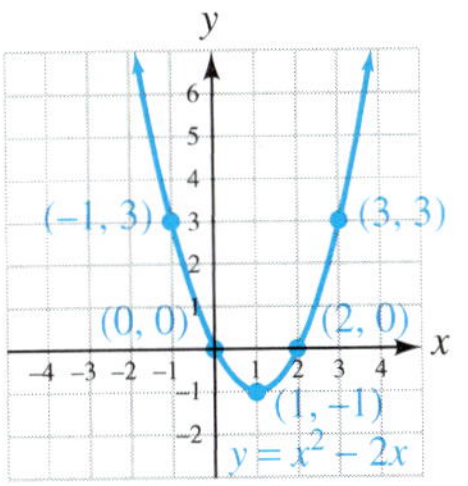

35.

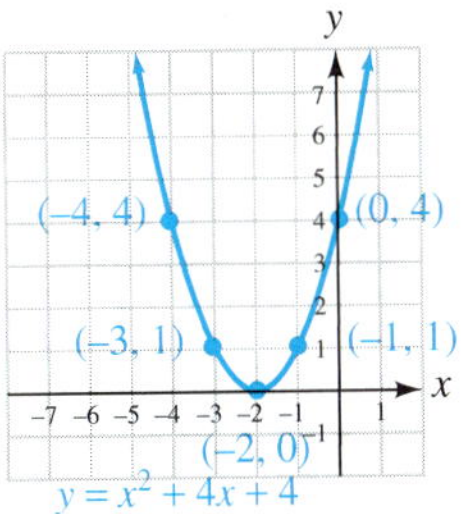

37.

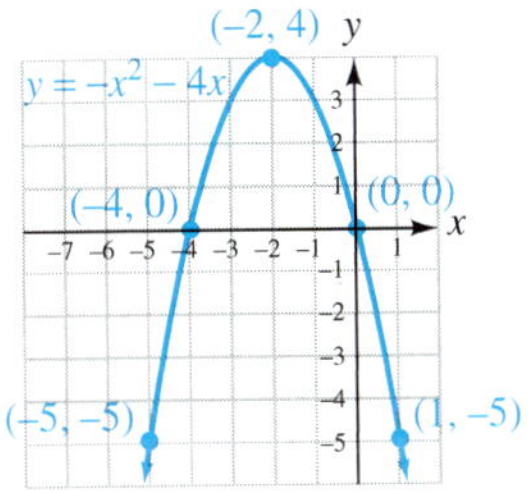

39.

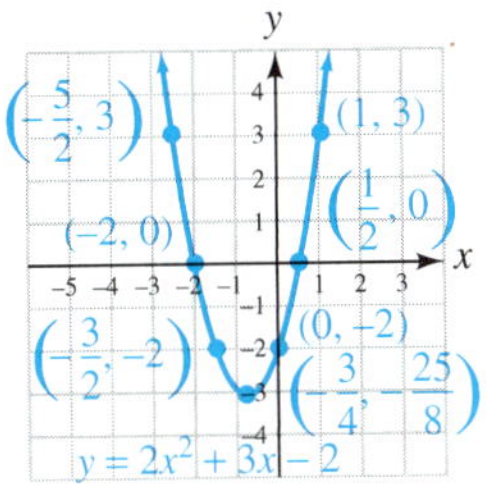

41.

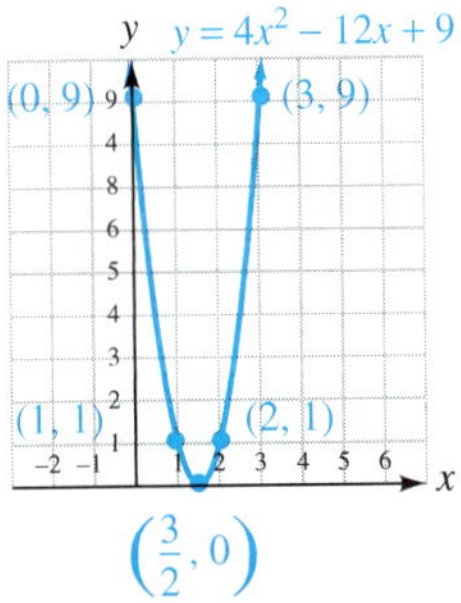

43. Irrational x-intercepts

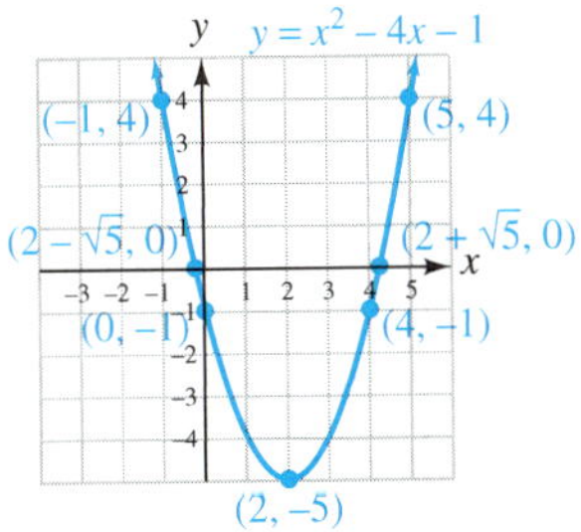

45. Irrational x-intercepts

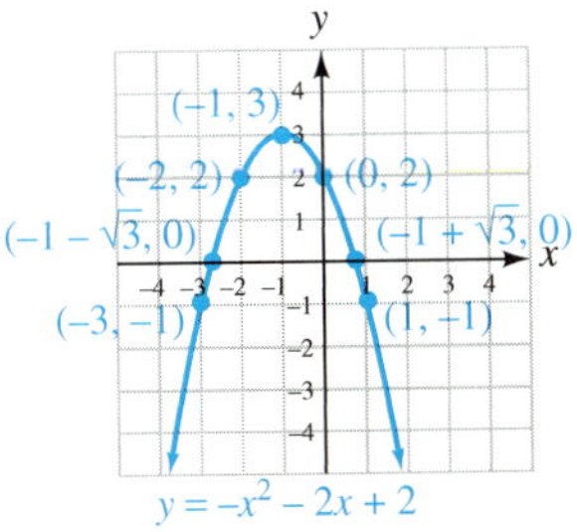

47. Irrational x-intercepts

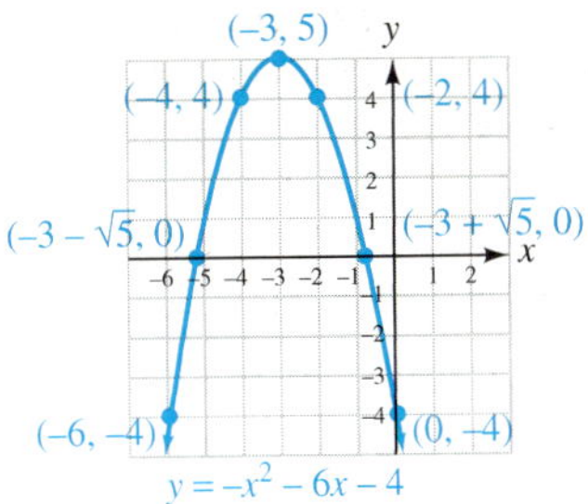

49. No x-intercepts

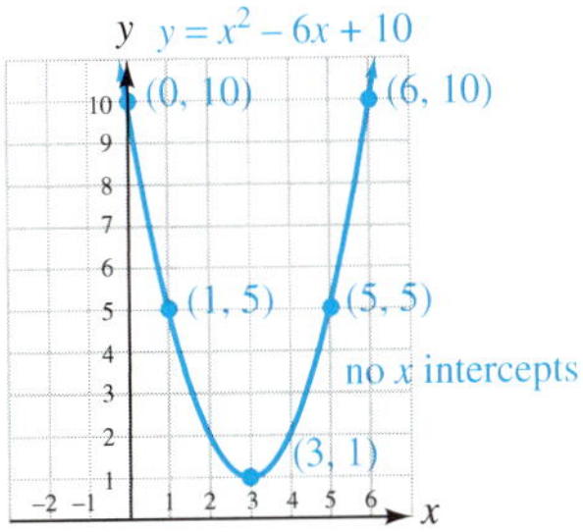

51. It is a vertical line through the body of the butterfly. **53.** The cost to manufacture a carburetor is lowest ($100) for a production run of 30 units. **55. a.** 14 ft **b.** 0.25 sec and 1.75 sec **c.** 18 ft; 1.0 sec **63.** $3\sqrt{2}$ **65.** $3z$

Chapter 9 Review (page 783)

1. ± 8 **2.** $\pm 2\sqrt{2}$ **3.** $\pm 5\sqrt{3}$ **4.** $-4, 6$ **5.** $\frac{8 \pm 2\sqrt{10}}{9}$ **6.** $\frac{7}{2}, \frac{1}{2}$ **7.** 13, 7 **8.** $\frac{-1 \pm \sqrt{6}}{3}$ **9.** ± 3.46 **10.** $-6.42, 8.42$ **11.** No real-number solutions **12.** No real-number solutions **13.** 3.0 sec **14.** $2\frac{1}{4}$ in. **15.** $x^2 + 4x + 4 = (x + 2)^2$ **16.** $t^2 - 5t + \frac{25}{4} = \left(t - \frac{5}{2}\right)^2$ **17.** 3, 5 **18.** 2, −7 **19.** $-1 \pm \sqrt{6}$ **20.** $\frac{4 \pm \sqrt{23}}{2}$ **21.** $\frac{1 \pm \sqrt{3}}{2}$ **22.** $-1, -\frac{2}{3}$ **23.** $-0.27, -3.73$ **24.** $-0.65, 7.65$ **25.** $x^2 + 2x + 5 = 0$; 1, 2, 5 **26.** $6x^2 - 2x - 1 = 0$; 6, −2, −1 **27.** 5, −3 **28.** $\frac{3}{2}, -\frac{1}{3}$ **29.** $1 \pm \sqrt{5}$ **30.** $3 \pm \sqrt{2}$ **31.** $\frac{-3 \pm \sqrt{21}}{6}$ **32.** $\frac{-1 \pm \sqrt{21}}{10}$ **33.** No real-number solutions **34.** $\frac{-3 \pm \sqrt{19}}{2}$ **35.** 0, −4 **36.** 1, −7 **37.** $\pm 3\sqrt{3}$ **38.** $\frac{3 \pm \sqrt{6}}{3}$ **39.** $\frac{-1 \pm \sqrt{11}}{2}$ **40.** 2 **41.** $\frac{13}{2}, -\frac{3}{2}$ **42.** No real-number solutions **43.** $\frac{-1 \pm \sqrt{7}}{3}$; $-1.22, 0.55$ **44.** 10 ft, 24 ft **45.** 18 sec **46.** 6 in. **47.** $5i$ **48.** $3i\sqrt{2}$ **49.** $-7i$ **50.** $\frac{3}{8}i$ **51.** Real numbers, imaginary numbers **52. a.** True **b.** True **c.** False **d.** False **53.** $3 - 6i$ **54.** $-1 + 7i$ **55.** $0 - 19i$ **56.** $0 + i$ **57.** $8 - 2i$ **58.** $3 - 5i$ **59.** $3 + 6i$ **60.** $9 + 7i$ **61.** $-\frac{5}{13} + \frac{12}{13}i$ **62.** $\frac{15}{26} - \frac{3}{26}i$ **63.** $0 \pm 3i$ **64.** $0 \pm \frac{4\sqrt{3}}{3}i$ **65.** $2 \pm 2i\sqrt{6}$ **66.** $-3 \pm 3i\sqrt{6}$ **67.** $-1 \pm i$ **68.** $\frac{3}{4} \pm \frac{\sqrt{7}}{4}i$ **69. a.** $(-3, 0), (1, 0)$ **b.** $(0, -3)$ **c.** $(-1, -4)$ **d.** A vertical line through $(-1, -4)$ **70.** $(-2, -3)$ **71.** (1, 5); upward **72.** (3, 16); downward **73.** $(-5, 0), (-1, 0)$; $(0, 5)$ **74.** No x-intercepts; (0, 3)

75.

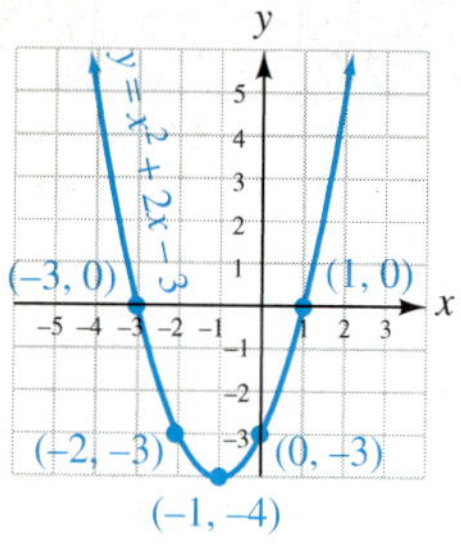

76.

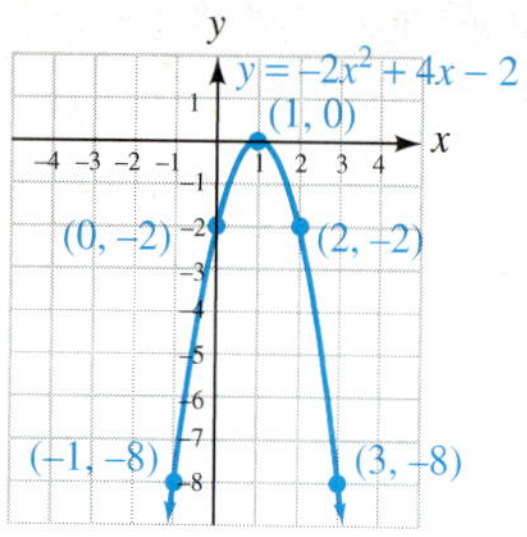

77. Irrational x-intercepts

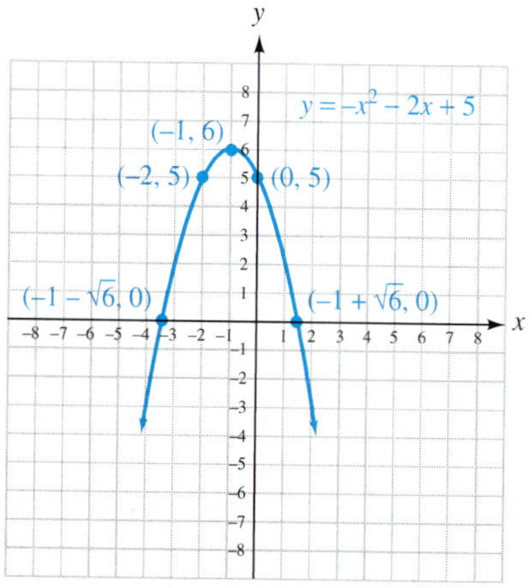

78. Irrational x-intercepts

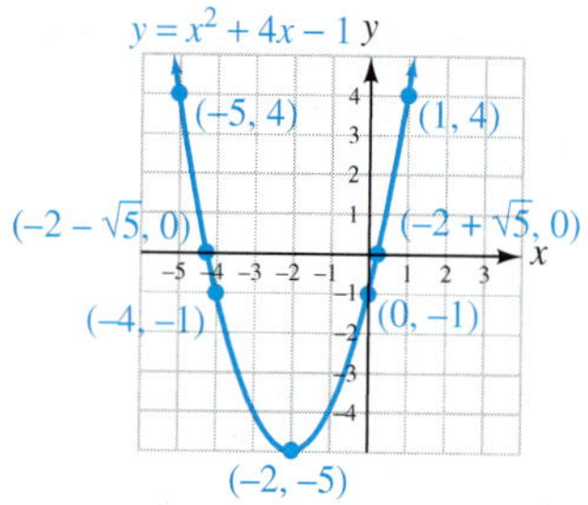

79. 2; −2, 1; 1; −3; no **80.** The maximum profit of $16,000 is obtained from the sale of 400 units.

Chapter 9 Test (page 790)

1. a. complex **b.** parabola **c.** quadratic **d.** imaginary **e.** leading **2.** $x = \pm\sqrt{5}$ **3.** $\pm\sqrt{17}$ **4.** $2 \pm \sqrt{3}$ **5.** $\pm\frac{5}{2}$ **6.** $-8 \pm 2\sqrt{6}$ **7.** $m = \pm\sqrt{-49}$, and $\sqrt{-49}$ is not a real number. **8.** 40 cm **9.** $x^2 - 14x + 49 = (x - 7)^2$ **10.** $a^2 - \frac{5}{3}a + \frac{25}{36} = \left(a - \frac{5}{6}\right)^2$ **11.** $-1 \pm \sqrt{5}$; −3.24, 1.24 **12.** $2, -\frac{1}{2}$ **13.** $-\frac{3}{2}, 4$ **14.** $\frac{-11 \pm \sqrt{61}}{10}$ **15.** $\frac{1 \pm \sqrt{7}}{3}$; −0.55, 1.22 **16.** 41 ft, 65 ft **17.** $2 \pm \sqrt{2}$ **18.** $1, -\frac{5}{3}$ **19.** $\pm 2\sqrt{6}$ **20.** $-\frac{1}{3} \pm \frac{\sqrt{2}}{3}i$ or $\frac{-1 \pm i\sqrt{2}}{3}$ **21.** $10i$ **22.** $-3i\sqrt{2}$ **23.** $1 + i$ **24.** $-1 + 12i$ **25.** $14 - 8i$ **26.** $\frac{5}{13} - \frac{12}{13}i$ **27.** The most air conditioners sold in a week (18) occurred when 3 ads were run. **28.** $<, >$

29.

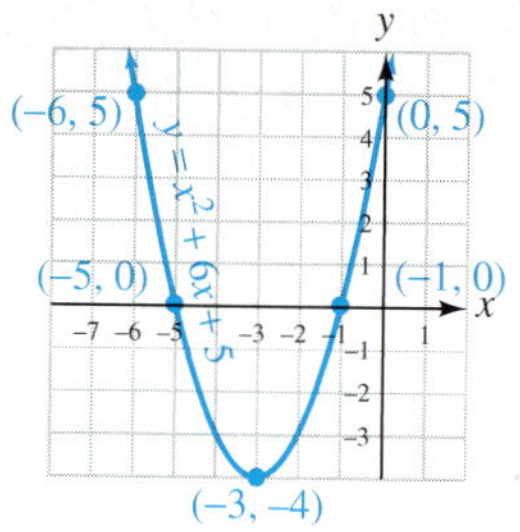

30.

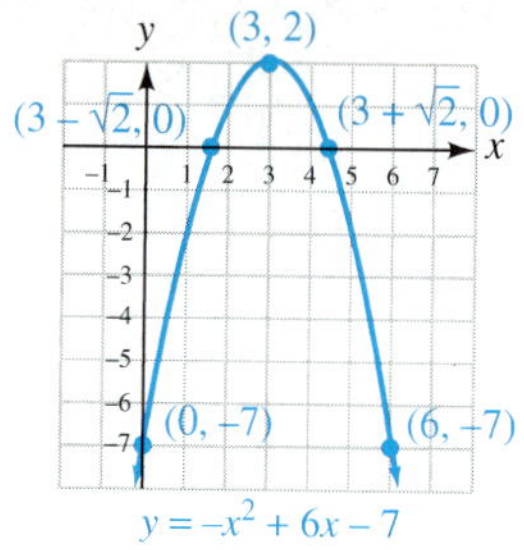

Cumulative Review Chapters 1–9 (page 791)

1. a. True **b.** True **c.** True **2.** 2 hours before salt is spread **3.** 36 **4.** 4 **5.** 65 **6.** $-2p + 54$ **7.** 12 **8.** $-\frac{3}{2}$ **9.** 8.9% **10.** $r = \frac{T - 2t}{2}$ **11.** $343,750 **12.** $8,000 at 7%, $20,000 at 10% **13.** $(-\infty, -2)$

14. Yes

15.

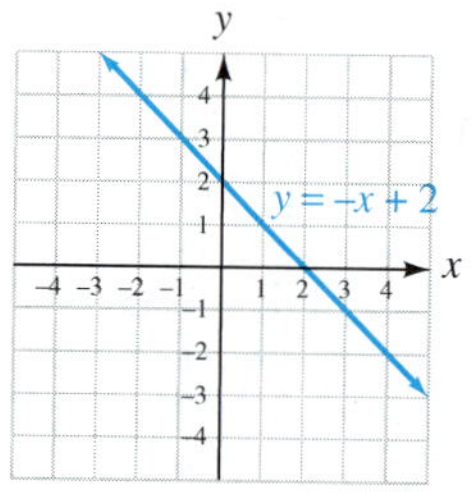

16.

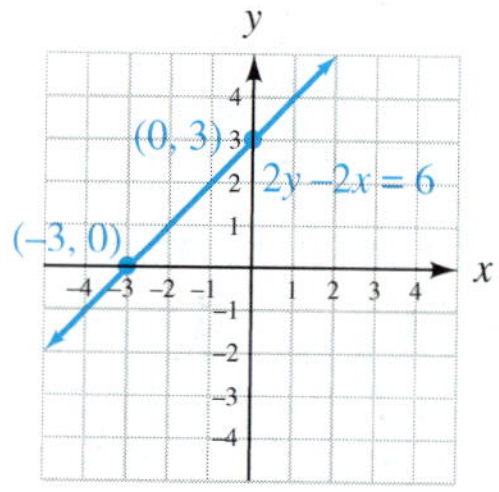

17.

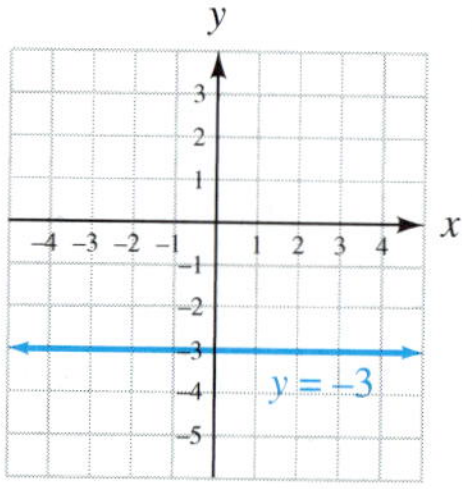

18.

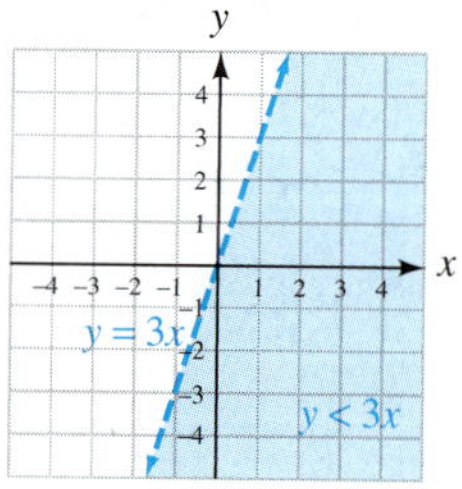

19. $\frac{3}{5}$ **20.** A decrease of 750,000 viewers per year **21.** $-\frac{4}{5}$ **22.** $y = -2x + 1$ **23.** Perpendicular **24.** $y = \frac{1}{4}x - 1$ **25.**

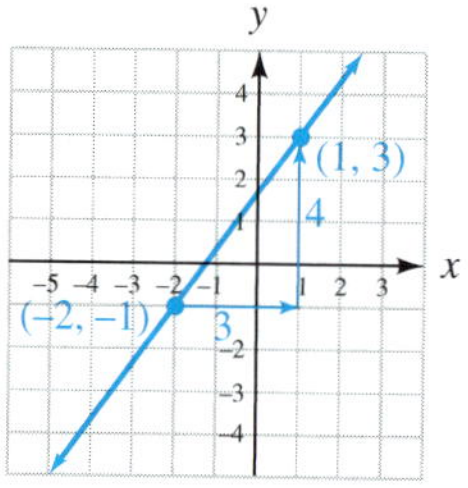

26. −8 **27.** Domain: {−4, 1, 4, 5}; range: {−3, 2, 8} **28.** Yes

29. (−2, 3)

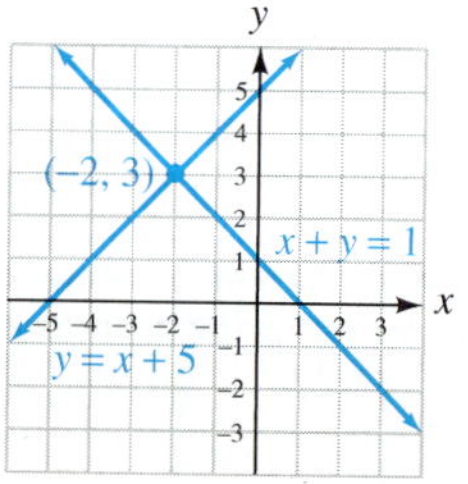

30. (−3, −1) **31.** (−1, 2) **32.** 550 mph **33.** 36 lb of hard candy, 12 lb soft candy

34.

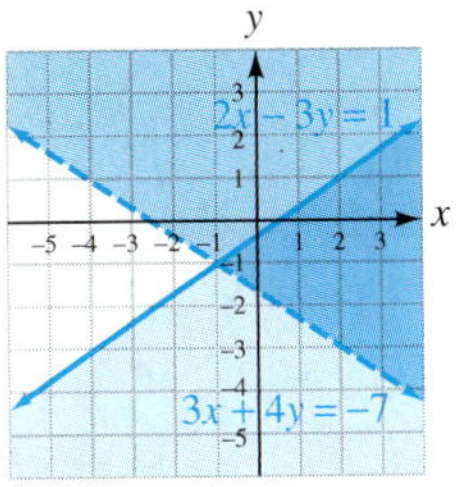

35. y^9 **36.** $\frac{b^6}{27a^3}$ **37.** 2
38. $\frac{x^{10}z^4}{9y^6}$ **39.** 2,600,000
40. 7.3×10^{-4}

41.

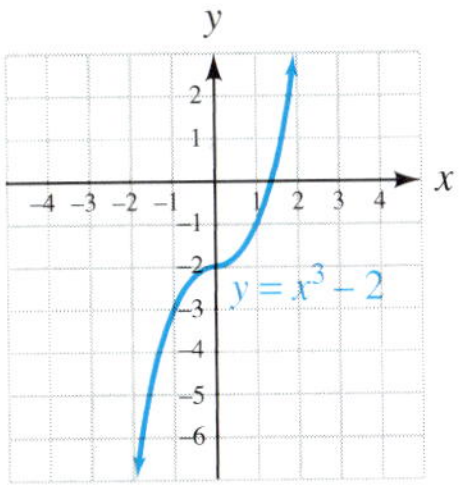

42. $6x^3 + 4x$
43. $6x^3 + 8x^2 + 3x - 72$
44. $-6a^5$ **45.** $6b^2 + 5b - 4$
46. $6x^3 - 7x^2y + y^3$
47. $4x^2 + 20xy + 25y^2$
48. $81m^4 - 1$ **49.** $2a - \frac{3}{2}b + \frac{1}{2a}$
50. $2x + 1$
51. $6a(a - 2a^2b + 6b)$
52. $(x + y)(2 + a)$
53. $(x + 2)(x - 8)$ **54.** $3y^3(5y - 2)(2y + 5)$
55. $(t^2 + 4)(t + 2)(t - 2)$ **56.** $(b + 5)(b^2 - 5b + 25)$
57. $0, -\frac{8}{3}$ **58.** $\frac{2}{3}, -\frac{1}{5}$ **59.** 5 in. **60.** -8 **61.** $\frac{3(x + 3)}{x + 6}$

62. -1 **63.** $\frac{x - 3}{x - 5}$ **64.** $\frac{1}{s - 5}$ **65.** $\frac{x^2 + 4x + 1}{x^2y}$ **66.** $\frac{x^2 + 5x}{x^2 - 4}$
67. $\frac{9m^2}{2}$ **68.** $\frac{9y + 5}{4 - y}$ **69.** 3 **70.** 1 **71.** $a = \frac{b}{b - 1}$
72. $2\frac{6}{11}$ days **73.** 875 days **74.** 39 **75.** 9 **76.** 1.2 rpm
77. $10x$ **78.** $-3b\sqrt{2b}$ **79.** $9\sqrt{6}$ **80.** $-1 - 2\sqrt{2}$
81. $\frac{4\sqrt{10}}{5}$ **82.** $\frac{3\sqrt{2} + \sqrt{2a}}{9 - a}$ **83.** 4 **84.** $-2, -1$ **85.** $\frac{3m}{2n^2}$
86. 2 **87.** 125 **88.** $\frac{1}{16}$ **89.** $\pm 5\sqrt{3}$ **90.** $\frac{-5 \pm 6\sqrt{2}}{6}$
91. 21.2 in. **92.** $-2, -6$ **93.** $\frac{1 \pm \sqrt{33}}{8}$; $-0.59, 0.84$
94. 83 ft × 155 ft **95.** $7i$ **96.** $3i\sqrt{6}$ **97.** $1 + 5i$
98. $16 - 2i$ **99.** $6 - 17i$ **100.** $1 - i$ **101.** $0 \pm 4i$
102. $2 \pm i$ **103.**

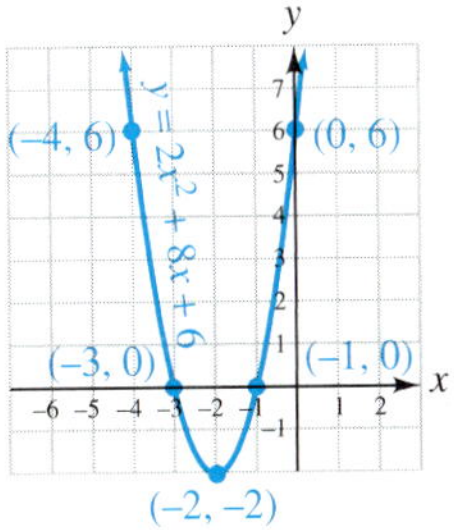

104. 4,000 rpm

INDEX

PERIMETER AND AREA FORMULAS

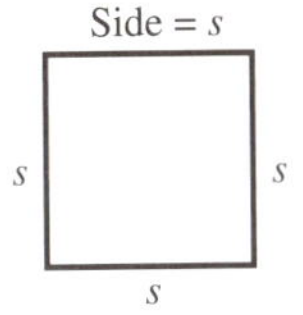

Square
$P = 4s$
$A = s^2$

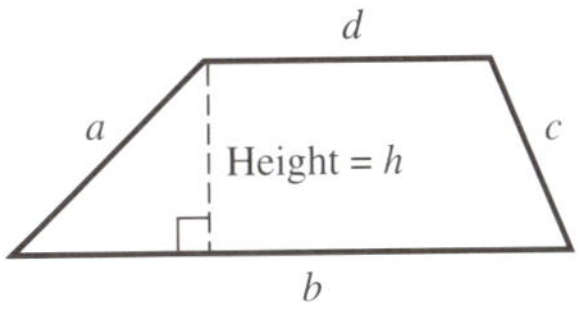

Trapezoid
$P = a + b + c + d$
$A = \frac{1}{2}h(b + d)$

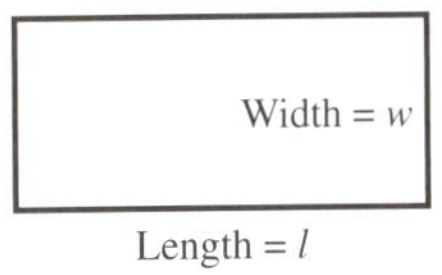

Rectangle
$P = 2l + 2w$
$A = lw$

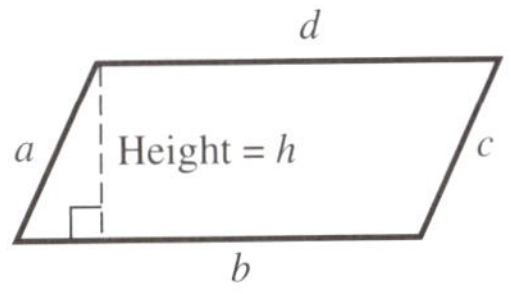

Parallelogram
$P = a + b + c + d$
$A = bh$

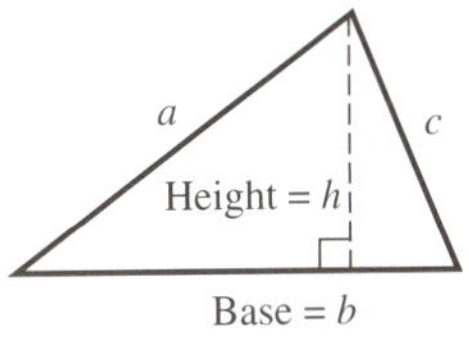

Triangle
$P = a + b + c$
$A = \frac{1}{2}bh$

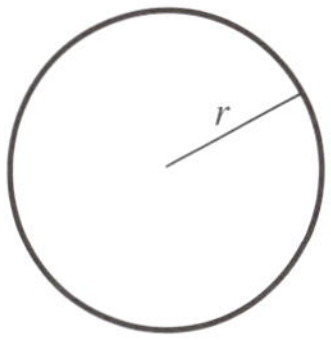

Circle
$C = 2\pi r$ or $C = \pi D$
where $\pi \cong 3.14$
$A = \pi r^2$

VOLUME FORMULAS

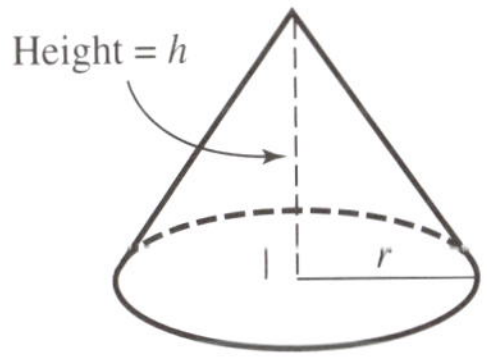

Cone
$V = \frac{1}{3}\pi r^2 h$

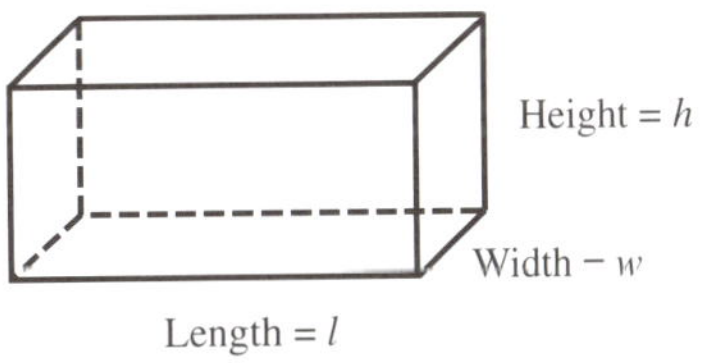

Rectangular solid
$V = lwh$

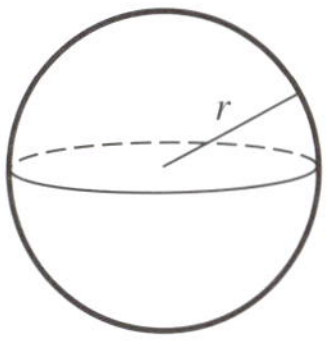

Sphere
$V = \frac{4}{3}\pi r^3$

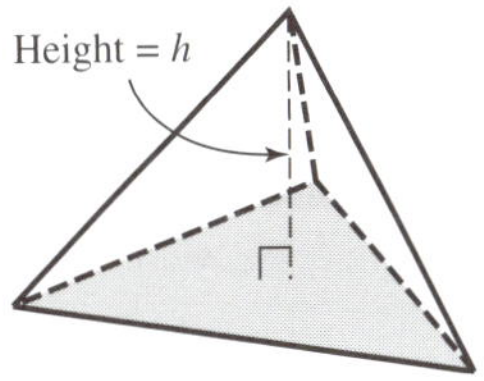

Pyramid
$V = \frac{1}{3}Bh$*
**B* represents the area of the base.

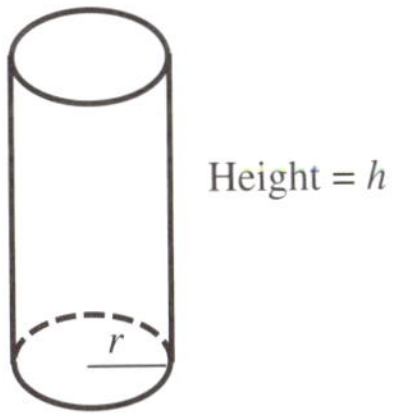

Cylinder
$V = \pi r^2 h$

ENHANCED WEBASSIGN

The Start Smart Guide *for Students*

CENGAGE Learning™

Australia • Brazil • Japan • Korea • Mexico • Singapore • Spain • United Kingdom • United States

CENGAGE Learning™

Enhanced WebAssign: The Start Smart Guide for Students

Acquisitions Editor: Gary Whalen

Copyeditor: Deborah Todd

Editorial Assistant: Lynh Pham

Cover Design: Fabio Fernandes

WebAssign
Centennial Campus
730 Varsity Drive
Raleigh, NC 27606
Web: http://webassign.net
Tel: (800) 955-8275 or (919) 829-8181
Fax: (919) 829-1516
E-mail: info@webassign.net

For product information and technology assistance, contact us at **Cengage Learning Customer & Sales Support, 1-800-354-9706**

For permission to use material from this text or product, submit all requests online at **www.cengage.com/permissions**
Further permissions questions can be emailed to **permissionrequest@cengage.com**

ISBN-13: 978-0-495-38479-3

ISBN-10: 0-495-38479-8

Cengage Learning is a leading provider of customized learning solutions with office locations around the globe, including Singapore, the United Kingdom, Australia, Mexico, Brazil, and Japan. Locate your local office at: **www.cengage.com/global**

Cengage Learning products are represented in Canada by Nelson Education, Ltd.

To learn more about Cengage Learning, visit **www.cengage.com**

Purchase any of our products at your local college store or at our preferred online store **www.ichapters.com**

Printed in the United States of America
6 7 8 9 10 11 10 09 08

Contents

WebAssign works with any recent browser and computer. Some assignments may require plugins like Java, Flash, Shockwave, or Adobe Reader.

For technical support go to http://webassign.net/student.html or email support@webassign.net.

GETTING STARTED

Welcome to Enhanced WebAssign, the integrated, online learning system that gives you 24/7 access to your math, physics, astronomy, chemistry, biology, and statistics assignments.

Now, you can do homework, take quizzes and exams, and receive your scores and graded assignments from any computer with an Internet connection and web browser, any time of the day or night.

Note: As a live, web-based program, Enhanced WebAssign is updated regularly with new features and improvements. Please refer to WebAssign's online Help for the most current information.

Technical Startup Tips

Before you start, please note the following important points:

- Most standard web connections should work with WebAssign. We recommend using Firefox 1.0 or later, or Internet Explorer 5.5 or later. *We do not recommend the AOL browser.*
- You can use a 56 KBPS modem, broadband, or school network connection.
- Your browser needs to have both JavaScript and Java enabled.
- *You cannot skip the login page.* WebAssign must know it is you before delivering your assignments.

Note: If you'd like to bookmark WebAssign on your computer, we recommend that you bookmark **https://www.webassign.net/login.html** or the appropriate address for your school.

Login to WebAssign

In order to access WebAssign your instructor will provide you with login information or a Class Key. Login information will consist of a username, institution code, and an initial password. The Class Key will allow you to self-register and create your own login. You will need to remember the username and initial password you set after self-registering.

Please note that Class Keys are not the same as access codes. See pages 8–9 for instructions on registering your access code number. You will need to login first before you are able to register an access code.

➢ To get started

1. If you are using a shared computer, completely exit any browsers that are already open.
2. Open a new web browser and go to https://www.webassign.net/login.html, or the web address provided by your instructor.

   If your instructor has provided you with a **Username, Institution** (school code), and **Password,** continue with step 3. If you have been provided with a **Class Key** (usually your institution name and a series of numbers), then skip to step 5.
3. Enter your **Username**, **Institution** (school code), and **Password** *provided by your instructor.*

   Institution

   If you do not know your **Institution,** you can search for it by clicking **(what's this?)** above the **Institution** entry box.

   In the **What's My Institution Code** pop-up window, enter your school name and click **go!**. The **Institution Search Results** table will give you choices of the School Names that most closely match your entry, and the **Institution Code** that you should enter in the **Institution** entry box on the **WebAssign Login** screen.

   Password

   If you have forgotten or do not know your **Password,** click **(Reset Password)** above the **Password** entry box, and follow the directions on the **WebAssign New Password Request** screen. You will need to submit your username, institution code, and the email address on file in your WebAssign account. If you are unsure of your username or listed email address, please check with your instructor. WebAssign cannot reset your username or password.

4. Click **Log In**.
5. If your instructor gave you a **Class Key,** you will use it to create your account. Click the **I have a Class Key** button. You will need to use this key only once when you register.
6. Enter the Class Key code in the field provided and click **Submit**. If your Class Key is recognized, you will be given fields for creating your username and password and for entering basic student information.
7. Enter a username in the field provided and then click **Check Availability** to determine whether or not your username is already in use. If it is, an available alternate username will be suggested. Remember your username because you will use it every time you login to WebAssign.
8. Enter and then re-enter a password. Remember your password because you will use it every time you login to WebAssign.
9. Under **Student Information** enter your first and last name, email address, and student ID.
10. Click **Create My Account**.
11. If you see confirmation of your account creation, you will now be able to login to WebAssign. Click **Log in now**.

Note: Before starting WebAssign on a shared computer, always exit any browsers and restart your browser application. *If you simply close the browser window or open a new window, login information contained in an encrypted key may not be yours.*

Logout

When you are finished with your work, click the **Logout** link in the upper right corner of your Home page, and *exit the browser completely* to avoid the possibility of someone else accessing your work.

Your Enhanced WebAssign Home Page

Your personalized Home page is your hub for referencing and managing all of your Enhanced WebAssign assignments.

Using Access Codes

Some classes require an **access code** for admission. Please remember:

- An **access code** is *not* the same as a Class Key or a login password.
- An **access code** is good for *one class only* unless the textbook includes a two-term **access code**.
- An **access code** is an alphanumeric code that is *usually* packaged with your textbook. It can begin with 2 or 3 letters, followed by an alphanumeric code, or it can have a longer prefix such as **BCEnhanced-S** followed by four sets of four characters.
- If your textbook did not include an **access code**, you can buy one at your bookstore, or from your personalized Home page by clicking the **Purchase an access code online** button.

➢ To enter an Access Code

1. Under **WebAssign Notices**, select the proper prefix from the **Choose your access code prefix** pull-down menu.

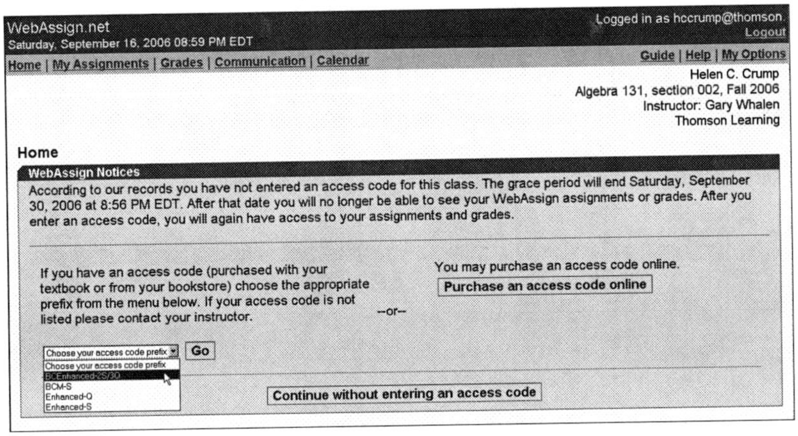

WebAssign notices

2. Click **Go**.
3. In the entry boxes, type in your access code *exactly* as it appears on your card. (When you purchase online, the access code is entered automatically.)

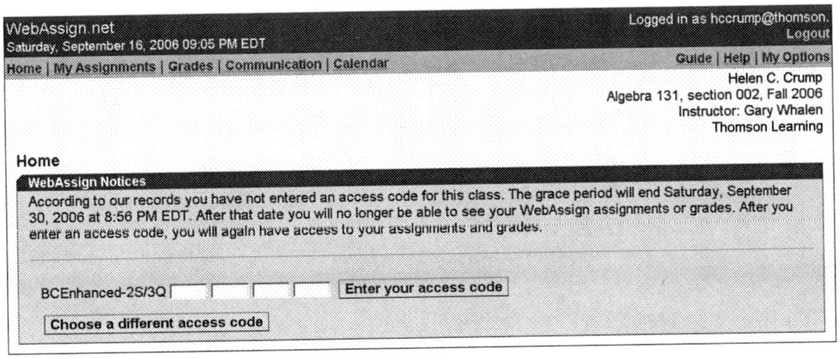

Access code entry

4. Click **Enter your access code.**

   If you have chosen the wrong prefix from the previous screen, you can click the **Choose a different access code** button to try again.

   If your **access code** is a valid unused code, you will receive a message that you have successfully entered the code for the class. Click the **Home** or **My Assignments** button to proceed.

Customizing Your Home Page

Your instructor has initial control over what you see on your Home page to make sure that you have all of the information you need. Your instructor might also set controls so that you can further personalize this page by moving or hiding certain modules.

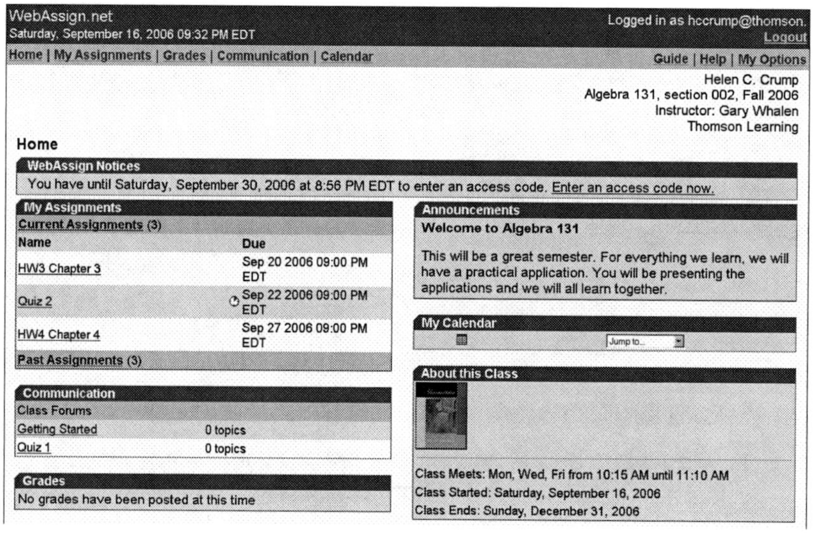

Student Home page

If your instructor has allowed you to personalize your Home page, each module will have markings like this:

Calendar module

To move a module

On the module's heading line, click an up, down, or sideways arrow (indicated by white triangles) until the module is where you'd like it placed on the page.

To minimize a module

On the module's heading line, click the underscore.

To hide a module

On the module's heading line, click the x.

Changing Your Password

For your personal security, it's a good idea to change the initial password provided by your instructor.

➢ To change your password

1. Click the **My Options** link in the upper right of your Home page.
2. In the **My Options** pop-up window, under the **Personal Info** tab:

   Enter your *new* password in the **Change Password** entry box next to **(enter new password)**, then

   Reenter your new password *exactly* the same in the entry box next to **(reenter for confirmation)**.
3. Enter your *current* password in the entry box under **If you made any changes above, enter your current password here and then click save:**, located at the bottom of the pop-up window.
4. Click the **Save** button in the bottom right corner of the pop-up window.

   If the change was successful, you will see the message **Your password has been changed**.

   Note: Passwords are case-sensitive. This means that if you capitalize any of the letters, you must remember to capitalize them the same way each time you sign in to Enhanced WebAssign.

Changing Your Email Address

If your instructor provided you with an email address, you can easily change it to your own personal email address any time.

➢ To change your email address

1. Click the **My Options** link in the upper right of your Home page.

2. In the **My Options** pop-up window, under the **Personal Info** tab, enter your *valid* email address in the **Email Address** box.
3. Enter your current password in the entry box under **If you made any changes above enter your current password here and then click save:**, located at the bottom of the pop-up screen.
4. Click the **Save** button in the bottom right corner of the pop-up window.

   A confirmation email will be sent to your new email address.

Once you receive the confirmation email, you must click the link in the email to successfully complete and activate this change.

WORKING WITH ASSIGNMENTS

The courses that have been set up for you by your instructor(s) appear on your Enhanced WebAssign personalized Home page. If you have more than one course, simply select the course you want to work with from the pull-down menu.

Assignment Summary

There are two ways to get a quick summary of your assignments. On the Home page:

- Click the **My Assignments** link in the upper left *menu bar*, *or*
- Click the **Current Assignments** link in the **My Assignments** *module* on the Home page.

Accessing an Assignment

Once your assignments are displayed on your Home page, simply click the name of the assignment you'd like to begin.

- If you have previously submitted an assignment, you will see your most recent responses, if your instructor allows this feature.
- If you have already submitted the assignment, there will usually be a link to **Review All Submissions** on the page, if your instructor has allowed it.

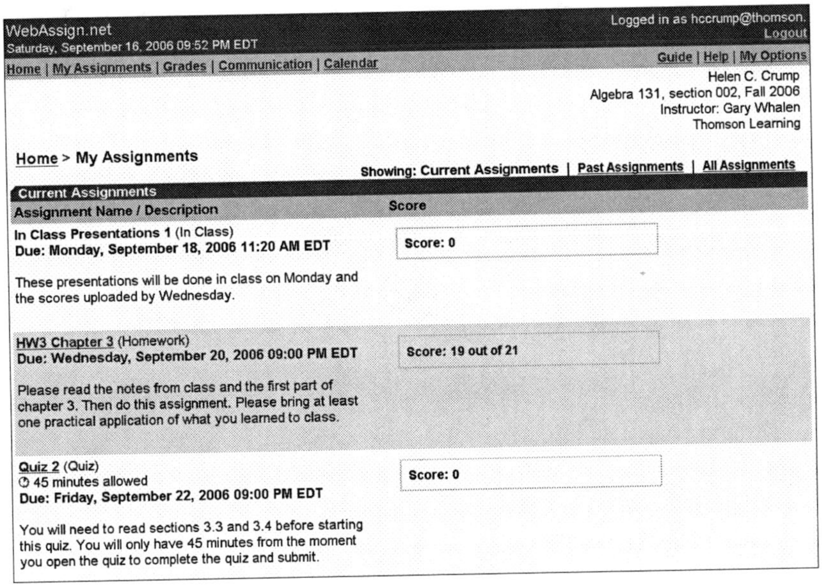

Assignment summary

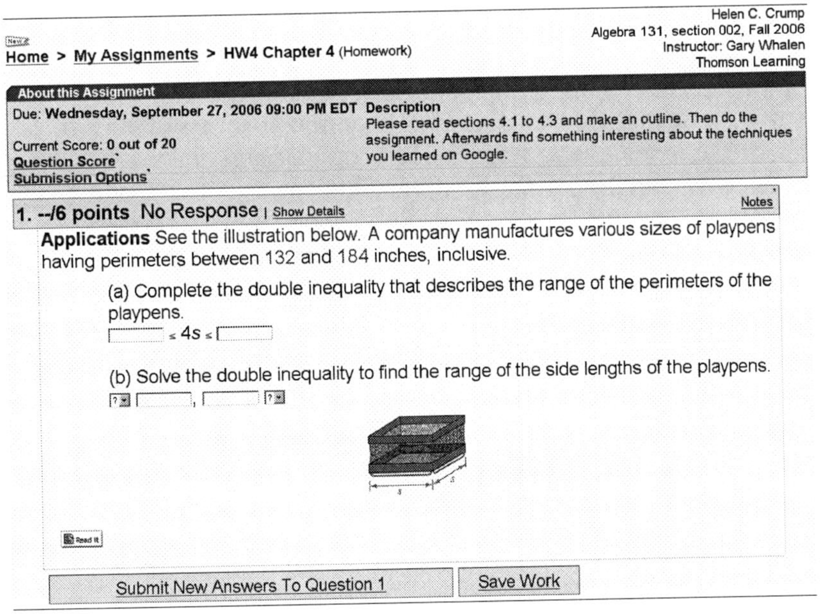

Math assignment

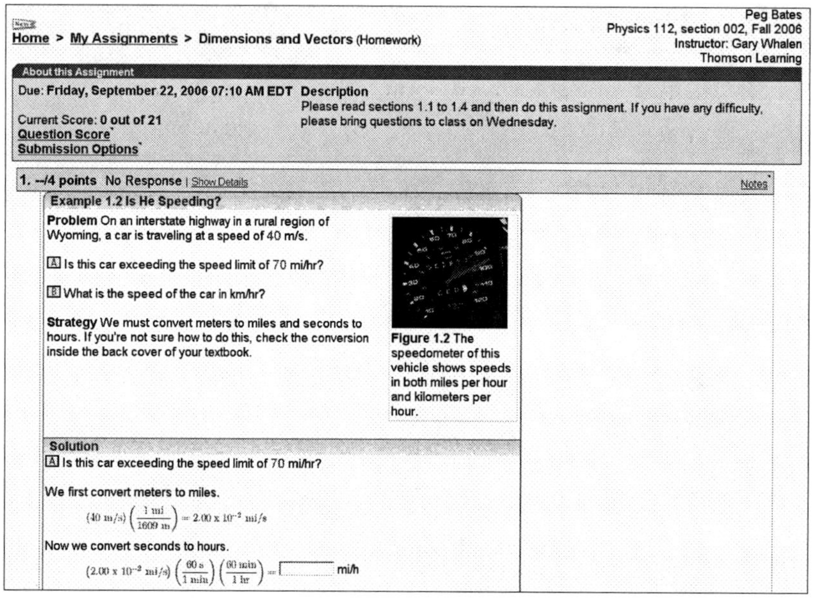

Physics assignment

Using the Assignment Page

When you click on an assignment name, your assignment will load. Within the **About this Assignment** page are links to valuable information about your assignment's score, submission options, and saving your work in progress. Within each question, there might also be "enhanced" action links to useful tutorial material such as book content, videos, animations, active figures, simulations, and practice problems. The links available may vary from one assignment to another.

Actions

Click a button or link to take one of the following actions:

Current Score

This gives you a quick look at your current score versus the maximum possible score.

Question Score

This gives you a pop-up window showing your score for each question.

Submission Options

This gives you a pop-up window explaining how you can submit the assignment and whether it can be submitted by question part, by whole question, or by the whole assignment.

Submissions Made

This shows you the number of submissions you've made. This information is only displayed on assignments that require submission of the entire assignment.

Notes

This feature gives you a pop-up window with a text box in which you can enter and save notes or show your work with a particular question.

Submit New Answers To Question

Use this button when you're ready to submit your answer for the question. This feature allows you to answer just the parts you want scored. If you leave any part of a question unanswered, the submission *will not* be recorded for that part.

Submit Whole Question

Use this button to submit your answer(s) for the entire question. If you leave any part of a question unanswered, the submission *will* be recorded as if the entire question has been answered, and graded as such.

Save Work

This button allows you to save the work you've done so far on a particular question, but does not submit that question for grading.

View Saved Work

Located in the question's header line, this allows you to view work that you previously saved for that question.

Show Details

Located in the question's header line, this link shows your score on each part of the question, how many points each part of the question is worth, and how many submissions are allowed for each part if you can submit each part separately.

Submit All New Answers

This submits all of your new answers for all of the questions in the assignment.

Save All Work

This allows you to save all the work you've done on all of the questions in the assignment, but does not submit your work for grading.

Ask Your Teacher

This feature allows you to send a question about the assignment to your instructor.

Extension Request

This allows you to submit a request to your instructor for an extension of time on an assignment.

Home

This link takes you to your personalized Home page.

My Assignments

This link takes you to your assignments page.

Open Math Palette

This opens a tool to use in writing answers that require math notation.

Read it

This links to question-specific textbook material in PDF form.

Practice Another Version

This provides you with an alternate version of the assigned problem. Within the pop-up window you will be able to answer the practice problem and have that answer checked. You will also be able to practice additional versions of your assigned problem.

Practice it

This links to a practice problem or set of practice problems in a pop-up window. No grade is recorded on the work you do on practice problems.

See it

This links to a tutorial video.

Hint

This links to a pop-up window with helpful hints in case you get stuck on a question.

Hint: Active Figure

This links to an animated simulation to help you better understand the concepts being covered.

Note: Your instructor has the ability to turn on/off many of the options listed above.

ANSWERING QUESTIONS

Enhanced WebAssign uses a variety of question types that you're probably already familiar with using, such as multiple choice, true/false, free response, etc.

Always be sure to pay close attention to any instructions within the question regarding how you are supposed to submit your answers.

Numerical Questions

There are a few key points to keep in mind when working on numerical questions:

- ❍ Numbers can be entered in both scientific notation and numerical expressions, such as fractions.
- ❍ WebAssign uses the standard scientific notation "E" or "e" for "times 10 raised to the power." (Note: both uppercase E and lowercase e are acceptable in WebAssign.) For example, 1e3 is the scientific notation for 1000.
- ❍ Numerical answers may *not* contain commas (,) or equal signs (=).
- ❍ Numerical answers may only contain:
  - Numbers
  - E or e for scientific notation
  - Mathematical operators +, -, *, /
- ❍ Numerical answers within 1% of the actual answer are counted as correct, unless your instructor chooses a different tolerance. This is to account for rounding errors in calculations. In general, enter three significant figures for numerical answers.

➢ Example: Numerical Question

Let's suppose you're presented a question to which your answer is the fraction "one over sixty-four." Following are examples of Correct and Incorrect answer formats:

Correct Answers

Any of these formats would be correct:

1/64

0.015625

0.0156

.0156

1.5625E-2

Incorrect Answers

These formats would be graded as incorrect:

O.015625	The first character is the letter "O"
0. 015625	There is an improper space in the answer
1.5625 E-2	There is an improper space using E notation
l/64	The first character is lowercase letter "L"
5,400	There is a comma in the answer
1234.5=1230	There is an equal sign in the answer

Numerical Questions with Units

Some Enhanced WebAssign questions require a number and a unit, and this is generally, although not always, indicated in the instructions in the question.

You will know that a unit is expected when there is no unit after the answer box.

When you are expected to enter units and do not, you will get an error message telling you that units are required.

Note: Whether omission of the unit counts as a submission depends on the submission options chosen by the instructor.

Find the horizontal and vertical components of the d = 130 m displacement of a superhero who flies from the top of a tall building following the path shown in Fig. P1.40 where θ = 35.0°.

x = 106 ✗ Units are required for this answer.

y = -74.6 m ✓

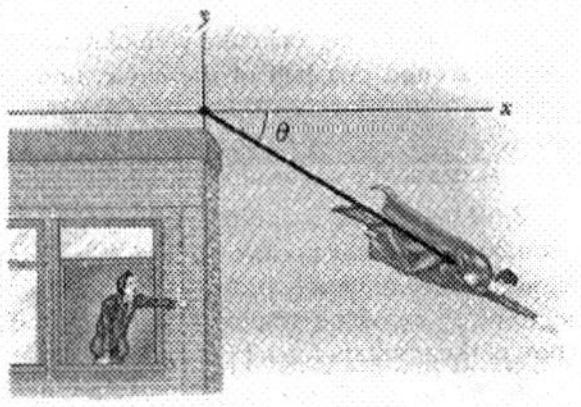

Figure P1.40

Numerical with units

The easiest units to use in this question are m, but the answer converted to yd would also be scored correct.

Numerical Questions with Significant Figures

Some numerical questions require a specific number of significant figures (sig figs) in your answer. If a question checks sig figs, you will see a sig fig icon next to the answer box.

If you enter the correct value with the wrong number of sig figs, you will not receive credit, but you will receive a hint that your number does not have the correct number of sig figs. The sig fig icon 4.0✓ is also a link to the rules used for sig figs in WebAssign.

Carry out the following arithmetic operations. (Use the correct number of significant figures.)

(a) the sum of the measured values 760., 37.2, 0.81, and 2.2

4.0✓ 8e2 ✗ Check the number of significant figures.

(b) the product 3.4 × 3.563

4.0✓ 12 ✓

(c) the product 5.7 × π

4.0✓ 18 ✓

Check for significant figures

Math Notation: Using the Math Palette

In many math questions, Enhanced WebAssign gives you an answer box with a **Math Palette** button. The **Math Palette** provides easy input of math answers, even the more complicated ones.

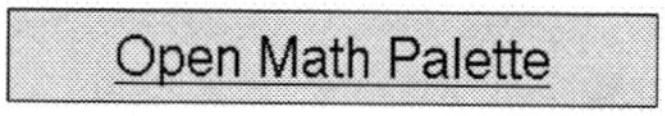

Math Palette button

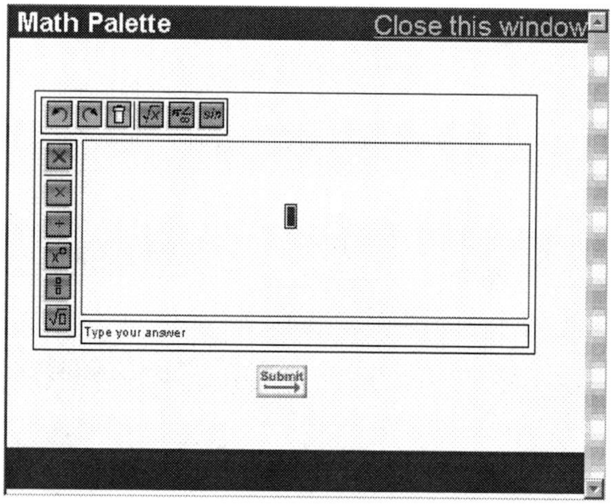

Math Palette tool

Top Symbols

The **Math Palette** has a toolbar of symbols on top that, when selected, give you a drop-down menu with more symbols from which to choose.

Side Symbols

The buttons on the side are single input buttons for frequently used operations.

The yellow buttons are editing buttons:

 Arrow keys are for undo and redo

 Trashcan is to clear your input and start over with an empty field

 The red "x" is used as a short cut for the answer "no solution."

After using the **Math Palette** to write your answer, click the **Submit** button. Your answer will appear in the appropriate boxed area with the question. Your answer will be graded once you actually submit your answers for grading.

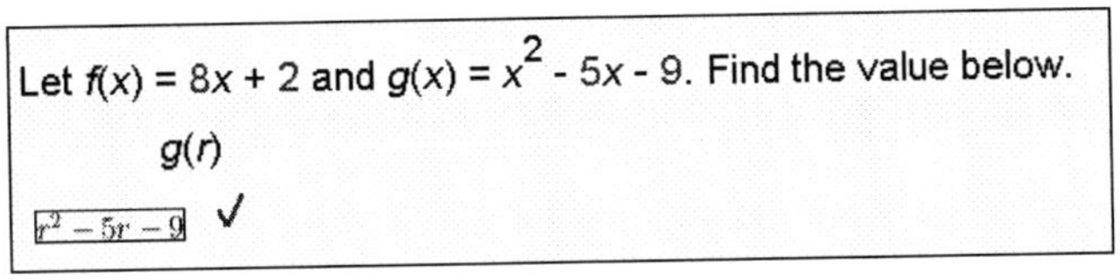

After using Math Palette

Math Notation: Using the Keyboard

If you use your keyboard to enter math notation (calculator notation), *you must use the exact variables specified in the questions.*

The order is not important, as long as it is mathematically correct.

➢ Example: Math Notation Using Keyboard

In the example below, the keyboard is used to enter the answer in the answer field.

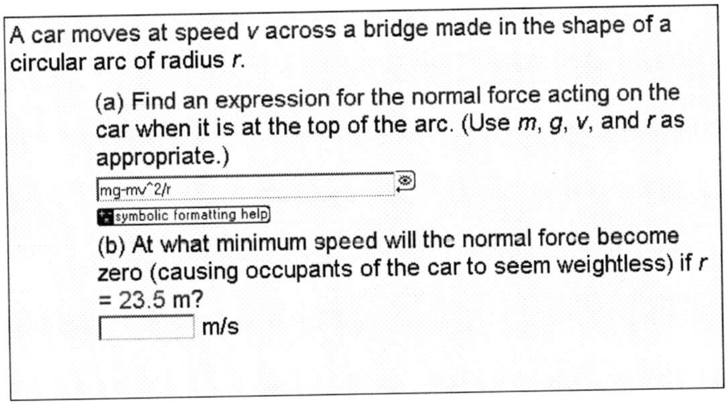

Symbolic question

Expression Preview

Clicking the eye button allows you to preview the expression you've entered in calculator notation.

Use this preview feature to help determine if you have properly placed your parentheses.

Symbolic Formatting Help

If you're unsure about how to symbolically enter your answer properly, use the **symbolic formatting help** button to display allowed notation.

Allowed notation for symbolic formatting

+ for addition	x+1
- for subtraction	x-1, or –x
* or nothing for multiplication	4*x, or 4x
/ for division	x/4
** or ^ for exponential	x**3, or x^3
( ) where necessary to group terms	4/(x+1), or 3(x+1)
abs( ) to take the absolute value of a variable or expression	abs(-5) = 5
sin, cos, tan, sec, csc, cot, asin, acos, atan functions (angle x expressed in radians)	sin(2x)
sqrt( ) for square root of an expression	sqrt(x/5)
x^ (1/n) for the nth root of a number	x^ (1/3), or (x-3)^ (1/5)
pi for 3.14159...	2 pi x
e for scientific notation	1e3 = 1000
ln( ) for natural log	ln(x)
exp( ) for "e to the power of"	exp(x) = e^x

Using the GraphPad

Introduction

The Enhanced WebAssign GraphPad lets you graph one or more mathematical elements directly on a set of coordinate axes. Your graph is then scored automatically when you submit the assignment for grading.

The GraphPad currently supports points, rays, segments, lines, circles, and parabolas. Inequalities can also be indicated by filling one or more areas.

GraphPad Interface Overview

The middle of GraphPad is the drawing area. It contains labeled coordinate axes, which may have different axis scales and extents depending on the nature of the question you are working on.

On the left side of GraphPad is the list of Tools that lets you create graph objects and select objects to edit.

The bottom of the GraphPad is the Object Properties toolbar, which becomes active when you have a graph element selected. This toolbar shows you all the details about the selected graph object and also lets you edit properties of that.

On the right side of GraphPad is the list of Actions that lets you create fills and delete objects from your graph.

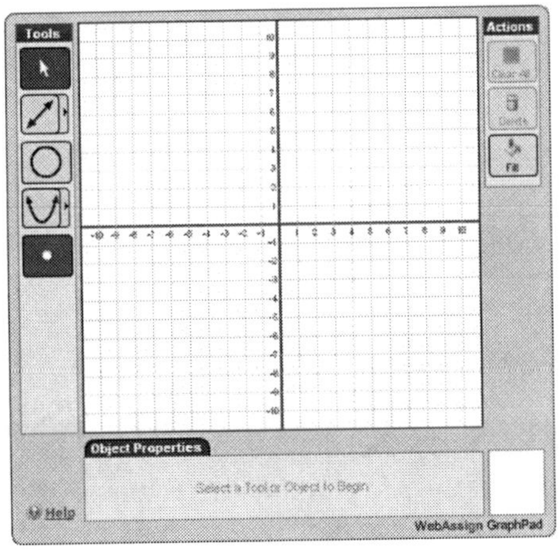

Drawing Graph Objects

To draw a line, first click on the line button in the Tools area. The line button will highlight to blue, and then you can place two points (which are on that line) by clicking twice within the drawing area. Don't worry if you don't place the line exactly where you want it initially; you can move these points around before submitting for grading. The arrows on the end of the line indicate that the line goes off to infinity on both ends.

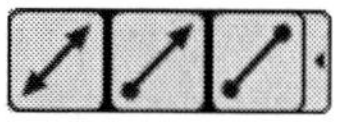

To draw a ray or a segment, first click the small arrow on the right side of the line button to open the selection of line-type tools. Choose ray or segment, and then place it by clicking twice within the drawing area. For rays, the first point is the endpoint of the ray. The arrow on the other end of the ray indicates that it goes off to infinity on that end.

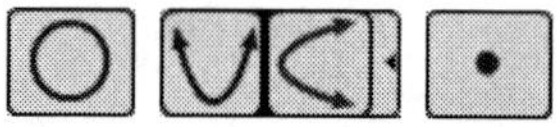

Circles, points, and parabolas can be drawn in the same manner. Circles are drawn by placing a point at the center first, then a point on the radius of the circle. Parabolas are drawn by placing the vertex first, then a point on the parabola. Parabolas can be horizontal or vertical. Points are even easier—just click the point button and then click where you want the point to appear.

Selecting Graph Objects

To edit a graph object you have drawn, that object must be "selected" as the active object. (When you first draw an object, it is created in the selected state.) When a graph element is "selected", the color of the line changes and two "handles" are visible. The handles are the two square points you clicked to create the object. To select an object, click on the object's line. To deselect the object, click on the object's line, a blank area on the drawing area, or a drawing tool.

Not Selected Selected

Moving and Editing Graph Objects

Once an object is selected, you can modify it by using your mouse or the keyboard. As you move it, you'll notice that you cannot move the handles off the drawing area.

To move an object with the mouse, click and drag the object's line. Or click and drag one of the handles to move just that handle.

On the keyboard, the arrow keys also move the selected object around by one unit.

As you move the object or handle you'll see that the Object Properties toolbar changes to remain up to date.

You can also use the coordinate boxes in the Object Properties toolbar to edit the coordinates of the handles directly. Use this method to enter decimal or fractional coordinates.

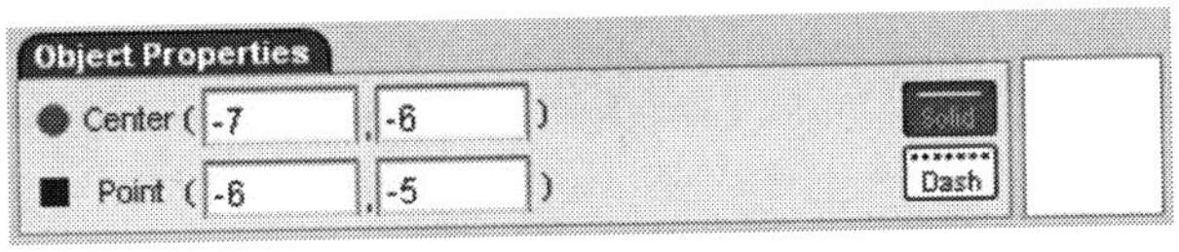

Using Fractions or Decimals as Coordinates

To draw an object with handle coordinates that are fractions or decimals, you must use the Object Properties toolbar. Draw the desired object anywhere on the drawing area, then use the coordinate boxes in the Object Properties toolbar to change the endpoint(s) to the desired value. To enter a fraction just type "3/4," for example.

Note: The points and lines you draw must be exactly correct when you submit for grading. This means you should not round any of your values—if you want a point at 11/3, you should enter 11/3 in the coordinate box rather than 3.667. Also, mixed fractions are not acceptable entries. This means that 3 2/3 is an incorrect entry.

Endpoints—Closed or Open?

If the selected object is a segment or ray, the Endpoint controls in the Object Properties toolbar can be clicked to toggle the endpoint from closed to open.

As a shortcut, you can also toggle an endpoint by clicking on the endpoint when the ray or segment is in the unselected state.

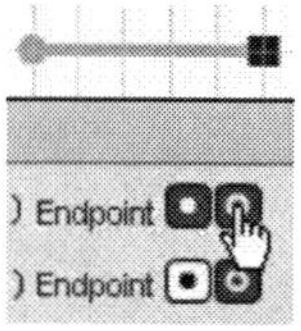

Graph Objects—Solid or Dashed?

For any selected object other than a point, the Solid/Dash buttons in the Object Properties toolbar can be used to make the object solid or dashed. To change graph objects to solid or dashed (for inequalities, for example), select the object and click the Solid or Dash button.

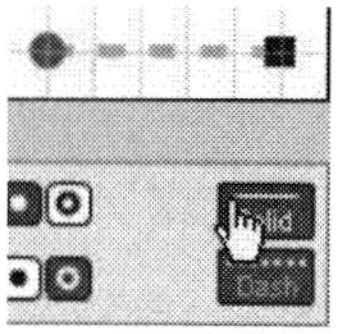

Specifying a Region with a Fill for Inequalities

To graph an inequality, you must specify a region on the graph. To do this, first draw the line(s), circle(s), or other object(s) that will define the region you want to represent your answer. Be sure to specify the objects as either solid or dashed, according to the inequality you are graphing! Then choose the fill button in the Actions area, and click inside the region that you want filled.

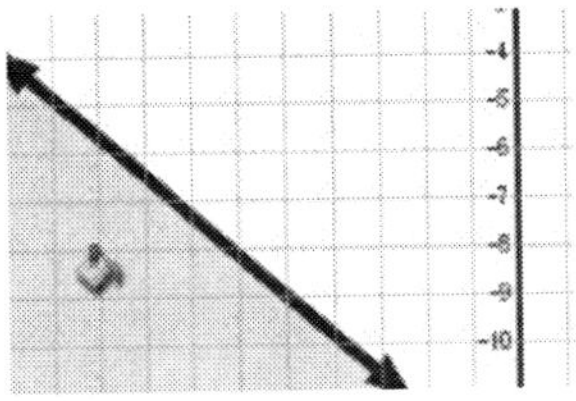

If you decide you wanted the fill in a different area, you can use the fill tool again to undo and then redo the fill in a different location. Choose the fill tool, click the filled region that you want to unfill, and then click the region that you do want to fill.

Erasing One Graph Object

To erase a single graph object, first select that element in the drawing area, then click the Delete icon in the Actions area or press the Delete key on your keyboard.

Erasing Everything on Your Graph

The Clear All button in the Actions area will erase all of your graph objects. (If the drawing area is already empty, the Clear All button is disabled.)

Example: Graphing Question

Let's suppose you're asked to graph the inequality $y > 5x + \frac{1}{5}$, and you want to use the points $\left(0, \frac{1}{5}\right)$ and $\left(1, 5\frac{1}{5}\right)$. You would first place any line on the drawing area.

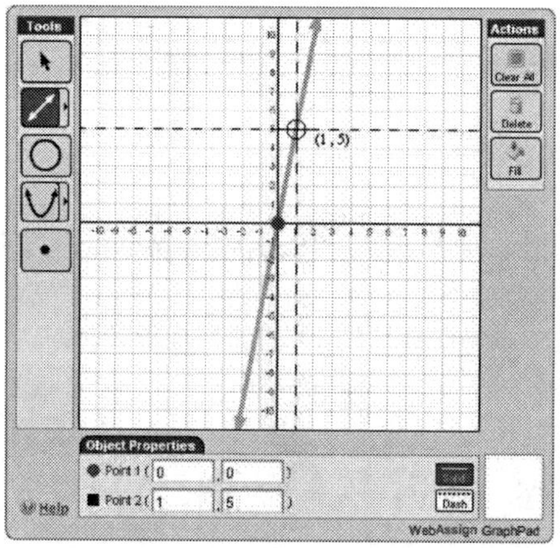

Then, adjust the points using the Object Properties Boxes.

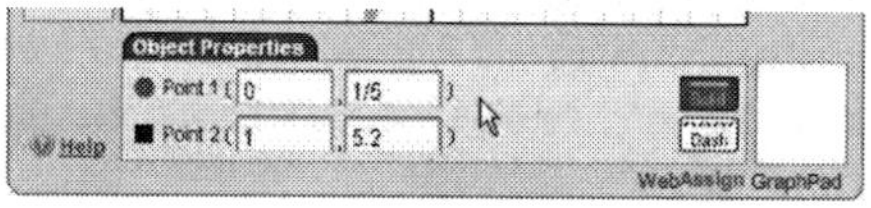

Next, you would define the line as dashed since the inequality does not include the values on the line.

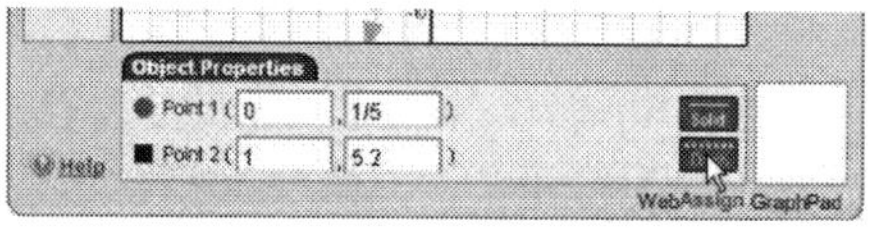

Finally, you would select the Fill Tool and click on the desired region to complete the graph.

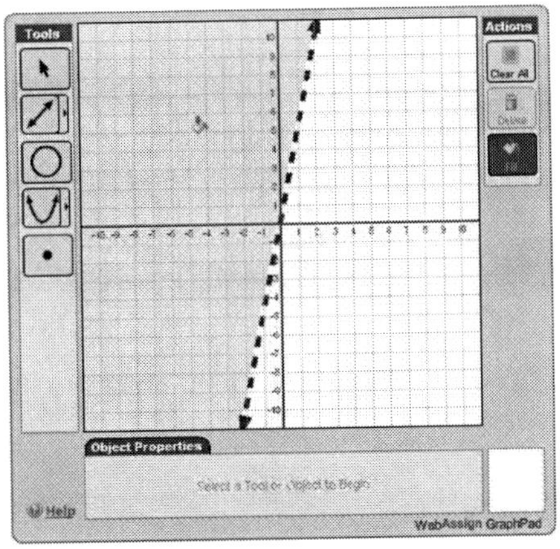

Additional Features

Calendar

The Calendar link presents you with a calendar showing all of your assignments on their due dates. You can also click on any date and enter your own personal events.

Communication

The Communication link gives you access to **Private Messages** and course **Forums**, if your instructor has enabled these features.

Forums

The **Forums** are for discussions with all the members of your class. Your instructor can create forums, and you can create topics within a forum or contribute to a current topic.

Private Messages

Private Messages are for communication between you and your instructor. If your instructor has enabled private messages, click the **New Message** link to send your instructor a message.

GRADES

The **Grades** link at the top of all your WebAssign pages gives you access to the raw scores and grades that your instructor posts. This page may also include statistics on the whole class, and a histogram of scores for each category of assignment and each individual assignment. It may have your individual average for each category of assignment, as well as the score on each of your assignments.

Your instructor will let you know what Scores and Grades will be posted in your course.

If your instructor has enabled all of the options, your display will be similar to the one below.

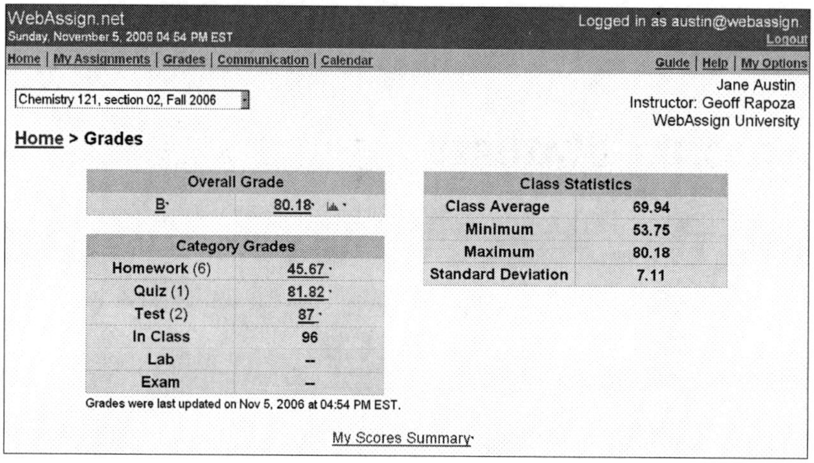

Grades

Overall Grade

This score is calculated from the various categories of assignments, for example, **Homework, Test**, **In Class**, **Quiz**, **Lab**, and **Exam**. Your instructor may have different categories.

Category Grades

The **Category Grades** give the contribution to your overall grade from each of the categories. If you click a grade that is a link, you will get a pop-up window explaining how the number was calculated.

Class Statistics

Class Statistics shows the averages, minimum scores, maximum scores, and standard deviation of the class at large.

My Scores Summary

This link presents a pop-up window with a summary of your raw scores and the class statistics on each assignment, if your teacher has posted these.

My Scores Summary — Close this window

Scores for Jane Austin
Class: Chemistry 121, section 02

Raw Scores

	My Scores			Class Statistics				
Category / Assignment	Score	Max Possible	%	Average	Min	Max	St Dev	
Homework (6)	49	81	60.5	10.2	0	49	11.6	
Quiz (1)	9	11	81.8	6.93	0	9	3.10	
Test (2)	174	200	87	174	109	197	21.4	
Exam (1)	0	10	0	0	0	0	0	
In Class (1)	96	100	96	88.3	74	99	7.32	
Homework for 2nd 9 (1)	4	4	100	–	–	–	–	–

Collapse All | Expand All

My Scores summary

Technical Tips

Enhanced WebAssign relies on web browsers and other related technology that can lead to occasional technical issues. The following technical tips can help you avoid some common problems.

Cookies

Allow your browser to accept cookies.

WebAssign will work if you set your browser to not accept cookies; however, if an encrypted cookie is not saved to your

For technical support go to http://webassign.net/student.html or email support@webassign.net.

computer during your session, you may be asked to login again more frequently. Once you logout and exit your browser, the cookie is deleted.

Login and Credit

If you see an assignment that does not have your name at the top, you have not logged in properly.

You will not receive credit for any work you do on an assignment if your name is not associated with it. If you find yourself in the midst of this situation, make notes of your solution(s) and start over. Be aware that any randomized values in the questions will probably change.

Logout When You Finish Your Session

If someone wants to use your computer for WebAssign, logout and exit the browser before relinquishing control.

Otherwise, the work you have just completed may be written over by the next user.

Server

Although it is very rare, the WebAssign server may occasionally be unavailable.

If the WebAssign server is unavailable, instructors will provide instructions for submitting your assignments—possibly including new due dates. The policy for handling server problems will vary from instructor to instructor.

Use the Latest Browser Software

Use the latest version of Firefox, Mozilla, Netscape, or Internet Explorer browsers.

Older versions of browsers may not be supported by WebAssign.

For technical support go to http://webassign.net/student.html or email support@webassign.net.